Florian Luca, Melvyn B. Nathanson, Jaroslav Nešetřil, Bruce Landman,
Richard J. Nowakowski, Aaron Robertson (Eds.)
Integers

Integers

Annual Volume 2013

Edited by
Florian Luca
Melvyn B. Nathanson
Jaroslav Nešetřil
Bruce Landman
Richard J. Nowakowski
Aaron Robertson

DE GRUYTER

Mathematics Subject Classification 2010
05A, 05C55, 05C65, 05D, 11A, 11B, 11K, 11N, 11P, 11Y, 91A46

Editors

Florian Luca
Universidad Nacional Autónoma de Mexico
Fundación Marcos Moshinsky
Circuito Exterior C.U., Apdo. Postal
70-543, Mexico D.F. 04510, Mexico
fluca@matmor.unam.mx

Melvyn B. Nathanson
Lehman College (CUNY)
Department of Mathematics
250 Bedford Park Blvd W
Bronx NY 10468, USA
melvyn.nathanson@lehman.cuny.edu

Jaroslav Nešetřil
Charles University
Department of applied Mathematics
Malostranske nam. 25,
118 00 Prague, Czech Republic
nesetril@kam.ms.mff.cuni.cz

Bruce M. Landman
University of West Georgia
Department of Mathematics
1601 Maple Street, Carrollton
GA 30118, USA
landman@westga.edu

Richard J. Nowakowski
Dalhousie University
Department Mathematics and Statistics
Chase Building, Halifax,
Nova Scotia B3H 3J5, Canada
r.nowakowski@dal.ca

Aaron Robertson
Colgate University
Department of Mathematics
13 Oak Drive, Hamilton
NY 13346, USA
arobertson@colgate.edu

ISBN 978-3-11-029811-6
e-ISBN 978-3-11-029816-1
Set-ISBN 978-3-11-029817-8

Library of Congress Cataloging-in-Publication Data
A CIP catalog record for this book has been applied for at the Library of Congress.

Bibliographic information published by the Deutsche Nationalbibliothek
The Deutsche Nationalbibliothek lists this publication in the Deutsche Nationalbibliografie; detailed bibliographic data are available in the Internet at http://dnb.dnb.de.

Printing and binding: Hubert & Co. GmbH & Co. KG, Göttingen
♾ Printed on acid-free paper
Printed in Germany

MIX
Papier aus verantwortungsvollen Quellen
FSC
www.fsc.org
FSC® C016439

www.degruyter.com

PREFACE TO VOLUME 13 OF INTEGERS

This book contains all of the articles published in Volume 13 of **Integers**, a refereed electronic journal devoted to research in combinatorial number theory and related parts of mathematics. This volume includes contributions to many subjects, including additive number theory, multiplicative number theory, sequences and sets, extremal combinatorics, Ramsey theory, the theory of partitions, elementary number theory, classical combinatorial problems, and probabilistic number theory. The volume also includes papers in the area of combinatorial games, demarcated by the use of a `Gxx` numbering system.

The editors of **Integers** are grateful to all of the authors who have contributed to this volume. We also express our appreciation to the anonymous referees whose hard work is essential to the success of a mathematics research journal.

– *The Editors*

CONTENTS

SIMPLE ARITHMETICAL CRITERIA FOR IRREDUCIBILITY OF POLYNOMIALS WITH INTEGER COEFFICIENTS

Natalio H. Guersenzvaig
Dept. of Mathematics, Universidad CAECE (Retired), Buenos Aires, Argentina
nguersenz@fibertel.com.ar

Received: 4/13/12, Revised: 11/17/12, Accepted: 1/7/13, Published: 1/25/13

Abstract

A simultaneous generalization of well-known arithmetical criteria for irreducibility in $\mathbb{Q}[X]$ of polynomials in $\mathbb{Z}[X]$, including a classical result of G. Pólya and G. Szegö, will be achieved via a general framework for that family of irreducibility criteria. A further generalization, which for any $f(X) \in \mathbb{Z}[X]$ provides an arithmetical function whose values are upper bounds for the number of irreducible factors of $f(X)$ in $\mathbb{Q}[X]$, will also be established.

1. Introduction

Let $Z \in \{\mathbb{Z}, \mathbb{Q}\}$ and let $\mathcal{U} = \{\pm 1\}$ or $\mathcal{U} = \mathbb{Z}$ according to whether $Z = \mathbb{Z}$ or $Z = \mathbb{Q}$, respectively. Let $f(X)$ denote an arbitrary polynomial in $\mathbb{Z}[X] \setminus \mathcal{U}$. We first remind the reader that $f(X)$ is called *reducible (irreducible) in* $Z[X]$, if there are (there are not, respectively) polynomials $g(X)$, $h(X)$ in $Z[X] \setminus \mathcal{U}$ satisfying $f(X) = g(X)h(X)$.

In this paper we focus our attention on a particular type of irreducibility criteria in $Z[X]$. Such results could be called *simple arithmetical criteria*, because they provide sufficient conditions for the irreducibility of $f(X)$ that only depend on the nature of the value $f(q)$ in a unique and conveniently chosen integer q. When $|f(q)| > 1$ we can write, possibly in several ways, $f(q) = \pm\, dp^e$, where d, p and e are positive integers with p prime. The most important criteria of this kind were established, in chronological order, by P. Stäckel (case $d = e = 1$), G. Pólya and G. Szegö (case $d = e = 1$), O. Ore (case $e = 1$), L. Weisner (general case), and M. Filaseta (case $e = 1$). A useful review of the known irreducibility criteria before 1935 is given in [6]; other results can be found in [13], and [18]. In spite of the different formulations of these results, including a generalization by the author of Pólya-Szegö's criterion, a careful scrutiny of their hypotheses allowed us to detect that besides the technical condition $p \nmid f'(q))^{e-1}$ introduced by Weisner in [21], which allows us to treat p^e like a prime number, similar sets of hypotheses are satisfied by ρ, q and d, where ρ denotes a real parameter which is used to locate the zeros of $f(X)$ in the complex plane. As a result of such observations a general framework for expanded versions of these criteria will be established. Perhaps unexpectedly, an even more general result can be proved. Indeed, our generalization will be extended (in the last section of this paper) via a theorem that for any "f-admissible" triple (ρ, q, d) provides integer upper bounds for N_f, the number of irreducible factors of $f(X)$ in $\mathbb{Q}[X]$.

2. A General Framework for Simple Arithmetical Criteria of Irreducibility in $\mathbb{Z}[X]$

First, summarizing our analysis of the sets of hypotheses of the aforementioned irreducibility criteria, we introduce the notion of "admissible triple."

Definition 2.1. Let S be a nonempty subset of $\mathbb{R} \times \mathbb{Z}$ and let $\mathcal{Z}$ be a mapping defined in $S \times \mathbb{N}$ whose values $\mathcal{Z}(\rho, q, d)$ are subsets of $\mathbb{C}$. Let $\mathcal{F}$ be a function defined in $(\mathbb{Z}[X] \setminus \{0\}) \times S$ with values $\mathcal{F}(f, \rho, q)$ in the real interval $[1, +\infty)$. Let $f(X)$ be any nonzero polynomial in $\mathbb{Z}[X]$.

The triple $(S, \mathcal{Z}, \mathcal{F})$ will be called *f-admissible*, if for any $(\rho, q, d) \in S \times \mathbb{N}$ such that

$$d | f(q),\ q \notin \mathcal{Z}(\rho, q, d) \text{ and the zeros of } f(X) \text{ belong to } \mathcal{Z}(\rho, q, d)$$

the following condition is satisfied:

(**Condition $\mathcal{A}$**) for any polynomial $g(X) \in \mathbb{Z}[X]$ of positive degree dividing $f(X)$,

$$g(q) | d \implies |g(q)| > \mathcal{F}(f, \rho, q).$$

Most of the above criteria are related to the irreducibility of $f(X)$ in $\mathbb{Q}[X]$. In order to also include in our framework sufficient conditions for the irreducibility of $f(X)$ in $\mathbb{Z}[X]$ we recall that the greatest common divisor of the coefficients of $f(X)$, say $\mathrm{c}(f)$, is called *content* of $f(X)$ and that $f(X)$ is called *primitive* if $\mathrm{c}(f) = 1$. Notice now that the given definitions of irreducibility guarantee that $f(X)$ is irreducible in $\mathbb{Z}[X]$ if and only if

$$\text{either } f(X) = \pm p, \text{ or } f(X) \text{ is primitive and irreducible in } \mathbb{Q}[X]. \tag{1}$$

Our theoretical framework also includes sufficient conditions for the irreducibility of $f(X)$ in $\mathbb{Z}[X]$ without invoking explicitly the primitivity of $f(X)$, as described below.

Theorem 2.2. ((S,$\mathcal{Z}$,$\mathcal{F}$)–Irreducibility Criterion) *Let $f(X)$ be an arbitrary nonzero polynomial in $\mathbb{Z}[X]$ and let $(S,\mathcal{Z},\mathcal{F})$ be any f-admissible triple. Let $(\rho, q, d) \in S \times \mathbb{N}$ such that $d | f(q)$, $q \notin \mathcal{Z}(\rho, q, d)$ and the zeros of $f(X)$ belong to $\mathcal{Z}(\rho, q, d)$. Suppose also that*

$$d \leq \mathcal{F}(f, \rho, q), \quad f(q) = \pm\, dp^e \ \textit{and} \ p \nmid (f'(q))^{e-1},$$

where p and e are positive integers with p prime. Then

$$f(X) \textit{ is irreducible in } \mathbb{Q}[X] \iff f(X) \textit{ has positive degree} \iff p \nmid \mathrm{c}(f). \qquad (2)$$

Moreover,

$$f(X) \textit{ is irreducible in } \mathbb{Z}[X] \iff \gcd(d, \mathrm{c}(f)) = 1. \qquad (3)$$

Proof. First we will prove (2). Suppose $f(X) = g(X)h(X)$ with $g(X)$ and $h(X)$ in $\mathbb{Z}[X]$. Since the definition of irreducibility in $\mathbb{Q}[X]$ requires that $f(X)$ has positive degree, to complete the proof of the first equivalence we only need to show that one of the polynomials $g(X)$, $h(X)$ has degree zero.

From $g(q)h(q) = \pm\, dp^e$ it easily follows that there are nonnegative integers d_1, d_2, e_1, e_2, with $d = d_1 d_2$ and $e = e_1 + e_2$, such that $|g(q)| = d_1 p^{e_1}$ and $|h(q)| = d_2 p^{e_2}$. Certainly $e_1 e_2 = 0$ if $e = 1$. We also have $e_1 e_2 = 0$ if $e > 1$, because otherwise from $f'(q) = g'(q)h(q) + g(q)h'(q)$ we would obtain $p|(f'(q))^{e-1}$, a contradiction. Hence we can assume $e_1 = 0$, that is, $|g(q)| = d_1$. Then $g(X)$ is a constant polynomial, because otherwise (Condition $\mathcal{A}$) yields $|g(q)| > \mathcal{F}(f, \rho, q)$ against $d_1 \leq d \leq \mathcal{F}(f, \rho, q)$.

On the other hand, it is clear that $p|\mathrm{c}(f)$ if $f(X)$ is a constant polynomial. To prove the converse statement we assume $p|\mathrm{c}(f)$. This clearly implies $p|f'(q)$, which combined with the hypothesis $p \nmid (f'(q))^{e-1}$ ensures $e = 1$. Hence we get that $f(X)/p$ is a divisor of $f(X)$ satisfying $|f(q)/q| = d$. Then $f(X)$ is a constant polynomial, because otherwise (Condition $\mathcal{A}$) yields the contradiction $d > \mathcal{F}(f, \rho, q)$.

Finally, to prove (3), note that (2) guarantees that (1) is equivalent to

$$\text{either } f(X) = \pm\, p, \text{ or } f(X) \text{ has positive degree and } \gcd(d, \mathrm{c}(f)) = 1,$$

and hence, via $f(q) = \pm\, dp^e$, to $\gcd(d, \mathrm{c}(f)) = 1$. □

Next we will show that Weisner's hypothesis is unnecessary if d is appropriately chosen. In fact, a precise result can be established via the following definition:

Definition 2.3 Let a, b be arbitrary positive integers. The *b-part of a*, say $\delta(a, b)$, is defined by $\delta(a, b) = d_0 \cdots d_{k-1}$, where $d_0 = 1$ and

$$d_j = \gcd\left(\frac{a}{d_0 \cdots d_{j-1}}, b\right) \text{ for } j = 1, \ldots, k-1.$$

Here k denotes the smallest positive integer with $d_k = 1$.

It can be readily seen that we have defined $\delta(a,b)$ so that the following is true:

$$\delta(a,b) = \min\{c \in \mathbb{N} : c|a \text{ and } \gcd(a/c,b) = 1\}.$$

Now we can easily derive of Theorem 2.2 the following result.

Corollary 2.4. *Let $f(X)$ be an arbitrary primitive polynomial in $\mathbb{Z}[X]$ of positive degree and let $(S, \mathcal{Z}, \mathcal{F})$ be any f-admissible triple. Let $(\rho, q, d) \in S \times \mathbb{N}$ such that $d|f(q)$, $\delta(|f(q)|, |f'(q)|)|d$, $q \notin \mathcal{Z}(\rho, q, d)$ and the zeros of $f(X)$ belong to $\mathcal{Z}(\rho, q, d)$. Suppose also that $d \leq \mathcal{F}(f, \rho, q)$ and $f(q) = \pm dp^e$, where p and e are positive integers with p prime. Then $f(X)$ is irreducible in $\mathbb{Q}[X]$.*

Proof. The aforementioned property of $\delta(|f(q)|, |f'(q)|)$ and $\delta(|f(q)|, |f'(q)|)|d$ guarantees $\gcd(p^e, f'(q)) = \gcd(f(q)/d, f'(q)) = 1$, that is, $p \nmid f'(q)$. Then, as all hypotheses of Theorem 2.2 that are required to prove that $f(X)$ is irreducible in $\mathbb{Q}[X]$ are fulfilled, the proof is complete. □

In the next section, for a better understanding of Definition 2.1, the irreducibility criteria listed above will be conveniently reformulated. The admissible triples corresponding to the more general criteria will be presented in Section 5, after establishing our generalizations of the Pólya-Szegö criterion.

3. A Review of Simple Arithmetical Criteria and Related Facts

There are irreducible polynomials in $\mathbb{Z}[X]$ that can not represent infinitely many primes over the integers. For example, as an extreme case, the polynomial $f(X) = X^2 + X + 4$ is irreducible in $\mathbb{Z}[X]$ but $|f(q)| = q(q+1)+4$ is an even integer greater than 3 for any integer q. We recall now a conjecture of V. Bouniakowsky (see cite 5, p. 333) that would generalize the celebrated Dirichlet's Theorem on primes in arithmetic progressions (1837). This conjecture, which has not yet been proved or refuted, can be stated as follows.

Bouniakowsky's Conjecture. (1857) For any $f(X) \in \mathbb{Z}[X]$ of positive degree m that is irreducible in $\mathbb{Z}[X]$ there exist infinitely many integers q such that $f(q)/d(f)$ is a prime number, where $d(f)$ denotes the greatest common divisor of all the values of $f(X)$ over the integers.

Remark 3.1. (I) Bouniakowsky also gave a complicated procedure to compute $d(f)$. A better one was given in 1896 by K. Hensel (see [5], pp. 332, 334]) who proved, using the well-known Newton's Formula

$$f(q+k) = \sum_{j=0}^{\min\{k,m\}} \binom{k}{j} \Delta^j f(q),\ k=0,1,\ldots, \text{ where } \Delta^j f(q) = \sum_{i=0}^{j} (-1)^{j-i} \binom{j}{i} f(q+i),$$

that $d(f)$ is the greatest common divisor of the values of $f(X)$ in any $m+1$ consecutive integers, say $q, q+1, \ldots, q+m$.

(II) A stronger conjecture formulated in 1962 by Bateman and Horn (see [1]) would imply in the case $d(f) = 1$ (see [12]) that

$$\pi(n, f) \underset{n \to \infty}{\sim} \frac{\mathrm{C}(f)}{m} \frac{n}{\log n}.$$

Here, $\pi(n, f)$ denotes the number of integers q with $1 < q < n$ for which $|f(q)|$ is prime and

$$\mathrm{C}(f) = \prod_{p \text{ prime}} \frac{p - w(p)}{p - 1},$$

where $w(p)$ stands for the number of solutions of the congruence $f(x) \equiv 0 \pmod{p}$.

The converse of the case $d(f) = 1$ of the Bouniakowsky Conjecture is an easy consequence of the following theorem of P. Stäckel (see [20], Satz 1).

Stäckel's Theorem 1. (1918) *A reducible polynomial $f(X) \in \mathbb{Z}[X]$ of degree $m \geq 2$ can represent at most $2m$ prime numbers over the integers, and as soon as the absolute values of the integer q exceeds a certain limit, $f(q)$ will represent only composite numbers.*

Remark 3.2. In a relatively recent work ([4]) it is proved that $2m$ can be replaced by $m+2$ if $m \notin \{4, 5\}$ and that there exist examples with m+1 instead of $m+2$. For m=4 or 5 the maximum possible value is 8.

Subsequently, Stäckel used a remark of O. Gmelin concerning the Fundamental Theorem of Algebra to establish the following result about the polynomials that represent prime numbers (see [20], Satz 7).

Stäckel's Theorem 7. (1918) *Let $f(X) \in \mathbb{Z}[X]$ and let $S = 1 + A$, where A denotes the maximum of the absolute values of the coefficients of $f(X)$. If there is an integer q with $|q| > S$ for which $f(q)$ is a prime number, then $f(X)$ is irreducible in $\mathbb{Z}[X]$.*

Remark 3.3. It should be noted that S constitutes the greatest possible value of a well-known Cauchy's upper bound for the absolute values of the zeros of $f(X)$, namely, $\rho_0 = 1 + (A/|a_n|)$, where a_n denotes the leading coefficient of $f(X)$ (see [11], Theorem (27, 2)). Previously, without using the Fundamental Theorem of Algebra, Stäckel gave for the case $n = \deg(f) \geq 2$ (see [20], Satz 6) the value

$$S = 2 + \frac{An!(n-1)^{(n^2+n+2)/2}}{1! \cdots (n-1)!}.$$

Stäckel's Theorem 7 was improved by G. Pólya and G. Szegö through the following result (see [15], **127**, pp. 137, 350-351, or [17], **127**, pp. 130, 330).

Pólya-Szegö's Theorem. (1925) *Let $f(X)$ be an arbitrary polynomial in $\mathbb{Z}[X]$. Assume that there exists an integer q such that the zeros of $f(X)$ lie in the half* *plane* $\Re(z) < q - \frac{1}{2}$, $f(q-1) \neq 0$ *and* $f(q) = p$, *where* p *is a prime number. Then* $f(X)$ *is irreducible in* $\mathbb{Z}[X]$.

Nine years later O. Ore gave a different generalization of Stäckel's Theorem 7 establishing sufficient conditions for the irreducibility in $\mathbb{Q}[X]$ of the polynomials in $\mathbb{Z}[X]$ that represent multiples of primes (see [14], Satz 5).

Ore's Theorem. (1934) *Let* $f(X)$ *be an arbitrary polynomial in* $\mathbb{Z}[X]$ *of degree* $n \geq 2$. *Assume that there exist a real number* ρ, *an integer* q *and a positive integer* d *such that* $\rho \geq 1$, $d|f(q)$ *and the zeros of* $f(X)$ *are outside of the disk* $|z-q| \leq \rho$. *Suppose also that* $d \leq \rho$ *and* $f(q) = dp$, *where* p *is a prime number. Then* $f(X)$ *is irreducible in* $\mathbb{Q}[X]$.

Remark 3.4. An equivalent result that does not use the real parameter ρ can be established replacing the hypotheses

$$\textit{the zeros of } f(X) \textit{ are outside of the disk } |z-q| \leq \rho, \quad d \leq \rho$$

by a single statement, namely,

$$\textit{the zeros of } f(X) \textit{ are outside of the disk } |z-q| \leq d.$$

In the last paragraph of [14] Ore says that the sufficient conditions of irreducibility of the previous theorem can be extended in distinct directions. This task was accomplished, almost simultaneously, by L. Weisner who established necessary conditions for the reducibility of the polynomials that represent multiples of prime powers (see [21]). In particular, besides establishing the case $q = 0$ of Ore's Theorem, Weisner derives from its first two theorems a result that we state as an irreducibility criterion in the following equivalent way, where $\mathbb{Z}_{(1)}[X] = \mathbb{Z}[X] \setminus \{-1, 1\}$ and $\mathbb{Z}_{(2)}[X]$ stands for the set of polynomials in $\mathbb{Z}_{(1)}[X]$ without rational roots.

Weisner's Theorem 3. (1934) *Let* $v \in \{1, 2\}$ *and let* $f(X)$ *be an arbitrary polynomial in* $\mathbb{Z}_{(v)}[X]$ *of degree* $n \geq 2$ *and leading coefficient* a_n. *Assume that there exist a real number* ρ, *an integer* q *and a positive integer* d *such that* $\rho \geq 1$, $|q| \geq \rho + 1$, $d|f(q)$ *and the zeros of* $f(X)$ *are in the disk* $|z| < \rho$. *Suppose also that*

$$\text{(A) } d \leq (|q| - \rho)^v \quad \textit{or} \quad \text{(B) } p^e \geq |a_n|(|q| + \rho)^{n-v},$$

$$f(q) = \pm\, dp^e \quad \textit{and} \quad p \nmid (f'(q))^{e-1},$$

where p *and* e *are positive integers with* p *prime. Then* $f(X)$ *is irreducible in* $\mathbb{Q}[X]$.

Remark 3.5. (I) Case $v = e = 1$ of part (A) was also established by Ore in the last paragraph of [14] and rediscovered several times in the last forty years; see, for example, [2], Theorem 1 and [9], Theorem 1.

(II) Since $p^e \geq |a|(|q| + \rho)^{n-v}$ and $d \leq |f(q)|/(|a_n|(|q| + \rho)^{n-v})$ are equivalent inequalities, we can rewrite

$$\text{(A)}\, d \leq (|q| - \rho)^v \ \ or \ \ \text{(B)}\, p^e \geq |a_n|(|q| + \rho)^{n-v}$$

as

$$d \leq \max\left\{(|q| - \rho)^v,\ |f(q)|/(|a_n|(|q| + \rho)^{n-v})\right\}.$$

More recently, using a well-known bound for the absolute value of the zeros with positive real part of the polynomials in $\mathbb{R}[X]$ with nonnegative coefficients (say $|z| < \rho_1$; see, for example, [16], (**24**)), M. Filaseta improves a previous result of J. Brillhart, M. Filaseta and A. Odlizko (see [3], Theorem 4) establishing the following theorem (see [7], Theorem 4]).

Filaseta's Theorem 1. (1982) *Let $f(X) = \sum_{j=0}^{m} a_j X^j$ be any polynomial in $\mathbb{Z}_{(2)}[X]$ with $a_m > 0$ and $a_{m-1} \geq 0$. Let $\rho_1 = (1 + \sqrt{4M+1})/2$, where $M = \max_{k=0,1,\ldots,m-2} |a_k/a_m|$. Let q be any integer and let d be any positive integer such that $q \geq \rho_1 + 1$ and $d|f(q)$. Suppose also that $d \leq (q - \rho_1)^2$ and $f(q) = dp$, where p is a prime number. Then $f(X)$ is irreducible in $\mathbb{Q}[X]$.*

Filaseta also established the following result concerning the construction of irreducible polynomials from multiples of primes (see [7], Theorem 2; case $d = 1$ can be found in Corollary 2 of [3]; see also [19]).

Filaseta's Theorem 2. (1982) *Let p, d, q be positive integers, with p prime and $dp \geq q > d$. Let $f(X)$ denote the polynomial in $\mathbb{Z}[X]$ which is obtained replacing q by X in the representation of dp in base q. Then $f(X)$ is irreducible in $\mathbb{Q}[X]$.*

Remark 3.6. Other important results of Filaseta for polynomials in $\mathbb{Z}[X]$ with nonnegative coefficients, also related to the case $f(q) = dp$, can be found in [8]. In particular, the preceding theorem is improved by Theorem 5 which admits much larger coefficients of relatively low order. It should be also noted that to prove this theorem the case $\rho = \sqrt{d}$ of Ore's Theorem (which is proved in Lemma 1) is used.

4. Generalizations of the Pólya-Szegö Theorem

In this section we generalize Pólya-Szegö's Theorem, which according to the best knowledge of the author has not been done yet. We first present a result similar to Ore's Theorem (see Remark 3.4). Indeed, Pólya-Szegö's Theorem is equivalent to the case $d = 1$ of the following result (see Remark 2.3).

Theorem 4.1. *Let $f(X)$ be an arbitrary polynomial in $\mathbb{Z}_{(1)}[X]$. Assume that there exist an integer q and a positive integer d such that $d|f(q)$, $\gcd(d, c(f)) = 1$ and the zeros of $f(X)$ lie in the punctured half plane $\Re(z) < q - \frac{d}{2}$, $z \neq q - d$. Suppose also that $f(q) = dp$, where p is a prime number. Then $f(X)$ is irreducible in $\mathbb{Z}[X]$.*

Remark 4.2. We can not conclude in all cases that $f(X)$ is irreducible in $\mathbb{Z}[X]$ if the fraction $d/2$ is replaced by d/k, where k is any real number greater than 2. In this situation we have, for example, the reducible polynomial

$$f(X) = ((X-q)^{2n+1}+p)(p(X-q)+d),$$

where q, n, p and d are arbitrary positive integers with p prime and $d<p<k$. Indeed, $f(X)$ is primitive and has its zeros in the half plane $\Re(z) \le q - \frac{d}{p} < q - \frac{d}{k}$, also satisfying $f(q-d) = d(p+(-d)^{2n+1})(1-p) \ne 0$ and $f(q)=dp$.

A generalization of the above theorem using Weisner's condition can be established. Indeed, Theorem 4.1 is equivalent to the case $v=e=1$ of the following result.

Theorem 4.3. (Generalized Pólya-Szegö's Theorem (1)) *Let $v \in \{1,2\}$ and let $f(X)$ be an arbitrary polynomial in $\mathbb{Z}_{(v)}[X]$. Assume that there exist an integer q and a positive integer d such that $d|f(q)$, $\gcd(d, c(f)) = 1$ and the zeros of $f(X)$ are in the punctured half plane*

$$\Re(z) < q - \min\left\{\tfrac{d}{2},\ \sqrt[v]{d}\right\},\ z \ne q-d.$$

Suppose also that $f(q) = \pm\, dp^e$ and $p \nmid (f'(q))^{e-1}$, where p and e are positive integers with p prime. Then $f(X)$ is irreducible in $\mathbb{Z}[X]$.

Now, since for any real number ρ,

$$\rho \le q - \min\left\{\frac{d}{2},\ \sqrt[v]{d}\right\} \text{ if and only if } d \le \max\{2(q-\rho),\ (q-\rho)^v\},$$

we state our final generalization of the Pólya-Szegö Theorem rewriting Theorem 4.3 in the following way, which became the model for Theorem 2.2.

Theorem 4.4. (Generalized Pólya-Szegö's Theorem (2)) *Let $v \in \{1,2\}$ and let $f(X)$ be an arbitrary polynomial in $\mathbb{Z}_{(v)}[X]$. Assume that there exist a real number ρ, an integer $q \ge \rho + \frac{1}{2}$ and a positive integer d such that $d|f(q)$ and the zeros of $f(X)$ are in the punctured half plane $\Re(z) < \rho$, $z \ne q-d$. Suppose also that*

$$d \le \max\{2(q-\rho),\ (q-\rho)^v\},\ f(q) = \pm\, dp^e,\ \text{and } p \nmid (f'(q))^{e-1},$$

where p and e are positive integers with p prime. Then

$$f(X) \text{ is irreducible in } \mathbb{Q}[X] \iff f(X) \text{ has positive degree.} \qquad (2^*)$$

Moreover,

$$f(X) \text{ is irreducible in } \mathbb{Z}[X] \iff \gcd(d, c(f)) = 1. \qquad (3^*)$$

Remark 4.5. Because $|z| < \rho$ implies $\Re(z) < \rho$, Weisner's Theorem 3 (A) can also be derived of Theorem 4.4 (the proof of the case $q < 0$ requires to use $f^*(X) = f(-X)$ and $q^* = -q$ instead of $f(X)$ and q, respectively). It is also clear that Filaseta's Theorem 1 is included in the case $\rho = \rho_1$, $e = 1$, $v = 2$ of Theorem 4.4.

A proof of Theorem 4.4 will be given in the next section. Now we use it to build irreducible polynomials from multiples of prime powers. Given any integer $q \geq 2$ we shall say that a nonzero polynomial in $\mathbb{Z}[X]$ is a *q-polynomial* if all its coefficients are in the interval $0 \leq x < q$. The following lemma, which will be proved in Appendix A, gives a very simple upper bound for the real part of the zeros of any q-polynomial.

Lemma 4.6. *For every integer $q \geq 2$, the zeros of any q-polynomial in $\mathbb{Z}[X]$ lie in the half plane $\Re(z) < \sqrt{q}$.*

A proof of the following theorem will be given in Appendix B.

Theorem 4.7. *Let a, d, p and e be arbitrary positive integers, with $a \geq 2$ and p prime, such that $d < q < dp^e$, where $q = pa$. Let $f(X)$ denote the polynomial in $\mathbb{Z}[X]$ which is obtained replacing q by X in the representation of dp^e in base q. Then*

$$f(X) \text{ is irreducible in } \mathbb{Q}[X] \text{ if } q \notin \bigcup_{1 \leq j < dp^{e-1}} \left(\frac{dp^e}{pj+1}, \frac{dp^e}{pj} \right]. \tag{4}$$

In particular,

$$f(X) \text{ is irreducible in } \mathbb{Q}[X] \text{ if } a^2 | (p-1) \text{ and } a \nmid d. \tag{5}$$

Furthermore,

$$\text{if } a^2 | (p-1) \text{ and } \gcd(d, [d/a]q) = 1, \text{ then } f(X) \text{ is irreducible in } \mathbb{Z}[X]. \tag{6}$$

Thus, for example, considering the simplest case of the above theorem, namely, $a = 2$, $p = 5$ and $d = 1$, we get that all polynomials that are obtained from the integer powers of 5, that is, 5, $2X + 5$, $X^2 + 2X + 5$, $6X^2 + 2X + 5, \ldots$, are irreducible in $\mathbb{Z}[X]$.

5. Main Admissible Triples

In this section we present the admissible triples associated with the missing generalization of Ore's Theorem, Weisner's Theorem 3 (according to (II) of Remark 3.5) and Theorem 4.4, our final generalization of the Pólya and Szegö Theorem. By comparing the definitions below with the corresponding statements it can be readily seen that these theorems are immediate consequences of Theorem 2.2 and the following result.

Theorem 5.1. *Let $v \in \{1, 2\}$ and let $f(X)$ be an arbitrary polynomial in $\mathbb{Z}_{(v)}[X]$. Assume that $f(X)$ has degree n and leading coefficient a_n. Then the following triples* *are f-admissible:*

$$(S,\mathcal{Z},\mathcal{F})_v^{\mathrm{Ore}} : \begin{cases} S = \{(\rho,q) \in \mathbb{R}\times\mathbb{Z} : \rho \geq 1\} \\ \mathcal{Z}(\rho,q,d) = \begin{cases} \{z \in \mathbb{C} : |q-z| > \rho\} & \text{if } v=1 \\ \{z \in \mathbb{C}\setminus\mathbb{Q} : |q-z| > \rho\} & \text{if } v=2 \end{cases} \\ \mathcal{F}(f,\rho,q) = \rho^v, \end{cases}$$

$$(S,\mathcal{Z},\mathcal{F})_v^{\mathrm{Weisner}} : \begin{cases} S = \{(\rho,q) \in \mathbb{R}\times\mathbb{Z} : \rho \geq 1, |q| \geq \rho+1\} \\ \mathcal{Z}(\rho,q,d) = \begin{cases} \{z \in \mathbb{C} : |z| < \rho\} & \text{if } v=1 \\ \{z \in \mathbb{C}\setminus\mathbb{Q} : |z| < \rho\} & \text{if } v=2 \end{cases} \\ \mathcal{F}(f,\rho,q) = \max\left\{(|q|-\rho)^v, |f(q)|/(|a_n|(|q|+\rho)^{n-v})\right\}, \end{cases}$$

$$(S,\mathcal{Z},\mathcal{F})_v^{\mathrm{P\&S}} : \begin{cases} S = \left\{(\rho,q) \in \mathbb{R}\times\mathbb{Z} : q \geq \rho+\frac{1}{2}\right\} \\ \mathcal{Z}(\rho,q,d) = \begin{cases} \{z \in \mathbb{C}\setminus\{q-d\} : \Re(z)<\rho\} & \text{if } v=1,\ d \leq 2(q-\rho) \\ \{z \in \mathbb{C}\setminus\mathbb{Z} : \Re(z) < \rho\} & \text{if } v=1,\ d > 2(q-\rho) \\ \{z \in \mathbb{C}\setminus\mathbb{Q} : \Re(z) < \rho\} & \text{if } v=2 \end{cases} \\ \mathcal{F}(f,\rho,q) = \max\{2(q-\rho),(q-\rho)^v\}. \end{cases}$$

Our proof that the triple $(S,\mathcal{Z},\mathcal{F})_v^{\mathrm{P\&S}}$ is f-admissible depends on the following lemma (which constitutes the core of the original proof of Pólya-Szegö's Theorem; see [17], **127**, p. 330).

Lemma 5.2. *Let $G(X)$ be an arbitrary nonzero polynomial in $\mathbb{Z}[X]$. Let $\rho \in \mathbb{R}$ such that the zeros of $G(X)$ are in the half plane $\Re(z) < \rho$ and let $q \in \mathbb{Z}$, with $q \geq \rho + \frac{1}{2}$, such that $G(q-1) \neq 0$ and $G(q) = \pm 1$. Then $G(X) = \pm 1$.*

Proof. Clearly we only need to prove that $G(X)$ is a constant polynomial. On the contrary suppose that $G(X)$ has positive degree. It easily follows from our hypotheses that the zeros of $G(X+q-\frac{1}{2})$ lie in the half plane $\Re(z) < 0$, so all its significant coefficients have the same sign. Therefore $|G(-x+q-\frac{1}{2})| < |G(x+q-\frac{1}{2})|$ for any positive real number x. Hence, letting $x = 1/2$, we get the contradiction $1 \leq |G(q-1)| < |G(q)| = 1$. □

We can now give a more simple proof of Theorem 5.1.

Proof. Note first that the triples previously defined satisfy the basic specifications of Definition 2.1. Let $(\rho,q,d) \in S\times\mathbb{N}$ with $d|f(q)$ such that the zeros of $f(X)$ belong to $\mathcal{Z}(\rho,q,d)$, and let $g(X)$ be an arbitrary polynomial of positive degree in $\mathbb{Z}[X]$ that divides $f(X)$. Assume that $g(X)$ has degree m (so that $m \geq v$), leading coefficient c_m, and zeros $z_1,\ldots,z_m$. To prove that $g(X)$ satisfies (Condition $\mathcal{A}$) we will show that in each case at least one of the two statements $|g(q)| > \mathcal{F}(f,\rho,q)$ and $g(q) \nmid d$ holds.

In the first place we will prove that $|g(q)| > \mathcal{F}(f,\rho,q)$ for the first two triples. For $(S,\mathcal{Z},\mathcal{F})_v^{\mathrm{Ore}}$ we have,

$$|g(q)| = |c_m| \prod_{j=1}^{m} |q - z_j| \geq \prod_{j=1}^{m} |q - z_j| > \rho^m \geq \rho^v.$$

Considering $(S, \mathcal{Z}, \mathcal{F})_v^{\text{Weisner}}$, in the case $(|q| - \rho)^v > |f(q)|/(a|(|q| + \rho)^{n-v})$ we have,

$$|g(q)| = |c_m| \prod_{j=1}^{m} |q - z_j| \geq \prod_{j=1}^{m} ||q| - |z_j|| > (|q| - \rho)^m \geq (|q| - \rho)^v.$$

For the remaining case, assuming that $h(X) := f(X)/g(X)$ has leading coefficient b and roots $z_{m+1}, \ldots, z_n$ we have,

$$\frac{|f(q)|}{|a_n|(|q|+\rho)^{n-v}} \leq \frac{|g(q)||h(q)|}{|b|(|q|+\rho)^{n-v}} = \frac{|g(q)|\prod_{j=m+1}^{n}|q - z_j|}{(|q|+\rho)^{n-v}} < \frac{|g(q)|}{(|q|+\rho)^{m-v}} \leq |g(q)|.$$

Then it only remains to consider the triple $(S, \mathcal{Z}, \mathcal{F})_v^{\text{P\&S}}$. In the first place we prove that $g(q) \nmid d$ if $d/2 \leq q - \rho$.

On the contrary, suppose $d/2 \leq q - \rho$ and $g(q)|d$. Let $G(X) = g(dX + q)/|g(q)|$. From $g(q)|d$ we get $G(X) \in \mathbb{Z}[X]$. Clearly $G(0) = \pm 1$. Note also that for any complex root z of $G(X)$ we have $d\Re(z) + q < \rho$, so $\Re(z) < -(q - \rho)/d \leq -1/2$. On the other hand, $f(q - d) \neq 0$ implies $G(-1) = g(q - d)/|g(q)| \neq 0$. Therefore, since Lemma 5.2 applies to $G(X)$ with $\rho = -1/2$ and $q = 0$, we have $G(X) = \pm 1$ which means that $g(X)$ is a constant polynomial, a contradiction.

Now assume $d/2 > q - \rho$. We shall prove that $|g(q)| > \max\{2(q - \rho), (q - \rho)^v\}$. Notice that

$$\max\{2(q - \rho), (q - \rho)^v\} = \begin{cases} (q - \rho)^2 & \text{if } v = 2 \text{ and } q - \rho > 2 \\ (q - \rho)^2 = 2(q - \rho) & \text{if } v = 2 \text{ and } q - \rho = 2 \\ 2(q - \rho) & \text{if } v = 1 \text{ or } q - \rho < 2. \end{cases}$$

In the first place we suppose $m = 1$, so $v = 1$. In this situation we have $|c_1| \geq 2$, because otherwise $g(X)$ has an integer root against the definition of $\mathcal{Z}(\rho, q, d)$. Therefore, $|g(q)| = |c_1||q - z_1| > 2|q - \rho| = 2(q - \rho) = \max\{2(q - \rho), (q - \rho)^v\}$.

We assume then that $m \geq 2$. Note that for any real $t \geq \rho$ all the significant coefficients of $g(X + t)$ have the same sign. Hence, letting $t = \rho$, $X = q - t$ we easily get,

$$|g(q)| > |c_m|(q - \rho)^m \geq (q - \rho)^m \geq \begin{cases} (q - \rho)^2 & \text{if } v = 2 \text{ and } q - \rho > 2 \\ (q - \rho)^2 = 2(q - \rho) & \text{if } q - \rho = 2 \\ 2(q - \rho) & \text{if } v = 1 \text{ and } q - \rho > 2, \end{cases}$$

which proves that $|g(q)| > \max\{2(q - \rho), (q - \rho)^v\}$ in the case $q - \rho \geq 2$.

Thus it only remains to show that $|g(q)| > 2(q-\rho)$ when $q-\rho < 2$. As $q-\rho \geq 1/2$ and

$$\left[\tfrac{1}{2}, 2\right) = \left[\tfrac{1}{2}, 1\right) \cup \left[1, \tfrac{3}{2}\right) \cup \left[\tfrac{3}{2}, 2\right),$$

it will be sufficient to prove that for $k = 1, 2, 3$,

$$q - \rho \in [k/2, (k+1)/2) \text{ implies } |g(q)| > k.$$

This holds for $k=1$ because otherwise Lemma 5.2 applies to $G(X)=g(X)$ giving the contradiction $g(X)=\pm 1$. Then suppose $k\in\{2,3\}$. Notice that $g^{(j)}(q-1)/j!$ and $2^{m-j}g^{(j)}\left(q-\frac{3}{2}\right)/j!$ are integer numbers for $j=0,1,\ldots,m$. Consequently,

$$
\begin{aligned}
|g(q)| &= \sum_{j=0}^{m}\left|\frac{g^{(j)}\left(q-\frac{k}{2}\right)}{j!}\right|\left(\frac{k}{2}\right)^{j} \\
&= \begin{cases} \sum_{j=0}^{m}\left|\dfrac{g^{(j)}(q-1)}{j!}\right| & \text{if } k=2 \\ \dfrac{1}{2^m}\sum_{j=0}^{m}\left|\dfrac{2^{m-j}g^{(j)}\left(q-\frac{3}{2}\right)}{j!}\right| 3^j & \text{if } k=3 \end{cases} \\
&\geq \begin{cases} \sum_{j=0}^{m} 1 = m+1 > 2 & \text{if } k=2 \\ \dfrac{1}{2^m}\sum_{j=0}^{m} 3^j = \left(\dfrac{3}{2}\right)^{m+1} - \dfrac{1}{2^{m+1}} \geq \left(\dfrac{3}{2}\right)^{3} - \dfrac{1}{8} > 3 & \text{if } k=3, \end{cases}
\end{aligned}
$$

as we wanted to show. □

6. On the Number of Irreducible Factors of the Polynomials in $\mathbb{Z}[X]$

In this section we will extend Theorem 2.2 via a theorem that for any polynomial $f(X)\in\mathbb{Z}[X]$ of positive degree and any "f-admissible triple" (ρ,q,d) yields an integer upper bound for N_f, the number of irreducible factors of $f(X)$ in $\mathbb{Q}[X]$ (multiplicities counted). Better bounds for the number of distinct irreducible factors of $f(X)$ in $\mathbb{Q}[X]$ will be obtained under certain additional hypotheses which guarantee that $f(X)$ is *square free*, which means that there does not exist a polynomial $k(X)\in\mathbb{Z}[X]$ of positive degree such that $k^2(X)|f(X)$.

Remark 6.1. In general, since $f(X)$ is square free if and only if $f(X)$ and $f'(X)$ have no common divisors of positive degree in $\mathbb{Z}[X]$, to obtain better estimates of the number of distinct irreducible factors of $f(X)$ in $\mathbb{Q}[X]$ we should replace $f(X)$ by its *square free part*, say $f_{\mathrm{sqf}}(X)$, defined by

$$
f_{\mathrm{sqf}}(X)=\frac{f(X)}{\gcd(f(X),f'(X))},
$$

where gcd stands for the greatest common divisor in $\mathbb{Z}[X]$ (an algorithm to compute it can be found in [13]). Such denomination is justified by the fact that $f_{\mathrm{sqf}}(X)$ is the product of the different irreducible factors of $f(X)$ in $\mathbb{Z}[X]$ of positive degree, so that $f_{\mathrm{sqf}}(X)$ is also primitive. Furthermore, in order to also diminish the computational cost of factoring large values of $|f(q)|$, it is convenient to work separately with the nonconstant components $P_1(X),\ldots,P_n(X)$ of the so-called *square free factorization* of $f(X)$,

$$f(X) = P_1(X)P_2^2(X)\cdots P_n^n(X),\ n = \deg(f),$$

which are square free and pairwise coprime polynomials in $\mathbb{Z}[X]$ (so $f_{\text{sqf}}(X) = P_1(X)P_2(X)\cdots P_n(X)$). This can be done using, for example, the well-known Tobey-Horowitz algorithm, which is described together with a new procedure in [10].

Some definitions are needed to establish the bounds for N_f associated to a suitable triple (ρ, q, d). First we define two closely related counting functions which will be used to estimate the maximum number of irreducible factors of positive degree that can have a divisor of $f(X)$ in $\mathbb{Z}[X]$, say $g(X)$, satisfying $g(q)|d$.

Definition 6.2. Let d be a positive integer and let x be a real number, $x \geq 1$. Let $\Delta_x(d) = \Delta_x^*(d) = 0$ if $d \leq x$. Otherwise,

(a) $\Delta_x(d)$ denotes the largest positive integer k for which we can write $d = d_1 \cdots d_k$ with $d_j > x$ for $j = 1, \ldots, k$;

(b) $\Delta_x^*(d)$ denotes the largest positive integer k for which we can write $d = d_1 \cdots d_k$ with $d_1, \ldots, d_k$ pairwise coprime and $d_j > x$ for $j = 1, \ldots, k$.

Remark 6.3. Given the factorization $1 < d = p_1^{e_1} \cdots\ p_t^{e_t}$, with $p_1^{e} < \cdots < p_t^{e_t}$, it is clear that $\Delta_x^*(d) \leq t$ and $\Delta_x(d) \leq e_1 + \cdots + e_t$. It can also be shown without difficulty that $\Delta_x(d) < \log_x d$ if $x > 1$. Hence, for any $f(X) \in \mathbb{Z}_{(v)}[X]$ with $v \in \{1, 2\}$ and any of the f-admissible triples considered in Theorem 5.1, from the case $d = |f(q)|$, $x = \mathcal{F}(f, \rho, q)$ it can be readily deduced that for any q sufficiently large, $\Delta_{\mathcal{F}(f,\rho,q)}(|f(q)|) \leq \deg(f)/v$.

Next we present a quantitative version of Weisner's condition $p \nmid (f'(q))^{e-1}$, which will be used to estimate the maximum number of irreducible factors of positive degree that can have a divisor of $f(X)$ in $\mathbb{Z}[X]$, say $h(X)$, satisfying $h(q)|(f(q)/d)$.

Let $\Pi(a)$ denote the set of positive prime divisors of an arbitrary nonzero integer a. For each $p \in \Pi(a)$ let $e_p = e_p(a)$ be the largest positive integer k satisfying $p^k|a$. As usual, a is called *square free* if $e_p = 1$ for each $p \in \Pi(a)$. The *square free part of* a, say sqf(a), is defined as the product of the primes $p \in \Pi(a)$ with $e_p = 1$.

Definition 6.4. Let $f(X)$ be an arbitrary primitive polynomial in $\mathbb{Z}[X]$, and let q, d be any integers such that $f(q) \neq 0$ and $d|f(q)$. For each $p \in \Pi(f(q)/d)$ we define:

$$r_p(f, q) = \min\left\{ k \in \mathbb{N} : p \nmid \left(\frac{f^{(k)}(q)}{k!}\right)^{e_p - 1} \right\}.$$

It should be noted that $r_p(f, q)$ is a well defined integer belonging to the set $\{1, \ldots, \deg(f)\}$. Indeed, $r_p(f, q) = 1$ if $e_p = 1$, and for $e_p \geq 2$ the supposition $p|f^{(k)}(q)/k!$ for $k = 1, \ldots, n - \deg(f)$ implies $p|f(X)$, via

$$f(X) = f(q) + f'(q)(X - q) + \cdots + \frac{f^{(n)}(q)}{n!}(X - q)^n,$$

which contradicts that $f(X)$ is primitive.

Now, to shorten the proof of the main theorem of this section, we establish a technical lemma which actually constitutes the part of such theorem that does not depend on admissibility.

Lemma 6.5. *Let $h(X)$ be a primitive polynomial in $\mathbb{Z}[X]$ and let q be any integer with $h(q) \neq 0$. Suppose that d_h is a positive divisor of $h(q)$ such that there is no polynomial $k(X)$ in $\mathbb{Z}[X]$ of positive degree satisfying $k(X)|h(X)$ and $k(q)|d_h$. Then*

$$N_h \leq \sum_{p\in\Pi(h(q)/d_h)} r_p(h,q). \tag{7}$$

Furthermore,

$$\sum_{p\in\Pi(h(q)/d_h)} r_p(h,q) = \#(\Pi(h(q)/d_h)) \iff \gcd(h(q)/d_h, h'(q))|\mathrm{sqf}(h(q)/d_h). \tag{8}$$

Proof. Let $\Pi = \Pi(h(q)/d_h)$ and, for each $p \in \Pi$, let $e_p = e_p(h(q)/d_h)$, $r_p = r_p(h,q)$.

Proof of (7). In case $|h(q)| = 1$ we have $N_h = 0 = \#(\Pi) = \sum_{p\in\Pi} r_p$. Now assume $|h(q)| \neq 1$. Thus $h(X)$ has positive degree and $t := \#(\Pi) \geq 1$. Let $\Pi = \{p_1, \ldots, p_t\}$, $e_j = e_{p_j}$ and $r_j = r_{p_j}$ for $j = 1, \ldots, t$. Thus we can write $|h(q)| = d_h p_1^{e_1} \ldots p_t^{e_t}$ and $h(X) = h_1(X) \cdots h_{N_h}(X)$, where each $h_j(X)$ has positive degree and is irreducible in $\mathbb{Z}[X]$. Because $h_j(q) \nmid d_h$ for $j = 1, \ldots, N_h$ we can assume $|h_j(q)| = d_j p_j^*$, where the d_j's are positive integers with $d_1 \cdots d_{N_h} = d_h$ and the p_j^*'s are integers greater than 1 with $p_1^* \cdots p_{N_h}^* = p_1^{e_1} \ldots p_t^{e_t}$.

We now define two polynomial sequences in $\mathbb{Z}[X]$, say $P_0(X), \ldots, P_t(X)$ and $H_1(X), \ldots, H_t(X)$. The $P_j(X)$'s are recursively defined starting with $P_0(X) = 1$. For $1 \leq j \leq t$ the polynomial $P_j(X)$ is defined as the product of all polynomials $h_k(X)$ with $k \in \{1, \ldots, N_h\}$ that are relatively prime to $P_0(X) \cdots P_{j-1}(X)$ and satisfy $p_j | p_k^*$. Notice that $h(X) = P_1(X) \cdots P_t(X)$. On the other hand we define $H_j(X)$ for $j \in \{1, \ldots, t\}$ as the product of all polynomials $h_k(X)$ with $k \in \{1, \ldots, N_h\}$ that satisfy $p_j | p_k^*$. It is clear that each $P_j(X)$ divides $H_j(X)$. Therefore, letting $N_j = N_{H_j}$ it will be enough to prove that $N_j \leq r_j$, $j = 1, \ldots, t$.

Let $j \in \{1, \ldots, t\}$. We can write $|H_j(q)| = \delta p_j^{e_j}$ for some $\delta \mid d_h \prod_{\substack{1\leq i\leq t \\ i\neq j}} p_i^{e_i}$. From $|h(q)|/d_h = p_1^{e_1} \ldots p_t^{e_t}$ and the definition of $H_j(X)$ we easily get $N_j \leq e_j$. Hence, to prove $N_j \leq r_j$, we can suppose $r_j < e_j$. Assume $H_j(X) = Q_1(X) \cdots Q_{N_j}(X)$, where each $Q_k(X)$ is an irreducible polynomial in $\mathbb{Z}[X]$. By definition of H_j we can write $|Q_k(q)|$ in the form $|Q_k(q)| = \delta_k p_j^{\epsilon_k}$, where ϵ_k, δ_k are positive integers with $e_j = \epsilon_1 + \cdots + \epsilon_{N_j}$ and $\delta = \delta_1 \cdots \delta_{N_j}$. Let $G_j(X) = h(X)/H_j(X)$ and let $r_{jk} = r_{p_j}(Q_k, q)$ for $k = 1, \ldots, N_j$. Hence, $h(X) = G_j(X)Q_1(X) \cdots Q_{N_j}(X)$.

From Leibniz's Formula for the successive derivatives of a product of polynomials we obtain

$$\frac{h^{(k)}(q)}{k!} = \sum_{\substack{k_0+k_1+\cdots+k_m=k \\ \text{each } k_i \geq 0}} \frac{G_j^{(k_0)}(q)}{k_0!} \frac{Q_1^{(k_1)}(q)}{k_1!} \cdots \frac{Q_{N_j}^{(k_m)}(q)}{k_m!}, \quad k = 1, 2, \ldots .$$

In the case $k < r_{j1} + \cdots + r_{jN_j}$, from $e_j > 1$ and Definition 6.4 it easily follows that for every term of the sum on the right there exists $i \in \{1, \ldots, N_j\}$ such that $k_i = 0$. In other words,

$$k < r_{j1} + \cdots + r_{jN_j} \implies p_j \left| \left(\frac{h^{(k)}(q)}{k!} \right)^{e_j - 1} \right.,$$

from which follows immediately $r_j \geq r_{j1} + \cdots + r_{jN_j} \geq N_j$.

Proof of (8). From Definition 6.4 it can be easily deduced that for each $p \in \Pi$,

$$r_p = 1 \iff p \nmid (h'(q)^{e_p - 1},$$
$$p \nmid (h'(q))^{e_p - 1} \iff \gcd(p^{e_p}, h'(q)) | \mathrm{sqf}(p^{e_p}).$$

Now, since each $r_p \geq 1$, the equivalence follows immediately from the basic equality $\gcd(h(q)/d_h, h'(q)) = \prod_{p \in \Pi} \gcd(p^{e_p}, h'(q))$. □

The following theorem contains our main result about N_f and some of its consequences. For the sake of simplicity and without lost of generality we will limit ourselves to consider primitive polynomials (the general case requires the use of the so-called *primitive part of* $f(X)$, that is, of the primitive polynomial $f_{\mathrm{pr}}(X) := f(X)/\mathrm{c}(f)$).

Theorem 6.6. *Let $f(X)$ be an arbitrary primitive polynomial in $\mathbb{Z}[X]$ of positive degree, and let $(S, \mathcal{Z}, \mathcal{F})$ be any f-admissible triple. Let $(\rho, q, d) \in S \times \mathbb{N}$ such that $d | f(q)$, $q \notin \mathcal{Z}(\rho, q, d)$ and the zeros of $f(X)$ belong to $\mathcal{Z}(\rho, q, d)$. We have*

$$N_f \leq \Delta_{\mathcal{F}(f,\rho,q)}(d) + \sum_{p \in \Pi(f(q)/d)} r_p(f, q). \tag{9}$$

Suppose also that $\gcd(f(q)/d, f'(q)) | \mathrm{sqf}(f(q)/d)$. *Then*

$$N_f \leq \Delta_{\mathcal{F}(f,\rho,q)}(d) + \#(\Pi(f(q)/d)). \tag{10}$$

Moreover,

if $\gcd(f(q), f'(q)) | \mathrm{sqf}(f(q))$, *then* (11)

$$f(X) \text{ is square free and } N_f \leq \Delta^*_{\mathcal{F}(f,\rho,q)}(d) + \#(\Pi(f(q)/d));$$

if $d \leq \mathcal{F}(f, \rho, q)$ and $f(q)/d$ is squarefree, then (12)

$$f(X) \text{ is square free and } N_f \leq \#(\Pi(f(q)/d)).$$

Proof. Let $\Pi=\Pi(f(q)/d)$ and let $e_p=e_p(f(q)/d)$, $r_p=r_p(f,q)$ for each $p\in\Pi$. Let $g(X)$ be a divisor of $f(X)$ in $\mathbb{Z}[X]$ with the highest possible degree satisfying $g(q)|d$.

Proof of (9). Letting $h(X)=f(X)/g(X)$ and $d_h=d/|g(q)|$ we get at once that $h(X)$ is primitive, $|h(q)|/d_h=|f(q)|/d$ and y $N_f=N_g+N_h$. Therefore, it will be sufficient to prove

$$\text{(i) } N_g\le\Delta_{\mathcal{F}(f,\rho,q)}(d) \text{ and (ii) } N_h\le\sum_{p\in\Pi}r_p.$$

Proof of (i). In case that $g(X)=\pm1$ we have $N_g=0\le\Delta_{\mathcal{F}(f,\rho,q)}(d)$. Assume that $g(X)$ has positive degree. Therefore we can write $g(X)=g_1(X)\cdots g_{N_g}(X)$, where $N_g\ge1$ and each $g_j(X)$ is a polynomial in $\mathbb{Z}[X]$ of positive degree that is irreducible in $\mathbb{Z}[X]$. From $g_j(X)|g(X)$ and $g(q)|d$ we obtain $g_j(X)|f(X)$ and $g_j(q)|d$ for $j=1,\dots,N_g$. Hence, since $g(q)|d$, (Condition $\mathcal{A}$) yields

$$|g_j(q)|>\mathcal{F}(f,\rho,q),\ j=1,\dots,N_g,$$

Consequently, as $d=|g_1(q)|\cdots|g_{N_g}(q)|d_h$, (i) follows via Definition 6.2 (a).

Proof of (ii). The given definition of $g(X)$ guarantees that there is no polynomial $k(X)$ in $\mathbb{Z}[X]$ of positive degree with $k(X)|h(X)$ and $k(q)|d_h$. Therefore (ii) follows directly from (7) of Lemma 6.5.

Proof of (10). Considering the case $h(X)=f(X)$, $d_h=d$ of Lemma 6.5, we get from (8) that $\gcd(f(q)/d,f'(q))|\mathrm{sqf}(f(q)/d)$ and $\sum_{p\in\Pi}r_p=\#(\Pi)$ are equivalent statements. Now (10) is an immediate consequence of (9).

Proof of (11). Assume $\gcd(f(q),f'(q))|\mathrm{sqf}(f(q))$. Hence it readily follows

$$\gcd(d,f'(q))|\mathrm{sqf}(d) \text{ and } \gcd(f(q)/d,f'(q))|\mathrm{sqf}(f(q)/d).$$

Therefore, taking into account what has already been proved, it will be sufficient to show that

(iii) $f(X)$ is square free and (iv) $\gcd(d,f'(q))|\mathrm{sqf}(d)$ implies $N_g\le\Delta^*_{\mathcal{F}(f,\rho,q)}(d)$.

Proof of (iii). Suppose that $k_1(X)\in\mathbb{Z}[X]$ satisfies $k_1^2(X)|f(X)$, say $f(X)=k_1^2(X)k_2(X)$ with $k_2(X)\in\mathbb{Z}[X]$. Note first that $|k_1(q)|=1$. Indeed, otherwise there is a prime p dividing $k_1(q)$ and $f'(q)=2k_1'(q)k_1(q)k_2(q)+k_1^2(q)k_2'(q)$, which together with $p^2|f(q)$ contradicts our assumption. From the proof of (ii) above it follows directly that $k_1(X)$ is relatively prime to $h(X)$, so that $k_1(X)|g(X)$. Therefore $k_1(X)$ is a constant polynomial, because otherwise, since $(S,\mathcal{Z},\mathcal{F})$ is f-admissible we would have $1=|k(q)|>\mathcal{F}(f,\rho,q)$ against the definition of $\mathcal{F}$.

Proof of (iv). Suppose that $k_1(X),k_2(X)$ are polynomials in $\mathbb{Z}[X]$ of positive degree with $g(X)=k_1(X)k_2(X)$. From Definition 6.2 (b) and the above proof of (i) we only need to show that $k_1(q)$ and $k_2(q))$ are coprime integers. On the contrary, suppose that a prime p divides $\gcd(k_1(q),k_2(q))$. From $f(X)=g(X)h(X)$ it follows that

$$f'(q)=k_1'(q)k_2(q)h(q)+k_1(q)k_2'(q)h(q)+k_1(q)k_2(q)h'(q),$$

whence we get $p|f'(q)$. Therefore $p|\gcd(d, f'(q))$, which together with $p^2|d$ contradicts $\gcd(d, f'(q))|\text{sqf}(d)$.

Proof of (12). Assume $d \leq \mathcal{F}(f,\rho,q)$ and that $f(q)/d$ is square free. These assumptions guarantee $\Delta_{\mathcal{F}(f,\rho,q)}(d)=0$ and $\gcd(f(q)/d, f'|\text{sqf}(f(q)/d)$ respectively, so (10) yields $N_f \leq \#(\Pi)$. In order to prove that $f(X)$ is square free suppose that $k(X)$ is a polynomial in $\mathbb{Z}[X]$ satisfying $k^2(X)|f(X)$. Certainly we can write $k(q) = ab$, with $a|d$ and $b|(f(q)/d)$. Since $k^2(q) = a^2b^2$ and $f(q)/d$ is square free, we have $ab|d$, that is, $k(q)|d$. Therefore $k(X)$ is a constant polynomial, because otherwise, as $(S,\mathcal{Z},\mathcal{F})$ is f-admissible we would have $|k(q)| > \mathcal{F}(f,\rho,q)$ against $d \leq \mathcal{F}(f,\rho,q)$. □

At this point it can be readily seen that the case $f(X)$ primitive of positive degree of Theorem 2.2 is equivalent to the case $\Delta_{\mathcal{F}(f,\rho,q)}(d) = 0$, $\#(\Pi(f(q)/d) = 1$ of (10). Furthermore, replacing here the hypothesis $\gcd(f(q)/d, f'(q))|\text{sqf}(f(q)/d)$ by $\delta(|f(q)|, |f'(q)|)|d$ (see Definition 2.3) we get a direct generalization of Corollary 2.4. Consequently, the case $d = |f(q)|$ of Theorem 6.6 yields (see Remark 6.3)

$$(9^*)\, N_f \leq \Delta_{\mathcal{F}(f,\rho,q)}(|f(q)|) \text{ and } (11^*)\, N_f \leq \Delta^*_{\mathcal{F}(f,\rho,q)}(|f(q)|),$$

which in general improve the estimations for N_f established in (9) and (11). It is also clear that (12) generalizes the case $e = 1$ of Theorem 2.2. As an application we will prove the following extension of Filaseta's Theorem 2.

Corollary 6.7. *Let t be a positive integer and let $p_1, \ldots, p_t$ be distinct prime numbers. Let q, a be positive integers with $ap_1 \cdots p_t \geq q > a$. Let $f(X)$ denote the polynomial in $\mathbb{Z}[X]$ which is obtained replacing q by X in the representation of $ap_1 \cdots p_t$ in base q. Then*

$$f(X) \text{ is squarefree and } N_f \leq t. \tag{13}$$

Proof. Let $d = a/\gcd(a, \text{c}(f))$. As $f(X)$ has nonnegative coefficients the hypothesis $ap_1 \cdots p_t \geq q > a \geq d$ ensures that $f(X)$ has positive degree and $f(q-d) \neq 0$. Hence, from Lemma 4.6, we can use in Theorem 6.6 the triple $(S,\mathcal{Z},\mathcal{F})^{\text{P\&S}}_{v=1}$ with $\rho = \sqrt{q}$. Note also that $q - 1 < 2(q - \sqrt{q})$, so that $d \leq a \leq q - 1 < 2(q - \sqrt{q}) = \mathcal{F}(f,\rho,q)$.

At this point it should be noticed that the aforementioned properties of $f(X)$ are also satisfied by $f_{\text{pr}}(X) = f(X)/\text{c}(f)$, the primitive part of $f(X)$. Furthermore, since $c = \text{c}(f)/\gcd(a, \text{c}(f))$ is relatively prime to a and therefore to d, we have that c divides $p_1 \cdots p_t$ from which it follows $f_{\text{pr}}(q) = ap_1 \cdots p_t/\text{c}(f) = d(p_1 \cdots p_t/c)$, which ensures that $f_{\text{pr}}(q)/d$ is square free. Thus, since all the conditions required in Theorem 6.6 to prove (12) are satisfied with $f_{\text{pr}}(X)$ instead of $f(X)$, and the equality above also implies $\#(\Pi(f_{\text{pr}}(q))/d) \leq t$, we get that $f_{\text{pr}}(X)$ is square free and $N_{f_{\text{pr}}} \leq t$. Hence, as $N_f = N_{f_{\text{pr}}}$ and $f(X)$ is square free if and only if $f_{\text{pr}}(X)$ is, the proof of (13) is complete. □

References

[1] P. T. Bateman and R. Horn, *A heuristic asymptotic formula concerning the distribution of prime numbers*, Math Comp. **16** (1962), 363-367.

[2] J. Brillhart, *Note on irreducibility testing*, Math. Comp. **35**, (1980) 1379–1381.

[3] J. Brillhart, M. Filaseta M. and A. Odlizko, *On an irreducibility theorem of A. Cohn*, Canadian J. Math. **5** (1981), 1055–1059.

[4] Y. Chen, G. Kun, G. Pete and I. Z. Ruzsa, *Prime values of reducible polynomials, II*, Acta Arithmetica **104** (2002), no. 2, 117–127.

[5] L. E. Dickson, History of the Theory of Numbers, Vol 1, Chelsea, 1971, 332–334.

[6] H. L. Dorwarth, *Irreducibility of polynomials*, Amer. Math. Monthly **42** (1935), no. 6, 369–381.

[7] M. Filaseta, *A Further generalization of an irreducibility theorem of A. Cohn*, Canadian J. Math. **6**, (1982), 1390–1395.

[8] M. Filaseta, *Irreducibility criteria for polynomials with non-negative coefficients*, Canadian J. Math. **XV**, (1988), 339–351.

[9] K. Girstmair, *On an irreducibility criterion of M. Ram Murty*, Amer. Math. Monthly **112** (2005), 269–270.

[10] N. H. Guersenzvaig and F. Szechtman, *Roots multiplicity and square free factorization of polynomials using companion matrices*, Linear algebra and its Applications **436** (2012), 3160–3164.

[11] M. Marden, The Geometry of the Zeros of a Polynomial in a Complex Variable, AMS, 1949, p. 96.

[12] K. S. McCurley, *Prime values of polynomials and irreducibility testing*, Bull. Amer. Math. Society **11** (1984), 155–158.

[13] M. Mignotte and D. Stefanescu, Polynomials. An Algorithmic approach, Springer (1999), 22–23, 58–63.

[14] O. Ore, *Einige Bemerkungen über Irreduzibilität*, Jahresbericht der Deutschen Mathematiker-Vereinigung **44** (1934), 147–151.

[15] G. Pólya und G. Szegö, Aufgaben und Lehrsätze aus der Analysis II, Springer-Verlag, 1925, 137, 350–351.

[16] G. Pólya and G. Szegö, Problems and Theorems in Analysis I, Springer-Verlag, 1976, 107, 301.

[17] G. Pólya and G. Szegö, Problems and Theorems in Analysis II, Springer-Verlag, 1976, 130, 133, 330.

[18] V. V. Prasolov, Polynomials, Springer, 2004, pp. 47–74.

[19] M. Ram Murty, *Prime numbers and irreducible polynomials*, Amer. Math. Monthly **109** (2002), 452-458.

[20] P. Stäckel, *Arithmetischen Eigenschaften ganzer Funktionen*, Journal für Mathemathik **148** (1918), 101–112.

[21] L. Weisner, *Criteria for the irreducibility of polynomials*, Bulletin of the Amer. Math. Society **40** (1934), 864–870.

Appendix A

Here we prove Lemma 4.6. A straightforward calculation shows that $\sqrt{3}(1+\sqrt{4m+1})/4 \le \sqrt{m+1}$ for any positive integer m (equality holds if and only if $m=2$). Thus Lemma 4.6 is an immediate consequence of the case $M \le q-1$ of the result below.

Lemma 4.6*. *Let $f(X)=\sum_{k=0}^{n} a_k X^k \in \mathbb{C}[X]$, $\Re(a_n)\ge 1$ and $\Re(a_{n-k})\ge 0$ for $k=1,2,3$. Let $M=\max_{k=0,1,\dots,n-2}|a_k/\Re(a_n)|$. Then the zeros of $f(X)$ lie in the half plane*

$$\Re(z) < \frac{\sqrt{3}(1+\sqrt{4M+1})}{4}.$$

Proof. Let w be any complex number with $|w|>1$. For $k=1,\dots,n$ we have,

$$\begin{aligned}\left|\frac{f(w)}{w^n}\right| &\ge \left|a_n+\frac{a_{n-1}}{w}+\cdots+\frac{a_{n-k}}{w^k}\right| - \frac{|a_{n-k-1}|}{|w|^{k+1}} - \cdots - \frac{|a_0|}{|w|^n}\\ &> \Re\left(a_n+\frac{a_{n-1}}{w}+\cdots+\frac{a_{n-k}}{w^k}\right) - \frac{M\Re(a_n)}{|w|^{k+1}-|w|^k}. \end{aligned} \tag{14}$$

Now we assume $\Re(w)>0$. Certainly

$$\Re\left(\frac{1}{w}\right) = \frac{\Re(w)}{|w|} > 0. \tag{15}$$

Hence, letting $k=1$ in (14), from $\Re(a_n)\ge 1$ and $\Re(a_{n-1})\ge 0$ we get,

$$\begin{aligned}\left|\frac{|w|^2-|w|}{w^n}\right| |f(w)| &> |w|^2-|w|-M)\\ &= \left(|w|-\frac{1+\sqrt{4M+1}}{2}\right)\left(|w|-\frac{1-\sqrt{4M+1}}{2}\right).\end{aligned}$$

In this way we arrive to a well-known fact, namely, the roots of $f(X)$ with positive real part are in the disk

$$|w| < \frac{1+\sqrt{4M+1}}{2}. \tag{16}$$

Now assume $\Re(w)\ge B(M)$, where $B(M)=\sqrt{3}(1+\sqrt{4M+1})/4$. From (16) we obtain

$$\cos(\arg(w)) = \frac{\Re(w)}{|w|} > \frac{B(M)}{\frac{1+\sqrt{4M+1}}{2}} = \frac{\sqrt{3}}{2}.$$

Consequently, $\arg(w)<\pi/6$, whence $\Re(1/w^k)=\cos(k\arg(w))/|w|^k>0$ for $k=2,3$. Thus, letting $k=3$ in (14), from (15), $\Re(a_{n-2})\ge 0$ and $\Re(a_{n-3})\ge 0$ we get

$$\left|\frac{|w|^4-|w|^3}{w^n}\right| |f(w)| > |w|^4-|w|^3-M.$$

The function h defined over $\mathbb{R}$ by $h(x) = x^4 - x^3 - M$ is strictly increasing in the interval $(3/4, +\infty)$. On the other hand, it can be readily verified that $-M = -4B^2(M)/3 + 4B(M)/\sqrt{3}$. Then, since $|w| \geq B(M) \geq \sqrt{3}/2 > 3/4$ and $3X^3 - 3X^2 - 4X + 4\sqrt{3}$ has no positive real zeros, we have

$$\left|\frac{|w|^4 - |w|^3}{w^n}\right| |f(w)| > h(|w|) \geq h(B(M)) = B^4(M) - B^3(M) - M$$
$$= B(M)(3B^3(M) - 3B^2(M) - 4B(M) + 4\sqrt{3})/3 > 0$$
$$> 0.$$

Hence, as $f(w) \neq 0$ for $\Re(w) \geq B(M)$, the proof is complete. □

Appendix B

Here we shall prove Theorem 4.7.

Proof. Let $(b_m \ldots b_1 b_0)_{(q)}$ denote the base q representation of $b = dp^e$ (with $b_m \neq 0$), which means (as usual $[x]$ stands for the integer part of any real number x)

$$b = \sum_{0 \leq k \leq m} b_k q^k, \text{ with } b_k = [b/q^k] - q[b/q^{k+1}] \text{ for } k=0,1,\ldots,m = [\log b/\log q].$$

Obviously, we have $f(q) = dp^e$. Our assumption $d < q < dp^e$ guarantees that $f(X)$ has positive degree and $f(q-d) \neq 0$. On the other hand, Lemma 4.6 ensures that the zeros of $f(X)$ are in the half plane $\Re(z) < \sqrt{q}$. Now, since $q > \sqrt{q} + \frac{1}{2}$, from (2^*) of the case $\rho = \sqrt{q}$ of Theorem 4.4 it follows immediately that for proving that $f(X)$ is irreducible in $\mathbb{Q}[X]$ only remains to show that $p \nmid (f'(q))^{e-1}$. To this end, note first that from the above definition of $f(X)$ and $q = pa$ are easily deduced the following equivalences:

$$p \nmid f'(q) \iff p \nmid [dp^{e-1}/a] \iff p \nmid b_1. \tag{17}$$

Proof of (4). It can be readily shown that for arbitrary positive integers n, k and j, with $k, j \in \{1, \ldots, n\}$,

$$[n/k] = j \iff n/(j+1) < k \leq n/j.$$

Replacing n, k and j by dp^e, q and jp, respectively, for the case $e > 1$ we obtain

$$[dp^e/q] = jp \iff dp^e/(jp+1) < q \leq dp^e/jp \text{ for } j = 1, 2, \ldots, dp^{e-1}.$$

Equivalently, we can write

$$p \nmid [dp^e/q] \iff q \notin \bigcup_{1 \leq j < dp^{e-1}} \left(\frac{dp^e}{pj+1}, \frac{dp^e}{pj}\right],$$

which togheter with the first equivalence of (17) completes the proof of (4).

From now on r denotes the remainder of dividing d by a.

Proof of (5). Suppose that $a^2|(p-1)$ and $r \neq 0$. From (17) it follows that we only need to show that $p \nmid b_1^{e-1}$, so here we also assume $e \geq 2$. First we will prove, inductively, that there is a positive integer n such that the base q representation of p^e has the form $p^e = (c_n \dots c_0)_{(q)}$ with $(c_1, c_0) = (a^*, p)$, where $a^* = (p-1)/a$.

When $e = 2$ we have $p^2 = p(aa^* + 1) = a^*pa + p = (a^*p)_{(q)}$. Now suppose $2 \leq i < e$ and $p^i = (c_n \dots c_0)_{(q)}$ with $(c_1, c_0) = (a^*, p)$. Therefore, since $p^{i+1} \equiv p(a^*q + p) \pmod{q^2}$, we just need to prove that the representation of $p(a^*q+p)$ in base q has the form $(sa^*p)_{(q)}$. From $a^2|(p-1)$ we get $a^* = sa$ with $s = (p-1)/a^2 < q$, so $p(a^*q+p) = s(pa)^2 + p^2 = s(pa)^2 + a^*(pa) + p$, as we wanted to show.

Therefore, assuming that the base q representation of p^e has the form $p^e = (c_n \dots c_0)_{(q)}$ with $(c_1, c_0) = (a^*, p)$, we have $dp^e \equiv d(a^*q + p) \pmod{q^2}$. As before we only need to consider the representation of $d(a^*q+p)$ in base q. Let $j = [d/a]$. From $d = ja + r$ we get

$$\begin{aligned} d(a^*q+p) &= da^*q + dp = da^*q + jq + rp \\ &= (da^* + j)q + rp = (jaa^* + ra^* + j)q + rp \\ &= (jp + ra^*)q + rp. \end{aligned}$$

Notice that $jp + ra^* = (pd - r)/a < pd/a < p^2 < q^2$. Consequently, there are integers s_1, s_2 in the interval $[0, q)$ such that

$$jp + ra^* = s_2q + s_1.$$

From $0 < r < a$, $a|a^*$ and $a^* < p$ it follows that both a^* and r are relatively prime to p. Then $p \nmid s_1$, because otherwise we have the contradiction $p|ra^*$. Finally, since $rp < q$, the equality

$$d(a^*q+p) = s_2q^2 + s_1q + rp$$

yields $b_0 = rp$, and hence $s_1 = b_1$ as we needed to show.

Proof of (6). Assume that $a^2|(p-1)$ and $\gcd(d, [d/a]q) = 1$. From $a \geq 2$ and

$$\gcd(d, rp) = \gcd(d, (d-r)p) = \gcd(d, [d/a]q) = 1 \tag{18}$$

it easily follows $r \neq 0$, so the previous proof and (17) guarantee that $p \nmid (f'(q))^{e-1}$. We have also proved above that $b_0 = rp$ when $e \geq 2$. Since $dp = [d/a]q + rp$, such equality holds as well for $e = 1$. Thus, since (18) implies $\gcd(d, \mathrm{c}(f)) = 1$, in accordance with (3*) of Theorem 4.4 we can conclude that $f(X)$ is irreducible in $\mathbb{Z}[X]$. □

A SHORT PROOF OF A RESULT OF GICA AND LUCA

Mickey Polasub
Nuanchan, Beungkum, Bangkok, Thailand
mickey.polasub@gmail.com

Received: 10/13/12, Accepted: 1/8/13, Published: 1/25/13

Abstract

We provide a short proof of a generalization of a recent result of Gica and Luca on the diophantine equation $2^x = y^2 + z^2(x^2 - 2)$.

– *Dedicated to the memory of Preechar Polasub.*

1. Introduction

In a recent paper, Gica and Luca [2], in connection with a result of Lee [3] on the class number one problem for quadratic fields of the shape $\mathbb{Q}(\sqrt{n^2 \pm 2})$, were led to consider the diophantine equation

$$2^x = y^2 + z^2(x^2 - 2), \tag{1}$$

where x, y and z are positive integers. This equation has many known solutions, including $(x, y, z) = (3, 1, 1), (5, 3, 1), (7, 9, 1)$, and infinite families of solutions obtained by choosing, say,

$$x = 2^\alpha + 1, \; y = 2^{2^{\alpha-1}-\alpha}\left(2^\alpha - 1\right), \; z = 2^{2^{\alpha-1}-\alpha}$$

or

$$x = 2^\alpha - 1, \; y = 2^{2^{\alpha-1}-\alpha-1}\left(2^\alpha + 1\right), \; z = 2^{2^{\alpha-1}-\alpha-1}.$$

Gica and Luca prove that there are only the three known solutions with $z = 1$. Their argument relies fundamentally upon lower bounds for linear forms in p-adic logarithms. Our goal in this short note is to give a quick proof of a stronger result, which immediately generalizes to partially resolve a conjecture of Gica and Luca on equation (1). Apparently, this strengthening does not follow from the techniques of [2].

Suppose we have a solution to (1). Then, working 2-adically, we readily obtain that x is odd, say $x = 2x_0 + 1$. It follows that

$$0 < \sqrt{2} - \frac{y}{2^{x_0}} = \frac{z^2(x^2-2)}{2^{x_0}\left(2^{x_0+1/2} + y\right)} < \frac{z^2(x_0^2 + x_0)}{2^{2x_0 - 3/2}}.$$

On the other hand, from Corollary 1.6 of Bauer and Bennett [1], either $x_0 \in \{3,7,8\}$, or

$$\sqrt{2} - \frac{y}{2^{x_0}} > 2^{-1.48x_0}.$$

In the latter case, if $z = 1$, we thus have

$$2^{0.52x_0} < 2^{1.5}\left(x_0^2 + x_0\right)$$

and so, using calculus, $x_0 \leq 19$. A quick check yields the main result of [2].

The same argument with a little more work, applying Corollary 1.6 of [1] to either

$$\sqrt{2} - \frac{y}{2^{x_0}} \quad \text{or} \quad \sqrt{2} - \frac{zx}{2^{x_0}},$$

implies the following :

Theorem. For any solution to equation (1) in positive integers x, y and z, we either have $(x,y,z) \in \{(3,1,1),(5,3,1),(7,9,1),(9,14,2),(13,3,7)\}$ or may conclude that

$$\min\{y,z\} > 2^{x/8}.$$

Gica and Luca conjecture that equation (1) has only the solutions $(x,y,z) = (3,1,1),(5,3,1),(7,9,1)$ and $(13,3,7)$ in odd integers x, y and z with $x^2 - 2$ prime. The above result confirms this in case either y or z are "small". Indeed, an almost immediate consequence of this theorem (together with routine computation) is that (1) has only the following solutions with $z < 10^8$:

$(x,y,z) = (3,1,1),(5,3,1),(7,9,1),(9,14,2),(11,12,4),(13,3,7),(15,136,8),$
$(17,240,16),(21,1324,28),(23,2496,64),(25,5568,64),(27,11456,64),$
$(31,33792,1024),(31,29502,1154),(33,71300,1796),(33,63488,2048),$
$(37,318048,5152),(39,741152,544),(39,376832,16384),(45,1934080,124672),$
$(45,655360,131072),(47,8639880,173048),(51,23078252,813316),$
$(51,7995392,917504),(55,189629256,151672),(63,2181038080,33554432),$
$(63,2309764600,31307768),(63,2043658240,35667968),(65,4490035200,62947328),$
$(65,4227858432,67108864),(65,3949423632,71012336),(69,24270711104,16066112).$

References

[1] M. Bauer and M. Bennett, *Application of the hypergeometric method to the generalized Ramanujan-Nagell equation*, Ramanujan J. 6 (2002), 209–270.

[2] A. Gica and F. Luca, *On the diophantine equation* $2^x = x^2 + y^2 - 2$, Funct. Approx. Comment. Math. 46 (2012), 109–116.

[3] J. Lee, *The complete determination of wide Richaud-Degert types which are not* 5 *modulo* 8 *with class number one*, Acta Arith. 140 (2009), 229–234.

COUNTING HERON TRIANGLES WITH CONSTRAINTS

Pantelimon Stănică
Applied Mathematics, Naval Postgraduate School, Monterey, California
pstanica@nps.edu

Santanu Sarkar
Applied Statistics Unit, Indian Statistical Institute, Kolkata, India
sarkar.santanu.bir@gmail.com

Sourav Sen Gupta
Applied Statistics Unit, Indian Statistical Institute, Kolkata, India
sg.sourav@gmail.com

Subhamoy Maitra
Applied Statistics Unit, Indian Statistical Institute, Kolkata, India
subho@isical.ac.in

Nirupam Kar
Aptivaa Consulting Solutions P. Ltd., Mumbai, India
nirupamkar@gmail.com

Received: 3/26/12, Revised: 10/12/12, Accepted: 1/13/13, Published: 1/25/13

Abstract

Heron triangles have the property that all three of their sides as well as their area are positive integers. In this paper, we give some estimates for the number of Heron triangles with two of their sides fixed. We provide a general bound on this count $H(a, b)$, where the sides a, b are fixed positive integers, and the estimate here is better than the one of Ionascu, Luca and Stănică for the general situation of fixed sides a, b. In the case of primes sides p, q, there is an additional hypothesis which helps us to drop the upper bounds on $H(p, q)$. In particular, we prove that $H(p, q)$ is less than or equal to 1 when $p - q \equiv 2 \pmod 4$. We also provide a count for the number of Heron triangles with a fixed height (there exists only one such when the height is prime). Moreover, we study the decomposability property of a Heron triangle into two similar ones, and provide some cases when a Heron triangle is not decomposable.

1. Introduction

In the field of two-dimensional Euclidean geometry, a *Heron triangle* has the property that its three sides, as well as its area are positive integers. Many interesting questions can be raised about these triangles, and there has been a plethora of research regarding several properties of the Heron triangles.

Ionascu, Luca and Stănică [10] found an upper bound for the number of Heron triangles with two fixed sides. They also found sharper upper bounds for the number of Heron triangles with two fixed prime sides. We improve upon the general bounds for two fixed integer sides, as well as give tight bounds for the case where the two fixed sides are primes. We also prove certain upper bounds for the number of Heron triangles for special cases involving the fixed sides; namely, fixed sides that are prime squares, twin primes, or Sophie Germain primes. We also find an example of a Heron triangle whose sides are all perfect squares (question raised in [13]), namely, the triangle of sides $[1853^2; 4380^2; 4427^2]$ and of area 32918611718880. We present the first instance of such a triangle in this paper.

Further, we study the decomposability of Heron triangles into two smaller Heron triangles. It is known from [5] that any Heron triangle is radially decomposable. We show that any isosceles Heron triangle is decomposable, and prove a few results regarding the non-decomposability of certain Heron triangles.

Throughout this paper, we denote the three sides of a general triangle by lower case letters, like a, b, c, and the corresponding vertices by capital letters, like A, B, C. By abuse of notation, we use the same capital letters for the angles, as well as for the corresponding vertices of the triangle. The semi-perimeter $(a+b+c)/2$ is denoted by s and hence the area is given by $\Delta = \sqrt{s(s-a)(s-b)(s-c)}$ (Heron formula).

2. Heron Triangles with Two Fixed Sides

Let $H(a,b)$ be the number of Heron triangles whose two sides a, b are fixed. In [10], a general upper bound for $H(a,b)$ has been proposed, as follows.

Proposition 1 ([10])**.** *If $a \leq b$ are fixed, then $0 \leq H(a,b) \leq \min\{2a-1, 4\left(\tau(ab)\right)^2\}$, where $\tau(n)$ represents the number of positive divisors of an integer n.*

We first find a better bound on $H(a,b)$. Later in this section, we also study the case with two fixed prime sides, and once again improve the corresponding bounds proved in [10].

2.1. Counting Heron Triangles with Fixed Integer Sides

Since a triangle can be uniquely determined by the length of two sides and the angle between them, we simply find the maximum number of all possible values of the angle between the two fixed sides such that the triangle is a Heron triangle; this number will give an upper bound for $H(a,b)$.

Let the prime factorization of an integer n be

$$n = 2^{a_0} \prod_{i=1}^{s} p_i^{a_i} \prod_{j=1}^{r} q_j^{e_j} \tag{1}$$

with primes $p_i \equiv 3 \pmod 4$ and $q_j \equiv 1 \pmod 4$ for all i, j.

We shall require the following known results (see for example [1, Chapter XIV], with a correction in [15]) on primitive Pythagorean triples involving the prime factorization of n as above.

Lemma 2. *Assume that the prime factorization of a positive integer n is as in (1). Then:*

1. *n can be represented as the sum of two positive coprime squares if and only if $a_i = 0$ for all $i = 0, 1, \ldots, s$.*

2. *Assuming that $a_i = 0$ for all $i = 0, 1, \ldots, s$, the number of representations of n as the sum of two positive coprime squares, ignoring signs and order, is given by $2^{w(n)-1}$, where $w(n)$ denotes the number of distinct primes factors of n which are $\equiv 1 \pmod 4$.*

Lemma 3. *Assume that the prime factorization of a positive integer n is as in (1). Then the number of primitive Pythagorean triples (u, w, v) with hypotenuse v dividing n is given by $T(n) = \frac{1}{2}(\prod_{j=1}^{r}(2e_j + 1) - 1)$.*

Proof. For each factor v of n, the number of primitive Pythagorean triples with v as hypotenuse is given by the number of representations of v as a sum of two positive coprime squares $m^2 + n^2$. From Lemma 2, we know that v can be represented as $m^2 + n^2$ with $\gcd(m, n) = 1$ if and only if all of its prime factors are $\equiv 1 \pmod 4$, and the number of such representations is $2^{w(v)-1}$, where $w(v)$ denotes the number of distinct prime factors of v.

Thus, the number of primitive Pythagorean triples (u, w, v) with hypotenuse v dividing n is given by

$$T(n) = \sum_{j=1}^{r} 2^{j-1} \sigma_j(\alpha_1, \alpha_2, \ldots, \alpha_r),$$

where $\sigma_j(\alpha_1, \alpha_2, \ldots, \alpha_r)$, the j-th elementary symmetric polynomial, and $0 \le \alpha_j \le e_j$ for $1 \le j \le r$. The term $\sigma_j(\alpha_1, \alpha_2, \ldots, \alpha_r)$ denotes the number of ways in which a factor v of n can be constructed using j distinct prime factors of n, each $\equiv 1 \pmod 4$. Using known properties of elementary polynomials, the expression for

$T(n)$ further simplifies to:

$$\begin{aligned} T(n) &= \frac{1}{2}\sum_{j=1}^{r} 2^j \sigma_j(\alpha_1, \alpha_2, \ldots, \alpha_r) \\ &= \frac{1}{2}\left(\sum_{j=0}^{r} 2^j \sigma_j(\alpha_1, \alpha_2, \ldots, \alpha_r) - 1\right) \\ &= \frac{1}{2}\left(\prod_{j=1}^{r}(2e_j+1) - 1\right), \end{aligned}$$

as we have $0 \le \alpha_j \le e_j$ for all $1 \le j \le r$. Hence the result. □

Assuming the previous notation, we present our first result, which follows an idea of [10, Theorem 2.3].

Theorem 4. *Let a, b be two fixed integers, and let ab be factored as in* (1)*. Then*

$$H(a,b) \le \frac{3+(-1)^{ab}}{2}\left(\prod_{j=1}^{r}(2e_j+1) - 1\right).$$

Proof. Let C be the angle between the sides a, b, and so, $\triangle = \frac{1}{2}ab\sin C$. Then, $\sin C = \frac{2\triangle}{ab} = \frac{u}{v}$ for some integers u, v with $\gcd(u,v) = 1$, which implies that the integers u, v are two components from a primitive Pythagorean triple with v as hypotenuse. Thus we have $v = m^2 + n^2$, $u \in \{m^2 - n^2, 2mn\}$ for $\gcd(m,n) = 1$, $m > n$.

Also note that $\frac{2\triangle}{ab} = \frac{u}{v}$ implies $v|ab$, and for each factor v of ab, there are two options $\{m^2 - n^2, 2mn\}$ for u. Thus, the number of possible values of $\sin C$ is bounded above by $2T(ab)$, where $T(ab) = \frac{1}{2}(\prod_{j=1}^{r}(2e_j+1) - 1)$ denotes the number of primitive Pythagorean triples with hypotenuse v dividing ab, derived from Lemma 3.

In case ab is odd, so is $v = m^2 + n^2$. In this case, $\frac{2\triangle}{ab} = \frac{u}{v}$ implies that $2|u$, and thus for each factor v of ab, the only possible choice for u is $2mn$ (as $m^2 - n^2$ is odd). Therefore, for ab odd, the number of possible values of $\sin C$ is bounded above by $T(ab)$, where $T(ab) = \frac{1}{2}(\prod_{j=1}^{r}(2e_j+1) - 1)$, derived from Lemma 3.

For each possible value of $\sin C$, there are at most two possible values of C, and thus we have $H(a,b) \le 2T(ab)$ if ab is odd, and $H(a,b) \le 4T(ab)$ if ab is even, from which we derive the result. □

Theorem 4 immediately offers us an interesting observation regarding a special class of fixed sides (a, b).

Corollary 5. *If all the prime factors of ab are $\equiv 3 \pmod 4$, then $H(a,b) = 0$.*

Proof. If all prime factors of ab are $\equiv 3 \pmod 4$, the prime factorization of ab is $\prod_{i=1}^{s} p_i^{a_i}$, as in (1), with $e_j = 0$ for all $1 \le j \le r$. In this case, $T(ab) = \frac{1}{2}(\prod_{j=1}^{r}(2e_j+1)-1) = 0$, and hence $H(a,b) = 0$ from Theorem 4. □

At this point, let us refer to the results available in [10]. When two fixed sides are primes, $p, q\ (\neq 2)$ say, one can get the following bounds from [10, Theorem 2.4]:

$$H(p,q) \text{ is } \begin{cases} = 0 & \text{if both } p \text{ and } q \text{ are} \equiv 3 \pmod 4, \\ = 2 & \text{if } p = q \equiv 1 \pmod 4, \\ \le 2 & \text{if } p \neq q \text{ and exactly one of } p \text{ and } q \text{ is} \equiv 3 \pmod 4, \\ \le 5 & \text{if } p \neq q \text{ and both } p \text{ and } q \text{ are} \equiv 1 \pmod 4. \end{cases} \quad (2)$$

From Theorem 4, we immediately obtain the same bounds in two of the cases.

Corollary 6. *Let $p, q\ (\neq 2)$ be two fixed prime sides of a triangle. Then:*

$$H(p,q) \text{ is } \begin{cases} = 0 & \text{if both } p \text{ and } q \text{ are} \equiv 3 \pmod 4, \\ \le 2 & \text{if } p \neq q \text{ and exactly one of } p \text{ and } q \text{ is} \equiv 3 \pmod 4. \end{cases}$$

Proof. In the first case, $H(p,q) = 0$ from Corollary 5. In the second case, $T(pq) = \frac{1}{2}((2 \times 1 + 1) - 1) = 1$, and thus $H(p,q) \le 2T(p,q) = 2$ from Theorem 4, as pq is obviously odd. □

For the remaining two cases, we obtain the following from Theorem 4.

Corollary 7. *Let $p, q\ (\neq 2)$ be two fixed prime sides of a triangle. Then:*

$$H(p,q) \text{ is } \begin{cases} \le 4 & \text{if } p = q \equiv 1 \pmod 4, \\ \le 8 & \text{if } p \neq q \text{ and both } p \text{ and } q \text{ are} \equiv 1 \pmod 4. \end{cases}$$

Proof. In both the cases, pq is odd, and thus we have $H(p,q) \le 2T(pq)$ from Theorem 4. In the first case, $T(pq) = T(p^2) = \frac{1}{2}((2 \times 2 + 1) - 1) = 2$, and thus $H(p,q) \le 2T(p,q) = 4$. In the second case, $T(pq) = \frac{1}{2}((2 \times 1 + 1)(2 \times 1 + 1) - 1) = 4$, and thus $H(p,q) \le 2T(p,q) = 8$. □

Corollary 7 is slightly weaker than [10], but it has the advantage that it follows easily from a more general result (Theorem 4), whereas the result from [10] had to be dealt with as a separate problem, just for the primes. In the following section, we discuss further improvements in the bounds for prime sides.

Let us compare our result with that of [10] by displaying some examples in Table 1. Note that in the work of Ionascu, Luca and Stănică [10], $H(a,b) \le \min\{2a-1, 4(\tau(ab))^2\}$ (column 7 of Table 1), as stated in Proposition 1, as well. The bound on $H(a,b)$ that we propose in this section (column 4 in Table 1) is much closer to the actual value (column 3 in Table 1), and it is clear that our result is sharper than the existing bound, when a, b are composite integers.

a	b	$H(a,b)$ (actual)	Upper bound from Theorem 4	Upper bound of [10]		
				$2a-1$	$4\,(\tau(ab))^2$	$\min\{2a-1, 4(\tau(ab))^2\}$
17	27	0	2	33	256	33
51	52	5	16	101	2304	101
65	87	4	26	129	1024	129
125	125	6	12	249	196	196
305	377	4	80	609	1024	609
714	728	5	16	1427	57600	1427
1189	1275	8	134	2377	9216	2377
3034	3434	7	160	6067	9216	6067
7089	7228	5	16	14177	20736	14177
81713	49274	0	16	98547	256	256

Table 1: Comparison of the bounds on $H(a,b)$ when a, b are composite integers.

2.2. Counting Heron Triangles with Fixed Prime Sides

In this section, we present certain results, which under an additional hypothesis slightly improves the bounds of $H(a,b)$ from [10, Theorem 2.4].

Theorem 8. *Let p, q ($\neq 2$) be two fixed prime sides of a triangle. Then:*

$$H(p,q) \text{ is } \begin{cases} = 0 & \text{if both } p \text{ and } q \text{ are} \equiv 3 \pmod 4,\\ = 2 & \text{if } p = q \equiv 1 \pmod 4,\\ = 0 & \text{if } p < q \text{ with } p \equiv 1 \pmod 4 \text{ and } q \equiv 3 \pmod 4,\\ \leq 1 & \text{if } p > q \text{ with } p \equiv 1 \pmod 4 \text{ and } q \equiv 3 \pmod 4,\\ \leq 5 & \text{if } p \neq q \text{ and both } p \text{ and } q \text{ are} \equiv 1 \pmod 4. \end{cases}$$

Moreover, when both p and q are $\equiv 1 \pmod 4$ with $q > p$ and $(t+1)(q-tp)$ is not a perfect square, where $t = \lfloor q/p \rfloor$, then $H(p,q) \leq 4$.

Proof. We consider each case, separately (some are included in [10], certainly, but we include them nonetheless, since we provide simpler proofs).

`Case I.` Both p and q are $\equiv 3 \pmod 4$. This follows from Corollary 5.

`Case II.` $p = q \equiv 1 \pmod 4$. When $p = q \equiv 1 \pmod 4$ (and third side $2w$), we get $H(p,q) \leq 4$ as an immediate consequence from Theorem 4, but an exact count is easy to obtain. The area is $\Delta = w\sqrt{p^2 - w^2}$, and so, p is part of a Pythagorean triple, that is, there exist integers m, n with $p = m^2 + n^2$. By Lemma 2, since $p \equiv 1 \pmod 4$, there is only one such representation (excluding order, say $m > n$), and so, either $w = 2mn$, or $w = m^2 - n^2$, which implies that $H(p,q) = 2$, in this case.

For the next three cases, recall that the third side of the triangle is even, say $2w$. Using Heron's formula,

$$\begin{aligned} 4\Delta &= \sqrt{(p+q+2w)(p+q-2w)(2w+p-q)(2w-p+q)} \\ &= \sqrt{((p+q)^2 - 4w^2)(4w^2 - (p-q)^2)}. \end{aligned} \tag{3}$$

`Case III.` $p < q$ with $p \equiv 1 \pmod 4$ and $q \equiv 3 \pmod 4$. As $p < q$, let us write $q - p = 4k + 2$, and suppose that $2w$ is the third side. Certainly, by the triangle inequality, $(q-p)/2 = 2k+1 < w < (q+p)/2 = p + 2k + 1$. From (3), we obtain

$$\begin{aligned}\Delta &= \sqrt{(p+2k+1+w)(p+2k+1-w)(w+2k+1)(w-2k-1)}\\ &= \sqrt{((p+2k+1)^2-w^2)(w^2-(2k+1)^2)}.\end{aligned}$$

We consider several cases. If there exists a prime $r \mid \gcd(p+2k+1-w, w-2k-1)$, then $r = p$, and so, $p \mid w - 2k - 1 < p$, a contradiction. If there exists a prime $r \mid \gcd(p+2k+1-w, w+2k+1)$, then $r = p+4k+2 = q$, and so, $q \mid p+2k+1-w < p < q$, a contradiction. Moreover, if there exists a prime $r \mid \gcd(p+2k+1+w, w-2k-1)$, then $r = p + 4k + 2 = q$, and so $q \mid w - 2k - 1 < p$, a contradiction. It remains to consider the case of $r \mid \gcd(p + 2k + 1 + w, w + 2k + 1)$, which implies that $r = p$. Combining all cases, we obtain that, for some integers m, n, we have

$$(p+2k+1)^2 - w^2 = p^\ell m^2 \text{ and } w^2 - (2k+1)^2 = p^\ell n^2.$$

Add the previous two equations and obtain $pq = p^\ell(m^2+n^2)$, and so $\ell \in \{0,1\}$. If $\ell = 0$, then $pq = m^2 + n^2$, and by Lemma 2, there are no representations of pq as a sum of squares, as $pq \equiv 3 \pmod 4$. If $\ell = 1$, then we get that $q = m^2 + n^2$, which is also a contradiction by Lemma 2. Thus, we get that $H(p,q) = 0$.

`Case IV.` $p > q$ with $p \equiv 1 \pmod 4$ and $q \equiv 3 \pmod 4$. In this case, we write $p - q - 4k + 2$, and proceed almost identically as in the previous case to obtain that, for some integers m, n, we have

$$(q+2k+1)^2 - w^2 = q^\ell m^2 \text{ and } w^2 - (2k+1)^2 = q^\ell n^2.$$

Add the previous two equations and obtain $pq = q^\ell(m^2+n^2)$, and so $\ell \in \{0,1\}$. If $\ell = 0$, then $pq = m^2 + n^2$, and by Lemma 2, there are no representations of pq as a sum of squares, as $pq \equiv 3 \pmod 4$. If $\ell = 1$, then we get that $p = m^2 + n^2$, which can be written in only one way, according to Lemma 2.

Now since $w^2 - (2k+1)^2 = qn^2$, we get

$$w = \pm(2k+1) + tq = \pm\frac{p-q}{2} + tq,$$

for some integer t. We also know that $2w < p + q$ and $p < 2w + q$, and thus w can not be of the form $\frac{p-q}{2} + tq$. Hence w must be of the form $-\frac{p-q}{2} + tq$. Moreover, since $2w < p + q$ and $p < 2w + q$, the only option for t is $\lfloor p/q \rfloor$. Thus, considering all cases discussed till now, we get that $H(p,q) \le 1$.

`Case V.` $p \neq q$ and both p and q are $\equiv 1 \pmod 4$. Assume $p < q$ and write $q - p = 4k$, and $2w$ for the third side of the Heron triangle. Certainly, by the triangle inequality, $(q-p)/2 = 2k < w < (q+p)/2 = p+2k$. From (3), we obtain

$$\begin{aligned}\Delta &= \sqrt{(p+2k+w)(p+2k-w)(w+2k)(w-2k)}\\ &= \sqrt{((p+2k)^2-w^2)(w^2-(2k)^2)}.\end{aligned}$$

We consider several subcases. If there exists a prime $r|\gcd(p+2k-w, w-2k)$, then $r=p$, and so, $p|w-2k<p$, a contradiction. If there exists a prime $r|\gcd(p+2k-w, w+2k)$, then $r=p+4k=q$, and so, $q|p+2k-w<p<q$, a contradiction. Moreover, if there exists a prime $r|\gcd(p+2k+w, w-2k)$, then $r=p+4k=q$, and so $q|w-2k<p$, a contradiction. It remains to consider the case of $r|\gcd(p+2k+w, w+2k)$, which implies that $r=p$. Combining all cases, we obtain that, for some integers m, n, we have

$$(p+2k)^2 - w^2 = p^\ell m^2 \text{ and } w^2-(2k)^2 = p^\ell n^2.$$

Add the previous two equations and obtain $pq = p^\ell(m^2+n^2)$, and so $\ell \in \{0,1\}$. If $\ell = 0$, then $pq = m^2+n^2$, and by Lemma 2, there are only two representations of pq as a sum of squares.

If $\ell=1$, then we get that q can be written as $q=m^2+n^2$ in only one way, disregarding order. In this situation $w^2 \equiv (2k)^2 \pmod p$, and thus w is of the form $w = \pm 2k+tp$ for some integer t. Since we know $2w < p+q$ and $q < 2w+p$, it can not be of the form $w=2k+tp$. Hence we must have $w=-2k+tp$, and as we know $2w<p+q$ and $q<2w+p$, the only option for t is $\lfloor q/p \rfloor$. Considering all cases, we get that $H(p,q) \le 2\times 2+1 = 5$.

The last claim follows easily, since if we assume that $(t+1)(q-tp)$ is not a perfect square, for $t = \lfloor q/p \rfloor$, then the two displayed identities from above (for $\ell=1$) would imply that $(p+2k)^2-(tp-2k)^2 = pm^2$, which is equivalent to $(t+1)(4k-(t-1)p) = m^2$, that is, $(t+1)(q-tp) = m^2$, a contradiction. Thus the count for the $l=1$ case will not appear in this case and the bound will be $2\times 2 = 4$. □

A natural question is whether the range of values in the last three inequalities of Theorem 8 is completely covered. Given the extensive computations we performed, *we conjecture that $H(p,q)$ never attains the values* 4, 5 (under $p \ne q$ with $p \equiv 1 \pmod 4$ and $q \equiv 1 \pmod 4$). However, we can certainly show that the following values are attained:

1. $H(p,q)$ is 0 or 1, when $p>q$ with $p\equiv 1 \pmod 4$ and $q \equiv 3 \pmod 4$ and

2. $H(p,q)$ is 0 or 1 or 2 or 3, when $p\ne q$ with $p \equiv 1 \pmod 4$ and $q\equiv 1 \pmod 4$.

In Table 2, we provide examples that indeed all such cases are possible. It does require further study for properties of the primes to exactly identify the corresponding values of $H(p,q)$, which we leave as an open problem.

In the following section, we take a look at the bounds on $H(a,b)$ where the sides a, b are fixed primes with special properties.

Case	p	q	$H(p,q)$	Third side
1	13	3	0	-
	5	3	1	4
2	17	5	0	-
	13	5	1	12
	37	13	2	30,40
	4241	2729	3	1530,1850, 6888

Table 2: $H(p,q)$ for different values of p and q.

2.3. Counting Heron Triangles with Special Prime Sides

The case where the two sides a, b are Sophie Germain primes comes directly from Theorem 8, as follows.

Corollary 9. *Let $p, q = 2p+1$ be Sophie Germain primes. Then $H(p,q) = 0$.*

Proof. We consider two cases. First suppose that $p \equiv 3 \pmod 4$. Then $q = 2p+1 \equiv 3 \pmod 4$, and by Case I of Theorem 8, we have $H(p,q) = 0$. Now suppose $p \equiv 1 \pmod 4$. Then we have $q = 2p+1 \equiv 3 \pmod 4$, and by Case III of Theorem 8, we immediately obtain $H(p,q) = 0$. □

We obtain a similar result for Mersenne primes.

Corollary 10. *Let p, q be two Mersenne prime numbers. Then $H(p,q) = 0$.*

Proof. Suppose that the sides p, q are Mersenne primes, $p = 2^x - 1, q = 2^y - 1$ (for some primes $x \le y$). Then, both p and q are $\equiv 3 \pmod 4$ and so, by Case I of Theorem 8, we have $H(a,b) = 0$. □

In case of twin primes, we not only obtain a bound on $H(p,q)$, but can also estimate the third side in each case. The result is as follows.

Theorem 11. *Let $p, q = p+2$ be twin primes. Then $H(p,q) \le 1$. Moreover, $H(p,q) = 1$ if and only if $p-2$ is a perfect square, and if that is so, the third side of the triangle must be $2p-2$, and one must have $p \equiv 11 \pmod{12}$, $p \not\equiv 7 \pmod 8$.*

Proof. Recall that the third side of the triangle is even, say $2w$. Also, by the triangle inequality, we have $w < p+1$. Using Heron's formula and (3),

$$\Delta = \sqrt{(p+1+w)(p+1-w)(w+1)(w-1)} = \sqrt{((p+1)^2 - w^2)(w^2-1)}.$$

Suppose that $r \ne 2$ is a prime. As $\gcd(w+1, w-1)|2$, one cannot have $r|\gcd(w+1, w-1)$. If $r|\gcd(p+1-w, w-1)$, then $r = p$, which implies that $p|w-1 < p$, a contradiction. If $r|\gcd(p+1+w, w-1)$, then $r = p+2 = q$, which implies that $q|w-1 < p$, a contradiction. If $r|\gcd(p+1-w, w+1)$, then $r = p+2 = q$, which implies that $q|w+1 < p+2 = q$, again a contradiction. If $r|\gcd(p+1+w, w+1)$, then $r = p$, which implies that $p|w+1 \le p+1$, that is $w+1 = p$. To summarize, if there exists a prime $r \neq 2$ such that $r|w^2-1$, then either $w+1 = p$, or either of $w+1$ and $w-1$ is a whole square (note that $w+1$ and $w-1$ cannot be simultaneously integer squares).

In the first case, where $w+1 = p$, we have $H(a,b) \le 1$. In fact, we can say more. Since $w+1 = p$, then $\Delta = 2p\sqrt{p-2}$. Writing $\Delta = 2p\Delta'$, we get $p = \Delta'^2 + 2$. Thus, $p \equiv 3 \pmod 4$ and since p is the smaller prime in a twin prime pair, it must also satisfy $p \equiv 5 \pmod 6$, and so $p \equiv 11 \pmod{12}$. From Gauss' sum of three squares function formula applied to $p = \Delta'^2 + 1^2 + 1^2$ (or simply by looking at $\Delta' \pmod 4$), we see that we also need $p \not\equiv 7 \pmod 8$.

In the second case, assume that the odd prime r divides $w-1$, and $w+1 = 2^e$ for some positive integer e. From our discussion, we must have $w - 1 = m^2$, for some integer m. It easily follows that $e = 1$, and so $w = 1$. In that case, $4\Delta = \sqrt{2p(p+2)} \notin \mathbb{Z}$, since p is odd. Next, assume that $w - 1 = 2^f$ for some positive integer f, and there is an odd prime $r|w+1$, that is, $w+1 = n^2$ (for some integer n), as per our earlier discussion. Then, we easily obtain $w = 3$, and so, $2\Delta = \sqrt{2(p-2)(p+4)} \notin \mathbb{Z}$, since the prime p is odd. Thus, $H(p,q) = 0$ in both these cases.

The only remaining case is when $w+1 = 2^e$ and $w-1 = 2^f$ simultaneously for positive integer e, f. In this case, one obtains $w = 3$ once again, and thus, $H(p,q) = 0$ in this case. □

Example 12. We can give several examples of Heron triangles based on twin primes with parameters $(p, q, 2p-2, \Delta)$, satisfying the conditions of our theorem: $(3,5,4,6)$, $(11,13,20,66)$, $(227,229,452,6810)$.

It may be interesting to investigate primes with other special properties as well.

2.4. Heron Triangles with Square Sides

In this section, we first answer affirmatively an open question of Sastry [13], that there exists a primitive (co-prime sides) Heron triangle with square sides. For that, we ran experiments (using a "bounded" approach) in GNU/Linux environment using C with GMP, and stopped when we obtained the following example.

Example 13. There exists a Heron triangle with square sides, namely

$$[a, b, c, \triangle] = [1853^2, 4380^2, 4427^2, 32918611718880].$$

Now, we turn our attention to general Heron triangles with square sides, and prove the following.

Proposition 14. *There is no isosceles Heron triangles with square sides.*

Proof. Let the sides of the triangle be a^2, a^2, b^2, and the height h corresponding to b^2. Since the semi-perimeter of a Heron triangle is an integer, then b is even, say $b = 2b_1$. By Pythagoras' theorem, $a^4 = h^2 + 4b_1^2$, and since in a Heron triangle, the heights are always rational numbers, we obtain that h is an integer. As $a^2, h, 2b_1^2$ form a Pythagorean triple, we can have the following two cases.

Case 1: $2b_1^2 = 2mn, h = m^2 - n^2, a^2 = m^2 + n^2$ for some integers m, n with $\gcd(m, n) = 1$. So, $b_1^2 = mn$. Now since m, n are co-prime to each other, $m = x^2, n = y^2$ for some integers x, y. Therefore we get $a^2 = x^4 + y^4$, which does not have any integer solution, by a known consequence of Fermat's Last Theorem.

Case 2: $2b_1^2 = m^2 - n^2, h = 2mn, a^2 = m^2 + n^2$ for some integers m, n with $\gcd(m, n) = 1$. As, $m^2 - n^2$ is even and $\gcd(m, n) = 1$, m, n are both odd. So, $m^2 + n^2 \equiv 2 \pmod 8$. But, for any integer a, $a^2 \not\equiv 2 \pmod 8$, and hence we get a contradiction. □

3. Heron Triangles with Other Constraints

In this section, rather than fixing the sides, we impose constraints on other properties of the triangle, namely one of the heights of the triangle, and the property of decomposability.

3.1. Counting Heron Triangles with Fixed Height

We first fix one of the heights of a Heron triangle, which we assume integer, unless otherwise specified. It is an easy exercise to show that if the Heron triangle contains more than one integer height it cannot be primitive (that is, $\gcd(a, b, c) = 1$). We consider non-Pythagorean Heron triangles with fixed height h (of corresponding vertex A) to obtain the following results.

Theorem 15. *The following statements are true:*

1. *Let h be an integer. Then h is the height of a Heron triangle if and only if $h > 2$.*

2. *For a fixed prime height h, there exists only one non-Pythagorean Heron triangle which has $b = c = \frac{h^2+1}{2}$ and $a = h^2 - 1$.*

3. *For a fixed height $h = 2^{\alpha_0} \prod_{i=1}^{k} p_i^{\alpha_i}$, where $p_i > 2$ for all $i \geq 1$, there exist $\frac{1}{4}\left(|2\alpha_0 - 1| \prod_{i=1}^{k}(2\alpha_i + 1) - 1\right)^2$ many non-Pythagorean Heron triangles.*

Proof. Assume that the height of length h is the one corresponding to the side a and to the angle A.

Claim 1: Assume that h is rational and $h \leq 2$; recall that the semi-perimeter $s = (a+b+c)/2$. Square the area and impose the condition on the height to get $s(s-a)(s-b)(s-c) = \frac{a^2h^2}{4} \leq a^2$. Label $x = s-a, y = s-b, z = s-c$. Using the triangle inequality, we get $x+y+z = s > a$ and so, $xyz < a$. Therefore, $x+y+z > xyz$, which we rewrite as $\frac{1}{xy} + \frac{1}{xz} + \frac{1}{yz} > 1$. If $\min\{x,y,z\} \geq 2$, then $1 < \frac{1}{xy} + \frac{1}{xz} + \frac{1}{yz} \leq \frac{3}{4}$, which is a contradiction. Thus, $\min\{x,y,z\} = 1$. Assume $x = 1$ (all the other cases are similar). Then, $\frac{1}{y} + \frac{1}{z} + \frac{1}{yz} > 1$, and so, $\min\{y,z\} \leq 2$.
Case 1 Let $\min\{y,z\} = 1$, say $y = 1$. Since $x = y = 1$, then $s = a+1 = b+1$, and $a = b$. It follows from $s = a + \frac{c}{2} = a+1$ that $c = 2$. We get a triangle of sides $(a,a,2)$, which cannot be Heron for any integer $a \geq 2$.
Case 2: Let $\min\{y,z\} = 2$, say $y = 2$. Then, $\frac{1}{z} + \frac{1}{2z} > \frac{1}{2}$, and so, $z \leq 2$. We have two possibilities, namely $(x,y,z) \in \{(1,1,1),(1,1,2)\}$, and $(a,b,c) \in \{(2,2,2),(3,3,2)\}$. However, neither of these triangles has integer area.

We show the converse under the assumption that the height h is an integer. Certainly, a Heron triangle of height 2 does not exist, since 2 must be part of a Pythagorean triple and that is impossible. If $h = 2k$, then $h^2 = (k^2+1)^2 - (k^2-1)^2$; if $h = 2k+1$, then $h^2 = \left((k+1)^2 + k^2\right)^2 - (2k(k+1))^2$. Thus, we can always write $h^2 = x^2 - y^2$ (for some integers x, y), and so, one may construct an isosceles Heron triangle of height h by taking $b = c = x$, $a = 2y$ (since h is integer and the base is an even integer, the area is an integer).

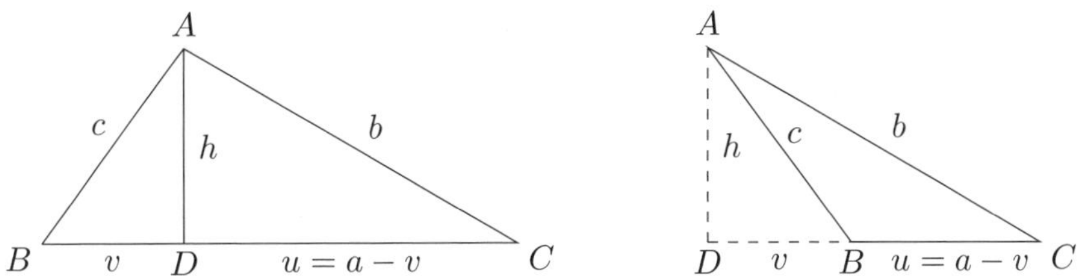

Figure 1: Heron triangle with fixed height h.

Claim 2: Suppose that a Heron triangle (as in Figure 1) has sides of lengths a, b, c and the prime number height h corresponding to A. In $\triangle ABD$, we have $h^2+v^2 = c^2$, and so, $h^2 = (c+v)(c-v)$. As h is a prime and $v \neq 0$, we get $c+v = h^2$ and $c - v = 1$. Therefore $c = \frac{h^2+1}{2}$ and $v = \frac{h^2-1}{2}$. Similarly, in $\triangle ACD$, we have $b+(a-v) = h^2, b-(a-v) = 1$, and so $b = \frac{h^2+1}{2}$ and $a = h^2 - 1$. Hence the claim.

Claim 3: Now, let us consider the general case $h = 2^{\alpha_0} \prod_{i=1}^{k} p_i^{\alpha_i}$, where $p_i > 2$ for all $i \geq 1$. In $\triangle ABD$, by Pythagoras' rule, we have $h^2 + v^2 = c^2$, and so, $(c+v)(c-v) = h^2 = 2^{2\alpha_0} \prod_{i=1}^{k} p_i^{2\alpha_i}$. The choices for the pair $(c+v, c-v)$ for which c, v are positive integers must have both $(c+v), (c-v)$ odd or both even.

The number of such distinct choices is

$$X = \frac{1}{2}\left(|2\alpha_0 - 1|\prod_{i=1}^{k}(2\alpha_i + 1) - 1\right).$$

Once v is determined, in $\triangle ACD$, the number of choices for the pair $(b+a-v, b-a+v)$ is, as before, $X = \frac{1}{2}\left(|2\alpha_0 - 1|\prod_{i=1}^{k}(2\alpha_i + 1) - 1\right)$, which will determine a, b.

Note that the choices of sides (c, v) in $\triangle ABD$ and that of sides $(b, a-v)$ of $\triangle ACD$ come from the same list, and each choice generates two non-Pythagorean Heron triangles with height h, as shown in Figure 1, if and only if $v \neq a - v$. In case the choices are the same, i.e., $v = a - v$, then it will generate only one triangle, namely, the first one shown in Figure 1. Thus, for $X = \frac{1}{2}\left(|2\alpha_0 - 1|\prod_{i=1}^{k}(2\alpha_i + 1) - 1\right)$ many distinct choices in the list, we will have $2 \times \binom{X}{2} + 1 \times X = (X^2 - X) + X = X^2$ many distinct non-Pythagorean Heron triangles with fixed height h. Hence the total number of Heron triangles having fixed height h is

$$X^2 = \frac{1}{4}\left(|2\alpha_0 - 1|\prod_{i=1}^{k}(2\alpha_i + 1) - 1\right)^2.$$

Hence the result. □

In view of Claim 1 in Theorem 15, it might be tempting to propose that, for all rationals $h > 2$, a Heron triangle having height h exists. However, that is not true. Take for example, $h = 5/2$. Assume that such a Heron triangle exists, with sides a, b, c ($u + v = a$, with $u, v \in \mathbb{Q}$, since $h \in \mathbb{Q}$), as in Figure 1. Then we get $5^2 = (2c)^2 - (2u)^2$, with $u' := 2u \in \mathbb{Z}$. Thus, $2c - u' \in \{1, 5, 25\}$, respectively, $2c + u' \in \{25, 5, 1\}$. However, none of the obtained systems have integers c, u' as solutions.

In case of Claim 3, to take an example, one may consider the case $h = 10$. From Theorem 15, we obtain the number of distinct non-Pythagorean Heron triangles as

$$\frac{1}{4}\left(|2 \times 1 - 1|(2 \times 1 + 1) - 1\right)^2 = \frac{1}{4} \times 2^2 = 1.$$

It can be verified that the only non-Pythagorean Heron triangle with height $h = 10$ has sides $(26, 48, 26)$.

3.2. Decomposable Heron Triangles

It is known that any Heron triangle is radially decomposable, that is, it can be subdivided into n isosceles Heron triangles each composed of two circum-radii and one side of the n-gon [5]. In this section we investigate the decomposability of a Heron triangle into two Heron triangles as in Figure 2 (we say, on the side BC). We first consider the case when the Heron triangle is isosceles. In this setting, we can prove that a Heron triangle is always decomposable.

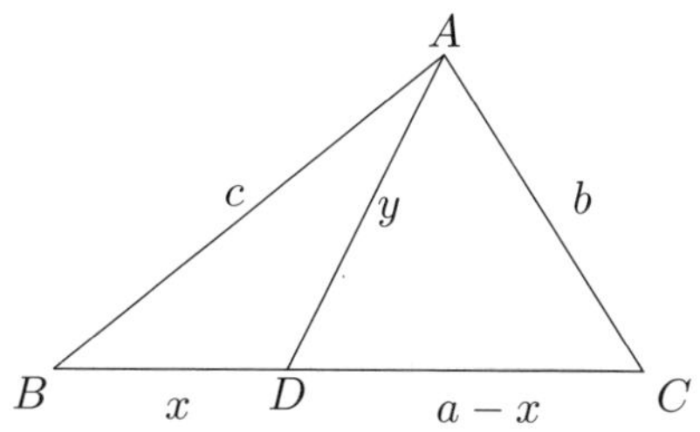

Figure 2: Heron triangle decomposed into two Heron triangles.

Proposition 16. *An isosceles Heron triangle of sides a, a, b can always be decomposed along the height corresponding to b.*

Proof. Since our triangle is a Heron triangle, its semi-perimeter $s = \frac{2a+b}{2}$ is an integer. Hence, b must be even. So, $b_1 = \frac{b}{2}$ is an integer. As $\triangle = \frac{bh}{2}$ is the area of the triangle, which is an integer, the height h must be a rational number. Also $h^2 = a^2 - b_1^2$ is an integer. Hence, h must be an integer.

Now, to prove the proposed statement, we need to show that $\frac{b_1 h}{2}$ is an integer. Note that we have the equation $a^2 = h^2 + b_1^2$ and hence one of h, b_1 must be even (being part of a Pythagorean triple). So, $\frac{b_1 h}{2}$ is indeed an integer, hence the result. □

We can give many examples of such triangles, by simply concatenating two copies of the same Pythagorean triangle. The parameters of the new triangle, which can be decomposed along the height $2mn$, will then be $(a, b, c, \Delta) = (m^2 + n^2, m^2 + n^2, 2(m^2 - n^2), 2mn(m^2 - n^2))$, for any integers $m > n$.

Regarding indecomposability, we can prove the following results.

Proposition 17. *Let a, b, c be the lengths of three sides of a Heron triangle, such that a is prime and $a \nmid b + c$. Then one cannot decompose $\triangle ABC$ into two Heron triangles $\triangle ABD$ and $\triangle ADC$.*

Proof. Let the side BC of $\triangle ABC$ be decomposed into BD and DC where the length of BD is the integer x. Let the length of AD be y. We refer to Figure 2 for an illustration of such a case.

Now, from $\triangle ABC$, we have $\cos B = \frac{c^2+a^2-b^2}{2ac}$, and from $\triangle ABD$, we have $\cos B = \frac{c^2+x^2-y^2}{2cx}$. So, we get $\frac{c^2+a^2-b^2}{2ac} = \frac{c^2+x^2-y^2}{2cx}$, from which we obtain

$$y^2 = c^2 + x^2 - ax - \frac{x(c-b)(c+b)}{a}.$$

Since a is prime, $a \nmid (b + c)$, and $x < a$, we obtain that a must be a divisor of $c - b$. If $b \neq c$, then we must have $a \leq c - b$, which contradicts the triangle inequality. If

$b = c$, then a would have to be even (and prime), that is, $a = 2$. Then $x = 2 - x = 1$, and y would be a non integer height. The proof is done. □

An example of such a Heron triangle has parameters $(a, b, c, \Delta) = (13, 14, 15, 84)$ (where 13 does not divide 14+15), which cannot be decomposed along the base of length 13 (but in this case, it can be decomposed along the height corresponding to the side 14). Another example has parameters $(5, 29, 30, 72)$, and it satisfies our conditions on two sides $5 \nmid 29 + 30$, $29 \nmid 5 + 30$).

From Proposition 17, we can derive the following two observations.

Remark 18. The above proof shows in fact that if the Heron triangle is isosceles, with $b = c$, then, we only need the base a to be prime, and not the divisibility condition.

Corollary 19. *Let $\triangle ABC$ be a Heron triangle of sides a, b, c and area Δ, such that a is a prime number and $a \nmid \Delta$. Then one cannot decompose $\triangle ABC$ into two Heron triangles $\triangle ABD$ and $\triangle ADC$.*

Proof. Suppose that one can decompose $\triangle ABC$ into two Heron triangles $\triangle ABD$ and $\triangle ADC$. Then, by Proposition 17, we must have $a|b + c$, and hence $a^2|(b + c)^2 - a^2 = (b + c + a)(b + c - a)$. Now, the area Δ satisfies

$$16\Delta^2 = (a + b + c)(a + c - b)(a + b - c)(b + c - a),$$

and hence $a^2|16\Delta^2$, which implies $a|\Delta$, since $a > 2$ is prime (no side of a Heron triangle can be 2). That is a contradiction and the proof is done. □

From Proposition 17, it is clear that $a|b+c$ is a necessary condition for decomposability of $\triangle ABC$ on the side BC. However, this condition is not at all a sufficient condition for such a decomposition. In this direction, we prove the following result.

Proposition 20. *Suppose that a Heron triangle $\triangle ABC$ (as in Figure 2), of sides a, b, c is decomposed into two triangles $\triangle ABD$ and $\triangle ADC$, with $BD = x$. Assume that c, x are odd. If $a|(b+c)$ and $a + (c - b)\frac{c+b}{a} \equiv 4 \pmod 8$, then $\triangle ABD$ can not be a Heron triangle.*

Proof. Suppose that $\triangle ABC$ could be decomposed into $\triangle ABD$ and $\triangle ADC$, both of which are Heron triangles. In such a case, the relation

$$y^2 = c^2 + x^2 - \frac{x}{a} \cdot (a^2 + c^2 - b^2) \qquad (4)$$

holds with x, y integers. Note that $\triangle ABD$ is a Heron triangle and two of its sides $AB = c, BD = x$ are odd. Hence, $AD = y$ must be even, and thus $y^2 \equiv 0, 4 \pmod 8$ in the LHS of (4). On the RHS of (4), $c^2 + x^2 \equiv 2 \pmod 8$ as a, x are odd. Thus, for the relation to be true, we require $\frac{x}{a}(a^2 + c^2 - b^2)$ to be equal to

$\pm 2 \pmod 8$. Since x is odd, one requires $\frac{1}{a}(a^2+c^2-b^2) \equiv \pm 2 \pmod 8$, which contradicts the given condition $\frac{1}{a}(a^2+c^2-b^2) \equiv 4 \pmod 8$. Hence $\triangle ABD$ can not be a Heron triangle. □

4. Conclusion

In this paper, we estimate the number $H(a, b)$ of Heron triangles with two fixed sides a, b. We also investigate $H(p, q)$ when the sides p, q are fixed primes, and provide slightly better results compared to [10]. In particular, we prove that $H(p, q)$ is less than or equal to 1 when $p - q \equiv 2 \pmod 4$. We also count Heron triangles with a fixed height and provide an estimate of the number of Heron triangles with a fixed prime height. Moreover, we study the decomposability property of a Heron triangle into two similar ones, and provide some cases where a Heron triangle is not decomposable.

Acknowledgment. We sincerely thank the anonymous reviewer for detailed and pertinent comments that substantially improved the technical and the editorial quality of the paper.

References

[1] A.H. Beiler, *Recreations in the Theory of Numbers: The Queen of Mathematics Entertains*, New York: Dover, 1966.

[2] K. D. Boklan, *Heronian Triangles: Problem* 11134, Amer. Math. Monthly 112 (2005), 180; with a solution by K.D. Boklan, M.R. Avidon, in The American Mathematical Monthly, Vol. 114, No. 3 (2007), 261–262.

[3] R. H. Buchholz, *On triangles with rational altitudes, angle bisectors or medians*, Doctoral Dissertation. Newcastle, Australia: Newcastle University, 1989.

[4] R. H. Buchholz and J. A. MacDougall, *Heron Quadrilaterals with Sides in Arithmetic or Geometric Progression*, Bull. Australian Math. Soc. 59 (1999), 263–269.

[5] R. H. Buchholz and J. A. MacDougall, *Cyclic polygons with rational sides and area*, J. Number Theory 128 (2008), 17–48.

[6] G. Campbell and E. H. Goins, *Heron triangles, Diophantine problems and elliptic curves*, preprint; available at `http://www.swarthmore.edu/NatSci/gcampbe1/papers/heron-Campbell-Goins.pdf`

[7] L. E. Dickson, *History of the Theory of Numbers*, volume 2, Chelsea, 1952.

[8] N. J. Fine, *On rational triangles*, American Math. Monthly. 83 (1976), 517–521.

[9] R. C. Fleenor, *Heronian triangles with consecutive integer sides*, J. Recreational Math. 28 (1987), 113–115.

[10] E. J. Ionascu, F. Luca and P. Stănică, *Heron triangle with two fixed sides*, J. Number Theory 126 (2007), 52–67.

[11] J. A. MacDougall, *Heron triangles with sides in arithmetic progression*, J. Recreational Math. 31 (2003), 189–196.

[12] D. Rusin, *Rational triangles with equal area*, New York J. Math. 4 (1998), 1–15.

[13] K. R. S. Sastry, *A Heron difference*, Canadian Math. Soc. (2001), 22–26.

[14] J. B. Tunnell, *A classical Diophantine problem and modular forms of weight* 3/2, Inventiones Math. 72 (1983), 323–334.

[15] E. W. Weisstein, *Pythagorean Triple*, From MathWorld–A Wolfram Web Resource, `http://mathworld.wolfram.com/PythagoreanTriple.html`.

RAMSEY TYPE RESULTS ON THE SOLVABILITY OF CERTAIN EQUATION IN $\mathbb{Z}_M$

Péter Pál Pach[1]
Dept. of Algebra & Number Theory, Eötvös Loránd University, Budapest, Hungary
and
Department of Computer Science and Information Theory, Budapest University of Technology and Economics, Budapest,, Hungary

ppp24@cs.elte.hu,ppp@cs.bme.hu

Received: 6/15/12, Revised: 9/26/12, Accepted: 1/13/13, Published: 1/25/13

Abstract

Csikvári, Gyarmati and Sárközy asked whether there exist Ramsey type theorems for the equations $a + b = cd$ and $ab + 1 = cd$ in $\mathbb{Z}_m$ for large enough m. In this paper it is proved that for any r-coloring of $\mathbb{Z}_m$ the more general equation $a_1 + \cdots + a_n = cd$ has a nontrivial monochromatic solution. Furthermore, an example is presented which shows that the corresponding statement does not hold for the equation $ab + 1 = cd$. We reformulate this problem with an additional natural condition, and give a partial positive answer.

1. Introduction

Sárközy [10], [11] proved that if $\mathcal{A}, \mathcal{B}, \mathcal{C}, \mathcal{D}$ are "large enough" subsets of $\mathbb{Z}_p$, then the equations

$$a + b = cd \tag{1}$$

and

$$ab + 1 = cd \tag{2}$$

can be solved with $a \in \mathcal{A}, b \in \mathcal{B}, c \in \mathcal{C}, d \in \mathcal{D}$. Gyarmati and Sárközy [5] generalized these results on the solvability of (1) and (2) to finite fields. Moreover, there are several papers written on the solvability of equations similar to (1) and (2) over a finite field, especially over $\mathbb{Z}_p$. (See for example, [3], [4].) It is natural to consider the solvability of these equations in $\mathbb{Z}_m$, as well ([8]). However, in [1] and [5] the authors note that for composite m no density-type theorem can be proved for equations (1) and (2) in $\mathbb{Z}_m$, which shows that $\mathbb{Z}_p$ and $\mathbb{Z}_m$ behave differently. Furthermore, it is

[1] *Key words and phrases*: algebraic equation, Ramsey type, solution set

asked whether there exist Ramsey type results: Is it true that for every r-coloring of $\mathbb{Z}_m$ equation (1) (or (2)) has a monochromatic solution, if r, the number of colors, is fixed and $m > N(r)$?

Problem 1. Are there Ramsey type results on the solvability of (1), resp. (2), in $\mathbb{Z}_m$?

Hindman answered the analogue of this question over $\mathbb{N}$ positively ([6]). He showed that for every r-coloring of $\mathbb{N}$ the equation $a_1 + \cdots + a_n = b_1 \ldots b_n$ has a solution where not only the numbers $a_1, \ldots, a_n, b_1, \ldots, b_n$, but also the sums $\sum_{i \in I} a_i$ (where $\emptyset \neq I \subseteq \{1, \ldots, n\}$) and products $\prod_{j \in J} b_j$ (where $\emptyset \neq J \subseteq \{1, \ldots, n\}$) are all distinct (except $\sum_{i=1}^{n} a_i$ and $\prod_{j=1}^{n} b_j$), and all of these sums and products have the same color.

In this paper we consider Problem 1 in $\mathbb{Z}_m$. First note that in the case of equation (1) trivial monochromatic solutions like $0+0 = 0 \cdot 0$ or $2+2 = 2 \cdot 2$ exist, naturally these have to be excluded. This kind of solution, where $a = b = c = d$ is called trivial. In Section 2 we prove that a nontrivial monochromatic solution of (1) always exists. On the other hand in Section 3 a counterexample is presented in the case of equation (2), namely we show a coloring of $\mathbb{Z}_m$ for infinitely many m such that (2) does not have a monochromatic solution. Therefore, instead of $m > N(r)$ the condition $p(m) > N(r)$ (where $p(m)$ denotes the smallest prime divisor of m) has to be assumed, otherwise no Ramsey type result exists. Finally, we show that the answer is affirmative to this modified question in the special case when m is a squarefree number satisfying $r \sum_{p \mid m} \frac{1}{p^{1/4}} \leq \frac{1}{\sqrt{10}}$. To avoid confusion, throughout the paper the notion $(a)_m$ is going to be used for the modulo m residue class of $a \in \mathbb{Z}$ if more than one moduli are used.

2. The Equation $a_1 + \cdots + a_n = cd$

In this section the equation $a + b = cd$, and more generally, the equation $a_1 + \cdots + a_n = cd$ will be studied. The case of prime moduli is well-known by the following theorem of Sárközy:

Theorem A (Sárközy, [10]). *If p is a prime, $\mathcal{A}, \mathcal{B}, \mathcal{C}, \mathcal{D} \subseteq \mathbb{Z}_p$, $|\mathcal{A}||\mathcal{B}||\mathcal{C}||\mathcal{D}| > p^3$, then equation* (1) *has a solution in $\mathbb{Z}_p$ satisfying $a \in \mathcal{A}, b \in \mathcal{B}, c \in \mathcal{C}, d \in \mathcal{D}$.*

In Theorem A the prime p cannot be replaced by an arbitrary $m \in \mathbb{N}$. Moreover, there is no density theorem for equation (1) in $\mathbb{Z}_m$ for arbitrary m, that is, there exists a constant $c > 0$ such that for infinitely many m there exists a set $\mathcal{A} \subseteq \mathbb{Z}_m$ having at least cm elements such that (1) does not have a solution in $\mathcal{A}$.

Example 2. Let $4|m$ and $\mathcal{A} = \{3, 7, 11, \ldots, m-1\} \subseteq \mathbb{Z}_m$. The size of $\mathcal{A}$ is $\frac{m}{4}$. If $a, b, c, d \in \mathcal{A}$, then $a + b \equiv 2 \pmod 4$, $cd \equiv 1 \pmod 4$, hence (1) does not have a solution in $\mathcal{A}$.

Now our aim is to prove that while there is no density theorem, a Ramsey type result exists for the equation $a + b = cd$ over $\mathbb{Z}_m$. Note that in general there are many trivial solutions. First we have to determine all the trivial solutions, and to do this we have to solve the congruence $a^2 \equiv 2a \pmod m$. Let $m = \prod_{i=1}^{r} p_i^{\alpha_i}$ be the canonical form of the number m. By the Chinese Remainder Theorem, it is enough to determine the trivial solutions in $\mathbb{Z}_{p_i^{\alpha_i}}$. Let us denote the number of solutions of the congruence $a^2 \equiv 2a \pmod{p^\alpha}$ by $s(p^\alpha)$. The following cases have to be considered:

- $p > 2$: the congruence $a^2 \equiv 2a \pmod{p^\alpha}$ has 2 solutions, namely $a \equiv 0$ and $a \equiv 2$, hence $s(p^\alpha) = 2$.
- $p^\alpha = 2$: $a \equiv 0$ is the only solution: $s(2) = 1$.
- $p^\alpha = 4$: the 2 solutions are $a \equiv 0$ and $a \equiv 2$, so $s(4) = 2$.
- $p = 2, \alpha \geq 3$: there are four solutions: $a \equiv 0, 2, 2^{\alpha-1}, 2^{\alpha-1} + 2$, hence $s(2^\alpha) = 4$.

By the Chinese Remainder Theorem, the congruence $a + b \equiv cd \pmod m$ has $\prod_{i=1}^{r} s(p_i^{\alpha_i})$ trivial solutions.

Naturally, our goal is to prove that there exists a nontrivial solution of (1), as well. To see this we will show that even the more general equation

$$a_1 + \cdots + a_n = cd \tag{3}$$

always has a monochromatic solution such that $a_1, \ldots, a_n, c, d \in \mathbb{Z}_m$ are pairwise distinct. These solutions, where $a_1, \ldots, a_n, c, d \in \mathbb{Z}_m$ are pairwise distinct, will be called primitive. The proof of this result is based on the following version of Rado's theorem ([7], Theorem 9.4):

Rado's Theorem. *Let $v \geq 2$. Let $c_i \in \mathbb{Z} \setminus \{0\}$, $1 \leq i \leq v$ be constants such that there exists a nonempty set $D \subseteq \{i : 1 \leq i \leq v\}$ with $\sum_{i \in D} c_i = 0$. If there exist distinct (not necessarily positive) integers y_i such that $\sum_{i=1}^{v} c_i y_i = 0$, then for every natural number r there exists some t such that for every r-coloring of the set $\{1, 2, \ldots, t\}$ the equation*

$$c_1 x_1 + \cdots + c_v x_v = 0$$

has a monochromatic solution $b_1, \ldots, b_v$ in $\{1, 2, \ldots, t\}$, where the b_i-s are distinct.

For more details on Rado's theorems, see [2], [7] and [9]. The following observation is also needed:

Lemma 3. *Let $T \in \mathbb{N}$ and $N = T^T$. If $m > N$, then m has a prime power divisor greater than T.*

Proof. For the sake of contradiction, suppose the contrary. Then each prime divisor of m is at most T, therefore m is the product of at most T prime powers. Since each prime power divisor is at most T, we have that $m \leq T^T$, which contradicts our assumption. □

Theorem 4. *For every $n, r \in \mathbb{N}$ there exists some $N = N(n, r)$ such that for every $N < m \in \mathbb{N}$ and every r-coloring of $\mathbb{Z}_m$, equation (3) has a primitive monochromatic solution in $\mathbb{Z}_m$.*

Proof. First assume that $n \geq 2$. Let $\alpha_i = (1-n) + 2(i-1)$ (for $i = 1, \ldots, n$), $\gamma = n, \delta = -n$. Note that the numbers $\alpha_1, \ldots, \alpha_n, \gamma, \delta$ are distinct integers and $\alpha_1 + \cdots + \alpha_n - n\gamma - n\delta = 0$. Therefore, the equation $\alpha_1 + \cdots + \alpha_n - n\gamma - n\delta = 0$ has a solution in $\mathbb{Z}$ where the α_i, γ, δ are distinct. Moreover, the sum of the coefficients of $\alpha_1, \ldots, \alpha_n, \gamma$ is $1 + \cdots + 1 - n = 0$, and thus the equation $\alpha_1 + \cdots + \alpha_n - n\gamma - n\delta = 0$ satisfies the conditions of Rado's theorem, so the equation has a primitive monochromatic solution in $\{1, 2, \ldots, K\}$ for every r-coloring of $\{1, 2, \ldots, K\}$, if K is large enough, say $K \geq K_0$. Let $C = \max(K_0, r^4(n+2)^4)$.

Take an arbitrary r-coloring of $\mathbb{Z}_m$. By applying Lemma 3 with $T = C^3$ we obtain that if $m > N = T^T$, then m has a prime power divisor greater than T.

Now we prove that $N = T^T$ satisfies the condition of the theorem. In the proof we distinguish two cases: the prime power divisor guaranteed by Lemma 3 is itself a prime or it is not.

As the first case suppose that $p > r^4(n+2)^4$ is a prime divisor of m such that $p^2 \nmid m$. Therefore, p and m/p are coprime, since $p \nmid m/p$. For $1 \leq i \leq p$ define the mod m residue class $(x_i)_m$ by the congruences $x_i \equiv i \pmod{p}$ and $x_i \equiv 0 \pmod{m/p}$. Now, we define an r-coloring of $\mathbb{Z}_p$ depending on the given r-coloring of $\mathbb{Z}_m$ in the following way: For $1 \leq i \leq p$ let the color of $(i)_p \in \mathbb{Z}_p$ be the color of $(x_i)_m$. Note that $\mathbb{Z}_p$ is colored by r colors, so we can choose (at least) $\frac{p}{r}$ elements having the same color. Let us denote the set of these (at least) $\frac{p}{r}$ elements by $\mathcal{S}$. Now we partition $\mathcal{S} \subseteq \mathbb{Z}_p$ into $n+2$ disjoint sets $\mathcal{S}_1, \ldots, \mathcal{S}_{n+2} \subseteq \mathcal{S}$ such that the size of any two of them differ by at most 1. Since $p \geq r^4(n+2)^4 \geq 2r(n+2)$, each of the sets $\mathcal{S}_i$ has size at least $\lfloor \frac{p}{r(n+2)} \rfloor \geq \frac{p}{2r(n+2)}$. Now let $\mathcal{A}, \mathcal{B}, \mathcal{C}, \mathcal{D} \subseteq \mathbb{Z}_p$ be defined in the following way: $\mathcal{A} = \mathcal{S}_1$, $\mathcal{B} = \mathcal{S}_2 + \cdots + \mathcal{S}_n = \{s_2 + \cdots + s_n | s_2 \in \mathcal{S}_2, \ldots, s_n \in \mathcal{S}_n\}$, $\mathcal{C} = \mathcal{S}_{n+1}$, $\mathcal{D} = \mathcal{S}_{n+2}$. By $p > r^4(n+2)^4$ we obtain that $|\mathcal{A}||\mathcal{B}||\mathcal{C}||\mathcal{D}| \geq \left(\frac{p}{r(n+2)}\right)^4 > p^3$, so Theorem A can be applied, which yields that there exist $a \in \mathcal{A}, b \in \mathcal{B}, c \in \mathcal{C}, d \in \mathcal{D}$ such that $a + b = cd$ in $\mathbb{Z}_p$. As $b \in \mathcal{B}$, we have $b = a_2 + \cdots + a_n$ for some $a_i \in \mathcal{S}_i$.

Let $a_1 = a$. Therefore, there exist $a_1, \ldots, a_n, c, d \in \{1, 2, \ldots, p\}$ such that the corresponding mod p residue classes have the same color, and the congruence

$$a_1 + \cdots + a_n \equiv cd \pmod{p}$$

holds. The elements $(x_{a_1})_m, \ldots, (x_{a_n})_m, (x_c)_m, (x_d)_m \in \mathbb{Z}_m$ have the same color as $(a_1)_p, \ldots, (a_n)_p, (c)_p, (d)_p \in \mathbb{Z}_p$; moreover, they are distinct, since $(a_1)_p, \ldots, (a_n)_p, (c)_p, (d)_p \in \mathbb{Z}_p$ are distinct as well. Furthermore, $(x_{a_1})_m, \ldots, (x_{a_n})_m, (x_c)_m, (x_d)_m$ is a solution of (3), since

- $(x_{a_1})_{m/p} \ldots (x_{a_n})_{m/p} \equiv (x_c)_{m/p}(x_d)_{m/p} \pmod{m/p}$, as $x_i \equiv 0 \pmod{m/p}$ for every i, and
- $(x_{a_1})_p \ldots (x_{a_n})_p \equiv (x_c)_p(x_d)_p \pmod{p}$, as $x_i \equiv i \pmod{p}$ for every i and $a_1 \ldots a_n \equiv cd \pmod{p}$.

Hence, $(x_{a_1})_m, \ldots, (x_{a_n})_m, (x_c)_m, (x_d)_m$ form a primitive monochromatic solution of (3) in $\mathbb{Z}_m$.

As the second case, assume that for a prime power (but not prime) $p^t \geq C^3$ we have $p^t | m$, where $t \geq 2$ and t is the largest integer such that $p^t | m$. Let $t_0 = \lfloor t/2 \rfloor$. As $t_0 \geq t/3$, we have $p^{t_0} \geq C$. We show that a monochromatic solution of the equation $a_1 + \cdots + a_n \equiv cd \pmod{m}$ can be found among the residue classes of the form $(y \cdot \frac{m}{p^{t_0}} + n)_m$. Note that the congruence

$$\left(\alpha_1 \cdot \frac{m}{p^{t_0}} + n\right)_m + \cdots + \left(\alpha_n \cdot \frac{m}{p^{t_0}} + n\right)_m \equiv \left(\gamma \cdot \frac{m}{p^{t_0}} + n\right)_m \left(\delta \cdot \frac{m}{p^{t_0}} + n\right)_m \pmod{m} \quad (4)$$

is equivalent to

$$\alpha_1 + \cdots + \alpha_n \equiv n\gamma + n\delta \pmod{p^{t_0}}.$$

As the next step we define an r-coloring of $\mathbb{N}$ depending on the given r-coloring of $\mathbb{Z}_m$. Let the color of $y \in \mathbb{N}$ be the color of $(y \cdot \frac{m}{p^{t_0}} + n)_m \in \mathbb{Z}_m$.

Since $C \geq K_0$, Rado's theorem implies that there exist distinct integers $\alpha_1, \ldots, \alpha_n, \gamma, \delta \in \{1, 2, \ldots, C\}$ having the same color and satisfying $\alpha_1 + \cdots + \alpha_n - n\gamma - n\delta = 0$. The residue classes $a_i = (\alpha_i \cdot \frac{m}{p^{t_0}} + n)_m, c = (\gamma \cdot \frac{m}{p^{t_0}} + n)_m, d = (\delta \cdot \frac{m}{p^{t_0}} + n)_m$ give a solution of (3), moreover they are distinct, since $\alpha_1, \ldots, \alpha_n, \gamma, \delta \in \{1, 2, \ldots, C\}$ are distinct and $p^{t_0} \geq C$.

Finally, let us examine the case when $n = 1$. By Rado's theorem for every $r \in \mathbb{N}$ there exists some $M = M(r)$ such that for every r-coloring of $\mathbb{N}$ the equation $\alpha = \gamma + \delta$ has a primitive monochromatic solution in $\{1, \ldots, M\}$. Suppose that $2^M < m$ and take an arbitrary r-coloring of $\mathbb{Z}_m$. Define a coloring of $\mathbb{N}$ in the following way: Let the color of $a \in \mathbb{N}$ be the color of $(2^a)_m$ in $\mathbb{Z}_m$. Rado's theorem yields that there exist three distinct positive integers $\alpha, \gamma, \delta \in \{1, \dots, M\}$ having the same color such that $\alpha = \gamma + \delta$. Then $a = (2^\alpha)_m, c = (2^\gamma)_m, d = (2^\delta)_m$ is a primitive monochromatic solution of $a = cd$ in $\mathbb{Z}_m$.

Hence, we showed that if $m > N = T^T$, then (3) has a nontrivial monochromatic solution in $\mathbb{Z}_m$. □

3. The Equation $ab + 1 = cd$

In this section equation (2) will be studied. First, we will show that if m has a small prime divisor, then there is no Ramsey type theorem on the solvability of $ab + 1 = cd$ in $\mathbb{Z}_m$ in the classical sense: If we fix the number of colors r and m is large enough, then a monochromatic solution need not exist.

Example 5. Let $p|m$ and the color of $(x)_m \in \mathbb{Z}_m$ be the mod p residue class containing x. If $(a)_m, (b)_m, (c)_m, (d)_m \in \mathbb{Z}_m$ have the same color, then $ab \equiv cd \pmod{p}$, so $ab + 1 \neq cd$ in $\mathbb{Z}_m$.

In this example we colored $\mathbb{Z}_m$ by p colors, where $p|m$, and there is no monochromatic solution of the equation $ab+1 = cd$, which shows that the least prime divisor of m, denoted by $p(m)$, has to be greater than the number of colors. To exclude counterexamples of this kind we reformulate the problem in the following way:

Problem 6. Are there Ramsey type results on the solvability of $ab+1 = cd$ in $\mathbb{Z}_m$ if r, the number of colors is fixed and $p(m)$ is large enough in terms of r?

We give a partial positive answer to this question, namely we show that the answer is affirmative, if m is squarefree and

$$r \sum_{p|m} \frac{1}{p^{1/4}} \leq \frac{1}{\sqrt{10}}.$$

To prove this result the following theorem of Sárközy is needed:

Theorem B (Sárközy, [11]). *If p is a prime, $\mathcal{A}, \mathcal{B}, \mathcal{C}, \mathcal{D} \subseteq \mathbb{Z}_p$, $|\mathcal{A}||\mathcal{B}||\mathcal{C}||\mathcal{D}| > 100p^3$, then the equation $ab + 1 = cd$ has a solution in $\mathbb{Z}_p$ satisfying $a \in \mathcal{A}, b \in \mathcal{B}, c \in \mathcal{C}, d \in \mathcal{D}$.*

Now we are ready to solve Problem 6 under a certain condition.

Theorem 7. *Let $m = p_1 \dots p_s$ be the product of s different primes. Let $\mathcal{A} \subseteq \mathbb{Z}_m$ and $\alpha = \frac{|\mathcal{A}|}{m}$. If $\sum_{j=1}^{s} \frac{1}{p_j^{1/4}} \leq \frac{\alpha}{\sqrt{10}}$, then there exist $a, b, c, d \in \mathcal{A}$ satisfying the equation $ab + 1 = cd$.*

Proof. The main idea of the proof is to solve the congruence system $ab + 1 \equiv cd \pmod{p_i}$ (for $1 \le i \le s$) step by step. Our aim is to obtain a solution finally where $(a)_m, (b)_m, (c)_m, (d)_m$ lie in $\mathcal{A}$. As the first step we show that the following statement holds: Let $m = m_1 m_2 \dots m_s$, where $m_1, m_2, \dots, m_s$ are pairwise coprime. Let $\mathcal{A} \subseteq \mathbb{Z}_m = \mathbb{Z}_{m_1} \times \dots \times \mathbb{Z}_{m_s}$, $\alpha = \frac{|\mathcal{A}|}{m}$ and $\alpha_1, \dots, \alpha_s \ge 0$ satisfying $\alpha_1 + \dots + \alpha_s \le \alpha$. Then there exist sets $\mathcal{A}_1 \subseteq \mathbb{Z}_{m_1}$, $\mathcal{A}_j(a_1, \dots, a_{j-1}) \subseteq \mathbb{Z}_{m_j}$ (for every $a_1 \in \mathcal{A}_1$, $a_2 \in \mathcal{A}_2(a_1)$, and so on, $a_{j-1} \in \mathcal{A}_{j-1}(a_1, \dots, a_{j-2})$) satisfying the following conditions:

- $|\mathcal{A}_1| \ge \alpha_1 m_1$
- For every $2 \le j \le r$, for every $a_1 \in \mathcal{A}_1$, for every $a_2 \in \mathcal{A}_2(a_1)$, for every $a_3 \in \mathcal{A}_3(a_1, a_2)$ and so on, for every $a_{j-1} \in \mathcal{A}_{j-1}(a_1, \dots, a_{j-2})$ the set $\mathcal{A}_j(a_1, \dots, a_{j-1})$ has at least $\alpha_j m_j$ elements.
- If $a_1 \in \mathcal{A}_1, a_2 \in \mathcal{A}_2(a_1), \dots, a_s \in \mathcal{A}_s(a_1, \dots, a_{s-1})$, then $(a_1, \dots, a_s) \in \mathcal{A}$.

So $\mathcal{A}_j(a_1, \dots, a_{j-1}) \subseteq \mathbb{Z}_{m_j}$ contains at least $\alpha_j m_j$ possible continuations of the vector $(a_1, \dots, a_{j-1}) \in \mathbb{Z}_{m_1} \times \dots \times \mathbb{Z}_{m_{j-1}}$. More precisely, we could add at least $\alpha_j m_j$ elements $a_j \in \mathbb{Z}_{m_j}$ as the j-th coordinate to the vector $(a_1, \dots, a_{j-1})$ such that after the s-th step we have vectors belonging to $\mathcal{A} \subseteq \mathbb{Z}_{m_1} \times \dots \times \mathbb{Z}_{m_s}$.

We prove this assertion by induction on s. For $s = 1$ the statement holds trivially. Let $s = 2$. Let $\mathcal{A}_2(a_1) = \{a_2 \in \mathbb{Z}_{m_2} : (a_1, a_2) \in \mathcal{A}\}$. Then let $\mathcal{A}_1 = \{a_1 \in \mathbb{Z}_{m_1} : |\mathcal{A}_2(a_1)| \ge \alpha_2 m_2\}$. As $\mathcal{A} = \bigcup_{a_1 \in \mathcal{A}_1} \{a_1\} \times \mathcal{A}_2(a_1) \cup \bigcup_{a_1 \in \mathbb{Z}_{m_1} \setminus \mathcal{A}_1} \{a_1\} \times \mathcal{A}_2(a_1)$, we have

$$\alpha m_1 m_2 = |\mathcal{A}| \le |\mathcal{A}_1| m_2 + (m_1 - |\mathcal{A}_1|)\alpha_2 m_2 \le |\mathcal{A}_1| m_2 + \alpha_2 m_1 m_2.$$

Thus the size of $\mathcal{A}_1$ is at least $\alpha_1 m_1$, as needed. Applying this repeatedly we get that the statement is true for every $s > 2$ as well.

This implies that in $\mathbb{Z}_{m_1} \times \dots \times \mathbb{Z}_{m_s}$ at least $\alpha_1 m_1$ first coordinates can be chosen, the set $\mathcal{A}_1$ contains them. For every $a_1 \in \mathcal{A}_1$, $\alpha_2 m_2$ second coordinates can be chosen, the set $\mathcal{A}_2(a_1)$ contains them. And so on. Finally, $\alpha_s m_s$ s-th coordinates can be chosen in such a way that all of the elements $(a_1, a_2, \dots, a_s)$ obtained in $\mathbb{Z}_{m_1} \times \dots \times \mathbb{Z}_{m_s}$ lie in $\mathcal{A}$.

As the second step let $m = p_1 \dots p_s$. As $\mathbb{Z}_m = \mathbb{Z}_{p_1} \times \dots \times \mathbb{Z}_{p_s}$ by the Chinese Remainder theorem, the modulo m residue class of a can be identified by an ordered s-tuple where the jth coordinate is the mod p_j residue of the residue class of a: $a \leftrightarrow (a_1, \dots, a_s)$, where $(a)_{p_j} = (a_j)_{p_j}$ for every $1 \le j \le s$. Solving the equation $ab + 1 = cd$ in $\mathcal{A} \subseteq \mathbb{Z}_m$ is equivalent to solve the system of equations $a_i b_i + 1 = c_i d_i$ in $\mathbb{Z}_{p_i}$ (where $1 \le i \le s$) in such a way that $(a_1, \dots, a_s), (b_1, \dots, b_s), (c_1, \dots, c_s), (d_1, \dots, d_s) \in \mathcal{A}$. We have just proved that for every $\alpha_1, \dots, \alpha_s \ge 0$ satisfying $\alpha_1 + \dots + \alpha_s \le \alpha$ subsets $\mathcal{A}_j(a_1, \dots, a_{j-1}) \subseteq \mathbb{Z}_{p_j}$ can be chosen which satisfy the following conditions:

- $|\mathcal{A}_1| \geq \alpha_1 p_1$.
- For every $2 \leq j \leq s$, for every $a_1 \in \mathcal{A}_1$, for every $a_2 \in \mathcal{A}_2(a_1)$, for every $a_3 \in \mathcal{A}_3(a_1, a_2)$ and so on, for every $a_{j-1} \in \mathcal{A}_{j-1}(a_1, \dots, a_{j-2})$ the set $\mathcal{A}_j(a_1, \dots, a_{j-1})$ has at least $\alpha_j p_j$ elements,
- If $a_1 \in \mathcal{A}_1, a_2 \in \mathcal{A}_2(a_1), \dots, a_s \in \mathcal{A}_s(a_1, \dots, a_{s-1})$, then $(a_1, \dots, a_s) \in \mathcal{A}$.

As a next step we are going to apply Theorem B repeatedly. In order to do this the inequalities $(\alpha_j p_j)^4 \geq 100 p_j^3$ $(1 \leq j \leq s)$ have to hold. Therefore, let $\alpha_j = \frac{\sqrt{10}}{p_j^{1/4}}$ (for every $1 \leq j \leq s$). Now note that $\sum_{j=1}^{s} \alpha_j = \sqrt{10} \sum_{j=1}^{s} \frac{1}{p_j^{1/4}} \leq \alpha$, so the previous statement can be applied, and the sets $\mathcal{A}_j(a_1, \dots, a_{j-1})$ can be chosen. As $|\mathcal{A}_1| \geq \alpha_1 p_1$, Theorem B yields that the equation $a_1 b_1 + 1 = c_1 d_1$ (in $\mathbb{Z}_{p_1}$) can be solved in $\mathcal{A}_1$. Fix this solution. Since each of the sets $\mathcal{A}_2(a_1), \mathcal{A}_2(b_1), \mathcal{A}_2(c_1), \mathcal{A}_2(d_1)$ has cardinality at least $\alpha_2 p_2$, the equation $a_2 b_2 + 1 = c_2 d_2$ (in $\mathbb{Z}_{p_2}$) has a solution such that $a_2 \in \mathcal{A}_2(a_1), b_2 \in \mathcal{A}_2(b_1), c_2 \in \mathcal{A}_2(c_1), d_2 \in \mathcal{A}_2(d_1)$. In the general step $a_1, \dots, a_j, b_1, \dots, b_j, c_1, \dots, c_j, d_1, \dots, d_j$ are already fixed. Since each of the sets $\mathcal{A}_{j+1}(a_1, \dots, a_j)$, $\mathcal{A}_{j+1}(b_1, \dots, b_j)$, $\mathcal{A}_{j+1}(c_1, \dots, c_j)$, $\mathcal{A}_{j+1}(d_1, \dots, d_j)$ has cardinality at least $\alpha_{j+1} p_{j+1}$, the equation $a_{j+1} b_{j+1} + 1 = c_{j+1} d_{j+1}$ (in $\mathbb{Z}_{p_{j+1}}$) has a solution such that $a_{j+1} \in \mathcal{A}_{j+1}(a_1, \dots, a_j)$, $b_{j+1} \in \mathcal{A}_{j+1}(b_1, \dots, b_j)$, $c_{j+1} \in \mathcal{A}_{j+1}(c_1, \dots, c_j)$, $d_{j+1} \in \mathcal{A}_{j+1}(d_1, \dots, d_j)$. At the end, since each of the sets $\mathcal{A}_s(a_1, \dots, a_{s-1})$, $\mathcal{A}_s(b_1, \dots, b_{s-1})$, $\mathcal{A}_s(c_1, \dots, c_{s-1})$, $\mathcal{A}_s(d_1, \dots, d_{s-1})$ has cardinality at least $\alpha_s p_s$, the equation $a_s b_s + 1 = c_s d_s$ (in $\mathbb{Z}_{p_s}$) has a solution such that $a_s \in \mathcal{A}_s(a_1, \dots, a_{s-1})$, $b_s \in \mathcal{A}_s(b_1, \dots, b_{s-1})$, $c_s \in \mathcal{A}_s(c_1, \dots, c_{s-1})$, $d_s \in \mathcal{A}_s(d_1, \dots, d_{s-1})$.

Therefore, for $a = (a_1, \dots, a_s), b = (b_1, \dots, b_s), c = (c_1, \dots, c_s), d = (d_1, \dots, d_s) \in \mathcal{A}$ we have $ab + 1 = cd$ (in $\mathbb{Z}_m$).

□

Corollary 8. *Let $m = p_1 \dots p_s$ be the product of s different primes. If $r \sum_{j=1}^{s} \frac{1}{p_j^{1/4}} \leq \frac{1}{\sqrt{10}}$, then for every r-coloring of $\mathbb{Z}_m$ the equation $ab+1 = cd$ has a monochromatic solution.*

4. Acknowledgements

The author is thankful to A. Sárközy for turning his attention to this topic and for the useful hints.

References

[1] P. Csikvári, K. Gyarmati, A. Sárközy: *Density and Ramsey type results on algebraic equations with restricted solution sets*, Combinatorica, to appear.

[2] R. L. Graham, B. L. Rotschild, J. H. Spencer: *Ramsey Theory*. Wiley-Interscience Series in Discrete Mathematics and Optimization. John Wiley & Sons, 1990.

[3] K. Gyarmati: *On a problem of Diophantus*, Acta Arith. 97 (2001), 53-65.

[4] K. Gyarmati, A. Sárközy: *Equations in finite fields with restricted solution sets, I. (Character sums)*, Acta Math. Hungar. 118 (2008), 129-148.

[5] K. Gyarmati, A. Sárközy: *Equations in finite fields with restricted solution sets, II (Algebraic equations)*, Acta Math. Hungar. 119 (2008), 259-280.

[6] N. Hindman: *Monochromatic sums equal to products in* $\mathbb{N}$, Integers 11A (2011), Article 10, 1-10.

[7] B. M. Landman, A. Robertson: *Ramsey Theory on the Integers*, American Mathematical Society, 2003.

[8] C. Pomerance, A. Schinzel: *Multiplicative properties of sets of residues*, Moscow J. Combin. and Number Theory 1 (2011), 52–66.

[9] R. Rado: *Studien zur Kombinatorik*, Math. Z. 36 (1) (1933), 424-470.

[10] A. Sárközy: *On sums and products of residues modulo p*, Acta Arith. 118 (2005), 403-409.

[11] A. Sárközy: *On products and shifted products of residues modulo p*, Integers 8 (2008), A9, 8pp.

A COMBINATORIAL PROOF OF S-ADICITY FOR SEQUENCES WITH LINEAR COMPLEXITY

Julien Leroy[1]
LAMFA, CNRS UMR 6410, Univ. de Picardie Jules Verne, Amiens, France
julien.leroy@uni.lu

Gwenaël Richomme
LIRMM (CNRS, Univ. Montpellier 2), Montpellier, France
and
Univ. Montpellier 3, Montpellier, France
gwenael.richomme@lirmm.fr

Received: 5/14/12, Revised: 12/3/12, Accepted: 1/21/13, Published: 1/25/13

Abstract

Using Rauzy graphs, Ferenczi proved that if a symbolic dynamical system has linear complexity then it is S-adic. Being more specific, the result can also be proved for infinite words. We provide a new proof of this latter result using the notion of return words to a set of words.

1. Introduction

Since Morse and Hedlund's work [18], due to their numerous properties and applications, Sturmian words were given a lot of attention (see for instance surveys and studies in [1, 3, 8, 15, 20]). In this paper, we are concerned with two of these properties. First, Sturmian words are examples of words with linear complexity [18]. Recall that the complexity of an infinite word is the function that associates to each integer n the number of factors (blocks of consecutive letters) of length n that the infinite word contains. Secondly, Sturmian words have a multi-scale property since each Sturmian word is the image of another Sturmian word by a morphism (a map, also called substitution, defined on letters and compatible with the concatenation operation) taken from a finite set S_{sturm} consisting of 4 morphisms [17]. More precisely, given a Sturmian word $\mathbf{s}$, one can find a sequence $(f_i)_{i\geq 0}$ of morphisms of S_{Sturm} and a letter $(a_i)_{i\geq 0}$, such that $\mathbf{s} = \lim_{n\to\infty} f_0 \cdots f_n(a_{n+1}^{\omega})$. Sturmian words are then called S_{Sturm}-adic. More generally, an infinite word $\mathbf{w}$ is said to be S-adic

[1]This paper was partially written while the first author was invited to Montpellier in June 2011 thanks to funding from Groupe AMIS from Montpellier 3.

when S is a finite set of morphisms and the previous equality holds for $\mathbf{w}$ instead of $\mathbf{s}$ and for a suitable sequence of morphisms of S.

Since Pansiot's work [19], it is known that fixed points of morphisms that are S-adic words for a singleton set S do not necessarily have linear complexity. Also, Cassaigne showed [21] that any word $\mathbf{w} = (\mathbf{w}_i)_{i\geq 0}$ (where each $\mathbf{w}_i$ is a letter) over an alphabet A is S-adic, with $S = \{\sigma_a, \varphi_a \mid a \in A\}$, where, using a letter α that does not belong to A, the morphisms σ_a and φ_a are defined by, for all letters a, b in A, $\sigma_a(\alpha) = a$, $\varphi_a(\alpha) = \alpha a$, $\sigma_a(b) = b = \varphi_a(b)$: $\mathbf{w} = \lim_{n\to\infty} \sigma_{\mathbf{w}_0}\varphi_{\mathbf{w}_1}\cdots\varphi_{\mathbf{w}_n}(\alpha^\omega)$.

Nevertheless, it can be observed that (as far as we know) all infinite words with linear complexity that have been studied are S-adic for a suitable (finite) set S of morphisms. Moreover, an important result due to Ferenczi in the context of symbolic dynamical systems can be reformulated in the context of infinite words as follows.

Theorem 1.1. *(*[12], see also [13]*) If an aperiodic and uniformly recurrent word has linear complexity then, for a finite set S of non-erasing morphisms, it is an everywhere growing S-adic sequence.*

The *everywhere growing property*[2] means that there exists a sequence $(\sigma_n)_{n\in\mathbb{N}} \in S^{\mathbb{N}}$ of morphisms of S generating the considered infinite word such that for all letters a in A, the length of $\sigma_0\sigma_1\cdots\sigma_n(a)$ tends to infinity with n. This property is not verified by Cassaigne's morphisms. The previous result of Ferenczi and other considerations (as for instance, the existence of an S-adic characterization of the family of linearly recurrent words [10]) are the origin of the S-adic conjecture which states the existence of a strong definition of S-adicity that would be equivalent to having linear complexity for a symbolic dynamical system or an infinite word (see [20, Section 12.1]). Unfortunately the everywhere growing property is not sufficient to solve this conjecture. Some examples [11] tend to show that the strong expected definition (if it exists) should concern not only the set S of morphisms but also the sequence of these morphisms used to define the infinite word.

Ferenczi's proof of Theorem 1.1 generalizes an idea of Arnoux and Rauzy [2]; it is based on the study of Rauzy graphs, a tool independently introduced by Rauzy [23] and Boshernitzan [5] and now often used to study factor complexity of words. The Rauzy graph G_n of order n of an infinite word $\mathbf{w}$ is the directed graph whose vertices are the factors of length n of $\mathbf{w}$ and such that any factor $u = u_1\cdots u_{n+1}$ of length $n+1$ of $\mathbf{w}$ induces an edge in G_n from $u_1\cdots u_n$ to $u_2\cdots u_{n+1}$. This edge is labelled by the pair (u_1, u_{n+1}). The label of a path in the graph is the concatenation of labels of edges along the path.

Leroy [13] revisited Ferenczi's proof in order to provide more information on the morphisms arising in the S-adic representation. He succeeded in going deeper into

[2]This property was called *ω-growth property* in [13]. However, when S is a singleton, the terminology *everywhere growing* is widely used so it is natural to keep working with it.

the study of underlying Rauzy graphs. He observed, without going further, that some important paths (called segments) in the Rauzy graphs used by Ferenczi are related to the notion of return words to a set. More precisely, the first (resp. the second) component of their labels corresponds to left (resp. right) return words to the set of left special factors. In this paper, we use this notion of return words to prove Theorem 1.1. Even though it is an adaptation of Ferenczi's, we hope that our more compact proof will open new ideas to tackle the S-adic conjecture.

An interest of our approach may be to provide non-contextual definitions of short or long return words before the proof when their corresponding terms (short or long segments) in the context of Rauzy graphs are defined in Ferenczi's proof. Furthermore, this concept of return words to a set W generalizes the one of Rauzy graphs since it does not require that the words in W share the same length.

The next two sections are preliminaries for our proof in Section 4. They allow us to present these notions of short and long return words in a general context independently of the complexity of the infinite words considered.

2. Return Words to Special Words

The aim of this section is to recall the main combinatorial notions used in our proof. This is done through the development of an example that will be used in Section 3 to illustrate the existence of short return words that are not constant return words.

2.1. Words and Morphisms

We assume that readers are familiar with combinatorics on words; for basic (possibly omitted) definitions, we follow [4, 14, 15].

Given an *alphabet* A, that is, a finite set of symbols called *letters*, we let A^* denote the set of all finite words over A (that is, the set of all finite sequences of elements of A). As usual, the *concatenation* of two words u and v is simply denoted uv. It is well known that the set A^* equipped with the concatenation operation is a free monoid with neutral element ε, the *empty word*.

Given two alphabets A and B, a free monoid morphism, or simply *morphism* σ, is a map from A^* to B^* such that $\sigma(uv) = \sigma(u)\sigma(v)$ for all words u and v over A (observe that this implies $\sigma(\varepsilon) = \varepsilon$). It is well known that a morphism is completely determined by the images of the letters.

From now on, denoting $\Gamma = \{e_0, e_1, e_2, e_3, b_1, b_2, b_3, b_4, a_1, a_2, 1, 2, c\}$ and $\Sigma = \{e_1, e_2, e_3, b, a, 1, 2, c\}$, we consider the following two morphisms ν and ψ defined respectively from Γ^* to Γ^* and from Γ^* to Σ^* by

$$\begin{array}{rclcrcl}
\nu(e_0) &=& e_0\ 1\ \ b_4\ 1\ \ 1\ \ e_0 & & \psi(e_0) &=& e_1\\
\nu(e_1) &=& e_1\ c\ \ c\ \ b_1\ 1\ \ e_1 & & \psi(e_1) &=& e_1\\
\nu(e_2) &=& e_2\ c\ \ c\ \ b_2\ 2\ \ e_2 & & \psi(e_2) &=& e_2\\
\nu(e_3) &=& e_3\ c\ \ c\ \ b_3\ 2\ \ e_3 & & \psi(e_3) &=& e_3\\
\nu(b_1) &=& c\ \ c\ \ c\ \ e_1\ a_1\ b_1 & & \psi(b_1) &=& b\\
\nu(b_2) &=& c\ \ c\ \ c\ \ e_2\ a_2\ b_2 & & \psi(b_2) &=& b\\
\nu(b_3) &=& c\ \ c\ \ c\ \ c\ \ e_3\ b_3 & & \psi(b_3) &=& b\\
\nu(b_4) &=& c\ \ c\ \ c\ \ c\ \ c\ \ b_4 & & \psi(b_4) &=& b\\
\nu(1) &=& c\ \ c\ \ c\ \ c\ \ b_1\ 1 & & \psi(1) &=& 1\\
\nu(2) &=& c\ \ c\ \ c\ \ c\ \ b_1\ 2 & & \psi(2) &=& 2\\
\nu(a_1) &=& c\ \ a_1\ c\ \ c\ \ e_0\ e_1 & & \psi(a_1) &=& a\\
\nu(a_2) &=& c\ \ a_2\ c\ \ c\ \ e_2\ e_3 & & \psi(a_2) &=& a\\
\nu(c) &=& c\ \ c\ \ c\ \ c\ \ a_1\ a_2 & & \psi(c) &=& c
\end{array}$$

One can observe that ν is *uniform*, that is, the images of the letters have the same length (let us recall that the *length* of a word u, denoted by $|u|$, is the number of letters of u). The morphism ν is also *primitive*, that is, there exists an integer k (here $k = 5$) such that for all letters x, y in Γ, the letter y occurs in $\nu^k(x)$.

Note also that the morphisms ν and $\psi \circ \nu$ are injective. Moreover, they are synchronizing morphisms; a morphism f from A^* to B^* is *synchronizing* if for all letters x, y, z in the alphabet A and words p, s over A, the equality $f(x)f(y) = pf(z)s$ implies $p = \varepsilon$ or $s = \varepsilon$.

2.2. Infinite Words and Factors

A (right) *infinite word* (or *sequence*) over an alphabet A is an element of A^ω.

When a morphism is not *erasing*, that is, the images of letters are never the empty word, the notion of morphism extends naturally to infinite words. From now on, $\mathbf{u}$ will be the fixed point of ν ($\mathbf{u} = \nu(\mathbf{u})$) starting with the letter c.

A word v is a *factor* of a finite or infinite word w over an alphabet A if it occurs in w or equivalently, if $w = xvy$ for some words x, y in $A^* \cup A^\omega$. When $x = \varepsilon$ (resp. $y = \varepsilon$), v is a *prefix* (resp. *suffix*) of w. We let $\mathrm{F}(w)$ denote the set of all factors of w.

Observe that for all $n \geq 0$, words $\nu^n(1)$ and $\nu^n(2)$ are factors of $\mathbf{u}$ (let us recall that, given a morphism f and an integer $n \geq 0$, f^n is the identity if $n = 0$, and is $f^{n-1} \circ f$ otherwise).

Fact 2.1. *Let $n \geq 0$ be an integer and v be a factor of $\mathbf{u}$. If $\psi(v) = \psi(\nu^n(1))$ then $v = \nu^n(1)$. If $\psi(v) = \psi(\nu^n(2))$ then $v = \nu^n(2)$.*

Proof. Let $i \in \{1, 2\}$. The result is immediate for $n = 0$ from the definitions of ν and ψ. Assume $n \geq 1$. Since $\psi \circ \nu$ is synchronizing and $\psi(v)$ is a factor of $\psi(\mathbf{u}) = \psi \circ \nu(\mathbf{u})$, there must exist a word v' such that $\psi(v) = \psi \circ \nu(v')$. The injectivity of $\psi \circ \nu$ implies $v' = \nu^{n-1}(i)$. Observe that for all letters α in Γ and for all words x in Γ^*, the equality $\psi(x) = \psi \circ \nu(\alpha)$ implies $x = \psi(\alpha)$. This extends to α in Γ^*. Hence $v = \nu^n(i)$. □

As with any fixed point of a primitive morphism [22], the word $\mathbf{u}$ is an example of a *uniformly recurrent* infinite word, *i.e.*, its factors occur infinitely often and with bounded gaps.

The complexity function $p_{\mathbf{w}}$ of an infinite word $\mathbf{w}$ is the function that counts the number of factors of each length (See [7] for a recent survey on this function). An infinite word $\mathbf{w}$ is said to have *linear complexity* if there is a constant C such that $p_{\mathbf{w}}(n) \leq Cn$ for all $n \geq 1$. Again, as with any fixed point of a primitive morphism [22], the word $\mathbf{u}$ is an example of an infinite word with linear complexity.

2.3. Special Factors

Let us recall that a factor u of an infinite word $\mathbf{w}$ over an alphabet A is *left special* if there exist at least two different letters a and b in A such that the words au and bu are factors of $\mathbf{w}$.

For all $n \geq 1$, observing that the words $\nu^n(1)$ and $\nu^n(2)$ differ only by their last letters, we let v_n denote their longest common prefix; $v_0 = \varepsilon$, $v_n = \nu(v_{n-1})ccccb_1$ for $n \geq 1$. We also let u_n denote the word $\psi(v_n)$.

Fact 2.2. *For all $n \geq 1$, the words abu_n, bu_n2 and u_n1e_1 are examples of left special factors of $\psi(\mathbf{u})$.*

Proof. As the words b_11e_1, b_22e_2, $11e_0$, and b_32e_3 are factors of $\mathbf{u}$, for all $n \geq 1$, the following words are also factors of $\mathbf{u}$: $e_1a_1b_1v_n1e_1c$, $e_2a_2b_2v_n2e_2c$, $1v_n1e_01$, $e_3b_3v_n2e_3$.

Thus, the words $e_1abu_n1e_1c$, $e_2abu_n2e_2c$, $1u_n1e_11$, $e_3bu_n2e_3$ are factors of $\psi(\mathbf{u})$. Hence the fact. □

As a consequence of the previous fact, the word $\psi(\nu(u))$ has arbitrarily large special factors. Thus it is *aperiodic*, that is, denoting by x_i the letter in position i of $\psi(\nu(u))$, there does not exist any positive integers m, k such that, for all $i \geq m$, $x_i = x_{i+k}$.

We will also need the next property.

Lemma 2.3. *For all $n \geq 1$, the following words are factors of $\psi(\mathbf{u})$ but are not left special factors of $\psi(\mathbf{u})$: bu_n1, u_n1e_1c, u_n1c, u_n1e_11, u_n2e_2, u_n2e_3.*

Proof. First, let $x = bu_n1 = b\psi(\nu^n(1))$ and assume that, for a letter α, αx is a factor of $\psi(\mathbf{u})$. As $n \geq 1$, u_n starts with ccc. This implies that αb must be a suffix of a word in $\psi(\nu(\Gamma))$ and so $\alpha \in \{a, e_3\}$. If $\alpha = e_3$, then $\mathbf{u}$ must contain as a factor the word $e_3b_3v_n1 = e_3b_3\nu^n(1)$ (see Fact 2.1). But one can observe that $\mathbf{u}$ contains no factor of the form $b_3\nu^n(1)$. Thus $\alpha = a$ and bu_n1 is not a left special factor of $\mathbf{u}$.

Now let $x = u_n 1 e_1 c = \psi(\nu^n(1))e_1 c$. If αx is a factor of $\psi(\mathbf{u})$ for a letter α, then there exists a letter β such that $\beta v_n 1 e_1 c$ is a factor of $\mathbf{u}$ with $\alpha = \psi(\beta)$ (we use here Fact 2.1 and the fact that $e_0 c$ is not a prefix of a word in $\nu(\Gamma)$). We let the reader verify (for instance by induction) that if, for an integer ℓ and a letter δ, $\delta\nu^\ell(1)e_1$ is a factor of $\mathbf{u}$, then $\delta = b_1$. Hence $\beta = b_1$ and $\alpha = b$.

The proof for other words is similar. Let us just state the following facts (with the same notation as above).

$$\begin{array}{llll} x = u_n 1 c & \Rightarrow \ \beta \in \{b_1, b_4\} & \text{and} & \alpha = b \\ x = u_n 1 e_1 1 & \Rightarrow \ \beta = b_1 & \text{and} & \alpha = b \\ x = u_n 2 e_2 & \Rightarrow \ \beta \in \{b_1, b_2\} & \text{and} & \alpha = b \\ x = u_n 2 e_3 & \Rightarrow \ \beta \in \{b_1, b_3\} & \text{and} & \alpha = b\,. \end{array}$$

□

2.4. Return Words

Given a set L of factors of an infinite word $\mathbf{w}$, a word r is called a (left) *return word* to L in $\mathbf{w}$ if there exist two words u and v in L such that rv is a factor of $\mathbf{w}$, rv admits u as a prefix and rv contains no other occurrences of words of L. Clearly if a set L contains a prefix of $\mathbf{w}$ and $\mathbf{w}$ is *recurrent* (all of its factors occur infinitely often), then $\mathbf{w}$ can be decomposed over the set of left return words to L.

In this paper, given an infinite word $\mathbf{w}$, we will mostly consider the set $\mathcal{R}_n(\mathbf{w})$ of return words to the set $\mathrm{LS}_n(\mathbf{w})$ composed of the prefix of length $n \geq 1$ of $\mathbf{w}$ and of all left special factors of length n of $\mathbf{w}$. At the end of this section, we show that a, b and ab are elements of $\mathcal{R}_n(\mathbf{u})$ (where $\mathbf{u}$ is the word defined in the previous sections) with a more precise description of the values of n for which this membership holds.

Lemma 2.4. *For all $n \geq 1$, $\{a, b\} \subseteq \mathcal{R}_n(\mathbf{u})$.*
For all $n \geq 0$, $ab \in \mathcal{R}_{|u_n|+2}(\mathbf{u}) = \mathcal{R}_{6^n+1}(\mathbf{u})$.

Proof. Let $n \geq 1$, $|u_n| = 6^n - 1 \geq n$. Let p_i be the prefix of length i of u_n where $0 \leq i \leq |u_n|$. From Fact 2.2, we know that abp_n is a factor of $\psi(\mathbf{u})$ and that the words abp_{n-2} (when $n \geq 2$), bp_{n-1} and p_n are all left special factors (and all of length n). As p_{n-2} is a prefix of p_{n-1}, itself a prefix of p_n, a and b belong to $\mathcal{R}_n(\mathbf{u})$.

The word $abu_n 1 e_1$ is also a factor of $\psi(\mathbf{u})$ with abu_n and $u_n 1 e_1$ left special factors. As, by Lemma 2.3, $bu_n 1$ is not a left special factor, $ab \in \mathcal{R}_{|u_n|+2}(\mathbf{u}) = \mathcal{R}_{6^n+1}(\mathbf{u})$. □

The last part of Lemma 2.4 admits a converse.

Lemma 2.5. *Let $n \geq 2$ be an integer. If $ab \in \mathcal{R}_n(\mathbf{u})$ then $n = 6^k + 1$ for some $k \geq 0$.*

Although this lemma will allow the important Remark 3.1, we provide its proof in an appendix (see Section 5) as it is rather long, technical, and a bit far from the main aim of the paper.

3. Short, Constant, and Long Return Words

In this section, we consider particular return words that will be useful in the next section where we present our proof of Theorem 1.1. Note that, as already said in the introduction, the terms short and long were already used in Ferenczi's paper [12] but, even though our concepts are related, they are differently defined. For instance the fact that the length of long return words tends to infinity is a consequence of the definition, not part of it.

It is important to observe that all results in this section are independent of the complexity of the words considered.

3.1. Definitions

We will need to distinguish various types of return words to the set $\mathrm{LS}_n(\mathbf{w})$. A factor u of $\mathbf{w}$ is called a *short return word* of $\mathbf{w}$ if for all $n \geq 0$, there exists an integer $N \geq n$ such that u is an element of $\mathcal{R}_N(\mathbf{w})$. When a factor u is a return word of $\mathbf{w}$ (*i.e.* belongs to $\mathcal{R}_n(\mathbf{w})$ for some integer n) without being a short return word, we said that u is a *long return word.* Amongst short return words, we say that a factor u of $\mathbf{w}$ is a *constant return word* of $\mathbf{w}$ if there exists an integer $N \geq 0$ such that for all $n \geq N$, u is an element of $\mathcal{R}_n(\mathbf{w})$. For $n \geq 0$, we let respectively $\mathcal{R}_n^{(L)}(\mathbf{w})$, $\mathcal{R}_n^{(S)}(\mathbf{w})$ and $\mathcal{R}_n^{(C)}(\mathbf{w})$ denote the set of long return words to $\mathrm{LS}_n(\mathbf{w})$, the set of short return words to $\mathrm{LS}_n(\mathbf{w})$ and the set of constant return words to $\mathrm{LS}_n(\mathbf{w})$.

Remark 3.1. Our example $\mathbf{u}$ studied in Section 2 shows that not all short return words are constant return words. More precisely Lemma 2.4 and Lemma 2.5 prove that a, b and ab are examples of short return words with a and b constant return words and ab not a constant return word. Indeed, for all n we have $a, b \in \mathcal{R}_n^{(C)}(\psi(\mathbf{u})) \subseteq \mathcal{R}_n^{(S)}(\psi(\mathbf{u}))$; $ab \in \bigcup_{i=1}^n \mathcal{R}_n^{(S)}(\psi(\mathbf{u})) \setminus \bigcup_{i=1}^n \mathcal{R}_n^{(C)}(\psi(\mathbf{u}))$.

As c, cc and ccc are left special factors of $\psi(\mathbf{u})$, and as $cccc$ is a factor of $\psi(\mathbf{u})$, c is a return word of $\psi(\mathbf{u})$, and more precisely, $c \in \mathcal{R}_1(\psi(\mathbf{u})) \cap \mathcal{R}_2(\psi(\mathbf{u})) \cap \mathcal{R}_3(\psi(\mathbf{u}))$. Observe that for all words x of length at least four such that cx is a factor of $\psi(\mathbf{u})$, x is always preceded in $\psi(\mathbf{u})$ by the letter c. Thus, for all $n \geq 4$, c is not a return word to $\mathrm{LS}_n(\psi(\mathbf{u}))$; $c \in (\mathcal{R}_1^{(L)}(\psi(\mathbf{u})) \cap \mathcal{R}_2^{(L)}(\psi(\mathbf{u})) \cap \mathcal{R}_3^{(L)}(\psi(\mathbf{u}))) \setminus \mathcal{R}_4^{(L)}(\psi(\mathbf{u}))$.

One would like to have examples of short or constant return words for an infinite word defined over a smaller alphabet than Σ. We let the reader check that this can be obtained by recoding $\psi(\mathbf{u})$ for instance with the following morphism f defined from Σ to $\{a, b\}$ by $f(a) = a^8b$, $f(b) = a^7b^2$, $f(e_1) = a^6b^3$, $f(e_2) = a^5b^4$, $f(e_3) = a^4b^5$, $f(1) = a^3b^6$, $f(2) = a^2b^7$, $f(c) = ab^8$.

We will see that for infinite words with linear complexity the set of short return words is finite. Thus it is convenient to let $\mathcal{R}^{(S)}(\mathbf{w})$ (resp. $\mathcal{R}^{(C)}(\mathbf{w})$) denote the set $\bigcup_{n\geq 1} \mathcal{R}_n^{(S)}(\mathbf{w})$ (resp. $\bigcup_{n\geq 1} \mathcal{R}_n^{(C)}(\mathbf{w})$).

3.2. Coherent Decompositions

From our example, we can observe that the sets $\mathcal{R}_n(\mathbf{w})$ may not be codes (a *code* is a set of words such that any word w in C^* has a unique decomposition over C). This leads us to introduce the notion of coherent decomposition.

A sequence $(r_k)_{k\in\mathbb{N}} \in (\mathcal{R}_n(\mathbf{w}))^{\mathbb{N}}$ such that $\mathbf{w} = r_0r_1r_2\cdots$ is called a *decomposition* of $\mathbf{w}$. The unique decomposition $(r_k)_{k\in\mathbb{N}}$ of $\mathbf{w}$ such that a factor of $\mathbf{w}$ belongs to $\mathrm{LS}_n(\mathbf{w})$ if and only if it is the prefix of length n of $r_ir_{i+1}\cdots$ for some $i \geq 0$ is called *the coherent decomposition* of $\mathbf{w}$ of *level* n.

Let $s_1, \ldots, s_k$ be words. Let n be an integer and let $\mathbf{w}$ be an infinite word. We say that $(s_1, \ldots, s_k)$ is a factor of the coherent decomposition of $\mathbf{w}$ of level n if there exists an integer $m \geq 0$ such that for all i in $\{1, \ldots, k\}$, $s_i = r_{m+i-1}$. Note that if $(s_1, \ldots, s_k)$ is a factor of the coherent decomposition of $\mathbf{w}$ of level n, then $\{s_1, \ldots, s_k\} \subseteq \mathcal{R}_n(\mathbf{w})$.

The next result shows a connection between the coherent decompositions of $\mathbf{w}$ for two successive levels.

Lemma 3.2. *Let $(r_i)_{i\geq 0}$ and $(r'_i)_{i\geq 0}$ be the coherent decompositions of an infinite word $\mathbf{w}$ respectively of level n and level $n+1$ (for an integer $n \geq 1$). There exists a strictly increasing sequence of integers $(i_j)_{j\geq 0}$ such that $i_0 = 0$ and for all $j \geq 0$, $r'_j = r_{i_j} \cdots r_{i_{j+1}-1}$.*

Proof. This is a direct consequence of the definition of coherent decompositions and the observation that the prefix of length n of any element of $\mathrm{LS}_{n+1}(\mathbf{w})$ is an element of $\mathrm{LS}_n(\mathbf{w})$. □

3.3. Short Return Words Are Concatenations of Constant Return Words

In this section, we prove the following proposition that claims in a more precise way that any short return word of an infinite word $\mathbf{w}$ is a concatenation of constant return words to $\mathrm{LS}_n(\mathbf{w})$.

Proposition 3.3. *Given an infinite word $\mathbf{w}$, for any short return word s of $\mathbf{w}$, there exist infinitely many integers n such that, for the coherent decomposition $(r_i)_{i\geq 0}$ of level n of $\mathbf{w}$, $s = r_i \cdots r_j$ for some integers i and j such that all words r_ℓ, $i \leq \ell \leq j$, are constant return words.*

Actually Proposition 3.3 is a particular case of the following technical lemma (Proposition 3.3 corresponds to the case $k = 1$ of Lemma 3.4).

Lemma 3.4. *Let* $\mathbf{w}$ *be an infinite word and let* $s_1, \ldots, s_k$ *be words such that* $(s_1, \ldots, s_k)$ *is a factor of the coherent decomposition of* $\mathbf{w}$ *of level* n *for infinitely many integers* n *(this implies in particular that all words* $s_1, \ldots, s_k$ *are short return words). There exist integers* $i_0, i_1, \ldots, i_k$ *with* $i_0 = 0 < i_1 < \ldots < i_k$ *and there exist constant return words* $r_1, \ldots, r_{i_k}$ *such that* $(r_1, \ldots, r_{i_k})$ *is a factor of the coherent decomposition of* $\mathbf{w}$ *of level* m *for infinitely many integers* m*, and for all* $j \in \{1, \ldots, k\}$*,* $s_j = r_{i_{j-1}+1} \cdots r_{i_j}$*.*

Proof. We start by defining a partial order on tuples of words. Given any words $t_1, \ldots, t_\ell, s_1, \ldots, s_k$ such that $t_1 \ldots t_\ell = s_1 \ldots s_k$, we let $(t_1, \ldots, t_\ell) < (s_1, \ldots, s_k)$ denote the fact that there exists an integer j in $\{1, \ldots, \min(\ell, k)\}$ such that for all $i = 1, \ldots, j-1$, $t_i = s_i$ and $|t_j| < |s_j|$. Observe that, given a word s, this partial order is a total order on the set of all tuples $(s_1, \ldots, s_k)$ such that $s = s_1 \ldots s_k$. This allows us to prove Lemma 3.4 by induction.

Let $s_1, \ldots, s_k$ and $\mathbf{w}$ be as in the lemma. Let $\mathcal{D}$ be the set of all integers n such that $(s_1, \ldots, s_k)$ is a factor of the coherent decomposition of $\mathbf{w}$ of level n. By hypothesis, this set is infinite.

First, consider the case where for all j, s_j is of length 1. Let j, $1 \leq j \leq k$. For all n in $\mathcal{D}$, by definition of this set, there exists a word u_n of length n such that both u_n and the prefix of length n of $s_j u_n$ belong to $\mathrm{LS}_n(\mathbf{w})$. As the prefix of length $n-1$ of any element of $\mathrm{LS}_n(\mathbf{w})$ is an element of $\mathrm{LS}_{n-1}(\mathbf{w})$, we deduce that for all $n \geq 1$, there exists a word v_n of length n such that both v_n and the prefix of length n of $s_j v_n$ is an element of $\mathrm{LS}_n(\mathbf{w})$. As $\mathcal{D}$ is infinite, this shows that s_j is a constant return word.

We now turn to the case where there exists a word s_j (with $j \in \{1, \ldots, k\}$) which is not a constant return word. It follows that there exists an infinite subset $\mathcal{D}'$ of $\mathcal{D}$ such that for all n in $\mathcal{D}'$, $(s_1, \ldots, s_k)$ is not a factor of the coherent decomposition of $\mathbf{w}$ of level $n-1$, whereas $(s_1, \ldots, s_k)$ is a factor of the coherent decomposition of $\mathbf{w}$ of level n. By Lemma 3.2, there exist some words $t_1(n)$, $\ldots, t_{\ell(n)}(n)$ and some integers $i_0 = 0 < i_1 < \ldots < i_k = \ell(n)$ such that for all $j = 1, \ldots, k$, we have $s_j = t_{i_{j-1}+1}(n) \cdots t_{i_j}(n)$ and $(t_1(n), \ldots, t_{\ell(n)}(n))$ is a factor of the coherent decomposition of $\mathbf{w}$ of level $n-1$: each word s_j is a composition of return words to $\mathrm{LS}_{n-1}(\mathbf{w})$, and at least one is the concatenation of at least two such words. Moreover, we have $(t_1(n), \ldots, t_{\ell(n)}(n)) < (s_1, \ldots, s_k)$. Even though the words $t_1(n), \ldots, t_{\ell(n)}(n)$ depend on n, as the length of $|t_1(n) \cdots t_{\ell(n)}(n)| = |s_1 \ldots s_k|$ is fixed, there must exist an integer ℓ, some words $t_1, \ldots, t_\ell$, and an infinite set $\mathcal{D}''$ such that for all n in $\mathcal{D}''$, $\ell(n) = \ell$ and for all $i = 1, \ldots, \ell$, $t_i(n) = t_i$. The lemma ends by induction applied to $(t_1, \ldots, t_\ell)$. □

3.4. More on Long Return Words

The next result justifies *a posteriori* the terminology *long* for long return words.

Proposition 3.5. *For an infinite word* $\mathbf{w}$*, if* $\mathcal{R}_n^{(L)}(\mathbf{w}) \neq \emptyset$ *for infinitely many integers* n*, then*

$$\lim_{n\to\infty} \min\{|r| \mid r \in \mathcal{R}_n^{(L)}(\mathbf{w})\} = +\infty.$$

Proof. This result is a direct consequence of the definitions of short and long return words.

Assume by contradiction that $\lim_{n\to\infty} \min\{|r| \mid r \in \mathcal{R}_n^{(L)}(\mathbf{w})\}$ is finite. Let $\mathcal{D}$ be an infinite subset of $\mathbb{N}$ such that $\mathcal{R}_n^{(L)}(\mathbf{w}) \neq \emptyset$ for all n in $\mathcal{D}$. For all $i \in \mathcal{D}$, let r_i be a word with minimal length in $\mathcal{R}_i^{(L)}(\mathbf{w})$. As the length of r_i is bounded, the set $\{r_i \mid i \in \mathcal{D}\}$ is finite. In particular, there exists a word r such that $r \in \mathcal{R}_i^{(L)}(\mathbf{w})$ for all i in an infinite subset of $\mathcal{D}$. Thus, r is a short return word of $\mathbf{w}$, which is a contradiction as long return words cannot be short by definition. □

3.5. Return Words to a Word

In almost the entire paper we are concerned with the set of return words to the set of left special words of an infinite word. The next important and well-known result uses the notion of return word to a single word.

A nonempty word r is a *return word* to a factor u of an infinite word $\mathbf{w}$ if ru is a factor of $\mathbf{w}$ starting with u and containing exactly two occurrences of u as a factor (one as a prefix and one as a suffix).

Proposition 3.6. *Let* $\mathbf{w}$ *be an aperiodic and uniformly recurrent infinite word. We have*

$$\lim_{n\to+\infty} \min_{\substack{u \text{ factor of } \mathbf{w} \\ |u|=n}} \{|r| \mid r \text{ return word to } u\} = +\infty.$$

Although the previous result is certainly well known (see for instance [9] for a proof of a similar result), we provide a short proof in order to get a self-contained paper. Let us recall one fundamental lemma of combinatorics on words (see [16] and for instance [14, Prop. 1.3.4]).

Lemma 3.7. *For two nonempty words* x *and* y*, there exists a word* z *such that* $xz = zy$ *if and only if there exist two words* u *and* v *such that* $x = uv$*,* $z \in (uv)^*u$ *and* $y = vu$*.*

Proof of Proposition 3.6. Assume by contradiction that the limit in Proposition 3.6 is finite. This implies that there exists a word r which is the return word to infinitely many words $(u_n)_{n\geq 0}$. This means that there exist some words $(x_n)_{n\geq 0}$ such that $u_n x_n = r u_n$. By Lemma 3.7, this implies that r^k is a factor of $\mathbf{w}$ for arbitrarily large k, which is a contradiction with the fact that $\mathbf{w}$ is aperiodic and uniformly recurrent. □

4. S-adicity of Infinite Words with Linear Complexity

In this section, we provide our proof of Theorem 1.1. Each subsection corresponds to a step or a key element of the proof. We first clarify the notion of S-adicity. Let A be an alphabet. Recall that with the product topology, the set A^ω of infinite words over A is a compact metric space. As usual, for a finite word u, u^ω denotes the infinite word obtained by concatenating infinitely many copies of u.

Let $\mathbf{w}$ be an infinite word over A. An *adic representation* of $\mathbf{w}$ is given by a sequence $(A_n)_{n\in\mathbb{N}}$ of alphabets, a sequence $(\sigma_n : A_{n+1}^* \to A_n^*)_{n\in\mathbb{N}}$ of morphisms and a sequence $(a_n)_{n\in\mathbb{N}}$ of letters such that $a_i \in A_i$ for all $i \geq 0$, $A_0 = A$, $\lim_{n\to+\infty} |\sigma_0\sigma_1\cdots\sigma_n(a_{n+1})| = +\infty$ and

$$\mathbf{w} = \lim_{n\to+\infty} \sigma_0\sigma_1\cdots\sigma_n(a_{n+1}^\omega).$$

When all morphisms σ_n, $n \in \mathbb{N}$, belong to a given *finite* set S of morphisms, we say that $\mathbf{w}$ is S-adic (often omitting the sequence of morphisms and letters).

4.1. The Number of Left Special Factors Is Bounded

The starting point of the proof of Theorem 1.1 is the following breakthrough result of Cassaigne result on complexity.

Theorem 4.1. [6] *An infinite word* $\mathbf{w}$ *has linear complexity if and only if the first difference of its complexity* $(p_{\mathbf{w}}(n+1) - p_{\mathbf{w}}(n))$ *is bounded.*

For a finite word u and an infinite word $\mathbf{w}$, let $\delta_{\mathbf{w}}^- u$ denote the number of letters a such that au is a factor of $\mathbf{w}$. Recall that a word u is a left special factor of $\mathbf{w}$ if and only if $\delta_{\mathbf{w}}^- u \geq 2$. Since Rauzy's work [23], it is well known that the first difference of complexity is related to special factors by the following Equation (1). For a recurrent infinite word $\mathbf{w}$,

$$p_{\mathbf{w}}(n+1) - p_{\mathbf{w}}(n) = \sum_{\substack{|u|=n,\\ u \text{ left special factor of } \mathbf{w}}} (\delta_{\mathbf{w}}^- u - 1). \tag{1}$$

Thus Theorem 4.1 has the following corollary (let us recall from Section 2.4 that $\mathrm{LS}_n(\mathbf{w})$ is the set consisting of the prefix of length n and the left special factors of length n of $\mathbf{w}$).

Corollary 4.2. *A recurrent infinite word* $\mathbf{w}$ *has linear complexity if and only if the sets* $\mathrm{LS}_n(\mathbf{w})$, $n \geq 0$, *have bounded cardinality.*

4.2. The Number of Return Words Is Bounded

Let us recall that $\mathcal{R}_n(\mathbf{w})$ is the set of return words to $\mathrm{LS}_n(\mathbf{w})$ where $\mathbf{w}$ is an infinite word and $n \geq 1$ is an integer.

Proposition 4.3. *If* $\mathbf{w}$ *is a recurrent infinite word with linear complexity, then the cardinality of the sets* $\mathcal{R}_n(\mathbf{w})$ *is bounded.*

Proof. Let r be a return word to the set $\mathrm{LS}_n(\mathbf{w})$. By definition of this set, there exists a word u in $\mathrm{LS}_n(\mathbf{w})$ (u is of length n and is a left special factor or a prefix of $\mathbf{w}$) such that ru is a factor of $\mathbf{w}$. Moreover for each factor sp of ru of length n with $1 \leq |s| < |r|$, s suffix of r and p prefix of u, there exists a unique letter γ such that γsp is a factor of $\mathbf{w}$. Hence r is entirely determined by the pair (u, α) with α its last letter (but the pair may not be unique). Thus, $\mathrm{Card}(\mathcal{R}_n(\mathbf{w})) \leq \sum_{u \in \mathrm{LS}_n(\mathbf{w})} \delta^- u = p(n+1) - p(n) + \mathrm{Card}(\mathrm{LS}_n(\mathbf{w}))$. The lemma follows from Theorem 4.1 and Corollary 4.2. □

The reader will note that the cardinality of a set X is denoted by $\mathrm{Card}(X)$.

4.3. The Number of Constant Return Words Is Finite

Let us recall from Section 3.1 that $\mathcal{R}^{(C)}(\mathbf{w}) = \cup_{n \geq 1} \mathcal{R}_n^{(C)}(\mathbf{w})$ is the set of all constant return words to the sets $\mathrm{LS}_n(\mathbf{w})$.

Proposition 4.4. *For any recurrent infinite word* $\mathbf{w}$ *with linear complexity, the set* $\mathcal{R}^{(C)}(\mathbf{w})$ *is finite.*

Proof. This is a direct consequence of Proposition 4.3. Let us consider $K = \max_{n \geq 1} \mathrm{Card}(R_n(\mathbf{w}))$. Assume by contradiction that $R^{(C)}(\mathbf{w})$ is not finite. Let $r_1, \ldots, r_{K+1}$ be distinct elements of $R^{(C)}(\mathbf{w})$. There are some integers $N_1, \ldots, N_{K+1}$ such that for all i in $\{1, \ldots, K+1\}$ and for all $n \geq N_i$, r_i belongs to $R_n^{(C)}(\mathbf{w})$. Thus, for $n \geq \max\{N_i \mid 1 \leq i \leq K+1\}$, we have $\{r_1, \ldots, r_{K+1}\} \subseteq R_n(\mathbf{w})$, which is a contradiction with $\mathrm{Card}(R_n(\mathbf{w})) \leq K$. □

4.4. The Number of Short Return Words Is Finite

Let us recall from Section 3.1 that $\mathcal{R}^{(S)}(\mathbf{w}) = \cup_{n \geq 1} \mathcal{R}_n^{(S)}(\mathbf{w})$ is the set of all short return words to the sets $\mathrm{LS}_n(\mathbf{w})$.

Proposition 4.5. *For any uniformly recurrent infinite word* $\mathbf{w}$ *with linear complexity, the set* $\mathcal{R}^{(S)}(\mathbf{w})$ *is finite.*

This result is a consequence of the following technical lemma and Proposition 3.3.

Lemma 4.6. *Let* $\mathbf{w}$ *be an aperiodic and uniformly recurrent infinite word with linear complexity. Let* $(L_i)_{i\geq 1}$ *be a family of sets of words such that* $L_i \subseteq \mathcal{R}_i(\mathbf{w})$ *for all* $i \geq 1$ *and* $R = \cup_{i\geq 1} L_i$ *is finite.*

There exist two integers K *and* N *such that for all* $n \geq N$ *and for any coherent decomposition* $(r_i)_{i\in\mathbb{N}}$ *of* $\mathbf{w}$ *into words of* $\mathcal{R}_n(\mathbf{w})$*, the number of consecutive elements of* R *in* $(r_i)_{i\in\mathbb{N}}$ *is bounded by* K*.*

In particular, the concatenation of such return words in a coherent decomposition of level greater than N *has length bounded by* $K\ell$ *where* ℓ *is the maximal length of words in* R*.*

Proof. We let ℓ be, as defined in the lemma, the maximal length of words in R. By Corollary 4.2 and Proposition 4.3, there exists an integer K such that for all n, $\mathrm{Card}(\mathcal{R}_n(\mathbf{w}) \times \mathrm{LS}_n(\mathbf{w})) \leq K$. By Proposition 3.6, there exists an integer N such that for all factors u of length $n \geq N$ of $\mathbf{w}$, all return words to u have length greater than $K\ell$. Finally, consider an integer $n \geq N$ and let $(r_i)_{i\geq 0}$ be a coherent decomposition of $\mathbf{w}$ of level n.

Assume that there exist $K+1$ consecutive elements of L_n in the considered decomposition of $\mathbf{w}$, say $r_m, \ldots, r_{m+K}$ for an integer $m \geq 1$. For $i \geq 0$, let l_i be the prefix of length n of the word $r_{i+1}r_{i+2}\cdots$. From the choice of K, there exist two integers i and j such that $m \leq i < j \leq m+K$ and $(r_i, l_i) = (r_j, l_j)$. Consequently, the word $r_{i+1}\cdots r_j$ is a return word to l_i and $|r_{i+1}\cdots r_j| \leq K\ell$, which contradicts the choice of N. □

Proof of Proposition 4.5. By Proposition 3.3, any short return word is the concatenation of constant return words $r_1, \ldots, r_k$ such that $(r_1, \ldots, r_k)$ is a factor of infinitely many coherent decompositions of $\mathbf{w}$. By Proposition 4.4, the set $\mathcal{R}^{(C)}(\mathbf{w})$ is finite. We can then apply Lemma 4.6 with $L_n = \mathcal{R}^{(C)}(\mathbf{w}) \cap \mathcal{R}_n(\mathbf{w})$ for all $n \geq 0$ to deduce that short return words have bounded length. Consequently, the set of short return words is finite. □

4.5. About Long Return Words

As an immediate consequence of Lemma 4.6, the next result states that any sufficiently long factor of a coherent decomposition (of sufficiently large level n) contains a long return word.

Lemma 4.7. *Let* $\mathbf{w}$ *be a uniformly recurrent infinite word with linear complexity. There exist some integers* K *and* N *such that for all* $n \geq N$*, if* $k \geq 1$ *and* $(r_1, \ldots, r_k)$ *is a factor of the coherent decomposition of* $\mathbf{w}$ *of level* n *with* $|r_1 \cdots r_k| \geq K$*, then at least one of the words* r_i $(1 \leq i \leq k)$ *belongs to* $\mathcal{R}_n^{(L)}(\mathbf{w})$*.*

Proof. By Lemma 4.6 (taking $L_i = \mathcal{R}_i^{(S)}(\mathbf{w})$ for all $i \geq 0$), there exist some integers K_1 and N such that for all $n \geq N$, at most K_1 short return words can occur successively in the coherent decomposition of $\mathbf{w}$ of level n. Let $\ell = \max\{r \mid r \in \mathcal{R}^{(S)}(\mathbf{w})\}$. Lemma 4.7 follows with $K = K_1\ell + 1$. □

Remark 4.8. With the same notation as in Lemma 4.7, we let M denote the smallest integer such that $M \geq N$ and $\min\{|u| \mid u \in \mathcal{R}_M^{(L)}(\mathbf{w})\} \geq K$. Due to Lemma 3.2, we have $\min\{|u| \mid u \in \mathcal{R}_n^{(L)}(\mathbf{w})\} \geq K$ for all $n \geq M$.

4.6. Morphisms

Up to the end of Section 4, $\mathbf{w}$ always denotes a recurrent infinite word with linear complexity. For all $n \geq 0$, let $\mathcal{U}_n(\mathbf{w})$ be the set of tuples $((r_1, \ell_1), \ldots, (r_k, \ell_k))$, $k \geq 1$, such that

- there exists an integer j, $1 \leq j \leq k$, such that $r_j \in \mathcal{R}_n^{(L)}(\mathbf{w})$ and for all $i \neq j$, $r_i \in \mathcal{R}_n^{(S)}(\mathbf{w})$ (by abuse of notation, we write $(r_1, \ldots, r_k) \in (\mathcal{R}_n^{(S)}(\mathbf{w}))^*\mathcal{R}_n^{(L)}(\mathbf{w})(\mathcal{R}_n^{(S)}(\mathbf{w}))^*$);
- with $(s_i)_{i\geq 0}$ the coherent decomposition of $\mathbf{w}$ of level n, there exists an integer i such that $(r_1, \ldots, r_k) = (s_i, \ldots, s_{i+k-1})$ (*i.e.* $(r_1, \ldots, r_k)$ is a factor of the coherent decomposition of $\mathbf{w}$ of level n), and for all $j = 1, \ldots, k$, ℓ_j is the prefix of length n of $s_{i+j}\cdots$ ($\ell_j \in \mathrm{LS}_n(\mathbf{w})$).

Lemma 4.9. *For any uniformly recurrent infinite word* $\mathbf{w}$ *with linear complexity, the cardinality of the sets* $(\mathcal{U}_n(\mathbf{w}))_{n\geq 0}$ *is ultimately bounded, i.e., there exist some integers* K *and* N *such that for all* $n \geq N$, $\mathrm{Card}(\mathcal{U}_n(\mathbf{w})) \leq K$.

Proof. By Lemma 4.6 (taking $L_i = \mathcal{R}_i^{(S)}(\mathbf{w})$ for all $i \geq 0$), there exist some integers K and N such that for all $n \geq N$, at most K short return words can occur successively in the coherent decomposition of $\mathbf{w}$ of level n. Note that, by definition of $\mathcal{U}_n(\mathbf{w})$, for any of its elements $((r_1, \ell_1), \ldots, (r_k, \ell_k))$, $r_1 \ldots r_k\ell_k$ is a factor of $\mathbf{w}$ and for all $j = 1, \ldots, k-1$, ℓ_j is the prefix of length n of $r_{j+1} \ldots r_k\ell_k$ (ℓ_j is uniquely determined by r_{j+1}, …, r_k and ℓ_k). By Proposition 4.5 and Proposition 4.3, it follows that, for all $n \geq N$, the cardinality of $\mathcal{U}_n(\mathbf{w})$ admits the upper bound $(\mathrm{Card}(\mathcal{R}_{(S)}(\mathbf{w})))^{2(K+1)}K'\mathrm{Card}(\mathrm{LS}_n(\mathbf{w}))$ with $K' = \max_{n\geq 0} \mathrm{Card}(\mathcal{R}_n^{(L)}(\mathbf{w}))$. □

Let K and M be integers as in Lemma 4.7 and Remark 4.8. Let $n \geq M$ and let $((r_1, \ell_1), \ldots, (r_k, \ell_k))$ be an element of $\mathcal{U}_{n+1}(\mathbf{w})$. Let $(s_i)_{i\geq 0}$ be the coherent decomposition of $\mathbf{w}$ of level n. By Lemma 3.2, there exists a strictly increasing sequence of integers $(i_j)_{1\leq j\leq k+1}$ such that for all $j \geq 0$, $r_j = s_{i_j} \cdots s_{i_{j+1}-1}$ and ℓ_k is a prefix of $s_{i_k+1} \cdots$. By the choice of M, as $|s_{i_1} \cdots s_{i_{k+1}-1}| = |r_1 \cdots r_k| \geq K$, at least one of the words s_j is a long return word to $\mathrm{LS}_n(\mathbf{w})$. Let k' be the number of integers i between i_1 and $i_{k+1} - 1$ such that $s_i \in \mathcal{R}_n^{(L)}(\mathbf{w})$ and let $j_1 < j_2 < \ldots < j_{k'}$ be these integers. Let ℓ'_j be the prefix of length n of $s_{j+1}s_{j+2}\cdots$. We have just described a way to associate to each element $((r_1, \ell_1), \ldots, (r_k, \ell_k))$

of $\mathcal{U}_{n+1}(\mathbf{w})$ a sequence $((s_{i_1},\ell'_{i_1}),\dots,(s_{j_1},\ell'_{j_1}))$, $((s_{j_1+1},\ell'_{j_1+1}),\dots,(s_{j_2},\ell'_{j_2}))$, $\dots$, $((s_{j_{k'-2}+1},\ell'_{j_{k'-2}+1}),\dots,(s_{j_{k'-1}},\ell'_{j_{k'-1}}))$, $((s_{j_{k'-1}+1},\ell'_{j_{k'-1}+1}),\ \dots,(s_{i_k},\ell'_{i_k}))$ of elements of $\mathcal{U}_n(\mathbf{w})$.

For $n \geq M$, let $\mathcal{A}_n = \{0,1,\dots,\mathrm{Card}(\mathcal{U}_n)(\mathbf{w})-1\}$ and let Θ_n be a bijection from $\mathcal{A}_n$ to $\mathcal{U}_n(\mathbf{w})$. Now we define our morphisms from $\mathcal{A}^*_{n+1}$ to $\mathcal{A}^*_n$ by $\sigma_n(a) = a_1 \dots a_k$ whenever $\Theta_n(a)$ can be associated as previously to $\Theta_n(a_1)\dots\Theta_n(a_k)$.

We also let τ be the morphism from $\mathcal{A}^*_n$ to A^* (the alphabet of $\mathbf{w}$) defined for all $a \in \mathcal{A}_n$ by $\tau(a) = u$ if $\Theta_n(a) = ((r_1,\ell_1),\dots,(r_k,\ell_k))$ and $u = r_1 \cdots r_k$.

Of course our morphisms depend on our construction. Other morphisms can be suitable.

4.7. The Number of Morphisms Is Finite

We continue with the notation introduced in the previous section. Let $S = \{\sigma_n \mid n \geq M\}$.

Fact 4.10. *The set S is finite.*

Proof. We start with a technical observation. Let (r,l) be an element of $\mathcal{R}_{n+1}(\mathbf{w}) \times \mathrm{LS}_{n+1}(\mathbf{w})$ such that rl is a factor of $\mathbf{w}$. Following Lemma 3.2, there exist (r_1,l_1), $(r_2,l_2),\dots,(r_k,l_k)$, elements of $\mathcal{R}_n(\mathbf{w}) \times \mathrm{LS}_n(\mathbf{w})$, such that $rl_{[1,n]} = r_1r_2\cdots r_kl_k$ (where $l_k = l_{[1,n]}$ denotes the prefix of length n of l) and l_i is a prefix of $r_{i+1}\dots r_kl_k$ for all i in $\{1,\dots,k\}$. We are going to show that, for i,j in $\{2,\dots,k-1\}$, $i \neq j$, we have $(r_i,l_i) \neq (r_j,l_j)$ and, consequently, $k < 2 + \mathrm{Card}(\mathcal{R}_n(\mathbf{w})) \times \mathrm{Card}(\mathrm{LS}_n(\mathbf{w}))$.

Let $l' \in \mathrm{LS}_{n+1}(\mathbf{w})$ be the prefix of rl of length $n+1$. Assume by contradiction that $(r_i,l_i) = (r_j,l_j)$ with $1 < i < j < k$. Since there are no left special factors of length $n+1$ in rl (except its prefix l' and its suffix l), for all integers $h \in \{1,\dots,i\}$, the words $r_h \cdots r_il_i$ can be uniquely extended to the left in $L(\mathbf{w})$ by r_{h-1}. As $r_il_i = r_jl_j$, this implies that $r_1\dots r_il_i$ is also a suffix of $r_1\dots r_jl_j$. Thus l' has an internal occurrence in rl, which is a contradiction.

We now turn to the proof of the fact. We need to go back to the construction of the morphisms and so we let $((r_1,\ell_1),\dots,(r_k,\ell_k))$ and integers $(i_j)_{1\leq j\leq k+1}$ be as in this construction. The previous observation shows that for all j, we have $1 \leq j \leq k$, $i_{j+1} - i_j \leq 2 + \mathrm{Card}(\mathcal{R}_n(\mathbf{w})) \times \mathrm{Card}(\mathrm{LS}_n(\mathbf{w}))$, and so $i_{k+1} - i_1 \leq k(2 + \mathrm{Card}(\mathcal{R}_n(\mathbf{w})) \times \mathrm{Card}(\mathrm{LS}_n(\mathbf{w})))$. By Corollary 4.2, Proposition 4.3 and Lemma 4.9, there exists an integer K_1 such that for all $n \geq M$, $\mathrm{Card}(\mathcal{R}_n(\mathbf{w})) \leq K_1$, $\mathrm{Card}(\mathrm{LS}_n(\mathbf{w})) \leq K_1$ and $k \leq K_1$. As the length of $\sigma_n(\Theta^{-1}_{n+1}((r_1,\ell_1),\dots,(r_k,\ell_k)))$ is bounded by $i_{k+1} - i_1$, it is also bounded above by $K_1(2 + K_1^2)$.

In other words, there exists an integer K, such that for all $n \geq M$ and for all letters $a \in \mathcal{A}_{n+1}$, $|\sigma_n(a)| \leq K$. As also $\mathrm{Card}(\mathcal{A}_n) \leq K$, the set S is finite. □

4.8. Everywhere Growing Property

We continue with the notation introduced in the two previous sections.

Fact 4.11. *For any sequence $(a_n \in \mathcal{A}_n)_{n \geq M}$, we have*

$$\lim_{n \to \infty} |\tau\sigma_M \dots \sigma_n(a_{n+1})| = +\infty .$$

Proof. By construction, for any $n \geq M$, $a_{n+1} = \Theta_{n+1}^{-1}((r_1, \ell_1), \dots, (r_k, \ell_k))$ for some (r_1, ℓ_1), ..., (r_k, ℓ_k) in $\mathcal{U}_{n+1}(\mathbf{w})$ and it can be verified by induction that $|\tau\sigma_M \dots \sigma_n(a_{n+1})| = |r_1 \cdots r_k|$. By Proposition 3.5, the length of minimal elements of $\mathcal{R}_n^{(L)}(\mathbf{w})$ grows to infinity. So the length of minimal elements of $\mathcal{U}_n(\mathbf{w})$ also grows to infinity. The lemma follows. □

4.9. Proof of Theorem 1.1

Let $n \geq 0$. Let $(r_i)_{i \geq 0}$ be the coherent decomposition of $\mathbf{w}$ over $\mathcal{R}_n(\mathbf{w})$ of level n, and for all $i \geq 0$, let ℓ_i be the prefix of $r_i \cdots$. For $n \geq M$, we know that at least one of the elements of $\mathcal{R}_n(\mathbf{w})$ belongs to $\mathcal{R}_n^{(L)}(\mathbf{w})$. Let j be the smallest integer such that $r_j \in \mathcal{R}_n^{(L)}(\mathbf{w})$ and let $a_n = \Theta_n^{-1}((r_0, \ell_0), \dots, (r_j, \ell_j))$. By definition, for $n \geq M$, $\tau\sigma_M \dots \sigma_n(a_{n+1})$ is a prefix of $\mathbf{w}$.

By Fact 4.11, it follows that

$$\mathbf{w} = \lim_{n \to +\infty} \tau\sigma_M \cdots \sigma_n(a_{n+1}^\omega). \tag{2}$$

This, with Fact 4.11, ends the proof of Theorem 1.1.

Acknowledgement The authors would like to thank Fabien Durand for many fruitful discussions concerning return words and S-adicity. Many thanks also to Narad Rampersad for his careful reading and to the referees for their valuable comments.

References

[1] J.-P. Allouche and J. Shallit. *Automatic sequences: Theory, applications, generalizations.* Cambridge University Press, Cambridge, 2003.

[2] P. Arnoux and G. Rauzy. Représentation géométrique de suites de complexité $2n + 1$. *Bull. Soc. Math. France*, 119:199–215, 1991.

[3] J. Berstel, A. Lauve, C. Reutenauer, and F. Saliola. *Combinatorics on Words: Christoffel Words and Repetitions in Words*, volume 27 of *CRM Monograph Series.* American Mathematical Society, 2008.

[4] V. Berthé and M. Rigo, editors. *Combinatorics, Automata and Number Theory*, volume 135 of *Encyclopedia of Mathematics and its Applications*. Cambridge University Press, Cambridge, 2010.

[5] M. Boshernitzan. A condition for minimal interval exchange maps to be uniquely ergodic. *Duke Math. J.*, 52(3):723–752, 1985.

[6] J. Cassaigne. Special factors of sequences with linear subword complexity. In *Developments in language theory, II (Magdeburg, 1995)*, pages 25–34. World Sci. Publ., River Edge, NJ, 1996.

[7] J. Cassaigne and F. Nicolas. Factor complexity. Chapter 3 in [4].

[8] A. de Luca. Sturmian words: structure, combinatorics, and their arithmetics. *Theoretical Computer Science*, 183:45–82, 1997.

[9] F. Durand. A characterization of substitutive sequences using return words. *Discrete Math.*, 179:89–101, 1998.

[10] F. Durand. Linearly recurrent subshifts have a finite number of non-periodic subshift factors. *Ergodic Theory Dynam. Systems*, 20:1061–1078, 2000.

[11] F. Durand, J. Leroy, and G. Richomme. Do properties of an S-adic representation determine factor complexity? To appear in *J. Integer Seq.* ArXiv e-prints:1208.6376v1.

[12] S. Ferenczi. Rank and symbolic complexity. *Ergodic Theory Dynam. Systems*, 16:663–682, 1996.

[13] J. Leroy. Some improvements of the S-adic conjecture. *Advances in Applied Mathematics*, 48(1):79 – 98, 2012.

[14] M. Lothaire. *Combinatorics on Words*, volume 17 of *Encyclopedia of Mathematics and its Applications*. Addison-Wesley, 1983. Reprinted in the *Cambridge Mathematical Library*, Cambridge University Press, UK, 1997.

[15] M. Lothaire. *Algebraic combinatorics on words*, volume 90 of *Encyclopedia of Mathematics and its Applications*. Cambridge University Press, Cambridge, 2002.

[16] R. C. Lyndon and M. P. Schützenberger. The equation $a^M = b^N c^P$ in a free group. *Michigan Math. J.*, 9:289–298, 1962.

[17] F. Mignosi and P. Séébold. Morphismes sturmiens et règles de Rauzy. *J. Théor. Nombres Bordeaux*, 5:221–233, 1993.

[18] M. Morse and G. A. Hedlund. Symbolic dynamics II. Sturmian trajectories. *Amer. J. Math.*, 62:1–42, 1940.

[19] J.-J. Pansiot. Complexité des facteurs des mots infinis engendrés par morphismes itérés. In *Automata, languages and programming (Antwerp, 1984)*, volume 172 of *Lecture Notes in Comput. Sci.*, pages 380–389. Springer, Berlin, 1984.

[20] N. Pytheas Fogg. *Substitutions in dynamics, arithmetics and combinatorics*, volume 1794 of *Lecture Notes in Mathematics*. Springer-Verlag, Berlin, 2002. Edited by V. Berthé, S. Ferenczi, C. Mauduit and A. Siegel.

[21] N. Pytheas Fogg, Terminologie S-adique et propriétés, Preprint available at `http://tinyurl.com/8opdb8s`, 2011.

[22] M. Queffélec. *Substitution dynamical systems—spectral analysis*, volume 1294 of *Lecture Notes in Mathematics*. Springer-Verlag, Berlin, 1987.

[23] G. Rauzy. Suites à termes dans un alphabet fini. *Séminaire de Théorie des nombres de Bordeaux*, exposé 25:1–16, 1982-1983.

5. Appendix: Proof of Lemma 2.5

This entire section is dedicated to the proof of Lemma 2.5. Two words x and y are said to be *prefix-comparable* if x is a prefix of y or y is a prefix of x.

Fact 5.1. *For any morphism f and any prefix-comparable words x and y, $f(x)$ and $f(y)$ are also prefix-comparable.*

Fact 5.2. *If two words x and y are prefix-comparable and z is a prefix of y, then x and z are also prefix-comparable.*

In what follows, let us recall that $\mathbf{u}$, introduced in Section 2, is the fixed point of ν starting with the letter c.

Lemma 5.3. *Any factor of $\mathbf{u}$ of length at least two and starting with b_1 is prefix-comparable to one of the words $b_1\nu^\ell(1)e_1c$, $b_1\nu^\ell(1)c$, $b_1\nu^\ell(1)e_01$, $b_1\nu^\ell(2)e_2$ and $b_1\nu^\ell(2)e_3$ for some $\ell \geq 0$.*

Proof. We proceed by induction. For factors of length two, three or four, the result is true with $\ell \in \{0, 1\}$ since

$$b_1\Gamma \cap F(\mathbf{u}) = \{b_12, b_1c\},$$

$$b_1\Gamma^2 \cap F(\mathbf{u}) = \{b_11e_1, b_1cc, b_11c, b_11e_0, b_12e_2, b_12e_3\},$$
$$b_1\Gamma^3 \cap F(\mathbf{u}) \subseteq \{b_11c, b_12e_2, b_12e_3\}\Gamma \cup \{b_11e_1c, b_11e_01, b_1ccc\}.$$

Let b_1x be a factor of length at least five of $\mathbf{u}$. The result is immediately true if x starts with $1c$, $2e_2$, $2e_3$, $1e_1c$ or $1e_01$, (with $\ell = 0$). Thus, assume that x starts with ccc. By the definition of ν, this implies the existence of a factor y of $\mathbf{u}$ such that b_1y is a factor of $\mathbf{u}$, x is a prefix of $\nu(y)$ and $|y| < |x|$. By the inductive hypothesis, b_1y is prefix-comparable to one of the words $b_1\nu^\ell(1)e_1c$, $b_1\nu^\ell(1)c$, $b_1\nu^\ell(1)e_01$, $b_1\nu^\ell(2)e_2$ or $b_1\nu^\ell(2)e_3$ for some $\ell \geq 0$. So, by Fact 5.1 and Fact 5.2, b_1x is prefix-comparable to one of the words $b_1\nu^{\ell+1}(1)e_1c$, $b_1\nu^{\ell+1}(1)c$, $b_1\nu^{\ell+1}(1)e_01$, $b_1\nu^{\ell+1}(2)e_2$ or $b_1\nu^{\ell+1}(2)e_3$. □

Corollary 5.4. *Any factor of $\mathbf{u}$ of length at least three and starting with a_1b_1 is prefix-comparable to one of the words $a_1b_1\nu^\ell(1)e_1c$, $a_1b_1\nu^\ell(1)c$, $a_1b_1\nu^\ell(1)e_01$, $a_1b_1\nu^\ell(2)e_2$ or $a_1b_1\nu^\ell(2)e_3$ for some $\ell \geq 1$.*

Proof. After Lemma 5.3, it is sufficient to observe that the words $a_1b_11e_1c$, a_1b_11c, $a_1b_11e_0$, $a_1b_12e_2$ and $a_1b_12e_3$ are not factors of words in $\nu(\Gamma^*)$. □

Lemma 5.5. *Any factor a_2b_2m of $\mathbf{u}$ of length at least three is prefix-comparable to $a_2b_2\nu^\ell(2)e_2$ for some $\ell \geq 1$.*

Proof. The proof is similar to that of Corollary 5.4. □

Proof of Lemma 2.5. Let $n \geq 2$ be an integer such that $ab \in \mathcal{R}_n(\mathbf{u})$. This means that there exists a word v and two letters x, y such that $|v| = n - 2$, $abvxy$ is a factor of $\psi(\mathbf{u})$, both words abv and vxy are left special factors of $\psi(\mathbf{u})$ and bvx is not a left special factor of $\psi(\mathbf{u})$. We aim to prove that, for some integer k, $v = u_k$ (and so $n = 6^k + 1$), $x = 1$ and $y = e_1$ (see before Fact 2.2 for the definition of words u_k).

By construction of $\psi(\mathbf{u})$, two cases can hold (maybe not exclusively).

- Case I: There exists a word w_1 and two letters x_1, y_1 such that $abvxy = \psi(a_1b_1w_1x_1y_1)$ and $a_1b_1w_1x_1y_1$ is a factor of $\mathbf{u}$.

- Case II: There exists a word w_2 and two letters x_2, y_2 such that $abvxy = \psi(a_2b_2w_2x_2y_2)$ and $a_2b_2w_2x_2y_2$ is a factor of $\mathbf{u}$.

First, we prove that Case II is impossible. By Lemma 5.5, the word $a_2b_2w_2x_2y_2$ is prefix-comparable to $a_2b_2\nu^k(2)e_2$ for an integer k. If $|a_2b_2w_2x_2y_2| \leq |a_2b_2\nu^k(2)e_2|$, then $b_2w_2x_2$ is a prefix of b_2v_k2. Thus, the non-left special word $bvx = \psi(bw_2x_2)$ is a prefix of bu_k2. This contradicts the fact that bu_k2 is a left special factor of $\psi(\mathbf{u})$ (the words $abu_k2 = \psi(a_2b_2v_k2)$ and $e_3bu_k2 = \psi(e_3b_3v_k2)$ are factors of $\psi(\mathbf{u})$). Hence $|a_2b_2w_2x_2y_2| > |a_2b_2\nu^k(2)e_2|$ and $\nu^k(2)e_2$ is a prefix of $w_2x_2y_2$. Consequently, the non-left special factor u_k2e_2 (see Lemma 2.3) is a prefix of the left special factor $vxy = \psi(w_2x_2y_2)$, which is a contradiction.

Therefore, Case I holds. By Corollary 5.4, the word $a_1b_1w_1x_1y_1$ is prefix-comparable to one of the words $a_1b_1\nu^k(1)e_1c$, $a_1b_1\nu^k(1)c$, $a_1b_1\nu^k(1)e_01$, $a_1b_1\nu^k(2)e_2$ and $a_1b_1\nu^k(2)e_3$ for some $k \geq 1$.

If $|a_1b_1w_1x_1y_1| \leq |a_1b_1\nu^k(1)|$, then b_1w_1x is a prefix of b_1v_k. Thus the non-left special factor $bvx = \psi(b_1w_1x)$ of $\psi(\mathbf{u})$ is a prefix of the left special factor bu_k, which is a contradiction.

If $|a_1b_1w_1x_1y_1| \geq |a_1b_1\nu^k(1)|+2$, then one of the words $a_1b_1\nu^k(1)e_1c$, $a_1b_1\nu^k(1)c$, $a_1b_1\nu^k(1)e_01$, $a_1b_1\nu^k(2)e_2$ or $a_1b_1\nu^k(2)e_3$ is a prefix of $a_1b_1w_1x_1y_1$. Thus u_k1e_1c, u_k1c, u_k1e_11, u_k2e_2 or u_k2e_3 is a prefix of the left special factor vxy of $\psi(\mathbf{u})$. This contradicts Lemma 2.3, which states that these five words are not left special.

Thus, $|a_1b_1w_1x_1y_1| = |a_1b_1\nu^k(1)| + 1$. Consequently, we have $w_1 = v_k$ and $x_1y_1 \in \{1e_1, 1c, 1e_0, 2e_2, 2e_3\}$, that is, $xy \in \{1e_1, 1c, 2e_2, 2e_3\}$. Noting that vxy is left special whereas by Lemma 2.3, the words u_k1c, u_k2e_2 and u_k2e_3 are not left special, we deduce that $xy = 1e_1$. This ends the proof of Lemma 2.5. □

ON INTERPOLATING POWER SERIES

James Guyker
Department of Mathematics, SUNY College at Buffalo, Buffalo, New York
guykerj@buffalostate.edu

Received: 1/24/12, Revised: 12/3/12, Accepted: 1/21/13, Published: 1/25/13

Abstract

We derive a simple error estimate for equally spaced, polynomial interpolation of power series that does not require the uniform bounds on derivatives of the Cauchy remainder. The key steps are expressing Newton coefficients in terms of Stirling numbers $S(i,\, j)$ of the second kind and applying the concavity of $\ln S(i,\, j)$.

1. Introduction

Let f be in $C[a,b]$ and let $P_n f$ denote the unique polynomial of degree at most n that interpolates f at $n+1$ equally spaced nodes. By a classical result [3, Theorem 4.3.1] of interpolation theory, $P_n f$ converges uniformly to f if f can be extended analytically to a certain region of the complex plane that contains the interior of a lemniscate formed from disks centered at a and b with radii greater than $b-a$. We show that more is true for real functions: if f is a Taylor series about a or b, then there is a simple bound on the uniform error that implies uniform convergence when the derived series f' is in $C[a,b]$. We expand the Newton coefficients of $P_n f$ in terms of Stirling numbers $S(i,\, j)$ of the second kind and use the well known log-concavity property of L. H. Harper and E. H. Lieb ([7], [9]):

$$\textit{for each } i > 1, \textit{ the ratio } \frac{S(i,j+1)}{S(i,j)} \textit{ is strictly decreasing.} \tag{1}$$

2. The Main Result

By the transformation $x \mapsto \frac{x-a}{b-a}$, we may assume that the underlying interval is $[0,1]$. An analog of Abel's partial summation formula [1, vol. I, Theorem 10.16] will be used in our arguments:

$$\sum_{i=m}^{n} a_i b_i = \left(\sum_{i=m}^{n} a_i\right) b_m + \sum_{j=m}^{n-1}\left(\sum_{i=j+1}^{n} a_i\right)(b_{j+1}-b_j). \qquad (2)$$

For example, by Abel's theorem [2, p. 325] a power series $\sum_{i\geq 0} a_i x^i$ is in $C[0,1]$ if and only if the series $\sum a_i$ converges. In this case, since for fixed $x<1$, $b_i = x^i$ decreases monotonically to zero, we have the error estimate

$$\left\|\sum_{i>n} a_i x^i\right\|_\infty \leq 2\epsilon_n(<a_i>)$$

by (2), where

$$\epsilon_n(<a_i>) := \max\left\{\left|\sum_{i>k} a_i\right| : k \geq n\right\}$$

converges to zero. A similar estimate that is a refinement of [5, Theorem 1] holds for polynomial interpolation:

Theorem 1. *Let* f *be in* $C[0,1]$ *and let* $P_n f$ *be its interpolating polynomial at the nodes* $0, 1/n, 2/n, \ldots, 1$. *If either* $f(x)$ *or the reflection* $f(1-x)$ *of* $f(x)$ about $x=\frac{1}{2}$ *is represented by a power series* $\sum a_i x^i$ *such that* $\sum a_i$ *converges, then* $P_n f$ satisfies

$$\|f - P_n f\|_\infty \leq (2n+1)\epsilon_n(<a_i>). \qquad (3)$$

In particular, $P_n f$ *converges to* f *uniformly whenever* $\sum i a_i$ *converges.*

Proof. Stirling numbers of the second kind are the coefficients in the formula that converts powers to binomial coefficients: for $i>0$

$$x^i = \sum_{j=1}^{i} S(i,j) j! \binom{x}{j} \qquad (4)$$

where $j!\binom{x}{j} = x(x-1)\cdots(x-j+1)$ is a factorial polynomial. Since $x^i = x\,x^{i-1}$ and $\binom{x}{j} = \binom{x-1}{j-1} + \binom{x-1}{j}$, we have the usual recurrence relation

$$S(i,j) = S(i-1,j-1) + jS(i-1,j)$$

with

$$S(i,1) = S(i,i) = 1 \text{ and } S(i,j) = 0 \text{ for } j > i.$$

Hence (4) may be easily solved for $S(i,\, j)$ by induction on row i:

$$S(i,j) = \frac{1}{j!}\sum_{k=1}^{j}(-1)^{j-k}k^i\binom{j}{k}. \tag{5}$$

By Lagrange's form [1, vol. II, Theorem 15.2] for P_nf, we have that for linear combinations, $P_n(\alpha f + \beta g) = \alpha P_nf + \beta P_ng$; and if $f_m \longrightarrow f$ pointwise, then $P_nf_m \longrightarrow P_nf$. Moreover by (4), for $i > 0$,

$$P_nx^i = \sum_{j=1}^{n}\frac{S(i,j)}{n^i}j!\binom{nx}{j} = \sum_{j=1}^{n}\frac{S(i,j)}{n^{i-j}}x\left(x-\frac{1}{n}\right)\cdots\left(x-\frac{j-1}{n}\right)$$

in Newton's form [1, vol. II, Theorem 15.5]. Therefore, if $\sum a_i$ converges, then

$$P_n\left(\sum_{i\ge 0}a_ix^i\right) = a_0 + \sum_{j=1}^{n}\left(\sum_{i>0}\frac{S(i,j)}{n^i}a_i\right)j!\binom{nx}{j}.$$

Suppose first that $f(x) = \sum a_ix^i$ where $\sum a_i$ converges. If $f_n(x) := \sum_{i=0}^{n}a_ix^i$, then $f_n = P_nf_n$ by uniqueness and

$$\begin{aligned}\|f-P_nf\|_\infty &\le \|f-f_n\|_\infty + \|P_n(f-f_n)\|_\infty \\ &\le 2\epsilon_n(<a_i>) + \left\|\sum_{j=1}^{n}\left(\sum_{i>n}\frac{S(i,j)}{n^i}a_i\right)j!\binom{nx}{j}\right\|_\infty.\end{aligned}$$

Clearly $\left|\binom{nx}{j}\right| \le \binom{n}{j}$ if $j-1 \le nx \le n$. Suppose that $k-1 \le nx < k$ for some integer k in $[1, j-1]$. Then $j!\left|\binom{nx}{j}\right| = (\Pi_{i=0}^{k-1}|nx-i|)(\Pi_{i=k}^{j-1}|nx-i|) \le k!(j-k)!$ where $k! \le \frac{n!}{(n-k)!}$ and $(j-k)! \le \frac{(n-k)!}{(n-j)!}$. It follows that $j!\left|\binom{nx}{j}\right| \le j!\binom{n}{j}$ for all x in $[0,1]$. Thus, if

$$b_{ij} := \frac{S(i,j)}{n^i}j!\binom{n}{j},$$

then

$$\|f-P_nf\|_\infty \le 2\epsilon_n(<a_i>) + \sum_{j=1}^{n}\left|\sum_{i>n}a_ib_{ij}\right| \tag{6}$$

where $< b_{ij} : j = 1, \ldots, n >$ is a probability vector for each $i > n$: $b_{ij} \geq 0$, and by (5) and the identity $\binom{n}{j}\binom{j}{k} = \binom{n}{k}\binom{n-k}{j-k}$,

$$\sum_{j=1}^{n} b_{ij} = \sum_{k=1}^{n} \left(\frac{k}{n}\right)^i \left[\sum_{j=1}^{n} (-1)^{j-k} \binom{n}{j}\binom{j}{k}\right] = \sum_{k=1}^{n} \left(\frac{k}{n}\right)^i \binom{n}{k} \delta_{nk} = 1$$

where δ_{nk} is the Kronecker delta.

Consider the term $j = n$ in (6). If $n = 1$, then $b_{in} = 1$. And if $n > 1$, then the sequence b_{in} $(i > n)$ increases to 1 since

i. $b_{i+1,n} > b_{in}$ if and only if $S(i+1, n) > n\, S(i, n)$. But $S(i+1, n) = S(i, n-1) + n\, S(i, n) > n\, S(i, n)$.

ii. By (5),

$$b_{in} = \sum_{k=1}^{n} (-1)^{n-k} \left(\frac{k}{n}\right)^i \binom{n}{k} \longrightarrow 1.$$

Therefore by (2), for all n,

$$\left|\sum_{i>n} a_i b_{in}\right| \leq \epsilon_n(<a_i>). \tag{7}$$

Next, suppose that $j < n$. We show that $\{b_{ij} : i > n\}$ is unimodal: There exists some integer $i(j) > n$ such that for $i > n$,

$$b_{i+1,j} \leq b_{ij} \text{ ifandonlyif } i \geq i(j). \tag{8}$$

As above, (8) is equivalent to

$$S(i+1, j) \leq nS(i, j) \text{ ifandonlyif } i \geq i(j).$$

Clearly $i(1) = n + 1$ so assume that $j > 1$. We first verify that the sequence $\ln S(i, j)$ $(i \geq j)$ is concave, i.e., the ratio $\frac{S(i+1,j)}{S(i,j)}$ is strictly decreasing. It is easy to check the case $j = 2$ with (5). Then for $j > 2$, the following inequalities are equivalent:

$$\frac{S(i+2,j)}{S(i+1,j)} = \frac{S(i+1,j-1)}{S(i+1,j)} + j < \frac{S(i+1,j)}{S(i,j)} = \frac{S(i,j-1)}{S(i,j)} + j$$

$$\frac{S(i+1,j-1)}{S(i+1,j)} = \frac{S(i,j-2) + (j-1)S(i,j-1)}{S(i,j-1) + jS(i,j)} < \frac{S(i,j-1)}{S(i,j)}$$

$$\frac{S(i,j)}{S(i,j-1)} - \frac{S(i,j)}{S(i,j-2)} < \frac{S(i,j-1)}{S(i,j-2)}.$$

The last inequality is true by (1). Thus by (5), the sequence $\frac{S(i+1,j)/j^i}{S(i,j)/j^i}$ strictly decreases to $j < n$, and therefore (8) holds.

Also by (5),

$$b_{ij} = \sum_{k=1}^{j} (-1)^{j-k} \left(\frac{k}{n}\right)^i \binom{n}{j} \binom{j}{k}$$

converges to zero. Hence by (2), for $j = 1, \ldots, n-1$,

$$\left| \sum_{i>n} a_i b_{ij} \right| \leq 2b_{i(j),j}\epsilon_n(<a_i>) \leq 2\epsilon_n(<a_i>)$$

and (3) now follows from (6) and (7).

Finally suppose that $g(x) := f(1-x) = \sum_{i \geq 0} a_i x^i$ where $\sum a_i$ converges. The polynomial $(P_n f)(1-x)$ is of degree at most n and interpolates g at the nodes $0, \frac{1}{n}, \frac{2}{n}, \ldots, 1$. Hence by uniqueness, $(P_n f)(1-x) = (P_n g)(x)$ and we have that for x in $[0,1]$,

$$| f(x) - (P_n f)(x)| \leq \max\{|g(x) - (P_n g)(x)| : 0 \leq \mathrm{x} \leq 1\} = \|g - P_n g\|_\infty .$$

Thus (3) is a consequence of the first case.

Assume now that $\sum i a_i$ converges. By (2), $\sum a_i$ converges and

$$\begin{aligned} \epsilon_n(<a_i>) &= \max\left\{ \left| \sum_{i>k} \frac{1}{i}(i a_i) \right| : k \geq n \right\} \\ &\leq \max\left\{ \frac{2}{k+1}\epsilon_k(<i a_i>) : k \geq n \right\} \leq \frac{2}{n+1}\epsilon_n(<i a_i>). \end{aligned}$$

Hence

$$(2n+1)\epsilon_n(<a_i>) \leq 4\epsilon_n(<i a_i>) - \epsilon_n(<a_i>)$$

and therefore $P_n f \longrightarrow f$ uniformly by (3). □

3. Examples

Runge's famous example $f(x) = \frac{1}{x^2+1}$ on $[-5,5]$ is *not* the uniform limit of $P_n f$ ([4], [8, Sec. 3.4]). However, Theorem 1 implies uniform convergence on $[0,1]$:

Example 2. For $f(x) = \frac{1}{x^2+1}$ in $C[0,1]$, we have that

$$f(1-x) = \frac{1}{2i}\left[\frac{1}{x-(1+i)} - \frac{1}{x-(1-i)}\right]$$

$$= \frac{i}{4}\left[(1-i)\sum_{k\geq 0}\left(\frac{x}{1+i}\right)^k - (1+i)\sum_{k\geq 0}\left(\frac{x}{1-i}\right)^k\right] = \sum_{k\geq 0}\frac{\sin[(k+1)\pi/4]}{2^{(k+1)/2}}x^k.$$

Therefore,

$$\|f - P_n f\|_\infty \leq \frac{2n+1}{(\sqrt{2}-1)(\sqrt{2})^{n+1}} \longrightarrow 0.$$

Uniform convergence of $P_n f$ holds for functions f in the large class of Taylor series about 0 or 1 with absolutely summable coefficients [6]. This also follows from the proof above since if $\sum_{i>n}|a_i| < \infty$, then in (6),

$$\sum_{j=1}^{n}\left|\sum_{i>n} a_i b_{ij}\right| \leq \sum_{i>n}|a_i|\left(\sum_{j=1}^{n} b_{ij}\right) = \sum_{i>n}|a_i|.$$

Moreover, $\|f - P_n f\|_\infty \leq 2\sum_{i>n}|a_i|$ in this case since $\left\|\sum_{i>n} a_i x^i\right\|_\infty \leq \sum_{i>n}|a_i|$. Thus we have an improvement for Example 2:

$$\|f - P_n f\|_\infty \leq \frac{1}{(\sqrt{2}-1)(\sqrt{2})^{n-1}}.$$

Example 3. The coefficients of $f(x) = \sum_{i\geq 2}\frac{(-1)^i x^i}{i\ln i}$ are not absolutely summable by the integral test. However, f and f' are in $C[0,1]$ by the alternating series test and therefore

$$\|f - P_n f\|_\infty \leq \frac{2n+1}{(n+1)\ln(n+1)}.$$

References

[1] T. M. Apostol, *Calculus*, 2nd ed., vols. I, II, Blaisdell, Waltham, MA, 1967, 1969.

[2] R. G. Bartle, *The Elements of Real Analysis*, 2nd ed., Wiley, New York, 1976.

[3] P. J. Davis, *Interpolation and Approximation*, Blaisdell, New York, 1963.

[4] J. F. Epperson, *On the Runge example*, Amer. Math. Monthly **94** (1987), 329-341.

[5] J. Guyker, *On the uniform convergence of interpolating polynomials*, Appl. Math. Comput. **210** (2009), no. 1, 11-17.

[6] J. Guyker, *Uniform polynomial approximation*, Far East J. Math. Sci. (FJMS) **39** (2010), no. 2, 243-254.

[7] L. H. Harper, *Stirling behavior is asymptotically normal*, Ann. Math. Stat. **31** (1967), 410-414.

[8] E. Isaacson and H. Keller, *Analysis of Numerical Methods*, Wiley, New York, 1966.

[9] E. H. Lieb, *Concavity properties and a generating function for Stirling numbers*, J. Combinatorial Theory **5** (1968), 203-206.

CAPTURED-REVERSIBLE MOVES AND STAR DECOMPOSITION DOMINATION IN HEX

Philip Henderson[1]
pthender@gmail.com

Ryan B. Hayward[2]
Department of Computing Science, University of Alberta, Edmonton AB, Canada
hayward@ualberta.ca

Received: 4/16/10, Revised: 8/17/12, Accepted: 1/15/13, Published: 1/25/13

Abstract

By applying the combinatorial game theory notions of dominated move and reversible move, and by exploiting graph-theoretic properties of Hex board decompositions, we identify two new types of inferior Hex move.

1. Introduction

Hex is the classic two-person alternate-turn board game invented by Piet Hein [10] and John Nash [14, 15, 16].

The players are Left and Right. The board is an $m \times n$ array of hexagonal cells; usually $m = n$, in which case the board has the shape of a rhombus. The color black and two opposite sides of the board are assigned to Left; the color white and the other two sides of the board are assigned to Right. A move consists of coloring an uncolored cell. The game ends when a player completes a path of their color joining their two sides; this player is the winner.

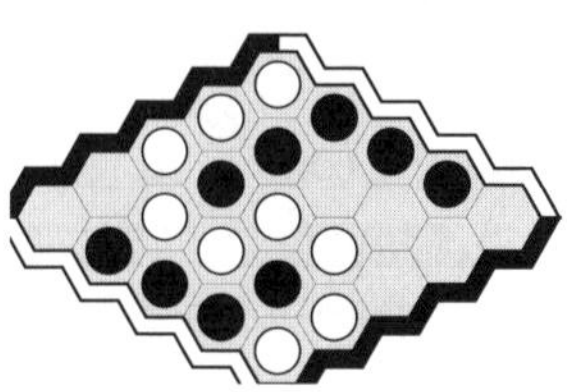

Figure 1: A game won by Right (white).

[1]supported by NSERC and iCORE
[2]supported by NSERC

Hex has several elegant properties:

- Hex is *monotonic*, or *regular* [20]: it is never disadvantageous to move,
- Hex never ends in a draw [5],
- Hex on an $n \times n$ board is a first-player win [2, 16],
- solving Hex — determing the winner of an abitrary position — is PSPACE-complete[17].

The last result hints that there might be no efficient algorithm to find a winning move whenever one exists. However, there are efficient methods that can identify some kinds of inferior move, and pruning such moves often improves the efficiency of finding a winning move [2, 4, 8, 9, 18].

Hex is not a combinatorial game in the strictest sense, since a Hex game ends when a player connects their two sides, whereas a combinatorial game ends when a player has no legal moves. But many combinatorial game theory (CGT) concepts — e.g., dominated and reversible moves — can be applied to Hex. Indeed, Hex is often included when CGT is discussed in a broad context, for example in Albert, Nowakowski and Wolfe's *Lessons in Play* [1].

In this paper[3], we consider inferior Hex moves in the context of CGT outcome classes. In the process, we identify two new kinds of inferior Hex cell that allow efficient pruning of the corresponding move.

In the rest of this section we present our notation. In §2 we reformulate previous Hex inferior cell analysis in terms of CGT. In §3 we identify a new class of Hex reversible moves. In §4 we identify a new class of Hex dominated moves.

1.1. Notation

Definition 1.1. A *(Hex) position* is defined by specifying the color status — uncolored, black, or white — of each board cell.

Definition 1.2. A *(Hex) state* is defined by a position and the player to move next. Given a position H and player to move next X, we denote the associated state as $H[X]$.

In CGT, the *outcome (class)* of a combinatorial game G is the result that can be achieved with perfect play. A game G is *positive* if Left wins regardless of who moves next; *negative* if Right wins regardless of who moves next; *fuzzy* if the first player wins; *zero* if the second player wins. In these four cases, we write $G > 0$, $G < 0$, $G \,||\, 0$, $G = 0$ respectively. A state is *Left-win* if Left wins and *Right-win* if Right wins.

Hex has no draws, so:

[3]See [19] for further discussion on modelling Hex with CGT.

Observation 1.3. *A Hex state is Left-win or Right-win.*

Hex is monotonic, so a position cannot have outcome zero, so:

Observation 1.4. *A Hex position is positive, negative, or fuzzy.*

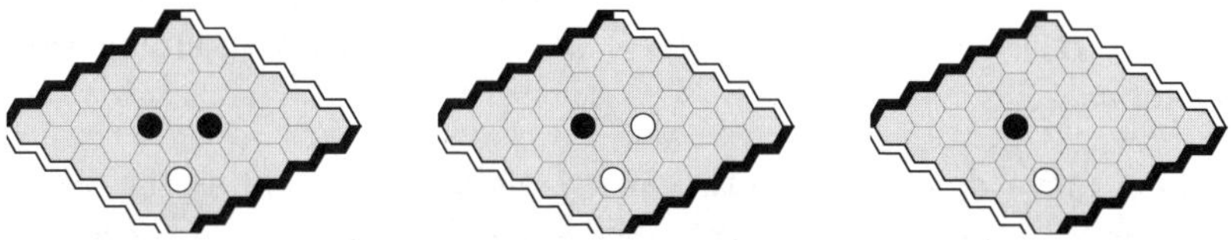

Figure 2: Three Hex positions: positive, negative, fuzzy.

In this paper, we consider (in)equality among states and positions not with respect to general CGT game values but only with respect to CGT outcome classes. Following CGT convention, we compare outcome classes of different states or positions from the perspective of Left.

Definition 1.5. For Hex states S_1, S_2, we write $S_1 \geq S_2$ if S_1 is at least as good for Left as S_2, namely if Left wins S_1 whenever Left wins S_2. Similarly, for Hex positions H_1, H_2, we write $H_1 \geq H_2$ if $H_1[L] \geq H_2[L]$ and $H_1[R] \geq H_2[R]$.

Corollary 1.6. *For Hex states S and S', $S \geq S'$ if and only if S is Left-win and/or S' is Right-win. For positions H and H', $H \geq H'$ if and only if at least one of these conditions holds: $H > 0$, $H' < 0$, or $H \,||\, 0$ and $H' \,||\, 0$.*

Definition 1.7. Hex positions (or states) A and B are *equal* if the associated sets of uncolored/black/white color assignments are equal (and the player to move is equal), in which case we write $A = B$. Hex positions (or states) A and B are *equivalent* if they have the same outcomes, in which case we write $A \equiv B$.

Definition 1.8. An *L-move* is a move by the Hex player Left, and an *L(c)-move* is an L-move to an uncolored cell c. For a position H with an uncolored cell c, a set of uncolored cells C, and a set of colored cells D:

- $H + L(c)$ is the position obtained from H by coloring c black,
- $H + L(C)$ is the position obtained from H by coloring all cells in C black,
- $H - D$ is the position obtained from H by uncoloring all cells in D.

These terms are defined similarly for Right and white.

Caveat: in CGT, "+" usually indicates the sum of games; in this paper, "+" is used only as defined above.

Definition 1.9. For a Hex position H, $\chi_U(H)$, $\chi_L(H)$, $\chi_R(H)$ denote the sets of all cells of H that are respectively uncolored, black, or white. For a set D of cells of a Hex position H, $\chi_U(H, D)$, $\chi_L(H, D)$, $\chi_R(H, D)$ denote the sets of all cells of D that are respectively uncolored, black, or white.

Definition 1.10. For Hex positions H_1 and H_2, H_2 is a *continuation* of H_1 if $\chi_L(H_1) \subseteq \chi_L(H_2)$ and $\chi_R(H_1) \subseteq \chi_R(H_2)$. A continuation with no uncolored cells is a *completion.*

In CGT a move *dominates* another if it is at least as good as the other, while a move *reverses* a previous move if it renders the previous move useless. In Hex, we define these notions in terms of the associated uncolored cells and restrict them according to outcome class.

Definition 1.11. For a Hex position H with uncolored cells c_1 and c_2,

- c_1 *Left-dominates* c_2 if $H + L(c_1) \geq H + L(c_2)$,
- c_1 is *Left-reversible* if $H \geq H + L(c_1) + R(c_2)$, in which case c_2 is a *Right-reverser* of c_1.

As explained in *Winning Ways for your Mathematical Plays* by Berlekamp, Conway, and Guy, when determining a game's value, dominated moves can be pruned (as long as one dominating move remains) and reversible moves can be bypassed, and — in its simplest form — a combinatorial game has no dominated or reversible moves [3].

The same result holds for outcome classes: when determining a game's outcome class, dominated moves can be pruned (as long as one dominating move remains) and reversible moves can be bypassed; this follows by simple case analysis from the minimax calculation of outcome classes.

We are interested in simplifying Hex positions. In the next section we rephrase some previous Hex results in terms of dominated and reversible moves.

2. Previously Known Inferior Cell Analysis

Following observations by Beck et al. [2, pp. 327-339] and Schensted and Titus [18], Hayward and van Rijswijck defined a class of provably useless Hex cells, called dead cells:

Definition 2.1. ([9]) For a Hex position H, an uncolored cell c is *live* if H has a completion H' in which changing c's color changes the winner of H'; otherwise, c is *dead.*

Observation 2.2. *([4]) For a position H, a cell c is live if and only if c is in a set S of uncolored cells of H such that some coloring of S yields the winning condition, but no coloring of a proper subset of S yields the winning condition.*

Thus determining whether a cell is live reduces to determining in a graph whether a given vertex is on a minimal path joining two other given vertices. This problem is NP-complete for general graphs, although its complexity on graphs that arise from Hex positions is unknown [4].

Observation 2.3. *([9]) Coloring a dead cell does not change a position's outcome.*

Observation 2.4. *A dead cell remains dead in all continuations in which it is uncolored.*

Some dead cells can be recognized by matching patterns of neighboring cells. For example, for each pattern in Figure 3, the uncolored cell is dead [8].

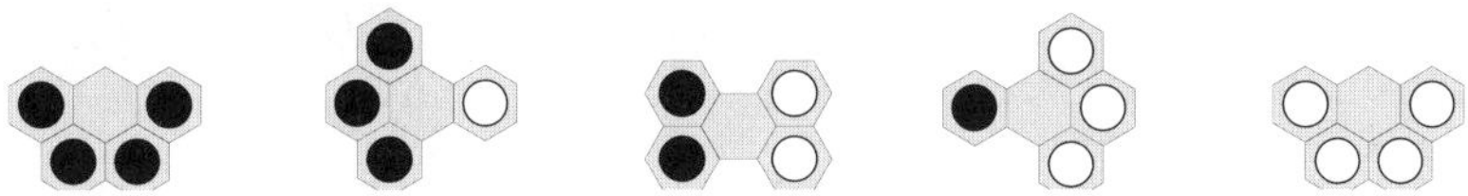

Figure 3: Some dead patterns. For any containing position, the uncolored cell is dead: coloring the cell cannot change the position's outcome.

In CGT terms, Hex is a hot game — for any position H with uncolored cell c, $H + L(c) \geq H$ — so it is never disadvantageous to have the next move. However, a move to a dead cell is equivalent to a pass move, so:

Observation 2.5. *([9]) A Hex player with a winning strategy has a winning strategy with no move to a dead cell.*

Definition 2.6. ([13]) For a Hex position and a player, a move to a cell c is *vulnerable* if the opponent has a move to a cell k that makes that cell dead; k is a *killer* of c.

E.g., in a position H, an uncolored cell c is Left-vulnerable if there is an uncolored cell k such that c is dead in $H + R(k)$.

We now redefine vulnerability in CGT terms:

Definition 2.7. For a Hex position and a player, a move to a cell c is *dead-reversible* if it is vulnerable.

Our first result is that dead-reversible cells are reversible:

Lemma 2.8. *Let H be a Hex position with a cell c that is Left-dead-reversible to killer k. Then c is Left-reversible in H, with Right-reverser k.*

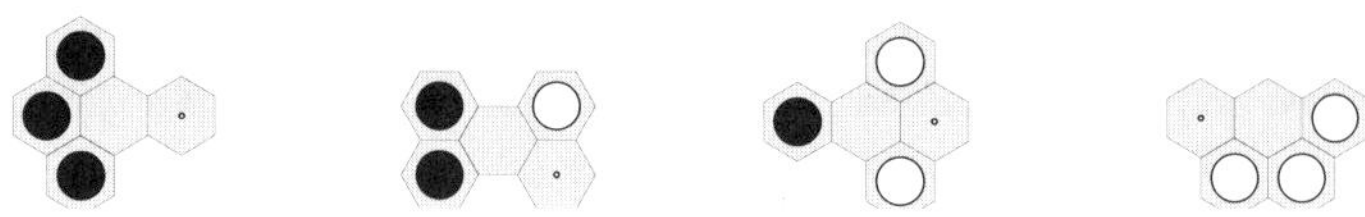

Figure 4: Some Left-dead-reversible patterns. For any containing position, the empty cell is killed by white-coloring the dotted cell.

Proof. $H \geq H + R(k)$ by monotonicity. $H + R(k) \equiv H + L(c) + R(k)$ by Observation 2.3. Thus $H \geq H + L(c) + R(k)$. □

Recall Observation 2.4: in a continuation, a dead cell remains dead for as long as it is uncolored. By contrast, a dead-reversible cell need not remain dead-reversible:

Observation 2.9. *Let H_1 be a Hex position with a Left-dead-reversible cell c and Right-reverser k, and let H_2 be a continuation of H_1 obtained by Right-coloring some cell(s) other than c. Then in H_2, c is either dead or Left-dead-reversible with Right-reverser k.*

The following rephrases a result of Hayward and van Rijswijck.

Theorem 2.10. *([9]) A Hex player with a winning strategy has a winning strategy with no move to a dead or dead-reversible cell.* □

Thus, dead-reversible Hex moves can be pruned. This is stronger than what is guaranteed by CGT, namely that reversible moves can be bypassed.

As noted in Observation 2.3, dead cells can be assigned to either player without changing a position's outcome. We now identify a class of cells that can be assigned to one particular player without changing a position's outcome.

Definition 2.11. For a position H, a set U of uncolored cells is *Left-captured* if Left has a second-player strategy on U such that, for each leaf position F in the strategy tree, each cell in $\chi_R(F, U)$ is dead in position $F - \chi_R(F, U)$. In H, any cell in such a set U is *Left-captured.*

Notice that, for each leaf position F in Definition 2.11, it follows by Observation 2.3 that

$$F \equiv F - \chi_R(F, U) \equiv (F - \chi_R(F, U)) + L(\chi_R(F, U)).$$

In other words, if Right ever plays in a Left-captured set, then Left has a replying strategy that guarantees no possible benefit to Right.

Consider any pattern in Figure 5. Left has a second-player strategy on the uncolored cell pair — if Right colors one, Left colors the other — that kills the cell just colored by Right. Thus the uncolored cell pair is Left-captured.

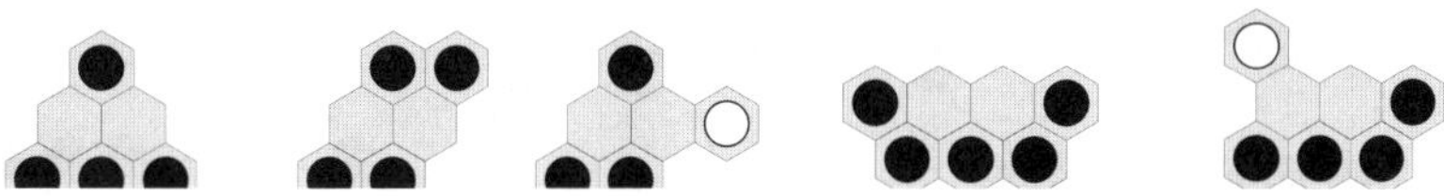

Figure 5: Some Left-captured patterns. For any containing position, black-coloring the pattern's uncolored cells cannot alter the position's outcome.

Observation 2.12. *Let H be a position with a Left-captured set S. Then S remains Left-captured in any continuation of H in which S remains uncolored.*

The observation holds because the uncolored cells outside S neither affect the capturing strategy nor revive any cell that is dead in a leaf position. Thus if H is a Hex position with a Left-captured set S, then $H \equiv H + L(S)$. Moreover, combining a captured set strategy for S with a winning strategy on the reduced board $H + L(S)$ yields a winning strategy in the original position [7].

Our second result shows that playing in one's own captured set is equivalent to a pass move:

Lemma 2.13. *Let H be a position with a cell c that is both Left-captured in H and a winning move in $H[L]$. Then $H > 0$.*

Proof. Let c be in a set F of Left-captured cells of H. Then $H \equiv H + L(F)$, and $H + L(F) \geq H + L(c) \geq H$ by monotonicity, so $H \equiv H + L(c)$. But c is a winning move in $H[L]$, so $(H + L(c))[R]$ is Left-win, so $H + L(c) \equiv H > 0$. □

Definition 2.14. An uncolored cell set F of a position H is *Left-fillin* if F partitions into cell sets $F_1, \ldots, F_t$ such that, for $1 \leq j \leq t$, each F_j is dead or Left-captured in position $H + L(F_1) + \ldots + L(F_{j-1})$.

This follows by induction on the fillin partition:

Observation 2.15. *For a position H with Left-fillin F, $H \equiv H + L(F)$.*

Definition 2.16. For uncolored cells c and f of a position H, c *Left-fillin-dominates* f if f is in some Left-fillin set F of $H + L(c)$.

This follows by monotonicity [6]:

Observation 2.17. *For a position H with uncolored cell c such that $H + L(c)$ has Left-fillin F, the cell c Left-dominates all cells f in F; namely,*

$$H + L(c) \equiv H + L(c) + L(F) \geq H + L(f).$$

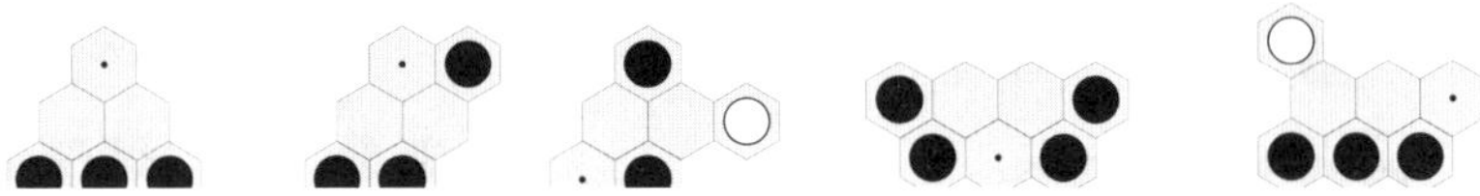

Figure 6: Some Left-domination patterns. For any containing position, black-coloring a dotted cell fillin-dominates black-coloring either of the other two uncolored cells.

In Hex, another form of domination arises when a cell is on all induced winning paths of another cell [11, 13]. When we wish to distinguish between these two forms of domination, we call the former *fillin domination* and the latter *induced path domination.* In this paper we use only fillin domination.

Thus far we have discussed inferior cells that can be pruned. Now we examine certain decompositions of Hex positions.

Definition 2.18. A *chain* is a maximal set of connected same-colored cells. A *Left chain* is a maximal set of connected black cells.

Definition 2.19. Two opposite-colored chains C_1, C_2 *touch* if there are cells c_1, c_2 in C_1, C_2 respectively such that c_1 and c_2 are adjacent or form an opposite-colored bridge pattern as shown in Figure 7.

During a Hex game, chains can form that decompose the board.

Figure 7: A bridge and a half-bridge between opposite-coloured cells. In the former, regarding the empty cells as non-adjacent enables decompositions.

Lemma 2.20. *Let H be a Hex position with an opposite-colored half-bridge composed of cells c_1, c_2, c_3 that are respectively black, white, uncolored, and let C_1, C_2 be the chains containing c_1, c_2 respectively. Then C_1 and C_2 touch in H.*

Proof. Let c_4 be the unique cell in H adjacent to c_1, c_2, and c_3. If c_4 is black then it is in C_1, and c_4 in C_1 is adjacent to c_2 in C_2. Similarly, the conclusion holds if c_4 is white. Lastly, if c_4 is uncolored, then C_1 and C_2 touch via the opposite-colored bridge $\{c_1 \dots c_4\}$. □

Definition 2.21. A *Left-splitting decomposition* is a Left chain that touches both of Right's sides.

In a position with such a decomposition, if Left has a winning strategy, then Left has a connection strategy for each of the two subgames (connect the chain to one Left side; connect the chain to the other Left side), and combining these two strategies yields a winning Left strategy for the whole board [12].

Definition 2.22. A *four-sided decomposition* is a 4-cycle of consecutively touching chains. The *carrier* of this decomposition is the set of uncolored cells bounded by the four chains and their touching points.

If Left has a second-player strategy within the carrier of a four-sided decomposition that guarantees connection of the decomposition's two bounding Left chains, then the decomposition carrier is Left-captured [12].

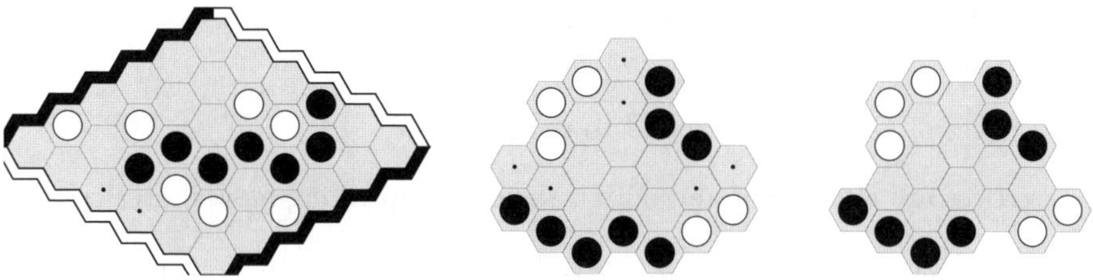

Figure 8: A Left-split decomposition, a four-sided decomposition, and the same four-sided decomposition using half-bridges. Dotted cells show opposite-coloured bridges of the decomposition. In the rightmost figure, all empty cells are black-captured.

3. Captured-Reversible Moves

Just as 'dead' leads to 'dead-reversible', so 'captured' leads to 'captured-reversible'. The following definition may seem counter-intuitive, since the *opponent's* move — not the player's — yields the *player's* fillin.

Definition 3.1. A cell c in position H is *Left-captured-reversible* if there is a cell r such that $H + R(r)$ has Left-fillin F containing c.

Lemma 3.2. *Let cell c be Left-captured-reversible in Hex position H. Then cell c is Left-reversible in H.*

Proof. By Definition 3.1, some Right-move r in H yields Left-fillin F containing c. By monotonicity, $H + L(F) + R(r) \geq H + L(c) + R(r) \geq H + R(r)$. By Observation 2.15, $H + R(r) \equiv H + L(F) + R(r)$, so $H + R(r) \equiv H + L(c) + R(r)$. By monotonicity, $H \geq H + R(r)$, so $H \geq H + R(r) \equiv H + L(c) + R(r)$, satisfying the definition of Left-reversible. □

Captured-reversible cells are reversible, so they can be bypassed. However, we would like to prune them from consideration completely, as is done with dead-reversible cells. It is an open question whether all capture-reversible cells can be pruned while preserving a position's outcome. However, we now show sufficient conditions for such pruning.

Definition 3.3. For a Left-captured-reversible move m with Right-reverser r and Left-fillin F, we call F a *Left-captured-reversible carrier* of m. With respect to carriers F_1, F_2 and Right-reversers r_1, r_2 of Left-captured-reversible moves m_1, m_2, we say that m_1 and m_2 *interfere* if r_1 is in F_2 or r_2 is in F_1. The *Left-captured-reversible graph* $G_{\gamma(H,L)}$ of position H is defined as follows:

- for each Left-captured-reversible move m_j in H, select a Right-reverser r_j and carrier F_j,
- vertices of $G_{\gamma(H,L)}$ correspond to the moves m_j,
- vertices are adjacent if and only if their corresponding moves interfere.

An independent vertex set in $G_{\gamma(H,L)}$ is called an *independent Left-captured-reversible set* in position H.

Lemma 3.4. *Let H_1 be a position with an uncolored cell c and an independent Left-captured-reversible set $I_1 = \{m_1, \ldots, m_n\}$, where each captured-reversible cell m_j has selected Right-reverser r_j and carrier F_j. Then in position $H_2 = H_1 + R(c)$ the set $I_2 = \{m_j \in I_1 : c \notin \{r_j\} \cup F_j\}$ is an independent Left-captured-reversible set.*

Proof. Each m_j in I_2 is Left-captured-reversible in H_2 since r_j remains a legal move for Right, and F_j remains fillin for any continuation of $H_1 + R(r_j)$ in which all cells in F_j are uncolored. In defining the Left-captured-reversible graph $G_{\gamma(H_2,L)}$, we can select the same reversers and carriers for all cells in I_2 to guarantee independence. □

Theorem 3.5. *Let H be a Hex position with a set of dead cells D, a set of Left-dead-reversible cells V, and an independent Left-captured-reversible set I. If Left wins $H[L]$, then either Left has a winning move not in D or V or I, or Left wins $H[R]$.*

Proof. Proof by contradiction. Let H be a counterexample with the smallest number of uncolored cells. Thus, Left has a winning move in $H[L]$, but each such move is in D or V or I, and Right wins $H[R]$. For $H[L]$, let $W = \{m_1, \ldots, m_n\}$ be the set of Left-winning moves not in D or V. Pruning dead and dead-reversible moves cannot eliminate all winning moves (Theorem 2.10), so W is a nonempty subset of I.

For some $m_j \in W$, let $T_j = H + R(r_j)$, where r_j is the Right-reverser of m_j. By the definition of captured-reversible, it follows that $T_j \equiv H + L(m_j) + R(r_j)$: see the proof of Lemma 3.2. Left wins $(H + L(m_j))[R]$, so Left wins $(H + L(m_j) + R(r_j))[L] \equiv T_j[L]$. By monotonicity $H \geq T_j$, so Right wins $T_j[R]$.

Since T_j has fewer uncolored cells than H, it is not a counterexample to this theorem. Thus in position T_j, for any dead cell set D_j, Left-dead-reversible cell set V_j, and independent Left-captured-reversible set I_j, Left has a winning move not in D_j or V_j or I_j.

Dead cells are dead in all continuations (Observation 2.4), and Left-dead-reversible cells are dead or Left-dead-reversible in all continuations in which only Right-colored cells are added (Observation 2.9). Thus we can select D_j, V_j such that $D_j \cup V_j \supseteq (D \cup V) \setminus \{r_j\}$. Also, we can select I_j to be the set of cells in I whose Right-reverser is not r_j. By Lemma 3.4, I_j is an independent Left-captured-reversible set in T_j. Thus some winning Left-move m of $T_j[L]$ is not in D_j or V_j or I_j.

Since $T_j = H + R(r_j)$, m is also winning in $H[L]$. Thus, by our assumption, m is in $(D \cup V \cup I) \setminus (D_j \cup V_j \cup I_j \cup \{r_j\}) \subseteq I \setminus I_j$. Thus m is Left-captured-reversible with Right-reverser r_j in H, meaning that m is Left-captured in T_j. By Lemma 2.13, $T_j > 0$, contradicting the fact that R wins $T_j[R]$. □

If Left wins $H[R]$, then any legal move in state $H[L]$ is Left-winning. Thus we can apply Theorem 3.5 as follows: given a Hex position in which we are trying to find a Left-winning move, we can identify dead cells, Left-dead-reversible cells, and an independent Left-captured-reversible set, and prune all these inferior cells from consideration with the caveat that we consider at least one legal move.

Using known captured patterns allows us to identify captured-reversible patterns.

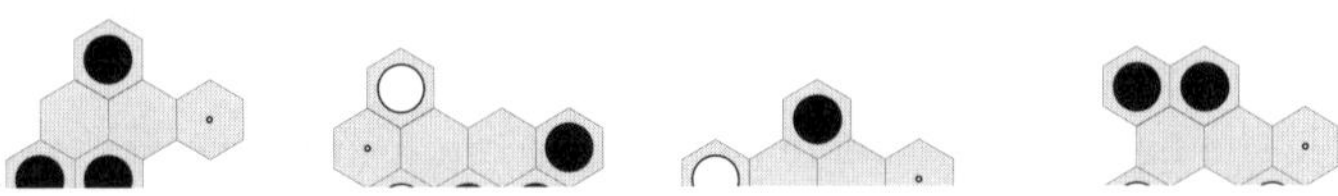

Figure 9: Some Left-captured-reversible patterns. For any containing position, white-coloring the dotted cell black-captures the other two uncolored cells.

4. Star Decomposition Domination

As mentioned in §2, the carrier C of a four-sided decomposition is Left-captured if Left has a second-player connection strategy joining the two bounding Left-chains within C. Thus if a Left-move m creates such a four-sided decomposition, then m fillin-dominates all moves in C.

Since Hex has no draws, it follows that every four-sided decomposition falls into one of three cases: the carrier is Left-captured, the carrier is Right-captured, or each player has a first-player strategy to connect their bounding chains within the carrier. In this last case, each player has a move available that captures the carrier for themselves and — by fillin domination — no other move in the carrier need be considered. Due to the strategic resemblance to the nimber $* = \{0|0\}$, we call such four-sided decompositions *star decompositions*.

Definition 4.1. A four-sided decomposition is a *star decomposition* if each player has a first-player connection strategy to join their two bounding chains within the carrier.

Unlike the situation with captured-carrier decompositions, a move that creates a star decomposition need not dominate all cells inside the carrier. However, some domination can be deduced by determining if coloring cells inside the carrier does not alter the decomposition's outcome.

Theorem 4.2. *Let H be a position such that a Left-move m yields a star decomposition with carrier C. Let $D \subseteq C$ be a set of cells on which Right has a first-player strategy to connect his two bounding chains. Then in H, m Left-dominates every cell in $C \setminus D$.*

Proof. If we can show that $H + L(m) \equiv H + L(m) + L(C \setminus D)$, then the result follows immediately by monotonicity. Since the cells in $C \setminus D$ are within the star decomposition, they can only affect Right's strategy within the decomposition.

If Left is the first to play inside the star decomposition carrier, then C becomes Left-captured, so the Left-coloring of cells $C \setminus D$ does not alter the position's outcome as they would be assigned to Left in either case.

If Right is the first to play inside the star decomposition carrier, then D becomes Right-captured since Left's additional cells do not obstruct Right's connection strategy on D. Let X be a continuation of $H + L(m)$ such that $\chi_U(X) \supseteq C$. In the position $X + R(D)$, the cells $C \setminus D$ are dead. So $X + L(C \setminus D) + R(D) \equiv X + R(D) \equiv X + R(C)$ since dead cells can be assigned any color without altering a position's outcome. Thus once again the Left-coloring of cells $C \setminus D$ does not alter the position's outcome. □

In other words, Right's star decomposition strategy is not adversely affected by Left-coloring $C \setminus D$.

Corollary 4.3. *Let H be a Hex position such that a Left-move m creates a star decomposition with carrier C. Let $I \subseteq C$ be the set of cells intersecting all of Right's* *first-player strategies to connect his two bounding chains within C. Then in H, m Left-dominates every cell in $C \setminus I$.*

Proof. Repeatedly apply Theorem 4.2 to every such Right first-player strategy. □

Star decompositions often allow the pruning of moves that cannot be pruned by the other techniques mentioned in this paper.

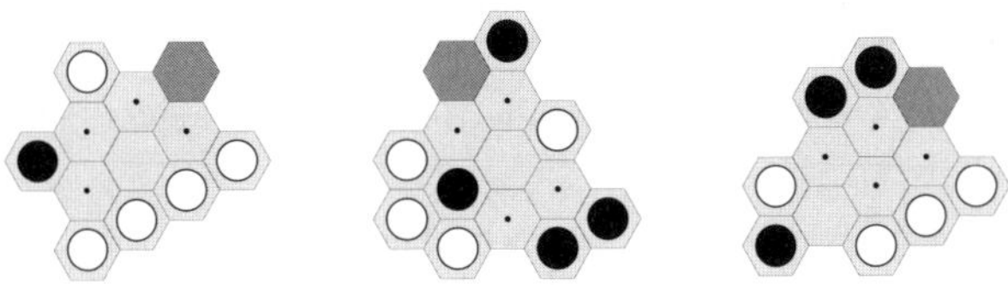

Figure 10: Star decomposition domination. For any containing position, black-coloring a pattern's shaded cell forms a star decomposition and dominates black-coloring any of the pattern's dotted cells.

5. Conclusions

By examining previous inferior cell analysis in terms of CGT, we have identified two new classes of inferior cells for the game of Hex. It remains an open question whether captured-reversible cells can be unconditionally pruned. Also, it would be interesting to know whether decomposition domination exists in other combinatorial games.

Acknowledgements. We thank Broderick Arneson, Martin Müller for helpful comments on an early version of this work, and David Spies for a detailed critique of a later version.

References

[1] Michael H. Albert, Richard J. Nowakowski, and David Wolfe. *Lessons in Play.* AK Peters, 2007.

[2] Anatole Beck, Michael N. Bleicher, and Donald W. Crowe. *Excursions into Mathematics,* Worth, New York, 1969.

[3] Elwyn Berlekamp, John H. Conway, and Richard K. Guy. *Winning Ways for Your Mathematical Plays,* Volumes 1–4, A.K. Peters, 2nd edition, 2000.

[4] Yngvi Björnsson, Ryan Hayward, Michael Johanson, and Jack van Rijswijck. Dead cell analysis in Hex and the Shannon game. In Adrian Bondy, Jean Fonlupt, Jean-Luc Fouquet, Jean-Claude Fournier, and Jorge L. Ramirez Alfonsin, editors, *Graph Theory in Paris: Proceedings of a Conference in Memory of Claude Berge*, pages 45–60. Birkhäuser, 2007.

[5] David Gale. The game of Hex and the Brouwer fixed point theorem. *Amer. Math. Monthly* **86(10)** 1979, 818–827.

[6] Ryan Hayward, Yngvi Björnsson, Michael Johanson, Morgan Kan, Nathan Po, and Jack van Rijswijck. Solving 7×7 Hex with domination, fill-in, and virtual connections. *Theoretical Computer Science* **349(2)** (2005), 123–139.

[7] Ryan B. Hayward. A note on domination in Hex. Technical report, University of Alberta, 2003.

[8] Ryan B. Hayward, Yngvi Björnsson, Michael Johanson, Morgan Kan, Nathan Po, and Jack van Rijswijck. Solving 7×7 Hex: Virtual connections and game-state reduction. In H. Jaap van den Herik, Hiroyuki Iida, and Ernst A. Heinz, editors, *Advances in Computer Games*, volume 263 of *International Federation for Information Processing*, pages 261–278. Kluwer Academic Publishers, Boston, 2003.

[9] Ryan B. Hayward and Jack van Rijswijck. Hex and combinatorics. *Discrete Math.* **306(19–20)** (2006), 2515–2528.

[10] Piet Hein. Vil de laere Polygon? *Politiken*, December 26 1942.

[11] Philip Henderson. *Playing and Solving the Game of Hex.* PhD thesis, University of Alberta, Edmonton, Alberta, Canada, 2010.

[12] Philip Henderson, Broderick Arneson, and Ryan B. Hayward. Solving 8x8 Hex. In Craig Boutilier, editor, *IJCAI*, pages 505–510, 2009.

[13] Philip Henderson and Ryan B. Hayward. Probing the 4-3-2 edge template in Hex. In H. Jaap van den Herik, Xinhe Xu, Zongmin Ma, and Mark H.M. Winands, editors, *Computers and Games*, volume 5131 of *Lecture Notes in Computer Science*, pages 229–240. Springer, 2008.

[14] Harold William Kuhn and Sylvia Nasar, editors. *The Essential John Nash.* Princeton University Press, 2002.

[15] Sylvia Nasar. *A Beautiful Mind: A Biography of John Forbes Nash, Jr.* Simon and Schuster, 1998.

[16] John Nash. Some games and machines for playing them. Technical Report D-1164, RAND, February 1952.

[17] Stefan Reisch. Hex ist PSPACE-vollständig. *Acta Informatica* **15** (1981), 167–191.

[18] Craige Schensted and Charles Titus. *Mudcrack Y and Poly-Y.* Neo Press, Peaks Island, Maine, 1975.

[19] Jack van Rijswijck. *Set Colouring Games.* PhD thesis, University of Alberta, Edmonton, Alberta, Canada, 2006.

[20] Yōhei Yamasaki. Theory of division games. *Publications of the Research Institute for Mathematical Sciences* **14(2)** (1978), 337–358.

ON BOUNDS FOR TWO DAVENPORT-TYPE CONSTANTS

H. G. Grundman
Department of Mathematics, Bryn Mawr College, Bryn Mawr, Pennsylvania
grundman@brynmawr.edu

C. S. Owens
Department of Mathematics, Bryn Mawr College, Bryn Mawr, Pennsylvania
catherineowens90@gmail.com

Received: 6/17/12, Revised: 12/9/12, Accepted: 1/27/13, Published: 2/8/13

Abstract

Let G be an additive abelian group of finite order n and let A be a non-empty set of integers. The Davenport constant of G with weight A, $D_A(G)$, is the smallest $k \in \mathbb{Z}^+$ such that for any sequence $x_1, \ldots, x_k$ of elements in G, there exists a non-empty subsequence $x_{j_1}, \ldots, x_{j_r}$ and corresponding weights $a_1, \ldots, a_r \in A$ such that $\sum_{i=1}^{r} a_i x_{j_i} = 0$. Similarly, $E_A(G)$ is the smallest positive integer k such that for any sequence $x_1, \ldots, x_k$ of elements in G there exists a non-empty subsequence of exactly n terms, $x_{j_1}, \ldots, x_{j_n}$, and corresponding weights $a_1, \ldots, a_n \in A$ such that $\sum_{i=1}^{n} a_i x_{j_i} = 0$. We consider these constants when $G = \mathbb{Z}_n$ and $A = \{b^2 | b \in \mathbb{Z}_n^*\}$, proving lower bounds for each.

1. Introduction

Let G be an additive abelian group of finite order n. The Davenport constant of G, $D(G)$, is the smallest $k \in \mathbb{Z}^+$ such that for any sequence $x_1, \ldots, x_k$ of elements in G, there exists a non-empty subsequence $x_{j_1}, \ldots, x_{j_r}$ such that $\sum_{i=1}^{r} x_{j_i} = 0$. Let A be a non-empty set of integers. The Davenport constant of G with weight A, $D_A(G)$, is the smallest $k \in \mathbb{Z}^+$ such that for any sequence $x_1, \ldots, x_k$ of elements in G, there exists a non-empty subsequence $x_{j_1}, \ldots, x_{j_r}$ and corresponding weights $a_1, \ldots, a_r \in A$ such that $\sum_{i=1}^{r} a_i x_{j_i} = 0$. Similarly, $E_A(G)$ is the smallest positive integer k such that for any sequence $x_1, \ldots, x_k$ of elements in G there exists a non-empty subsequence of exactly n terms, $x_{j_1}, \ldots, x_{j_n}$, and corresponding weights $a_1, \ldots, a_n \in A$ such that $\sum_{i=1}^{n} a_i x_{j_i} = 0$.

In 2008, Adhikari, David, and Urroz [1] considered the case where G is $\mathbb{Z}_n$, the cyclic group of order n, and A is the set of quadratic residues modulo n,

$$A = A_n = \{b^2 | b \in \mathbb{Z}_n^*\}, \tag{1}$$

proving a collection of bounds for each of these constants. Unfortunately, the first theorem in that paper holds only for odd integers. In this work, we provide some counter-examples in the even case, then state and prove a corrected version of the theorem, explaining the error made in the original proof.

Fix $n \geq 2$, let $G = \mathbb{Z}_n$, and let $A = A_n$, as defined in (1). Let $\Omega(n)$ denote the number of prime factors of n counting multiplicity and let $\Omega_o(n)$ denote the number of odd prime factors of n counting multiplicity.

In [1, Theorem 1], it is claimed that

$$D_A(\mathbb{Z}_n) \geq 2\Omega(n) + 1 \quad \text{and} \quad E_A(\mathbb{Z}_n) \geq 2\Omega(n) + n. \tag{2}$$

Theorem 1. *The bounds in (2) are incorrect for even n. For example,*

$$\begin{array}{ll} D_A(\mathbb{Z}_2) = 2 < 3 = 2\Omega(2) + 1, & E_A(\mathbb{Z}_2) = 3 < 4 = 2\Omega(2) + n, \\ D_A(\mathbb{Z}_4) = 4 < 5 = 2\Omega(2) + 1, & E_A(\mathbb{Z}_4) = 7 < 8 = 2\Omega(2) + n, \\ D_A(\mathbb{Z}_{10}) = 4 < 5 = 2\Omega(2) + 1, & E_A(\mathbb{Z}_{10}) = 13 < 14 = 2\Omega(2) + n. \end{array}$$

Proof. First note that $A_2 = A_4 = \{1\}$ and so, for $n = 2$ or 4, $D_A(\mathbb{Z}_n) = D(\mathbb{Z}_n) = n$, from which the first two examples follow. For $n = 10$, we have $A = A_{10} = \{1, -1\}$, and so, from [2, Lemma 2.1], it follows that $D_A(10) \leq \lfloor \log_2 10 \rfloor + 1 = 4 < 5$. The remaining results follow, for $A = \{1\}$, from $E_A(G) = D_A(G) + n - 1$, which was proved in [3] and, for $A = \{-1, 1\}$, from $E_A(G) = n + \lfloor \log_2 n \rfloor$, which was proved in [2]. □

2. Corrected Version of the Theorem

We now state and prove our corrected version of the theorem. The proof follows closely the proof of the original theorem in [1].

Theorem 2. *For $n \geq 2$, $D_A(\mathbb{Z}_n) \geq 2\Omega_o(n) + 1$ and $E_A(\mathbb{Z}_n) \geq 2\Omega_o(n) + n$.*

Proof. Given $n \geq 2$, let $n = 2^{\alpha_0} p_1^{\alpha_1} \cdots p_r^{\alpha_r}$, with $\alpha_0 \geq 0$ and $\alpha_i \geq 1$ for $i \geq 1$. To prove the first inequality, it suffices to produce a sequence of $2\Omega_o(n) = 2(\alpha_1 + \alpha_2 + \cdots + \alpha_r)$ terms with no non-zero weighted zero-sum subsequence.

For each $1 \leq i \leq r$, fix $v_i \in \mathbb{Z}_n$ such that, modulo p_i, $v_i \notin A_{p_i} \cup \{0\}$. (Note that, since $p_i > 2$, $A_{p_i} \cup \{0\} \subsetneq \mathbb{Z}_{p_i}$, while $A_2 \cup \{0\} = \mathbb{Z}_2$. This is precisely the problem invalidating the proof giving in [1]: it was not possible for a v_2 to exist satisfying the given conditions.)

For $1 \leq i \leq r$ and $0 \leq j_i \leq \alpha_i - 1$, define $x_{i,j_i} = np_i^{j_i - \alpha_i}$ and $y_{i,j_i} = -v_i x_{i,j_i}$. Let S be the $2\Omega_o(n)$-term sequence:

$$x_{1,0}, y_{1,0}, x_{1,1}, y_{1,1}, \ldots, x_{1,\alpha_1 - 1}, y_{1,\alpha_1 - 1}, x_{2,0}, \ldots, y_{2,\alpha_2 - 1}, \ldots, x_{r,\alpha_r - 1}, y_{r,\alpha_r - 1}.$$

Suppose that S has a non-empty weighted zero-sum subsequence. Then there exist $s_{i,j_i}, t_{i,j_i} \in A_n \cup \{0\}$, not all zero, such that

$$\sum_{i,j_i} (s_{i,j_i} x_{i,j_i} + t_{i,j_i} y_{i,j_i}) = 0. \tag{3}$$

Fix an arbitrary k, $1 \leq k \leq r$ and notice that for $(i, j_i) \neq (k, 0)$, $p_k | x_{i,j_i}$ and $p_k | y_{i,j_i}$. So reducing equation (3) modulo p_k yields

$$s_{k,0}x_{k,0} + t_{k,0}y_{k,0} \equiv 0 \pmod{p_k}. \tag{4}$$

Since $x_{k,0}$ is a unit modulo p_k, the congruence simplifies to

$$s_{k,0} \equiv v_k t_{k,0} \pmod{p_k}. \tag{5}$$

Suppose that $s_{k,0} \neq 0$. Then, recalling that $s_{k,0}$, $t_{k,0} \in A_n \cup \{0\}$, it follows that $s_{k,0} \not\equiv 0 \pmod{p_k}$, and so $t_{k,0} \neq 0$. Thus, there exist units, u_1, $u_2 \in \mathbb{Z}_n$ such that ${u_1}^2 = s_{k,0}$ and $u_2^2 = t_{k,0}$. But then, by (5), $v_k \equiv (u_1 u_2^{-1})^2 \pmod{p_k}$, which is a contradiction, since $v_k \notin A_{p_k}$. Thus $s_{k,0} = 0$ and so $v_k t_{k,0} \equiv 0 \pmod{p_k}$. Since v_k is defined to be non-zero modulo p_k, $t_{k,0} \equiv 0 \pmod{p_k}$, and thus $t_{k,0} = 0$.

Now, fix ℓ, $0 < \ell < \alpha_k$, and assume by induction that for all $j_k < \ell$, $s_{k,j_k} = t_{k,j_k} = 0$. Reducing equation (3) modulo $p_k^{\ell+1}$, yields

$$s_{k,j_k}x_{k,j_k} + t_{k,j_k}y_{k,j_k} \equiv 0 \pmod{p_k^{\ell+1}}.$$

Dividing through by p_k^ℓ, we find that

$$s_{k,\ell}\frac{x_{k,\ell}}{p_k^\ell} + t_{k\ell}\frac{y_{k,\ell}}{p_k^\ell} \equiv 0 \pmod{p_k}.$$

So $\frac{x_{k,\ell}}{p_k^\ell}(s_{k,\ell} - v_k t_{k,\ell}) \equiv 0 \pmod{p_k}$. Since $\frac{x_{k,\ell}}{{p_k}^\ell}$ is a unit modulo p_k, $s_{k,\ell} \equiv v_k t_{k,\ell} \pmod{p_k}$. Using the same arguments as above, $s_{k,\ell} = t_{k,\ell} = 0$. Hence by induction, for all j_k, $s_{k,j_k} = t_{k,j_k} = 0$. Since k was arbitrary, we have that for all i, j_i, $s_{i,j_i} = t_{i,j_i} = 0$, which is a contradiction.

Hence, S is a sequence of length $2\Omega_o(n)$ that does not have a non-empty weighted zero-sum subsequence. Therefore, $D_A(n) \geq 2\Omega_o(n) + 1$, as desired.

Finally, to prove the bound on $E_A(n)$, let T be a sequence of length $D_A(n)-1$ with no weighted zero-sum subsequence. Let T' be the sequence obtained by appending $n-1$ zeros to T. Then T' is a sequence of $D_A(n) + n - 2$ terms with no zero-sum subsequence of exactly n terms. Thus, $E_A(n) > D_A(n) + n - 2$, and so $E_A(n) \geq 2\Omega_o(n) + n$. □

References

[1] S. D. Adhikari, C. David, J. J. Urroz. "Generalizations of some zero-sum theorems." *Integers* **8**, #A52 (2008).

[2] S. D. Adhikari, Y. G. Chen, J. B. Friedlander, S. V. Konyagin, F. Pappalardi. "Contributions to zero-sum problems." *Discrete Math.* **306**, 1-10 (2006)

[3] W. D. Gao. "A combinatorial problem on finite abelian groups." *J. Number Theory* **58** 100-103 (1996)

ON THE DIOPHANTINE EQUATION $X^2 - KXY + Y^2 + LX = 0$

Li Feng
School of Mathematics, South China Normal University, Guangzhou, P.R.CHINA
fengli05061012@163.com

Pingzhi Yuan
School of Mathematics, South China Normal University, Guangzhou, P.R.CHINA
yuanpz@scnu.edu.cn

Yongzhong Hu
Department of Mathematics, Foshan University, Foshan, P.R.CHINA
fsyzhujx@pub.foshan.gd.cn

Received: 1/6/12, Revised: 8/4/12, Accepted: 12/27/12, Published: 2/8/13

Abstract

For any given positive integer l, we prove that there are only finitely many integers k such that the Diophantine equation $x^2 - kxy + y^2 + lx = 0$ has an infinite number of positive integer solutions (x, y). Moreover, we determine all integers k such that the Diophantine equation $x^2 - kxy + y^2 + lx = 0$, $1 \leq l \leq 33$, has an infinite number of positive integer solutions (x, y).

1. Introduction

In [2], Marlewski and Zarzycki proved that the Diophantine equation

$$x^2 - kxy + y^2 + x = 0 \tag{1}$$

has an infinite number of positive integer solutions (x, y) if and only if $k = 3$. Some computer experiments suggest that for many integers k there are infinitely many positive integer solutions, so they hoped that it is possible to characterize positive integer solutions of the equation $x^2 - kxy + y^2 + lx = 0$ in the general case.

Recently, Yuan and Hu [5] showed that the equation

$$x^2 - kxy + y^2 + 2x = 0 \tag{2}$$

has an infinite number of positive integer solutions (x, y) if and only if $k = 3, 4$; and the equation

$$x^2 - kxy + y^2 + 4x = 0 \tag{3}$$

has an infinite number of positive integer solutions (x, y) if and only if $k = 3, 4, 6$.

The main purpose of the present paper is to determine integers k such that the equation

$$x^2 - kxy + y^2 + lx = 0, \tag{4}$$

where l is a given positive integer, has an infinite number of positive integer solutions (x, y).

In this paper, we use a completely different method to deal with this problem. The main result is as follows.

Theorem 1. *For any given positive integer l, there are only finite many integers k such that equation* (4) *has an infinite number of positive integer solutions* (x, y).

For positive integers l with $1 \le l \le 33$, by the method indicated in the proof of the main theorem, we compute and list all (k, l) such that equation (4) has infinitely many positive integer solutions (x, y) (see the table at the end of Section 4).

2. Lemmas

In this section, we will present the lemmas that will be needed in the proof of the main theorems.

Lemma 2. ([2, Theorem 1]) *If positive integers x, y, k satisfy equation* (1), *then there exist positive integers c, e such that $x = c^2$, $y = ce$, and* $\gcd(c, e) = 1$.

Lemma 3. *If positive integers x, y, k satisfy equation* (4) *with* $\gcd(x, y, l) = 1$, *then there exist positive integers c, e such that $x = c^2$, $y = ce$, and* $\gcd(c, e) = 1$.

Proof. It follows from (4) that if p is a prime number, then $p\,|\,x$ implies $p\,|\,y$. In particular, l is prime to p. Let $x = p^\mu x_1$ and $y = p^\nu y_1$ with $\gcd(p, x_1 y_1) = 1$. Substituting these values of x and y into equation (4) we have

$$p^{2\nu} y_1^2 = p^\mu (p^\nu k x_1 y_1 - p^\mu x_1^2 - l x_1),$$

which implies that $\mu = 2\nu$ since $\gcd(l x_1, p) = 1$. This means that $x = c^2$, $y = ce$, and $\gcd(c, e) = 1$. □

To prove our results, we need some results on continued fractions.

Definition 4. *The fraction*

$$a_0 + \cfrac{1}{a_1 + \cfrac{1}{a_2 + \cfrac{1}{a_3 + \ddots + \cfrac{1}{a_N}}}}$$

is called a finite continued fraction, and denoted by $[a_0, a_1, \ldots, a_N]$. *It is called a continued fraction when* $N = +\infty$. *For simplicity, usually it is denoted by* $[a_0, a_1, a_2, \ldots, a_N, \ldots]$. *We call* $\alpha_n = [a_n, a_{n+1}, \ldots]$ *the* $(n+1)$*-st complete quotient of the continued fraction* $\alpha = [a_0, a_1, \ldots, a_n, \ldots]$.

Lemma 5. ([1, Theorem 10.8.1]) *Let* d *be a positive integer which is not a square. Then the* $(n+1)$*-st complete quotient* α_n *of the continued fraction* $\alpha = \sqrt{d}$ *is of the form*

$$\frac{\sqrt{d}+P_n}{Q_n}, \qquad P_n^2 \equiv d \pmod{Q_n},$$

where P_n *and* Q_n *are positive integers.*

Lemma 6. ([1, Theorem 10.8.2]) *Let* d, P_n, *and* Q_n *be as in Lemma 5. Then the quadratic equation*

$$x^2 - dy^2 = (-1)^n Q_n$$

has a positive integer solution (x, y). *If* $l \neq (-1)^n Q_n$ *and* $|l| < \sqrt{d}$, *then the Diophantine equation*

$$x^2 - dy^2 = l$$

has no integer solutions (x, y).

The proof of the following lemma is well-known, so we omit the proof here.

Lemma 7. *Let* $d > 1$ *be a positive integer which is not a square and* $c \neq 0$ *a given integer. If the Diophantine equation*

$$x^2 - dy^2 = c \tag{5}$$

has a positive integer solution (x, y) *with* $\gcd(x, y) = 1$, *then equation (5) has infinite many positive integer solutions* (x, y) *with* $\gcd(x, y) = 1$.

Lemma 8. ([3, Theorem 108a]) *Let* N, D *be positive integers and* D *not a square. Suppose that* $x_0 + y_0\sqrt{D}$ *is the fundamental solution of the Pell equation* $x^2 - Dy^2 = 1$ *and the equation*

$$u^2 - Dv^2 = -N, \ u, \, v \in \mathbb{Z}, \tag{6}$$

where $\gcd(u, v) = 1$, *is solvable. Then (6) has a solution* $u_0 + v_0\sqrt{D}$ *with the following property;*

$$0 < v_0 \leq \frac{y_0\sqrt{N}}{\sqrt{2(x_0 - 1)}}, \quad 0 \leq u_0 \leq \sqrt{\frac{1}{2}(x_0 - 1)N}. \tag{7}$$

Lemma 9. ([4, Theorem 2]) *Let* N, D *be odd positive integers with* D *non-square. Suppose that the equation*

$$x^2 - Dy^2 = 4, \quad \gcd(x, y) = 1 \tag{8}$$

is solvable and let $x_0 + y_0\sqrt{D}$ *be the least solution. If the equation*

$$u^2 - Dv^2 = -4N, \; u, \, v \in \mathbb{Z}, \tag{9}$$

where $\gcd(u, v)\,|2$, *is solvable, then (9) has a solution* $u_0 + v_0\sqrt{D}$ *with the following property:*

$$0 < v_0 \le \frac{y_0\sqrt{N}}{\sqrt{(x_0 - 2)}}, \quad 0 \le u_0 \le \sqrt{(x_0 - 2)N}. \tag{10}$$

3. Proof of Theorem 1

Proof. Since $\gcd(x, y, l) = 1$, by Lemma 3 we have $x = c^2$, $y = ce$, $\gcd(c, e) = 1$. Substituting these values of x and y into equation (4), we have $c^2 - kce + e^2 + l = 0$. It follows that

$$(2c - ke)^2 - (k^2 - 4)e^2 = -4l. \tag{11}$$

`Case (i).` $2\,|k$. Let $k = 2k_1$. Then from (11), we have

$$(c - k_1e)^2 - (k_1^2 - 1)e^2 = -l.$$

Since

$$\sqrt{k_1^2 - 1} = [k_1 - 1, \overline{1, 2k_1 - 2}],$$

where ($k_1 \ge 2$, $k_1 \in \mathbb{Z}^+$), we therefore have $Q_{2t-1} = 2k_1 - 2$, $Q_{2t} = 1$, $t > 0$. By Lemma 6, if equation (11) has a positive integer solution (c, e) and $|l| < k_1$, then we have

$$-l = (-1)^{2t-1}Q_{2t-1} = -2k_1 + 2,$$

which is possible only when $l = k - 2$.

`Case (ii).` $2 \nmid k$ and $2\,|e$. Let $e = 2e_1$. Then from (11), we have

$$(c - ke_1)^2 - (k^2 - 4)e_1^2 = -l.$$

Since

$$\sqrt{k^2 - 4} = [k - 1, \overline{1, (k-3)/2, 2, (k-3)/2, 1, 2k - 1}],$$

where $k \ge 5$ is odd, a short computation shows that

$$Q_{6t+1} = 2k - 5, \; Q_{6t+2} = 4, \; Q_{6t+3} = k - 2,$$

$$Q_{6t+4} = 4, \; Q_{6t+5} = 2k - 5, \; Q_{6t+6} = 1, \; t \ge 0.$$

By Lemma 6, if equation (11) has a positive integer solution (c, e) and $|l| < k$, then we have

$$-l = (-1)^n Q_n = -2k + 5, 4, -k + 2, 1,$$

which is possible only when $l = k-2, -4, -1$.

(iii) $2 \nmid ke$. Then from (11), we have $(2c-ke)^2 - (k^2-4)e^2 = -4l$. Since

$$\sqrt{k^2-4} = [k-1, \overline{1, (k-3)/2, 2, (k-3)/2, 1, 2k-1}],$$

where $k \geq 5$ is odd, similarly, we have

$$Q_{6t+1} = 2k-5,\ Q_{6t+2} = 4,\ Q_{6t+3} = k-2,$$

$$Q_{6t+4} = 4,\ Q_{6t+5} = 2k-5,\ Q_{6t+6} = 1,\ t \geq 0.$$

By Lemma 6, if equation (11) has a positive integer solution (c, e) and $4|l| < k$, then we have

$$-4l = (-1)^n Q_n = -2k+5, 4, -k+2, 1,$$

which is possible only when $l = -1$.

To sum up, we derive that if equation (11) has a positive integer solution (c, e), then $k < 4l$, which implies the theorem. □

Remark. If $l = dl_1$, d, $l_1 \in \mathbb{Z}$ and the equation $x^2 - kxy + y^2 + l_1 x = 0$ has infinitely many positive integer solutions (x_n, y_n), $n = 1, 2, \ldots$, then the equation

$$u^2 - kuv + v^2 + lu = 0$$

has infinitely many positive integer solutions (u_n, v_n), $n = 1, 2, \ldots$, where $u_n = dx_n$, $v_n = dy_n$.

In view of the arguments in the proof of Theorem 1, the Remark and Lemmas 6-8, for a given positive integer l, to find all possible integers k such that the equation $x^2 - kxy + y^2 + lx = 0$ has infinitely many positive integer solutions (x, y), we only need to find all $k_1 = k/2$ such that the equation

$$x^2 - (k_1^2 - 1)y^2 = -l,\ \gcd(x, y) = 1$$

has a positive integer solution (x, y) with $1 < k_1 \leq l \leq 33$, $y \leq \sqrt{33/2(x_0-1)} < 5$; and to find all integers k such that the equation

$$x^2 - (k^2-4)y^2 = -4l,\ \gcd(x, y)\,|2$$

has a positive integer solution (x, y) with $1 < k \leq 4l \leq 132$, $y \leq \sqrt{132/(x_0-2)} < 7$, $x_0 \geq 5$. This can be easily done by MATHLAB. The following is the list of the computation.

l	k	l	k
2	$3,4$	18	$3,4,5,7,8,11,20$
3	$3,4,5$	19	$3,9,12,21$
4	$3,4,6$	20	$3,4,5,6,7,10,12,22$
5	$3,5,7$	21	$3,4,5,6,9,13,23$
6	$3,4,5,8$	22	$3,4,7,8,13,24$
7	$3,6,9$	23	$3,4,6,10,11,14,25$
8	$3,4,6,10$	24	$3,4,5,6,8,10,14,26$
9	$3,4,5,7,11$	25	$3,5,7,15,27$
10	$3,4,5,7,12$	26	$3,4,9,12,15,28$
11	$3,4,7,8,13$	27	$3,4,5,7,11,13,16,29$
12	$3,4,5,6,8,14$	28	$3,4,6,8,9,16,30$
13	$3,9,15$	29	$3,7,11,13,17,31$
14	$3,4,6,8,9,16$	30	$3,4,5,7,8,10,12,17,32$
15	$3,4,5,7,8,10,17$	31	$3,6,12,18,33$
16	$3,4,6,10,18$	32	$3,4,6,10,14,18,34$
17	$3,5,9,11,19$	33	$3,4,5,7,8,13,19,35$

4. The Equation $2x^2 - kxy + y^2 + x = 0$

In this section, we consider a variation of the above problem. We ask the same question for the equation

$$2x^2 - kxy + y^2 + x = 0, \tag{12}$$

i.e., for which k, equation (12) has infinitely many positive integer solutions (x, y). To our surprise, it is much difficult to completely solve this question.

Lemma 10. *If positive integers x, y, k satisfy equation (12), then there exist positive integers c, e such that $x = c^2$, $y = ce$, and $\gcd(c, e) = 1$.*

Proof. It follows from (12) that if p is a prime number, then $p \mid x$ implies $p \mid y$. Let $x = p^{\mu}x_1$ and $y = p^{\nu}y_1$ with $\gcd(p, x_1y_1) = 1$. Substituting these values of x and y into equation (12) we have

$$p^{2\nu}y_1^2 = p^{\mu}(p^{\nu}kx_1y_1 - p^{\mu}x_1^2 - x_1),$$

which implies that $\mu = 2\nu$ since $\gcd(x_1, p) = 1$. This means that $x = c^2$, $y = ce$, and $\gcd(c, e) = 1$. □

Theorem 11. *Equation (12) has infinitely many positive integer solutions (x, y) if and only if the equation $x^2 - (k^2 - 8)y^2 = -1$ has a positive integer solution.*

Proof. If (x, y) is a positive integer solution of equation (12), by Lemma 10, we have $x = c^2$, $y = ce$, $\gcd(c, e) = 1$. Substituting these values of x and y into equation (12), we have

$$2c^2 - kce + e^2 + 1 = 0. \tag{13}$$

We divide the proof into two cases.

`Case 1` $2 \mid k$. Let $k = 2k_1$. Then from (13), we have

$$(e - k_1c)^2 - (k_1^2 - 2)c^2 = -1.$$

Since

$$\sqrt{k_1^2 - 2} = [k_1 - 1, \overline{1, k_1 - 2, 1, 2k_1 - 2}],$$

where ($k_1 \geq 2, k_1 \in \mathbb{Z}^+$), so we have $Q_{2t} = 2k_1 - 3, Q_{2t} = 1$ or $2, t > 0$. By Lemma 6, equation (12) has no solutions in this case.

`Case 2` $2 \nmid k$. From (13) we have $2 \nmid e$. We divide the proof into two subcases.

(i) $2 \mid c$. Let $c = 2c_1$. Then from (13), we have

$$(e - kc_1)^2 - (k^2 - 8)c_1^2 = -1. \tag{14}$$

Therefore, in this case, equation (12) has a positive integer solution (x, y) if and only if the equation $x^2 - (k^2 - 8)y^2 = -1$ has a positive integer solution.

(ii) $2 \nmid kec$. From (13), we have $(2e - kc)^2 - (k^2 - 8)c^2 = -4$, and we derive that $1 - 1 \equiv 4 \pmod 8$ by taking modulo 8, which is a contradiction.

In conclusion, by Lemma 6 and the above arguments, equation (12) has infinitely many positive integer solutions (x, y) if and only if the equation $x^2 - (k^2 - 8)y^2 = -1$ has a positive integer solution. This completes the proof of Theorem 11. □

Finally, we propose a conjecture which is closely related to equation (12). It is generally believed that there are infinitely many primes of the form $u^2 - 8$, so we believe that the following conjecture is true.

Conjecture 12. *There are infinitely many positive integers k such that the equation $x^2 - (k^2 - 8)y^2 = -1$ has a positive integer solution (x, y).*

If the above conjecture is true, then by Theorem 11, there are infinitely many positive integers k such that equation (12) has infinitely many positive integer solutions (x, y).

Acknowledgments. The first and the second authors are supported by the Guangdong Provincial Natural Science Foundation (No.S2012010009942) and NSF of China (No. 11271142).

References

[1] L. K. Hua, *Introduction to Number Theory*, Springer-Verlag, Berlin, 1982.

[2] A. Marlewski and P. Marzycki, *Infinitely many positive solutions of the Diophantine equation* $x^2 - kxy + y^2 + x = 0$, Comp. Math. Appl. **47** (2004), 115-121.

[3] T. Nagell, *Introduction to Number Theory*, Chelsea, 1981.

[4] B. Stolt, *On the Diophantine equation* $u^2 - Dv^2 = \pm 4N$, Ark. Mat. **2** (1952), 1-23.

[5] P. Yuan and Y. Hu, *On the Diophantine equation* $x^2 - kxy + y^2 + lx = 0$, $l \in \{1, 2, 4\}$, Comp. Math. Appl. **61** (2011), 573-577.

ON A SEQUENCE OF POLYNOMIALS WITH HYPOTHETICALLY INTEGER COEFFICIENTS

Vladimir Shevelev
Dept. of Mathematics, Ben-Gurion University of the Negev, Beersheva, Israel
shevelev@bgu.ac.il

Peter J. C. Moses
Moparmatic Company, Astwood Bank, Nr. Redditch, Worcestershire, England
mows@mopar.freeserve.co.uk

Received: 3/31/12, Revised: 12/11/12, Accepted: 2/3/13, Published: 2/8/13

Abstract

The first author introduced a sequence of polynomials defined recursively. One of the main results of this study is proof of the integrality of its coefficients.

1. Introduction

In point of fact, there are only a few examples of sequences known where the question of the integrality of the terms is a difficult problem. In 1989, Somos [9] posed a problem on the integrality of sequences depending on parameter $k \geq 4$ which are defined by the recursion

$$a_n = \frac{\sum_{j=1}^{\lfloor \frac{k}{2} \rfloor} a_{n-j} a_{n-(k-j)}}{a_{n-k}}, \quad n \geq k \geq 4,$$

with the initial conditions $a_i = 1, \ \ i = 1, \dots, k-1$.
Gale [3] proved the integrality of Somos sequences when $k = 4$ and 5, attributing a proof to Malouf [4]. Hickerson and Stanley (see [6]) independently proved the integrality of the $k = 6$ case in unpublished work and Fomin and Zelevinsky (2002) gave the first published proof. Finally, Lotto (1990) gave an unpublished proof for the $k = 7$ case. These are sequences A006720-A006723 in [8]. It is interesting that, for $k \geq 8$, the property of integrality disappears (see sequence A030127 in [8]). In connection with this, note that in the so-called Göbel's sequence ([11]) defined by the recursion

$$x_n = \frac{1}{n}(1 + \sum_{i=0}^{n-1} x_i^2), \ \ n \geq 1, \ \ x_0 = 1,$$

the first non-integer term is $x_{43} = 5.4093 \times 10^{178485291567}$.

In this paper we study the Shevelev sequence of polynomials $\{P_n(x)\}_{n\geq 1}$ that are defined by the following recursion: $P_1 = 1$, $P_2 = 1$, and, for $n \geq 2$,

$$4(2x+n)P_{n+1}(x) = 2(x+n)P_n(x)+$$
$$(2x+n)P_n(x+1) + (4x+n)l_n(x), \text{ if } n \text{ is odd}, \quad (1)$$

$$4P_{n+1}(x) = 4(x+n)P_n(x)+$$
$$2(2x+n+1)P_n(x+1) + (4x+n)l_{n-1}(x), \text{ if } n \text{ is even}, \quad (2)$$

where

$$l_n(x) = (x + \frac{n-1}{2})(x + \frac{n-3}{2}) \cdots (x+1).$$

The first few polynomials are ([8], sequence A174531):

$$P_1 = 1,$$
$$P_2 = 1,$$
$$P_3 = 3x + 4,$$
$$P_4 = 2x + 4,$$
$$P_5 = 5x^2 + 25x + 32,$$
$$P_6 = 3x^2 + 19x + 32,$$
$$P_7 = 7x^3 + 77x^2 + 294x + 384,$$
$$P_8 = 4x^3 + 52x^2 + 240x + 384,$$
$$P_9 = 9x^4 + 174x^3 + 1323x^2 + 4614x + 6144,$$
$$P_{10} = 5x^4 + 110x^3 + 967x^2 + 3934x + 6144,$$
$$P_{11} = 11x^5 + 330x^4 + 4169x^3 + 27258x^2 + 90992x + 122880,$$
$$P_{12} = 6x^5 + 200x^4 + 2842x^3 + 21040x^2 + 79832x + 122880.$$

According to our observations, the following conjectures are natural.
1) The coefficients of all the polynomials are integers. Moreover, the greatest common divisor of all coefficients is $n/rad(n)$, where $rad(n) = \prod_{p|n} p$;
2) $P_n(0) = 4^{\lfloor \frac{n-1}{2} \rfloor} \lfloor \frac{n-1}{2} \rfloor !$;
3) For even n, $P_n(1) = (2^n - 1)(\frac{n}{2})!/(n+1)$, and for odd n, $P_n(1) = (2^n - 1)(\frac{n-1}{2})!$;
4) $P_n(x)$ has a real rational root if and only if either $n = 3$ or $n \equiv 0 \pmod 4$. In the latter case, such a unique root is $-\frac{n}{2}$;

5) Coefficients of x^k increase when k decreases;
6) If n is even, then the coefficients of P_n do not exceed the corresponding coefficients of P_{n-1} and the equality holds only for the last ones; moreover, the ratios of coefficients of x^k of polynomials P_{n-1} and P_n monotonically decrease to 1 when k decreases;
7) All coefficients of P_n, except of the last one, are multiples of n if and only if n is prime.
The main results of our paper consist of the following two theorems.

Theorem 1. (Explicit formula for $P_n(k)$) *For an integer k we have*

$$P_n(k) = \begin{cases} (\binom{(n-1)/2+k-1}{k-1}/\binom{n+2k-2}{k-1})(\frac{n-1}{2})!\ T_n(k), & \text{if } n \geq 1 \text{ is odd} \\ (\binom{n/2+k-1}{k}/\binom{n+2k-1}{k})(\frac{n}{2}-1)!\ T_n(k), & \text{if } n \geq 2 \text{ is even,} \end{cases} \tag{3}$$

$$= 2^{-(\lfloor \frac{n}{2} \rfloor + k - 1)} \frac{(n+k-1)!}{(2\lfloor \frac{n}{2} \rfloor + 2k - 1)!!} T_n(k), \tag{4}$$

where

$$T_n(k) = \sum_{i=1}^{n} 2^{i-1} \binom{n+2k-i-1}{k-1}. \tag{5}$$

Using Theorem 1, we prove Conjectures (2), (3) and the following main result.

Theorem 2. *For $n \geq 1$, $P_n(x)$ is a polynomial of degree $\lfloor \frac{n-1}{2} \rfloor$ with integer coefficients.*

Nevertheless, the subtle second part of Conjecture (1) remains open.

2. Representation of $P_n(k)$ Via a Polynomial in n of Degree $k-1$ with Integer Coefficients

Theorem 3. *For integers $k \geq 1,\ n \geq 1$, the following recursion holds*

$$P_n(k) = c_n(k)\Big(2^{n+k-1} - \frac{R_k(n)}{(2k-2)!!}\Big), \tag{6}$$

where $R_k(n)$ is a polynomial in n of degree $k-1$ with integer coefficients and

$$c_n(k) = \begin{cases} (\frac{n-1}{2})! \prod_{i=1}^{k-1} \frac{n+i}{n+2i}, & \text{if } n \text{ is odd,} \\ \frac{1}{2}(\frac{n}{2}-1)! \prod_{i=0}^{k-1} \frac{n+i}{n+2i+1}, & \text{if } n \text{ is even .} \end{cases} \tag{7}$$

Proof. Write (3) and (4) in the form

$$P_n(k+1) = -\frac{2f}{g}P_n(k) + 4P_{n+1}(k) - \frac{h}{g}\left(\frac{n-1}{2}\right)!\binom{\frac{g-1}{2}}{k}, \quad \textit{if } n \equiv 1 \pmod 2; \tag{8}$$

$$P_n(k+1) = -\frac{2f}{g+1}P_n(k) + \frac{2}{g+1}P_{n+1}(k) - \frac{h}{2(g+1)}\left(\frac{n}{2}-1\right)!\binom{\frac{g}{2}-1}{k} \tag{9}$$

$$\textit{if } n \equiv 0 \pmod 2,$$

where $f = n+k,\ \ g = n+2k,\ \ h = n+4k.$

Let n be odd. We use induction over k. For $k=1$, (6) gives

$$R_1(n) = 2^n - \frac{P_n(1)}{c_n(1)} = Const(k). \tag{10}$$

Thus the base of induction is valid. Suppose the theorem is true for some value of k. Then, using this supposition and (6) to (9), we have

$$P_n(k+1) =$$

$$-\frac{2f}{g}\left(\frac{n-1}{2}\right)!\Big(2^{n+k-1} - \frac{R_k(n)}{(2k-2)!!}\Big)\prod_{i=1}^{k-1}\frac{n+i}{n+2i}+$$

$$2\left(\frac{n-1}{2}\right)!\Big(2^{n+k} - \frac{R_k(n+1)}{(2k-2)!!}\Big)\prod_{i=0}^{k-1}\frac{n+i+1}{n+2i+2}-$$

$$\frac{h}{g}\left(\frac{n-1}{2}\right)!\frac{\frac{g-1}{2}\frac{g-3}{2}\cdots\frac{n+1}{2}}{k!}.$$

Note that

$$\frac{f}{g}\prod_{i=0}^{k-1}\frac{n+i}{n+2i} = \prod_{j=1}^{k}\frac{n+j}{n+2j} = \prod_{i=0}^{k-1}\frac{n+i+1}{n+2i+2}.$$

Therefore,

$$P_n(k+1) = \left(\frac{n-1}{2}\right)!\Big(-2^{n+k} + \frac{2R_k(n)}{(2k-2)!!} + 2^{n+k+1}-$$

$$\frac{2R_k(n+1)}{(2k-2)!!} - \frac{h}{g}\frac{\frac{g-1}{2}\frac{g-3}{2}\cdots\frac{n+1}{2}}{k!}\prod_{j=1}^{k}\frac{n+2j}{n+j}\Big)\prod_{j=1}^{k}\frac{n+j}{n+2j}.$$

Here we note that

$$(g-1)(g-3)\cdots(n+1)\prod_{j=1}^{k}\frac{n+2j}{n+j} = (n+2k)_k,$$

where $(x)_k$ is a falling factorial. Hence

$$P_n(k+1) = c_n(k+1)\Big(2^{n+k} - 2\frac{R_k(n+1)-R_k(n)}{(2k-2)!!} -$$

$$\frac{4k+n}{(2k)!!}(n+2k-1)_{k-1}\Big) = c_n(k+1)\Big(2^{n+k} - \frac{R_{k+1}(n)}{(2k)!!}\Big),$$

where

$$R_{k+1}(n) = 4k(R_k(n+1)-R_k(n)) + (4k+n)(n+2k-1)_{k-1}. \tag{11}$$

Since, by the inductive supposition, $R_k(n)$ is a polynomial of degree $k-1$ with integer coefficients, then, by (11), $R_{k+1}(n)$ is a polynomial of degree k with integer coefficients. Note that the case of even n is considered quite analogously, obtaining the *same* formula (11). □

In (6) and (7), put $n = 1$. Then, for $k \geq 1$ we have

$$\Big(2^k - \frac{R_k(1)}{(2k-2)!!}\Big)\frac{k!}{(2k-1)!!} = 1,$$

from which

$$R_k(1) = (k-1)!\Big(2^{2k-1} - \binom{2k-1}{k}\Big). \tag{12}$$

In particular, $R_1(1) = 1$ and, since $R_1(n)$ is of degree 0, $R_1(n) = 1$. Further, we find polynomials $R_k(n)$ using the recursion (11). The first polynomials $R_k(n)$ are

$$R_1(n) = 1,$$

$$R_2(n) = n+4,$$
$$R_3(n) = n^2+11n+32,$$
$$R_4(n) = n^3+21n^2+152n+384,$$
$$R_5(n) = n^4+34n^3+443n^2+2642n+6144,$$
$$R_6(n) = n^5+50n^4+1015n^3+10510n^2+55864n+122880.$$

3. Proof of Conjectures (2) and (3)

We start with the proof of Conjecture (3) for $P_n(1)$.

Proof. Note that, since $R_1(n) = 1$, from (10) we find

$$P_n(1) = c_n(1)(2^n - 1). \tag{13}$$

Besides, by (7), we have

$$c_n(1) = \begin{cases} (\frac{n-1}{2})!, & if \ n \ is \ odd, \\ \\ \frac{1}{2}(\frac{n}{2} - 1)! \frac{n}{n+1} = (\frac{n}{2})!/(n+1), & if \ n \ is \ even, \end{cases} \tag{14}$$

and Conjecture (3) follows. □

Let us now prove Conjecture (2).

Proof. Note that (8) and (9), as in (1) and (2), are valid for every nonnegative k. For $k = 0$ and odd $n \geq 1$, (8) gives

$$P_n(1) = -2P_n(0) + 4P_{n+1}(0) - (\frac{n-1}{2})!,$$

or, using (13) and (14), we have

$$4P_{n+1}(0) - 2P_n(0) = 2^n(\frac{n-1}{2})!$$

Analogously, for $k = 0$ and even $n \geq 1$, from (9), (13) and (14) we find

$$P_{n+1}(0) - nP_n(0) = 2^{n-1}(\frac{n}{2})!$$

Thus

$$P_{n+1}(0) = \begin{cases} \frac{1}{2}P_n(0) + 2^{n-2}(\frac{n-1}{2})!, & \text{if } n \text{ is odd,} \\ \\ nP_n(0) + 2^{n-1}(\frac{n}{2})!, & \text{if } n \text{ is even} \end{cases}$$

with $P_1(0) = 1, \ P_2(0) = 1$. Since the difference equation

$$y(n+1) = \begin{cases} \frac{1}{2}y(n) + 2^{n-2}(\frac{n-1}{2})!, & \text{if } n \text{ is odd,} \\ \\ ny(n) + 2^{n-1}(\frac{n}{2})!, & \text{if } n \text{ is even} \end{cases}$$

with the initial values $y(1) = 1, \ y(2) = 1$ has an unique solution, it is sufficient to verify that $y(n) = P_n(0) = 4^{\lfloor \frac{n-1}{2} \rfloor} \lfloor \frac{n-1}{2} \rfloor!$ is a solution. □

4. Explicit Formula for $R_k(n)$

Since from (11)

$$4kR_k(n+1) = 4kR_k(n) + R_{k+1}(n) - (4k+n)(n+2k-1)_{k-1}, \tag{15}$$

we have a recursion in n for $R_k(n)$ given by (12) and (15).
Our aim in this section is to find a generalization of (12) for an arbitrary integer $n \geq 1$. Note that we can write (12) in the form

$$R_k(1) = 2(k-1)!4^{k-1} - \frac{(2k-1)!}{k!}.$$

Using (15) and (12), after some transformations, we find

$$R_k(2) = 2^2(k-1)!4^{k-1} - 2\frac{(2k-1)!}{k!} - \frac{(2k)!}{(k+1)!}.$$

The regularity is fixed in the following theorem.

Theorem 4. *For integer $k \geq 1$, $n \geq 1$, we have*

$$R_k(n) = 2^n(k-1)!4^{k-1} - \sum_{i=1}^{n} 2^{n-i}\frac{(2k+i-2)!}{(k+i-1)!}. \tag{16}$$

Proof. Taking into account that $\frac{(2k+i-2)!}{(k+i-1)!} = \binom{2k+i-2}{k-1}(k-1)!$, we prove (16) in the following equivalent form:

$$R_k(n) = 2^n(k-1)!\left(4^{k-1} - \sum_{i=1}^{n} 2^{-i}\binom{2k+i-2}{k-1}\right). \tag{17}$$

We use induction over n. Suppose that (17) is valid for some value of n and an arbitrary integer $k \geq 1$. By (15), we have

$$R_k(n+1) = 2^n(k-1)!\left(4^{k-1} - \sum_{i=1}^{n} 2^{-i}\binom{2k+i-2}{k-1}\right) +$$

$$2^{n-2}(k-1)!\left(4^k - \sum_{i=1}^{n} 2^{-i}\binom{2k+i}{k}\right) - \frac{4k+n}{4k}(n+2k-1)_{k-1} =$$

$$2^n(k-1)!\left(4^{k-1} - \sum_{i=1}^{n} 2^{-i}\binom{2k+i-2}{k-1}\right) + 2^n(k-1)!\left(4^{k-1} - \sum_{i=1}^{n} 2^{-i-2}\binom{2k+i}{k}\right) -$$

$$\frac{(n+2k-1)!}{(n+k)!} - \frac{n}{4k}\frac{(n+2k-1)!}{(n+k)!}.$$

Thus we should prove the identity

$$2^{n+1}(k-1)!4^{k-1} - 2^n(k-1)!\sum_{i=1}^{n} 2^{-i}\binom{2k+i-2}{k-1} -$$

$$2^{n-2}(k-1)!\sum_{i=1}^{n} 2^{-i}\binom{2k+i}{k} - \frac{n+4k}{4k}\frac{(n+2k-1)!}{(n+k)!} =$$

$$2^{n+1}(k-1)!\Big(4^{k-1} - \sum_{i=1}^{n+1} 2^{-i}\binom{2k+i-2}{k-1}\Big),$$

which is easily reduced to the identity

$$4\sum_{i=1}^{n} 2^{-i}\binom{2k+i-2}{k-1} - \sum_{i=1}^{n} 2^{-i}\binom{2k+i}{k} =$$

$$2^{-n}\frac{n+4k}{4k}\frac{(n+2k-1)!}{(n+k)!} - 4\cdot 2^{-n}\binom{2k+n-1}{k-1}.$$

Note that, the right hand part is $\frac{n}{k2^n}\binom{2k+n-1}{k-1}$. Therefore, it is left to prove the identity

$$4\sum_{i=1}^{n} 2^{-i}\binom{2k+i-2}{k-1} - \sum_{i=1}^{n} 2^{-i}\binom{2k+i}{k} = \frac{n}{2^n k}\binom{2k+n-1}{k-1}.$$

Since this is trivially satisfied for $n = 0$, it is sufficient to verify the equality of the first differences of the left and the right hand parts, which is reduced to the identity

$$2(n+2k-1)\binom{2k+n-2}{k-1} = n\binom{2k+n-1}{k-1} + k\binom{2k+n}{k},$$

which is verified directly. □

5. Proof of Theorem 1

Now we are able to prove Theorem 1.

Proof. According to (5), we have

$$T_n(k) = \sum_{i=1}^{n} 2^{i-1}\binom{n+2k-i-1}{k-1} = \sum_{j=1}^{n} 2^{n-j}\binom{2k+j-2}{k-1}. \tag{18}$$

Hence, by (17), we find

$$R_k(n) = 2^n(k-1)!(4^{k-1} - 2^{-n}T_n(k)) =$$

$$(k-1)!(2^{n+2k-2} - T_n(k)). \tag{19}$$

Now from (6) and (19) we have

$$P_n(k) = 2^{-(k-1)} c_n(k) T_n(k). \tag{20}$$

Let n be odd. Note that, by (7),

$$2^{-(k-1)} c_n(k) =$$

$$2^{-(k-1)} (\frac{n-1}{2})! \frac{(n+k-1)(n+k-2)\cdots(n+1)}{(n+2k-2)(n+2k-4)\cdots(n+2)} =$$

$$2^{-(k-1)} (\frac{n-1}{2})! \frac{(n+k-1)!n!!}{n!(n+2k-2)!!}. \tag{21}$$

Taking into account that

$$n!! = \frac{n!}{(n-1)!!} = \frac{n!}{2^{\frac{n-1}{2}} (\frac{n-1}{2})!}, \tag{22}$$

we find from (21)

$$2^{-(k-1)} c_n(k) = \frac{(n+k-1)!(\frac{n-1}{2}+k-1)!}{(n+2k-2)!} =$$

$$\frac{(\frac{n-1}{2}+k-1)!}{(k-1)!\binom{n+2k-2}{k-1}} = \frac{\binom{\frac{n-1}{2}+k-1}{k-1}}{\binom{n+2k-2}{k-1}} (\frac{n-1}{2})!$$

and (3) follows from (20). Furthermore, since by (22) $\frac{n!!(\frac{n-1}{2})!}{n!} = 2^{-\frac{n-1}{2}}$, from (20) and (21) we find

$$P_n(k) = 2^{-(\frac{n-1}{2}+k-1)} \frac{(n+k-1)!}{(n+2k-2)!!} T_n(k)$$

corresponds to (4) in the case of odd n. The case of even n is considered quite analogously. □

6. Bisection of Sequence $\{P_n(x)\}$

Note that $T_n(k)$, (5), has rather a simple structure, which allows us to find different relations for it. Using (3) and (4), we are able to find recursion relations for $P_n(x)$ which are simpler than the basis recursion (1) and (2). We start with the following simple recursions for $T_n(k)$.

Lemma 5.

$$T_n(k) - 2T_{n-1}(k) = \binom{n+2k-2}{k-1}, \quad k \geq 1; \tag{23}$$

$$T_n(k) - 4T_{n-2}(k) = \binom{n+2k-2}{k-1} + 2\binom{n+2k-3}{k-1}, \quad k \geq 2. \tag{24}$$

Proof. By (5), we have

$$T_n(k) - 2T_{n-1}(k) =$$

$$\sum_{i=1}^{n} 2^{i-1}\binom{n+2k-i-1}{k-1} - \sum_{j=1}^{n-1} 2^{j}\binom{n+2k-j-2}{k-1} =$$

$$\sum_{i=1}^{n} 2^{i-1}\binom{n+2k-i-1}{k-1} - \sum_{i=2}^{n} 2^{i-1}\binom{n+2k-i-1}{k-1}$$

and (23) follows; (24) is a simple corollary of (23). □

Theorem 6. (Bisection) *If $n \geq 3$ is odd, then*

$$(2x+n-2)P_n(x) = 2(x+n-1)(x+n-2)P_{n-2}(x) +$$

$$(4x+3n-4)(x+\frac{n-1}{2}-1)(x+\frac{n-1}{2}-2)\cdots x; \tag{25}$$

if $n \geq 4$ is even, then

$$(2x+n-1)P_n(x) = 2(x+n-1)(x+n-2)P_{n-2}(x) +$$

$$\frac{1}{2}(4x+3n-4)(x+\frac{n-2}{2}-1)(x+\frac{n-2}{2}-2)\cdots x. \tag{26}$$

Proof. According to (3), we have

$$T_n(k) = \begin{cases} \binom{n+2k-2}{k-1}/(\binom{(n-1)/2+k-1}{k-1}(\frac{n-1}{2})!)\ P_n(k), & \text{if } n \text{ is odd,} \\ \binom{n+2k-1}{k}/(\binom{n/2+k-1}{k}(\frac{n}{2}-1)!)\ P_n(k), & \text{if } n \text{ is even.} \end{cases} \tag{27}$$

Substituting this to (24), after simple transformations, we obtain (25) and (26), where k is replaced by arbitrary x. □

Note that from (25) and (26), using a simple induction, we conclude that, for even $n \geq 4$, $P_n(x)$ is a polynomial of degree $\frac{n-2}{2}$, while, for odd $n \geq 3$, $P_n(x)$ is a polynomial of degree $\frac{n-1}{2}$. However, the structure of formulas (25) and (26) does not allow us to prove that all coefficients of $P_n(x)$ are integer. This will be done in the following section by the discovery of the special relationships with the required structure.

7. Proof of Theorem 2

Lemma 7. *For $n \geq 1$, we have*

$$T_n(k) - T_{n-2}(k+1) = \binom{n+2k-1}{k}. \tag{28}$$

Proof. By (18), we should prove that

$$\binom{2k+n-1}{k} = T_n(k) - T_{n-2}(k+1) =$$

$$\sum_{j=1}^{n} 2^{n-j}\binom{2k+j-2}{k-1} - \sum_{j=1}^{n-2} 2^{n-j-2}\binom{2k+j}{k} =$$

$$\sum_{j=1}^{n} 2^{n-j}\binom{2k+j-2}{k-1} - \sum_{i=1}^{n} 2^{n-i}\binom{2k+i-2}{k} +$$

$$2^{n-1}\binom{2k-1}{k} + 2^{n-2}\binom{2k}{k},$$

or

$$\sum_{j=1}^{n} 2^{-j}\left(\binom{2k+j-2}{k-1} - \binom{2k+j-2}{k}\right) =$$

$$2^{-n}\binom{2k+n-1}{k} - \frac{1}{2}\binom{2k-1}{k} - \frac{1}{4}\binom{2k}{k}. \tag{29}$$

It is verified directly that (29) is valid for $n = 1$. Therefore, it is sufficient to verify that the first differences over n of the left hand side and the right hand side coincide. The corresponding identity

$$2^{-n}\left(\binom{2k+n-2}{k-1} - \binom{2k+n-2}{k}\right) =$$

$$2^{-n}\binom{2k+n-1}{k} - 2^{-n+1}\binom{2k+n-2}{k}$$

reduces to the equality $\binom{2k+n-2}{k-1} + \binom{2k+n-2}{k} = \binom{2k+n-1}{k}$.

Now we are able to complete proof of Theorem 2. Considering even $n \geq 4$, by (27), we obtain the following relation for $P_n(k)$ corresponding to (28):

$$P_n(x) = (n+x-1)P_{n-2}(x+1)+$$

$$(x+\frac{n}{2}-1)(x+\frac{n}{2}-2)\cdots(x+1). \tag{30}$$

On the other hand, using (23), for odd $n \geq 3$, we obtain the following relation

$$P_n(x) = 2(x+n-1)P_{n-1}(x)+$$

$$(x + \frac{n-1}{2} - 1)(x + \frac{n-1}{2} - 2)\cdots x. \tag{31}$$

From (30), by simple induction, we see that, for even $n \geq 4$, $P_n(x)$ is a polynomial with integer coefficients. Then from (31) we find that $P_n(x)$, for odd n, is also a polynomial with integer coefficients. □

8. Other Relations

Together with (25), (26), (30) and (31) there exist many other relations for $P_n(x)$. All of them are corollaries of the corresponding relations for $T_n(k)$. Below we give a few pairs of some such relations.

As we saw, for odd $n \geq 3$, (31) follows from (23). Let us consider even $n \geq 4$. Then we obtain the second component of the following recursion

$$P_n(x) = \begin{cases} 2(x+n-1)P_{n-1}(x)+ \\ ((x+n-1)P_{n-1}(x)+ \end{cases}$$

$$\begin{cases} (x+\frac{n-1}{2}-1)(x+\frac{n-1}{2}-2)\cdots x, \text{ if } n \geq 3 \text{ is odd,} \\ (x+\frac{n}{2}-1)(x+\frac{n}{2}-2)\cdots x)/(2x+n-1), \text{ if } n \geq 4 \text{ is even.} \end{cases}$$

Lemma 8. *For $n \geq 1$, $k \geq 1$, we have*

$$T_n(k+1) = 4T_n(k) - \frac{n}{k}\binom{n+2k-1}{k-1}. \tag{32}$$

Proof. By (24) and (28), we have

$$T_n(k+1) = T_{n+2}(k) - \binom{n+2k+1}{k} =$$

$$4T_n(k) + \binom{n+2k}{k-1} + 2\binom{n+2k-1}{k-1} - \binom{n+2k+1}{k}.$$

It is left to note that

$$\binom{n+2k}{k-1} + 2\binom{n+2k-1}{k-1} - \binom{n+2k+1}{k} = -\frac{n}{k}\binom{n+2k-1}{k-1}.$$

□

From Lemma 8 and (27) we find the following recursion

$$\begin{cases} (2x+n)P_n(x+1) = 2(x+n)P_n(x) - \\ \\ (2x+n+1)P_n(x+1) = 2(x+n)P_n(x) - \end{cases}$$

$$\begin{cases} n(x+\frac{n-1}{2})(x+\frac{n-1}{2}-1)\cdots(x+1), \text{ if } n \geq 3 \text{ is odd}, \\ \\ \frac{n}{2}(x+\frac{n}{2}-1)(x+\frac{n}{2}-2)\cdots(x+1), \text{ if } n \geq 4 \text{ is even}. \end{cases}$$

Lemma 9. *For* $n \geq 2,\ k \geq 1$, *we have*

$$(n+k-1)(T_n(k) - 4T_n(k-1)) = n(T_{n-1}(k) - 2T_n(k-1)). \tag{33}$$

Proof. By (32),

$$T_n(k) - 4T_n(k-1) = -\frac{n}{k-1}\binom{n+2k-3}{k-2}. \tag{34}$$

By (23),

$$T_n(k-1) = 2T_{n-1}(k-1) + \binom{n+2k-4}{k-2}.$$

Therefore,

$$T_{n-1}(k) - 2T_n(k-1) = T_{n-1}(k) - 4T_{n-1}(k-1) - 2\binom{n+2k-4}{k-2}.$$

Using again (32), we find

$$T_{n-1}(k) - 2T_n(k-1) = -(\frac{n-1}{k-1}+2)\binom{n+2k-4}{k-2}. \tag{35}$$

Now the lemma follows from (34) and (35) since $(n+k-1)\binom{n+2k-3}{k-2} = (n+2k-3)\binom{n+2k-4}{k-2}$. □

Going from (33) to the corresponding formula for $P_n(x)$ in the case of odd $n \geq 3$ unexpectedly leads to a very simple homogeneous relation

$$P_n(x) = P_n(x-1) + nP_{n-1}(x) \tag{36}$$

which we use in Sections 9 and 12. The corresponding relation for even $n \geq 4$ is

$$(2x+n-1)P_n(x) = (2x+n-2)P_n(x-1) + \frac{n}{2}P_{n-1}(x). \tag{37}$$

Lemma 10. *For $n \geq 1, \ \ k \geq 2$, we have*

$$2T_n(k) - T_{n-1}(k+1) = \binom{n+2k-1}{k}.$$

Proof. By (23), we have

$$2T_n(k) - T_{n-1}(k+1) = 4T_{n-1}(k) + 2\binom{n+2k-2}{k-1} - T_{n-1}(k+1).$$

Furthermore, by (28),

$$T_{n-1}(k+1) = T_{n+1}(k) - \binom{n+2k}{k}.$$

Hence,

$$2T_n(k) - T_{n-1}(k+1) =$$
$$4T_{n-1}(k) - T_{n+1}(k) + 2\binom{n+2k-2}{k-1} + \binom{n+2k}{k}. \tag{38}$$

Finally, by (24),

$$T_{n+1}(k) - 4T_{n-1}(k) = \binom{n+2k-1}{k-1} + 2\binom{n+2k-2}{k-1}$$

and the lemma follows from (38). □

Using Lemma 10 and (27), for even $n \geq 4$, we find

$$2P_n(x) = P_{n-1}(x+1) + (x + \frac{n}{2} - 1)(x + \frac{n}{2} - 2)\cdots(x+1), \tag{39}$$

while, for odd $n \geq 3$,

$$P_n(x) = (2x+n)P_{n-1}(x+1) + (x + \frac{n-1}{2})(x + \frac{n-1}{2} - 1)\cdots(x+1).$$

Proposition 11. *For odd $n \geq 3$, we have $P_n(k) \equiv P_n(0) \pmod{n}$.*

Proof. From (36) we find that $\sum_{i=1}^{k} P_{n-1}(i) = (P_n(k) - P_n(0))/n$, and the proposition follows. □

9. On the Coefficients of $P_n(x)$

Using formulas (25) and (26), we give a recursion for the calculation of the coefficients of $P_n(x)$ with a fixed parity of n. Let

$$P_n(x) = a_0(n)x^m + a_1(n)x^{m-1} + \cdots + a_{m-1}(n)x + a_m(n),$$

where $m = \lfloor \frac{n-1}{2} \rfloor$. We prove the following.

Theorem 12. *For $n \geq 1$, we have*

$$a_0(n) = \begin{cases} n, & \text{if } n \text{ is odd} \\ \frac{n}{2}, & \text{if } n \text{ is even;} \end{cases} \tag{40}$$

$$a_1(n) = \begin{cases} \frac{1}{24}(7n^3 - 12n^2 + 5n) \\ \frac{1}{48}(7n^3 - 18n^2 + 8n) \end{cases} = \begin{cases} \frac{1}{24}n(n-1)(7n-5), & \text{if } n \text{ is odd,} \\ \frac{1}{48}n(n-2)(7n-4), & \text{if } n \text{ is even.} \end{cases} \tag{41}$$

In general, for a fixed i, $a_i(n) = U_i(n)$, if n is odd, and $a_i(n) = V_i(n)$, if n is even, where U_i and V_i are polynomials in n of degree $2i+1$.

Proof. `Case 1)`. Let n be even. Then, using (26), for integer x and $m = \frac{n-2}{2}$, we have

$$\begin{aligned} &(2x+n-1)(a_0(n)x^m + a_1(n)x^{m-1} + \cdots) = \\ &2(x+n-1)(x+n-2)(a_0(n-2)x^{m-1} + a_1(n-2)x^{m-2} + \cdots) + \\ &\frac{1}{2}(\frac{n-2}{2})!(4x+3n-4)\binom{x-1+\frac{n-2}{2}}{\frac{n-2}{2}}. \end{aligned} \tag{42}$$

Comparing the coefficient of x^{m+1} on both sides, we find

$$a_0(n) = a_0(n-2) + 1, \quad n \geq 4, \quad a_0(4) = 2.$$

Thus $a_0(6) = 3, a_0(8) = 4, \ldots, a_0(n) = n/2$.

Furthermore, comparing the coefficient of x^m on both sides of (42), we have

$$\begin{aligned} &2a_1(n) + (n-1)a_0(n) = 2a_1(n-2) + 2(2n-3)a_0(n-2) + \\ &Coef[x^m](\frac{1}{2}(4x+3n-4)(x+\frac{n-4}{2})(x+\frac{n-6}{2})\cdots(x+1)x). \end{aligned} \tag{43}$$

Note that

$$\begin{aligned} &Coef[x^m](\frac{1}{2}(4x+3n-4)(x+\frac{n-4}{2})(x+\frac{n-6}{2})\cdots(x+1)x) = \\ &\frac{3n-4}{2} + 2(\frac{n-4}{2} + \frac{n-6}{2} + \cdots + 1) = \\ &\frac{3n-4}{2} + \sum_{i=2}^{m}(n-2i) = \frac{n^2}{4}. \end{aligned}$$

Therefore, by (43),

$$a_1(n) - a_1(n-2) = \frac{(2n-3)(n-2)}{2} - \frac{(n-1)n}{4} + \frac{n^2}{8} = \frac{7n^2-26n+24}{8}.$$

Hence

$$a_1(n) = \sum_{i=4,6,\cdots,n} (a_1(i) - a_1(i-2)) = \frac{1}{8} \sum_{i=4,6,\ldots,n} (7i^2 - 26i + 24) =$$

$$\frac{1}{2}\sum_{j=2}^{n/2}(7j^2 - 13j + 6) = \frac{1}{48}(7n^3 - 18n^2 + 8n).$$

Finally, comparing the coefficient of x^{m-i} on both sides of (42), we find

$$2a_{i+1}(n) + (n-1)a_i(n) = 2a_{i+1}(n-2)+$$

$$2(2n-3)a_i(n-2) + 2(n-1)(n-2)a_{i-1}(n-2)+$$

$$\frac{1}{2}Coef[x^{m-i}]((4x+3n-4)(x+\frac{n-4}{2})(x+\frac{n-6}{2})\cdots(x+1)(x)). \quad (44)$$

Note that, polynomial $(4x+3n-4)(x+\frac{n-4}{2})(x+\frac{n-6}{2})\cdots(x+1)x$ has degree $m+1$. Therefore, in order to calculate $Coef[x^{m-i}]$ in (44), we should choose, in all possible ways, in $m-i$ brackets (from $m+1$ ones) $x's$, and for the other $i+1$ brackets we choose linear forms of n. Thus $\frac{1}{2}Coef[x^{m-i}]$ in (44) is a polynomial $r_i(n)$ of degree $i+1$. Further we use induction over i with the formulas (40) and (41) as the inductive base. Write (44) in the form

$$2(a_{i+1}(n) - a_{i+1}(n-2)) =$$

$$2(2n-3)a_i(n-2) - (n-1)a_i(n) + 2(n-1)(n-2)a_{i-1}(n-2) + r_i(n). \quad (45)$$

By the inductive supposition, $a_{i-1}(n)$ and $a_i(n)$ are polynomials of degree $2i-1$ and $2i+1$ respectively. Thus $a_{i+1}(n) - a_{i+1}(n-2)$ is a polynomial of degree $2i+2$. This means that $a_{i+1}(n)$ is a polynomial of degree $2i+3$.

`Case 2).` Let n be odd. By (25), for integer x and $m = \frac{n-1}{2}$, we have

$$(2x+n-2)(a_0(n)x^m + a_1(n)x^{m-1} + \cdots) =$$

$$2(x+n-1)(x+n-2)(a_0(n-2)x^{m-1} + a_1(n-2)x^{m-2} + \cdots)+$$

$$(\frac{n-1}{2})!(4x+3n-4)\binom{x+\frac{n-3}{2}}{\frac{n-1}{2}}. \quad (46)$$

Hence, comparing the coefficient of x^{m+1} on both sides, we find

$$a_0(n) = a_0(n-2) + 2, \ \ n \geq 3, \ \ a_0(1) = 1.$$

Thus $a_0(3) = 3, a_0(5) = 5, \ldots, a_0(n) = n$.
Furthermore, comparing the coefficient of x^m on both sides of (46), using the same arguments as in 1), we have

$$a_1(n) = a_1(n-2) + \frac{7n^2 - 22n + 19}{4}, \quad n \geq 3, \quad a_1(1) = 0.$$

Since $a_1(n) = \sum_{i=3,5,\ldots,n}(a_1(i) - a_1(i-2))$, we find

$$a_1(n) = \frac{1}{4} \sum_{i=3,5,\ldots,n} (7i^2 - 22i + 19) = \frac{1}{24}(7n^3 - 12n^2 + 5n).$$

Finally, comparing the coefficient of x^{m-i} on both sides of (46), we find

$$\begin{aligned} &2(a_{i+1}(n) - a_{i+1}(n-2)) = \\ &2(2n-3)a_i(n-2) - (n-2)a_i(n) + \\ &2(n-1)(n-2)a_{i-1}(n-2) + s_i(n), \end{aligned} \tag{47}$$

where

$$s_i(n) = Coef[x^{m-i}]((4x + 3n - 4)(x + \frac{n-3}{2})(x + \frac{n-5}{2}) \cdots (x+1)x)$$

and, as in 1), the statement is proved by induction over i. □

A few such polynomials are the following:
For odd n :

$$U_0(n) = n,$$

$$U_1(n) = \frac{1}{24}(n-1)n(7n-5),$$

$$U_2(n) = \frac{1}{640}(n-3)(n-1)n(29n^2 - 44n + 7),$$

$$U_3(n) = \frac{1}{322560}(n-5)(n-3)(n-1)n(1581n^3 - 3775n^2 + 1587n + 223);$$

For even n :

$$V_0(n) = \frac{1}{2}n,$$

$$V_1(n) = \frac{1}{48}(n-2)n(7n-4),$$

$$V_2(n) = \frac{1}{3840}(n-4)(n-2)n(87n^2 - 98n + 16),$$

$$V_3(n) = \frac{1}{645120}(n-6)(n-4)(n-2)n(1581n^3 - 2686n^2 + 936n + 64).$$

Proposition 13.

$$a_i(n) \equiv \begin{cases} r_i(n), & \text{if } n \text{ is even,} \\ \\ s_i(n), & \text{if } n \text{ is odd.} \end{cases} \pmod 2$$

Proof. The proposition follows from (45), (47) and Theorem 2. □

Finally, note that, from (36) and (37) the following homogeneous recursions for the coefficients of $P_n(x)$ follow.

Theorem 14. *For odd $n \geq 3$ and $i \geq 0$,*

$$(m-i)a_i(n) = na_i(n-1) + \sum_{j=0}^{i-1}(-1)^{i-j+1}\binom{m-j}{m-i-1}a_j(n).$$

For even $n \geq 4$ and $i \geq 0$,

$$(n-2i-1)a_i(n) = \frac{n}{2}a_i(n-1)+$$

$$2\sum_{j=0}^{i-1}(-1)^{i-j+1}\left(m\binom{m-j}{m-i} - \binom{m-j}{m-i-1}\right)a_j(n).$$

10. Arithmetic Proof of the Integrality $P_n(x)$ in Integer Points

From Theorem 2 we conclude that the polynomial, $P_n(x)$, takes integer values for integer $x = k$. Here we give an independent arithmetic proof of this fact using the explicit expression (3). It is well known (cf. [5], Section 8, Problem 87) that, if a polynomial $P(x)$ of degree m takes integer values for $x = 0, 1, \ldots, m$, then it takes integer values for every integer x. Since, as we proved at the end of Section 6, $\deg P_n(k) = \lfloor\frac{n-1}{2}\rfloor$, we suppose that $0 \leq k \leq \lfloor\frac{n-1}{2}\rfloor$. Moreover, from the results of Section 3, $P_n(0)$ and $P_n(1)$ are integers. (In the case when $n+1$ is an odd prime, $P_n(1) = (2^n-1)(\frac{n}{2})!/(n+1)$ is integer, since $2^n - 1 \equiv 0 \pmod{n+1}$, while in the case when $n+1$ is an odd composite number, no divisor exceeds $\frac{n+1}{3}$, therefore, $(\frac{n}{2})! \equiv 0 \pmod{n+1}$.) Thus we can suppose that

$$2 \leq k \leq \lfloor\frac{n-1}{2}\rfloor. \tag{48}$$

Suppose that n is *even* (the case of odd n is considered quite analogously). Let p be a prime. Denote the maximal power of p dividing n by $[n]_p$. We say that, for integer $l,\ h$, the fraction $\frac{l}{h}$ is *p-integer,* if $[l]_p - [h]_p \geq 0$.

A) Firstly, we show that, for $n \geq 4$, $P_n(k)$ is 2-integer. Indeed, $2k+n-1$ is odd, while $4k+3n-4$ is even. Therefore, by (26), using a trivial induction, we see that $P_n(k)$ is 2-integer.

Further we use the explicit formula (3) of Theorem 1.

B) Let p be an odd prime divisor of $\binom{n+2k-2}{k-1}$ which does not coincide with any factor of the product $(n+2k-1)(n+2k-2)\cdots(n+k)$. Thus p could divide one or several *composite* factors of this product. Therefore, the following condition holds

$$3 \leq p \leq \frac{n+2k-1}{3}. \tag{49}$$

Let us show that

$$a(n;k) = \frac{\binom{\frac{n}{2}+k-1}{k}}{\binom{n+2k-1}{k}}(\frac{n-2}{2})! =$$

$$2^{-k}\frac{(n+2k-2)(n+2k-4)\cdots n}{(n+2k-1)(n+2k-2)\cdots(n+k)}(\frac{n-2}{2})!$$

is p-integer and, consequently, $P_n(k)$ is p-integer.

Let $k \geq 3$ be even. Then, after a simplification, we have

$$2^k a(n;k) = \frac{(n+k-2)(n+k-4)\cdots n}{(n+2k-1)(n+2k-3)\cdots(n+k+1)}(\frac{n-2}{2})!,$$

or

$$2^{\frac{k}{2}} a(n;k) = \frac{(\frac{n+k-2}{2})!}{(n+2k-1)(n+2k-3)\cdots(n+k+1)} \tag{50}$$

We distinguish several cases.

`Case a).` For $t \geq 2$, let p^t divide at least one factor of the denominator. Then $p \leq (n+2k-1)^{\frac{1}{t}}$ Let us show that $p \leq \frac{n+k-2}{2t}$. We should show that $n+2k-1 \leq (\frac{n+k-2}{2t})^t$, or, since, by (48), $k \leq \frac{n-2}{2}$, it is sufficient to show that $\frac{3}{2}(n+k-2) \leq (\frac{n+k-2}{2t})^t$, or $(2t)^{\frac{t}{t-1}} \leq (\frac{2}{3})^{\frac{1}{t-1}}(n+k-2)$. Since $(\frac{2}{3})^{\frac{1}{t-1}} \geq \frac{2}{3}$, it is sufficient to prove that $(2t)^{\frac{t}{t-1}} \leq \frac{2}{3}(n+k-2)$. Note that $e^t < p^t \leq n+2k-2$, $t \leq \ln(n+2k-2)$. Therefore we find $(2t)^{\frac{t}{t-1}} \leq (2\ln(n+2k-2))^2$. Furthermore, note that, if $n \geq 152$, then $\ln^2 n < \frac{n}{6}$. Thus $(2t)^{\frac{t}{t-1}} \leq \frac{2}{3}(n+k-2)$. It is left to add that up to $n=161$ we verified that the polynomials $P_n(k)$ have integer coefficients and, consequently, is integer-valued.

`Case b).` Let p divide only one factor of the denominator. Then, in view of (48) and (49), $p \leq \frac{n+2k-1}{3} \leq \frac{n+k-2}{2}$ and, by (50), $a(n;k)$ is p-integer.

`Case c).` Let p divide exactly l factors of the denominator. Then

$$p \leq \frac{(n+2k-1)-(n+k+1)}{l} = \frac{k-2}{l},$$

and, since, by (48), $n \geq 2k+2$, we conclude that $\frac{n+k-2}{2} \geq \frac{3k}{2} \geq k-2 \geq lp$. Hence, by (50), $a(n;k)$ is p-integer.
It is left to notice that the case of odd k is considered quite analogously.
C) Suppose that, as in B), $k \geq 2$ is even. Let p be an odd prime divisor of $\binom{n+2k-1}{k}$ which coincides with some factor of the product $(n+2k-1)(n+2k-3)\cdots(n+k+1)$. In this case the fraction (50) is not integer. Thus in order to prove that $P_n(k)$ is p-integer, we should prove that $T_n(k)$ (18) is p-integer. By the condition, p has form

$$p = n+2k-1-2r, \;\; 0 \leq r \leq \frac{k-2}{2}. \tag{51}$$

According to (18) and (51), we should prove that

$$\sum_{j=0}^{n-1} 2^j \binom{n+2k-j-2}{k-1} =$$

$$\sum_{j=0}^{n-1} 2^j \binom{p+2r-1-j}{k-1} \equiv 0 \pmod{p}, \tag{52}$$

or

$$A(n,r,k) :=$$

$$\sum_{j=0}^{n-1} 2^j (j-(2r-1))(j+1-(2r-1))\cdots(j+k-2-(2r-1)) \equiv 0 \pmod{p}.$$

Note that, since $n-2r = p-2k+1$, we have

$$\sum_{j=0}^{n-1} x^{j+k-2-(2r-1)} =$$

$$(x^{n+k-2r-1} - x^{k-2r-1})(x-1)^{-1} = (x^{p-k} - x^{k-2r-1})(x-1)^{-1}.$$

Therefore,

$$A(n,r,k) = 2^{2r} \sum_{j=0}^{n-1} (x^{j+k-2-(2r-1)})^{(k-1)} \mid_{x=2} =$$

$$2^{2r}((x^{p-k} - x^{k-2r-1})(x-1)^{-1})^{(k-1)} \mid_{x=2} .$$

Thus we should prove that

$$((x^{p-k} - x^{k-2r-1})(x-1)^{-1})^{(k-1)} \mid_{x=2} \equiv 0 \pmod{p},$$

or, using the Leibnitz formula,

$$\sum_{j=0}^{k-1} (-1)^{k-j-1} \binom{k-1}{j} (k-j-1)!(p-k)(p-k-1)\cdots(p-k-j+1)2^{p-k-j} \equiv$$

$$\sum_{j=0}^{k-1}(-1)^{k-j-1}\binom{k-1}{j}(k-j-1)!(k-2r-1)(k-2r-2)\cdots(k-2r-j)2^{k-2r-j-1} \pmod{p}.$$

Since $2^{p-1} \equiv 1 \pmod{p}$, we should prove the identity

$$\sum_{j=0}^{k-1}(-1)^{k-j-1}\binom{k-1}{k-j-1}(k-j-1)!(p-k)(p-k-1)\cdots(p-k-j+1)\mid_{p=0} 2^{-k-j+1} =$$

$$\sum_{j=0}^{k-1}(-1)^{k-j-1}\binom{k-1}{k-j-1}(k-j-1)!(k-2r-1)(k-2r-2)\cdots(k-2r-j)2^{k-2r-j-1},$$

or, after simple transformations, the identity

$$\sum_{j=0}^{k-1}\binom{k+j-1}{j}2^{-j} = 2^{2k-2r-2}\sum_{j=0}^{k-1}(-1)^j\binom{k-2r-1}{j}2^{-j}. \tag{53}$$

It is known ((([7], Ch.1, problem 7), that

$$\sum_{i=0}^{n}\binom{2n-i}{n}2^{i-n} = 2^n.$$

Putting $n-i=j$, we have

$$\sum_{j=0}^{n}\binom{n+j}{n}2^{-j} = \sum_{j=0}^{n}\binom{n+j}{j}2^{-j} = 2^n.$$

Therefore, the left hand side of (53) is 2^{k-1} and it is left to prove that

$$\sum_{j=0}^{k-1}(-1)^j\binom{k-2r-1}{j}2^{k-j} = 2^{2r+1}.$$

We have

$$\sum_{j=0}^{k-1}(-1)^j\binom{k-2r-1}{j}2^{k-j} = \sum_{j=0}^{k-2r-1}(-1)^j\binom{k-2r-1}{j}2^{k-j} =$$

$$2^{2r+1}\sum_{j=0}^{k-2r-1}(-1)^j\binom{k-2r-1}{j}2^{k-2r-1-j} = 2^{2r+1}(2-1)^{k-2r-1} = 2^{2r+1}$$

and we are done. The case of odd $k \geq 3$ is considered quite analogously. So, formulas (50) and (51) take the form

$$2^{\frac{k-1}{2}}a(n;k) = \frac{(\frac{n+k-1}{2})!}{(n+2k-1)(n+2k-3)\cdots(n+k)},$$

$$p = n+2k-2r-1,\ 0 \leq r \leq \frac{k-1}{2},$$

and, for odd k, the proof reduces to the same congruence (52).

11. Representation of $P_n(x)$ in Basis $\{\binom{x}{i}\}$

The structure of the explicit formula (3) allows us to conjecture that the coefficients of $P_n(x)$ in basis $\{\binom{x}{i}\}$ have simpler properties. A process of expansion of a polynomial $P(x)$ in the binomial basis is indicated in [5] in a solution of Problem 85: "Functions 1, x, $x^2, \ldots, x^n$ one can consecutively express in the form of linear combinations with the constant coefficients of 1, $\frac{x}{1}$, $\frac{x(x-1)}{2}, \ldots, \frac{x(x-1)\cdots(x-n+1)}{n!}$." Therefore,

$$P(x) = b_0\binom{x}{m} + b_1\binom{x}{m-1} + \cdots + b_{m-1}\binom{x}{1} + b_m,$$

where $b_0,\ b_1, \ldots, b_m$ are defined from the equations

$$P(0) = b_m,$$

$$P(1) = b_m + \binom{1}{1}b_{m-1},$$

$$P(2) = b_m + \binom{2}{1}b_{m-1} + \binom{2}{2}b_{m-2},$$

$$\vdots$$

$$P(m) = b_m + \binom{m}{1}b_{m-1} + \cdots + \binom{m}{m}b_0.$$

This process can be simplified in the following way. In the identity

$$n^x = (1 + (n-1))^x =$$

$$1 + (n-1)\binom{x}{1} + (n-1)^2\binom{x}{2} + \cdots + (n-1)^x\binom{x}{x} =$$

$$n^0 + (n-n^0)\binom{x}{1} + (n-n^0)^2\binom{x}{2} + \cdots + (n-n^0)^x\binom{x}{x}$$

we can evidently replace powers $n^j,\ j = 0, \ldots, x$, by the arbitrary numbers $a_j,\ j = 0, \ldots, x$. Thus we have a general identity

$$a_x = a_0 + (a_1 - a_0)\binom{x}{1} + (a_2 - 2a_1 + a_0)\binom{x}{2} + (a_3 - 3a_2 + 3a_1 - a_0)\binom{x}{3} + \cdots +$$

$$(a_x - \binom{x}{1}a_{x-1} + \binom{x}{2}a_{x-2} - \cdots + (-1)^x\binom{x}{x}a_0)\binom{x}{x}.$$

Essentially, we quickly obtained a special case of the so-called "Newton's forward difference formula" (cf. [10]). Here, put $a_j = P(j),\ j = 0, \ldots, m$, and, firstly, consider values $0 \le x \le m$. Since $\binom{x}{l} = 0$ for $l > m$, we obtain the required representation under the condition $0 \le x \le m$:

$$P(x) = P(0) + (P(1) - P(0))\binom{x}{1} +$$

$$(P(2) - 2P(1) + P(0))\binom{x}{2} + \cdots + (P(m) - \binom{m}{1}P(m-1) +$$

$$\binom{m}{2}P(m-2) - \cdots + (-1)^m\binom{m}{m}P(0))\binom{x}{m}. \tag{54}$$

It is left to note that, since a polynomial of degree m is fully defined by its values in $m+1$ points $0, 1, \ldots, m$, then (54) is the required representation for all x.

So, for the polynomials $\{P_n(x)\}$, we have

$$P_1 = 1,$$

$$P_2 = 1,$$

$$P_3 = 3\binom{x}{1} + 4,$$

$$P_4 = 2\binom{x}{1} + 4,$$

$$P_5 = 10\binom{x}{2} + 30\binom{x}{1} + 32,$$

$$P_6 = 6\binom{x}{2} + 22\binom{x}{1} + 32,$$

$$P_7 = 42\binom{x}{3} + 196\binom{x}{2} + 378\binom{x}{1} + 384,$$

$$P_8 = 24\binom{x}{3} + 128\binom{x}{2} + 296\binom{x}{1} + 384,$$

$$P_9 = 216\binom{x}{4} + 1368\binom{x}{3} + 3816\binom{x}{2} + 6120\binom{x}{1} + 6144,$$

$$P_{10} = 120\binom{x}{4} + 840\binom{x}{3} + 2664\binom{x}{2} + 5016\binom{x}{1} + 6144,$$

$$P_{11} = 1320\binom{x}{5} + 10560\binom{x}{4} + 38544\binom{x}{3} + 84480\binom{x}{2} + 122760\binom{x}{1} + 122880,$$

$$P_{12} = 760\binom{x}{5} + 6240\binom{x}{4} + 25152\binom{x}{3} + 62112\binom{x}{2} + 103920\binom{x}{1} + 122880.$$

12. On Coefficients of $P_n(x)$ in Basis $\{\binom{x}{i}\}$

Let

$$P_n(x) = b_0(n)\binom{x}{m} + b_1(n)\binom{x}{m-1} + \cdots + b_{m-1}(n)\binom{x}{1} + b_m(n),$$

where $m = \lfloor \frac{n-1}{2} \rfloor$.
Since, for integer k, we have the explicit formula for $P_n(k)$, (3), then, according to (54), we have the following explicit formula for $b_i(n),\ \ i = 0, \dots, m:$

$$b_i(n) = \sum_{k=0}^{m-i} (-1)^{m-i-k} \binom{m-i}{k} P_n(k). \tag{55}$$

Let

$$P_n(x) = \sum_{j=0}^{m} a_j(n) x^{m-j}.$$

Then

$$b_i(n) = \sum_{j=0}^{m} a_j(n) \sum_{k=0}^{m-i} (-1)^{m-i-k} k^{m-j} \binom{m-i}{k}. \tag{56}$$

Since the l-th difference of $f(x)$ is (cf. [1], formula 25.1.1)

$$\Delta^l f(x) = \sum_{k=0}^{l} (-1)^{l-k} \binom{l}{k} f(x+k),$$

one can write (56) in the form

$$b_i(n) = \sum_{j=0}^{m} a_j(n) \Delta^{m-i} x^{m-j} \mid_{x=0} .$$

Here the summands corresponding to $j > i$, evidently, equal 0. Therefore, we have

$$b_i(n) = \sum_{j=0}^{i} a_j(n) \Delta^{m-i} x^{m-j} \mid_{x=0} . \tag{57}$$

Theorem 15. *For $n \geq 1$, we have*

$$b_0(n) = \begin{cases} n(\frac{n-1}{2})!, & \text{if } n \text{ is odd,} \\ (\frac{n}{2})!, & \text{if } n \text{ is even;} \end{cases} \tag{58}$$

$$b_1(n) = \begin{cases} \frac{1}{6}n(5n-7)(\frac{n-1}{2})!, & \text{if } n \text{ is odd,} \\ \frac{1}{6}(5n-8)(\frac{n}{2})!, & \text{if } n \text{ is even.} \end{cases} \tag{59}$$

In general, for a fixed i, $b_i(n) = (m-i)!Y_i(n)$, if n is odd, and $b_i(n) = (m-i)!Z_i(n)$, if n is even, where Y_i and Z_i are polynomials in n of degree $2i+1$.

Proof. Note that the Stirling number of the second kind $S(n,m)$ is connected with the m-th difference of $\Delta^m x^n\,|_{x=0}$ in the following way (see [1], formulas 24.1.4):

$$S(n,m)m! = \sum_{k=0}^{m}(-1)^{m-k}\binom{m}{k}k^n = \Delta^m x^n\,|_{x=0}\,. \tag{60}$$

In particular, since $S(m,m) = 1$, $S(m+1,m) = \binom{m+1}{2}$, we have

$$\Delta^m x^m\,|_{x=0} = m!$$

and

$$\Delta^m x^{m+1}\,|_{x=0} = \frac{m}{2}(m+1)!\,.$$

Therefore, by (57),

$$b_0(n) = m!a_0(n),$$

$$b_1(n) = \frac{m-1}{2}m!a_0(n) + (m-1)!a_1(n),$$

and, by (40) and (41) (where $m = \lfloor\frac{n-1}{2}\rfloor$), we find formulas (58) and (59).

Further, we need the following lemma.

Lemma 16. *$S(n+k,n)$ is a polynomial in n of degree $2k$.*

Proof. For $k \geq 1$, let

$$Q_k(n) = S(n+k,n).$$

Note that, since $S(n,n) = 1$, we have $Q_0(n) = 1$. Further, since $S(n,0) = \delta_{n,0}$, for $k \geq 1$, $Q_k(0) = 0$. From the main recursion for $S(n,m)$ which is $S(n,m) = mS(n-1,m) + S(n-1,m-1)$, we have

$$Q_k(n) - Q_k(n-1) = nQ_{k-1}(n).$$

Also, in view of $Q_k(0) = 0$, we find the recursion

$$Q_0(n) = 1, \quad Q_k(n) = \sum_{i=1}^{n} iQ_{k-1}(i). \tag{61}$$

Using a simple induction, from (61) we obtain the lemma. □

Remark 17. The first few polynomials $\{Q_k(n)\}$ are:

$$Q_0 = 1,$$
$$Q_1 = \frac{1}{2}n(n+1),$$
$$Q_2 = \frac{1}{24}n(n+1)(n+2)(3n+1),$$
$$Q_3 = \frac{1}{48}n^2(n+1)^2(n+2)(n+3),$$
$$Q_4 = \frac{1}{5760}n(n+1)(n+2)(n+3)(n+4)(15n^3+30n^2+5n-2).$$

It can be proven that the sequence of denominators coincides with $A053657$ [8], such that the denominator of $Q_k(n)$ is $\prod p^{\sum_{j\geq 0}\lfloor \frac{k}{(p-1)p^j}\rfloor}$, where the product is over all primes.

Note that from (57) and (60) we find

$$b_i(n) = (m-i)!\sum_{j=0}^{i} a_j(n)S(m-j,m-i), \quad m = \lfloor\frac{n-1}{2}\rfloor.$$

Since, by Lemma 10, $S(m-j,m-i)$ is a polynomial in n of degree $2((m-j)-(m-i)) = 2(i-j)$, while, by Theorem 12, $a_j(n)$ is a polynomial of degree $2j+1$, it follows that $a_j(n)S(m-j,m-i)$ is a polynomial of degree $2i+1$. Thus $\sum_{j=0}^{i} a_j(n)S(m-j,m-i)$ is a polynomial of degree $2i+1$. This completes the proof. □

The first polynomials $Y_i(n)$, $Z_i(n)$ are

$$Y_0 = n,$$
$$Y_1 = \frac{1}{12}(n-1)n(5n-7),$$
$$Y_2 = \frac{1}{480}(n-3)(n-1)n(43n^2-168n+149),$$
$$Y_3 = \frac{1}{13440}(n-5)(n-3)(n-1)n(177n^3-1319n^2+3063n-2161);$$
$$Z_0 = \frac{n}{2},$$
$$Z_1 = \frac{1}{24}(n-2)n(5n-8),$$
$$Z_2 = \frac{1}{960}(n-4)(n-2)n(43n^2-182n+184),$$
$$Z_3 = \frac{1}{26880}(n-6)(n-4)(n-2)n(3n-8)(59n^2-306n+352)).$$

Finally, we prove the following attractive result.

Theorem 18. 1) *For odd* n, $b_j(n)/n, j = 0, \dots, m-1$, *are integer. Moreover, for* $n \geq 3$,

$$b_i(n) = n(b_i(n-1) + b_{i-1}(n-1)), \quad i = 1, \dots, m-1.$$

2) *For even* $n \geq 4$,

$$2b_i(n) = b_i(n-1) + b_{i-1}(n-1) + m!\binom{m}{i}, \quad i = 1, \dots, m-1. \tag{62}$$

Proof. 1) According to (55), we should prove that for odd $n \geq 3$,

$$\sum_{k=0}^{m-i}(-1)^{m-i-k}\binom{m-i}{k}P_n(k) =$$

$$n\Big(\sum_{k=0}^{m-i}(-1)^{m-i-k}\binom{m-i}{k}P_{n-1}(k)+$$

$$\sum_{k=0}^{m-i-1}(-1)^{m-i-k-1}\binom{m-i-1}{k}P_{n-1}(k)\Big), \quad i = 1, 2, \dots, m-1,$$

or, putting $m - i = t$,

$$\sum_{k=0}^{t}(-1)^k\binom{t}{k}P_n(k) = n\Big(\sum_{k=0}^{t}(-1)^k\binom{t}{k}P_{n-1}(k)-$$

$$\sum_{k=0}^{t}(-1)^k\binom{t-1}{k}P_{n-1}(k)\Big), \quad t = 1, 2, \dots, m-1,$$

or, finally, for $t = 1, \dots, \frac{n-3}{2}$,

$$\sum_{k=1}^{t}(-1)^{k-1}\Big(\binom{t}{k}P_n(k) - \binom{t-1}{k-1}nP_{n-1}(k)\Big) = P_n(0). \tag{63}$$

To prove (63), note that, by (36), $nP_{n-1}(k) = P_n(k) - P_n(k-1)$. Hence,

$$\binom{t}{k}P_n(k) - \binom{t-1}{k-1}nP_{n-1}(k) =$$

$$P_n(k)\Big(\binom{t}{k} - \binom{t-1}{k-1}\Big) + \binom{t-1}{k-1}P_n(k-1) =$$

$$\binom{t-1}{k}P_n(k) + \binom{t-1}{k-1}P_n(k-1).$$

Thus the summands of (63) are

$$(-1)^{k-1}\Big(\binom{t}{k}P_n(k) - \binom{t-1}{k-1}nP_{n-1}(k)\Big) =$$

$$(-1)^{k-1}\binom{t-1}{k}P_n(k)-(-1)^{k-2}\binom{t-1}{k-1}P_n(k-1),$$

and the summing gives

$$\sum_{k=1}^{t}(-1)^{k-1}\left(\binom{t}{k}P_n(k)-\binom{t-1}{k-1}nP_{n-1}(k)\right)=$$

$$(-1)^{k-1}\binom{t-1}{k}P_n(k)\mid_{k=t}-(-1)^{k-2}\binom{t-1}{k-1}P_n(k-1)\mid_{k=1}=P_n(0).$$

2) Analogously, the proof of (62) reduces to the proof of the following equality for $t=1,2,\dots,m-1$:

$$\sum_{k=0}^{t+1}(-1)^k\left(2\binom{t}{k}P_n(k)+\binom{t}{k-1}P_{n-1}(k)\right)=(-1)^t m!\binom{m}{t}. \tag{64}$$

Note that, by (39),

$$(-1)^k\left(2\binom{t}{k}P_n(k)+\binom{t}{k-1}P_{n-1}(k)\right)=$$

$$(-1)^k\binom{t}{k}P_{n-1}(k+1)-$$

$$(-1)^{k-1}\binom{t}{k-1}P_{n-1}(k)+(-1)^k\binom{t}{k}\binom{k+m}{m}m! \tag{65}$$

Since

$$\sum_{k=0}^{t+1}\left((-1)^k\binom{t}{k}P_{n-1}(k+1)-(-1)^{k-1}\binom{t}{k-1}P_{n-1}(k)\right)=$$

$$(-1)^k\binom{t}{k}P_{n-1}(k)\mid_{k=t+1}-(-1)^{k-1}\binom{t}{k-1}P_{n-1}(k)\mid_{k=0}=0,$$

by (64) and (65), the proof reduces to the known combinatorial identity

$$\sum_{k=0}^{t}(-1)^k\binom{t}{k}\binom{k+m}{m}=(-1)^t\binom{m}{t},\ \ t=1,\dots,m-1$$

(see [7], Ch.1, formula (8) with $p=0$ up to the notations). □

References

[1] M. Abramowitz and I. A. Stegun (Eds.), Bernoulli and Euler Polynomials and the Euler-Maclaurin Formula in *Handbook of Mathematical Functions with Formulas, Graphs, and Mathematical Tables, 9th printing,* New York: Dover, pp. 804-806, 1972.

[2] S. Fomin and A. Zelevinsky, The Laurent phenomenon, *Adv. Appl. Math.,* **28** (2002), 19–24.

[3] D. Gale, Mathematical entertainments: the strange and surprising saga of the Somos sequences, *Math. Intel.* **13** (1991), 40-42.

[4] J. L. Malouf, An integer sequence from a rational recursion, *Discrete Math.,* **110** (1992), 257–261.

[5] G. Polya and G. Szegö, *Problems and theorems in analysis,* Vol. 2, Springer-Verlag, 1976.

[6] J. Propp and E. W. Weisstein, "Somos sequence," From *MathWorld* –A Wolfram Web Resource, `http://mathworld.wolfram.com/GoebelsSequence.html`.

[7] J. Riordan, *Combinatorial Identities,* Wiley & Sons, 1968.

[8] N. J. A. Sloane, *The On-Line Encyclopedia of Integer Sequences,* `http://oeis.org`.

[9] M. Somos, Problem 1470, Crux Mathematicorum, **15** (1989), 208.

[10] E. W. Weisstein, "Newton's forward difference formula," From *MathWorld* –A Wolfram Web Resource, `http://mathworld.wolfram.com/NewtonsForwardDifferenceFormula.html`.

[11] E. W. Weisstein, "Göbel's sequence," From *MathWorld* –A Wolfram Web Resource, `http://mathworld.wolfram.com/GoebelsSequence.html`.

INTEGERS: 13 (2013) #A10

ON DIVISIBILITY PROPERTIES OF SOME DIFFERENCES OF THE CENTRAL BINOMIAL COEFFICIENTS AND CATALAN NUMBERS

Tamás Lengyel
Department of Mathematics, Occidental College, Los Angeles, California
lengyel@oxy.edu

Received: 5/1/12, Revised: 10/16/12, Accepted: 2/24/13, Published: 3/15/13

Abstract

We discuss divisibility properties of some differences of the central binomial coefficients and Catalan numbers. The main tool is the application of various congruences modulo high prime powers for binomial coefficients combined with some recurrences relevant to these combinatorial quantities.

1. Introduction

The differences of certain combinatorial quantities exhibit interesting divisibility properties. For example, we can consider differences of central binomial coefficients

$$c_n = \binom{2n}{n}, n \geq 0,$$

and Catalan numbers

$$C_n = \frac{1}{n+1}\binom{2n}{n}, n \geq 0.$$

We will need some basic notation. Let n and k be positive integers, p be a prime, $d_p(k)$ and $\nu_p(k)$ denote the sum of digits in the base p representation of k and the highest power of p dividing k, respectively. The latter one is often referred to as the p-adic order of k. For the rational n/k we set $\nu_p(n/k) = \nu_p(n) - \nu_p(k)$.

Our main results concern the p-adic order of the differences of central binomial coefficients $c_{ap^{n+1}+b} - c_{ap^n+b}$ (cf. Theorems 2.1 and 2.4), Catalan numbers $C_{ap^{n+1}+b} - C_{ap^n+b}$ (cf. Theorems 2.2 and 2.11, and Remark 4.1) with a prime p, $(a,p) = 1$, and $n \geq n_0$ for some integer $n_0 \geq 0$. These results are essential in obtaining the p-adic order of the differences of certain Motzkin numbers, more precisely $M_{ap^{n+1}+b} - M_{ap^n+b}$ with a prime p and different settings of a and b that are discussed in [8] and [9]. Of course, results involving the exact orders of differences or lower bounds on them can be easily rephrased in terms of super congruences for the underlying quantities.

Section 2 collects some of the main results while Section 3 is devoted to some known results and their direct consequences regarding congruential and p-adic properties of the binomial coefficients, e.g., Corollary 3.9. Section 4 contains the proofs of the main results of Section 2 and presents some interesting congruences for central binomial coefficients (cf. Corollary 4.1 improving Corollary 3.5) and Catalan numbers (cf. Theorem 2.12). In Section 5, we derive and prove some important p-adic properties of the differences of certain harmonic numbers stated in Section 2. These properties are important in the actual use of Theorem 2.4.

2. Main Results

We state some of the main results of the paper.

Theorem 2.1. *For $p = 2$ and a odd, we have that*

$$\nu_2(c_{a2^{n+1}} - c_{a2^n}) = 3(n+1) + \nu_2\left(\binom{2a}{a}\right) - 1 = 3(n+1) + d_2(a) - 1, n \geq 1.$$

For $p = 3$ and $(a, 3) = 1$,

$$\nu_3(c_{a3^{n+1}} - c_{a3^n}) = 3(n+1) + \nu_3\left(\binom{2a}{a}\right) - 1, n \geq 0.$$

Let B_n denote the nth Bernoulli number. For any prime $p \geq 5$, $(a,p) = 1$, and $\nu_p(B_{p-3}) = 0$ or -1, we have that

$$\nu_p(c_{ap^{n+1}} - c_{ap^n}) = 3(n+1) + \nu_p\left(\binom{2a}{a}\right) + \nu_p(B_{p-3}), n \geq 0.$$

Note. The term $\nu_p(\binom{2a}{a})$ contributes to the p-adic orders above exactly if at least one of the p-ary digits of a is at least as large as $p/2$.

Remark 2.1. *It is well known that $\nu_p(B_n) \geq -1$ by the von Staudt–Clausen theorem. If the prime p divides the numerator of B_{p-3}, i.e., $\nu_p(B_{p-3}) \geq 1$, or equivalently $\binom{2p}{p} \equiv 2 \bmod p^4$, then it is sometimes called a Wolstenholme prime [2]. The only known Wolstenholme primes up to 10^9 are $p = 16843$ and 2124679. For such primes Theorem 2.1 is inconclusive and gives only the lower bound $3(n+1) + \nu_p(\binom{2a}{a}) + 1$ on the p-adic order.*

For the p-adic order of the differences of Catalan numbers we obtain the following theorem.

Theorem 2.2. *For any prime $p \geq 2$ and $(a,p)=1$, we have*

$$\nu_p(C_{ap^{n+1}} - C_{ap^n}) = n + \nu_p\left(\binom{2a}{a}\right), n \geq 1.$$

The case with $n=0$ is slightly different and included in

Lemma 2.3. *For $n \geq 1$ and any prime $p \geq 2$, we have $\nu_p(C_{ap^n}) = \nu_p(c_{ap^n}) = \nu_p(\binom{2a}{a})$ and $\nu_p(C_{ap} - C_a) = \nu_p(C_a)$ if in addition $(a,p)=1$.*

The nature of the p-adic order in Theorem 2.1 changes as we introduce an additive term b, and we get a result similar to Theorem 2.2.

Theorem 2.4. *For $p=2$, a odd, and $n \geq n_0 = 1$ we have*

$$\nu_2(c_{a2^{n+1}+1} - c_{a2^n+1}) = n + \nu_2\left(\binom{2a}{a}\right) + 1,$$

and in general, for $b \geq 1$ and $n \geq n_0 = \lfloor \log_2 2b \rfloor$

$$\begin{aligned}\nu_2(c_{a2^{n+1}+b} - c_{a2^n+b}) &= n + \nu_2\left(\binom{2a}{a}\right) + \nu_2(f(b)) \\ &= n + d_2(a) + d_2(b) - \lfloor \log_2 b \rfloor,\end{aligned}$$

where $f(b) = 2\binom{2b}{b}(H_{2b} - H_b)$ with $H_n = \sum_{j=1}^n 1/j$ being the nth harmonic number.

For any prime $p \geq 3$, $(a,p)=1$, and $b \geq 1$ we have that

$$\nu_p(c_{ap^{n+1}+b} - c_{ap^n+b}) = n + \nu_p\left(\binom{2a}{a}\right) + \nu_p(f(b)),$$

for $n \geq n_0 = \max\{\nu_p(f(b)) + 2r - \nu_p(\binom{2b}{b}) + 1, r+1\} = \max\{\nu_p(2(H_{2b} - H_b)) + 2r + 1, r+1\}$ and $r = \lfloor \log_p 2b \rfloor$.

In general, for any prime $p \geq 3$, $(a,p)=1, b \geq 1$, and $n > \lfloor \log_p 2b \rfloor$, we have

$$\nu_p(c_{ap^{n+1}+b} - c_{ap^n+b}) \geq n + \nu_p\left(\binom{2a}{a}\right) + \nu_p\left(\binom{2b}{b}\right) - \lfloor \log_p 2b \rfloor$$

and for $p=2$, a odd, and $n \geq \lfloor \log_2 2b \rfloor$

$$\nu_2(c_{a2^{n+1}+b} - c_{a2^n+b}) \geq n + \nu_2\left(\binom{2a}{a}\right) + \nu_2\left(\binom{2b}{b}\right) - \lfloor \log_2 b \rfloor.$$

We can make some useful statements on the magnitude of $\nu_p(f(b))$. By taking the common denominator in $H_{2b} - H_b$, we note that $\nu_p(f(b)) \geq \nu_p(\binom{2b}{b}) - \lfloor \log_p 2b \rfloor$ for $p \geq 3$ and $\nu_2(f(b)) = d_2(b) - \lfloor \log_2 b \rfloor$. This implies that $\nu_p(f(b)) \geq 0$ for $1 \leq b \leq (p-1)/2$. In this range the binomial factor of $f(b)$ can be dropped, leaving only $\nu_p(H_{2b} - H_b)$. It appears that $\nu_p(f(b)) = 0$ in many cases, however $f(2) = 7$ and $f(15) = 450351518582/2145$; thus, $\nu_p(f(2)) = 1$ and $\nu_p(f(15)) = 2$ with $p=7$. Also note that for $p \geq 3$ we have $\nu_p(f(b)) = 0$ if $b+1 \leq p \leq 2b$ and $\nu_p(f(p)) = -1$. In fact, a much stronger statement about $\nu_p(f(b))$ is given in

Theorem 2.5. *For $p = 2$ and $c \geq 1$, we have $\nu_2(f(c)) = d_2(c) - \lfloor \log_2 c \rfloor$. For $p \geq 3$, we have $\nu_p(f(p^k)) = -k, k \geq 1$, and in general, for $c \geq 1$ and $k \geq 0$, we have $\nu_p(f(cp^k)) = -k + \nu_p(f(c))$ provided that $\nu_p(H_{2c} - H_c) \leq 0$.*

We observe that $\nu_p(f(b))$; and therefore, the p-adic order of the difference $c_{ap^{n+1}+b} - c_{ap^n+b}$ changes only slightly for bs that are small relative to p.

Clearly, $\nu_2(f(b)) \leq 1$ and equality holds only if $b+1$ is a power of two. We observe that $\nu_3(f(b)) \leq 0$ and equality holds exactly if $b = 1$ or all ternary digits of b are at least one and $b \equiv 2 \bmod 3$, i.e., the least significant digit is two, as these facts follow from Theorem 2.8. Clearly, the study of $\nu_p(f(b))$ requires the understanding of the behavior of $\nu_p(H_{2b} - H_b)$. The next lemma and theorem are straightforward.

Lemma 2.6. *For any prime p and integer $b \geq 1$, we have $\nu_p(H_{2b} - H_b) \geq -\lfloor \log_p 2b \rfloor$.*

This follows immediately as the exponent in the largest power of p can not exceed $\lfloor \log_p 2b \rfloor$. Also note the following

Theorem 2.7. *For the positive integers $b < a$, $H_a - H_b$ is never an integer.*

We include the standard proof of this theorem which, with some tweaking, leads to the proof of the following theorem. Indeed, for the exact orders, we get

Theorem 2.8. *For $p = 2$ and 3, we have $\nu_p(H_{2b} - H_b) = -\lfloor \log_p 2b \rfloor$. For $p = 5$ we have*

$$\nu_5(H_{2b} - H_b) = -\lfloor \log_5 2b \rfloor + \chi_{\exists m: \frac{3}{2}\cdot 5^m < b < 2\cdot 5^m} \tag{2.1}$$

with the indicator variable $\chi_{\exists m: \frac{3}{2}\cdot 5^m < b < 2\cdot 5^m}$ which is 1 if for b there exists an m so that $\frac{3}{2} \cdot 5^m < b < 2 \cdot 5^m$ and otherwise, it is 0.

For primes larger than 5 it seems more complicated to establish the exact order of $\nu_p(H_{2b} - H_b)$. Also note that somewhat related investigations have been initiated by [4] in which the set $J(p) = \{b \mid \nu_p(H_b) \geq 1\}$ is analyzed and determined for some primes including $3, 5, 13, 17, 23$, and 67, e.g., $J(5) = \{4, 20, 24\}$, and $\{p-1, p(p-1), p^2-1\} \subseteq J(p)$ for $p > 3$. It has been conjectured that the set $J(p)$ is finite for all primes.

We also note that we could not find any b for which $\nu_p(f(b))$ exceeded two and thus suggest

Conjecture 2.9. *We have $\nu_p(f(b)) \leq 2$ for $p \geq 5$ and $b \geq 1$.*

In terms of the harmonic numbers, Conjecture 2.9 can be rephrased as

Conjecture 2.10. *We have $\nu_p(H_{2b} - H_b) \leq 2 - \nu_p(\binom{2b}{b})$ for $p \geq 5$ and $b \geq 1$.*

Finally, for the Catalan numbers we obtain

Theorem 2.11. *For $p = 2$, a odd, and $n \geq n_0 = 2$ we have*

$$\nu_2(C_{a2^{n+1}+1} - C_{a2^n+1}) = n + \nu_2\left(\binom{2a}{a}\right) - 1,$$

and in general, for $b \geq 1$ and $n \geq n_0 = \lfloor \log_2 2b \rfloor + 1$

$$\nu_2(C_{a2^{n+1}+b} - C_{a2^n+b}) = n + \nu_2\left(\binom{2a}{a}\right) + \nu_2(g(b))$$
$$= n + d_2(a) + d_2(b) - \lceil \log_2(b+2) \rceil - \nu_2(b+1) + 1$$

where $g(b) = 2\binom{2b}{b}(b+1)^{-1}(H_{2b} - H_b - 1/(2(b+1))) = 2C_b(H_{2b} - H_b - 1/(2(b+1))) = (f(b) - C_b)/(b+1)$ with $H_n = \sum_{j=1}^{n} 1/j$.

For any prime $p \geq 3$, $(a,p) = 1$, and $b \geq 1$ we have that

$$\nu_p(C_{ap^{n+1}+b} - C_{ap^n+b}) = n + \nu_p\left(\binom{2a}{a}\right) + \nu_p(g(b)),$$

with $n \geq n_0 = \max\{\nu_p(g(b)) + 2r - \nu_p(C_b) + 1, r+1\} = \max\{\nu_p(2(H_{2b} - H_b - 1/(2(b+1)))) + 2r + 1, r+1\}$ and $r = \lfloor \log_p 2b \rfloor$.

In general, for any prime $p \geq 2$, $(a,p) = 1, b \geq 1$, and $n > \lfloor \log_p 2b \rfloor$, we have

$$\nu_p(C_{ap^{n+1}+b} - C_{ap^n+b}) \geq n + \nu_p\left(\binom{2a}{a}\right) + \nu_p\left(\binom{2b}{b}\right) - \lfloor \log_p 2b \rfloor - \nu_p(b+1).$$

Note. Clearly, $\nu_p(g(b)) \geq 0$ for $1 \leq b \leq (p-1)/2$. We note that in general, for $b \geq 1$ we have $\nu_p(g(b)) \geq \nu_p(\binom{2b}{b}) - \lfloor \log_p 2b \rfloor - \nu_p(b+1)$ if $p \geq 2$ while $\nu_2(g(b)) = d_2(b) - \lceil \log_2(b+2) \rceil - \nu_2(b+1) + 1 = d_2(b+1) - \lceil \log_2(b+2) \rceil$ if $p = 2$.

Remark 2.2. *The combination of Theorems 2.2 and 2.11 proves some kind of generalization of the observation in [11] that for any $n \geq 2$ the remainders $C_{2^{n+m-1}-1} \bmod 2^n$ are equal for each $m \geq 0$. The latter fact has been proven in [10] and [13] and extended in [14] recently.*

Our result focuses on $C_{ap^m+b} \bmod p^n$, with any large enough m, for non-negative values of b only.

Theorem 2.12. *For any prime $p \geq 2$, $(a,p) = 1, b \geq 0$, we have that $C_{ap^m+b} \bmod p^n$ is constant for $m \geq n + \nu_p(b+1) + \max\{0, \lfloor \log_p 2b \rfloor\}, n \geq 1$.*

3. Preparation

We note that there are many places in the literature where relevant divisibility and congruential properties of the binomial coefficients are discussed. Excellent surveys can be found in [6] and [12].

The following three theorems comprise the most basic facts regarding divisibility and congruence properties of the binomial coefficients. We assume that $0 \le k \le n$.

Theorem 3.1 (Kummer, 1852). *The power of a prime p that divides the binomial coefficient $\binom{n}{k}$ is given by the number of carries when we add k and $n-k$ in base p.*

Theorem 3.2 (Legendre, 1830). *We have* $\nu_p(\binom{n}{k}) = \frac{n-d_p(n)}{p-1} - \frac{k-d_p(k)}{p-1} - \frac{n-k-d_p(n-k)}{p-1} = \frac{d_p(k)+d_p(n-k)-d_p(n)}{p-1}$. *In particular,* $\nu_2(\binom{n}{k}) = d_2(k) + d_2(n-k) - d_2(n)$ *represents the carry count in the addition of k and $n-k$ in base 2.*

From now on M and N will denote integers such that $0 \le M \le N$.

Theorem 3.3 (Lucas, 1877). *Let* $N = (n_d, \ldots, n_1, n_0)_p = n_0 + n_1 p + \cdots + n_d p^d$ *and* $M = m_0 + m_1 p + \cdots + m_d p^d$ *with* $0 \le n_i, m_i \le p-1$ *for each i, be the base p representations of N and M, respectively. Then*

$$\binom{N}{M} \equiv \binom{n_0}{m_0}\binom{n_1}{m_1}\cdots\binom{n_d}{m_d} \bmod p.$$

Lucas' theorem has some remarkable extensions.

Theorem 3.4 (Anton, 1869, Stickelberger, 1890, Hensel, 1902). *Let* $N = (n_d, \ldots, n_1, n_0)_p = n_0 + n_1 p + \cdots + n_d p^d, M = m_0 + m_1 p + \cdots + m_d p^d$ *and* $R = N \quad M = r_0 + r_1 p + \cdots + r_d p^d$ *with* $0 \le n_i, m_i, r_i \le p-1$ *for each i, be the base p representations of* N, M*, and* $R = N - M$*, respectively. Then with* $q = \nu_p(\binom{N}{M})$,

$$(-1)^q \frac{1}{p^q}\binom{N}{M} \equiv \left(\frac{n_0!}{m_0!r_0!}\right)\left(\frac{n_1!}{m_1!r_1!}\right)\cdots\left(\frac{n_d!}{m_d!r_d!}\right) \bmod p.$$

An immediate consequence of this theorem is

Corollary 3.5. *For* $(a,p) = 1, b \ge 1, n \ge 1$, *with* $q = \nu_p(\binom{2a}{a})$, $t = \nu_p(\binom{2b}{b})$, *and* $t \le \lfloor \log_p 2b \rfloor < n$, *we have that*

$$c_{ap^n+b} = \binom{2ap^n + 2b}{ap^n + b} \equiv \binom{2a}{a}\binom{2b}{b} \bmod p^{q+t+1}.$$

This congruence also holds for $b = 0$ *with* $n \ge 0$.

Davis and Webb (1990) and Granville (1995) have independently generalized Lucas' theorem and its extension Theorem 3.4. Here we use the latter version whose strength is in the choice of modulus which can be arbitrarily large. For a given integer n and prime p, we define $(n!)_p = n!/(p^{\lfloor n/p \rfloor}\lfloor n/p \rfloor!)$ to be the product of positive integers not exceeding n and not divisible by p, and which is closely related to the p-adic Morita gamma function.

Theorem 3.6. (Granville, 1995 in [6]) *Let* $N = (n_d, \dots, n_1, n_0)_p = n_0 + n_1 p + \dots + n_d p^d, M = m_0 + m_1 p + \dots + m_d p^d$ *and* $R = N - M = r_0 + r_1 p + \dots + r_d p^d$ *with* $0 \le n_i, m_i, r_i \le p-1$ *for each* i*, be the base* p *representations of* N, M*, and* $R = N - M$*, respectively. Let* $N_j = n_j + n_{j+1}p + \dots + n_{j+k-1}p^{k-1}$ *for each* $j \ge 0$*, i.e., the least positive residue of* $\lfloor N/p^j \rfloor \bmod p^k$ *with some integer* $k \ge 1$*; also make the corresponding definitions for* M_j *and* R_j*. Let* ε_j *be the number of carries when adding* M *and* R *on and beyond the* j*th digit. Then with* $q = \varepsilon_0 = \nu_p(\binom{N}{M})$*,*

$$\frac{1}{p^q}\binom{N}{M} \equiv (\pm 1)^{\varepsilon_{k-1}} \left(\frac{(N_0!)_p}{(M_0!)_p (R_0!)_p}\right)\left(\frac{(N_1!)_p}{(M_1!)_p (R_1!)_p}\right) \cdots \left(\frac{(N_d!)_p}{(M_d!)_p (R_d!)_p}\right) \bmod p^k$$

where ± 1 *is* -1 *except if* $p = 2$ *and* $k \ge 3$.

We also need another related theorem.

Theorem 3.7. (Granville, 1995, (39) in [6]) *We have*

$$(p^{K+1}!)_p \equiv (p^K!)_p^p \bmod p^{3K+1}, \quad \text{for } p \ge 5, \text{and}$$
$$(p^{K+1}!)_p \equiv (p^K!)_p^p \bmod p^{3K-1}, \quad \text{for } p = 2,3 \text{ except if } p^K = 2.$$

The following generalization of the Jacobstahl–Kazandzidis [2] congruences will be helpful in proving Theorem 2.1 and congruence (4.5).

Theorem 3.8. (Corollary 11.6.22 [2]) *Let* M *and* N *such that* $0 \le M \le N$ *and* p *prime. We have*

$$\binom{pN}{pM} \equiv \begin{cases} \left(1 - \frac{B_{p-3}}{3} p^3 NM(N-M)\right)\binom{N}{M} & \bmod p^4 NM(N-M)\binom{N}{M}, \text{ if } p \ge 5, \\ (1 + 45NM(N-M))\binom{N}{M} & \bmod p^4 NM(N-M)\binom{N}{M}, \text{ if } p = 3, \\ (-1)^{M(N-M)} P(N,M)\binom{N}{M} & \bmod p^4 NM(N-M)\binom{N}{M}, \text{ if } p = 2, \end{cases}$$

where $P(N,M) = 1 + 6NM(N-M) - 4NM(N-M)(N^2 - NM + M^2) + 2(NM(N-M))^2$.

We note that a stronger version of Corollary 3.5 also follows by Theorem 3.8.

Corollary 3.9. *For* $(a,p) = 1, b \ge 1, n \ge 1$*, with* $q = \nu_p(\binom{2a}{a})$*,* $t = \nu_p(\binom{2b}{b})$*, and* $\lfloor \log_p 2b \rfloor + h(p) < n$*, we have that*

$$c_{ap^n+b} = \binom{2ap^n + 2b}{ap^n + b} \equiv \binom{2a}{a}\binom{2b}{b} \bmod p^{q+t+h(p)},$$

where

$$h(p) = \begin{cases} 3 + \nu_p(B_{p-3}), & \text{if } p \ge 5, \\ 2, & \text{if } p = 3, \\ 1, & \text{if } p = 2. \end{cases}$$

The congruence also holds for $b = 0$ *with* $n \ge 1$*. The exponents of* p *are best possible in the sense that* $\nu_p(\binom{2ap^n+2b}{ap^n+b} - \binom{2a}{a}\binom{2b}{b}) = q + t + h(p)$.

4. Proofs and More Results

This section contains the proofs of the main theorems (Theorems 2.1, 2.2, 2.4, and 2.11) and present some additional results.

To shed light on the nature of the problems at hand, we note that by Legendre's theorem $\nu_2(c_k) = d_2(k)$, i.e., $\nu_2(c_{2^{n+1}}) = \nu_2(c_{2^n}) = 1$. In a similar fashion,

$$\nu_2(C_k) = d_2(k) - \nu_2(k+1) = d_2(k+1) - 1, \tag{4.1}$$

i.e., $\nu_2(C_{2^{n+1}}) = \nu_2(C_{2^n}) = 1$. It also follows that C_k is odd if and only if $k = 2^q - 1$ for some integer $q \geq 0$.

A natural way to start proving Theorem 2.1 is to apply a generalization of Lucas' theorem. However, it turns out that it will provide only lower bounds on $\nu_p(c_{ap^{n+1}} - c_{ap^n})$ although it will suffice to prove Theorem 2.2. For example, with $p = 2, a = 1, q = 1, K = n+1$, and $k = 3(n+1)$, and combining Theorems 3.6 and 3.7, it follows that

$$\nu_2(c_{2^{n+1}} - c_{2^n}) \geq 3(n+1), n \geq 1.$$

In fact, we set $N = 2^{n+2}$ and $M = 2^{n+1}$ and obtain that

$$\frac{1}{2}\binom{2^{n+2}}{2^{n+1}} \equiv \frac{(2^{n+2}!)_2}{(2^{n+1}!)_2^2} \prod_{j=1}^{n+1} \left(\frac{(2^{n+2-j}!)_2}{(2^{n+1-j}!)_2^2} \right) \bmod 2^{3(n+1)}$$

and

$$\frac{1}{2}\binom{2^{n+1}}{2^n} \equiv \prod_{j=1}^{n+1} \left(\frac{(2^{n+2-j}!)_2}{(2^{n+1-j}!)_2^2} \right) \bmod 2^{3(n+1)}.$$

From here on, o_i, e_i, and $a_i, i \geq 1$, denote odd, even, or arbitrary non-negative integers, respectively, whose actual values are of no significance. By Theorem 3.7, we get

$$c_{2^{n+1}} - c_{2^n} = \binom{2^{n+2}}{2^{n+1}} - \binom{2^{n+1}}{2^n} \equiv 2\left(\frac{(2^{n+2}!)_2}{(2^{n+1}!)_2^2} - 1 \right) o_1 \equiv$$
$$\equiv 2 \cdot 2^{3(n+1)-1} a_1 \bmod 2^{3(n+1)+1}. \tag{4.2}$$

It turns out that a_1 is always an odd number as it can be derived by Theorem 3.8. In fact, this will be already sufficient to obtain a full

Proof of Theorem 2.1. We set $N = 2ap^n$ and $M = ap^n$ and proceed as follows. If $p \geq 5$ then we get that

$$\binom{pN}{pM} - \binom{N}{M} \equiv -\frac{B_{p-3}}{3} p^3 NM(N-M) \binom{N}{M} \bmod p^4 NM(N-M) \binom{N}{M};$$

thus, $\nu_p(c_{ap^{n+1}} - c_{ap^n}) = 3(n+1) + \nu_p(\binom{2a}{a}) + \nu_p(B_{p-3})$ by the assumption on $\nu_p(B_{p-3})$ and $\nu_p(\binom{N}{M}) = \nu_p(\binom{2a}{a})$.

If $p = 3$ then

$$\binom{pN}{pM} - \binom{N}{M} \equiv 5 \cdot 3^2 \cdot NM(N-M)\binom{N}{M} \bmod p^4 NM(N-M)\binom{N}{M};$$

thus, $\nu_3(c_{a3^{n+1}} - c_{a3^n}) = 3(n+1) - 1 + \nu_p(\binom{2a}{a})$.

If $p = 2$ and $n \geq 1$ then

$$\binom{pN}{pM} - \binom{N}{M} \equiv 2 \cdot 3 \cdot NM(N-M)\binom{N}{M} \bmod p^4 NM(N-M)\binom{N}{M};$$

thus, $\nu_2(c_{a2^{n+1}} - c_{a2^n}) = 1 + (n+1) + n + n + \nu_2(\binom{2a}{a}) = 3(n+1) - 1 + d_2(a)$. Note that here we need $n \geq 1$ otherwise the factor $(-1)^{M(N-M)}$ evaluates to -1 in Theorem 3.8. □

Proof of Theorem 2.2. For $p = 2$ we can prove the statement by Theorem 3.8 but in fact, Theorem 3.7 suffices. As in (4.2), we get

$$\begin{aligned} C_{2^{n+1}} - C_{2^n} &= \frac{c_{2^{n+1}}}{2^{n+1}+1} - \frac{c_{2^n}}{2^n+1} \equiv \left(\frac{1}{2^{n+1}+1} 2 \frac{(2^{n+2}!)_2}{(2^{n+1}!)_2^2} - \frac{1}{2^n+1} 2\right) o_1 \\ &\equiv 2\left(\frac{1 + a_2 2^{3(n+1)-1}}{2^{n+1}+1} - \frac{1}{2^n+1}\right) o_1 \equiv o_3 2^{n+1} \bmod 2^{3(n+1)}. \end{aligned}$$

If $p \geq 3$ then we cannot apply Theorem 3.7. However, by taking all factors in the numerator and denominator modulo p^{n+1}, we get that

$$\frac{((2ap^{n+1})!)_p}{((ap^{n+1})!)_p^2} \equiv 1 \bmod p^{n+1}.$$

By the repeated application of the definition of $(m!)_p$, we get that

$$\binom{2ap^n}{ap^n} = \frac{((2ap^n)!)_p}{((ap^n)!)_p^2} \binom{2ap^{n-1}}{ap^{n-1}} = \ldots = \prod_{j=1}^{n} \frac{((2ap^j)!)_p}{((ap^j)!)_p^2} \cdot \binom{2a}{a},$$

and thus,

$$\binom{2ap^n}{ap^n} = f^* p^q \tag{4.3}$$

with $q = \nu_p(\binom{2a}{a})$ and

$$f^* = \frac{\binom{2a}{a}}{p^q} \prod_{j=1}^{n} \frac{((2ap^j)!)_p}{((ap^j)!)_p^2}.$$

(This also follows by Theorem 3.6 applied with a sufficiently large k.) We set $f' = a(1-p)f^*$ and observe that $\nu_p(f^*) = \nu_p(f') = 0$ and

$$\begin{aligned} C_{ap^{n+1}} - C_{ap^n} &= \frac{c_{ap^{n+1}}}{ap^{n+1}+1} - \frac{c_{ap^n}}{ap^n+1} \equiv p^q\left(\frac{1}{ap^{n+1}+1}\frac{((2ap^{n+1})!)_p}{((ap^{n+1})!)_p^2} - \frac{1}{ap^n+1}\right)f^* \\ &\equiv p^q\left(\frac{1+a_3p^{n+1}}{ap^{n+1}+1} - \frac{1}{ap^n+1}\right)f^* \equiv \frac{a(1-p)+a_4p}{(ap^{n+1}+1)(ap^n+1)}f^*p^{n+q} \\ &\equiv f'p^{n+q} \bmod p^{n+1+q}, \end{aligned} \tag{4.4}$$

and the proof is complete. □

Remark 4.1. *Theorems 2.1 and 2.2 can be viewed as corollaries of Theorem 3.8. In fact, by using Theorem 3.8 we can improve (4.4) if we replace the term a_3p^{n+1} by a_5p^{3n} (although a_5p^{2n} suffices) and we obtain*

$$C_{ap^{n+1}} - C_{ap^n} \equiv f'p^{n+q} \bmod p^{2n+q}. \tag{4.5}$$

Proof of Theorem 2.4. First we prove the lower bounds then we derive the exact orders by applying a recurrence. We observe for $M = N/2$ that

$$\binom{pN+2b}{pM+b} = \binom{pN}{pM}\frac{(pN+1)(pN+2)\cdots(pN+2b)}{\left((pM+1)(pM+2)\cdots(pM+b)\right)^2} \tag{4.6}$$

and

$$\binom{N+2b}{M+b} = \binom{N}{M}\frac{(N+1)(N+2)\cdots(N+2b)}{\left((M+1)(M+2)\cdots(M+b)\right)^2}. \tag{4.7}$$

We use the setting $N = 2ap^n$ and $M = ap^n$ from the proof of Theorem 2.1 and take the factors of the second factor modulo p^n on the right-hand side in the above equations (4.6) and (4.7) for $2 \le 2b < p$ (and thus, $r = \lfloor \log_p 2b \rfloor = 0$ and $t = \nu_p(\binom{2b}{b}) = 0$). By Theorem 2.1, with some f_i so that $\nu_p(f_i) \ge 0$ for each $i \ge 1$, and $q = \nu_p(\binom{N}{M}) = \nu_p(\binom{2a}{a})$, we get that

$$\begin{aligned} c_{ap^{n+1}+b} - c_{ap^n+b} &= \binom{pN+2b}{pM+b} - \binom{N+2b}{M+b} \\ &= \binom{pN}{pM}\binom{2b}{b}\left(1+f_1p^n\right) - \binom{N}{M}\binom{2b}{b}\left(1+f_2p^n\right) \\ &= \binom{N}{M}\left(1+f_3p^{3n}\right)\binom{2b}{b}\left(1+f_1p^n\right) - \binom{N}{M}\binom{2b}{b}\left(1+f_2p^n\right) \\ &\equiv 0 \bmod p^{n+q} \end{aligned}$$

which already guarantees that $\nu_p(c_{ap^{n+1}+b} - c_{ap^n+b}) \ge n + \nu_p(\binom{2a}{a})$ for $1 \le b \le (p-1)/2$.

For larger values of b, with some modifications that also allow us to handle terms of the second factor in (4.6) and (4.7) that are multiples of p, we get that

$$\begin{aligned} c_{ap^{n+1}+b} - c_{ap^n+b} &= \binom{pN+2b}{pM+b} - \binom{N+2b}{M+b} \\ &= \binom{N}{M}\left(1+f_3p^{3n}\right)\binom{2b}{b}\left(1+f_4p^{n-r}\right) \\ &- \binom{N}{M}\binom{2b}{b}\left(1+f_5p^{n-r}\right) \\ &\equiv 0 \bmod p^{n+q+t-r} \end{aligned} \tag{4.8}$$

with $r = \lfloor \log_p 2b \rfloor < n$. Indeed, if $i : 1 \leq i \leq 2b$, $s(i) = \nu_p(i) \leq r < n$, and $t = \nu_p(\binom{2b}{b})$ then e.g., with some f_i' and f_i" so that $\nu_p(f_i'), \nu_p(f_i") \geq 0$ for each $i \geq 1$, we obtain that

$$\frac{(N+1)(N+2)\cdots(N+2b)}{((M+1)(M+2)\cdots(M+b))^2} = \frac{\prod_{i=1}^{2b} p^{s(i)}(Np^{-s(i)}+ip^{-s(i)})}{(\prod_{i=1}^{b} p^{s(i)}(Mp^{-s(i)}+ip^{-s(i)}))^2} =$$

$$p^t\prod_{i=1}^{b}(1+f_i'p^{n-s(i)})\frac{\prod_{i=b+1}^{2b}(f_i"p^{n-s(i)}+ip^{-s(i)})}{\prod_{i=1}^{b}(f_i"p^{n-s(i)}+ip^{-s(i)})} = p^t\frac{\binom{2b}{b}}{p^t}(1+f_6p^{n-r}) \tag{4.9}$$

with r being the maximum value of $s(i), 1 \leq i \leq 2b$.

It follows that

$$\nu_p(c_{ap^{n+1}+b} - c_{ap^n+b}) \geq n + \nu_p\left(\binom{2a}{a}\right) + \nu_p\left(\binom{2b}{b}\right) - \lfloor \log_p 2b \rfloor$$

for $b \geq 1$. Meanwhile we also proved another useful version of Corollary 3.5 which involves n in the exponent of the modulus.

Corollary 4.1. *For* $(a,p) = 1, b \geq 1, n \geq 1$, *with* $q = \nu_p(\binom{2a}{a})$, $t = \nu_p(\binom{2b}{b})$, *and* $r = \lfloor \log_p 2b \rfloor < n$, *we have that*

$$c_{ap^n+b} = \binom{2ap^n+2b}{ap^n+b} \equiv \binom{2ap^n}{ap^n}\binom{2b}{b} \bmod p^{n+q+t-r}.$$

To find the exact order $\nu_p(\Delta_b)$ of $\Delta_b = c_{ap^{n+1}+b} - c_{ap^n+b}$, another approach is suggested by the part of the statement that the p-adic order only slightly depends on b in the given range. We set

$$\begin{aligned}\Delta_{b+1} &= c_{ap^{n+1}+b+1} - c_{ap^n+b+1} = \binom{2ap^{n+1}+2b+2}{ap^{n+1}+b+1} - \binom{2ap^n+2b+2}{ap^n+b+1}\\ &= \binom{2ap^{n+1}+2b}{ap^{n+1}+b}\frac{(2ap^{n+1}+2b+1)(2ap^{n+1}+2b+2)}{(ap^{n+1}+b+1)^2}\\ &- \binom{2ap^n+2b}{ap^n+b}\frac{(2ap^n+2b+1)(2ap^n+2b+2)}{(ap^n+b+1)^2}\\ &= 2\Delta_b\frac{2ap^{n+1}+2b+1}{ap^{n+1}+b+1} + \binom{2ap^n+2b}{ap^n+b}\frac{2a(p-1)p^n}{(ap^{n+1}+b+1)(ap^n+b+1)}\end{aligned}$$

which yields

$$\frac{\Delta_{b+1}}{2^{b+1}} = \frac{\Delta_b}{2^b}\frac{2ap^{n+1}+2b+1}{ap^{n+1}+b+1} + \binom{2ap^n+2b}{ap^n+b}\frac{ap^n(p-1)}{2^b(ap^{n+1}+b+1)(ap^n+b+1)}. \tag{4.10}$$

In order to determine $\nu_p(\Delta_b)$, we first deal with the case $1 \le b \le (p-1)/2+1$. The case with $p = 2$, a odd, and $b = 1$ immediately follows by setting b to 0 in (4.10) and using Theorem 2.1 on $\nu_2(\Delta_0)$.

For $p \ge 3$, Theorem 2.1 already yields $\nu_p(\Delta_0)$, $\nu_p(\Delta_1) = n + \nu_p(\binom{2a}{a})$, and thus, $\nu_p(\Delta_b) \ge n + \nu_p(\binom{2a}{a})$. Also, if we can justify that $\nu_p(\Delta_{(p-1)/2}) = n + \nu_p(\binom{2a}{a})$ then identity (4.10) implies that $\nu_p(\Delta_{(p-1)/2+1}) = n + \nu_p(\binom{2a}{a})$, too, since now $\nu_p(2b+1) = 1$. The latter comment will take care of the largest value of b in the given range.

By the recurrence (4.10), we obtain that

$$\frac{\Delta_{b+1}}{2^{b+1}} = \sum_{i=0}^{b}\binom{2ap^n+2i}{ap^n+i}\frac{ap^n(p-1)}{2^i(ap^{n+1}+i+1)(ap^n+i+1)}\prod_{j=i+1}^{b}\frac{2ap^{n+1}+2j+1}{ap^{n+1}+j+1} \tag{4.11}$$

$$+\,\Delta_0\prod_{j=0}^{b}\frac{2ap^{n+1}+2j+1}{ap^{n+1}+j+1}.$$

We start with the first summand on the right-hand side. The first factor $\binom{2ap^n+2i}{ap^n+i}, 0 \le i \le b$, can be handled by Corollary 4.1, and it contributes a constant factor $\binom{2ap^n}{ap^n}$ with p-adic order $q = \nu_p(\binom{2ap^n}{ap^n}) = \nu_p(\binom{2a}{a})$, and a factor $\binom{2i}{i} \bmod p^m$ with $m = n + t - r \le n$ (since $t \le r$) and any large enough n so that $\max\{0, \nu_p(f(b))\} < m \le n$. (We can choose $m = 1$ if $\nu_p(f(b)) \le 0$, which in fact means $\nu_p(f(b)) = 0$ in the range $1 \le b \le (p-1)/2$.)

The second factor contributes an extra p-adic order n. We factor out the constant factors (i.e., those that are independent of i) including $a(p-1)$ and the one with p at the exponent $n+q$. After this, we take everything modulo p^m and get

$$\sum_{i=0}^{b}\binom{2i}{i}\frac{1}{2^i(i+1)^2}\prod_{j=i+1}^{b}\frac{2j+1}{j+1}=\frac{\binom{2b+2}{b+1}}{2^{b+1}}\sum_{i=0}^{b}\frac{1}{(2i+1)(i+1)}$$
$$=\frac{\binom{2b+2}{b+1}}{2^b}(H_{2b+2}-H_{b+1}). \tag{4.12}$$

Finally, the second summand on the right-hand side of (4.11) has a p-adic order larger than the other terms by Theorem 2.1. We cross multiply with 2^{b+1} to complete the proof for $\nu_p(\Delta_{b+1})$.

Note that n can be selected so that $n \geq n_0 = \nu_p(f(b))+1$ to get $\nu_p(\Delta_b)$.

From now on in this section $f_k, k \geq 7$, with $\nu_p(f_k) \geq 0$, may depend on i and j. For larger values of b, i.e., if $b \geq (p-1)/2+2$, the equation (4.12) turns into

$$\sum_{i=0}^{b}\binom{2i}{i}\left(1+f_7p^{n-r}\right)\frac{1}{2^i(i+1)^2}\left(1+f_8p^{n-r}\right)\prod_{j=i+1}^{b}\frac{2j+1}{j+1}\left(1+f_9p^{n-r}\right)$$
$$=\frac{\binom{2b+2}{b+1}}{2^{b+1}}\sum_{i=0}^{b}\left(\frac{1}{(2i+1)(i+1)}\left(1+f_{10}p^{n-r}\right)\right)$$
$$=\frac{\binom{2b+2}{b+1}}{2^b}(H_{2b+2}-H_{b+1})+f_{11}p^{n-r-r'+\nu_p\left(\binom{2b+2}{b+1}\right)} \tag{4.13}$$

with $r' = \lfloor \log_p(2b+1)\rfloor$ via Corollary 4.1 and in a similar fashion to the case with smaller bs and using that $\nu_p((2i+1)(i+1)) \leq \lfloor \log_p(2b+1)\rfloor = r'$ for $2i+1 \leq 2b+1$. Therefore, $n \geq n_0 = \max\{\nu_p(f(b))+2r-\nu_p(\binom{2b}{b})+1, r+1\}$ suffices in order to get $\nu_p(\Delta_b)$.

Note that in (4.13) we used the fact that we could write e.g.,

$$\frac{1}{ap^n+i+1}=\frac{1}{p^{s(i+1)}(ap^{n-s(i+1)}+(i+1)p^{-s(i+1)})}$$
$$=\frac{1}{p^{s(i+1)}}\left(\frac{1}{(i+1)p^{-s(i+1)}}+f_{12}p^{n-s(i+1)}\right)$$
$$=\frac{1}{(i+1)}\left(1+f_{13}p^{n-s(i+1)}\right)$$

which guarantees that in the range $0 \leq i \leq b$, $b \geq 1$, we get

$$\frac{1}{ap^n+i+1}=\frac{1}{(i+1)}\left(1+f_{14}p^{n-r}\right).$$

Finally, for $p = 2$ and $1 \le b < 2^n$, i.e., $n \ge \lfloor \log_2 2b \rfloor$, equations (4.11), (4.12), (4.13) and the above argument with some modifications yield that

$$\nu_2(\Delta_b) = n + \nu_2\left(\binom{2a}{a}\right) + \nu_2(f(b)) = n + d_2(a) + d_2(b) - \lfloor \log_2 b \rfloor.$$

□

We can modify the above proof and obtain the

Proof of Theorem 2.11. The proof is similar to that of Theorem 2.4. We set $N = 2ap^n$ and $M = ap^n$ again and rewrite (4.8) for $b \ge 1$,

$$\begin{aligned}
\Delta_b^* - \frac{C_{ap^{n+1}+b}}{ap^{n+1}+b+1} - \frac{C_{ap^n+b}}{ap^n+b+1} &= \frac{\binom{pN+2b}{pM+b}}{pM+b+1} - \frac{\binom{N+2b}{M+b}}{M+b+1} \\
&= \binom{N}{M}\left(1+f_3p^{3n}\right)\binom{2b}{b}\left(1+f_4p^{n-r}\right)\frac{1}{b+1}\left(1+f_{15}p^{n-s(b+1)}\right) \\
&\quad - \binom{N}{M}\binom{2b}{b}\left(1+f_5p^{n-r}\right)\frac{1}{b+1}\left(1+f_{16}p^{n-s(b+1)}\right) \\
&\equiv 0 \bmod p^{n+q+t-r-\nu_p(b+1)}
\end{aligned}$$

which guarantees the inequality on the order $\nu_p(\Delta_b^*)$.

Now we also have

$$\begin{aligned}
\Delta_{b+1}^* &= \frac{C_{ap^{n+1}+b+1}}{ap^{n+1}+b+2} - \frac{C_{ap^n+b+1}}{ap^n+b+2} = \frac{\binom{2ap^{n+1}+2b+2}{ap^{n+1}+b+1}}{ap^{n+1}+b+2} - \frac{\binom{2ap^n+2b+2}{ap^n+b+1}}{ap^n+b+2} \\
&= \binom{2ap^{n+1}+2b}{ap^{n+1}+b}\frac{(2ap^{n+1}+2b+1)(2ap^{n+1}+2b+2)}{(ap^{n+1}+b+1)^2(ap^{n+1}+b+2)} \\
&\quad - \binom{2ap^n+2b}{ap^n+b}\frac{(2ap^n+2b+1)(2ap^n+2b+2)}{(ap^n+b+1)^2(ap^n+b+2)} \\
&= 2\Delta_b^*\frac{2ap^{n+1}+2b+1}{ap^{n+1}+b+2} + \frac{\binom{2ap^n+2b}{ap^n+b}}{ap^n+b+1}\frac{6a(p-1)p^n}{(ap^{n+1}+b+2)(ap^n+b+2)}.
\end{aligned}$$

It follows that

$$\frac{\Delta_{b+1}^*}{2^{b+1}} = \frac{\Delta_b^*}{2^b}\frac{2ap^{n+1}+2b+1}{ap^{n+1}+b+2} + \frac{\binom{2ap^n+2b}{ap^n+b}}{ap^n+b+1}\frac{3ap^n(p-1)}{2^b(ap^{n+1}+b+2)(ap^n+b+2)} \tag{4.14}$$

which can be rewritten as

$$\frac{\Delta_{b+1}^*}{2^{b+1}} = \sum_{i=0}^{b} \frac{\binom{2ap^n+2i}{ap^n+i}}{ap^n+i+1} \frac{3ap^n(p-1)}{2^i(ap^{n+1}+i+2)(ap^n+i+2)} \prod_{j=i+1}^{b} \frac{2ap^{n+1}+2j+1}{ap^{n+1}+j+2} \tag{4.15}$$

$$+ \Delta_0^* \prod_{j=0}^{b} \frac{2ap^{n+1}+2j+1}{ap^{n+1}+j+2}.$$

We set $q = \nu_p(\binom{2ap^k}{ap^k}) = \nu_p(\binom{2a}{a}), k \geq 0$. Since $\nu_p(\Delta_0^*) = n+q$ by (4.4), the terms in (4.15) are of similar magnitude and it requires a more careful analysis than the treatment of (4.11) did.

We start with the second summand on the right-hand side of (4.15) and obtain

$$\Delta_0^* \equiv p^{n+q} a(1-p) f^* \bmod p^{2n+q}$$

by the congruence (4.5). Now we use Corollary 4.1 and equation (4.3).

If $p \neq 2$ then

$$\Delta_{b+1}^* = 2^{b+1} p^{n+q} a(p-1) f^* \left(3 \sum_{i=0}^{b} \left(\frac{\binom{2i}{i}}{i+1} \frac{1}{2^i(i+2)^2} \prod_{j=i+1}^{b} \frac{2j+1}{j+2} \left(1 + f_{17} p^{n-r}\right)\right)\right.$$
$$\left. - \prod_{j=0}^{b} \frac{2j+1}{j+2} \left(1 + f_{18} p^{n-r}\right)\right).$$

With some simplifications we obtain

$$\Delta_{b+1}^* = p^{n+q} a(p-1) f^* \frac{\binom{2b+2}{b+1}}{b+2} \left(\left(3 \sum_{i=0}^{b} \frac{1}{(i+2)(2i+1)}\right) - 1 \right) + f_{19} p^{2n+q-r-r'+\nu_p(C_{b+1})} \tag{4.16}$$

with $r' = \lfloor \log_p(2b+1) \rfloor$, and we are done since the product of the last two factors of the first term on the right-hand side of (4.16) simplifies to $g(b+1) = 2\binom{2b+2}{b+1}(b+2)^{-1}\left(H_{2b+2} - H_{b+1} - 1/(2(b+2))\right)$.

Similarly to the proof of Theorem 2.4, although now using the extra factor 3 in (4.16), we also get that $n \geq n_0 = \max\{\nu_p(g(b)) + 2r - \nu_p(C_b) + 1, r+1\}$ suffices in order to get $\nu_p(\Delta_b^*)$.

Note that the case with $p = 2$, a odd, and $b = 1$ immediately follows by setting b to 0 in (4.14) and $g(b+1)$ which is $g(1) = 1/2$. The 2-adic error term in (4.16) comes with the modified exponent $2n + q - 2$.

The general case with $p = 2$ and $n \geq \lfloor \log_2 2b \rfloor + 1$ follows in a similar manner. □

5. Other Proofs

Now we prove some p-adic properties of the differences of the harmonic numbers $H_{2b} - H_b$ (cf. Theorems 2.5, 2.7, and 2.8) and a related quantity (cf. (5.3)). Although these properties are not essential to other parts of the paper, they are important in the actual use of Theorem 2.4.

The proof of Theorem 2.5. For $p = 2$, the definition obviously yields that $\nu_2(f(c)) = 1 + d_2(c) - \lfloor \log_2 2c \rfloor$. If $p \geq 3$ then we proceed with induction on k. Note that

$$f(dp) = 2\binom{2dp}{dp}(H_{2dp} - H_{dp}).$$

This implies that

$$f(dp) = 2\binom{2dp}{dp}\left(\frac{1}{p}(H_{2d} - H_d) + \sum_{\substack{i=dp+1\\(i,p)=1}}^{2dp} \frac{1}{i}\right), \tag{5.1}$$

and clearly,

$$\nu_p\left(\sum_{\substack{i=dp+1\\(i,p)=1}}^{2dp} \frac{1}{i}\right) \geq 0.$$

For $k = 1$, we set $d = 1$ and the statement follows. For $k \geq 2$ we have two cases.
Since $\nu_p(\binom{2dp}{dp}/\binom{2d}{d}) = 0$, by identity (5.1) we get that

$$\nu_p(f(dp)) = \nu_p\left(\binom{2dp}{dp}\right) + \nu_p\left(\frac{1}{p}(H_{2d} - H_d)\right) = \nu_p(f(d)) - 1 \tag{5.2}$$

provided that $\nu_p(H_{2d} - H_d) \leq 0$. By the inductive hypothesis, with $d = p^{k-1}$, we have that $\nu_p(f(p^{k-1})) = -(k-1), k \geq 2$, which yields $\nu_p(f(p^k)) = -k$ by recurrence (5.2). Similarly, with $d = cp^{k-1}, c \geq 1$ and $k \geq 1$, we obtain $\nu_p(f(cp^k)) = -k + \nu_p(f(c))$ for $k \geq 0$ provided that $\nu_p(H_{2c} - H_c) \leq 0$. □

Note that finer details of divisibility by p can be recovered from

$$\nu_p\left(H_{(a+1)p^n} - H_{ap^n} - \frac{1}{p}\left(H_{(a+1)p^{n-1}} - H_{ap^{n-1}}\right)\right) \geq 2n,$$

for $p \geq 5$ prime, $n \geq 1$, and $a \geq 0$, e.g., if $a = 1$, cf. [7, p245: Solution to Problem 1997, B3]. In fact, more can be said according to Theorem 5.2 as stated in Theorem 5.1.

Theorem 5.1. *For $p \geq 3$ prime, $a \geq 0$, and $n \geq 1$, we have that*

$$\nu_p\left(H_{(a+1)p^n} - H_{ap^n} - \frac{1}{p}\left(H_{(a+1)p^{n-1}} - H_{ap^{n-1}}\right)\right) = 2n + \nu_p(2a+1) - \chi_{p=3}. \tag{5.3}$$

This follows by

Theorem 5.2. (Theorem 5.2 in [1]) *Let p be an odd prime. Then there is a sequence $c_k \in \mathbb{Q}_p$ such that, for all $m \geq 1$,*

$$H_{pm} - \frac{1}{p}H_m = \sum_{k=1}^{\infty} c_k p^{2k} m^{2k},$$

where the series converges in p-adic norm. The c_k are p-adic integers unless $(p-1)|2k$ or $p|k$. In general, $\nu_p(c_k) = -1 + \nu_p(1/k)$ if $(p-1)|2k$, and $\nu_p(c_k) = \nu_p(1/k)$ otherwise.

Proof of Theorem 5.1. By setting $m = bp^{n-1}, n \geq 1$, and $b \geq 1$ in Theorem 5.2, we obtain that

$$\nu_p(H_{bp^n} - \frac{1}{p}H_{bp^{n-1}}) = 2n + 2\nu_p(b) - \chi_{p=3}$$

and (5.3), too. □

Before we turn to the proof of Theorem 2.8, as a starting point, we include

The proof of Theorem 2.7 (attributed to J. Kürschák, cf. [1]). Let $m \geq 1$ be the largest integer so that 2^m divides k for some k with $b+1 \leq k \leq a$. We prove that there is a unique integer $k, b+1 \leq k \leq a$, for which $\nu_2(k) = m$ and thus, $\nu_2(H_a - H_b) = -m < 0$ already guaranteeing the statement. Assume to the contrary that there are two such integers: $2^m o_1 < 2^m o_2$ with odd integers o_1 and o_2. Clearly, $2^m(o_1+1)$ falls between $b+1$ and a and it has 2-adic order at least $m+1$ which contradicts the definition of m. □

The proof of Theorem 2.8. The case of $p = 2$ is included in Theorem 2.5. If $p = 3$ then we take the highest power of three not exceeding $2b$, i.e., let $3^m < 2b$. We prove that there is a unique term in the sum $\sum_{k=b+1}^{2b} \frac{1}{k}$ with k of 3-adic order $m = \lfloor \log_3 2b \rfloor$. In fact, for this term we have either $b+1 \leq 3^m < 2b \leq 2 \cdot 3^m$ or $3^m < b+1 \leq 2 \cdot 3^m \leq 2b < 3^{m+1}$.

If $p = 5$ then we take the highest power of five not exceeding $2b$, i.e., let $5^m < 2b$. We set $m = \lfloor \log_5 2b \rfloor$. We have the following cases:

$$\begin{array}{llll}
(1) & b+1 \leq & 5^m & < 2b < 2 \cdot 5^m, \\
(2) & 5^m < b+1 \leq 2 \cdot 5^m & & \leq 2b < 3 \cdot 5^m, \\
(3) & 5^m < b+1 \leq 2 \cdot 5^m < 3 \cdot 5^m & & < 2b < 4 \cdot 5^m, \\
(4) & 2 \cdot 5^m < b+1 \leq 3 \cdot 5^m < 4 \cdot 5^m & & \leq 2b < 5^{m+1}.
\end{array}$$

Clearly, in Cases (1), (2), and (4), we have $\nu_5(H_{2b} - H_b) = -\lfloor \log_5 2b \rfloor$ and thus, (2.1) holds.

In case (3) we have $m \geq 1$ and we face the problem that $\frac{1}{2\cdot 5^m} + \frac{1}{3\cdot 5^m} = \frac{1}{6\cdot 5^{m-1}}$ and other terms might also have the 5-adic order $-(m-1)$. Note, however, that this is exactly the case when there exists an m so that $\frac{3}{2}\cdot 5^m < b < 2\cdot 5^m$. We can finish the proof by proceeding with induction on m. The cases with $m = 1$ and 2 are easily tested.

We assume that $m \geq 3$ and write

$$H_{2b} - H_b = \sum_{k=b+1}^{2b} \frac{1}{k} = \sum_{\substack{k=b+1\\(k,5)=1}}^{2b} \frac{1}{k} + \frac{1}{5}\sum_{k=b'}^{b"} \frac{1}{k} \tag{5.4}$$

with $b' = \lceil (b+1)/5 \rceil$ and $b" = \lfloor 2b/5 \rfloor$. Clearly, $\lceil \frac{3}{2}\cdot 5^m \rceil \equiv 3 \bmod 5$ with $m \geq 1$. We have two cases.

Case 1. If $b \equiv 0, 1$ or $2 \bmod 5$ then $b" = 2b' - 2$ and $\frac{3}{2}\cdot 5^{m-1} < b' - 1 < 2\cdot 5^{m-1}$. Therefore,

$$\begin{aligned}\nu_5\left(\frac{1}{5}\sum_{k=b'}^{b"}\frac{1}{k}\right) &= \nu_5\left(\frac{1}{5}\sum_{k=(b'-1)+1}^{2(b'-1)}\frac{1}{k}\right) = -1 - \lfloor \log_5 2(b'-1)\rfloor + 1\\ &= -1 - (m-1) + 1 = -m + 1 = -\lfloor \log_5 2b \rfloor + 1\end{aligned}$$

by the induction hypothesis. Therefore, (5.4) guarantees the statement.

Case 2. If $b \equiv 3$ or $4 \bmod 5$ then we have two subcases.

Case 2.1. This case deals with b values within a distance of 2 of the boundaries in (3). If $b = \lceil \frac{3}{2}\cdot 5^m \rceil$ or $\lceil \frac{3}{2}\cdot 5^m \rceil + 1$ then $b' = \lceil \frac{3}{2}\cdot 5^{m-1} \rceil$, $b" = 2b' - 1$, and $\frac{3}{2}\cdot 5^{m-1} < b' < 2\cdot 5^{m-1}$ as well as $2b' \equiv 1 \bmod 5$ and hence, $\nu_5(2b') = 0$. Thus, we have

$$\begin{aligned}\nu_5\left(\frac{1}{5}\sum_{k=b'}^{b"}\frac{1}{k}\right) &= \nu_5\left(\frac{1}{5}\sum_{k=b'+1}^{2b'}\frac{1}{k} + \frac{1}{5}\frac{1}{b'} - \frac{1}{5}\frac{1}{2b'}\right) = \nu_5\left(\frac{1}{5}\sum_{k=b'+1}^{2b'}\frac{1}{k} + \frac{1}{5}\frac{1}{2b'}\right)\\ &= -1 - \lfloor \log_5 2b' \rfloor + 1 = -1 - (m-1) + 1 = -m + 1 = -\lfloor \log_5 2b \rfloor + 1\end{aligned}$$

by the induction hypothesis. In a similar fashion, if $b = 2\cdot 5^m - 1$ or $2\cdot 5^m - 2$ then $b' = 2\cdot 5^{m-1}$, $b" = 2b' - 1$, and $\frac{3}{2}\cdot 5^{m-1} < b' - 1 < 2\cdot 5^{m-1}$ as well as $2b' - 1 \equiv 4 \bmod 5$ and thus, $\nu_5(2b' - 1) = 0$. We have

$$\begin{aligned}\nu_5\left(\frac{1}{5}\sum_{k=b'}^{b"}\frac{1}{k}\right) &= \nu_5\left(\frac{1}{5}\sum_{k=(b'-1)+1}^{2(b'-1)}\frac{1}{k} + \frac{1}{5}\frac{1}{2b'-1}\right) = -1 - \lfloor \log_5 2(b'-1)\rfloor + 1\\ &= -1 - (m-1) + 1 = -m + 1 = -\lfloor \log_5 2b \rfloor + 1\end{aligned}$$

by the induction hypothesis.

`Case 2.2.` If $\lceil \frac{3}{2}\cdot 5^m\rceil + 1 < b < 2\cdot 5^m - 2$ then $b" = 2b' - 1$ and $\frac{3}{2}\cdot 5^{m-1} + 1 < b' < 2\cdot 5^{m-1}$, so both $b' - 1$ and b' will be in the right range to apply the induction hypothesis. We either have $\nu_5(2b'-1) = 0$ or $\nu_5(2b') = 0$. We proceed as in Case 2.1. In the former case, we get that

$$\nu_5\left(\frac{1}{5}\sum_{k=b'}^{b"}\frac{1}{k}\right) = \nu_5\left(\frac{1}{5}\sum_{k=(b'-1)+1}^{2(b'-1)}\frac{1}{k} + \frac{1}{5}\frac{1}{2b'-1}\right) = -1 - \lfloor \log_5 2(b'-1)\rfloor + 1$$
$$= -1 - (m-1) + 1 = -m + 1 = -\lfloor \log_5 2b\rfloor + 1$$

while in the latter case,

$$\nu_5\left(\frac{1}{5}\sum_{k=b'}^{b"}\frac{1}{k}\right) = \nu_5\left(\frac{1}{5}\sum_{k=b'+1}^{2b'}\frac{1}{k} + \frac{1}{5}\frac{1}{b'} - \frac{1}{5}\frac{1}{2b'}\right) = \nu_5\left(\frac{1}{5}\sum_{k=b'+1}^{2b'}\frac{1}{k} + \frac{1}{5}\frac{1}{2b'}\right)$$
$$= -1 - \lfloor \log_5 2b'\rfloor + 1 = -1 - (m-1) + 1 = -m + 1 = -\lfloor \log_5 2b\rfloor + 1$$

by the induction hypothesis.

The proof is complete by (5.4). □

Acknowledgements The author wishes to thank the referee and Gregory P. Tollisen for helpful comments and suggestions that improved the organization of the paper.

References

[1] D. W. Boyd, A p-adic study of the partial sums of the harmonic series, *Experiment. Math.* **3** (1994), 287–302.

[2] H. Cohen, Number Theory, vol 2: Analytic and Modern Tools, Graduate Texts in Mathematics, Springer, 2007.

[3] Emeric Deutsch and Bruce E. Sagan, Congruences for Catalan and Motzkin numbers and related sequences, *J. Number Theory* **117** (2006), 191–215.

[4] A. Eswarathasan and E. Levine, p-integral harmonic sums, *Discrete Math.* **91** (1991), 249–257.

[5] S.-P. Eu, S.-C. Liu, and Y.-N. Yeh, Catalan and Motzkin numbers modulo 4 and 8, *European J. Combin.* **29** (2008), 1449-1466.

[6] A. Granville, Arithmetic properties of binomial coefficients. I. Binomial coefficients modulo prime powers, in *Organic mathematics* (Burnaby, BC, 1995), volume 20 of CMS Conf. Proc., 253-276, Amer. Math. Soc., Providence, RI, 1997. Electronic version (a dynamic survey): `http://www.dms.umontreal.ca/~andrew/Binomial/`.

[7] K. S. Kedlaya, B. Poonen, and R. Vakil, The William Lowell Putnam Mathematical Competition, 1985–2000, Mathematical Association of America, Washington, DC, 2002.

[8] T. Lengyel, On divisibility properties of some differences of Motzkin numbers, manuscript, 2012, submitted to the XVth International Conference on Fibonacci Numbers and Their Applications, June 25-30, 2012, Eger, Hungary.

[9] T. Lengyel, Exact p-adic orders for differences of Motzkin numbers, manuscript, 2012.

[10] H.-Y. Lin, Odd Catalan numbers modulo 2^k, *INTEGERS* **11** (2011), #A55, 1–5.

[11] Shu-Chung Liu and Jean C.-C. Yeh, Catalan Numbers Modulo 2^k, *J. Integer Sequences* **13** (2010), Article 10.5.4.

[12] Andrew D. Loveless, A congruence for products of binomial coefficients modulo a composite, *INTEGERS* **7** (2007), #A44, 1–9.

[13] F. Luca and P. T. Young, On the binary expansion of the odd Catalan numbers, *Aportaciones Matematicas, Investigacion* **20** (2011), 185–190.

[14] G. Xin and J-F. Xu, A short approach to Catalan numbers modulo 2^r, *Elec. J. Combinatorics* **18** (2011), #P177, 1–12.

GENERALIZED BINOMIAL EXPANSIONS AND BERNOULLI POLYNOMIALS

Hieu D. Nguyen
Department of Mathematics, Rowan University, Glassboro, New Jersey
nguyen@rowan.edu

Received: 5/2/12, Revised: 1/4/13, Accepted: 2/28/13, Published: 3/15/13

Abstract

We investigate generalized binomial expansions that arise from two-dimensional sequences satisfying a broad generalization of the triangular recurrence for binomial coefficients. In particular, we present a new combinatorial formula for such sequences in terms of a 'shift by rank' quasi-expansion based on ordered set partitions. As an application, we give a new proof of Dilcher's formula for expressing generalized Bernoulli polynomials in terms of classical Bernoulli polynomials.

1. Introduction

The binomial expansion formula

$$(a+b)^n = \sum_{k=0}^{n} \binom{n}{k} a^k b^{n-k}, \tag{1}$$

where $\binom{n}{k}$ denotes the binomial coefficients given by the formula

$$\binom{n}{k} = \frac{n!}{k!(n-k)!} \quad (0 \le k \le n), \tag{2}$$

is well known to all mathematicians. In this paper, we consider instead a generalized binomial expansion of two sequences $a(k)$ and $b(k)$ where the first few cases are given by the quasi-expansion that we shall refer to as 'shift by rank':

$$\begin{aligned}
(a(k)+b(k))^0 &:= 1 \\
(a(k)+b(k))^1 &:= a(k)+b(k) \\
(a(k)+b(k))^2 &:= a(k)^2 + a(k-1)b(k) + b(k)a(k) + b(k-1)b(k) \\
(a(k)+b(k))^3 &:= a(k)^3 + b(k)a(k)^2 + a(k-1)b(k)a(k) + a(k-1)^2b(k) \\
&\quad +b(k-1)b(k)a(k) + b(k-1)a(k-1)b(k) \\
&\quad +a(k-2)b(k-1)b(k) + b(k-2)b(k-1)b(k).
\end{aligned} \tag{3}$$

Can the reader guess the correct expansion for the next case, $(a(k)+b(k))^4$, and more generally the formula for $(a(k)+b(k))^n$ where n is any non-negative integer? Observe that if $a(k)$ and $b(k)$ are constant sequences, then the pattern reduces to that for binomial coefficients.

As we shall explain in this paper, the 'shift-by-rank' expansion above appears in a formula for two-dimensional sequences $x(n,k)$ satisfying the generalized triangular recurrence

$$x(n,k) = a(n,k)x(n-1,k) + b(n,k)x(n-1,k-1) \tag{4}$$

where $a(n,k)$ and $b(n,k)$ are arbitrary two-dimensional sequences. If $a(n,k)$ and $b(n,k)$ are constant and equal to one, then (4) reduces to the classic triangular recurrence satisfied by the binomial coefficients:

$$\binom{n}{k} = \binom{n-1}{k} + \binom{n-1}{k-1}. \tag{5}$$

This recurrence is commonly illustrated by arranging the binomial coefficients into the much-celebrated figure known as Pascal's triangle:

$$\begin{array}{ccccccccc} &&&&1&&&& \\ &&&1&&1&&& \\ &&1&&2&&1&& \\ &1&&3&&3&&1& \\ 1&&4&&6&&4&&1 \\ &&&&\cdots&&&& \end{array} \tag{6}$$

Moreover, the recurrence (5) uniquely defines the binomial coefficients if we initialize the values on the boundary of Pascal's triangle to be 1, i.e.

$$\binom{n}{0} = \binom{n}{n} = 1, \tag{7}$$

since all other entries can then be generated by this recurrence.

Of course, the mathematical literature contains many extensions of (5) which are special cases of (4) (see [1], [2], and [5]). For example, the unsigned Stirling numbers of the first kind are defined by the recurrence

$$\left[\begin{matrix} n \\ k \end{matrix}\right] = (n-1)\left[\begin{matrix} n-1 \\ k \end{matrix}\right] + \left[\begin{matrix} n-1 \\ k-1 \end{matrix}\right] \tag{8}$$

and Stirling numbers of the second kind are defined by

$$\left\{\begin{matrix} n \\ k \end{matrix}\right\} = k\left\{\begin{matrix} n-1 \\ k \end{matrix}\right\} + \left\{\begin{matrix} n-1 \\ k-1 \end{matrix}\right\}. \tag{9}$$

An even more interesting recurrence is one satisfied by the generalized Bernoulli polynomials $B_k^{(n)}$ of order n:

$$B_k^{(n)}(z) = \left(1 - \frac{k}{n-1}\right) B_k^{(n-1)}(z) + k\left(\frac{z}{n-1} - 1\right) B_{k-1}^{(n-1)}(z). \tag{10}$$

We shall consider this recurrence further in Section 3.

Our approach to studying $x(n,k)$ relies on a different initialization of Pascal's triangle, which will appear quite natural if we realize it as an infinite rectangular array (also referred to as Pascal's matrix or square; see [3] and [6]), where infinitely many 0's are appended to both ends of each row as shown in (11):

$$\begin{array}{lccccccccl} & & (k=-1) & (k=0) & (k=1) & (k=2) & (k=3) & \dots & & \\ (n=0) & \dots & 0 & 1 & 0 & 0 & 0 & 0 & \dots & \\ (n=1) & \dots & 0 & 1 & 1 & 0 & 0 & 0 & \dots & \\ (n=2) & \dots & 0 & 1 & 2 & 1 & 0 & 0 & \dots & \\ (n=3) & \dots & 0 & 1 & 3 & 3 & 1 & 0 & \dots & \\ \dots & & & & & & & & & \end{array} \qquad (11)$$

In this case, it suffices to initialize only those values along the first row $(n=0)$ as follows:

$$\binom{0}{k} \equiv \delta_{0,k}, \quad k \in \mathbb{Z} \qquad (12)$$

where

$$\delta_{i,j} = \begin{cases} 1 & \text{if } i = j \\ 0 & \text{if } i \neq j \end{cases} \qquad (13)$$

is Kronecker's delta. All other binomial coefficients in this array can then be generated from the triangular recurrence as before.

As an example, let $x(n,k)$ be any two-dimensional sequence which satisfies the classical triangular recurrence, i.e.

$$x(n,k) = x(n-1,k) + x(n-1,k-1), \qquad (14)$$

but whose values along row $n = 0$ are prescribed by an arbitrary bi-directional sequence $f(k)$, i.e.

$$x(0,k) = f(k), \quad k \in \mathbb{Z}. \qquad (15)$$

It is straightforward to show that $x(n,k)$ satisfies the following convolution formula:

$$x(n,k) = \sum_{m=0}^{n} \binom{n}{m} f(k-m). \qquad (16)$$

This demonstrates that binomial coefficients can be considered as fundamental building blocks for generating all two-dimensional sequences which satisfy the triangular recurrence.

In this paper we shall describe the fundamental building blocks for two-dimensional sequences $x(n,k)$ satisfying the generalized triangular recurrence (4). This will require us to introduce the notion of rank for comparing the relative size of an integer with respect to a given subset. As a result, we obtain what seems to be a new combinatorial formula for $x(n,k)$ (Theorem 1) involving ordered set partitions; an extensive search of the literature, including the standard reference [1], which mentions (4) but only discusses special cases, did not reveal any formula similar to ours. As an application, we give a different proof of Dilcher's formula ([1], Theorem 2), which expresses generalized Bernoulli polynomials in terms of classical Bernoulli polynomials. We conclude by presenting a generalization of our results to three-term recurrences.

2. Generalized Triangular Recurrences

We begin our study of generalized triangular recurrences by first considering the simple generalization

$$x(n,k) = ax(n-1,k) + bx(n-1,k-1), \tag{17}$$

where a and b are constants. Cadogan has proven in [2] that

$$x(n,k) = \sum_{m=0}^{n} \binom{n}{m} a^{n-m} b^m f(k-m) \tag{18}$$

which shows that the coefficients in (18) can be generated by expanding the binomial $(ax+by)^n$.

As a next step, let us assume that the more general triangular recurrence holds:

$$x(n,k) = a(n)x(n-1,k) + b(n)x(n-1,k-1), \tag{19}$$

where $a(n)$ and $b(n)$ are one-dimensional sequences. In this case, the situation becomes much more interesting. To gain some intuition, let's write out $x(n,k)$ explicitly for the first several values of n:

$$\begin{aligned}
x(0,k) &= f(k)\\
x(1,k) &= a(1)f(k) + b(1)f(k-1)\\
x(2,k) &= a(1)a(2)f(k) + [b(1)a(2) + a(1)b(2)]f(k-1) + b(1)b(2)f(k-2)\\
x(3,k) &= a(1)a(2)a(3)f(k)\\
&\quad + [a(1)a(2)b(3) + a(1)b(2)a(3) + b(1)a(2)a(3)]f(k-1)\\
&\quad + [a(1)b(2)b(3) + b(1)a(2)b(3) + b(1)b(2)a(3)]f(k-2)\\
&\quad + b(1)b(2)b(3)f(k-3).
\end{aligned} \tag{20}$$

Guided by the combinatorial definition of binomial coefficients as counting the number of subsets of a given set, we look to describe the coefficients of f in (20) by ordered partitions of $\{1,2,3\}$.

Definition 1. Let $A_n = \{1, 2, ..., n\}$. We define $P = (\sigma_1, \sigma_2)$ to be a *2-block ordered partition* of A_n if σ_1 and σ_2 are disjoint sets whose union is A_n. Morever, let $\mathcal{P}_2(A_n)$ denote the set of all 2-block ordered partitions of A_n.

Observe that in the definition above we allow either σ_1 or σ_2 to be empty (but not both), which differs from the standard definition of set partitions (see [8]). Also, the correspondence $(\sigma_1, \sigma_2) \leftrightarrow \sigma_2$ gives a bijection between $\mathcal{P}_2(A_n)$ and the power set of A_n, i.e., the set of all subsets of A_n, since σ_1 and σ_2 are complements of each other in A_n. Thus, we can express any 2-block ordered partition as $(\bar{\sigma}, \sigma)$, where σ is a subset of A_n. Moreover, $\mathcal{P}_2(A_n)$ has cardinality 2^n.

Definition 2. Given $(\bar{\sigma}, \sigma) \in \mathcal{P}_2(A_n)$ where the elements of σ and $\bar{\sigma}$ are denoted explicitly by $\{i_1, i_2, ..., i_m\}$ and $\{j_1, j_2, ..., j_{n-m}\}$, respectively, we define the product

$$\pi_{a,b}(\bar{\sigma}, \sigma) \equiv a(j_1)a(j_2)\cdots a(j_{n-m})b(i_1)b(i_2)\cdots b(i_m). \tag{21}$$

It follows from this definition that $x(n, k)$ can be described by the formula

$$x(n,k) = \sum_{(\bar{\sigma},\sigma)\in\mathcal{P}_2(A_n)} \pi_{a,b}(\bar{\sigma}, \sigma) f(k - |\sigma|), \tag{22}$$

or equivalently,

$$x(n,k) = \sum_{m=0}^{n} \left(\sum_{\sigma\in A_n(m)} \pi_{a,b}(\bar{\sigma}, \sigma) \right) f(k-m). \tag{23}$$

where $A_n(m)$ denotes the set of m-element subsets of A_n. We shall prove a more general formula later on.

Let us now assume that $x(n, k)$ satisfies the recurrence

$$x(n, k) = a(k)x(n - 1, k) + b(k)x(n - 1, k - 1) \tag{24}$$

where $a(k)$ and $b(k)$ are again one-dimensional sequences as in (19), but this time dependent on the second index k instead of the first index n. How does this affect our formula for $x(n, k)$? Again, we write out $x(n, k)$ explicitly for the first several values of n:

$$\begin{aligned} x(0,k) &= f(k) \\ x(1,k) &= a(k)f(k) + b(k)f(k-1) \\ x(2,k) &= a(k)a(k)f(k) + [a(k-1)b(k) + b(k)a(k)]f(k-1) \\ &\quad + vb(k-1)b(k)f(k-2) \\ x(3,k) &= a(k)a(k)a(k)f(k) + [b(k)a(k)a(k) \\ &\quad + a(k-1)b(k)a(k) + a(k-1)a(k-1)b(k)]f(k-1) \\ &\quad + [b(k-1)b(k)a(k) + b(k-1)a(k-1)b(k) \\ &\quad + a(k-2)b(k-1)b(k)]f(k-2) + b(k-2)b(k-1)b(k)f(k-3). \end{aligned} \tag{25}$$

The pattern in this case appears more complicated than the previous case due to shifts in certain indices; in fact, (25) is essentially the 'shift and rank' expansion (modulo f) discussed at the beginning of this paper. To shed some light on this pattern, we first fix an ordering for how the factors a and b should appear in each coefficient: indices should be written left to right in ascending order and if two factors have the same index, then they are written left to right in reverse-alphabetical order. For example, the expansion in (25) is written using this ordering.

Next, we compare the shifts of each index with the corresponding positions of the factor a and b. In particular, if we denote by s_j the amount of shift in the index of the j-th factor of a given coefficient of f, then what pattern does s_j satisfy? The answer is that the shift in the index of a given factor depends on the number of factors involving b that are higher in position, i.e., further to the right.

We formalize this pattern by introducing a rank function to indicate the relative size of an integer with respect to a given subset.

Definition 3. Let $\sigma = \{i_1, i_2, ..., i_m\} \in A_n(m)$. We define the *rank* of a positive integer j relative to σ, and denote it by $R_\sigma(j)$, to be the number of elements in σ that are greater than j, i.e.

$$R_\sigma(j) = |\{i \in \sigma : i > j\}|$$

.

For example, if $\sigma = \{1, 2, 4\} \in A_5(3)$, then $R_\sigma(3) = 1$ since there is only one element in σ, namely 4, that is greater than 3. We also have $R_\sigma(5) = 0$. In general, $R_\sigma(n) = 0$ for every subset $\sigma \in A_n$. Then given $(\bar{\sigma}, \sigma) \in \mathcal{P}_2(A_n)$, we define the product (analogous to (21))

$$\pi_{a,b}(\bar{\sigma}, \sigma; k) = \prod_{s=1}^{n-m} a(k - R_\sigma(j_s)) \prod_{r=1}^{m} b(k - R_\sigma(i_r)). \tag{26}$$

As a result, we are now able to express $x(n, k)$ by the *shift-by-rank* expansion formula

$$x(n, k) = \sum_{m=0}^{n} \left(\sum_{\sigma \in A_n(m)} \pi_{a,b}(\bar{\sigma}, \sigma; k) \right) f(k - m). \tag{27}$$

This brings us to the last case where we assume that $x(n, k)$ satisfies the most general recurrence given by (17) in terms of arbitrary two-dimensional sequences $a(n, k)$ and $b(n, k)$. Again, given $(\bar{\sigma}, \sigma) \in \mathcal{P}_2(A_n)$, we define the product

$$\pi_{a,b}(\bar{\sigma}, \sigma; k) = \prod_{s=1}^{n-m} a(j_s, k - R_\sigma(j_s)) \prod_{r=1}^{m} b(i_r, k - R_\sigma(i_r)). \tag{28}$$

By combining the results of the two previous cases, we obtain the following main theorem, which we prove rigorously.

Theorem 4. *Suppose a two-dimensional sequence $x(n,k)$ satisfies the generalized triangular recurrence*

$$x(n,k) = a(n,k)x(n-1,k) + b(n,k)x(n-1,k-1)$$
$$x(0,k) = f(k)$$

where n is a non-negative integer and $k \in \mathbb{Z}$. Then

$$x(n,k) = \sum_{m=0}^{n} \left(\sum_{\sigma \in A_n(m)} \pi_{a,b}(\bar{\sigma}, \sigma; k) \right) f(k-m). \tag{29}$$

Proof. For $n = 0$, formula (29) reduces to $x(0,k) = f(k)$. By induction on n, we have

$$\begin{aligned} x(n+1,k) &= a(n+1,k)x(n,k) + b(n+1,k)x(n,k-1) \\ &= a(n+1,k) \textstyle\sum_{m=0}^{n} \left(\sum_{\sigma \in A_n(m)} \pi_{a,b}(\bar{\sigma}, \sigma; k) \right) f(k-m) \\ &\quad + b(n+1,k) \textstyle\sum_{m=0}^{n} \left(\sum_{\sigma \in A_n(m)} \pi_{a,b}(\bar{\sigma}, \sigma; k-1) \right) f(k-1-m) \\ &= \textstyle\sum_{m=0}^{n} \left(\sum_{\sigma \in A_n(m)} a(n+1,k)\pi_{a,b}(\bar{\sigma}, \sigma; k) \right) f(k-m) \\ &\quad + \textstyle\sum_{m=0}^{n} \left(\sum_{\sigma \in A_n(m)} b(n+1,k)\pi_{a,b}(\bar{\sigma}, \sigma; k-1) \right) f(k-1-m). \end{aligned}$$

Next, we simplify the inner summation in each of double summations above as follows. First, assume $\sigma = \{i_1, i_2, ..., i_m\} \in A_n(m)$ so that $\bar{\sigma} = \{j_1, j_2, ..., j_{n-m}\} \in A_n(n-m)$. Then observe that we can view σ as a subset of $A_{n+1}(m)$; moreover, the complement of σ in $A_{n+1}(m)$ equals $\bar{\sigma} \cup \{n+1\}$. Set $j_{n-m+1} = n+1$ and recall that $R_\sigma(n+1) = 0$. It follows that the inner summation of the first term can be rewritten as

$$\begin{aligned} &\sum_{\sigma \in A_n(m)} a(n+1,k)\pi_{a,b}(\bar{\sigma}, \sigma; k) \\ &\qquad = \textstyle\sum_{\sigma \in A_n(m)} a(n+1,k) \prod_{s=1}^{n-m} a(j_s, k - R_\sigma(j_s)) \prod_{r=1}^{m} b(i_r, k - R_\sigma(i_r)) \\ &\qquad = \textstyle\sum_{\substack{\sigma \in A_{n+1}(m) \\ n+1 \notin \sigma}} \prod_{s=1}^{n+1-m} a(j_s, k - R_\sigma(j_s)) \prod_{r=1}^{m} b(i_r, k - R_\sigma(i_r)) \\ &\qquad = \textstyle\sum_{\substack{\sigma \in A_{n+1}(m) \\ n+1 \notin \sigma}} \pi_{a,b}(\bar{\sigma}, \sigma; k). \end{aligned}$$

As for the second term, since $R_{\sigma'}(j) = R_\sigma(j) + 1$ for $\sigma \in A_n(m)$, where $\sigma' = \sigma \cup \{n+1\}$ and $1 \le j < n+1$, we can simplify it as

$$\sum_{\sigma\in A_n(m)} b(n+1,k)\pi_{a,b}(\bar{\sigma},\sigma;k-1)$$

$$= \sum_{\sigma\in A_n(m)} b(n+1,k)\prod_{s=1}^{n-m} a(j_s,k-1-R_\sigma(j_s))\prod_{r=1}^{m} b(i_r,k-1-R_\sigma(i_r))$$

$$= \sum_{\substack{\sigma'=\sigma\cup\{n+1\}\\ \sigma\in A_n(m)}} \prod_{s=1}^{n-m} a(j_s,k-R_{\sigma'}(j_s))\prod_{r=1}^{m+1} b(i_r,k-R_{\sigma'}(i_r))$$

$$= \sum_{\substack{\sigma'\in A_{n+1}(m+1)\\ n+1\in\sigma'}} \pi_{a,b}(\bar{\sigma}',\sigma';k).$$

It follows that

$$x(n+1,k) = \sum_{m=0}^{n}\left(\sum_{\substack{\sigma\in A_{n+1}(m)\\ n+1\notin\sigma}} \pi_{a,b}(\bar{\sigma},\sigma;k)\right) f(k-m)$$

$$+\sum_{m=0}^{n}\left(\sum_{\substack{\sigma'\in A_{n+1}(m+1)\\ n+1\in\sigma'}} \pi_{a,b}(\bar{\sigma}',\sigma';k)\right) f(k-(m+1)).$$

But the two summations above can be viewed as a partition of A_{n+1} into those subsets σ' that contain the integer $n+1$ and those subsets σ that do not. Then by re-indexing the second summation on the right hand side above and making the limits of summation the same for both, which involve inserting empty cases, we can combine both summations into one as follows:

$$x(n+1,k) = \sum_{m=0}^{n+1}\left(\sum_{\substack{\sigma\in A_{n+1}(m)\\ n+1\notin\sigma}} \pi_{a,b}(\bar{\sigma},\sigma;k)\right) f(k-m)$$

$$+\sum_{m=0}^{n+1}\left(\sum_{\substack{\sigma'\in A_{n+1}(m)\\ n+1\in\sigma'}} \pi_{a,b}(\bar{\sigma}',\sigma';k)\right) f(k-m)$$

$$= \sum_{m=0}^{n+1}\left(\sum_{\sigma\in A_{n+1}(m)} \pi_{a,b}(\bar{\sigma},\sigma;k)\right) f(k-m).$$

This completes the proof. □

3. Application to Bernoulli Polynomials

As an application of Theorem 1, we prove Dilcher's formula ([1]) for expressing generalized Bernoulli polynomials in terms of classical Bernoulli polynomials $B_n(z)$. The latter arises in many important applications involving special functions such as the Riemann zeta function.

Let p be a positive integer. Then the generalized Bernoulli polynomials $B_n^{(p)}(z)$ of order p are defined by the generating function

$$\frac{t^p e^{zt}}{(e^t-1)^p} = \sum_{n=0}^{\infty} B_n^{(p)}(z)\frac{t^n}{n!}. \tag{30}$$

If $p = 1$, then $B_n^p(z)$ reduces to the classical Bernoulli polynomials $B_n(z)$. Since

$$\frac{t^p e^{zt}}{(e^t-1)^p} = \prod_{q=1}^{p}\left(\frac{te^{z_q t}}{e^t-1}\right) = \prod_{q=1}^{p}\left(\sum_{n=0}^{\infty} B_n(z_q)\frac{t^n}{n!}\right) \tag{31}$$

where $z = z_1 + ... + z_p$, it follows from equating (30) and (31) that

$$S_{n,p}(z) \equiv \sum_{\substack{i_1\geq 0, i_2\geq 0,\dots,i_p\geq 0\\ i_1+i_2+\dots+i_p=n}} \frac{n!}{i_1!\cdots i_p!}B_{i_1}(z_1)\cdots B_{i_p}(z_p) = B_n^{(p)}(z). \tag{32}$$

In [3], Nörlund proved that the generalized Bernoulli polynomials satisfy the generalized triangular recurrence

$$B_n^{(p)}(z) = \left(1-\frac{n}{p-1}\right)B_n^{(p-1)}(z) + n\left(\frac{z}{p-1}-1\right)B_{n-1}^{(p-1)}(z) \tag{33}$$

for $n > p \geq 1$. By applying Theorem 1 to this recurrence, we obtain the following corollary.

Corollary 5. *The generalized Bernoulli polynomials are given in terms of the classical Bernoulli polynomials by*

$$B_n^{(p)}(z) = \sum_{k=0}^{p-1}\left(\sum_{\sigma\in A_{p-1}(k)} \pi_{a,b}(\bar{\sigma},\sigma;n)\right)B_{n-k}(z). \tag{34}$$

Proof. Formula (34) follows immediately from (29) by setting $x(p,n) = B_n^{(p+1)}(z)$, $a(p,n) = 1 - \frac{n}{p}$ and $b(p,n) = n\left(\frac{z}{p}-1\right)$. □

The following lemma describing certain properties of the rank function will be useful in demonstrating that formula (34) reduces to Dilcher's formula, which we show afterwards.

Lemma 6. *Let* $\sigma = \{i_1, i_2, ..., i_m\} \in A_n(m)$ *with* $i_1 < i_2 < ... < i_m$ *and* $\bar{\sigma} = \{j_1, j_2, ..., j_{n-m}\} \in A_n(n-m)$ *with* $j_1 < j_2 < ... < j_{n-m}$. *Then*

$$\{R_\sigma(i_r)\}_{r=1}^{m} = \{0, ..., m-1\} \tag{35}$$

and

$$\{R_\sigma(j_s) + j_s\}_{s=1}^{n-m} = \{m+1, ..., n\}. \tag{36}$$

Proof. Since $i_1 < i_2 < ... < i_k$ we have $R_\sigma(i_r) = m - r$ for $r = 1, ..., m$ and so

$$\{R_\sigma(i_r)\}_{r=1}^m = \{0, ..., m-1\}.$$

This proves (35). To prove (36), we first show that $R_\sigma(j_s) + j_s$ is bounded between $m+1$ and n for $s = 1, ..., n-m$. This is because $R_\sigma(j_s) \geq m - j_s + 1$ and so

$$R_\sigma(j_s) + j_s \geq (m - j_s + 1) + j_s = m + 1.$$

On the other hand, we have $R_\sigma(j_s) \leq n - j_s$ and so

$$R_\sigma(j_s) + j_s \leq n - j_s + j_s = n.$$

Next, we claim that the values $R_\sigma(j_s) + j_s$ are all distinct for $1 \leq s \leq n-m$ and in particular increasing in s, i.e., $R_\sigma(j_r) + j_r < R_\sigma(j_s) + j_s$ for $r < s$. To prove this, we set $k = j_s - j_r > 0$ (recall that $j_r < j_s$). Since $R_\sigma(j_r) < R_\sigma(j_s) + k$, it follows that

$$R_\sigma(j_r) + j_r < R_\sigma(j_s) + j_r + k = R_\sigma(j_s) + j_s$$

which proves our claim. Thus, (36) holds. □

Let $s(n,k)$ denote the Stirling numbers of the first kind, which are defined by the generating function

$$z(z-1)(z-2)\cdots(z-n+1) = \sum_{m=0}^{n} s(n,m) z^m. \tag{37}$$

In particular, we have

$$s(n,m) = \sum_{\{i_1, i_2, ..., i_m\} \in A_n(m)} (-1)^m i_1 i_2 \cdots i_m. \tag{38}$$

We are now ready to prove Dilcher's formula.

Theorem 7. (Dilcher [1]) *For positive integers n and p with $n > p \geq 1$,*

$$S_{n,p}(z) = (-1)^{p-1} p \binom{n}{p} \sum_{k=0}^{p-1} \frac{(-1)^k}{n-k} \left(\sum_{m=0}^{k} \binom{q-1}{m} s(p,q) z^m \right) B_{n-k}(z) \tag{39}$$

where $q = p - k + m$.

Proof. Because of (32) it suffices to show that $B_n^{(p)}(x)$, given by (34), equals the right-hand side of (39). Towards this end, suppose $\sigma \in A_{p-1}(k)$ with $\sigma = \{i_1, i_2, ..., i_k\}$ and $\bar{\sigma} = \{j_1, j_2, ..., j_{p-k-1}\}$. Then since $a(p,n) = 1 - \frac{n}{p}$ and $b(p,n) = n\left(\frac{z}{p} - 1\right)$, we have

$$
\begin{aligned}
\pi_{a,b}(\bar{\sigma},\sigma;n) &= \prod_{s=1}^{p-k-1} a(j_s, n-R_\sigma(j_s)) \prod_{r=1}^{k} b(i_r, n-R_\sigma(i_r)) \\
&= \prod_{s=1}^{p-k-1} \left(1 - \frac{n-R_\sigma(j_s)}{j_s}\right) \prod_{r=1}^{k} [(n-R_\sigma(i_r))\left(\frac{z}{i_r}-1\right)] \\
&= \frac{(-1)^{p-k-1}}{(p-1)!} \prod_{s=1}^{p-k-1} (n-R_\sigma(j_s)-j_s) \prod_{r=1}^{k}(n-R_\sigma(i_r)) \prod_{r=1}^{k}(z-i_r)
\end{aligned}
$$

where we have used the fact that $i_1 \cdots i_k j_1 \cdots j_{p-k-1} = (p-1)!$. By the previous lemma, the products involving R_σ can be combined into

$$
\prod_{s=1}^{p-k-1} (n-R_{\bar{\sigma}}(j_s)-j_s) \prod_{r=1}^{k}(n-R_\sigma(i_r)) = \prod_{\substack{r=0 \\ r\neq k}}^{p-1} (n-r) = \frac{(n)_p}{n-k}
$$

which yields

$$
\pi_{a,b}(\bar{\sigma},\sigma;n) = \frac{(-1)^{p-k-1}}{(p-1)!} \cdot \frac{(n)_p}{n-k} \cdot \prod_{r=1}^{k}(z-i_r).
$$

It follows that

$$
\sum_{\sigma\in A_{p-1}(k)} \pi_{a,b}(\bar{\sigma},\sigma;n) = (-1)^{p-k-1}\frac{p}{n-k}\binom{n}{p} \sum_{\sigma\in A_{p-1}(k)} \prod_{r=1}^{k}(z-i_r).
$$

Now, write the polynomial $P_\sigma(z) \equiv \prod_{r=1}^{k}(z-i_r)$, which has degree k, in the form

$$
P_\sigma(z) = c_k(\sigma)z^k + c_{k-1}(\sigma)z^{k-1} + \ldots + c_0(\sigma).
$$

Here each coefficient $c_m(\sigma)$ corresponding to z^m is a sum of terms of the form $(-1)^{k-m} i_{r_1} i_{r_2} \ldots i_{r_{k-m}}$, where $\{i_{r_1}, \ldots, i_{r_{k-m}}\}$ is an $(k-m)$-element subset of σ. Since σ is a k-element subset of $\{1, \ldots, p-1\}$, we claim that there are exactly $\binom{p-1-k+m}{m}$ subsets $\sigma \in A_{p-1}(k)$ that contain $\{i_{r_1}, \ldots, i_{r_{k-m}}\}$. This is because, assuming the elements $\{i_{r_1}, \ldots, i_{r_{k-m}}\}$ have already been chosen, we can fill in the remaining m elements of σ by choosing them from the $(p-1-k+m)$ un-chosen elements. It follows that

$$
\begin{aligned}
\sum_{\sigma\in A_{p-1}(k)} P_\sigma(z) &= \sum_{m=0}^{k} \binom{q-1}{m} \left(\sum_{\{r_1,\ldots,r_{k-m}\}\in A_{p-1}(k-m)} (-1)^{k-m} i_{r_1} i_{r_2} \ldots i_{r_{k-m}} \right) z^m \\
&= \sum_{m=0}^{k} \binom{q-1}{m} s(p,q) z^m
\end{aligned}
$$

where $q = p - k + m$. Thus,

$$\begin{aligned} S_{n,p}(x) &= B_n^{(p)}(x) \\ &= \sum_{k=0}^{p-1} \left(\sum_{\sigma \in A_{p-1}(k)} \pi_{a,b}(\bar{\sigma}, \sigma; n) \right) B_{n-k}(z) \\ &= \sum_{k=0}^{p-1} \left((-1)^{p-k-1} \frac{p}{n-k} \binom{n}{p} \sum_{\sigma \in A_{p-1}(k)} P_\sigma(z) \right) B_{n-k}(z) \\ &= (-1)^{p-1} p \binom{n}{p} \sum_{k=0}^{p-1} \frac{(-1)^k}{n-k} \left(\sum_{m=0}^{k} \binom{q-1}{m} s(p,q) z^m \right) B_{n-k}(z) \end{aligned}$$

as desired. □

4. Generalization to Three-Term Recurrences

Our results for two-term triangular recurrences can be generalized to three-term recurrences of the form

$$x(n,k) = a(n,k)x(n-1,k) + b(n,k)x(n-1,k-1) + c(n,k)x(n-1,k-2) \quad (40)$$

where $a(n,k)$, $b(n,k)$ and $c(n,k)$ are given arbitrary two-dimensional sequences. Towards this end, let $\sigma = \{i_1, i_2, ..., i_{m_1}\} \in A_n(m_1)$ and $\tau = \{j_1, j_2, ..., j_{m_2}\} \in A_n(m_2)$ be two disjoint subsets of $\{1, 2, ..., n\}$. We define the *rank* of a positive integer j with respect to (σ, τ) to be

$$R_{\sigma,\tau}(j) = |\{i \in \sigma : j < i\}| + 2\,|\{i \in \tau : j < i\}|. \quad (41)$$

Let $\mathcal{P}_3(A_n)$ denote the set of 3-block ordered partitions of the form $(\sigma_1, \sigma_2, \sigma_3)$, where σ_1, σ_2, and σ_3 are mutually disjoint and whose union equals A_n. Write out the elements of these three sets explicitly as

$$\begin{aligned} \sigma_1 &= \{j_1, j_2, ..., j_{n-m_1-m_2}\} \\ \sigma_2 &= \{i_1, i_2, ..., i_{m_2}\} \\ \sigma_3 &= \{h_1, h_2, ..., h_{m_1}\} \end{aligned}$$

and define the product

$$\pi_{a,b,c}(\sigma_1, \sigma_2, \sigma_3; k) =$$

$$\prod_{s=1}^{m_3} a(j_s, k - R_{\sigma_2,\sigma_3}(j_s)) \prod_{r=1}^{m_2} b(i_r, k - R_{\sigma_2,\sigma_3}(i_r)) \prod_{q=1}^{m_1} c(h_q, k - R_{\sigma_2,\sigma_3}(h_q)) \quad (42)$$

where $m_3 = n - m_1 - m_2$.

We conjecture the following formula for $x(n,k)$, which generalizes (29) to three-term recurrences:

$$x(n,k) = \sum_{(\sigma_1,\sigma_2,\sigma_3)\in\mathcal{P}_3(A_n)} \pi_{a,b,c}(\sigma_1,\sigma_2,\sigma_3;k) f(k - |\sigma_2| - 2|\sigma_3|). \qquad (43)$$

Formulas (29) and (43) can of course be naturally extended to higher order n-term recurrences for $n \geq 4$.

Acknowledgement The author wishes to thank the referee for his useful suggestions to improve the paper.

References

[1] B. A. Bondarenko, *Generalized Pascal Triangles and Pyramids: Their Fractals, Graphs, and Applications* (Translated by Richard C. Bollinger), Fibonacci Association, 1993, available at: http://www.fq.math.ca/pascal.html

[2] C. Cadogan, Some Generalizations of the Pascal Triangle, *Mathematics Magazine* **45** (1972), No. 3, 158-162.

[3] G. S. Call and D. J. Velleman, Pascal's matrices, *Amer. Math. Monthly* **100** (1993), 372-376.

[4] K. Dilcher, Sums of Products of Bernoulli Numbers, *J. Number Theory* **60** (1996), 23-41.

[5] M. Klika, Zeilenmaxima in verallgemeinerten Pascalschen Dreiecken, *J. Reine Angew. Math.* **274-275** (1975), 27-37.

[6] D. Moran, Pascal Matrices, *Mathematics Magazine* **40** (1967), No. 1, 12-14.

[7] N. E. Nörlund, *Differenzenrechnung*, Springer, 1924.

[8] G. C. Rota, The Number of Partitions of a Set, *Amer. Math. Monthly* **71** (1964), No. 5, 498-504.

THE DISTRIBUTION OF SOLUTIONS TO $XY = N$ (MOD A) WITH AN APPLICATION TO FACTORING INTEGERS

Michael O. Rubinstein [1]
Pure Mathematics, University of Waterloo, Waterloo, Ontario, Canada
mrubinst@uwaterloo.ca

Received: 3/6/09, Revised: 9/21/12, Accepted: 3/6/13, Published: 3/15/13

Abstract
We consider the uniform distribution of solutions (x, y) to $xy = N \mod a$, and obtain a bound on the second moment of the number of solutions in squares of length approximately $a^{1/2}$. We use this to study a new factoring algorithm that factors $N = UV$ provably in $O(N^{1/3+\epsilon})$ time, and discuss the potential for improving the runtime to sub-exponential.

1. Introduction

Let $\gcd(a, N) = 1$. A classic application of Kloosterman sums shows that the points $(x, y) \mod a$ satisfying $xy = N \mod a$ become uniformly distributed in the square of side length a as $a \to \infty$. In this paper we investigate an application of this fact to the problem of factoring integers. We give a new method to factor the integer N which beats trial division, and prove that it runs in time $O(N^{1/3+\epsilon})$. Unlike many existing factoring algorithms that attempt to factor N by considering various congruences modulo N, the method presented here revisits trial division and exploits information available on a failed trial division, namely it considers the remainder of N when divided by an integer a, and also by its neighbor, $a - 1$.

While the complexity of our method is not exciting, considering the existence of several probabilistic sub-exponential factoring algorithms, the runtime here is provable. For comparison, the best known *provable* factoring algorithm, Pollard-Strassen, runs in time $O(N^{1/4+\epsilon})$. Shank's class group method runs in time $O(N^{1/5+\epsilon})$ assuming the GRH. It has a comparable runtime to Lehman's method [7], but the method is different. Our algorithm is described in Section 2.

Furthermore, proving this runtime requires understanding the finer distribution of solutions to $xy = N \mod a$, and our results in this regards are interesting in their own right. We discuss the problem on uniform distribution in Sections 4 and 5.

Finally, all existing sub-exponential factoring algorithms have grown out of much weaker exponential algorithms, and we hope that the factoring ideas presented here

[1]Support for work on this paper was provided by an NSERC Discovery Grant

will be improved. In Section 3 we discuss some needed improvements to achieve a better runtime.

We have not implemented the algorithms described in this paper. The purpose of this paper is to present a new approach to factoring integers and analyze its runtime.

2. Algorithm – Hide and Seek

Let N be a positive integer that we wish to factor. Say $N = UV$ where U and V are positive integers, not necessarily prime, with $1 < U \leq V$. For simplicity, assume $V < 2U$, so that $V < (2N)^{1/2}$. The general case, without this restriction, will be handled at the end of this section.

The idea behind the algorithm is to perform trial division of N by a couple of integers, and to use information about the remainder to determine the factors U and V.

Let a be a positive integer, $1 < a < N$. By the division algorithm, write

$$
\begin{aligned}
U &= u_1 a + u_0, \quad \text{with } 0 \leq u_0 < a \\
V &= v_1 a + v_0, \quad \text{with } 0 \leq v_0 < a.
\end{aligned} \tag{1}
$$

Assume that u_0 is relatively prime to a, and likewise for v_0, since otherwise we easily extract a factor of N by taking $\gcd(a, N)$. If, for a given a, we can determine u_0, u_1, v_0, v_1 then we have found U and V.

Consider $N = u_0 v_0 \mod a$. One cannot simply determine u_0 and v_0 from the value of $N \mod a$, because $\phi(a)$ pairs of integers $(x, y) \mod a$ satisfy $xy = N \mod a$ (if $x = m u_0 \mod a$, then $y = m^{-1} v_0 \mod a$, where $\gcd(m, a) = 1$).

However, say a is large, $a \geq \lceil (2N)^{1/3} \rceil > V^{2/3}$, so that v_1 and u_1 are comparatively small, $u_1, v_1 \leq V^{1/3}$, i.e., both are $< a^{1/2}$. If we consider $N \mod a - \delta$

$$
N = UV = (u_1 \delta + u_0)(v_1 \delta + v_0) \mod a - \delta, \tag{2}
$$

for $\delta = 0, 1$, we get, as solutions (x, y) to $xy = N \mod a - \delta$, two nearby points, (u_0, v_0) and $(u_0 + u_1, v_0 + v_1)$, whose coordinates are within $a^{1/2}$ of one another. This pair of points is just one pair amongst the many pairs of solutions to the above equations, for $\delta = 0, 1$. However, the fact that the solutions are nearby reduces the amount of checking that we need to do in order to find the pair of points, (u_0, v_0) and $(u_0 + u_1, v_0 + v_1)$, that we seek.

Figures 2 and 2 illustrate this fact, for $N = 1910861 = 1061 \times 1801$, and $a = 157$. Thus, $U = 1061$, $u_0 = 119$, $u_1 = 8$, and $V = 1801$, $v_0 = 74$, $v_1 = 15$. Rather than just depict the solutions to $xy = N \mod a - \delta$, for $\delta = 0, 1$, we also plot the solutions for $\delta = 2, 3$ (though our algorithm below only makes use of solutions for

$\delta = 0, 1$). Plotting four sets of solutions, for $\delta = 0, 1, 2, 3$ makes it easier for the human eye to tell the points $(u_0, v_0) = (119, 74)$, $(u_0 + u_1, v_0 + v_1) = (125, 85)$, $(u_0 + 2u_1, v_0 + 2v_1) = (131, 96)$, and $(u_0 + 3u_1, v_0 + 3v_1) = (137, 107)$ from the random coincidences of nearby points as these all lie equally spaced apart and on one line.

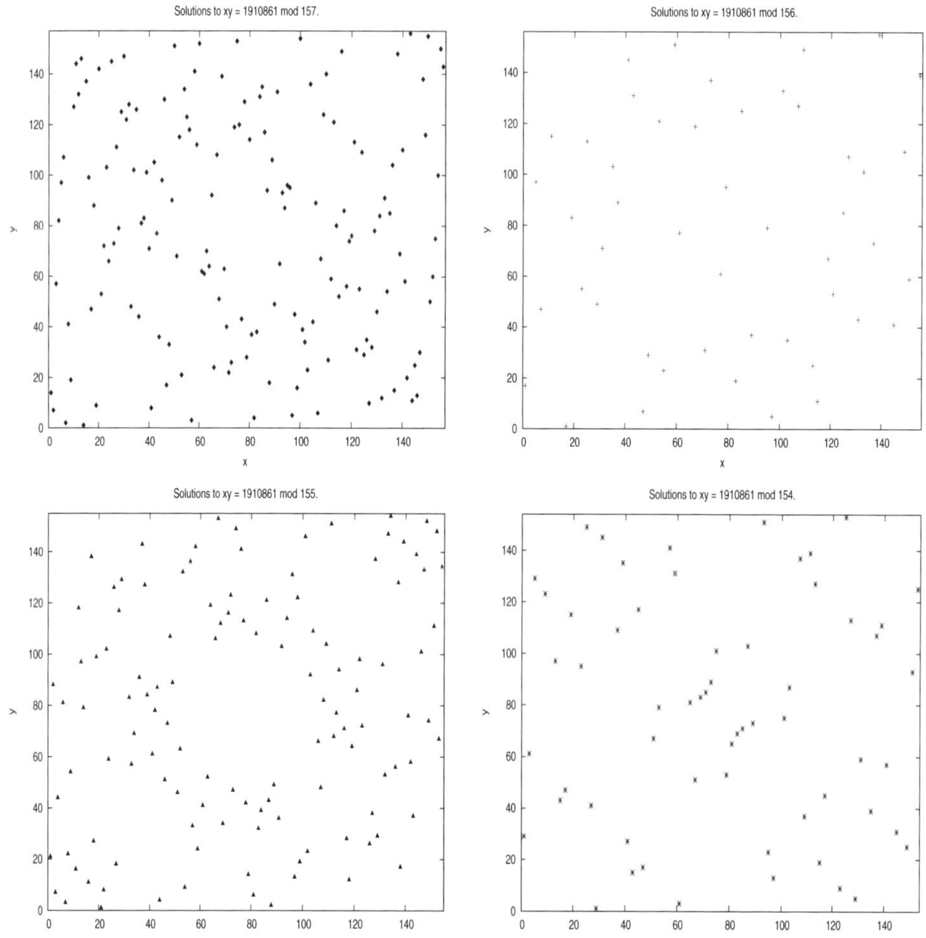

Figure 1: The solutions (x, y) to $xy = 1910861 \mod 157 - \delta$, for $\delta = 0, 1, 2, 3$

So, we can set, say, $a = \lceil (2N)^{1/3} \rceil$, and partition the Cartesian plane into squares of side length $a^{1/2}$, each square being of the form $\{(x, y) \in \mathbb{R}^2 | ma^{1/2} \leq x < (m+1)a^{1/2}, na^{1/2} \leq y < (n+1)a^{1/2}\}$, where $m, n \in \mathbb{Z}$.

We then list *all* $\phi(a)$ pairs of integers (x, y), with $1 \leq x, y \leq a$, that satisfy $xy = N \mod a$, throwing them into our squares of side lengths $a^{1/2}$. We can

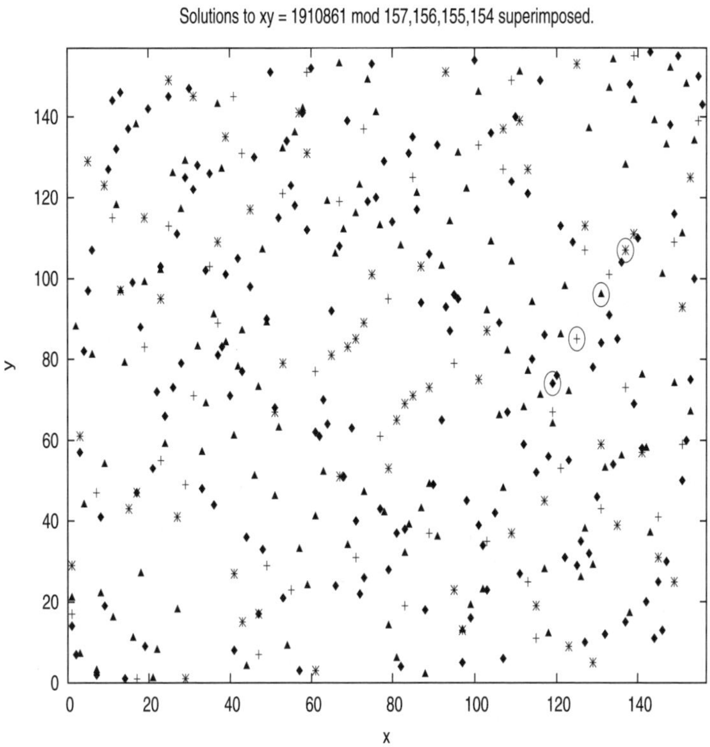

Figure 2: The solutions (x, y) to $xy = 1910861 \mod 157 - \delta$, for $\delta = 0, 1, 2, 3$, superimposed. The four circled points are $(u_0, v_0) = (119, 74)$, $(u_0 + u_1, v_0 + v_1) = (125, 85)$, $(u_0 + 2u_1, v_0 + 2v_1) = (131, 96)$, and $(u_0 + 3u_1, v_0 + 3v_1) = (137, 107)$.

assume that $\gcd(a, N) = 1$, because, otherwise we easily extract a factor of N.

We can compute all inverses mod a, and hence all $(x, y) = (x, x^{-1}N) \mod a$ in $O(a)$ operations mod a. To compute all inverses, start with $m = 2$, multiply mod a by m until we arrive at 1, or hit a residue class already encountered (in which case m is not invertible). Then, take the first residue not yet encountered and repeat the previous step until all residue classes are exhausted.

Having produced all solutions for the modulus a, we then repeat the process for the modulus $a - 1$. For each solution (x_1, y_1) to $xy = N \mod a - 1$, we determine which $a^{1/2} \times a^{1/2}$ square it falls within, and consider all nearby (with each coordinate within $a^{1/2}$, wrapping to the opposite side of the larger $a \times a$ square if needed) solutions (x_0, y_0) to $xy = N \mod a$ from our list of stored solutions. We set $\mu_0 = x_0$, $\nu_0 = y_0$, $\mu_1 = x_1 - \mu_0$, $\nu_1 = y_1 - \nu_0$, and check whether $(\mu_1 a + \mu_0)(\nu_1 a + \nu_0) = N$. If so, we have determined a non-trivial factor of N and quit.

How much work does comparing pairs of points (x_0, y_0) and (x_1, y_1) entail? There are $\phi(a-1)$ solutions to $xy = N \mod a-1$, and, typically, we expect there to be only a handful of solutions to $xy = N \mod a$ whose coordinates are each within $a^{1/2}$. Each such pair of solutions gives us candidate values μ_0, ν_0 and μ_1, ν_1 for u_0, v_0 and u_1, v_1, and we check to see whether they produce $N = (u_1 a + u_0)(v_1 a + v_0)$. On average, each $a^{1/2} \times a^{1/2}$ square contains $O(1)$ points, the overall time to check all squares and points is roughly predicted to be $O(a)$. In Section 4 we obtain a runtime bound of $O(a^{1+\epsilon})$. This algorithm terminates successfully when the true points (u_0, v_0) and $(u_0 + u_1, v_0 + v_1)$ are found. Since $a = O(N^{1/3})$ this gives a running time that is provably $O(N^{1/3+\epsilon})$.

The idea that lies behind the algorithm suggests the name 'Hide and Seek'. The solutions that we seek (u_0, v_0) and $(u_0 + u_1, v_0 + v_1)$ are hiding amongst many solutions in the large $a \times a$ square, but, like children who have hidden next to one another while playing the game Hide and Seek, they have become easier to spot.

We summarize the above in the following algorithm.

Algorithm 2.1. (Hide and Seek) *Let $N = UV$ be a positive integer, and assume that $1 < U \leq V < 2U$, with $U, V \in \mathbb{Z}$. Thus $V < (2N)^{1/2}$. For given positive integers N, r, define*

$$H_{N,a} = \{(x, y) \mid xy = N \mod a,\ 0 \leq x, y < a\}. \tag{3}$$

Step 1 *Set $a = \lceil (2N)^{1/3} \rceil$.*

Step 2 *Use the Euclidean algorithm to compute* $\gcd(N, a - \delta)$ *for $\delta = 0, 1$. If either gcd is > 1 then we have determined a non-trivial factor of N and quit.*

Step 3 *Compute and store in an array all $\phi(a)$ points of $H_{N,a}$. This can be done using $O(a)$ arithmetic operations mod a as described above.*

Step 4 *For $0 \leq m, n < a^{1/2}$, initialize a doubly indexed array, 'Bin'. Each element, Bin[m,n], will contain a list of points and be used to partition $H_{N,a}$. Each is initially set to empty.*

Step 5 *Partition the elements of $H_{N,a}$ according to squares of side length $a^{1/2}$ by computing, for each $(x, y) \in H_{N,a}$, the values $m = \lfloor x/a^{1/2} \rfloor$ and $n = \lfloor y/a^{1/2} \rfloor$, and appending the point (x, y) to Bin[m,n].*

Step 6 *Compute the $\phi(a-1)$ elements of $H_{N,a-1}$. For each $(x_1, y_1) \in H_{N,a-1}$:*

Step 6a *Determine which bin it corresponds to by computing $m = \lfloor x_1/a^{1/2} \rfloor$ and $n = \lfloor y_1/a^{1/2} \rfloor$.*

Step 6b *Loop through the nearby points (x_0, y_0) of $H_{N,a}$ whose coordinates lie, left and downwards, within $a^{1/2}$. Typically, this entails examining the four bins Bin[$m - \epsilon_1, n - \epsilon_2$], where $\epsilon_1, \epsilon_2 \in \{0, 1\}$. However, slight care* *is needed when crossing over an edge of the $a \times a$ square- one should wrap to the opposite side of the square.*

Step 6c *Set $\mu_0 = x_0$, $\nu_0 = y_0$, $\mu_1 = x_1 - \mu_0$, $\nu_1 = y_1 - \nu_0$, and check whether $(\mu_1 a + \mu_0)(\nu_1 a + \nu_0) = N$. If so, we have determined a non-trivial factor of N and quit.*

The storage requirement of $O(N^{1/3})$ can be improved to $O(N^{1/6})$ by generating the solutions (x, y) to $xy = N \mod a-\delta$ lying in one vertical strip of width $O(a^{1/2})$ at a time (easy to do since we can choose x as we please, which then determines y). In general, we are then no longer free to generate all modular inverses at once, and must compute inverses in intervals of size $a^{1/2}$, one at a time, at a cost, using the Euclidean algorithm, of $O(a^{\epsilon})$ per inverse.

A slight improvement on the runtime of the algorithm can be achieved my modifying the choice of a so that $\phi(a)$ or $\phi(a-1)$ are comparatively small. This can be achieved by selecting an a or $a-1$ with smaller distinct prime factors.

2.1. Variant: $1 < U \leq V < N$ Without Restriction

Say $U = N^{\alpha}$, $V = N^{1-\alpha}$, with $1/3 < \alpha \leq 1/2$. We may assume that $\alpha > 1/3$, for, if not, we can find U by performing $O(N^{1/3})$ trial divisions.

Let $a = \lceil 2N^{1/3} \rceil$ (we do, here, mean $2N^{1/3}$, rather than $(2N)^{1/3}$ of the previous section, as explained below).

Instead of working with small squares of side length $a^{1/2}$, partition the $a \times a$ square into rectangles of width w and height h, with $wh = N^{1/3}$. We would like to select w roughly equal to $N^{\alpha-1/3}$ and hence $h = N^{1/3}/w$ roughly equal to $N^{2/3-\alpha}$.

These rough values of w and h are needed to make sure that, using the same notation as before, (u_0, v_0) and $(u_0 + u_1, v_0 + v_1)$ are in the same, or neighboring, rectangles. More precisely, say $N^{\alpha-1/3} < w \leq 2N^{\alpha-1/3}$. Then $h = N^{1/3}/w \geq N^{2/3-\alpha}/2$. Then, in (1), $u_1 = \lfloor U/a \rfloor \leq N^{\alpha}/\lceil 2N^{1/3} \rceil \leq N^{\alpha-1/3}/2 < w$, and $v_1 = \lfloor V/a \rfloor \leq N^{1-\alpha}/\lceil 2N^{1/3} \rceil \leq N^{2/3-\alpha}/2 \leq h$. Thus the x-coordinates of (u_0, v_0) and $(u_0 + u_1, v_0 + v_1)$ are $< w$ apart and the y-coordinates are $\leq h$ apart.

Since we do not, a priori know α, we cannot simply set w and h. Instead, we use an exponentially increasing set of w's, for example starting with $w = 2$, and, repeatedly applying the above procedure, each time doubling the size of w, until $w > N^{\alpha-1/3}$ and one successfully factors N.

The area of each rectangle is $N^{1/3}$, and of the $a \times a$ square is approximately $N^{2/3}$, so there are $O(N^{1/3})$ rectangles (at the top and right edges these will typically be truncated), and, on average, each contains $O(1)$ solutions to $xy = N \mod a - \delta$. Running through each rectangle and its immediate neighbors, checking all pairs of points in these rectangles suggests $O(N^{1/3})$ operations are needed for a particular choice of w and h. Since we might have to repeat this a few times, doubling the size of w, the overall running time gets multiplied by $O(\log N)$ which is $O(N^{\epsilon})$.

In Section 5, a running time equal to $O(N^{1/3+\epsilon})$ is proven.

The steps described in this section are summarized below.

Algorithm 2.2. *Let $N = UV$ be a positive integer, and assume that $1 < U \leq V < N$, with $U, V \in \mathbb{Z}$.*

Step 1 *Carry out trial division on N up to $N^{1/3}$. If a non-trivial factor of N is found, quit.*

Step 2 *Set $a = \lceil 2N^{1/3} \rceil$.*

Step 3 *Use the Euclidean algorithm to compute $\gcd(N, a - \delta)$ for $\delta = 0, 1$. If either gcd is > 1 then we have determined a non-trivial factor of N and quit.*

Step 4 *Compute and store, in two arrays, all the points of $H_{N,a}$ and $H_{N,a-1}$.*

Step 5 *Set $j = 0$. While we have not succeeded in finding a non-trivial factor of N:*

- *Step 5a* *Increment j by 1 and set $w = 2^j$ and $h = N^{1/3}/w$.*
- *Step 5b* *For $0 \leq m < a/w$ and $0 \leq n < a/h$, initialize a doubly indexed array, 'Bin', whose elements, Bin[m,n], will contain lists of points and be used to partition $H_{N,a}$. Each bin is initially set to empty.*
- *Step 5c* *Partition the elements of $H_{N,a}$ according to rectangles of width w and height h by computing, for each $(x, y) \in H_{N,a}$, the values $m = \lfloor x/w \rfloor$ and $n = \lfloor y/h \rfloor$, and appending the point (x, y) to Bin[m,n].*
- *Step 5d* *For each $(x_1, y_1) \in H_{N,a-1}$:*
  - *Step 5d1* *Determine which bin it corresponds to by computing $m = \lfloor x_1/w \rfloor$ and $n = \lfloor y_1/h \rfloor$.*
  - *Step 5d2* *Loop through the nearby points (x_0, y_0) of $H_{N,a}$ whose coordinates lie, left and downwards, within w and h respectively. Typically, this entails examining the four bins Bin[$m - \epsilon_1, n - \epsilon_2$], where $\epsilon_1, \epsilon_2 \in \{0, 1\}$. However, slight care is needed when crossing over an edge of the $a \times a$ square- one should wrap to the opposite side of the square.*
  - *Step 5d3* *Set $\mu_0 = x_0$, $\nu_0 = y_0$, $\mu_1 = x_1 - \mu_0$, $\nu_1 = y_1 - \nu_0$, and check whether $(\mu_1 a + \mu_0)(\nu_1 a + \nu_0) = N$. If so, we have determined a non-trivial factor of N and quit.*
- *Step 5e* *Free up the memory used by 'Bin'.*

3. Towards a Subexponential Bound

The above algorithm exploits the fact that when a is large, and δ is small, the points with coordinates $(U, V) \mod a - \delta$ are close to one another. In fact they lie equally spaced on a line with common horizontal difference u_1, and vertical difference v_1.

An obvious thing to try is to reduce the size of a. However, as a decreases, u_1 and v_1 increase so that not only do the points (u_0, v_0) and $(u_0 + u_1, v_0 + v_1)$ move far apart, the latter point soon falls far outside the square of side length a.

To fix this, one can view (1) as the base a expansion of U and V. When a is smaller, one could instead use a polynomial expansion

$$\begin{aligned} U &= u_{d_1}a^{d_1} + \ldots + u_1 a + u_0, \quad 0 \le u_i < a \\ V &= v_{d_2}a^{d_2} + \ldots + v_1 a + v_0, \quad 0 \le v_i < a, \end{aligned} \tag{4}$$

with $u_{d_1} \neq 0$ and $v_{d_1} \neq 0$. For simplicity in what follows, assume that the degrees of both polynomials are equal, $d_1 = d_2 = d$, so that both U and V satisfy $a^d \le U, V < a^{d+1}$.

A polynomial of degree d is determined uniquely by $d+1$ values. Imitating the approach in Section 1, we evaluate $N \mod a - \delta$ for $d+1$ values of δ. A natural choice might be $\delta = 0, \pm 1, \pm 2, \ldots$, but, to keep our polynomial values positive, we consider non-negative values of δ, and, for good measure, take extra values, $\delta = 0, 1, 2, \ldots 2d$ (by extra, we mean $\delta \le 2d$ rather than $d\ leqd$). Now,

$$N = UV = (u_d\delta^d + \ldots + u_1\delta + u_0)(v_d\delta^d + \ldots + v_1\delta + v_0) \mod a - \delta. \tag{5}$$

Since $0 \le u_j < a$, we have

$$u_d\delta^d + \ldots + u_1\delta + u_0 < a\lambda(d, \delta) \tag{6}$$

where

$$\lambda(d, \delta) = \delta^d + \delta^{d-1} + \ldots + 1 = (\delta^{d+1} - 1)/(\delta - 1) \sim \delta^d, \quad \text{as } \delta \to \infty. \tag{7}$$

and similarly for the v_j's.

For each δ one lists all solutions (x, y) to

$$xy = N \mod a - \delta \tag{8}$$

$$0 < x, y < a\lambda(d, \delta). \tag{9}$$

The number of points (x, y) for a given δ is $\phi(a - \delta)$ per $a \times a$ square, and hence, overall, equals

$$\phi(a - \delta)\lambda(d, \delta)^2 = O(a(2d)^{2d}). \tag{10}$$

We are again assuming that $\gcd(a - \delta, N) = 1$, otherwise one easily pulls out a factor of N.

We need a method to recognize the solutions that we seek $(u_d\delta^d + \ldots + u_0, v_d\delta^d + \ldots + v_0)$ hiding amongst all the (x, y)'s. This leads to the question:

Let $X > 0$ and let $S_0, S_1, \ldots, S_{2d}$ be $2d + 1$ sets of points $\in \mathbb{Z}^2$ all of whose coordinates are positive and $\le X$. Assume that amongst these points there exists

$2d+1$ points, one from each S_δ, whose coordinates are described by polynomials $u(\delta), v(\delta) \in \mathbb{Z}[\delta]$ of degree d. More precisely, for each $0 \le \delta \le 2d$ there exists a point $(x_\delta, y_\delta) \in S_\delta$ such that

$$\begin{aligned} x_\delta &= u(\delta) = u_d\delta^d + \ldots + u_0 \\ y_\delta &= v(\delta) = v_d\delta^d + \ldots + v_0. \end{aligned} \tag{11}$$

Can one find these $2d+1$ points much more efficiently than by exhaustively searching through all possible $2d+1$ tuples of points? For example, can one find these points in time $O(X^\alpha d^{\beta d})$ for some $\alpha, \beta > 0$?

In our application, $X = O(a(2d)^{2d})$. Since $N = UV$ and $a^d \le U < V < a^{d+1}$, we have $a < N^{1/(2d)}$. Assuming that there is an $O(X^\alpha d^{\beta d})$ time algorithm for finding points with polynomial coordinates, on taking d proportionate to

$$\left(\frac{\log N}{\log\log N}\right)^{1/2} \tag{12}$$

one gets a factoring algorithm requiring

$$\exp\left(\gamma(\log N \log\log N)^{1/2}\right) \tag{13}$$

time and storage, for some $\gamma > 0$.

One can cut back a bit on the search space, by noting, for example, that the coefficients of $u(\delta)$ and $v(\delta)$ are integers (this imposes a divisibility restriction on finite differences between points lying on the polynomial), and, in our particular application, that the coefficients are non-negative and bounded, and this restricts the rate of growth of the polynomials. However, to get down to a running time polynomial in X, one needs to do much better.

4. Uniform Distribution

Let $\gcd(a, N) = 1$. A classic application of Kloosterman sums shows that the points $(x, y) \mod a$ satisfying $xy = N \mod a$ become uniformly distributed in the square of side length a as $a \to \infty$. While the tools used in this section are fairly standard, they will also be applied in the next section to estimate the running time of the Hide and Seek algorithm. Similar theorems can be found in the literature ([1], [2], [3], [4], [8], [9], [11]), often with restrictions to prime values of a or to $N = 1$.

Consider the following identity which detects pairs of integers (x, y) such that $xy = N \mod a$:

$$\frac{1}{a}\sum_{k=0}^{a-1} e\left(\frac{k}{a}(y - \bar{x}N)\right) = \begin{cases} 1 & \text{if } xy = N \mod a \\ 0 & \text{otherwise} \end{cases} \tag{14}$$

where $e(z) = \exp(2\pi iz)$, and where $\bar{x}$ stands for any integer congruent to x^{-1} mod a, if the inverse exists. Recall that we have assumed $\gcd(a, N) = 1$ so that any solution to $xy = N \mod a$ must have $\gcd(x, a) = 1$. Thus, for such solutions, $x^{-1} \mod a$ exists.

Let R be the rectangle bounded horizontally by $x_1, x_2 \in \mathbb{Z}$ and vertically by $y_1, y_2 \in \mathbb{Z}$, where $0 \leq x_1 < x_2 \leq a$ and $0 \leq y_1 < y_2 \leq a$:

$$R = R(x_1, x_2, y_1, y_2) = \{(x, y) \in \mathbb{Z}^2 | x_1 \leq x < x_2, y_1 \leq y < y_2\}. \tag{15}$$

Let $c_R(N, a)$ denote the number of pairs of integers (x, y) that lie in the rectangle R, and satisfy $xy = N \mod a$:

$$c_R(N, a) = \sum_{\substack{(x,y)\in R \\ xy=N \mod a}} 1. \tag{16}$$

The identity above gives

$$c_R(N, a) = \frac{1}{a} \sum_{k=0}^{a-1} \sum_{\substack{(x,y)\in R \\ \gcd(x,a)=1}} e\left(\frac{k}{a}(y - \bar{x}N)\right). \tag{17}$$

Notice that we only need to restrict x to $\gcd(x, a) = 1$ and that y runs over all residues in $y_1 \leq y < y_2$. This will allow us to deal with the sum over y as a geometric series.

The $k = 0$ term provides the main contribution while the other terms can be estimated using bounds for Kloosterman sums. We require two lemmas. The first considers the main contribution, and the second bounds the remaining terms.

Lemma 4.1. *The $k = 0$ term in (20) equals*

$$\frac{area(R)}{a^2}\phi(a) + O(a^\epsilon) \tag{18}$$

for any $\epsilon > 0$.

Proof. The $k = 0$ term is

$$\frac{1}{a} \sum_{\substack{(x,y)\in R \\ \gcd(x,a)=1}} 1 = \frac{(y_2 - y_1)}{a} \sum_{\substack{x_1\leq x<x_2 \\ \gcd(x,a)=1}} 1. \tag{19}$$

Using the Mobius function we have

$$\begin{aligned} \sum_{\substack{x_1\leq x<x_2 \\ \gcd(x,a)=1}} 1 &= \sum_{x_1\leq x<x_2} \sum_{d|\gcd(x,a)} \mu(d) = \sum_{d|a} \mu(d) \sum_{x_1/d\leq x<x_2/d} 1 \\ &= \sum_{d|a} \mu(d)((x_2 - x_1)/d + O(1)) = (x_2 - x_1) \prod_{p|a}(1 - 1/p) + O(\tau(a)) \end{aligned} \tag{20}$$

where $\tau(a)$ equals the number of divisors of a and is $O(a^\epsilon)$ for any $\epsilon > 0$. This implies that the $k = 0$ contribution to $c_R(N, a)$ equals

$$\frac{\text{area}(R)}{a^2}\phi(a) + O((y_2 - y_1)a^{-1+\epsilon}) \tag{21}$$

which gives the lemma. □

The next lemma bounds the contribution of the $k \geq 1$ terms in (20).

Lemma 4.2. *For any $\epsilon > 0$ we have*

$$\frac{1}{a}\sum_{k=1}^{a-1} \sum_{\substack{(x,y)\in R \\ \gcd(x,a)=1}} e\left(\frac{k}{a}(y - \bar{x}N)\right) = O(a^{1/2+\epsilon}). \tag{22}$$

Proof. One can separate the sum over y and evaluate it as a geometric series obtaining for the left-hand side above

$$\frac{1}{a}\sum_{k=1}^{a-1} \frac{e\left(\frac{k}{a}y_2\right) - e\left(\frac{k}{a}y_1\right)}{e\left(\frac{k}{a}\right) - 1} \sum_{\substack{x_1 \leq x < x2 \\ \gcd(x,a)=1}} e\left(\frac{-k}{a}\bar{x}N\right). \tag{23}$$

Taking absolute values we get an upper bound of

$$\frac{1}{a}\sum_{k=1}^{a-1} \frac{\left|\sin\left(\frac{\pi k}{a}(y_2 - y_1)\right)\right|}{\left|\sin\left(\frac{\pi k}{a}\right)\right|} \left|\sum_{\substack{x_1 \leq x < x2 \\ \gcd(x,a)=1}} e\left(\frac{-k}{a}xN\right)\right|. \tag{24}$$

Next, notice that the terms k and $a - k$ give the same contribution, so we may restrict our attention to just the terms $1 \leq k \leq (a-1)/2$. If $a-1$ is odd, the middle term is left out at a cost of $O(1)$, and the bound becomes

$$\frac{2}{a}\sum_{1\leq k\leq (a-1)/2} \frac{\left|\sin\left(\frac{\pi k}{a}(y_2 - y_1)\right)\right|}{\left|\sin\left(\frac{\pi k}{a}\right)\right|} \left|\sum_{\substack{x_1 \leq x < x2 \\ \gcd(x,a)=1}} e\left(\frac{-k}{a}\bar{x}N\right)\right| + O(1). \tag{25}$$

The second sum above over x can be expressed in terms of Kloosterman sums, and using estimates for Kloosterman sums one has

$$\sum_{\substack{x_1 \leq x < x2 \\ \gcd(x,a)=1}} e\left(\frac{-k}{a}\bar{x}N\right) = O(a^{1/2+\epsilon}\gcd(k, a)^{1/2}). \tag{26}$$

For a proof, see Lemma 4 on page 36 of Hooley's book [5] where a proof is given (his r corresponds to our a, and his l is $-kN$. Also recall that we are assuming $\gcd(N, a) = 1$ so that N does not appear in the gcd of the O term).

Furthermore, using the Taylor expansion of $\sin(x)$ one obtains the two inequalities

$$\begin{aligned} \sin(x) &\le \min(x,1), \quad x \ge 0, \\ 1/\sin(x) &< 2/x, \quad 0 < x < \pi/2. \end{aligned} \tag{27}$$

For the second inequality, use $0 < x/2 < x - x^3/3! < \sin(x)$ in the stated interval.

Applying (31) and (30) gives an upper bound for (29) of

$$O\left(a^{-1/2+\epsilon} \sum_{1\le k\le (a-1)/2} \min(\frac{\pi k}{a}(y_2-y_1),1)\frac{2a}{\pi k}\gcd(k,a)^{1/2}+1\right). \tag{28}$$

Breaking up the sum into $1 \le k \le a/(\pi(y_2-y_1))$ and $a/(\pi(y_2-y_1)) < k \le (a-1)/2$, the sum over k in the O term equals

$$2(y_2-y_1)\sum_{1\le k\le a/(\pi(y_2-y_1))}\gcd(k,a)^{1/2}+\frac{2a}{\pi}\sum_{a/(\pi(y_2-y_1))<k\le (a-1)/2}\gcd(k,a)^{1/2}/k. \tag{29}$$

Both kinds of sums can be easily handled (the first can also be found in Hooley). Let $X > 0$. Then,

$$\sum_{1\le k\le X}\gcd(k,a)^{1/2} \le \sum_{d|a} d^{1/2}\sum_{\substack{1\le k\le X\\ d|k}} 1 \le X\sum_{d|a} d^{-1/2} = O(Xa^\epsilon). \tag{30}$$

Next, let $0 < X_1 < X_2$. Then

$$\begin{aligned} \sum_{X_1<k\le X_2}\gcd(k,a)^{1/2}/k &\le \sum_{d|a} d^{1/2}\sum_{\substack{X_1<k\le X_2\\ d|k}} 1/k \\ &= O\left(\log(X_2-X_1+2)\sum_{d|a} d^{-1/2}\right) \end{aligned} \tag{31}$$

which equals

$$O(\log(X_2-X_1+2)a^\epsilon). \tag{32}$$

Applying (35) and (38) to (34), we have that (33) is

$$O(a^{1/2+\epsilon}), \tag{33}$$

completing the proof. □

These two lemmas together give the following theorem.

Theorem 4.3. *Let* $\gcd(N,a) = 1$ *and* R *as described in (18). Then,* $c_R(N,a)$*, the number of solutions* (x,y) *to* $xy = N \mod a$ *with* (x,y) *lying in the rectangle* R*, is equal to*

$$\frac{area(R)}{a^2}\phi(a) + O(a^{1/2+\epsilon}) \tag{34}$$

for any $\epsilon > 0$.

This theorem shows that the points (x, y) satisfying $xy = N \mod a$ are uniformly dense in the sense that the rectangle R contains its fair share of solutions, so long as the area of R is of larger size than $a^{3/2+\epsilon}$.

For example, if R is a square, it needs to have side length at least $a^{3/4+\epsilon}$ to contain its fair share of points. This is considerably larger than the side length of $a^{1/2}$ that is used in the algorithm of Section 1.

The papers of Shparlinski ([8], [9]) contain many references to the problem of uniform distribution and discusses improved results on average over N.

5. Second Moment and Running Time

We now examine the assertion made in Section 1 that $O(a^{1+\epsilon})$ time is needed to scan across all a squares of side length $a^{1/2}$ and their immediate neighbors, comparing all pairs of points contained in said squares.

The running time of the algorithm in Section 2.1, i.e., in the case $U \leq V < 2U$ depends on how the solutions to $xy = N \mod a$ and $x'y' = N \mod (a-1)$ are distributed amongst the small squares of side length $a^{1/2}$. In Section 5.1 we will consider the running time of the variant in Section 2.1 which is used for the general situation $1 < U \leq V < N$.

Let S denote one such square, i.e., of side length $a^{1/2}$. Then the running time needed to examine just the square S, looking at all pairs of points (x, y), (x', y') in S is $O(c_S(N, a)c_S(N, a-1))$, which, by the arithmetic geometric inequality is $O(c_S(N, a)^2 + c_S(N, a-1)^2)$. The algorithm also requires us to compare points in neighboring squares, say S_1 and S_2, which, similarly, takes $O(c_{S_1}(N, a)^2 + c_{S_2}(N, a-1)^2)$ time. Hence, the overall running time to compare pairs of points is

$$O\left(\sum_S c_S(N, a)^2 + c_S(N, a-1)^2\right), \tag{35}$$

the sum being over the roughly a squares of side length $a^{1/2}$ that partition the $a \times a$ square $\{(x, y) \in \mathbb{Z}^2 | 0 \leq x, y < a\}$ (at the top and right edges we get rectangles, unless $a^{1/2}$ is an integer).

Consider now the contribution from the points $\mod a$:

$$\sum_S c_S(N, a)^2. \tag{36}$$

For convenience, rather than deal with squares S of side length $a^{1/2}$, we will estimate (42) by making a small adjustment and partitioning the $a \times a$ square into squares of side length

$$b = \lceil a^{1/2} \rceil. \tag{37}$$

We also assume that $\gcd(b,a)=1$. If not, replace b with $b+1$ until this condition holds. By equation (23), this will not take long to occur, so that, for any $\epsilon>0$, $b=a^{1/2}+O(a^{\epsilon})$.

Thus, consider the squares

$$B_{ij}=\{(x,y)\in\mathbb{Z}^2|ib\le x<(i+1)b, jb\le y<(j+1)b\} \tag{38}$$

with $0\le i,j<a/b-1$.

Since $b\nmid a$, these will not entirely cover the $a\times a$ square, but the number of points $(x,y)\in\mathbb{Z}^2$ satisfying $xy=N \mod a$ that are neglected at the right most and top portions of the $a\times a$ square is, by (23), $O(\phi(a)b/a)$, and these therefore contribute $O(\phi(a)^2b^2/a^2)=O(\phi(a))$ to (42).

The points $(x,y)\in\mathbb{Z}^2$ belonging to an $a^{1/2}\times a^{1/2}$ square S are contained entirely in at most four squares, say $B_{i_1j_1}, B_{i_2j_2}, B_{i_1j_3}, B_{i_4j_4}$, of side length b. Therefore,

$$c_S(N,a)^2\le(c_{B_{i_1j_1}}(N,a)+c_{B_{i_2j_2}}(N,a)+c_{B_{i_3j_3}}(N,a)+c_{B_{i_4j_4}}(N,a))^2 \tag{39}$$

which, by the Cauchy-Schwartz inequality is

$$\le 4(c_{B_{i_1j_1}}(N,a)^2+c_{B_{i_2j_2}}(N,a)^2+c_{B_{i_3j_3}}(N,a)^2+c_{B_{i_4j_4}}(N,a)^2). \tag{40}$$

Since each B square overlaps with $O(1)$ S squares, we thus have that

$$\sum_S c_S(N,a)^2=O\left(\phi(a)+\sum_B c_B(N,a)^2\right), \tag{41}$$

the $\phi(a)$ accounting for the contribution from the neglected portion at the right most and top portions of the $a\times a$ square.

A similar consideration for the points satisfying $xy=N \mod (a-1)$, partitioning the larger $a\times a$ square into squares D of side length d, where d is the smallest integer greater than $\lceil (a-1)^{1/2}\rceil$ which is coprime to a, gives the same kind of sum

$$\sum_S c_S(N,a-1)^2=O\left(\phi(a-1)+\sum_D c_D(N,a-1)^2\right). \tag{42}$$

Therefore, we need to estimate the second moment

$$\sum_B c_B(N,a)^2 \tag{43}$$

where B ranges over all $\lfloor a/b\rfloor^2$ squares of the form (44). To prove that the running time of the hide and seek algorithm of Section 1 is $O(N^{1/3+\epsilon})$ we need to prove that (49) is $O(a^{1+\epsilon})$.

Theorem 5.1. *Let*

$$P = \{B_{ij}\}_{0 \leq i,j < a/b-1} \tag{44}$$

Then

$$\sum_{B \in P} c_B(N,a)^2 = O(a^{1+\epsilon}). \tag{45}$$

Proof. Rather than look at just the $a \times a$ square, it is helpful to consider the $ba \times ba$ square $\{(x,y) \in \mathbb{Z}^2 | 0 \leq x, y < ba\}$. The advantage of looking at the larger square will become apparent when we turn to the discrete Fourier transform, and will be summing over all the ath roots of unity.

This larger square can be partitioned into b^2 squares of side length a. Because the solutions to $xy = N \mod a$ repeat $\mod a$, we can count each $c_B(N,a)^2$ once per $a \times a$ square, by summing $c_{B'}(N,a)^2 = c_B(N,a)^2$ over all b^2 translates $B' = B + (r_1 a, r_2 a)$ of B, with $0 \leq r_1, r_2 < b$.

On the other hand, we can also partition the $ba \times ba$ square into a^2 squares of side length b:

$$P_2 = \{B_{ij}\}_{0 \leq i,j \leq a-1} \tag{46}$$

with B_{ij} given by (44).

Each translate of a B square, $B' = B + (r_1 a, r_2 a)$, is covered by at most four $B_{ij} \in P_2$, and each $B_{ij} \in P_2$ overlaps at most four such translates of B.

Hence, applying the Cauchy-Schwartz inequality as before,

$$b^2 \sum_{B \in P} c_B(N,a)^2 = O\left(\sum_{B \in P_2} c_B(N,a)^2 \right). \tag{47}$$

To study $c_B(N,a)^2$ we multiply equation (20) by its conjugate, giving

$$c_B(N,a)^2 = \frac{1}{a^2} \sum_{0 \leq k_1, k_2 \leq a-1} \sum_{\substack{(x_1,y_1) \in B \\ \gcd(x_1,a)=1}} \sum_{\substack{(x_2,y_2) \in B \\ \gcd(x_2,a)=1}} e\left(\frac{k_1}{a}(y_1 - \bar{x_1}N) - \frac{k_2}{a}(y_2 - \bar{x_2}N) \right). \tag{48}$$

Next, sum over all $B_{ij} \in P_2$, and break up each sum over $(x,y) \in B_{ij}$ into a double sum $ib \leq x < (i+1)b$, $jb \leq y < (j+1)b$,

$$\sum_{B \in P_2} c_B(N,a)^2 = \frac{1}{a^2} \sum_{0 \leq k_1, k_2 \leq a-1} \left(\sum_{i=0}^{a-1} \sum_{\substack{ib \leq x_1, x_2 < (i+1)b \\ \gcd(x_1,a)=\gcd(x_2,a)=1}} e\left(-\frac{N}{a}(k_1 \bar{x_1} - k_2 \bar{x_2}) \right) \right) \times \left(\sum_{j=0}^{a-1} \sum_{jb \leq y_1, y_2 < (j+1)b} e\left(\frac{k_1 y_1 - k_2 y_2}{a} \right) \right). \tag{49}$$

Now, the inner most sum,

$$\sum_{jb \le y_1, y_2 < (j+1)b} e\left(\frac{k_1 y_1 - k_2 y_2}{a}\right), \tag{50}$$

is a product of two geometric series and equals

$$e((k_1 - k_2)jb/a)\frac{e(k_1 b/a) - 1}{e(k_1/a) - 1}\frac{e(-k_2 b/a) - 1}{e(-k_2/a) - 1}. \tag{51}$$

We understand $\frac{e(kb/a)-1}{e(k/a)-1}$ to equal b if $k = 0 \mod a$. Summing (58) over $0 \le j \le a-1$ gives

$$\begin{cases} a\left|\frac{e(k_1 b/a)-1}{e(k_1/a)-1}\right|^2 & \text{if } k_1 = k_2 \mod a \\ 0 & \text{otherwise} \end{cases}$$

(recall that we have chosen b so that $\gcd(b, a) = 1$). Therefore, only the terms with $k_1 = k_2$ contribute to (56) and it equals

$$\frac{1}{a}\sum_{k=0}^{a-1}\left(\sum_{i=0}^{a-1}\sum_{\substack{ib \le x_1, x_2 < (i+1)b \\ \gcd(x_1,a)=\gcd(x_2,a)=1}} e\left(-\frac{N}{a}(k(\bar{x_1} - \bar{x_2}))\right)\right)\left|\frac{e(kb/a) - 1}{e(k/a) - 1}\right|^2. \tag{52}$$

The $k = 0$ term gives, on separating the sum over x_1 and x_2,

$$\frac{b^2}{a}\sum_{i=0}^{a-1}\left(\sum_{\substack{ib \le x < (i+1)b \\ \gcd(x,a)=1}} 1\right)^2 \tag{53}$$

which, by (23) and using $b \sim a^{1/2}$, equals

$$b^2(\phi(a)b/a + O(a^{\epsilon}))^2 = O(\phi(a)^2). \tag{54}$$

Next, we deal with the terms $1 \le k \le a-1$. The sum over i in (60) equals

$$\sum_{i=0}^{a-1}\sum_{ib \le x_1, x_2 < (i+1)b} A_{x_1,x_2}(-Nk), \tag{55}$$

where

$$A_{x_1,x_2}(t) = \begin{cases} 0 & \text{if } \gcd(x_1 x_2, a) > 1 \\ e(t(\bar{x_1} - \bar{x_2})/a) & \text{otherwise.} \end{cases}$$

To analyze this sum, we use the two dimensional discrete Fourier transform

$$\hat{A}_{m_1,m_2}(t) = \sum_{0 \le x_1, x_2 \le a-1} A_{x_1,x_2}(t)e\left(-\frac{m_1 x_1 + m_2 x_2}{a}\right), \tag{56}$$

so that

$$A_{x_1,x_2}(t) = \frac{1}{a^2} \sum_{0\le m_1,m_2\le a-1} \hat{A}_{m_1,m_2}(t) e\left(\frac{m_1x_1+m_2x_2}{a}\right), \tag{57}$$

and (63) equals, on changing order of summation,

$$\frac{1}{a^2} \sum_{0\le m_1,m_2\le a-1} \hat{A}_{m_1,m_2}(-Nk) \left(\sum_{0\le i\le a-1} \sum_{ib\le x_1,x_2<(i+1)b} e\left(\frac{m_1x_1+m_2x_2}{a}\right)\right). \tag{58}$$

The bracketed sum over i is similar to the sum over j worked out above and equals

$$\begin{cases} a\left|\frac{e(m_1b/a)-1}{e(m_1/a)-1}\right|^2 & \text{if } m_2 = -m_1 \mod a \\ 0 & \text{otherwise.} \end{cases}$$

Therefore, (67) equals

$$\frac{1}{a}\sum_{m=0}^{a-1} \hat{A}_{m,a-m}(-Nk) \left|\frac{e(mb/a)-1}{e(m/a)-1}\right|^2. \tag{59}$$

So, (60), and hence (56), equals

$$\frac{1}{a^2}\sum_{k=0}^{a-1}\sum_{m=0}^{a-1} \hat{A}_{m,a-m}(-Nk) \left|\frac{e(mb/a)-1}{e(m/a)-1}\right|^2 \left|\frac{e(kb/a)-1}{e(k/a)-1}\right|^2. \tag{60}$$

But,

$$\begin{aligned} \hat{A}_{m,a-m}(-Nk) &= \sum_{\substack{0\le x_1,x_2\le a-1\\ \gcd(x_1x_2,a)=1}} e\left(-\frac{Nk}{a}(\bar{x_1}-\bar{x_2})\right) e\left(-\frac{mx_1-mx_2}{a}\right) \\ &= \left|\sum_{\substack{0\le x\le a-1\\ \gcd(x,a)=1}} e\left(-\frac{Nk\bar{x}+mx}{a}\right)\right|^2. \end{aligned} \tag{61}$$

However, the sum on the right-hand side is a Kloosterman sum,

$$\sum_{\substack{0\le x\le a-1\\ \gcd(x,a)=1}} e\left(-\frac{Nk\bar{x}+mx}{a}\right) = S(-m,-Nk,a), \tag{62}$$

and are known [10] [6] to satisfy the bound

$$|S(-m,-Nk,a)| \le \tau(a)\gcd(m,k,a)^{1/2}a^{1/2} = O(a^{1/2+\epsilon}\gcd(k,a)^{1/2}) \tag{63}$$

(recall we are assuming that $\gcd(N,a) = 1$ so that N does not appear on the right-hand side of this inequality). Applying this bound to $\hat{A}_{m,a-m}(-Nk)$, shows that (70) is

$$O\left(\frac{a^\epsilon}{a}\sum_{k=1}^{a-1}\sum_{m=0}^{a-1}\gcd(k,a)\left|\frac{e(mb/a)-1}{e(m/a)-1}\right|^2\left|\frac{e(kb/a)-1}{e(k/a)-1}\right|^2+\phi(a)^2\right). \quad (64)$$

The $\phi(a)^2$ terms comes from the $k = 0$ contribution, (62). We must isolate this term, otherwise the estimate below will be too large.

Separating sums gives

$$O\left(\frac{a^\epsilon}{a}\left(\sum_{k=1}^{a-1}\gcd(k,a)\left|\frac{e(kb/a)-1}{e(k/a)-1}\right|^2\right)\left(\sum_{m=0}^{a-1}\left|\frac{e(mb/a)-1}{e(m/a)-1}\right|^2\right)+\phi(a)^2\right). \quad (65)$$

Both sums can be bounded using the same approach as for (28) in the previous section, namely: combining terms k and $a-k$ (similarly for the m sum, but taking the $m = 0$ term alone), breaking up the sum into the terms with $k \le a/(\pi b) \sim a^{1/2}/\pi$ (respectively, m), applying inequalities (31), estimating the resulting sums, using $b \sim a^{1/2}$, we find, for any $\epsilon > 0$, that (76) equals

$$O(b^2 a^{1+\epsilon}). \quad (66)$$

We have thus estimated the sum that appears on the right-hand side of (53). The sum that we wish to bound appears on the left-hand side of (53) but with an extra factor of b^2. Hence, dividing the above by b^2 gives $O(a^{1+\epsilon})$ for the sum in theorem. □

Remark: In certain cases, such as when $a = p^2$, with p prime, one can improve the above estimate for the second moment to $O(a)$ by taking $b = p$ and, for $x = jp + l$, with $\gcd(l,p) = 1$, using $\bar{x} = \bar{l}^2(l - jp)$.

5.1. Running Time of the Variant for $1 < U \le V < N$

Instead of partitioning the $a \times a$ square into smaller squares of side length $b \sim a^{1/2}$, we partition it into rectangles R of width $w < a$ and height $h < a$, where $w, h \in \mathbb{Z}$ and, for convenience, $\gcd(w,a) = \gcd(h,a) = 1$.

We partition the $a \times a$ square and also the larger $wa \times ha$ rectangle into smaller rectangles R:

$$\begin{aligned} R &= R_{ij} = \{(x,y) \in \mathbb{Z}^2 | iw \le x < (i+1)w, jh \le y < (j+1)h\} \\ Q &= \{R_{ij}\}_{\substack{0\le i<a/w-1 \\ 0\le j<a/h-1}} \\ Q_2 &= \{R_{ij}\}_{0\le i,j\le a-1}. \end{aligned}$$

As in Section 4.1, we have

$$wh \sum_{R \in Q} c_R(N,a)^2 = O\left(\sum_{R \in Q_2} c_R(N,a)^2 \right). \tag{67}$$

with wh appearing on the left-hand side since the large $wa \times ha$ rectangle has that many copies of the $a \times a$ square.

Using the discrete Fourier transform, as before,

$$\sum_{R \in Q_2} c_R(N,a)^2 = \frac{1}{a^2} \sum_{k=0}^{a-1} \sum_{m=0}^{a-1} |S(-m,-Nk,a)|^2 \left| \frac{e(mw/a)-1}{e(m/a)-1} \right|^2 \left| \frac{e(kh/a)-1}{e(k/a)-1} \right|^2. \tag{68}$$

This useful identity expresses the second moment for the larger $wa \times ha$ rectangle as a sum involving Kloosterman sums.

The $k = 0$ term can be estimated as in (62) and asymptotically equals

$$\frac{h^2 w^2}{a^2} \phi(a)^2. \tag{69}$$

For the $k \geq 1$ terms, we use bound (74) to estimate the Kloosterman sums and separate the double sum above to get a contribution of

$$O\left(\frac{a^\epsilon}{a} \left(\sum_{k=1}^{a-1} \gcd(k,a) \left| \frac{e(kh/a)-1}{e(k/a)-1} \right|^2 \right) \left(\sum_{m=0}^{a-1} \left| \frac{e(mw/a)-1}{e(m/a)-1} \right|^2 \right) \right). \tag{70}$$

The first sum is estimated to equal $O(\tau(a)ah)$ while the second sum is $O(aw)$, giving, for $k \geq 1$, a contribution of

$$O(a^{1+\epsilon} wh) \tag{71}$$

for any $\epsilon > 0$. Putting (83) and (85) together, then dividing the left-hand side of (81) by wh gives the following estimate for the second moment:

Theorem 5.2. *Let $1 < w, h < a$, with $\gcd(w,a) = \gcd(h,a) = 1$. Then, using the notation above, we have an estimate for the second moment that depends on the area wh of the rectangles R:*

$$\sum_{R \in Q} c_R(N,a)^2 = \begin{cases} O(a^{1+\epsilon}) & \text{if } wh = O(a^{1+\epsilon}), \\ O(wh\phi(a)^2/a^2) & \text{if } wh \gg a^\lambda \text{ for some } \lambda > 1. \end{cases} \tag{72}$$

Remark: if $\gcd(w,a) = \gcd(h,a) = 1$ does not hold, one can bound the left-hand side of (86) by comparing with the same kind of sum, but where w and h are incremented, as before, by at most $O(a^\epsilon)$ until this gcd condition holds. So long as $w, h \gg a^\epsilon$ to begin with, the estimates in the above theorem are unaffected.

In Section 1.2, our choice of w and h has $wh = O(a)$, and the estimate for the second moment is thus $O(a^{1+\epsilon})$, as in the previous section.

The second estimate of the theorem (not relevant for our particular application), $O(wh\phi(a)^2/a^2)$, can probably be turned into an asymptotic formula and a central limit theorem proven. This will remain an inquiry for the future.

Acknowledgements I wish to thank Andrew Granville, Jeffrey Lagarias, Carl Pomerance, and Matthew Young for helpful feedback, and the referee for making a number of suggested improvements to the presentation and description of the algorithms.

References

[1] J. Beck, and M. R. Khan, *On the uniform distribution of inverses modulo n*, Periodica Mathematica Hungarica **44** (2002), 147–155.

[2] F. P. Boca, C. Cobeli and A. Zaharescu, *Distribution of Lattice Points Visible from the Origin*, Commun. Math. Phys. **213** (2000), 433–470.

[3] A. Fujii, and Y.Kitaoka, *On plain lattice points whose coordinates are reciprocals modulo a prime*, Nagoya Math. J. **147** (1997), 137–146.

[4] D.R. Heath-Brown, *Arithmetic applications of Kloosterman sums*, Nieuw Arch. Wiskd. (5) **1** (2000), no. 4, 380–384.

[5] C. Hooley, *Applications of sieve methods to the theory of numbers*, Cambridge University Press (1976).

[6] H. Iwaniec, E. Kowalski, *Analytic Number Theory*, Colloquium Publications **53**, American Math. Soc., Providence, RI, 2004, Chapter 11.

[7] R.S. Lehman, *Factoring large integers*, Math. Comp. 28 (1974), 637–646.

[8] I. Shparlinski, *Distribution of points on modular hyperbolas*, Sailing on the Sea of Number Theory: Proc. 4th China-Japan Seminar on Number Theory, Weihai, 2006, World Scientific, 2007, 155–189.

[9] I. Shparlinski, *Modular hyperbolas*, `http : //arxiv.org/abs/1103.2879`.

[10] A. Weil, *On some exponential sums*, Proc. Nat. Acad. Sci. USA **34** (1948), 204–207.

[11] W. Zhang, *On the distribution of inverses modulo n*, Journal of Number theory **61** (1996), 301–310.

DISTRIBUTION LAWS OF PAIRS OF DIVISORS

S. Nyandwi
Université du Burundi, Département de Mathématiques, Bujumbura, Burundi
nservat@yahoo.fr

A. Smati
Université de Limoges, UMR-CNRS 7252, Mathématiques, Limoges, France
smati@unilim.fr

Received: 8/17/11, Revised: 12/17/12, Accepted: 3/8/13, Published: 3/15/13

Abstract
In this paper we study the distribution of pairs (d_1, d_2) of positive integers such that the product d_1d_2 divides a given integer n from a probabilistic point of view. The number of these pairs, denoted by $\tau_3(n)$, is equal to the number of ways to write n as a product of three positive integers. To these pairs we associate a random vector taking the values $((\log d_1)/(\log n), (\log d_2)/(\log n))$ with uniform probability $1/\tau_3(n)$ and its distribution function F_n. We show that the mean of F_n uniformly converges to the distribution function of the Beta two-dimensional law (Dirichlet law). Our study generalizes a work done by Deshouillers, Dress and Tenenbaum in the case of the divisors of an integer where they showed that the average distribution of divisors of a given integer follows the arcsine law.

1. Introduction

In order to study the distribution of divisors of a given integer n, Deshouillers, Dress and Tenenbaum [1], introduce the random variable D_n which takes the values $(\log d)/(\log n)$ as d runs through all divisors of n with uniform probability $1/\tau_2(n)$, where $\tau_2(n)$ is the number of divisors of n, and its distribution function $G_n(u) := Prob(D_n \leqslant u)$, $u \in [0, 1]$. The sequence $(G_n)_n$ does not converge pointwise in $[0, 1]$, they studied its mean value and showed that

$$\frac{1}{x}\sum_{n\leqslant x} G_n(u) = \frac{1}{x}\sum_{n\leqslant x} Prob(D_n \leqslant u) = \frac{2}{\pi}\arcsin(\sqrt{u}) + O\left(\frac{1}{\sqrt{\log x}}\right),$$

uniformly for $x \geqslant 2$ and $u \in [0, 1]$. Moreover, the order of the remainder term's magnitude is optimal if the uniformity in $[0, 1]$ is required. The method is based on the sums estimation $\Sigma_{n\leqslant x}1/\tau_2(kn)$; see Théorème T of [1] and also II.5 of [2]. In the present work we are interested in the distribution of pairs (d_1, d_2) of positive integers such that the product d_1d_2 divides n. The number of these pairs is equal to the number of ways to write n as a product of three positive integers, which will be denoted as $\tau_3(n)$. We consider the random vector

$$(X_n, Y_n) : \{(d_1, d_2) : d_1d_2|n\} \longrightarrow [0,1] \times [0,1],$$

which takes the values $((\log d_1)/(\log n), (\log d_2)/(\log n))$ with uniform probability equal to $1/\tau_3(n)$ and its distribution function, given by

$$F_n(u,v) := Prob(X_n \leqslant u, Y_n \leqslant v) = \frac{1}{\tau_3(n)} \sum_{qm|n, q\leqslant n^u, m\leqslant n^v} 1\,.$$

The sequence $(F_n)_n$ does not converge pointwise on $[0,1] \times [0,1]$, as can be easily seen by observing that for a fixed (u_0, v_0), $\frac{1}{3} < u_0 < \frac{2}{3}$ and $\frac{1}{3} < v_0 < \frac{2}{3}$, the subsequences $(F_p(u_0, v_0))_p$ and $(F_{p^3}(u_0, v_0))_{p^3}$ with p as a prime number, do not converge to the same limit. We will study the convegence of the mean of $(F_n)_n$:

$$\frac{1}{x}\sum_{n\leqslant x} F_n(u,v) = \frac{1}{x}\sum_{n\leqslant x} Prob(X_n \leqslant u, Y_n \leqslant v) = \frac{1}{x}\sum_{n\leqslant x}\frac{1}{\tau_3(n)} \sum_{qm|n, q\leqslant n^u, m\leqslant n^v} 1,$$

which gives the average distribution of solutions of the equation $xyz = n$ in integers $x \geqslant 1, y \geqslant 1, z \geqslant 1$. In the sequel, we will use the notation:

$$S(x;u,v) := \sum_{n\leqslant x}\frac{1}{\tau_3(n)} \sum_{qm|n, q\leqslant n^u, m\leqslant n^v} 1\,. \tag{1}$$

2. Statement of the Theorem

Denote by Γ the Euler gamma function and for $a, b \in]0, +\infty[$ let

$$B(a,b) = \int_0^1 y^{a-1}(1-y)^{b-1}dy = \frac{\Gamma(a)\Gamma(b)}{\Gamma(a+b)}.$$

Set

$$T_1 = \{(u,v) \in [0,1] \times [0,1] : u+v < 1\}; T_2 = \{(u,v) \in [0,1] \times [0,1] : u+v \geqslant 1\}\,.$$

The following theorem shows that the mean of the distribution function defined above uniformly converges in T_1 to the distribution function of the Beta two-dimensional law which has parameters $1/3, 1/3, 1/3$ and uniformly converges in T_2 to a sum of distribution functions of the Beta-dimensional laws which has parameters $2/3, 1/3$.

Theorem 2.1. *1. Uniformly for $x > 1$ and $(u,v) \in T_1$, we have*

$$\frac{1}{x}\sum_{n\leqslant x} Prob(X_n \leqslant u, Y_n \leqslant v) = \frac{1}{\Gamma^3(\frac{1}{3})}\int_0^u\int_0^v y^{-\frac{2}{3}}z^{-\frac{2}{3}}(1-y-z)^{-\frac{2}{3}}dydz + O\left(\frac{1}{\sqrt[3]{\log x}}\right).$$

2. Uniformly for $x > 1$ and $(u,v) \in T_2$, we have

$$\frac{1}{x}\sum_{n\leqslant x} Prob(X_n \leqslant u, Y_n \leqslant v) = -1 + \frac{1}{B(\frac{2}{3},\frac{1}{3})}\int_0^u y^{-\frac{1}{3}}(1-y)^{-\frac{2}{3}}\,dy + \frac{1}{B(\frac{2}{3},\frac{1}{3})}\int_0^v y^{-\frac{1}{3}}(1-y)^{-\frac{2}{3}}\,dy + O\left(\frac{1}{\sqrt[3]{\log x}}\right).$$

Remark 2.2. The remainder term $O\left(\frac{1}{\sqrt[3]{\log x}}\right)$ in Theorem 2.1 is optimal if uniformity in (u,v) is required. Indeed, by using partial summation and lemma 3.1 below, we can show that for $0 \leqslant v < (\log 2)/(\log x)$

$$\frac{1}{x}\sum_{n\leqslant x} Prob(X_n \leqslant \frac{1}{2}, Y_n \leqslant v) \sim \frac{c_2\Gamma(\frac{1}{3})}{\Gamma(\frac{2}{3})}\frac{1}{\sqrt[3]{\log x}}, \quad (x\to+\infty),$$

where c_2 is a constant defined in (4) below.
We also note that the transition from the first formula to the second in Theorem 2.1 is regular. Indeed, we can show that for (u,v) such $u+v=1$,

$$-1 + \frac{1}{B(\frac{2}{3},\frac{1}{3})}\int_0^u y^{-\frac{1}{3}}(1-y)^{-\frac{2}{3}}\,dy + \frac{1}{B(\frac{2}{3},\frac{1}{3})}\int_0^{1-u} y^{-\frac{1}{3}}(1-y)^{-\frac{2}{3}}\,dy$$
$$= \frac{1}{\Gamma^3(\frac{1}{3})}\int_0^u\int_0^{1-u} y^{-\frac{2}{3}}z^{-\frac{2}{3}}(1-y-z)^{-\frac{2}{3}}dydz$$

For $x \geqslant 2$, we set

$$\epsilon_x := \frac{\log 2}{\log x}, \quad \epsilon''_x := (\frac{\log 2}{\log x})^{\eta}, \tag{2}$$

where $0 < \eta < 1/3$ is an arbitrary fixed number.

$$\begin{aligned} T_x &:= \{(u,v) \in [0,1]\times[0,1] : u+v \leqslant 1-\epsilon''_x\}, \\ \overline{T}_x &:= \{(u,v) \in [0,1]\times[0,1] : 1-\epsilon''_x < u+v < 1\}. \end{aligned} \tag{3}$$

For technical reasons, we divide the proof of Theorem 2.1 into two parts. In the first part we prove the first formula for (u,v) in T_x and in the second one we prove the same formula for (u,v) in $\overline{T}_x$ and we also prove the second formula for (u,v) in T_2. However, the two parts use the same ideas. In Section 3, we will give some necessary lemmas. In Section 4, we will give full proof of Theorem 2.1 for (u,v) in T_x. In Section 5, to avoid repetitions, we will just describe the proof of Theorem 2.1 for (u,v) in $\overline{T}_x$ and for (u,v) in T_2 without details. All notations introduced here will be retained throughout the rest of the article.

3. Lemmas

We introduce two multiplicative functions h_1 et h_2 that will be used in the sequel. For a prime power they are defined by

$$h_1(p^r) = \left(\sum_{j\geqslant 0} \frac{1}{p^j\tau_3(p^{j+r})}\right)\left(\sum_{j\geqslant 0} \frac{1}{p^j\tau_3(p^j)}\right)^{-1}$$

and

$$h_2(p^r) = \left(\sum_{j\geqslant 0} \frac{h_1(p^{j+r})}{p^j}\right)\left(\sum_{j\geqslant 0} \frac{h_1(p^j)}{p^j}\right)^{-1}.$$

We also set

$$c_1 := \prod_p (1-\frac{1}{p})^{\frac{1}{3}} \sum_{j\geqslant 0} \frac{p^{-j}}{C_{j+2}^2};\ c_2 := \prod_p (1-\frac{1}{p})^{\frac{1}{3}} \sum_{j\geqslant 0} \frac{h_1(p^j)}{p^j};$$

$$c_3 := \prod_p (1-\frac{1}{p})^{\frac{1}{3}} \sum_{j\geqslant 0} \frac{h_2(p^j)}{p^j}. \quad (4)$$

Lemma 3.1. *The following hold:*

1. *For every $\theta \in]0,+\infty[$ and every integer $d \geqslant 1$, there is a positive constant $M_{\theta,d} := (1/3+\theta)^{\omega(d)}$, where $\omega(d)$ is the number of prime divisors of d, such that uniformly for any real number $x \geqslant 2$ and any integer $d \geqslant 1$ we have*

$$\sum_{n\leqslant x} \frac{1}{\tau_3(dn)} = \frac{c_1 h_1(d)}{\Gamma(\frac{1}{3})} \frac{x}{\log^{\frac{2}{3}}(x)} + O\left(\frac{M_{\theta,d}\, x}{\log^{\frac{5}{3}}(x)}\right);$$

$$\sum_{n\leqslant x} h_1(dn) = \frac{c_2 h_2(d)}{\Gamma(\frac{1}{3})} \frac{x}{\log^{\frac{2}{3}}(x)} + O\left(\frac{M_{\theta,d}\, x}{\log^{\frac{5}{3}}(x)}\right).$$

2. *For every $\theta \in]0,+\infty[$, and $i = 1, 2$, we uniformly have for $x \geqslant 2$,*

$$\sum_{n\leqslant x} h_2(n) = \frac{c_3}{\Gamma(\frac{1}{3})} \frac{x}{\log^{\frac{2}{3}}(x)} + O\left(\frac{x}{\log^{\frac{5}{3}}(x)}\right);\sum_{n\leqslant x} \frac{h_i(n)}{n} = O\left(\log^{\frac{1}{3}}(x)\right);$$

$$\sum_{n\leqslant x} \frac{M_{\theta,n}}{n} = O\left(\log^{\frac{1}{3}+\theta}(x)\right).$$

Proof. The lemma is an immediate consequence of Théorème T of [1]. □

Lemma 3.2. *The following two equalities hold:*
1. *for any $x \in]0,1[$, $\sum_{r\geqslant 0}\sum_{j\geqslant 0}\sum_{\ell\geqslant 0} \frac{x^{r+j+\ell}}{C_{r+j+\ell+2}^{r+j+\ell}} = \frac{1}{1-x}$;*
2. *$c_1c_2c_3 = 1$.*

Proof. 1. We clearly have

$$\frac{x^{r+j+\ell}}{C_{r+j+\ell+2}^{r+j+\ell}} = \frac{2x^{r+j+\ell}}{r+j+\ell+1} - \frac{2x^{r+j+\ell}}{r+j+\ell+2} = \frac{2}{x}\int_0^x t^{r+j+\ell}dt - \frac{2}{x^2}\int_0^x t^{r+j+\ell+1}dt\,.$$

Then

$$\sum_{r\geqslant 0}\sum_{j\geqslant 0}\sum_{\ell\geqslant 0}\frac{x^{r+j+\ell}}{C_{r+j+\ell+2}^{r+j+\ell}} = \frac{2}{x}\int_0^x \frac{1}{(1-t)^3}dt - \frac{2}{x^2}\int_0^x \frac{t}{(1-t)^3}dt = \frac{1}{1-x}\cdot$$

2. From definitions, we have

$$c_1c_2c_3 = \prod_p \left(1-\tfrac{1}{p}\right)\left(\sum_{r\geqslant 0} p^{-r}\sum_{j\geqslant 0}\tfrac{p^{-j}}{C_{j+r+2}^{j+r}}\right)\left(\sum_{j\geqslant 0} p^{-j}\sum_{\ell\geqslant 0}\tfrac{p^{-\ell}}{C_{j+\ell+2}^{j+\ell}}\right)^{-1}\times$$

$$\left(\sum_{r\geqslant 0}\sum_{j\geqslant 0}\sum_{\ell\geqslant 0}\tfrac{p^{-r-j-\ell}}{C_{r+j+\ell+2}^{r+j+\ell}}\right) = \prod_p \left(1-\tfrac{1}{p}\right)\left(\sum_{r\geqslant 0}\sum_{j\geqslant 0}\sum_{\ell\geqslant 0}\tfrac{(p^{-r-j-\ell}}{C_{r+j+\ell+2}^{r+j+\ell}}\right),$$

which immediately yields $c_1c_2c_3 = 1$ by the formula proved in 1. □

Lemma 3.3. *1. For $x \geqslant 2$, let $\epsilon_x := (\log 2)/(\log x)$ and $(u,v) \in [0,1]^2$ be such that $\epsilon_x \leqslant u+v \leqslant 1-\epsilon_x$. For $2 \leqslant t \leqslant x^u$, set*

$$I(t,x,v) := \int_{\epsilon_x}^{v} z^{-\frac{2}{3}}(1-\frac{\log t}{\log x}-z)^{-\frac{2}{3}}dz,\ J(x,u,v) := \int_2^{x^u}(\log t)^{-\frac{2}{3}}\frac{\partial}{\partial t}I(t,x,v)\,dt\,.$$

Then, we uniformly have $J(x,u,v) = O(1)$.

2. Let

$$I(t,x,v) := \int_v^{1-\epsilon_x-\frac{\log t}{\log x}} z^{-\frac{2}{3}}(1-\frac{\log t}{\log x}-z)^{-\frac{2}{3}}dz.$$

Uniformly for $\epsilon_x \leqslant v \leqslant 1$ and $x \geqslant 2$, we have

$$J_1(x,v) := \int_2^{\frac{x}{2}} \log^{-5/3}(t)\,I(t,x,v)\frac{dt}{t} = O(1),$$

and

$$J(x,v) := \int_2^{\frac{x}{2}} \log^{-2/3}(t)\,\frac{\partial}{\partial t}I(x,t,v)\,dt = O(1).$$

Proof. 1. We have $\frac{\partial}{\partial t}I(t,x,v) = \frac{2}{3t\log x}\int_{\epsilon_x}^{v} z^{-\frac{2}{3}}(1-\frac{\log t}{\log x}-z)^{-\frac{5}{3}}dz$. By a change of variable $y = \log t/\log x$, we obtain $J(x,u,v) = \frac{2}{3(\log x)^{2/3}}\int_{\epsilon_x}^{u}\int_{\epsilon_x}^{v} y^{-\frac{2}{3}}z^{-2/3}(1-y-z)^{-\frac{5}{3}}dzdy$. When $y \to 0$ (resp. $y \to 1$), we have $z \to 0$ or $z \to 1$ (resp. $z \to 0$), as $\epsilon_x \leqslant u+v \leqslant 1-\epsilon_x$. The integrand is therefore equivalent to $y^{-\frac{2}{3}}z^{-\frac{2}{3}}$ or to $y^{-\frac{2}{3}}(1-z)^{-5/3}$ (resp. to $z^{-\frac{2}{3}}(1-y)^{-5/3}$). An easy calculation yields $J(x,u,v) = O(1)$.
2. The proof is similar to 1. □

Lemma 3.4. *1. For $x \geqslant 2$, let $\epsilon''_x := (\log 2)^\eta/(\log x)^\eta$, $0 < \eta < 1/3$. For $u \in [\epsilon_x, 1-\epsilon''_x]$, $v \in [\sqrt{\epsilon_x}, 1-\epsilon''_x]$, $u+v \leqslant 1-\epsilon''_x$, and $2 \leqslant q \leqslant x^u$, we uniformly have*

$$\widehat{I}(x,v) := \int_{x^{\sqrt{\epsilon_x}}}^{x^v} \log^{-\frac{5}{3}}(t)\, \log^{-\frac{2}{3}}(\frac{x/q}{t})\, \frac{dt}{t} = O\left((\log x)^{-1+\frac{4}{3}\eta}\right),$$

and

$$\widehat{\widehat{I}}(x,v) := \int_{x^{\sqrt{\epsilon_x}}}^{x^v} \log^{-\frac{2}{3}}(t)\, \log^{-\frac{5}{3}}(\frac{x/q}{t})\, \frac{dt}{t} = O\left((\log x)^{-1+\frac{4}{3}\eta}\right).$$

2. For $\epsilon_x \leqslant u \leqslant 1-\epsilon''_x$, $2 \leqslant m \leqslant x^{\sqrt{\epsilon_x}}$ and $u+v \leqslant 1-\epsilon''_x$, we uniformly have

$$\overline{I}(x,u) := \int_{2}^{x^u} \log^{-\frac{5}{3}}(t)\, \log^{-\frac{2}{3}}(\frac{x/m}{t})\, \frac{dt}{t} = O\left((\log x)^{-\frac{2}{3}+\frac{2}{3}\eta}\right),$$

and

$$\overline{\overline{I}}(x,u) := \int_{2}^{x^u} \log^{-\frac{2}{5}}(t)\, \log^{-\frac{5}{3}}(\frac{x/m}{t})\, \frac{dt}{t} = O\left((\log x)^{-\frac{2}{3}+\frac{2}{3}\eta}\right).$$

3. For $\epsilon_x \leqslant v \leqslant 1-\epsilon''_x$ and $2 \leqslant q \leqslant \sqrt{x}$, we uniformly have

$$\int_{x^v}^{\frac{x}{2q}} \log^{-\frac{2}{3}+\theta}(t) \log^{-\frac{5}{3}}(x/qt) \frac{dt}{t} = O\left((\log x)^{-\frac{2}{3}+\theta}\right).$$

4. For $\epsilon_x \leqslant v \leqslant 1-\epsilon_x$ and $2 \leqslant q \leqslant x^{1-v}$, we uniformly have

$$\int_{x^v}^{x/2q} \log^{-5/3}(t)\, \log^{-2/3}(\frac{x/q}{t})\, \frac{dt}{t} = O\left((\log x)^{-\frac{4}{3}+\frac{4}{3}\eta}\right).$$

Proof. The proofs of the four statements are similar. Let us prove the first one. We write

$$\widehat{I}(x,v) = \log^{-2/3}(x/q) \int_{x^{\sqrt{\epsilon_x}}}^{x^v} \log^{-5/3}(t)\, (1 - \frac{\log t}{\log(x/q)})^{-2/3}\, \frac{dt}{t},$$

and by the change of variable $y = 1 - \frac{\log t}{\log(x/q)}$, we get

$$\begin{aligned}
\widehat{I}(x,v) &= \log^{-4/3}(x/q) \int_{1-\frac{v\log x}{\log(x/q)}}^{1-\frac{\sqrt{\epsilon_x}\log x}{\log(x/q)}} y^{-2/3}(1-y)^{-5/3}\, dy \\
&\leqslant \log^{-4/3}(x/q)(1 - \tfrac{v\log x}{\log(x/q)})^{-2/3} \int_{1-\frac{v\log x}{\log(x/q)}}^{1-\frac{\sqrt{\epsilon_x}\log x}{\log(x/q)}} (1-y)^{-5/3}\, dy \\
&\leqslant \tfrac{3}{2} \log^{-4/3}(x/q)(\tfrac{1-u-v}{1-u})^{-2/3}(\tfrac{\sqrt{\epsilon_x}\log x}{\log(x/2)})^{-2/3} \ll \log^{-1+(4/3)\eta}(x).
\end{aligned}$$

□

4. Proof of Theorem 2.1 for (u, v) in T_x

Recall that the notation $\overline{T}_x$ has been introduced in (3), ϵ_x and ϵ''_x in (2) and $S(x; u, v)$ in (1). First, we note that Theorem 2.1 is obvious for x bounded. From now on we suppose that x is sufficently large. We divide T_x into two zones: $[0, \epsilon_x] \times [0, 1] \cup [0, 1] \times [0, \epsilon_x]$, and $\mathcal{D}_0 := \{(u, v) \in [\epsilon_x, 1 - \epsilon''_x]^2, u + v \leqslant 1 - \epsilon''_x\}$. In the first zone, we show that $S(x, u, v)$ has the same order of magnitude as the remainder term (see Lemma 4.1 below). In order to study the sum $S(x; u, v)$ in the second zone, we decompose it as follows: $S(x, u, v) = S_1(x, u, v) - S_2(x, u, v) - S_3(x, u, v) - S_4(x, u, v)$, with

$$S_1(x, u, v) := \sum_{\substack{q \leqslant x^u, m \leqslant x^v \\ d \leqslant \frac{x}{qm}}} \frac{1}{\tau_3(qmd)}; \quad S_2(x, u, v) := \sum_{\substack{n^u \leqslant q \leqslant x^u, n^v < m \leqslant x^v \\ n = qmd \leqslant x}} \frac{1}{\tau_3(qmd)};$$

$$S_3(x, u, v) := \sum_{\substack{q \leqslant n^u, n^v < m \leqslant x^v \\ n = dmq \leqslant x}} \frac{1}{\tau_3(qmd)}; \quad S_4(x, u, v) := \sum_{\substack{n^u \leqslant q \leqslant x^u, m \leqslant n^v \\ n = qmd \leqslant x}} \frac{1}{\tau_3(qmd)}\cdot$$

We then show that $S_2(x, u, v)$, $S_3(x, u, v)$ and $S_4(x, u, v)$ have the same order of magnitude as the remainder term (see Lemma 4.2 below) and that $S_1(x, u, v)$ provides the main term (see Lemma 4.3 below).

Lemma 4.1. *Uniformly for $x \geqslant 2$ and $(u, v) \in [0, 1] \times [0, \epsilon_x[\cup [0, \epsilon_x[\times [0, 1]$, we have*

$$S(x, u, v) = O\left(\frac{x}{\sqrt[3]{\log x}}\right)\cdot$$

Proof. By symmetry, it suffices to prove the lemma for $(u, v) \in [0, 1] \times [0, \epsilon_x[$. We have

$$S(x, u, v) \leqslant \sum_{q \leqslant x^u} \sum_{d \leqslant x/q} \frac{1}{\tau_3(dq)} \leqslant \sum_{q \leqslant x} \sum_{d \leqslant x/q} \frac{1}{\tau_3(dq)}\cdot$$

The condition $dq \leqslant x$ implies that $d \leqslant \sqrt{x}$ or $q \leqslant \sqrt{x}$, we clearly have

$$\sum_{q \leqslant x} \sum_{d \leqslant x/q} \frac{1}{\tau_3(dq)} \ll \sum_{q \leqslant \sqrt{x}} \sum_{d \leqslant x/q} \frac{1}{\tau_3(dq)} =: L\,.$$

Applying Lemma 3.1 and observing that $q \leqslant \sqrt{x}$, we get

$$L \ll x \log^{-2/3}(x) \sum_{q \leqslant \sqrt{x}} \frac{h_1(q)}{q} \ll \frac{x}{\sqrt[3]{\log x}}\cdot$$

□

Lemma 4.2. *Let $\epsilon_x = \log 2/\log x$. Uniformly for $x \geqslant 2$ and $(u, v) \in [\epsilon_x, 1]^2$, we have*

$$S_i := S_i(x, u, v) = O\left(x/\sqrt[3]{\log x}\right), \quad (i = 2, 3, 4)\cdot$$

Proof. Let $M_2 := \min(M_3, M_4)$, with $M_3 = \frac{x^{1-v}}{q}$, $M_4 = \frac{x^{1-u}}{m}$ and

$$\widetilde{S_3} = \widetilde{S_3}(x,u,v) := \sum_{\substack{q\leqslant x^u, m\leqslant x^v \\ d\leqslant M_3}} \frac{1}{\tau_3(qmd)}, \quad \widetilde{S_4} = \widetilde{S_4}(x,u,v) := \sum_{\substack{q\leqslant x^u, m\leqslant x^v \\ d\leqslant M_4}} \frac{1}{\tau_3(qmd)}\cdot$$

We clearly have

$$S_2 \leqslant \sum_{\substack{q\leqslant x^u, m\leqslant x^v \\ d\leqslant M_2}} \frac{1}{\tau_3(qmd)} \leqslant \min(\widetilde{S_3}, \widetilde{S_4}),\ S_3 \leqslant \widetilde{S_3},\ S_4 \leqslant \widetilde{S_4},\ \widetilde{S_3}(x,u,v) = \widetilde{S_4}(x,v,u)\cdot$$

Then to prove the lemma, it suffices to prove that $\widetilde{S_4} = O\left(x/\sqrt[3]{\log x}\right)$ uniformly in $[\epsilon_x, 1]^2$. Set $\epsilon'_x := \sqrt{\log 2/\log x}$. We will first give the estimate for $(u,v) \in [\epsilon_x, \epsilon'_x]^2$. By Lemma 3.1, we get

$$\begin{aligned} \widetilde{S_4} &\ll x^{1-u} \textstyle\sum_{q\leqslant x^u, m\leqslant x^v} \frac{h_1(qm)}{m} \log^{-2/3}\left(\frac{x^{1-u}}{m}\right) \\ &\ll x(u(1-u-v))^{-2/3} \log^{-4/3}(x) \textstyle\sum_{m\leqslant x^v} \frac{h_2(m)}{m} \ll \frac{x}{\sqrt[3]{\log x}}\cdot \end{aligned}$$

Let us now estimate $\widetilde{S_4}$ for (u,v) in $[\epsilon'_x, 1] \times [\epsilon_x, 1]$. We distinguish two cases:
1st case, suppose that $u + v \leqslant 1 - \epsilon'_x$. Lemma 3.1 gives

$$\widetilde{S_4} \ll x(u(1-u-v))^{-2/3} \log^{-4/3}(x) \sum_{m\leqslant x^v} \frac{h_2(m)}{m} \ll \frac{x}{\sqrt[3]{\log x}},$$

as $u^{-2/3} \leqslant (\epsilon'_x)^{-2/3} \ll \log^{1/3}(x)$ and $(1-u-v)^{-2/3} \ll (\epsilon'_x)^{-2/3} \ll \log^{1/3}(x)$.
2nd case, suppose that $u + v > 1 - \epsilon'_x$. Then, we can write $\widetilde{S_4} = T_1 + T_2$, with

$$T_1 := \sum_{m\leqslant x^{1-u-\epsilon'_x}, q\leqslant x^u, d\leqslant M_4} \frac{1}{\tau_3(qmd)},\ T_2 := \sum_{x^{1-u-\epsilon'_x} < m\leqslant x^v, q\leqslant x^u, d\leqslant M_4} \frac{1}{\tau_3(qmd)}\cdot$$

Consider T_1. By Lemma 3.1 we obtain, as before,

$$\begin{aligned} T_1 &\ll \textstyle\sum_{m\leqslant x^{1-u-\epsilon'_x}, q\leqslant x^u} x^{1-u} \frac{h_1(qm)}{m} \log^{-2/3}(x^{1-u}/m) \\ &\ll xu^{-2/3} \log^{-2/3}(x) (\epsilon'_x)^{-2/3} \log^{-2/3}(x) \log^{1/3}(x) \ll \frac{x}{\sqrt[3]{\log x}}\cdot \end{aligned}$$

Consider now T_2. The condition $d \leqslant M_4$ that is $dm \leqslant x^{1-u}$ implies that $d \leqslant \sqrt{x^{1-u}}$ or $m \leqslant \sqrt{x^{1-u}}$. We therefore have, by symmetry, $T_2 \leqslant T_3 + 2\,T_4$ with

$$T_3 := \sum_{q\leqslant x^u, m\leqslant\sqrt{x^{1-u}}, d\leqslant\sqrt{x^{1-u}}} \frac{1}{\tau_3(qmd)},\ T_4 := \sum_{q\leqslant x^u, m\leqslant\sqrt{x^{1-u}}, \sqrt{x^{1-u}} < d\leqslant \frac{x^{1-u}}{m}} \frac{1}{\tau_3(qmd)}\cdot$$

In order to evaluate T_3 and T_4 we consider three cases:
1st case. We suppose that $\epsilon'_x \leqslant u \leqslant 1 - \epsilon'_x$. By Lemma 3.1, we obtain

$$T_3 \ll x^{\frac{1-u}{2}} x^u x^{\frac{1-u}{2}} \log^{-2/3}(x^{\frac{1-u}{2}}) \log^{-2/3}(x^u) \log^{-2/3}(x^{\frac{1-u}{2}}) \ll \frac{x}{\log x},$$

and

$$T_4 \ll x^{1-u} \log^{-2/3}(x^{\frac{1-u}{2}}) x^u \log^{-2/3}(x^u) \sum_{m \leqslant \sqrt{x^{1-u}}} \frac{h_2(m)}{m} \ll \frac{x}{\sqrt[3]{\log x}}.$$

2nd case. We suppose that $1 - \epsilon'_x < u \leqslant 1 - \epsilon_x$. We obtain

$$T_3 \ll x^{\frac{1-u}{2}} x^u x^{\frac{1-u}{2}} \log^{-2/3}(x^{\frac{1-u}{2}}) \log^{-2/3}(x^u) \log^{-2/3}(x^{\frac{1-u}{2}}) \ll \frac{x}{(\log x)^{2/3}},$$

and

$$T_4 \ll x^{1-u} \log^{-2/3}(x^{\frac{1-u}{2}}) x^u \log^{-2/3}(x^u) \sum_{m \leqslant \sqrt{x^{1-u}}} \frac{h_2(m)}{m} \ll \frac{x}{\sqrt[3]{\log x}}.$$

3rd case. We suppose that $1 - \epsilon_x < u \leqslant 1$. We have, $1 \leqslant d \leqslant x^{\frac{1-u}{2}} < x^{\frac{\epsilon_x}{2}} = \sqrt{2}$ then $d = 1$. Similarly $m = 1$. Hence

$$T_3 = T_4 = \sum_{q \leqslant x} \frac{1}{\tau_3(q)} \ll x(\log x)^{-2/3}.$$

To complete the proof, it remains to estimate $\widetilde{S_4}$ for (u, v) in $[\epsilon_x, \epsilon'_x] \times [\epsilon'_x, 1]$. Using the notations T_3 and T_4 above, we have $\widetilde{S_4} \leqslant T_3 + 2T_4$ and by Lemma 3.1, we easily see that $T_3 \ll x/\log^{4/3} x$ and $T_4 \ll x/\sqrt[3]{\log x}$ for (u, v) in $[\epsilon_x, \epsilon'_x] \times [\epsilon'_x, 1]$. □

Lemma 4.3. *Uniformly for $(u, v) \in \mathcal{D}_0$ and $x \geqslant 2$, we have*

$$S_1 = \frac{x}{\Gamma^3(\frac{1}{3})} \int_0^u y^{-2/3}(1-y)^{-1/3} \int_0^{\frac{v}{1-y}} t^{-2/3}(1-t)^{-2/3}\, dt\, dy + O\left(\frac{x}{\sqrt[3]{\log x}}\right).$$

Proof. Write the decomposition $S_1 = S_{1,1} + S_{1,2}$ with

$$S_{1,1} := \sum_{q \leqslant x^u} \sum_{x^{\sqrt{\epsilon_x}} < m \leqslant x^v} \sum_{d \leqslant \frac{x}{qm}} \frac{1}{\tau_3(qmd)}, \quad S_{1,2} := \sum_{q \leqslant x^u} \sum_{m \leqslant x^{\sqrt{\epsilon_x}}} \sum_{d \leqslant \frac{x}{qm}} \frac{1}{\tau_3(qmd)}.$$

We will estimate these two quantities.

`1. Estimation of` $S_{1,1}$. The application of Lemma 3.1(1) gives

$$\sum_{d \leqslant \frac{x}{qm}} \frac{1}{\tau_3(qmd)} = x \frac{c_1}{\Gamma(\frac{1}{3})} \frac{h_1(qm)}{qm} \log^{-\frac{2}{3}}(\frac{x}{qm}) + O\left(M_{qm,\theta} \frac{x}{qm} \log^{-\frac{5}{3}}(\frac{x}{qm})\right). \quad (5)$$

Consider the remainder term. Recall that $M_{qm,\theta} = (\frac{1}{3} + \theta)^{\omega(qm)}$, $\theta > 0$. Hence

$$\sum_{q \leqslant x^u} \sum_{x^{\sqrt{\epsilon_x}} < m \leqslant x^v} \frac{M_{qm,\theta}}{qm} \leqslant \sum_{q \leqslant x^u} \frac{(\frac{1}{3} + \theta)^{\omega(q)}}{q} \sum_{m \leqslant x^v} \frac{(\frac{1}{3} + \theta)^{\omega(m)}}{m} \ll \log^{2(\theta + \frac{1}{3})}(x),$$

by Lemma 3.1(2) and $\log^{-5/3}(x/qm) \leqslant \log^{-5/3}(x^{1-u-v}) \leqslant \log^{-5/3}(x^{\epsilon''_x}) \ll \log^{-5/3+(5/3)\eta}(x)$. From the choices $0 < \theta < (1-(5/2)\eta)/3$ and $\eta < 2/5$, it follows that

$$O\left(x \sum_{q \leqslant x^u} \sum_{x^{\sqrt{\epsilon_x}} < m \leqslant x^v} \frac{M_{qm,\theta}}{qm} \log^{-\frac{5}{3}}(\frac{x}{qm})\right) = O\left(\frac{x}{\sqrt[3]{\log x}}\right). \qquad (6)$$

Now, we consider the main term in (5). By partial summation and Lemma 3.1(1) we have

$$\begin{aligned} \textstyle\sum_{x^{\sqrt{\epsilon_x}} < m \leqslant x^v} \frac{h_1(qm)}{m} \log^{-\frac{2}{3}}(\frac{x}{qm}) &= -\textstyle\int_{x^{\sqrt{\epsilon_x}}}^{x^v} \left(\sum_{m \leqslant t} h_1(qm)\right) d(\frac{\log^{-\frac{2}{3}}(\frac{x/q}{t})}{t}) \\ &+ O\left(\log^{-1+\frac{2}{3}\eta}(x)\, h_2(q)\right)\cdot \end{aligned} \qquad (7)$$

Taking the sum over q, and using Lemma 3.1(2), we see that the remainder term is

$$O\left(\log^{-1+\frac{2}{3}\eta}(x) \sum_{q \leqslant x^u} \frac{h_2(q)}{q}\right) = O\left(\frac{1}{\sqrt[3]{\log x}}\right)\cdot \qquad (8)$$

Now, consider the first term in the second member of (7). By Lemma 3.1(1), we get

$$\frac{c_2 h_2(q)}{\Gamma(\frac{1}{3})} \int_{x^{\sqrt{\epsilon_x}}}^{x^v} \log^{-2/3}(t) \log^{-2/3}(\frac{x/q}{t}) \frac{dt}{t} + O(h_2(q)\widehat{\widehat{I}}(x,v)) + O(M_{q,\theta'}\widehat{I}(x,v)), \quad (9)$$

where $\widehat{I}(x,v)$ and $\widehat{\widehat{I}}(x,v)$ are defined in Lemma 3.4 (1). Summing over q and using Lemma 3.4 (1) and Lemma 3.1 (2) in each of the two remainder terms, we get

$$\begin{aligned} &O\left((\log x)^{-1+\frac{4}{3}\eta} \textstyle\sum_{q \leqslant x^u} \frac{M_{q,\theta'}}{q}\right) = O\left(\frac{1}{\sqrt[3]{\log x}}\right); \\ &O\left(\widehat{I}(x,v) \textstyle\sum_{q \leqslant x^u} \frac{h_2(q)}{q}\right) = O\left(\frac{1}{\sqrt[3]{\log x}}\right)\cdot \end{aligned} \qquad (10)$$

In the first term of (9), by the changes of de variables $y = \frac{\log t}{\log x/q}$ and $z = y(1 - \frac{\log q}{\log x})$, and summation over q, we get

$$\frac{c_2}{\Gamma(\frac{1}{3})} \log^{-\frac{1}{3}}(x) \sum_{q \leqslant x^u} \frac{h_2(q)}{q} \int_{\sqrt{\epsilon_x}}^{v} z^{-\frac{2}{3}}(1 - \frac{\log q}{\log x} - z)^{-\frac{2}{3}} dz. \qquad (11)$$

It remains to study the quantity

$$\begin{aligned} K &:= \textstyle\sum_{q \leqslant x^u} \frac{h_2(q)}{q} \int_{\sqrt{\epsilon_x}}^{v} z^{-\frac{2}{3}}(1 - \frac{\log q}{\log x} - z)^{-\frac{2}{3}} dz \\ &= \textstyle\sum_{2 \leqslant q \leqslant x^u} \frac{h_2(q)}{q} \int_{\sqrt{\epsilon_x}}^{v} z^{-\frac{2}{3}}(1 - \frac{\log q}{\log x} - z)^{-\frac{2}{3}} dz + O(1). \end{aligned} \qquad (12)$$

By partial summation, Lemma 3.1 (2) and Lemma 3.3 (1) with the notations therein, we get

$$\begin{aligned} K &= -\int_2^{x^u}\left(\sum_{q\leqslant t} h_2(q)\right) d\left(\frac{1}{t}\int_{\sqrt{\epsilon_x}}^{v} z^{-\frac{2}{3}}(1-\frac{\log t}{\log x}-z)^{-\frac{2}{3}}dz\right)+O(1)\\ &= \frac{c_3}{\Gamma(\frac{1}{3})}\int_2^{x^u}\log^{-\frac{2}{3}} t\left(\frac{I(t,x,v)}{t}-\frac{\partial}{\partial t}I(t,x,v)\right)dt+O(1)\\ &= \frac{c_3}{\Gamma(\frac{1}{3})}\int_2^{x^u}\log^{-\frac{2}{3}}(t)\,\frac{I(t,x,v)}{t}\,dt+O\left(J(x,u,v)\right))+O(1)\\ &= \frac{c_3}{\Gamma(\frac{1}{3})}\int_2^{x^u}\log^{-\frac{2}{3}}(t)\,\frac{I(t,x,v)}{t}\,dt+O(1)\cdot \end{aligned}$$

Now, the changes of variables $y=\frac{\log t}{\log x}$, and $t=\frac{z}{1-y}$ yield

$$\begin{aligned} K &= \frac{c_3}{\Gamma(\frac{1}{3})}\log^{1/3}(x)\int_{\epsilon_x}^{u} y^{-2/3}(1-y)^{-1/3}\int_{\frac{\sqrt{\epsilon_x}}{1-y}}^{\frac{v}{1-y}} t^{-2/3}(1-t)^{-2/3}\,dt\,dy\\ &\quad +O(1). \end{aligned} \tag{13}$$

Finally by collecting successively (5), (6),(7), (8), (9), (10), (11), (12), (13) and using the equality $c_1c_2c_3=1$ established in Lemma 3.2 (2) we get

$$S_{1,1}=\frac{x}{\Gamma^3(\frac{1}{3})}\int_{\epsilon_x}^{u} y^{-2/3}(1-y)^{-1/3}\int_{\frac{\sqrt{\epsilon_x}}{1-y}}^{\frac{v}{1-y}} t^{-2/3}(1-t)^{-2/3}\,dt\,dy+O\left(\frac{x}{\sqrt[3]{\log x}}\right). \tag{14}$$

2. `Estimation of` $S_{1,2}$. We invert the first two sums that we estimate:

$$S_{1,2}=\sum_{m\leqslant x^{\sqrt{\epsilon_x}}}\sum_{q\leqslant x^u}\sum_{d\leqslant\frac{x}{qm}}\frac{1}{\tau_3(qmd)}\cdot$$

The method described above works. To study various remainder terms, we use Lemma 3.4 (2). We obtain in each case $O\left(x/\sqrt[3]{\log x}\right)$. we obtain

$$\begin{aligned} S_{1,2} &= \frac{x}{\Gamma^3(\frac{1}{3})}\int_{\epsilon_x}^{\sqrt{\epsilon_x}} y^{-2/3}(1-y)^{-1/3}\int_{\frac{\sqrt{\epsilon_x}}{1-y}}^{\frac{u}{1-y}} t^{-2/3}(1-t)^{-2/3}\,dt\,dy+O\left(\frac{x}{\sqrt[3]{\log x}}\right)\\ &= \frac{x}{\Gamma^3(\frac{1}{3})}\int_{\sqrt{\epsilon_x}}^{u} y^{-2/3}(1-y)^{-1/3}\int_{\frac{\epsilon_x}{1-y}}^{\frac{\sqrt{\epsilon_x}}{1-y}} t^{-2/3}(1-t)^{-2/3}\,dt\,dy+O\left(\frac{x}{\sqrt[3]{\log x}}\right)\cdot \end{aligned} \tag{15}$$

The last equality is obtained by a change of variables.

Finally, we obtain the required formula for $S_1=S_{1,1}+S_{1,2}$ from (14) and (15). This completes the proof of Lemma 4.3. □

5. Description of the Proof of Theorem 2.1 for (u,v) in $\overline{T}_x$ and for (u,v) in T_2

Recall that the notation $\overline{T}_x$ has been introduced in (3), ϵ_x and ϵ''_x in (2) and $S(x;u,v)$ in (1). Also $T_2=\{(u,v)\in[0,1]\times[0,1]:u+v\geqslant 1\}$. In $\overline{T}_x\cup T_2$, we will use the expression of $S(x,u,v)$ given in the following lemma which reduces the estimate of $S(x,u,v)$ to that of the three quantities Σ_i, $1 \leqslant i \leqslant 3$, with

$$\Sigma_1 = \sum_{2\leqslant q\leqslant x; x^v<m\leqslant x; dmq\leqslant x} \frac{1}{\tau_3(dmq)}; \quad \Sigma_2 = \sum_{x^u<q\leqslant x; 2\leqslant m\leqslant x; dmq\leqslant x} \frac{1}{\tau_3(dmq)};$$

$$\Sigma_3 = \sum_{x^u<q\leqslant x; x^v<m\leqslant x; dmq\leqslant x} \frac{1}{\tau_3(dmq)}.$$

Lemma 5.1. *For every $(u,v) \in [\epsilon_x, 1]^2$ and $x \geqslant 2$, we uniformly have*

$$S(x,u,v) = [x] - \Sigma_1 - \Sigma_2 + \Sigma_3 + O\left(\frac{x}{\sqrt[3]{\log x}}\right).$$

By following the proof of Theorem 2.1 in T_x, we can prove the following two lemmas:

Lemma 5.2. *Uniformly for $\epsilon_x \leqslant v \leqslant 1$, $\epsilon_x \leqslant u \leqslant 1$, and $x \geqslant 2$, we have,*

$$\begin{aligned}\Sigma_1 &= \tfrac{x}{\Gamma^3(\frac{1}{3})} \int_0^{1-v} y^{-\frac{2}{3}}(1-y)^{-\frac{1}{3}} \int_{\frac{v}{1-y}}^1 s^{-\frac{2}{3}}(1-s)^{-\frac{2}{3}}\, dy\, ds + O\left(\tfrac{x}{\sqrt[3]{\log x}}\right)\\ &= x - \tfrac{x}{B(\frac{2}{3},\frac{1}{3})} \int_0^v y^{-\frac{1}{3}}(1-y)^{-\frac{2}{3}}\, dy + O\left(\tfrac{x}{\sqrt[3]{\log x}}\right),\end{aligned}$$

and

$$\begin{aligned}\Sigma_2 &= \tfrac{x}{\Gamma^3(\frac{1}{3})} \int_0^{1-u} y^{-\frac{2}{3}}(1-y)^{-\frac{1}{3}} \int_{\frac{u}{1-y}}^1 s^{-\frac{2}{3}}(1-s)^{-\frac{2}{3}}\, dy\, ds + O\left(\tfrac{x}{\sqrt[3]{\log x}}\right)\\ &= x - \tfrac{x}{B(\frac{2}{3},\frac{1}{3})} \int_0^u y^{-\frac{1}{3}}(1-y)^{-\frac{2}{3}}\, dy + O\left(\tfrac{x}{\sqrt[3]{\log x}}\right).\end{aligned}$$

Remark 5.3. We trivially have $\Sigma_3 = 0$ if $u + v \geqslant 1$. So, the formula of Theorem 2.1 for $(u,v) \in T_2$ results from the estimates of Σ_1 and Σ_2, given in Lemma 5.2.

Lemma 5.4. *Uniformly for $(u,v) \in \overline{T}_x$, with $u + v < 1$ and $x \geqslant 2$, we have*

$$\begin{aligned}\Sigma_3 &= x - \tfrac{x}{B(\frac{2}{3},\frac{1}{3})} \int_0^u y^{-\frac{1}{3}}(1-y)^{-\frac{2}{3}}\, dy - \tfrac{x}{B(\frac{2}{3},\frac{1}{3})} \int_0^v y^{-\frac{1}{3}}(1-y)^{-\frac{2}{3}}\, dy\\ &+ \tfrac{x}{\Gamma^3(\frac{1}{3})} \int_0^u y^{-2/3}(1-y)^{-1/3} \int_0^{\frac{v}{1-y}} t^{-2/3}(1-t)^{-2/3}\, dt\, dy + O\left(\tfrac{x}{\sqrt[3]{\log x}}\right).\end{aligned}$$

The Proof of Theorem 2.1 in $\overline{T}_x$ and in T_2 results from Lemmas 5.2 and 5.3 and Remark 5.1. Note that by a change of variable $z = (1-y)t$, we have

$$\frac{1}{\Gamma^3(\frac{1}{3})} \int_0^u y^{-2/3}(1-y)^{-1/3} \int_0^{\frac{v}{1-y}} t^{-2/3}(1-t)^{-2/3}\, dt\, dy$$
$$= \frac{1}{\Gamma^3(\frac{1}{3})} \int_0^u \int_0^v y^{-\frac{2}{3}} z^{-\frac{2}{3}} (1-y-z)^{-\frac{2}{3}}\, dy\, dz.$$

Acknowledgements. We are grateful to the referee for her or his interesting remarks and advice.

References

[1] J.-M. Deshouillers, F. Dress, G. Tenenbaum, Lois de répartition des diviseurs, I, Acta Arith. 34 (1979), 273-285.

[2] G.Tenenbaum, Introduction à la théorie analytique et probabiliste des nombres, Troisième édition, coll. Échelles, Belin, Paris 2008.

SHARP THRESHOLD ASYMPTOTICS FOR THE EMERGENCE OF ADDITIVE BASES

Anant Godbole
Department of Mathematics and Statistics, East Tennessee State University, Johnson City, Tennessee

Chang Mou Lim
Department of Mathematics, University of Chicago, Chicago, Illinois

Vince Lyzinski
Department of Applied Mathematics and Statistics, The Johns Hopkins University, Baltimore, Maryland
lyzinski@ams.jhu.edu

Nicholas George Triantafillou
Department of Mathematics, University of Michigan, Ann Arbor, Michigan

Received: 7/5/12, Revised: 1/7/13, Accepted: 3/11/13, Published: 3/15/13

Abstract

A set $\mathcal{A} \subseteq [n] \cup \{0\}$ is said to be a 2-additive basis for $[n]$ if each $j \in [n]$ can be written as $j = x + y$, $x, y \in \mathcal{A}$, $x \leq y$. If we pick each integer in $[n] \cup \{0\}$ independently with probability $p = p_n \to 0$, thus getting a random set $\mathcal{A}$, what is the probability that we have obtained a 2-additive basis? We address this question when the target sum-set is $[(1-\alpha)n, (1+\alpha)n]$ (or equivalently $[\alpha n, (2-\alpha)n]$) for some $0 < \alpha < 1$. We use a delicate application of Janson's correlation inequalities in conjuction with the Stein-Chen method of Poisson approximation to tease out a very sharp threshold for the emergence of a 2-additive basis. Generalizations to k-additive bases are then given.

1. Introduction

In 1956, Erdős [3] answered a question posed in 1932 by Sidon by proving that there exists an infinite sequences of natural numbers $\mathcal{S}$ and constants c_1 and c_2 such that for large n,

$$c_1 \log n \leq r_2(n) \leq c_2 \log n, \tag{1}$$

where, for $k \geq 2$, $r_k(n)$ is the *number of ways* of representing the integer n as the sum of k elements from $\mathcal{S}$, a so-called *asymptotic basis* of order k. The result was generalized in the 1990 work of Erdős and Tetali [4] which established that there exists an infinite sequence $\mathcal{S}$ for which (1) was true for each fixed $k \geq 2$, i.e., for each large n,

$$r_k(n) = \Theta(\log n). \tag{2}$$

To achieve this result for fixed k, Erdős and Tetali constructed a random sequence $\mathcal{S}$ of natural numbers by including z in $\mathcal{S}$ with probability

$$p(z) = \begin{cases} C\frac{(\log z)^{1/k}}{z^{(k-1)/k}}, & \text{if } z > z_0 \\ 0 & \text{otherwise} \end{cases}$$

where C is a determined constant and z_0 is the smallest constant such that

$$C\frac{(\log z_0)^{1/k}}{z_0^{(k-1)/k}} \leq 1/2.$$

They then showed that (2) holds a.s. for large n (implying that $\mathcal{S}$ is an asymptotic basis of order k).

A natural finite variant of the above problem concerns the notion of k-additive bases.

Definition 1. With $[n] := \{1, 2, \ldots, n\}$, a set $\mathcal{A} \subseteq [n] \cup \{0\}$ is said to be a *k-additive basis* for $[n]$ if each $j \in [n]$ can be written as $j = x_1 + x_2 + \ldots + x_k$, $x_i \in \mathcal{A}$, $i = 1, \ldots, k$.

Note that this definition allows for $x_i = x_j$ $i \neq j$. In [4], Erdős and Tetali showed that in some probability space, almost all infinite sequences $\mathcal{S}$ satisfy (2) and are asymptotic bases of order k. It is natural to then ask, for finite $\mathcal{A}$, how small $\mathcal{A}$ can be while still being a k-additive basis. For $k = 2$, if $\mathcal{A}$ is a 2-additive basis we must clearly have $\binom{|\mathcal{A}|+1}{2} \geq n$, so that $|\mathcal{A}| \geq \sqrt{2n}(1 + o(1)) = 1.4142(1 + o(1))\sqrt{n}$. Extensive work, using predominantly analytic techniques, has been done to improve this trivial lower bound to $|\mathcal{A}| \geq 1.428\sqrt{n}$ ([9]); $|\mathcal{A}| \geq 1.445\sqrt{n}$ ([6]); $|\mathcal{A}| \geq 1.459\sqrt{n}$ ([13]);, and $|\mathcal{A}| \geq 1.463\sqrt{n}$ ([5]). The best upper bound appears to be $|\mathcal{A}| \lesssim 1.871\sqrt{n}(1 + o(1))$ ([10], [7]).

In this paper, we will use a probability model in which each integer in $[n] \cup \{0\}$ is chosen to be in $\mathcal{A}$ with equal (and low) probability $p = p_n$. We will then give sharp bounds on the probability that the random set $\mathcal{A}$ is a k-additive basis. It is evident that smaller numbers must be present in an additive basis, since, e.g., the only way to represent 1 in a 2-additive basis is as 1+0, and so in the random model edge effects come into play. Therefore, the random ensemble is unlikely to form an additive basis unless we adopt a different approach.

Definition 2. A set $\mathcal{A} \subseteq [n] \cup \{0\}$ is said to be a *truncated k-additive basis* for $[n]$ if each $j \in [\alpha n, (k-\alpha)n]$ can be written as $j = x_1 + x_2 + \ldots + x_k$, $x_i \in \mathcal{A}$, $i = 1, \ldots, k; \alpha \in (0,1)$.

Although our model differs from that of Erdős and Tetali (our set is constructed using constant probability p as opposed to the $p(z)$ used in [4]), the output threshold probabilities for the size of a truncated (or modular) k-additive basis end up being remarkably close to the input probabilities used by Tetali-Erdős to construct their asymptotic bases. We will stress the similarities between our results as appropriate. It also bears mentioning that the threshold size for our random $\mathcal{A}$ to be a 2-additive basis is, up to a logarithmic factor, similar to the bounds outlined above which were derived using predominantly analytic techniques.

Our work is organized as follows: We present threshold results on truncated 2-additive bases in Section 2, using the Stein-Chen method of Poisson approximation (see [2]). This method allows one to not just find the limiting probability that $\mathcal{A}$ forms an additive basis, but also to approximate the probability distribution $\mathcal{L}(X)$ of the random variable X, defined as the number of integers j that cannot be expressed as a sum of elements in $\mathcal{A}$. In Section 3, we consider similar questions for truncated k-additive bases; where we make use of the Janson exponential inequalities ([1], [8]) to estimate some critical baseline quantities that were calculated exactly in Section 2 for $k = 2$.

Remark 3. An alternate way of dealing with the boundary effects encountered in finite additive bases is to define *modular k-additive bases.* A set $\mathcal{A} \subseteq [n-1] \cup \{0\}$ is said to be a modular k-additive basis for $[n]$ if each $j \in [n-1] \cup \{0\}$ can be written as $j = x_1 + x_2 + \ldots + x_k \pmod{n}$, $x_i \in \mathcal{A}$, $i = 1, \ldots, k$. Definitive results on the emergence of modular additive bases have been proved in the paper of Sandor using the method of Brun's sieve [12], and in the papers by Yadin [14] and Yu [13] using Janson's correlation inequalities. We have recovered these results using the implementation of the Janson's inequality and the Stein-Chen method outlined below though details are omitted for the sake of brevity.

We believe that the truncated basis is the more natural finite variant of the problem considered by Erdős and Tetali in [4]. They were concerned with constructing a basis with $r_k(n) = \Theta(\log n)$ for all integers greater than a fixed but arbitrary N. This allowed them great flexibility in choosing the threshold N to be large enough to achieve the desired behavior. A natural question is how small N can be while still maintaining an additive basis with $r_k(n) = \Theta(\log n)$, which is a natural analogue to the truncated basis question explored below. Also, in a sequel to this paper, we will recover the logarithmic behavior of r_k in the truncated basis setting.

Remark 4. Throughout the rest of the paper, we suppress the descriptor "additive," referring simply to "truncated k-bases." Here and in the sequal, we write $f(n) \cong g(n)$ to signify that $\frac{f(n)}{g(n)} = 1 + o(1)$. Also, we write $Y \sim \text{Po}(\lambda)$ to signify that Y is a random variable distributed as Poisson(λ).

2. 2-Additive Bases

We begin by investigating truncated 2-bases in our random model outlined above. We first select each integer from $\{0, 1, \ldots, n\}$ independently with probability $p = p_n$. We thus obtain a random set $\mathcal{A}$, and denote by X the number of integers in $[\alpha n, (2-\alpha)n]$ that cannot be written in the form $x_1 + x_2$, $x_i \in \mathcal{A}$. Evidently, $\mathcal{A}$ is a truncated 2-basis if and only if $X = 0$. We can write $X = \sum_{j=\lceil \alpha n \rceil}^{\lfloor (2-\alpha)n \rfloor} I_j$, where I_j equals one or zero according as the integer j cannot or can be represented as a 2-sum of elements in $\mathcal{A}$. To simplify the notation a bit, we will write simply $\sum_{j=\alpha n}^{(2-\alpha)n} I_j$ for $\sum_{j=\lceil \alpha n \rceil}^{\lfloor (2-\alpha)n \rfloor} I_j$ in the sequel. The main results of this section are summarized in the following theorem:

Theorem 5. *Pick each integer in* $[n] \cup \{0\}$ *independently with probability* $p = p_n \to 0$ *thus getting a random set* $\mathcal{A}$.

(i) *Let* $Y \sim \mathrm{Po}(\lambda = \mathbb{E}(X))$. *Let* $\delta > 0$; *then for all*

$$p \geq \sqrt{\frac{(\frac{1}{\alpha} + \delta)\log n}{n}}, \tag{3}$$

we have that $d_{TV}(X, Y) \to 0$ $(n \to \infty)$, *where*

$$d_{TV}(X, Y) := \sup_{A \subseteq \mathbb{Z}^+} |\mathbb{P}(X \in A) - \mathbb{P}(Y \in A)|.$$

(ii) *With*

$$p = \sqrt{\frac{\frac{2}{\alpha}\log n - \frac{2}{\alpha}\log\log n + A_n}{n}}, \tag{4}$$

$$A_n \to \infty \Rightarrow \mathbb{P}(\mathcal{A} \text{ is a truncated } 2 - \text{basis}) \to 1 \ (n \to \infty);$$

$$A_n \to -\infty \Rightarrow \mathbb{P}(\mathcal{A} \text{ is a truncated } 2 - \text{basis}) \to 0 \ (n \to \infty);$$

and

$$A_n \to A \Rightarrow \mathbb{P}(\mathcal{A} \text{ is a truncated } 2 - \text{basis}) \to \exp\left\{-2\alpha e^{-\alpha A/2}\right\} \ (n \to \infty).$$

Proof. Here and throughout the paper, let $\mathcal{S}$ denote the random sumset generated by $\mathcal{A}$. The first step in the proof will be to calculate the precise asymptotics of $\mathbb{E}(X)$:

Proposition 6. *With X as above,*

$$\mathbb{E}(X) = \sum_{j=\alpha n}^{(2-\alpha)n} \mathbb{P}(I_j = 1) \cong 2 \sum_{j=\alpha n}^{n} (1-p^2)^{j/2} \cong \frac{4}{p^2}\exp\{-np^2\alpha/2\}.$$

Proof. This is a critical computation and needs to be justified for the entire range of p's that we encounter, in particular for all p satisfying (3). The first direction is easy:

$$\begin{aligned} 2\sum_{j=\alpha n}^{n}(1-p^2)^{j/2} &\leq 2\sum_{j=\alpha n}^{\infty}(1-p^2)^{j/2} \\ &\leq \frac{2(1-p^2)^{\alpha n/2}}{1-\sqrt{1-p^2}} \\ &\leq \frac{4}{p^2}(1-p^2)^{\alpha n/2} \\ &\leq \frac{4}{p^2}\exp\{-np^2\alpha/2\}, \end{aligned}$$

using the inequalities $1 - x \leq e^{-x}$ and $\sqrt{1-p^2} \leq 1 - \frac{p^2}{2}$. For the other direction, we have

$$\begin{aligned} 2\sum_{j=\alpha n}^{n}(1-p^2)^{j/2} &\geq \frac{2(1-p^2)^{\alpha n/2} - 2(1-p^2)^{n/2}}{1-\sqrt{1-p^2}} \\ &= \frac{2(1-p^2)^{\alpha n/2}(1-(1-p^2)^{(1-\alpha)n/2})}{1-\sqrt{1-p^2}} \\ &\cong \frac{2(1-p^2)^{\alpha n/2}\left(1-e^{-np^2(1-\alpha)/2}\right)}{\frac{p^2}{2}} \\ &\cong \frac{4}{p^2}(1-p^2)^{\alpha n/2} \\ &\cong \frac{4}{p^2}\exp\{-np^2\alpha/2\}, \end{aligned}$$

as desired. □

With p defined as in (4) we see that $\mathbb{E}(X) = (1+o(1))(2\alpha \cdot \exp\{-\alpha A_n/2\})$. As is often the case with threshold phenomena, Markov's inequality can be used to easily establish the first part of (ii), as we have $\mathbb{P}(X \geq 1) \leq \mathbb{E}(X) \to 0$ provided that $A_n \to \infty$ in (4).

To establish the second and third statements in part (ii) of the theorem, we go beyond estimating the point probability $\mathbb{P}(X = 0)$, using instead the Stein-Chen method of Poisson approximation [2] to establish a total variation approximation for the distribution $\mathcal{L}(X)$ of X. If the distribution of X is approximately Poisson, then we will have $e^{-\lambda} - \varepsilon_n \le \mathbb{P}(X = 0) \le e^{-\lambda} + \varepsilon_n$, where $\lambda = \mathbb{E}(X)$ is the mean of the approximating Poisson variable and ε_n is the total variation error bound for the approximation. We have that $\mathbb{E}(X) \to \infty$ if $A_n \to -\infty$; $\mathbb{E}(X) \to 0$ if $A_n \to \infty$; and $\mathbb{E}(X) \to 2\alpha e^{-\alpha A/2}$ if $A_n \to A$. Thus, if we can show that ε_n tends to zero in a window around

$$p = \sqrt{\frac{\frac{2}{\alpha}\log n}{n}},$$

we will have that $\mathbb{P}(X = 0) \to 0$ if $A_n \to -\infty$; $\mathbb{P}(X = 0) \to 1$ if $A_n \to \infty$; $\mathbb{P}(X = 0) \to \exp\{-2\alpha e^{-\alpha A/2}\}$ if $A_n \to A$.

With $Y \sim \mathrm{Po}(\lambda = \mathbb{E}(X))$, and throughout writing $d_{\mathrm{TV}}(A, B)$ instead of the more appropriate $d_{\mathrm{TV}}(\mathcal{L}(A), \mathcal{L}(B))$, we seek to bound $d_{TV}(X, Y)$. Following [2], we first need to determine, for each j separately, an auxiliary sequence of variables J_{j_i} defined on the same probability space with the property that

$$\mathcal{L}(J_{j_1}, J_{j_2}, \ldots) = \mathcal{L}(I_1, I_2, \ldots | I_j = 1). \tag{5}$$

Our explicitly constructed coupling of the J_{j_i}'s is as follows: If $I_j = 1$, set $J_{j_i} = I_i$ for all $i \ne j$. If $I_j = 0$, for all pairs $x_1, x_2 \in \mathcal{A}$ such that $x_1 \ne x_2$ and $x_1 + x_2 = j$, remove x_1 from $\mathcal{A}$ with probability $\frac{p(1-p)}{1-p^2}$, remove x_2 from $\mathcal{A}$ with probability $\frac{p(1-p)}{1-p^2}$, and remove both x_1 and x_2 from $\mathcal{A}$ with probability $\frac{p^2+1-2p}{1-p^2}$. If $x \in \mathcal{A}$ and $x + x = j$, then we remove x from $\mathcal{A}$ with probability 1. Finally, define $J_{j_i} = 1$ if $i \notin \mathcal{S}$ after the above coupling is implemented. It is clear that (5) is satisfied since we have "de-selected" offending integers based on the conditional probability of one or both integers in a pair being absent, given that both are not present. The total variation bounds derived in [2] are expressed in terms of the probability that the coupled indicator variables are different after the coupling is implemented; i.e., $I_i = 1, J_{j_i} = 0$ or $I_i = 0, J_{j_i} = 1$. Now, $\mathbb{P}(I_i = 1, J_{j_i} = 0) = 0$, since if integer i is not present in $\mathcal{S}$, it cannot magically appear after some integers have been de-selected.

The formula we need thus reduces to

$$d_{TV}(X, Y) \le \frac{1 - e^{-\lambda}}{\lambda} \sum_j \left(\mathbb{P}^2(I_j = 1) + \mathbb{P}(I_j = 1) \cdot \sum_{i \ne j} \mathbb{P}(I_i = 0, J_{j_i} = 1) \right). \tag{6}$$

There are two terms above. The first,

$$\frac{1 - e^{-\lambda}}{\lambda} \sum_j \mathbb{P}^2(I_j = 1) \le \frac{1}{\lambda} \max_i \mathbb{P}(I_i = 1) \sum_i \mathbb{P}(I_j = 1) = \max_i \mathbb{P}(I_i = 1),$$

and thus is bounded above by $\mathbb{P}(I_{\lceil \alpha n \rceil} = 1) \cong (1 - p^2)^{\alpha n/2}$. This bound tends to zero if $1/\sqrt{n} = o(p)$, a condition that is satisfied for p satisfying (3).

To bound the second term in (6), we begin by noting

$$\frac{1-e^{-\lambda}}{\lambda}\sum_j\left[\mathbb{P}(I_j=1)\sum_{i\neq j}\mathbb{P}(I_i=0,J_{j_i}=1)\right]\leq\max_j\sum_{i\neq j}\mathbb{P}(I_i=0,J_{j_i}=1).$$

We further bound this second term by conditioning on the number of 2-sums of i that are present pre-coupling, denoting this number by B_i. Denote the sumset of $\mathcal{A}$ post-coupling by $\mathcal{S}^*$. We see that

$$\mathbb{P}(I_i=0,J_{j_i}=1)=\sum_{k=1}^{\lceil\frac{i+1}{2}\rceil}\mathbb{P}(i\notin\mathcal{S}^*|B_i=k)\cdot\mathbb{P}(B_i=k).$$

Note that if there exists x such that $i=2x$, then

$$\begin{aligned}\mathbb{P}(B_i=k)=&\binom{\lceil\frac{i+1}{2}\rceil-1}{k}p^{2k}(1-p^2)^{\lceil\frac{i+1}{2}\rceil-k-1}(1-p)\\&+\binom{\lceil\frac{i+1}{2}\rceil-1}{k-1}p^{2k-2}(1-p^2)^{\lceil\frac{i+1}{2}\rceil-k}p,\end{aligned}$$

otherwise (when i is odd and $f(i):=\lceil\frac{i+1}{2}\rceil$ to simplify the notation)

$$\begin{aligned}&\mathbb{P}(B_i=k)=\binom{f(i)}{k}p^{2k}(1-p^2)^{f(i)-k}\\&=\binom{f(i)-1}{k}p^{2k}(1-p^2)^{f(i)-k}+\binom{f(i)-1}{k-1}p^{2k}(1-p^2)^{f(i)-k}\\&\leq\binom{f(i)-1}{k}p^{2k}(1-p^2)^{f(i)-k}+\binom{f(i)-1}{k-1}p^{2k-2}(1-p^2)^{f(i)-k}p\\&\cong\binom{f(i)-1}{k}p^{2k}(1-p^2)^{f(i)-1-k}(1-p)+\binom{f(i)-1}{k-1}p^{2k-2}(1-p^2)^{f(i)-k}p.\end{aligned}$$

Our next step is to bound $\mathbb{P}(i\notin\mathcal{S}^*|B_i=k)$ via

$$\mathbb{P}(i\notin\mathcal{S}^*|B_i=k)\leq\left(\mathbb{P}(\text{a given 2-sum of } i \text{ removed}|B_i=k)\right)^k.$$

To bound the above term, we will assume all elements of the 2-sums of i are part of 2-sums of j and thus have positive probability of being removed by the coupling process. Fix x,y $x\neq y$, such that $x+y=i$. For ease of notation, define P_x to be the event that x is part of a pre-coupling 2-sum of j, and analogously define P_y. Further define R_x to be the event x is removed from $\mathcal{A}$ by the coupling, and analogously define R_y. Consider the following simple calculations:

$$\mathbb{P}(P_x,\ P_y^c,\ R_x)=p(1-p)\left(1-\frac{p(1-p)}{1-p^2}\right)\leq p,$$

since the only undesirable outcome is if only the "other" component of the 2-sum of 0 is removed in lieu of x;

$$\mathbb{P}(P_x,\ P_y,\ R_x,\ R_y^c) = p^2\frac{p(1-p)}{1-p^2}\left(1-\frac{p(1-p)}{1-p^2}\right) \le p^3;$$

and

$$\mathbb{P}(P_x,\ P_y,\ R_x,\ R_y) = p^2\left(1-\frac{p(1-p)}{1-p^2}\right)^2 \le p^2.$$

In the case where $x = y$, the only case we would need to consider is

$$\mathbb{P}(P_x,\ R_x) = p\left(1-\frac{p(1-p)}{1-p^2}\right) \le p.$$

From this, we see that

$$\mathbb{P}(\text{a given 2-sum of } i \text{ removed} \mid B_i = k) \le 2p + 2p^3 + p^2.$$

Let $C(p) := 2p + 2p^3 + p^2$. Our above calculations yield that

$$\max_j \sum_{i\neq j} \mathbb{P}(I_i = 0, J_{j_i} = 1) \le \Sigma_1 + \Sigma_2, \tag{7}$$

where

$$\Sigma_1 := \max_j \sum_{i\neq j} \sum_{k=1}^{\lceil\frac{i+1}{2}\rceil} \binom{\lceil\frac{i+1}{2}\rceil - 1}{k} p^{2k}(1-p^2)^{\lceil\frac{i+1}{2}\rceil-k-1}(1-p)(C(p))^k,$$

and

$$\Sigma_2 := \max_j \sum_{i\neq j} \sum_{k=1}^{\lceil\frac{i+1}{2}\rceil} \binom{\lceil\frac{i+1}{2}\rceil - 1}{k-1} p^{2k-2}(1-p^2)^{\lceil\frac{i+1}{2}\rceil-k}p(C(p))^k.$$

Now,

$$\begin{aligned}
\Sigma_1 &= \max_j \sum_{i\neq j}(1-p)\left[(1-p^2+p^2C(p))^{\lceil\frac{i+1}{2}\rceil-1} - (1-p^2)^{\lceil\frac{i+1}{2}\rceil-1}\right] \\
&\le (1-p)\max_j \sum_{i\neq j}\left(\left\lceil\frac{i+1}{2}\right\rceil - 1\right)C(p)p^2(1-p^2+C(p)p^2)^{\lceil\frac{i+1}{2}\rceil-2} \\
&= O\left(np^3\sum_i (1-p^2+C(p)p^2)^{\frac{i}{2}}\right) \\
&= O\left(n^2p^3e^{-\alpha np^2/2}\right),
\end{aligned} \tag{8}$$

and hence $\Sigma_1 \to 0$ if p satisfies (3). For Σ_2, we have

$$\begin{aligned}
\Sigma_2 &= C(p)p \max_j \sum_{i\neq j} \sum_{k=1}^{\lceil \frac{i+1}{2}\rceil} \binom{\lceil \frac{i+1}{2}\rceil - 1}{k-1} p^{2k-2}(1-p^2)^{\lceil \frac{i+1}{2}\rceil - k}(C(p))^{k-1} \\
&= C(p)p \max_j \sum_{i\neq j} (1-p^2+C(p)p^2)^{\lceil \frac{i+1}{2}\rceil - 1} \\
&\cong C(p)p \max_j \sum_{i\neq j} e^{-\lceil i/2\rceil p^2} \\
&= O(np^2 e^{-\alpha np^2/2}) \\
&= o(n^2p^3e^{-\alpha np^2/2}). \qquad (9)
\end{aligned}$$

Applying (8) and (9) to (7), we see that the second term in (6) tends to 0 if p satisfies (3), and hence $d_{TV}(X,Y) \to 0$ if p satisfies (3). For p in this range

$$|\mathbb{P}(X=0) - e^{-\lambda}| \to 0$$

which finishes the proof. Note that if we consider a p not in this range, the result will hold by monotonicity. □

Remark 7. We have so far dealt with random sets of fixed expected size, and now indicate briefly how we can easily transition to the case of random sets of fixed size. This can be done for all the results in this paper, but we indicate the method in the context of Theorem 5: Choose one of $\binom{n+1}{|\mathcal{A}|}$ families randomly, and suppose $|\mathcal{A}| = \sqrt{Kn\log n}; K > \frac{2}{\alpha}$ (this corresponds to the expected size of $\mathcal{A}$ with $p = \sqrt{K\log n/n}$). Then we reconcile the two models as follows with $p = \sqrt{K\log n/n}$:

$$\begin{aligned}
&\mathbb{P}(\mathcal{A} \text{ is not a basis}||\mathcal{A}| = \sqrt{Kn\log n}) \\
&\leq \mathbb{P}(\mathcal{A} \text{ is not a basis}||\mathcal{A}| \leq \sqrt{Kn\log n}) \\
&\leq \mathbb{P}(\mathcal{A} \text{ is not a basis})/\mathbb{P}(|\mathcal{A}| \leq \sqrt{Kn\log n}) \\
&\cong 2\mathbb{P}(\mathcal{A} \text{ is not a basis}) \to 0,
\end{aligned}$$

by Theorem 5 and the central limit theorem. A similar argument holds if $K < \frac{2}{\alpha}$, or even if $|\mathcal{A}| = \sqrt{\frac{2}{\alpha}n\log n - \frac{2}{\alpha}n\log\log n + nA_n}$, where $|A_n| \to \infty$. It follows that we can easily go back and forth from the independent model to the fixed set size model except possibly when we are at the threshold, i.e., when

$$p = \sqrt{\frac{\frac{2}{\alpha}\log n - \frac{2}{\alpha}\log\log n + A_n}{n}},$$

with $A_n \to A \in \mathbb{R}$.

3. Truncated k-Bases

Following the same setup as before, we wish to represent each $j \in [\alpha n, (k-\alpha)n]$ as $j = x_1 + \ldots + x_k$ for $x_i \in \mathcal{A}$, where $x_1 \leq x_2 \leq \ldots \leq x_k$. Again, for each j, let I_j equal 1 if j cannot be expressed as a k-sum of elements of $\mathcal{A}$ and 0 otherwise. We set $X := \sum_{j=\alpha n}^{(k-\alpha)n} I_j$, and note that $X = 0 \Leftrightarrow \mathcal{A}$ is a truncated k-basis. Our main theorem is as follows:

Theorem 8. *With X defined as above, if we choose elements of $\{0\} \cup [n]$ to be in $\mathcal{A}$ with probability*

$$p = \sqrt[k]{\frac{K \log n - K \log\log n + A_n}{n^{k-1}}},$$

where $|A_n| = o(\log\log n)$ and $K = K_{\alpha,k} = \frac{k!(k-1)!}{\alpha^{k-1}}$, then

$$\mathbb{P}(\mathcal{A} \text{ is a truncated } k-\text{basis}) \to \begin{cases} 0 & \text{if } A_n \to -\infty \\ 1 & \text{if } A_n \to \infty \\ \exp\{-\frac{2\alpha}{k-1}e^{-A/K}\} & \text{if } A_n \to A \in \mathbb{R} \end{cases}.$$

Also for

$$p = \sqrt[k]{\frac{\beta K \log n}{n^{k-1}}}, \tag{10}$$

$\beta > (k-1)/k$, and $Y \sim \mathrm{Po}(\mathbb{E}(X))$,

$$d_{TV}(X, Y) \to 0 \text{ as } n \to \infty.$$

Before proving this theorem, we need some preliminary work. Let S_j be the set of all unordered k-tuples of nonnegative integers in $\{0\} \cup [n]$ that sum to j.

Claim 9. *For $j \in [\alpha n, n]$,*

$$|S_j| = j^{k-1}/[(k-1)!k!] + O(j^{k-2}).$$

Proof. The number of ordered k-tuples of nonnegative integers that sum to j is $\binom{j+k-1}{j} \cong \frac{j^{k-1}}{(k-1)!}$. All such tuples will be composed entirely of numbers in $\{0\} \cup [j]$, and at most $\binom{k}{2} \cdot n \cdot \binom{j+k-3}{j} = O(j^{k-2})$ of these contain a number repeated once or more often. We can disregard these in the asymptotic analysis, and consider the remaining unordered and ordered tuples. Each remaining unordered tuple appears $k!$ times among the remaining ordered tuples, giving us the desired first order asymptotics. □

Claim 10. *If $1 \leq j \leq kn/2$, then $|S_j| = |S_{kn-j}|$.*

Proof. There is a bijection between k-tuples in S_j for $1 \leq j \leq kn/2$ and those in S_j for $j \geq kn/2$ given by $\{x_1, x_2, \ldots, x_k\} \leftrightarrow \{n - x_1, n - x_2, \ldots, n - x_k\}$. □

Lastly, we have:

Claim 11. *For* $j \in [n+1,(k-1)n]$,

$$|S_j| \geq \frac{n^{k-1}}{(k-1)!k!} + O(n^{k-2}).$$

Proof. We will use the fact that $|S_j|$, i.e., the number of partitions of j of size at most k each part of which is less than or equal to n, is the coefficient of q^j in the q-binomial coefficient

$$\binom{n+k}{k}_q := \frac{(1-q)(1-q^2)\cdots(1-q^{n+k})}{(1-q)(1-q^2)\cdots(1-q^n)(1-q)(1-q^2)\cdots(1-q^k)}.$$

It is well known, see for example [11], that $\binom{n+k}{k}_q = \sum a_i q^i$ is a polynomial in q and that the coefficients are unimodal, namely $a_{j-1} < a_j$ for $j \leq nk/2$. Claim 9 yields that

$$a_n = |S_n| = \frac{n^{k-1}}{(k-1)!k!} + O(n^{k-2}),$$

and the proof now follows directly from Claim 10. □

We next begin our analysis of $\mathbb{E}(X)$ with a preliminary claim.

Claim 12. *With* S_j *defined as above,*

$$\mathbb{E}(X) = \sum_{j=\alpha n}^{(k-\alpha)n} \exp\left(-|T_j|p^k(1+o(1))\right),$$

where T_j*, the set of* k*-tuples of distinct elements that add to* j*, satisfies*

$$|T_j| \begin{cases} = j^{k-1}/(k-1)!k!, & \text{if } j \in [\alpha n, n]. \\ \geq n^{k-1}/(k-1)!k! & \text{if } j \in [n,(k-1)n]. \\ = (kn-j)^{k-1}/(k-1)!k! & \text{if } j \in [(k-1)n,(k-\alpha)n]. \end{cases}$$

Proof. Fix j, and let $\{B_{(j,i)}\}_{i=1}^{|S_j|}$ be the event that the ith unordered k-tuple of elements of $\{0\} \cup [n]$ that add to j is in $\mathcal{A}$. Then by Janson's inequality, we have

$$\begin{aligned}
\mathbb{P}(I_j = 1) &= \mathbb{P}(\cap_i B^C_{(j,i)}) \\
&\geq (1-p)^{O(1)}(1-p^2)^{O(j)}(1-p^3)^{O(j^2)}\ldots(1-p^{k-1})^{O(j^{k-2})}(1-p^k)^{|T_j|} \\
&\geq \exp(-\frac{O(1)p}{1-p} - \frac{O(j)p^2}{1-p^2} - \ldots \frac{|T_j|p^k}{1-p^k}) \\
&= \exp\Bigg(- |T_j|p^k \left(\frac{1}{1-p^k} + \frac{O(j^{k-2})}{(1-p^{k-1})|T_j|p} + \ldots + \frac{O(1)}{(1-p)|T_j|p^{k-1}}\right)\Bigg) \\
&= \exp\{-|T_j|p^k(1+o(1))\}
\end{aligned}$$

as for $x \in (0,1)$, $1-x \geq e^{-x/(1-x)}$, and with our choice of p, $np \to \infty$ and for all j, we have $j = O(n)$.

For an upper bound, Janson's inequality yields

$$P(I_j = 1) \leq \exp\left(-|T_j|p^{k-1}(1+o(1)) + \Delta_j/2\right).$$

Now $\Delta_j = O(n^{2k-3}p^{2k-1})$ since the maximal contribution to Δ_j is when the two tuples have distinct elements and intersect in a single element; as for all values of p in our window, $np \to \infty$. Thus

$$\begin{aligned} P(I_j = 1) &\leq \exp\left(-|T_j|p^{k-1} + O(n^{2k-3}p^{2k-1})\right) \\ &= \exp\left(-|T_j|p^{k}(1 + O(n^{k-2}p^{k-1}))\right). \end{aligned}$$

With $p = o(n^{-(k-2)/(k-1)})$, a condition satisfied by all of the $p's$ in our theorem, $n^{k-2}p^{k-1} \to 0$, finishing the proof. □

We are now ready to begin the proof of Theorem 8.

Proof. We now have that (with $(1+o(1))$ terms suppressed

$$\mathbb{E}(X) = 2\sum_{j=\alpha n}^{n} \exp\{-p^k j^{k-1}/(k-1)!k!\} + \sum_{j=n}^{(k-1)n} \exp\{-p^k|T_j|\}. \tag{11}$$

The first summation, which we will call Σ_1 (correspondingly calling the second sum Σ_2) can be bounded above as follows, recalling the definition of K from the statement of the theorem, and setting $B = \alpha^{k-2}/k!(k-2)!$:

$$\begin{aligned} \Sigma_1 &\leq 2\sum_{j=\alpha n}^{\infty} \exp\{-p^k j^{k-1}/(k-1)!k!\} \\ &\leq 2\exp\{-n^{k-1}p^k/K\}\left(\sum_{j=0}^{\infty}\left(\exp\{-(k-1)(n\alpha)^{k-2}p^k/(k-1)!k!\}\right)^j\right) \\ &= \frac{2\exp\{-n^{k-1}p^k/K\}}{1-\exp\{-Bn^{k-2}p^k\}} \\ &\leq \frac{2\exp\{-n^{k-1}p^k/K\}}{Bn^{k-2}p^k}(1+Bn^{k-2}p^k) \\ &= \frac{2\exp\{-n^{k-1}p^k/K\}}{Bn^{k-2}p^k}(1+o(1)) \end{aligned} \tag{12}$$

where we used the facts that $\frac{x}{x+1} \leq 1-e^{-x}$ in the fourth line and, in the final line of the display, that for all choices of p considered in the theorem, $n^{k-2}p^k \to 0$.

Finally, the geometric bound in the second line follows from the fact that the ratio of consecutive terms in the sum satisfies

$$\exp\{-\frac{p^k}{(k-1)!k!}[(j+1)^{k-1}-j^{k-1}]\} \geq \exp\{-\frac{p^k}{(k-1)!k!}(k-1)(\alpha n)^{k-2}\}.$$

For Σ_2 we have

$$\begin{aligned} \Sigma_2 &= \sum_{j=n}^{(k-1)n} \exp\{-p^k|T_j|\} \\ &\leq (k-2)n\exp\{-p^k n^{k-1}/(k-1)!k!\} \to 0, \end{aligned} \tag{13}$$

if p satisfies

$$p \geq \left(\frac{D\log n}{n^{k-1}}\right)^{1/k}; D > (k-1)!k!.$$

A lower bound on Σ_1 (and thus on $\mathbb{E}(X)$) is obtained by using elementary integration by parts to derive a tight estimate for the integral

$$\int_x^\infty e^{-t^{k-1}}dt; x \to \infty,$$

in much the same way that Gaussian tails are analyzed (which is the $k=3$ case.) We have

$$\begin{aligned} \mathbb{E}(X) &\geq 2\sum_{j=\alpha n}^{n} \exp\{-p^k j^{k-1}/(k-1)!k!\} \\ &\cong 2\int_{\alpha n}^{n} \exp\{-p^k x^{k-1}/(k-1)!k!\}dx \\ &= 2\int_{\alpha n}^{\infty} \exp\{-p^k x^{k-1}/(k-1)!k!\}dx \\ &\quad -2\int_{n}^{\infty} \exp\{-p^k x^{k-1}/(k-1)!k!\}dx. \end{aligned} \tag{14}$$

Setting, for $t > 0$, $\Psi(t,k) = \int_t^\infty \exp\{-Cx^{k-1}\}dx$, we see that

$$\begin{aligned} \Psi(t,k) &\leq \frac{1}{t^{k-2}}\int_t^\infty x^{k-2}\exp\{-Cx^{k-1}\}dx \\ &= \frac{1}{C(k-1)t^{k-2}}\int_{Ct^{k-1}}^\infty e^{-u}du \\ &= \frac{1}{C(k-1)t^{k-2}}\exp\{-Ct^{k-1}\}, \end{aligned} \tag{15}$$

and, for another two constants $E, E' > 0$,

$$\begin{aligned}
\Psi(t,k) &= \int_t^\infty \frac{x^{k-2}}{x^{k-2}} \exp\{-Cx^{k-1}\}dx \\
&= \frac{1}{C(k-1)t^{k-2}} \exp\{-Ct^{k-1}\} - \frac{E}{C}\int_t^\infty \frac{1}{x^{k-1}} \exp\{-Cx^{k-1}\}dx \\
&\geq \frac{1}{C(k-1)t^{k-2}} \exp\{-Ct^{k-1}\} - \frac{E'}{C^2}\frac{1}{t^{2k-3}} \exp\{-Ct^{k-1}\} \qquad (16)
\end{aligned}$$

using (15). Thus, since in our context $C^2t^{2k-3} \gg Ct^{k-2}$, (12) and (16) combine to give

$$\Sigma_1 \cong \frac{2(k-1)!k!}{(k-1)\alpha^{k-2}n^{k-2}p^k} \exp\{-\alpha^{k-1}n^{k-1}p^k/(k-1)!k!\}.$$

It is easy to verify from (13) that $\Sigma_2 = o(1)\Sigma_1$, and thus

$$\mathbb{E}(X) = \Sigma_1(1+o(1)) \cong \frac{2(k-2)!k!}{\alpha^{k-2}n^{k-2}p^k} \exp\{-\alpha^{k-1}n^{k-1}p^k/(k-1)!k!\}.$$

Let $\lambda = \mathbb{E}(X)$ and let $Y \sim \text{Po}(\lambda)$. First we note that $\mathbb{E}(X)$ tends to zero, infinity, or $\frac{2\alpha}{k-1}e^{-A/K}$ if p is as stated as in the theorem with $A_n \to \infty$, $A_n \to -\infty$ or $A_n \to A$ respectively. To complete the proof of Theorem 8, we use Poisson approximation to show that the total variation distance between X and Y converges to 0, so that, in particular, $\mathbb{P}(X=0) \to e^{-\lambda}$.

Following [2], we first need to determine, for each j separately, an auxiliary sequence of variables J_{j_i} defined on the same probability space with the property that

$$\mathcal{L}(J_{j_1}, J_{j_2}, \ldots) = \mathcal{L}(I_1, I_2, \ldots | I_j = 1). \qquad (17)$$

Such a coupling can *probably* be described explicitly as we did for $k=2$ but there is no need to do so: It is clear that the more integers in $\mathcal{A}$, the higher the probability that I_j is 0, as integer j is more likely to be representable as a k-sum. Therefore if we let

$$Y_i = \begin{cases} 1 & \text{if} \quad i \in \mathcal{A} \\ 0 & \text{else} \end{cases}$$

then the I_j's are decreasing functions of the baseline i.i.d. random variables $\{Y_i\}$ that have distribution $\mathbb{P}(Y_i = 1) = p; \mathbb{P}(Y_i = 0) = 1-p$, and a monotone coupling satisfying (17) exists. We can then apply Theorem 2.C of [2] with $\Gamma_\alpha^0 = \Gamma_\alpha^- = \emptyset$,

$$d_{TV}(X,Y) \leq \frac{1-e^{-\lambda}}{\lambda}\left(\sum_j \mathbb{P}^2(I_j=1) + \sum_j\sum_\ell(\mathbb{E}(I_jI_\ell) - \mathbb{E}(I_j)\mathbb{E}(I_\ell))\right). \qquad (18)$$

Note that the above is just a variation of the bound used in the proof of Theorem 5. Now the first term in (18) is bound by

$$\begin{aligned}\frac{1-e^{-\lambda}}{\lambda}\sum_j \mathbb{P}^2(I_j=1) &\le \max_j \mathbb{P}(I_j=1)\\ &\le \max_j \exp\{-|T_j|p^k(1+o(1))\}\\ &\le e^{-\delta n^{k-1}p^k(1+o(1))},\end{aligned}$$

where δ is a constant not depending on p or n, and this term converges to 0 if $p=\sqrt[k]{\frac{G\log n}{n^{k-1}}}$ for any constant $G>0$. For the double sum in (18), we start by using the estimates

$$\mathbb{P}(I_i=1)\ge \exp\{-|S_i|p^k(1+O(1/np))\}$$

for $i=j,\ell$. For the cross product term $\mathbb{E}(I_jI_\ell)$, for fixed j,ℓ, note that $\mathbb{P}(I_jI_\ell=1)=\mathbb{P}(S=0)$, where

$$S=\sum_{r=j,\ell}\sum_{s=1}^{k}\sum_{\mathbf{a}:=\{a_1,\ldots,a_s\}:\sum a_i=r} J_{r,s,\mathbf{a}},$$

and where $J_{\mathbf{a}}:=J_{r,s,\mathbf{a}}$ equals one if the s integers in $\mathbf{a}$ that sum to r (with possible repetition) are all selected to be in $\mathcal{A}$. We use Janson's inequality to get the bound

$$\mathbb{E}(I_jI_\ell)\le e^{-\mathbb{E}(S)+\Delta}, \tag{19}$$

where

$$\Delta=\sum_{\{r,l,\mathbf{a}\}}\sum_{\{s,m,\mathbf{b}\}\sim\{r,l,\mathbf{a}\}}\mathbb{E}(J_{\mathbf{a}}J_{\mathbf{b}}),$$

and $\{r,l,\mathbf{a}\}\sim\{s,m,\mathbf{b}\}$ if $\mathbf{a}\cap\mathbf{b}\ne\emptyset; \{r,l,\mathbf{a}\}\ne\{s,m,\mathbf{b}\}$. Using a worst case scenario estimate for Δ, we see

$$\Delta=O(n^{2k-3}p^{2k-1}).$$

As in the proofs of Claims 9-11, we see

$$\mathbb{E}(S)\ge(|S_j|+|S_\ell|)p^k+p^{k-1}O(n^{k-2})\ge(|S_j|+|S_\ell|)p^k(1+O(1/np)).$$

Equation (19) now yields

$$\mathbb{E}(I_jI_\ell)\le\exp\{-(|S_j|+|S_\ell|)p^k(1+O(1/np))+O(n^{2k-3}p^{2k-1})\},$$

and we then conclude that

$$\begin{aligned}&\sum_j\sum_{\ell\ne j}(\mathbb{E}(I_jI_\ell)-\mathbb{E}(I_j)\mathbb{E}(I_\ell))\\ &\le\sum_j\sum_{\ell\ne j}\exp\{-(|S_j|+|S_\ell|)p^k(1+O(1/np))\}\left(e^{O(n^{2k-3}p^{2k-1})}-1\right)\\ &=O(n^{2k-3}p^{2k-1}\lambda^2),\end{aligned}$$

so that

$$\frac{\sum_j \sum_{\ell \neq j} (\mathbb{E}(I_j I_\ell) - \mathbb{E}(I_j)\mathbb{E}(I_\ell))}{\lambda} = O(n^{2k-3} p^{2k-1} \lambda) \to 0$$

if $\lambda \ll 1/n^{2k-3}p^{2k-1}$, i.e., if

$$\frac{2(k-2)!k!n^{k-1}p^{k-1}}{\alpha^{k-2}} \exp\{-\alpha^{k-1}n^{k-1}p^k/(k-1)!k!\} \to 0,$$

i.e., if

$$p = \sqrt[k]{\frac{[\beta K] \log n}{n^{k-1}}},$$

for $\beta > (k-1)/k$. Applying this to (18) yields that $d_{\mathrm{TV}}(X, Y) \to 0$ as $n \to \infty$ for all p satisfying (10). □

Open Problems. In both the modular and truncated cases, the representation function question has yet to be addressed. We have preliminary results in this direction and will be publishing them in a future paper.

Acknowledgments. The research of AG, CML, and NGT was supported by NSF grant 1004624. The collaboration between AG and VL was made possible due to the support of the Acheson J. Duncan Fund for the Advancement of Research in Statistics. VL was supported by U.S. Department of Education GAANN grant P200A090128. A preliminary version of these results was presented by AG at the INTEGERS 2009 conference. We also want to acknowledge the helpful comments and suggestions of an anonymous referee.

References

[1] N. Alon and J. Spencer (1992). *The Probabilistic Method.* Wiley, New York.

[2] A. Barbour, L. Holst, and S. Janson (1992). *Poisson Approximation.* Oxford University Press.

[3] P. Erdős (1956). Problems and results in additive number theory, in *Colloque sur la Théorie des Nombres* (CBRM), Bruxelles, 127–137.

[4] P. Erdős and P. Tetali (1990). Representations of integers as the sum of k terms, *Rand. Structures Algorithms* **1**, 245–261.

[5] K. Ford (2008). Unpublished work.

[6] C. Güntürk and M. Nathanson (2006). A new upper bound for finite additive bases, *Acta Arith.* **124**, 235–255.

[7] N. Hämmerer and G. Hofmeister (1976). Zu einer Vermutung von Rohrbach, *J. Reine Angew. Math.* **286-287**, 239–247.

[8] S. Janson (1990). Poisson approximation for large deviations, *Rand. Structures Algorithms* **1**, 221–229.

[9] L. Moser (1960). On the representation of $\{1, 2, \ldots, n\}$ by sums, *Acta Arith.* **6**, 11–13.

[10] A. Mrose (1979). Untere Schranken für die Reichweiten von Extremalbasen fester Ordnung, *Abh. Math. Sem. Univ. Hamburg* **48**, 118–124.

[11] K. M. O'Hara (1990). Unimodality of Gaussian coefficients: a constructive proof, *J. Combin. Number Theory, Series A*.**43**, 29–52.

[12] C. Sandor (2007). Random B_h sets and additive bases in $\mathbb{Z}_n$, *Integers* **7**, Paper #A32.

[13] G. Yu (2009). Upper bounds for finitely additive 2-bases, *Proc. Amer. Math. Soc.* **137**, 11–18.

[14] A. Yadin (2009). When Do Random Subsets Decompose a Finite Group? *Israel J. Math.* **174**, 203–219.

ON AN ANALOGUE TO THE LUCAS-LEHMER-RIESEL TEST USING ELLIPTIC CURVES

Markus Kirschmer
Lehrstuhl D für Mathematik, RWTH Aachen University, Aachen, Germany
markus.kirschmer@math.rwth-aachen.de

Michael H. Mertens
Lehrstuhl D für Mathematik, RWTH Aachen University, Aachen, Germany
michael.helmut.mertens@rwth-aachen.de

Received: 5/17/11, Revised: 2/17/12, Accepted: 3/24/13, Published: 3/29/13

Abstract

Following an idea of B. H. Gross, who presented an elliptic curve test for Mersenne primes $M_p = 2^p - 1$, we propose a similar test with elliptic curves for generalized Thabit primes $K(h,n) := h \cdot 2^n - 1$ for any positive odd number h and any integer $n > \log_2(h) + 2$.

1. Introduction

In 1876, Édouard Lucas invented an efficient primality test for Mersenne numbers $M_p = 2^p - 1$ for prime numbers p. The test uses a recursion defined by

$$L_0 = 4, \quad L_{k+1} = L_k^2 - 2.$$

He showed that a given Mersenne number M_p for $p \geq 3$ is prime if and only if $L_{p-2} \equiv 0 \pmod{M_p}$ (cf. [6]). The general idea of the proof given in [9] is to interpret the recursion as squaring a point on the algebraic torus over $\mathbb{Q}$ associated to the quadratic field $\mathbb{Q}(\sqrt{3})$. This test was optimized by Derrick Henry Lehmer in 1935 (cf. [5]) and therefore it is known as Lucas-Lehmer test.

In 1969, H. Riesel proposed a test for so called generalized Thabit numbers

$$K(h,n) = h \cdot 2^n - 1 \text{ for } h \in \mathbb{N} \text{ odd and } n \in \mathbb{N}$$

using the Lucasian sequence

$$L_{k+1} := L_k^2 - 2$$

where the first value L_0 of the recursion depends on h and n (cf. [8]). Some special cases of his generalization, especially for $3 \nmid h$, had already been found by Lucas and Lehmer themselves. This test is referred to as the Lucas-Lehmer-Riesel test. The idea of using elliptic curves for primality tests is due to Hendrik Lenstra. He invented a general primality test for integers. In 2005, Benedict H. Gross used the ideas of Lucas and Lenstra to create an elliptic curve test especially for Mersenne primes. It is based on doubling the point $P = (-2, 4)$ on the rational elliptic curve with Weierstrass equation $y^2 = x^3 - 12x$. In [2], R. Denomme and G. Savin developed similar primality tests for Fermat numbers $F_n = 2^{2^n} + 1$ and numbers of the form $2^{2^n} - 2^{2^{n-1}} + 1$ and $3^{2^n} - 3^{2^{n-1}} + 1$. These tests are also based on doubling some suitable point on a twisted rational elliptic curve.

In this paper, we will extend the algorithm of Gross to all generalized Thabit numbers $K(h, n)$ with $n > \log_2(h) + 2$. If h is divisible by 3, we will usually not be able to work with a global point on some rational elliptic curve. Instead, we will be using quadratic twists of rational elliptic curves E_ε as it was done in [2]. The curves E_ε we are using have Weierstrass equations of the form $y^2 = x^3 - \varepsilon x$, where $\varepsilon \in \mathbb{Z}$ depends on h and n. We also give an algorithm to find appropriate values for ε.

All the primality tests mentioned above are based on the following idea. Let N be some integer such that the factorization of a sufficiently large factor of $N \pm 1$ is known. These algorithms then construct an element g in some suitable group (e.g., the reduction of some rational elliptic curve modulo N or $(\mathbb{Z}_F/N\mathbb{Z}_F)^*$ where $\mathbb{Z}_F$ denotes the ring of integers in some algebraic number field F). Then N is prime if and only if the order of g is sufficiently large. The survey article [7] by Carl Pomerance contains some more primality tests based on this idea.

The paper is organized as follows. A short proof of the Lucas-Lehmer-Riesel test is given in Section 2. In Section 3, we recall division polynomials and in Section 4, we summarize some properties of elliptic curves of the form E_ε. In Section 5, we prove our primality test for generalized Thabit numbers. Finally in Section 6, we compare the efficiency of the mentioned primality tests: the Lucas-Lehmer test to Gross' test using elliptic curves for Mersenne numbers (cf. [3]) and the Lucas-Lehmer-Riesel test to our test for generalized Thabit numbers.

Notation By $\mathbb{P}$ we denote the set of positive prime numbers. For positive integers m, n let $\left(\frac{m}{n}\right)$ denote the Jacobi-Symbol. In particular, if n is prime, then $\left(\frac{m}{n}\right)$ coincides with the Legendre-Symbol. For an elliptic curve E over a field K we set $E(L)$ the group of L-rational points of E for any extension field L of K and $\mathcal{O}$ as the point at infinity, the zero-element of E as an abelian group. If $K = \mathbb{Q}$ and $p \in \mathbb{Z}$ is prime, we denote the reduction of E modulo p by $\widehat{E}$ and we set $E(p) := \widehat{E}(\mathbb{F}_p)$. If $\mathcal{O} \neq P = (x_0, y_0)$ is a point of an elliptic curve E, we denote by $x(P) := x_0$ the projection to the first affine coordinate of P.

2. The Lucas-Lehmer-Riesel Test

At first we want to give a short proof of Riesel's original primality test for numbers of the form $K(h,n)$ (see [8]). For this section, let $d \geq 2$ be a square-free integer, $F := \mathbb{Q}(\sqrt{d})$ the corresponding real quadratic number field and $\mathbb{Z}_F$ its ring of integers. Further, let $\bar{\ }\colon F \to F$ denote the nontrivial Galois automorphism of F. The following theorem is well known from the theory of quadratic number fields, but essential for the further discussion of the Lucas-Lehmer-Riesel test. Thus we give a short proof here.

Theorem 2.1. (Fermat) *Let $\alpha \in \mathbb{Z}_F$ and let $p \in \mathbb{P}$ be an odd prime such that $\alpha\mathbb{Z}_F + p\mathbb{Z}_F = \mathbb{Z}_F$.*

1. *If $\left(\frac{d}{p}\right) = 1$ then $\alpha^{p-1} \equiv 1 \pmod{p\mathbb{Z}_F}$.*
2. *If $\left(\frac{d}{p}\right) = -1$ then $\alpha^{p+1} \equiv \alpha\bar{\alpha} \pmod{p\mathbb{Z}_F}$.*

Proof. Let $\varphi\colon \mathbb{Z}_F \to \mathbb{Z}_F/p\mathbb{Z}_F$ be the canonical epimorphism. If $\left(\frac{d}{p}\right) = 1$ then $\mathbb{Z}_F/p\mathbb{Z}_F \cong \mathbb{F}_p \times \mathbb{F}_p$ since p splits in $\mathbb{Z}_F$. By assumption, $\varphi(\alpha)$ is a unit, thus $\varphi(\alpha)^{p-1} = 1$. If $\left(\frac{d}{p}\right) = -1$ then p is inert in $\mathbb{Z}_F$. Thus $\mathbb{Z}_F/p\mathbb{Z}_F \cong \mathbb{F}_{p^2}$ and $\varphi(\bar{x}) = \varphi(x)^p$ for all $x \in \mathbb{Z}_F$. Hence $\varphi(\alpha^{p+1}) = \varphi(\alpha\bar{\alpha})$. □

From this fact, we can now derive the Lucas-Lehmer-Riesel test:

Proposition 2.2. *Suppose $p \in \mathbb{P}$ is an odd prime with $\left(\frac{d}{p}\right) = -1$. If there are $a, b, r \in \mathbb{Z}$ such that*

$$\alpha := \frac{(a+b\sqrt{d})^2}{r} \in \mathbb{Z}_F \text{ and } \left(\frac{r}{p}\right) \cdot \frac{a^2-b^2d}{r} = -1$$

then

$$\alpha^{(p+1)/2} \equiv -1 \pmod{p\mathbb{Z}_F}.$$

Proof. According to Theorem 2.1 we obtain modulo $p\mathbb{Z}_F$

$$\alpha^{(p+1)/2} = \frac{(a+b\sqrt{d})^{p+1}}{r^{(p+1)/2}} \overset{2.1}{\equiv} \frac{(a+b\sqrt{d})(a-b\sqrt{d})}{r^{(p-1)/2} \cdot r} \equiv \frac{a^2-b^2d}{r} \cdot \left(\frac{r}{p}\right) = -1. \qquad \square$$

Proposition 2.3. *Let $n \geq 2$ and let $h < 2^n$ be an odd number. Further let $K := K(h,n) = h \cdot 2^n - 1$ denote the corresponding generalized Thabit number. If $\gcd(K,d) = 1$ and if there are $a, b \in \mathbb{Z}$ such that*

$$\alpha := \frac{(a+b\sqrt{d})^2}{|a^2-b^2d|} \in \mathbb{Z}_F \quad \text{and} \quad \alpha^{(K+1)/2} \equiv -1 \pmod{K\mathbb{Z}_F}$$

then K is prime.

Proof. Let p be some prime factor of $K(h,n)$. Then $\alpha^{(K+1)/2} \equiv -1 \pmod{p\mathbb{Z}_F}$. Let $\varphi\colon \mathbb{Z}_F \to \mathbb{Z}_F/p\mathbb{Z}_F$ be the canonical epimorphism and let

$$k = \min\{t \in \mathbb{N} \mid \alpha^t \equiv -1 \pmod{p\mathbb{Z}_F}\}\,.$$

The order of $\varphi(\alpha)$ in $(\mathbb{Z}/p\mathbb{Z}_F)^*$ is $2k$ since p is odd. Further, $\varphi(\alpha^{(K+1)/2-k}) = 1$ implies that $h \cdot 2^{n-1} = (K+1)/2 = k + 2kr = k(1+2r)$ for some $r \in \mathbb{Z}$. In particular, k is divisible by 2^{n-1}. Let $\varepsilon := \left(\frac{d}{p}\right)$. Then $\varepsilon \neq 0$ since $\gcd(K,d) = 1$. Further, $\alpha\overline{\alpha} = 1$, so α is a unit in $\mathbb{Z}_F$. Thus, if $\varepsilon = 1$ then Theorem 2.1 shows that

$$\alpha^{(p-1)/2} = \frac{(a+b\sqrt{d})^{p-1}}{|a^2-b^2d|^{(p-1)/2}} \equiv \left(\frac{|a^2-b^2d|}{p}\right) \pmod{p\mathbb{Z}_F}$$

and if $\varepsilon = -1$ then

$$\alpha^{(p+1)/2} \equiv \frac{a^2-b^2d}{|a^2-b^2d|}\left(\frac{|a^2-b^2d|}{p}\right) \pmod{p\mathbb{Z}_F}\,.$$

In particular, $\alpha^{p-\varepsilon} \equiv 1 \pmod{p\mathbb{Z}_F}$. Thus we have $p+1 \geq p-\varepsilon \geq 2k \geq 2^n$. Furthermore, K is not a square since $K \equiv -1 \pmod 4$. So if K is not a prime, then K has two distinct prime factors p and q. We may assume that $p < q$. Then

$$K \geq p\cdot q \geq p(p+2) \geq (2^n-1)(2^n+1) = 2^n\cdot 2^n - 1 \gneq h\cdot 2^n - 1 = K$$

gives the desired contradiction. Hence K must be prime. □

Corollary 2.4 (Lucas-Lehmer-Riesel test). *Let $n \geq 2$ and let $h < 2^n$ be an odd integer. Suppose there exist $a, b \in \mathbb{Z}$ and some square-free integer $d \geq 2$ such that*

$$\alpha = \tfrac{(a+b\sqrt{d})^2}{r} \in \mathbb{Z}_{\mathbb{Q}(\sqrt{d})} \text{ and } \left(\tfrac{d}{K(h,n)}\right) = \left(\tfrac{r}{K(h,n)}\right)\cdot\tfrac{a^2-b^2d}{r} = -1$$

where $r = |a^2-b^2d|$.
Then the generalized Thabit number $K(h,n)$ is prime if and only if

$$L_{n-2} \equiv 0 \pmod{K(h,n)}$$

where $L_0 = \alpha^h + \overline{\alpha}^h \in \mathbb{Z}$ and $L_{s+1} := L_s^2 - 2$ for $s \geq 0$.

Proof. Let $F = \mathbb{Q}(\sqrt{d})$. Since $\alpha\overline{\alpha} = 1$ we have $\alpha^{-1} = \overline{\alpha}$. Hence it follows from

$$(\alpha^{h\cdot 2^s} + \alpha^{-h\cdot 2^s})^2 = \alpha^{h\cdot 2^{s+1}} + \alpha^{-h\cdot 2^{s+1}} + 2$$

that $L_s = \alpha^{h\cdot 2^s} + \alpha^{-h\cdot 2^s}$ for all $s \geq 0$. If $K(h,n)$ is prime, then Proposition 2.2 implies

$$\begin{aligned} L_{n-2} &= \alpha^{h\cdot 2^{n-2}} + \alpha^{-h\cdot 2^{n-2}} \\ &= \alpha^{-h\cdot 2^{n-2}}(\alpha^{h\cdot 2^{n-1}} + 1) \equiv 0 \pmod{K(h,n)\mathbb{Z}_F}\,. \end{aligned}$$

Since $L_{n-2} \in \mathbb{Z}$ this shows that $L_{n-2} \equiv 0 \pmod{K(h,n)}$ as claimed. Conversely, by Proposition 2.3 and the above calculation we see that the given conditions are also sufficient for the primality of $K(h,n)$. □

Given a generalized Thabit number $K(h,n)$, a triple (a,b,d) as in the Corollary above can be found by inspecting the fundamental units of various real quadratic number fields, see [8] for details.

Remark 2.5. Suppose the notation of Corollary 2.4. Then the Lucas-Lehmer-Riesel test can be summed up as follows. Let $\varphi\colon \mathbb{Z}_F \to \mathbb{Z}_F/K(h,n)\mathbb{Z}_F$ denote the canonical epimorphism and set $\beta = \varphi(\alpha)$. Then $K(h,n)$ is prime if and only if β^h generates a cyclic subgroup of $(\mathbb{Z}_F/K(h,n)\mathbb{Z}_F)^*$ of order 2^n.

3. Division Polynomials

In this section, let E be an elliptic curve over a field K with $\operatorname{char}(K) \neq 2,3$ given by a Weierstrass equation of the form

$$y^2 = x^3 + ax + b, \quad a,b \in K.$$

Definition 3.1. The *division polynomials* $\psi_m \in \mathbb{Z}[x,y,a,b]$ of the elliptic curve E are given via the recursion

$$\begin{aligned}
\psi_0 &:= 0, \qquad \psi_1 := 1, \qquad \psi_2 := 2y,\\
\psi_3 &:= 3x^4 + 6ax^2 + 12bx - a^2,\\
\psi_4 &:= 4y(x^6 + 5ax^4 + 20bx^3 - 5a^2x^2 - 4abx - 8b^2 - a^3),\\
\psi_{2m+1} &:= \psi_{m+2}\psi_m^3 - \psi_{m-1}\psi_{m+1}^3 \quad (m \geq 2),\\
2y\psi_{2m} &:= \psi_m(\psi_{m+2}\psi_{m-1}^2 - \psi_{m-2}\psi_{m+1}^2) \quad (m \geq 3).
\end{aligned}$$

These polynomials have the following property.

Lemma 3.2. *Let m be a positive integer.*

1. *If one identifies y^2 with $x^3 + ax + b$, then ψ_{2m+1}, ψ_{2m}/y and ψ_{2m}^2 can be viewed as polynomials in $\mathbb{Z}[x,a,b]$.*

2. *Let L/K be a field extension and $P = (x,y)$ be a point on $E(L)$. If $m \cdot P \neq \mathcal{O}$ then $\psi_m^2(x) \neq 0$ and*

$$m \cdot P = \left(x - \frac{\psi_{m-1}(x)\psi_{m+1}(x)}{\psi_m^2(x)}, \frac{\psi_{2m}(x,y)}{2\psi_m^4(x)} \right).$$

In particular, $m \cdot P = (f_m(x,a,b), g_m(x,a,b)y)$ for some rational functions $f_m, g_m \in \mathbb{Q}(x,a,b)$.

Proof. See for example [4, Chapter 2, §§1-2]. □

4. Some Properties of the Curves E_ε and Their Quadratic Twists

In the following, let E_ε^t be the (twisted) elliptic curve over $\mathbb{Q}$ given by

$$ty^2 = x^3 - \varepsilon x, \qquad \varepsilon, t \in \mathbb{Z} \text{ with } t \neq 0.$$

When $t = 1$, we will usually omit the superscript t. Then

$$\tau_{\varepsilon,t}\colon E_\varepsilon^t \to E_{\varepsilon t^2},\ (x,y) \mapsto (xt, yt^2) \tag{1}$$

is an isogeny. Further, E_ε^t defined over $\mathbb{Q}(i)$ has complex multiplication by the ring of Gaussian integers $\mathbb{Z}[i]$:

$$[i] : E_\varepsilon^t(\mathbb{Q}(i)) \to E_\varepsilon^t(\mathbb{Q}(i)),\ (x,y) \mapsto (-x, i \cdot y)$$

and the discriminant of E_ε^t is

$$\Delta(E_\varepsilon^t) = 2^6 \cdot \varepsilon^3 \cdot t^6.$$

This leads to the following lemma.

Lemma 4.1. *Let $q \in \mathbb{P}$ be a prime not dividing εt such that $q \equiv -1 \pmod 4$.*

1. *The reduction of E_ε^t modulo q is supersingular and the group $E_\varepsilon^t(q)$ has order $q+1$.*

2. *If ε is not a square modulo q, then the group $E_\varepsilon^t(q)$ is cyclic and the only point of order 2 in $E_\varepsilon^t(q)$ is $(0,0)$.*

Proof. Because of the isogeny $\tau_{\varepsilon,t}$ from equation (1), we may assume that $t = 1$. Let i be a primitive fourth root of unity in $\overline{\mathbb{F}}_q$. Then the reduction $\widehat{E}_\varepsilon$ of E_ε modulo q has complex multiplication by $[i]$ as well. Let Φ be the q-th power Frobenius endomorphism and $P = (x,y)$ a point of E_ε. We have

$$(\Phi \circ [i])(P) = (-x^q, i^q y^q) = (-x^q, -i \cdot y^q) \quad \text{and} \quad ([i] \circ \Phi)(P) = (-x^q, iy^q).$$

Therefore the endomorphism ring $\mathrm{End}(\widehat{E}_\varepsilon)$ is not commutative, hence $\widehat{E}_\varepsilon$ is supersingular by [10, Theorem V.3.1]. Thus the abelian group $E_\varepsilon(q)$ has order $q+1$ (cf. [10, exercise 5.10]). This proves the first claim.

Let $[q+1]$ denote the isogeny of E_ε which multiplies each point on E_ε by $q+1$. The kernel of $[q+1]$ is isomorphic to $\mathbb{Z}/(q+1)\mathbb{Z} \times \mathbb{Z}/(q+1)\mathbb{Z}$ by [10, Corollary III.6.4]. Since $\#E_\varepsilon(q) = q+1$ we have $E_\varepsilon(q) \leq \ker([q+1])$. In particular, $E_\varepsilon(q) \cong (\mathbb{Z}/d_1\mathbb{Z}) \times (\mathbb{Z}/d_2\mathbb{Z})$ where $d_1 \mid d_2$ and $d_1 \cdot d_2 = q+1$. Furthermore $d_1 \mid (q-1)$ since the Weil-pairing is surjective and Galois-invariant (cf. [10, Proposition III.8.1]). Thus $d_1 \mid \gcd(q-1, q+1) = 2$, so either $E_\varepsilon(q)$ is cyclic or all 2-torsion points of $\widehat{E}_\varepsilon$ are $\mathbb{F}_q$-rational, which means all roots of $x^3 - \varepsilon x = x \cdot (x^2 - \varepsilon)$ lie in $\mathbb{F}_q$. But this is impossible because ε is not a square in $\mathbb{F}_q$. Hence the group $E_\varepsilon(q)$ must be cyclic. □

Lemma 4.2. *Let $q \in \mathbb{P}$ be a prime not dividing $2\varepsilon t$ and let $P = (x_0, y_0) \in E_\varepsilon^t(q)$. If $x_0 t$ is not a square in $\mathbb{F}_q$, then P is not divisible by 2 in $E_\varepsilon^t(q)$.*

Proof. It suffices to show that $P' := \tau_{\varepsilon,t}(P) = (x_0 t, y_0 t^2)$ is not divisible by 2 in $E_{\varepsilon t^2}(q)$. Let $Q = (x, y)$ be a point in $E_{\varepsilon t^2}(q)$ such that $2 \cdot Q \neq \mathcal{O}$. Then we have

$$x(2 \cdot Q) = \frac{(x^2 + \varepsilon t^2)^2}{4 \cdot y^2},$$

which is a square in $\mathbb{F}_q$. Since $x_0 t = x(P')$ is not, there cannot be any point $Q \in E_{\varepsilon t^2}(q)$ such that $2 \cdot Q = P'$. □

5. Generalized Thabit Numbers

5.1. Choosing the Curve

For our primality test for generalized Thabit numbers $K(h, n) = h \cdot 2^n - 1$ in Theorem 5.5 we need a (twisted) rational elliptic curve E such that

$$\#E(K(h, n)) = K(h, n) + 1 = h \cdot 2^n$$

and a point that generates the Sylow-2-subgroup of $E(K(h, n))$ provided that $K(h, n)$ is prime.

Proposition 5.1. *Let $n \geq 2$ and $h \in \mathbb{N}$ be odd. Suppose there exists a pair $(\varepsilon, x_0) \in \mathbb{Z}^2$ such that*

$$\left(\frac{\varepsilon}{K(h,n)}\right) = \left(\frac{x_0^2 - \varepsilon}{K(h,n)}\right) = -1 \,.$$

Write $x_0(x_0^2 - \varepsilon) = t y_0^2$ for some $t, y_0 \in \mathbb{Z}$. Then $P = (x_0, y_0) \in E_\varepsilon^t(\mathbb{Q})$. If $K(h, n)$ is prime then

1. *$K(h, n)$ does not divide x_0.*

2. *The reduction of E_ε^t modulo $K(h, n)$ is supersingular and $E_\varepsilon^t(K(h, n))$ is cyclic of order $h \cdot 2^n$.*

3. *The reduction of P modulo $K(h, n)$ is not divisible by 2 in $E_\varepsilon^t(K(h, n))$ and $h \cdot P$ generates the Sylow-2-subgroup of $E_\varepsilon^t(K(h, n))$.*

Proof. If $K(h, n)$ would divide x_0, then

$$-1 = \left(\frac{x_0^2 - \varepsilon}{K(h,n)}\right) = \left(\frac{-\varepsilon}{K(h,n)}\right) = \left(\frac{-1}{K(h,n)}\right)\left(\frac{\varepsilon}{K(h,n)}\right) = (-1)^2$$

which is impossible. In particular, $K(h,n)$ does not divide $x_0(x_0^2-\varepsilon) = ty_0^2$. Hence the second statement immediately follows from Lemma 4.1.
Further, $(x_0t)(x_0^2-\varepsilon) = t^2y_0^2$ is a square modulo $K(h,n)$ but $x^2-\varepsilon$ is not. So x_0t is also not a square. By Lemma 4.2 the point P is not divisible by 2 in $E_\varepsilon^t(K(h,n))$. Thus $h \cdot P$ generates the Sylow-2-subgroup of $E_\varepsilon^t(K(h,n))$. □

Note that different factorizations of $x_0(x_0^2-\varepsilon)$ into $t \cdot y_0^2$ yield isomorphic curves.

If h is not divisible by 3 then one can easily find pairs (ε, x_0) satisfying the conditions in Proposition 5.1 for all generalized Thabit primes of the form $K(h,n)$ by quadratic reciprocity.

Example 5.2. Suppose $n \geq 3$ and let $h \in \mathbb{N}$ be odd and not divisible by 3. Then either $K(h,n)$ is divisible by 3 or $(\varepsilon, x_0) := (12, -2)$ satisfies the conditions in Proposition 5.1. Moreover, $x_0^3 - \varepsilon x_0 = 4^2$. Thus we can work with the global point $P = (-2, 4)$ on the elliptic curve E_{12}. This is the setup used by B. Gross in [3] for Mersenne numbers $M_p = K(1,p)$ where $p \in \mathbb{P}$.

If h is divisible by 3 then it is not possible to find such a pair (ε, x_0) which is independent from h and n. However, we have the following lemma.

Lemma 5.3. *Let $h \in \mathbb{N}$ be odd and let $n \geq 2$.*

1. *There exists a minimal prime $p \in \mathbb{P}$ such that $\left(\frac{p}{K(h,n)}\right) \neq 1$.*

2. *Either p is a proper divisor of $K(h,n)$ or $(\varepsilon, x_0) := (p, 1)$ satisfies the conditions in Proposition 5.1.*

Proof. The integer $K(h,n)$ is not a square in $\mathbb{Z}$ since $K(h,n) \equiv -1 \pmod 4$. Thus the existence of a minimal number $p \in \mathbb{N}$ satisfying $\left(\frac{p}{K(h,n)}\right) \neq 1$ follows from the Chinese Remainder Theorem. Since the Jacobi symbol is multiplicative in the first component, any nontrivial factorization of p would yield a smaller number d with $\left(\frac{d}{K(h,n)}\right) \neq 1$, which is a contradiction to the minimality of p. Hence p must be prime.
Moreover, since $1 < p < K(h,n)$ we know that either p is a proper divisor of $K(h,n)$ or $\left(\frac{p}{K(h,n)}\right) = -1$. In the latter case, the choice of p implies $\left(\frac{p-1}{K(h,n)}\right) = 1$ and therefore

$$\left(\frac{x_0^2-\varepsilon}{K(h,n)}\right) = \left(\frac{-1}{K(h,n)}\right)\left(\frac{p-1}{K(h,n)}\right) = -1 \cdot 1 = -1$$

as claimed. □

If $n \geq 3$ and $h \in \mathbb{N}$ is odd but not divisible by 3 then

$$\min\left\{q \in \mathbb{P} \mid \left(\frac{q}{K(h,n)}\right) \neq 1\right\} = 3\,.$$

Hence, if we plug $(\varepsilon, x_0) = (3,1)$ into Proposition 5.1, we end up with the point $P = (1,1)$ on the curve E_3^{-2}. Under the isogeny $\tau_{3,-2}$ from equation (1) this point corresponds to the point $(-2,4)$ on the curve E_{12}. So Example 5.2 is just a special case of Lemma 5.3.

The curve E_{12} has rank 1. Similarly, the rational curves E_1^{30} and the one given by $7y^2 = x^3 + 1$ also have rank 1. These latter curves were used in [2] for primality tests of Fermat numbers and numbers of the form $2^{2^\ell} - 2^{2^\ell - 1} + 1$ respectively. So one might wonder what the ranks of the rational curves E_p^{1-p} are.

Proposition 5.4. *Let p be a prime. Then the rank of the rational elliptic curve E_p^{1-p} is at least 1.*

Proof. For $p = 2$ one can check the rank explicitly. So suppose $p \geq 3$ and let m be the product of all primes $\leq p$. Then by the Chinese Remainder Theorem, there exists a positive odd number h such that $p = \min\{q \in \mathbb{P} \mid \left(\frac{q}{K(h,3)}\right) \neq 1\}$. For $k \in \mathbb{N}$ let $a_k = K(h + km, 3) = k \cdot (8m) + (8h - 1)$. Then $p = \min\{q \in \mathbb{P} \mid \left(\frac{q}{a_k}\right) \neq 1\}$ for all k. The choice of p implies $\gcd(8m, 8h-1) = 1$. Thus the sequence (a_k) contains infinitely many primes by Dirichlet's theorem on arithmetic progressions. But then Lemma 5.3 and Proposition 5.1 imply that the rank of the twisted rational curve E_p^{1-p} cannot be 0. □

It turns out that the smallest prime p such that E_p^{1-p} has rank 2 is $p = 7$.

5.2. An Elliptic Curve Primality Test for Generalized Thabit Numbers

Now we are able to prove the main theorem of this paper, which gives a primality test for numbers of the form $K(h,n)$.

Let (ε, x_0) be a pair of integers that satisfies the conditions of Proposition 5.1. See Example 5.2 or Lemma 5.3 on how to find such a pair. Write $x_0^3 - \varepsilon x_0 = t y_0^2$ for some $y_0, t \in \mathbb{Z}$ and set $P = (x_0, y_0) \in E_\varepsilon^t(\mathbb{Q})$.

Now we recursively define a sequence of rational numbers as follows:

$$T_0 := x(h \cdot P) \text{ and } T_{k+1} := \frac{(T_k^2 + \varepsilon)^2}{4T_k(T_k^2 - \varepsilon)} \qquad (2)$$

Then $T_k = x(h2^k \cdot P)$ for all $k \geq 0$. Also note that the initial value T_0 depends only on x_0 and ε. Moreover, it can be computed without knowing y_0 and t as explained in Remark 5.7 below.

Theorem 5.5. *Let $h \in \mathbb{N}$ be odd and $n > \log_2(h) + 2$ be some integer. Suppose $(\varepsilon, x_0) \in \mathbb{Z}^2$ satisfies the conditions of Proposition 5.1 and let (T_k) be the sequence defined in equation (2). Then the number $K(h,n) = h \cdot 2^n - 1$ is prime if and only if the following three conditions are met:*

- $\gcd(K(h,n), x_0) = 1$.
- $T_k(T_k^2 - \varepsilon)$ *is a unit in* $\mathbb{Z}/K(h,n)\mathbb{Z}$ *for all* $0 \le k \le n-2$.
- $T_{n-1} \equiv 0 \pmod{K(h,n)}$.

Proof. Write $x_0^3 - \varepsilon x_0 = t y_0^2$ for some $y_0, t \in \mathbb{Z}$. Further let $Q = h \cdot P$ where P denotes the point $(x_0, y_0) \in E_\varepsilon^t(\mathbb{Q})$.
First, suppose that $K(h,n)$ is prime. Then $\gcd(x_0, K(h,n)) = 1$ by Proposition 5.1. This proposition also shows that the point Q generates the Sylow-2-subgroup of the cyclic group $E_\varepsilon^t(K(h,n)) \cong C_h \times C_{2^n}$. Hence Q has order 2^n, which implies $2^{n-1} \cdot Q = (0,0)$ is the unique point of order 2. In particular, $2^k \cdot Q \ne (0,0)$ for all $k \in \{0, \ldots, n-2\}$. The duplication formula shows that $T_k \equiv x(2^k \cdot Q) \pmod{K(h,n)}$, so we have that $T_{n-1} \equiv 0 \pmod{K(h,n)}$ and $T_k \not\equiv 0 \pmod{K(h,n)}$ for $0 \le k \le n-2$. Therefore T_k is invertible in $\mathbb{Z}/K(h,n)\mathbb{Z}$ for all $0 \le k \le n-2$. The same holds for $T_k^2 - \varepsilon$ since ε is not a square modulo $K(h,n)$.
Conversely, suppose that $K(h,n)$ is composite and satisfies the three conditions above. Let $q \le \sqrt{K(h,n)}$ be the smallest positive prime divisor of $K(h,n)$. Note that by assumption $q \ne 2$ is coprime to x_0, $x_0^2 - \varepsilon$ and ε. So q does also not divide $x_0(x_0^2 - \varepsilon) = t y_0^2$. In particular, $\gcd(q, 2\varepsilon t) = 1$. Thus E_ε^t has good reduction modulo q. By assumption, $2^{n-1} \cdot Q$ has order 2 in $E_\varepsilon^t(q)$, so Q has order 2^n in $E_\varepsilon^t(q)$. Since $(0,0) \in E_\varepsilon^t(q)$ is the only point with x-coordinate 0, we get the trivial bound $\#E_\varepsilon^t(q) \le 2q$. But then

$$2^n \le \#E_\varepsilon^t(q) \le 2q \le 2\sqrt{K(h,n)} = 2^{\frac{n+\log_2(h)}{2}+1},$$

which is impossible by the choice of n. Hence q does not exist. □

If $h = 1$ and $(\varepsilon, x_0) = (12, -2)$ then the above primality criterion is the same as the one found by Gross in [3] for Mersenne numbers $M_p = K(1,p)$. Also note that if one chooses (ε, x_0) as in Example 5.2 or Lemma 5.3 then $x_0 \in \{1, -2\}$. Hence the condition $\gcd(K(h,n), x_0) = 1$ is always satisfied.

Remark 5.6. The primality test given in Theorem 5.5 is based on the following idea. The candidate $K(h,n)$ is prime if and only if the reduction of $h \cdot P$ modulo $K(h,n)$ generates a cyclic subgroup of order 2^n. So it is completely analogous to the Lucas-Lehmer-Riesel test (see Remark 2.5).

We close this section with a remark concerning the implementation of the above primality criterion.

Remark 5.7. If $K(h,n)$ is prime then the order of $P = (x_0, y_0) \in E_\varepsilon^t(\mathbb{Q})$ is at least $2^n > h$ (see Lemma 4.1). So in particular, $m \cdot P \ne \mathcal{O}$ for any $1 \le m \le h$.

Let $1 \leq m \leq h$ and $\alpha \in \mathbb{C}$ such that $\alpha^2 = t$. Then Lemma 3.2 and the change of coordinates

$$E^t_\varepsilon(\mathbb{C}) \to E_\varepsilon(\mathbb{C}),\ (x, y) \mapsto (x, \alpha y)$$

show that $m \cdot P = (f_m, g_m y_0)$ for some $f_m, g_m \in \mathbb{Q}$ that depend only on ε, x_0 and $ty_0^2 = x_0^3 - \varepsilon x_0$. Thus without actually knowing t or y_0, one can compute $T_0 = x(h \cdot P) = f_h$ using "square and multiply" by adding at most $2\log_2(h)$ points.

Moreover, these calculations can be done directly in $\mathbb{Z}/K(h,n)\mathbb{Z}$. This has the advantage that if a necessary computation cannot be performed in $\mathbb{Z}/K(h,n)\mathbb{Z}$ (i.e., some nonzero element cannot be inverted) then $K(h,n)$ is immediately proven not to be prime. The same holds for the computation of $T_1, T_2, \ldots$.

6. Efficiency

We have implemented our primality test for generalized Thabit numbers as well as the Lucas-Lehmer-Riesel test in Magma (see [1]) to compare the efficiency of both tests. All tests were performed on a Core i7, 940 running at 2.93 GHz.

In our primality test it is necessary to divide in the residue class ring $\mathbb{Z}/K(h,n)\mathbb{Z}$. Thus on the one hand the calculations in each step of the iteration are more complex than the calculations in the Lucas-Lehmer-Riesel test (four multiplications and one division versus one multiplication in the ring $\mathbb{Z}/K(h,n)\mathbb{Z}$) but on the other hand we know that if the necessary division is not possible, then $K(h,n)$ cannot be prime. So if $K(h,n)$ is composite, it might happen that the algorithm aborts without computing all values $T_0, T_1, \ldots, T_{n-1}$ modulo $K(h,n)$.

For example, $K(3, 100\,008)$ is composite and the Lucas-Lehmer-Riesel test as well as Magma's build-in primality test both take about 4 minutes to verify that. Our elliptic curve test (using $(\varepsilon, x_0) = (5,1)$) recognizes this in virtually no time since already $T_4 \pmod{K(h,n)}$ does not exist.

To see how often such premature aborts occur, we have run the Lucas-Lehmer-test and our primality test using 27 different pairs (ε, x_0) for Mersenne primes in Magma to compute a list of all prime exponents $3 \leq p \leq 10\,000$, such that the Mersenne number $M_p = K(1,p) = 2^p - 1$ is prime. For some pairs (ε, x_0), there was no premature abort at all, for others there were up to five. Table 1 gives the chosen values ε (where $\varepsilon = 0$ represents the Lucas-Lehmer-test), the chosen initial values $T_0 = x_0$, the exponents p that yield an premature abort, and the time in seconds required for the test.

We see that it takes the elliptic curve tests about 50 times as long as the Lucas-Lehmer-test to do the task. This is easily explained by the fact that there were never more than four exponents out of 1 228 (i.e., the number of odd primes $\leq$ 10 000), which yielded an premature abort. So the possible advantage of the elliptic curve tests towards the Lucas-Lehmer-test does not play any significant role for the efficiency on average.

Similarly, we considered generalized Thabit numbers. We compared the Lucas-Lehmer-Riesel test (LLR) to Theorem 5.5 (where the pair (ε, x_0) was chosen as in Lemma 5.3) and measured the time needed to compute all exponents $\log_2(h) + 2 < n \leq 3\,000$ such that $K(h,n)$ is prime for all odd numbers $h \in \{1, \dots 99\}$ using both tests. Additionally we calculated the ratio ρ of premature aborts that occurred using the algorithm from Theorem 5.5 to the number of composite generalized Thabit numbers in the tested range. The results are given in Table 2.

Note that $n \geq 3$, i.e., $\left(\frac{2}{K(h,n)}\right) = 1$. So if $K(h,n)$ is divisible by 3, then Lemma 5.3 will immediately detect this. Hence we excluded all pairs (h,n) from the test for which Lemma 5.3 does find a nontrivial divisor since this would definitely favor our algorithm. Moreover, any serious implementation of a primality test would test for small prime factors anyway.

By our experiments we conclude that our test is obviously less efficient than the Lucas-Lehmer-Riesel test. Although in some cases, where there are very many premature aborts (for example for $h = 11$ or $h = 71$) we are very close to its efficiency. We also see, that the value of ρ does indeed play a significant role for the velocity of our test. But still there are not enough premature aborts to compensate the fact that the calculations in each iteration step are much more expensive for our test compared to the Lucas-Lehmer-Riesel test.

Acknowledgments The authors would like to thank the anonymous referee and the editors for several helpful comments and suggestions.

References

[1] Wieb Bosma, John Cannon, and Catherine Playoust. The Magma algebra system. I The user language. *J. Symbolic Comput.*, 24:235–265, 1997.

[2] Robert Denomme and Gordan Savin. Elliptic curve primality tests for Fermat and related primes. *J, Number Theory*, 128:2398–2412, 2008.

[3] Benedict H. Gross. An elliptic curve test for Mersenne primes. *J, Number Theory*, 110:114–119, 2005.

[4] Serge Lang. *Elliptic Curves - Diophantine Analysis*. Springer-Verlag, 1978.

[5] Derrick Henry Lehmer. On Lucas's test for the primality of Mersenne's numbers. *J. London Math. Soc.*, 10:162–165, 1935.

[6] Édouard Lucas. Nouveaux théorèmes d'Arithmétique supérieur. *C.R. Acad. Sci. Paris*, 83:1286–1288, 1876.

[7] Carl Pomerance. Primality testing: Variations on a theme of Lucas. *Congr. Numer.*, 201:301–312, 2010.

[8] Hans Riesel. Lucasian Criteria for the Primality of $\mathcal{N} = h \cdot 2^n - 1$. *Math. of Comp.*, 23(108):869–875, 1969.

[9] Michael Rosen. A proof of the Lucas-Lehmer test. *Amer. Math. Monthly*, 95:855–856, 1995.

[10] Joseph H. Silverman. *The Arithmetic of Elliptic Curves*. Springer-Verlag, 1986.

7. Tables

ε	$T_0 = x_0$	Abort at	Time in sec.
0	-	-	71
-242	1	-	3544
-242	25	47, 191, 397, 1013	3550
-242	29	113, 4649	3550
-242	101	11	3548
-242	115	11, 23, 191, 5717, 6491	3537
-50	2	11, 37, 191	3542
-50	26	-	3541
-50	46	11, 23, 47, 1321	3543
-2	1	-	3540
-2	2	-	3539
-2	5	11, 37, 191	3540
-2	19	11, 23, 179	3540
-2	22	-	3539
-2	71	191, 9791	3529
3	1	23	3535
3	3	23	3534
3	27	11, 23, 47, 191, 743	3536
6	2	11, 37, 47, 191	3537
6	3	11, 37, 47, 191	3537
6	9	-	3536
6	33	23	3537
6	123	431, 3023	3535
27	5	37	3536
54	2	-	3534
75	9	37	3537
75	33	191, 4871	3536
243	1	11, 23, 47, 191, 743	3537

Table 1: Efficiency of the tests for Mersenne numbers

h	n for which $K(h,n)$ is prime	ρ (in %)	LLR (in sec)	Thm 5.5 (in sec)
1	3, 5, 7, 13, 17, 19, 31, 61, 89, 107, 127, 521, 607, 1279, 2203, 2281	66.3	8	67
3	4, 6, 7, 11, 18, 34, 38, 43, 55, 64, 76, 94, 103, 143, 206, 216, 306, 324, 391, 458, 470, 827, 1274	34.1	21	208
5	8, 10, 12, 14, 18, 32, 48, 54, 72, 148, 184, 248, 270, 274, 420, 1340, 1438, 1522, 1638, 1754, 1884, 2014, 2170, 2548, 2622, 2652, 2704	14.1	14	185
7	5, 9, 17, 21, 29, 45, 177	26.7	14	157
9	7, 13, 15, 21, 43, 63, 99, 109, 159, 211, 309, 343, 415, 469, 781, 871, 939, 1551	52.2	20	141
11	26, 50, 54, 126, 134, 246, 354, 362, 950, 1310, 2498	76.9	14	50
13	7, 23, 287, 291, 795, 2203	31.9	14	148
15	10, 14, 17, 31, 41, 73, 80, 82, 116, 125, 145, 157, 172, 202, 224, 266, 289, 293, 463, 1004, 1246, 2066, 2431, 2705	39.3	18	157
17	16, 20, 36, 54, 60, 96, 124, 150, 252, 356, 460, 612, 654, 664, 698, 702, 972, 1188, 1312	21.2	14	172
19	21, 41, 49, 89, 133, 141, 165, 189, 293, 305, 395, 651, 665, 771, 801, 923, 953	27.2	14	158
21	7, 10, 13, 18, 27, 37, 51, 74, 157, 271, 458, 530, 891, 1723, 1793, 1849, 1986, 2191, 2869	37.6	19	180
23	12, 46, 72, 244, 264, 544, 888, 1146	46.4	14	116
25	9, 11, 17, 23, 35, 39, 75, 105, 107, 155, 161, 215, 225, 335, 635, 651, 687, 1479, 1515, 1953, 2435, 2963	49.4	14	110
27	8, 10, 14, 28, 37, 38, 70, 121, 122, 160, 170, 253, 329, 362, 454, 485, 500, 574, 892, 962, 1213, 1580, 2642, 2708	43.8	20	172
29	16, 76, 148, 184	36.7	14	135
31	7, 11, 13, 23, 33, 35, 37, 47, 115, 205, 235, 271, 409, 739, 837, 887, 2189	31.9	14	146
33	8, 10, 22, 35, 42, 43, 46, 56, 91, 102, 106, 142, 190, 208, 266, 330, 360, 382, 462, 503, 815, 1038, 1651, 1855, 1858, 1992, 2232	25.6	22	243
35	10, 20, 44, 114, 146, 156, 174, 260, 306, 380, 654, 686, 702, 814, 906, 1196, 1316	47.7	14	116
37	2553	51.3	14	105

h	n for which $K(h,n)$ is prime	ρ (in %)	LLR (in sec)	Thm 5.5 (in sec)
39	24, 105, 153, 188, 605, 795, 813, 839, 2135, 2619	64.4	19	102
41	10, 14, 18, 50, 114, 122, 294, 362, 554, 582, 638, 758	57.8	14	91
43	31, 67, 251, 767, 1171, 1643	74.4	14	56
45	8, 9, 14, 15, 16, 22, 28, 29, 36, 37, 54, 59, 85, 93, 117, 119, 161, 189, 193, 256, 308, 322, 327, 411, 466, 577, 591, 902, 928, 946, 1162, 1428, 1708, 1724, 2063, 2922, 2951	43.9	28	226
47	14, 70, 78, 1374, 1824, 2158, 2654	40.2	14	129
49	9, 13, 15, 29, 33, 39, 55, 81, 95, 205, 279, 581, 807, 813, 2551, 2565	20.5	14	170
51	9, 10, 19, 22, 57, 69, 97, 141, 169, 171, 195, 238, 735, 885, 1287, 1365, 2026, 2211, 2361, 2889	63.4	18	102
53	8, 42, 50, 62, 362, 488, 642, 846, 2870	55.5	14	96
55	15, 33, 41, 57, 69, 75, 77, 131, 133, 153, 247, 305, 351, 409, 471, 1251, 1259, 2253, 2411, 2425, 2699	31.9	14	147
57	8, 10, 20, 22, 25, 26, 32, 44, 62, 77, 158, 317, 500, 713, 1657, 1790, 2761, 2794	37.5	19	178
59	12, 16, 72, 160, 256, 916, 1216, 1840	24.6	14	161
61	9, 13, 17, 19, 25, 39, 63, 67, 75, 119, 147, 225, 419, 715, 895, 1025, 1103, 1179, 1345, 1829	34.8	14	141
63	8, 11, 14, 16, 28, 32, 39, 66, 68, 91, 98, 116, 126, 164, 191, 298, 323, 443, 714, 758, 759, 1059, 1168, 1511, 1792, 2116	45.7	21	175
65	12, 22, 28, 52, 78, 94, 124, 162, 174, 192, 204, 304, 376, 808, 930, 972, 1714, 1776, 2176, 2568	44.6	15	122
67	9, 21, 45, 65, 77, 273, 677, 1049, 1721	36.5	14	138
69	9, 11, 13, 17, 19, 23, 29, 37, 49, 61, 79, 99, 121, 133, 141, 164, 173, 181, 185, 193, 233, 299, 313, 351, 377, 540, 569, 909, 1057, 1081, 1189, 1679, 2043, 2641	20.8	22	260
71	14, 410, 1390	72.4	14	61
73	11, 19, 71, 79, 131, 1167, 1191	30.6	14	151
75	18, 19, 20, 22, 28, 29, 39, 43, 49, 75, 85, 92, 111, 126, 136, 159, 162, 237, 349, 381, 767, 969, 1247, 1893, 1951, 2363, 2657	45.1	28	221

h	n for which $K(h,n)$ is prime	ρ (in %)	LLR (in sec)	Thm 5.5 (in sec)
77	14, 26, 58, 60, 64, 100, 122, 212, 566, 638, 1214, 2080	15.8	14	182
79	15, 43, 57, 61, 75, 145, 217, 247, 2803	55.1	14	98
81	11, 17, 21, 27, 81, 101, 107, 327, 383, 387, 941, 1665	55.3	19	132
83	10, 14, 18, 22, 24, 26, 28, 36, 42, 58, 64, 78, 158, 198, 206, 424, 550, 676, 904, 1276, 1374, 1536, 1642, 2124, 2796	29.0	14	154
85	11, 71, 113, 115, 355, 473, 563, 883, 1235	68.7	14	69
87	9, 10, 12, 22, 29, 32, 50, 57, 69, 81, 122, 138, 200, 296, 514, 656, 682, 778, 881, 1422, 1494, 1857	35.7	20	200
89	12, 24, 48, 52, 64, 84, 96, 1272, 2028	53.3	14	102
91	9, 13, 15, 17, 19, 23, 47, 57, 67, 73, 77, 81, 83, 191, 301, 321, 435, 867, 869, 917	20.3	14	172
93	10, 15, 18, 19, 24, 27, 39, 60, 84, 111, 171, 192, 222, 639, 954, 2400, 2587	45.6	19	153
95	26, 32, 66, 128, 170, 288, 320, 470, 1278, 1296, 1316, 1536, 1608	58.2	14	91
97	9, 45, 177, 585, 1409, 2617	34.0	14	142
99	11, 19, 25, 28, 35, 65, 79, 212, 271, 361, 461, 1237, 1297, 1577, 1747, 1901, 1943, 2741	31.1	19	197

Table 2: Efficiency of the tests for Thabit numbers

RESTRICTED INTEGER PARTITION FUNCTIONS

Noga Alon [1]
Sackler School of Mathematics and Blavatnik School of Computer Science, Tel Aviv University, Tel Aviv, Israel
nogaa@tau.ac.il

Received: 7/13/12, Accepted: 3/24/13, Published: 3/29/13

Abstract

For two sets A and M of positive integers and for a positive integer n, let $p(n, A, M)$ denote the number of partitions of n with parts in A and multiplicities in M, that is, the number of representations of n in the form $n = \sum_{a \in A} m_a a$ where $m_a \in M \cup \{0\}$ for all a, and all numbers m_a but finitely many are 0. It is shown that there are infinite sets A and M so that $p(n, A, M) = 1$ for every positive integer n. This settles (in a strong form) a problem of Canfield and Wilf. It is also shown that there is an infinite set M and constants c and n_0 so that for $A = \{k!\}_{k \geq 1}$ or for $A = \{k^k\}_{k \geq 1}$, $0 < p(n, A, M) \leq n^c$ for all $n > n_0$. This answers a question of Ljujić and Nathanson.

1. Introduction

For two sets A and M of positive integers and for a positive integer n, let $p(n, A, M)$ denote the number of representations of n in the form

$$n = \sum_{a \in A} m_a a \tag{1}$$

where $m_a \in M \cup \{0\}$ for all a, and all numbers m_a but finitely many are 0. We say that the function $p = p(n, A, M)$ has polynomial growth if there exists an absolute constant c and an integer n_0 so that $p(n, A, M) \leq n^c$ for all $n > n_0$. Canfield and Wilf [2] raised the following question.

Question 1 *(Canfield and Wilf [2])* Do there exist two **infinite** sets A and M so that $p(n, A, M) > 0$ for all sufficiently large n and yet p has polynomial growth ?

Ljujić and Nathanson [4] showed, among other things, that this cannot be the case if the set A has at least $\delta \log n$ members in $[n] = \{1, 2, \ldots, n\}$ for all sufficiently

[1] Research supported in part by an ERC Advanced grant, by a USA-Israeli BSF grant and by the Israeli I-Core program.

large n, where $\delta > 0$ is any positive constant, and asked the following two more specific questions.

Question 2 *(Ljujić and Nathanson [4]).* Let $A = \{k!\}_{k=1}^{\infty}$. Does there exist an infinite set M of positive integers so that $p(n, A, M) > 0$ for all sufficiently large n, and p has polynomial growth ?

Question 3 *(Ljujić and Nathanson [4]).* Let $A = \{k^k\}_{k=1}^{\infty}$. Does there exist an infinite set M of positive integers so that $p(n, A, M) > 0$ for all sufficiently large n and p has polynomial growth ?

In this note we prove that the answer to all three questions above is positive.

Our first result is simple, and shows that the answer to the first question is positive in the strongest possible way: there are infinite sets A and M so that $p(n, A, M) = 1$ for all n. This is stated in the following Theorem.

Theorem 1.1. *There are two infinite sequences of positive integers A and M so that $p(n, A, M) = 1$ for every positive integer n.*

The proof is by an explicit construction, which is general enough to give examples with sets A that grow at any desired rate which is faster than exponential. This rate of growth is tight, by the observation of Ljujić and Nathanson mentioned above.

We also prove the following result, which settles Questions 2 and 3.

Theorem 1.2. *Let A be an infinite set of positive integers, and suppose that $1 \in A$ and*

$$|A \cap [n]| = (1 + o(1)) \frac{\log n}{\log \log n},$$

where the $o(1)$ term tends to 0 as n tends to infinity. Then there exists n_0 and an infinite set M of positive integers so that $0 < p(n, A, M) < n^{8+o(1)}$ for all $n > n_0$.

The upper estimate $n^{8+o(1)}$ can be improved and we make no attempt to optimize it here. The proof of this theorem is probabilistic and can be easily modified to provide the existence of such sets M for many other sparse infinite sets A.

The rest of this note is organized as follows. In Section 2 we present the simple proof of Theorem 1.1. In Section 3 we describe the probabilistic construction of the set M used in the proof of Theorem 1.2. The proof it satisfies the required properties and hence establishes Theorem 1.2 is described in Section 4. Section 5 contains some concluding remarks.

All logarithms throughout the paper are in base 2, unless otherwise specified.

2. The Proof of Theorem 1.1

Proof. Let $N \cup \{0\} = B \cup C$, $B \cap C = \emptyset$ be a partition of the set of all non-negative integers into two disjoint infinite sets B and C. Let

$$D = \{\sum_{i \in B'} 2^i \ : \ B' \subset B\}$$

be the set of all sums of powers of two in which each exponent lies in B. Similarly, put

$$E = \{\sum_{j \in C'} 2^j \ : \ C' \subset C\}.$$

This is the set of all sums of powers of two in which each exponent is a member of C. Note that both D and E contain 0, as the sets B' and C' in their definition may be taken to be empty. Since every non-negative integer has a unique binary representation, it is clear that each such integer has a unique representation as a sum of an element of D and an element of E.

Define, now,

$$A = \{2^d \ : \ d \in D\},$$

and

$$M = \{\sum_{e \in E'} 2^e \ : \ E' \subset E\}.$$

Clearly both A and M are infinite. We claim that every positive integer has a unique representation of the form (1) with these A and M, that is, $p(n, A, M) = 1$ for all $n \geq 1$. Indeed, a general expression of the form $\sum_{a \in A} m_a a$ with $m_a \in M$ for all a satisfies

$$\sum_{a \in A} m_a a = \sum_{d \in D} \sum_{e \in E'_d} 2^{e+d}.$$

Let $n = 2^{t_1} + 2^{t_2} + \ldots + 2^{t_r}$ be the binary representation of n, with $0 \leq t_1 < t_2 < \ldots < t_r$. By the fact that each nonnegative integer has a unique representation as a sum of an element of D and an element of E, there are, for each $1 \leq j \leq r$, unique $d_i \in D$ and $e_i \in E$ so that $t_i = d_i + e_i$. The elements d_i are not necessarily distinct. Let $D' = \{d_1, d_2 \ldots, d_r\}$ be the set of all distinct ones. For each $d \in D'$ define $m_{2^d} = \sum_{i: d_i = d} 2^{e_i}$ and observe that

$$n = \sum_{d \in D} m_{2^d} \cdot 2^d.$$

Therefore n has a representation of the form (1). It is not difficult to check that this representation is unique, that is, the elements d and m_{2^d} in the expression above can be reconstructed in the unique way described above from the binary representation of n. This shows that indeed $p(n, A, M) = 1$ for all n, completing the proof of Theorem 1.1. □

Note that by choosing the set B so that $|B \cap \{0, 1, \ldots, r\}| = r - g(r)$, where $g(r)$ is an arbitrary monotone function of r that tends to infinity, and $r - g(r)$ also tends to infinity in a monotone way, we get that for every integer $r \geq 1$:

$$|A \cap [2^{2^r}]| = |D \cap [2^r]| = |B \cap \{r\}| + 2^{|B \cap \{0,1,\ldots,r-1\}|} = 2^{r-1-g(r-1)} + O(1).$$

As $r = \log \log n$ this shows that $|A \cap [n]|$ is

$$\Theta(\frac{\log n}{2^{g(\log \log n)}}) + O(1),$$

showing that by an appropriate choice of g we can get sets A whose growth function is at least $\log n / w(n)$ for an arbitrary slowly growing function $w(n)$.

3. The Probabilistic Construction

The set M defined for the proof of Theorem 1.2 is a union of the form $M = \cup_{a \in A} M_a$. Each set M_a is a union $M_a = \cup_{i:2^i \geq a} M_{a,i}$. Each of the sets $M_{a,i}$ is a random subset of $[2^i]$ obtained by picking each number in $[2^i]$, randomly and independently, to be a member of $M_{a,i}$ with probability $\frac{i^6}{2^i}$ (if this ratio exceeds 1 then all members of $[2^i]$ lie in $M_{a,i}$; clearly this happens only for finitely many values of i.)

We claim that M satisfies the required properties with high probability (that is, with probability that tends to 1 as n_0 tends to infinity). This is proved in the next section. The fact that $p = p(n, A, M)$ has polynomial growth is simple: we show that with high probability the set M is sparse enough to ensure that the total number of expressions of the form (1) that are at most n does not exceed $n^{8+o(1)}$. The fact that with high probability $p(n, A, M) > 0$ for all sufficiently large n is more complicated and requires some work. It turns out to be convenient to restrict attention to expressions (1) of a special form that enable us to control their behaviour in an effective manner. We can then apply the Janson Inequality (c.f., [1], Chapter 8) to derive the required result.

4. The Proof of Theorem 1.2

4.1. Polynomial Growth

Let A be as in Theorem 1.2, and let $M = \cup_{a \in A} \cup_{i:2^i \geq a} M_{a,i}$ be as in the previous section. In this subsection we show that with high probability the function $p(n, A, M)$ has polynomial growth.

Lemma 4.1. *For every $\epsilon > 0$ there exists an $n_0 = n_0(\epsilon)$ so that with probability at least $1 - \epsilon$, M is infinite and $|M \cap [n]| < \log^8 n - 1$ for all $n > n_0$.*

Proof. It is obvious that M is infinite with probability 1. We proceed to prove the main part of the lemma.

Let m be a sufficiently large integer. If $2^i < m$ then m cannot lie in $M_{a,i}$. If $2^i \geq m$, then the probability that $m \in M_{a,i}$ is $\frac{i^6}{2^i}$. Since for large i, $\frac{(i+1)^6}{2^{i+1}} < \frac{2}{3}\frac{i^6}{2^i}$ it follows that for $a \leq m$ the probability $Pr[m \in M_a = \cup_{i:2^i \geq m} M_{a,i}]$ is smaller than $3\frac{\log^6 m}{m}$. Similarly, for $a > m$, $Pr[m \in M_a] < 3\frac{\log^6 a}{a}$. Summing over all $a \in A$, $a > m$ and using the fact that A is sparse we conclude that $Pr[m \in \cup_{a>m} M_a] < o(\frac{\log^7 m}{m})$. Since there are $o(\log m)$ members of A that are smaller than m we also get that (since m is large) $Pr[m \in \cup_{a \leq m} M_a] < 0.9\frac{\log^7 m}{m}$. Altogether, the probability that m lies in $M = \cup_a M_a$ does not exceed $\frac{\log^7 m}{m}$.

It follows that the expected value of $|M \cap [n]|$ is smaller than $O(1) + \frac{1}{8}\log^8 n < 0.5\log^8 n$ (where the $O(1)$ term is added to account for the small values of m). As this cardinality is a sum of independent random variables we can apply the Chernoff-Hoeffding Inequality (c.f., e.g., [1], Appendix A) and conclude that the probability that $|M \cap [n]|$ is at least $\log^8 n - 1$ is at most $e^{-\Omega(\log^8 n)}$. For sufficiently large n_0 the sum $\sum_{n \geq n_0} e^{-\Omega(\log^8 n)} < \epsilon$, completing the proof of the lemma. □

Corollary 4.2. *Let A and M be as above. For any $\epsilon > 0$ there is an $n_0 = n_0(\epsilon)$ so that with probability at least $1 - \epsilon$, M is infinite and $\sum_{n \leq m} p(n, A, M) \leq m^{8+o(1)}$ for all $m > n_0$.*

Proof. Let ϵ and $n_0 = n_0(\epsilon)$ be as in Lemma 4.1 and suppose that M satisfies the assertion of the lemma (this happens with probability at least $1 - \epsilon$). Then, the number of representations of the form (1) of integers n that do not exceed m is at most $|(M \cap [m]) \cup \{0\}|^{|A \cap [m]|}$, as all coefficients m_a and all numbers a with a nonzero coefficient must be at most m. This expression is at most $(\log^8 m)^{(1+o(1))\log m/\log\log m} = m^{8+o(1)}$, as needed. □

4.2. Representing All Large Integers

In this subsection we prove that with high probability every sufficiently large number n has a representation of the form (1) where A and M are as above. To do so, it is convenient to insist on a representation of a special form, which we proceed to define. Put $A' = \{a \in A : a \leq n^{1/3}\}$. Let $q = |A'|$, $A' = \{a_1, a_2, \ldots, a_q\}$ where $1 = a_1 < a_2 < \ldots < a_q$. Thus $q = (\frac{1}{3} + o(1))\frac{\log n}{\log\log n}$. Let i_1 be the smallest integer t so that $2^t \geq n$. For $2 \leq j \leq q$, let i_j be the smallest integer t so that $2^t \geq \frac{n}{a_j \log n}$. Therefore, $n \leq 2^{i_1} < 2n$ and $\frac{n}{a_j \log n} \leq 2^{i_j} < \frac{2n}{a_j \log n}$ for all $2 \leq j \leq q$. Since $a_j \leq n^{1/3}$ for all $a_j \in A'$, it follows that $i_j > \frac{1}{2}\log n$ for all admissible j.

We say that n has a special representation as a partition with parts in A and multiplicities in M (for short: n has a special representation) if there is a representation of the form $n = \sum_{j=1}^{q} m_j a_j$ where $m_j \in M_{a_j,i_j}$ for all $1 \leq j \leq q$. Recall that here $a_j \leq n^{1/3}$ for all $j \leq q$.

Lemma 4.3. *For all sufficiently large n, the probability that n does not have a special representation is at most $e^{-\Omega(\log^5 n)}$. Therefore, for any $\epsilon > 0$ there is an $n_0 = n_0(\epsilon)$ such that the probability that there is a special representation for every integer $n > n_0$ is at least $1 - \epsilon$.*

The assertion of Theorem 1.2 follows from Corollary 4.2 (with $\epsilon < 1/2$) and Lemma 4.3 (with $\epsilon < 1/2$) that supply the existence of an infinite set M satisfying the conclusions of the theorem.

In the proof of Lemma 4.3 we apply the Janson Inequality (c.f. [1], Chapter 8). We first state it to set the required notation. Let X be a finite set, and let $R \subset X$ be a random subset of X obtained by picking each element $r \in X$ to be a member of R, randomly and independently, with probability p_r. Let $\mathcal{C} = \{C_i\}_{i \in I}$ be a collection of subsets of X, let B_i denote the event that $C_i \subset R$, and let $i \sim j$ denote the fact that $i \neq j$ and $C_i \cap C_j \neq \emptyset$. Let $\mu = \sum_{i \in I} Pr[B_i] = \sum_{i \in I} \prod_{j \in C_i} p_j$ be the expected number of events B_i that hold, and define $\Delta = \sum_{i,j \in I, i \sim j} Pr[B_i \cap B_j]$ where the sum is computed over ordered pairs. The inequality we need is the following.

Lemma 4.4. (The Janson Inequality) *In the notation above, if $\Delta \leq D$ with $D \geq \mu$ then the probability that no event B_i holds is at most $e^{-\mu^2/2D}$.*

Note that the above statement is an immediate consequence of the two Janson inequalities described in [1], Chapter 8. If $\Delta < \mu$ than the above statement follows from the first inequality, whereas if $\Delta \geq \mu$ then it follows from the second.

We can now prove Lemma 4.3.

Proof. We apply the Janson inequality as stated in Lemma 4.4 above. The set X here is a disjoint union of the sets $X_j = [2^{i_j}]$, $1 \leq j \leq q$, and each element of X_j is chosen with probability $\frac{i_j^6}{2^{i_j}}$. The sets C_i are indexed by sequences $(m_1, m_2, \ldots, m_q)$ with $m_j \in X_j$ which satisfy

$$\sum_{j=1}^{q} m_j a_j = n. \tag{2}$$

If $(m_1, m_2, \ldots, m_q)$ is the i-th such sequence, then the set C_i is defined by the rule $C_i \cap X_j = \{m_j\}$. Note that there are exactly $\prod_{j=2}^{q} 2^{i_j}$ such sets, as for any choice of $m_j \in X_j$, $2 \leq j \leq q$ there is a unique choice of $m_1 \in X_1$ so that (2) holds. Therefore, in the notation above

$$\mu = \prod_{j=2}^{q} 2^{i_j} \prod_{j=1}^{q} \frac{i_j^6}{2^{i_j}} = \frac{1}{2^{i_1}} \prod_{j=2}^{q} i_j^6 = n^{1-o(1)},$$

where the last equality follows from the fact that $\frac{1}{2}\log n \leq i_j \leq \log n$ for all $2 \leq j \leq q$ and the fact that $q = (\frac{1}{3} + o(1))\frac{\log n}{\log \log n}$.

We proceed with the estimation of the quantity Δ that appears in the inequality. This is the sum, over all ordered pairs of sequences $m_j^{(1)}$ and $m_j^{(2)}$ with $m_j^{(1)}, m_j^{(2)} \in X_j$, where $\sum m_j^{(1)} a_j = \sum m_j^{(2)} a_j = n$ and for at least one r, $m_r^{(1)} = m_r^{(2)}$, of the probability that both $m_j^{(1)}$ and $m_j^{(2)}$ belong to M_{a_j,i_j} for all j.

Write $\Delta = \sum_{\ell=1}^{q} \Delta_\ell$, where Δ_ℓ is the sum of these probabilities over all pairs $m_j^{(1)}$ and $m_j^{(2)}$ as above for which $\ell = \min\{j : m_j^{(1)} \neq m_j^{(2)}\}$.

Claim 1: We have

$$\Delta_1 \leq \frac{1}{2^{i_1}} \prod_{j=1}^{q} i_j^6 \sum_{\emptyset \neq I \subset \{2,3,\ldots,q\}} \frac{i_1^6}{2^{i_1}} \prod_{j>1, j \notin I} i_j^6.$$

Indeed, for each choice of the sequence $m_j^{(1)}$ the contribution to Δ_1 arises from sequences $m_j^{(2)}$ for which $m_r^{(1)} = m_r^{(2)}$ for all r in some nonempty subset I of $\{2,3,\ldots,q\}$. There are $\prod_{j=2}^{q} 2^{i_j}$ choices for the sequence $m_j^{(1)}$ (as the value of $m_1^{(1)}$ is determined by the value of the sum $\sum_{j=1}^{q} m_j^{(1)} a_j$). The probability that each $m_j^{(1)}$ lies in M_{a_j,i_j} is $\frac{i_j^6}{2^{i_j}}$. For each fixed choice of $m_j^{(1)}$ and for each nonempty subset I as above, there are $\prod_{j>1, j \notin I}(2^{i_j} - 1)$ possibilities to choose the numbers $m_j^{(2)}, j > 1, j \notin I$, and the value of $m_1^{(2)}$ is determined (and has to differ from $m_j^{(1)}$, which is another reason the above is an upper estimate for Δ_1 and not a precise computation). The probability that $m_j^{(2)} \in M_{a_j,i_j}$ for all these values of j is $\frac{i_1^6}{2^{i_1}} \prod_{j>1, j \notin I} \frac{i_j^6}{2^{i_j}}$, implying Claim 1.

Plugging the value of μ in Claim 1, we conclude that

$$\Delta_1 \leq \mu^2 \sum_{\emptyset \neq I \subset \{2,3,\ldots,q\}} \prod_{j \in I} \frac{1}{i_j^6} = \mu^2 ([\prod_{j=2}^{q} (1 + \frac{1}{i_j^6})] - 1) \leq \mu^2 \frac{1}{\log^5 n}. \tag{3}$$

The final inequality above follows from the fact that $i_j > 0.5 \log n$ for all j and that $q = o(\log n)$.

Claim 2: For any $2 \leq \ell \leq q$,

$$\Delta_\ell \leq \frac{1}{2^{i_1}} \prod_{j=1}^{q} i_j^6 \cdot \frac{i_\ell^6}{2^{i_\ell}} \sum_{I \subset \{\ell+1,\ell+2,\ldots,q\}} \prod_{r \in \{\ell+1,\ell+2,\ldots,q\} - I} i_r^6.$$

The proof is similar to that of Claim 1 with a few modifications. For each choice of the sequence $m_j^{(1)}$ the contribution to Δ_ℓ arises from sequences $m_j^{(2)}$ for which $m_r^{(1)} = m_r^{(2)}$ for all $r < \ell$, $m_\ell^{(1)} \neq m_\ell^{(2)}$, and there is a possibly empty subset I of $\{\ell+1, \ell+2, \ldots, q\}$ so that $m_r^{(1)} = m_r^{(2)}$ for all $r \in I$ (and for no other r in $\{\ell+1, \ell+2, \ldots, q\}$.) As in the proof of Claim 1 there are $\prod_{j=2}^{q} 2^{i_j}$ choices for the sequence $m_j^{(1)}$, and the probability that each $m_j^{(1)}$ lies in M_{a_j,i_j} is $\frac{i_j^6}{2^{i_j}}$. For each fixed choice of $m_j^{(1)}$ and for each subset I as above, there are $\prod_{r\in\{\ell+1,\ell+2,\ldots,q\}-I}(2^{i_r}-1)$ possible choices for $m_r^{(2)}$, $r \in \{\ell+1, \ell+2, \ldots, q\} - I$, and the probability that each of those lies in the corresponding M_{a_r,i_r} is $\frac{i_r^6}{2^{i_r}}$. The product of these two terms is at most $\prod_{r\in\{\ell+1,\ell+2,\ldots,q\}-I} i_r^6$. Finally, the value of $m_\ell^{(2)}$ is determined by the values of all other $m_j^{(2)}$ and by the fact that $\sum_{j=1}^q m_j^{(2)} a_j = n$. (Note that this value has to lie in $[2^{i_\ell}]$, otherwise we do not get any contribution here. This is fine, as we are only upper bounding Δ_ℓ.) Finally, the probability that $m_\ell^{(2)} \in M_{a_\ell,i_\ell}$ is $\frac{i_\ell^6}{2^{i_\ell}}$. This completes the explanation for the estimate in Claim 2.

Plugging the value of μ we get, by Claim 2, that for any $2 \le \ell \le q$

$$\Delta_\ell \le \mu^2 \frac{2^{i_1}}{2^{i_\ell}} \frac{1}{i_1^6 i_2^6 \cdots i_{\ell-1}^6} \sum_{I\subset\{\ell+1,\ell+2,\ldots,q\}} \prod_{r\in I} \frac{1}{i_r^6}$$

$$\le \mu^2 \frac{2^{i_1}}{2^{i_\ell}} \frac{1}{i_1^6 i_2^6 \cdots i_{\ell-1}^6} \prod_{r=\ell+1}^{q} \left(1+\frac{1}{i_r^6}\right) \le 4\mu^2 a_\ell \log n \left(\frac{2^6}{\log^6 n}\right)^{\ell-1}.$$

In the last inequality we used the fact that

$$\frac{2^{i_1}}{2^{i_\ell}} \le 2a_\ell \log n \quad \text{and} \quad \prod_{r=\ell+1}^{q} \left(1+\frac{1}{i_r^6}\right) < 2.$$

Since $a_2 = O(1)$ we conclude that $\Delta_2 \le O(\frac{\mu^2}{\log^5 n})$. For any $\ell \ge 3$ we use the fact that $a_\ell < \ell^{(1+o(1))\ell} \le (\log n)^{(1+o(1))\ell}$ to conclude that $\Delta_\ell \le O(\frac{\mu^2}{\log^{(5-o(1))\ell-7} n}) \le O(\frac{\mu^2}{\log^7 n})$.

Summing over all values of ℓ we conclude that $\Delta = O(\frac{\mu^2}{\log^5 n})$. The assertion of Lemma 4.3 thus follows from Lemma 4.4. As mentioned after the statement of this lemma, this also completes the proof of Theorem 1.2. □

5. Concluding Remarks

- Since the proof of Theorem 1.2 is probabilistic and the probabilistic estimates are strong enough it follows that for any finite collection of sequences A_j, each satisfying the assumptions of Theorem 1.2, there is a sequence of multiplicities M that is good for each of them, where the number n_0 here depends on all sequences A_j.
- The proof of Theorem 1.2 can be easily modified to work for any sequence A that grows to infinity at least as fast as $k^{\Omega(k)}$ and satisfies $gcd(A) = 1$.

- Erdős and Turán [3] asked if for any asymptotic basis of order 2 of the positive integers (that is, a set A of positive integers so that each sufficiently large integer has a representation as a sum of two elements of A), there must be, for any constant t, integers that have more than t such representations. Theorem 1.1 shows that a natural analogous statement does not hold for partition functions with restricted multiplicities.

Acknowledgment Part of this work was carried out during a visit at the Ecole Polytechnique in Palaiseau, France. I would like to thank Julia Wolf for her hospitality during this visit. I would also like to thank Mel Nathanson for communicating the Canfield-Wilf problem, thank him and Julia for helpful comments, and thank an anonymous referee for useful suggestions.

References

[1] N. Alon and J. Spencer, The Probabilistic Method (3rd edition), Wiley Interscience, 2008.

[2] E. R. Canfield and H. S. Wilf, On the growth of restricted integer partition functions, arXiv: 1009.4404, 2010.

[3] P. Erdős and P. Turán, On a problem of Sidon in additive number theory and some related problems, J. London Math. Soc. 16 (1941), 212-215; *Addendum* (by P. Erdős), ibid. 19 (1944), 208.

[4] Z. Ljujić and M. Nathanson, On a partition problem of Canfield and Wilf, Integers, to appear. (See also: arXiv:1101.0599, 2011).

BRUN MEETS SELMER

Fritz Schweiger
FB Mathematik, University of Salzburg, Salzburg, Austria
fritz.schweiger@sbg.ac.at

Received: 5/12/12, Revised: 1/24/12, Accepted: 3/24/13, Published: 3/29/13

Abstract

The most famous 2-dimensional continued fraction algorithm is the Jacobi algorithm. However, Brun and Selmer algorithms are also interesting 2-dimensional subtractive algorithms. Schratzberger shows that all these three algorithms are deeply related by a process similar to insertion and extension for continued fractions. In this note the basic ergodic properties of two mixtures of both maps are explored. Furthermore a digression to a quite different map is made which exhibits an "exotic" invariant measure.

1. Introduction

The most famous 2-dimensional continued fraction algorithm is the Jacobi algorithm. Last years saw an increasing interest in other 2-dimensional algorithms (see [9], chapters 6 and 7, and [2]). The Brun and the Selmer algorithms are remarkable examples of this type. In the first section we give a short description of both algorithms and look shortly on the flip-flop map built on both maps. It generalizes the 1-dimensional map

$$x \mapsto \frac{x}{1-x}, \; 0 \le x \le \frac{1}{2}$$

$$x \mapsto \frac{1-x}{x}, \; \frac{1}{2} \le x \le 1$$

to the set

$$B := \{(x_1, x_2) : \quad 0 \le x_2 \le x_1 \le 1\}.$$

The jump map (see [9], chapter 3) which avoids the critical point $(0,0)$ leads to Garrity's triangle sequence (Assaf et al. [1]). The next section is devoted to the study of the composition of the Brun and the Selmer map. The set

$$D^- := \{(x_1, x_2) \in B : x_1 + x_2 \le 1\}$$

is transient for the Selmer map and therefore the study of its ergodic behaviour concentrates on the set

$$D^+ := \{(x_1, x_2) \in B : x_1 + x_2 \geq 1\}.$$

The Brun map expands this set D^+ onto the full set B. Therefore, the study of the interplay of these different dynamics may be of some interest.
In the last section a digression to a different map is made which exhibits an "exotic" invariant measure. "Exotic" means that it is possible to construct a fractal like set with positive Lebesgue measure and an invariant density.

2. The Brun, the Selmer Algorithm, and the Flip-flop Map

The Brun algorithm $T : B \to B$ is given by the matrices of its inverse branches

$$M_\alpha = \begin{pmatrix} 1 & 1 & 0 \\ 0 & 1 & 0 \\ 0 & 0 & 1 \end{pmatrix}, M_\beta = \begin{pmatrix} 1 & 1 & 0 \\ 1 & 0 & 0 \\ 0 & 0 & 1 \end{pmatrix}, M_\gamma = \begin{pmatrix} 1 & 0 & 1 \\ 1 & 0 & 0 \\ 0 & 1 & 0 \end{pmatrix}$$

which correspond to a partition of B into three cells $B(\alpha) = M_\alpha B, B(\beta) = M_\beta B$, and $B(\gamma) = M_\gamma B = D^+$ (see Figure 1).

The Selmer algorithm $S : B \to B$ is defined by the matrices of its inverse branches

$$M_0 = \begin{pmatrix} 1 & 0 & 1 \\ 0 & 1 & 0 \\ 0 & 0 & 1 \end{pmatrix}, M_1 = \begin{pmatrix} 0 & 1 & 1 \\ 1 & 0 & 0 \\ 0 & 0 & 1 \end{pmatrix}, M_2 = \begin{pmatrix} 0 & 1 & 1 \\ 1 & 0 & 0 \\ 0 & 1 & 0 \end{pmatrix}.$$

There is an important difference to be observed. $M_0 B$ is the triangle $B(0) = D^-$ with vertices $[1, 0, 0], [1, 1, 0]$ and $[2, 1, 1]$ but M_1 and M_2 are restricted to the triangle D^+. Then $M_1 D^+$ is the triangle $B(1)$ with vertices $[1, 1, 0], [2, 2, 1]$ and $[2, 1, 1]$. $M_2 D^+$ is the triangle $B(2)$ with vertices $[1, 1, 1], [2, 2, 1]$ and $[2, 1, 1]$ (see Figure 2). The flip-flop map uses the matrices M_0 and M_γ. It gives the (forward) map

$$F(x_1, x_2) = \left(\frac{x_1}{1 - x_2}, \frac{x_2}{1 - x_2} \right); \, x \in B(0),$$

$$F(x_1, x_2) = \left(\frac{x_2}{x_1}, \frac{1 - x_1}{x_1} \right); \, x \in B(\gamma).$$

Although Pipping used a kind of mixture of both algorithms [6] this kind of a flip-flop between both algorithms seems not to be investigated. We show that this algorithm admits a σ-finite invariant measure but is related to Garrity's triangle sequence.

A product of n matrices M_η, $\eta \in \{0, \gamma\}$ gives a matrix $((B_{ij}^{(n)}))$, $0 \le i, j \le 2$ and the Jacobian of an inverse branch after n steps is given by

$$\omega(\eta_1, \ldots, \eta_n; x) = \frac{1}{(B_{00}^{(n)} + B_{01}^{(n)} x_1 + B_{02}^{(n)} x_2)^3}.$$

Therefore the measure of a cylinder of rank n is given by

$$\lambda(B(\eta_1, ..., \eta_n)) = \frac{1}{2B_{00}^{(n)}(B_{00}^{(n)} + B_{01}^{(n)})(B_{00}^{(n)} + B_{01}^{(n)} + B_{02}^{(n)})}.$$

Theorem 1: *The function*

$$h(x_1, x_2) = \frac{1}{x_1 x_2}$$

is the density of a σ-finite invariant measure.
This assertion is easily verified.

If we consider the jump map over the cylinder $B(0)$ we obtain a map with matrices

$$\begin{pmatrix} 1 & 0 & k \\ 0 & 1 & 0 \\ 0 & 0 & 1 \end{pmatrix} \begin{pmatrix} 1 & 0 & 1 \\ 1 & 0 & 0 \\ 0 & 1 & 0 \end{pmatrix} = \begin{pmatrix} 1 & k & 1 \\ 1 & 0 & 0 \\ 0 & 1 & 0 \end{pmatrix}$$

This algorithm is Garrity's triangle sequence (see e. g. [1, 4, 10]). Therefore the map F is ergodic. Since the segment $\overline{(0,0)(1,0)}$ is pointwise invariant it is no surprise that this algorithm does not converge everywhere. If $p^{(s)} = p(k_1, ..., k_s)$ and $q^{(s)} = q(k_1, ..., k_s)$ are the vertices of the cylinder $B(k_1, ..., k_s)$ such that $F^s p^{(s)} = (0, 0)$ and $F^s q^{(s)} = (1, 0)$ then the segments $\overline{p(k_1, ..., k_s, k_{s+1}), q(k_1, ..., k_s, k_{s+1})}$ converge to the segment $\overline{p(k_1, ..., k_s), q(k_1, ..., k_s)}$ as $k_{s+1} \to \infty$. Then we choose a sequence $(k_1, k_2, k_3, ...)$ such that

$$\frac{d(p(k_1, ..., k_s, k_{s+1}), q(k_1, ..., k_s, k_{s+1})}{d(p(k_1, ..., k_s), q(k_1, ..., k_s))} > \frac{k_s}{1 + k_s}$$

and the infinite product $\prod_s \frac{k_s}{1+k_s}$ converges. More details can be found in Assaf et al. [1].

3. The Composition of Both Maps

We now consider the mixed map $(S \circ T)x = T(Sx)$. Since $SB(1) = SB(2) = D^+ = B(\gamma)$ the map $S \circ T$ can be described by the five matrices

$$M_{0\alpha} = \begin{pmatrix} 1 & 1 & 1 \\ 0 & 1 & 0 \\ 0 & 0 & 1 \end{pmatrix}, M_{0\beta} = \begin{pmatrix} 1 & 1 & 1 \\ 1 & 0 & 0 \\ 0 & 0 & 1 \end{pmatrix}, M_{0\gamma} = \begin{pmatrix} 1 & 1 & 1 \\ 1 & 0 & 0 \\ 0 & 1 & 0 \end{pmatrix},$$

$$M_{1\gamma} = \begin{pmatrix} 1 & 1 & 0 \\ 1 & 0 & 1 \\ 0 & 1 & 0 \end{pmatrix}, M_{2\gamma} = \begin{pmatrix} 1 & 1 & 0 \\ 1 & 0 & 1 \\ 1 & 0 & 0 \end{pmatrix},$$

These five matrices give a partition of B into five cylinders (see Figure 3).

Lemma 1: *The set*

$$E = \{x : (S \circ T)^j x \in B(0\alpha) \cup B(0\beta) \text{ for all } j \geq 0\}$$

has measure $\lambda(E) = 0$.

Proof. The product of N matrices $M_{0\alpha}$ and $M_{0\beta}$ has the form

$$M^{(N)} = \begin{pmatrix} B_{00}^{(N)} & B_{01}^{(N)} & B_{02}^{(N)} \\ B_{10}^{(N)} & B_{11}^{(N)} & B_{12}^{(N)} \\ 0 & 0 & 1 \end{pmatrix}.$$

Therefore $x = (x_1, x_2)$ is mapped onto

$$(x_1^{(N)}, x_2^{(N)}) = \left(\frac{B_{10}^{(N)} + B_{11}^{(N)} x_1 + B_{12}^{(N)} x_2}{B_{00}^{(N)} + B_{01}^{(N)} x_1 + B_{02}^{(N)} x_2}, \frac{x_2}{B_{00}^{(N)} + B_{01}^{(N)} x_1 + B_{02}^{(N)} x_2} \right).$$

This implies $\lim_{N\to\infty} x_2^{(N)} = 0$. □

Lemma 2: *We have* $B_{02}^{(N)} \leq B_{00}^{(N)} + B_{01}^{(N)}$.

Proof. For $N = 1$ this is verified by inspection. Then we use induction. Let 0α or 0β be the N-th digit. Then

$$B_{02}^{(N+1)} = B_{00}^{(N)} + B_{02}^{(N)} \leq B_{00}^{(N)} + B_{00}^{(N)} + B_{01}^{(N)} = B_{00}^{(N+1)} + B_{01}^{(N+1)}.$$

If $\varepsilon_N \in \{0\gamma, 1\gamma, 2\gamma\}$ the assertion is immediate.
Now we consider the jump transformation $R : B \to B$ which leaves out the digits 0α and 0β. This means we define

$$Rx := (S \circ T)^n x$$

if $x \in B(\varepsilon_1, \ldots, \varepsilon_n)$, $\varepsilon_1, \ldots, \varepsilon_{n-1} \in \{0\alpha, 0\beta\}$ but $\varepsilon_n \in \{0\gamma, 1\gamma, 2\gamma\}$. Lemma 1 implies that R is defined almost everywhere. □

Lemma 3: *R satisfies a Rényi condition.*

Proof. Let

$$\omega(\varepsilon_1, \ldots, \varepsilon_N; x) = \frac{1}{(B_{00}^{(N)} + B_{01}^{(N)} x_1 + B_{02}^{(N)} x_2)^3}$$

be the Jacobian of an inverse branch of R. We have to compare $B_{00}^{(N)}$ with $B_{00}^{(N)} + B_{01}^{(N)} + B_{02}^{(N)}$. Since $\varepsilon_N \in \{0\gamma, 1\gamma, 2\gamma\}$ we see that

$$B_{00}^{(N)} \geq B_{00}^{(N-1)} + B_{01}^{(N-1)}$$

but

$$B_{00}^{(N)} + B_{01}^{(N)} + B_{02}^{(N)} \leq 3B_{00}^{(N-1)} + 2B_{01}^{(N-1)} + B_{02}^{(N-1)} \leq 4B_{00}^{(N-1)} + 3B_{01}^{(N-1)}$$

by Lemma 2. □

Lemma 4: *If the sequence* $(\varepsilon_1, \varepsilon_2, \varepsilon_3, \ldots)$ *contains one of the digits* $0\gamma, 1\gamma$, *or* 2γ *infinitely often then* $\lim\limits_{n\to\infty}$ *diam* $B(\varepsilon_1, \ldots, \varepsilon_n) = 0$.

Proof. We describe the vertices of the cylinders we consider as the pictures of points in projective coordinates (see Figure 4) and suppress the upper index of the relevant matrix

$$\beta = \beta(\varepsilon_1, \ldots, \varepsilon_n) = \begin{pmatrix} B_{00} & B_{01} & B_{02} \\ B_{10} & B_{11} & B_{12} \\ B_{20} & B_{21} & B_{22} \end{pmatrix}.$$

We look for triangles which lie inside the triangle $B(\varepsilon_1, \ldots, \varepsilon_n)$ and contain the triangle $B(\varepsilon_1, \ldots, \varepsilon_n, \varepsilon_{n+1})$ or in some cases the triangle $B(\varepsilon_1, \ldots, \varepsilon_n, \varepsilon_{n+1}, \varepsilon_{n+2})$. If the points $[a, b, c]$, $[a', b', c']$, and $[a'', b'', c'']$ are collinear such that

$$\lambda[a, b, c] + [a', b', c'] = [a'', b'', c'']$$

we will estimate the ratio

$$\frac{d(\beta[a, b, c], \beta[a', b', c'])}{d(\beta[a, b, c], \beta[a'', b'', c''])} = \frac{B_{00}a'' + B_{01}b'' + B_{02}c''}{B_{00}a' + B_{01}b' + B_{02}c'}.$$

We further use that for $\alpha < \delta$ the function $f(t) = \frac{\alpha+t}{\delta+t}$ is increasing on $0 \leq t$.

$\boxed{\varepsilon_{n+1} = 0\alpha}$

$$\frac{d(\beta[1, 0, 0], \beta[2, 1, 0])}{d(\beta[1, 0, 0], \beta[1, 1, 0])} = \frac{B_{00} + B_{01}}{2B_{00} + B_{01}}.$$

$$\frac{d(\beta[1, 0, 0], \beta[3, 1, 1])}{d(\beta[1, 0, 0], \beta[1, 1, 1])} = \frac{B_{00} + B_{01} + B_{02}}{3B_{00} + B_{01} + B_{02}} \leq \frac{B_{00} + B_{01}}{2B_{00} + B_{01}}.$$

Since the periodic point $\overline{0\alpha}$ shrinks to the point $(0,0)$ we can additionally assume that $\varepsilon_n \in \{0\beta, 0\gamma, 1\gamma, 2\gamma\}$. Then the recursion relations show $B_{01} \leq 2B_{00}$ and we obtain

$$\frac{B_{00}+B_{01}}{2B_{00}+B_{01}} \leq \frac{3}{4}.$$

$\boxed{\varepsilon_{n+1} = 2\gamma}$

In a similar way as before we find the ratios

$$\frac{d(\beta[1,1,1],\beta[2,2,1])}{d(\beta[1,1,1],\beta[1,1,0])} = \frac{B_{00}+B_{01}}{2B_{00}+2B_{01}+B_{02}} \leq \frac{1}{2}.$$

$$\frac{d(\beta[1,1,1],\beta[2,1,1])}{d(\beta[1,1,1],\beta[1,0,0])} = \frac{B_{00}}{2B_{00}+B_{01}+B_{02}} \leq \frac{1}{2}.$$

$\boxed{\varepsilon_{n+1} = 0\beta}$

Here we use the additional points $\beta[3,2,1]$ and $\beta[2,1,1]$ which lie outside on the line which joins $\beta[1,1,0]$ and$\beta[2,1,1]$.

$\boxed{\varepsilon_{n+2} = 0\beta, 0\gamma, 1\gamma}$

$$\frac{d(\beta[2,1,0],\beta[3,2,0])}{d(\beta[2,1,0],\beta[1,1,0])} = \frac{B_{00}+B_{01}}{3B_{00}+2B_{01}} \leq \frac{1}{2}.$$

$$\frac{d(\beta[2,1,0],\beta[5,3,1])}{d(\beta[2,1,0],\beta[3,2,1])} = \frac{3B_{00}+2B_{01}+B_{02}}{5B_{00}+3B_{01}+B_{02}} \leq \frac{3}{4}.$$

$$\frac{d(\beta[2,1,0],\beta[4,2,1])}{d(\beta[2,1,0],\beta[2,1,1])} = \frac{2B_{00}+B_{01}+B_{02}}{4B_{00}+2B_{01}+B_{02}} \leq \frac{2}{3}.$$

$$\frac{d(\beta[2,1,0],\beta[5,2,1])}{d(\beta[2,1,0],\beta[3,1,1])} = \frac{3B_{00}+B_{01}+B_{02}}{5B_{00}+2B_{01}+B_{02}} \leq \frac{2}{3}.$$

$\boxed{\varepsilon_{n+1} = 0\gamma}$

Here we use the additional points $\beta[3,2,0]$, $[2,1,0]$, and $\beta[1,0,0]$.

$\boxed{\varepsilon_{n+2} = 0\gamma, 1\gamma}$

$$\frac{d(\beta[2,1,1],\beta[5,3,1])}{d(\beta[2,1,1],\beta[3,2,0])} = \frac{3B_{00}+2B_{01}}{5B_{00}+3B_{01}+B_{02}} \leq \frac{2}{3}.$$

$$\frac{d(\beta[2,1,1],\beta[4,2,1])}{d(\beta[2,1,1],\beta[2,1,0])} = \frac{2B_{00}+B_{01}}{4B_{00}+2B_{01}+B_{02}} \leq \frac{1}{2}.$$

$$\frac{d(\beta[2,1,1],\beta[5,2,2])}{d(\beta[2,1,1],\beta[1,0,0])} = \frac{B_{00}}{5B_{00}+2B_{01}+2B_{02}} \le \frac{1}{5}.$$

$\boxed{\varepsilon_{n+1} = 1\gamma}$

Only the case $\overline{1\gamma}$ remains; however, the sequence of associated triangles shrinks to the point $(\lambda-1, \lambda^2-\lambda-1)$, where $\lambda > 1$ is the greatest root of $\lambda^3 = \lambda^2+2\lambda-1$. $\square$

Lemmas 1-4 provide the necessary machinery to deduce the following:

Theorem 2: *$S \circ T$ is ergodic and admits a σ-finite invariant measure $\mu \sim \lambda$.*

Remark: The map $(T \circ S)(x) = S(Tx)$ divides B into nine cells. Since $S \circ (T \circ S) = (S \circ T) \circ S$ their ergodic behaviors are equivalent.

4. A Split Algorithm

The next algorithm is not directly related to the Brun or the Selmer algorithm but shows that the "exotic" behaviour which was first detected with the Parry-Daniels map is quite common (see [5]).
The starting point are the three matrices

$$\beta(1) = \begin{pmatrix} 1 & 0 & 0 \\ 1 & 1 & 0 \\ 1 & 0 & 1 \end{pmatrix}, \beta(2) = \begin{pmatrix} 1 & 0 & 1 \\ 1 & 1 & 0 \\ 1 & 0 & 0 \end{pmatrix}, \beta(3) = \begin{pmatrix} 1 & 1 & 0 \\ 1 & 0 & 0 \\ 1 & 0 & 1 \end{pmatrix}.$$

These matrices form a 2-dimensional continued fraction on the basic set $(\mathbb{R}^+)^2$ with the three inverse branches

$$\begin{aligned} V(1)(u,v) &= (1+u, 1+v) \\ V(2)(u,v) &= \left(\frac{1+u}{1+v}, \frac{1}{1+v}\right) \\ V(3)(u,v) &= \left(\frac{1}{1+u}, \frac{1+v}{1+u}\right) \end{aligned}$$

and the basic partition is

$$B(1) = \{(u,v) : 1 \le u, 1 \le v\}$$

$$B(2) = \{(u,v) : 0 \le v \le u, v \le 1\}$$

$$B(3) = \{(u,v) : 0 \le u \le v, u \le 1\}.$$

The dual map is given given as

$$V^{\#}(1)(x,y) = \left(\frac{x}{1+x+y}, \frac{y}{1+x+y}\right)$$

$$
\begin{aligned}
V^{\#}(2)(x,y) &= (\frac{x}{1+x+y}, \frac{1}{1+x+y}) \\
V^{\#}(3)(x,y) &= (\frac{1}{1+x+y}, \frac{y}{1+x+y})
\end{aligned}
$$

which may be compared with the 2-dimensional Farey-Brocot algorithm which was considered in Schweiger [10]. This algorithm sits on a set E with $\lambda(E) = 0$ but the function

$$g(x_1, x_2) = \frac{1}{x_1 x_2}$$

behaves formally as an invariant density. It would be nice to explore if in some limiting sense the integral

$$\int_E \frac{dx_1 dx_2}{x_1 x_2}$$

is finite.

Let

$$E_{12} = \{(u,v) : T^s(u,v) \in B(1) \cup B(2), s \geq 0\}$$

and

$$E_{13} = \{(u,v) : T^s(u,v) \in B(1) \cup B(3), s \geq 0\}.$$

We will show that $\lambda(E_{12}) = \lambda(E_{13}) > 0$ and calculate an invariant density for the map T restricted to E_{12}.

We consider the first return map on the set on the set $B(2)$ of the restriction of T to E_{12}. This map is given as $R(u,v) = T^k(u,v)$ if $(u,v) \in B(2), T^j(u,v) \in B(1), 1 \leq j \leq k-1, T^k(u,v) \in B(2)$. The associated matrices are given as

$$\beta(2)\beta(1)^k =: \gamma(a) = \begin{pmatrix} a & 0 & 1 \\ a & 1 & 0 \\ 1 & 0 & 0 \end{pmatrix}$$

where $a = k+1$. These matrices are related to continued fractions! If

$$\gamma(a_1)...\gamma(a_s) = \begin{pmatrix} q_s & 0 & q_{s-1} \\ r_s & 1 & r_{s-1} \\ p_s & 0 & p_{s-1} \end{pmatrix}$$

then as usual $q_s = a_s q_{s-1} + q_{s-2}$, $p_s = a_s p_{s-1} + p_{s-2}$ but $r_s = a_s r_{s-1} + r_{s-2} + a_s$. The last recursion can be written as $r_s + 1 = a_s(r_{s-1} + 1) + r_{s-2} + 1$ which shows that $q_s \leq r_s \leq 2q_s$.

Theorem 3: $\lambda(E_{12}) > 0$.

Proof. We transport the map T into the triangle with vertices $(0,0)$, $(1,0)$, and $(0,1)$ by using the map $\psi(u,v) = (\frac{u}{1+u+v}, \frac{v}{1+u+v})$. The quotient of the measure of the cylinder $B(a_1, ..., a_s)$ and the length of the associated continued fraction interval $I(a_1, ...a_s)$ is bounded from below. Therefore we find $\lambda(E_{12}) > 0$. □

Theorem 4:*Let* $\theta = [a_1, a_2, ...]$ *be a regular continued fraction and define* $\Gamma(\theta) = \sum_{n=0}^{\infty}(\prod_{j=0}^{n} T^j\theta)a_{n+1}$. *Then the function*

$$h(u, v) = \frac{1}{(1 + v)(u - \Gamma(v))}$$

is an invariant density for the map T *restricted to the set* E_{12}.

Proof. We first remark

$$\Gamma(\theta) = \Gamma(\frac{1}{a + \theta})(a + \theta) - a.$$

Then we calculate

$$\sum_{a=1}^{\infty} h(\frac{a+u}{a+v}, \frac{1}{a+v})\frac{1}{(a+v)^3} = \sum_{a=1}^{\infty} \frac{1}{(a+1+v)(a+v)(a+u-\Gamma((a+v)^{-1})(a+v))}$$

$$= \frac{1}{u - \Gamma(v)} \sum_{a=1}^{\infty} \frac{1}{(a+v)(a+1+v)} = \frac{1}{(1+v)(u-\Gamma(v))}.$$

Remark: The dual map defined by

$$\gamma^{\#}(a) = \begin{pmatrix} a & a & 1 \\ 0 & 1 & 0 \\ 1 & 0 & 0 \end{pmatrix}$$

formally has the invariant density

$$f(x_1, x_2) = \frac{1}{x_1} \int_0^1 \frac{dv}{(1 + x_1\Gamma(v) + x_2 v)^2}.$$

We verify this by direct calculation:

$$\sum_{a=1}^{\infty} f(\frac{x_1}{a + ax_1 + x_2}, \frac{1}{a + ax_1 + x_2})\frac{1}{(a + ax_1 + x_2)^3}$$

$$= \frac{1}{x_1} \sum_{a=1}^{\infty} \int_0^1 \frac{dv}{(a + (a + \Gamma(v))x_1 + x_2 + v)^2}$$

$$= \frac{1}{x_1} \sum_{a=1}^{\infty} \int_{\frac{1}{a+1}}^{\frac{1}{a}} \frac{dw}{(1 + \Gamma(w))x_1 + x_2 w)^2} = F(x_1, x_2).$$

This follows from $w = \frac{1}{a+v}$ and the equation $\Gamma(v) + a = \Gamma(w)(a + v)$. □

Acknowledgement The author wants to express his sincere thanks to the referee whose critical remarks helped to improve the present paper.

References

[1] Assaf, S.; Li-Chung Chen; Cheslack-Postava, T.; Diesl, A.; Garrity, T.; Lepinsky, M.; Schuyler, A.: A dual approach to triangle sequences: a multidimensional continued fraction algorithm. *Integers* 5 (2005).

[2] Bryuno, A.D. ; Parusnikov, A. D.: Comparison of various generalizations of continued fractions *Math. Notes* 61 (1997), 278-286.

[3] Iosifescu, M. and Kraaikamp, C.: *Metrical Theory of Continued Fractions.* Dordrecht - Boston - London: Kluwer Academic Publishers, 2002.

[4] Messaoudi, A. & Nogueira, A. & Schweiger, F.: Ergodic properties of triangle partitions. *Monatsh. Math.* 157 (2009), 253-299.

[5] Nogueira, A.: The three-dimensional Poincaré continued fraction algorithm. *Israel J. Math.* **90** (1995), 373–401.

[6] Pipping, N.: Über eine Verallgemeinerung des Euklidischen Algorithmus. *Acta Acad. Åbo* ser B **1**:7 (1922), 1-14.

[7] Schratzberger, B.: *S-expansions in dimension two.* J. Theor. Nombres Bordeaux, 16 no. 3 (2004), p. 705-732.

[8] Schratzberger, B.: *On the singularisation of the two-dimensional Jacobi-Perron algorithm.* J. Experiment. Math. 16, Issue 4 (2007), 441-454.

[9] Schweiger, F.: *Multidimensional Continued Fractions*, Oxford: Oxford University Press, 2000.

[10] Schweiger, F.: A 2-dimensional algorithm related to the Farey-Brocot sequence. *Int. J. Number Theory* 8 (2012), 149-160.

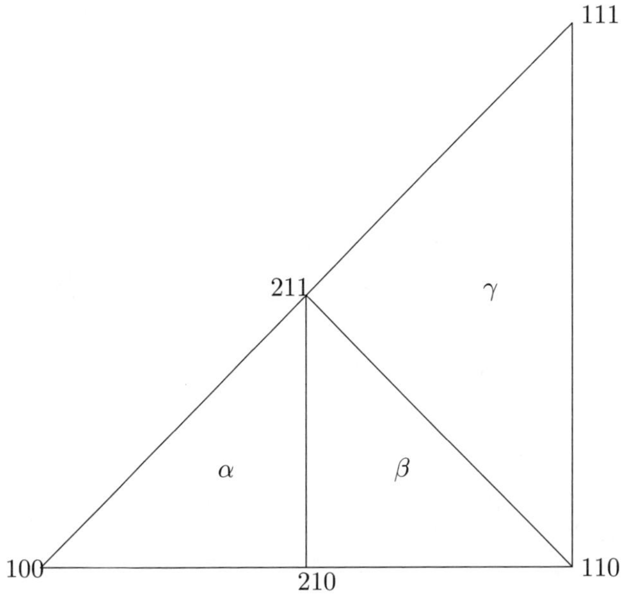

Figure 1

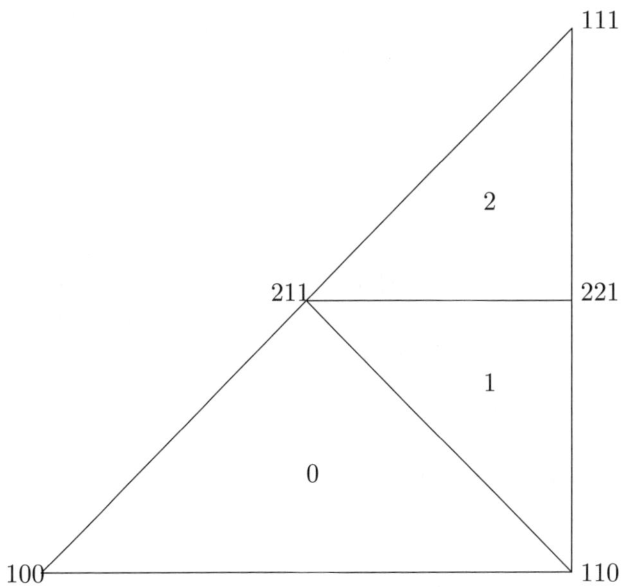

Figure 2

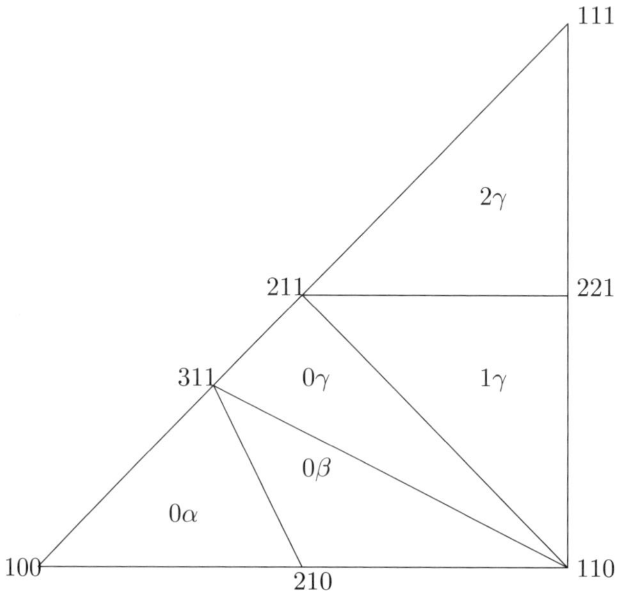

Figure 3

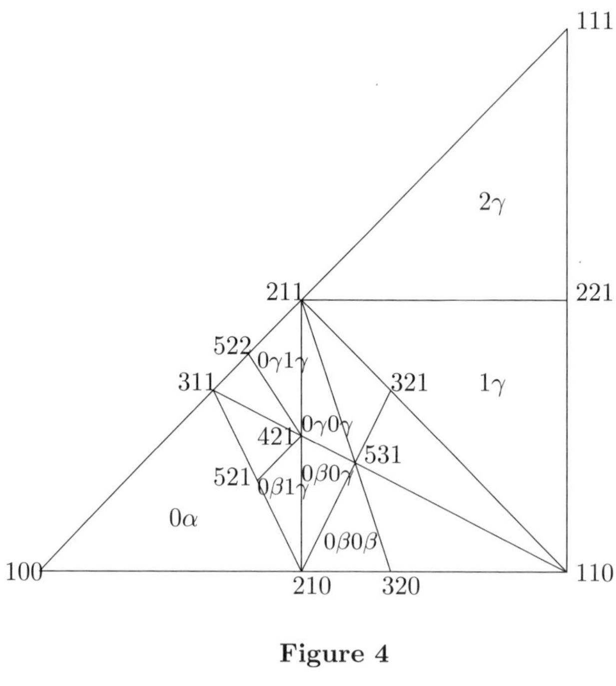

Figure 4

THICK SUBSETS THAT DO NOT CONTAIN ARITHMETIC PROGRESSIONS

Kevin O'Bryant[1]
Department of Mathematics, City University of New York, College of Staten Island and The Graduate Center, New York
obryant@gmail.com

Received: 10/25/11, Accepted: 3/25/13, Published: 3/29/13

Abstract

We adapt the construction of subsets of $\{1, 2, \ldots, N\}$ that contain no k-term arithmetic progressions to give a relatively thick subset of an arbitrary set of N integers. Particular examples include a thick subset of $\{1, 4, 9, \ldots, N^2\}$ that does not contain a 3-term AP, and a positive relative density subset of a random set (contained in $\{1, 2, \ldots, n\}$ and having density $cn^{-1/(k-1)}$) that is free of k-term APs.

1. Introduction

For a finite set $\mathcal{N}$ (whose cardinality we denote by N), set $r_k(\mathcal{N})$ to be the largest possible size of a subset of $\mathcal{N}$ that contains no k-term arithmetic progressions (k-APs). A little-known result of Komlós, Sulyok, and Szemerédi [8] implies that

$$r_k(\mathcal{N}) \geq C\, r_k([N]), \tag{1}$$

for an explicit positive constant C. Abbott [2] reports that their proof gives $C = 2^{-15}$, and indicates some refinements that yield $C = 1/34$.

In this work, we focus on the situation when $\mathcal{N}$ itself has few solutions and we can give bounds on $r_k(\mathcal{N})$ that are much stronger than those implied by Eq. (1) and the currently best bounds on $r_k([N])$. In particular, we adapt the Behrend-type construction [9] of subsets of $[N]$ without k-APs to arbitrary finite sets $\mathcal{N}$. We draw particular attention to subsets of the squares and to subsets of random sets. As the statement of our theorem requires some notation and terminology, we first give two corollaries.

Our first corollary brings attention to the fact that while the squares contain many 3-APs, they also contain unusually large subsets that do not. Here and

[1]The work was supported in part by a grant for The City University of New York PSC-CUNY Research Award Program.

throughout this paper, $\exp(x) = 2^x$ and $\log x = \log_2(x)$. For comparison, $r_3([N]) \geq N \exp(-2\sqrt{2}\sqrt{\log N} + \frac{1}{4}\log\log N)$.

Corollary 1. *There is an absolute constant $C > 0$ such that for every N there is a subset of $\{1, 4, 9, \ldots, N^2\}$ with cardinality at least*

$$C \cdot N \cdot \exp\left(-2\sqrt{2}\sqrt{\log\log N} + \frac{1}{4}\log\log\log N\right)$$

that does not contain any 3-term arithmetic progressions.

Our second corollary identifies sets that have subsets with no k-APs and with *positive* relative density .

Corollary 2. *For every real ψ and integer $k \geq 3$, there is a real $\delta > 0$ such that every sufficiently large $\mathcal{N} \subseteq \mathbb{Z}$ that has fewer than $\psi|\mathcal{N}|$ arithmetic progressions of length k contains a subset that is free of k-term arithmetic progressions and has relative density at least δ. In particular, for each $\delta > 0$, if n is sufficiently large and $\mathcal{N} \subseteq \{1, 2, \ldots, n\}$ is formed by including each number independently with probability $cn^{-1/(k-1)} > 0$, then with high probability $\mathcal{N}$ contains a subset A with relative density δ and no k-term arithmetic progressions.*

The structure of the proof requires us to consider a generalization of arithmetic progressions. A k-term D-progression is a nonconstant sequence $a_1, \ldots, a_k$ whose $(D+1)$-st differences are all zero:

$$\sum_{i=0}^{D+1} (-1)^i \binom{D+1}{i} a_{i+v} = 0, \qquad (1 \leq v \leq k - D - 1).$$

Equivalently, $a_1, \ldots, a_k$ is a k-term D-progression if there is a nonconstant polynomial $Q(j)$ with degree at most D and $Q(i) = a_i$ for $i \in [k]$. Clarifying examples of 5-term 2-progressions of integers are $1, 2, 3, 4, 5$ (from $Q(j) = j$), and $4, 1, 0, 1, 4$ (from $Q(j) = (j-3)^2$), and $1, 3, 6, 10, 15$ (from $Q(j) = \frac{1}{2}j + \frac{1}{2}j^2$). Let $Q(j) = \sum_{i=0}^{D'} q_i j^i$ be a polynomial with degree $D' \geq 1$, so that $Q(1), Q(2), \ldots, Q(k)$ is a k-term D-progression for all $D \geq D'$. The quantity $D'!q_{D'}$, which is necessarily nonzero, is called the difference of the sequence, and $(D', Q(1), D'!q_{D'})$ is the type of the sequence. Note that different progressions can have the same type: both $1, 4, 9, 16, 25$ and $1, 5, 11, 19, 29$ have type $(2, 1, 2)$. For any set $\mathcal{N}$, we let $\text{TYPE}_{k,D}(\mathcal{N})$ be the number of types of k-term D-progressions contained in $\mathcal{N}$. The proof of [9]*Lemma 4 shows that $\text{TYPE}_{k,D}(\mathcal{N}) \ll |\mathcal{N}| \operatorname{diam}(\mathcal{N})$. Since the type of a k-term D-progression is determined by its first $D+1$ elements, we also have $\text{TYPE}_{k,D}(\mathcal{N}) \leq N^{D+1}$. We define

$$r_{k,D}(\mathcal{N}) := \max_{A \subseteq \mathcal{N}} \{|A| \colon A \text{ does not contain any } k\text{-term } D\text{-progressions}\}$$

and recall the lower bound proved in [9]:

$$\frac{r_{k,D}([N])}{N} \geq C \exp\left(-n2^{(n-1)/2} D^{(n-1)/n} \sqrt[n]{\log N} + \frac{1}{2n} \log\log N\right). \quad (2)$$

We can now state our main theorem.

Theorem 1. *Let $k \geq 3, n \geq 2, D \geq 1$ be integers satisfying $k > 2^{n-1}D$. Let $\Psi(N)$ be any function that is at least 2. There is a constant $C = C(k, D, \Psi)$ such that for all $\mathcal{N} \subseteq \mathbb{Z}$ with $\textsc{Type}_{k,D}(\mathcal{N}) \leq N\Psi(N)$ (where $N := |\mathcal{N}|$)*

$$\frac{r_{k,D}(\mathcal{N})}{N} \geq C \exp\left(-n2^{(n-1)/2} D^{(n-1)/n} \sqrt[n]{\log \Psi(N)} + \frac{1}{2n} \log\log \Psi(N)\right).$$

Corollary 2 is now straightforward: set $D = 1$ and $\Psi(N) = \max\{\psi, 2\}$ and take

$$\delta = \exp\left(-n2^{(n-1)/2} \sqrt[n]{\log C} + \frac{1}{2n} \log\log C\right),$$

to arrive at the first sentence. Considering the random set $\mathcal{N}$ described in the second sentence of Corollary 2, for each pair $(a, a+d)$ of elements of $\mathcal{N}$ the likelihood of the next $k-2$ elements $a+2d, \ldots, a+(k-1)d$ of the arithmetic progression being in $\mathcal{N}$ is $(cn^{-1/(k-1)})^{k-2}$. Consequently, the expected number of k-term arithmetic progressions in $\mathcal{N}$ is

$$\binom{n}{2} (n^{-1/(k-1)})^{k-2} \leq \frac{c^{k-2}}{2} n^{k/(k-1)},$$

and the expected size of $\mathcal{N}$ is $N = n \cdot cn^{-1/(k-1)} = cn^{k/(k-1)}$. We can take $\Psi(N)$ to be a constant with high probability, and so Corollary 2 follows from Theorem 1.

Corollary 1 is only a bit more involved. It is known (perhaps since Fermat, see [4, 5, 10, 3, 6, 7, 1] for a history and for the results we use here) that while the squares do not contain any 4-term arithmetic progressions, the 3-term arithmetic progressions a^2, b^2, c^2 are parameterized by

$$a = u(2st - s^2 + t^2), b = u(s^2 + t^2), c = u(2st + s^2 - t^2),$$

with $s, t, u \geq 1$ and $\gcd(s,t) = 1$. Merely observing that $s, t, u \geq 1, b \leq N$ yields that there are fewer than $2\pi N \log N$ triples (s, t, u) with a, b, c in $[N]$, i.e.,

$$\textsc{Type}_{3,1}(\{1, 4, 9, \ldots, N^2\}) \leq 2\pi N \log N.$$

Now, setting $k = 1, n = 2, D = 1, \Psi(N) = 2\pi \log N$ in Theorem 1 produces Corollary 1.

Section 2 gives a short outline of the construction behind Theorem 1, which is given in greater detail in Section 3. We conclude in Section 4 with some unresolved questions.

2. Overview of Construction Proving Theorem 1

Throughout this work we fix three integers, $k \geq 3$, $n \geq 2$, $D \geq 1$, that satisfy $k > 2^{n-1}D$; in other words, one may take $n = \lceil \log(k/D) \rceil$.

In this section, we outline the construction, suppressing as much technical detail as possible. In the following sections, all definitions are made precisely and all arguments are given full rigor.

Fix $\Psi(N)$, and take $\mathcal{N} \subseteq \mathbb{Z}$ with $|\mathcal{N}| = N$, and so that $\mathcal{N}$ contains less than $N\Psi(N)$ types of k-term D-progressions. The parameters N_0, d, δ are chosen at the end for optimal effect.

Let $A_0 = R_{k,2D}(N_0)$ be a subset of $[N_0]$ without k-term $2D$-progressions, and

$$|A_0| = r_{k,2D}(N_0).$$

Consider $\overline{\omega}, \overline{\alpha}$ in $\mathbb{T}^d$ (we average over all choices of $\overline{\omega}, \overline{\alpha}$ later in the argument), and set

$$A := \{a \in \mathcal{N} \colon a\overline{\omega} + \overline{\alpha} \bmod \overline{1} = \langle x_1, \ldots, x_d \rangle, |x_i| < 2^{-D-1}, \sum x_i^2 \in \textsc{Annuli}\},$$

where $\textsc{Annuli}$ is a union of thin annuli in $\mathbb{R}^d$ with thickness δ whose radii are affinely related to elements of A_0. Set

$$T := \{a \in A \colon \text{there is a } k\text{-term } D\text{-progression in } A \text{ starting at } a\}.$$

Then $A \setminus T$ is free of k-term D-progressions, and so $r_{k,D}(\mathcal{N}) \geq |A \setminus T| = |A| - |T|$, and more usefully

$$r_{k,D}(\mathcal{N}) \geq \mathbb{E}_{\overline{\omega},\overline{\alpha}}\left[|A|\right] - \mathbb{E}_{\overline{\omega},\overline{\alpha}}\left[|T|\right],$$

with the expectation referring to choosing $\overline{\omega}, \overline{\alpha}$ uniformly from the torus $\mathbb{T}^d$. We have

$$\mathbb{E}_{\overline{\omega},\overline{\alpha}}\left[|A|\right] = \mathbb{E}_{\overline{\omega}}\left[\mathbb{E}_{\overline{\alpha}}\left[|A|\right]\right] = \mathbb{E}_{\overline{\omega}}\left[N\,\mathbf{vol}(\textsc{Annuli})\right] = N\,\mathbf{vol}(\textsc{Annuli}).$$

We also have

$$\mathbb{E}_{\overline{\omega},\overline{\alpha}}\left[|T|\right] \leq \mathbb{E}_{\overline{\omega},\overline{\alpha}}\left[\sum E(D',a,b)\right] = \sum \mathbb{E}_{\overline{\omega},\overline{\alpha}}\left[E(D',a,b)\right]$$

where $E(D',a,b)$ is 1 if A contains a progression of type (D',a,b), and is 0 otherwise, and the summation has $\textsc{Type}_{k,D}(\mathcal{N})$ summands. Using the assumption that A_0 is free of k-term $2D$-progressions, we are able to bound

$$\mathbb{E}_{\overline{\omega},\overline{\alpha}}\left[E(D',a,b)\right]$$

efficiently in terms of the volume of $\textsc{Annuli}$ and the volume of a small sphere. We arrive at

$$\mathbb{E}_{\overline{\omega},\overline{\alpha}}\left[|T|\right] \leq \textsc{Type}_{k,D}(\mathcal{N})\,\mathbf{vol}(\textsc{Annuli})\,\mathbf{vol}(\textsc{Ball}),$$

which gives us a lower bound on $r_{k,D}(\mathcal{N})$ in terms of Ψ, N_0, d, δ and A_0. The work [9] gives a lower bound on the size of A_0, and optimization of the remaining parameters yields the result.

3. Proof of Theorem 1

The open interval $(a-b, a+b)$ of real numbers is denoted $a \pm b$. The interval $[1, N] \cap \mathbb{Z}$ of natural numbers is denoted $[N]$. The box $(\pm 2^{-D-1})^d$, which has Lebesgue measure 2^{-dD}, is denoted $\textsc{Box}_D$. We define $\textsc{Box}_0 = [-1/2, 1/2)^d$.

Although we make no use of this until the very end of the argument, we set

$$d := \left\lfloor 2^{n/2} \left(\frac{\log \Psi(N)}{D} \right)^{1/(n+1)} \right\rfloor.$$

Given $\overline{x} \in \mathbb{R}^d$, we denote the unique element $\overline{y}$ of $\textsc{Box}_0$ with $\overline{x} - \overline{y} \in \mathbb{Z}^d$ as $\overline{x} \bmod \overline{1}$.

A point $\overline{x} = \langle X_1, \dots, X_d \rangle$ chosen uniformly from $\textsc{Box}_D$ has components X_i independent and uniformly distributed in $(-2^{-D-1}, 2^{-D-1})$. Therefore, $\|\overline{x}\|_2^2 = \sum_{i=1}^d X_i^2$ is the sum of d iidrvs, and is consequently normally distributed as $d \to \infty$. Further, $\|\overline{x}\|_2^2$ has mean $\mu := 2^{-2D} d/12$ and variance $\sigma^2 := 2^{-4D} d/180$.

Let A_0 be a subset of $[N_0]$ with cardinality $r_{k,2D}([N_0])$ that does not contain any k-term $2D$-progression, and assume $2\delta N_0 \le 2^{-2D}$. We define $\textsc{Annuli}$ in the following manner:

$$\textsc{Annuli} := \left\{ \overline{x} \in \textsc{Box}_D \colon \frac{\|\overline{x}\|_2^2 - \mu}{\sigma} \in \bigcup_{a \in A_0} \left(z - \frac{a-1}{N_0} \pm \delta \right) \right\},$$

where $z \in \mu \pm \sigma$ is chosen to maximize the volume of $\textsc{Annuli}$. Geometrically, $\textsc{Annuli}$ is the union of $|A|$ spherical shells, intersected with $\textsc{Box}_D$. From [9]*Lemma 3, the Barry-Esseen central limit theorem and the pigeonhole principle yield:

Lemma 1 (Annuli has large volume). *If d is sufficiently large, $A_0 \subseteq [N_0]$, and $2\delta \le 1/n$, then the volume of* $\textsc{Annuli}$ *is at least* $\frac{2}{5} 2^{-dD} |A_0| \delta$.

Set

$$A := A(\overline{\omega}, \overline{\alpha}) = \{ n \in \mathcal{N} \colon n\overline{\omega} + \overline{\alpha} \bmod \overline{1} \in \textsc{Annuli} \},$$

which we will show is typically (with respect to $\overline{\omega}, \overline{\alpha}$ being chosen uniformly from $\textsc{Box}_0$) a set with many elements and few types of D-progressions. After removing one element from A for each type of progression it contains, we will be left with a set that has large size and no k-term D-progressions.

Define $T := T(\overline{\omega}, \overline{\alpha})$ to be the set

$$\left\{ a \in \mathcal{N} \colon \begin{array}{c} \exists b \in \mathbb{R}, D' \in [D] \text{ such that } A(\overline{\omega}, \overline{\alpha}) \text{ contains} \\ \text{a } k\text{-term progression of type } (D', a, b) \end{array} \right\},$$

which is contained in $A(\overline{\omega}, \overline{\alpha})$. Observe that $A \setminus T$ is a subset of $\mathcal{N}$ and contains no k-term D-progressions, and consequently $r_{k,D}(\mathcal{N}) \geq |A \setminus T| = |A| - |T|$ for every $\overline{\omega}, \overline{\alpha}$. In particular,

$$r_{k,D}(\mathcal{N}) \geq \mathbb{E}_{\overline{\omega},\overline{\alpha}}\left[|A \setminus T|\right] = \mathbb{E}_{\overline{\omega},\overline{\alpha}}\left[|A| - |T|\right] = \mathbb{E}_{\overline{\omega},\overline{\alpha}}\left[|A|\right] - \mathbb{E}_{\overline{\omega},\overline{\alpha}}\left[|T|\right]. \qquad (3)$$

First, we note that

$$\mathbb{E}_{\overline{\omega},\overline{\alpha}}\left[|A|\right] = \sum_{n \in \mathcal{N}} \mathbb{P}_{\overline{\omega},\overline{\alpha}}\left[n \in A\right] = \sum_{n \in \mathcal{N}} \mathbb{P}_{\overline{\alpha}}\left[n \in A\right] = N\,\mathbf{vol}(\textsc{Annuli}). \qquad (4)$$

Let $E(D', a, b)$ be 1 if A contains a k-term progression of type (D', a, b), and $E(D', a, b) = 0$ otherwise. We have

$$|T| \leq \sum_{(D',a,b)} E(D', a, b),$$

where the sum extends over all types (D', a, b) for which $D' \in [D]$ and there is a k-term D'-progression of that type contained in $\mathcal{N}$; by definition there are $AP_{k,D}(\mathcal{N})$ such types.

Suppose that A has a k-term progression of type (D', a, b), with $D' \in [D]$. Let p be a degree D' polynomial with lead term $p_{D'} = b/D'! \neq 0$, and $p(1), \ldots, p(k)$ a D'-progression contained in A. Then

$$\overline{x}_i := p(i)\,\overline{\omega} + \overline{\alpha} \bmod \overline{1} \in \textsc{Annuli} \subseteq \textsc{Box}_D\,.$$

We now pull a lemma from [9]*Lemma 2.

Lemma 2. *Suppose that $p(j)$ is a polynomial with degree D', with D'-th coefficient p'_D, and set $\overline{x}_j := \overline{\omega}\,p(j) + \overline{\alpha} \bmod \overline{1}$. If $\overline{x}_1, \overline{x}_2, \ldots, \overline{x}_k$ are in $\textsc{Box}_D$ and $k \geq D+2$, then there is vector polynomial $\overline{P}(j) = \sum_{i=0}^{D'} \overline{P}_i j^i$ with $\overline{P}(j) = \overline{x}_j$ for $j \in [k]$, and $D'!\overline{P}_{D'} = \overline{\omega}\,D'!p_{D'} \bmod \overline{1}$.*

Thus, the $\overline{x}_i$ are a D'-progression in $\mathbb{R}^d$, say $\overline{P}(j) = \sum_{i=0}^{D'} \overline{P}_i j^i$ has $\overline{P}(j) = \overline{x}_j$ and $D'!\overline{P}_{D'} = D'!p_{D'}\,\overline{\omega} \bmod \overline{1} = b\overline{\omega} \bmod \overline{1}$. Recalling that z was chosen in the definition of $\textsc{Annuli}$, by elementary algebra

$$Q(j) := \frac{\|\overline{P}(j)\|_2^2 - \mu}{\sigma} - z$$

is a degree $2D'$ polynomial in j (with real coefficients), and since $\overline{P}(j) = \overline{x}_j \in \textsc{Annuli}$ for $j \in [k]$, we know that

$$Q(j) \in \bigcup_{a \in A_0} \left(-\frac{a-1}{N_0} \pm \delta\right)$$

for all $j \in [k]$, and also $Q(1), \ldots, Q(k)$ is a $2D'$-progression. Define the real numbers $a_j \in A_0$, $\epsilon_j \in \pm\delta$ by

$$Q(j) = -\frac{a_j - 1}{N_0} + \epsilon_j.$$

For a finite sequence $(a_i)_{i=1}^k$, we define the forward difference $\Delta(a_i)$ to be the slightly shorter finite sequence $(a_{v+1} - a_v)_{v=1}^{k-1}$. The formula for repeated differencing is

$$\Delta^m(a_i) = \left(\sum_{i=0}^{m} \binom{m}{i} (-1)^i a_{i+v}\right)_{v=1}^{k-m}.$$

We note that a nonconstant sequence (a_i) with at least $2D+1$ terms is a $2D$-progression if and only if $\Delta^{2D+1}(a_i)$ is a sequence of zeros. If $a_i = p(i)$, with p a polynomial with degree $2D$ and lead term $p_{2D} \neq 0$, then $\Delta^{2D}(a_i) = ((2D)!p_{2D})$, a nonzero-constant sequence. Note also that Δ is a linear operator. Finally, we make use of the fact, provable by induction for $1 \leq m \leq k$, that

$$|\Delta^m(a_i)| \leq 2^{m-1}\left(\max_i a_i - \min_i a_i\right).$$

We need to handle two cases separately: either the sequence (a_i) is constant or it is not. Suppose first that it is not constant. Since $a_i \in A_0$, a set without k-term $2D$-progressions, we know that $\Delta^{2D+1}(a_i) \neq (0)$, and since (a_i) is a sequence of integers, for some v

$$|\Delta^{2D+1}(a_i)(v)| \geq 1.$$

Consider:

$$(0) = \Delta^{2D+1}(Q(i)) = \frac{1}{N_0}\Delta^{2D+1}(a_i) + \Delta^{2D+1}(\epsilon_i),$$

whence

$$|\Delta^{2D+1}(\epsilon_i)(v)| = \frac{1}{N_0}|\Delta^{2D+1}(a_i)(v)| \geq \frac{1}{N_0}.$$

Since $|\epsilon_i| < \delta$, we find that

$$|\Delta^{2D+1}(\epsilon_i)(v)| = \left|\sum_{i=0}^{2D+1} \binom{2D+1}{i} (-1)^i \epsilon_{i+v}\right| < 2^{2D+1}\delta,$$

and since we assumed that $2\delta N_0 \leq 2^{-2D}$, we arrive at the impossibility

$$\frac{1}{N_0} \leq |\Delta^{2D+1}(\epsilon_i)(v)| < 2^{2D+1}\delta \leq 2^{2D} \cdot \frac{2^{-2D}}{N_0} = \frac{1}{N_0}.$$

Now assume that (a_i) is a constant sequence, say $a := a_i$, so that

$$Q(j) \in -\frac{a-1}{N_0} \pm \delta$$

for all $j \in [k]$. This translates to

$$\|\overline{P}(j)\|_2^2 \in \mu - (z - \frac{a-1}{N_0})\sigma \pm \delta\sigma.$$

Clearly a degree $2D'$ polynomial, such as $\|\overline{P}(j)\|_2^2$, cannot have the same value at $2D'+1$ different arguments; we pull now another lemma from [9]*Lemma 1 that quantifies this.

Lemma 3. *Let δ, r be real numbers with $0 \le \delta \le r$, and let k, D be integers with $D \ge 1, k \ge 2D+1$. If $\overline{P}(j)$ is a polynomial with degree D, and $r-\delta \le \|\overline{P}(j)\|_2^2 \le r+\delta$ for $j \in [k]$, then the lead coefficient of $\overline{P}$ has norm at most $2^D\ (2D)!^{-1/2}\ \sqrt{\delta}$.*

Using Lemma 3, the lead coefficient $\overline{P}_{D'}$ of $\overline{P}(j)$ satisfies

$$\|D'!\overline{P}_{D'}\|_2 \le D'!\, 2^{D'}(2D')!^{-1/2}\sqrt{\delta\sigma} \le \sqrt{F\sigma\delta},$$

where F is an explicit constant. We have deduced that $E(D', a, b) = 1$ only if

$$a\,\overline{\omega} + \overline{\alpha} \bmod 1 \in \text{ANNULI} \quad \text{and} \quad \|b\,\overline{\omega} \bmod 1\|_2 \le \sqrt{F\sigma\delta}.$$

Since $\overline{\alpha}$ is chosen uniformly from BOX_0, we notice that

$$\mathbb{P}_{\overline{\alpha}}\,[a\,\overline{\omega} + \overline{\alpha} \bmod 1 \in \text{ANNULI}] = \mathbf{vol}\,\text{ANNULI},$$

independent of $\overline{\omega}$. Also, we notice that the event $\{\|b\,\overline{\omega} \bmod 1\|_2 \le \sqrt{F\sigma\delta}\}$ is independent of $\overline{\alpha}$, and that since b is an integer, $\overline{\omega} \bmod \overline{1}$ and $b\,\overline{\omega} \bmod \overline{1}$ are identically distributed. Therefore, the event $\{\|b\,\overline{\omega} \bmod 1\|_2 \le \sqrt{F\sigma\delta}\}$ has probability at most

$$\mathbf{vol}\,\text{BALL}(\sqrt{F\sigma\delta}) = \frac{2\pi^{d/2}(\sqrt{F\sigma\delta})^d}{\Gamma(d/2)d},$$

where $\text{BALL}(x)$ is the d-dimensional ball in $\mathbb{R}^d$ with radius x. It follows that

$$\mathbb{P}_{\overline{\omega},\overline{\alpha}}\,[E(D', a, b) = 1] \le \mathbf{vol}\,\text{ANNULI} \cdot \mathbf{vol}\,\text{BALL}(\sqrt{F\sigma\delta}),$$

and so

$$\mathbb{E}_{\overline{\omega},\overline{\alpha}}\,[|T|] \le \text{TYPE}_{k,D}(\mathcal{N})\,\mathbf{vol}\,\text{ANNULI} \cdot \mathbf{vol}\,\text{BALL}(\sqrt{F\sigma\delta}). \tag{5}$$

Equations (3), (4), and (5) now give us

$$\frac{r_{k,D}(N)}{N} \ge \mathbf{vol}(\text{ANNULI})\left(1 - \frac{\text{TYPE}_{k,D}(\mathcal{N})}{N}\,\mathbf{vol}\,\text{BALL}(\sqrt{F\sigma\delta})\right).$$

Setting

$$\delta = \frac{2ed}{\pi F\sigma}\left(\frac{d}{d+2}\right)^{2/d}\frac{\Gamma(d/2)^{2/d}}{2ed}\left(\frac{\text{TYPE}_{k,D}(\mathcal{N})}{N}\right)^{-2/d} \sim C\frac{d^{1/2}}{\Psi(N)^{2/d}}$$

we observe that

$$1-\frac{\textsc{Type}_{k,D}(\mathcal{N})}{N}\,\mathbf{vol}\,\textsc{Ball}(\sqrt{F\sigma\delta})=\frac{d}{d+2}\sim 1.$$

Now,

$$\begin{aligned}\frac{r_{k,D}(\mathcal{N})}{N}&\geq \mathbf{vol}\,\textsc{Annuli}\,\frac{d}{d+2}\\&\gg 2^{-dD}\,\delta|A_0|\\&\gg 2^{-dD}d^{1/2}\Psi(N)^{-2/d}|A_0|\\&=C\exp\left(-dD-\frac{2}{d}\log\Psi(N)+\frac{1}{2}\log d+\log|A_0|\right).\end{aligned}$$

Recall that we set

$$d:=\left\lfloor 2^{n/2}\left(\frac{\log\Psi(N)}{D}\right)^{1/(n+1)}\right\rfloor.$$

If $2D<k\leq 4D$, we take $N_0=1$ and $A_0=\{1\}$ to complete the proof. If $k>4D$, we set

$$N_0:=C\frac{\Psi(N)^{2/d}}{d^{1/2}},$$

and use the bound

$$|A_0|=r_{k,2D}(N_0)\geq CN_0\exp\left(-n2^{(n-1)/2}(2D)^{(n-1)/n}(\log N_0)^{1/n}+\frac{1}{2n}\log\log N_0\right),$$

proved in [9], to complete the proof.

4. Unanswered Questions

Kolountzakis (personal communication) asks whether

$$r_{3,1}([N])=\min\{r_{3,1}(\mathcal{N})\colon \mathcal{N}\subseteq\mathbb{Z},|\mathcal{N}|=N\}.$$

More generally, which set $\mathcal{N}$ (for fixed k,D,N) minimizes $r_{k,D}(\mathcal{N})$? It is not even clear to this author which set maximizes $\textsc{Type}_{k,D}(\mathcal{N})$, nor even what that maximum is, although the interval $[N]$ is the natural suspect and has $\textsc{Type}_{k,D}([N])\leq 2^{D+1}N^2$.

We doubt that there is a subset of the squares with positive relative density that does not contain any 3-term arithmetic progressions, but have not been able to prove such. We note that there are 4-term 2-progressions of positive cubes: $3^3,16^3,22^3,27^3$ is the image of $0,1,2,3$ under $Q(x)=\frac{2483}{2}x^2+\frac{5655}{2}x+27$. For which k,D,p are there k-term D-progressions of perfect p-th powers, and when they exist how many types are there?

References

[1] McRae, Graeme, What's the longest arithmetic progression of perfect squares?, www.2000clicks.com/MathHelp/PuzzleSequenceOfSquares.aspx, 2010.

[2] Abbott, H. L., Sidon sets, Canad. Math. Bull **33** (1990), 335-341.

[3] Brown, Tom C., Freedman, Allen R., and Shiue, Peter Jau-Shyong, Progressions of squares, Australas. J. Combin. **27** (2003), 187-192.

[4] Conrad, Keith, Arithmetic progressions of three squares, www.math.uconn.edu/ kconrad/blurbs/ugradnumthy/3squarearithprog.pdf, 2008.

[5] Conrad, Keith, Arithmetic progressions of four squares, www.math.uconn.edu/ kconrad/blurbs/ugradnumthy/4squarearithprog.pdf, 2007.

[6] Fogarty, Kenneth and O'Sullivan, Cormac, Arithmetic progressions with three parts in prescribed ratio and a challenge of Fermat, Math. Mag. **77** (2004), 283-292.

[7] Khan, M. A. and Kwong, Harris, Arithmetic progressions with square entries, Fibonacci Quart. **43** (2005), 98-103.

[8] Komlós, J., Sulyok, M. and Szemeredi, E., Linear problems in combinatorial number theory, Acta Math. Acad. Sci. Hungar. **26** (1975), 113-121.

[9] O'Bryant, Kevin, Sets of integers that do not contain long arithmetic progressions, preprint, arXiv:0811.3057.

[10] Van der Poorten, Alf, Fermat's four squares theorem, www.maths.mq.edu.au/ alf/SomeRecentPapers/183.pdf, 2007.

COFINITE INDUCED SUBGRAPHS OF IMPARTIAL COMBINATORIAL GAMES: AN ANALYSIS OF CIS-NIM

Scott M. Garrabrant[1]
Pitzer College, Claremont, California
coscott@math.ucla.edu

Eric J. Friedman
International Computer Science Institute, Berkeley, California

Adam Scott Landsberg
W.M. Keck Science Department, Claremont McKenna, Pitzer, and Scripps Colleges, Claremont, California

Received: 5/20/12, Revised: 1/11/13, Accepted: 3/20/13, Published: 3/29/13

Abstract

Given an impartial combinatorial game G, we create a class of related games (CIS-G) by specifying a finite set of positions in G and forbidding players from moving to those positions (leaving all other game rules unchanged). Such modifications amount to taking *cofinite induced subgraphs* (CIS) of the original game graph. Some recent numerical/heuristic work has suggested that the underlying structure and behavior of such "CIS-games" can shed new light on, and bears interesting relationships with, the original games from which they are derived. In this paper we present an analytical treatment of the cofinite induced subgraphs associated with the game of (three-heap) Nim. This constitutes one of the simplest nontrivial cases of a CIS game. Our main finding is that although the structure of the winning strategies in games of CIS-Nim can differ greatly from that of Nim, CIS-Nim games inherit a type of period-two scale invariance from the original game of Nim.

1. Introduction

Questions surrounding the underlying structure of the N- and P-positions in impartial combinatorial games (and associated issues of complexity and optimal strategies) continue to pose substantive challenges to researchers in the field. For some impartial games, the N- and P-positions form readily characterizable patterns (such as in Nim, as shown by Bouton's analysis [4]), while for others the structure is much

[1]Current address: Department of Mathematics, UCLA, Los Angeles, CA 90095, USA

more complex and appears to be resilient against standard analytical treatments (such as the game of Chomp). Indeed, a great deal of work has been devoted to understanding and characterizing the N- and P- positions in a variety of different games (see, e.g., [7, 2, 3, 12, 13]). However, rather than considering an individual impartial game in isolation, recent work [10, 11, 8] suggests that new and sometimes surprising insights can be had by considering a given game *within the context of a family of 'closely related' games.* In particular, for a given impartial game, the idea is to construct a set of similar games whose game graphs are all 'close' to that of the original game (in some suitably defined metric). One then examines how the underlying structure of the N- and P- positions in this associated family of games compares to that of the original. Indeed, this is the premise behind the earlier notion of "generic games" first introduced in [11]: Given an impartial combinatorial game G, one can create slightly perturbed versions of the original game by selecting a finite number of P-positions in G and declaring them to be automatic N positions. The class of games formed by arbitrary perturbations of this type has been dubbed the "generic" form of the game G. The generic forms of Chomp, Nim, and Wythoff's games have been previously investigated using a combination of numerical methods and (nonrigorous) renormalization techniques from physics [10, 11]. It has been observed that in some cases (e.g., three-row Chomp) the original game and its associated family of generic games all share a similar underlying structure, which in turn has yielded a novel geometric characterization of Chomp's N- and P-positions. In other cases (e.g., three-heap Nim) it has been found that the family of generic games appears to have a rather different underlying structure from the original game. (See also [9] for an alternative discussion of perturbed games.)

The present work on cofinite induced subgraph (CIS) games is a formalization and extension of some of this earlier work on generic games. Using three-heap Nim as a case study, it provides a new approach which not only yields novel results but for the first time allows *rigorous* mathematical statements to be made about the structure of the N and P positions in this family of Nim-like games. In particular, Figure 1b illustrates the structure of the P-positions in ordinary (three-heap) Nim, while Figure 1a shows an example of a CIS-Nim game (these figures will be discussed more fully later). Despite the striking structural differences between the two, we prove that the overall structure of P-positions in CIS-Nim exhibits the same 'period-two scale invariance' (to be defined more precisely later) as Nim. This work constitutes the first formal proofs regarding the properties of CIS-Nim and its relationship to Nim – relationships that were conjectured to exist based on nonrigorous techniques from physics but never formally proven. Moreover, the proofs themselves, although geared for CIS-Nim, provide more general insights into other impartial games and suggest a means of determining which structural properties of a game's P-positions are unstable and dependent on its specific end-game positions, and which properties are stable and independent of the details of the end game.

2. Background

2.1. Game Graphs

Impartial combinatorial games are often represented as directed graphs called "game graphs" wherein the vertices of the game graph represent the possible positions of the game and there is a directed edge from vertex u to vertex v if and only if there is a legal move from u to v. In this case, we will call u a *parent* of v and v a *child* of u. Starting from any vertex in the game graph, the two players will alternate in moving along any directed edge from the current vertex to another vertex. We will only consider games under normal play, meaning that if the current position has zero out-degree, the player whose turn it is to move has no legal options and is declared the loser. (This is in contrast with misère play, in which a player with no legal move is declared the winner.) All game graphs of impartial combinatorial games are acyclic and have the property that from any given position there are only finitely many positions which are reachable using any sequence of moves. However, since we will think of these games generally and not limit ourselves to a single starting position, the game graphs we consider will not necessarily be finite. In fact, all of the game graphs discussed in this paper will have infinitely many vertices.

2.2. P- and N- Positions

It follows from Zermelo's theorem [15] that from any position either the next player to move can guarantee herself a win under optimal play, or the previous player can guarantee himself a win. Any position in which the *next* player to move can force a win is known as an "N-position," while if the *previous* player can force the win the position is called a "P-position." This partition of positions into P-positions and N-positions has the property that no P-position has a P-position child and every N-position has at least one P-position child. Further, this is the only partition which satisfies this property. Most importantly, knowledge of the N- and P-positions of a game defines an optimal strategy for the game: A player at an N-position need only move his/her opponent to a P-position whenever possible to guarantee a win. This means that given a game, a primary goal is to determine the unique partition of positions into P-positions and N-positions.

More generally, Sprague-Grundy theory asserts that it is possible to assign a non-negative integer value to each position (i.e., a 0 to P-positions and positive-integer values to N-positions) such that simultaneous play of multiple games can be reduced to an equivalent game of Nim [12, 13].

3. Cofinite Induced Subgraph Games

Once the the positions of a game graph are partitioned into P-positions and N-positions, one interesting question is to ask how stable this partition is to minor perturbations made in the game graph. One simple way of making perturbations in a game graph is to remove some finite number of vertices, resulting in an cofinite induced subgraph of the original game graph.

Definition 3.1. Let G be a game graph, and let F be a finite set of vertices in G, called the set of "forbidden positions". Let $G - F$ denote the game whose game graph is the induced subgraph formed by removing from G the vertices in F and all edges incident to vertices in F. For a given game G, "Cofinite Induced Subgraph G" or "CIS-G" will refer to the general class of games of the form $G - F$ for any F.

Loosely speaking, the game $G - F$ is effectively G, except that players are forbidden from moving to any position in F.

4. Nim and CIS-Nim

4.1. Nim

Nim [4] is a game played with multiple heaps of beans. Two players alternate taking any positive number of beans from any one heap. When all heaps are empty, the player whose turn it is to play has no move and is therefore declared the loser. We will restrict our attention to games of Nim with three heaps and we will consider the three heaps of beans unlabeled so that the positions in this game can be thought of as three-element multisets of non-negative integers, where the three numbers represent the number of beans in the three heaps. The children of a position $\{x, y, z\}$, are all positions of the form $\{x', y, z\}$ with $x' < x$, $\{x, y', z\}$ with $y' < y$, or $\{x, y, z'\}$ with $z' < z$. From now on, unless otherwise stated, the word "Nim" will refer to three-heap Nim with unlabeled heaps.

4.2. CIS-Nim

Our goal is to analyze the class of Cofinite Induced Subgraph Nim, or CIS-Nim. Figures 1.a and 1.b show the structure of the P-positions in the standard game of Nim and in another instance of Cofinite Induced Subgraph Nim, or CIS-Nim. These structures are remarkably different. The data suggests that the structure of any instance of CIS-Nim looks like one of these two. Most games look similar to Figure 1.a. However, in special cases where none of the forbidden positions are P-positions of Nim or the forbidden positions are set up to correct any errors they introduce, the structure will look similar to 1.b. This is because games like Nim are unstable special cases in the generic class of CIS-Nim Games [11, 8].

(a) The P-positions of Nim-$\{\{1,1,0\}\}$ of the form $\{x,y,z\}$ with $x,y < 5000$.

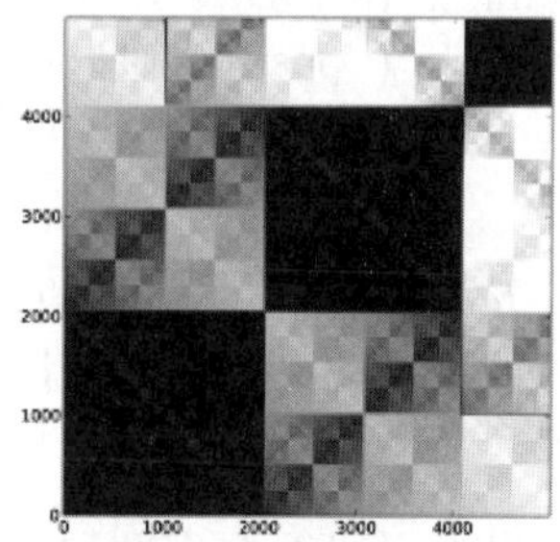

(b) The P-positions of Nim of the form $\{x,y,z\}$ with $x,y < 5000$.

(c) The P-positions of Nim-$\{\{1,1,0\}\}$ of the form $\{x,y,z\}$ with $x,y < 2500$.

Figure 1: The structure of P-positions in Nim and Nim-$\{\{1,1,0\}\}$. Point (x,y) is given with a color representing the unique z such that $\{x,y,z\}$ is a P-position. Lighter shades of gray represent larger z values, and are normalized based on the largest value of z in each figure.

There is one significant property of Nim which also holds for all instances of CIS-Nim: The structure of P-positions is invariant up to scaling by a factor of two. Figures 1.a and 1.c demonstrate this period-two scale invariance.

A somewhat weaker but more formal way to state this period-two scale invariance is that given any instance, Nim $- F$ of CIS-Nim, if we let $\pi(n)$ denote the number of P-positions of the form $\{x, y, z\}$, with x, y, and z all less than n, then for any positive integer n, $\lim_{k\to\infty} \frac{\pi(n2^k)}{(n2^k)^2}$ converges to a nonzero constant. In the case of Nim, this can be shown directly using Bouton's well-known analytical solution to Nim [4]. The primary result of this paper is a proof that this period-two scale invariance holds for any game of CIS-Nim.

5. Basic Properties of CIS-Nim

Before we can prove the period-two scale invariance, we will have to establish some basic properties of CIS-Nim games. In Figure 1, the point (x, y) is given a color representing z, where z is the unique P-position of the form $\{x, y, z\}$. These figures are only well-defined because such a unique P-position of the form $\{x, y, z\}$ is known to always exist.

Theorem 5.1. *Given any instance Nim $-$ F of CIS-Nim, for any nonnegative integers x and y there is a unique z such that $\{x, y, z\}$ is a P-position in Nim $- F$. This value of z satisfies the inequality $z \leq x + y + |F|$.*

Proof. To show uniqueness, assume by way of contradiction that there existed two P-positions $\{x, y, z_1\}$ and $\{x, y, z_2\}$. Without loss of generality, assume that $z_1 < z_2$. This means that $\{x, y, z_1\}$ is a P-position child of P-position $\{x, y, z_2\}$, contradicting the fact that no P-position has a P-position child.

Next, assume by way of contradiction that every position of the form $\{x, y, z\}$ with $z \leq x+y+|F|$ is an N-position. There are $x+y+|F|+1$ values of z satisfying this condition, and for all but at most $|F|$ of them, $\{x, y, z\}$ is valid position in $Nim - F$. There are at least $x + y + 1$ N-positions of this form and each of these positions, $\{x, y, z\}$, therefore has a P-position child. This child cannot be of the form $\{x, y, z'\}$ with $z' < z$, so it must be of the form $\{x', y, z\}$ with $x' < x$ or $\{x, y', z\}$ with $y' < y$. There are x different pairs of the form (x', y) with $x' < x$ and we know that for each of these pairs, there is at most one value of z such that $\{x', y, z\}$ is a P-position. Similarly, there are y different pairs of the form (x, y') with $y' < y$ and we know that for each of these pairs, there is at most one value of z such that $\{x, y', z\}$ is a P-position. There are $x + y + 1$ N-positions of the form $\{x, y, z\}$. At most x of these positions can have a P-position child of the form $\{x', y, z\}$, and at most y of them can have a P-position child of the form $\{x, y', z\}$.

Therefore, at least one of them has no P-position child, contradicting the fact that every N-position has a P-position child. Therefore, there is at least one P-position of the form $\{x, y, z\}$ with $z \leq x + y + |F|$. This means that there is exactly one P-position of the form $\{x, y, z\}$, and it satisfies the inequality $z \leq x + y + |F|$. □

The bound of $z \leq x + y + |F|$ given in Theorem 5.1 is only necessary for small valued P-positions. For all but finitely many P-positions near $\{0, 0, 0\}$, we can improve this bound to $z \leq x + y$.

Definition 5.2. Given an instance Nim $- F$ of CIS-Nim, let F_{max} equal the largest element (largest number of beans in a single heap) of any position in F.

Theorem 5.3. *Given any instance Nim $-F$ of CIS-Nim, if $\{x, y, z\}$ is a P-position with $z > 2F_{max} + |F|$, then $z \leq x + y$*

Proof. Since $\{x, y, z\}$ is a P-position, we know from Theorem 5.1, that $z \leq x + y + |F|$. Therefore, $2F_{max} < z - |F| \leq x + y$, so either $F_{max} < x$ or $F_{max} < y$. Therefore, each of the z positions of the form $\{x, y, z'\}$ with $z' < z$ has an element greater than F_{max}, and is therefore not in F. Since all of these positions are also children of the P-position, $\{x, y, z\}$, we know that each of these positions are actually N-positions. Therefore, each of these z positions has a distinct P-position child of the form $\{x', y, z'\}$ or $\{x, y', z'\}$ with $x' < x$, $y' < y$, and $z' < z$. Similarly to in Theorem 5.1, there can be at most $x + y$ such P-positions, so $z \leq x + y$. □

Corollary 5.4. *Given any instance Nim $- F$ of CIS-Nim, for all $n > 2F_{max} + |F|$, $\{n, n, 0\}$ is a P-position.*

Proof. If $\{n, n, 0\}$ were not a P-position, it would have a P-position child of the form $\{n, n', 0\}$ with $n' < n$. In this case, $n > 2F_{max} + |F|$, but $n > n' + 0$, contradicting Theorem 5.3. □

In the game of Nim, $\{n, n, 0\}$ is a P-position for all n. This means that if we only consider positions of the form $\{x, y, 0\}$, the structures of P-positions in Nim and in generic games of CIS-Nim agree on all but finitely many small valued positions. Positions of the form $\{x, y, 0\}$, are effectively positions in two-heap Nim games, so Corollary 5.4 tells that two-heap Nim is stable in that large valued P-positions are unaffected by removal of small valued positions.

It turns out that if we fix the size of any one heap, the structure of P-positions is eventually additively periodic. This is an generalization of Corollary 5.4 which shows that if we fix one heap to be of size 0, the structure of P-positions is additively periodic with period 1.

Claim 5.5. *For any x, there exists a p and a q such that for any $y > q$, $\{x, y, z\}$ is a P-position if and only if $\{x, y + p, z + p\}$ is a P-position.*

We will not prove this claim, as it is technical and unnecessary for our main result. However, the proof is almost identical to an argument given L. Abrams and D. S. Cowen-Morton for a game with similar structure [1].

6. Period-Two Scale Invariance in CIS-Nim

The primary result of this paper is the proof of the following theorem, which is a formulation of the observation that the overall structure of P-positions in any game of CIS-Nim is invariant under scaling by a factor of two.

Theorem 6.1 (Period-Two Scale Invariance). *Given any instance $Nim - F$ of CIS-Nim, let $\pi(n)$ denote the number of P-positions (assuming unlabeled heaps) in $Nim - F$ of the form $\{x, y, z\}$, with x, y, and z all less than n. For any positive integer n, $\lim_{k\to\infty} \frac{\pi(n2^k)}{(n2^k)^2}$ converges to a nonzero constant.*

We note that an analog of this result holds for the much simpler case of ordinary Nim (see Figure 2). In the case of ordinary Nim, it is possible to give an explicit formula for $\pi(x)$ as $\frac{3x^2-6xy+4y^2+3x+2}{6}$, where y is the greatest power of 2 less than or equal to x. This formula comes from intersecting the region $x \le y \le z < n$ with the set of P-positions found in Bouton's original analysis [4]. (Specifically, note that if we label the piles, the number of P-positions with all three heaps of size less than x is $y^2 + 3(x - y)^2$. The full formula comes from modifying this to account for the fact that the heaps are unlabeled.) Notice that this implies that for ordinary Nim, $\lim_{k\to\infty} \frac{\pi(2^k)}{(2^k)^2} = \frac{1}{6}$, while $\lim_{k\to\infty} \frac{\pi(3\cdot 2^k)}{(3\cdot 2^k)^2} = \frac{7}{54}$. These limiting values are consistent with the main result, Theorem 6.1.

To prove the main theorem, we will first need to prove several Lemmas.

6.1. The Set S

To start, we will need to think about this problem in terms of a new set S, which encodes much of the information about the structure of the P-positions as a set of ordered pairs.

Definition 6.2. Given any instance $\mathrm{Nim} - F$ of CIS-Nim, let S be the infinite set of ordered pairs of integers such that $(x, y) \in S$ if and only if there exists a z such that $z < y < x$ and $\{x, y, z\}$ is a P-position in $\mathrm{Nim} - F$.

Definition 6.3. Given any instance $\mathrm{Nim} - F$ of CIS-Nim, for any nonnegative integers x and y, let $r(x, y)$ be the number of elements of S of the form (x', y) with $x' \ge x$. Let $b(x, y)$ be the number of elements of S of the form (x, y') with $y' \le y$.

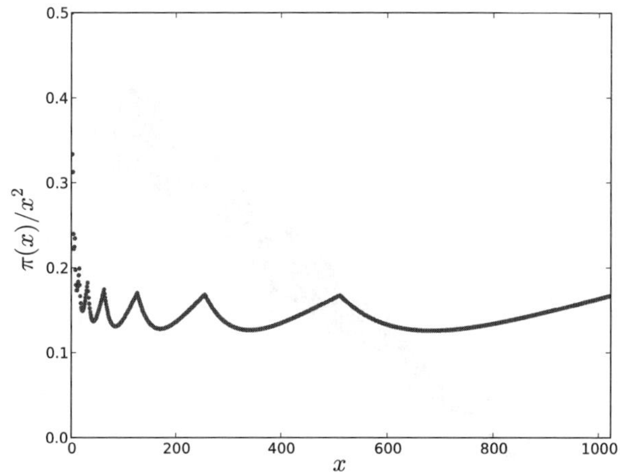

Figure 2: Plot of $\pi(x)/x^2$ vs. x for ordinary Nim, illustrating a period-two scale invariance.

It will be helpful to visualize S as a subset of the plane, as shown in Figure 3. With this visualization in mind, the definitions of $r(x, y)$ and $b(x, y)$ are very natural as the number of points directly to the right or equal to (x, y) and the number of points to below or equal to (x, y) respectively. Notice that there are points $(x, y) \notin S$ with $x > y$ for which $b(x, y)$ is positive. We will refer to such points as "holes." The following two Lemmas will prove some properties of $r(x, y)$ and $b(x, y)$. Lemma 6.4 captures the essence of why CIS-Nim follows the period-two scale invariance. In fact, if not for the existence of holes, the period-two scale invariance would follow almost directly from Lemma 6.4. However, holes do exist, which is why we will need Lemma 6.5 which places limitations on the ways which holes can show up.

Lemma 6.4. *Given any instance Nim* $-$ *F of CIS-Nim, for all* $x > 4F_{max} + 3|F|$, $r(x, x) + 2b(x, x) + 1 = x$.

Proof. For each of the x values of y satisfying $y < x$, there exists a unique z, such that $\{x, y, z\}$ is a P-position. For each value of y, this unique z will satisfy exactly one of the following: $z > x$, $z = x$, $x > z > y$, $z = y$, or $z < y$. The number of values of y which satisfy $z > x$ is exactly the number of P-positions of the form $\{z, x, y\}$ with $z > x > y$, which is $r(x, x)$. The number of values of y which satisfy $z = x$ is 1, since $\{x, x, 0\}$ is the only P-position of the form $\{x, z, y\}$ with $x = z$. The number of values of y which satisfy $x > z > y$ is exactly the number of P-positions of the form $\{x, z, y\}$ with $x > z > y$, which is $b(x, x)$. If there were a P-position of the form $\{y, z, x\}$ with $y = z$, then $\{y, y, 0\}$ would not

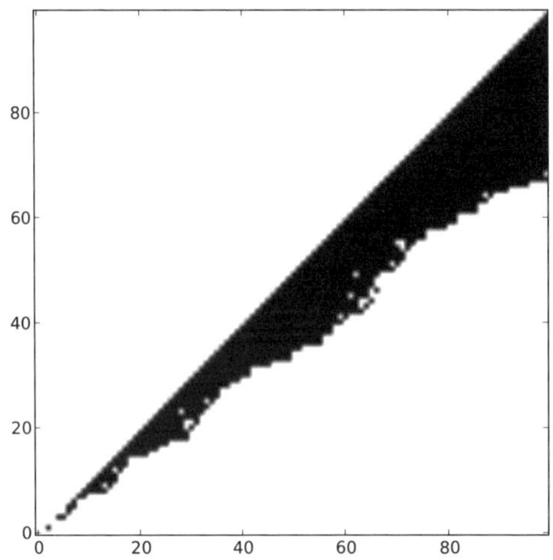

Figure 3: The set of all points $(x, y) \in S$ with $x, y < 100$ for the game Nim-$\{\{1, 1, 0\}\}$

be a P-position, which implies by Corollary 5.4 that $y \leq 2F_{max} + |F|$. By Theorem 5.1, this would imply that $x \leq (2F_{max} + |F|) + (2F_{max} + |F|) + |F| = 4F_{max} + 3|F|$, a contradiction. Therefore, the number of values of y which satisfy $z = y$ is 0. The number of values of y which satisfy $z < y$ is exactly the number of P-positions of the form $\{x, y, z\}$ with $x > y > z$, which is $b(x, x)$. Adding all these together, we get $r(x, x) + 2b(x, x) + 1$, and we know that the total number of values of y less than x is exactly x, so $r(x, x) + 2b(x, x) + 1 = x$. □

Lemma 6.5. *Given any instance Nim* $-$ *F of CIS-Nim, for all* $x > y > 4F_{max} + 3|F|$, *if* $(x, y) \notin S$ *then* $b(x, y) \geq r(x, y)$.

Proof. There is an element of S of the form (x', y) with $x' > x$ for each P-position of the form $\{x', y, z\}$ with $z < y < x < x'$. No two of these positions can have the same value for z, since then they would have two of the three elements in common, so there would be a move from one to the other. There are therefore $r(x, y)$ distinct values of z for which there is a P-position of the form $\{x', y, z\}$ with $z < y < x < x'$. For each of these values of z, $\{x, y, z\}$ cannot be a P-position, since it has a P-position parent. It therefore must have a P-position child. This child cannot be of the form $\{x'', y, z\}$ with $x'' < x$ because then it would also be a child of $\{x', y, z\}$. This child cannot be of the form $\{x, y, z'\}$ since then it would satisfy $x > y > z'$, contradicting the fact that $(x, y) \notin S$. Therefore, for each of the $r(x, y)$ values of z, there is a P-position of the form $\{x, y', z\}$ with $y' < y$. If two of these P-positions, $\{x, y_1', z_1\}$ and $\{x, y_2', z_2\}$ were the same, then $y_1' = z_2$ and $y_2' = z_1$. Because z is one of our $r(x, y)$ values, we know that there exist x_1 and x_2 greater than x such that $\{x_1, y, z_1\}$ and $\{x_2, y, z_2\}$ are both P-positions. This would mean that $\{y, z_1, z_2\}$ would have three P-position parents, $\{x_1, y, z_1\}$, $\{x_2, y, z_2\}$, and $\{x, z_1, z_2\}$. In order to form a P-position by changing any of the three elements of $\{y, z_1, z_2\}$, we would have to increase that element. Therefore, we cannot decrease one element of $\{y, z_1, z_2\}$ to form a P-position, so $\{y, z_1, z_2\}$ has no P-position children, making it a P-position, which contradicts the fact that $\{x, z_1, z_2\}$ is also a P-position. Therefore, each of the $r(x, y)$ P-positions of the form $\{x, y', z\}$ with $y' < y$ are unique, and each one has one coordinate equal to x, and the other two less than y. For each of these P-positions, y' and z must be distinct, since otherwise, $4F_{max} + 3|F| < x \leq y' + z = 2z$, so $z < 2F_{max} + |F|$, implying that $\{0, z, z\}$ is a P-position child of $\{x, y', z\}$. Therefore, either (x, y') or (x, z) is in S, so each of the positions will contribute to S an ordered pair of the form (x, y'') with $y'' < y$, which will contribute 1 to $b(x, y)$. Therefore, $b(x, y) \geq r(x, y)$. □

6.2. The Sets $U_{x,y}$, $\bar{U}_{x,y}$, and S_n

In this section, we define a sequence S_n of sets. These sets, and the intermediate sets, $U_{x,y}$ and $\bar{U}_{x,y}$ which are used to define S_n encode information about S. In particular, the S_n should be thought of as increasingly accurate approximations of S which are defined to be free of holes. The next three lemmas are building up to proving that for and $n < m$, it is possible to get from S_n to S_m by changing points in a very limited way. This information about how to construct S_m from S_n will be useful for the next section, where we will prove several important properties of the sequence S_n.

The first sets we will need to define on our way to S_n are the $U_{x,y}$. The set $U_{x,y}$ is very similar to the set of all points (x', y') in S with $y' < x$. In fact, these two sets have the same size. However, some of the points are moved so that $U_{x,y}$ is free of holes.

Definition 6.6. Given any instance Nim $- F$ of CIS-Nim, and given any $x > 4F_{max} + 3|F|$ and $y \leq x$:

Let $A_{x,y}$ be the set of all ordered pairs (x', y'), such that $0 \leq y' < x' \leq x$ and $b(x', x') \geq x' - y'$.

Let $B_{x,y}$ be the set of all ordered pairs (x, y'), such that $0 \leq y' < y$ and $b(x, y-1) \geq y - y'$.

Let $C_{x,y}$ be the set of all ordered pairs (x', y'), such that $y \leq y' < x \leq x'$ and $r(x, y') > x' - x$.

Let $D_{x,y}$ be the set of all ordered pairs (x', y'), such that $0 \leq y' < y \leq x < x'$ and $r(x+1, y') > x' - (x+1)$.

Let $U_{x,y} = A_{x,y} \cup B_{x,y} \cup C_{x,y} \cup D_{x,y}$.

The next lemma will describe the map necessary to get from $U_{x,y}$ to $U_{x,y+1}$. This will be extended in the following two lemmas to describe the map necessary to get from S_n to S_m.

Lemma 6.7. *Given any instance Nim* $-$ *F of CIS-Nim, and any* $x > 4F_{max} + 3|F|$ *and* $0 \leq y < x$, *there exists a bijection from* $U_{x,y}$ *to* $U_{x,y+1}$ *which either fixes all elements or fixes all but one element and sends* $(x, y - b(x,y))$ *to* $(x + r(x,y), y)$. *This bijection will be the identity if and only if* $(x,y) \in S$ *or* $b(x,y) = 0$.

Proof. We will partition the set of all points (x, y) with $y' < x'$ into 11 regions. We will show that $U_{x,y}$ and $U_{x,y+1}$ agree for most of these regions. We will see that they do not always agree for regions 3 and 9, but we will show that the way the points in which regions 3 and 9 may differ will exactly follow the statement of the lemma.

Region 1 ($x' < x$): $(x', y') \in U_{x,y}$ if and only if $(x', y') \in A_{x,y}$ if and only if $b(x', x') \geq x' - y'$ if and only if $(x', y') \in A_{x,y+1}$ if and only if $(x', y') \in U_{x,y+1}$.

Region 2 ($x' = x$ and $y' < y - b(x,y)$): Since $b(x, y-1) \leq b(x,y) < y - y'$, we know that $(x', y') \notin B_{x,y}$ so $(x', y') \notin U_{x,y}$. Similarly, since $b(x, (y+1)-1) \leq y - y' < (y+1) - y'$, we know that $(x', y') \notin B_{x,y+1}$ so $(x', y') \notin U_{x,y+1}$.

Region 3 ($x' = x$ and $y' = y - b(x,y) < y$): This case must further be divided into two cases:

Case 1 ($(x,y) \in S$): This means that $b(x,y) = b(x, y-1) + 1$. Therefore, $(x', y') \in U_{x,y}$ if and only if $(x', y') \in B_{x,y}$ if and only $b(x, y-1) \geq y - y'$ if and only if $b(x, (y+1)-1) \geq (y+1) - y'$ if and only if $(x', y') \in B_{x,y+1}$ if and only if $(x', y') \in U_{x,y+1}$.

Case 2 ($(x,y) \notin S$): This means that $b(x,y) = b(x, y-1)$. Therefore, $b(x, y-1) = b(x,y) = y - y'$, so $(x', y') \in B_{x,y}$, so $(x', y') \in U_{x,y}$. On the other hand, $b(x,y) = y - y' < y + 1 - y'$, so $(x', y') \notin B_{x,y+1}$ and $(x', y') \notin U_{x,y+1}$.

Region 4 ($x' = x$ and $y - b(x,y) < y' < y$): Since $b(x,y) > y - y'$, we know that $b(x, y-1) \geq y - y'$, which implies that $(x', y') \in B_{x,y}$ so $(x', y') \in U_{x,y}$. Similarly, since $b(x, (y+1)-1) > y - y'$ we know that $b(x, (y+1)-1) \geq (y+1) - y'$ which implies that $(x', y') \in B_{x,y+1}$ so $(x', y') \in U_{x,y+1}$.

Region 5 ($x' = x < x + r(x,y)$ and $y' = y$): Since $r(x, y') > x' - x$, we know that $(x', y') \in C_{x,y}$, so $(x', y') \in U_{x,y}$. If $(x,y) \in S$, then $b(x, (y+1)-1) \geq 1 = (y+1) - y'$. Otherwise, $(x,y) \notin S$, so by Lemma 6.5, $b(x, (y+1)-1) \geq r(x,y) \geq 1 = (y+1) - y'$. Either way, $(x', y') \in B_{x,y+1}$, so $(x', y') \in U_{x,y+1}$.

Region 6 ($x' \geq x$ and $y < y'$): $(x', y') \in U_{x,y}$ if and only if $(x', y') \in C_{x,y}$ if and only if $r(x, y') > x' - x$ if and only if $(x', y') \in C_{x,y+1}$ if and only if $(x', y') \in U_{x,y+1}$.

Region 7 ($x' > x$ and $y' < y$): $(x', y') \in U_{x,y}$ if and only if $(x', y') \in D_{x,y}$ if and only if $r(x+1, y') > x' - (x+1)$ if and only if $(x', y') \in D_{x,y+1}$ if and only if $(x', y') \in U_{x,y+1}$.

Region 8 ($x < x' < x + r(x, y)$ and $y' = y$): Since $r(x, y') > x' - x$, we know that $(x', y') \in C_{x,y}$, so $(x', y') \in U_{x,y}$. Since $r(x+1, y') \geq r(x+1, y') - 1 > x' - (x+1)$, we know that $(x', y') \in D_{x,y+1}$, so $(x', y') \in U_{x,y+1}$.

Region 9 ($x' = x + r(x, y)$ and $y' = y$): This region must be further divided into three cases:

Case 1 ($(x, y) \in S$): Since $r(x, y') = r(x, y) = x' - x$, we know that $(x', y') \notin C_{x,y}$, so $(x', y') \notin U_{x,y}$. Since $(x, y) \in S$, $r(x, y) > 0$, so we know that $x' > x$. Since $(x, y) \in S$, $r(x+1, y') = r(x, y) - 1 = x' - (x+1)$, so $(x', y') \notin D_{x,y+1}$, so $(x', y') \notin U_{x,y+1}$.

Case 2 ($b(x, y) = 0$): Since $r(x, y') = r(x, y) = x' - x$, we know that $(x', y') \notin C_{x,y}$, so $(x', y') \notin U_{x,y}$. By Lemma 6.5, since $b(x, y) = 0$, we know that $r(x, y) = 0$, so $x' = x$. Also, $b(x, (y+1) - 1) = 0 < 1 = (y+1) - y'$, which implies that $(x', y') \notin B_{x,y}$, so $(x', y') \notin U_{x,y}$.

Case 3 ($(x, y) \notin S$ and $b(x, y) > 0$): Since $r(x, y') = r(x, y) = x' - x$, we know that $(x', y') \notin C_{x,y}$, so $(x', y') \notin U_{x,y}$. If $r(x, y) = 0$, then $x' = x$, which since $b(x, (y+1) - 1) \geq 1 = (y+1) - y'$, which implies that $(x', y') \in B_{x,y+1}$. Otherwise, $r(x, y) > 0$, and $x' > x$. Since $(x, y) \notin S$, $r(x+1, y') = r(x, y') > x + r(x, y') - (x+1) > x' - (x+1)$, which implies that $(x', y') \in D_{x,y+1}$. Either way, $(x', y') \in U_{x,y+1}$.

Region 10 ($x + r(x, y) < x'$ and $y' = y$): Since $r(x, y') \leq x' - x$, we know that $(x', y') \notin C_{x,y}$, so $(x', y') \notin U_{x,y}$. Since $r(x+1, y') \leq r(x, y') \leq x' - (x+1)$, we know that $(x', y') \notin D_{x,y+1}$, so $(x', y') \notin U_{x,y+1}$.

Region 11 ($y' \geq x$): All of $A_{x,y}, B_{x,y}, C_{x,y}, D_{x,y}, A_{x,y+1}, B_{x,y+1}, C_{x,y+1}$, and $D_{x,y+1}$ are defined to not allow any points in this region, so no points in this region are in either $U_{x,y}$ or $U_{x,y+1}$.

Notice that all points are in $U_{x,y}$ if and only if they are in $U_{x,y+1}$, except for those contained in region 3 case 2, and region 9 case 3. Further, if $b(x, y) > 0$ and $b(x, y) \notin S$, then both of these cases contain exactly one point, and otherwise they contain no points. Therefore, if $b(x, y) = 0$ or $(x, y) \in S$, then $U_{x,y} = U_{x,y+1}$, and the identity map is a bijection from $U_{x,y}$ to $U_{x,y+1}$. Otherwise, $U_{x,y}$ and $U_{x,y+1}$ are identical, except for the fact that $U_{x,y}$ contains $(x, y - b(x, y))$ while $U_{x,y+1}$

contains $(x + r(x, y), y)$. In this case the map which fixes all but one element and sends $(x, y - b(x, y))$ to $(x + r(x, y), y)$ is a bijection from $U_{x,y}$ to $U_{x,y}$.

Therefore, for any $x > 4F_{max} + 3|F|$ and $0 \leq y < x$, there exists a bijection from $U_{x,y}$ to $U_{x,y+1}$ which either fixes all elements or fixes all but one element and sends $(x, y - b(x, y))$ to $(x + r(x, y), y)$. Further, this bijection will be the identity if and only if $(x, y) \in S$ or $b(x, y) = 0$. □

Next, we will need to introduce the sets $\bar{U}_{x,y}$ which are an extention of the sets $U_{x,y}$. In fact, $\bar{U}_{x,y}$ is defined from $\bar{U}_{x,y}$ by adding infinitely many points so that $\bar{U}_{x,y}$ satisfies relations similar to those shown to be satisfied for S in Lenna 6.4. Finally, the S_n are just the sets of the form $\bar{U}_{n,0}$.

Definition 6.8. Given any instance Nim$-F$ of CIS-Nim, and any $x > 4F_{max}+3|F|$ and $0 \leq y < x$, let $\bar{U}_{x,y}$ be the unique set of ordered pairs (x', y') with the following two properties:

Property 1: For all (x', y') with $y' < x$, $(x', y') \in \bar{U}_{x,y}$ if and only if $(x', y') \in U_{x,y}$.

Property 2: For all (x', y') with $y' \geq x$, $(x', y') \in \bar{U}_{x,y}$ if and only if $y' < x' \leq 2y'$ and $(y', \left\lfloor \frac{x'}{2} \right\rfloor) \notin \bar{U}_{x,y}$.

Let S_n be a sequence of sets of ordered pairs defined by $S_n = \bar{U}_{n,0}$.

Remark 6.9. We know $\bar{U}_{x,y}$ exists and is unique since this definition comes with a natural way to determine whether or not (x', y') is in $\bar{U}_{x,y}$ as a function of the set of all points in $\bar{U}_{x,y}$ of the form (x'', y'') with $x'' < x'$.

We will now extend the result of the previous lemma to these infinite sets, $\bar{U}_{x,y}$ and $\bar{U}_{x,y+1}$. We find that when a single point is moved in the map from $U_{x,y}$ to $U_{x,y+1}$, this will cause infinitely many points to move in the map from $\bar{U}_{x,y}$ to $\bar{U}_{x,y+1}$. However, we will show that all of the points moved in this way will move in the same general direction. In particular, if the map sends (x_1, y_1) to (x_2, y_2), then $x_1 - y_1 \geq x_2 - y_2$ and $x_1 - 2y_1 \geq x_2 - 2y_2$.

Lemma 6.10. *Given any instance Nim$-F$ of CIS-Nim, and any $x > 4F_{max}+3|F|$ and $0 \leq y < x$, there exists a bijection ϕ from $\bar{U}_{x,y}$ to $\bar{U}_{x,y+1}$ such that if $\phi(x_1, y_1) = (x_2, y_2)$, then $x_1 - y_1 \geq x_2 - y_2$ and $x_1 - 2y_1 \geq x_2 - 2y_2$.*

Proof. Let ϕ' be the map constructed in Lemma 6.7. If ϕ' fixes all elements, then $U_{x,y} = U_{x,y+1}$, so $\bar{U}_{x,y} = \bar{U}_{x,y+1}$. In this case, the identity map sends $\bar{U}_{x,y}$ to $\bar{U}_{x,y+1}$ and clearly satisfies the necessary relations.

Otherwise, ϕ' sends $(x, y - b(x, y))$ to $(x + r(x, y), y)$, and $(x, y) \notin S$. Consider the map ϕ from $\bar{U}_{x,y}$ to $\bar{U}_{x,y+1}$ defined so that

$$(2^n x + k_1, 2^n(y - b(x, y)) + k_2) \mapsto (2^n(x + r(x, y)) + k_1, 2^n y + k_2)$$

and

$$(2^{n+1}y+k_3, 2^n(x+r(x,y))+k_1) \mapsto (2^{n+1}(y-b(x,y))+k_3, 2^nx+k_1)$$

for all $n \geq 0$, $0 \leq k_1, k_2 < 2^n$, and $0 \leq k_3 < 2^{n+1}$. Let ϕ fix all other elements of $\bar{U}_{x,y}$.

First, notice that in this case, $x < 2(y-b(x,y)-1)$. This is because since $(x,y) \notin S$, there are $b(x,y)$ values of $y' < y$ with $(x,y') \in S$, so there exists at least one $(x,y') \in S$ with $y' < y-b(x,y)$. There is therefore a P-position of the form $\{x,y',k\}$ with $k < y' < x$. If it were true that $x \geq 2(y-b(x,y)-1)$, then $x \geq 2y' > y'+k$, which since $x \geq 4F_{max}+3|F|$, contradicts Theorem 5.3. Therefore, $x < 2(y-b(x,y)-1)$.

It is easy to verify that this implies that the ordered pairs, (x',y') of the form $(2^nx+k_1, 2^n(y-b(x,y))+k_2)$, $(2^n(x+r(x,y))+k_1, 2^ny+k_2)$, $(2^{n+1}y+k_3, 2^n(x+r(x,y))+k_1)$, or $(2^{n+1}(y-b(x,y))+k_3, 2^nx+k_1)$ all satisfy $x' \leq 2y'$.

We want to show that the ordered pairs in $\bar{U}_{x,y}$ but not in $\bar{U}_{x,y+1}$ are exactly those of the form $(2^nx+k_1, 2^n(y-b(x,y))+k_2)$ or $(2^{n+1}y+k_3, 2^n(x+r(x,y))+k_1)$, and the ordered pairs in $\bar{U}_{x,y+1}$ but not in $\bar{U}_{x,y}$ are exactly those of the form $(2^n(x+r(x,y))+k_1, 2^ny+k_2)$ or $(2^{n+1}(y-b(x,y))+k_3, 2^nx+k_1)$. We will show that this is true for all ordered pairs (x',y') with $y' < m$ by induction on m, and it will follow that it holds for all ordered pairs.

Base Case: ($m = x$): In this case, a point (x',y') with $y' < m$ is in $\bar{U}_{x,y}$ if and only if it is in $U_{x,y}$, and a point (x',y') with $y' < m$ is in $\bar{U}_{x,y+1}$ if and only if it is in $U_{x,y+1}$. This means that the only point which is in $\bar{U}_{x,y}$ but not $\bar{U}_{x,y+1}$ is $(x, y-b(x,y))$, which since $2(y-b(x,y)) \geq x = m$ is also the only point (x',y') of the form $(2^nx+k_1, 2^n(y-b(x,y))+k_2)$ or $(2^{n+1}y+k_3, 2^n(x+r(x,y))+k_1)$ with $y < x = m$.

Similarly, the only point which is in $\bar{U}_{x,y+1}$ but not $\bar{U}_{x,y}$ is $(x+r(x,y), y)$, which since $2y \geq 2(y-b(x,y)) \geq x = m$ is also the only point (x',y') of the form $(2^n(x+r(x,y))+k_1, 2^ny+k_2)$ or $(2^{n+1}(y-b(x,y))+k_3, 2^nx+k_1)$ with $y < x = m$.

Inductive Hypothesis: The ordered pairs of the form (x',y') with $y' < m$ in $\bar{U}_{x,y}$ but not in $\bar{U}_{x,y+1}$ are exactly those of the form $(2^nx+k_1, 2^n(y-b(x,y))+k_2)$ or $(2^{n+1}y+k_3, 2^n(x+r(x,y))+k_1)$, and the ordered pairs (x',y') with $y' < m$ in $\bar{U}_{x,y+1}$ but not in $\bar{U}_{x,y}$ are exactly those of the form $(2^n(x+r(x,y))+k_1, 2^ny+k_2)$ or $(2^{n+1}(y-b(x,y))+k_3, 2^nx+k_1)$.

Inductive Step: Consider some arbitrary ordered pair (x',y') with $y' = m$. If $x' \geq 2y'$, then (x',y') is in neither $\bar{U}_{x,y}$ nor $\bar{U}_{x,y+1}$, and (x',y') is not of the form $(2^nx+k_1, 2^n(y-b(x,y))+k_2)$, $(2^{n+1}y+k_3, 2^n(x+r(x,y))+k_1)$, $(2^n(x+r(x,y))+k_1, 2^ny+k_2)$ or $(2^{n+1}(y-b(x,y))+k_3, 2^nx+k_1)$.

Otherwise by definition, we know that (x', y') is in $\bar{U}_{x,y}$ but not in $\bar{U}_{x,y+1}$ if and only if $(y', \left\lfloor \frac{x'}{2} \right\rfloor)$ is in $\bar{U}_{x,y}$ but not in $\bar{U}_{x,y+1}$, which, since $\left\lfloor \frac{x'}{2} \right\rfloor < y'$, is true if and only if $(y', \left\lfloor \frac{x'}{2} \right\rfloor)$ is of the form $(2^n(x + r(x,y)) + k_1, 2^n y + k_2)$ or $(2^{n+1}(y - b(x,y)) + k_3, 2^n x + k_1)$. Notice that $(y', \left\lfloor \frac{x'}{2} \right\rfloor)$ is of the form $(2^n(x + r(x,y)) + k_1, 2^n y + k_2)$ if and only if (x', y') is of the form $(2^{n+1}y + k_3, 2^n(x + r(x,y)) + k_1)$, and $(y', \left\lfloor \frac{x'}{2} \right\rfloor)$ is of the form $(2^{n+1}(y - b(x,y)) + k_3, 2^n x + k_1)$ if and only if (x', y') is of the form $(2^n x + k_1, 2^n(y - b(x,y)) + k_2)$. Therefore, (x', y') is in $\bar{U}_{x,y}$ but not in $\bar{U}_{x,y+1}$ if and only if it is of the form $(2^n x + k_1, 2^n(y - b(x,y)) + k_2)$ or $(2^{n+1}y + k_3, 2^n(x + r(x,y)) + k_1)$.

A similar argument shows that (x', y') is in $\bar{U}_{x,y+1}$ but not in $\bar{U}_{x,y}$ if and only if it is of the form $(2^n(x + r(x,y)) + k_1, 2^n y + k_2)$ or $(2^{n+1}(y - b(x,y)) + k_3, 2^n x + k_1)$.

It is easy to verify that ordered pairs of the form $(2^n x + k_1, 2^n(y - b(x,y)) + k_2)$, $(2^n(x + r(x,y)) + k_1, 2^n y + k_2)$, $(2^{n+1}y + k_3, 2^n(x + r(x,y)) + k_1)$, or $(2^{n+1}(y - b(x,y)) + k_3, 2^n x + k_1)$ are all distinct, which is the last thing we need to see that ϕ sends the points in $\bar{U}_{x,y}$ but not in $\bar{U}_{x,y+1}$ bijectively to the points in $\bar{U}_{x,y+1}$ but not in $\bar{U}_{x,y}$, and fixes all other points in $\bar{U}_{x,y}$, so ϕ is a bijection from $\bar{U}_{x,y}$ to $\bar{U}_{x,y+1}$.

Further,

$$2^n x + k_1 - 2^n(y - b(x,y)) + k_2 \geq 2^n(x + r(x,y)) + 2^n y + k_2$$

and

$$2^{n+1}y + k_3 - (2^n(x + r(x,y)) + k_1) \geq 2^{n+1}(y - b(x,y)) + k_3 - (2^n x + k_1),$$

so if $\phi(x_1, y_1) = (x_2, y_2)$, then $x_1 - y_1 \geq x_2 - y_2$. Similarly,

$$2^n x + k_1 - 2(2^n(y - b(x,y)) + k_2) \geq 2^n(x + r(x,y)) + k_1 - 2(2^n y + k_2)$$

and

$$2^{n+1}y + k_3 - 2(2^n(x + r(x,y)) + k_1) \geq 2^{n+1}(y - b(x,y)) + k_3 - 2(2^n x + k_1),$$

so if $\phi(x_1, y_1) = (x_2, y_2)$, then $x_1 - 2y_1 \geq x_2 - 2y_2$.

Therefore, ϕ is a bijection from $\bar{U}_{x,y}$ to $\bar{U}_{x,y+1}$ such that if $\phi(x_1, y_1) = (x_2, y_2)$, then $x_1 - y_1 \geq x_2 - y_2$ and $x_1 - 2y_1 \geq x_2 - 2y_2$. □

Finally, we will now show that the same properties relating $\bar{U}_{x,y}$ and $\bar{U}_{x,y+1}$ will also relate S_n and S_m. We will do this by showing that $\bar{U}_{x,x}$ is the same as $\bar{U}_{x+1,0}$. This will allow us to compose the maps described in Lemma 6.10 to get from $\bar{U}_{x,0}$ to $\bar{U}_{x,x} = \bar{U}_{x+1,0}$. We will then be able to compose these maps to get from $S_n = \bar{U}_{n,0}$ to $S_m = \bar{U}_{m,0}$.

Lemma 6.11. *Given any instance Nim−F of CIS-Nim, and any* $m > n > 4F_{max} + 3|F|$*, there exists a bijection* $\phi_{n,m}$ *from* S_n *to* S_m *such that if* $\phi_{n,m}(x_1, y_1) = (x_2, y_2)$*, then* $x_1 - y_1 \geq x_2 - y_2$ *and* $x_1 - 2y_1 \geq x_2 - 2y_2$*.*

Proof. It follows directly from the definitions that $A_{x,x} \cup B_{x,x} = A_{x+1,0}$, and that $B_{x+1,0}$, $C_{x,x}$, and $D_{x+1,0}$ are all empty. Also, the set of all points (x', y') in $C_{x+1,0}$ with $y' < x$ is exactly $D_{x,x}$. Therefore, $U_{x,x}$ is the set of all points (x', y') in $U_{x+1,0}$ with $y' < x$.

Given a point (x', y') with $y' = x$ and $y' < x' \leq 2y'$, this point is in $\bar{U}_{x+1,0}$ if and only if it is in $U_{x+1,0}$ if and only if it is in $C_{x+1,0}$ if and only if $r(x, y') > x' - x$.

On the other hand, this point is in $\bar{U}_{x,x}$ if and only if $(y', \left\lfloor \frac{x'}{2} \right\rfloor) \notin U_{x,x}$ if and only if $(y', \left\lfloor \frac{x'}{2} \right\rfloor) \notin B_{x,x}$ if and only if $b(x, x-1) < x - \left\lfloor \frac{x'}{2} \right\rfloor$.

From Lemma 6.4, we know that $r(x, y') > x' - x$ if and only if $x - 1 - 2b(x, x) > x' - x$ if and only if $2b(x, x-1) \leq 2x - x' - 2$ if and only if $b(x, x-1) \leq x - 1 - \left\lfloor \frac{x'}{2} \right\rfloor$ if and only if $b(x, x-1) < x - \left\lfloor \frac{x'}{2} \right\rfloor$. Therefore, the point is in $\bar{U}_{x,x}$ if and only if it is in $\bar{U}_{x+1,0}$.

Finally, notice that points with $y' \geq x'$ are clearly in neither set, and points with $x' > 2y'$ are not in $\bar{U}_{x,x}$ by definition, and not in $\bar{U}_{x+1,0}$, since otherwise $r(x, x)$ would be greater than x, making it impossible for $r(x, x) + 2b(x, x) + 1$ to equal x. Therefore, a point with $y' \leq x$ is in $\bar{U}_{x+1,0}$ if and only if it is in $\bar{U}_{x,x}$. Therefore, $\bar{U}_{x,x}$ satisfies property 1 for $\bar{U}_{x+1,0}$, and clearly property 2 for $\bar{U}_{x,x}$ is stronger than property 2 for $\bar{U}_{x+1,0}$. Therefore, $\bar{U}_{x,x}$ satisfies both property 1 and property 2 for $\bar{U}_{x+1,0}$, so $\bar{U}_{x,x} = \bar{U}_{x+1,0}$.

We know from Lemma 6.7 that for all $y < x$ there is a bijection from $\bar{U}_{x,y}$ to $\bar{U}_{x,y+1}$ with all the properties described in Lemma 6.7. There is therefore a bijection from $\bar{U}_{x,0}$ to $\bar{U}_{x,x}$ which can be expressed as a composition of the bijections described in Lemma 6.7. Since $\bar{U}_{x,x} = \bar{U}_{x+1,0}$, this means there is a bijection from $S_x = \bar{U}_{x,0}$ to $S_{x+1} = \bar{U}_{x+1,0}$ which can be expressed as a composition of the bijections described in Lemma 6.10. By composing these bijections, we get that for any $n < m$ there is a bijection from S_n to S_m which can be expressed as a composition of the bijections described in Lemma 6.7.

Clearly, any composition of functions described in Lemma 6.7 also satisfy the same relation. We therefore constructed a bijection $\phi_{n,m}$ such that if $\phi_{n,m}(x_1, y_1) = (x_2, y_2)$, then $x_1 - y_1 \geq x_2 - y_2$ and $x_1 - 2y_1 \geq x_2 - 2y_2$. □

6.3. Properties of S_n

In this section we will define two functions, $g(n, m)$ and $h(n, m)$ which will contain information about S_n. We will use g as a potential function that will limit how much $h(n, n)$ will be able to change as n increases. This will ultimately allow us to prove that $\lim_{k\to\infty} \frac{h(n2^k, n2^k)}{4^k}$ converges. This will be helpful, because $h(n,n)$ will allow us to approximately construct $\pi(n)$ and show that $\lim_{k\to\infty} \frac{\pi(n2^k)}{(n2^k)^2}$ also converges, and prove that CIS-Nim exhibits a period-two scale invariance.

Definition 6.12. Given any instance Nim $- F$ of CIS-Nim, and positive integers $m > 4F_{max} + 3|F|$ and n, let R_n denote the set of all (x,y) with $y < n \le x \le 2y$, and let $h(m,n) = |R_n \cap S_m|$.

Lemma 6.13. *Given any instance Nim $- F$ of CIS-Nim, for all $n > 4F_{max} + 3|F|$ and for all non-negative integers k, $h(n, n2^k) = 4^k h(n,n)$.*

Proof. We will proceed by induction on k.

Base Case ($k = 0$): $h(n, n2^k) = h(n,n) = 4^k h(n,n)$

Inductive Hypothesis: $h(n, n2^{k-1}) = 4^{k-1} h(n,n)$

Inductive Step: Consider the map $\psi : R_{n2^k} \to R_{n2^{k-1}}$ defined by $(x,y) \mapsto (\lfloor \frac{x}{2} \rfloor, \lfloor \frac{y}{2} \rfloor)$. Notice that $y < n2^k \le x \le 2y$ if and only if $\left\lfloor \frac{x}{y} \right\rfloor < n2^{k-1} \le \lfloor \frac{x}{2} \rfloor \le 2 \lfloor \frac{y}{2} \rfloor$, so this map is well defined and surjective. Further, if $(x,y) \in R_{n2^k}$, then $y < n2^k \le x \le 2y$, so by the definition of S_n, we know that

$$(x,y) \in S_n \Leftrightarrow (y, \left\lfloor \frac{x}{2} \right\rfloor) \notin S_n \Leftrightarrow (\left\lfloor \frac{x}{2} \right\rfloor, \left\lfloor \frac{y}{2} \right\rfloor) \in S_n.$$

Therefore, for any $(x,y) \in R_{n2^k}$ we get that $(x,y) \in S_n$ if and only if $\psi(x,y) \in S_n$. Therefore, ψ maps $R_{n2^k} \cap S_n$ onto $R_{n2^{k-1}} \cap S_n$. Further, there are exactly four points, $(2x, 2y)$, $(2x+1, 2y)$, $(2x, 2y+1)$, and $(2x+1, 2y+1)$, which map to the point (x,y). Therefore, ψ maps four points in $R_{n2^k} \cap S_n$ onto each point in $R_{n2^k} \cap S_n$, so $|R_{n2^k} \cap S_n| = 4|R_{n2^{k-1}} \cap S_n|$. Therefore, $h(n, n2^k) = 4h(n, n2^{k-1}) = 4(4^{k-1} h(n,n)) = 4^k h(n,n)$.

□

We have shown that $h(n, n2^k) = 4^k h(n,n)$. We would like to relate $h(n2^k, n2^k)$ to $h(n, n2^k)$, which will allow us to relate $h(n2^k, n2^k)$ to $h(n,n)$. To do this, we will have to limit how much $h(m, n2^k)$ can change as m changes from n to $n2^k$. We will need $g(m,n)$, which will serve as a potential function limiting how much $h(m,n)$ can change as we repeatedly double m.

Definition 6.14. Given any instance Nim $- F$ of CIS-Nim, and positive integers $m > 4F_{max} + 3|F|$ and n, let T_n be set of ordered pairs of the form (x,y), with $2y - x \le n$. Let $f((x,y), n) = n + 2x - 3y + 2$, and let $g(m,n)$ denote the sum over all pairs $(x,y) \in T_n \cap S_m$ of $f((x,y), n)$.

Definition 6.15. Given any instance Nim $-$ F of CIS-Nim, if $r(x,x) < x-1$ for all but finitely many x, we will say the game is "well-behaved." In this case, let c_1 be the least natural number, such that $c_1 \geq 4F_{max} + 3|F|$ and $r(x,x) < x-1$ for all $x > c_1$.

It turns out that all the interesting games of CIS-Nim are well-behaved. However, some special cases such as Nim are not well-behaved. Games which are not well-behaved are much easier to analyze. We will continue our analysis in Lemmas 6.16 and 6.17 only considering well-behaved games. Then, in Lemma 6.18, we will show that the result of 6.17 also holds for games which are not well-behaved.

We are going to use g as a potential function to limit how much h will be able to change. The following lemma will provide an upper bound for $g(c_1, 2^k)$ in terms of k, which will give us our initial finite potential.

Lemma 6.16. *Given any well-behaved instance Nim $-$ F of CIS-Nim, there exists a constant c_2 such that for all k, $\frac{g(c_1,2^k)}{8^k} \leq c_2$.*

Proof. Consider the map $\psi : T_{2^k} \to T_{2^{k-1}}$ defined by $(x,y) \mapsto \left(\left\lfloor \frac{x}{2} \right\rfloor, \left\lfloor \frac{y}{2} \right\rfloor\right)$. Given any point, $(x,y) \in T_{2^k}$, we know that

$$2\left\lfloor \frac{y}{2} \right\rfloor - \left\lfloor \frac{x}{2} \right\rfloor \leq 2\frac{y}{2} + \left\lceil \frac{-x}{2} \right\rceil = \left\lceil \frac{2y-x}{2} \right\rceil \leq \left\lceil \frac{2^k}{2} \right\rceil = 2^{k-1}.$$

Therefore, $\psi(x,y) \in T_{2^{k-1}}$.

Similarly, from the definition of S_{c_1}, we know that for any point $(x,y) \in S_{c_1}$ with $\left\lfloor \frac{x}{2} \right\rfloor > c_1$, $(y, \left\lfloor \frac{x}{2} \right\rfloor) \notin S_{c_1}$, so $(\left\lfloor \frac{x}{2} \right\rfloor, \left\lfloor \frac{y}{2} \right\rfloor) \in S_{c_1}$. Therefore, $\psi(x,y) \in S_{c_1}$.

For every point $(x,y) \in T_{2^k}$,

$$f((x,y),2^k) = 2^k + 2x - 3y + 2 \leq 2(2^{k-1}) + 4\left\lfloor \frac{x}{2} \right\rfloor - 6\left\lfloor \frac{y}{2} \right\rfloor + 4 = 2f(\psi(x,y), 2^{k-1}).$$

Combining these three facts, we get that ψ maps each point in $T_{2^k} \cap S_{c_1}$ with $\left\lfloor \frac{x}{2} \right\rfloor > c_1$ to a point in $T_{2^{k-1}} \cap S_{c_1}$. This map clearly sends at most four points to any given point, and $f((x,y),2^k) \leq 2f(\psi(x,y),2^{k-1})$. Therefore, the sum over all elements (x,y) in $T_{2^k} \cap S_{c_1}$ with $\left\lfloor \frac{x}{2} \right\rfloor > c_1$ of $f((x,y),2^k)$ is at most four times the sum over all elements $(x,y) \in T_{2^{k-1}} \cap S_{c_1}$ of $2f((x,y),2^{k-1})$. This value equals $8g(c_1, 2^{k-1})$.

There are at most $2c_1 + 2$ values of x with $\left\lfloor \frac{x}{2} \right\rfloor \leq c_1$, and for each of these values, at most $2c_1 + 2$ values of y with $y < x$. Therefore, there are at most $(2c_1+2)^2$ points $(x,y) \in T_{2^k} \cap S_{c_1}$ with $\left\lfloor \frac{x}{2} \right\rfloor \leq c_1$. For each of these points, $f((x,y),2^k) \leq 2^k + 2(2c_1+2) + 2$. Therefore, the sum over all elements (x,y) in $T_{2^k} \cap S_{c_1}$ with $\left\lfloor \frac{x}{2} \right\rfloor \leq c_1$ of $f((x,y),2^k)$ is at most $(2^k + 2(2c_1+2)+2)(2c_1+2)^2$. Combining this with the last result gives us that the sum over all elements (x,y) in $T_{2^k} \cap S_{c_1}$ of $f((x,y),2^k)$ is at most $8g(c_1,2^{k-1}) + (2^k + 4c_1 + 6)(2c_1+2)^2$.

Therefore, $g(c_1, 2^k) \leq 8g(c_1, 2^{k-1}) + (2^k + 4c_1 + 6)(2c_1 + 2)^2$, which implies that

$$\frac{g(c_1, 2^k)}{8^k} \leq \frac{g(c_1, 2^{k-1})}{8^{k-1}} + \frac{(2^k + 4c_1 + 6)(2c_1 + 2)^2}{8^{k-1}}.$$

Therefore,

$$\frac{g(c_1, 2^k)}{8^k} \leq g(c_1, 1) + \sum_{i=1}^{\infty} \frac{(2^i + 4c_1 + 6)(2c_1 + 2)^2}{8^{i-1}}.$$

Because the game is well-behaved, $g(c_1, 1)$ is a finite constant. We also know $\sum_{i=1}^{\infty} \frac{(2^i + 4c_1 + 6)(2c_1 + 2)^2}{8^{i-1}} = \sum_{i=1}^{\infty} \frac{(8)(2c_1 + 2)^2}{4^i} + \frac{8(4c_1 + 6)(2c_1 + 2)^2}{8^i}$ is the sum of two geometric series with ratio less than one, and therefore converges to a finite constant. Therefore, $\frac{g(c_1, 2^k)}{8^k}$ is bounded above by some finite constant c_2. □

The following lemma is a key part of proving the period-two scale invariance. After this result, all that will remain are a few technical details. We will prove that $\lim_{m \to \infty} \frac{h(n2^m, n2^m)}{4^m}$ converges, which will later be modified to a similar statement about π. The general strategy is to assume by way of contradiction that it does not converge, and therefore must contain infinitely many points above and below a interval of positive length. This means that $\frac{h(n2^m, n2^m)}{4^m}$ must increase and decrease by a fixed amount infinitely many times. We will use g as a potential function, and show that as h changes, g must decrease by some fixed amount. This means that g must decrease infinitely, but g will start at a finite potential, and will remain nonnegative, causing a contradiction.

Lemma 6.17. *Given any well-behaved instance Nim − F of CIS-Nim, for any positive integer n, the limit $\lim_{m \to \infty} \frac{h(n2^m, n2^m)}{4^m}$ converges.*

Proof. Let $\zeta_m = \frac{h(n2^m, n2^m)}{4^m}$ and assume for the purpose of contradiction that ζ_m does not converge. For every point $(x, y) \in R_{n2^m}$, we know that $y < n2^m$, and $x < 2y < n2^{m+1}$. There are only $8(n2^m)^2$, such points, so $h(n2^m, n2^m)$ is bounded above by $8n^2 4^m$, and ζ_m is bounded above by $8n^2$ Therefore, ζ_m is a sequence in the compact interval $[0, 8n^2]$, which does not converge. There therefore exists some real numbers $q < r$, such that infinitely many terms of ζ_m are less than q and infinitely many terms of ζ_m are greater than r. There therefore exists some subsequence ζ_{s_m} such that ζ_{s_m} is less than q for all odd m and greater than r for all even m. Fix some even m, and for convenience of notation let $t(m) = n2^{s_m}$. From Lemma 6.13 we know that for any $k \geq s_{m+1}$,

$$h(t(m), n2^k) = h(t(m), t(m))4^{k - s_m} > r4^k,$$

and

$$h(t(m+1), n2^k) = h(t(m+1), t(m+1))4^{k - s_{m+1}} < q4^k.$$

We know that $|S_{t(m)} \cap R_{n2^k}| > r4^k$ and $|S_{t(m+1)} \cap R_{n2^k}| < q4^k$. Therefore, there are at least $4^k(r-q)$ points $(x,y) \in S_{t(m)}$, such that $(x,y) \in R_{n2^k}$ but $\phi_{t(m),t(m+1)}(x,y) \notin R_{n2^k}$. Given any constant, d, at most $dn2^k$ of these $4^k(r-q)$ points can satisfy the relation $x < n2^k + d$, and at most $dn2^k$ of them can satisfy the relation $y \geq n2^k - d$. Now notice that if (x_1, y_1) is one of the at least $4^k(r-q) - 2dn2^k$ points remaining, and $\phi_{t(m),t(m)}(x_1, y_1) = (x_2, y_2)$, then either $x_2 \leq n2^k < x_1 - d$ or $y_2 > n2^k \geq y_1 + d$.

If $x_2 < x_1 - d$, then since we also know that $x_1 - 2y_1 \geq x_2 - 2y_2$, algebra shows that $(2x_1 - 3y_1) - (2x_2 - 3y_2) \geq \frac{d}{2}$. On the other hand, if $y_2 \geq y_1 + d$, then since we also know that $x_1 - y_1 \geq x_2 - y_2$ algebra shows that $(2x_1 - 3y_1) - (2x_2 - 3y_2) \geq d$. Either way, $(n2^k + 2x_1 - 3y_1) - (n2^k + 2x_2 - 3y_2) \geq \frac{d}{2}$, so $f((x_2, y_2), n2^k) \leq f((x_1, y_1), n2^k) - \frac{d}{2}$. Therefore, for at least $4^k(r-q) - 2dn2^k$ points, (x,y), we have the relation $f(\phi_{t(m),t(m+1)}(x,y), n2^k) \leq f((x,y), n2^k) - \frac{d}{2}$. Further, for all of these points, we have know that $y \leq n2^k - d$ and $x \geq n2^k + d$, so $f((x,y), n2^k) \geq n2^k + 2(n2^k + d) - 3(n2^k - d) = 5d \geq \frac{d}{2}$. Each of these $4^k(r-q) - 2dn2^k$ contribute to $g(t(m), n2^k)$, and whether or not they contribute to $g(t(m+1), n2^k)$, the contribution for each point is reduced by at least $\frac{d}{2}$. It is easy to see that $f(\phi_{t(m),t(m+1)}(x,y), n2^k) \leq f((x,y), n2^k)$ for all (x,y), so every point which contributes to $g(t(m+1), n2^k)$ will contribute at least as much to $g(t(m), n2^k)$. It is also easy to see that no point can contribute negatively to $g(t(m), n2^k)$. Therefore, the contribution of at least $4^k(r-q) - 2dn2^k$ decreases by at least $\frac{d}{2}$ and no point increases its contribution, so we know that

$$g(t(m+1), n2^k) \leq g(t(m), n2^k) - \frac{d}{2}(4^k(r-q) - 2dn2^k).$$

In particular, if we let $d = \frac{2^k(r-q)}{4n}$, we get that

$$g(t(m+1), n2^k) \leq g(t(m), n2^k) - \frac{8^k(r-q)^2}{16n}.$$

It is easy to see that $f(\phi_{t(m+1),t(m+2)}(x,y), n2^k) \leq f((x,y), n2^k)$ for all (x,y), so every point which contributes to $g(t(m+2), n2^k)$ will contribute at least as much to $g(t(m+1), n2^k)$. It is also easy to see that no point can contribute negatively to $g(t(m+1), n2^k)$. Therefore, $g(t(m+2), n2^k) \leq g(t(m+1), n2^k)$, and since m was an arbitrary even integer, we can compose this relation multiple times, to get that for any m and any $k > s_m$

$$g(t(m), n2^k) \leq g(t(0), n2^k) - m\frac{8^k(r-q)^2}{16n}.$$

We know that $g(t(m), n2^k) \geq 0$, and that

$$g(t(0), n2^k) \leq g(c_1, n2^k) \leq g(c_1, 2^{k+\lceil \log_2(n) \rceil}) \leq 8^{k+\lceil \log_2(n) \rceil} c_2.$$

Combining these three relations, we get that

$$m\frac{8^k(r-q)^2}{16n} \le 8^{k+\lceil \log_2(n)\rceil}c_2 \le 8^{k+\log_2(n)+1}c_2 = 8^{k+1}n^3c_2.$$

Now, finally, if we choose m such that $m > \frac{128n^4c_2}{(r-q)^2}$, we get that

$$8^{k+1}n^3c_2 = \frac{128n^4c_2}{(r-q)^2}\frac{8^k(r-q)^2}{16n} < m\frac{8^k(r-q)^2}{16n} \le 8^{k+1}n^3c_2.$$

This is a contradiction, implying that ζ_m does converge. □

Now, we will show that we can reach the same conclusion for games that are not well-behaved that we just reached for well-behaved games.

Lemma 6.18. *Given any instance $Nim-F$ of CIS-Nim which is not well-behaved, for any positive integer n, the limit $\lim_{k\to\infty}\frac{h(n2^k,n2^k)}{4^k}$ converges.*

Proof. The methods here will be very different. Games which are not well-behaved are much easier to analyze, so we will be able to describe the S_n in great detail, and the fact that $\lim_{k\to\infty}\frac{h(n2^k,n2^k)}{4^k}$ converges will follow directly.

We know that for infinitely many values of $m > 4F_{max}+3|F|+1$, $r(m,m) \ge m-1$. It is not possible to have $r(m,m) > m-1$, since $r(m,m)+2b(m,m)+1 = m$. Therefore, for infinitely many values of $m > 4F_{max}+3|F|+1$, $r(m,m) = m-1$. For any such m, we know that $r(m,m)+2b(m,m)+1 = m$, so $b(m,m) = 0$. This means that for all $y < m$, $(m,y) \notin S$ and $b(m,y) = 0$. Therefore, $r(m,y) = 0$, which means that for any $y < m \le x$, $(x,y) \notin S$. In particular, this means that $r(m-1,m-1) = r(m,m-1) = 0$, so $2b(m-1,m-1)+0+1 = m-1$, so $b(m-1,m-1) = \frac{m-2}{2}$.

Notice that if there were some $m' \le \frac{m-1}{2}$, with $(m-1,m') \in S$, then there would be a P-position of the form $\{m-1,m',m''\}$ with $m'' < m'$. In this case, $m-1 > 4F_{max}+3|F| \ge 2F_{max}+|F|$ and

$$m'+m'' < 2m' \le 2(\frac{m-1}{2}) = m-1,$$

contradicting Theorem 5.3. Therefore, for all $(m-1,m') \in S$, $m' > \frac{m-1}{2}$. This means that the $\frac{m-2}{2}$ values of m' with $(m-1,m') \in S$ are exactly the integers from $\frac{m}{2}$ to $m-2$ inclusive. Therefore, $(m-1,\frac{m}{2}) \in S$.

Now, we want to show that $(m',\frac{m}{2}) \in S$ for all $\frac{m}{2} < m' < m-1$. Assume for the purpose of contradiction that there exists some $\frac{m}{2} < m' < m-1$, such that $(m',\frac{m}{2}) \notin S$, and consider the greatest such m'. $r(m',\frac{m}{2}) = m-1-m'$. Therefore, by Lemma 6.5, $b(m',\frac{m}{2}) \ge m-1-m'$, so there exists some $m'' \le \frac{m}{2}-b(m',\frac{m}{2})-1 =$

$m' - \frac{m}{2}$ with $(m', m'') \in S$. There would therefore exist a P-position of the form $\{m', m'', m'''\}$ with $m'' > m'''$. Notice that $m' > \frac{m}{2} > 2F_{max} + |F|$ and

$$m'' + m''' < 2m'' \leq 2(m' - \frac{m}{2}) = 2m' - m < m' + (m' - 1) - m < m',$$

contradicting Theorem 5.3. Therefore, $(m', \frac{m}{2}) \in S$ for all $\frac{m}{2} < m' \leq m - 1$, so $r(\frac{m}{2}, \frac{m}{2}) \geq m - 1 - \frac{m}{2}$, so $r(\frac{m}{2}, \frac{m}{2}) = \frac{m}{2} - 1$.

For any point m satisfying $r(m, m) = m$, we know the greatest value $m' < m$ satisfying $r(m', m') = m' - 1$ is $\frac{m}{2}$. This also tells us that the least value $m' > m$ satisfying $r(m', m') = m' - 1$ is $2m$. Therefore, if we let m be the least value greater than $4F_{max} + 3|F| + 1$ with $r(m, m) = m - 1$, then for any $m' > 4F_{max} + 3|F| + 1$, $r(m', m') = m' - 1$ if and only if $m' = m2^k$ for some nonnegative integer k.

We also know that for any $y < m2^k \leq x$, $(x, y) \notin S$, which implies that $r(y, y) \leq m2^k - y - 1$, and $b(x, x) \leq x - m2^k$. Therefore, given any $m2^{k-1} \leq m' < m2^k$, $r(m', m') \leq m2^k - m' - 1$ and $b(m', m') \leq m' - m2^{k-1}$. Notice that $r(m', m') + 2b(m', m') + 1 = m'$ is satisfied if and only if both of these inequalities are tight, so given any $m2^{k-1} \leq m' < m2^k$, $r(m', m') = m2^k - m' - 1$ and $b(m', m') = m' - m2^{k-1}$.

This means that all of the points (x, y) which we have not already determined to not be in S must be in S. Therefore, for any $x > 4F_{max} + 3|F| + 1, (x, y) \in S$ if and only if there exists a k, such that $m2^{k-1} \leq x, y < m2^k$. This in particular means that there is no (x, y) with $x > 4F_{max} + 3|F| + 1$, such that $(x, y) \notin S$ but $b(x, y) > 0$. This means that all of the functions defined in Lemmas 6.7, 6.10, and 6.11 are the identity, which in particular means that if we set k' to be the least nonnegative integer such that $n2^{k'} > 4F_{max} + 3|F| + 1$, then $S_{n2^k} = S_{n2^{k'}}$ for all $k \geq k'$. Therefore,

$$\lim_{k\to\infty} \frac{h(n2^k, n2^k)}{4^k} = \lim_{k\to\infty} \frac{h(n2^{k'}, n2^k)}{4^k} = \lim_{k\to\infty} \frac{h(n2^{k'}, n2^{k'})}{4^{k'}} = \frac{h(n2^{k'}, n2^{k'})}{4^{k'}}.$$

Therefore, $\lim_{k\to\infty} \frac{h(n2^k, n2^k)}{4^k}$ converges. □

6.4. Proof of the Period-Two Scale Invariance

We now have all the lemmas necessary to complete the proof of the period-two scale invariance.

Theorem 6.1 (Period-Two Scale Invariance). *Given any instance Nim − F of CIS-Nim, let $\pi(n)$ denote the number of P-positions in Nim − F of the form $\{x, y, z\}$, with x, y, and z all less than n. For any positive integer n, $\lim_{k\to\infty} \frac{\pi(n2^k)}{(n2^k)^2}$ converges to a nonzero constant.*

Proof. All that needs to be done to complete the proof is to convert the result from Lemmas 6.17 and 6.18 from a statement about $h(m,m)$ to an analogous statement about $\pi(n)$.

Notice that $h(m,m)$ is the number of $(x,y) \in S_m$ with $y < m \leq x \leq 2y$. From the definition of S_m, this is the number of ordered pairs (x,y) with $y < m \leq x \leq 2y$ and $r(m,y) > x - m$.

Notice that if there existed a point (x,y) with $y < m \leq x$ and $x \geq 2y$ such that $r(m,y) > x - m$, then there must be greater than $2y + 1 - m$ values of $x' \geq m$ with $(x', y) \in S$, then there must be at least 1 value of $x' \geq 2y$ with $(x', y) \in S$. However, this would mean that there would be a P-position of the form $\{x,y,z\}$ with $x \geq 2y$ and $x > y > z$. However, this means that $x \geq 2y > y + z$, which, since $x \geq m > 4F_{max} + 3|F|$, contradicts Theorem 5.3. This means that the $x \leq 2y$ condition is unnecessary, so $h(m,m)$ is the number of ordered pairs (x,y) with $y < m \leq x$ and $r(m,y) > x - m$.

Notice that for each y, there are exactly $r(m,y)$ values of x with $m \leq x$ and $r(m,y) > x - m$. Therefore, $h(m,m) = \sum_{y=0}^{m-1} r(m,y)$, which is exactly the number of ordered pairs $(x,y) \in S$ with $y < m \leq x$, or equivalently the number of P-positions of the form $\{x,y,z\}$ with $x \geq m > y > z$.

For each ordered pair (y,z) with $z < y < m$, let x be the unique value such that $\{x,y,z\}$ is a P-position. There are exactly $\frac{m^2-m}{2}$ such ordered pairs, and exactly $h(m,m)$ of them satisfy the relation $x \geq m$. Therefore, the remaining $\frac{m^2-m}{2} - h(m,m)$ of them satisfy the relation $x < m$. Let V_m denote this set of $\frac{m^2-m}{2} - h(m,m)$ points with $x < m$ and $z < y < m$. Let $\pi_3(m)$ be the number of P-positions of the form $\{x,y,z\}$ with x, y, and z distinct and less than m. Let $\pi_2(m)$ be the number of P-positions of the form $\{x,x,y\}$ with x and y distinct and less than m. Let $\pi_1(m)$ be the number of P-positions of the form $\{x,x,x\}$ with $x < m$. Clearly, $\pi_3(m) + \pi_2(m) + \pi_1(m) = \pi(m)$.

Notice that given a P-position of the form $\{x,y,z\}$ with $m > x > y > z$, by definition, (x,y), (x,z), and (y,z) are in V_m. Given a P-position of the form $\{x,x,y\}$ with x and y distinct and less than m, clearly exactly one of (x,y) and (y,x) is in V_m. Also, all points in V_m clearly fall into one of two cases. Therefore,

$$\frac{m^2 - m}{2} - h(m,m) = 3\pi_3(m) + \pi_2(m).$$

Therefore,

$$6\pi(m) = m^2 - m - 2h(m,m) + 4\pi_2(m) + 6\pi_1(m).$$

Therefore,

$$\frac{\pi(n2^k)}{(n2^k)^2} = \frac{n^2 4^k - n2^k - 2h(n2^k, n2^k) + 4\pi_2(n2^k) + 6\pi_1(n2^k)}{6(4^k)n^2},$$

so
$$\frac{\pi(n2^k)}{(n2^k)^2} = \frac{1}{6} - \frac{1}{3n^2}\frac{h(n2^k, n2^k)}{4^k} - \frac{1}{6(2^k)n} + \frac{2\pi_2(n2^k)}{3(4^k)n^2} + \frac{\pi_1(n2^k)}{4^k n^2}.$$

Notice that $\frac{1}{6}$ is a constant, $-\frac{1}{3n^2}\frac{h(n2^k,n2^k)}{4^k}$ converges as k goes to infinity by Lemmas 6.17 and 6.18, and $-\frac{1}{6(2^k)n}$ converges to 0 as k goes to infinity. For any x, there is only one P-position of the form $\{x, x, y\}$, and at most one P-position of the form $\{x, x, x\}$. Therefore, $\pi_2(n2^k)$ and $\pi_1(n2^k)$ are both less than or equal to $n2^k$, so $\frac{2\pi_2(n2^k)}{3(4^k)n^2}$ and $\frac{\pi_1(n2^k)}{4^k n^2}$ both converge to 0 as k goes to infinity. Therefore, $\frac{\pi(n2^k)}{(n2^k)^2}$ converges as k goes to infinity.

Further, we know that $\frac{\pi(n2^k)}{(n2^k)^2}$ does not converge to 0, since $\pi(n2^k)$ is at least one sixth the number of ordered triples (x, y, z) with x, y, z distinct less than $n2^k$ such that $\{x, y, z\}$ is a P-position. For each pair $y, z < \frac{n2^k-|F|}{2}$, there exists an ordered triple (x, y, z) with $x \leq y + z + |F| < n2^k$ and $y, z < n2^k$, such that $\{x, y, z\}$ is a P-position. There are $\frac{(n2^k-|F|)^2}{4}$ such pairs, so $\pi(n2^k) \geq \frac{(n2^k-|F|)^2}{24}$. Therefore, $\frac{\pi(n2^k)}{(n2^k)^2} \geq \frac{(n2^k-|F|)^2}{24(n2^k)^2}$, which converges to $\frac{1}{24}$ as k goes to infinity.

Therefore, for any positive integer n, $\lim_{k\to\infty} \frac{\pi(n2^k)}{(n2^k)^2}$ converges to a nonzero constant. □

7. Concluding Remarks

We have thus shown here that the class of combinatorial games CIS-Nim obeys a form of scale invariance (period-two scale invariance). The existence of such scaling properties in combinatorial games had been previously hinted at using renormalization techniques adapted from physics. However, such techniques were nonrigorous in nature; the present work is the first formal characterization of scaling in this context. Additionally, it has been demonstrated that certain properties of combinatorial games persist under perturbations (the perturbations here being defined by the forbidden set F), and hence are 'generic' in the sense of dynamical systems theory.

That said, the version of the period-two scale invariance proven in this paper was not the strongest version possible. A much stronger version, which is also appears to be true, would allow more general regions than the set of points $\{x, y, z\}$ with $\{x, y, z\} < n$. We therefore conjecture a stronger version of the period-two scale invariance:

Conjecture 7.1. *Given any instance Nim $-$ F of CIS-Nim and any open set $S \subseteq \mathbb{R}^3$, let $\pi(R, k)$ be the number of P-positions of the form $\{x2^k, y2^k, z2^k\}$, with $x, y, z \in \mathbb{Q}$ and $(x, y, z) \in S$. Then $\lim_{k\to\infty} \frac{\pi(R,k)}{4^k}$ converges.*

Alternatively, we could make this statement stronger by considering more general versions of the game of Nim. This can be done by considering the piles in Nim to be labeled, so the positions are ordered triples. This would allow for non-symmetric forbidden sets. We could also consider Nim played with an arbitrary number of piles. We conjecture that that this generalization will also preserve the period-two scale invariance.

Conjecture 7.2. *Let m be any positive integer, let F be a set of positions in m-Heap Nim with labeled piles. Given any open set $S \subseteq \mathbb{R}^m$, let $\pi(R,k)$ be the number of P-positions of m-Heap Nim $-$ F of the form $2^k v$ with $v \in S$. Then $\lim_{k\to\infty} \frac{\pi(R,k)}{2^{(n-1)k}}$ converges.*

Finally, this general notion of a Cofinite Induced Subgraph Games can be applied to any other impartial combinatorial games. Since other games do not necessarily satisfy a period-two scale invariance, this result will not generalize to the CIS version of most of these other games. However, by analyzing the cofinite induced subgraphs of a game graph, we learn which properties of the structure of the P-positions are unstable and dependent on a finite set of end game positions, and which properties stable and inevitable regardless of the details of the end game.

Acknowledgments. ASL's research has been supported in part through a W.M. Keck Foundation research grant. EJF's research has been supported in part by the NSF under grant CDI-0835706.

References

[1] L. Abrams and D. S. Cowen-Morton. Periodicity and other structure in a colorful family of nim-like arrays. *The Electron. J. Comb.* 17(1), 2010.

[2] M. Albert, R. Nowakowski, and D. Wolfe. *Lessons in Play: An Introduction to the Combinatorial Theory of Games.* AK Peters, 2007.

[3] Michael H. Albert and Richard J. Nowakowski, Eds. *Games of No Chance 3.* Cambridge University Press, 2009.

[4] C. L. Bouton. Nim, a game with a complete mathematical theory. *Annals Math.* 3(1/4):35–39, 1901 - 1902.

[5] Steven Byrnes. Poset game periodicity. *Integers* 3, 2003, #G3.

[6] J. H. Conway. *On Numbers and Games.* A. K. Peters, Wellesley, Massachusetts, 2nd edition, 2001.

[7] J. H. Conway E. R. Berlekamp and R. K. Guy. *Winning Ways For Your Mathematical Plays*, volume 1. A. K. Peters, Natick, Massachusetts, 2nd edition, 2001.

[8] Rebecca E. Morrison, Eric J. Friedman, and Adam S. Landsberg, Combinatorial games with a pass: A dynamical systems approach, *Chaos* 21, 2011, 14 pp.

[9] A. S. Fraenkel and M. Ozery. Adjoining to wythoff's game its p-positions as moves. *Theor. Comput. Sci.* 205 (1-2), 283, 1998.

[10] E. J. Friedman and A. S. Landsberg. Nonlinear dynamics in combinatorial games: Renormalizing chomp. *Chaos* 17(2), 2007.

[11] E. J. Friedman and A. S. Landsberg. On the geometry of combinatorial games: A renormalization approach in *Games of No Chance 3*, Michael H. Albert and Richard J. Nowakowski, Eds. Cambridge University Press, 2009.

[12] P. Grundy. Mathematics and games *Eureka* 2:6-8, 1939.

[13] R. Sprague. Uber mathematische kampfspiele *Tohoku Mathematical Journal* 41:438-444, 1936.

[14] Doron Zeilberger. Chomp, recurrences and chaos(?). *Journal of Difference Equations and Applications* 10(13-15), 2004.

[15] Ernst Zermelo. Uber eine anwendung der mengenlehre auf die theorie *Proc. Fifth Congress Mathematicians*, (Cambridge 1912), Cambridge University Press, 1913.

PIRATES AND TREASURE

Fraser Stewart
Department of Mathematics and Statistics, Xi'An Jiaotong University, Xi'An, Shaanxi, China
fraseridstewart@gmail.com

Received: 8/14/12, Accepted: 3/23/13, Published: 3/29/13

Abstract

In this paper we introduce a new game; in this game there are two players, who play as rival pirate gangs. The goal is to gather more treasure than your rival. The game is played on a graph and a player gathers treasure by moving to an unvisited vertex. At the end of the game, the player with the most treasure wins. We will show that this game is NP-hard, and we will also look at the structure of this game under the disjunctive sum. We will show that there are cases where this game behaves like a normal play game, and cases where it behaves like a misère play game. We then leave an open problem about scoring play games in general.

1. Introduction

Coin-sliding games have been studied for many years, one of the best known is the game Geography. This is a simple game that parents often tell their children to play during long car journeys. The idea is that a person says the name of a country, and the next person must name a country whose first letter is the same as the last letter of the country just named. For example, Britain, Norway, Yugoslavia, America, Argentina, Australia and so on.

The generalised version of this game is played on a directed graph, and players take it in turns to move a coin to a previously unvisited neighbour. The game ends when a player cannot move, and the last player to move is the winner.

Another game, simply titled "The Coin-Sliding Game", was introduced in a paper by Moews [14]. This is a game where the players have coins of various values, that are placed on a vertical strip. The player then chooses to either move a coin down the strip, or remove one of his opponents coins.

The players collect coins that are removed from the strip, either by sliding them off it, or from removing them. At the end of the game the players add the values of the coins that they have collected, and the player who has the most wins. In his paper, Moews gave the solution to this game.

Our game is played on an undirected graph, where players have multiple coins placed on various vertices of the graph. These coins represent their pirate ships. The remaining vertices are given a numerical value, and players gather points (treasure) by moving onto those vertices. The player who gathers the most points (treasure) wins.

The formal rules are given as follows:

1. The game is played on a finite simple graph, defined arbitrarily before the game begins. Left has n ships, and Right has m ships.
2. Each ship has a pre-defined starting vertex.
3. Every node is numbered to indicate how much treasure there is at that node, the players' starting vertices are not numbered.
4. On a player's turn he moves to an adjacent, unvisited vertex. The number of points he gets, corresponds to the number on that vertex. A player may not move a previously visited vertex, including the starting vertices.
5. The game ends when it is a player's turn, and he is not adjacent to an unvisited vertex.
6. The player who gathers the most treasure wins.

In this paper, first we will be examining the complexity of this game, and then showing that there are variations of it which are comparable with normal and misère play.

1.1. Scoring Play Combinatorial Game Theory

Scoring play combinatorial games have not been studied anywhere near as much as their normal, and misere play counterparts. The first papers were written by Milnor and Hanner [7, 13]. These were then followed up with papers with Ettinger [3, 4], the first of which was published, while the second remains unpublished. Johnson then did some follow up work subsequently [9].

All of them studied well-tempered scoring play games, that is games where the game always lasts a fixed number of moves. However, in 2011 Stewart introduced the most general theory for scoring play games [15]. This work was done entirely independently of Milnor, Hanner and Johnson, and is based on the theories of Berlkeamp, Conway and Guy [1, 2].

The idea behind this theory is very simple, consider the game tree given in Figure 1.

On a typical game tree as shown in Figure 1, the nodes represent the positions of a game, and edges represent possible moves for both players from those positions. Left sloping edges are Left's moves, and right sloping edges are Right's moves.

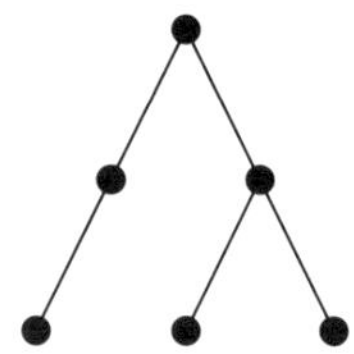

Figure 1: A typical game tree.

A scoring play game tree is exactly the same, but for one difference, the nodes now have numbers on them which represent the score associated with that position. The score is the difference between Left's total points, and Right's total points, at that point in the game.

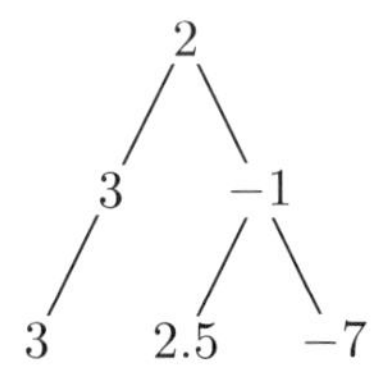

Figure 2: A scoring play game tree.

Formally scoring play games are defined as follows.

Definition 1. A scoring play game $G = \{G^L|G^S|G^R\}$, where G^L and G^R are sets of games and $G^S \in \mathbb{R}$, the base case for the recursion is any game G where $G^L = G^R = \emptyset$. $G^L = \{\text{All games that Left can move to from } G\}$,
$G^R = \{\text{All games that Right can move to from } G\}$, and for all G there is an $S = (P, Q)$ where P and Q are the number of points that Left and Right have on G respectively. Then $G^S = P - Q$, and for all $g^L \in G^L$, $g^R \in G^R$, there is a $p^L, p^R \in \mathbb{R}$ such that $g^{LS} = G^S + p^L$ and $g^{RS} = G^S + p^R$.

By convention, we will take G^S to be 0, unless stated otherwise. This is simply to give games a "default" setting, i.e., if we don't know what G^S is, then it is natural to simply let it be 0. We also write $\{.|G^S|.\}$ as G^S, e.g. $\{\{.|3|.\}|4|\{.|2|.\}\}$ would be written as $\{3|4|2\}$. This simply for convenience and ease of reading.

For these games we also need the idea of a "final score". That is the best possible score that both players can get when they move first. Formally, this is defined as follows.

Definition 2. We define the following:

- G_F^{SL} is called the Left final score, and is the maximum score –when Left moves first on G– at a terminal position on the game tree of G, if both Left and Right play perfectly.
- G_F^{SR} is called the Right final score, and is the minimum score –when Right moves first on G– at a terminal position on the game tree of G, if both Left and Right play perfectly.

Our game Pirates and Treasure clearly falls under this theory, so we will be referencing it –and using it– throughout the paper.

The paper "Scoring Play Combinatorial Game Theory" [15], discusses the structure of these games under the disjunctive sum, which is defined below. In this paper it is shown that these games do not form a group, there is no non-trivial identity, and almost no games that can be compared in the usual sense. However, these games are partially ordered under the disjunctive sum, and do form equivalence classes with a canonical form. The games can also be reduced using the usual rules of domination and reversibility.

Definition 3. The disjunctive sum is defined as follows:

$$G +_\ell H = \{G^L +_\ell H, G +_\ell H^L | G^S + H^S | G^R +_\ell H, G +_\ell H^R\},$$

where $G^S + H^S$ is the normal addition of two real numbers.

We abuse notation by letting G^L and G^R represent the set of options and the individual options themselves. The reader will also notice that we have used $+_\ell$ and $+$. This is to distinguish between the addition of games (the disjunctive sum), and the addition of scores.

Definition 4. We define the following:

- $-G = \{-G^R| - G^S| - G^L\}$.
- For any two games G and H, $G = H$ if $G +_\ell X$ has the same outcome as $H +_\ell X$ for all games X.
- For any two games G and H, $G \geq H$ if $H +_\ell X \in O$ implies $G +_\ell X \in O$, where $O = L_\geq$, $R_\geq$, $L_>$ or $R_>$, for all games X.
- For any two games G and H, $G \leq H$ if $H +_\ell X \in O$ implies $G +_\ell X \in O$, where $O = L_\leq$, $R_\leq$, $L_<$ or $R_<$, for all games X.
- $G \cong H$ means G and H have identical game trees.
- $G \approx H$ means G and H have the same outcome.

The reason that we need this, is because our game naturally splits up into multiple smaller components that are played independently of one another. So, the disjunctive sum is the natural operator to use when analysing this game.

Finally we need to define the outcome classes of games. Before we can define what the outcome classes are precisely, we first need the following definition.

Definition 5.

$L_> = \{G|G_F^{SL} > 0\}$, $L_< = \{G|G_F^{SL} < 0\}$, $L_= = \{G|G_F^{SL} = 0\}$.

$R_> = \{G|G_F^{SR} > 0\}$, $R_< = \{G|G_F^{SR} < 0\}$, $R_= = \{G|G_F^{SR} = 0\}$.

$L_\geq = L_> \cup L_=$, $L_\leq = L_< \cup L_=$, $R_\geq = R_> \cup R_=$, $L_\leq = R_< \cup R_=$.

Next we can use this to give the definition of outcome classes for scoring play games. Note that scoring play games, unlike normal and misère play games have five outcome classes.

Definition 6. The outcome classes of scoring games are defined as follows:

- $\mathcal{L} = (L_> \cap R_>) \cup (L_> \cap R_=) \cup (L_= \cap R_>)$
- $\mathcal{R} = (L_< \cap R_<) \cup (L_< \cap R_=) \cup (L_= \cap R_<)$
- $\mathcal{N} = L_> \cap R_<$
- $\mathcal{P} = L_< \cap R_>$
- $\mathcal{T} = L_= \cap R_=$

1.2. An Example

In this section we will give an example of Pirates and Treasure, so that the reader has a better idea for how this game is played. Consider the game shown in Figure 3, this Figure shows a typical Pirates and Treasure position, as well as, a sequence of moves that the players could make.

In the diagrams L represents Left's current position, R Right's current position, and the numbers represent the amount of treasure at that vertex. The number in brackets will represent the current score, we also change numbered nodes to non-numbered nodes once they have been visited. This indicates that the pirate has gathered all of the available treasure at that particular node.

In this particular example, $G_F^{SL} = G_F^{SR} = 2$. Note that the winner is not related to who moves last. If Left moves first then Right moves last, but Left wins. If Right moves first then Left moves last, but again, Left wins.

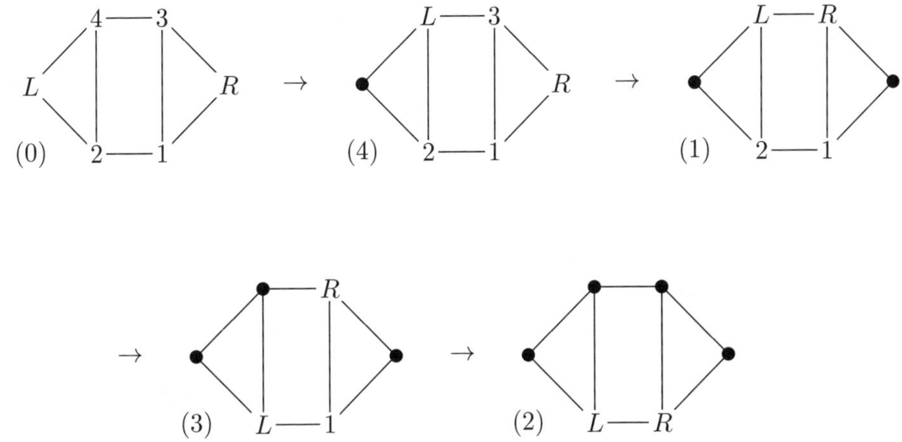

Figure 3: An example of Pirates and Treasure

2. Complexity

As always when studying a new game –or problem– like this, the very first question we ask as a matter-of-course is "how hard is it?". This is a very important question, and as we will show it is NP-hard to determine the final score of this game, and it remains NP-hard for various types of graphs.

Problem: **Hamiltonian Path**

Instance: A Graph $G = (V, E)$.

Question: Does G contain a Hamiltonian Path?

This problem has been shown to be NP-complete [10]. We will be doing a reduction from Hamiltonian Path to our problem.

We define our problem as follows:

Problem: **Pirates and Treasure**

Instance: A Graph $G = (V, E)$, weight $w(v) \in \mathbb{Z}^+$ for each $v \in V$, specified vertices L and R.

Question: Can Left win moving first on G?

Theorem 7. *Pirates and treasure is NP-hard.*

Proof. To do this reduction we first start with a graph G and pick a vertex L on G. The vertex L will be the starting position of player Left. We then give all remaining vertices on G value 1.

We add a path P to G, such that L is one of the end vertices. We then choose the vertex R on P that is adjacent to L, and let it be the starting position of player Right. P is chosen such that $|P| = |V|$. The reason we choose it this way is to ensure that Left can win only by visiting every vertex of the graph G. That is, Right has $|V| - 2$ (subtracting the vertices L and R) vertices he can visit, while Left has at most $|V| - 1$ (subtracting the vertex L).

By the rules of our game, Left cannot move on P, since the vertex R was once occupied by Right. Likewise, Right cannot move onto G since he must move to vertex L, and this was once occupied by Left.

If there is no hamiltonian path then Left can only visit at most $|V| - 2$ of the vertices on G, meaning that Right is guaranteed to tie. If there is a hamiltonian path, then Left will make the final move of the game and visit all $|V| - 1$ vertices, while Right will only have visited $|V| - 2$ vertices.

Therefore, Left can win moving first, if and only if, there is a hamiltonian path on G and the theorem is proven.

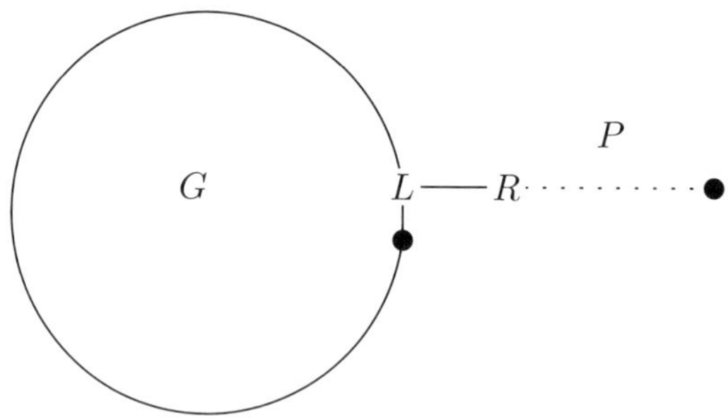

Figure 4: Reduction from Hamiltonian Path.

□

The problem Hamiltonian path remains NP-complete if G is planar, cubic, 3-connected, or has no face with fewer than 5 edges [6]. It also remains NP-complete if G is bipartite [11], or a grid graph [8].

Definition 8. A grid graph is the graph whose vertices correspond to the points in the plane with integer coordinates, x-coordinates being in the range $1, \ldots, n$, y-coordinates being in the range $1, \ldots, m$, and two vertices are connected by an edge whenever the corresponding points are at distance 1.

Another way of thinking about this, is that a grid graph is simply an induced subgraph of a 2-dimensional grid.

Theorem 9. *Pirates and Treasure is NP-hard if G is either a planar or, a grid graph.*

Proof. The proof of this is almost identical to Theorem 7. Adding a path to the outer face of a planar graph, is still a planar graph. Likewise for a grid-graph. Therefore, we can use the same reduction method, and the theorem is proven. □

For the remaining types of graphs, we must use a different proof technique. The reason is that if we add a path, to say, a cubic graph, then the resulting graph is no longer cubic. If we want to say that the game remains NP-hard for cubic graphs, say, then the graph we use for our reduction must also be cubic. So we make the following conjecture.

Conjecture 10. Pirates and Treasure is NP-hard if G is either cubic, or 3-connected.

Since it is unlikely that this game is in NP, and this game is clearly in PSpace, we also make this conjecture.

Conjecture 11. Pirates and Treasure is PSpace-complete.

3. The Game in General

Due to the fact that scoring games are not as nicely behaved as normal play games, we must devise a new technique for studying them. The technique that we propose is to restrict our set to only include those scoring games that represent a position of the game we wish to analyse.

In some cases this makes the problem much simpler, as was shown by Stewart in his paper on impartial scoring play games [16]. In this paper he looked only at the set of impartial games, and in doing was able to devise a general strategy for solving any scoring play octal game. Here we will attempt to do the same thing, but for Pirates and Treasure.

The most obvious question to ask is, "will playing greedily always work?". The answer to that is "no", as demonstrated in the following example.

Example 12. Consider the game in Figure 5.

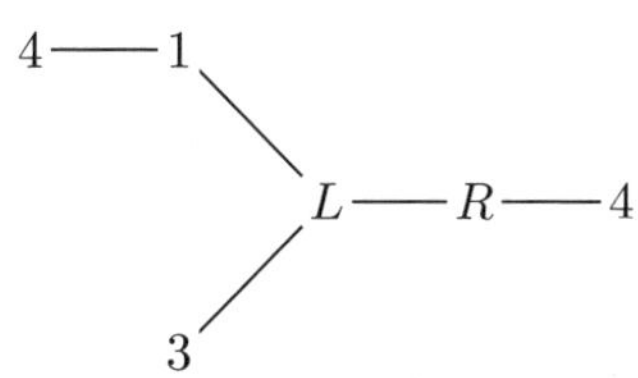

Figure 5: Playing greedily is not the best strategy.

If Left were playing greedily he would move to the neighbour with value 3. However, if he does so Right will move and get 4 points and therefore Left will lose.

Left's best strategy is to move to the neighbour with value 1, Right still moves and gets 4 points, but then Left can move again, get 4 points and win. So playing greedily certainly does not always work.

Definition 13. $\mathcal{PT} = \{G|G$ represents a position of Pirates and Treasure$\}$

Theorem 14. *For all* $G \in \mathcal{PT}$, *if* $G \not\approx 0$ *then* $G \neq 0$.

Proof. First let P be a single edge, with one Right piece on it (note that $P = \{.|a|b\}$), and let Left move first on $G +_\ell P$, where G is any graph that has at least one Left piece. The case where Right moves first will follow by symmetry. Note, $P_F^{SL} = a$, and we let $a \geq 0$.

We will label the vertices of P, p_1 and p_2. We place the Right piece of p_1 and give p_2 a value that is larger than the sum of all the values on G. Since G is a finite graph, we can always do this.

Left moving first, must move on G, since he has no move on P. Right simply moves to p_2 and wins. Therefore $(G +_\ell P)_F^{SL} < 0$, i.e., $G +_\ell P \not\approx P$ and the theorem is proven. □

This means that it is highly unlikely that we will be able to find any general technique for solving different variations of this game.

3.1. Comparison with Normal Play

In this section we are aiming to show that it is possible to find a variation of Pirates and Treasure that behaves very similarly to a normal play game. That is, best strategy under normal play, corresponds to best strategy under scoring play. First, Table 1 shows the sums of the four outcome classes under normal play.

$G + H$	$G \in \mathcal{P}$	$G \in \mathcal{L}$	$G \in \mathcal{R}$	$G \in \mathcal{N}$
$H \in \mathcal{P}$	$\mathcal{P}$	$\mathcal{L}$	$\mathcal{R}$	$\mathcal{N}$
$H \in \mathcal{L}$	$\mathcal{L}$	$\mathcal{L}$	$\mathcal{L}, \mathcal{R}, \mathcal{N}, \mathcal{P}$	$\mathcal{L}, \mathcal{N}$
$H \in \mathcal{R}$	$\mathcal{R}$	$\mathcal{L}, \mathcal{R}, \mathcal{N}, \mathcal{P}$	$\mathcal{R}$	$\mathcal{R}, \mathcal{N}$
$H \in \mathcal{N}$	$\mathcal{N}$	$\mathcal{L}, \mathcal{N}$	$\mathcal{R}, \mathcal{N}$	$\mathcal{L}, \mathcal{R}, \mathcal{N}, \mathcal{P}$

Table 1: Outcome Class Table for Normal Play Games

Our aim, is to find a variation of Pirates and Treasure that gives an outcome class table which looks like that one. By doing this, we are effectively demonstrating that there is a non-trivial subset of scoring play games that exhibit the same "nice" behavior of normal play games.

First we define the games $\mathcal{PT}_x$.

Definition 15. $\mathcal{PT}_x$, is a subset of $\mathcal{PT}$, where every node on the graph has value $x \in \mathbb{R}$, $x > 0$, and for all $G \in \mathcal{PT}_x$, $G^S = 0$.

What the following theorems will show is that $\mathcal{PT}_x$ exhibits behavior that is almost identical to normal play. In all diagrams R represents a Right piece, and L represents a Left piece.

Theorem 16. *If $G \in \mathcal{PT}_x$, then G belongs to either $\mathcal{L}$, $\mathcal{R}$, $\mathcal{N}$ or $\mathcal{T}$, i.e., there are no $\mathcal{P}$ positions.*

Proof. Whenever a player moves they gain x points. So the first player to move will have an x point advantage over the second player. Since all nodes have value x, the most the second player can do is bring the game back to a tie. Therefore when the game ends, either the first player has won, or the game is a tie.

For a position G to be in $\mathcal{P}$, the second player has to be able to win outright. But, this is impossible, therefore there are no $\mathcal{P}$ positions, and the theorem is proven. □

To show the similarities between this particular variation, and normal play, we will be looking at the outcome class table. This is given in the following theorem.

Theorem 17. *The outcome class table for $\mathcal{PT}_x$ is given as follows:*

$G +_\ell H$	$G \in \mathcal{T}$	$G \in \mathcal{L}$	$G \in \mathcal{R}$	$G \in \mathcal{N}$
$H \in \mathcal{T}$	$\mathcal{T}$	$\mathcal{L}$	$\mathcal{R}$	$\mathcal{N}$
$H \in \mathcal{L}$	$\mathcal{L}$	$\mathcal{L}$	$\mathcal{L}, \mathcal{R}, \mathcal{N}, \mathcal{T}$	$\mathcal{L}, \mathcal{N}$
$H \in \mathcal{R}$	$\mathcal{R}$	$\mathcal{L}, \mathcal{R}, \mathcal{N}, \mathcal{T}$	$\mathcal{R}$	$\mathcal{R}, \mathcal{N}$
$H \in \mathcal{N}$	$\mathcal{N}$	$\mathcal{L}, \mathcal{N}$	$\mathcal{R}, \mathcal{N}$	$\mathcal{L}, \mathcal{R}, \mathcal{N}, \mathcal{T}$

Table 2: Outcome Class Table for PT_x

Proof. The proof of this will be split into five cases, and the remaining cases follow by symmetry.

`Case 1`: $G \in \mathcal{X}$, $H \in \mathcal{T}$ implies $G +_\ell H \in \mathcal{X}$, where $\mathcal{X} = \mathcal{L}, \mathcal{R}, \mathcal{N}$ or $\mathcal{T}$.

Since $H \in \mathcal{T}$, this implies that the second player to move, must be the last player to move. First assume that Left wins moving first on G, since the case where Right wins moving first follows by symmetry. Again, Left winning on G implies that he also moves last on G.

So when they play $G +_\ell H$, Left will choose his winning move on G, and move to $G^L +_\ell H$. If Right also moves on G, i.e., moves to $G^{LR} +_\ell H$, Left will respond on G. When G is over, they must play H, which ends in a tie, and therefore Left wins.

If Right chooses to move to $G^L +_\ell H^R$, then Left will respond by moving to $G^L +_\ell H^{RL}$. Since H ends in a tie, we know that Left must move last on H, and therefore can force Right to move first on G^L, which he loses. So if Left wins moving first on G, then Left wins moving first on $G +_\ell H$.

We know that neither Left nor Right can win moving second on any game G, therefore the final case to consider is Left ties moving second on G. Left can tie $G +_\ell H$, by simply following Right's moves, i.e., Left moves on the same component as Right.

Since Right cannot win moving first or either G or H, Right will choose to move to $G^R +_\ell H$, otherwise he may give Left an opportunity to win. If Left chooses to move to $G^R +_\ell H^L$, then Right can still force a tie by playing to $G^R +_\ell H^{LR}$. In other words, Left cannot change the parity of $G +_\ell H$, and therefore the best he can do is tie moving second on $G +_\ell H$.

Case 2: G and $H \in \mathcal{L}$ implies $G +_\ell H \in \mathcal{L}$.

Left playing first on $G +_\ell H$, simply makes his winning move on G or H. Then, whichever component Right chooses to move on, Left will also move on. Since we know that Left can move last on both G and H, he is guaranteed to keep his advantage over Right and therefore win, moving first.

When Right moves first, Left can at least tie by playing the same strategy as before (i.e., moving on the same component as Right). Therefore $(G +_\ell H)_F^{SL} > 0$ and $(G +_\ell H)_F^{SR} \geq 0$, and we conclude that $G +_\ell H \in \mathcal{L}$.

Case 3: $G \in \mathcal{L}$, $H \in \mathcal{R}$ implies $G +_\ell H \in \mathcal{X}$, where $\mathcal{X} = \mathcal{L}, \mathcal{R}, \mathcal{N}$ or $\mathcal{T}$.

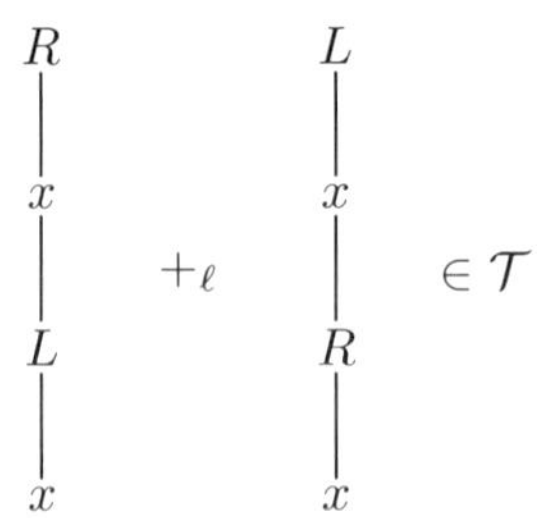

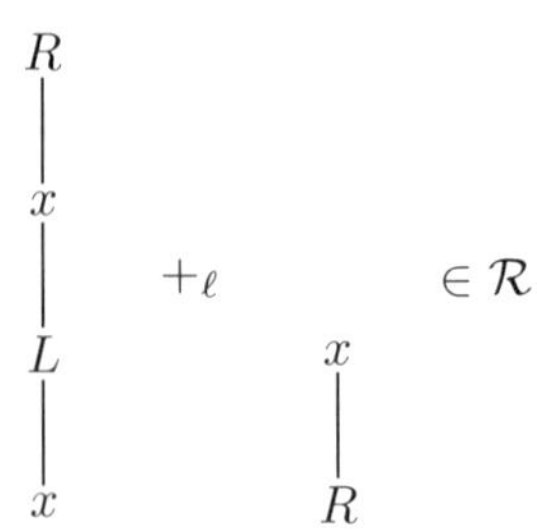

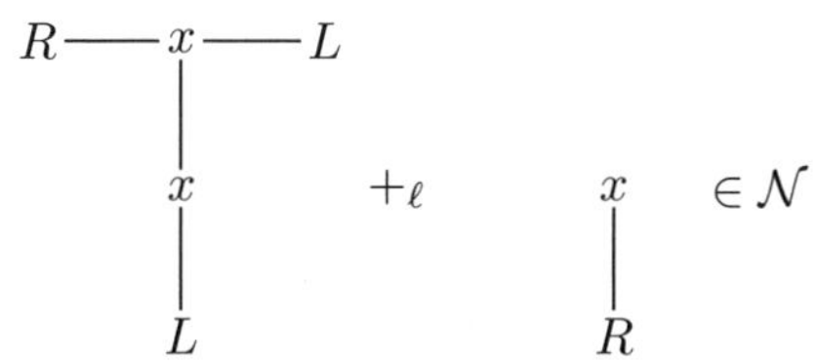

The case $G \in \mathcal{L}$, $H \in \mathcal{R}$ implies $G +_\ell H \in \mathcal{L}$ follows by symmetry.

Case 4: $G \in \mathcal{L}$, $H \in \mathcal{N}$ implies $G +_\ell H \in \mathcal{L}$ or $\mathcal{N}$.

If $G \in \mathcal{L}$ and $H \in \mathcal{N}$, then Left moving first on $G +_\ell H$, can win by moving $G +_\ell H^L$. The reason is that he will gain an x point advantage over Right, and whichever component Right moves on, Left will move on. Since he moves last on both G and H, playing this strategy will guarantee that he maintains his x point advantage over Right.

Since Left can always win moving first on $G +_\ell H$, whenever $G \in \mathcal{L}$ and $H \in \mathcal{N}$, then we conclude that $G +_\ell H$ cannot be in $\mathcal{R}$ or $\mathcal{T}$. To complete the proof, we simply give an example of $G +_\ell H \in \mathcal{L}$ and $G +_\ell H \in \mathcal{N}$.

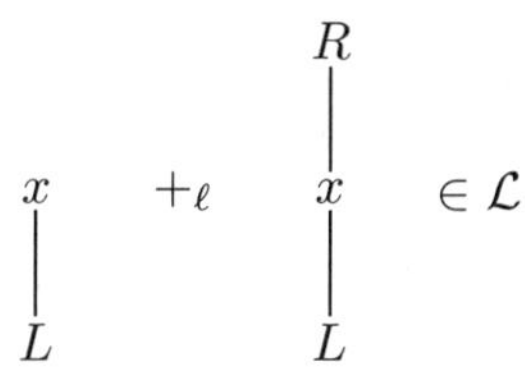

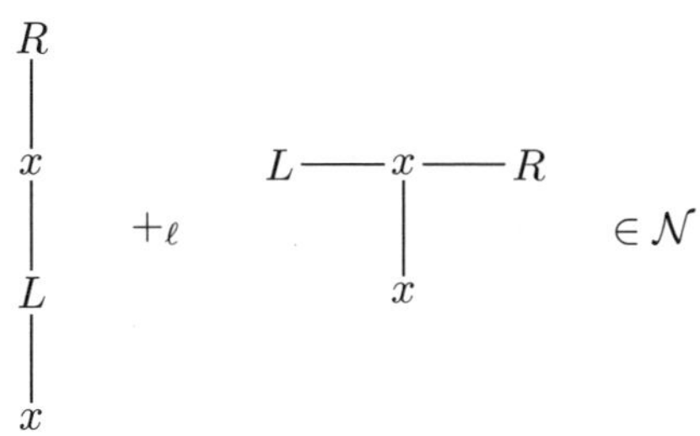

Case 5: G and $H \in \mathcal{N}$ implies $G +_\ell H \in \mathcal{X}$, where $\mathcal{X} = \mathcal{L}, \mathcal{R}, \mathcal{N}$ or $\mathcal{T}$.

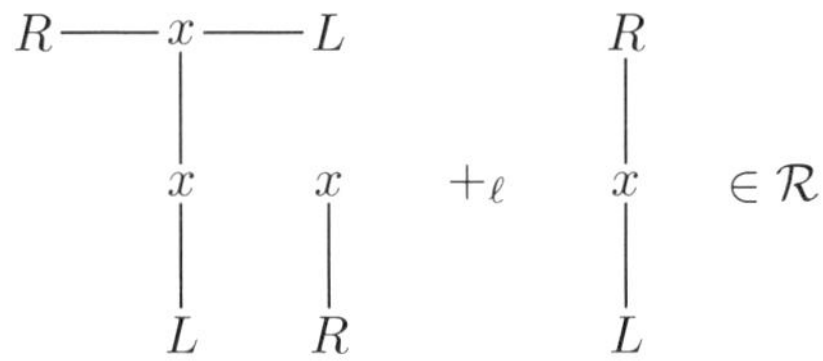

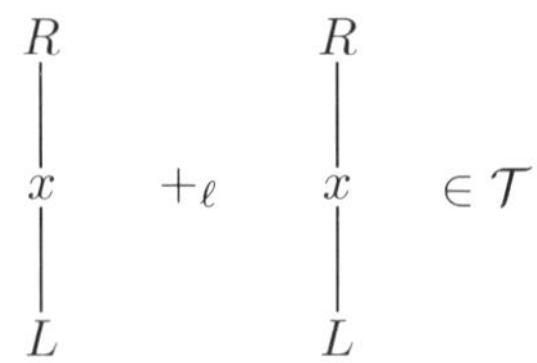

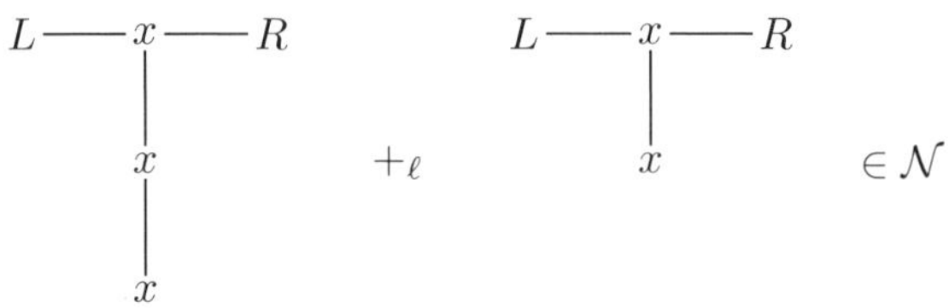

The case $G \in \mathcal{L}$, $H \in \mathcal{R}$ implies $G +_\ell H \in \mathcal{L}$ follows by symmetry. All remaining cases follow by symmetry, and therefore the theorem is proven. □

Theorem 18. *We have $G +_\ell (-G) \in \mathcal{T}$ for all $G \in \mathcal{PT}_x$.*

Proof. To prove this, first consider Left moving first, since Right moving first will follow by symmetry. Right can tie $G +_\ell (-G)$, simply by playing the "tweedle-dum, tweedle-dee" strategy, i.e., whichever move Left makes, Right makes the identical move in the opposite component. Since this will allow Right to move last on $G +_\ell (-G)$, and therefore tie the game.

We know that Right cannot win moving second, because from Theorem 16, there are no $\mathcal{P}$ positions. Therefore Right's strategy can guarantee him a tie, and $G +_\ell (-G) \in \mathcal{T}$. The theorem is proven. □

The natural question to ask is, "are the sets $\mathcal{PT}_x$ groups?" The answer to that is "no." The reason is that the games G^L are G^R are not in the set. We demand that every game $G \in \mathcal{PT}_x$ have $G^S = 0$. But, $g^{LS} = x$ and $g^{RS} = -x$, for all $g^L \in G^L$ and $g^R \in G^R$. So when a player moves, he is moving to something outside of the set.

However, what we have shown is that this particular variant of Pirates and Treasures behaves very similarly to a normal play game. In fact the similarity goes further than that. What we will attempt to convince the reader, although we have no proof of this, is that a winning a strategy for Pirates and Treasure under normal play, is identical to a "non-losing" strategy of a game in $\mathcal{PT}_x$.

Consider the game in Figure 6. If we played it under normal play, this game has value $\{-1, 0|1\} = \{0|1\} = \frac{1}{2}$. Now consider the game in Figure 7. Under normal play Right's best move, moving first, is to move the Right piece on the left hand graph down adjacent to the Left piece. This is exactly the same for scoring play. Likewise, Left's best move –moving first– is to move his piece up, so it is adjacent to the Right piece, under both normal and scoring play.

Figure 6: One half?

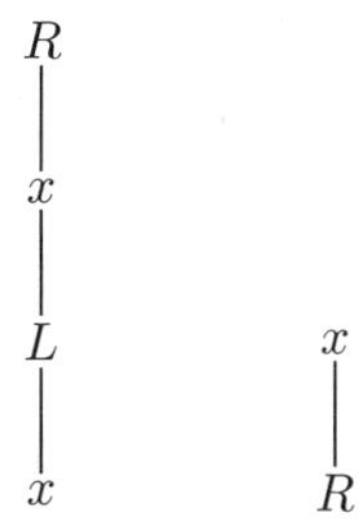

Figure 7: An $\mathcal{R}$ position, under normal and scoring play.

In fact, if the reader is particularly vigilant, and checks the best strategies in the examples given in the proof of Theorem 17, he/she will find that they are exactly the same as the best strategies under normal play. So what we have shown is that there is a non-trivial subset of scoring play games, that behaves very similarly to normal play games.

3.1.1. A Few Notes

There are several things we need to note about this idea. The first is that this will not work on the set $\mathcal{PT}_x \cup \mathcal{PT}_y$, where $x \neq y$. It will also not work if $G^S \neq 0$, i.e., before the players have moved the score is something other than zero. Finally, if we let $x = 0$, then this game is trivial since every position is a $\mathcal{T}$ position.

The reader may feel that what we did, is simply changed the rules to "last player to move wins." However, we did not not change the rules of the game; all we did was look at a particular case of the game. The real question is "are there any other scoring games that fall into this set?" So we leave the following open problem.

Problem 19. Can we define, and classify, the set of scoring play games that behave like a normal play game? If yes, which games lie in this set?

3.2. Comparison With Misère Play

In this paper we do not intend to say much about a comparison with misère play. All we really are doing is giving examples to show that there are variations where the winning strategy under scoring play is identical to the winning strategy under misère play.

To show that it is very similar to misère play is much harder, given that the general structure of misère games is not as "nice" as normal play. All we will be doing is demonstrating that the last player to move does not win, i.e., he loses or ties. Therefore, both players are trying *not* to move last, just like a misère play game.

Definition 20. $\mathcal{PT}_{-x}$, is a subset of $\mathcal{PT}$, where every node on the graph has value $x \in \mathbb{R}$, $x > 0$, and for all $G \in \mathcal{PT}_{-x}$, $G^S = 0$.

Conjecture 21. For all $G \in \mathcal{PT}_{-x}$, if $G \not\approx 0$ then $G \neq 0$.

The reason we make this a conjecture, is because we were unable to prove it. There is a good reason for that. In [12], the authors used the game G in Figure 8 to prove a similar theorem for misère games.

The idea is that if we play $G + H$, where $H^L \neq \emptyset$, then Right moving first will move on G. Left has to move on H, as he has no move on G^R, Right moves again on G. We then let the string on Left moves on G be longer than the depth of H, and Left will be forced to move last.

While this worked for misère games, it would not work for Pirates and Treasure. The problem is that there is no position in Pirates and Treasure, with a game tree that has that general shape. Either a player can move from the start of the game, or he cannot move at all. This makes proving the conjecture considerably more challenging, and it will be beyond the scope of this paper.

We will also not be examining the outcome class table for these games either. As far as a comparison to misère play, it is known that outcome of any two games

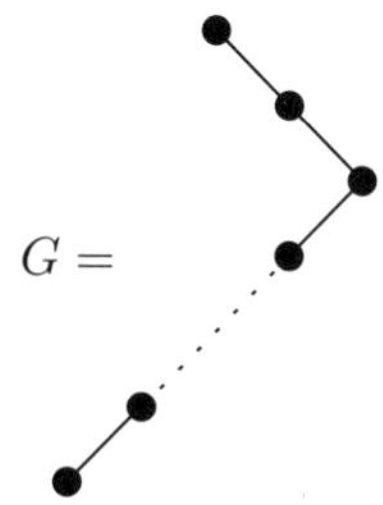

Figure 8: Game to show $G \neq 0$ under misère rules.

G and H are not related to their sum under misère play. Therefore, there is really very little that the outcome class Table would tell us.

If were were playing this game under misère rules, not every position is represented. We know that moving last is bad, so showing that the outcome class Table is also "bad" is not particularly interesting. So, this is why we will not be looking at it in this paper.

To finish this section we will simply look at an example, and show that the winning strategy under scoring play corresponds to the winning strategy under misère play, and conjecture that this is always the case for these variations. Again, this is not true if $G^S \neq 0$, or if we examine a set $\mathcal{PT}_{-x} \cup \mathcal{PT}_{-y}$, where $x \neq y$.

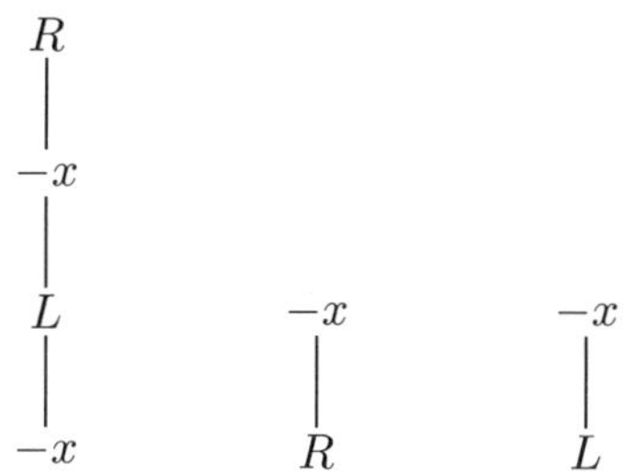

Figure 9: A $\mathcal{T}$ position under misère play.

Consider the game in Figure 9. If we played this game under misère rules, then Left's best move –moving first– would be to slide his piece, on the left hand graph, down to the lower vertex. This is exactly the same as his best move under scoring play rules. Likewise Right's best move, under misère rules is either to move his piece on the left hand graph down, or move his piece in the center graph up. Again, these correspond to best strategy under scoring rules.

Also note that the difference in playing this game with negative values and positive values on the vertices, is very similar to the difference between playing it under misère and normal play rules.

We finish with one further problem:

Problem 22. If there is a variation of a scoring game that behaves like a normal play game, will using negative values for the points players gain or lose on their turns give a game that behaves like a misère play? If no, can we classify which games have this property, and which do not?

4. Conclusion

We have introduced a new game, and shown that there are variations of this game that exhibit behavior that is very similar to normal play, and variations that exhibit behaviour that is very similar to misère play. This means that there are non-trivial subsets of scoring play games that have this same behaviour. The question is whether we can define these subsets precisely, and determine which scoring games lie in them, and which do not? We hope that we have convinced the reader that this is an interesting problem to pursue.

References

[1] E. Berlekamp, J. Conway, R. Guy, *Winning Ways for your Mathematical Plays*, Volumes 1-4, A.K. Peters (2002).

[2] J. Conway, *On Numbers and Games*, A.K. Peters (2000).

[3] J. M. Ettinger, *On the Semi-group of Positional Games*. Online at `http://citeseerx.ist.psu.edu/viewdoc/summary?doi=10.1.1.37.5278`, 1996

[4] J. M. Ettinger, *A Metric for Positional Games*, Theoretical Computer Science (Math. Games), 230 (2000) 207-219.

[5] M. R. Garey, D. S. Johnson, *COMPUTERS AND INTRACTABILITY A Guide to the Theory of NP-Completeness*, Freeman (1979).

[6] M. R. Garey, D. S. Johnson, R. E. Tarjan, *The Planar Hamiltonian Circuit Problem is NP-Complete*, SIAM J. Comput. 5 (1976),704-714.

[7] O. Hanner, *Mean Play of Sums of Positional Games*, Pacific Journal of Mathematics, 9(1) (1959), 81-99.

[8] A. Itai, C. H. Papadimitriou, J. L. Szwarcfiter, *Hamilton Paths in Grid Graphs*, SIAM J. Comput., 11:4 (1982), 676-686.

[9] W. Johnson, *The Combinatorial Game Theory of Well-Tempered Scoring Games*. Online at `http://arxiv.org/abs/1112.3610`

[10] R. M. Karp, *Reducibility Among Combinatorial Problems*, Complexity of Computer Computations, Plenum Press, New York, pg 85-103, 1972.

[11] M. Krishnamoorthy, *An NP-Hard Problem in Bipartite Graphs*, SIGACT News 7:1 (1975), 26.

[12] G. A. Mesdal, P. Ottaway, *Simplification of Partizan Games in Misère Play*, Integers 7 (2007)

[13] J. W. Milnor, *Sums of Positional Games*, Contributions to the Theory of Games II, 28. Annals of Mathematics Studies, pg 291-301, 1953.

[14] D. Moews, *Infinitesimals and Coin-Sliding*, Games of No Chance, 1996.

[15] F. Stewart, *Scoring Play Combinatorial Games*, To Appear, Games of No Chance 5. Online at `http://arxiv.org/abs/1202.4653`

[16] F. Stewart, *Impartial Scoring Play Games*, To Appear, Games of No Chance 5. Online at `http://arxiv.org/abs/1202.4655`

TAKING THE CONVOLUTED OUT OF BERNOULLI CONVOLUTIONS: A DISCRETE APPROACH

Neil J. Calkin
Dept. of Mathematical Sciences, Clemson University, Clemson, South Carolina
calkin@ces.clemson.edu

Julia Davis
Dillsburg, Pennsylvania
juliablackburn87@gmail.com

Michelle Delcourt[1]
Dept. of Mathematics, University of Illinois at Urbana-Champaign, Urbana, Illinois
delcour2@illinois.edu

Zebediah Engberg
Department of Mathematics, Dartmouth College, Hanover, New Hampshire
zeb@dartmouth.edu

Jobby Jacob
School of Math. Sciences, Rochester Institute of Technology, Rochester, New York
jxjsma@rit.edu

Kevin James
Department of Math. Sciences, Clemson University, Clemson, South Carolina
kevja@clemson.edu

Received: 7/7/12, Accepted: 4/5/13, Published: 4/10/13

Abstract

In this paper we consider a discrete version of the Bernoulli convolution problem traditionally studied via functional analysis. We discuss several innovative algorithms for computing the sequences with this new approach. In particular, these algorithms assist us in gathering data regarding the maximum values. By looking at a family of associated polynomials, we gain insight on the local behavior of the sequence itself. This work was completed as part of the Clemson University REU, an NSF funded program[2].

[1]NSF Graduate Research Fellow DGE 1144245
[2]This research was supported by NSF grant DMS-0552799

1. Introduction

The classic Bernoulli convolution problem in analysis has an elegant discrete analogue as follows. Define two maps $\text{dup}_n, \text{shf}_n : \mathbb{R}^n \longrightarrow \mathbb{R}^{3n}$ defined by

$$\text{dup}_n : (a_1, a_2, ..., a_{n-1}, a_n) \longmapsto (a_1, a_1, a_2, a_2, ..., a_{n-1}, a_{n-1}, a_n, a_n, \overbrace{0, ..., 0}^{n}) \quad (1)$$

$$\text{shf}_n : (a_1, a_2, ..., a_{n-1}, a_n) \longmapsto (\overbrace{0, ..., 0}^{n}, a_1, a_1, a_2, a_2, ..., a_{n-1}, a_{n-1}, a_n, a_n). \quad (2)$$

Consider the finite sequences recursively given by $B_0 = (1)$ and $B_{n+1} = \text{dup}_{3^n}(B_n) + \text{shf}_{3^n}(B_n)$. B_n is called the nth *level Bernoulli sequence.* The names "dup" and "shf" reference the duplication and shifting of the coordinates.

In this paper, we are primarily interested in the rate at which the maximum value of B_n is growing with n. We develop two independent algorithms to do this. The first gives a recursive method for computing the entire sequence B_n when n is small. By encoding the sequence as coefficients of a polynomial, the second gives a method for computing specified entries of B_n when n is larger. In addition, we provide numerical data concerning B_n.

1.1. Motivation

Classically, Bernoulli convolutions have been studied as a problem in functional analysis. A Bernoulli convolution is obtained as an infinite convolution of Bernoulli measures [1]. The Bernoulli measure, denoted by $b(X)$, is the measure corresponding to the discrete probability density function on the real line with value $1/2$ at 1 and -1. The Bernoulli convolution with parameter q for $0 < q < 1$ is the measure

$$\mu_q(X) = b(X) * b(X/q) * b(X/q^2) * \dots .$$

This measure was first studied by Jessen and Wintner [4]. They showed that μ_q is continuous for any q.

A different perspective on Bernoulli convolutions is obtained through a functional equation. For $0 < q < 1$, consider the functional equation

$$F(t) = \frac{1}{2}F\left(\frac{t-1}{q}\right) + \frac{1}{2}F\left(\frac{t+1}{q}\right) \quad (3)$$

for t on the interval $I_q := [-1/(1-q), 1/(1-q)]$. It can be shown that there is a unique bounded solution $F_q(t)$ to the above equation. Moreover, $F_q(t)$ is the distribution function of μ_q, that is $F_q(t) = \mu_q((-\infty, t])$. For an introduction to this, refer to Chapter 5 in *Experimental Mathematics in Action* [1]. For a more in depth analysis, refer to *Sixty years of Bernoulli convolutions* [6].

Jessen and Winter, in [4], showed that $F_q(t)$ is either absolutely continuous or purely singular. The major question regarding the solutions of (3) is to determine the values of q which make $F_q(t)$ absolutely continuous. If $F_q(t)$ is absolutely continuous, rather than considering the function $F_q(t)$, one may consider its derivative $f_q(t) := F_q'(t)$. Upon differentiating, the functional equation for $F_q(t)$ yields the following equation:

$$f(t) = \frac{1}{2q} f\left(\frac{t-1}{q}\right) + \frac{1}{2q} f\left(\frac{t+1}{q}\right). \tag{4}$$

The existence of an absolutely continuous solution $F_q(t)$ to (3) is equivalent to the the existence of an $L^1(I_q)$ solution $f_q(t)$ to (4).

When $0 < q < 1/2$, Kershner and Wintner [5] proved that $F_q(t)$ is always singular. For these values of q, the solution $F_q(t)$ is an example of a *Cantor function*, a function that is constant almost everywhere. It can be shown that if $q = 1/2$, then the solution $F_q(t)$ is absolutely continuous.

The case when $q > 1/2$, however, is significantly harder and more interesting. In 1939, Erdős [3] showed that $F_q(t)$ is again singular for q of the form $q = 1/\theta$ with θ a Pisot number. There is little else that is known for other values of $q > 1/2$. One interesting result due to Solomyak [7] is that almost every $q > 1/2$ yields a solution $F_q(t)$ that is absolutely continuous. Hence it is surprising that no actual example of such a q is known. Specifically, the case when $q = 2/3$ remains a mystery.

In [1], Girgensohn asks the question of computing $f_q(t)$ for various values of q. The author starts an arbitrary initial function $f^0(t) \in L^1(I_q)$ and iterates the transform

$$T_q : f(t) \longmapsto \frac{1}{2q} f\left(\frac{t-1}{q}\right) + \frac{1}{2q} f\left(\frac{t+1}{q}\right) \tag{5}$$

to gain a sequence of functions $f^0, f^1, f^2, \ldots.$ He shows that if this sequence of functions converges to a bounded function, then it converges to the unique solution of (4).

Calkin [2] specifically looked at the above process for $q = 2/3$. Rather than working on the interval I_q, we shift the entire interval to $[0, 1]$ for simplicity. The transform T_q now becomes the map $T : L^1([0,1]) \longrightarrow L^1([0,1])$ where

$$T : f(x) \longmapsto \frac{3}{4} f\left(\frac{3x}{2}\right) + \frac{3}{4} f\left(\frac{3x-1}{2}\right). \tag{6}$$

Geometrically, this transform (6) takes two scaled copies of $f(x)$: one on the interval $[0, 2/3]$ and the other on $[1/3, 1]$, and adds them. The scaling factor of $3/4$ gives us that

$$\int_0^1 f(x)dx = \int_0^1 Tf(x)dx.$$

In other words, the average value of $Tf(x)$ equals that of $f(x)$. In this setting, the question now reads: starting with the constant function $f^0(x) = 1$, does the iteration determined by the transform in (6) converge to a bounded function?

1.2. Discrete Version

Rather than viewing T as a transform on $[0,1]$, we consider the discrete analogue mentioned earlier. The sequences $\text{dup}(B_n)$ and $\text{shf}(B_n)$ defined in (1) and (2) are analogous to the two shifted copies of the function on $[0,1]$ in (6). We start with the sequence $B_0 = (1)$. We recursively generate sequences given by

$$B_{n+1} = \text{dup}_{3^n}(B_n) + \text{shf}_{3^n}(B_n). \tag{7}$$

We are interested in the global behavior of B_n. In particular, we investigate the growth rate of the maximum element of B_n as n grows large.

Before we discuss our results, we present some notations that are useful throughout this paper. We call B_n the *Bernoulli sequence on level n.* We refer to the map $(\text{dup}_n + \text{shf}_n) : \mathbb{R}^n \longrightarrow \mathbb{R}^{3n}$ seen in (7) as the process of *duplicate, shift, add* or DSA for short. We usually write the nth level Bernoulli sequence as $B_n = (b_0, b_1, ..., b_t)$ where $t = 3^n - 1$. The fact that B_n has a total of 3^n terms follows directly from the definition of dup_n and shf_n in (1) and (2). We define $m_n := \max(B_n)$.

```
1 → 1 1
      1 1
    -----
    1 2 1

1 2 1 → 1 1 2 2 1 1
              1 1 2 2 1 1
        -----------------
        1 1 2 3 2 3 2 1 1

1 1 2 3 2 3 2 1 1 → 1 1 1 1 2 2 3 3 2 2 3 3 2 2 1 1 1 1
                                      1 1 1 1 2 2 3 3 2 2 3 3 2 2 1 1 1 1
                    -----------------------------------------------------
                    1 1 1 1 2 2 3 3 2 3 4 4 3 4 3 4 4 3 2 3 3 2 2 1 1 1 1
```

Figure 1: This shows the process of DSA at three low levels. The figure demonstrates computation of the Bernoulli sequence on level $n+1$ from the Bernoulli sequence on level n for $n = 0, 1$ and 2 using the definition.

It is immediate from the definition that the maximum value of B_n must occur on the middle third of the Bernoulli sequence. Furthermore, because the sequence is palindromic, the maximum must occur on the first half of the middle third.

The following observation can be proved using induction on n and the definition of B_n in (7).

Observation 1.1. *The mean of the elements of* B_n *is* $(4/3)^n$.

The major question that we considered is whether m_n also grows like $(4/3)^n$. A positive result here is equivalent to the existence of a bounded solution to the functional equation (4) for $q = 2/3$. $F_q(t)$ is absolutely continuous at $q = 2/3$ if and only if two conditions hold: first, $m_n = O((4/3)^n)$, and secondly, the iterations of the step functions is pointwise continuous. Beginning with a computational approach, we address this question and other related questions throughout this paper.

2. Recursive Attempts at Computing the Bernoulli Sequence

2.1. The Naive Method: DSA

The process of *duplicate, shift, add* gives a naive method for computing Bernoulli sequences. Despite the simplicity in describing this process, DSA is not computationally feasible for large values of n. The primary shortfall of the DSA method is that each Bernoulli level B_n has 3^n terms—each successive computation takes roughly three times as long as the previous. Likewise, at each successive level we must keep track of three times as many entries as on the previous level. Using the DSA method without modifications, we have been able to compute B_n for $n = 1, 2, ..., 20$. However, the fact that one must know all 3^n entries of level n to compute level $n + 1$ limits the practicality of this algorithm.

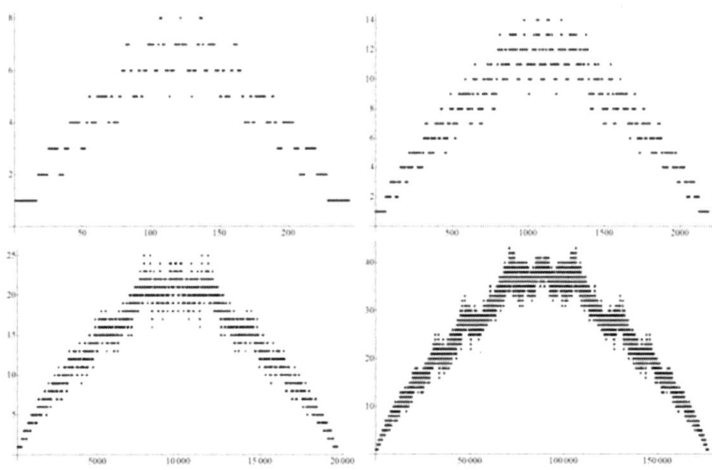

Figure 2: This shows some of the low level Bernoulli sequences generated using DSA. In the above plots, the horizontal axis gives the index and the vertical axis gives the entry corresponding to that index in the Bernoulli sequence. Moving clockwise from the top left, we see B_n for $n = 5, 7, 11$ and 9.

2.2. The Improvement: DEM

We now consider an alternate approach that addresses some of the issues arising from the DSA algorithm. The process we call *double, enlarge, merge*, abbreviated DEM, is a way of encoding the Bernoulli sequence B_n as a sequence of length $2(2^n - 1)$. The advantage with DEM is that this auxiliary sequence is in the order of 2^n as opposed to the 3^n size increase required for the DSA process. The DEM algorithm is based on the observation that in a given Bernoulli sequence, many individual entries are consecutively repeated. Rather than keeping consecutive repeats, we only keep the index where the Bernoulli sequence either increases or decreases. For this encoding to be meaningful, it is important to note that when comparing two consecutive entries in a Bernoulli sequence, the difference between these entries will never be greater than one in absolute value. This can be proved inductively from the definition in (7).

Given the Bernoulli sequence $B_n = (b_0, b_1, ..., b_t)$ with $t = 3^n - 1$, the DEM representation is $(d_1, ..., d_r)$ where d_i is the index of the ith jump in B_n up to a sign (by jump, we mean two consecutive entries which are not equal). Suppose the ith jump occurs at index j in B_n, so b_j and b_{j+1} are different. Then

$$d_i = \begin{cases} j+1 & \text{if } b_j < b_{j+1} \\ -(j+1) & \text{if } b_j > b_{j+1}. \end{cases}$$

The DEM representation of $B_2 = (1, 1, 2, 3, 2, 3, 2, 1, 1)$ is $(2, 3, -4, 5, -6, -7)$.

We now describe how to translate the DSA method to this new representation. The process of *duplicate* becomes *double*: each element from the original list is multiplied by two. The process of *shift* becomes *enlarge*: each element from the doubled list is modified by 3^n as follows: if the element is positive, we add 3^n, for negative elements we subtract 3^n. The process of add translates to *merge*: we discard the original sequence and consider the new lists attained in the *double* and *enlarge* processes. We then add two additional elements 3^n and $-2(3^n)$ to these lists, respectively. Finally, we merge sort the elements of the two lists according to their absolute value to obtain the DEM representation of the next Bernoulli level.

2.3. Data

Using the DEM method, we are able to compute the Bernoulli sequence B_n up to level $n = 27$. For each of these levels, the maximum value is of particular interest. Typically, the maximum is attained many times on a given level. Although we found no clear pattern regarding the location of the maximum values, we do see obvious patterns in the convergence of the maximums themselves.

Below we provide a table detailing the data we collected using the DSA and DEM algorithms.

Level n	m_n	$m_n(3/4)^n$	Level n	m_n	$m_n(3/4)^n$
0	1	1	14	99	1.763976853
1	2	1.5	15	131	1.750613395
2	3	1.6875	16	176	1.763976853
3	4	1.6875	17	232	1.743931662
4	6	1.8984375	18	309	1.742052425
5	8	1.8984375	19	410	1.733595860
6	11	1.957763672	20	545	1.728310507
7	14	1.868774414	21	728	1.731481719
8	18	1.802032471	22	962	1.716022061
9	25	1.877117157	23	1283	1.716468012
10	33	1.858345985	24	1705	1.710782128
11	43	1.816110849	25	2266	1.705263476
12	56	1.773875713	26	3024	1.706768563
13	75	1.781794801	27	4025	1.703805423

Table 1: m_n denotes the maximum value at each level. The maximums m_n appear to grow like $(4/3)^n$. The fact that $m_n(3/4)^n$ seems to be converging, albeit slowly, supports this claim.

3. A Polynomial Approach to DSA

3.1. Translating DSA Into a Polynomial Recursion

By encoding these sequences as coefficients of polynomials, the process of *duplicate, shift, add* gives a particularly nice recursive relation among the polynomials. Let $B_n = (b_0, b_1, ..., b_t)$ be the Bernoulli sequence on level n where $t = 3^n - 1$. Consider the polynomial $p_n(x) := b_0 + b_1 x + ... + b_t x^t$.

We describe how these polynomials behave under the process of DSA. We see that the duplication $b_0, b_0, b_1, b_1, ..., b_r, b_r$ corresponds to the polynomial $(1+x)p_n(x^2)$. Shifting the sequence 3^n places to the right corresponds to multiplication by x^{3^n}. By adding the duplicate and the shift of the sequence, we arrive at the sequence on the next level. This yields the recursive relation

$$p_{n+1}(x) = (1+x)p_n(x^2)\left(1 + x^{3^n}\right) \tag{8}$$

with $p_0(x) = 1$.

3.2. Explicit Formula for $p_n(x)$

This formula (8) allows us to explicitly solve for $p_n(x)$.

Theorem 3.1. *For $n \geq 1$, the polynomials $p_n(x)$ satisfy*

$$p_n(x) = \prod_{i=0}^{n-1}\left(1 + x^{2^i}\right)\prod_{j=0}^{n-1}\left(1 + x^{2^{n-1}(3/2)^j}\right). \tag{9}$$

Proof. We proceed by induction on n. We know $p_1(x) = 1 + 2x + x^2$, and thus the formula holds for the case $n = 1$. Assume the formula holds for $p_n(x)$. We have

$$\begin{aligned}
p_{n+1}(x) &= (1+x)p_n(x^2)\left(1 + x^{3^n}\right) \\
&= (1+x)\prod_{i=0}^{n-1}\left(1 + (x^2)^{2^i}\right)\prod_{j=0}^{n-1}\left(1 + (x^2)^{2^{n-1}(3/2)^j}\right)\left(1 + x^{3^n}\right) \\
&= (1+x)\prod_{i=0}^{n-1}\left(1 + x^{2^{i+1}}\right)\prod_{j=0}^{n-1}\left(1 + x^{2^n(3/2)^j}\right)\left(1 + x^{2^n(3/2)^n}\right) \\
&= \prod_{i=0}^{n}\left(1 + x^{2^i}\right)\prod_{j=0}^{n}\left(1 + x^{2^n(3/2)^j}\right).
\end{aligned}$$

□

3.3. A Bound on the Coefficients

By factoring p_n, we can obtain a bound on how fast the coefficients grow with the level n.

Theorem 3.2. *We have $m_n = O((\sqrt{2})^n)$.*

Proof. Define polynomials q_n, r_n, s_n by

$$q_n(x) = \prod_{i=0}^{n-1}\left(1 + x^{2^i}\right) \qquad s_n(x) = \prod_{\substack{1 \leq j \leq n-1 \\ j \text{ odd}}}\left(1 + x^{2^{n-1}(3/2)^j}\right)$$

$$r_n(x) = \prod_{\substack{1 \leq j \leq n-1 \\ j \text{ even}}}\left(1 + x^{2^{n-1}(3/2)^j}\right) = \prod_{j=1}^{\lfloor (n-1)/2 \rfloor}\left(1 + x^{2^{n-1}(9/4)^j}\right).$$

We see that

$$p_n(x) = q_n(x)\left(1 + x^{2^{n-1}}\right)r_n(x)s_n(x).$$

Consider the polynomial $q_n(x)r_n(x)$. Because $9/4 > 2$, we are left with distinct powers of x when we expand $q_n(x)r_n(x)$. In other words, the coefficients of $q_n(x)r_n(x)$ are all either 0 or 1. Hence the coefficients of $q_n(x)r_n(x)(1 + x^{2^{n-1}})$ are all either 0, 1, or 2. In particular, the coefficients are bounded. On the other hand, there are at most $n/2$ terms in the product defining $s_n(x)$. Thus there are at most $2^{n/2}$ nonzero terms in the polynomial $s_n(x)$. Also, the coefficients of $s_n(x)$ are bounded as in the case of $r_n(x)$. Therefore the coefficients of $p_n(x)$ are all $O(2^{n/2}) = O((\sqrt{2})^n)$. □

3.4. An Algorithmic Implementation: PIP

The fact that the Bernoulli sequence can be realized as the coefficients of an explicitly defined polynomial provides us with an algorithm for computing specified values on high levels. The algorithms DSA and DEM are useful for computing entire levels, though this task becomes impossible for large n due to the recursive nature of the algorithm. The following algorithm, which we dub PIP, allows us to compute values on much higher levels. The algorithm is non-recursive—we need no lower levels to compute entries of B_n. The name PIP stands for *polynomial isolated point*; this algorithm computes the entry corresponding to a given index on a given level.

Our algorithm is based on the following idea. Suppose $S = \{a_1, ..., a_n\}$ is a sequence of positive integers. Consider the polynomial

$$f(x) = \prod_{i=1}^{n}(1 + x^{a_i}) = \sum_{j=1}^{m} \alpha_j x^j.$$

Then α_j is the number of ways that j can be written as a sum of distinct elements from S [8]. This idea is applicable to the coefficients of our polynomial because our polynomial is a product of terms of the form $(1 + x^a)$. Applying this idea to our situation, we find that the coefficient b_j of x^j in $p_n(x)$ (which is the jth entry on the nth Bernoulli sequence) is precisely the number of ways that j can be written as a sum of distinct terms in the sequence

$$S = \{1, 2, 4, ..., 2^{n-2}, 2^{n-1}, 2^{n-1}, 2^{n-2}3, 2^{n-3}3^2, ..., 2^2 3^{n-3}, 2^1 3^{n-2}, 3^{n-1}\}. \qquad (10)$$

The values in the sequence S are just the exponents of the factors of $p_n(x)$ in (9).

We now outline an algorithm that can be used to calculate the entry b_j for a fixed level n. Let $S = \{a_1, ..., a_n\}$ where each $a_i > 0$. Let $\mathrm{N}_S(k)$ denote the number of ways k can be written as a sum of elements from S. Our entire algorithm is based on the following observations:

- For any $i \in \{1, ..., n\}$, we have

$$\mathrm{N}_S(k) = \mathrm{N}_{S\setminus\{a_i\}}(k) + \mathrm{N}_{S\setminus\{a_i\}}(k - a_i). \qquad (11)$$

- If $k > \sum_{s \in S} s$, then $\mathrm{N}_S(k) = 0$.

- If $k < 0$, then $\mathrm{N}_S(k) = 0$.

These three relations provide an algorithm for computing the jth entry on a given Bernoulli level. However, our sequence S in (10) takes on a particularly nice form—taking advantage of this we can increase the efficiency of our algorithm. Let

S be as in (10). Suppose $0 < k < 2^{n-1}$. Any element of S containing a positive power of 3 is strictly larger than k itself, so these terms become useless when writing k as a sum of elements in S. Hence, we are left with the distinct powers of 2 in S. But k can now be written uniquely; this is simply the binary expansion of k. So $N_S(k) = 1$ in this case.

Based on the above ideas, we implement a tree branching algorithm to compute $N_S(k)$ for various values of k and n. See Figure 3 for a specific example of the PIP algorithm.

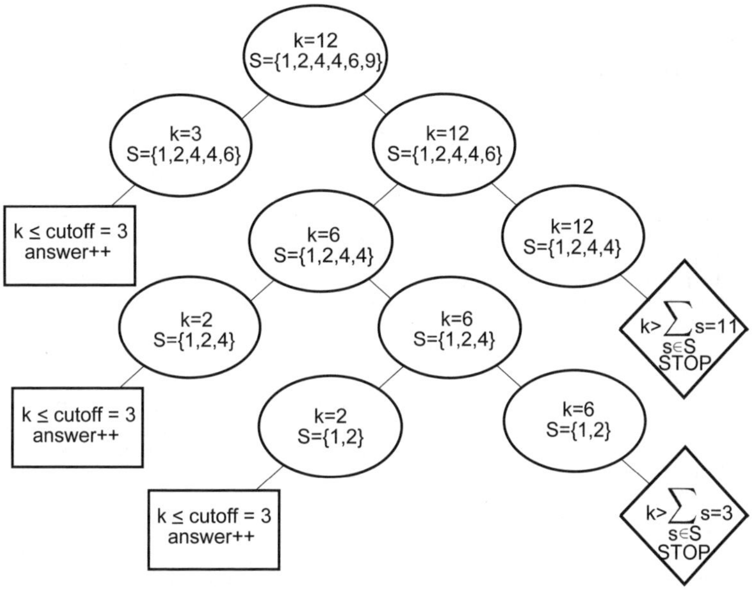

Figure 3: The above flowchart illustrates the PIP algorithm. It shows the computation of $\mathrm{N}_S(k)$ for $k = 12$ and $S = \{1, 2, 4, 4, 6, 9\}$. The boxes containing the word "cutoff" account for the fact that if $0 < k < 2^{n-1}$, then there is a unique way to write k as a sum of elements in S (the binary expansion of k). As the diagram suggests, $\mathrm{N}_S(k) = 3$, corresponding to the fact that there are three boxes containing the word answer++. Hence the 12th entry on the 3rd Bernoulli level is 3.

3.5. Data

Our PIP algorithm is used to compute isolated points on B_n for a given n. It is infeasible to compute entire levels using PIP, so we are not able to calculate the global maximum of a given level with this method. Instead, PIP provides us with local information on the Bernoulli sequence. In particular, the PIP algorithm gives evidence that the Bernoulli sequence B_n is converging uniformly as n grows large.

Furthermore, we can gather local data on much higher levels than we were able to with DSA or DEM.

We now describe one way in which PIP can be used to gather local information on the Bernoulli sequence. Fix α in $[0,1]$. Consider the index $k = \lceil \alpha(3^n - 1) \rceil$. We now use PIP to compute the entry b_k at index k for levels $n = 1, 2, ..., 50$ (we have used PIP to compute individual entries on levels as high as $n = 70$). Finally, we take the quotient of b_k by $(4/3)^n$. This data is presented in Table 2.

Level	0.36	0.38	0.4	0.42	0.44	0.46	0.48	0.5
1	1.500000	1.500000	1.500000	1.500000	1.500000	1.500000	1.500000	1.500000
2	1.687500	1.687500	1.687500	1.687500	1.687500	1.125000	1.125000	1.125000
3	1.265625	1.687500	1.687500	1.687500	1.687500	1.265625	1.265625	1.687500
4	1.265625	1.582031	1.582031	1.265625	1.582031	1.265625	1.582031	1.898438
5	1.186523	1.186523	1.661133	1.423828	1.898438	1.423828	1.423828	1.898438
6	1.245850	1.423828	1.779785	1.423828	1.779785	1.423828	1.423828	1.779785
7	1.201355	1.334839	1.735291	1.601807	1.601807	1.334839	1.601807	1.601807
8	1.201355	1.301468	1.802032	1.701920	1.501694	1.401581	1.501694	1.802032
9	1.201355	1.351524	1.877117	1.802032	1.501694	1.351524	1.576778	1.802032
10	1.182584	1.407838	1.858346	1.689405	1.464151	1.520465	1.520465	1.576778
11	1.309289	1.435995	1.816111	1.562700	1.393759	1.520465	1.647170	1.689405
12	1.330407	1.393759	1.742199	1.615494	1.488789	1.520465	1.647170	1.710523
13	1.354164	1.401679	1.710523	1.591737	1.472950	1.520465	1.591737	1.520465
14	1.336346	1.425436	1.674887	1.567979	1.496708	1.478890	1.621433	1.639251
15	1.296256	1.469981	1.643706	1.576888	1.496708	1.496708	1.630342	1.630342
16	1.312960	1.433231	1.603615	1.573548	1.533457	1.493367	1.583570	1.583570
17	1.315466	1.450771	1.578559	1.563525	1.525940	1.510906	1.563525	1.623661
18	1.324862	1.460167	1.578559	1.544733	1.516544	1.482718	1.572921	1.668762
19	1.314996	1.429160	1.594063	1.530638	1.530638	1.496812	1.577149	1.640574
20	1.319224	1.436559	1.585606	1.525353	1.534867	1.474614	1.569750	1.617318
21	1.305747	1.446073	1.591156	1.534074	1.541209	1.479370	1.560236	1.584020
22	1.309314	1.455586	1.589372	1.539425	1.539425	1.478776	1.566182	1.641102
23	1.311098	1.446221	1.581345	1.547898	1.547898	1.485019	1.577331	1.650913
24	1.312436	1.453914	1.585358	1.551243	1.535189	1.483012	1.578334	1.609440
25	1.310178	1.455419	1.592382	1.563785	1.537446	1.478748	1.576579	1.610443
26	1.304910	1.448270	1.603482	1.563409	1.533495	1.479877	1.581470	1.609690
27	1.308438	1.448129	1.609832	1.559881	1.526864	1.478607	1.577237	1.601789
28	1.308967	1.444848	1.615017	1.556601	1.532790	1.485486	1.579142	1.612160
29	1.315555	1.448896	1.616049	1.554379	1.529853	1.476517	1.576046	1.618192
30	1.308114	1.451872	1.616346	1.554914	1.527055	1.471516	1.577594	1.610810
31	1.304274	1.449729	1.617150	1.550718	1.531565	1.470757	1.577639	1.608846
32	1.304073	1.447017	1.616983	1.552593	1.536219	1.473034	1.575094	1.610252
33	1.307213	1.447419	1.616631	1.551990	1.531498	1.474767	1.572030	1.613467
34	1.308757	1.447589	1.615463	1.549579	1.532515	1.474315	1.575458	1.615463
35	1.308474	1.449637	1.616141	1.549989	1.529139	1.473750	1.574780	1.617582
36	1.308125	1.449245	1.617667	1.550127	1.530548	1.474640	1.574123	1.616396
37	1.307886	1.450556	1.617897	1.552057	1.530150	1.474251	1.574417	1.617445
38	1.307892	1.450008	1.619012	1.551682	1.529727	1.473464	1.574763	1.617743
39	1.307544	1.450535	1.619714	1.550310	1.529070	1.473370	1.575921	1.617233
40	1.307356	1.450019	1.620015	1.550705	1.529959	1.475150	1.575666	1.616817
41	1.308606	1.449664	1.620206	1.550265	1.529659	1.475490	1.574831	1.616616
42	1.308272	1.448889	1.620105	1.550073	1.529771	1.476246	1.575167	1.617084
43	1.309044	1.449437	1.619960	1.550729	1.529682	1.477035	1.574522	1.616727
44	1.309736	1.449494	1.619917	1.550938	1.529485	1.475599	1.574488	1.617941
45	1.308585	1.449791	1.619751	1.550583	1.529578	1.475369	1.574587	1.617180
46	1.308839	1.449973	1.619832	1.550568	1.529498	1.476223	1.574642	1.615671
47	1.309522	1.450225	1.619787	1.550845	1.529697	1.476398	1.574457	1.617048
48	1.309481	1.449946	1.619833	1.550384	1.529741	1.476457	1.574743	1.616217
49	1.309458	1.450029	1.619892	1.550488	1.529521	1.476353	1.575004	1.616225
50	1.309438	1.450046	1.619848	1.550631	1.529865	1.476265	1.574653	1.616802

Table 2: We consider various values for α (appearing in the top row), and compute associated entry of the Bernoulli sequence. We chose these specific α values because the only portion of the sequence of interest is the first half of the middle third.

4. Conclusion

Studying Bernoulli convolutions through a discrete lens sheds much new insight on the subject. Many of our algorithms would not have been discovered without combinatorial thinking—for example, the PIP algorithm is made possible by the fact that the coefficients of a polynomial can be thought of as counting the number of representations of integers as sums from a certain sequence. The discrete point of view is a very simple way to think about Bernoulli convolutions (the *duplicate, shift, add* method could be explained to a small child), but a computer has trouble computing more than a handful of Bernoulli sequences. In particular, studying Bernoulli convolutions via combinatorics has led to the discovery and development of two elegant algorithms (DEM and PIP). Using these algorithms we were able to generate the entire Bernoulli sequences at many new levels (up to 27) and also were also able to calculate individual entries of B_n at levels as high as 70.

We conjecture that $m_n = O((4/3)^n)$. Our two algorithms provide much data to support this claim.

References

[1] D. H. Bailey, J. M. Borwein, N. J. Calkin, R. Girgensohn, D. R. Luke, and V. H. Moll. *Experimental mathematics in action.* A K Peters Ltd., Wellesley, MA, 2007.

[2] N. Calkin. Counting kings, collectings coupons, and other applications of linear algebra to combinatorics. *Clemson University REU Colloquium Lecture*, 2008.

[3] P. Erdős. On a family of symmetric Bernoulli convolutions. *Amer. J. Math.*, 61:974–976, 1939.

[4] B. Jessen and A. Wintner. Distribution functions and the Riemann zeta function. *Trans. Amer. Math. Soc.*, 38(1):48–88, 1935.

[5] R. Kershner and A. Wintner. On symmetric Bernoulli Convolutions. *Amer. J. Math.*, 57(3):541–548, 1935.

[6] Y. Peres, W. Schlag, and B. Solomyak. Sixty years of Bernoulli convolutions. In *Fractal geometry and stochastics, II (Greifswald/Koserow, 1998)*, volume 46 of *Progr. Probab.*, pages 39–65. Birkhäuser, Basel, 2000.

[7] B. Solomyak. On the random series $\sum \pm\lambda^n$ (an Erdős problem). *Ann. of Math. (2)*, 142(3):611–625, 1995.

[8] H. S. Wilf. *generatingfunctionology.* A K Peters Ltd., Wellesley, MA, third edition, 2006.

SOME RESULTS ON BALANCING, COBALANCING, (a, b)-TYPE BALANCING, AND (a, b)-TYPE COBALANCING NUMBERS

Bouroubi Sadek[1]
Department of Operational Research, Faculty of Mathematics, USTHB University, Algiers, Algeria
bouroubis@yahoo.fr and sbouroubi@usthb.dz

Debbache Ali[2]
Dept. of Algebra, Faculty of Mathematics, USTHB University, Algiers, Algeria
a_debbache2003@yahoo.fr

Received: 4/16/12, Revised: 12/2/12, Accepted: 4/7/13, Published: 4/10/13

Abstract

In this paper, we present new results on balancing, cobalancing, (a, b)-type balancing and (a, b)-type cobalancing numbers as well as establish some new identities.

1. Introduction and Notation

A positive integer n is called by Behera et al. a balancing number [1], if there exists a positive integer r, which is called the balancer of n, such that:

$$1 + 2 + \cdots + (n - 1) = (n + 1) + (n + 2) + \cdots + (n + r). \tag{1}$$

Panda [4] sets $n = 1$ as the first balancing number and $r = 0$ as its corresponding balancer. Panda et al. [5] define cobalancing numbers as the solutions to the diophantine equation:

$$1 + 2 + \cdots + n = (n + 1) + (n + 2) + \cdots + (n + r), \tag{2}$$

where r is the cobalancer of n.

[1]Supported by L'IFORCE Laboratory
[2]Supported by Laboratory of Algebra

Throughout this paper, we denote by B_m, R_m, b_m and r_m, the mth balancing number, the mth balancer, the mth cobalancing number and the mth cobalancer, respectively. These numbers have already been extensively investigated in several papers.

2. Background

The present work is strongly connected to the theory of diophantine equations and more specifically, to the integer solutions of the following equation in two variables:

$$x^2 - 2y^2 = u^2 - 2v^2, \tag{3}$$

where u and v are integers. Note that for $u = \pm 1$ and $v = 0$, Equation (3) is Pell's equation. It is well known, that the form $x^2 - 2y^2$ is irreducible over the field $\mathbb{Q}$ of rational numbers, but in the extension field $\mathbb{Q}(\sqrt{2})$ it can be factored as a product of linear factors $(x + y\sqrt{2})(x - y\sqrt{2})$. Using the norm concept for the extension field $\mathbb{Q}(\sqrt{2})$, Equation (3) which has $\xi = u + v\sqrt{2}$ as solution, can be written in the form:

$$N(x + y\sqrt{2}) = N(\xi). \tag{4}$$

It is easily checked that the set of all numbers of the form $x + y\sqrt{2}$, where x and y are integers, form a ring, which is denoted $\mathbb{Z}[\sqrt{2}]$. The subset of units of $\mathbb{Z}[\sqrt{2}]$, which we denote $\mathcal{U}$ forms a group. It is easy to show that $\alpha \in \mathcal{U}$ if and only if $N(\alpha) = \pm 1$ [2]. Applying Dirichlet's Theorem of units via subtle calculations, we can show that $\mathcal{U} = \{\pm \left(1 + \sqrt{2}\right)^m,\ m \in \mathbb{Z}\}$. Since

$$N\left(\left(1 + \sqrt{2}\right)^m\right) = N\left(\left(1 + \sqrt{2}\right)\right)^m = (-1)^m, \tag{5}$$

we obtain

$$N(\alpha) = +1 \Leftrightarrow \alpha = \left(1 + \sqrt{2}\right)^{2m},\ m \in \mathbb{Z}, \tag{6}$$

and

$$N(\alpha) = -1 \Leftrightarrow \alpha = \left(1 + \sqrt{2}\right)^{2m+1},\ m \in \mathbb{Z}. \tag{7}$$

For any $\alpha \in \mathcal{U}$ with $N(\alpha) = 1$, Equation (4) becomes

$$N(x + y\sqrt{2}) = N(\alpha\xi).$$

Thus, all integral solutions of Equation (3) have take the form:

$$x + y\sqrt{2} = \xi \left(1 + \sqrt{2}\right)^{2m},\ m \in \mathbb{Z}. \tag{8}$$

3. Preliminary Results

From (1) we have

$$r^2 + (2n+1)\,r - n\,(n-1) = 0. \tag{9}$$

The discriminant Δ of Equation (9) with respect to r is $\Delta = 8n^2 + 1$. Then

$$r = \frac{-(2n+1) + \sqrt{8n^2+1}}{2}. \tag{10}$$

Since r is a positive integer, $8n^2+1$ is a perfect square, i.e., $8n^2+1 = u^2$, with u odd. Therefore

$$2n^2 = \left(\frac{u-1}{2}\right)\left(\frac{u+1}{2}\right). \tag{11}$$

Letting $A = \dfrac{u-1}{2}$, we get from (10) and (11)

$$r = A - n, \tag{12}$$

and

$$n^2 = \frac{A\,(A+1)}{2} = 1 + \cdots + A. \tag{13}$$

Consequently, n^2 is a triangle number (see also [1]).

`Case 1.` If A is even, then from (13) we have $n^2 = \frac{A}{2}\,(A+1)$. Letting $a = \frac{A}{2}$, we get

$$n^2 = a\,(2a+1). \tag{14}$$

Since a and $2a+1$ are coprime, they are both necessarily perfect squares. Hence, from (12) and (14), we get

$$\begin{aligned} a &= d^2, \\ r &= 2d^2 - n, \\ n &= d\sqrt{2d^2+1}. \end{aligned} \tag{15}$$

`Case 2.` If A is odd, we obtain from (13) that $n^2 = \left(\frac{A+1}{2}\right) A\cdot$ Letting $a = \frac{A+1}{2}$, we get

$$n^2 = a\,(2a-1). \tag{16}$$

Since a and $2a-1$ are coprime, they are necessarily both perfect squares. Hence, from (12) and (16), we get

$$\begin{aligned} a &= d^2, \\ r &= 2d^2 - n - 1, \\ n &= d\sqrt{2d^2-1}. \end{aligned} \tag{17}$$

Now we are in a position to formulate our result as follows:

Theorem 1. *Let n be a positive integer. The number n is a balancing number if and only if there exists a proper divisor d of n (except for $n = 1$) for which $2d^2 + 1$ or $2d^2 - 1$ is a perfect square. The pair (n, r) of each balancing with its cobalancer is then explicitly given by*

$$(n,r) = \begin{cases} \left(d\sqrt{2d^2+1},\ \ 2d^2 - n\right) & \text{if} \quad 2d^2+1 \text{ is a perfect square,} \\ \left(d\sqrt{2d^2-1},\ \ 2d^2 - n - 1\right) & \text{if} \quad 2d^2-1 \text{ is a perfect square.} \end{cases}$$

Table 1 summarizes the 10 first balancing numbers based on Theorem 1.

d	$2d^2-1$	$2d^2+1$	n	r
1	1		$1\sqrt{1} = 1$	0
2		9	$2\sqrt{9} = 6$	2
5	49		$5\sqrt{49} = 35$	14
12		289	$12\sqrt{289} = 204$	84
29	1681		$29\sqrt{1681} = 1189$	492
70		9801	$70\sqrt{9801} = 6930$	2870
169	57121		$169\sqrt{57121} = 40391$	16730
408		332929	$408\sqrt{332929} = 235416$	97512
985	1940449		$985\sqrt{1940449} = 1372105$	568344
2378		11309769	$2378\sqrt{11309769} = 7997214$	3312554

Table 1.

Remark 1. Theorem 1 proves that no prime number could be a balancing number. This result was also obtained by Panda et al., who showed that $B_m = P_m Q_m$, where P_m and Q_m are the mth Pell number and the mth associated Pell number respectively [6].

4. An Explicit Formula for Balancing Numbers and Some New Identities

A quick glance at Table 1 seems to indicate that the balancing numbers are alternatively odd and even (see also [8]), while the balancer numbers are even. In the present section we prove this indication in a more explicit form. Indeed, from (15) and (17), we have both,

$$\left(\frac{n}{d}\right)^2 - 2d^2 = 1, \tag{18}$$

and

$$\left(\frac{n}{d}\right)^2 - 2d^2 = -1. \tag{19}$$

Letting $x = \frac{n}{d}$ and $y = d$, Equations (18) and (19) become the Pell equations

$$x^2 - 2y^2 = 1, \tag{20}$$

and

$$x^2 - 2y^2 = -1, \tag{21}$$

respectively. According to (6) and (7), all the solutions to Equations (20) and (21) are given by

$$\begin{aligned} x + \sqrt{2}y &= \left(1+\sqrt{2}\right)^{2m} \\ &= \sum_{i=0}^{2m} \binom{2m}{i} 2^{i/2} \\ &= \left(\sum_{i=0}^{m} \binom{2m}{2i} 2^i\right) + \sqrt{2}\left(\sum_{i=0}^{m-1} \binom{2m}{2i+1} 2^i\right), \end{aligned} \tag{22}$$

and

$$\begin{aligned} x + \sqrt{2}y &= \left(1+\sqrt{2}\right)^{m} \\ &= \sum_{i=0}^{m} \binom{m}{i} 2^{i/2} \\ &= \left(\sum_{i=0}^{\lfloor m/2 \rfloor} \binom{m}{2i} 2^i\right) + \sqrt{2}\left(\sum_{i=0}^{\lfloor (m-1)/2 \rfloor} \binom{m}{2i+1} 2^i\right), \end{aligned}$$

respectively, with m a positive integer.
Substituting x by $\dfrac{n}{d}$ and d by y, we get after identification

$$B_{2m-1} = n = yx = \left(\sum_{i=0}^{m-1} \binom{2m-1}{2i+1} 2^i\right)\left(\sum_{i=0}^{m-1} \binom{2m-1}{2i} 2^i\right),$$

and

$$B_{2m} = n = yx = \left(\sum_{i=0}^{m-1} \binom{2m}{2i+1} 2^i\right)\left(\sum_{i=0}^{m} \binom{2m}{2i} 2^i\right).$$

for $m \geq 1$.

Since both $\sum_{i=0}^{m-1} \binom{2m-1}{2i+1} 2^i$ and $\sum_{i=0}^{m-1} \binom{2m-1}{2i} 2^i$ are odd, the balancing numbers of the subsequence $\{B_{2m-1}\}_{m\geq 1}$ are odd as well. Similarly, since $\sum_{i=0}^{m-1} \binom{2m}{2i+1} 2^i$ is even, the balancing numbers of the subsequence $\{B_{2m}\}_{m\geq 1}$ are even. Hence, according to Theorem 1, we have proved the following theorem.

Theorem 2. *For any positive integer $m \geq 1$, (B_{2m}, R_{2m}) is an even-even pair and (B_{2m-1}, R_{2m-1}) is an odd-even pair and we have*

$$
\begin{aligned}
B_{2m-1} &= \left(\sum_{i=0}^{m-1} \binom{2m-1}{2i+1} 2^i\right)\left(\sum_{i=0}^{m-1} \binom{2m-1}{2i} 2^i\right), \\
R_{2m-1} &= 2\left(\sum_{i=0}^{m-1} \binom{2m-1}{2i+1} 2^i\right)^2 - B_{2m-1} - 1,
\end{aligned}
$$

and

$$
\begin{aligned}
B_{2m} &= \left(\sum_{i=0}^{m-1} \binom{2m}{2i+1} 2^i\right)\left(\sum_{i=0}^{m} \binom{2m}{2i} 2^i\right), \\
R_{2m} &= 2\left(\sum_{i=0}^{m-1} \binom{2m}{2i+1} 2^i\right)^2 - B_{2m}.
\end{aligned}
$$

Now let us rewrite Equation (9) as $(2(r+n)+1)^2 - 2(2n)^2 = 1$. Letting $x = 2(r+n)+1$ and $y = 2n$, we find Pell's equation (20) again. By identification, according to (22), we get

$$
\begin{aligned}
n &= \frac{y}{2} \\
&= \frac{1}{2}\sum_{i=0}^{m-1} \binom{2m}{2i+1} 2^i \\
&= m + \sum_{i=0}^{m-2} \binom{2m}{2i+3} 2^i \\
&= \sum_{i=-1}^{m-2} \binom{2m}{2i+3} 2^i,
\end{aligned}
\tag{23}
$$

and since $x = 2r + y + 1$, we get

$$
\begin{aligned}
r &= \frac{x-y-1}{2} \\
&= -n + \frac{x-1}{2} \\
&= -n + \frac{1}{2}\left(-1 + \sum_{i=0}^{m} \binom{2m}{2i} 2^i\right) \\
&= -n + \sum_{i=1}^{m} \binom{2m}{2i} 2^{i-1} \\
&= -n + \sum_{i=0}^{m-1} \binom{2m}{2i+2} 2^i.
\end{aligned}
\tag{24}
$$

We have thus proved, via the above discussion, the following theorem.

Theorem 3. *For $m \geq 1$, the balancing number B_m and its balancer number R_m are given by*

$$B_m = \sum_{i=-1}^{m-2} \binom{2m}{2i+3} 2^i \quad \text{and} \quad R_m = -B_m + \sum_{i=0}^{m-1} \binom{2m}{2i+2} 2^i.$$

The following identities on binomial coefficients are a direct consequence of both Theorem 2 and Theorem 3.

Corollary 2. *For $m \geq 1$, we have*

$$\sum_{i=-1}^{2m-2} \binom{4m}{2i+3} 2^i = \left(\sum_{i=0}^{m-1} \binom{2m}{2i+1} 2^i \right) \left(\sum_{i=0}^{m} \binom{2m}{2i} 2^i \right),$$

$$\sum_{i=-1}^{2m-3} \binom{4m-2}{2i+3} 2^i = \left(\sum_{i=0}^{m-1} \binom{2m-1}{2i+1} 2^i \right) \left(\sum_{i=0}^{m-1} \binom{2m-1}{2i} 2^i \right),$$

$$\sum_{i=0}^{2m-1} \binom{4m}{2i+2} 2^i = 2 \left(\sum_{i=0}^{m-1} \binom{2m}{2i+1} 2^i \right)^2 = \left(\sum_{i=0}^{m} \binom{2m}{2i} 2^i \right)^2 - 1,$$

$$\sum_{i=0}^{2m-2} \binom{4m-2}{2i+2} 2^i = 2 \left(\sum_{i=0}^{m-1} \binom{2m-1}{2i+1} 2^i \right)^2 - 1 = \left(\sum_{i=0}^{m-1} \binom{2m-1}{2i} 2^i \right)^2.$$

Remark 2. In [8], Ray establishes an other interesting formula for B_m using the generating function $g(z) = \dfrac{z}{1-6z+z^2}$. He gets

$$B_m = \sum_{i=0}^{\lfloor \frac{m-1}{2} \rfloor} (-1)^i \binom{m-i-1}{i} 6^{m-2i-1}.$$

From this Remark and Theorem 3, we obtain the new identity in the following Corollary.

Corollary 3. *For $m \geq 1$, we have*

$$\sum_{i=-1}^{m-2} \binom{2m}{2i+3} 2^i = \sum_{i=0}^{\lfloor \frac{m-1}{2} \rfloor} (-1)^i \binom{m-i-1}{i} 6^{m-2i-1}.$$

5. An Explicit Formula for Cobalancing Numbers

From (2), we have

$$r^2 + (2n+1)\,r - n\,(n+1) = 0, \tag{25}$$

which, when solved for r gives

$$r = \frac{-(2n+1) + \sqrt{8n^2+8n+1}}{2}. \tag{26}$$

Since r is positive, $8n^2 + 8n + 1$ is a perfect square, i.e.,

$$8n^2 + 8n + 1 = u^2, \text{ with } u \text{ odd}. \tag{27}$$

Therefore,

$$2n\,(n+1) = \left(\frac{u-1}{2}\right)\left(\frac{u+1}{2}\right). \tag{28}$$

Letting $A = \dfrac{u-1}{2}$, we get from (26) and (28)

$$r = A - n,$$

and

$$n\,(n+1) = \frac{A\,(A+1)}{2} = 1 + \cdots + A. \tag{29}$$

Consequently, $n\,(n+1)$ is a triangle number (see also [8]).

Letting $x = 2\,(n-r) + 1$ and $y = 2r$, Equation (25) leads again to the above Pell's equation (20). It follows from (23) and (24), that

$$r = \frac{y}{2} = B_m,$$

and

$$\begin{aligned} n &= \frac{x+y-1}{2} \\ &= \frac{x-y-1}{2} + y \\ &= R_m + 2r. \end{aligned}$$

The above discussion proves the following theorem.

Theorem 4. *For $m \geq 1$, the cobalacing number b_m and its cobalancer r_m are given by: $b_m = 2B_{m-1} + R_{m-1}$ and $r_m = B_{m-1}$, with $B_0 = R_0 = 0$.*

m	$b_m = 2B_{m-1} + R_{m-1}$	$r_m = B_{m-1}$
1	0	0
2	2	1
3	14	6
4	84	35
5	492	204
6	2870	1189
7	16730	6930
8	97512	40391
9	568344	235416
10	3312554	1372105

Table 2.

Table 2 summarizes the 10 first cobalancing numbers with there cobalancers, based on Table 1 and Theorem 4.
The following corollary is a direct consequence of Theorem 3 and Theorem 4.

Corollary 4. *For $m \geq 1$, we have*

$$b_{m+1} = \sum_{i=1}^{2m} \binom{2m}{i} 2^{\lfloor \frac{i-2}{2} \rfloor} \quad \textit{and} \quad r_{m+1} = \sum_{i=-1}^{m-2} \binom{2m}{2i+3} 2^i.$$

An immediate consequence of Theorems 2 and 4 is the following (see also [5]).

Corollary 5. *Every cobalancing number is even. Thus, no odd prime number could be a cobalancing number.*

6. New Formulas for (a, b)-Type Balancing and (a, b)-Type Cobalancing Numbers

Panda [7] defines sequence balancing and sequence cobalancing numbers as follows:

Definition 1. Let $\{u_n\}_{n\geq 1}$ be a sequence of real numbers. The number u_n is called a sequence balancing number if there exists a natural number r such that

$$u_1 + u_2 + \cdots + u_{n-1} = u_{n+1} + u_{n+2} + \cdots + u_{n+r}.$$

Similarly, the number u_n is called a sequence cobalancing number if

$$u_1 + u_2 + \cdots + u_n = u_{n+1} + u_{n+2} + \cdots + u_{n+r},$$

for some natural number r.

Kovács et al. [3] extend the concept of balancing numbers to arithmetic progressions as follows:

Definition 2. Let a, b be nonnegative coprime integers. If for some positive integers n and r, we have

$$(a+b)+\cdots+(a(n-1)+b)=(a(n+1)+b)+\cdots+(a(n+r)+b), \tag{30}$$

then we say that $an+b$ is an (a,b)-type balancing number.

Similarly, $an+b$ is an (a,b)-type cobalancing number if

$$(a+b)+\cdots+(an+b)=(a(n+1)+b)+\cdots+(a(n+r)+b), \tag{31}$$

for some natural number r.

Let $B_m^{(a,b)}$, $R_m^{(a,b)}$, $b_m^{(a,b)}$ and $r_m^{(a,b)}$ denote the mth (a,b)-type balancing number, the mth (a,b)-type cobalancing number, the mth (a,b)-type balancer and the mth (a,b)-type cobalancer, respectively.

6.0.1. (a,b)-Type Balancing Numbers

From (30), we have

$$an(n-1)+2b(n-1)-2arn-ar(r+1)-2br=0,$$

which, via straightforward calculations, is equivalent to

$$(2a(n-r-1)+a+2b)^2-2(a(2r+1))^2=(a+2b)^2-2a^2. \tag{32}$$

Letting $x=2a(n-r-1)+a+2b$, $y=a(2r+1)$, $u=a+2b$ and $v=a$, Equation (25) becomes:

$$x^2-2y^2=u^2-2v^2, \tag{33}$$

which has from (8), the integral solutions in the form:

$$x+y\sqrt{2}=\left(u+v\sqrt{2}\right)\left(1+\sqrt{2}\right)^{2m},\ m\geq 0. \tag{34}$$

From (22), we obtain

$$x+y\sqrt{2}=\left(u\sum_{i=0}^{m}\binom{2m}{2i}2^i+2v\sum_{i=0}^{m-1}\binom{2m}{2i+1}2^i\right)$$
$$+\sqrt{2}\left(v\sum_{i=0}^{m}\binom{2m}{2i}2^i+u\sum_{i=0}^{m-1}\binom{2m}{2i+1}2^i\right).$$

After identification, we get

$$2a\left(n-r-1\right)+a+2b=\left(a+2b\right)\sum_{i=0}^{m}\binom{2m}{2i}2^{i}+2a\sum_{i=0}^{m-1}\binom{2m}{2i+1}2^{i},$$

and

$$a\left(2r+1\right)=a\sum_{i=0}^{m}\binom{2m}{2i}2^{i}+\left(a+2b\right)\sum_{i=0}^{m-1}\binom{2m}{2i+1}2^{i}.$$

Therefore

$$n=1+r+\frac{a+2b}{a}\sum_{i=0}^{m-1}\binom{2m}{2i+2}2^{i}+\sum_{i=0}^{m-1}\binom{2m}{2i+1}2^{i},$$

and

$$r=\sum_{i=0}^{m-1}\binom{2m}{2i+2}2^{i}+\frac{a+2b}{a}\sum_{i=-1}^{m-2}\binom{2m}{2i+3}2^{i}.$$

From Theorem 3 and Theorem 4, we obtain

$$\begin{aligned} n &= 1+r+\frac{a+2b}{a}\left(B_m+R_m\right)+2B_m \\ &= 1+r+\frac{a+2b}{a}\left(b_{m+1}-r_{m+1}\right)+2r_{m+1} \\ &= 1+r+\frac{a-2b}{a}\,r_{m+1}+\frac{a+2b}{a}\,b_{m+1} \\ &= 1+r+r_{m+1}+b_{m+1}+\frac{2b}{a}\left(b_{m+1}-r_{m+1}\right), \end{aligned}$$

and

$$\begin{aligned} r &= B_m+R_m+\frac{a+2b}{a}\,B_m \\ &= b_{m+1}-r_{m+1}+\frac{a+2b}{a}\,r_{m+1} \\ &= b_{m+1}+\frac{2b}{a}\,r_{m+1}. \end{aligned}$$

Since n and r are positive integers and a and b are coprime, $2b_{m+1}$ and $2r_{m+1}$ should be both divisible by a. This discussion proves the following theorem.

Theorem 5. *Let* $\left(b^{/a}_{\varphi(m)}, r^{/a}_{\varphi(m)}\right)$ *denote the* m^{th} *pair of cobalacing number and its cobalancer such that* $2b^{/a}_{\varphi(m)}$ *and* $2r^{/a}_{\varphi(m)}$ *are both divisible by* a*. Then we have*

$$B_m^{(a,b)}=1+r^{/a}_{\varphi(m)}+\frac{2\left(a+b\right)}{a}\,b^{/a}_{\varphi(m)},$$

and

$$R_m^{(a,b)} = b_{\varphi(m)}^{/a} + \frac{2b}{a}\, r_{\varphi(m)}^{/a}.$$

Example 1. Let $a = 9$. The first pair $\left(b_{\varphi(1)}^{/9}, r_{\varphi(1)}^{/9}\right)$ of cobalancing number and its cobalancer both divisible by 9 is $(b_1, r_1) = (0, 0)$. Hence

$$B_1^{(9,b)} = 1 \text{ and } R_1^{(9,b)} = 0.$$

According to Corollary 4 and using Maple, the second pair $\left(b_{\varphi(2)}^{/9}, r_{\varphi(2)}^{/9}\right)$ of cobalancing number and its cobalancer both divisible by 9 is $(b_{13}, r_{13}) = (655869060, 271669860)$. Thus

$$B_2^{(9,b)} = 1 + r_{13} + \frac{2\,(9+b)}{9}\, b_{13} = 1583407981 + 145748680\, b,$$

and

$$R_2^{(9,b)} = b_{13} + \frac{2b}{9}\, r_{13} = 655869060 + 60371080\, b.$$

6.1. (a, b)-Type Cobalancing Numbers

From (31), we have

$$an\,(n+1) + 2bn - 2arn - ar\,(r+1) - 2br = 0,$$

which, via straightforward calculations, is equivalent to

$$(a\,(2n - 2r + 1) + 2b)^2 - 2\,(2ar)^2 = (a + 2b)^2\,. \tag{35}$$

Then, from (8) and (22), we obtain

$$a\,(2n - 2r + 1) + 2b = (a + 2b) \sum_{i=0}^{m} \binom{2m}{2i} 2^i,$$

i.e.,

$$\begin{aligned} n &= r + \frac{a+2b}{2a} \sum_{i=1}^{m} \binom{2m}{2i} 2^i \\ &= r + \frac{a+2b}{a} \sum_{i=0}^{m-1} \binom{2m}{2i+2} 2^i, \end{aligned}$$

and

$$\begin{aligned} r &= \frac{a+2b}{2a}\sum_{i=0}^{m-1}\binom{2m}{2i+1}2^i \\ &= \frac{a+2b}{a}\sum_{i=-1}^{m-2}\binom{2m}{2i+3}2^i. \end{aligned}$$

Hence, from Corollary 4, Theorem 3 and Theorem 4, we get

$$r = \frac{a+2b}{a}\, r_{m+1},$$

and

$$\begin{aligned} n &= r + \frac{a+2b}{a}\,(R_m + B_m) \\ &= \frac{a+2b}{a}\, r_{m+1} + \frac{a+2b}{a}\,(b_{m+1} - r_{m+1}) \\ &= \frac{a+2b}{a}\, b_{m+1}. \end{aligned}$$

Since n and r are positive integers and a and b are coprime, then $2b_{m+1}$ and $2r_{m+1}$ should be both divisible by a. Hence we have proved the following theorem.

Theorem 6. *Let* $\left(b^{/a}_{\varphi(m)}, r^{/a}_{\varphi(m)}\right)$ *denote the* m^{th} *pair of cobalacing number and its cobalancer such that* $2b^{/a}_{\varphi(m)}$ *and* $2r^{/a}_{\varphi(m)}$ *are both divisible by* a*. Then we have*

$$b^{(a,b)}_m = \frac{a+2b}{a}\, b^{/a}_{\varphi(m)} \quad \textit{and} \quad r^{(a,b)}_m = \frac{a+2b}{a}\, r^{/a}_{\varphi(m)}.$$

Example 2. For $a = 9$, we have $b^{(9,b)}_1 = r^{(9,b)}_1 = 0$, and according to Example 1, we get

$$b^{(9,b)}_2 = \frac{9+2b}{9}\, b_{13} = 72874340\,(9+2b)\,,$$

and

$$r^{(9,b)}_2 = \frac{9+2b}{9}\, r_{13} = 30185540\,(9+2b)\,.$$

Acknowledgements. The authors would like to thank the anonymous referee for his valuable comments which have improved the quality of the manuscript. We also thank Professor Hannoun Noureddine for his help in improving the English.

References

[1] A. Behera and G. K. Panda, *On the square roots of triangular numbers*, Fibonacci Quart. 37 (1999), 98–105.

[2] Z. I. Borevich and I. R. Shafarevich, Number Theory, Academic Press, 1966.

[3] T. Kovács, K. Liptai, P. Olajos, *On (a,b)-balancing numbers*, Publ. Math. Debrecen (accepted).

[4] G. K. Panda, *Some fascinating properties of balancing numbers*, to appear in "Applications of Fibonacci Numbers" Vol. 10, Kluwer Academic Pub., 2006.

[5] G. K. Panda and P. K. Ray, *Cobalancing numbers and cobalancers*, Int. J. Math. Math. Sciences, 8 (2005), 1189–1200.

[6] G. K. Panda and P. K. Ray, *Some links of balancing and cobalancing numbers with Pell and associated Pell numbers*, Bull. Inst. Math. Acad. Sin. (N.S.) 6(1), (2011), 41–72.

[7] G. K. Panda, *Sequence balancing and cobalancing numbers*, Fibonacci Quart. 45 (2007), 265–271.

[8] P. K. Ray, Balancing and cobalancing numbers, Ph. D. Thesis, Submitted to National Institute of Technology, Rourkela, India, August, 2009.

VALUATIONS AND COMBINATORICS OF TRUNCATED EXPONENTIAL SUMS

T. Amdeberhan
Department of Mathematics, Tulane University, New Orleans, Louisiana
tamdeber@tulane.edu

D. Callan
Department of Statistics, University of Wisconsin-Madison, Madison, Wisconsin
callan@stat.wisc.edu

V. Moll
Department of Mathematics, Tulane University, New Orleans, Louisiana
vhm@tulane.edu

Received: 6/24/12, Accepted: 4/18/13, Published: 4/25/13

Abstract

A conjecture of G. McGarvey for the 2-adic valuation of the Schenker sums is established. These sums are $n!$ times the sum of the first $n+1$ terms of the series for e^n. A certain analytic expression for the p-adic valuation of these sums is provided for a class of primes. Some combinatorial interpretations (using rooted trees) are furnished for identities that arose along the way.

1. Introduction

Let $0 \neq x \in \mathbb{Q}$. The Fundamental Theorem of Arithmetic implies the prime factorization $|x| = \prod_p p^{n_p}$ where the product is over all primes and for some $n_p \in \mathbb{Z}$ (all but finitely many being zero). The *p-adic valuation* of x, denoted $\nu_p(x)$, is the exponent n_p in the power of p in the above factorization. For example, $\nu_2(2^k) = k$ and $\nu_2(2^k - 1) = 0$. By convention, $\nu_p(0) = +\infty$.

Given a sequence of positive integers a_n and a prime p, determining a closed form for the sequence of p-adic valuations $\nu_p(a_n)$ often presents interesting challenges. Legendre's classical formula for the factorials

$$\nu_p(n!) = \sum_{j=0}^{\infty} \left\lfloor \frac{n}{p^j} \right\rfloor \tag{1.1}$$

appears in elementary textbooks. If $n \in \mathbb{N}$ is expanded in base p and $s_p(n)$ denotes the sum of its p-ary digits, then the alternative form

$$\nu_p(n!) = \frac{n - s_p(n)}{p-1} \tag{1.2}$$

follows directly from (1.1).

The presence of a compact formula, such as (1.2), facilitates the analysis of arithmetical properties of a given sequence a_n. For instance, it follows directly from (1.2) that

$$\nu_p\binom{2n}{n} = \frac{2s_p(n) - s_p(2n)}{p-1} \tag{1.3}$$

and in particular, for $p = 2$, this yields

$$\nu_2\binom{2n}{n} = s_2(n), \tag{1.4}$$

in view of $s_2(2n) = s_2(n)$. This provides an elementary proof that the central binomial coefficients $\binom{2n}{n}$ are always even, and exactly divisible by 2 if and only if n is a power of 2.

Introduce the sequence of positive integers

$$a_n = \sum_{k=0}^{n} \frac{n!}{k!} n^k. \tag{1.5}$$

One immediately recognizes that $\frac{a_n}{n!}$ equals the nth partial sum of the exponential e^n. The sequence a_n appeared in a paper by S. Ramanujan [10] where he proposes the following problem:

Show that

$$\frac{1}{2}e^n = 1 + \frac{n}{1!} + \frac{n^2}{2!} + \cdots + \frac{n^n}{n!}\theta, \tag{1.6}$$

for some θ in the range between $\frac{1}{2}$ and $\frac{1}{3}$.

The relation (1.6) may be expressed in the form

$$a_n = \frac{1}{2}n!e^n + (1-\theta)n^n. \tag{1.7}$$

The sequence $\{a_n\}$ resurfaced in Exercise 1.2.11.3.18 of [8] in an urn problem,

There are n balls in an urn. How many selections with replacement are made, on average, if we stop when we reach a ball already selected?

with answer a_n/n^n. In relation to this question, D. Knuth introduces the functions

$$Q(n) = 1 + \frac{n-1}{n} + \frac{(n-1)(n-2)}{n^2} + \cdots \text{ and } R(n) = 1 + \frac{n}{n+1} + \frac{n^2}{(n+1)(n+2)} + \cdots,$$

with $Q(n)+R(n) = n!e^n/n^n$. To derive asymptotics of the function $Q(n)$, Ramanujan resorts to the integral representation

$$Q(n) = \int_0^\infty e^{-x}\left(1+\frac{x}{n}\right)^{n-1} dx. \tag{1.8}$$

More details on an asymptotic analysis of the sequence a_n can be found in [2] and [5].

The sequence a_n is listed as A063170 on OEIS and the name *Schenker sum* is given to it. The comments there include the integral representation

$$a_n = \int_0^\infty e^{-x}(x+n)^n \, dx, \tag{1.9}$$

due to M. Somos and the following conjecture by G. McGarvey for the 2-adic valuation of a_n.

Conjecture 1.1. *For $n \in \mathbb{N}$, we have*

$$\nu_2(a_n) = \begin{cases} 1 & \text{if } n \text{ is odd} \\ n - s_2(n) & \text{if } n \text{ is even.} \end{cases} \tag{1.10}$$

A primary focus of this paper is to establish the above conjecture and extend the discussion to odd primes.

2. The Proof

The proof starts with an elementary observation.

Lemma 2.1. *Suppose $A(x)$ is a polynomial with integer coefficients. Assume every coefficient is divisible by r. Then, the integer*

$$\int_0^\infty A(x)e^{-x} \, dx \tag{2.1}$$

is divisible by r.

Proof. Write $A(x) = a_0 + a_1x + \cdots + a_nx^n$ and observe that

$$\int_0^\infty A(x)e^{-x} \, dx = \sum_{j=0}^{n} a_j \, j! \tag{2.2}$$

is clearly divisible by r. □

The previous result shows that if $A(x) \equiv B(x) \bmod r$, then

$$\int_0^\infty A(x)e^{-x}\,dx \equiv \int_0^\infty B(x)e^{-x}\,dx \bmod r. \tag{2.3}$$

For the proof of the conjecture, the integral representation (1.9) will be useful. The process consists of two cases based on the parity of n.

`Case 1`: Suppose n is odd, say $n = 2m+1$. Now write $n = 1 + 2n_1 + \cdots + 2^r n_r$ in base 2 and raise $n + x \equiv 1 + x \bmod 2$ to the n-th power to produce

$$(n+x)^n \equiv (x+1)\prod_{i=1}^{r}(1+x)^{2^{k_i}} \equiv (1+x)\prod_{i=1}^{r}(1+x^{2^{k_i}}) \equiv 1 + x + O(x^w) \bmod 2,$$

with $w \geq 2$. Fermat's little theorem was employed in the second congruence. Then

$$\begin{aligned} a_n &= \int_0^\infty (n+x)^n e^{-x}\,dx \\ &\equiv \int_0^\infty (1 + x + O(x^w))\, e^{-x}\,dx \bmod 2 \\ &\equiv \int_0^\infty (1+x)e^{-x}\,dx = 2 \equiv 0 \bmod 2. \end{aligned}$$

It follows that a_n is even. But a_n is not divisible by 4. Indeed, if m is even

$$a_n \equiv \int_0^\infty (1+x)e^{-x}\,dx = 2 \equiv 2 \bmod 4 \tag{2.4}$$

and for m odd,

$$a_n \equiv \int_0^\infty (3+x)^3 e^{-x}\,dx = 78 \equiv 2 \bmod 4. \tag{2.5}$$

This proves the conjecture when n odd.

`Case 2`: Suppose n is even, say $n = 2m$. Then

$$\begin{aligned} a_{2m} &= \int_0^\infty (2m+x)^{2m} e^{-x}\,dx \qquad (2.6) \\ &= \sum_{k=0}^{2m} \binom{2m}{k} (2m)^{2m-k} \int_0^\infty x^k e^{-k}\,dx \\ &= \sum_{k=0}^{2m} \binom{2m}{k} (2m)^{2m-k}\, k!. \end{aligned}$$

Let t_k be the summand in the last sum. Then $2mt_{k+1} = (2m-k)t_k$ and if $j = 2m - k$, this becomes

$$2mt_{2m-j+1} = jt_{2m-j}. \tag{2.7}$$

This recurrence is now utilized in expressing the coefficients t_i in terms of t_{2m} and also in analyzing the 2-adic valuation of each term in the sum for a_{2m}. For example, $j = 1$ yields $t_{2m-1} = 2mt_{2m}$, therefore

$$\nu_2(t_{2m-1}) = 1 + \nu_2(m) + \nu_2(t_{2m}) > \nu_2(t_{2m}). \tag{2.8}$$

Similarly, $j = 2$ yields $t_{2m-2} = 2m^2 t_{2m}$ from which it follows that $\nu_2(t_{2m-2}) > \nu_2(t_{2m})$ and $j = 3$ gives the relation $4m^3 t_{2m} = 3t_{2m-3}$ and $\nu_2(t_{2m-3}) > \nu_2(t_{2m})$ is obtained. In general

Lemma 2.2. *For $1 \leq j \leq 2m$, the inequality $\nu_2(t_{2m-j}) > \nu_2(t_{2m})$ holds.*

Proof. Define $u_j = t_{2m-j}$. Then (2.7) gives

$$2mu_{j-1} = ju_j. \tag{2.9}$$

From here it follows that

$$u_j = \frac{2m}{j} u_{j-1} = \frac{2m}{j} \cdot \frac{2m}{j-1} u_{j-2} \tag{2.10}$$

and iterating produces

$$u_j = \frac{(2m)^j}{j!} t_{2m}. \tag{2.11}$$

Now write

$$j! = 2^{\nu_2(j!)} O_*(j) = 2^{j - s_2(j)} O_*(j), \tag{2.12}$$

with $O_*(j)$ representing an odd number, to obtain

$$O_*(j) u_j = 2^{s_2(j)} m^j t_{2m}. \tag{2.13}$$

This gives

$$\nu_2(u_j) = s_2(j) + j\nu_2(m) + \nu_2(t_{2m}) > \nu_2(t_{2m}), \tag{2.14}$$

completing the proof as required. □

Note 2.3. Lemma 2.2 implies $\nu_2(a_{2m}) = \nu_2(t_{2m}) = \nu_2(n!) = n - s_2(n)$. This completes the analysis of Case 2 and establishes Conjecture 1.1.

3. The p-Adic Valuations for p an Odd Prime

In view of the results established in the previous section, it is natural to consider the question of what happens when p is an odd prime, i.e., is there a simple expression for $\nu_p(a_n)$ when $p \neq 2$ is a prime? The present section gives partial answers to this problem.

Proposition 3.1. *Let p be an odd prime and assume $n = pm$ for some $m \in \mathbb{N}$. Then*

$$\nu_p(a_n) = \frac{n - s_p(n)}{p-1}. \tag{3.1}$$

Proof. Consider the integral expression

$$a_{pm} = \sum_{k=0}^{pm} \binom{pm}{k} (pm)^{pm-k} \int_0^{\infty} x^k e^{-x}\, dx = \sum_{k=0}^{pm} \binom{pm}{k} (pm)^{pm-k} k! \tag{3.2}$$

and let

$$t_{m,p}(k) = \binom{pm}{k} (pm)^{pm-k}\, k! \tag{3.3}$$

be the summand in (3.2). Observe that $t_{m,p}(mp) = (pm)!$. Pursuant, the case $p = 2$, suppose that

$$\nu_p(t_{m,p}(k)) > \nu_p(t_{m,p}(pm)) = \nu_p(n!). \tag{3.4}$$

Then

$$\nu_p(a_{pm}) = \nu_p(n!) = \frac{n - s_p(n)}{p-1}, \tag{3.5}$$

as claimed.

The proof of (3.4) begins with the computation of the ratio of two consecutive terms $t_{m,p}$ to produce the relation

$$pm\, t_{m,p}(k+1) = (pm - k) t_{m,p}(k). \tag{3.6}$$

The proof then proceeds as in the case $p = 2$. □

The next result is a crucial reduction towards the modular arithmetic employed in the computation of $\nu_p(a_n)$.

Proposition 3.2. *Let p be a prime and $n = pm + r$ with $0 < r < p$. Then $p|a_n$ if and only if $p|a_r$.*

Proof. The reduction

$$(x+n)^n \equiv (x+r)^{pm} (x+r)^r \equiv (x^{pm} + r^{pm})(x+r)^r \equiv (x^{pm} + r^m)(x+r)^r \bmod p,$$

is due the fact that p divides $\binom{pm}{k}$ for any $0 < k < pm$. This implies

$$\begin{aligned}
a_n &= \int_0^\infty (x+n)^n e^{-x}\,dx \\
&\equiv \int_0^\infty (x^{pm}+r^m)(x+r)^r e^{-x}\,dx \\
&= \sum_{j=0}^{r} \binom{r}{j} r^{r-j} \int_0^\infty \left(x^{pm}+r^m\right) x^j e^{-x}\,dx \\
&= \sum_{j=0}^{r} \binom{r}{j} r^{r-j} \left[(pm+j)! + r^m j!\right] \\
&\equiv \sum_{j=0}^{r} \binom{r}{j} r^{m+r-j}\, j! \\
&\equiv \sum_{j=0}^{r} \binom{r}{j} r^{m+j}\, (r-j)! \\
&\equiv r^m \sum_{j=0}^{r} \frac{r!}{j!} r^j \\
&\equiv r^m a_r \bmod p.
\end{aligned}$$

The assertion follows. □

Before embarking on the more general study, it is worthwhile to consider some toy examples (small primes). The reader will hopefully find these illustrative of the potential subtleties and obstacles.

Example 3.3. Let $p = 3$. Proposition 3.1 gives

$$\nu_3(a_{3n}) = \frac{1}{2}(3n - s_3(n)). \tag{3.7}$$

The remaining two cases are established by Proposition 3.2. Assume $n = 3m + r$ with $r = 1, 2$. Then $3|a_n$ if and only if $3|a_r$. Neither $a_1 = 2$ nor $a_2 = 10$ are divisible by 3, and therefore 3 does not divide a_n.

In summary,

$$\nu_3(a_n) = \begin{cases} \frac{1}{2}(n - s_3(n)) & \text{if } n \equiv 0 \bmod 3 \\ 0 & \text{if } n \not\equiv 0 \bmod 3. \end{cases} \tag{3.8}$$

Example 3.4. Let $p = 5$. This brings in the first difficult problem. Start with the simpler cases. Proposition 3.1 ensures that

$$\nu_5(a_{5n}) = \frac{1}{4}(n - s_5(n)). \tag{3.9}$$

By Proposition 3.2 and since none of the numbers $a_1 = 2$, $a_3 = 78$, $a_4 = 824$ is divisible by 5, the following holds

$$\nu_5(a_n) = 0 \quad \text{if } n \equiv 1,\, 3,\, 4 \bmod 5. \tag{3.10}$$

The remaining case $\nu_5(a_{5n+2})$ requires a closer look. A preliminary discussion is presented in the next section.

Example 3.5. Let $p = 7$. Because the first six numbers $a_1 = 2$, $a_3 = 78$, $a_4 = 824$, $a_5 = 10970$, $a_6 = 176112$ are not divisible by 7, it follows that

$$\nu_7(a_n) = \begin{cases} \frac{1}{6}(n - s_7(n)) & \text{if } n \equiv 0 \bmod 7 \\ 0 & \text{if } n \not\equiv 0 \bmod 7. \end{cases} \tag{3.11}$$

A direct computation of the values of a_j modulo 11 shows that a_j is not divisible by 11 for $1 \leq j < 11$. Therefore

$$\nu_{11}(a_n) = \begin{cases} \frac{1}{10}(n - s_{11}(n)) & \text{if } n \equiv 0 \bmod 11 \\ 0 & \text{if } n \not\equiv 0 \bmod 11. \end{cases} \tag{3.12}$$

The case $p = 13$ is similar to $p = 5$ since 13 divides $a_3 = 78$.

4. Schenker Primes

The results established in the previous sections determine the valuation $\nu_p(a_n)$ for a class of prime numbers. The primes not completely covered by those methods are fall under a special category as defined below.

Definition 4.1. A prime p is called a *Schenker prime* if p divides a_r for some value r in the range $1 \leq r \leq p-1$.

The result is summarized in the next theorem.

Theorem 4.2. *Let p be a prime and assume that p is not a Schenker prime. Then*

$$\nu_p(a_n) = \begin{cases} \frac{1}{p-1}(n - s_p(n)) & \text{if } n \equiv 0 \bmod p \\ 0 & \text{if } n \not\equiv 0 \bmod p. \end{cases} \tag{4.1}$$

Example 4.3. The prime $p = 17$ is not a Schenker prime. The factorization of the numbers a_r, for $1 \leq r \leq 16$ is

$a_1 = 2$	$a_2 = 2 \cdot 5$
$a_3 = 2 \cdot 3 \cdot 13$	$a_4 = 2^3 \cdot 103$
$a_5 = 2 \cdot 5 \cdot 1097$	$a_6 = 2^4 \cdot 3^2 \cdot 1223$
$a_7 = 2 \cdot 5 \cdot 7 \cdot 41 \cdot 1153$	$a_8 = 2^7 \cdot 556403$
$a_9 = 2 \cdot 3^4 \cdot 149 \cdot 163 \cdot 439$	$a_{10} = 2^8 \cdot 5^2 \cdot 7281587$
$a_{11} = 2 \cdot 11 \cdot 9431 \cdot 6672571$	$a_{12} = 2^{10} \cdot 3^5 \cdot 5^3 \cdot 1443613$
$a_{13} = 2 \cdot 13 \cdot 179 \cdot 339211523363$	$a_{14} = 2^{11} \cdot 7^2 \cdot 595953719897$
$a_{15} = 2 \cdot 3^6 \cdot 5^3 \cdot 317 \cdot 13103$	$a_{16} = 2^{15} \cdot 13 \cdot 179 \cdot 116371 \cdot 11858447$

The prime $p = 17$ does not appear in any of these factorizations confirming that it is not a Schenker prime. In accord with Theorem 4.2, the 17-adic valuation of the sequence a_n is explicit:

$$\nu_{17}(a_n) = \begin{cases} \frac{1}{16}(n - s_{17}(n)) & \text{if } n \equiv 0 \bmod 17 \\ 0 & \text{if } n \not\equiv 0 \bmod 17. \end{cases} \tag{4.2}$$

Example 4.4. The prime 5 is a Schenker prime because 5 divides $a_2 = 10$. Similarly 37 is a Schenker prime since 37 divides a_{25}. The list of all Schenker primes up to 200 is

$$\{5, 13, 23, 31, 37, 41, 43, 47, 53, 59, 61, 71, 79, 101, 103, \\ 107, 109, 127, 137, 149, 157, 163, 173, 179, 181, 191, 197, 199\}. \tag{4.3}$$

Note 4.5. The valuation $\nu_5(a_n)$ is not obvious or as simple, so finding an analytic/explicit formula for it stands as an open question. *The description given below is purely experimental and no proofs are available at the moment.* The only rigorous result is Example 3.4, which determines the value of $\nu_5(a_n)$ except for indices congruent to 2 modulo 5.

The indices of the form $5n + 2$ are first divided according to the parity of n modulo 5. Symbolic computations show that

$$\nu_5(a_{5n+2}) = 1 \text{ for } n \not\equiv 2 \bmod 5. \tag{4.4}$$

Therefore it is now required to consider indices of the form

$$m_1 = 5(5n + 2) + 2 = 5^2 n + 5 \cdot 2 + 2. \tag{4.5}$$

Then it is observed that

$$\nu_5(a_{5^2n+5\cdot 2+2}) = 2 \text{ for } n \not\equiv 0 \bmod 5, \tag{4.6}$$

leading to indices of the form

$$m_2 = 5^3 n + 5 \cdot 2 + 2. \tag{4.7}$$

Continuing this process, it is then observed that

$$\nu_5(a_{5^3n+5\cdot 2+2}) = 3 \text{ for } n \not\equiv 4 \bmod 5, \tag{4.8}$$

leading to indices of the form

$$m_3 = 5^4 n + 5^3 \cdot 4 + 5^2 \cdot 0 + 5^1 \cdot 2 + 2, \tag{4.9}$$

and also

$$\nu_5(a_{5^4n+5^3\cdot 2+5\cdot 2+2}) = 4 \text{ for } n \not\equiv 4 \bmod 5, \tag{4.10}$$

leading to

$$m_4 = 5^5 n + 5^4 \cdot 4 + 5^3 \cdot 4 + 5^2 \cdot 0 + 5^1 \cdot 2 + 2. \tag{4.11}$$

This process can be described in terms of the expansion of the index n in base 5 in the form

$$n = x_0 + x_1 \cdot 5 + x_2 \cdot 5^2 + x_3 \cdot 5^3 + x_4 \cdot 5^4 + \cdots \tag{4.12}$$

The results of Example 3.4 for $\nu_5(a_n)$ are

$$x_0 = \begin{cases} 0 & \nu_5(a_n) = \frac{1}{4}(n - s_5(n)) \\ 1,\, 3,\, 4 & \nu_5(a_n) = 0 \\ 2 & \nu_5(a_n) \text{ depends on } x_1. \end{cases} \tag{4.13}$$

The next steps are

$$x_0 = 2 \text{ and } x_1 = \begin{cases} \neq 2 & \nu_5(a_n) = 1 \\ 2 & \text{depends on } x_2, \end{cases} \tag{4.14}$$

and

$$x_0 = 2,\, x_1 = 2 \text{ and } x_2 = \begin{cases} \neq 0 & \nu_5(a_n) = 2 \\ 0 & \text{depends on } x_3, \end{cases} \tag{4.15}$$

and

$$x_0 = 2,\, x_1 = 2,\, x_2 = 0 \text{ and } x_3 = \begin{cases} \neq 4 & \nu_5(a_n) = 3 \\ 4 & \text{depends on } x_4. \end{cases} \tag{4.16}$$

The next conjecture has been verified numerically, for the prime $p = 5$, up to depth/level 10.

Conjecture 4.6. *Assume the valuation $\nu_5(a_n)$ is not determined by the first r digits of n; that is $x_0, x_1, \cdots, x_{r-1}$ do not determine $\nu_5(a_n)$. Then, among the 5 possible values for x_r, there is a single value for which the valuation is not determined by $x_0, x_1, \cdots, x_{r-1}, x_r$.*

Note 4.7. Denote by d_j the j-th exceptional digit in Conjecture 4.6. The list of these digits begins with

$$d_0 = 2,\, d_1 = 2,\, d_2 = 0,\, d_3 = 4,\, d_4 = 4. \tag{4.17}$$

A similar conjecture has been proposed in [1] and [3] for the p-adic valuation of Stirling numbers of the second kind. Conjecture 4.6 is rephrased using *valuation trees.*

Tree construction. The tree starts with a top vertex v_0 labeled n that represents all of $\mathbb{N}$. This top vertex forms the *first level* of the tree. The expansion of n in base 5 in (4.12) is employed in the description of this tree.

From the top vertex, form the *second level* consisting of 5 vertices connected to v_0. Each vertex corresponds to a value of x_1 in the expansion of n in base 5. The figure shows three types of vertices: those with $x_0 = 0$ for which $\nu_5(a_n) = \frac{1}{4}(n - s_5(n))$ (shown to the left of the tree), those with $x_0 \neq 0, 2$ for which $\nu_5(a_n) = 0$ (shown at the center) and finally those vertices with $x_0 = 2$ for which the valuation $\nu_5(a_n)$ is not determined by x_0. In this form, each vertex represents a subset of $\mathbb{N}$ determined by some property of the digits x_i. Each vertex has a symbol indicating the type of digit x_i it represents (to be more precise all the properties determining this subset is obtained by reading the path from the top vertex to the vertex in question) and also the valuation $\nu_5(a_n)$ for those indices n associated to the vertex.

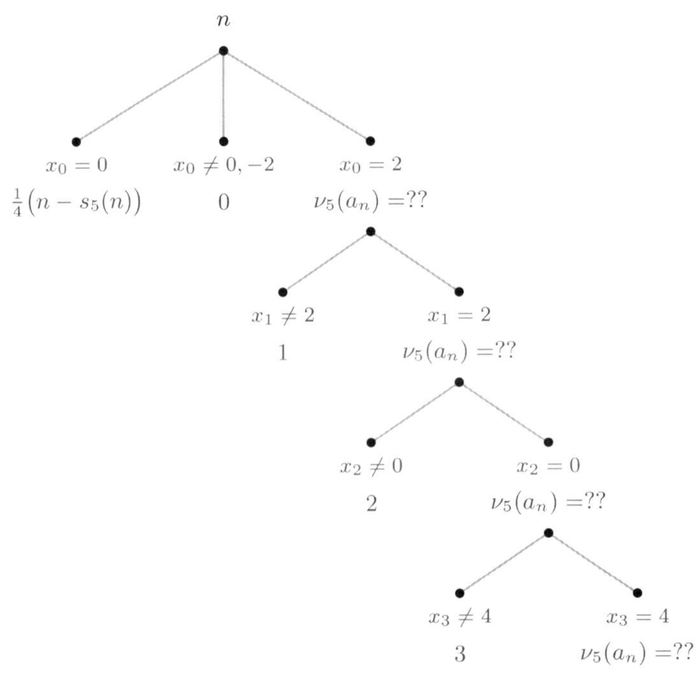

The valuation tree for $p = 5$

The discussion that follows excludes the vertex corresponding to $x_0 = 0$. The valuation for the indices corresponding to this vertex are determined by Proposition 3.1.

Definition 4.8. A vertex is called *terminal* if the valuation is the same for all indices associated to the vertex.

Example 4.9. All indices n associated to the vertex corresponding to $x_0 = 1$ have valuation $\nu_5(a_n) = 0$; that is $\nu_5(a_{5n+1}) = 0$. Therefore this vertex is terminal. On the other hand, if $n = 7$ then

$$\nu_5(a_7) = \nu_5(3309110) = 1 \tag{4.18}$$

and

$$\nu_5(a_{17}) = \nu_5(48458665918962686950l0) = 3. \tag{4.19}$$

Both indices 7 and 17 are associated to the vertex with $x_0 = 2$ and they have different valuation. Therefore this is not a terminal vertex.

Note 4.10. The first level consists of the vertex with $x_0 = 0$, excluded from this discussion, the three vertices with $x_0 = 1, 3, 4$ (shown as one single vertex in the tree), and the vertex with $x_0 = 2$. This last vertex produces 5 new ones that form the second level. These five vertices correspond to indices with $x_0 = 2$ and $0 \leq x_1 \leq 4$. Each of them have a set of indices attached to them, for instance $x_1 = 2$ correspond to indices of the form $n = x_0 + 5x_1 + 5^2m = 2 + 5 \cdot 2 + 5^2m = 12 + 25m$. This describes the construction of the valuation tree: non-terminal vertices produce 5 new vertices at the next level.

Definition 4.11. The tree constructed above, extended naturally by simply replacing 5 by a prime p, is called *the valuation tree for* p.

The structure of this valuation tree described in the next conjecture generalizes Conjecture 4.6.

Conjecture 4.12. *Assume* p *is a Schenker prime. Then each level of the valuation tree for* p *contains a single non-terminal vertex.*

5. The Combinatorics of a_n

The arithmetic properties of the sequence a_n discussed in the earlier sections are based on the integral representation (1.9). In this section, the Abel-type identity Theorem 5.1 gives an alternative binomial representation of a_n. There is an extensive literature on Abel's identity and its numerous variants (see, for example, [9] and its references). Here, we give two short direct proofs, one analytic and one bijective.

Theorem 5.1. *The following identity provides two different formulation for the sequence* a_n*:*

$$\sum_{k=0}^{n} \frac{n!}{k!} n^k = \sum_{k=0}^{n} \binom{n}{k} k^k (n-k)^{n-k}. \tag{5.1}$$

Proof. Define

$$\begin{aligned} A_n(t) &= t\sum_{k=0}^{n}\binom{n}{k}(t+k)^{k-1}(n-k)^{n-k} \\ B_n(t) &= \sum_{k=0}^{n}\frac{n!}{(n-k)!}(t+n)^{n-k}, \\ C_n(t) &= \sum_{k=0}^{n}\binom{n}{k}(t+k)^{k}(n-k)^{n-k}. \end{aligned} \tag{5.2}$$

The relation $(t+k)^k = t(t+k)^{k-1} + k(t+k)^{k-1}$ gives

$$C_n(t) = A_n(t) + nC_{n-1}(t+1). \tag{5.3}$$

The value

$$A_n(t) = (t+n)^n \tag{5.4}$$

follows directly from Abel's identity

$$\sum_{k=0}^{n}\binom{n}{k}(t+k)^{k-1}(s-k)^{n-k} = \frac{(t+s)^n}{t}, \tag{5.5}$$

Then

$$C_n(t) = (t+n)^n + nC_{n-1}(t+1), \tag{5.6}$$

and it is easy to check that B_n also satisfies this recurrence. Since both B_n and C_n have the same initial conditions, it follows that $B_n(t) = C_n(t)$. The stated result now comes from $B_n(0) = C_n(0)$. □

Note 5.2. A nice proof of Abel's identity (5.5) appears in [4]. A nice combinatorial interpretation may be found in [6, 7] with the following picturesque formulation. Whereas the binomial identity $(t+s)^n = \sum_{k=0}^{n}\binom{n}{k}t^k s^{n-k}$ counts functions $f : [n] \to [t+s]$ by the number of elements that *map directly to* $[t]$, that is, by number of elements $i \in [n]$ for which $f(i) \in [t]$, (5.5) counts these same functions by the number of elements that *ultimately map to* $[s+1, s+t]$, that is, by number of elements $i \in [n]$ for which $\underbrace{f \circ f \circ \ldots \circ f}_{m}(i) \in [s+1, s+t]$ for some $m \geq 1$ (assuming $s \geq n$ so that all summands are nonnegative).

The identity $B_n(t) = C_n(t)$ implies another well known identity.

Corollary 5.3.

$$n! = \sum_{r=0}^{n}(-1)^r\binom{n}{r}(n-r)^n. \tag{5.7}$$

Proof. Matching powers of t in $B_n(t-n) = C_n(t-n)$ gives

$$\frac{n!}{k!} = (-1)^k \sum_{r=k}^{n} (-1)^r \binom{n}{r} \binom{r}{k} (n-r)^{n-k} \tag{5.8}$$

and the special case $k = 0$ gives the result. □

Note 5.4. An elementary combinatorial proof of (5.7) is obtained by counting all the $n!$ bijective functions on a set of n elements. The right-hand side employs the inclusion-exclusion principle by excluding maps according to the number of elements missed in the range.

Note 5.5. Theorem 5.1, after canceling some equal terms, is equivalent to the identity,

$$\sum_{k=2}^{n} \frac{n!}{(n-k)!} n^{n-k} = \sum_{k=1}^{n-1} \binom{n}{k} k^k (n-k)^{n-k} \tag{5.9}$$

for which we now give a combinatorial interpretation.

We will show that (5.9) counts a class of rooted trees in two different ways. Let us say a vertex in a rooted tree is a *descendant* of an edge in the tree if the path from the vertex to the root includes the edge. Define an *ev-tree* to be a rooted vertex-labeled tree on $[n]$ *with a highlighted edge e and a marked descendant v of e*, as illustrated below with $n = 9$. Call the (unique) path starting at edge e and ending at vertex v the *critical path* of an ev-tree.

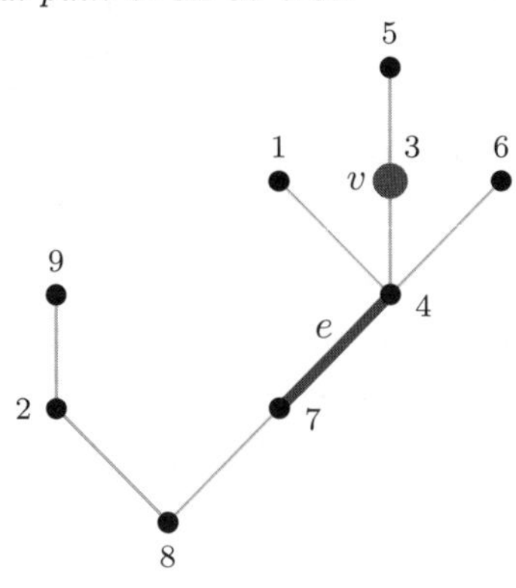

An ev-tree

In the example, $e = 74$ (the bolded edge) and $v = 3$ (the large point). The descendants of e are $4, 1, 3, 6, 5$ and the critical path is $7 \to 4 \to 3$.

The left side of (5.9) counts ev-trees by the length k of the critical path as follows. Choose the k vertices that occur on the critical path—$\binom{n}{k}$ choices. Form a forest of trees on $[n]$ rooted at these k vertices—kn^{n-k-1} choices [11, Proposition 5.3.2]. Put a cycle structure on the k roots—$(k-1)!$ choices. Mark one of the vertices in the forest—n choices. Turn the cycle of roots into a path, $r_1 \to r_2 \to \cdots \to r_k$, by starting at the root of the tree containing the marked vertex. Ignoring the orientation of edges in this path, we now have a tree rooted at the marked vertex. Take e to be the edge r_1r_2 and v to be the vertex r_k. This is the desired ev-tree and by construction, there are $\binom{n}{k} \cdot kn^{n-k-1} \cdot (k-1)! \cdot n = \frac{n!}{(n-k)!}n^{n-k}$ of them.

On the other hand, the right side of (5.9) counts ev-trees by the number k of descendants of e as follows. Choose the descendants of e—$\binom{n}{k}$ choices—and form a rooted tree on these vertices with one vertex colored blue—k^k choices, because Cayley's formula says there are k^{k-1} rooted trees. Similarly, form a rooted tree on the remaining $n-k$ vertices with one vertex colored blue—$(n-k)^{n-k}$ choices. Now join the two blue vertices with a blue edge and change the root of the first tree to a red vertex. The result will form an ev-tree by taking the blue edge as e and the red vertex as v.

Actually, identity (5.9) can be sharpened. Every term on the left side is obviously divisible by n and it is a fact, not quite so obvious, that every term on the right is also divisible by n. So we can divide by n to get another integer identity,

$$\sum_{k=2}^{n} \frac{(n-1)!}{(n-k)!}n^{n-k} = \sum_{k=1}^{n-1} \frac{1}{n}\binom{n}{k}k^k(n-k)^{n-k}. \tag{5.10}$$

Basically the same interpretation works for (5.10): simply observe that incrementing the vertex labels, $i \to i+1 \pmod{n}$, in an ev-tree leaves the statistics "number of descendants of the highlighted edge" and "length of the critical path" invariant, and partitions the class of ev-trees on $[n]$ into orbits, each of which has size n. So just pick out the ev-tree in each orbit whose root is, say, 1. Note that this argument provides a combinatorial proof that the summand on the right side of (5.10) is an integer.

Note 5.6. The sums in (5.10) give $(a_n)_{n\geq 1} = (1, 8, 78, 944, \ldots)$, A000435, "the sequence that started it all". A comment on A000435 by Geoffrey Critzer says that a_n, for $n > 1$, is the number of connected endofunctions on $[n]$ with no fixed points, that is, functions $f : [n] \to [n]$ with only one orbit of periodic points (connected) whose length is ≥ 2 (no fixed points). In fact, ev-trees with root 1 are just another way of looking at these endofunctions.

Acknowledgements. The last author wishes to thank the partial support of NSF-DMS 1112656.

References

[1] T. Amdeberhan, D. Manna, and V. Moll. The 2-adic valuation of Stirling numbers. *Experiment. Math.*, 17:69–82, 2008.

[2] B. Berndt. *Ramanujan's Notebooks, Part I.* Springer-Verlag, New York, 1985.

[3] A. Berribeztia, L. Medina, A. Moll, V. Moll, and L. Noble. The p-adic valuation of Stirling numbers. *J. Algebra Number Theory Academia*, 1:1–30, 2010.

[4] Shalosh B. Ekhad and J. E. Majewicz. A short WZ-style proof of Abel's identity. *Elec. Jour. Comb.*, 3:R16, 1996.

[5] P. Flajolet, P. Grabner, P. Kirschenhofer, and H. Prodinger. On Ramanujan's Q-function. *J. Comput. Appl. Math.*, 58:103–116, 1995.

[6] J. Françon. Preuves combinatoires des identités d'Abel. *Discrete Math.*, 8:331–343, 1974.

[7] S. Getu and L. W. Shapiro. A natural proof of Abel's identity. In *Proceedings of the Eleventh Southeastern Conference on Combinatorics, Graph Theory and Computing, Vol. I.* Congr. Numer., Vol. 28, pages 447–452, 1980.

[8] D. E. Knuth. *Art of Computer Programming*, Vol. 1. Addison-Wesley, 3rd edition, 1997.

[9] J. Pitman. Forest volume decompositions and abel-cayley-hurwitz multinomial expansions. *J. Comb. Theory, Series A*, 98:175–191, 2002.

[10] S. Ramanujan. Question 294. *J. Indian Math. Soc.*, 3:128, 1911.

[11] R. Stanley. *Enumerative Combinatorics*, Vol. 2. Cambridge University Press, 1999.

STRICT SCHUR NUMBERS

Tanbir Ahmed
Department of Computer Science and Software Engineering
Concordia University, Montréal, Canada
ta_ahmed@cs.concordia.ca

Michael G. Eldredge
University of Idaho, Moscow, Idaho
michael.eldredge@vandals.uidaho.edu

Jonathan J. Marler
University of Idaho, Moscow, Idaho
marler8997@vandals.uidaho.edu

Hunter S. Snevily
Department of Mathematics, University of Idaho, Moscow, Idaho
snevily.uidaho.edu

Received: 2/6/9, Revised: 9/11/12, Accepted: 4/18/13, Published: 4/25/13

Abstract

Let $S(h,k)$ be the least positive integer such that any 2-coloring of the interval $[1, S(h,k)]$ must admit either

(i) a monochromatic solution to $x_1 + \ldots + x_{h-1} = x_h$ with $x_1 < x_2 < \ldots < x_h$ or

(ii) a monochromatic solution to $x_1 + \ldots + x_{k-1} = x_k$ with $x_1 < x_2 < \ldots < x_k$.

We prove $S(3,3) = 9$, $S(3,4) = 16$, and for all $k \geqslant 5$,

$$S(3,k) = \begin{cases} 3k^2/2 - 7k/2 + 3 & \text{if } k \equiv 0,1 \pmod 4, \\ 3k^2/2 - 7k/2 + 4 & \text{if } k \equiv 2,3 \pmod 4. \end{cases}$$

1. Introduction

Let $\mathbb{N}$ denote the set of positive integers and $[a,b] = \{n \in \mathbb{N} : a \leqslant n \leqslant b\}$. A mapping $\chi : [a,b] \to [1,t]$ is called a t-coloring of $[a,b]$. Let L_m denote the system of inequalities given by

$$x_1 + x_2 + \ldots, x_{m-1} = x_m, \qquad x_1 < x_2 < \cdots < x_m.$$

A solution $n_1, n_2, \ldots, n_m$ to L_m is *monochromatic* if $\chi(n_i) = \chi(n_j)$ for $1 \leqslant i, j \leqslant m$. Henceforth, we assume a *two-coloring* ($t = 2$) of the interval and denote each color by *red* and *blue*. Furthermore, a monochromatic solution to L_m such that $\chi(n_1) = \chi(n_2) = \ldots = \chi(n_m) = red$ will be called a "*red* solution," and likewise for a "*blue* solution." Lastly, we define $S(h, k)$ to be the least positive integer such that every coloring of the interval $[1, S(h, k)]$ by the colors *red* and *blue* contains either a *red* solution to L_k or a *blue* solution to L_h.

In the following proofs, we show that $S(3, 3) = 9$, $S(3, 4) = 16$, and for all $k \geqslant 5$,

$$S(3,k) = \begin{cases} 3k^2/2 - 7k/2 + 3 & \text{if } k \equiv 0,1 \pmod 4, \\ 3k^2/2 - 7k/2 + 4 & \text{if } k \equiv 2,3 \pmod 4. \end{cases}$$

2. The Lower Bound

Lemma 1. (Lower Bound) *For $k \geqslant 3$,*

$$S(3,k) \geqslant N = \begin{cases} 3k^2/2 - 7k/2 + 3 & \textit{if } k \equiv 0,1 \pmod 4, \\ 3k^2/2 - 7k/2 + 4 & \textit{if } k \equiv 2,3 \pmod 4. \end{cases}$$

Proof. Consider a coloring of $\chi : [1, N-1] \to \{blue, red\}$ defined as follows. For $n \in [1, N-1]$, let

$$\chi(n) = \begin{cases} blue & \text{if } n \equiv 1 \pmod 2 \text{ and } n \leqslant k(k-1)/2, \\ blue & \text{if } n \equiv 0 \pmod 2 \text{ and } n \geqslant k(k-1), \\ red & \text{otherwise.} \end{cases}$$

We claim this coloring has no *blue* solution to L_3 and no *red* solution to L_k.

Suppose $n_1+n_2 = n_3$, where $n_1 < n_2 < n_3$, is a *blue* solution to L_3 on the interval $[1, N-1]$. Suppose $n_2 \equiv 1 \pmod 2$. Then $n_1 < n_2 \leqslant k(k-1)/2$, which implies $n_1 \equiv 1 \pmod 2$ and $n_3 < k(k-1)$. Since $n_3 = n_1 + n_2 \equiv (1+1) \pmod 2 \equiv 0 \pmod 2$, we must have that $n_3 \geqslant k(k-1)$, which is a contradiction. Therefore, $n_2 \equiv 0 \pmod 2$.

Hence, $n_3 > n_2 \geqslant k(k-1)$, which implies $n_3 \equiv 0 \pmod 2$, which then implies $n_1 \equiv 0 \pmod 2$. Therefore, $n_2 > n_1 \geqslant k(k-1)$, which implies $n_3 = n_1 + n_2 \geqslant k(k-1) + (k(k-1)+2) > N-1$, another contradiction implying no such *blue* solution to L_3 exists.

Next, suppose $n_1 + n_2 + \cdots + n_{k-1} = n_k$, where $n_1 < n_2 < \cdots < n_k$, is a *red* solution to L_k on the interval $[1, N-1]$. Let q denote the minimal sum of $k-2$ *red* numbers. Clearly, $q = \sum_{i=1}^{k-2} 2i = k^2 - 3k + 2$.

If $n_{k-1} \leqslant k(k-1)/2$, then $n_i \equiv 0 \pmod 2$ for $i \in [1, k-1]$. This implies $n_k \equiv 0 \pmod 2$, which implies $n_k < k(k-1)$, but this is a contradiction since $k(k-1) > n_k \geqslant q + 2(k-1) = k(k-1)$. Therefore, $n_{k-1} > k(k-1)/2$.

If $k \equiv 0, 1 \pmod 4$, then $n_k > q + k(k-1)/2 = 3k^2/2 - 7k/2 + 2 = N - 1$, a contradiction that implies $k \equiv 2, 3 \pmod 4$.

If $n_{k-1} = k(k-1)/2 + 1$, then $n_i \equiv 0 \pmod 2$ for all $i \in [1, k-1]$, which implies $n_k \equiv 0 \pmod 2$. Since n_k is *red*, $n_k < k(k-1)$, which is a contradiction since $n_k \geqslant q + k(k-1)/2 + 1$. Therefore, $n_{k-1} > k(k-1)/2 + 1$, which implies $n_k \geqslant q + k(k-1)/2 + 2 = 3k^2/2 - 7k/2 + 4 = N$. □

3. The Upper Bound

Throughout this section, let p denote the sum of the first k *red* numbers and let r_i and b_i denote the ith *red* and *blue* numbers, respectively. Then, $r_i < r_j$ and $b_i < b_j$, for all $i < j$.

Lemma 2. *For $n \geqslant 3$, if at least $n+1$ numbers in the interval $[1, 2n]$ are colored blue, then the only coloring that avoids a blue solution to L_3 is given by*

$$\chi(x) = \begin{cases} red & \text{if } x \in [1, n-1], \\ blue & \text{if } x \in [n, 2n]. \end{cases}$$

Proof. Since the case $n = 3$ is trivial, assume $n > 3$.

Case 1. $\chi(2n) = red$. We proceed via induction on n. For some $n > 3$, assume the claim holds for $n-1$. To avoid a *blue* solution on the interval $[1, 2(n-1)]$, we must have $\chi(x) = blue$ for all $x \in [(n-1), 2(n-1)]$ and $\chi(x) = red$ for all $x \in [1, n-2]$. Since we need another *blue* in the interval $[1, 2n]$, the number $(2n-1)$ must be *blue*, but then $(n-1) + n = (2n-1)$ is a *blue* solution to L_3.

Case 2. $\chi(2n) = blue$. By the pigeonhole principle, $\chi(n) = blue$, since otherwise one of the pairs $\{x, 2n-x\}$ with $1 \leqslant x < n$ would be all *blue* giving us the *blue* solution $(x) + (2n-x) = 2n$. Now suppose $\chi(1) = blue$, which implies the pair $\{n-1, n+1\}$ is all *red*. Hence, some other pair $\{x, 2n-x\}$ with $1 \leqslant x < n-1$, is all *blue* and we get a contradiction. Therefore, $\chi(1) = red$; hence $\chi(2n-1) = blue$, otherwise some other pair $\{x, 2n-x\}$ with $2 \leqslant x \leqslant n-2$ would be all *blue*, giving us a contradiction as before. But then we must have $\chi(n-1) = red$ since $(n-1)+n = (2n-1)$; hence $\chi(n+1) = blue$. But then we must have $\chi(n-2) = red$ since $(n-2)+(n+1) = (2n-1)$; hence $\chi(n+2) = blue$. Continuing in this manner, we get the desired coloring. □

Corollary 1. *To avoid a blue solution to L_3, $r_i \leqslant 2i+1$ for all i.*

Proof. The claim is easily proven for r_1 and r_2. Suppose $r_i > 2i+1$ for some $i \geqslant 3$. This would imply at least $i+1$ numbers are colored *blue* in the interval $[1, 2i]$. Applying Lemma 2 gives us that $\chi(i) = \chi(i+1) = blue$. Since there are $2i+1-(i-1) = i+2$ *blue* integers in $[1, 2i+1]$, $\chi(2i+1) = blue$ as well, but this yields the *blue* solution $i + (i+1) = 2i+1$. □

Corollary 2. *There exists at least k integers in the interval $[1, N]$ colored red, that is, p exists.*

Proof. If p does not exist, then by Corollary 1, $r_{k-1} \leqslant 2k - 1$, which implies that all numbers in $[2k, N]$ are colored *blue*, while the Pigeonhole Principle ensures that some number in $[1, k]$ is also colored *blue*, say $x \in \{1, 2, \ldots, k\}$. But then $x + 2k = x + 2k \leqslant N$ is a *blue* solution for $k \geqslant 4$, a contradiction. The case $k = 3$ can be done separately by hand varying this argument and handling several cases. □

Corollary 3. *Let $i > b_1$ and $i \geqslant 3$. To avoid a blue solution to L_3, $r_i \leqslant 2i$.*

Proof. Suppose $r_i > 2i$. Then Corollary 1 implies $r_i = 2i + 1$. Therefore, the interval $[1, 2i + 1]$ must contain exactly $i + 1$ *blue* numbers. Since $i \geqslant 3$, Lemma 2 implies that the interval $[1, i - 1]$ is all *red*, but this contradicts the hypothesis $b_1 < i$. □

Lemma 3. (P Lemma) *The following hold:*

(*i*) *If $b_1 = 1$, then $p \leqslant k^2 + k - 12 + r_1 + r_2 + r_3$,*

(*ii*) *If $1 < b_1 \leqslant k$, then $p \leqslant k^2 + k + 1 - b_1(b_1 - 1)/2$,*

(*iii*) *If $b_1 > k$, then $p = k(k + 1)/2$.*

Proof. (*i*) If $b_1 = 1$, then by Corollary 3

$$\begin{aligned} p = r_1 + r_2 + r_3 + \sum_{i=4}^{k} r_i &\leqslant r_1 + r_2 + r_3 + \sum_{i=4}^{k} 2i \\ &= k^2 + k - 12 + r_1 + r_2 + r_3. \end{aligned}$$

(*ii*) If $1 < b_1 \leqslant k$, then by Corollaries 1 and 3

$$\begin{aligned} p = \sum_{i=1}^{b_1-1} r_i + r_{b_1} + \sum_{i=b_1+1}^{k} r_i &\leqslant \sum_{i=1}^{b_1-1} i + (2b_1 + 1) + \sum_{i=b_1+1}^{k} r_i \\ &\leqslant \sum_{i=1}^{b_1-1} i + (2b_1 + 1) + \sum_{i=b_1+1}^{k} 2i \\ &= k^2 + k + 1 - b_1(b_1 - 1)/2. \end{aligned}$$

(*iii*) If $b_1 > k$, then $p = \sum_{i=1}^{k} r_i = \sum_{i=1}^{k} i = k(k + 1)/2$.

□

Given a valid colouring, the upper bound for p can be improved by modifying Lemma 3. For example, for $k \geqslant 6$, if $r_1 = 1$, $b_1 = 2$, $r_2 = 3$, and $r_3 = 4$, then

$$p \leqslant k^2 + k + 1 - b_1(b_1 - 1)/2 - (5 - r_2) - (6 - r_3) = k^2 + k - 4.$$

Fact 1. *If $k \geqslant 6$, then $k^2 + k - 5 \leqslant N$.*

Corollary 4. *If $p - r_j \leqslant N$ for some $j \in [1, k]$, then to avoid a red solution to L_k, $\chi(p - r_i) =$ blue for all $i \in [j, k]$.*

Proof. If $\chi(p - r_i) =$ *red* for some $i \in [j, k]$, then we get the *red* solution to L_k

$$r_1 + r_2 + \ldots + r_{i-1} + r_{i+1} + r_{i+2} + \ldots + r_k = p - r_i,$$

where $p - r_i \leqslant p - r_j \leqslant N$ (by hypothesis). Hence, $\chi(p - r_i) =$ *blue* for all $i \in [j, k]$. □

Corollary 5. *If $k \geqslant 6$ and $b_1 > 1$, then $p - r_i \leqslant N$ for all r_i.*

Proof. We have $p - r_i \leqslant p - 1$. In view of Fact 1, if $p \leqslant k^2 + k - 4$, then $p - r_i \leqslant N$ for all r_i. If $b_1 = 2$, then modifying Lemma 3, we get the following cases:

#	1	2	3	4	5	6	7	8	9	10	11	12	n s.t. $p \leqslant n$
1.	r_1	b_1	b_2	b_3	r_2	r_3	r_4	b_4	b_5	r_5	r_6	r_7	k^2+k-4
2.	r_1	b_1	b_2	b_3	r_2	r_3	r_4	b_4	r_5	r_6			k^2+k-4
3.	r_1	b_1	b_2	b_3	r_2	r_3	r_4	r_5	b_4	b_5	r_6		k^2+k-4
4.	r_1	b_1	b_2	b_3	r_2	r_3	r_4	r_5	b_4	r_6			k^2+k-5
5.	r_1	b_1	b_2	b_3	r_2	r_3	r_4	r_5	r_6				k^2+k-6
6.	r_1	b_1	b_2	r_2	r_3	b_3	b_4	r_4	r_5	r_6			k^2+k-5
7.	r_1	b_1	b_2	r_2	r_3	b_3	r_4	r_5					k^2+k-5
8.	r_1	b_1	b_2	r_2	r_3	r_4							k^2+k-4
9.	r_1	b_1	r_2	b_2	b_3	r_3	r_4	b_4	r_5				k^2+k-4
10.	r_1	b_1	r_2	b_2	b_3	r_3	r_4	r_5					k^2+k-5
11.	r_1	b_1	r_2	b_2	r_3	r_4							k^2+k-5
12.	r_1	b_1	r_2	r_3									k^2+k-4

Similarly, if $b_1 = 3$ then modifying Lemma 3, we get the following cases:

#	1	2	3	4	5	6	7	8	9	10	n s.t. $p \leqslant n$
1.	r_1	r_2	b_1	b_2	b_3	b_4	r_3	r_4	r_5	r_6	k^2+k-5
2.	r_1	r_2	b_1	b_2	b_3	r_3	r_4				k^2+k-4
3.	r_1	r_2	b_1	b_2	r_3						k^2+k-4
4.	r_1	r_2	b_1	r_3							k^2+k-5

For $4 \leqslant b_1 \leqslant k$, by Lemma 3 we have $p \leqslant k^2 + k - 5$. For $b_1 > k$, $p = k(k+1)/2$ by part (iii) of Lemma 3. For $k \geqslant 6$ and $b_1 > 1$, we have $p \leqslant k^2 + k - 4$, and hence, using Fact 1

$$p - r_i \leqslant p - 1 \leqslant (k^2 + k - 4) - 1 = k^2 + k - 5 \leqslant N$$

for all r_i with $i \in [1, k]$. □

Remark 1. Combining Corollaries 4 and 5, we see that, for $k \geqslant 6$ and $b_1 > 1$, $\chi(p - r_j) = \textit{blue}$ for all $j \in [1, k]$.

Lemma 4. (Upper Bound) *For* $k \geqslant 6$*,*

$$S(3, k) \leqslant N = \begin{cases} 3k^2/2 - 7k/2 + 3 & \text{if } k \equiv 0, 1 \pmod 4, \\ 3k^2/2 - 7k/2 + 4 & \text{if } k \equiv 2, 3 \pmod 4. \end{cases}$$

Proof. Suppose to the contrary that N is not an upper bound for $k \geqslant 6$. This occurs if and only if there exists a coloring of $[1, N]$ without a *blue* solution to L_3 and a *red* solution to L_k. Consider the following two cases:

(1) $\chi(1) = \textit{blue}$ (with $k \geqslant 6$). Suppose $\chi(2) = \textit{blue}$. Then $r_1 = 3$ and $r_2 \leqslant 5$ (by Corollary 1) to avoid *blue* solutions $1 + 2 = 3$ and $1 + 4 = 5$, respectively. Therefore, by Lemma 3 and Corollary 3, we have

$$p \leqslant k^2 + k - 4 + r_3 \leqslant k^2 + k + 2,$$

which implies (by Fact 1) that $p - r_i \leqslant k^2 + k - 5 \leqslant N$ for $r_i \geqslant 7$.

If $\chi(x) = \textit{blue}$ for some $x \in [6, 7]$, then this implies $\chi(x+1) = \chi(x+2) = \textit{red}$ to avoid the *blue* solutions $1+x = x+1$ and $2+x = x+2$. Corollary 4 implies $\chi(p - x - 1) = \chi(p - x - 2) = \textit{blue}$, and then $1 + (p - x - 2) = p - x - 1$ is a *blue* solution since $p - x - 2 \geqslant k(k + 1)/2 - 9 \geqslant 12 > 1$. Therefore, $\chi(6) = \chi(7) = \textit{red}$. Considering the present information, it can be shown that $p - r_i \leqslant N$ for $r_i \geqslant 6$. If $r_2 = 4$, then $p \leqslant k^2 + k + 1$, otherwise 7 being *red* means that the estimate for r_4 can be improved by one. Now, Corollary 4 gives $\chi(p - 6) = \chi(p - 7) = \textit{blue}$. Thus $1 + (p - 7) = p - 6$ is a *blue* solution in view of $p - 8 \geqslant k(k + 1)/2 - 8 \geqslant 13 > 1$. So we conclude that $\chi(2) = \textit{red}$.

If $b_2 \geqslant 5$, then $r_2 = 3$, $r_3 = 4$, and, by Lemma 3 and Fact 1,

$$p - r_1 \leqslant k^2 + k - 12 + r_2 + r_3 = k^2 + k - 5 \leqslant N,$$

which leads to a contradiction since Corollary 4 gives us the *blue* solution $1 + (p - 4) = p - 3$.

If $b_2 = 4$, then $r_2 = 3$ and $r_3 = 5$, and by Lemma 3 and Fact 1

$$p - r_2 \leqslant k^2 + k - 12 + r_1 + r_3 = k^2 + k - 5 \leqslant N.$$

By Corollary 4, $\chi(p-3) = \chi(p-5) = \textit{blue}$. To avoid a *blue* solution $1+(p-6) = p - 5$, we need $\chi(p - 6) = \textit{red}$, but by Corollary 4, this implies $\chi(6) = \textit{blue}$.

In that case, to avoid the *blue* solution $1 + 6 = 7$, we need $\chi(7) = red$, but that yields the *blue* solution $4 + (p - 7) = p - 3$.

Now, suppose $b_3 > r_3$. If $b_2 = 3$, then $r_2 = 4$, $r_3 = 5$ (since $b_3 > r_3$), and by Lemma 3 and Fact 1, $p - r_2 \leqslant k^2 + k - 5 \leqslant N$, but that yields the *blue* solution $1 + (p - 5) = p - 4$ (by Corollary 4).

Therefore, $b_3 < r_3$, which implies the interval $[3, 5]$ has two *blue* numbers. Since these *blue* numbers cannot be adjacent, the only valid coloring is $\chi(3) = \chi(5) = blue$ and $\chi(4) = \chi(6) = red$. With this coloring, Corollary 1, Lemma 3, Corollary 4, and Fact 1 imply that $\chi(p - r_i) = blue$, for $i \in [3, k]$.

To avoid the *blue* solution $1+(p-r_{i+1}) = p-r_i$, we must have $r_{i+1} > r_i+1$, that is, $\chi(r_i+1) = blue$, for all $i \in [3, k-1]$. Thus $\chi(7) = blue$. Also, to avoid the *blue* solution $1 + b_i = b_{i+1}$, we must have $\chi(b_i + 1) = red$, for all $i > 1$. Thus $\chi(8) = red$, which implies $\chi(9) = blue$, which implies $\chi(10) = red$, and continuing in this manner, we get that, for all $x \in [1, 2k]$,

$$\chi(x) = \begin{cases} red & \text{if } x \equiv 0 \pmod 2, \\ blue & \text{if } x \equiv 1 \pmod 2. \end{cases}$$

Furthermore, for all $x \in [1, 2k - 3]$ with $\chi(x) = blue$, we must have $\chi(x + (2k - 1)) = red$, otherwise we get the *blue* solution $x + (2k - 1) = x + 2k - 1$. This implies $\chi(y) = red$ for all $y \in [2k, 4k - 4]$ with $y \equiv 0 \pmod 2$.

Using the block of even *red* numbers, we can extend the *blue* interval. Clearly, the sum of any $k - 1$ *red* numbers which is less than N must be *blue*. The maximal sum of $k-1$ *red* numbers from the block is $\sum_{i=0}^{k-2}((4k-4)-2i) = 3k^2 - 5k + 2$, which is clearly greater than N. Furthermore, the minimal sum of $k - 1$ *red* numbers from the block is $\sum_{i=1}^{k-1} 2i = k^2 - k$. Since we can always replace a *red* number in the minimal sum by an adjacent even number which is also *red*, and the maximal sum is greater than N, we get that all even numbers greater than or equal to $k^2 - k$ must be *blue*. This yields the extended coloring

$$\chi(x) = \begin{cases} red & \text{if } x \equiv 0 \pmod 2 \text{ and } x \in [2, 4k - 4], \\ blue & \text{if } x \equiv 1 \pmod 2 \text{ and } x \in [1, 2k - 1], \\ blue & \text{if } x \equiv 0 \pmod 2 \text{ and } x \in [k^2 - k, N]. \end{cases}$$

It can easily be shown that $N \equiv 1 \pmod 2$, implying $\chi(N - 1) = blue$. Since $\chi(1) = \chi(3) = blue$, we must have $\chi(N) = \chi(N - 2) = \chi(N - 4) = red$. Let q be the sum of first $k - 2$ *red* numbers. Then $q = k^2 - 3k + 2$. To avoid a *red* solution to L_k, we must have

$$\chi(N - q) = \chi(N - 2 - q) = \chi(N - 4 - q) = blue,$$

since $N-4-q > r_{k-2} = 2(k-2)$.

If $k \equiv 0,1 \pmod 4$, then we get the *blue* solution

$$(N-q)+(N-2-q) = 2(3k^2/2 - 7k/2 + 3) - 2(k^2-3k+2) - 2 = k^2 - k.$$

Likewise, if $k \equiv 2,3 \pmod 4$, then we get the *blue* solution

$$(N-q)+(N-4-q) = 2(3k^2/2 - 7k/2 + 4) - 2(k^2-3k+2) - 4 = k^2 - k.$$

Therefore, $\chi(1) \neq$ *blue*.

(2) $\chi(1) =$ *red* (with $k \geqslant 6$). Since $k \geqslant 6$ and $b_1 > 1$, by Remark 1, we have $\chi(p-r_i) =$ *blue* for all $i \in [1,k]$. Let a be the minimum *red* number such that $\chi(a-1) =$ *blue*. It can be shown that a exists. Suppose a does not exist, that is, x (say) *red* numbers are followed by $N-x$ *blue* numbers. If $x \geqslant k(k-1)/2$, then we have a *red* solution $1+2+\cdots+(k-1) = k(k-1)/2$, or else we have a *blue* solution $k(k-1)/2 + (k(k-1)/2+1) = k^2-k+1 \leqslant N$.

If $a < r_k$, then $\chi(p-a) =$ *blue*, which gives us a potential *blue* solution $(a-1)+(p-a) = p-1$. In order for it to be a valid solution, we must have that $a-1 \neq p-a$. However, since $a = r_i$ for some $i \in [1,k]$, this has already been proven in Corollary 4 ($p - r_i > r_k$ for all $i \in [1,k]$).

The argument of the previous paragraph yields a contradiction unless $p - r_k = p-a = a-1$, in which case $k+(k+1) = 2k+1 \leqslant \sum_{i=1}^{k-1} i = p - r_k = p-a = a-1$, from which it is clear in view of the definition of a that $k+(k+1) = 2k+1$ is a *blue* solution.

Suppose $b_1 \leqslant 3k/2$. To avoid the *blue* solution, $b_1 + (b_1+1) = 2b_1+1$, either b_1+1 or $2b_1+1$ must be *red*, which implies $a \leqslant 2b_1+1 \leqslant 3k+1$. Now consider,

$$\begin{aligned}(p-r_k)-1+a &= \sum_{i=1}^{k-1} i + (a-1) = k(k-1)/2 + (a-1) \\ &\leqslant k(k-1)/2 + (3k+1) - 1 = (k^2+5k)/2 \leqslant N.\end{aligned}$$

Since $\chi(a) =$ *red*, to avoid the *red* solution $a + \sum_{i=2}^{k-1} i = (p-r_k)-1+a$, we have $\chi(p-r_k+a-1) =$ *blue*, which yields the potential *blue* solution

$$(p-r_k)+(a-1) = (p-r_k)-1+a.$$

To be a valid solution, we must have $a-1 \neq p-r_k$. If $a-1 = p-r_k$, then $p-r_k+1 = a$, which implies $\chi(p-r_k+1) =$ *red*. However, this is a contradiction since $\chi(p-r_{k-1}) =$ *blue* and $r_{k-1} = r_k - 1$.

Therefore, $b_1 > 3k/2$, which implies $\chi(x) =$ *red* for all $x \in [1, 3k/2]$. Using this *red* interval, we can create another *blue* interval. The minimum sum of

$k-1$ *red* numbers in this interval is $k(k-1)/2$. If k is even, then the maximal sum is k^2-1, and if k is odd, then since only integers are colored, we only know that the interval $[1,(3k-1)/2]$ is colored all *red*, in which case, the maximal sum of $k-1$ *red* integers is $k^2-k/2-1/2$.

Since every integer in this new interval can be represented by a sum of $k-1$ *red* numbers, the interval $[k(k-1)/2, k^2-k/2-1/2]$ must be colored *blue* to avoid a *red* solution. Since k^2-k+1 is in the *blue* interval, we have the *blue* solution $k(k-1)/2+(k(k-1)/2+1)=k^2-k+1$.

Hence, for $k \geqslant 6$, every coloring of $[1,N]$ has a *blue* solution to L_3 or a *red* solution to L_k. □

4. The Cases $3 \leqslant k \leqslant 5$

In this section, we formally prove the exact values of $S(3,3)$ and $S(3,4)$, and provide the computer proof for the exact value of $S(3,5)$.

Lemma 5. $S(3,3)=9$.

Proof. Let $\chi(1)=\chi(2)=\chi(4)=\chi(8)=$ *red* and $\chi(3)=\chi(5)=\chi(6)=\chi(7)=$ *blue*. This coloring has no *red* or *blue* solution to L_3. Therefore, $S(3,3)>8$.

Suppose to the contrary that $S(3,3)>9$. Without loss of generality, let *blue* be the color used 5 or more times from 1 to 9. If $\chi(9)=$ *red*, then Lemma 2 gives us the *red* solution $1+2=3$. Therefore, $\chi(9)=$ *blue*. We are left with two cases.

`Case 1.` $\chi(8)=$ *blue*. To avoid the *blue* solution $1+8=9$, we have $\chi(1)=$ *red*. If $\chi(5)=$ *blue*, then we must have $\chi(3)=$ *red* (to avoid the *blue* solution $3+5=8$) and $\chi(4)=$ *red* (to avoid the *blue* solution $4+5=9$). But then we have the *red* solution $1+3=4$. Therefore, $\chi(5)=$ *red*. To avoid the *red* solutions $1+4=5$ and $1+5=6$, we must have $\chi(4)=\chi(6)=$ *blue*. Then $\chi(2)=$ *red* (to avoid the *blue* solution $2+4=6$), which implies $\chi(3)=$ *blue* (to avoid the *red* solution $1+2=3$), which gives the *blue* solution $3+6=9$.

`Case 2.` $\chi(8)=$ *red*. If $\chi(7)=$ *red*, then Lemma 2 gives us the *red* solution $1+7=8$. Therefore, $\chi(7)=$ *blue*, which leads to a contradiction after a chain of implications:

$\chi(2)=$ *red* (to avoid the *blue* solution $2+7=9$),
$\chi(6)=$ *blue* (to avoid the *red* solution $2+6=8$),
$\chi(1)=$ *red* (to avoid the *blue* solution $1+6=7$),
$\chi(3)=$ *blue* (to avoid the *red* solution $1+2=3$),
and hence the *blue* solution $3+6=9$. □

Lemma 6. $S(3,4)=16$.

Proof. For all $x \in [1, 15]$, let $x \in [6, 12]$ be *blue* and x be *red* otherwise. This coloring has no *blue* solution to L_3 and no *red* solution to L_4. Therefore, $S(3, 4) > 15$.

Suppose to the contrary that $S(3, 4) > 16$. Then suppose $\chi(1) = blue$. Corollary 3 implies $r_i \leqslant 2i$, for all $i \geqslant 3$.

Since $r_1+r_2+r_4 \leqslant 3+5+8 = 16$, we have $\chi(r_1+r_2+r_3) = \chi(r_1+r_2+r_4) = blue$. If $r_4 > r_3 + 2$, we get the *blue* solution $1 + (r_3 + 1) = r_3 + 2$, and if $r_4 = r_3 + 1$, we get the *blue* solution $1 + (r_1 + r_2 + r_3) = r_1 + r_2 + r_4$. Hence, $r_4 = r_3 + 2$. To avoid the *blue* solution $2 + (r_1 + r_2 + r_3) = r_1 + r_2 + r_4$, we must have $\chi(2) = red$, that is, $r_1 = 2$, which implies $r_1 + r_3 + r_4 \leqslant 2+6+8 = 16$. Thus $\chi(r_1 + r_3 + r_4) = blue$.

Applying the same reasoning to r_3 as we did to r_4, we get that $r_3 = r_2 + 2$. Then to avoid the *blue* solution $4 + (r_1 + r_2 + r_3) = r_1 + r_3 + r_4$, we must have $\chi(4) = red$. If $\chi(3) = red$, then $r_2 = 3$, and so $r_3 = 4$. But $r_3 \neq r_2 + 1$. Therefore, $\chi(3) = blue$, which implies $r_2 = 4$, and so $r_3 = 6$ and $r_4 = 8$. Then we must have $\chi(5) = \chi(7) = \chi(12) = blue$, but then we get the *blue* solution $5 + 7 = 12$. Therefore, $\chi(1) = red$.

Corollary 1 gives us that $r_2 = 2, 3, 4$, or 5. We handle these four cases separately.

`Case 1.` $r_2 = 5$. This implies $\chi(2) = \chi(3) = \chi(4) = blue$. Therefore, $\chi(6) = red$ and $\chi(7) = red$ to avoid the *blue* solutions $2 + 4 = 6$ and $3 + 4 = 7$, respectively. Hence, $\chi(12) = blue$ and $\chi(14) = blue$ to avoid the *red* solutions $1 + 5 + 6 = 12$ and $1 + 6 + 7 = 14$, respectively. But, then we get the *blue* solution $2 + 12 = 14$.

`Case 2.` $r_2 = 4$. This implies $\chi(2) = \chi(3) = blue$. Therefore, $\chi(5) = red$ (to avoid the *blue* solution $2 + 3 = 5$), which implies $\chi(10) = blue$ (to avoid the *red* solution $1 + 4 + 5 = 10$). Therefore, $\chi(7) = red$ and $\chi(12) = red$ to avoid the *blue* solutions $3 + 7 = 10$ and $2 + 10 = 12$, respectively. But then we get the *red* solution $1 + 4 + 7 = 12$.

`Case 3.` $r_2 = 3$. This implies $\chi(2) = blue$. If $r_4 = 9$, then Lemma 2 implies $\chi(2) = red$. Thus $r_4 \leqslant 8$.

If $r_3 \geqslant 6$, then $\chi(4) = \chi(5) = blue$, which implies $r_3 = 6$ (to avoid the *blue* solution $2 + 4 = 6$). Therefore, $\chi(7) = red$ (to avoid the *blue* solution $2 + 5 = 7$), which implies $\chi(14) = blue$ (to avoid the *red* solution $1+6+7 = 14$) and $\chi(16) = blue$ (to avoid the *red* solution $3 + 6 + 7 = 16$). But then we get the *blue* solution $2 + 14 = 16$. Therefore, $r_3 \leqslant 5$.

This implies $3+r_3+r_4 \leqslant 16$, which gives us that $\chi(1+r_3+r_4) = \chi(3+r_3+r_4) = blue$, but then we get the *blue* solution $2 + (1 + r_3 + r_4) = 3 + r_3 + r_4$.

`Case 4.` $r_2 = 2$. Suppose $\chi(7) = red$. Then $\chi(4) = blue$ (to avoid the *red* solution $1 + 2 + 4 = 7$) and $\chi(10) = blue$ (to avoid the *red* solution $1 + 2 + 7 = 10$). This implies $\chi(6) = red$ and $\chi(14) = red$ to avoid the *blue* solutions $4 + 6 = 10$ and $4 + 10 = 14$, respectively. But then we get the *red* solution $1 + 6 + 7 = 14$. Therefore, $\chi(7) = blue$.

Suppose $\chi(3) = blue$. Then $\chi(4) = red$ and $\chi(10) = red$ to avoid the *blue* solutions $3+4=7$ and $3+7=10$, respectively. This implies $\chi(13) = blue$ and $\chi(16) = blue$ to avoid the *red* solutions $1+2+10=13$ and $2+4+10=16$, respectively. Therefore, $\chi(3) = red$, which leads to a contradiction after a chain of implications:

$\chi(6) = blue$ (to avoid the *red* solution $1+2+3=6$),
$\chi(13) = red$ (to avoid the *blue* solution $6+7=13$),
$\chi(9) = blue$ (to avoid the *red* solution $1+3+9=13$),
$\chi(16) = red$ (to avoid the *blue* solution $7+9=16$),
and hence the *red* solution $1+2+13=16$. □

4.1. Computer Assisted Proof for $S(3,5)$

Let us write a coloring of $[1,n]$ as a bit-string of length n where the i-th bit is zero if $\chi(i) = blue$, and one if $\chi(i) = red$.

4.1.1. S(3,5)=23

By Lemma 1, the lower bound is $S(3,5) > 22$. We consider all of the ten colorings of $[1,22]$ (obtained by computer search) without a *blue* solution to L_3 and a *red* solution to L_5.

1. For each of the following four colorings

   0010110111111111111110,

   0010110111111011111110,

   0010110111111011011110, and

   0010110111101111011110,

   if $\chi(23) = blue$, then we have a *blue* solution $1+22=23$ to L_3; and

   if $\chi(23) = red$, then we have a *red* solution $3+5+6+9=23$ to L_5.

2. For each of the following four colorings

   0010111011111111111101,

   0010111011111110111101,

   0010111011111101111101, and

   0010111011111011111101,

   if $\chi(23) = blue$, then we have a *blue* solution $2+21=23$ to L_3; and

   if $\chi(23) = red$, then we have a *red* solution $3+5+6+9=23$ to L_5.

3. For each of the following two colorings

   0101010101111111101010, and

   0101010101111111111010,

   if $\chi(23) = blue$, then we have a *blue* solution $1 + 22 = 23$ to L_3; and

   if $\chi(23) = red$, then we have a *red* solution $2 + 4 + 6 + 11 = 23$ to L_5.

Therefore, $S(3, 5) = 23$.

Acknowledgements We sincerely would like to thank the anonymous referee for the helpful comments and detailed list of suggestions on how to improve this paper. We would also like to thank the Managing Editor Bruce Landman for his patience as the paper went several rounds during the review process.

References

[1] Bialostocki, A. and Schaal, D. On a variation of Schur numbers, *Graphs Combin.*, **16(2)** (2000), 139–147.

[2] Landman, B. and Robertson, A. Ramsey Theory on the Integers, *Student mathematical library*, American Mathematical Society, Providence, RI, 2004.

[3] Martinelli, B. and Schaal, D. On generalized Schur numbers for $x_1 + x_2 + c = kx_3$, *Ars Combin.* **85** (2007), 33–42.

ON q-ANALOG OF WOLSTENHOLME TYPE CONGRUENCES FOR MULTIPLE HARMONIC SUMS

Jianqiang Zhao
Department of Mathematics, Eckerd College, St. Petersburg, Florida
zhaoj8888@gmail.com

Received: 9/25/12 , Accepted: 4/20/13, Published: 4/25/13

Abstract

Multiple harmonic sums are iterated generalizations of harmonic sums. Recently Dilcher has considered congruences involving q-analogs of these sums in depth one. In this paper we shall study the homogeneous case for arbitrary depth by using generating functions and shuffle relations of the q-analog of multiple harmonic sums. At the end, we also consider some non-homogeneous cases.

1. Introduction

In [8] Shi and Pan extended Andrews' result [1] on the q-analog of Wolstenholme Theorem to the following two cases: for all prime $p \geq 5$

$$\sum_{j=1}^{p-1} \frac{1}{[j]_q} \equiv \frac{p-1}{2}(1-q) + \frac{p^2-1}{24}(1-q)^2[p]_q \pmod{[p]_q^2}, \tag{1}$$

$$\sum_{j=1}^{p-1} \frac{1}{[j]_q^2} \equiv -\frac{(p-1)(p-5)}{12}(1-q)^2 \pmod{[p]_q}, \tag{2}$$

where $[n]_q = (1-q^n)/(1-q)$ for any $n \in \mathbb{N}$ and $q \neq 1$. This type of congruences is considered in the polynomial ring $\mathbb{Z}[q]$ throughout this paper. Notice that the modulus $[p]_q$ is an irreducible polynomial in q when p is a prime. In [3] Dilcher generalized the above two congruences further to sums of the form $\sum_{j=1}^{p-1} \frac{1}{[j]_q^n}$ and $\sum_{j=1}^{p-1} \frac{q^n}{[j]_q^n}$ for all positive integers n in terms of certain determinants of binomial coefficients. However, his modulus is always $[p]_q$. He also expressed these congruences using Bernoulli numbers, Bernoulli numbers of the second kind, and Stirling numbers of the first kind, which we briefly recall now.

The well-known Bernoulli numbers are defined by the following generating series:

$$\frac{x}{e^x-1} = \sum_{n=0}^{\infty} B_n \frac{x^n}{n!} = 1 - \frac{1}{2}\frac{x}{1!} + \frac{1}{6}\frac{x^2}{2!} - \frac{1}{30}\frac{x^4}{4!} + \cdots .$$

On the other hand, the Bernoulli numbers of the second kind are defined by the power series (cf. [7, p. 114]).

$$\frac{x}{\log(1+x)} = \sum_{n=0}^{\infty} b_n \frac{x^n}{n!} = 1 + \frac{1}{2}\frac{x}{1!} - \frac{1}{6}\frac{x^2}{2!} + \frac{1}{4}\frac{x^3}{3!} - \frac{19}{24}\frac{x^4}{4!} + \cdots .$$

This is a little different from the definition of $\tilde{b}_n$ in [3], which is changed to b_n later in the same paper. Finally, the Stirling numbers of the first kind $s(n,j)$ are defined by

$$x(x-1)(x-2)\cdots(x-n+1) = \sum_{j=0}^{n} s(n,j)x^j.$$

Define

$$K_n(p) := (-1)^{n-1}\frac{b_n}{n!} - \frac{(-1)^n}{(n-1)!}\sum_{j=1}^{[n/2]} \frac{B_{2j}}{2j} s(n-1, 2j-1)p^{2j}. \tag{3}$$

By [3, Thm. 1, (6.5) and Thm. 4] and [4, Thm. 3.1] one gets:

Theorem 1.1. *If $p > 3$ is a prime, then for all integers $n > 1$ we have*

$$\sum_{j=1}^{p-1} \frac{q^j}{[j]_q^n} \equiv K_n(p)(1-q)^n \pmod{[p]_q}.$$

We will need the following easy generalization of this theorem.

Theorem 1.2. *If $p > 3$ is a prime, then for every integer $n > t \geq 1$ we have*

$$\sum_{j=1}^{p-1} \frac{q^{tj}}{[j]_q^n} \equiv (1-q)^n \sum_{i=0}^{t-1} \binom{t-1}{i} (-1)^i K_{n-i}(p) \pmod{[p]_q}. \tag{4}$$

Moreover,

$$\sum_{j=1}^{p-1} \frac{1}{[j]_q^n} \equiv (1-q)^n \left(\frac{p-1}{2} + \sum_{j=2}^{n} K_j(p) \right) \pmod{[p]_q}. \tag{5}$$

Proof. If $t > 1$ it is clear that

$$q^{tj} = q^j\big(1-(1-q^j)\big)^{t-1} = q^j \sum_{i=0}^{t-1} \binom{t-1}{i} (-1)^i (1-q^j)^i.$$

So (4) follows from Theorem 1.1 immediately. Congruence (5) is a variation of [3, (5.11)]. □

All the sums in Theorem 1.1 and 1.2 are special cases of the q-analog of multiple harmonic sums. The congruence properties of the classical multiple harmonic sums (MHS for short) are systematically investigated in [10]. In this paper we shall study their q-analogs which are natural generalizations of the congruences obtained by Shi and Pan [8] and Dilcher [3].

Similar to its classical case (cf. [10]) a q-analog of multiple harmonic sum (q-MHS for short) is defined as follows. For $\mathbf{s} := (s_1, \ldots, s_\ell) \in \mathbb{N}^\ell$, $\mathbf{t} := (t_1, \ldots, t_\ell) \in \mathbb{N}^\ell$ and $n \in \mathbb{Z}_{\geq 0}$ set

$$H_q^{(\mathbf{t})}(\mathbf{s}; n) := \sum_{1 \leq k_1 < \cdots < k_\ell \leq n} \frac{q^{k_1 t_1 + \cdots + k_\ell t_\ell}}{[k_1]_q^{s_1} \cdots [k_\ell]_q^{s_\ell}}, \qquad H_q^{*(\mathbf{t})}(\mathbf{s}; n) = H_q^{(\mathbf{t})}(\mathbf{s}; n)/(1-q)^{w(\mathbf{s})}, \tag{6}$$

where $w(\mathbf{s}) := s_1 + \cdots + s_\ell$ is the *weight*, $\ell(\mathbf{s}) := \ell$ the *depth* and $\mathbf{t}$ the *modifier*. For trivial modifier we set

$$H_q(\mathbf{s}; n) := H_q^{(0,\ldots,0)}(\mathbf{s}; n), \qquad H_q^*(\mathbf{s}; n) = H_q(\mathbf{s}; n)/(1-q)^{w(\mathbf{s})}.$$

Note that in [3] $\tilde{H}_q(s; p-1) := H_q^{(1)}(s; p-1)$ are studied in some detail and are related to $H_q(s; p-1)$. Also note that $H_q^{(s_1-1,\ldots,s_\ell-1)}(\mathbf{s}; n)$ are the partial sums of the most convenient form of q-multiple zeta functions (see [9]).

In this paper we mainly consider q-MHS with the trivial modifier. By convention we set $H_q^{(\mathbf{t})}(\mathbf{s}; r) = 0$ for $r = 0, \ldots, \ell(\mathbf{s}) - 1$, and $H_q^{(\mathbf{t})}(\emptyset; n) = 1$. To save space, for an ordered set $(e_1, \ldots, e_t)$ we denote by $\{e_1, \ldots, e_t\}^d$ the ordered set formed by repeating $(e_1, \ldots, e_t)$ d times. For example, $H_q(\{s\}^\ell; n)$ will be called a *homogeneous* sum.

Throughout the paper, we use short-hand $H_q(\mathbf{s})$ to denote $H_q(\mathbf{s}; p-1)$ for some fixed prime p.

2. Homogeneous q-MHS

It is extremely beneficial to study the so-called stuffle (or quasi-shuffle) relations among MHS (see, for e.g., [10]). The same mechanism works equally well for q-MHS.

Recall that for any two ordered sets $(r_1, \ldots, r_t)$ and $(r_{t+1}, \ldots, r_n)$ the shuffle operation is defined by

$$\operatorname{Shfl}\big((r_1, \ldots, r_t), (r_{t+1}, \ldots, r_n)\big) := \bigcup_{\substack{\sigma \text{ permutes } \{1,\ldots,n\}, \\ \sigma^{-1}(1) < \cdots < \sigma^{-1}(t), \\ \sigma^{-1}(t+1) < \cdots < \sigma^{-1}(n)}} \big(r_{\sigma(1)}, \ldots, r_{\sigma(n)}\big).$$

Fix a positive integer s. For any $k = 1, \ldots, \ell - 1$, by stuffle relation we have

$$H_q^*\big((\ell-k)s\big) \cdot H_q^*\big(\{s\}^k\big) = \sum_{\mathbf{s} \in \operatorname{Shfl}\left(\{(\ell-k)s\}, \{s\}^k\right)} H_q^*(\mathbf{s}) + \sum_{\mathbf{s} \in \operatorname{Shfl}\left(\{(\ell-k+1)s\}, \{s\}^{k-1}\right)} H_q^*(\mathbf{s}).$$

Applying $\sum_{k=1}^{\ell-1}(-1)^{\ell-k-1}$ on both sides we get

$$H_q^*(\{s\}^\ell) = \frac{1}{\ell}\sum_{k=0}^{\ell-1}(-1)^{\ell-k-1}H_q^*((\ell-k)s)\cdot H_q^*(\{s\}^k). \tag{7}$$

Theorem 2.1. *Let s be a positive integer and let $\eta_s = \exp(2\pi\sqrt{-1}/s)$ be the sth primitive root of unity. Then*

$$\sum_{\ell=0}^\infty H_q^*(\{s\}^\ell)x^\ell \equiv \frac{(-1)^s}{p^s x}\prod_{n=0}^{s-1}\Big(1-(1-\eta_s^n(-x)^{1/s})^p\Big) \pmod{[p]_q}.$$

Proof. Let $\zeta = \exp(2\pi\sqrt{-1}/p)$ be the primitive pth root of unity and set

$$P_n = \sum_{j=1}^{p-1}\frac{1}{(1-\zeta^j)^n}. \tag{8}$$

It is easy to see that $H_q^*(n) \equiv P_n \pmod{[p]_q}$. By using partial fractions Dilcher [4, (4.2)] obtained essentially the following generating function of P_n:

$$g(x) := \sum_{n=0}^\infty P_n x^n = -\frac{px(x-1)^{p-1}}{1-(1-x)^p}. \tag{9}$$

Let $a_\ell = H_q^*(\{s\}^\ell)$ for all $\ell \geq 0$. Let $w(x) = \sum_{\ell=0}^\infty a_\ell x^\ell$ be its the generating function. By (7) we get

$$w(x) = \sum_{\ell=0}^\infty a_\ell x^\ell \equiv 1 + \sum_{\ell=1}^\infty \frac{1}{\ell}\sum_{k=0}^{\ell-1}(-1)^{\ell-k-1}P_{(\ell-k)s}a_k x^\ell \pmod{[p]_q}.$$

Differentiating both sides and changing index $\ell \to \ell+1$ we get, modulo $[p]_q$,

$$w'(x) \equiv \sum_{\ell=0}^\infty\sum_{k=0}^{\ell}(-1)^{\ell-k}P_{(\ell-k+1)s}a_k x^\ell \equiv \sum_{k=0}^\infty\sum_{\ell=k}^\infty(-1)^{\ell-k}P_{(\ell-k+1)s}a_k x^\ell.$$

Changing index $\ell \to \ell+k$ and then exchanging the order of summation we get

$$\begin{aligned} w'(x) &\equiv w(x)\sum_{\ell=0}^\infty P_{(\ell+1)s}(-x)^\ell \equiv \frac{w(x)}{-x}\left(\sum_{\ell=0}^\infty P_{\ell s}(-x)^\ell + 1\right) \\ &\equiv \frac{w(x)}{-sx}\left(s + \sum_{n=0}^{s-1}\sum_{\ell=0}^\infty P_\ell(\eta_s^n(-x)^{1/s})^\ell\right) \\ &\equiv \frac{w(x)}{-sx}\Big(s + \sum_{n=0}^{s-1} g\big(\eta_s^n(-x)^{1/s}\big)\Big) \\ &\equiv \frac{w(x)}{-sx}\left(s - \sum_{n=0}^{s-1}\frac{p\eta^n(-x)^{1/s}(\eta^n(-x)^{1/s}-1)^{p-1}}{1-(1-\eta_s^n(-x)^{1/s})^p}\right). \end{aligned}$$

Here $\eta_s = \exp(2\pi\sqrt{-1}/s)$ is the sth primitive root of unity. Thus

$$(\ln w(x))' = \left(-(\ln x)' + \sum_{n=0}^{s-1} \frac{(1-(1-\eta^n(-x)^{1/s})^p)'}{1-(1-\eta_s^n(-x)^{1/s})^p}\right).$$

Therefore by comparing the constant term we get

$$w(x) \equiv \frac{(-1)^s}{p^s x} \prod_{n=0}^{s-1} \left(1-(1-\eta_s^n(-x)^{1/s})^p\right) \pmod{[p]_q}$$

as desired. □

Corollary 2.2. *For every positive integer $\ell < p$ we have*

$$H_q(\{1\}^\ell) \equiv \frac{1}{\ell+1}\binom{p-1}{\ell} \cdot (1-q)^\ell \qquad \pmod{[p]_q}.$$

Proof. By the theorem we get

$$\begin{aligned}\sum_{\ell=0}^{\infty} H_q^*(\{1\}^\ell)x^\ell \equiv& \frac{(1+x)^p-1}{px} \\ \equiv& \frac{1}{px}\sum_{\ell=0}^{\infty}\binom{p}{\ell+1}x^{\ell+1} \equiv \sum_{\ell=0}^{\infty}\frac{1}{\ell+1}\binom{p-1}{\ell}x^\ell \pmod{[p]_q}.\end{aligned}$$

The corollary follows immediately. □

Corollary 2.3. *For every positive integer $\ell < p$ we have*

$$H_q(\{2\}^\ell) \equiv (-1)^\ell \frac{2\cdot \ell!}{(2\ell+2)!}\binom{p-1}{\ell} \cdot F_{2,\ell}(p)\cdot(1-q)^{2\ell} \pmod{[p]_q},$$

where $F_{2,\ell}(p)$ is a monic polynomial in p of degree ℓ.

Proof. By Theorem 2.1 we have modulo $[p]_q$

$$\begin{aligned}\sum_{\ell=0}^{\infty} H_q^*(\{2\}^\ell)x^\ell \equiv& \frac{1}{p^2x}\left(1-(1-\sqrt{-1}\sqrt{x})^p\right)\left(1-(1+\sqrt{-1}\sqrt{x})^p\right) \\ \equiv& \frac{1}{p^2x}\left|\sum_{j=1}^{(p-1)/2}\binom{p}{2j}(-1)^jx^j + \sqrt{-1}\sqrt{x}\sum_{j=0}^{(p-1)/2}\binom{p}{2j+1}(-1)^jx^j\right|^2,\end{aligned}$$

which easily yields

$$H_q^*(\{2\}^\ell) \equiv \frac{(-1)^\ell}{p^2}\left\{\sum_{\substack{j+k=\ell \\ 0\le j,k<p/2}}\binom{p}{2j+1}\binom{p}{2k+1} - \sum_{\substack{j+k=\ell+1 \\ 1\le j,k<p/2}}\binom{p}{2j}\binom{p}{2k}\right\}.$$

In the first sum above if $j+k=\ell+1$ and $1\le j,k<p/2$ then we may assume $j>\ell/2$. Then $(\ell+1)!\binom{p}{\ell+1}$ is a factor of $(2j+1)!\binom{p}{2j+1}$ as a polynomial of p, and so is $\ell!\binom{p-1}{\ell}$. Similarly we can see that $\ell!\binom{p-1}{\ell}$ is a factor of the second sum.

In order to determine the leading coefficient we set

$$C_1(x)=\sum_{j=0}^{\ell}\frac{(2\ell+2)!\,x^{2j+1}}{(2j+1)!(2l-2j+1)!}=\frac{(x+1)^{2\ell+2}-(x-1)^{2\ell+2}}{2},$$
$$C_2(x)=\sum_{j=0}^{\ell+1}\frac{(2\ell+2)!\,x^{2j}}{(2j)!(2l-2j+2)!}=\frac{(x+1)^{2\ell+2}+(x-1)^{2\ell+2}}{2}.$$

Hence

$$\sum_{\substack{j+k=\ell\\0\le j,k<p/2}}\frac{1}{(2j+1)!(2k+1)!}-\sum_{\substack{j+k=\ell+1\\1\le j,k<p/2}}\frac{1}{(2j)!(2k)!}$$
$$=\frac{C_1(1)-(C_2(1)-2)}{(2\ell+2)!}=\frac{2}{(2\ell+2)!}.$$

This finishes the proof of the corollary. □

Corollary 2.4. *Let ℓ be a positive integer. Set $\delta_\ell=(1+(-1)^\ell)$ and $L=3\ell+3$. Then for every prime $p\ge L$ we have modulo $[p]_q$*

$$H_q(\{3\}^\ell)\equiv\begin{cases}\dfrac{-3\cdot\ell!}{(3\ell+1)!}\dbinom{p-1}{\ell}\cdot F_{3,\ell}(p)\cdot(1-q)^{3\ell} & ,\ \text{if } \ell \text{ is odd,}\\[2ex] \dfrac{6\cdot\ell!}{(3\ell+3)!}\dbinom{p-1}{\ell}\cdot F_{3,\ell}(p)\cdot(1-q)^{3\ell} & ,\ \text{if } \ell \text{ is even,}\end{cases}\tag{10}$$

where $F_{3,\ell}(p)$ is a monic polynomial in p of degree $2\ell-1$ if ℓ is odd and of degree 2ℓ if ℓ is even.

Proof. Let $\eta=\exp(2\pi i/3)$. Then $\eta^2+\eta+1=0$. By Theorem 2.1 we have

$$\sum_{\ell=0}^{\infty}H_q^*(\{3\}^\ell)x^\ell\equiv\frac{-1}{p^3x}\prod_{a=0}^{2}\Big(1-(1-\eta^a\sqrt[3]{-x})^p\Big).\tag{11}$$

We now use two ways to expand this. Set $y=\sqrt[3]{-x}$. First, the product on the right hand side of (11) can be expressed as

$$\begin{aligned}
&1-\sum_{a=0}^{2}(1-\eta^a y)^p+\sum_{a=0}^{2}(1-\eta^a y)^p(1-\eta^{a+1}y)^p-\prod_{a=0}^{2}(1-\eta^a y)^p\\
=&1-\sum_{j=0}^{p}\binom{p}{j}\sum_{a=0}^{2}\eta^{aj}y^j+\sum_{a=0}^{2}(1+\eta^a y+\eta^{a+1}y^2)^p-(1+x)^p\\
=&1-3\sum_{j=0}^{[p/3]}\binom{p}{3j}x^j+3\sum_{\substack{j,k\ge 0,j+k<p\\ 2j+k\equiv 0(3)}}\frac{p!\,(-x)^{(j+2k)/3}}{j!k!(p-j-k)!}-(1+x)^p.
\end{aligned}$$

Thus for $\ell>0$ we get

$$H_q^*(\{3\}^\ell)\equiv\frac{1}{p^3}\left\{3\delta_\ell\binom{p}{L}+(-1)^\ell\cdot 3\sum_{k\ge 1}\binom{p}{L-k}\binom{L-k}{k}+\binom{p}{\ell+1}\right\}$$

Note that if ℓ is odd then the degree of the polynomial is reduced to $3\ell-1$ with leading coefficient given by

$$(-1)^\ell\cdot 3\frac{1}{(L-1)!}\binom{L-1}{1}=\frac{-3}{(L-2)!}=\frac{-3}{(3\ell+1)!}$$

as we wanted.

Now to prove $\ell!\binom{p}{\ell}$ is a factor we use the following expansion of (11):

$$\sum_{\ell=0}^{\infty}\frac{1}{p^3x}\sum_{j,k,n\ge 1}(-1)^{j+k+n}\binom{p}{j}\binom{p}{k}\binom{p}{n}x^{(j+k+n)/3}\eta^{k+2n}.$$

Thus

$$H_q^*(\{3\}^\ell)\equiv\frac{1}{p^3}\sum_{\substack{1\le j,k,n\le p\\ j+k+n=3\ell+3}}(-1)^{j+k+n}\binom{p}{j}\binom{p}{k}\binom{p}{n}\eta^{k+2n}\pmod{[p]_q}.$$

Notice that $j+k+n=3\ell+3$ implies one of the indices, say j, is at least $\ell+1$. Then clearly $\binom{p}{j}$ contains $\ell!\binom{p}{\ell}$ as a factor, therefore so does $H_q^*(\{3\}^\ell)\pmod{[p]_q}$. This completes the proof of the corollary. □

3. Some Non-Homogeneous q-MHS Congruences

In this section we consider some non-homogeneous q-MHS of depth two with modifiers of special type.

Theorem 3.1. *Let m, n be two positive integers. For every prime p we have*

$$H_q^{(m,n)}(2m,2n) \equiv \frac{1}{2}\{f(m;p)f(n;p) - f(m+n;p)\} \pmod{[p]_q}.$$

where

$$f(N;p) = (1-q)^{2N}\sum_{i=0}^{N-1}\binom{N-1}{i}(-1)^i K_{2N-i}(p)$$

Proof. By definition and substitution $i \to p-i$ and $j \to p-j$ we have

$$\begin{aligned}
H_q^{*(m,n)}(2m,2n) &= \sum_{1\le i<j<p} \frac{q^{mi+nj}}{(1-q^i)^{2m}(1-q^j)^{2n}} \\
&= \sum_{1\le j<i<p} \frac{q^{pm+pn-mi-nj}}{(1-q^{p-i})^{2m}(1-q^{p-j})^{2n}} \\
&\equiv \sum_{1\le j<i<p} \frac{q^{mi+nj}}{(q^i-q^p)^{2m}(q^j-q^p)^{2n}} && \pmod{[p]_q} \\
&\equiv \sum_{1\le j<i<p} \frac{q^{mi+nj}}{(1-p^i)^{2m}(1-p^j)^{2n}} && \pmod{[p]_q} \\
&\equiv H_q^{*(n,m)}(2n,2m) && \pmod{[p]_q} \qquad (12)
\end{aligned}$$

By shuffle relation we have

$$H_q^{*(m)}(2m)H_q^{*(n)}(2n) = H_q^{*(m,n)}(2m,2n) + H_q^{*(n,m)}(2n,2m) + H_q^{*(m+n)}(2m+2n).$$

Together with (12) this yields

$$2H_q^{*(m,n)}(2m,2n) \equiv H_q^{*(m)}(2m)H_q^{*(n)}(2n) - H_q^{*(m+n)}(2m+2n) \pmod{[p]_q}.$$

Our theorem follows from (4) quickly. □

In the study of q-multiple zeta functions the following function appears naturally (see [9, (47)] or [2, Theorem 1]):

$$\varphi_q(n) = \sum_{k=1}^{\infty}(k-1)\frac{q^{(n-1)k}}{[k]_q^n} = \sum_{k=1}^{\infty}\frac{kq^{(n-1)k}}{[k]_q^n} - \zeta_q(n),$$

where $\zeta_q(n) = \sum_{k=1}^{\infty} \frac{q^{(n-1)k}}{[k]_q^n}$ is the q-Riemann zeta value defined by Kaneko et al. in [5]. Using the results we have obtained so far in this paper we discover a congruence related to the partial sums of $\varphi_q(2)$.

Proposition 3.2. *For every prime p we have*

$$\sum_{k=1}^{p-1}\frac{kq^k}{[k]_q^2} \equiv -\frac{p(p-1)(p+1)}{24}(1-q)^2 \pmod{[p]_q}. \qquad (13)$$

Proof. To save space, all congruences are modulo $[p]_q$ throughout this proof.

We can check the congruence (13) for $p=2$ and $p=3$ easily by hand. Now we assume $p \geq 5$. By definition we have

$$H_q^*(2,1) = \sum_{1\leq i<j<p} \frac{1}{(1-q^i)^2(1-q^j)}.$$

With the substitutions $i \to p-i$ and $j \to p-j$ we get

$$\begin{aligned}
-H_q^*(2,1) &= -\sum_{1\leq j<i<p} \frac{q^{2i}\cdot q^j}{(q^i-q^p)^2(q^j-q^p)} \\
&\equiv -\sum_{1\leq j<i<p} \frac{q^{2i}\cdot q^j}{(q^i-1)^2(q^j-1)} \\
&\equiv -\sum_{1\leq j<i<p} \frac{(q^i-1)^2+2(q^i-1)+1}{(q^i-1)^2}\cdot\frac{1-q^j-1}{1-q^j} \\
&\equiv H_q^*(1,2) - 2H_q^*(1,1) + \sum_{k=1}^{p-1}\frac{p-3+k}{1-q^k} - \sum_{k=1}^{p-1}\frac{k-1}{(1-q^k)^2} - \binom{p-1}{2} \\
&\equiv H_q^*(1,2) - 2H_q^*(1,1) + (p-3)H_q^*(1) + H_q^*(2) \\
&\qquad\qquad - \binom{p-1}{2} - \sum_{k=1}^{p-1}\frac{kq^k}{(1-q^k)^2}.
\end{aligned}$$

Notice that we have the stuffle relations

$$\begin{aligned}
H_q^*(2,1) + H_q^*(1,2) &= H_q^*(1)H_q^*(2) - H_q^*(3), \\
2H_q^*(1,1) &= H_q^*(1)^2 - H_q^*(2).
\end{aligned}$$

Hence

$$\sum_{k=1}^{p-1}\frac{kq^k}{(1-q^k)^2} \equiv (H_q^*(1)+2)H_q^*(2) - H_q^*(3) - H_q^*(1)^2 + (p-3)H_q^*(1) - \binom{p-1}{2}.$$

Notice that by [3, Theorem 2]

$$H_q^*(3) \equiv -\frac{(p-1)(p-3)}{8}. \tag{14}$$

The proposition now follows from (1) and (2) immediately. □

4. A Congruence of Lehmer Type

Instead of the harmonic sums up to $(p-1)$-st term Lehmer also studied the following type of congruence (see [6]): for every odd prime p

$$\sum_{j=1}^{(p-1)/2} \frac{1}{j} \equiv -2q_p(2) + q_p(2)^2 p \pmod{p^2},$$

where $q_p(2) = (2^{p-1} - 1)/p$ is the Fermat quotient. It is also easy to see that for every positive integer n and prime $p > 2n+1$

$$\sum_{j=1}^{(p-1)/2} \frac{1}{j^{2n}} \equiv 0 \pmod{p}.$$

As a q-analog of the above we have

Theorem 4.1. *Let n be a positive integer. For every odd prime p we have*

$$H_q^{(n)}(2n;(p-1)/2) \equiv \frac{1}{2}(1-q)^{2n} \sum_{j=0}^{n-1} \binom{n-1}{j} (-1)^j K_{2n-j}(p) \qquad \pmod{[p]_q}.$$

Proof. By definition and substitution $i \to p - i$ we have

$$\begin{aligned} H_q^{*(n)}(2n) =& H_q^{*(n)}(2n;(p-1)/2) + \sum_{1 \le i \le (p-1)/2} \frac{q^{n(p-i)}}{(1-q^{p-i})^{2n}} \\ \equiv& 2H_q^{*(n)}(2n;(p-1)/2) \pmod{[p]_q} \end{aligned}$$

which yields the theorem by (4) easily. □

To conclude the paper we remark that the congruence for general q-MHS should involve some type of q-analog of Bernoulli numbers and Euler numbers similar to the classical cases treated in [10]. We hope to return to this theme in the future.

Acknowledgement. This work is partly supported by the Faculty Development Fund of Eckerd College and the NSF grant DMS1162116.

References

[1] G.E. Andrews, *q-analogs of the binomial coefficient congruences of Babbage, Wolstenholme and Glaisher*, Discrete Math., **204**(1) (1999), 15-25.

[2] D.M. Bradley, *A q-analog of Euler's decomposition formula for the double zeta function*, Int. J. Math. Math. Sci., Vol. 2005, Issue 21, 2005, pp. 3453–3458.

[3] K. Dilcher, *Determinant expressions for q-harmonic congruences and degenerate Bernoulli numbers*, Electron. J. Combin., **15**(1) (2008), R63.

[4] F. T. Howard, *Explicit formulas for degenerate Bernoulli numbers*, Discrete Math. **162** (1996), 175–185.

[5] M. Kaneko, N. Kurokawa, and M. Wakayama, *A variation of Euler's approach to values of the Riemann zeta function*, Kyushu J. Math. **57** (2003), 175–192.

[6] E. Lehmer, *On congruences involving Bernoulli numbers and the quotients of Fermat and Wilson*, Ann. Math. **39** (1938), 350–360.

[7] S. Roman, The Umbral Calculus, New York: Academic Press, 1984.

[8] L.-L. Shi and H. Pan, *A q-analogue of Wolstenholme's harmonic series congruence*, Amer. Math. Monthly, **114**(6) (2007), 529-531.

[9] J. Zhao, *Multiple q-zeta functions and multiple q-polylogarithms*, Ramanujan J., **14** (2)(2007), 189–221. arXiv: math/0304448

[10] J. Zhao, *Wolstenholme type theorem for multiple harmonic sums*, Int. J. Number Theory, **4**(1) (2008), 73–106.

ON THE FOURIER TRANSFORM OF THE GREATEST COMMON DIVISOR

Peter H. van der Kamp
Dept. of Mathematics and Statistics, La Trobe University, Victoria, Australia
P.vanderkamp@latrobe.edu.au

Received: 5/23/12, Revised: 12/5/12, Accepted: 5/4/13, Published: 5/10/13

Abstract

The discrete Fourier transform of the greatest common divisor

$$\widehat{\mathrm{id}}[a](m) = \sum_{k=1}^{m} \gcd(k,m)\alpha_m^{ka},$$

with α_m a primitive m-th root of unity, is a multiplicative function that generalizes both the gcd-sum function and Euler's totient function. On the one hand it is the Dirichlet convolution of the identity with Ramanujan's sum, $\widehat{\mathrm{id}}[a] = \mathrm{id} * c_\bullet(a)$, and on the other hand it can be written as a generalized convolution product, $\widehat{\mathrm{id}}[a] = \mathrm{id} *_a \phi$. We show that $\widehat{\mathrm{id}}[a](m)$ counts the number of elements in the set of ordered pairs (i,j) such that $i \cdot j \equiv a \bmod m$. Furthermore we generalize a dozen known identities for the totient function, to identities which involve the discrete Fourier transform of the greatest common divisor, including its partial sums, and its Lambert series.

1. Introduction

In [15] discrete Fourier transforms of functions of the greatest common divisor were studied, i.e.,

$$\widehat{h}[a](m) := \sum_{k=1}^{m} h(\gcd(k,m))\alpha_m^{ka},$$

where α_m is a primitive m-th root of unity. The main result in that paper is the identity[1] $\widehat{h}[a] = h * c_\bullet(a)$, where

$$c_m(a) := \sum_{\substack{k=1 \\ \gcd(k,m)=1}}^{m} \alpha_m^{ka} \tag{1}$$

[1]Similar results in the context of r-even function were obtained earlier; see [10] for details.

is Ramanujan's sum, and $*$ denotes Dirichlet convolution, i.e.,

$$\widehat{h}[a](m) = \sum_{d|m} h(\frac{m}{d})c_d(a). \tag{2}$$

Ramanujan's sum generalizes both Euler's totient function $\phi = c_\bullet(0)$ and the Möbius function $\mu = c_\bullet(1)$. Thus, identity (2) generalizes the formula

$$\sum_{k=1}^{m} h(\gcd(k,m)) = (h * \phi)(m), \tag{3}$$

already known to Cesàro in 1885. The formula (2) shows that $\widehat{h}[a]$ is multiplicative if h is multiplicative (because $c_\bullet(a)$ is multiplicative and Dirichlet convolution preserves multiplicativity).

Here we will take $h = \mathrm{id}$ to be the identity function (of the greatest common divisor) and study its Fourier transform. Obviously, as $\mathrm{id}(n) := n$ is multiplicative, the function $\widehat{\mathrm{id}}[a]$ is multiplicative, for all a. Two special cases are well-known. Taking $a = 0$ we have $\widehat{\mathrm{id}}[0] = \mathcal{P}$, where

$$\mathcal{P}(m) := \sum_{k=1}^{m} \gcd(k,m) \tag{4}$$

is known as Pillai's arithmetical function or the gcd-sum function [4]. Secondly, by taking $a = 1$ in (2), we find that $\widehat{\mathrm{id}}[1] = \mathrm{id} * \mu$ equals ϕ, by Möbius inversion of Euler's identity $\phi * u = \mathrm{id}$, where $u = \mu^{-1}$ is the unit function defined by $u(m) := 1$.

Let $\mathcal{F}_a^m$ denote the set of ordered pairs of congruence classes (i,j) such that $i \cdot j \equiv a \bmod m$, the set of *factorizations of a modulo m*. We claim that $\widehat{\mathrm{id}}[a](m)$ counts its number of elements. Let us first consider two special cases.

$a = 0$ For given $i \in \{1, 2, \ldots, m\}$ the congruence $i \cdot j \equiv 0 \bmod m$ yields

$$\frac{i}{\gcd(i,m)} j \equiv 0 \mod \frac{m}{\gcd(i,m)},$$

which has a unique solution modulo $m/\gcd(i,m)$, and so there are $\gcd(i,m)$ solutions modulo m. Hence, the total number of elements in $\mathcal{F}_0^m$ is $\mathcal{P}(m)$.

$a = 1$ The totient function $\phi(m)$ counts the number of invertible congruence classes modulo m. As for every invertible congruence class i modulo m there is a unique $j = i^{-1} \bmod m$ such that $i \cdot j \equiv 1 \bmod m$, it counts the number of elements in the set $\mathcal{F}_1^m$.

To prove the general case we employ a Kluyver-like formula for $\widehat{\mathrm{id}}[a]$, that is, a formula similar to the formula for the Ramanujan sum function:

$$c_k(a) = \sum_{d|\gcd(a,k)} d\mu\left(\frac{k}{d}\right). \tag{5}$$

attributed to Kluyver [8]. Together the identities (2) and (5) imply, cf. Section 3,

$$\widehat{\mathrm{id}}[a](m) = \sum_{d|\gcd(a,m)} d\phi\left(\frac{m}{d}\right), \tag{6}$$

and we will show, in the next section, that the number $\#\mathcal{F}_a^m$ of factorizations of a mod m is given by the same sum.

The right-hand sides of (5) and (6) are particular instances of so-called generalized Ramanujan sums [1], and both formulas follow as consequence of a general formula for the Fourier coefficients of such sums [2, 3]. In Section 3 we provide simple proofs for some of the nice properties of these sums. In particular we interpret the sums as a generalization of Dirichlet convolution. This interpretation lies at the heart of many of the generalized totient identities we establish in Section 4.

2. The Number of Factorizations of a mod m

We are after the number of pairs (i, j) of congruence classes modulo m such that

$$i \cdot j \equiv a \mod m. \tag{7}$$

For given $i, m \in \mathbb{N}$, let g denote $\gcd(i, m)$. If the congruence $i \cdot j \equiv a \bmod m$ has a solution j, then $g \mid a$ and $j \equiv i^{-1}a/g$ is unique mod m/g, so mod m there are g solutions. This yields

$$\#\mathcal{F}_a^m = \sum_{\substack{i=1 \\ \gcd(i,m)|a}}^{m} \gcd(i, m),$$

which can be written as

$$\#\mathcal{F}_a^m = \sum_{d|a} \sum_{\substack{i=1 \\ \gcd(i,m)=d}}^{m} d. \tag{8}$$

If $d \nmid m$ then the sum

$$\sum_{\substack{i=1 \\ \gcd(i,m)=d}}^{m} 1$$

is empty. Now let $d \mid m$. The only integers i which contribute to the sum are the multiples of d, kd, where $\gcd(k, m/d) = 1$. There are exactly $\phi(m/d)$ of them. Therefore the right-hand sides of formulae (6) and (8) agree, and hence $\#\mathcal{F}_a^m = \widehat{\mathrm{id}}[a](m)$.

3. A Historical Remark and Generalized Ramanujan Sums

It is well-known that Ramanujan was not the first who considered the sum $c_m(a)$. Kluyver proved his formula (5) in 1906, twelve years before Ramanujan published the novel idea of expressing arithmetical functions in the form of a series $\sum_s a_s c_s(n)$ [11]. It is not well-known that Kluyver actually showed that $c_m(a)$ equals Von Sterneck's function, introduced in [13], i.e.,

$$c_m(a) = \frac{\mu\Big(\dfrac{m}{\gcd(a,m)}\Big)\phi(m)}{\phi\Big(\dfrac{m}{\gcd(a,m)}\Big)}. \tag{9}$$

This relation is referred to in the literature as Hölder's relation, cf. the remark on page 213 in [1]. However, Hölder published this relation (9) thirty years after Kluyver [7]. We refer to [1, Theorem 2], or [2, Theorem 8.8] for a generalization of (9). The so-called generalized Ramanujan sums,

$$f *_a g(m) := \sum_{d \mid \gcd(a,m)} f(d) g\Big(\frac{m}{d}\Big), \tag{10}$$

were introduced in [1]. The notation $*_a$ is new, the sums are denoted $S(a;m)$ in [1], $s_m(a)$ in [2], and $S_{f,h}(a,m)$ in [3]. In the context of r-even functions [10] the sums are denoted $S_{f,g}(a)$, and considered as sequences of m-even functions, with argument a. We consider the sums as a sequence of functions with argument m, labeled by a. We call $f *_a g$ the **a-convolution of f and g**.

The concept of a-convolution provides a generalization of Dirichlet convolution as $f *_0 g = f * g$. As we will see below shortly, the function $f *_a g$ is multiplicative (for all a) if f and g are, and the following inter-associative property holds, cf. [3, Theorem 4]:

$$(f *_a g) * h = f *_a (g * h). \tag{11}$$

We also adopt the notation $f_a := \mathrm{id} *_a f$, and call this **the Kluyver, or a-extension of f**. Thus, we have $f_0 = \mathrm{id} * f$, $f_1 = f$, and formulas (5) and (6) become $c_m(a) = \mu_a(m)$, and $\widehat{\mathrm{id}}[a] = \phi_a$, respectively.

In terms of the Iverson bracket, with P a logical statement,

$$[P] := \begin{cases} 1 & \text{if } P, \\ 0 & \text{if not } P, \end{cases}$$

the identity function I is given by $I(k) := [k = 1]$. It is the identity function in the algebra where convolution is the multiplication, i.e., $I * f = f * I = f$. Let us consider the function $f *_a I$. It is

$$f *_a I(k) = \sum_{d \mid \gcd(a,k)} f(d)[d = k] = [k \mid a] f(k).$$

Since the function $k \to [k \mid a]$ is multiplicative, the function $f *_a I$ is multiplicative if f is multiplicative. We may write, cf. [3, eq. (9)],

$$f *_a g(m) = \sum_{d|m} [d \mid a] f(d) g(\frac{m}{d}) = (f *_a I) * g(m), \tag{12}$$

which shows that $f *_a g$ is multiplicative if f and g are. Using equation (12), the inter-associativity property (11) follows easily from the associativity of Dirichlet convolution.

Proof. We have $(f *_a g) * h = ((f *_a I) * g) * h = (f *_a I) * (g * h) = f *_a (g * h)$. □

We note that the a-convolution product is neither associative, nor commutative. The inter-associativity and the commutativity of Dirichlet convolution imply that

$$f_a * g = (f * g)_a = f * g_a. \tag{13}$$

We provide a simple proof for formula (6), which states that the Fourier transform of the greatest common divisor is the Kluyver extension of the totient function.

Proof. Using (2), (5) and (13) we get $\widehat{\mathrm{id}}[a] = \mathrm{id} * c_\bullet(a) = \mathrm{id} * \mu_a = (\mathrm{id} * \mu)_a = \phi_a$. □

Formula (6) also follows as a special case of the formula for the Fourier coefficients of a-convolutions,

$$f *_a g(m) = \sum_{k=1}^{m} q[k](m)\alpha_m^{ka}, \qquad q[k] := g *_k \frac{f}{\mathrm{id}}, \tag{14}$$

which was given in [1, 2]. Formula (14) combines with (2) and (5) to yield a formula for functions of the greatest common divisor, $\bar{h}[k] : m \to h(\gcd(k, m))$, namely

$$\bar{h}[k] = (h * \mu) *_k u. \tag{15}$$

Proof. The Fourier coefficients of $\widehat{h}[a](m)$ are $\bar{h}[k](m)$. But $\widehat{h}[a] = h * c_\bullet(a) = (\mathrm{id} *_a \mu) * h = \mathrm{id} *_a (\mu * h)$, and so, using (14), the Fourier coefficients are also given by $(h * \mu) *_k u(m)$. □

For a Dirichlet convolution with a Fourier transform of a function of the greatest common divisor we have

$$f * \widehat{g}[a] = \widehat{f * g}[a]. \tag{16}$$

Proof. We have $f * \widehat{g}[a] = f * (g * \mu_a) = (f * g) * \mu_a = \widehat{f * g}[a]$. □

Similarly, for an a-convolution with a Fourier transform of a function of the greatest common divisor, we have

$$f *_a \widehat{g}[b] = \widehat{f *_a g}[b]. \tag{17}$$

Proof. Indeed, $f *_a \widehat{g}[b] = f *_a (g * \mu_b) = (f *_a g) * \mu_b = \widehat{f *_a g}[b]$. □

4. Generalized Totient Identities

The totient function is an important function in number theory, and related fields of mathematics. It is extensively studied, connected to many other notions and functions, and there exist numerous generalization and extensions, cf. the chapter "The many facets of Euler's totient" in [12]. The Kluyver extension of the totient function is a very natural extension, and it is most surprising it has not been studied before. In this section we generalize a dozen known identities for the totient function ϕ, to identities which involve its Kluyver extension ϕ_a, a.k.a. the discrete Fourier transform of the greatest common divisor. This includes a generalization of Euler's identity, the partial sums of ϕ_a, and its Lambert series.

4.1. The Value of ϕ_a at Powers of Primes

We start by providing a formula for the value of ϕ_a at powers of primes. This depends only on the multiplicity of the prime in a. The formulae, with p prime,

$$\mathcal{P}(p^k) = (k+1)p^k - kp^{k-1}, \qquad \phi(p^k) = p^k - p^{k-1},$$

of which the first one is Theorem 2.2 in [4], generalize to

$$\phi_a(p^k) = \begin{cases} (p^k - p^{k-1})(l+1) & \text{if } l < k, \\ (k+1)p^k - kp^{k-1} & \text{if } l \geq k, \end{cases} \tag{18}$$

where l is the largest integer, or infinity, such that $p^l \mid a$.

Proof. We have

$$\begin{aligned} \phi_a(p^k) &= \sum_{d \mid \gcd(p^l, p^k)} d\phi(\frac{p^k}{d}) \\ &= \sum_{r=0}^{\min(l,k)} p^r \phi(p^{k-r}) \\ &= \begin{cases} \sum_{r=0}^{l} p^k - p^{k-1} & \text{if } l < k, \\ (\sum_{r=0}^{k-1} p^k - p^{k-1}) + p^k & \text{if } l \geq k, \end{cases} \end{aligned}$$

which equals (18). □

4.2. Partial Sums of ϕ_a/id

To generalize the totient identity

$$\sum_{k=1}^{n} \frac{\phi(k)}{k} = \sum_{k=1}^{n} \frac{\mu(k)}{k} \left\lfloor \frac{n}{k} \right\rfloor. \tag{19}$$

to an identity for ϕ_a we first establish

$$\sum_{k=1}^{n} \frac{f_0(k)}{k} = \sum_{k=1}^{n} \frac{f(k)}{k} \left\lfloor \frac{n}{k} \right\rfloor . \tag{20}$$

Proof. Since there are $\lfloor n/d \rfloor$ multiples of d in the range $[1, n]$ it follows that

$$\sum_{k=1}^{n} \frac{f * \mathrm{id}(k)}{k} = \sum_{k=1}^{n} \sum_{d|k} \frac{f(d)}{d} = \sum_{d=1}^{n} \frac{f(d)}{d} \left\lfloor \frac{n}{d} \right\rfloor .$$

□

As a corollary we obtain

$$\sum_{k=1}^{n} \frac{f *_a g_0(k)}{k} = \sum_{k=1}^{n} \frac{f *_a g(k)}{k} \left\lfloor \frac{n}{k} \right\rfloor . \tag{21}$$

Proof. Employing (11) we find

$$\sum_{k=1}^{n} \frac{f *_a (g * \mathrm{id})(k)}{k} = \sum_{k=1}^{n} \frac{(f *_a g) * \mathrm{id}(k)}{k} = \sum_{k=1}^{n} \frac{f *_a g(k)}{k} \left\lfloor \frac{n}{k} \right\rfloor .$$

□

Now taking $f = \mathrm{id}$ and $g = \mu$ in (21) we find

$$\sum_{k=1}^{n} \frac{\phi_a(k)}{k} = \sum_{k=1}^{n} \frac{c_k(a)}{k} \left\lfloor \frac{n}{k} \right\rfloor . \tag{22}$$

4.3. Partial Sums of $\mathcal{P}_a/\mathrm{id}$ Expressed in Terms of ϕ_a

Taking $f = \mathrm{id}$ and $g = \phi$ in (21) we find

$$\sum_{k=1}^{n} \frac{\mathcal{P}_a(k)}{k} = \sum_{k=1}^{n} \frac{\phi_a(k)}{k} \left\lfloor \frac{n}{k} \right\rfloor . \tag{23}$$

Note that by taking either $a = 0$ in (22), or $a = 1$ in (23), we find an identity involving the totient function and the gcd-sum function,

$$\sum_{k=1}^{n} \frac{\mathcal{P}(k)}{k} = \sum_{k=1}^{n} \frac{\phi(k)}{k} \left\lfloor \frac{n}{k} \right\rfloor . \tag{24}$$

4.4. Partial Sums of ϕ_a

To generalize the totient identity, with $n > 0$,

$$\sum_{k=1}^{n} \phi(k) = \frac{1}{2}\left(1 + \sum_{k=1}^{n} \mu(k) \left\lfloor \frac{n}{k} \right\rfloor^2\right), \tag{25}$$

we first establish

$$\sum_{k=1}^{n} f_0(k) = \frac{1}{2}\left(\sum_{k=1}^{n} f(k) \left\lfloor \frac{n}{k} \right\rfloor^2 + \sum_{k=1}^{n} f * u(k)\right). \tag{26}$$

Proof. By changing variables, $k = dl$, we find

$$\begin{aligned}\sum_{k=1}^{n}(2f * \mathrm{id} - f * u)(k) &= \sum_{k=1}^{n}\sum_{d|k} f(d)(\frac{2k}{d} - 1)\\ &= \sum_{d=1}^{n}\sum_{l=1}^{\lfloor n/d \rfloor} f(d)(2l-1)\\ &= \sum_{d=1}^{n} f(d)\left\lfloor \frac{n}{d} \right\rfloor^2.\end{aligned}$$

□

Note that this gives a nice proof of (25), taking $f = \mu$, as $\sum_{k=1}^{n} I(k) = [n > 0]$. When $f = \mu_a$, then (13) implies $f * \mathrm{id} = \phi_a$, and $f * u = I_a$, and therefore as a special case of (26) we obtain

$$\sum_{k=1}^{n} \phi_a(k) = \frac{1}{2}\left(\sum_{k|a} k[k \le n] + \sum_{k=1}^{n} c_k(a) \left\lfloor \frac{n}{k} \right\rfloor^2\right). \tag{27}$$

We remark that when $n \ge a$ we have $\sum_{k|a} k[k \le n] = \sigma(a)$, where $\sigma = \mathrm{id} * u$ is the sum of divisors function.

4.5. Generalization of Euler's Identity

Euler's identity, $\phi * u = \mathrm{id}$, generalizes to

$$\sum_{d|m} \phi_a(d) = \tau(\gcd(a, m))m, \tag{28}$$

where $\tau = u * u$ is the number of divisors function.

Proof. We have $\phi_a * u = (\phi * u)_a = \mathrm{id}_a$, where

$$\mathrm{id}_a(m) = \sum_{d|\gcd(a,m)} d\frac{m}{d} = m\tau(\gcd(a, m)). \tag{29}$$

□

4.6. Partial Sums of $\mathcal{P}_a$ Expressed in Terms of ϕ_a (and τ)

If $f = \phi_a$, then $f * \text{id} = \mathcal{P}_a$, and (26) becomes, using (28),

$$\sum_{k=1}^{n} \mathcal{P}_a(k) = \frac{1}{2}\Big(\sum_{k=1}^{n} \tau(\gcd(a,k))k + \sum_{k=1}^{n} \phi_a(k) \left\lfloor \frac{n}{k} \right\rfloor^2\Big). \tag{30}$$

4.7. Four Identities From Césaro

According to Dickson [5] the following three identities were obtained by Césaro:

$$\sum_{d|n} d\phi\big(\frac{n}{d}\big) = \mathcal{P}(n), \tag{31}$$

$$\sum_{d|n} \frac{d}{n}\phi(d) = \sum_{j=1}^{n} \frac{1}{\gcd(j,n)}, \tag{32}$$

$$\sum_{d|n} \phi(d)\phi\big(\frac{n}{d}\big) = \sum_{j=1}^{n} \phi(\gcd(j,n)). \tag{33}$$

Identity (31), which is Theorem 2.3 in [4], is obtained by taking $a = 0$ in (6), or $h = \text{id}$ in (3). It generalizes to

$$\sum_{d|n} d\phi_a\big(\frac{n}{d}\big) = \mathcal{P}_a(n). \tag{34}$$

Proof. Take $f = \phi$ and $g = \text{id}$ in (13). □

Identity (32) is obtained by taking $h = 1/\text{id}$ in (3) and generalizes to

$$\sum_{d|n} \frac{d}{n}\phi_a(d) = \sum_{j=1}^{n} \sum_{d|\gcd(a,n)} \frac{1}{\gcd\big(j, \frac{n}{d}\big)}. \tag{35}$$

Proof. Take $f = \phi$ and $g = 1/\text{id}$ in (13). □

Identity (33) is also a special case of (3), with $h = \phi$. It generalizes to

$$\sum_{d|n} \phi_a(d)\phi_b\big(\frac{n}{d}\big) = \sum_{j=1}^{n} \sum_{d|\gcd(a,n)} \phi_b\big(\gcd(j, \frac{n}{d})\big). \tag{36}$$

Proof. We have

$$(\text{id} *_a \phi) * (\text{id} *_b \phi) = \text{id} *_a (\text{id} *_b (\phi * \phi)) = \text{id} *_a (\text{id} *_b \widehat{\phi}[0]) = \text{id} *_a \widehat{\phi_b}[0],$$

and evaluation at m yields

$$\sum_{d|\gcd(a,m)} d\sum_{j=1}^{m/d}\phi_b(\gcd(j,\frac{m}{d})) = \sum_{d|\gcd(a,m)}\sum_{j=1}^{m}\phi_b(\gcd(j,\frac{m}{d})).$$

□

The more general identity (3) generalizes to

$$\sum_{k=1}^{m} h_a(\gcd(k,m)) = h * \phi_a(m). \tag{37}$$

4.8. Three Identities From Liouville

Dickson [5, p.285-286] states, amongst many others identities that were presented by Liouville in the series [9], the following identities:

$$\sum_{d|m}\phi(d)\tau(\frac{m}{d}) = \sigma(m), \tag{38}$$

$$\sum_{d|m}\phi(d)\sigma[n+1](\frac{m}{d}) = m\sigma[n](m), \tag{39}$$

$$\sum_{d|m}\phi(d)\tau\Big(\frac{m^2}{d^2}\Big) = \sum_{d|m} d\theta(\frac{m}{d}), \tag{40}$$

where $\sigma[n] = \mathrm{id}[n] * u$, $\mathrm{id}[n](m) := m^n$, and $\theta(m)$ is the number of decompositions of m into two relatively prime factors. All three are of the form $\phi * f = g$ and therefore they gain significance due to (3), though Liouville might not have been aware of this. For example, (3) and (38) combine to yield

$$\sum_{k=1}^{m}\tau(\gcd(k,m)) = \sigma(m).$$

The three identities are easily proven by substituting $\tau = u * u$, $\sigma[n] = \mathrm{id}[n] * u$, $\tau \circ \mathrm{id}[2] = \theta * u$, $\phi = \mu * \mathrm{id}$, and using $\mu * u = I$. They generalize to

$$\sum_{d|m}\phi_a(d)\tau(\frac{m}{d}) = \sigma_a(m), \tag{41}$$

$$\sum_{d|m}\phi_a(d)\sigma[n+1](\frac{m}{d}) = mu *_a \sigma[n](m), \tag{42}$$

$$\sum_{d|m}\phi_a(d)\tau\Big(\frac{m^2}{d^2}\Big) = \sum_{d|m} d\tau(\gcd(a,d))\theta(\frac{m}{d}). \tag{43}$$

These generalizations are proven using the same substitutions, together with (13), or for the latter identity, (11) and (29). We note that n need not be integer valued in (39) and (42).

4.9. One Identity from Dirichlet

Dickson [5] writes that Dirichlet [6], by taking partial sums on both sides of Euler's identity, obtained

$$\sum_{k=1}^{n} \left\lfloor \frac{n}{k} \right\rfloor \phi(k) = \binom{n+1}{2}.$$

By taking partial sums on both sides of equation (28) we obtain

$$\sum_{k=1}^{n} \left\lfloor \frac{n}{k} \right\rfloor \phi_a(k) = \sum_{d|a} d \binom{\lfloor \frac{n}{d} \rfloor + 1}{2}. \tag{44}$$

Proof. Summing the left-hand side of (28) over m yields

$$\sum_{m=1}^{n} \sum_{d|m} \phi_a(d) = \sum_{d=1}^{n} \left\lfloor \frac{n}{d} \right\rfloor \phi_a(d)$$

and summing the right-hand side of (28) over m yields

$$\sum_{m=1}^{n} \tau(\gcd(a,m))m = \sum_{m=1}^{n} \sum_{d|\gcd(a,m)} m = \sum_{d|a} \sum_{k=1}^{\lfloor n/d \rfloor} dk = \sum_{d|a} d \left\lfloor \frac{n}{d} \right\rfloor \left(\left\lfloor \frac{n}{d} \right\rfloor + 1 \right)/2.$$

□

4.10. The Lambert Series of ϕ_a

As shown by Liouville [9], cf. [5, p.120], the Lambert series of the totient function is given by

$$\sum_{m=1}^{\infty} \phi(m) \frac{x^m}{1-x^m} = \frac{x}{(1-x)^2}.$$

The Lambert series for ϕ_a is given by

$$\sum_{m=1}^{\infty} \phi_a(m) \frac{x^m}{1-x^m} = p[a](x) \frac{x}{(1-x^a)^2}, \tag{45}$$

where the coefficients of $p[a](x) = \sum_{k=1}^{2a} c[a](k)x^{k-1}$ are given by $c[a] = \mathrm{id}_a \circ t[a]$, and $t[a]$ is the piece-wise linear function $t[a](n) = a - |n-a|$.

Proof. Cesàro proved

$$\sum_{n=1}^{\infty} f(n) \frac{x^n}{1-x^n} = \sum_{n=1}^{\infty} x^n \sum_{d|n} f(d),$$

cf. [5] and exercise 31 to chapter 2 in [14]. By substituting (28) in this formula we find

$$\sum_{n=1}^{\infty} \phi_a(n) \frac{x^n}{1-x^n} = \sum_{n=1}^{\infty} x^n \tau(\gcd(a,n)) n.$$

Multiplying the right-hand side by $(1 - 2x^a + x^{2a})$ yields

$$\Big(\sum_{n=1}^{\infty} x^n \tau(\gcd(a,n)) n\Big) - 2\Big(\sum_{n=a+1}^{\infty} x^n \tau(\gcd(a,n))(n-a)\Big)$$
$$+ \Big(\sum_{n=1+2a}^{\infty} x^n \tau(\gcd(a,n))(n-2a)\Big) = \sum_{n=1}^{\infty} c[a](n) x^n,$$

where

$$c[a](n) = \begin{cases} \tau(\gcd(a,n))n & 0 < n \leq a, \\ \tau(\gcd(a,n))(n - 2(n-a)) = \tau(\gcd(a,n))(2a - n) & a < n \leq 2a, \\ \tau(\gcd(a,n))(n - 2(n-a) + n - 2a) = 0 & n > 2a. \end{cases}$$

Rewriting, using (29), the fact that $\gcd(a, a+k) = \gcd(a, a-k)$, and dividing by x, yields the result. □

The polynomials $p[a]$ seem to be irreducible over $\mathbb{Z}$ and their zeros are in some sense close to the a-th roots of unity; see Figures 4.10 and 4.10.

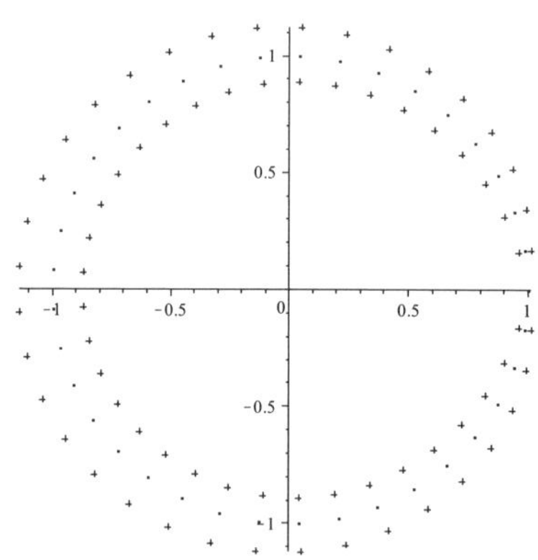

Figure 1: The roots of $p[37]$ are depicted as crosses and the 37th roots of unity as points. This figure shows that when a is prime the roots of $p[a]$ that are close to 1 are closer to a^{th} roots of unity.

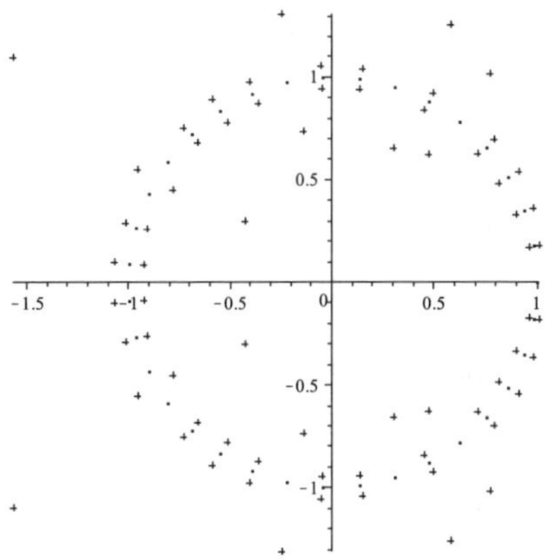

Figure 2: The roots of $p[35]$ are depicted as crosses and the 35th roots of unity as points. This figure shows that roots of $p[a]$ are closest to primitive a^{th} roots of unity.

4.11. A Series Related to the Lambert Series of ϕ_a

Liouville [9] also showed

$$\sum_{m=1}^{\infty} \phi(m) \frac{x^m}{1+x^m} = (1+x^2)\frac{x}{(1-x^2)^2}.$$

We show that

$$\sum_{m=1}^{\infty} \phi_a(m) \frac{x^m}{1+x^m} = q[a](x)\frac{x}{(1-x^{2a})^2}, \tag{46}$$

where $q[a](x) = \sum_{k=1}^{4a} b[a](k)x^{k-1}$, with $b[a] = h[a] \circ t[2a]$, and $h[a](k) = \mathrm{id}_a(k) - 2[2 \mid k]\mathrm{id}_a(k/2)$.

Proof. As the left-hand side of (46) is obtained from the left-hand side of (45) by subtracting twice the same series with x replaced by x^2, the same is true for the right-hand side. Thus it follows that $q[a](x) = p[a](x)(1+x^a)^2 - 2p[a](x^2)x$, and hence, that

$$b[a](k) = \begin{cases} \mathrm{id}_a(k) - 2[2 \mid k]\mathrm{id}_a(k/2) & k \le a, \\ \mathrm{id}_a(2a-k) + 2\mathrm{id}_a(k-a) - 2[2 \mid k]\mathrm{id}_a(k/2) & a < k \le 2a, \\ 2\mathrm{id}_a(3a-k) + \mathrm{id}_a(k-2a) - 2[2 \mid k]\mathrm{id}_a(2a-k/2) & 2a < k \le 3a, \\ \mathrm{id}_a(4a-k) - 2[2 \mid k]\mathrm{id}_a(2a-k/2) & 3a < k \le 4a. \end{cases}$$

The result follows due to the identities

$$\mathrm{id}_a(2a-k) + 2\mathrm{id}_a(k-a) = \mathrm{id}_a(k), \quad 2\mathrm{id}_a(3a-k) + \mathrm{id}_a(k-2a) = \mathrm{id}_a(4a-k),$$

which are easily verified using (29). □

We can express the functions $b[a]$ and $h[a]$ in terms of an interesting fractal function. Let a function κ of two variables be defined recursively by

$$\kappa[a](n) = \begin{cases} 0 & 2 \mid n, 2 \nmid a, \text{ or } n = 0, \\ \kappa[a/2](n/2) & 2 \mid n, 2 \mid a, \\ \tau(\gcd(a, n)) & 2 \nmid n. \end{cases} \tag{47}$$

The following properties are easily verified using the definition. We have

$$\kappa[a](2a + n) = \kappa[a](2a - n), \tag{48}$$

and, with $\gcd(a, b) = 1$,

$$\kappa[an](bn) = \begin{cases} 0 & 2 \mid b, \\ \alpha(n) & 2 \nmid b, \end{cases} \tag{49}$$

where α denotes the number of odd divisors function, i.e., for all k and odd m

$$\alpha(2^k m) = \tau(m). \tag{50}$$

Property (49) is a quite remarkable fractal property; from the origin in every direction we see either the zero sequence, or α, at different scales.

We claim that

$$h[a] = \kappa[a]\text{id} \tag{51}$$

follows from (47), (50), and (29). From (51) and (48) we obtain

$$b[a] = \kappa[a]t[2a]. \tag{52}$$

We now prove that $1 + x^2$ divides $q[a]$ when a is odd.

Proof. Noting that $b[a](2a + k) = b[a](2a - k)$ and, when $2 \nmid a$, $b[a](2k) = 0$, we therefore have

$$\begin{aligned} q[a](x) &= \sum_{n=1}^{2a} b[a](2n - 1)x^{2n-2} \\ &= \sum_{m=1}^{a} b[a](2a - 2m + 1)x^{2a-2m} + b[a](2a + 2m - 1)x^{2a+2m-2} \\ &= \sum_{m=1}^{a} b[a](2a - 2m + 1)x^{2a-2m}(1 + x^{4m-2}), \end{aligned}$$

which vanishes at the points where $x^2 = -1$. □

Apart from the factor $1 + x^2$ when a is odd, the polynomial $q[a]$ seems to be irreducible over $\mathbb{Z}$ and its zeros are in some sense close to the $(2a)^{\text{th}}$ roots of -1 or, to the $(a + 1)^{\text{st}}$ roots of unity, see Figure 4.11.

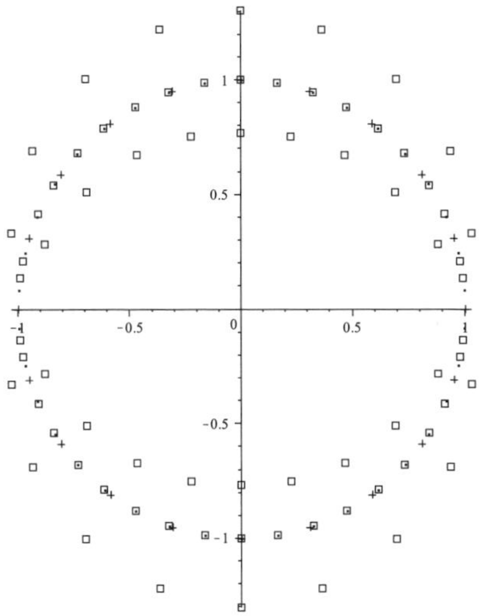

Figure 3: The roots of $q[19]$ are depicted as boxes, the 38th roots of -1 as points, and the 20th roots of unity as crosses.

4.12. A Perfect Square

Our last identity generalizes the faint fact that $\phi(1) = 1$. We have

$$\sum_{a=1}^{n} \phi_a(n) = n^2. \tag{53}$$

Proof. For any lattice point (i, j) in the square $[1, n] \times [1, n]$ the product $i \cdot j \bmod n$ is congruent to some a in the range $[1, n]$. □

Acknowledgment. This research has been funded by the Australian Research Council through the Centre of Excellence for Mathematics and Statistics of Complex Systems.

References

[1] D.R. Anderson and T.M. Apostol, *The evaluation of Ramanujan's sum and generalizations*, Duke Math. J. **20** (1953), 211–216.

[2] T.M. Apostol, Introduction to Analytic Number Theory, Springer-Verlag (1976).

[3] T.M. Apostol, *Arithmetical properties of generalized Ramanujan sums*, Pacific J. Math. 41 (1972), 281–293.

[4] K.A. Broughan, *The gcd-sum function*, Journal of Integer Sequences **4** (2001), Art. 01.2.2.

[5] L.E. Dickson, History of the theory of numbers, Carnegie Inst. (1919); reprinted by Chelsea (1952).

[6] G.L. Dirichlet, *Über die Bestimmung der mittleren Werte in der Zahlentheorie*, Abh. Akad. Wiss. Berlin (1849), 78–81; also in Werke, vol. 2 (1897), 60–64.

[7] O. Hölder, *Zur Theorie der Kreisteilungsgleichung* $K_m(x) = 0$, Prace Matematyczno Fizyczne **43** (1936), 13–23.

[8] J.C. Kluyver, *Some formulae concerning the integers less than n and prime to n*, in: KNAW, Proceedings 9 I, Amsterdam (1906), 408–414.

[9] J. Liouville, *Sur quelques séries et produits infinis*, J. Math. Pures Appl. **2** (1857), 433–440.

[10] L. Tóth and P. Haukkanen, *The discrete Fourier transform of r-even functions*, Acta Univ. Sapientiae Math. **3** (2011), 5–25.

[11] S. Ramanujan, *On Certain Trigonometric Sums and their Applications in the Theory of Numbers*, Transactions of the Cambridge Philosophical Society **22** (1918), 259–276; also in: Collected papers of Srinivasa Ramanujan, Ed. G.H. Hardy et al, Chelsea Publ. Comp. (1962), 179–199.

[12] J. Sandor and B. Crstici, Handbook of Number Theory II, Kluwer Acad. Publ. (2004).

[13] R. Daublebsky Von Sterneck, *Ein Analogon zur additiven Zahlentheorie*, Sitzber, Akad. Wiss. Wien, Math. Naturw. Klasse, vol. I ll (Abt. IIa) (1902), 1567–1601.

[14] H. S. Wilf, Generatingfunctionology, Academic Press, 2nd edition (1994).

[15] W. Schramm, *The Fourier transform of functions of the greatest common divisor*, Integers **8** (2008) A50.

WEIGHTED LONESUM MATRICES AND THEIR GENERATING FUNCTION

Ken Kamano
Department of Mathematics, Osaka Institute of Technology, Osaka, Japan
kamano@ge.oit.ac.jp

Yasuo Ohno[1]
Department of Mathematics, Kinki University, Osaka, Japan
ohno@math.kindai.ac.jp

Shuji Yamamoto[2]
Department of Mathematics, Faculty of Science and Technology, Keio University, Yokohama, Japan
yamashu@math.keio.ac.jp

Received: 7/9/12, Accepted: 4/21/13, Published: 5/10/13

Abstract

A lonesum matrix is a $(0,1)$-matrix uniquely determined by its column and row sums, and the sum of its all entries is called the "weight" of it. The generating function of numbers of weighted lonesum matrices of each weight is given. A certain explicit formula for the number of weighted lonesum matrices is also proved.

1. Introduction and Main Theorem

A matrix A is called a $(0,1)$-matrix if each of its entries is either zero or one, and a $(0,1)$-matrix A is called lonesum if it is uniquely determined by its column and row sums. For positive integers m and n, we denote by $L(m,n)$ the number of $m \times n$ lonesum matrices. Further, we define $L(m,0) = L(0,n) = 1$ for all non-negative integers m and n.

Brewbaker [1] proved that

$$L(m,n) = B_n^{(-m)} \tag{1}$$

[1]The author was supported in part by Grant-in-Aid for Scientific Research (C) (No. 23540036), Japan Society for the Promotion of Science.

[2]The author was supported in part by Grant-in-Aid for JSPS Fellows (No. 09J05093) and by Grant-in Aid for Young Scientists (S) (No. 21674001), Japan Society for the Promotion of Science.

for any non-negative integers m and n. Here, $B_n^{(k)}$ is the n-th *poly-Bernoulli number* of index k defined by

$$\frac{\mathrm{Li}_k(1-e^{-x})}{1-e^{-x}} = \sum_{n\geq 0} B_n^{(k)} \frac{x^n}{n!}, \tag{2}$$

where $\mathrm{Li}_k(x)$ denotes the k-th polylogarithm (cf. [2]). Shikata [5] gave another proof of equality (1). To obtain the equality, Shikata proved a generating function representation of $L(m,n)$ as follows:

$$\sum_{m\geq 0, n\geq 0} L(m,n)\frac{x^m}{m!}\frac{y^n}{n!} = (e^{-x}+e^{-y}-1)^{-1}. \tag{3}$$

For any lonesum matrix A, Nanbara [3] introduced the *weight* of A as the sum of all entries in A, and did numerical experiments to some extent. For positive integers m, n and a non-negative integer k, we denote by $L(m,n,k)$ the number of $m \times n$ lonesum matrices with weight k. Further, we define $L(m,0,0) = L(0,n,0) = 1$ and $L(m,0,k) = L(0,n,k) = 0$ for any non-negative integers m, n and positive integer k.

By definition, we have $L(m,n,k) = L(n,m,k)$ for $0 \leq k \leq mn$, and

$$L(m,n) = \sum_{k=0}^{mn} L(m,n,k).$$

The following is the main theorem of this paper, which gives a generating function of $L(m,n,k)$.

Theorem 1. *Let x, y and q be indeterminates satisfying*

$$qx = xq, \quad qy = yq \quad \text{and} \quad yx = qxy. \tag{4}$$

Then the following identity holds:

$$\sum_{m,n,k\geq 0} L(m,n,k)\frac{x^m}{m!}\frac{y^n}{n!}q^k = (e^{-x}+e^{-y}-1)^{-1}. \tag{5}$$

It is surprising that the right-hand sides of (3) and (5) coincide with each other. Hence, by setting $q = 1$ in the equality (5), we naturally obtain the original generating function (3) of $L(m,n)$.

2. Proof of Theorem 1

An $m \times n$ $(0,1)$-matrix $A = (a_{i,j})$ is called a *Ferrers matrix* if A satisfies the condition

$$\begin{aligned} a_{i,j} &\geq a_{i+1,j} \quad (1 \leq i < m-1,\ 1 \leq j \leq n), \\ a_{i,j} &\geq a_{i,j+1} \quad (1 \leq i \leq m,\ 1 \leq j < n-1). \end{aligned}$$

This condition means that all 1 entries of A are placed at the upper left of A (see Fig. 1 in the proof of Theorem 1).

The following result is due to Ryser [4] (see also [5, Theorem 3.2]).

Proposition 2. *Let A be an $m \times n$ $(0,1)$-matrix. Then the following conditions are equivalent:*

(i) A is a lonesum matrix.

(ii) A has no minor of the form $\begin{pmatrix} 1 & 0 \\ 0 & 1 \end{pmatrix}$ *or* $\begin{pmatrix} 0 & 1 \\ 1 & 0 \end{pmatrix}$.

(iii) A is obtained from a Ferrers matrix by permutations of columns and rows.

We can prove our main theorem from this proposition.

Proof of Theorem 1. For a Ferrers matrix $A = (a_{i,j})$, we define m_i and n_j $(1 \le i, j \le l)$ by the numbers of columns and rows as indicated in Figure 1. Here we have

$$m_1, \ldots, m_{l-1} > 0, \quad m_l \ge 0, \quad m_1 + \cdots + m_l = m,$$
$$n_1 \ge 0, \quad n_2, \ldots, n_l > 0, \quad n_1 + \cdots + n_l = n.$$

Then the weight of A is expressed as

$$m_1 n_1 + m_2(n_1 + n_2) + \cdots + m_l(n_1 + \cdots + n_l) = \sum_{l \ge i \ge j \ge 1} m_i n_j.$$

The number of lonesum matrices obtained from the matrix A is equal to

$$\frac{m!}{m_1! \cdots m_l!} \frac{n!}{n_1! \cdots n_l!}.$$

Therefore we have

$$\begin{aligned} &\sum_{m,n,k \ge 0} L(m,n,k) \frac{x^m}{m!} \frac{y^n}{n!} q^k \\ &= \sum_{l=1}^{\infty} \sum_{\substack{m_1, \ldots, m_{l-1} > 0, m_l \ge 0 \\ n_1 \ge 0, n_2, \ldots, n_l > 0}} \frac{x^{m_1 + \cdots + m_l}}{m_1! \cdots m_l!} \frac{y^{n_1 + \cdots + n_l}}{n_1! \cdots n_l!} q^{\sum_{i \ge j} m_i n_j}. \end{aligned} \tag{6}$$

By the commutation relations (4), we have

$$x^{m_1 + \cdots + m_l} y^{n_1 + \cdots + n_l} q^{\sum_{i \ge j} m_i n_j} = y^{n_1} x^{m_1} y^{n_2} x^{m_2} \cdots y^{n_l} x^{m_l}.$$

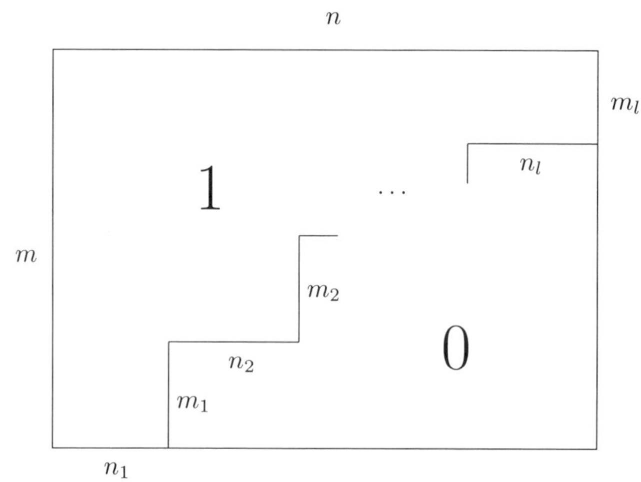

Figure 1: Ferrers matrix A

Hence the right-hand side of (2) equals

$$\sum_{l=1}^{\infty} \sum_{\substack{m_1,\ldots,m_{l-1}>0,\\ m_l\geq 0\\ n_1\geq 0, n_2,\ldots,n_l>0}} \frac{y^{n_1}}{n_1!}\frac{x^{m_1}}{m_1!}\frac{y^{n_2}}{n_2!}\frac{x^{m_2}}{m_2!}\cdots\frac{y^{n_l}}{n_l!}\frac{x^{m_l}}{m_l!} = \sum_{l=1}^{\infty} e^y((e^x-1)(e^y-1))^{l-1}e^x$$

$$= e^y\big\{1-(e^x-1)(e^y-1)\big\}^{-1}e^x$$
$$= \big\{e^{-x}(1-(e^x-1)(e^y-1))e^{-y}\big\}^{-1}$$
$$= (e^{-x}+e^{-y}-1)^{-1}.$$

Now the proof is complete. □

3. A Recurrence Relation and an Explicit Formula

In this section we investigate the following polynomial in q:

$$L_q(m,n) := \sum_{k=0}^{mn} L(m,n,k)q^k, \quad (m,n\geq 0).$$

When $q = 1$, we have

$$L_1(m,n) = \sum_{k=0}^{mn} L(m,n,k) = L(m,n).$$

This polynomial $L_q(m,n)$ satisfies the following recurrence relation:

Proposition 3. *For $(m,n) \neq (0,0)$, we have*

$$L_q(m,n) = -\sum_{k=0}^{m-1} \binom{m}{k} (-1)^{m-k} L_q(k,n) - \sum_{l=0}^{n-1} \binom{n}{l} (-q^m)^{n-l} L_q(m,l). \quad (7)$$

Proof. By Theorem 1, the identity

$$(e^{-x} + e^{-y} - 1) \sum_{m \geq 0, n \geq 0} L_q(m,n) \frac{x^m}{m!} \frac{y^n}{n!} = 1$$

holds. Since $e^{-y} x^m = x^m e^{-q^m y}$, we have

$$\sum_{m \geq 0, n \geq 0} L_q(m,n) \frac{x^m}{m!} (e^{-x} + e^{-q^m y}) \frac{y^n}{n!} = 1 + \sum_{m \geq 0, n \geq 0} L_q(m,n) \frac{x^m}{m!} \frac{y^n}{n!}.$$

Comparing the coefficients of both sides, we obtain that

$$\sum_{k=0}^{m} \binom{m}{k} (-1)^{m-k} L_q(k,n) + \sum_{l=0}^{n} \binom{n}{l} (-q^m)^{n-l} L_q(m,l) = L_q(m,n)$$

for $(m,n) \neq (0,0)$. The formula (6) immediately follows from this equation. □

Now we give an explicit formula for the polynomial $L_q(m,n)$. This formula can be proved by using Proposition 3, but we will prove it directly to keep the proof simple.

Theorem 4. *For integers $m \geq 0$ and $n \geq 0$, we have*

$$L_q(m,n) = \sum_{\lambda \vdash m} (-1)^{m+l(\lambda)} \binom{m}{\lambda} \Big(\sum_{i=0}^{l(\lambda)} q^{\lambda_1 + \cdots + \lambda_i} \Big)^n. \quad (8)$$

Here $\lambda \vdash m$ means that λ is a composition of m, i.e., $\lambda = (\lambda_1, \ldots, \lambda_l)$ is a sequence of positive integers such that $\lambda_1 + \cdots + \lambda_l = m$. For such λ, $l = l(\lambda)$ denotes its length and $\binom{m}{\lambda} = \frac{m!}{\lambda_1! \cdots \lambda_l!}$ the associated multinomial coefficient.

Proof. We denote the right-hand side of (7) by $T_q(m,n)$. By Theorem 1, it suffices to show that

$$\sum_{m \geq 0, n \geq 0} T_q(m,n) \frac{x^m}{m!} \frac{y^n}{n!} = (e^{-x} + e^{-y} - 1)^{-1} \quad (9)$$

under the commutation relations (4). We have

$$\begin{aligned}
&\sum_{n\geq 0} T_q(m,n)\frac{x^m}{m!}\frac{y^n}{n!}\\
&= \frac{x^m}{m!}\sum_{\lambda\vdash m}(-1)^{m+l(\lambda)}\binom{m}{\lambda}\sum_{n\geq 0}\left(\sum_{i=0}^{l(\lambda)} q^{\lambda_1+\cdots+\lambda_i}\right)^n\frac{y^n}{n!}\\
&= \frac{x^m}{m!}\sum_{\lambda\vdash m}(-1)^{m+l(\lambda)}\binom{m}{\lambda}e^{\sum_{i=0}^{l(\lambda)} q^{\lambda_1+\cdots+\lambda_i}y}\\
&= \sum_{\lambda\vdash m}(-1)^{l(\lambda)}\frac{(-x)^{\lambda_l}\cdots(-x)^{\lambda_1}}{\lambda_l!\cdots\lambda_1!}e^{\sum_{i=0}^{l(\lambda)} q^{\lambda_1+\cdots+\lambda_i}y}\\
&= \sum_{\lambda\vdash m}(-1)^{l(\lambda)}e^y\frac{(-x)^{\lambda_l}}{\lambda_l!}e^y\cdots e^y\frac{(-x)^{\lambda_1}}{\lambda_1!}e^y
\end{aligned}$$

(here we used the relation $(-x)^i e^{q^i y} = e^y(-x)^i$). Consequently we obtain that

$$\begin{aligned}
\sum_{m\geq 0,n\geq 0} T_q(m,n)\frac{x^m}{m!}\frac{y^n}{n!} &= \sum_{l\geq 0} e^y\overbrace{(1-e^{-x})e^y\cdots(1-e^{-x})e^y}^{l}\\
&= e^y\left\{1-(1-e^{-x})e^y\right\}^{-1}\\
&= \left\{e^{-y}-(1-e^{-x})\right\}^{-1}\\
&= (e^{-x}+e^{-y}-1)^{-1}
\end{aligned}$$

and this completes the proof. □

Examples. We have:

$$\begin{aligned}
L_q(1,n) &= (1+q)^n,\\
L_q(2,n) &= -(1+q^2)^n + 2(1+q+q^2)^n,\\
L_q(3,n) &= (1+q^3)^n - 3(1+q+q^3)^n - 3(1+q^2+q^3)^n + 6(1+q+q^2+q^3)^n.
\end{aligned}$$

By putting $q=-1$ in the cases $m=2$ and 3, we obtain the following alternating sum formulas:

Corollary 5. *For any integer $n \geq 1$, we have*

$$\sum_{k=0}^{2n} L(2,n,k)(-1)^k = 2-2^n, \tag{10}$$

$$\sum_{k=0}^{3n} L(3,n,k)(-1)^k = -3((-1)^n+1) = \begin{cases} 0 & \text{if } n \text{ is odd,}\\ -6 & \text{if } n \text{ is even.}\end{cases} \tag{11}$$

References

[1] C. Brewbaker, A combinatorial interpretation of the poly-Bernoulli numbers and two Fermat analogues, Integers **8** (2008), A02.

[2] M. Kaneko, Poly-Bernoulli numbers, J. Theor. Nombres Bordeaux **9** (1997), 221–228.

[3] Y. Nanbara, On poly-Bernoulli numbers of negative indices (in Japanese), Master's thesis, Kyushu University (2006).

[4] H. J. Ryser, Combinatorial properties of matrices of zeros and ones, Canad. J. Math. **9** (1957), 371–377.

[5] M. Shikata, Lonesum matrices and poly-Bernoulli numbers, Acta Human. Sci. Univ. Sangio Kyotien. Natur. Sci. Ser. **40** (2011), 1–12.

ON MULTIPLICATIVE SIDON SETS

David Wakeham
Department of Mathematics and Statistics, The University of Melbourne, Melbourne, Australia
d.wakeham@student.unimelb.edu.au

David R. Wood[1]
School of Mathematical Sciences, Monash University, Melbourne, Australia
david.wood@monash.edu

Received: 9/20/12, Accepted: 4/21/13, Published: 5/10/13

Abstract

Fix integers $b > a \geq 1$ with $g := \gcd(a, b)$. A set $S \subseteq \mathbb{N}$ is $\{a, b\}$*-multiplicative* if $ax \neq by$ for all $x, y \in S$. For all n, we determine an $\{a, b\}$-multiplicative set with maximum cardinality in $[n]$, and conclude that the maximum density of an $\{a, b\}$-multiplicative set is $\frac{b}{b+g}$. For $A, B \subseteq \mathbb{N}$, a set $S \subseteq \mathbb{N}$ is $\{A, B\}$-multiplicative if for all $a \in A$ and $b \in B$ and $x, y \in S$, the only solutions to $ax = by$ have $a = b$ and $x = y$. For $1 < a < b < c$ and a, b, c coprime, we give a $\mathcal{O}(1)$ time algorithm to approximate the maximum density of an $\{\{a\}, \{b, c\}\}$-multiplicative set to arbitrary given precision.

1. Introduction

Erdős [3], Erdős [4], Erdős [5] defined a set $S \subseteq \mathbb{N}$ to be *multiplicative Sidon*[2] if $ab = cd$ implies $\{a, b\} = \{c, d\}$ for all $a, b, c, d \in S$; see [9, 10, 11]. In a similar direction, Wang [14] defined a set $S \subseteq \mathbb{N}$ to be *double-free* if $x \neq 2y$ for all $x, y \in S$, and proved that the maximum density of a double-free set is $\frac{2}{3}$; see [1] for related results. Here $\mathbb{N} := \{1, 2, \dots\}$, $\mathbb{N}_0 = \mathbb{N} \cup \{0\}$, $[n] := \{1, 2, \dots, n\}$, and the *density* of $S \subseteq \mathbb{N}$ is

$$\lim_{n \to \infty} \frac{|S \cap [n]|}{n} .$$

Motivated by some questions in graph colouring, Pór and Wood [8] generalised the notion of double-free sets as follows. For $k \in \mathbb{N}$, a set $S \subseteq \mathbb{N}$ is k*-multiplicative*

[1]Research supported by the Australian Research Council.

[2]Additive Sidon sets have been more widely studied; see the classical papers [6, 12, 13] and the survey by O'Bryant [7].

(*Sidon*) if $ax = by$ implies $a = b$ and $x = y$ for all $a, b \in [k]$ and $x, y \in S$. Pór and Wood [8] proved that the maximum density of a k-multiplicative set is $\Theta(\frac{1}{\log k})$.

Here we study the following alternative generalization of double-free sets. For distinct $a, b \in \mathbb{N}$, a set $S \subseteq \mathbb{N}$ is *$\{a, b\}$-multiplicative* if $ax \neq by$ for all $x, y \in S$. Our first result is to determine the maximum density of an $\{a, b\}$-multiplicative set. Assume that $a < b$ throughout.

Say $x \in \mathbb{N}$ is an *i-th subpower* of b if $x = b^i y$ for some $y \not\equiv 0 \pmod b$. If x is an i-th subpower of b for some even/odd i then x is an *even/odd* subpower of b. The following table gives the even subpowers of $b \in \{2, 3, 4\}$ and the corresponding entry in *The On-Line Encyclopedia of Integer Sequences*.

$b = 2$	$1, 3, 4, 5, 7, 9, 11, 12, 13, 15, 16, 17, 19, 20, 21, 23, \ldots$	[A003159]
$b = 3$	$1, 2, 4, 5, 7, 8, 9, 10, 11, 13, 14, 16, 17, 18, 19, 20, 22, \ldots$	[A007417]
$b = 4$	$1, 2, 3, 5, 6, 7, 9, 10, 11, 13, 14, 15, 16, 17, 18, 19, 21, \ldots$	[A171948]

We prove the following result:

Theorem 1. *Fix integers $b > a \geqslant 1$. Let $g := \gcd(a, b)$. Then for every integer $n \in \mathbb{N}$, the even subpowers of $\frac{b}{g}$ in $[n]$ are an $\{a, b\}$-multiplicative set in $[n]$ with maximum cardinality. And the even subpowers of $\frac{b}{g}$ are an $\{a, b\}$-multiplicative set with density $\frac{b}{b+g}$, which is maximum.*

Note that if $g = a$ then $b \geqslant 2g$ and $b + g \leqslant \frac{3}{2}b$, and if $g < a$ then $a \geqslant 2g$ and $b + g \leqslant b + a < \frac{3}{2}b$. In both cases the density bound $\frac{b}{b+g}$ in Theorem 1 is at least $\frac{2}{3}$, which is the bound obtained by Wang [14] for the $a = 1$ and $b = 2$ case.

We propose a further generalization of double-free sets. Let $A, B \subseteq \mathbb{N}$. Say $S \subseteq \mathbb{N}$ is *$\{A, B\}$-multiplicative* if $ax = by$ implies $a = b$ and $x = y$ for all $a \in A$ and $b \in B$, and $x, y \in S$. One case is easily dealt with. If $B := \{b\}$ and b is coprime to each element of A, and there is some element $a \in A$ such that $a < b$, then, by the reasoning above, the even subpowers of b form an $\{A, B\}$-multiplicative set of (maximum) density $\frac{b}{b+1}$.

The simplest nontrivial case (not covered by Theorem 1) is $\{A, B\}$-multiplicativity for $A = \{a\}$, $B = \{b, c\}$, $1 < a < b < c$, with a, b, c pairwise coprime. We have the following theorem:

Theorem 2. *Consider $a, b, c \in \mathbb{N}$ pairwise coprime, with $1 < a < b < c$. For all fixed $\epsilon > 0$, there is a $\mathcal{O}(1)$ time algorithm that computes the maximum density of an $\{\{a\}, \{b, c\}\}$-multiplicative set to within ϵ.*

2. Proof of Theorem 1

First suppose that $\gcd(a, b) = 1$. Let T be the set of even subpowers of b. We now prove that T is an $\{a, b\}$-multiplicative set with maximum density. In fact, for all $[n]$, we prove that $T_n := T \cap [n]$ has maximum cardinality out of all $\{a, b\}$-multiplicative sets contained in $[n]$.

The key to our proof is to model the problem using a directed graph. Let G be the directed graph with $V(G) := [n]$ where $(x, y) \in E(G)$ whenever $bx = ay$ (implying $x < y$). Thus $S \subseteq [n]$ is $\{a, b\}$-multiplicative if and only if S is an independent set in G. If (x, y, z) is a directed path in G, then $x = \frac{a}{b}y$ and $z = \frac{b}{a}y$. Thus each vertex y has indegree and outdegree at most 1. Since $(x, y) \in E(G)$ implies $x < y$, G contains no directed cycles. Thus G is a collection of disjoint directed paths. Hence a maximum independent set in G is obtained by taking all the vertices at even distance from the source vertices[3], where a vertex y is a source (indegree 0) if and only if $\frac{a}{b}y$ is not an integer; that is, if $y \not\equiv 0 \pmod{b}$.

We now prove that the vertices at distance d from a source vertex are precisely the d-th subpowers of b. We proceed by induction on $d \geqslant 0$. Each vertex y of G has an incoming edge if and only if $\frac{a}{b}y \in \mathbb{N}$, which occurs if and only if $y \equiv 0 \pmod{b}$ since $\gcd(a, b) = 1$. Thus the source vertices of G are precisely the 0-th subpowers of b. This proves the $d = 0$ case of the induction hypothesis. Now consider a vertex y at distance d from a source vertex. Thus $y = \frac{b}{a}x$ for some vertex x at distance $d - 1$ from a source vertex. By induction, x is a $(d-1)$-th subpower of b. That is, $x = b^{d-1}z$ for some $z \not\equiv 0 \pmod{b}$. Thus $y = b^d\frac{z}{a}$, which, since $\gcd(a, b) = 1$, implies that $\frac{z}{a}$ is an integer. Hence $\frac{z}{a} \not\equiv 0 \pmod{b}$ and y is a d-th subpower of b, as claimed.

This proves that the even subpowers of b form a maximum independent set in G. That is, T_n is an $\{a, b\}$-multiplicative set of maximum cardinality in $[n]$. To illustrate this proof, the following table shows two examples of the graph G with $b = 3$. Observe that the i-th row consists of the i-th subpowers of 3 regardless of a.

$a = 1$ and $b = 3$									$a = 2$ and $b = 3$											
1	2	4	5	7	8	10	11	⋯	1	2	4	5	7	8	10	11	13	14	16	⋯
↓	↓	↓	↓	↓	↓	↓	↓			↓	↓			↓	↓			↓	↓	
3	6	12	15	21	24	30	33	⋯		3	6			12	15			21	24	⋯
↓	↓	↓	↓	↓	↓	↓	↓				↓			↓					↓	
9	18	36	45	63	72	90	99	⋯			9			18					36	⋯
↓	↓	↓	↓	↓	↓	↓	↓							↓					↓	
27	48	108	135	189	216	270	297	⋯						27					48	⋯
⋮	⋮	⋮	⋮	⋮	⋮	⋮	⋮												⋮	

[3]Note that this is not necessarily the only maximum independent set—for a path component with odd length, we may take the vertices at odd distance from the source of this path. This observation readily leads to a characterization of all maximum independent sets in G, and thus of all $\{a, b\}$-multiplicative sets in $[n]$ with maximum cardinality. Details omitted.

We now bound $|T_n|$ from above. Observe that

$$T_n = \left\{ b^{2i}y : 0 \leqslant i \leqslant \tfrac{1}{2}\log_b n,\ 1 \leqslant y \leqslant \frac{n}{b^{2i}},\ y \not\equiv 0 \pmod{b} \right\} .$$

Thus

$$\begin{aligned} |T_n| &\leqslant \sum_{i=0}^{\lfloor (\log_b n)/2 \rfloor} \left\lceil \frac{b-1}{b}\frac{n}{b^{2i}} \right\rceil \\ &\leqslant 1 + \tfrac{1}{2}(\log_b n) + \frac{(b-1)n}{b}\sum_{i\geqslant 0}\frac{1}{b^{2i}} \\ &\leqslant 1 + \tfrac{1}{2}(\log_b n) + \frac{(b-1)n}{b}\frac{b^2}{b^2-1} \\ &= 1 + \tfrac{1}{2}(\log_b n) + \frac{b}{b+1}n \ . \end{aligned}$$

We now bound $|T_n|$ from below. Observe that

$$T_n = [n] \setminus \left\{ b^{2i+1}y : 0 \leqslant i \leqslant \tfrac{1}{2}((\log_b n) - 1),\ 1 \leqslant y \leqslant \frac{n}{b^{2i+1}},\ y \not\equiv 0 \pmod{b} \right\} .$$

Thus

$$\begin{aligned} |T_n| &\geqslant n - \sum_{i=0}^{\lfloor ((\log_b n)-1)/2 \rfloor} \left\lceil \frac{b-1}{b}\frac{n}{b^{2i+1}} \right\rceil \\ &\geqslant n - \tfrac{1}{2}((\log_b n)+1) - \frac{(b-1)n}{b^2}\sum_{i\geqslant 0}\frac{1}{b^{2i}} \\ &\geqslant n - \tfrac{1}{2}((\log_b n)+1) - \frac{(b-1)n}{b^2}\frac{b^2}{b^2-1} \\ &= n - \tfrac{1}{2}((\log_b n)+1) - \frac{n}{b+1} \\ &= \frac{b}{b+1}n - \tfrac{1}{2}((\log_b n)+1) \ . \end{aligned}$$

These upper and lower bounds on $|T_n|$ imply that

$$|T_n| = \frac{b}{b+1}n + \Theta(\log_b n) \ .$$

Hence the density of T is $\frac{b}{b+1}$, and because T_n is optimal for each n, no $\{a,b\}$-multiplicative set has density greater than $\frac{b}{b+1}$.

We now drop the assumption that $\gcd(a,b) = 1$. Let $g := \gcd(a,b)$. Since $ax = by$ if and only if $\frac{a}{g}x = \frac{b}{g}y$, a set S is $\{a,b\}$-multiplicative if and only if S is $\{\frac{a}{g},\frac{b}{g}\}$-multiplicative. Since $\frac{b/g}{b/g+1} = \frac{b}{b+g}$, the theorem is proved.

3. Proof of Theorem 2

Fix $A = \{a\}$ and $B = \{b, c\}$, where $1 < a < b < c$, and a, b, c are pairwise coprime. For convenience, we use the infinite graph G with vertex set $\mathbb{N}$ and edge set

$$E(G) = \{\{x, y\} : bx = ay \text{ or } cx = ay, \text{ and } x, y \in \mathbb{N}\}.$$

Let G_n denote the subgraph of G induced by the vertex set $[n]$. Let δ be the maximum density of an $\{\{a\}, \{b, c\}\}$-multiplicative set. Then

$$\delta = \lim_{n\to\infty} \frac{\alpha(G_n)}{n} ,$$

where $\alpha(G_n)$ is the size of a maximum independent set in G_n.

The infinite graph G has components $C_{p,q}$ with vertex set

$$V(C_{p,q}) = \{a^{p-x-y} b^x c^y q : x, y \in \mathbb{N}_0\}$$

for all $p \in \mathbb{N}_0$, $q \in \mathbb{N}$, and q not divisible by a, b, or c. Note that each $C_{p,q}$ is finite. Define p as the *height* of the component, and subsets of constant $x + y$ as *rows*. Note that the maximum and minimum vertices in $C_{p,q}$ are $c^p q$ and $a^p q$ respectively. The first few components of G for $a = 2$, $b = 3$, and $c = 5$ are shown below:

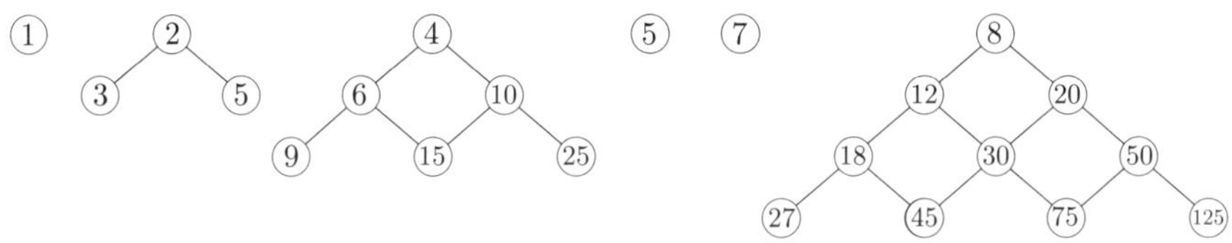

For a, b, c as above and fixed $\epsilon > 0$, let d be a non-negative integer $d \in \mathbb{N}_0$, to be specified later. Basically, d is a cutoff height which allows us to partition the components of G_n into three types, for any given $n \in \mathbb{N}$. The first are *complete* components $C_{p,q}$ where $n > c^p q$. The second are *small* incomplete components $S_{p,q}$ where $p \leqslant d$ and $a^p q \leqslant n < c^p q$. The third are *large* incomplete components $L_{p,q}$ with $p > d$ and $a^p q \leqslant n < c^p q$.

Let $\alpha_T(G_n)$ denote the size of a maximum independent set in the components of type T in G_n, for $T \in \{C, S, L\}$. We clearly have

$$\alpha(G_n) = \alpha_C(G_n) + \alpha_S(G_n) + \alpha_L(G_n) \ .$$

Thus,

$$\delta = \lim_{n\to\infty} \frac{\alpha_C(G_n)}{n} + \lim_{n\to\infty} \frac{\alpha_S(G_n)}{n} + \lim_{n\to\infty} \frac{\alpha_L(G_n)}{n} = \delta_C + \delta_S + \delta_L$$

where

$$\delta_T = \lim_{n\to\infty} \frac{\alpha_T(G_n)}{n}.$$

Below we show that these limits exist, and we determine δ_C and δ_S explicitly. Then we show that, for any $\epsilon > 0$, we can choose d so that $\delta_L < \epsilon$. Hence, we can calculate δ to arbitrary precision.

3.1. Complete Components

We require the following lemma about independent sets in grid-like graphs by Cassaigne and Zimmerman [2].

Lemma 1. *Define a graph H by $V(H) := \mathbb{N}_0 \times \mathbb{N}_0$ and*

$$E(H) := \{\{\mathbf{v}, \mathbf{w}\} : \mathbf{v}, \mathbf{w} \in V(H), |v_1 - w_1| + |v_2 - w_2| = 1\}.$$

Suppose that F is a finite subgraph of H such that $(x, y) \in V(F)$ implies $(x-1, y) \in V(F)$ unless $x = 0$, and $(x, y-1) \in V(F)$ unless $y = 0$. Then one of the sets

$$O := \{(x, y) \in V(F) : x + y \text{ is odd}\} \textit{ or}$$
$$E := \{(x, y) \in V(F) : x + y \text{ is even}\}$$

is a maximum independent set in F.

Now, consider a complete component $C_{p,1}$ of G_n. Note that every complete component $C_{p,q}$ of height q is isomorphic to $C_{p,1}$, and can be obtained by multiplying each vertex by q. Thus, we call $C_{p,q}$ a *q-copy* of $C_{p,1}$. In general, we use this terminology for isomorphic components of any type obtained by multiplying each vertex by q.

Observe that we can apply Lemma 1 to $C_{p,1}$, since it is isomorphic to a subgraph of H with the required properties. Define a function $\varphi : V(C_{p,1}) \to \mathbb{N}_0 \times \mathbb{N}_0$ by

$$\varphi(a^{p-x-y} b^x c^y) = (x, y).$$

If $a^{p-x-y} b^x c^y$ is adjacent to $a^{p-x'-y'} b^{x'} c^{y'}$, then $|x - x'| + |y - y'| = 1$ since they must differ by a factor of b/a or c/a. Thus, since φ is injective, it defines an isomorphism from $C_{p,1}$ to a subgraph of H. Assume $a^{p-x-y} b^x c^y \in V(C_{p,1})$. Then $a^{p-x-y+1} b^{x-1} c^y \in V(C_{p,1})$ unless $x = 0$, and similarly $a^{p-x-y+1} b^x c^{y-1} \in V(C_{p,1})$ unless $y = 0$. Under φ, these are clearly equivalent to the conditions required for Lemma 1.

Hence, by Lemma 1 and the definition of φ, a maximum independent set in $C_{p,1}$ is given by choosing all rows with $x + y$ even, or all rows with $x + y$ odd. In fact, it is clear that a maximum independent set is obtained by choosing the bottom row first, then alternating between remaining rows. Thus, if $p = 2i - 1$, then $\alpha(C_{p,1}) = i(i + 1)$. If $p = 2i$, then $\alpha(C_{p,1}) = (i + 1)^2$. Since the largest vertex in such a component is c^p, we must have $p \leqslant \log_c n$ for the component $C_{p,1}$ to be complete. Hence, the maximum height of a complete component is $\lfloor \log_c n \rfloor$.

Now we multiply by the number of components of height p that are complete. For a given p, we require $1 \leqslant q \leqslant nc^{-p}$. Since a, b, c are pairwise coprime, the density of numbers not divisible by a, b, or c is

$$\frac{(a-1)(b-1)(c-1)}{abc} ,$$

the number of components of height p in G_n is

$$\frac{(a-1)(b-1)(c-1)n}{c^p abc} + o(n) .$$

Let $M(n) = \frac{1}{2}\lfloor \log_c n \rfloor$. The total number of vertices in a maximum independent set in complete components is therefore

$$\alpha_C(G_n) = \frac{(a-1)(b-1)(c-1)n}{abc} \sum_{i=0}^{M(n)} \left[\frac{i(i+1)}{c^{2i-1}} + \frac{(i+1)^2}{c^{2i}} \right] + o(n) .$$

Thus, the density contribution is

$$\begin{aligned}
\delta_C = \lim_{n\to\infty} \frac{\alpha_C(G_n)}{n} &= \lim_{n\to\infty} \frac{1}{n} \cdot \frac{(a-1)(b-1)(c-1)n}{abc} \sum_{i=0}^{M(n)} \left[\frac{i(i+1)}{c^{2i-1}} + \frac{(i+1)^2}{c^{2i}} \right] \\
&= \frac{(a-1)(b-1)(c-1)}{abc} \sum_{i=0}^{\infty} \left[\frac{i(i+1)}{c^{2i-1}} + \frac{(i+1)^2}{c^{2i}} \right] \\
&= \frac{(a-1)(b-1)(c-1)}{abc} \cdot \frac{c^4}{(c-1)^3(c+1)} \\
&= \frac{(a-1)(b-1)c^3}{ab(c-1)^2(c+1)}.
\end{aligned}$$

3.2. Small Incomplete Components

Now we consider the small incomplete components. Let $C_{p,1}[r]$ be the subgraph of $C_{p,1}$ induced by $[r]$. Define

$$f(p, r) := \alpha(C_{p,1}[r])$$

for $r \in \mathbb{N}$. We can calculate all f for $p \leqslant d$ in $\mathcal{O}(c^d)$ time with a computer, again using Lemma 1. (In fact, these components have bounded size, so any exponential time maximum independent set algorithm runs in $O(1)$ time.) Note that $C_{p,q}[n]$ is a q-copy of $C_{p,1}[\lfloor n/q \rfloor]$, and therefore $\alpha(C_{p,q}[n]) = f(p, \lfloor n/q \rfloor)$. So we can find the size of maximum independent sets in the small components using the f's.

More precisely, given $p \leqslant d$ and n, for how many values of q is $C_{p,q}[n]$ a q-copy of $C_{p,1}[r]$, where $r = \lfloor n/q \rfloor$? First note that

$$\frac{n}{r+1} < q \leqslant \frac{n}{r} .$$

Thus, there are

$$\frac{(a-1)(b-1)(c-1)n}{abc}\left(\frac{1}{r}-\frac{1}{r+1}\right)+o(n)=\frac{(a-1)(b-1)(c-1)n}{abcr(r+1)}+o(n)$$

q-copies of $C_{p,1}[r]$. The only restriction on r is that $a^p \leqslant r \leqslant c^p - 1$. Hence, the size of a maximum independent set in components of type S is

$$\sum_{p=0}^{d}\sum_{r=a^p}^{c^p-1}\frac{(a-1)(b-1)(c-1)n}{abcr(r+1)}f(p,r)+o(n)\ .$$

As $n\to\infty$, the density contribution of small components is therefore

$$\begin{aligned}\delta_S &= \lim_{n\to\infty}\frac{1}{n}\sum_{p=0}^{d}\sum_{r=a^p}^{c^p-1}\frac{(a-1)(b-1)(c-1)n}{abcr(r+1)}f(p,r)\\ &= \sum_{p=0}^{d}\sum_{r=a^p}^{c^p-1}\frac{(a-1)(b-1)(c-1)}{abcr(r+1)}f(p,r).\end{aligned}$$

Since a, b and c are constants and d will be chosen so that it is bounded by a function of a, b and c (see the next section for details), δ_S can be computed in $\mathcal{O}(1)$ time.

3.3. Large Incomplete Components

Finally, we show that we can choose d so that the density of a maximum independent set in components of type L is less than ϵ. For large components,

$$p>d \quad\text{and}\quad a^p q\leqslant n<c^p q\ .$$

The latter implies $c^{-p}n < q \leqslant a^{-p}n$. From the density of q, the number of large incomplete components $L_{p,q}$ for a given $p > d$ is

$$\frac{(a-1)(b-1)(c-1)n}{abc}\left(\frac{1}{a^p}-\frac{1}{c^p}\right)+o(n)\ .$$

Since there are less than p^2 vertices in a component of height p,

$$\begin{aligned}\alpha_L(G_n) &\leqslant \sum_{p=d}^{\infty}p^2\cdot\frac{(a-1)(b-1)(c-1)n}{abc}\left(\frac{1}{a^p}-\frac{1}{c^p}\right)\\ &\leqslant \frac{(a-1)(b-1)(c-1)n}{abc}\sum_{p=d}^{\infty}\frac{p^2}{a^p}\\ &= \frac{(a-1)(b-1)(c-1)n}{abc}\cdot\frac{a^{1-d}((a-1)^2d^2+2(a-1)d+a+1)}{(a-1)^3}\end{aligned}$$

$$\leqslant \frac{(b-1)(c-1)n}{bc} \cdot a^{-d/2}$$

where the last inequality holds for $d \geqslant 22$. Define $\beta := (b-1)(c-1)/bc$. Hence,

$$\delta_L = \lim_{n\to\infty} \frac{\alpha_L(G_n)}{n} \leqslant \lim_{n\to\infty} \frac{1}{n} \cdot \beta n \cdot a^{-d/2} = \beta a^{-d/2} \ .$$

So, to obtain a precision of ϵ in the approximation $\delta \approx \delta_C + \delta_S$, we pick

$$d = \max\{2\log_a(\beta/\epsilon), 22\}$$

which is a function of a, b, c, and ϵ. This completes the proof of Theorem 2.

The following table gives approximate values of δ for small a, b, and c:

a	b	c	δ
2	3	5	0.7292
2	3	7	0.7407
2	5	7	0.8235
2	5	9	0.8187
2	7	9	0.8709
3	4	5	0.7093
3	4	7	0.7934
3	5	7	0.8239
3	5	8	0.8212
3	7	8	0.8727

These results were obtained by incrementing d and looking for convergence to 4 decimal places. We also approximated δ_S using a naive algorithm (based on Lemma 1) for large n. Numerical convergence occurred at values of d slightly lower than the bound given above.

References

[1] Jean-Paul Allouche, André Arnold, Jean Berstel, Srečko Brlek, William Jockusch, Simon Plouffe, and Bruce E. Sagan. A relative of the Thue-Morse sequence. *Discrete Math.*, 139(1–3):455–461, 1995. doi:10.1016/0012-365X(93)00147-W.

[2] Julien Cassaigne and Paul Zimmerman. Numerical evaluation of the strongly triple-free set constant, 1996. http://iml.univ-mrs.fr/~cassaign/publis/cz.ps.gz.

[3] Paul Erdős. On sequences of integers no one of which divides the product of two others and some related problems. *Izvestiya Naustno-Issl. Inst. Mat. i Meh. Tomsk*, 2:74–82, 1938. http://www.renyi.hu/~p_erdos/1938-07.pdf.

[4] Paul Erdős. On some applications of graph theory to number theoretic problems. *Publ. Ramanujan Inst. No.*, 1:131–136, 1968/1969. http://www.renyi.hu/~p_erdos/1968-09.pdf.

[5] Paul Erdős. Some applications of graph theory to number theory. In *The Many Facets of Graph Theory (Proc. Conf., Western Mich. Univ., Kalamazoo, Mich., 1968)*, pp. 77–82. Springer, Berlin, 1969. http://www.renyi.hu/~p_erdos/1969-14.pdf.

[6] Paul Erdős and Pál Turán. On a problem of Sidon in additive number theory, and on some related problems. *J. London Math. Soc.*, 16:212–215, 1941. doi:10.1112/jlms/s1-16.4.212.

[7] Kevin O'Bryant. A complete annotated bibliography of work related to Sidon sequences. *Electron. J. Combin.*, DS11, 2004. http://www.combinatorics.org/ojs/index.php/eljc/article/view/ds11.

[8] Attila Pór and David R. Wood. Colourings of the Cartesian product of graphs and multiplicative Sidon sets. *Combinatorica*, 29(4):449–466, 2009. doi:10.1007/s00493-009-2257-0.

[9] Imre Z. Ruzsa. Erdős and the integers. *J. Number Theory*, 79(1):115–163, 1999. doi:10.1006/jnth.1999.2395.

[10] Imre Z. Ruzsa. Additive and multiplicative Sidon sets. *Acta Math. Hungar.*, 112(4):345–354, 2006. doi:10.1007/s10474-006-0102-0.

[11] András Sárközy. Unsolved problems in number theory. *Period. Math. Hungar.*, 42(1-2):17–35, 2001. doi:10.1023/A:1015236305093.

[12] Simon Sidon. Ein Satz über trigonometrische Polynome und seine Anwendung in der Theorie der Fourier-Reihen. *Math. Ann.*, 106:536–539, 1932. MR:1512772.

[13] James Singer. A theorem in finite projective geometry and some applications to number theory. *Trans. Amer. Math. Soc.*, 43(3):377–385, 1938. doi:10.2307/1990067.

[14] Edward T. H. Wang. On double-free sets of integers. *Ars Combin.*, 28:97–100, 1989. MR:1039135.

SOME WEIGHTED SUMS OF PRODUCTS OF LUCAS SEQUENCES

Emrah Kılıç
Mathematics Department, TOBB University of Economics and Technology, Ankara, Turkey
ekilic@etu.edu.tr

Neşe Ömür
Mathematics Department, Kocaeli University, İzmit Kocaeli, Turkey
neseomur@kocaeli.edu.tr

Received: 8/16/12, Accepted: 4/21/13, Published: 5/10/13

Abstract

In this paper, we consider the weighted sums of products of Lucas sequences of the form

$$\sum_{k=0}^{n} \binom{n}{k} r_{mk} s_{m(tn+k)},$$

where r_n and s_n are the terms of Lucas sequences $\{U_n\}$ and $\{V_n\}$ for some positive integers t and m. By using generating function methods, we compute the weighted sums of products of Lucas sequences and show that these sums could be expressed via terms of the Lucas sequences.

1. Introduction

Define second order linear recurrences $\{U_n\}$ and $\{V_n\}$ for $n > 0$ as

$$\begin{aligned} U_n &= pU_{n-1} + U_{n-2}, \\ V_n &= pV_{n-1} + V_{n-2}, \end{aligned}$$

where $U_0 = 0,\ U_1 = 1$ and $V_0 = 2,\ V_1 = p$, respectively. If $p = 1$, then $U_n = F_n$ (nth Fibonacci number) and $V_n = L_n$ (nth Lucas number).

The Binet formulas of $\{U_n\}$ and $\{V_n\}$ are

$$U_n = \frac{\alpha^n - \beta^n}{\alpha - \beta} \text{ and } V_n = \alpha^n + \beta^n,$$

where $\alpha, \beta = \left(p \pm \sqrt{\Delta}\right)/2$ and $\Delta = p^2 + 4$.

Let $A(x)$ and $B(x)$ be the exponential generating functions of sequences $\{a_n\}$ and $\{b_n\}$, that is,

$$A(x) = \sum_{n\geq 0} a_n \frac{x^n}{n!} \text{ and } B(x) = \sum_{n\geq 0} b_n \frac{x^n}{n!}.$$

Then the convolution of them is given by

$$A(x) B(x) = \sum_{n\geq 0} \left(\sum_{k=0}^{n} \binom{n}{k} a_k b_{n-k} \right) \frac{x^n}{n!}.$$

Many authors have considered and computed many kinds of binomial sums as well as weighted binomial sums with terms of certain number sequences. As a consequence of convolution of two exponential generating functions, we have the following results from the literature (see [2]):

$$\sum_{i=0}^{n} \binom{n}{i} F_{mi} L_{mn-mi}, \quad \sum_{i=0}^{n} \binom{n}{i} F_{mi} F_{mn-mi}, \tag{1.1}$$

$$\sum_{i=0}^{n} \binom{n}{i} L_{mi} F_{mn-mi}, \quad \sum_{i=0}^{n} \binom{n}{i} L_{mi} L_{mn-mi}. \tag{1.2}$$

In this paper, motivated by (1.1) and (1.2), we consider the following new four kinds of weighted binomial sums with the product of terms of the sequences $\{U_n\}$ and $\{V_n\}$:

$$\sum_{i=0}^{n} \binom{n}{i} U_{mi} V_{m(kn+i)}, \quad \sum_{i=0}^{n} \binom{n}{i} V_{mi} U_{m(kn+i)},$$

$$\sum_{i=0}^{n} \binom{n}{i} U_{mi} U_{m(kn+i)}, \quad \sum_{i=0}^{n} \binom{n}{i} V_{mi} V_{m(kn+i)},$$

for some integers k and m. We consider the sums above and then show that the sums could nicely be expressed in terms of the terms of the sequences $\{U_n\}$ and $\{V_n\}$. Because of the indices of the terms in the sums, the convolution of exponential generating functions can not be used for computing these sums. Our approach for computing these kind of sums is mainly to use generating function methods and the Binet formula of the sequences. For computing weighted binomial sums with the product of terms of binary sequences and using generating functions in deriving combinatorial identities, we refer to [1, 3].

2. The Main Results

First we give a useful auxiliary lemma and its direct consequences. After this we give our main results. By the Binet formulas of $\{U_n\}$ and $\{V_n\}$, we have the following results without proof:

Lemma 1. *For odd m,*

$$1+\alpha^{2m}=\alpha^m U_m\sqrt{\Delta} \text{ and } \left(1+\beta^{2m}\right)=-\beta^m U_m\sqrt{\Delta},$$

and for even m,

$$1+\alpha^{2m}=\alpha^m V_m \text{ and } 1+\beta^{2m}=\beta^m V_m.$$

As straightforward consequences of Lemma 1, we have the following results.

Corollary 1. *Let m be a nonnegative odd integer. Then for $n>0$,*

$$\sum_{i=0}^{n}\binom{n}{i}U_{2mi}=U_m^n\begin{cases}\Delta^{\frac{n-1}{2}}V_{mn} & \text{if } n \text{ is odd,}\\ \Delta^{\frac{n}{2}}U_{mn} & \text{if } n \text{ is even.}\end{cases}$$

Proof. By Lemma 1, we write for odd n,

$$\left(1+\alpha^{2m}\right)^n=\alpha^{mn}U_m^n\Delta^{\frac{n}{2}} \text{ and } \left(1+\beta^{2m}\right)^n=-\beta^{mn}U_m^n\Delta^{\frac{n}{2}}$$

or

$$\sum_{i=0}^{n}\binom{n}{i}\alpha^{2mi}=\alpha^{mn}U_m^n\left(p^2+4\right)^{\frac{n}{2}} \text{ and } \sum_{i=0}^{n}\binom{n}{i}\beta^{2mi}=-\beta^{mn}U_m^n\left(p^2+4\right)^{\frac{n}{2}}$$

and so

$$\begin{aligned}\sum_{i=0}^{n}\binom{n}{i}\left(\frac{\alpha^{2mi}-\beta^{2mi}}{\alpha-\beta}\right) &= \Delta^{\frac{n-1}{2}}U_m^n\left(\alpha^{mn}+\beta^{mn}\right)\\ \sum_{i=0}^{n}\binom{n}{i}U_{2mi} &= \Delta^{\frac{n-1}{2}}U_m^nV_{mn},\end{aligned}$$

as claimed. For even n, consider

$$\left(1+\alpha^{2m}\right)^n=\alpha^{mn}U_m^n\Delta^{\frac{n}{2}} \text{and } \left(1+\beta^{2m}\right)^n=\beta^{mn}U_m^n\Delta^{\frac{n}{2}}$$

or

$$\sum_{i=0}^{n}\binom{n}{i}\alpha^{2mi}=\alpha^{mn}U_m^n\Delta^{\frac{n}{2}} \text{ and } \sum_{i=0}^{n}\binom{n}{i}\beta^{2mi}=\beta^{mn}U_m^n\Delta^{\frac{n}{2}}$$

and so

$$\begin{aligned}\sum_{i=0}^{n}\binom{n}{i}\left(\alpha^{2mi}-\beta^{2mi}\right) &= \Delta^{\frac{n}{2}}U_m^n\left(\alpha^{mn}-\beta^{mn}\right)\\ \sum_{i=0}^{n}\binom{n}{i}U_{2mi} &= \Delta^{\frac{n}{2}}U_m^nU_{mn},\end{aligned}$$

as claimed. □

For example, for odd $m > 0$ and $n > 0$, we have

$$\sum_{i=0}^{n} \binom{n}{i} F_{2mi} = F_m^n \begin{cases} 5^{\frac{n-1}{2}} L_{mn} & \text{if } n \text{ is odd,} \\ 5^{\frac{n}{2}} F_{mn} & \text{if } n \text{ is even,} \end{cases}$$

which can be found in [4].

Corollary 2. *Let m be a nonnegative even integer. Then for $n > 0$,*

$$\sum_{i=0}^{n} \binom{n}{i} U_{2mi} = V_m^n U_{mn},$$
$$\sum_{i=0}^{n} \binom{n}{i} V_{2mi} = V_m^n V_{mn}.$$

Proof. By Lemma 1, we have that for even m,

$$1 + \alpha^{2m} = \alpha^m V_m \text{ and } 1 + \beta^{2m} = \beta^m V_m$$

and write

$$\left(1 + \alpha^{2m}\right)^n = \alpha^{mn} V_m^n \text{ and } \left(1 + \beta^{2m}\right)^n = \beta^{mn} V_m^n,$$

which, by the binomial theorem, gives us

$$\sum_{i=0}^{n} \binom{n}{i} \alpha^{2mi} = \alpha^{mn} V_m^n \text{ and } \sum_{i=0}^{n} \binom{n}{i} \beta^{2mi} = \beta^{mn} V_m^n.$$

By subtracting these two equalities side by side and the Binet formula of $\{U_n\}$, we obtain

$$\sum_{i=0}^{n} \binom{n}{i} U_{2mi} = V_m^n U_{mn}.$$

By adding the above two equalities and the Binet formula of $\{V_n\}$, we obtain

$$\sum_{i=0}^{n} \binom{n}{i} V_{2mi} = V_m^n V_{mn},$$

as claimed. □

For even $m > 0$ and $n > 0$, we obtain

$$\sum_{i=0}^{n} \binom{n}{i} F_{2mi} = L_m^n F_{mn} \quad \text{and} \quad \sum_{i=0}^{n} \binom{n}{i} L_{2mi} = L_m^n L_{mn}.$$

Theorem 1. *Let k be a nonnegative integer. For odd m,*

$$\sum_{i=0}^{n}\binom{n}{i}U_{mi}V_{kmn+mi}=\Delta^{\lfloor\frac{n}{2}\rfloor}U_m^n\begin{cases}U_{(k+1)mn} & \text{if } n \text{ is even,}\\ V_{(k+1)mn} & \text{if } n \text{ is odd.}\end{cases} \tag{2.1}$$

For even m,

$$\sum_{i=0}^{n}\binom{n}{i}U_{mi}V_{kmn+mi}=V_m^nU_{mn(k+1)}-2^nU_{kmn}.$$

Proof. Multiplying the left-hand side of (2.1) by z^n and summing over n and by the Binet formulas of $\{U_n\}$ and $\{V_n\}$, we derive for odd m,

$$\begin{aligned}
&\sum_{n\geq 0}z^n\sum_{i=0}^{n}\binom{n}{i}U_{mi}V_{kmn+mi}\\
=\;&\frac{1}{\alpha-\beta}\sum_{n\geq 0}z^n\sum_{i=0}^{n}\binom{n}{i}\left(\left(\alpha^{kmn+2mi}-\beta^{kmn+2mi}\right)-\left(\alpha^{kmn}-\beta^{kmn}\right)(-1)^i\right)\\
=\;&\frac{1}{\alpha-\beta}\left(\sum_{i\geq 0}\alpha^{2mi}\sum_{n\geq 0}\binom{n}{i}\left(\alpha^{km}z\right)^n-\sum_{i\geq 0}\beta^{2mi}\sum_{n\geq 0}\binom{n}{i}\left(\beta^{km}z\right)^n\right)\\
&-\frac{1}{\alpha-\beta}\sum_{n\geq 0}\sum_{i=0}^{n}\binom{n}{i}\left(\alpha^{kmn}-\beta^{kmn}\right)(-1)^{mi}z^n\\
=\;&\frac{1}{\alpha-\beta}\sum_{i\geq 0}\left(\frac{\left(\alpha^{m(k+2)}z\right)^i}{\left(1-\left(\alpha^{km}z\right)\right)^{i+1}}-\frac{\left(\beta^{m(k+2)}z\right)^i}{\left(1-\left(\beta^{km}z\right)\right)^{i+1}}\right)\\
&-\frac{1}{\alpha-\beta}\left(\sum_{i\geq 0}\frac{\left((-1)^m\alpha^{km}z\right)^i}{\left(1-\left(\alpha^{km}z\right)\right)^{i+1}}-\sum_{i\geq 0}\frac{\left((-1)^m\beta^{km}z\right)^i}{\left(1-\left(\beta^{km}z\right)\right)^{i+1}}\right)\\
=\;&\frac{1}{\alpha-\beta}\left(\frac{1}{\left(1-\alpha^{km}z\right)}\frac{1}{1-\frac{\alpha^{m(k+2)}z}{1-\alpha^{km}z}}-\frac{1}{\left(1-\beta^{km}z\right)}\frac{1}{1-\frac{\beta^{m(k+2)}z}{1-\beta^{km}z}}\right)\\
&-\frac{1}{\alpha-\beta}\left(\frac{1}{1-\alpha^{km}\left(1+(-1)^m\right)z}-\frac{1}{1-\beta^{km}\left(1+(-1)^m\right)z}\right)\\
=\;&\frac{1}{\alpha-\beta}\left(\frac{1}{1-z\alpha^{km}\left(1+\alpha^{2m}\right)}-\frac{1}{1-z\beta^{km}\left(1+\beta^{2m}\right)}\right)-\frac{1}{\alpha-\beta}(1-1)\\
=\;&\frac{1}{\alpha-\beta}\sum_{n\geq 0}\left(\alpha^{kmn}\left(1+\alpha^{2m}\right)^n-\beta^{kmn}\left(1+\beta^{2m}\right)^n\right)z^n.
\end{aligned}$$

Therefore, we get the identity

$$\sum_{i=0}^{n}\binom{n}{i}U_{mi}V_{kmn+mi}=\sum_{i=0}^{n}\binom{n}{i}U_{2mi+kmn}.$$

Using Lemma 1, we write for even n,

$$\begin{aligned}
& \frac{1}{\alpha-\beta}\left(\alpha^{kmn}\left(1+\alpha^{2m}\right)^n-\beta^{kmn}\left(1+\beta^{2m}\right)^n\right) \\
= & \frac{1}{\alpha-\beta}\left(\alpha^{kmn}\left(\alpha^m U_m\sqrt{\Delta}\right)^n-\beta^{kmn}\left(-\beta^m U_m\sqrt{\Delta}\right)^n\right) \\
= & \frac{1}{\alpha-\beta}\left(\alpha^{kmn}\alpha^{mn}U_m^n\Delta^{\frac{n}{2}}-\beta^{kmn}\beta^{mn}U_m^n\Delta^{\frac{n}{2}}\right) \\
= & U_m^n\Delta^{\frac{n}{2}}U_{(k+1)mn}.
\end{aligned}$$

On the other hand, we get for odd n

$$\begin{aligned}
& \frac{1}{\alpha-\beta}\left(\alpha^{kmn}\left(1+\alpha^{2m}\right)^n-\beta^{kmn}\left(1+\beta^{2m}\right)^n\right) \\
= & \frac{1}{\alpha-\beta}\left(\alpha^{kmn}\left(\alpha^m U_m\sqrt{\Delta}\right)^n-\beta^{kmn}\left(-\beta^m U_m\sqrt{\Delta}\right)^n\right) \\
= & \frac{1}{\sqrt{p^2+4}}\left(\alpha^{kmn}\alpha^{mn}U_m^n\Delta^{\frac{n}{2}}+\beta^{kmn}\beta^{mn}U_m^n\Delta^{\frac{n}{2}}\right) \\
= & \alpha^{kmn}\alpha^{mn}U_m^n\Delta^{\frac{n-1}{2}}+\beta^{kmn}\beta^{mn}U_m^n\Delta^{\frac{n-1}{2}} \\
= & U_m^n\Delta^{\frac{n-1}{2}}V_{(k+1)mn},
\end{aligned}$$

as claimed. By combining the above two results, the proof is complete for the case m is odd.

Now we consider the case m is even:

$$\begin{aligned}
& \frac{1}{\alpha-\beta}\left(\sum_{i\geq 0}\alpha^{2mi}\sum_{n\geq 0}\binom{n}{i}\left(\alpha^{km}z\right)^n-\sum_{i\geq 0}\beta^{2mi}\sum_{n\geq 0}\binom{n}{i}\left(\beta^{km}z\right)^n\right) \\
& -\frac{1}{\alpha-\beta}\sum_{n\geq 0}\sum_{i=0}^{n}\binom{n}{i}\left(\alpha^{kmn}-\beta^{kmn}\right)z^n \\
= & \frac{1}{\alpha-\beta}\sum_{i\geq 0}\left(\frac{\left(\alpha^{m(k+2)}z\right)^i}{\left(1-\left(\alpha^{km}z\right)\right)^{i+1}}-\frac{\left(\beta^{m(k+2)}z\right)^i}{\left(1-\left(\beta^{km}z\right)\right)^{i+1}}\right) \\
& -\frac{1}{\alpha-\beta}\left(\sum_{i\geq 0}\frac{\left(\alpha^{km}z\right)^i}{\left(1-\left(\alpha^{km}z\right)\right)^{i+1}}-\sum_{i\geq 0}\frac{\left(\beta^{km}z\right)^i}{\left(1-\left(\beta^{km}z\right)\right)^{i+1}}\right) \\
= & \frac{1}{\alpha-\beta}\left(\frac{1}{\left(1-\alpha^{km}z\right)}\frac{1}{1-\frac{\alpha^{m(k+2)}z}{1-\alpha^{km}z}}-\frac{1}{\left(1-\beta^{km}z\right)}\frac{1}{1-\frac{\beta^{m(k+2)}z}{1-\beta^{km}z}}\right) \\
& -\frac{1}{\alpha-\beta}\left(\frac{1}{1-2\alpha^{km}z}-\frac{1}{1-2\beta^{km}z}\right)
\end{aligned}$$

$$
\begin{aligned}
&= \frac{1}{\alpha-\beta}\left(\frac{1}{1-z\alpha^{km}\left(1+\alpha^{2m}\right)}-\frac{1}{1-z\beta^{km}\left(1+\beta^{2m}\right)}\right)\\
&\quad -\frac{1}{\alpha-\beta}\left(\frac{1}{1-2\alpha^{km}z}-\frac{1}{1-2\beta^{km}z}\right)\\
&= \frac{1}{\alpha-\beta}\sum_{n\geq 0}\left(\alpha^{kmn}\left(1+\alpha^{2m}\right)^n-\beta^{kmn}\left(1+\beta^{2m}\right)^n-2^n\left(\alpha^{kmn}-\beta^{kmn}\right)\right)z^n
\end{aligned}
$$

By Lemma 1, we write

$$
\begin{aligned}
&\frac{1}{\alpha-\beta}\left(\alpha^{kmn}\left(\alpha^m V_m\right)^n-\beta^{kmn}\left(\beta^m V_m\right)^n-2^n\left(\alpha^{kmn}-\beta^{kmn}\right)\right)\\
=\ &\frac{1}{\alpha-\beta}\left(V_m^n\left(\alpha^{mn(k+1)}-\beta^{mn(k+1)}\right)-2^n\left(\alpha^{kmn}-\beta^{kmn}\right)\right)\\
=\ &V_m^n U_{mn(k+1)}-2^n U_{kmn},
\end{aligned}
$$

as claimed. □

Theorem 2. *Let k be a nonnegative integer. For odd m,*

$$
\sum_{i=0}^{n}\binom{n}{i}V_{mi}V_{kmn+mi}=\Delta^{\lfloor\frac{n+1}{2}\rfloor}U_m^n\begin{cases}V_{(k+1)mn} & \text{if } n \text{ is even,}\\ U_{(k+1)mn} & \text{if } n \text{ is odd.}\end{cases} \tag{2.2}
$$

For even m,

$$
\sum_{i=0}^{n}\binom{n}{i}V_{mi}V_{kmn+mi}=V_m^n V_{(k+1)mn}+2^n V_{kmn}.
$$

Proof. Multiplying the left-hand side of (2.2) by z^n and summing over n, we write

$$
\begin{aligned}
&\sum_{n\geq 0}\sum_{i=0}^{n}\binom{n}{i}V_{mi}V_{kmn+mi}z^n\\
=\ &\sum_{n\geq 0}\sum_{i=0}^{n}\binom{n}{i}\left(\alpha^{mi}+\beta^{mi}\right)\left(\alpha^{kmn+mi}+\beta^{kmn+mi}\right)z^n\\
=\ &\sum_{n\geq 0}\sum_{i=0}^{n}\binom{n}{i}\left(\alpha^{kmn+2im}+\beta^{kmn+2mi}+(-1)^{im}\left(\alpha^{kmn}+\beta^{kmn}\right)\right)z^n\\
=\ &\sum_{i\geq 0}\alpha^{2mi}\sum_{n\geq 0}\binom{n}{i}\left(\alpha^{km}z\right)^n+\sum_{i\geq 0}\beta^{2mi}\sum_{n\geq 0}\binom{n}{i}\left(\beta^{km}z\right)^n\\
&+\sum_{i\geq 0}(-1)^{im}\sum_{n\geq 0}\binom{n}{i}\left(\alpha^{kmn}+\beta^{kmn}\right)z^n\\
=\ &\sum_{i\geq 0}\left(\frac{\left(\alpha^{km+2m}z\right)^i}{\left(1-\alpha^{km}z\right)^{i+1}}+\frac{\left(\beta^{km+2m}z\right)^i}{\left(1-\beta^{km}z\right)^{i+1}}\right)
\end{aligned}
$$

$$
\begin{aligned}
& +\sum_{i\geq 0}\left(\frac{\left((-1)^m\alpha^{km}z\right)^i}{\left(1-\alpha^{km}z\right)^{i+1}}+\frac{\left((-1)^m\beta^{km}z\right)^i}{\left(1-\beta^{km}z\right)^{i+1}}\right) \\
= & \frac{1}{1-z\alpha^{km}\left(1+\alpha^{2m}\right)}+\frac{1}{1-z\beta^{km}\left(1+\beta^{2m}\right)} \\
& +\frac{1}{1-z\alpha^{km}\left(1+(-1)^m\right)}+\frac{1}{1-z\beta^{km}\left(1+(-1)^m\right)}.
\end{aligned}
$$

If m is even, then by Lemma 1, we write

$$
\begin{aligned}
& \sum_{n\geq 0}\sum_{i=0}^{n}\binom{n}{i}V_{mi}V_{kmn+mi}z^n \\
= & \frac{1}{1-z\alpha^{m(k+1)}V_m}+\frac{1}{1-z\beta^{m(k+1)}V_m}+\frac{1}{1-2z\alpha^{km}}+\frac{1}{1-2z\beta^{km}} \\
= & \sum_{n\geq 0}\left(\left(\alpha^{m(k+1)n}+\beta^{m(k+1)n}\right)V_m^n+2^n\left(\alpha^{kmn}+\beta^{kmn}\right)\right)z^n
\end{aligned}
$$

which gives us

$$
\sum_{i=0}^{n}\binom{n}{i}V_{mi}V_{kmn+mi}=V_m^nV_{(k+1)mn}+2^nV_{kmn}.
$$

If m is odd, then

$$
\begin{aligned}
& \sum_{n\geq 0}\sum_{i=0}^{n}\binom{n}{i}V_{mi}V_{kmn+mi}z^n \\
= & \frac{1}{1-z\alpha^{km}\left(\alpha^mU_m\sqrt{p^2+4}\right)}+\frac{1}{1-z\beta^{km}\left(-\beta^mU_m\sqrt{p^2+4}\right)} \\
= & \sum_{n\geq 0}\left(\left(\alpha^{(k+1)m}U_m\sqrt{p^2+4}\right)^n+\left(-\beta^{m(k+1)}U_m\sqrt{p^2+4}\right)^n\right)z^n.
\end{aligned}
$$

Now we consider two cases: first if n is odd, then we obtain

$$
\begin{aligned}
& \sum_{n\geq 0}\sum_{i=0}^{n}\binom{n}{i}V_{mi}V_{kmn+mi}z^n \\
= & \sum_{n\geq 0}\left(\alpha^{(k+1)mn}U_m^n\left(p^2+4\right)^{\frac{n}{2}}-\beta^{m(k+1)n}U_m^n\left(p^2+4\right)^{\frac{n}{2}}\right)z^n \\
= & \sum_{n\geq 0}\left(U_m^n\Delta^{\frac{n}{2}}\left(\alpha^{(k+1)mn}-\beta^{m(k+1)n}\right)\right)z^n \\
= & \sum_{n\geq 0}\left(U_m^n\Delta^{\frac{n-1}{2}}\left(\frac{\alpha^{(k+1)mn}-\beta^{m(k+1)n}}{\alpha-\beta}\right)\right)z^n \\
= & \sum_{n\geq 0}\left(U_m^n\Delta^{\frac{n-1}{2}}U_{(k+1)mn}\right)z^n.
\end{aligned}
$$

Second, if n is even, then we obtain

$$\begin{aligned}\sum_{n\geq 0}\sum_{i=0}^{n}\binom{n}{i}V_{mi}V_{kmn+mi}z^n &= \sum_{n\geq 0}\left(\alpha^{(k+1)mn}U_m^n\Delta^{\frac{n}{2}}+\beta^{(k+1)mn}U_m^n\Delta^{\frac{n}{2}}\right)z^n\\ &= \sum_{n\geq 0}\left(U_m^n\left(p^2+4\right)^{\frac{n}{2}}\left(\alpha^{(k+1)mn}+\beta^{(k+1)mn}\right)\right)z^n\\ &= \sum_{n\geq 0}U_m^n\Delta^{\frac{n}{2}}V_{(k+1)mn}z^n.\end{aligned}$$

By combining the last two results, we prove the claim for odd m. Thus the proof is complete. □

Similar to the proof methods of Theorems 1 and 2, we give the following results without proof.

Theorem 3. *Let k be a nonnegative integer. For odd m,*

$$\sum_{i=0}^{n}\binom{n}{i}V_{mi}U_{kmn+mi}=\Delta^{\lfloor\frac{n}{2}\rfloor}U_m^n\begin{cases}U_{(k+1)mn} & \text{if } n \text{ is even,}\\ V_{(k+1)mn} & \text{if } n \text{ is odd.}\end{cases}$$

For even m,

$$\sum_{i=0}^{n}\binom{n}{i}V_{mi}U_{kmn+mi}=V_m^nU_{mn(k+1)}+2^nU_{kmn}$$

Theorem 4. *Let k be a nonnegative integer. For odd m,*

$$\sum_{i=0}^{n}\binom{n}{i}U_{mi}U_{kmn+mi}=\Delta^{\lfloor\frac{n-1}{2}\rfloor}U_m^n\begin{cases}V_{(k+1)mn} & \text{if } n \text{ is even,}\\ U_{(k+1)mn} & \text{if } n \text{ is odd.}\end{cases}$$

For even m,

$$\sum_{i=0}^{n}\binom{n}{i}U_{mi}U_{kmn+mi}=\frac{1}{\Delta}\left(V_m^nV_{(k+1)mn}-2^nV_{kmn}\right).$$

References

[1] E. Kılıç, Y. Ulutas and N. Omur, Formulas for weighted binomial sums with the powers of terms of binary recurrences, *Miskolc Math. Notes* **13 (1)** (2012), 53-65.

[2] T. Koshy, Fibonacci and Lucas numbers with applications. Pure and Applied Mathematics, Wiley-Interscience, New York, 2001.

[3] P. Stanica, Generating functions, weighted and non-weighted sums for powers of second-order recurrence sequences, *Fibonacci Quarterly* **41.4** (2003), 321-333.

[4] S. Vajda, Fibonacci & Lucas numbers, and the golden section. John Wiley & Sons, Inc., New York, 1989.

INTEGERS: 13 (2013) #A28

ADDITIVE ENERGY AND THE FALCONER DISTANCE PROBLEM IN FINITE FIELDS

Doowon Koh
Department of Mathematics, Chungbuk National University, Cheongju city, Chungbuk-Do, Korea
koh131@chungbuk.ac.kr

Chun-Yen Shen
Department of Mathematics, Michigan State University, East Lansing, Michigan
shenc@umail.iu.edu

Received: 6/13/12, Accepted: 3/9/13, Published: 5/10/13

Abstract

We study the number of the vectors determined by two sets in d-dimensional vector spaces over finite fields. We observe that the lower bound of cardinality for the set of vectors can be given in view of an additive energy or the decay of the Fourier transform on given sets. As an application of our observation, we find sufficient conditions on sets where the Falconer distance conjecture for finite fields holds in two dimensions. Moreover, we give an alternative proof of the theorem, due to Iosevich and Rudnev, that any Salem set satisfies the Falconer distance conjecture for finite fields.

1. Introduction

Let $\mathbb{F}_q^d, d \geq 1$, be a d-dimensional vector space over the finite field $\mathbb{F}_q$ with q elements. Given $A, B \subset \mathbb{F}_q^d$, one may ask what is the cardinality of the set $A - B$, where the difference set $A - B$ is defined by

$$A - B = \{x - y \in \mathbb{F}_q^d : x \in A, y \in B\}.$$

It is clear that $|A-B| \geq \max\{|A|, |B|\}$. Here, and throughout the paper, we denote by $|E|$ the cardinality of the set E. However, taking $A = B = \mathbb{F}_q^s, 1 \leq s \leq d$, shows that the trivial estimate for $|A-B|$ is sharp in general, because $|A-B| = |\mathbb{F}_q^s| = q^s$. Moreover, if $s = d-1$, then the size of $A - B$ is much smaller than that of $\mathbb{F}_q^d$, although $|A||B| = q^{2d-2}$ is somewhat big. Therefore, it may be interesting to find some conditions on the sets $A, B \subset \mathbb{F}_q^d$ such that the cardinality of $A - B$ is much bigger than the trivial lower bound, $\max\{|A|, |B|\}$, of $|A - B|$, or the difference set

$A-B$ contains a positive proportion of all vectors in $\mathbb{F}_q^d$, that is $|A-B| \gtrsim |\mathbb{F}_q^d| = q^d$. Here, we recall that for $l, m > 0$, the expression $l \gtrsim m$ or $m \lesssim l$ means that there exists a constant $C > 0$ independent of q, the size of the underlying finite field $\mathbb{F}_q$, such that $Cl \geq m$. The problem to consider the size of difference sets is strongly motivated by the Falconer distance problem for finite fields, which was introduced by Iosevich and Rudnev [9]. In this paper, we shall make an effort to find the connection between the size of the difference set $A - B$ and the cardinality of the distance set determined by $A, B \subset \mathbb{F}_q^d$. As one of the main results, we shall give some examples for sets satisfying the Falconer distance conjecture for finite fields.

First, let us review the Falconer distance problem for the Euclidean case and the finite field case. In the Euclidean setting, the Falconer distance problem is to determine the Hausdorff dimensions of compact sets $E, F \subset \mathbb{R}^d, d \geq 2$, such that the Lebesgue measure of the distance set

$$\Delta(E, F) := \{|x - y| : x \in E, y \in F\}$$

is positive. In the case when $E = F$, Falconer [4] first addressed this problem and showed that if the Hausdorff dimension of the compact set E is greater than $(d + 1)/2$, then the Lebesgue measure of $\Delta(E, E)$ is positive. He also conjectured that every compact set with the Hausdorff dimension $> d/2$ yields a distance set with a positive Lebesgue measure. This is known as the Falconer distance conjecture which has not been solved in all dimensions. The best known result for this problem is due to Wolff [17] in two dimensions and Erdoğan [3] in all other dimensions. They proved that if the Hausdorff dimension of any compact set $E \subset \mathbb{R}^d$ is greater than $d/2 + 1/3$, then the Lebesgue measure of $\Delta(E, E)$ is positive. These results are a culmination of efforts going back to Falconer [4] in 1985 and Mattila [13] a few years later. The Falconer distance problem on generalized distances was also studied in [1], [6], [7], [8], and [10]. In addition, the chapter 13 in [14] is highly relevant to the topic we study in this paper.

In the Finite field setting, one can also study the Falconer distance problem. Given $A, B \subset \mathbb{F}_q^d, d \geq 2$, the distance set $\Delta(A, B)$ is given by

$$\Delta(A, B) = \{\|x - y\| \in \mathbb{F}_q : x \in A, y \in B\},$$

where $\|\alpha\| = \alpha_1^2 + \cdots + \alpha_d^2$ for $\alpha = (\alpha_1, \ldots, \alpha_d) \in \mathbb{F}_q^d$. It is clear that $|\Delta(A, B)| \leq q$, because the distance set is a subset of the finite field with q elements. In this setting, the Falconer distance problem is to determine the minimum value of $|A||B|$ such that $|\Delta(A, B)| \gtrsim q$. In the case when $A = B$, this problem was introduced by Iosevich and Rudnev [9] and they proved that if $A = B$ and $|A| \gtrsim q^{(d+1)/2}$, then $|\Delta(A, B)| \gtrsim q$. It turned out in [5] that if the dimension d is odd, then the theorem due to Alex and Rudnev gives the best possible result on the Falconer distance problem for finite fields. However, if the dimension d is even, then it has been believed that the aforementioned authors' result may be improved to the following conjecture.

Conjecture 1.1 (Iosevich and Rudnev [9]). *Let $K \subset \mathbb{F}_q^d$ with $d \geq 2$ even. If $|K| \geq Cq^{\frac{d}{2}}$, with $C > 0$ sufficiently large, then*

$$|\Delta(K,K)| \gtrsim q.$$

This conjecture has not been solved in all dimensions. The exponent $(d+1)/2$ obtained by Iosevich and Rudnev is currently the best known result for all dimensions except dimension two. In two dimensions, this exponent was improved by $4/3$ (see [2] or [11]). We may consider the following general version of Conjecture 1.1:

Conjecture 1.2. *Let $A, B \subset \mathbb{F}_q^d$ with $d \geq 2$ even. If $|A||B| \geq Cq^d$, with $C > 0$ large enough, then*

$$|\Delta(A,B)| \gtrsim q.$$

Theorem 2.1 in [15] due to Shparlinski implies that if $A, B \subset \mathbb{F}_q^d, d \geq 2$, and $|A||B| \gtrsim q^{d+1}$, then $|\Delta(A,B)| \gtrsim q$. This was improved by authors [11] who showed that if $|A||B| \gtrsim q^{8/3}$ for $A, B \subset \mathbb{F}_q^2$, then $|\Delta(A,B)| \gtrsim q$. For a variant of the Falconer distance problem for finite fields, see [16] and [12].

1.1. Purpose of This Paper

The goal of this paper is to find some sets $A, B \subset \mathbb{F}_q^d, d \geq 2$, for which Conjecture 1.2 holds. In general, it may not be easy to construct such examples, supporting the claim that Conjecture 1.2 holds. A well-known example is due to Iosevich and Rudnev [9] who showed that if $K \subset \mathbb{F}_q^d, d \geq 2$, is a Salem set and $|K| \gtrsim q^{d/2}$, then $|\Delta(K,K)| \gtrsim q$. Here, we recall that we say that $E \subset \mathbb{F}_q^d$ is a Salem set if for every $m \in \mathbb{F}_q^d \setminus \{(0,\ldots,0)\}$,

$$|\widehat{E}(m)| := \left| q^{-d} \sum_{x \in E} \chi(-x \cdot m) \right| \lesssim \frac{\sqrt{E}}{q^d},$$

where we denote by χ a nontrivial additive character of $\mathbb{F}_q$. They obtained this example by showing that the formula of $|\Delta(K,K)|$ is closely related to the decay of the Fourier transform on the set K. In this paper, we take a new approach to find such examples. First, we shall show that if $A, B \subset \mathbb{F}_q^d, d \geq 2$ and $|A - B| \gtrsim q^d$, then $|\Delta(A,B)| \gtrsim q$. Second, we find certain conditions on the set $A, B \subset \mathbb{F}_q^d$ such that $|A - B| \sim q^d$. Thus, estimating the size of the difference set $A - B$ makes an important role. For example, using our approach we can recover the example by Iosevich and Rudnev. Moreover, we can find a stronger result that if one of $A, B \subset \mathbb{F}_q^d$ is a Salem set and $|A||B| \gtrsim q^d$, then $A - B$ contains a positive proportion of all elements in $\mathbb{F}_q^d$. In particular, our method yields that if one of $A, B \subset \mathbb{F}_q^2$ intersects with $\sim q$ points in an algebraic curve which does not contain any line, and $|A||B| \gtrsim q^2$, then the sets A, B satisfies Conjecture 1.2 in two dimensions.

2. Cardinality of Difference Sets

In this section we introduce the formulas for the lower bound of difference sets. Such formulas are closely related to the additive energy

$$\Lambda(A,B) = |\{(x,y,z,w) \in A \times A \times B \times B : x - y + z - w = 0\}|.$$

To see this, for $c \in \mathbb{F}_q^d$, define

$$\begin{aligned}\nu(c) &= |\{(x,w) \in A \times B : x - w = c\}| \\ &= \sum_{\substack{x\in A, w \in B \\ : x-w=c}} 1.\end{aligned}$$

Applying the Cauchy-Schwarz inequality shows that if $A, B \subset \mathbb{F}_q^d, d \geq 2$, then

$$|A|^2|B|^2 = \left(\sum_{c \in A-B} \nu(c)\right)^2 \leq |A-B| \sum_{c \in \mathbb{F}_q^d} \nu^2(c).$$

Now, we observe that

$$\sum_{c\in\mathbb{F}_q^d} \nu^2(c) = \sum_{c\in\mathbb{F}_q^d} \left(\sum_{\substack{x\in A, w\in B \\ :x-w=c}} 1\right)\left(\sum_{\substack{y\in A, z\in B \\ :y-z=c}} 1\right) = \sum_{\substack{x,y\in A, w,z\in B \\ :x-w=y-z}} 1 = \Lambda(A,B).$$

It follows that

$$|A-B| \geq \frac{|A|^2|B|^2}{\Lambda(A,B)}. \tag{2.1}$$

Since $\Lambda(A,B) \leq \min\{|A|^2|B|, |A||B|^2\}$, it is clear that

$$|A-B| \geq \max\{|A|,|B|\},$$

which is in fact a trivial lower bound of $|A-B|$. However, if A and B are subspaces with $A = B$, then the trivial lower bound can not be improved. In this case, the difference set $A-B$ has much smaller cardinality than $|A||B|$. It therefore is natural to guess that if A and B do not contain big subspaces, then $|A-B|$ should be large. In this paper, we shall deal with this issue.

Recall that the Fourier transform on the set $E \subset \mathbb{F}_q^d$ is defined by

$$\widehat{E}(m) = \frac{1}{q^d} \sum_{x \in E} \chi(-x \cdot m) \quad \text{for} \quad m \in \mathbb{F}_q^d,$$

where χ denotes a nontrivial additive character of $\mathbb{F}_q$ and we write E for the characteristic function on the set E. Also recall that the orthogonality relation of the nontrivial additive character of $\mathbb{F}_q$ says that

$$\sum_{t \in \mathbb{F}_q} \chi(a \cdot t) = 0 \quad \text{for} \quad a \neq 0.$$

More generally we have

$$\sum_{x \in \mathbb{F}_q^d} \chi(m \cdot x) = 0 \quad \text{for} \quad m \neq (0, \cdots, 0).$$

The lower bound of $|A - B|$ can be also written in terms of the Fourier transforms on A and B. To see this, using the definition of the Fourier transform and the orthogonality relation of the nontrivial additive character of $\mathbb{F}_q$, observe that

$$\begin{aligned}
\Lambda(A, B) &= |\{(x, y, z, w) \in A \times A \times B \times B : x - y + z - w = 0\}| \\
&= \sum_{x,y,z,w \in \mathbb{F}_q^d} A(x)A(y)B(z)B(w)\delta_0(x - y + z - w) \\
&= \sum_{x,y,z,w \in \mathbb{F}_q^d} A(x)A(y)B(z)B(w) \left(q^{-d} \sum_{m \in \mathbb{F}_q^d} \chi(m \cdot (x - y + z - w)) \right) \\
&= q^{3d} \sum_{m \in \mathbb{F}_q^d} |\widehat{A}(m)|^2 |\widehat{B}(m)|^2,
\end{aligned}$$

where $\delta_0(\alpha) = 0$ for $\alpha \neq (0, \ldots, 0)$ and $\delta_0(\alpha) = 1$ for $\alpha = (0, \ldots, 0)$. Therefore, the formula (2.1) can be replaced by

$$|A - B| \geq \frac{|A|^2 |B|^2}{q^{3d} \sum_{m \in \mathbb{F}_q^d} |\widehat{A}(m)|^2 |\widehat{B}(m)|^2}. \tag{2.2}$$

This formula indicates that if the Fourier decay on A or B is good, then several kinds of vectors are contained in the difference set $A - B$. For example, if A or B is a Salem set such as the paraboloid or the sphere, then $|A - B|$ can be big and so a lot of distances can be determined by A, B.

3. Sets in $\mathbb{F}_q^2$ Satisfying the Falconer Distance Conjecture

In view of the sizes of difference sets, we shall find some sets $A, B \subset \mathbb{F}_q^2$ where the Falconer distance conjecture (Conjecture 1.2) holds. A core idea is due to the following fact.

Lemma 3.1. *Let $E \subset \mathbb{F}_q^2$. If $|E| \geq cq^2$ for some $0 < c \leq 1$, then we have*

$$|\{\|x\| \in \mathbb{F}_q : x \in E\}| \geq \frac{cq}{2},$$

where $\|x\| = x_1^2 + x_2^2$ for $x = (x_1, x_2) \in \mathbb{F}_q^2$. *Proof.* For each $a \in \mathbb{F}_q$, consider a vertical line $L_a = \{(a,t) \in \mathbb{F}_q^2 : t \in \mathbb{F}_q\}$. Since $|E| \geq cq^2$, it is clear from the pigeonhole principle that there exists a line L_b for some $b \in \mathbb{F}_q$ with $|E \cap L_b| \geq cq$. Thus, Lemma follows from the following observation that for the fixed $b \in \mathbb{F}_q$,

$$|\{b^2 + t^2 \in \mathbb{F}_q : (b,t) \in E \cap L_b\}| \geq \frac{cq}{2}.$$

□

If $|A - B| \gtrsim |A||B| \gtrsim q^2$, then Lemma 3.1 implies that $A, B \subset \mathbb{F}_q^2$ are the sets to satisfy the Falconer conjecture. Thus, the main task is to find sets A, B such that $|A - B|$ is extremely large. The following lemma tells us some properties of sets A, B where the size of $A - B$ can be large.

Lemma 3.2. *Let $B \subset \mathbb{F}_q^2$. Suppose that there exists a set $W \subset \mathbb{F}_q^2$ with $|W| \sim 1$ such that*

$$|B \cap (B + c)| \lesssim 1 \quad \text{for all} \quad c \in \mathbb{F}_q^2 \setminus W. \tag{3.1}$$

Then for any $A \subset \mathbb{F}_q^2$, we have

$$|A - B| \gtrsim \min(|A||B|, |B|^2).$$

Proof. From (2.1), it suffices to show that

$$\Lambda(A,B) = |\{(x,y,z,w) \in A \times A \times B \times B : x - y + z - w = 0\}| \lesssim |A||B| + |A|^2.$$

It follows that

$$\Lambda(A,B) = \sum_{x,y \in A} \left(\sum_{w,z \in B : z - w = y - x} 1 \right) = \sum_{x,y \in A} |B \cap (B + y - x)|$$

$$= \sum_{x,y \in A : y - x \notin W} |B \cap (B + y - x)| + \sum_{x,y \in A : y - x \in W} |B \cap (B + y - x)|$$

$$= \mathrm{I} + \mathrm{II}.$$

From the assumption (3.1), it is clear that $|\mathrm{I}| \lesssim |A|^2$. On the other hand, the value II can be estimated as follows.

$$\mathrm{II} = \sum_{\beta \in W} \sum_{x,y \in A : y - x = \beta} |B \cap (B + \beta)| \leq \sum_{\beta \in W} \sum_{x,y \in A : y - x = \beta} |B|.$$

Whenever we fix $x \in A$ and $\beta \in W$, there is at most one $y \in A$ such that $y - x = \beta$. We therefore see

$$\mathrm{II} \leq |W||A||B| \sim |A||B|.$$

Thus, we complete the proof. □

3.1. Examples of the Falconer Conjecture Sets in Two Dimensions

First recall that Bezout's theorem says that two algebraic curves of degrees d_1 and d_2 can not meet in more than $d_1 \cdot d_2$ points unless they have a component in common. As a direct application of Bezout's theorem, it can be shown that subsets of certain algebraic curves in two dimensions satisfy the condition in (3.1). This observation yields the following theorem.

Theorem 3.3. *Let $P(x) \in \mathbb{F}_q[x_1, x_2]$ be a polynomial which does not have any liner factor. Define an algebraic variety $V = \{x \in \mathbb{F}_q^2 : P(x) = 0\}$. If $B \subset V$, then for any $A \subset \mathbb{F}_q^2$, we have*

$$|A - B| \gtrsim \min(|A||B|, |B|^2).$$

Proof. First recall that we always assume that the degree of the polynomial is ~ 1. Thus, if $B \subset V$, then the pigeonhole principle implies that we can choose a subvariety V' of V and a set $B' \subset V'$ with $|B'| \sim |B|$. Therefore, we may assume that V is a variety generated by an irreducible polynomial with degree $k \geq 2$. Applying Bezout's theorem shows that for any $c \in \mathbb{F}_q^2 \setminus \{(0,0)\}$,

$$|V \cap (V + c)| \leq k^2 \lesssim 1.$$

Therefore, the proof is complete from Lemma 3.2. □

The following corollary follows immediately from Lemma 3.2 and Lemma 3.1.

Corollary 3.4. *Let $B \subset \mathbb{F}_q^2$ with $|B| \gtrsim q$. Suppose that $W \subset \mathbb{F}_q^2$ with $|W| \sim 1$, and $|B \cap (B + c)| \lesssim 1$ for any $c \in \mathbb{F}_q^2 \setminus W$. Then for any $A \subset \mathbb{F}_q^2$ with $|A| \gtrsim q$, we have*

$$|\Delta(A, B)| = |\{||x - y|| \in \mathbb{F}_q : x \in A, y \in B\}| \gtrsim q.$$

Notice that such sets A, B as in this corollary satisfy the Falconer distance conjecture. Moreover, the difference set $A - B$ contains a positive proportion of all elements in $\mathbb{F}_q^2$. As a consequence of Theorem 3.3 and Corollary 3.4, more concrete examples for the Falconer distance conjecture sets can be found.

Example 3.5. Let $A = \{(x_1, x_2) : x_1^2 + x_2^2 = a\}$ for some $a \neq 0$ as considered in [9]. Then it can be checked that $|A| \sim q$. From Theorem 3.3, this implies the difference set $A - A$ has cardinality $\sim q^2$ which in turns sharpens the result in [9]. In general, we may choose any polynomial $P \in \mathbb{F}_q[x_1, x_2]$ which does not contain any linear factor, and define a variety $V = \{x \in \mathbb{F}_q : P(x) = 0\}$. If $|V| \gtrsim q$, then choose a subset $B \subset V$ with $|B| \sim q$. Finally, choose any subset A of $\mathbb{F}_q^2$, whose cardinality is $\sim q$. Then the difference set $A - B$ contains the positive proportion of all elements in $\mathbb{F}_q^2$ and so $|\Delta(A, B)| \sim q$. Since $|A||B| \sim q^2$, the sets A, B satisfy the Falconer distance conjecture.

Observe that if both A and B contain many points in some lines L_1, L_2, respectively, then we can not proceed such steps as in the above example. For this reason, if sets A, B possess the structures like product sets, then it seems that two sets A, B determine the distance set $\Delta(A, B)$ with a small cardinality.

4. Salem Sets and Difference Sets

If the decay of the Fourier transform on $A, B \subset \mathbb{F}_q^d$ is known, then the formula (2.2) can be very useful to measure the lower bound of $|A - B|$. Here, we shall show that if one of A and B is a Salem set, then $|A - B|$ is so big that A, B satisfy the Falconer distance conjecture. We need the following lemma which shows the relation between the Fourier decay of sets and the size of difference sets.

Lemma 4.1. *Let $A, B \subset \mathbb{F}_q^d$. Suppose that for every $m \in \mathbb{F}_q^d \setminus \{(0, \dots, 0)\}$,*

$$|\widehat{B}(m)| \lesssim q^{\beta} \quad \text{for some} \quad \beta \in \mathbb{R}. \tag{4.1}$$

Then we have

$$|A - B| \gtrsim \min\left(q^d, \frac{|A||B|^2}{q^{2d+2\beta}}\right).$$

Proof. The proof is based on the formula (2.2) and discrete Fourier analysis. It follows that

$$q^{3d} \sum_{m \in \mathbb{F}_q^d} |\widehat{A}(m)|^2 |\widehat{B}(m)|^2$$

$$\leq q^{3d} |\widehat{A}(0, \dots, 0)|^2 |\widehat{B}(0, \dots, 0)|^2 + q^{3d} \left(\max_{m \in \mathbb{F}_q^d \setminus (0,\dots,0)} |\widehat{B}(m)|^2 \right) \sum_{m \in \mathbb{F}_q^d} |\widehat{A}(m)|^2$$

$$= \mathrm{I} + \mathrm{II}.$$

By the definition of the Fourier transform, it is clear that $\mathrm{I} = q^{-d}|A|^2|B|^2$. On the other hand, using the assumption (4.1) and the Plancherel theorem, we obtain that $\mathrm{II} \lesssim q^{2d+2\beta}|A|$. Thus, we have

$$q^{3d} \sum_{m \in \mathbb{F}_q^d} |\widehat{A}(m)|^2 |\widehat{B}(m)|^2 \lesssim q^{-d}|A|^2|B|^2 + q^{2d+2\beta}|A|.$$

Thus, Lemma 2.2 can be used to obtain that

$$|A-B| \gtrsim \frac{|A|^2|B|^2}{q^{-d}|A|^2|B|^2 + q^{2d+2\beta}|A|} \gtrsim \min\left(q^d, \frac{|A||B|^2}{q^{2d+2\beta}}\right),$$

which completes the proof. □

As mentioned in introduction, it is known that if $B \subset \mathbb{F}_q^d$ with $|B| \gtrsim q^{d/2}$ is a Salem set, then $|\Delta(B,B)| \gtrsim q$. Namely, the Salem set B satisfies the Falconer distance conjecture. In this case, we can state a strong fact that $B-B$ contains a positive proportion of all elements in $\mathbb{F}_q^d$. More precisely, we have the following theorem.

Theorem 4.2. *If $B \subset \mathbb{F}_q^d$ is a Salem set, then for any $A \subset \mathbb{F}_q^d$ with $|A||B| \gtrsim q^d$, we have*

$$|A-B| \gtrsim q^d.$$

Proof. Since $B \subset \mathbb{F}_q^d$ is a Salem set, taking $q^\beta = q^{-d}\sqrt{|B|}$ from Lemma 4.1 shows that

$$|A-B| \gtrsim \min\{q^d, |A||B|\}. \tag{4.2}$$

Since $|A||B| \gtrsim q^d$, the proof is complete. □

The following corollary follows immediately from above theorem and Lemma 3.1.

Corollary 4.3. *Let $A \subset \mathbb{F}_q^d$ is a Salem set. Then for any $B \subset \mathbb{F}_q^d$ with $|A||B| \gtrsim q^d$, we have*

$$|\Delta(A,B)| \gtrsim q.$$

In other words, the sets A, B satisfy the Falconer distance conjecture.

Acknowledgement. The first author's research was supported by the research grant of the Chungbuk National University in 2012 and Basic Science Research Program through the National Research Foundation of Korea(NRF) funded by the Ministry of Education, Science and Technology(2012010487).

References

[1] G. Arutyunyants and A. Iosevich, *Falconer conjecture, spherical averages and discrete analogs*, In Towards a theory of geometric graphs, 15-24, Contemp. Math. **342**, Amer. Math. Soc., Providence, (2004).

[2] J. Chapman, M. Erdoğan, D. Hart, A. Iosevich, and D. Koh, *Pinned distance sets, Wolff's exponent in finite fields and sum-product estimates*, Mathematische Zeitschrift, Volume 271, Issue 1 (2012), 63-93 .

[3] M. Erdoğan, *A bilinear Fourier extension theorem and applications to the distance set problem,* Internat. Math. Res. Notices 23 (2005), 1411-1425.

[4] K. Falconer, *On the Hausdorff dimensions of distance sets,* Mathematika, 32 (1985), 206–212.

[5] D. Hart, A. Iosevich, D. Koh and M. Rudnev, *Averages over hyperplanes, sum-product theory in vector spaces over finite fields and the Erdös-Falconer distance conjecture*, Trans. Amer. Math. Soc. Volume 363, June (2011), no.6, 3255-3275.

[6] S. Hofmann and A. Iosevich, *Circular averages and Falconer/Erdo"s distance conjecture in the plane for random metrics*, Proc. Amer. Math. Soc. 133 (2005) 133-143.

[7] A. Iosevich and I. Laba, *K-distance, Falconer conjecture, and discrete analogs*, Integers, Electronic Journal of Combinatorial Number Theory, Proceedings of the Integers Conference in honor of Tom Brown, (2005) 95–106.

[8] A. Iosevich and M. Rudnev, *On distance measures for well-distributed sets*, Journal of Discrete and Computational Geometry, **38**, (2007).

[9] A. Iosevich and M. Rudnev, *Erdös distance problem in vector spaces over finite fields*, Trans. Amer. Math. Soc. 359 (2007), 6127-6142.

[10] A. Iosevich and M. Rudnev, *Freiman's theorem, Fourier transform, and additive structure of measures*, Journal of the Australian Mathematical Society, **86**, (2009), 97–109.

[11] D. Koh and C. Shen, *Sharp extension theorems and Falconer distance problems for algebraic curves in two dimensional vector spaces over finite fields* , Revista Matematica Iberoamericana, Volume 28, issue 1 (2012).

[12] D. Koh and C. Shen, *The generalized Erdös-Falconer distance problems in vector spaces over finite fields*, preprint, arxiv.org.

[13] P. Mattila, *Spherical averages of Fourier transforms of measures with finite energy: dimension of intersections and distance sets,* Mathematika 34(1987), 207–228.

[14] P. Mattila, *Geometry of Sets and Measures in Euclidean Spaces: Fractals and Rectifiability,* Cambridge University Press, Feb 25, 1999.

[15] I. Shparlinski, *On the set of distance between two sets over finite fields*, International Journal of Mathematics and Mathematical Sciences Volume 2006, Article ID 59482, 1–5.

[16] V. Vu, *Sum-product estimates via directed expanders*, Math. Res. Lett. **15** (2008), 375–388.

[17] T. Wolff, *Decay of circular means of Fourier transforms of measures,* Internat. Math. Res. Notices 1999, 547–567.

A COMBINATORIAL INTERPRETATION OF THE CATALAN AND BELL NUMBER DIFFERENCE TABLES

Jocelyn Quaintance
Dept. of Mathematics, Rutgers University, Piscataway, New Jersey
quaintan@math.rutgers.edu

Harris Kwong
SUNY Fredonia, Dept. of Math. Sci., Fredonia, New York
kwong@fredonia.edu

Received: 5/10/12, Accepted: 3/11/13, Published: 5/10/13

Abstract
We study the recurrence relations and derive the generating functions of the entries along the rows and diagonals of the Catalan and Bell number difference tables.

– *In Memory of Professor Herb Wilf*

1. Introduction

One of the most famous sequences in all of enumerative combinatorics is the Catalan numbers

$$C_n = \frac{1}{n+1}\binom{2n}{n}.$$

They possess numerous fascinating properties and appear in many problems in mathematics and computer science; see, for example, [1, 3]. A glimpse of the old classic [2], and more recently, [4, 7], reveals there are hundreds of books and articles written about the sequence $\{C_n\}_{n=0}^{\infty}$. One of the reasons it occurs so frequently in enumeration is a well-known recurrence relation that it satisfies:

$$C_0 = 1, \qquad C_n = \sum_{k=1}^{n} C_{k-1}C_{n-k}, \quad n \geq 1.$$

Interestingly, C_n also satisfies several other less famous recurrence relations. The first is Touchard's formula

$$C_{n+1} = \sum_{k=0}^{\lfloor n/2 \rfloor} \binom{n}{2r} 2^{n-2r} C_r.$$

The second is an nth difference type recurrence formula

$$C_n = \sum_{r=1}^{\lfloor (n+1)/2 \rfloor} (-1)^{r-1} \binom{n-r+1}{r} C_{n-r} \tag{1}$$

They can be found on, for example, page 319 and 322 respectively of [4].

The structure of Equation (1) prompts us to study

$$K_n = \sum_{r=0}^{n} (-1)^{n-r} \binom{n}{r} C_r = \sum_{r=0}^{n} \frac{(-1)^{n-r}}{r+1} \binom{n}{r} \binom{2r}{r}. \tag{2}$$

In the OEIS [6], $\{K_n\}_{n=0}^{\infty}$ is sequence A005043. This sequence enumerates certain Motzkin and Dyck paths. We notice that $\{K_n\}_{n=0}^{\infty}$ is the leftmost diagonal of the Catalan number difference table (see Table 1). The difference table contains an infinite number of rows and diagonals. Only the first two diagonals and first four rows are found in the OEIS. Furthermore, the OEIS only provides combinatorial meaning for the two diagonals and first two rows. We wonder if there is a systematic interpretation of the combinatorial meaning for all the rows and diagonals of the Catalan difference table.

An answer can be found in the numeration of certain $n \times 1$ non-interlocking letter columns discussed in Chapter 4 of [5]. The purpose of this paper is to explain this association. During the process of exploring this connection we are able to derive recurrence formulas and generating functions for each of the rows and diagonals.

We close this section with a derivation of the generating function $K(t)$ of the sequence $\{K_n\}_{n=0}^{\infty}$. It is easy to show that

$$\frac{(-1)^k}{k+1} \binom{2k}{k} = 2 \cdot 4^k \binom{\frac{1}{2}}{k+1}.$$

Together with

$$\sum_{n=k}^{\infty} \binom{n}{k} (-t)^{n-k} = \sum_{r=0}^{\infty} \binom{k+r}{r} (-t)^r = \frac{1}{(1+t)^{k+1}},$$

we find

$$\begin{aligned} K(t) &= \sum_{n=0}^{\infty}\sum_{k=0}^{n}\frac{(-1)^{n-k}}{k+1}\binom{n}{k}\binom{2k}{k}t^n \\ &= 2\sum_{k=0}^{\infty}(-4t)^k\binom{\frac{1}{2}}{k+1}\sum_{n=k}^{\infty}\binom{n}{k}(-t)^{n-k} \\ &= -\frac{1}{2t}\sum_{k=0}^{\infty}\binom{\frac{1}{2}}{k+1}\left(\frac{-4t}{1+t}\right)^{k+1} \\ &= -\frac{1}{2t}\left[\left(1-\frac{4t}{1+t}\right)^{\frac{1}{2}}-1\right] \\ &= \frac{1}{2t}\left(1-\sqrt{\frac{1-3t}{1+t}}\right). \end{aligned}$$

As we shall see in Section 3, $K(t)$ plays an important role in finding the generating functions of the diagonals of the Catalan difference table.

2. Rows of the Catalan Difference Table

In the next two sections we shall combinatorially analyze the difference table of the Catalan numbers. Table 1 shows a portion of this infinite table. The top row consists of the Catalan numbers C_n, where $n \geq 0$. The main diagonal contains the numbers $1, 0, 1, 1, 3, 6, 15, 91, 252, 603, \ldots$. We shall show that this is precisely the sequence $\{K_n\}_{n=0}^{\infty}$.

1	1	2	5	14	42	132	429	1430	4862	16796	...
	0	1	3	9	28	90	297	1001	3432	11934	...
		1	2	6	19	62	207	704	2431	8502	...
			1	4	13	43	145	497	1727	6071	...
				3	9	30	102	352	1230	4344	...
					6	21	72	250	878	3114	...
						15	51	178	628	2236	...
							36	127	450	1608	...
								91	323	1158	...
									232	835	...
										603	...
											⋱

Table 1: A portion of the Catalan number difference table.

We use $c_{i,n}$, where $i \geq 0$ and $n \geq i$, to denote the entry in the ith row and nth column in this table. Obviously, $c_{0,n} = c_n$ are the Catalan numbers. To maintain consistency we will write c_n instead of C_n in the rest of the paper.

2.1. The First Row of the Difference Table

The top row of the difference table contains the Catalan numbers c_n, where $n \geq 0$. They enumerate many combinatorial objects. Standard interpretations include the following:

- The number of ways to triangulate the interior of a convex $(n+2)$-gon.
- The number of rooted binary trees of height n.
- The number of ways to parenthesize n pairs of left and right parentheses.

For our purpose we use an interpretation different from these three.

Consider an $n \times 1$ column of letters selected from an alphabet $\mathcal{A} = \{\alpha_i\}_{i \geq 1}$. For brevity we shall call it an ***n-column***. Denote entries in an n-column by $a_1 a_2 \dots a_n$. An n-column is said to be ***non-skipping*** if, for any i, all predecessors of a_i within $\mathcal{A}$ must have already appeared among $a_1 a_2 \dots a_{i-1}$. In other words, after $a_1 a_2 \dots a_i$ is generated, a_{i+1} is either a letter that has already appeared, or the next letter in $\mathcal{A}$ that has not appeared yet. Note that the definition implies that a non-skipping n-column always starts with $a_1 = \alpha_1$.

In addition, call a letter column ***interlocking*** if there exist $1 \leq i < j < k < \ell \leq n$ such that $a_i \neq a_j$, $a_i = a_k$, and $a_j = a_\ell$. In a way, we can say that the pattern $a_i a_j$, where $a_i \neq a_j$, repeats again later within the same column as $a_k a_\ell$, with the understanding that neither a_i and a_j nor a_k and a_ℓ need to be adjacent.

For demonstrative purposes we shall use the English letters. In Figure 1 both columns are non-skipping, but the one on the left is non-interlocking, while the one on the right is interlocking due to the repetition of the AB pattern.

A	A
B	B
A	C
C	A
A	C
A	B

Figure 1: Examples of non-interlocking and interlocking letter columns

We are interested in the enumeration of non-interlocking and non-skipping (NINS) n-columns. Notice that in an NINS n-column, if a_k is the last occurrence of α_1, then, because of the non-skipping property, we may assume that for some $t \geq 1$, all the letters $\alpha_1, \alpha_2, \ldots, \alpha_t$ appear within $a_1 a_2 \ldots a_k$. Now each of the letters $\alpha_2, \alpha_3, \ldots, \alpha_t$ cannot appear in the segment $a_{k+1}a_{k+2}\ldots a_n$, for otherwise, along with $a_k = \alpha_1$, one of the patterns $\alpha_1\alpha_i$, where $1 \leq i \leq t$, would have reappeared. This means only *new* letters $\alpha_{t+1}, \alpha_{t+2}, \ldots$ can appear in $a_{k+1}a_{k+2}\ldots a_n$. The non-skipping property also requires $a_{k+1} = \alpha_{t+1}$. In addition, the segment $a_{k+1}a_{k+2}\ldots a_n$ still obeys the NINS condition. Therefore, over all NINS n-columns with a_k as the last occurrence of α_1, the segments $a_{k+1}a_{k+2}\ldots a_n$ are in one-to-one correspondence with NINS $(n-k)$-columns. This important observation plays a key role in our analysis.

Theorem 2.1 *Let u_n denote the number of NINS n-columns. Define $u_0 = 1$. Then $u_n = c_n$.*

Proof. It is easy to verify that $u_1 = 1$, $u_2 = 2$, and $u_3 = 5$. Their respective letter columns are depicted in Figure 2.

A	A	A	A	A	A	A	A
	B	A	B	B	A	B	A
			C	B	B	A	A

Figure 2: NINS letter columns of size at most three.

Let a_k be the last occurrence of the letter $\alpha_1 = \mathrm{A}$ within an NINS n-column. Obviously, $1 \leq k \leq n$. Notice that $a_1 a_2 \ldots a_{k-1}$ is a NINS $(k-1)$-column. In this regard, defining $u_0 = 1$ counts the number of the null column. Our earlier remark asserts that all the letters from a_{k+1} to a_n must be new, and the segment $a_{k+1}a_{k+2}\ldots a_n$ can be reduced to an NINS $(n-k)$-column. Figure 3 lists the NINS 4-columns according to this classification.

A	A	A	A	A	A	A	A	A	A	A	A	A	A
B	B	B	B	B	A	A	B	A	B	B	A	B	A
C	C	B	C	B	B	B	A	A	C	B	B	A	A
D	C	C	B	B	C	B	C	B	A	A	A	A	A
$k=1$					$k=2$		$k=3$		$k=4$				

Figure 3: Classification of NINS 4-columns.

It follows that

$$u_0 = 1, \qquad u_n = \sum_{k=1}^{n} u_{k-1}u_{n-k}, \quad n \geq 1.$$

Since u_n satisfies the same initial condition and recurrence relation as the Catalan numbers, we find $u_n = c_n$. □

A closing remark: an alternate proof of Theorem 2.1 using parentheses can be found in Chapter 4 of [5].

2.2. The General Row of the Difference Table

The second row of the Catalan difference table forms the sequence $\{c_{1,n}\}_{n=1}^{\infty}$. By construction, we have $c_{1,n} = c_n - c_{n-1}$, which leads to our first result regarding $c_{1.n}$.

Theorem 2.2 *For $n \geq 2$, the number of NINS n-columns that starts with AB is $c_{1,n}$. Equivalently, $c_{1,n}$ is the number of NINS n-columns that do not start with AA.*

Proof. Without any restriction, there are c_n NINS columns of size $n \times 1$. Because of non-skipping property, any such column has to start with AA or AB. Hence we need to remove those that start with AA.

If an NINS n-column starts with AA, by removing the first A, we obtain an NINS $(n-1)$-column. Conversely, starting with any NINS $(n-1)$-column, by appending an A on top, we form an NINS n-column that starts with AA. Therefore there are c_{n-1} such columns, and $c_{1,n} = c_n - c_{n-1}$ counts the number of n-columns that do not start with AA (consequently they must start with AB). □

We say that a letter column has a ***repetition of type i***, or it has a ***repetition at position i***, if $a_i = a_{i+1}$. If $i = n$, an n-column has a repetition of type n if it ends with an A. Equivalently, we may define repetition of type i as having the property $a_i = a_{(i+1) \bmod n}$. Hence $c_{1,n}$ counts the number of NINS n-columns that do not have repetition at position 1. Notice that, in particular, $c_{1,1} = 0$ because the 1×1 letter column A has a type 1 repetition.

We can restate Theorem 2.2 as: for $n \geq 1$, the number $c_{1,n}$ enumerates NINS n-columns that do not have repetition at the first position. It turns out that this result also holds for other rows in the Catalan difference table.

Theorem 2.3 *For $j \geq 1$, the sequence $\{c_{j,n}\}_{n=j}^{\infty}$ counts the number of NINS n-columns that do not have any repetition at the first j positions.*

Proof. Induct on j. The case of $j = 1$ was proved in Theorem 2.2. Assume, for some $j \geq 2$, the sequence $\{c_{j-1,n}\}_{n=j-1}^{\infty}$ counts the number of NINS n-columns that do not have type i repetitions for $1 \leq i \leq j-1$. We now analyze the meaning of $\{c_{j,n}\}_{n=j}^{\infty}$, where $c_{j,n} = c_{j-1,n} - c_{j-1,n-1}$.

To account for all NINS n-columns that do not have repetition at the first j positions, we start with the $c_{j-1,n}$ columns that do not have repetition at the first $j-1$ positions, and remove from them those that do have a type j repetition.

The case of $n = j$ requires a separate treatment. Among the $c_{j-1,j}$ columns that do not have repetition at the first $j-1$ positions, some also have type j repetition, hence must be discarded from our consideration. By definition, these columns have $a_j = A$. Notice that no repetition at position $j-1$ implies $a_{j-1} \neq A$. Hence $a_1a_2 \dots a_{j-1}$ is an NINS $(j-1)$-column that does not have repetition at the first $j-1$ positions.

Conversely, by adding an A at the bottom of an NINS $(j-1)$-column that does not have repetition at its first $j-1$ positions will create a repetition of type j. Therefore, we need to throw out $c_{j-1,j-1}$ of the original $c_{j-1,j}$ columns. Hence the number of NINS n-columns that do not have repetition at the first j positions is $c_{j-1,j} - c_{j-1,j-1}$, which is precisely $c_{j,j}$.

Now we look at $n \geq j+1$. Any column of this type must have $a_{j-1} \neq a_j$ and $a_j = a_{j+1}$. By deleting a_j, we form a new NINS $(n-1)$-column $a_1a_2 \dots a_{j-1}a_{j+1} \dots a_n$, which does not have repetition at its first $j-1$ positions. Conversely, starting with NINS $(n-1)$-column $b_1b_2 \dots b_{n-1}$ that does not have repetition at its first $j-1$ positions, we can form an n-column $b_1b_2 \dots b_{j-1}b_jb_jb_{j+1} \dots b_{n-1}$ that does have a type j repetition. Therefore the number of forbidden columns is $c_{j-1,n-1}$.

We have proved that, for $n \geq j+1$, the number $c_{j,n} = c_{j-1,n} - c_{j-1,n-1}$ gives the number of NINS n-columns that do not have repetition at their first j positions. This completes the induction. □

The combinatorial meaning of $c_{j,n}$ yields the following recurrence relation, in which we adopt the usual convention that a summation is empty if its upper limit is less than its lower limit.

Theorem 2.4 *For* $n-1 \geq j \geq 1$,

$$c_{j,n} = c_{j-1,n-1} + \sum_{k=3}^{j} c_{k-1,k-1}c_{j-k,n-k} + \sum_{k=\max(3,j+1)}^{n} c_{n-k}c_{j,k-1}.$$

Proof. Consider any NINS n-column with no repetition at its first j positions. Let a_k be the last occurrence of A. Obviously $k \neq 2$. If $k = 1$, then $a_2a_3 \dots a_n$ is an NINS $(n-1)$-column over the alphabet $\mathcal{A} - \{A\} = \mathcal{A} - \{\alpha_1\}$ that does not have any repetition at its first $j-1$ positions. Hence there are $c_{j-1,n-1}$ such NINS columns.

For $3 \leq k \leq j$, the upper portion $a_1a_2 \dots a_{k-1}$ (recall that $a_{k-1} \neq A$) is an NINS $(k-1)$-column without repetition at its first $k-1$ positions. This upper portion can be formed in $c_{k-1,k-1}$ ways. The lower portion $a_{k+1}a_{k+2} \dots a_n$, as we had remarked before, is essentially an NINS $(n-k)$-column. Since this is part of a letter column that originally does not have repetition at positions 1 through j, we know that the lower portion does not have any repetition at its first $j-k$ positions. Hence it can be formed in $c_{j-k,n-k}$ ways. We have found that $\sum_{k=3}^{j} c_{k-1,k-1}c_{j-k,n-k}$ NINS columns of size $n \times 1$ meet our condition.

For each k that satisfies $\max(3, j+1) \leq k \leq n$, the same old familiar argument yields $c_{j,k-1}c_{n-k}$ columns. By combining the counts over all possible values of k, we obtain the desired recurrence relation. □

Here are a few examples:

$$\begin{aligned}
c_{1,n} &= c_{n-1} + \sum_{k=3}^{n} c_{n-k}c_{1,k-1}, \quad n \geq 2, \\
c_{2,n} &= c_{1,n-1} + \sum_{k=3}^{n} c_{n-k}c_{2,k-1}, \quad n \geq 3, \\
c_{3,n} &= c_{2,n-1} + c_{n-3} + \sum_{k=4}^{n} c_{n-k}c_{3,k-1}, \quad n \geq 4, \\
c_{4,n} &= c_{3,n-1} + c_{1,n-3} + c_{n-4} + \sum_{k=5}^{n} c_{n-k}c_{4,k-1}, \quad n \geq 5, \\
c_{5,n} &= c_{4,n-1} + c_{2,n-3} + c_{1,n-4} + 3c_{n-5} + \sum_{k=6}^{n} c_{n-k}c_{5,k-1}, \quad n \geq 6.
\end{aligned}$$

The generating function for the Catalan numbers is well-known [1, 2, 3, 7] to be

$$C(t) = \sum_{t=0}^{\infty} c_n t^n = \frac{1-\sqrt{1-4t}}{2t}.$$

The generating function for $\{c_{j,n}\}_{n=j}^{\infty}$ can be derived recursively.

Theorem 2.5 *Let* $C_j(t) = \sum_{n=j}^{\infty} c_{j,n}t^n$. *Then* $C_0(t) = C(t) = \frac{1-\sqrt{1-4t}}{2t}$, *and, for* $j \geq 1$,

$$C_j(t) = (1-t)C_{j-1}(t) - c_{j-1,j-1}t^{j-1}.$$

Proof. Since $c_{j,n} = c_{j-1,n} - c_{j-1,n-1}$, we obtain

$$\begin{aligned}
\sum_{n=j}^{\infty} c_{j,n}t^n &= \sum_{n=j}^{\infty} c_{j-1,n}t^n - t\sum_{n=j}^{\infty} c_{j-1,n-1}t^{n-1} \\
&= [C_{j-1}(t) - c_{j-1,j-1}t^{j-1}] - tC_{j-1}(t),
\end{aligned}$$

which simplifies to the stated result. □

For examples,

$$\begin{aligned}
C_1(t) &= (1-t)C_0(t) - 1 &&= (1-t)C(t) - 1, \\
C_2(t) &= (1-t)C_1(t) &&= (1-t)^2C(t) - (1-t), \\
C_3(t) &= (1-t)C_2(t) - t^2 &&= (1-t)^3C(t) - (1-2t+2t^2).
\end{aligned}$$

These examples suggests that it is possible to express $C_j(t)$ in terms of $C(t)$ explicitly.

Theorem 2.6 *For $j \geq 0$,*

$$C_j(t) = (1-t)^j C(t) - \sum_{k=0}^{j-1}\left[\sum_{i=0}^{k}(-1)^i\binom{j}{i}c_{k-i}\right]t^k.$$

Proof. For $n \geq j$, we know, from the construction of the difference table,

$$c_{j,n} = \sum_{i=0}^{j}(-1)^i\binom{j}{i}c_{n-i}. \tag{3}$$

Thus

$$\begin{aligned}
\sum_{n=j}^{\infty} c_{j,n}t^n &= \sum_{i=0}^{j}(-1)^i\binom{j}{i}t^i\sum_{n=j}^{\infty}c_{n-i}t^{n-i} \\
&= \sum_{i=0}^{j}(-1)^i\binom{j}{i}t^i\left[C(t) - \sum_{\ell=0}^{j-i-1}c_\ell t^\ell\right] \\
&= (1-t)^j C(t) - \sum_{i=0}^{j-1}\sum_{\ell=0}^{j-i-1}(-1)^i\binom{j}{i}c_\ell t^{i+\ell} \\
&= (1-t)^j C(t) - \sum_{k=0}^{j-1}\left[\sum_{i=0}^{k}(-1)^i\binom{j}{i}c_{k-i}\right]t^k,
\end{aligned}$$

which is what we want to prove. □

It is clear from the proof that a similar result also holds in *any* difference table.

3. Diagonals of the Catalan Difference Table

We now turn our attention to the sequences $\{\tilde{c}_{j,n}\}_{n=j}^{\infty}$, where $j = 0, 1, 2, \ldots$, formed by the diagonal entries in the Catalan difference table. Table 2 illustrates the meaning of the notation we use to identify the entries. It is clear that $\tilde{c}_{j,n} = c_{n-j,n}$. Hence $\tilde{c}_{j,n}$ counts the number of NINS n-columns that do not have repetition at their first $n-j$ positions.

3.1. The First Diagonal of the Catalan Difference Table

The entry $\tilde{c}_n = \tilde{c}_{0,n}$ on the first diagonal counts the number of $n \times 1$ NINS columns that do not have a repetition at their first n positions. Recall that this means

$\tilde{c}_{0,0}$	$\tilde{c}_{1,1}$	$\tilde{c}_{2,2}$	$\tilde{c}_{3,3}$	$\tilde{c}_{4,4}$	$\cdots$
	$\tilde{c}_{0,1}$	$\tilde{c}_{1,2}$	$\tilde{c}_{2,3}$	$\tilde{c}_{3,4}$	$\cdots$
		$\tilde{c}_{0,2}$	$\tilde{c}_{1,3}$	$\tilde{c}_{2,4}$	$\cdots$
			$\tilde{c}_{0,3}$	$\tilde{c}_{1,4}$	$\cdots$
				$\tilde{c}_{0,4}$	$\cdots$
					$\ddots$

Table 2: Renaming of the entries in the Catalan difference table.

these columns do not end with an A. Naturally we set $\tilde{c}_0 = 1$. The combinatorial interpretation also implies that $\tilde{c}_1 = 0$. For $n \geq 2$, we obtain the following recurrence relation.

Theorem 3.1 *For $n \geq 2$,*

$$\tilde{c}_n = \sum_{k=1}^{n-1} \tilde{c}_{k-1} \left(\tilde{c}_{n-k} + \tilde{c}_{n-k-1}\right).$$

Proof. Let a_k be the last occurrence of A. Since the column cannot end with an A, we find $1 \leq k \leq n-1$. In addition, $a_{k-1} \neq A$, hence $a_1 a_2 \dots a_{k-1}$ is an NINS $(k-1)$-column that does not have a repetition at its first $k-1$ positions. Thus it can be formed in $\tilde{c}_{k-1}$ ways. The remaining question is: in how many ways can $a_{k+1} a_{k+2} \dots a_n$ be formed?

Notice that all these letters are different from those used in the upper portion $a_1 a_2 \dots a_k$, and a_{k+1} is the next unused letter. The lower portion is essentially an NINS $(n-k)$-column. However, a_n may or may not be equal to a_{k+1}. If $a_n \neq a_{k+1}$, what we have is an NINS $(n-k)$-column that does not have a repetition at its first $n-k$ positions, hence it can be formed in $\tilde{c}_{n-k}$ ways. If $a_n = a_{k+1}$, then $a_{n-1} \neq a_{k+1}$. The removal of a_n will produce an NINS $(n-k-1)$-column that does not have a repetition at its first $n-k-1$ positions, hence the lower portion can be formed in $\tilde{c}_{n-k-1}$ in this case. Combining what we have found yields the given recurrence. □

This recurrence relation, along with the standard convolution argument, yields the generating function of the sequence $\{\tilde{c}_n\}_{n=0}^{\infty}$.

Theorem 3.2 *Let $\widetilde{C}(t) = \sum_{n=0}^{\infty} \tilde{c}_n t^n$, then*

$$\widetilde{C}(t) = \frac{(1+t) - \sqrt{1-2t-3t^2}}{2t(1+t)}.$$

Proof. Using the recurrence from Theorem 3.1, we find

$$\sum_{n=2}^{\infty} \tilde{c}_n t^n = t \sum_{n=2}^{\infty} \left(\sum_{k=1}^{n-1} \tilde{c}_{k-1} \tilde{c}_{n-k} \right) t^{n-1} + t^2 \sum_{n=2}^{\infty} \left(\sum_{k=1}^{n-1} \tilde{c}_{k-1} \tilde{c}_{n-k-1} \right) t^{n-2}.$$

In terms of $\widetilde{C}(t)$, we can rewrite it as

$$\widetilde{C}(t) - \tilde{c}_0 - \tilde{c}_1 t = t\widetilde{C}(t)[\widetilde{C}(t) - \tilde{c}_0] + t^2\widetilde{C}(t) \cdot \widetilde{C}(t).$$

Since $\tilde{c}_0 = 1$ and $\tilde{c}_1 = 0$, this reduces to the equation

$$(t + t^2)\widetilde{C}^2(t) - (1 + t)\widetilde{C}(t) + 1 = 0.$$

Solving for $\widetilde{C}(t)$ finishes the proof. □

Note that we can write

$$\widetilde{C}(t) = \frac{1}{2t}\left(1 - \frac{\sqrt{(1+t)(1-3t)}}{1+t}\right) = \frac{1}{2t}\left(1 - \sqrt{\frac{1-3t}{1+t}}\right),$$

which is precisely $K(t)$ in Section 1. This proves that $K_n = \tilde{c}_n$ for all $n \geq 0$.

3.2. Other Diagonals of the Catalan Difference Table

For $j \geq 1$, we find a recurrence for the sequence $\{\tilde{c}_{j,n}\}_{n=j}^{\infty}$.

Theorem 3.3 *For $n - 1 \geq j \geq 1$,*

$$\tilde{c}_{j,n} = \sum_{\ell=1}^{j} \tilde{c}_{j-\ell,j-\ell}\tilde{c}_{\ell-1,n-j+\ell-1} + \sum_{k=1}^{n-j} \tilde{c}_{k-1}\tilde{c}_{j,n-k}.$$

Proof. Let a_k be the last occurrence of A in an NINS n-column that does not have a repetition at its first $n - j$ positions. Then $1 \leq k \leq n$. Since $n - j \geq 1$, we know that $a_2 \neq A$. Hence, technically, $k \neq 2$. Nevertheless, we shall see that it will be taken care of, numerically, in the summation.

First consider $n - j + 1 \leq k \leq n$. In this case, the upper portion $a_1 a_2 \ldots a_{k-1}$ is an NINS $(k-1)$-column that does not have a repetition at the first $n - j$ positions, so it can be formed in $\tilde{c}_{k-1-(n-j),k-1}$ ways. The lower portion $a_{k+1}a_{k_2} \ldots a_n$ is essentially an NINS $(n - k)$-column with no restriction in regard to repetitions, hence it can formed in $c_{n-k} = \tilde{c}_{n-k,n-k}$ ways. This case produces

$$\sum_{k=n-j+1}^{n} \tilde{c}_{n-k,n-k}\tilde{c}_{k-1-(n-j),k-1} = \sum_{\ell=1}^{j} \tilde{c}_{j-\ell,j-\ell}\tilde{c}_{\ell-1,n-j+\ell-1}$$

columns.

For $1 \leq k \leq n - j$, the upper portion $a_1 a_2 \ldots a_{k-1}$ does not have a repetition in the first $k - 1$ positions, hence it can be formed in $\tilde{c}_{k-1}$ ways. Note that $\tilde{c}_1 = 0$, which explains why $k \neq 2$. The lower portion $a_{k+1}a_{k+2} \ldots a_n$ is essentially an NINS

$(n-k)$-column that does not have repetition in its first $n-j-k$ positions, it can be formed in $\tilde{c}_{j,n-k}$ ways. There are altogether

$$\sum_{k=1}^{n-j} \tilde{c}_{k-1}\tilde{c}_{j,n-k}$$

columns in this case. Combining this with the first case yields the stated recurrence relation. □

For $1 \le j \le 4$, we obtain these examples:

$$\begin{aligned}
\tilde{c}_{1,n} &= \tilde{c}_{n-1} + \sum_{k=1}^{n-1} \tilde{c}_{k-1}\tilde{c}_{1,n-k}, \quad n \ge 2, \\
\tilde{c}_{2,n} &= \tilde{c}_{n-2} + \tilde{c}_{1,n-1} + \sum_{k=1}^{n-2} \tilde{c}_{k-1}\tilde{c}_{2,n-k}, \quad n \ge 3, \\
\tilde{c}_{3,n} &= 2\tilde{c}_{n-3} + \tilde{c}_{1,n-2} + \tilde{c}_{2,n-1} + \sum_{k=1}^{n-3} \tilde{c}_{k-1}\tilde{c}_{3,n-k}, \quad n \ge 4, \\
\tilde{c}_{4,n} &= 5\tilde{c}_{n-4} + 2\tilde{c}_{1,n-3} + \tilde{c}_{2,n-2} + \tilde{c}_{3,n-1} + \sum_{k=1}^{n-4} \tilde{c}_{k-1}\tilde{c}_{4,n-k}, \quad n \ge 5.
\end{aligned}$$

The generating function $\widetilde{C}_j(t) = \sum_{n=j}^{\infty} \tilde{c}_{j,n}t^n$ can be expressed in terms of $\widetilde{C}(t)$, as follows.

Theorem 3.4 *For $j \ge 0$,*

$$\widetilde{C}_j(t) = (1+t)^j \widetilde{C}(t) - \sum_{k=0}^{j-1} \left[\sum_{i=0}^{k} \binom{j}{i} \tilde{c}_{k-i} \right] t^k.$$

Proof. The construction of the difference table ensures that for all $n \ge j \ge 1$,

$$\tilde{c}_{j,n} = \tilde{c}_{j-1,n} + \tilde{c}_{j-1,n-1}.$$

By applying it repeatedly, we obtain

$$\tilde{c}_{j,n} = \sum_{i=0}^{j} \binom{j}{i} \tilde{c}_{n-i}.$$

Note its similarity to Equation (3). The rest of the proof is almost identical to that of Theorem 2.6, and hence is omitted. □

To illustrate Theorem 3.4, we list below the results for $1 \leq j \leq 4$.

$$\begin{aligned}
\widetilde{C}_1(t) &= (1+t)\widetilde{C}(t) - 1, \\
\widetilde{C}_2(t) &= (1+t)^2\widetilde{C}(t) - (1+2t), \\
\widetilde{C}_3(t) &= (1+t)^3\widetilde{C}(t) - (1+3t+4t^2), \\
\widetilde{C}_4(t) &= (1+t)^4\widetilde{C}(t) - (1+4t+7t^2+9t^3).
\end{aligned}$$

Once again, we note that the same argument could be applied to the diagonal entries of *any* difference table to produce similar results.

4. The Bell Number Difference Table

Thus far, we have discussed the correlation between NINS n-columns and the Catalan number difference table. It is the non-interlocking property that gives rise to the convolution-type recurrence relation that is often found in Catalan numbers. What if we drop the non-interlocking property? The answer can be found in the Bell number B_n, which counts the number of partitions of an n-set.

Theorem 4.1 *The number of non-skipping n-columns, denoted NS n-columns, is B_n.*

Proof. Given any $n \times 1$ non-skipping letter column, the subsets containing subscripts whose respective positions hold the same letter form a partition of $[n] = \{1, 2, \ldots, n\}$. Conversely, given any partition of $[n]$, we can name the subsets $S_A, S_B, \ldots,$ such that their smallest elements are in ascending order. Next, let $a_i = k$ if $i \in S_k$. The result is an $n \times 1$ non-skipping letter column. This proves that the $n \times 1$ non-skipping columns are in one-to-one correspondence with the partitions of $[n]$, hence they are counted by the Bell number B_n. □

The Bell number B_n can be defined recursively as

$$B_0 = 1, \qquad B_n = \sum_{k=0}^{n-1} \binom{n-1}{k} B_k, \quad n \geq 1.$$

This enables us to obtain another proof.

Proof. (Alternate Proof) Let v_n denote the number of non-skipping n-columns. We adopt the convention that $v_0 = 1$, which can be considered as the number of null columns. For $n \geq 1$, assume that there are k occurrences of A other than the first one. Hence $0 \leq k \leq n-1$. The locations of these k occurrences can be selected in $\binom{n-1}{k}$ ways. Once the A's are placed, the remaining $n-k$ spaces essentially form an NS $(n-k)$-column that uses B, C, ... Therefore

$$v_n = \sum_{k=0}^{n-1} \binom{n-1}{k} v_{n-k}.$$

Since v_n and B_n obey the same recurrence and share the same initial condition, we conclude that $v_n = B_n$ for all $n \geq 0$. □

Since no interlocking occurs when $n \leq 3$, we find $B_1 = C_1 = 1$, $B_2 = C_2 = 3$, and $B_3 = C_3 = 5$. When $n = 4$ only one interlocking NS 4-column can be found, namely, ABAB. It follows that $B_4 = 15$, and $C_4 = 14$.

Naturally one may ask if there is any combinatorial meaning attached to the Bell number difference table (see Table 3). As in the case of the Catalan difference table, we denote the numbers in this table by $b_{j,n}$, where $n \geq j$, and let $b_{0,n} = b_n = B_n$.

1	1	2	5	15	52	203	877	4140	21147	115975	…
	0	1	3	10	37	151	674	3263	17007	94828	…
		1	2	7	27	114	523	2589	13744	77821	…
			1	5	20	87	409	2066	11155	64077	…
				4	15	67	322	1657	9089	52922	…
					11	52	255	1335	7432	43833	…
						41	203	1080	6097	36401	…
							162	877	5017	30304	…
								715	4140	25287	…
									3425	21147	…
										17722	…
											⋱

Table 3: A portion of the Bell number difference table.

4.1. Rows of the Bell Difference Table

The proofs of Theorems 2.2 and 2.3 do not rely on the non-interlocking property, hence we can apply them to non-skipping columns.

Theorem 4.2 *For $j \geq 1$ the sequence $\{b_{j,n}\}_{n=j}^{\infty}$ counts the number of NS n-columns that do not have any repetition at the first j positions.*

The argument for the recurrence relation that the $c_{j,n}$ satisfy does depend on the non-interlocking property. It appears that $b_{j,n}$ does not obey any simple recurrences.

Nonetheless, it is clear from the structure of the difference table that

$$b_{j,n} = \sum_{i=0}^{j}(-1)^i \binom{j}{i} b_{n-i}.$$

There is another way to connect $b_{j,n}$ to the Bell numbers.

Theorem 4.3 *For $n \geq 1$,*

$$b_{1,n+1} = \sum_{k=0}^{n-1} \binom{n-1}{k} b_{n-k} = \sum_{\ell=0}^{n-1} \binom{n-1}{\ell} b_{\ell+1}.$$

Proof. Once again let k count the number of A's that occur beyond the initial A in an NS $(n+1)$-column that begins with AB. Hence $0 \leq k \leq n-1$. The placement of these A's is enumerated by $\binom{n-1}{k}$. Once these A's are placed, the remaining $n-k$ spaces are completed with an arbitrary NS $(n-k)$-column over $\mathcal{A}-\{A\}$ which starts with B. The number of such letter columns is b_{n-k}. This completes the proof of the first equality. The second is obtained by a change in the index of summation. □

This result can be generalized.

Theorem 4.4 *For $n \geq j \geq 1$,*

$$b_{j,n+1} = \sum_{k=0}^{n-j} \binom{n-j}{k} b_{n-k} = \sum_{\ell=0}^{n-j} \binom{n-j}{\ell} b_{\ell+j}.$$

Proof. It suffices to prove the first equality. Induct on j. We have just proved the case of $j = 1$. So we may assume the result holds for $j-1$ for some $j \geq 2$. Then, using Pascal's identity, we find

$$\begin{aligned}
b_{j,n+1} &= b_{j-1,n+1} - b_{j-1,n} \\
&= \sum_{k=0}^{n-j+1} \binom{n-j+1}{k} b_{n-k} - \sum_{k=0}^{n-j} \binom{n-j}{k} b_{n-k-1} \\
&= \sum_{k=0}^{n-j+1} \binom{n-j+1}{k} b_{n-k} - \sum_{k=1}^{n-j+1} \binom{n-j}{k-1} b_{n-k} \\
&= b_n + \sum_{k=1}^{n-j+1} \left[\binom{n-j+1}{k} - \binom{n-j}{k-1} \right] b_{n-k} \\
&= \sum_{k=0}^{n-j} \binom{n-j}{k} b_{n-k},
\end{aligned}$$

which completes the induction. □

Corollary 4.5 *For $j \geq 1$, we have*

$$b_{j,j+1} = b_j, \qquad \textit{and} \qquad b_{j,j+2} = b_j + b_{j+1}.$$

For $j \geq 2$, we find

$$b_{j,j} = \sum_{k=0}^{j-2} (-1)^k b_{j-1-k}.$$

Proof. The first two results are direct consequences of Theorem 4.4. Apply

$$b_{j,j} = b_{j-1,j} - b_{j-1,j-1} = b_{j-1} - b_{j-1,j-1}$$

repeatedly to derive the last identity. □

The simplicity of Theorem 4.4 and Corollary 4.5 suggest that there may exist simple combinatorial proofs. We invite the readers to find them.

Bell numbers can be stated without using any recurrence. For instance, Dubinski's formula asserts that

$$B_n = \frac{1}{e} \sum_{k=0}^{\infty} \frac{k^n}{k!},$$

from which we obtain the following *ordinary* generating function

$$B(t) = \sum_{n=0}^{\infty} B_n t^n = \frac{1}{e} \sum_{k=0}^{\infty} \frac{1}{(1-kt)k!}.$$

Because of the remark following Theorem 2.6, we immediately obtain the next result.

Theorem 4.6 *Let $B_j(t) = \sum_{n=j}^{\infty} b_{j,n} t^n$. For $j \geq 0$.*

$$B_j(t) = (1-t)^j B(t) - \sum_{k=0}^{j-1} \left[\sum_{i=0}^{k} (-1)^i \binom{j}{i} b_{k-i} \right] t^k.$$

We are often more interested in the *exponential* generating function

$$\mathfrak{B}(t) = \sum_{n=0}^{\infty} b_n \frac{t^n}{n!} = \exp\left(e^t - 1\right).$$

For $j \geq 1$, the result for $b_{j,n}$ becomes rather easy if we study a non-standard form of exponential generating function:

$$\mathfrak{B}_j(t) = \sum_{n=j}^{\infty} b_{j,n+1} \frac{t^n}{(n-j)!}.$$

Although it may look odd, it is precisely its peculiar set up that allows us to apply Theorem 4.4 efficiently.

Theorem 4.7 *For $j \geq 1$, the "shifted" exponential generating function for the truncated sequence $\{b_{j,n+1}\}_{n=j}^{\infty}$ is*

$$\mathfrak{B}_j(t) = t^j e^t \mathfrak{B}^{(j)}(t),$$

where $\mathfrak{B}^{(j)}(t)$ denotes $D_t^j \mathfrak{B}(t)$.

Proof. Using Theorem 4.4 we find

$$\begin{aligned}
\mathfrak{B}_j(t) &= \sum_{n=j}^{\infty} \frac{t^n}{(n-j)!} \sum_{k=0}^{n-j} \binom{n-j}{k} b_{n-k} \\
&= t^j \sum_{k=0}^{\infty} \sum_{n=k+j}^{\infty} \frac{t^k}{k!} \cdot \frac{b_{n-k} t^{n-k-j}}{(n-k-j)!} \\
&= t^j \left(\sum_{k=0}^{\infty} \frac{t^k}{k!} \right) \left(\sum_{\ell=0}^{\infty} b_{j+\ell} \frac{t^\ell}{\ell!} \right),
\end{aligned}$$

from which the result follows. □

4.2. Diagonals of the Bell Difference Table

Following the same convention we adopted in the Catalan difference table, we name the diagonal entries in the Bell difference table $\tilde{b}_{j,n}$, where $n \geq j \geq 0$, and $\tilde{b}_{j,n} = b_{n-j,n}$. Combinatorially, $\tilde{b}_{j,n}$ counts the number of NS n-columns that do not have any repetition in its first $n-j$ positions.

Theorem 4.8 *For $n \geq j \geq 0$,*

$$\tilde{b}_{j,n} = \sum_{k=0}^{n-j} (-1)^k \binom{n-j}{k} b_{n-k}.$$

Proof. Analytically, since $\tilde{b}_{j,n} = b_{n-j,n}$, this is a direct consequence due to the construction of the difference table. What follows is a combinatorial proof. Let S denote the of set of all NS n-columns. For $1 \leq i \leq n-j$ define S_i to be the subset of S that contains all the NS n-columns with a repetition at position i. Obviously, $|S| = b_n$. Consider $S_{i_1} \cap S_{i_2} \cap \cdots \cap S_{i_k}$. We may assume the subscripts $i_1, i_2, \ldots, i_k$ form ℓ clusters of consecutive integers of sizes $m_1, m_2, \ldots, m_\ell$. For each q, the qth cluster of consecutive subscripts give rise to a block of $m_q + 1$ repeated letters in the n-columns. By deleting the first m_q occurrences of these repeated letters for each q, we obtain a NS column of size $n - \sum_{q=1}^{\ell} m_q = n - k$. Since the converse is also true, we see that $\left|S_{i_1} \cap S_{i_2} \cap \cdots \cap S_{i_k}\right| = n - k$ regardless of the choices of $i_1, i_2, \ldots, i_k$. The result follows from principle of inclusion-exclusion. □

Corollary 4.5 yields these special values for the first three diagonals:

$$\begin{aligned}\tilde{b}_{0,n} &= \sum_{k=0}^{n-2}(-1)^k b_{n-1-k}, \quad n \geq 2,\\ \tilde{b}_{1,n} &= b_{n-1}, \qquad\qquad n \geq 2,\\ \tilde{b}_{2,n} &= b_{n-2} + b_{n-1}, \quad n \geq 3.\end{aligned} \tag{4}$$

More generally, Theorem 4.4 becomes

Theorem 4.9 *For $n - 1 \geq j \geq 0$,*

$$\tilde{b}_{j+1,n+1} = \sum_{k=0}^{j} \binom{j}{k} b_{n-k} = \sum_{\ell=0}^{j} \binom{j}{\ell} b_{n+\ell-j}.$$

We can also compute the generating functions in the usual way.

Theorem 4.10 *Let $\widetilde{B}_j(t) = \sum_{n=j}^{\infty} \tilde{b}_{j,n} t^n$. Then $\widetilde{B}(t) = \widetilde{B}_0(t) = \frac{1+tB(t)}{1+t}$, and, for $j \geq 0$,*

$$\widetilde{B}_j(t) = (1+t)^j \widetilde{B}(t) - \sum_{k=0}^{j-1} \left[\sum_{i=0}^{k} \binom{j}{i} \tilde{b}_{k-i} \right] t^k.$$

Proof. It suffices to derive the generating function $\widetilde{B}(t) = \widetilde{B}_0(t)$. Define $b_{-1} = 1$ so that (4) can be rewritten as

$$\tilde{b}_{0,n} = \sum_{k=0}^{n} (-1)^k b_{n-1-k}, \qquad n \geq 0.$$

Then

$$\widetilde{B}(t) = \sum_{n=0}^{\infty} \sum_{k=0}^{n} (-1)^k b_{n-1-k} t^n = \left(\sum_{k=0}^{\infty} (-1)^k t^k \right) \left(\sum_{\ell=0}^{\infty} b_{\ell-1} t^\ell \right).$$

Hence $\widetilde{B}(t) = \frac{1+tB(t)}{1+t}$. □

For exponential generating functions, we study the shifted version

$$\widetilde{\mathfrak{B}}_j(t) = \sum_{n=j}^{\infty} \tilde{b}_{j,n} \frac{t^n}{(n-j)!}.$$

Theorem 4.11 *For $j \geq 0$, we find $\widetilde{\mathfrak{B}}_j(t) = t^j e^{-t} \mathfrak{B}^{(j)}(t)$. In particular, $\widetilde{\mathfrak{B}}_0(t) = e^{-t} \mathfrak{B}(t) = \exp(e^t - t - 1)$.*

Proof. Because of Theorem 4.8, we find

$$\begin{aligned}\widetilde{\mathfrak{B}}_j(t) = \sum_{n=j}^{\infty} \frac{t^n}{(n-j)!} \sum_{k=0}^{n-j} (-1)^k \binom{n-j}{k} b_{n-k} &= t^j \sum_{k=0}^{\infty} \sum_{n=k+j}^{\infty} \frac{(-t)^k}{k!} \cdot \frac{b_{n-k} t^{n-k-j}}{(n-k-j)!} \\ &= t^j \left(\sum_{k=0}^{\infty} \frac{(-t)^k}{k!} \right) \left(\sum_{\ell=0}^{\infty} b_{j+\ell} \frac{t^\ell}{\ell!} \right) \\ &= t^j e^{-t} \mathfrak{B}^{(j)}(t),\end{aligned}$$

from which the special case of $j = 0$ follows easily. □

5. Closing Remarks

Interested readers may want to study the difference tables of other famous sequences, especially those with well-known combinatorial meanings.

We could also extend the definition (2) to

$$K_{n,s,t} = \sum_{r=0}^{n} \frac{(-1)^{n-r}}{r+a} \binom{n}{r} \binom{2r+b}{r}.$$

Is there any combinatorial interpretation for this number?

We showed $K_n = \tilde{c}_n$ by proving that they share the same generating function. Is it possible to find a direct combinatorial proof? The algebraic structure of (2) suggests some kind of application of the principle of inclusion-exclusion may work.

We invite the readers to investigate these and other related problems.

References

[1] L. Comtet, *Advanced Combinatorics.* D. Reidel, Dordrecht, 1974.

[2] H. W. Gould, *Bell and Catalan Numbers: Research Bibliography of Two Special Number Sequences*, Revised Edition, Combinatorial Research Institute, Morgantown, WV, 1978, available at http://www.math.edu/~gould.

[3] D. E. Knuth, *The Art of Computer Programming*, Vol. 1, Addison-Wesley, Reading, MA, 1968.

[4] T. Koshy, *Catalan Numbers with Applications*, Oxford University Press, New York, 2009.

[5] J. Quaintance, "$m \times n$ Proper Arrays: Geometric and Algebraic Methods of Classification," Ph. D. thesis, University of Pittsburgh, Pittsburgh, PA, 2002.

[6] N. J. Sloane, The On-Line Encyclopedia of Integer Sequences, available at http://www.research.att.com/~njas/sequences.

[7] R. P. Stanley, Catalan Addendum, version 6, October 2008, available at http://www-math.mit.edu/~rstan/ec/catadd.pdf.

PISOT NUMBERS AND CHROMATIC ZEROS

Víctor F. Sirvent[1]
Departamento de Matemáticas, Universidad Simón Bolívar, Caracas, Venezuela
vsirvent@usb.ve

Received: 12/5/12, Accepted: 3/24/13, Published: 5/10/13

Abstract

In this article we show that Pisot numbers of even degree and their powers cannot be roots of chromatic polynomials. We also consider the family of smallest Pisot numbers of odd degree. We show that they cannot be roots of chromatic polynomials of connected graphs with a certain maximum number of vertices.

1. Introduction

An *algebraic integer* is a root of a monic polynomial with integer coefficients. A *Pisot number* is a real algebraic integer greater than 1 such that all its Galois conjugates are of norm smaller than 1. The *degree of the Pisot number* is the degree of its minimal polynomial. The Pisot numbers are also known as *Pisot-Vijayaraghavan numbers* or *PV numbers*. The best well-known is the golden mean, i.e. $(\sqrt{5}+1)/2$. The Pisot numbers form an infinite closed set ([6]). Due to their properties, these numbers play an important role in number theory (*cf.* [6, 7, 18]), harmonic analysis (*cf.* [6, 11, 16]), dynamical systems and ergodic theory (*cf.* [13, 14, 15, 17, 19]) and tilings (*cf.* [4, 12, 21, 22]).

One of the best known family of Pisot numbers is the so-called n-bonacci numbers, a generalization of the golden mean, i.e. the real root greater than 1, of the polynomial $x^n - x^{n-1} - \cdots - x - 1$ (*cf.* [8]).

The chromatic polynomial of a graph counts the number of its proper vertex colorings. More precisely, let G be a graph and t a positive integer. A *t-colouring* is a map from the set of vertices of G to $\{1, \ldots, t\}$ such that the images of adjacent vertices are different. Let $P_G(t)$ be the number of different t-colourings of G, it turns out that $P_G(t)$ is a polynomial in t ([5]), so we call $P_G(t)$ the *chromatic polynomial of G*. A root of the chromatic polynomial of G is called *chromatic zero* or *chromatic root*. Sokal ([20]) proved that the complex chromatic zeros are dense in the complex

[1]Webpage: http://www.ma.usb.ve/~vsirvent

plane. An important question is what kind of algebraic integers could be chromatic zeros, see [1] and references within, for different partial answers to this question. In particular, studies have been done concerning which algebraic numbers are not chromatic zeros. Alikhani and Peng ([2]) showed the golden mean is not a chromatic zero and in [3] they showed that n-bonacci numbers are not chromatic zeros when n is even. In the case of n odd, the authors showed that the n-bonacci number cannot be a chromatic zero of a connected graph having at most $4n+2$ vertices. In the present article we generalized those results to Pisot numbers. In particular we show in Theorem 1 that a Pisot number of even degree and its natural powers are not chromatic zeros. In Theorems 3 and 4, we consider some important families of Pisot numbers of odd degree. Using similar arguments to the proof of Theorem 5 in [3], we prove that those Pisot numbers could not be chromatic zeros of connected graphs having certain maximum number of vertices.

It is well-known that the chromatic zeros are contained in the complement of $(-\infty, 0) \cup (0,1) \cup (1, 32/27]$ ([10]). Since the smallest Pisot number (also known as the *plastic number*), the real root of $x^3 - x - 1$ (*cf.* [18]), is approximately 1.32, we ask when a Pisot number is a chromatic zero. In Theorems 3 and 4 we give a negative answer of this question for the smallest isolated Pisot numbers.

We consider the following families of polynomials:

1. $\Psi_n(x) := x^{2n+1} - x^{2n-1} - \cdots - x - 1$,
2. $\Phi_n(x) := x^{2n+1} - x^{2n} - x^{2n-2} - \cdots - x^2 - 1$,
3. $\Xi_n(x) := x^n(x^2 - x - 1) + x^2 - 1$,

where n is a positive integer. These polynomials are irreducible in $\mathbb{Z}[x]$ and have a real root greater than 1, which is a Pisot number (*cf.* [6]). Some of the Pisot numbers associated to these polynomials are well-known: $\Xi_1(x) = \Psi_1(x) = x^3 - x - 1$ is the minimal polynomial of the smallest Pisot number, ξ_1. The minimal polynomial of the second smallest Pisot number, $\xi_2 \approx 1.38$, is $\Xi_2(x) = x^4 - x^3 - 1$ (*cf.* [18]). We denote by ψ_n, ϕ_n and ξ_n the Pisot numbers associated to the polynomial $\Psi_n(x)$, $\Phi_n(x)$ and $\Xi_n(x)$, respectively. They were considered by C.L. Siegel in [18]. They are the smallest Pisot numbers and they are ordered as follows (*cf.* [6]):

$$\xi_1 = \psi_1 < \xi_2 < \xi_3 < \phi_1 < \xi_4 < \psi_2 < \xi_5 < \phi_2 < \xi_6 < \psi_3 < \cdots <$$
$$\cdots < \phi_n < \xi_{2n+2} < \psi_{n+1} < \xi_{2n+3} < \phi_{n+1} < \cdots < \frac{1+\sqrt{5}}{2}.$$

Moreover they are isolated points of the set of Pisot numbers and the golden mean is the smallest limit point of the set of Pisot numbers (*cf.* [6]).

We do not know if for all positive integers n and $l > 1$, the numbers ψ_n^l, ϕ_n^l are not chromatic zeros. And similarly for ξ_n^l, with n odd.

2. Results

Theorem 1. *A Pisot number of even degree is not a chromatic zero. Moreover the natural powers of a Pisot number of even degree are not chromatic zeros.*

Proof. Let θ be a Pisot number of even degree and $\Theta(x)$ its minimal polynomial. By definition, the degree of $\Theta(x)$ is even. So it has at least two real roots: one root is θ, and the other root θ' is inside the unit circle. So θ' is in the interval $(-1, 1)$. It cannot be zero due to the minimality of $\Theta(x)$. Therefore the root θ' is in $(-1, 0) \cup (0, 1)$ which is not possible for a chromatic polynomial. Hence $\Theta(x)$ is not a chromatic polynomial.

Let $C(x)$ be a polynomial with integer coefficients having θ as a root. So $\Theta(x)$ is a factor of $C(x)$, and hence $C(x)$ has a root in $(-1, 0) \cup (0, 1)$. Therefore $C(x)$ is not a chromatic polynomial and θ is not a chromatic zero.

Suppose that θ^n is a chromatic zero, for some positive integer n. So there exists a chromatic polynomial $P(x)$ such that θ^n is one of its roots. It follows that θ is a root of $Q(x) := P(x^n)$. Hence $\Theta(x)$ is a factor of $Q(x)$. We have seen in the first part of the proof that there exists $\theta' \in (-1, 0) \cup (0, 1)$, a root of $\Theta(x)$. Therefore θ' is a root of $Q(x)$, so θ'^n is a root of $P(x)$. Since $\theta'^n \in (-1, 0) \cup (0, 1)$, $P(x)$ could be a chromatic polynomial. We conclude that θ^n is not a chromatic zero. □

Theorem 2 ([9]). *Let G be a graph with n vertices and k connected components. Then the chromatic polynomial of G is of the form*

$$P_G(t) = a_n t^n + a_{n-1} t^{n-1} + \cdots + a_k t^k$$

where a_i are integers such that $a_n = 1$ and $(-1)^{n-i} a_i > 0$, for $k \leq i \leq n$. Moreover, if G has at least one edge, then 1 is a root of $P_G(t)$.

We prove the following theorem using arguments based on the proof of Theorem 5 in [3].

Theorem 3. *Let ψ_n be the Pisot number whose minimal polynomial is $\Psi_n(x)$. If $n \geq 2$ then ψ_n is not a chromatic zero of a connected graph with at most $4n + 1$ vertices.*

Proof. According to Theorem 2, the polynomial $\Psi_n(x)$ is not a chromatic polynomial. If ψ_n is a root of a chromatic polynomial $C(x) = \sum_{i=0}^{m} a_i x^i$, we have that $\Psi_n(x)$ is a proper factor of $C(x)$ and $m > 2n + 1$. So there exists a polynomial $D(x)$, such that $C(x) = \Psi_n(x) D(x)$ and

$$D(x) = b_{m-2n-1} x^{m-2n-1} + \cdots + b_1 x + b_0.$$

So when we develop the product $\Psi_n(x)D(x)$, we get the following equations when $m-2n-1 \le 2n-1$:

$$-b_0 = a_0, \quad \cdots \quad -b_{m-2n-1} - \cdots - b_1 - b_0 = a_{m-2n-1},$$

and for $m-2n-1 = 2n$:

$$-b_0 = a_0, \quad \cdots \quad -b_{m-2n-1} - \cdots - b_1 = a_{m-2n-1}.$$

Due to Theorem 2, $b_0 = a_0 = 0$, since 0 is a root of a chromatic polynomial. And by the same theorem, the number 1 is a root of $C(x)$. Since $\Psi_n(1) \neq 0$, we have $D(1) = 0$, so $b_{m-2n-1} + \cdots + b_1 + b_0 = 0$, i.e. $a_{m-2n-1} = 0$. This fact contradicts Theorem 2. Therefore ψ_n is not a chromatic zero of a connected graph with m vertices, where $m-2n-1 \le 2n$, i.e. $m \le 4n+1$. □

Theorem 4. *Let ξ_n and ϕ_n the Pisot numbers whose minimal polynomials are $\Xi_n(x)$ and $\Phi_n(x)$, respectively.*

(a) The plastic number, i.e. ξ_1, is not a chromatic zero of a connected graph with at most 7 vertices.

(b) The Pisot number ξ_3 is not a chromatic zero of a connected graph with at most 10 vertices.

(c) The Pisot number ξ_n, with $n \ge 5$ and odd, is not a chromatic zero of a connected graph with at most $n+5$ vertices.

(d) The Pisot number ϕ_n, with $n \ge 2$, is not a chromatic zero of a connected graph with at most $2n+4$ vertices.

Proof. Due to Theorem 2, $\Xi_1(x)$ is not a chromatic polynomial. We suppose that ξ_1 is a root of the chromatic polynomial $C(x) = \sum_{i=0}^{m} a_i x^i$, with $m > 3$, so $\Xi_1(x)$ is a factor of $C(x)$. Let $D(x)$ be as in the proof of Theorem 3, with $C(x) = D(x)\Xi_1(x)$. If $m-3 \le 4$, then we have the following equations:

$$-b_0 = a_0, \quad -b_1 - b_0 = a_1, \quad -b_1 - b_2 = a_2, \quad b_0 - b_2 - b_3 = a_3, \quad b_1 - b_3 - b_4 = a_4.$$

As in the proof of Theorem 3, $b_0 = a_0 = 0$, and $C(1) = D(1) = 0$, so

$$\begin{aligned} a_0 + \cdots + a_4 &= 0, \\ -b_0 + (-b_0 - b_1) + (-b_1 - b_2) + (b_0 - b_2 - b_3) + (b_1 - b_3 - b_4) &= 0, \\ -b_2 - b_3 &= 0, \\ a_3 &= 0. \end{aligned}$$

Hence $C(x)$ could not be a chromatic polynomial, by Theorem 2 . Therefore ξ_1 is not a chromatic zero of a connected graph having m vertices with $m \leq 7$. This completes the proof of statement (a).

The polynomial $\Xi_3(x)$ is not a chromatic polynomial, due to Theorem 2. Let $C(x) = \sum_{i=0}^{m} a_i x^i$, with $m > 5$, be a polynomial having ξ_3 as a root. We suppose that $C(x)$ is chromatic, so there exists $D(x) = \sum_{j=0}^{m-5} b_j x^j$ such that $C(x) = D(x)\Xi_3(x)$. If $m \leq 10$ we have the following equations:

$$\begin{array}{llll} -b_0 = a_0, \; -b_1 = a_1, & b_0 - b_2 = a_2, & & -b_0 + b_1 - b_3 = a_3, \\ -b_0 - b_1 + b_2 - b_4 = a_4, & b_0 - b_1 - b_2 + b_3 - b_5 = a_5. & & \end{array}$$

As in (a) $b_0 = a_0 = 0$ and $C(1) = D(1) = 0$, so

$$\begin{aligned} a_0 + \cdots + a_5 &= 0, \\ -b_1 - b_2 + (b_1 - b_3) + (-b_1 + b_2 - b_4) + (-b_1 - b_2 + b_3 - b_5) &= 0, \\ -b_1 + b_3 &= 0, \\ -a_3 &= 0. \end{aligned}$$

Hence $C(x)$ could not be a chromatic polynomial, by Theorem 2 . Therefore ξ_3 is not a chromatic zero of a connected graph having m vertices with $m \leq 10$. This completes the proof of statement (b).

The polynomial $\Xi_n(x)$ is not a chromatic polynomial, due to Theorem 2. Let n be an odd number and $n \geq 5$. Let $C(x) = \sum_{i=0}^{m} a_i x^i$, with $m > n+2$, be a polynomial having ξ_n as a root. We suppose that $C(x)$ is chromatic, so there exists $D(x) = \sum_{j=0}^{m-n-2} b_j x^j$ such that $C(x) = D(x)\Xi_n(x)$. If $m - n - 2 \leq 3$ we have the following equations:

$$-b_0 = a_0, \quad -b_1 = a_1, \quad b_0 - b_2 = a_2, \quad b_1 - b_3 = a_3.$$

As in (a), $b_0 = a_0 = 0$ and $C(1) = D(1) = 0$, so

$$\begin{aligned} a_0 + a_1 + a_2 + a_3 &= 0, \\ -b_0 - b_1 + b_0 - b_2 + b_1 - b_3 &= 0, \\ b_1 &= 0, \\ -a_1 &= 0. \end{aligned}$$

This contradicts the assumption that $C(x)$ is chromatic. Therefore ξ_n, with $n \geq 5$ is not a chromatic zero of a connected graph having m vertices with $m \leq n+5$. This completes the proof of statement (c).

Let $n \geq 2$. Due to Theorem 2, $\Phi_n(x)$ is not a chromatic polynomial. We suppose that ϕ_n is a root of the chromatic polynomial $C(x) = \sum_{i=0}^{m} a_i x^i$, with $m > 2n+1$, so

There exists $D(x) = \sum_{j=0}^{m-2n-1} b_j x^j$ such that $C(x) = D(x)\Phi_n(x)$. If $m-2n-1 \leq 3$ we have the following equations:

$$-b_0 = a_0, \quad -b_1 = a_1, \quad -b_0 - b_2 = a_2, \quad -b_1 - b_3 = a_3.$$

As in the previous proofs, $b_0 = a_0 = 0$ and $C(1) = D(1) = 0$, so

$$\begin{aligned} a_0 + a_1 + a_2 + a_3 &= 0, \\ -b_0 - b_1 - b_0 - b_2 - b_1 - b_3 &= 0, \\ -b_1 &= 0, \\ a_1 &= 0. \end{aligned}$$

Therefore ϕ_n, with $n \geq 2$, is not a chromatic zero of a connected graph having m vertices with $m \leq 2n + 4$. This completes the proof of statement (d). □

Open problem. We do not know if the numbers ψ_n^l, ϕ_n^l are chromatic zeros, for integers $n \geq 1$ and $l > 1$; and similarly for ξ_n^l, when n is odd.

References

[1] S. Alikhani, R. Hasni: *Algebraic integers as chromatic and domination roots,* Int. J. Combinatorics **2012** (2012), Article ID 780765, 8 pp.

[2] S. Alikhani, Y. H. Peng: *Chromatic zeros and the golden ratio*, Appl. Anal. Discrete Math. **3** (2009), 120-122.

[3] S. Alikhani, Y.H. Peng: *Chromatic zeros and generalized Fibonacci numbers,* Appl. Anal. Discrete Math. **3** (2009), 330-335.

[4] V. Berthé, A. Siegel: *Tilings associated with beta-numeration and substitutions,* Integers **5** (2005), #A02.

[5] G.D. Birkhoff: *A Determinant formula for the number of ways of coloring a map,* Ann. Math. **14**, 42–46, 1912.

[6] M. J. Bertin, A. Decomps-Guilloux, M. Grandet-Hugot, M. Pathiaux-Delefosse, J. P. Schreiber: *Pisot and Salem Numbers*, Basel, Birkhäuser, 1992.

[7] D.W. Boyd: *Pisot and Salem numbers in intervals of the real line,* Math. Comp. **32** (1978), 1244–1260.

[8] A. Brauer: *On algebraic equations with all but one root in the interior of the unit circle,* Math. Nachr. **4** (1951), 250–257.

[9] F. M. Dong, K. M. Koh, K. L. Teo: *Chromatic Polynomials and Chromaticity of Graphs*, World Scientific Publishing Co. Pte. Ltd. 2005.

[10] B. Jackson: *A zero free interval for chromatic polynomials of graphs,* Combin. Probab. Comput. **2** (1993), 325–336.

[11] Y. Meyer: *Algebraic Numbers and Harmonic Analysis*, North-Holland (1972).

[12] R.V. Moody (ed.): *The Mathematics of Long-Range Aperiodic Order*, Kluwer Acad. Publ. (1997).

[13] N. Pytheas Fogg: *Substitutions in Dynamics, Arithmetics and Combinatorics*, (edited by V. Berthé, S. Ferenczi, C. Mauduit, et al.), Lecture Notes in Mathematics **1794**, Springer, Berlin, 2002.

[14] M. Queffélec: *Substitution Dynamical Systems – Spectral Analysis*, Lecture Notes in Mathematics **1294**, Springer, Berlin, 1987.

[15] G. Rauzy: *Nombres algébriques et substitutions*, Bull. Soc. Math. France **110** (1982), 147–178.

[16] R. Salem: *Algebraic Numbers and Fourier Analysis*, Heath, 1963.

[17] K. Schmidt: *On periodic expansions of Pisot numbers and Salem numbers,* Bull. London Math. Soc. **12** (1980), 269-278.

[18] C.L. Siegel: *Algebraic numbers whose conjugates lie in the unit circle*, Duke Math. J. **11** (1944), 597-602.

[19] V.F. Sirvent, Y. Wang: *Self-affine tilings via substitution dynamical systems and Rauzy fractals,* Pacific Journal of Mathematics **206** (2002), 465–485.

[20] A. D. Sokal: *Chromatic roots are dense in the whole complex plane*, Combinatorics, Probability and Computing **13** (2004), 221–261.

[21] B. Solomyak: *Dynamics of self-similar tilings*, Ergodic Theory Dynam. Systems, **17** (1997), 695–738.

[22] W. Thurston: *Groups, Tilings and Finite State Automata*, Lecture Notes AMS, 1989.

COORDINATE SUM AND DIFFERENCE SETS OF d-DIMENSIONAL MODULAR HYPERBOLAS[1]

Amanda Bower
Department of Mathematics and Statistics, University of Michigan-Dearborn, Dearborn, Michigan
amandarg@umd.umich.edu

Ron Evans
Dept. of Mathematics, University of California San Diego, La Jolla, California
revans@ucsd.edu

Victor Luo
Department of Mathematics and Statistics, Williams College, Williamstown, MA
victor.d.luo@williams.edu

Steven J Miller
Department of Mathematics and Statistics, Williams College, Williamstown, MA
sjm1@williams.edu, Steven.Miller.MC.96@aya.yale.edu

Received: 12/20/13, Revised: 3/16/13, Accepted: 4/4/13, Published: 5/24/13

Abstract

Many problems in additive number theory, such as Fermat's last theorem and the twin prime conjecture, can be understood by examining sums or differences of a set with itself. A finite set $A \subset \mathbb{Z}$ is considered sum-dominant if $|A+A| > |A-A|$. If we consider all subsets of $\{0, 1, \dots, n-1\}$, as $n \to \infty$, it is natural to expect that almost all subsets should be difference-dominant, as addition is commutative but subtraction is not; however, Martin and O'Bryant in 2007 proved that a positive percentage are sum-dominant as $n \to \infty$. This motivates the study of "coordinate sum dominance." Given $V \subset (\mathbb{Z}/n\mathbb{Z})^2$, we call $S := \{x+y : (x,y) \in V\}$ a coordinate sumset and $D := \{x-y : (x,y) \in V\}$ a coordinate difference set, and we say V is coordinate sum dominant if $|S| > |D|$. An arithmetically interesting choice of V is $\bar{H}_2(a;n)$, which is the reduction modulo n of the modular hyperbola $H_2(a;n) := \{(x,y) : xy \equiv a \bmod n, 1 \le x, y < n\}$. In 2009, Eichhorn, Khan, Stein, and Yankov determined the sizes of S and D for $V = \bar{H}_2(1;n)$ and investigated conditions for coordinate sum dominance. We extend their results to reduced d-dimensional modular hyperbolas $\bar{H}_d(a;n)$ with a coprime to n.

[1]Keywords: Modular hyperbolas, coordinate sumset, coordinate difference set. MSC 2010 Subject Classification: 11P99, 14H99 (primary), 11T23 (secondary).

We thank Mizan R. Khan for introducing us to this problem and, along with the referee, for helpful comments on an earlier draft. The first and third named authors were supported by Williams College and NSF grant DMS0850577; the fourth named author was partially supported by NSF grant DMS0970067.

1. Introduction

Let $A \subset \mathbb{N} \cup \{0\}$. Two natural sets to study are

$$\begin{aligned} A+A &= \{x+y : x, y \in A\} \\ A-A &= \{x-y : x, y \in A\}. \end{aligned} \tag{1.1}$$

The former is called the sumset and the latter the difference set. Many problems in additive number theory can be understood in terms of sum and difference sets. For instance, the Goldbach conjecture says that the even numbers greater than 2 are a subset of $P+P$, where P is the set of primes. The twin prime conjecture states that there are infinitely many ways to write 2 as a difference of primes (and thus if P_N is the set of primes exceeding N, $P_N - P_N$ always contains 2). If we let A_n be the set of positive n^{th} powers, then Fermat's Last Theorem says $(A_n + A_n) \cap A_n = \emptyset$ for all $n > 2$.

Let $|S|$ denote the cardinality of a set S. A set A is sum dominant if $|A+A| > |A-A|$. We might expect that almost all sets are difference dominant since addition is commutative while subtraction is not. However, in 2007 Martin and O'Bryant [7] proved that a positive percentage of sets are sum dominant; i.e., if we look at all subsets of $\{0, 1, \dots, n-1\}$ then as $n \to \infty$ a positive percentage are sum dominant. One explanation is that choosing A uniformly from $\{0, 1, \dots, n-1\}$ is equivalent to taking each element from 0 to $n-1$ to be in A with probability $1/2$. By the Central Limit Theorem this implies that there are approximately $n/2$ elements in a typical A, yielding on the order of $n^2/4$ pairs whose sum must be one of $2n-1$ possible values. On average we thus have each possible value realized on the order of $n/8$ ways. It turns out most possible sums and differences are realized (the expected number of missing sums and differences are 10 and 6, respectively). Thus most sets are close to being balanced, and we just need a little assistance to push a set to being sum-dominant. This can be done by carefully controlling the fringes of A (the elements near 0 and $n-1$). Such constructions are the basis of numerous results in the field; see for example [6, 7, 8, 9, 15, 16].

This motivates the study of "coordinate sum dominance" on fringeless sets such as $(\mathbb{Z}/n\mathbb{Z})^2$. Given $V \subset (\mathbb{Z}/n\mathbb{Z})^2$, we call $S := \{x+y : (x,y) \in V\}$ a coordinate sumset and $D := \{x-y : (x,y) \in V\}$ a coordinate difference set, and we say V is coordinate sum dominant if $|S| > |D|$. An arithmetically interesting choice of V is $\bar{H}_2(a;n)$, which is the reduction modulo n of the modular hyperbola

$$H_2(a;n) := \{(x,y) : xy \equiv a \bmod n,\ 1 \le x, y < n\}, \tag{1.2}$$

where $(a,n) = 1$. Eichhorn, Khan, Stein, and Yankov [2] determined the cardinalities of S and D for $V = \bar{H}_2(1;n)$ and investigated conditions for coordinate sum dominance. See [10] for additional results on related problems in other modular settings.

The modular hyperbolas in (1.2) have very interesting structure, as is evidenced in Figure 1. See Figures 1 through 4 of [2] for additional examples.

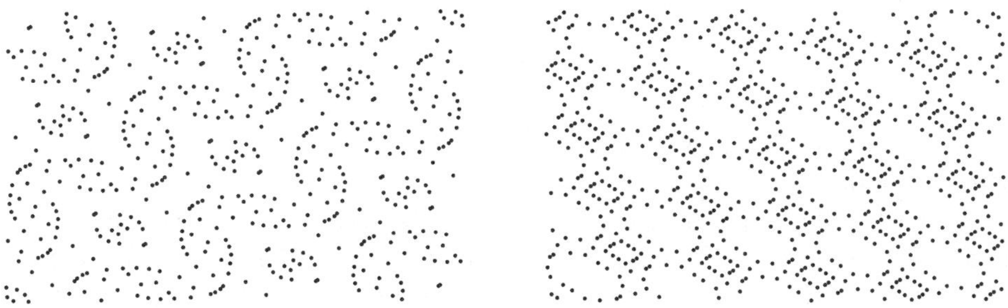

Figure 1: (a) Left: $H_2(51; 2^{10})$. (b) Right: $H_2(1325; 48^2)$.

In the sequel, coordinate sumsets will be the only type of sumset discussed. Hence we may drop the premodifier "coordinate" without fear of confusion. For a relatively prime to n, we define the sumset $S_2(a;n)$, the difference set $D_2(a;n)$, and their reduced counterparts as

$$\begin{aligned} S_2(a;n) &= \{x_1 + x_2 : (x_1,x_2) \in H_2(a;n)\} \\ D_2(a;n) &= \{x_1 - x_2 : (x_1,x_2) \in H_2(a;n)\} \\ \bar{S}_2(a;n) &= \{x_1 + x_2 \bmod n : (x_1,x_2) \in H_2(a;n)\} \\ \bar{D}_2(a;n) &= \{x_1 - x_2 \bmod n : (x_1,x_2) \in H_2(a;n)\}. \end{aligned} \tag{1.3}$$

From a geometric viewpoint, $\#S_2(a;n)$ counts the number of lines of slope -1 that intersect $H_2(a;n)$, and $\#D_2(a;n)$ counts the number of lines of slope 1 that intersect $H_2(a;n)$. When the ratio

$$c_2(a;n) := \#\bar{S}_2(a;n)/\#\bar{D}_2(a;n) \tag{1.4}$$

exceeds 1, we have sum-dominance of $\bar{H}_2(a;n)$.

A d-dimensional modular hyperbola is of the form

$$H_d(a;n) := \{(x_1,\ldots,x_d) : x_1 \cdots x_d \equiv a \bmod n, 1 \le x_1,\ldots,x_d < n\}, \tag{1.5}$$

where $(a,n) = 1$. We define the generalized signed sumset as

$$\bar{S}_d(m;a;n) = \{x_1 + \cdots + x_m - \cdots - x_d \bmod n : (x_1,\ldots,x_d) \in H_d(a;n)\}, \tag{1.6}$$

where m is the number of plus signs in $\pm x_1 \pm \cdots \pm x_d$. In particular, $\bar{S}_2(1;a;n) = \bar{D}_2(a;n)$ and $\bar{S}_2(2;a;n) = \bar{S}_2(a;n)$.

Modular hyperbolas have been extensively studied; see for example the recent survey by Shparlinski [11]. In particular, when a is not divisible by the prime p, Shparlinski and Winterhof [12] determined that the number of distances $|x-y|$ as (x,y) ranges over all points on the modular hyperbola $H_2(a;p)$ is

$$\frac{1}{4}\left(p+1+\left(\frac{a}{p}\right)\left(1+(-1)^{(p-1)/2}\right)\right). \tag{1.7}$$

In [13], they also found asymptotic formulas for the number of relatively prime points in $H_d(a;n)$.

The goal of this paper is to extend results of [2] to the general two-dimensional modular hyperbolas in (1.2), and to investigate the higher dimensional modular hyperbolas defined in (1.5) (this is a generalization of Question 24 of [11] to arbitrary dimensions and combinations). We prove explicit formulas for the cardinalities of the sumsets $\bar{S}_2(a;n)$ and difference sets $\bar{D}_2(a;n)$ (Theorems 3.3 and 3.6). This allows us to analyze the ratios $c_2(a;n)$ (Theorems 3.8 – 3.12), thus providing conditions on a and n for sum dominance and difference dominance of reduced modular hyperbolas $\bar{H}_2(a;n)$. For example, a special case of Theorem 3.9 shows that if $a=11$ and $n=3^t7^s$ with $t\geq 2$, then $c_2(a;n)>1$, i.e., we have sum dominance. A special case of Theorem 3.12 shows that when a is a fixed power of 4, we have sum dominance for more than 85% of those n relatively prime to a. For $d>2$ and positive integers n whose prime factors all exceed 7, we prove in Theorem 4.1 that $\#\bar{S}_d(m;a;n)=n$. This means that each such generalized sumset consists of all possible values mod n, i.e., all possible sums and differences occur.

2. Counting Preliminaries

In this section we present some counting results that are central to proving our main theorems. Many of these are natural generalizations of results from [2], so we refer the reader to the appendix for detailed proofs.

Throughout this paper, p always denotes a prime. The following proposition reduces the analysis of the cardinalities of $\bar{S}_d(m;a;n)$ to those of $\bar{S}_d(m;a;p^t)$, where p^t is a factor in the canonical factorization of n.

Proposition 2.1. *Let $n=\prod_{i=1}^k p_i^{e_i}$ be the factorization of n into distinct prime powers. Then*

$$\#\bar{S}_d(m;a;n) = \prod_{i=1}^k \#\bar{S}_d(m;a;p_i^{e_i}). \tag{2.1}$$

The proof is given in Appendix A.1.

Lemma 2.2 cuts our work in half, as once we understand the sumset we immediately have results for the corresponding difference set.

Lemma 2.2. *We have* $\bar{S}_2(a;n) = \bar{D}_2(-a;n)$.

Proof. We show $\bar{S}_2(a;n) \subseteq \bar{D}_2(-a;n)$; the reverse containment is handled similarly. Let $\tau \in \bar{S}_2(a;n)$. Then there exists $(x_0, y_0) \in H_2(a;n)$ such that $x_0 y_0 \equiv a \bmod n$ and $x_0 + y_0 \equiv \tau \bmod n$. Since $(x_0, n - y_0) \in H_2(-a;n)$ and $\tau \equiv x_0 - (n - y_0) \bmod n$, we see that $\tau \in \bar{D}_2(-a;n)$. □

Lemma 2.3. *We have* $(2k \bmod p^t) \in \bar{D}_2(a;p^t) \Leftrightarrow (k^2 + a)$ *is a square modulo* p^t. *The map* $f(k) = 2k \bmod p^t$ *defines a bijection*

$$f : \{k : k^2 + a \text{ is a square mod } p^t, 0 \leq k < p^t\} \to \bar{D}_2(a;p^t) \tag{2.2}$$

when $p > 2$. *If* $p = 2$, *then* f *defines a bijection*

$$f : \{k : k^2 + a \text{ is a square mod } 2^t, 0 \leq k < 2^{t-1}\} \to \bar{D}_2(a;2^t). \tag{2.3}$$

See Appendix A.2 for the proof. By Lemma 2.2, a similar result for S in place of D follows by replacing a by $-a$.

3. Cardinalities of $\bar{S}_2(a;p^t)$ and $\bar{D}_2(a;p^t)$

In this section we compute the cardinalities of $\bar{S}_2(a;p^t)$ and $\bar{D}_2(a;p^t)$. We then give conditions on a and n for sum dominance and difference dominance of $\bar{H}_2(a;n)$.

3.1. Case 1: $p = 2$

We isolate a useful result that we need for the proof of the next lemma. For a proof, see Proposition A.2 in the Appendix.

Proposition 3.1 (Gauss [4]). *For* $t \geq 1$, *any integer of the form* $4^k(8n+1)$ *is a square modulo* 2^t.

The next result is used in investigating some of the cases of Theorem 3.3.

Lemma 3.2. *Write* $m = 4b + r$ *with* $0 \leq r \leq 3$. *For* $t \geq 5$, $k^2 + 3 + 8m$ *is a square* mod 2^t *if and only if* $k \equiv \pm(4r+1) \bmod 16$.

Proof. We only prove the case $r = 0$, as the other proofs are similar. First assume that $k^2 + 3 + 8m$ is a square mod 2^t. Reducing mod 32, we have that $k^2 + 3 + 8m \equiv k^2 + 3 \bmod 32$, which implies that $k \equiv \pm 1 \bmod 16$. Conversely, assume that $k = 16l \pm 1$ for some $l \in Z$. Then $k^2 + 3 + 8m = (16l \pm 1)^2 + 3 + 8(4b) = 256l^2 \pm 32l + 4 + 32b = 4(8(8l^2 \pm l + b) + 1)$. Hence, by Proposition 3.1, $k^2 + 3 + 8m$ is a square mod 2^t. □

Theorem 3.3. *For $t \geq 5$,*

$$\#\bar{D}_2(a;2^t) = \begin{cases} \frac{2^{t-4}}{3} + \frac{(-1)^{t-1}}{3} + 3 & a \equiv 7 \bmod 8 \\ 2^{t-3} & a \equiv 1,5 \bmod 8 \\ 2^{t-4} & a \equiv 3 \bmod 8 \end{cases}$$

$$\#\bar{S}_2(a;2^t) = \begin{cases} \frac{2^{t-4}}{3} + \frac{(-1)^{t-1}}{3} + 3 & a \equiv 1 \bmod 8 \\ 2^{t-3} & a \equiv 3,7 \bmod 8 \\ 2^{t-4} & a \equiv 5 \bmod 8. \end{cases} \quad (3.1)$$

Moreover, $\#\bar{S}_2(a;16) = 2$ for all a, and when $t \leq 3$, we have $\#\bar{S}_2(a;2^t) = 1$ with the exception that $\#\bar{S}_2(a;8) = 2$ when $a \equiv 1 \bmod 4$.

Proof. The claim for $t \leq 4$ can be checked by direct calculation, so assume $t \geq 5$. By Lemma 2.2, it is enough to prove the claims about $\#\bar{D}_2(a;2^t)$ when $a \equiv 1,3,5 \bmod 8$, and about $\#\bar{S}_2(a;2^t)$ when $a \equiv 1 \bmod 8$.

We refer to the Appendix A.3 for the proofs of the results for $\bar{D}_2(a;2^t)$ when $a \equiv 1,5 \bmod 8$ and for $\bar{S}_2(a;2^t)$ when $a \equiv 1 \bmod 8$. It remains to prove the result for the difference set when $a \equiv 3 \bmod 8$. Write $a = 3 + 8m$. We consider only the case where $m \equiv 0 \bmod 4$, since the cases $m \equiv 1,2,3 \bmod 4$ are proved similarly and lead to the same result. By Lemma 3.2, we see that if $m \equiv 0 \bmod 4$, then

$$\begin{aligned} &\#\{k : k^2 + 3 + 8m \text{ is a square } \bmod 2^t,\ 0 \leq k < 2^{t-1}\} \\ &= \#\{1 + 16l : 0 \leq l < 2^{t-5}\} + \#\{15 + 16l : 0 \leq l < 2^{t-5}\} = 2^{t-4}. \end{aligned} \quad (3.2)$$

By Lemma 2.3, we know

$$\#\bar{D}(a;2^t) = \#\{k : k^2 + 3 + 8m \text{ is a square} \bmod 2^t,\ 0 \leq k < 2^{t-1}\}. \quad (3.3)$$

Hence $\#\bar{D}_2(a;2^t) = 2^{t-4}$. □

3.2. Case 2: $p > 2$

For this subsection, we adopt the following notation from [2]:

$$\begin{aligned} S_2'(a;p^t) &= \{k \bmod p^t : k^2 - a \text{ is a square} \bmod p^t,\ p \nmid (k^2 - a)\} \\ S_2''(a;p^t) &= \{k \bmod p^t : k^2 - a \text{ is a square} \bmod p^t,\ p | (k^2 - a)\}. \end{aligned} \quad (3.4)$$

By Lemma 2.3,

$$\#\bar{S}_2(a;p^t) = \#S_2'(a;p^t) + \#S_2''(a;p^t). \quad (3.5)$$

Lemma 3.4. *Let p be an odd prime. Then*

$$\#\bar{S}_2'(a;p^t) = \begin{cases} \frac{(p-1)p^{t-1}}{2} & \left(\frac{a}{p}\right) = -1 \\ \frac{(p-3)p^{t-1}}{2} & \left(\frac{a}{p}\right) = 1. \end{cases} \quad (3.6)$$

See Appendix A.4 for the proof.

Lemma 3.5. *Let p be an odd prime. If* $\left(\frac{a}{p}\right) = -1$, *then* $\#S_2''(a;p^t) = 0$ *and thus* $\#S_2(a;p^t) = \frac{\phi(p^t)}{2}$. *If* $\left(\frac{a}{p}\right) = 1$,

$$\#S_2''(a;p^t) \;=\; \frac{p^{t-1}}{p+1} + \frac{3}{2} + \frac{(-1)^{t-1}(p-1)}{2(p+1)}. \tag{3.7}$$

See Appendix A.5 for the proof.

Theorem 3.6. *For $t \geq 1$ and $p > 2$,*

$$\#\bar{S}_2(a,p^t) \;=\; \begin{cases} \frac{(p-3)p^{t-1}}{2} + \frac{p^{t-1}}{p+1} + \frac{3}{2} + \frac{(-1)^{t-1}(p-1)}{2(p+1)} & \left(\frac{a}{p}\right) = 1 \\ \frac{\phi(p^t)}{2} & \left(\frac{a}{p}\right) = -1 \end{cases}$$

$$\#\bar{D}_2(a,p^t) \;=\; \begin{cases} \frac{(p-3)p^{t-1}}{2} + \frac{p^{t-1}}{p+1} + \frac{3}{2} + \frac{(-1)^{t-1}(p-1)}{2(p+1)} & p \equiv 1 \bmod 4, \left(\frac{a}{p}\right) = 1 \\ \frac{\phi(p^t)}{2} & p \equiv 1 \bmod 4, \left(\frac{a}{p}\right) = -1 \\ \frac{(p-3)p^{t-1}}{2} + \frac{p^{t-1}}{p+1} + \frac{3}{2} + \frac{(-1)^{t-1}(p-1)}{2(p+1)} & p \equiv 3 \bmod 4, \left(\frac{a}{p}\right) = -1 \\ \frac{\phi(p^t)}{2} & p \equiv 3 \bmod 4, \left(\frac{a}{p}\right) = 1. \end{cases} \tag{3.8}$$

Proof. The result follows from Lemmas 3.4, 3.5, and 2.2. □

Corollary 3.7. *For $p \equiv 1 \bmod 4$, $c_2(a;p^k) = 1$.*

3.3. Ratios for $d = 2$

Now that we have explicit formulas for the cardinalities of the sum and difference sets, the next natural object to study is the ratio $c_2(a;n)$ of the size of the sumset to the size of the difference set. By Corollary 3.7, we only need to consider the prime factors of n which are congruent to 3 mod 4, since the primes which are congruent to 1 mod 4 do not change $c_2(a;n)$. When $p \equiv 3 \bmod 4$, it is sufficient to evaluate $c_2(a;p^t)$ in the case when $\left(\frac{a}{p}\right) = 1$, since $c_2(-a;p^t)$ is the reciprocal of $c_2(a;p^t)$.

Theorem 3.8. *For $p \equiv 3 \bmod 4$ and* $\left(\frac{a}{p}\right) = 1$,

$$c_2(a;p^t) \;=\; 1 - 2\sum_{i=0}^{[t/2]-1} \frac{1}{p^{2i+1}} + \frac{2}{\phi(p^t)}. \tag{3.9}$$

Proof. By Theorem 3.6,

$$\begin{aligned} c_2(a;p^t) \;&=\; \left(\frac{(p-3)p^{t-1}}{2} + \frac{p^{t-1}}{p+1} + \frac{3}{2} + \frac{(-1)^{t-1}(p-1)}{2(p+1)}\right)\frac{1}{\phi(p^t)/2} \\ &=\; \frac{p^2-2p-1}{p^2-1} + \frac{(-1)^{t-1}(p-1)+3p+3}{(p+1)\phi(p^t)}. \end{aligned} \tag{3.10}$$

Therefore

$$c_2(a;p^t) - 1 - \frac{2}{\phi(p^t)} = \frac{-2p}{p^2-1} + \frac{(-1)^{t-1}(p-1)+p+1}{(p+1)\phi(p^t)}$$
$$= \frac{-2p}{p^2-1} + 2\sum_{[t/2]}^{\infty} \frac{1}{p^{2i+1}} = -2\sum_{i=0}^{[t/2]-1} \frac{1}{p^{2i+1}}. \quad (3.11)$$

□

Theorem 3.9. *Let $p < q$ be primes, both congruent to* 3 mod 4, *and let $s, t \geq 2$. If a is a square* mod p, *then $c_2(a; p^t q^s) < 1$, so we have difference dominance. If a is not a square* mod p, *then $c_2(a; p^t q^s) > 1$, so we have sum dominance.*

Proof. It suffices to prove the first assertion, for then the second will follow by taking the reciprocal. By Theorem 3.8, $c_2(a;p^t) < 1$. If a is a square mod q, then also $c_2(a;q^s) < 1$, so that $c_2(a;p^tq^s) = c_2(a;p^t)c_2(a;q^s) < 1$, as desired. Finally, assume that a is not a square mod q. Then it remains to show that $c_2(-a;q^s) > c_2(a;p^t)$. By Theorem 3.8, $c_2(a;p^t)$ is monotone decreasing in t. Therefore it suffices to show that $\lim_{s\to\infty} c_2(-a;q^s) > c_2(a;p^2)$. This inequality is equivalent to $1-2q/(q^2-1) > 1-(2p-4)/(p^2-p)$, so we must show that $(p-2)/(p^2-p) > q/(q^2-1)$. Since the right member is a decreasing function of q, it suffices to prove this inequality when $q = p+4$, and this is easily accomplished. □

It is not hard to show that the conclusion of Theorem 3.9 still holds in the case $s = 1$, $t \geq 2$. However, the inequalities are reversed in the case $t = 1$, $s \geq 1$.

As the next three theorems are straightforward generalizations of results from [2], we omit the proofs.

Theorem 3.10. *Let $N_k = \prod_{i=1}^{k} p_i$, where p_i is the ith prime that is congruent to* 3 *modulo* 4. *Fix a perfect square a relatively prime to all of the p_i. Then*

$$c_2(a; N_k) \asymp \log\log N_k, \quad (3.12)$$

and for any $t \geq 2$,

$$c_2(a; N_k^t) \asymp (\log\log N_k)^{-1}. \quad (3.13)$$

Theorem 3.11. *Fix an integer a. Let n run through the positive integers relatively prime to a. Then*

(1) $$\frac{1}{\log\log n} \ll c_2(a;n) \ll \log\log n, \quad (3.14)$$

(2) $$\limsup_{n\to\infty} c_2(a;n) = \infty \quad \text{and} \quad \liminf_{n\to\infty} c_2(a;n) = 0, \quad (3.15)$$

(3) $$\limsup_{n\to\infty} \frac{\#S(a;n)}{\#D(a;n)} = \infty \quad \text{and} \quad \liminf_{n\to\infty} \frac{\#S(a;n)}{\#D(a;n)} = 0. \quad (3.16)$$

Theorem 3.12. *For a fixed nonzero integer a, let E_a denote the set of positive integers n relatively prime to a such that $\left(\frac{a}{p}\right) = 1$ for every prime $p \equiv 3 \bmod 4$ dividing n. Let $C_a(L) = \{n \in E_a : c_2(a;n) > L\}$. Define $E_a(x) = \{n \in E_a : n \le x\}$ and $C_a(L,x) = \{n \in C_a(L) : n \le x\}$. Then the lower density of $C_a(L)$ in E_a, defined by $\liminf \#C_a(L,x)/\#E_a(x)$, satisfies the inequality*

$$\liminf_{x\to\infty} \frac{\#C_a(1,x)}{\#E_a(x)} \ge K_a \prod \left(1 - \frac{1}{p^2}\right), \tag{3.17}$$

where the product is over all primes $p \equiv 3 \bmod 4$ for which $\left(\frac{a}{p}\right) = 1$, and where

$$K_a = \begin{cases} 1 & a \equiv 0 \bmod 2 \\ 63/64 & a \equiv 1 \bmod 8 \\ 31/32 & a \equiv 5 \bmod 8 \\ 15/16 & a \equiv 3 \bmod 4. \end{cases} \tag{3.18}$$

Furthermore, for any constant $L > 0$, the lower density of $C_a(L)$ in E_a is positive.

For example, if a is an odd power of 2, then the lower density in (3.17) exceeds 97%. Note that if the condition $\left(\frac{a}{p}\right) = 1$ is replaced by $\left(\frac{a}{p}\right) = -1$ throughout the statement of Theorem 3.12, then by Lemma 2.2, (3.17) holds with the inequality $c_2(a;n) > 1$ replaced by $c_2(a;n) < 1$.

4. Cardinality of $\bar{S}_d(m;a;n)$ for $d > 2$

We now turn our attention to modular hyperbolas with higher dimension ($d > 2$). Suppose that $p > 7$ for every prime p dividing n. Then Theorem 4.1 shows that the higher dimensional generalized sumsets $\bar{S}_d(m;a;n)$ all have cardinality n. In particular, this cardinality is the same for every value of m, i.e., there is no dependence on the number of plus and minus signs.

Theorem 4.1. *If the prime factors of n all exceed 7, then $\#\bar{S}_d(m;a;n) = n$.*

Proof. Let $q = p^t$ for a prime $p > 7$. By Proposition 2.1, it suffices to prove that $\#\bar{S}_d(m;a;q) = q$. We will show that for every a coprime to q and every $b \bmod q$, the system of congruences

$$\begin{aligned} x_1 + \cdots + x_d &\equiv b \bmod q \\ x_1 \cdots x_d &\equiv a \bmod q \end{aligned} \tag{4.1}$$

has a solution. This suffices, because x_i could be replaced by $q - x_i$ for any collection of subscripts i. If (4.1) can always be solved for $d = 3$, then it can always be solved for any $d > 3$, by setting $x_i = 1$ for $i > 3$. Thus assume that $d = 3$.

Solving (4.1) is equivalent to solving the congruence $xy(b-x-y) \equiv a \bmod q$ for $x, y \in (\mathbb{Z}/q\mathbb{Z})^*$. Replacing y by y^{-1} and then multiplying by y, we see that this is equivalent to solving

$$x^2 + x(y^{-1} - b) + ay \equiv 0 \bmod q. \tag{4.2}$$

The quadratic polynomial in x in (4.2) has discriminant

$$(-4ay^3 + b^2y^2 - 2by + 1)/y^2. \tag{4.3}$$

Let $R(y) \in (\mathbb{Z}/p\mathbb{Z})[y]$ denote the cubic polynomial in y obtained by reducing the numerator in (4.3) mod$p$. To solve (4.2), it remains to show that there exists $y \in (\mathbb{Z}/p\mathbb{Z})^*$ for which $R(y)$ is a non-zero square mod p; this is because a non-zero square mod p is also a square mod q (see Proposition A.2 in the Appendix).

Suppose for the purpose of contradiction that no term in the sum

$$\sum_{y=1}^{p-1} \left(\frac{R(y)}{p}\right) \tag{4.4}$$

is equal to 1. Then since $R(y)$ has at most 3 zeros in $(\mathbb{Z}/p\mathbb{Z})^*$, we have

$$S := \sum_{y=0}^{p-1} \left(\frac{R(y)}{p}\right) = w - p, \tag{4.5}$$

for some $w \in \{1, 2, 3, 4\}$.

Let D denote the discriminant of $R(y)$. Then $D \equiv 16a(b^3 - 27a) \bmod p$, and so D vanishes if and only if $a \equiv (b/3)^3 \bmod p$. When D vanishes, it follows that $b \in (\mathbb{Z}/p\mathbb{Z})^*$ and $y = 3/(4b)$ is a simple zero of $R(y)$. We conclude that $R(y)$ cannot equal a constant times the square of a polynomial in $(\mathbb{Z}/p\mathbb{Z})[y]$. Therefore (see equation (6.0.2) in [1]) we can apply Weil's bound to conclude that $|S| < 2\sqrt{p}$. Together with (4.5), this yields

$$p - 2\sqrt{p} < w \leq 4, \tag{4.6}$$

which contradicts the fact that $p > 7$. □

We remark that the conditions $p > 7$ cannot be weakened in Theorem 4.1. For example, (4.2) has no solution when $p = q = 2$, $b = 0$ and $a = 1$; when $p = q = 3$ and $b = a = 1$; when $p = q = 5$, $b = 1$ and $a = 2$; and when $p = q = 7$, $b = 0$ and $a = 3$.

5. Conclusion and Future Research

We generalized work of [2] on the modular hyperbola $H_2(1, n)$ by examining more general modular hyperbolas $H_d(a; n)$. The two-dimensional case ($d = 2$) provided interesting conditions on a and n for sum dominance and difference dominance. On the other hand, for higher dimensions ($d > 2$), all possible sums and differences are realized when the prime factors of n all exceed 7.

The following are some topics for future and ongoing research:

1. We can study the cardinality of sumsets and difference sets of the intersection of modular hyperbolas with other modular objects such as lower dimensional modular hyperbolas and modular ellipses. See [3] for work on the cardinality of the intersection of modular circles and $H_2(1;n)$.

2. Extend Theorem 3.9 by estimating $c_2(a;n)$ in cases where n has more than two prime factors of the form $4k+3$.

3. Extend Theorem 4.1 by finding the cardinality of the generalized higher dimensional sumsets in cases where $(n, 210) > 1$.

4. In higher dimensions ($d > 2$), nearly every sum and difference is realized for $\bar{H}_d(a;n)$. The situation becomes more interesting if we replace $\bar{H}_d(a;n)$ by a random subset chosen according to some probability distribution depending on d. If S and D denote the corresponding sumset and difference set, we can then compare the random variables $\#S$ and $\#D$.

A. Additional Proofs

The following proofs are a natural extension of the proofs given by [2], and are included for completeness.

A.1. Proof of Proposition 2.1

Proof of Proposition 2.1. Consider

$$g : \bar{S}_d(m;a;n) \longrightarrow \prod_{i=1}^{k} \bar{S}_d(m; a \bmod p_i^{e_i}; p_i^{e_i}) \tag{A.1}$$

defined by

$$g(x) = (x \bmod p_1^{e_1}, \dots, x \bmod p_k^{e_k}). \tag{A.2}$$

We claim g is a bijection.

To show g is injective, suppose $g(x) = g(y)$. Then we have $x \equiv y \bmod p_i^{e_i}$ for $i = 1, \dots, k$. Thus, by the Chinese Remainder Theorem, $x \equiv y \bmod n$, so g is injective.

To show g is surjective, let $(\alpha_1, \ldots, \alpha_k) \in \prod_{i=1}^k \bar{S}_d(m; a \bmod p_i^{e_i}; p_i^{e_i})$. Then, for each $i \in \{1, \ldots, k\}$, there exists $(x_1(i), \ldots, x_d(i)) \in H_d(a; p_i^{e_i})$ such that

$$x_1(i) + \cdots + x_m(i) - \cdots - x_d(i) \equiv \alpha_i \bmod p_i^{e_i}.$$

By the Chinese Remainder Theorem, for each fixed r with $1 \le r \le d$, the system of congruences

$$x \equiv x_r(i) \bmod p_i^{e_i}, \quad (1 \le i \le k) \tag{A.3}$$

has a unique solution $x_r \bmod n$. Since $x_1(i) \cdots x_d(i) \equiv a \bmod p_i^{e_i}$ for all $i \in \{1, \ldots, k\}$, we have $x_1 \cdots x_d \equiv a \bmod n$. Thus $g(x_1 + \cdots + x_m - \cdots - x_d \bmod n) = (\alpha_1, \ldots, \alpha_k)$, so g is a bijection, which completes the proof. □

A.2. Proof of Lemma 2.3

Before proving Lemma 2.3, we state a useful lemma that is a simple observation and immediate generalization of a result from [2].

Lemma A.1. *Let $(x_0, y_0) \in H_2(a; p^t)$. Then $x_0 - y_0 \equiv 2k \bmod p^t$ for some $k \in \mathbb{Z}$.*

Proof. If $p = 2$, then x_0 and y_0 are both odd since they are coprime to p^t, so their difference is even. If $p \neq 2$, then 2^{-1} exists mod p^t, so $x_0 - y_0 \equiv 2k \bmod p^t$ has a solution k. □

Proof of Lemma 2.3. Let $(x_0, y_0) \in \bar{H}_2(a; p^t)$ so that $x_0 - y_0 \in \bar{D}_2(a; p)$. When $x_0 - y_0 \equiv 2k \bmod p^t$, we have $k^2 + a \equiv (x_0 - k)^2 \bmod p^t$, so that $k^2 + a$ is a square mod p^t.

Conversely, suppose $k^2 + a$ is a square mod p^t. Then there exists $c \in \mathbb{Z}$ such that $c^2 - k^2 \equiv a \bmod p^t$. It follows that

$$(x_0, y_0) := ((c + k) \bmod p^t, (c - k) \bmod p^t) \in \bar{H}_2(a; p^t)$$

and $x_0 - y_0 \equiv 2k \bmod p^t$.

Next we show that f is a bijection. If $p > 2$, then the inverse of the function f is $f^{-1}(x) = 2^{-1}x \bmod p^t$. Now suppose $p = 2$. Clearly f is injective, since $2k \equiv 2j \bmod 2^t$ implies $k \equiv j \mod 2^{t-1}$. To show f is surjective, let $\tau \in \bar{D}_2(a; 2^t)$, so that there exists $(x_0, y_0) \in \bar{H}_2(a; 2^t)$ such that $x_0 - y_0 \equiv \tau \mod 2^t$. Then by Lemma A.1, $\tau \equiv 2k \mod 2^t$ for some $k \in \mathbb{Z}$ with $0 \le k < 2^{t-1}$, so $f(k) = \tau$. □

A.3. Proof of Theorem 3.3

To prove the required cases of Theorem 3.3, we will need the following two propositions. The first proposition is from [5] (see page 46). It gives us a quick way to count squares modulo prime powers.

Proposition A.2. *Let a be an integer not divisible by the prime p. Then we have*

1. *If $p \neq 2$ and the congruence $x^2 \equiv a \bmod p$ is solvable, then for every $t \geq 1$ the congruence $x^2 \equiv a \bmod p^t$ is solvable with precisely two distinct solutions.*
2. *If $p = 2$ and the congruence $x^2 \equiv a \bmod 2^3$ is solvable, then for every $t \geq 3$ the congruence $x^2 \equiv a \bmod 2^t$ is solvable with precisely four distinct solutions.*

Proposition A.3 (Stangl [14]). *Let p be an odd prime. Then*

$$\#\{k^2 \bmod p^t\} = \tfrac{p^{t+1}}{2(p+1)} + (-1)^{t-1}\tfrac{p-1}{4(p+1)} + \tfrac{3}{4}.$$

For the prime 2, we have

$$\#\{k^2 \bmod 2^t\} = \tfrac{2^{t-1}}{3} + \tfrac{(-1)^{t-1}}{6} + \tfrac{3}{2}.$$

Proof of Theorem 3.3 cases. We now prove the remaining cases of the theorem. `Case 1: Difference set for` $a \equiv 1 \bmod 8$. By Proposition 2.3 with $a = 8m+1$,

$$\#\bar{D}(a; 2^t) \;=\; \#\{k : k^2 + 1 + 8m \text{ is a square mod } 2^t, 0 \leq k < 2^{t-1}\}. \qquad \text{(A.4)}$$

We claim that

$$k^2 + 1 + 8m \text{ is a square mod } 2^t \;\Leftrightarrow\; k = 4l \text{ for some } l \in \mathbb{Z}. \qquad \text{(A.5)}$$

First assume that k^2+1+8m is a square mod 2^t. Then k^2+1 is a square mod 8, which yields $k \equiv 0, 4 \bmod 8$. Hence $k = 4l$ for some $l \in \mathbb{Z}$.

Conversely, assume that $k = 4l$ for some $l \in \mathbb{Z}$. We want to show that $(4l)^2+1+8m$ is a square mod 2^t. Reducing modulo 8 gives us $(4l)^2+1+8m \equiv 1 \bmod 8$, which is a square modulo 8. Hence, by the second part of Proposition A.2, $(4l)^2+1+8m$ is a square mod 2^t. Thus

$$\{k : k^2 + 1 + 8m \text{ is a square mod } 2^t, 0 \leq k < 2^{t-1}\} \;=\; \{4l : 0 \leq l < 2^{t-3}\}. \quad \text{(A.6)}$$

`Case 2: Difference set for` $a \equiv 5 \bmod 8$. We show that

$$k^2 + 5 + 8m \text{ is a square mod } 2^t \;\Leftrightarrow\; k = 2 + 4l \text{ for some } l \in \mathbb{Z}. \qquad \text{(A.7)}$$

First assume that k^2+5+8m is a square mod 2^t. Then k^2+5+8m is a square mod 8, which implies $k \equiv 2, 6 \bmod 8$ or $k = 2+4l$ for some $l \in \mathbb{Z}$.

Conversely, assume that $k = 2+4l$ for some $l \in \mathbb{Z}$. Reducing modulo 8 gives us $k^2+5+8m \equiv 1 \bmod 8$, which is a square modulo 8. Hence, by the second part of Proposition A.2, k^2+5+8m is a square mod 2^t. We conclude that

$$\{k : k^2{+}5{+}8m \text{ is a square mod } 2^t, 0 \leq k < 2^{t-1}\} \;=\; \{2{+}4l : 0 \leq l < 2^{t-3}\}. \quad \text{(A.8)}$$

Case 3: Sum set for $a \equiv 1 \bmod 8$. In view of Proposition A.3, it suffices to show that

$$\#\{k : k^2 - a \text{ is a square mod } 2^t, 0 \leq k < 2^{t-1}\} = 2\#\{k^2 \bmod 2^{t-4}\}. \qquad (A.9)$$

If $k^2 - a$ is a square mod 2^t then k must be odd, since $-a$ is not a square mod 4. The equality (A.9) is equivalent to

$$\#\{k : k^2 - a \text{ is a square mod } 2^t, 0 < k < 2^{t-2}, 2 \nmid k\} = \#\{k^2 \bmod 2^{t-4}\}, \quad (A.10)$$

since $k^2 - a \bmod 2^t$ has the same value when k is replaced by $2^{t-1} - k$. The left member of (A.10) equals the number of (distinct) squares modulo 2^t of the form $k^2 - a \bmod 2^t$. Any square divisible by 8 is also divisible by 16, so the left member of (A.10) also equals the number of squares modulo 2^{t-4} of the form $(k^2 - a)/16 \bmod 2^{t-4}$. It remains to show that every square modulo 2^{t-4} has the form $(k^2 - a)/16 \bmod 2^{t-4}$, i.e., that $16u^2 + a$ is a square modulo 2^t for every integer u. This follows from the second part of Proposition A.2. □

A.4. Proof of Lemma 3.4

Proof of Lemma 3.4. If $l \in S_2'(a; p^t)$, then $\left(\frac{l^2-a}{p}\right) = 1$. Thus

$$\begin{aligned}
\#S_2'(a;p^t) &= \frac{1}{2} \sum_{\substack{l=0 \\ (l^2-a,p)=1}}^{p^t-1} \left(\left(\frac{l^2-a}{p}\right)+1\right) \\
&= \frac{1}{2} \sum_{k=0}^{p^{t-1}-1} \sum_{\substack{l=0 \\ l^2 \neq a \bmod p}}^{p-1} \left(\left(\frac{(l+kp)^2-a}{p}\right)+1\right) \\
&= \left(\sum_{\substack{l=0 \\ l^2 \neq a \bmod p}}^{p-1} \left(\left(\frac{l^2-a}{p}\right)+1\right) \right) \frac{p^{t-1}}{2} \\
&= \left(-1 + \sum_{\substack{l=0 \\ l^2 \neq a \bmod p}}^{p-1} 1 \right) \frac{p^{t-1}}{2}, \qquad (A.11)
\end{aligned}$$

as (see for example page 63 of [5])

$$\sum_{i=0}^{p-1} \left(\frac{i^2-a}{p}\right) = -1. \qquad (A.12)$$

Substituting

$$\sum_{\substack{l=0 \\ l^2 \neq a \bmod p}}^{p-1} 1 = \begin{cases} p-2 & \left(\frac{a}{p}\right) = 1 \\ p & \left(\frac{a}{p}\right) = -1 \end{cases} \qquad (A.13)$$

into (A.11) gives us the desired result. □

A.5. Proof of Lemma 3.5

Proof. If $\left(\frac{a}{p}\right) = -1$, then the congruence $x^2 - a \equiv 0 \bmod p$ has no solutions, so $\#S_2''(a;p^t) = 0$. Now assume that $\left(\frac{a}{p}\right) = 1$. It is easily seen that $\#S_2''(a;p^t) = 2$ when $t = 1$ or $t = 2$, so let $t \geq 3$.

By Proposition A.3, it suffices to prove that

$$\#S_2''(a;p^t) = 2\#\{m^2 \bmod p^{t-2}\}. \tag{A.14}$$

Note that (A.14) is equivalent to

$$\#\{k : k^2 - a \text{ is a square mod } p^t, 0 < k < p^t/2, p \mid (k^2 - a)\} \;=\; \#\{m^2 \bmod p^{t-2}\}. \tag{A.15}$$

The left member of (A.15) also equals the number of squares modulo p^{t-2} of the form $(k^2 - a)/p^2 \bmod p^{t-2}$. It remains to show that every square modulo p^{t-2} has the form $(k^2 - a)/p^2 \bmod p^{t-2}$, i.e., that $p^2u^2 + a$ is a square modulo p^t for every integer u. This follows from the first part of Proposition A.2. □

References

[1] B. Berndt, R. Evans and K. Williams, *Gauss and Jacobi Sums*, Canadian Mathematical Society Series of Monographs and Advanced Texts, John Wiley & Sons, New York, 1998.

[2] D. Eichhorn, M. R. Khan, A. H. Stein and C. L. Yankov, *Sums and Differences of the Coordinates of Points on Modular Hyperbolas*, Integers **9** (2009), #A3.

[3] S. Hanrahan and M. R. Khan, *The cardinality of the value sets modulo n of $x^2 + x^{-2}$ and $x^2 + y^2$*, Involve **3** (2010), no. 2, 171–182.

[4] C.F. Gauss, *Disquisitiones Arithmeticae*, Art. 103.

[5] K. Ireland and M. Rosen, *A classical Introduction to Modern Number Theory*, second edition, Graduate Texts in Mathematics, Springer-Verlag, New York, 2010.

[6] G. Iyer, O. Lazarev, S. J. Miller and L. Zhang, *Generalized More Sums Than Differences Sets*, J. Number Theory **132** (2012), no. 5, 1054–1073.

[7] G. Martin and K. O'Bryant, *Many sets have more sums than differences*, in Additive Combinatorics, CRM Proc. Lecture Notes, vol. 43, AMS, Providence, RI, 2007, pp. 287-305.

[8] S. J. Miller, B. Orosz and D. Scheinerman, *Explicit constructions of infinite families of MSTD sets*, J. Number Theory **130** (2010) 1221–1233.

[9] S. J. Miller, S. Pegado and S. L. Robinson, *Explicit constructions of infinite families of generalized MSTD sets*, Integers **12** (2012), #A30.

[10] S. J. Miller and K. Vissuet, *Most sets are balanced in finite groups*, preprint.

[11] I. E. Shparlinski, *Modular hyperbolas*, Jap. J. Math **7** (2012), 235–294.

[12] I. E. Shparlinski and A. Winterhof, *On the number of distances between the coordinates of points on modular hyperbolas*, J. Number Theory **128** (2008), no. 5, 1224–1230.

[13] I. E. Shparlinski and A. Winterhof, *Visible points on multidimensional modular hyperbolas*, J. Number Theory **128** (2008), no. 9, 2695–2703.

[14] W. Stangl, *Counting squares in* $\mathbb{Z}_n$, Math. Mag. **69** (1996), 285–289.

[15] Y. Zhao, *Constructing MSTD Sets Using Bidirectional Ballot Sequences*, J. Number Theory **130** (2010), no. 5, 1212–1220.

[16] Y. Zhao, *Sets Characterized by Missing Sums and Differences*, J. Number Theory **131** (2011), 2107–2134.

AN INFINITE FAMILY OF RECURSIVE FORMULAS GENERATING POWER MOMENTS OF KLOOSTERMAN SUMS WITH TRACE ONE ARGUMENTS: $O(2N+1, 2^R)$ CASE

Dae San Kim[1]
Department of Mathematics, Sogang University, Seoul, Korea
dskim@sogong.ac.kr

Received: 7/14/12, Revised: 1/4/13, Accepted: 4/10/13, Published: 5/24/13

Abstract

In this paper, we construct an infinite family of binary linear codes associated with double cosets with respect to a certain maximal parabolic subgroup of the orthogonal group $O(2n+1, q)$. Here q is a power of two. Then we obtain an infinite family of recursive fomulas generating the odd power moments of Kloosterman sums with trace one arguments in terms of the frequencies of weights in the codes associated with those double cosets in $O(2n+1, q)$, and in the codes associated with similar double cosets in the symplectic group $Sp(2n, q)$. This is done via the Pless power moment identity and by utilizing the explicit expressions of exponential sums over those double cosets related to the evaluations of "Gauss sums" for the orthogonal group $O(2n+1, q)$.

1. Introduction

Let ψ be a nontrivial additive character of the finite field $\mathbb{F}_q$ with $q = p^r$ elements (p a prime). Then the Kloosterman sum $K(\psi; a)$ ([12]) is defined as

$$K(\psi; a) = \sum_{\alpha \in \mathbb{F}_q^*} \psi(\alpha + a\alpha^{-1}) \ (a \in \mathbb{F}_q^*).$$

For this, we have the Weil bound

$$|K(\psi; a)| \leq 2\sqrt{q}. \tag{1.1}$$

The Kloosterman sum was introduced in 1926([11]) to give an estimate for the Fourier coefficients of modular forms.

[1]This work was supported by National Research Foundation of Korea Grant funded by the Korean Government 2009-0072514.

For each nonnegative integer h, by $MK(\psi)^h$ we will denote the h-th moment of the Kloosterman sum $K(\psi; a)$. Namely, it is given by

$$MK(\psi)^h = \sum_{a \in \mathbb{F}_q^*} K(\psi; a)^h.$$

If $\psi = \lambda$ is the canonical additive character of $\mathbb{F}_q$, then $MK(\lambda)^h$ will be simply denoted by MK^h. Here we recall that $\lambda(x) = e^{2\pi i tr(x)/p}$ is the canonical additive character of $\mathbb{F}_q$, where $tr(x) = x + x^p + \cdots + x^{p^{r-1}}$ is the trace function $\mathbb{F}_q \to \mathbb{F}_p$.

Explicit computations on power moments of Kloosterman sums were begun with the paper [17] of Salié in 1931, where he showed, for any odd prime q,

$$MK^h = q^2 M_{h-1} - (q-1)^{h-1} + 2(-1)^{h-1} \ (h \geq 1).$$

Here $M_0 = 0$, and, for $h \in \mathbb{Z}_{>0}$,

$$M_h = |\{(\alpha_1, \cdots, \alpha_h) \in (\mathbb{F}_q^*)^h | \sum_{j=1}^{h} \alpha_j = 1 = \sum_{j=1}^{h} \alpha_j^{-1}\}|.$$

For $q = p$ odd prime, Salié obtained MK^1, MK^2, MK^3, MK^4 in [17] by determining M_1, M_2, M_3. On the other hand, MK^5 can be expressed in terms of the p-th eigenvalue for a weight 3 newform on $\Gamma_0(15)$(cf. [13], [16]). MK^6 can be expressed in terms of the p-th eigenvalue for a weight 4 newform on $\Gamma_0(6)$(cf. [3]). Also, based on numerical evidence, in [2] Evans was led to propose a conjecture which expresses MK^7 in terms of Hecke eigenvalues for a weight 3 newform on $\Gamma_0(525)$ with quartic nebentypus of conductor 105.

From now on, let us assume that $q = 2^r$. Carlitz[1] evaluated MK^h for $h \leq 4$. Recently, Moisio was able to find explicit expressions of MK^h, for $h \leq 10$ (cf.[15]). This was done, via Pless power moment identity, by connecting moments of Kloosterman sums and the frequencies of weights in the binary Zetterberg code of length $q+1$, which were known by the work of Schoof and Vlugt in [18].

In [7], the binary linear codes $C(SL(n, q))$ associated with finite special linear groups $SL(n, q)$ were constructed when n, q are both powers of two. Then we obtained a recursive formula for the power moments of multi-dimensional Kloosterman sums in terms of the frequencies of weights in $C(SL(n, q))$.

In order to describe our results, we introduce two incomplete power moments of Kloosterman sums, namely, the one with the sum over all a in $\mathbb{F}_q^*$ with $tr\, a$=0 and the other with the sum over all a in $\mathbb{F}_q^*$ with $tr\, a = 1$. For every nonnegative integer h, and ψ as before, we define

$$T_0K(\psi)^h = \sum_{a \in \mathbb{F}_q^*,\ tra=0} K(\psi; a)^h, \ T_1K(\psi)^h = \sum_{a \in \mathbb{F}_q^*,\ tra=1} K(\psi; a)^h, \qquad (1.2)$$

which will be respectively called the h-th moment of Kloosterman sums with "trace zero arguments" and those with "trace one arguments". Then, clearly we have

$$MK(\psi)^h = T_0K(\psi)^h + T_1K(\psi)^h. \tag{1.3}$$

If $\psi = \lambda$ is the canonical additive character of $\mathbb{F}_q$, then $T_0K(\lambda)^h$ and $T_1K(\lambda)^h$ will be respectively denoted by T_0K^h and T_1K^h, for brevity.

In this paper, we will show the main Theorem 1.1 giving an infinite family of recursive formulas generating the odd power moments of Kloosterman sums with trace one arguments. To do that, we construct binary linear codes $C(DC(n,q))$, associated with the double cosets $DC(n,q)=P\sigma_{n-1}P$, for the maximal parabolic subgroup $P=P(2n+1,q)$ of the orthogonal group $O(2n+1,q)$, and express those power moments in terms of the frequencies of weights in the codes $C(DC(n,q))$ and $C(\widehat{DC}(n,q))$. Here $C(\widehat{DC}(n,q))$ is a binary linear code constructed similarly from certain double cosets $\widehat{DC}(n,q)$ in the sympletic group $Sp(2n,q)$. Then, thanks to our previous results on the explicit expressions of exponential sums over those double cosets related to the evaluations of "Gauss sums" for the orthogonal group $O(2n+1,q)$ [10], we can express the weight of each codeword in the dual of the codes $C(DC(n,q))$ in terms of Kloosterman sums. Then our formulas will follow immediately from the Pless power moment identity. Analogously to these, in [8](resp. [9]), for q a power of three, two(resp. eight) infinite families of ternary linear codes associated with double cosets in the symplectic group $Sp(2n,q)$(resp. orthogonal group $O^-(2n,q)$) were constructed in order to generate one (resp. four)infinite families of recursive formulas for the power moments of Kloosterman sums with square arguments and for the even power moments of those in terms of the frequencies of weights in those codes. We emphasize here that there have been only a few recursive formulas generating power moments of Kloosterman sums including the one in [15].

Theorem 1.1 in the following(cf. (1.6)-(1.8)) is the main result of this paper. Henceforth, we agree that the binomial coefficient $\binom{b}{a} = 0$ if $a > b$ or $a < 0$. To simplify notations, we introduce the following ones which will be used throughout this paper at various places.

$$A(n,q) = q^{\frac{1}{4}(5n^2-1)} \begin{bmatrix} n \\ 1 \end{bmatrix}_q \Pi_{j=1}^{(n-1)/2}(q^{2j-1}-1), \tag{1.4}$$

$$B(n,q) = q^{\frac{1}{4}(n-1)^2}(q^n-1)\Pi_{j=1}^{(n-1)/2}(q^{2j}-1). \tag{1.5}$$

Here $\begin{bmatrix} n \\ 1 \end{bmatrix}_q = \frac{q^n-1}{q-1}$ is a q-binomial coefficient.

Theorem 1.1. *Let $q = 2^r$. Assume that n is any odd integer≥ 3, with all q, or n=1, with $q \geq 8$. Then, in the notations of (1.4) and (1.5), we have the following.*

For h=1,3,5,⋯,

$$T_1K^h = -\sum_{0\le l\le h-2,\ l\ odd} \binom{h}{l} B(n,q)^{h-l}T_1K^l + qA(n,q)^{-h}\sum_{j=0}^{min\{N(n,q),h\}} (-1)^j D_j(n,q)\sum_{t=j}^{h} t!S(h,t)2^{h-t-1}\binom{N(n,q)-j}{N(n,q)-t}, \tag{1.6}$$

where $N(n,q) = |DC(n,q)| = A(n,q)B(n,q)$, $D_j(n,q) = C_j(n,q) - \widehat{C}_j(n,q)$, *with* $\{C_j(n,q)\}_{j=0}^{N(n,q)}$, $\{\widehat{C}_j(n,q)\}_{j=0}^{N(n,q)}$ *respectively the weight distributions of the binary linear codes* $C(DC(n,q))$ *and* $C(\widehat{DC}(n,q))$ *given by: for* $j = 0,\cdots,N(n,q)$,

$$\begin{aligned} C_j(n,q) = &\sum \binom{q^{-1}A(n,q)(B(n,q)+1)}{\nu_1} \\ &\times \prod_{tr(\beta-1)^{-1}=0} \binom{q^{-1}A(n,q)(B(n,q)+q+1)}{\nu_\beta} \\ &\times \prod_{tr(\beta-1)^{-1}=1} \binom{q^{-1}A(n,q)(B(n,q)-q+1)}{\nu_\beta}, \end{aligned} \tag{1.7}$$

$$\begin{aligned} \widehat{C}_j(n,q) = &\sum \binom{q^{-1}A(n,q)(B(n,q)+1)}{\nu_0} \\ &\times \prod_{tr(\beta^{-1})=0} \binom{q^{-1}A(n,q)(B(n,q)+q+1)}{\nu_\beta} \\ &\times \prod_{tr(\beta^{-1})=1} \binom{q^{-1}A(n,q)(B(n,q)-q+1)}{\nu_\beta}. \end{aligned} \tag{1.8}$$

Here the first sum in (1.6) is 0 if $h = 1$ and the unspecified sums in (1.7) and (1.8) are over all the sets of nonnegative integers $\{\nu_\beta\}_{\beta\in\mathbb{F}_q}$ satisfying $\sum_{\beta\in\mathbb{F}_q}\nu_\beta = j$ and $\sum_{\beta\in\mathbb{F}_q}\nu_\beta\beta = 0$. In addition, $S(h,t)$ is the Stirling number of the second kind defined by

$$S(h,t) = \frac{1}{t!}\sum_{j=0}^{t}(-1)^{t-j}\binom{t}{j} j^h. \tag{1.9}$$

2. $O(2n+1,q)$

For more details about this section, one is referred to the paper [10]. Throughout this paper, the following notations will be used:

$$\begin{aligned} q &= 2^r \ (r \in \mathbb{Z}_{>0}), \\ \mathbb{F}_q &= \ the\ finite\ field\ with\ q\ elements, \\ TrA &= \ the\ trace\ of\ A\ for\ a\ square\ matrix\ A, \\ {}^tB &= \ the\ transpose\ of\ B\ for\ any\ matrix\ B. \end{aligned}$$

Let θ be the nondegenerate quadratic form on the vector space $\mathbb{F}_q^{(2n+1)\times 1}$ of all $(2n+1)\times 1$ column vectors over $\mathbb{F}_q$, given by

$$\theta(\sum_{i=1}^{2n+1} x_i e^i) = \sum_{i=1}^{n} x_i x_{n+i} + x_{2n+1}^2,$$

where $\{e^1 = {}^t[10\cdots 0], e^2 = {}^t[010\cdots 0], \cdots, e^{2n+1} = {}^t[0\cdots 01]\}$ is the standard basis of $\mathbb{F}_q^{(2n+1)\times 1}$.

The group $O(2n+1,q)$ of all isometries of $(\mathbb{F}_q^{(2n+1)\times 1}, \theta)$ consists of the matrices

$$\begin{bmatrix} A & B & 0 \\ C & D & 0 \\ g & h & 1 \end{bmatrix} \ (A, B, C, D\ n\times n, g, h\ 1\times n)$$

in $GL(2n+1,q)$ satisfying the relations:

$$\begin{aligned} &{}^tAC + {}^tgg\ is\ alternating \\ &{}^tBD + {}^thh\ is\ alternating \\ &{}^tAD + {}^tCB = 1_n. \end{aligned}$$

Here an $n \times n$ matrix (a_{ij}) is called alternating if

$$\begin{cases} a_{ii} &= 0, & for\ 1 \le i \le n, \\ a_{ij} &= -a_{ji} = a_{ji}, & for\ 1 \le i < j \le n. \end{cases}$$

Also, one observes, for example, that ${}^tAC + {}^tgg$ is alternating if and only if ${}^tAC = {}^tCA$ and $g = \sqrt{diag({}^tAC)}$, where $\sqrt{diag({}^tAC)}$ indicates the $1 \times n$ matrix $[\alpha_1, \cdots, \alpha_n]$ if the diagonal entries of ${}^tAC$ are given by

$$({}^tAC)_{11} = \alpha_1^2, \cdots, ({}^tAC)_{nn} = \alpha_n^2,\ for\ \alpha_i \in \mathbb{F}_q.$$

As is well known, there is an isomorphism of groups

$$\iota:\ O(2n+1,q) \to Sp(2n,q)\ (\begin{bmatrix} A & B & 0 \\ C & D & 0 \\ g & h & 1 \end{bmatrix} \mapsto \begin{bmatrix} A & B \\ C & D \end{bmatrix}). \qquad (2.1)$$

In particular, for any $w \in O(2n+1,q)$,

$$Trw = Tr\iota(w) + 1. \tag{2.2}$$

Let $P = P(2n+1,q)$ be the maximal parabolic subgroup of $O(2n+1,q)$ given by

$$P(2n+1,q) = \left\{ \begin{bmatrix} A & 0 & 0 \\ 0 & {}^tA^{-1} & 0 \\ 0 & 0 & 1 \end{bmatrix} \begin{bmatrix} 1_n & B & 0 \\ 0 & 1_n & 0 \\ 0 & h & 1 \end{bmatrix} \middle| \begin{array}{c} A \in GL(n,q) \\ B + {}^thh \textit{ is alternating} \end{array} \right\}.$$

The Bruhat decomposition of $O(2n+1,q)$ with respect to $P = P(2n+1,q)$ is

$$O(2n+1,q) = \coprod_{r=0}^{n} P\sigma_r P,$$

where

$$\sigma_r = \begin{bmatrix} 0 & 0 & 1_r & 0 & 0 \\ 0 & 1_{n-r} & 0 & 0 & 0 \\ 1_r & 0 & 0 & 0 & 0 \\ 0 & 0 & 0 & 1_{n-r} & 0 \\ 0 & 0 & 0 & 0 & 1 \end{bmatrix} \in O(2n+1,q).$$

The symplectic group $Sp(2n,q)$ over the field $\mathbb{F}_q$ is defined as:

$$Sp(2n,q) = \{w \in GL(2n,q) \mid {}^twJw = J\},$$

with

$$J = \begin{bmatrix} 0 & 1_n \\ 1_n & 0 \end{bmatrix}.$$

Let $P' = P'(2n,q)$ be the maximal parabolic subgroup of $Sp(2n,q)$ defined by:

$$P'(2n,q) = \left\{ \begin{bmatrix} A & 0 \\ 0 & {}^tA^{-1} \end{bmatrix} \begin{bmatrix} 1_n & B \\ 0 & 1_n \end{bmatrix} \middle| A \in GL(n,q), {}^tB = B \right\}.$$

Then, with respect to $P' = P'(2n,q)$, the Bruhat decomposition of $Sp(2n,q)$ is given by

$$Sp(2n,q) = \coprod_{r=0}^{n} P'\sigma'_r P',$$

where

$$\sigma_r^{'} = \begin{bmatrix} 0 & 0 & 1_r & 0 \\ 0 & 1_{n-r} & 0 & 0 \\ 1_r & 0 & 0 & 0 \\ 0 & 0 & 0 & 1_{n-r} \end{bmatrix} \in Sp(2n,q).$$

Put, for each r with $0 \leq r \leq n$,

$$A_r = \{w \in P(2n+1,q) \mid \sigma_r w \sigma_r^{-1} \in P(2n+1,q)\},$$
$$A_r^{'} = \{w \in P^{'}(2n,q) \mid \sigma_r^{'} w (\sigma_r^{'})^{-1} \in P^{'}(2n,q)\}.$$

Expressing them as the disjoint union of right cosets of maximal parabolic subgroups, the double cosets $P\sigma_r P$ and $P^{'}\sigma_r^{'}P^{'}$ can be written respectively as

$$P\sigma_r P = P\sigma_r(A_r \setminus P), \tag{2.3}$$

$$P^{'}\sigma_r^{'}P^{'} = P^{'}\sigma_r^{'}(A_r^{'} \setminus P^{'}). \tag{2.4}$$

The order of the general linear group $GL(n,q)$ is given by

$$g_n = \prod_{j=0}^{n-1}(q^n - q^j) = q^{\binom{n}{2}} \prod_{j=1}^{n}(q^j - 1).$$

For integers n,r with $0 \leq r \leq n$, the q-binomial coefficients are defined as:

$$\begin{bmatrix} n \\ r \end{bmatrix}_q = \prod_{j=0}^{r-1}(q^{n-j} - 1)/(q^{r-j} - 1).$$

The following results follow either from [10] or from [4] plus the observation that under the isomorphism ι in (2.1) P, A_r, σ_r are respectively mapped onto $P_,^{'}$ $A_r^{'}, \sigma_r^{'}$:

$$\begin{aligned}
|A_r| = |A_r^{'}| &= g_r g_{n-r} q^{\binom{n+1}{2}} q^{r(2n-3r-1)/2}, \\
|P(2n+1,q)| &= |P^{'}(2n,q)| = q^{\binom{n+1}{2}} g_n, \\
|A_r \setminus P(2n+1,q)| &= |A_r^{'} \setminus P^{'}(2n,q)| = q^{\binom{r+1}{2}} \begin{bmatrix} n \\ r \end{bmatrix}_q, \\
|P(2n+1,q)\sigma_r P(2n+1,q)| &= |P'(2n,q)\sigma_r^{'}P^{'}(2n,q)| \\
&= q^{n^2} \begin{bmatrix} n \\ r \end{bmatrix}_q q^{\binom{r}{2}} q^r \prod_{j=1}^{n}(q^j - 1) \\
&(= |P(2n+1,q)|^2 |A_r|^{-1} \\
&= |P^{'}(2n+q)|^2 |A_r^{'}|^{-1}).
\end{aligned} \tag{2.5}$$

In particular, with

$$DC(n,q) = P(2n+1,q)\sigma_{n-1}P(2n+1,q),$$

$$|DC(n,q)| = q^{\frac{1}{2}n(3n-1)} \begin{bmatrix} n \\ 1 \end{bmatrix}_q \prod_{j=1}^{n}(q^j - 1) = A(n,q)B(n,q) \quad (cf.\ (1.4),(1.5)). \tag{2.6}$$

3. Exponential Sums Over Double Cosets of $O(2n+1,q)$

The following notations will be employed throughout this paper:

$$tr(x) = x + x^2 + \cdots + x^{2^{r-1}} \text{ the trace function } \mathbb{F}_q \to \mathbb{F}_2,$$
$$\lambda(x) = (-1)^{tr(x)} \text{ the canonical additive character of } \mathbb{F}_q.$$

Then any nontrivial additive character ψ of $\mathbb{F}_q$ is given by $\psi(x) = \lambda(ax)$, for a unique $a \in \mathbb{F}_q^*$.

For any nontrivial additive character ψ of $\mathbb{F}_q$ and $a \in \mathbb{F}_q^*$, the Kloosterman sum $K_{GL(t,q)}(\psi; a)$ for $GL(t,q)$ is defined as

$$K_{GL(t,q)}(\psi; a) = \sum_{w \in GL(t,q)} \psi(Trw + aTrw^{-1}).$$

Notice that, for $t = 1$, $K_{GL(1,q)}(\psi; a)$ denotes the Kloosterman sum $K(\psi; a)$.

In [4], it is shown that $K_{GL(t,q)}(\psi; a)$ satisfies the following recursive relation: for integers $t \geq 2$, $a \in \mathbb{F}_q^*$,

$$K_{GL(t,q)}(\psi; a) = q^{t-1}K_{GL(t-1,q)}(\psi; a)K(\psi; a) + q^{2t-2}(q^{t-1} - 1)K_{GL(t-2,q)}(\psi; a),$$

where we understand that $K_{GL(0,q)}(\psi, a) = 1$.

From [4] and [10], we have (cf. (2.2)-(2.5)):

$$\begin{aligned}
&\sum_{w \in P\sigma_r P} \psi(Trw) \\
&\quad = |A_r \setminus P| \sum_{w \in P} \psi(Trw\sigma_r) \\
&\quad = \psi(1)|A_r^{'} \setminus P^{'}| \sum_{w \in P} \psi(Tr\iota(w)\sigma_r^{'}) \\
&\quad = \psi(1)|A_r^{'} \setminus P^{'}| \sum_{w \in P^{'}} \psi(Trw\sigma_r^{'}) \\
&\quad (= \psi(1) \sum_{w \in P^{'}\sigma_r^{'}P^{'}} \psi(Trw)) \\
&\quad = \psi(1)q^{\binom{n+1}{2}}|A_r^{'} \setminus P^{'}|q^{r(n-r)}a_r K_{GL(n-r,q)}(\psi; 1).
\end{aligned} \tag{3.1}$$

Here ψ is any nontrivial additive character of $\mathbb{F}_q$, $a_0 = 1$, and, for $r \in \mathbb{Z}_{>0}$, a_r denotes the number of all $r \times r$ nonsingular alternating matrices over $\mathbb{F}_q$, which is given by

$$a_r = \begin{cases} 0, & \text{if } r \text{ is odd,} \\ q^{\frac{r}{2}(\frac{r}{2}-1)} \prod_{j=1}^{\frac{r}{2}} (q^{2j-1}-1), & \text{if } r \text{ is even} \end{cases} \tag{3.2}$$

(cf.[4], Proposition 5.1).
Thus we see from (2.5), (3.1), and (3.2) that, for each r with $0 \le r \le n$,

$$\sum_{w \in P\sigma_r P} \psi(Trw) = \begin{cases} 0, & \text{if } r \text{ is odd,} \\ \psi(1) q^{\binom{n+1}{2}} q^{rn-\frac{1}{4}r^2} \begin{bmatrix} n \\ r \end{bmatrix}_q & \\ \qquad \times \prod_{j=1}^{r/2} (q^{2j-1}-1) K_{GL(n-r,q)}(\psi; 1), & \text{if } r \text{ is even.} \end{cases} \tag{3.3}$$

For our purposes, we need only one infinite family of exponential sums in (3.3) over $P(2n+1,q)\sigma_{n-1}P(2n+1,q) = DC(n,q)$, for $n = 1, 3, 5, \cdots$. So we state them separately as a theorem.

Theorem 3.1. *Let ψ be any nontrivial additive character of $\mathbb{F}_q$. Then in the notation of (1.4), we have*

$$\sum_{w \in DC(n,q)} \psi(Trw) = \psi(1) A(n,q) K(\psi; 1), for\ n = 1, 3, 5, \cdots. \tag{3.4}$$

Proposition 3.2. *([5]) For $n = 2^s (s \in \mathbb{Z}_{\ge 0})$, and λ the canonical additive character of $\mathbb{F}_q$,*

$$K(\lambda; a^n) = K(\lambda; a).$$

The next corollary follows from Theorem 3.1, Proposition 3.2 and a simple change of variables.

Corollary 3.3. *Let λ be the canonical additive character of $\mathbb{F}_q$, and let $a \in \mathbb{F}_q^*$. Then we have*

$$\sum_{w \in DC(n,q)} \lambda(aTrw) = \lambda(a) A(n,q) K(\lambda; a), for\ n = 1, 3, 5, \cdots \tag{3.5}$$

(cf. (1.4)).

Proposition 3.4. *([5]) Let λ be the canonical additive character of $\mathbb{F}_q$, $\beta \in \mathbb{F}_q$. Then*

$$\sum_{a \in \mathbb{F}_q^*} \lambda(-a\beta) K(\lambda; a) = \begin{cases} q\lambda(\beta^{-1}) + 1, & if\ \beta \neq 0, \\ 1, & if\ \beta = 0. \end{cases} \tag{3.6}$$

For any integer r with $0 \leq r \leq n$, and each $\beta \in \mathbb{F}_q$, we let

$$N_{P\sigma_r P}(\beta) = |\{w \in P\sigma_r P | Trw = \beta\}|.$$

Then it is easy to see that

$$qN_{P\sigma_r P}(\beta) = |P\sigma_r P| + \sum_{a \in \mathbb{F}_q^*} \lambda(-a\beta) \sum_{w \in P\sigma_r P} \lambda(aTrw). \tag{3.7}$$

For brevity, we write

$$n(\beta) = N_{DC(n,q)}(\beta). \tag{3.8}$$

Now, from (2.6) and (3.5)-(3.7), we have the following result.

Proposition 3.5. *With the notations in (1.4), (1.5), and (3.8), for* $n = 1, 3, 5, \cdots$,

$$n(\beta) = q^{-1}A(n,q)B(n,q) + q^{-1}A(n,q) \times \begin{cases} 1, & \beta = 1, \\ q+1, & tr(\beta-1)^{-1} = 0, \\ -q+1, & tr(\beta-1)^{-1} = 1. \end{cases} \tag{3.9}$$

Corollary 3.6. *For each odd* $n \geq 3$, *with all* q, $n(\beta) > 0$, *for all* β; *for* $n = 1$, *with all* q,

$$n(\beta) = \begin{cases} q, & \beta = 1, \\ 2q, & tr(\beta-1)^{-1} = 0, \\ 0, & tr(\beta-1)^{-1} = 1. \end{cases} \tag{3.10}$$

Proof. $n = 1$ case follows directly from (3.9). Let $n \geq 3$ be odd. Then, from (3.9), we see that, for any β, we have $n(\beta) \geq q^{-1}A(n,q)(B(n,q) - (q-1)) > 0$. □

4. Construction of Codes

Let

$$N(n,q) = |DC(n,q)| = A(n,q)B(n,q), \ for\ n = 1, 3, 5, \cdots \tag{4.1}$$

(cf. (1.4), (1.5), (2.6)).

Here we will construct one infinite family of binary linear codes $C(DC(n,q))$ of length $N(n,q)$ for all positive odd integers n and all q, associated with the double cosets $DC(n,q)$.

Let $g_1, g_2, \cdots, g_{N(n,q)}$ be a fixed ordering of the elements in $DC(n,q)$ ($n = 1, 3, 5, \cdots$). Then we put

$$v(n,q) = (Trg_1, Trg_2, \cdots, Trg_{N(n,q)}) \in \mathbb{F}_q^{N(n,q)}, \ for\ n = 1, 3, 5, \cdots.$$

Now, the binary linear code $C(DC(n,q))$ is defined as:

$$C(DC(n,q)) = \{u \in \mathbb{F}_2^{N(n,q)} | u \cdot v(n,q) = 0\}, \ for\ n = 1, 3, 5, \cdots, \tag{4.2}$$

where the dot denotes the usual inner product in $\mathbb{F}_q^{N(n,q)}$.

The following Delsarte's theorem is well-known.

Theorem 4.1. *([14]) Let B be a linear code over $\mathbb{F}_q$. Then*

$$(B|_{\mathbb{F}_2})^{\perp} = tr(B^{\perp}).$$

In view of this theorem, the dual $C(DC(n,q))^{\perp}$ of the code $C(DC(n,q))$ is given by

$$\begin{aligned} &C(DC(n,q))^{\perp} \\ &= \{c(a) = c(a;n,q) = (tr(aTrg_1)), \cdots, tr(aTrg_{N(n,q)})|a \in \mathbb{F}_q\} \end{aligned} \tag{4.3}$$

$(n = 1, 3, 5, \cdots)$.

Let $\mathbb{F}_2^+$, $\mathbb{F}_q^+$ denote the additive groups of the fields $\mathbb{F}_2$, $\mathbb{F}_q$, respectively. Then we have the following exact sequence of groups:

$$0 \to \mathbb{F}_2^+ \to \mathbb{F}_q^+ \to \Theta(\mathbb{F}_q) \to 0,$$

where the first map is the inclusion and the second one is the Artin-Schreier operator in characteristic two given by $\Theta(x) = x^2 + x$. So

$$\Theta(\mathbb{F}_q) = \{\alpha^2 + \alpha | \alpha \in \mathbb{F}_q\}, \ \textit{and} \ \ [\mathbb{F}_q^+ : \Theta(\mathbb{F}_q)] = 2.$$

Theorem 4.2. *([5]) Let λ be the canonical additive character of $\mathbb{F}_q$, and let $\beta \in \mathbb{F}_q^*$. Then*

$$\sum_{\alpha \in \mathbb{F}_q - \{0,1\}} \lambda(\frac{\beta}{\alpha^2 + \alpha}) = K(\lambda; \beta) - 1. \tag{4.4}$$

Theorem 4.3. *The map $\mathbb{F}_q \to C(DC(n,q))^{\perp}(a \mapsto c(a))$ is an $\mathbb{F}_2$-linear isomorphism for each odd integer $n \geq 1$and all q, except for $n = 1$ and $q = 4$.*

Proof. The map is clearly $\mathbb{F}_2$-linear and surjective. Let a be in the kernel of map. Then $tr(aTrg) = 0$, for all $g \in DC(n,q)$. If $n \geq 3$ is odd, then, by Corollary 3.6, $Tr : DC(n,q) \to \mathbb{F}_q$ is surjective and hence $tr(a\alpha) = 0$, for all $\alpha \in \mathbb{F}_q$. This implies that $a = 0$, since otherwise $tr : \mathbb{F}_q \to \mathbb{F}_2$ would be the zero map. Now, assume that $n = 1$. Then, by (3.10), $tr(a\beta) = 0$, for all $\beta \neq 1$, with $tr((\beta - 1)^{-1}) = 0$. Hilbert's theorem 90 says that $tr(\gamma) = 0 \Leftrightarrow \gamma = \alpha^2 + \alpha$, for some $\alpha \in \mathbb{F}_q$. This implies that $\lambda(a) \sum_{\alpha \in \mathbb{F}_q - \{0,1\}} \lambda(\frac{a}{\alpha^2 + \alpha}) = q - 2$. If $a \neq 0$, then, invoking (4.4) and the Weil bound (1.1), we would have

$$q - 2 = \lambda(a) \sum_{\alpha \in \mathbb{F}_q - \{0,1\}} \lambda(\frac{a}{\alpha^2 + \alpha}) = \pm(K(\lambda; a) - 1) \leq 2\sqrt{q} + 1.$$

For $q \geq 16$, this is impossible, since $x > 2\sqrt{x} + 3$, for $x \geq 16$. On the other hand, for $q = 2, 4, 8$, one easily checks from (3.10) that the kernel is trivial for $q = 2, 8$ and is $\mathbb{F}_2$, for $q = 4$. □

5. Power Moments of Kloosterman Sums With Trace One Arguments

Here we will be able to find, via the Pless power moment identity, an infinite family of recursive formulas generating the odd power moments of Kloosterman sums with trace one arguments over all $\mathbb{F}_q$ in terms of the frequencies of weights in $C(DC(n,q))$ and $C(\widehat{DC}(n,q))$, respectively.

Theorem 5.1. (Pless power moment identity, [14]) *Let B be a q-ary $[n,k]$ code, and let B_i(resp. $B_i^{\perp}$) denote the number of codewords of weight i in B(resp. in $B^{\perp}$). Then, for $h = 0, 1, 2, \cdots$,*

$$\sum_{j=0}^{n} j^h B_j = \sum_{j=0}^{min\{n,h\}} (-1)^j B_j^{\perp} \sum_{t=j}^{h} t! S(h,t) q^{k-t} (q-1)^{t-j} \binom{n-j}{n-t}, \tag{5.1}$$

where $S(h,t)$ is the Stirling number of the second kind defined in (1.9).

Lemma 5.2. *Let $c(a) = (tr(aTrg_1), \cdots, tr(aTrg_{N(n,q)})) \in C(DC(n,q))^{\perp}(n = 1, 3, 5, \cdots)$, for $a \in \mathbb{F}_q^*$. Then the Hamming weight $w(c(a))$ is expressed as follows:*

$$w(c(a)) = \frac{1}{2} A(n,q)(B(n,q) - \lambda(a)K(\lambda; a))(cf.\ (1.4),\ (1.5)). \tag{5.2}$$

Proof. Here we recall that the Hamming weight of the codeword $c(a)$ is just the number of nonzero coordinates.

$$w(c(a)) = \frac{1}{2} \sum_{j=1}^{N(n,q)} (1 - (-1)^{tr(aTrg_j)}) = \frac{1}{2}(N(n,q) - \sum_{w \in DC(n,q)} \lambda(aTrw)).$$

Our result now follows from (3.5) and (4.1). □

Let $u = (u_1, \cdots, u_{N_{N(n,q)}}) \in \mathbb{F}_2^{N(n,q)}$, with ν_β 1's in the coordinate palces where $Tr(g_j) = \beta$, for each $\beta \in \mathbb{F}_q$. Then from the definition of the codes $C(DC(n,q))$ (cf.(4.2)) that u is a codeword with weight j if and only if $\sum_{\beta \in \mathbb{F}_q} \nu_\beta = j$ and $\sum_{\beta \in \mathbb{F}_q} \nu_\beta \beta = 0$(an identity in $\mathbb{F}_q$). As there are $\prod_{\beta \in \mathbb{F}_q} \binom{n(\beta)}{\nu_\beta}$ (cf. (3.8)) many such codewords with weight j, we obtain the following result.

Proposition 5.3. *Let $\{C_j(n,q)\}_{j=0}^{N(n,q)}$ be the weight distribution of $C(DC(n,q))$ $(n = 1, 3, 5, \cdots)$. Then*

$$C_j(n,q) = \sum \prod_{\beta \in \mathbb{F}_q} \binom{n(\beta)}{\nu_\beta}, \tag{5.3}$$

where the sum is over all the sets of integers $\{\nu_\beta\}_{\beta \in \mathbb{F}_q}$ $(0 \le \nu_\beta \le n(\beta))$, satisfying

$$\sum_{\beta \in \mathbb{F}_q} \nu_\beta = j, \text{ and} \sum_{\beta \in \mathbb{F}_q} \nu_\beta \beta = 0. \tag{5.4}$$

Corollary 5.4. *Let* $\{C_j(n,q)\}_{j=0}^{N(n,q)}$ $(n = 1,3,5,\cdots)$ *be as above. Then we have*

$$C_j(n,q) = C_{N(n,q)-j}(n,q), \text{ for all } j, \text{ with } 0 \le j \le N(n,q).$$

Proof. Under the replacements $\nu_\beta \to n(\beta) - \nu_\beta$, for each $\beta \in \mathbb{F}_q$, the first equation in (5.4) is changed to $N(n,q) - j$, while the second one in there and the summand in (5.3) is left unchanged. Here the second sum in (5.4) is left unchanged, since $\sum_{\beta\in\mathbb{F}_q} n(\beta)\beta = 0$, as one can see by using the explicit expressions of $n(\beta)$ in (3.9) and (3.10). □

The formula appearing in the next theorem and stated in (1.7) follows from the formula in (5.3), using the explicit value of $n(\beta)$ in (3.9).

Theorem 5.5. *Let* $\{C_j(n,q)\}_{j=0}^{N(n,q)}$ *be the weight distribution of* $C(DC(n,q))$ $(n = 1,3,5,\cdots)$. *Then, for* $j = 0,\cdots,N(n,q)$,

$$\begin{aligned} C_j(n,q) = \sum \binom{q^{-1}A(n,q)(B(n,q)+1)}{\nu_1} \\ \times \prod_{tr(\beta-1)^{-1}=0} \binom{q^{-1}A(n,q)(B(n,q)+q+1)}{\nu_\beta} \\ \times \prod_{tr(\beta-1)^{-1}=1} \binom{q^{-1}A(n,q)(B(n,q)-q+1)}{\nu_\beta}, \end{aligned}$$

where the sum is over all the sets of nonnegative integers $\{\nu_\beta\}_{\beta\in\mathbb{F}_q}$ *satisfying* $\sum_{\beta\in\mathbb{F}_q} \nu_\beta = j$ *and* $\sum_{\beta\in\mathbb{F}_q} \nu_\beta\beta = 0$.

The recursive formula in the next theorem follows from the study of codes associated with the double cosets $\widehat{DC}(n,q) = P'(2n,q)\sigma'_{n-1}P'(2n,q)$ of the symplectic group $Sp(2n,q)$. It is slightly modified from its original version, which makes it more usable in below.

Theorem 5.6. *([6]) For each odd integer* $n \ge 3$, *with all* q, *or* $n = 1$, *with* $q \ge 8$,

$$\begin{aligned} &\frac{1}{2^h}A(n,q)^h \sum_{l=0}^{h}(-1)^l \binom{h}{l} B(n,q)^{h-l}MK^l \\ &= q \sum_{j=0}^{min\{N(n,q),h\}} (-1)^j \widehat{C}_j(n,q) \sum_{t=j}^{h} t!S(h,t)2^{-t}\binom{N(n,q)-j}{N(n,q)-t} \ (h = 1,2,\cdots), \end{aligned} \tag{5.5}$$

where $N(n,q) = A(n,q)B(n,q)$, *and* $\{\widehat{C}_j(n,q)\}_{j=0}^{N(n,q)}$ *is the weight distribution of*

$C(\widehat{DC}(n,q))$ *given by*

$$\widehat{C}_j(n,q) = \sum \binom{q^{-1}A(n,q)(B(n,q)+1)}{\nu_0}$$
$$\times \prod_{tr(\beta^{-1})=0} \binom{q^{-1}A(n,q)(B(n,q)+q+1)}{\nu_\beta}$$
$$\times \prod_{tr(\beta^{-1})=1} \binom{q^{-1}A(n,q)(B(n,q)-q+1)}{\nu_\beta}.$$

Here the sum is over all the sets of nonnegative integers $\{\nu_\beta\}_{\beta\in\mathbb{F}_q}$ satisfying $\sum_{\beta\in\mathbb{F}_q}\nu_\beta = j$ and $\sum_{\beta\in\mathbb{F}_q}\nu_\beta\beta = 0$. In addition, $S(h,t)$ is the Stirling number of the second kind as in (1.9).

From now on, we will assume that n is any odd integer ≥ 3, with all q, or $n=1$, with $q \geq 8$. Under these assumptions, each codeword in $C(DC(n,q))^{\perp}$ can be written as $c(a)$, for a unique $a \in \mathbb{F}_q$ (cf. Theorem 4.3, (4.3)) and Theorem 5.6 in the above can be applied.

Now, we apply the Pless power moment identity in (5.1) to $B = C(DC(n,q))^{\perp}$(and hence $B_j^{\perp} = C_j(n,q)$), in order to get the result in Theorem 1.1 (cf. (1.6)-(1.8)) about recursive formulas. Below, "the sum over $tra=0$(resp. $tra=1$)" will mean "the sum over all nonzero $a\in\mathbb{F}_q^*$, with $tra=0$ (resp. $tra=1$)." The left-hand side of that identity in (5.1) is equal to

$$\sum_{a\in\mathbb{F}_q^*} w(c(a))^h,$$

with $w(c(a))$ given by (5.2). We have

$$\sum_{a\in\mathbb{F}_q^*} w(c(a))^h = \frac{1}{2^h}A(n,q)^h \sum_{a\in\mathbb{F}_q^*}(B(n,q)-\lambda(a)K(\lambda;a))^h$$
$$= \frac{1}{2^h}A(n,q)^h \sum_{tra=0}(B(n,q)-K(\lambda;a))^h + \frac{1}{2^h}A(n,q)^h \sum_{tra=1}(B(n,q)+K(\lambda;a))^h \tag{5.6}$$

$$
\begin{aligned}
&= \frac{1}{2^h} A(n,q)^h \sum_{tra=0} \sum_{l=0}^{h} (-1)^l \binom{h}{l} B(n,q)^{h-l} K(\lambda;a)^l \\
&\quad + \frac{1}{2^h} A(n,q)^h \sum_{tra=1} \sum_{l=0}^{h} \binom{h}{l} B(n,q)^{h-l} K(\lambda;a)^l \\
&= \frac{1}{2^h} A(n,q)^h \sum_{l=0}^{h} (-1)^l \binom{h}{l} B(n,q)^{h-l} (MK^l - T_1 K^l)(\psi = \lambda \ case\ of\ (1.2),\ (1.3)) \\
&\quad + \frac{1}{2^h} A(n,q)^h \sum_{l=0}^{h} \binom{h}{l} B(n,q)^{h-l} T_1 K^l \\
&= \frac{1}{2^h} A(n,q)^h \sum_{l=0}^{h} (-1)^l \binom{h}{l} B(n,q)^{h-l} M K^l \\
&\quad + 2\frac{1}{2^h} A(n,q)^h \sum_{0 \le l \le h,\ l\ odd} \binom{h}{l} B(n,q)^{h-l} T_1 K^l \\
&= q \sum_{j=0}^{min\{N(n,q),h\}} (-1)^j \widehat{C}_j(n,q) \sum_{t=j}^{h} t! S(h,t) 2^{-t} \binom{N(n,q)-j}{N(n,q)-t}\ (cf.\ (5.5)) \\
&\quad + 2\frac{1}{2^h} A(n,q)^h \sum_{0 \le l \le h,\ l\ odd} \binom{h}{l} B(n,q)^{h-l} T_1 K^l.
\end{aligned}
$$

On the other hand, the right hand side of the identity in (5.1) is given by:

$$
q \sum_{j=0}^{min\{N(n,q),h\}} (-1)^j C_j(n,q) \sum_{t=j}^{h} t! S(h,t) 2^{-t} \binom{N(n,q)-j}{N(n,q)-t}. \tag{5.7}
$$

In (5.7), one has to note that $dim_{\mathbb{F}_2} C(DC(n,q))^{\perp} = r$. Our main result in (1.6) now follows by equating (5.6) and (5.7).

Acknowledgments I would like to thank the referee who read the manuscript carefully and made valuable comments and suggestions.

References

[1] L.Carlitz, *Gauss sums over finite fields of order* 2^n, Acta Arith. **15**(1969), 247–265.

[2] R.J.Evans, *Seventh power moments of Kloosterman sums*, Israel J. Math.**175** (2010), 349-362.

[3] K. Hulek, J. Spandaw, B. van Geemen and D.van van Straten, *The modulartiy of the Barth-Nieto quintic and its relatives*, Adv. Geom. **1** (2001), 263–289.

[4] D. S. Kim, *Gauss sums for symplectic groups over a finite field*, Mh. Math. **126** (1998), 55–71.

[5] D. S. Kim, *Codes associated with $O^+(2n, q)$ and power moments of Kloosterman sums*, Integers **11** (2011), A62, 19 pp.

[6] D. S. Kim, *Infinite families of recursive formulas generating power moments of Kloosterman sums: symplectic case*, submitted.

[7] D. S. Kim, *Codes associated with special linear groups and power moments of multi-dimensional Kloosterman sums*, Ann. Mat. Pura Appli. **190** (2011), 61-76.

[8] D. S. Kim, *Infinite families of recursive formulas generating power moments of ternary Kloosterman sums with square arguments arising from symplectic groups*, Adv. Math. Commun.**3** (2009), 167-178.

[9] D. S. Kim, *Infinite families of recursive formulas generating power moments of ternary Kloosterman sums with square arguments associated with $O^-(2n, q)$*, J. Korean Math. Soc.**48** (2011), 267-288.

[10] D. S. Kim and Y. H. Park, *Gauss sums for orthogonal groups over a finite field of characteristic two*, Acta Arith., **82** (1997), 331–357.

[11] H. D. Kloosterman, *On the representation of numbers in the form $ax^2 + by^2 + cz^2 + dt^2$*, Acta Math. **49** (1926), 407-464.

[12] R. Lidl and H. Niederreiter, *Finite Fields*, Encyclopedia Math. Appl. **20**, Cambridge University Pless, Cambridge, 1987.

[13] R. Livné, *Motivic orthogonal two-dimensional representations of $Gal(\overline{\mathbb{Q}}/\mathbb{Q})$*, Israel J. Math. **92** (1995), 149-156.

[14] F. J. MacWilliams and N. J. A. Sloane, *The Theory of Error Correcting Codes*, North-Holland, Amsterdam, 1998.

[15] M. Moisio, *The moments of a Kloosterman sum and the weight distribution of a Zetterberg-type binary cyclic code*, IEEE Trans. Inform. Theory. **53**(2007), 843–847.

[16] C. Peters, J. Top, and M. van der Vlugt, *The Hasse zeta function of a K3 surface related to the number of words of weight 5 in the Melas codes*, J. Reine Angew. Math. **432** (1992), 151-176.

[17] H. Salié, *Über die Kloostermanschen Summen $\mathcal{S}(u, v; q)$*, Math. Z. **34**(1931), 91-109.

[18] R. Schoof and M. van der Vlugt, *Hecke operators and the weight distributions of certain codes*, J. Combin. Theory Ser. A **57**(1991), 163-186.

THE DIOPHANTINE EQUATION $F_n^y + F_{n+1}^x = F_m^x$

Florian Luca[1]
Fundación Marcos Moshinsky, UNAM, Circuito Exterior, C.U., Apdo. Postal 70-543, Mexico D.F. 04510, Mexico
fluca@matmor.unam.mx

Roger Oyono
Équipe GAATI, Université de la Polynésie Française, BP 6570, 98702 Faa'a, Tahiti, French Polynesia
roger.oyono@upf.pf

Received: 8/16/12, Accepted: 4/9/13, Published: 6/14/13

Abstract

Here, we find all the solutions of the title Diophantine equation in positive integer variables (m, n, x, y), where F_k is the k–th term of the Fibonacci sequence.

1. Introduction

Let $(F_n)_{n\geq 0}$ be the Fibonacci sequence given by $F_0 = 0,\ F_1 = 1, F_{n+2} = F_{n+1} + F_n$ for all $n \geq 0$. The Diophantine equation

$$F_n^x + F_{n+1}^x = F_m \tag{1}$$

in positive integers (m, n, x) was studied in [7]. There, it was showed that there exists no solution other than $(m, n) = (3, 1)$ for which $1^x + 1^x = 2$ (valid for all positive integers x), and the solutions for $x = 1$ and $x = 2$ arising via the formulas $F_n + F_{n+1} = F_{n+2}$ and $F_n^2 + F_{n+1}^2 = F_{2n+1}$. Equation (1) was revisited in [6] under the more general form

$$F_n^x + F_{n+1}^x = F_m^y \tag{2}$$

in positive integers (m, n, x, y) and it was shown that the only solutions of equation (2) with $y > 1$ are $(m, n, x, y) = (3, 4, 1, 3),\ (4, 2, 3, 2)$. Here, we reverse the role of two exponents in equation (2) and study the equation

$$F_n^x + F_{n+1}^y = F_m^x \qquad \text{or} \qquad F_n^y + F_{n+1}^x = F_m^x \tag{3}$$

[1]Work supported in part by Projects PAPIIT IN104512, CONACyT 193539, CONACyT 163787 and a Marcos Moshinsky Fellowship.

in positive integers (m, n, x, y). Our result is the following.

Theorem 1. *The only positive integer solution* (m, n, x, y) *of one of equations* (3) *with* $n \geq 3$ *and* $x \neq y$ *is* $(5, 3, 2, 4)$ *for which* $F_3^4 + F_4^2 = F_5^2$.

We note that the solutions of equation (3) either with $n \in \{1, 2\}$ or $x = y$ are contained in the solutions of equation (2) and therefore are of no interest.

Before getting to the proof, we mention that similar looking equations have already been studied. For example, in [4], it was shown that the only solution in positive integers (k, ℓ, n, r) of the equation

$$F_1^k + F_2^k + \cdots + F_{n-1}^k = F_{n+1}^\ell + \cdots + F_{n+r}^\ell$$

is $(k, \ell, n, r) = (8, 2, 4, 3)$, while in [9], T. Miyazaki showed that the only positive integer solutions (x, y, z, n) of the equation

$$F_n^x + F_{n+1}^y = F_{2n+1}^z$$

are for $(x, y, z) = (2, 2, 1)$ (and for all positive integers n).

2. Preliminary Results

We write $(\alpha, \beta) = ((1 + \sqrt{5})/2, (1 - \sqrt{5})/2)$ and use the Binet formula

$$F_n = \frac{\alpha^n - \beta^n}{\alpha - \beta} \quad \text{valid for all} \quad n \geq 0. \tag{4}$$

We also use the inequality

$$\alpha^{n-2} \leq F_n \leq \alpha^{n-1} \quad \text{valid for all} \quad n \geq 1. \tag{5}$$

We will need the following elementary inequality.

Lemma 1. *For* $n \geq 3$, *we have* $F_n^5 \geq F_{n+1}^3$.

Proof. The inequality is clearly true for $n = 3$, so we assume that $n \geq 4$. Observe that $F_{n+1}/F_n \leq 5/3$, since the above inequality is equivalent to $3F_{n+1} \leq 5F_n$, or $3(F_n + F_{n-1}) \leq 5F_n$, or $3F_{n-1} \leq 2F_n$, further with $3F_{n-1} \leq 2(F_{n-1} + F_{n-2})$, or $F_{n-1} \leq 2F_{n-2}$, or $F_{n-2} + F_{n-3} \leq 2F_{n-2}$, or $F_{n-3} \leq F_{n-2}$, which is clearly true for $n \geq 4$. Thus,

$$\left(\frac{F_{n+1}}{F_n}\right)^3 \leq \left(\frac{5}{3}\right)^3 < 3^2 \leq F_n^2$$

for $n \geq 4$, which is equivalent to $F_{n+1}^3 \leq F_n^5$. □

We shall need a couple of results from the theory of lower bounds for nonzero linear forms in complex and p-adic logarithms which we now recall.

For an algebraic number η we write $h(\eta)$ for its logarithmic height whose formula is

$$h(\eta) = \frac{1}{d}\left(\log a_0 + \sum_{i=1}^{d} \log\left(\max\{|\eta^{(i)}|, 1\}\right)\right),$$

with d being the degree of η over $\mathbb{Q}$ and

$$f(X) = a_0 \prod_{i=1}^{d}(X - \eta^{(i)}) \in \mathbb{Z}[X] \tag{6}$$

being the minimal primitive polynomial over the integers having positive leading coefficient a_0 and η as a root.

With this notation, Matveev (see [8] or Theorem 9.4 in [1]) proved the following deep theorem:

Theorem 2. *Let $\mathbb{K}$ be a real number field of degree D over $\mathbb{Q}$, $\gamma_1, \ldots, \gamma_t$ be nonzero elements of $\mathbb{K}$, and $b_1, \ldots, b_t$ be nonzero integers. Put*

$$B \geq \max\{|b_1|, \ldots, |b_t|\},$$

and

$$\Lambda = \gamma_1^{b_1} \cdots \gamma_t^{b_t} - 1.$$

Let $A_1, \ldots, A_t$ be real numbers such that

$$A_i \geq \max\{Dh(\gamma_i), |\log \gamma_i|, 0.16\}, \quad i = 1, \ldots, t.$$

Then, assuming that $\Lambda \neq 0$, we have

$$|\Lambda| > \exp\left(-1.4 \times 30^{t+3} \times t^{4.5} \times D^2(1 + \log D)(1 + \log B)A_1 \cdots A_t\right).$$

We shall also need the rational case version of a linear form in p-adic logarithms proved by Kunrui Yu [10]. For a nonzero rational number r and a prime number p put $\mathrm{ord}_p(r)$ for the exponent of p in the factorization of r.

Theorem 3. *Let $\gamma_1, \ldots, \gamma_t$ be nonzero rational numbers and $b_1, \ldots, b_t$ be nonzero integers. Put*

$$B \geq \max\{|b_1|, \ldots, |b_t|, 3\},$$

and

$$\Lambda = \gamma_1^{b_1} \cdots \gamma_t^{b_t} - 1.$$

Let $A_1, \ldots, A_t$ be real numbers such that

$$A_i \geq \max\{h(\gamma_i), \log p\}, \quad i = 1, \ldots, t.$$

Then, assuming that $\Lambda \neq 0$, we have

$$\mathrm{ord}_p(\Lambda) < 19(20\sqrt{t+1})^{2t+2} \frac{p}{(\log p)^2} \log(e^5 t) A_1 \cdots A_t \log B.$$

3. The Proof of Theorem 1

3.1. Inequalities Among the Variables m, n and x, y

We start with the following lemma.

Lemma 2. *In any positive integer solution* (m, n, x, y) *of either one of equations* (3) *with* $n \geq 3$ *and* $x \neq y$, *we have:*

(i) $m > n$;

(ii) $x < y$;

(iii) $m \geq 5$;

(iv) $y(n+1) > (m-2)x$ *and* $(n-2)y < (m-1)x$.

Proof. (i) Either one of equations (3) implies that $F_m^x > F_n^x$, therefore $m > n$.

(ii) Let us now show that $x < y$. Assuming otherwise, we have that

$$F_m^x < F_n^x + F_{n+1}^x < (F_n + F_{n+1})^x = F_{n+2}^x,$$

therefore $m < n+2$. The case $m \in \{n, n+1\}$ is impossible because F_n and F_{n+1} are coprime, so we get $m < n$, contradicting (i). (iii) Since $m > n$ by (i) and the fact that F_n is coprime to F_{n+1}, we deduce in fact that $m > n+1$, and since $n \geq 3$, we get that $m \geq 5$.

(iv) This follows from (ii) and inequalities (5). More precisely,

$$\begin{aligned} \alpha^{(n+1)y} &> F_{n+2}^y = (F_n + F_{n+1})^y > \max\{F_n^x + F_{n+1}^y, F_n^y + F_{n+1}^x\} \\ &\geq F_m^x > \alpha^{(m-2)x}, \end{aligned}$$

implying the first inequality (iv), and

$$\alpha^{(n-2)y} < F_n^y < \min\{F_n^x + F_{n+1}^y, F_n^y + F_{n+1}^x\} \leq F_m^x < \alpha^{(m-1)x},$$

implying the second inequality (iv). □

3.2. Bounding y in Terms of m

Lemma 3. *Any positive integer solution* (m, n, x, y) *with* $n \geq 3$ *and* $x \neq y$ *of equation* (3) *satisfies one of the following inequalities*

(i) $y < 2 \times 10^{13} m^2 \log m$ *if* $y \leq 2x$;

(ii) $y < 10^{13} m \log m$ *if* $y > 2x$.

Proof. We distinguish the following two cases.

`Case 1.` $y \leq 2x$. In this case, we apply a linear form in 2-adic logarithms upon observing that exactly of F_n, F_{n+1}, F_m is even. The linear form is of the form

$$\Lambda = F_a^u F_b^{-v} - 1,$$

where a and b are distinct in $\{n, n+1, m\}$ such that F_a and F_b are odd, and u and v are in $\{2x, 2y\}$. In any case, if c is such that $\{a, b, c\} = \{n, n+1, m\}$ then it is always the case that F_c is even and $F_c^x \mid F_a^u - F_b^v$, therefore

$$\mathrm{ord}_2(\Lambda) \geq \mathrm{ord}_2(F_c^x) \geq x \geq y/2. \tag{7}$$

To get an upper bound on $\mathrm{ord}_2(\Lambda)$, we use Theorem 3. We take the parameters $t = 2$, $\gamma_1 = F_a$, $\gamma_2 = F_b$, $b_1 = u$, $b_2 = -v$. We can take $B = 2y$. Since $n + 1 < m$, by inequalities (5), we can take

$$A_1 = A_2 = m \log \alpha > \max\{\log F_a, \log F_b, \log 2\}.$$

Theorem 3 now gives

$$\mathrm{ord}_2(\Lambda) \leq 19(20\sqrt{3})^6 \left(\frac{2}{(\log 2)^2}\right) \log(2e^5)(m \log \alpha)^2 \log(2y), \tag{8}$$

which compared with (7) gives

$$\begin{aligned} 2y &\leq 4 \times 19 \times (20\sqrt{3})^6 \left(\frac{2}{(\log 2)^2}\right) \log(2e^5)(\log \alpha)^2 \log(2y) \\ &< 8 \times 10^{11} m^2 \log(2y). \end{aligned}$$

Using the fact that for $A > 3$ the inequality

$$t < A \log t \qquad \text{implies} \qquad t < 2A \log A$$

(with $A = 8 \times 10^{11} m^2$), we have

$$2y < 2 \times 8 \times 10^{11} m^2 (\log(8 \times 10^{11}) + 2 \log m) < 2 \times 8 \times 10^{11} m^2 (20 \log m),$$

therefore

$$y < 2 \times 10^{13} m^2 \log m, \tag{9}$$

which takes care of (i). In the above inequalities we also used the fact that

$$\log(8 \times 10^{11}) + 2 \log m < 20 \log m,$$

which holds because $m \geq 5$.

Case 2. $y > 2x$. In this case, we use a linear form in complex logarithms. This linear form is one of

$$\Lambda = F_m^x F_{n+1}^{-y} - 1 \qquad \text{or} \qquad F_m^x F_n^{-y} - 1$$

depending on whether we work with the left equation (3) or with the right equation (3), respectively. Clearly, $\Lambda > 0$. We first find an upper bound on Λ which follows from equation (3). In case of the left equation (3), we have

$$\Lambda = \frac{F_n^x}{F_{n+1}^y} < \frac{F_{n+1}^x}{F_{n+1}^y} < \frac{1}{F_{n+1}^{y/2}}. \tag{10}$$

In case of the right equation (3), we have, by Lemma 1,

$$\Lambda = \frac{F_{n+1}^x}{F_n^y} < \frac{F_{n+1}^x}{F_{n+1}^{3y/5}} < \frac{1}{F_{n+1}^{3y/5-y/2}} = \frac{1}{F_{n+1}^{y/10}}. \tag{11}$$

So, from (10) and (11), we get that the inequality

$$\Lambda < \frac{1}{F_{n+1}^{y/10}} \tag{12}$$

holds in all instances. We now find a lower bound on Λ by using Theorem 2. We take $t = 2$, $\gamma_1 = F_m$, $\gamma_2 = F_u$ with $u \in \{n, n+1\}$, $b_1 = x$, $b_2 = -y$. We take $\mathbb{K} = \mathbb{Q}$, so $D = 1$. We take $B = y$. By inequality (5), we can take $A_1 = m \log \alpha$ and $A_2 = \log F_{n+1}$. We then get that

$$\Lambda > \exp\left(-1.4 \times 30^5 \times 2^{4.5} \times (m \log \alpha) \times \log F_{n+1} \times (1 + \log y)\right),$$

which together with (12) gives

$$(y/10) \log F_{n+1} < 1.4 \times 30^5 \times 2^{4.5} \times (m \log \alpha) \times \log F_{n+1} \times (1 + \log y),$$

or

$$y < 14 \times 30^5 \times 2^{4.5} \times \log \alpha \times m \times (3 \log y) < 2 \times 10^{11} m \log y,$$

where we used the inequality $1 + \log y < 3 \log y$, which holds for all $y \geq 2$. Thus,

$$y < 4 \times 10^{11} m (\log(2 \times 10^{11}) + \log m) < 4 \times 10^{11} (20 \log m) < 10^{13} m \log m,$$

where we used the fact that $\log(2 \times 10^{11}) + \log m < 20 \log m$ for all $m \geq 5$. This takes care of (ii). □

3.3. Small m

Lemma 4. *If $(m, n, x, y) \neq (5, 3, 2, 4)$ is a positive integer solution of equation (3) with $n \geq 3$ and $x \neq y$, then $m \geq 1000$.*

Proof. Assume that we work with the left equation (3). Then

$$F_{n+1}^y = F_m^x - F_n^x. \tag{13}$$

Assume first that $y \geq 20$. Observe that from the above equation we get that $F_m - F_n$ is a divisor of F_{n+1}. Let $D_{m,n} = \gcd(F_m - F_n, F_{n+1})$. We first checked computationally that there is no pair (m, n) with $6 \leq n + 3 < m \leq 999$, such that $p^{20} \mid F_m - F_n$ for some prime factor p of $D_{m,n}$. It follows that all prime factors of $F_m - F_n$ appear in its factorization at powers smaller than 20. But if that is so, it should be the case that $D_{m,n}^{20}$ is divisible by $F_m - F_n$. We checked computationally that this is not the case for any such pair (m, n). The conclusion of this computation is that $m \in \{n + 2, n + 3\}$. Now

$$F_{n+2} - F_n = F_{n+1} \qquad \text{and} \qquad F_{n+3} - F_n = 2F_{n+1}.$$

Together with formula (8), we get that $F_m^x - F_n^x = F_{n+1}^y$ is divisible by exactly the same primes as $F_m - F_n$. By Carmichael's Primitive Divisor Theorem (see [3]) for Lucas sequences with coprime integer roots, we get that $x \leq 6$. So,

$$F_m^x - F_n^x < F_m^x \leq F_{n+3}^6 < (2F_{n+2})^6 < (4F_{n+1})^6 = 2^{12} F_{n+1}^6 < F_{n+1}^{20} \leq F_{n+1}^y,$$

a contradiction. This calculation shows that $1 \leq x < y \leq 19$. We tested the remaining range $1 \leq x < y \leq 19$ and $3 \leq n < m \leq 999$ by brute force and no solution came up. A similar argument works for the right equation (3) with one exception. Namely, in the case when $5 \leq n + 2 < m \leq 999$, by putting $D_{m,n} = \gcd(F_m - F_{n+1}, F_n)$ computations revealed that, as before, $p^{20} \nmid F_m - F_n$ for any prime $p \mid D_{m,n}$ and any such pair (m, n), but the pair $(m, n) = (14, 8)$ has the property that $D_{mn,}^{20}$ is a multiple of $F_m - F_n$ and is the only such pair. Namely, in this case $F_m - F_n = F_{14} - F_9 = 343 = 7^3$, and $F_8 = 21 = 3 \times 7$. In this last case however, again by Carmichael's Primitive Divisor Theorem, $F_{14}^x - F_9^x$ should have a prime factor $p \equiv 1 \pmod{x}$ if $x > 6$ which does not divide $F_{14} - F_8$, but this is not the case if $x > 6$ since $F_{14}^x - F_9^x = F_8^y = 3^y \times 7^y$. Hence, again $x \leq 6$, and we get a contradiction because $y \geq 20$. This shows that, as for the case of the left equation (3), we must have $1 \leq x < y \leq 19$. Again we tested this remaining range by brute force and only the solution $(5, 3, 2, 4)$ of the right equation (3) showed up. The lemma is therefore proved. □

3.4. Approximating F_m^x

From now on, we assume that $m \geq 1000$.

Lemma 5. *If (m,n,x,y) is a positive integer solution of equation (3) with $n \geq 3$ and $x \neq y$, then*

$$F_m^x = \frac{\alpha^{mx}}{5^{x/2}}\left(1+\zeta_{m,x}\right), \qquad \textit{where} \qquad |\zeta_{m,x}| < \frac{2}{\alpha^m}.$$

Proof. We use the Binet formula (4) to get

$$F_m^x = \frac{\alpha^{mx}}{5^{x/2}}\left(1-\left(\frac{\beta}{\alpha}\right)^m\right)^x = \frac{\alpha^{mx}}{5^{x/2}}\left(1-\frac{(-1)^m}{\alpha^{2m}}\right)^x. \tag{14}$$

Observe that, by Lemmas 2 and 3, we have

$$\frac{x}{\alpha^{2m}} < \frac{y}{\alpha^{2m}} < \frac{2\times 10^{13} m^2 \log m}{\alpha^{2m}} < \frac{1}{\alpha^m},$$

where the last inequality holds for all $m \geq 86$. Thus, if m is odd, then

$$\begin{aligned} 1 &< \left(1-\frac{(-1)^m}{\alpha^{2m}}\right)^x = \left(1+\frac{1}{\alpha^{2m}}\right)^x < \exp\left(\frac{x}{\alpha^{2m}}\right) \\ &< \exp\left(\frac{1}{\alpha^m}\right) < 1+\frac{2}{\alpha^m}, \end{aligned} \tag{15}$$

where we also used the fact that $\exp(t) < 1+2t$ if $t \in (0,1)$. Similarly, when m is even, using the fact that $1-t > \exp(-2t)$ holds for $t \in (0,1/2)$, we have

$$\begin{aligned} 1 &> \left(1-\frac{(-1)^m}{\alpha^{2m}}\right)^x = \left(1-\frac{1}{\alpha^{2m}}\right)^x > \exp\left(-\frac{2x}{\alpha^{2m}}\right) \\ &> \exp\left(-\frac{2}{\alpha^m}\right) > 1-\frac{2}{\alpha^m}. \end{aligned} \tag{16}$$

From estimates (15) and (16), we deduce that in both cases m odd and m even we have

$$\left(1+\frac{(-1)^m}{\alpha^{2m}}\right)^x = 1+\zeta_{m,x}, \qquad \text{with} \qquad |\zeta_{m,x}| < \frac{2}{\alpha^m},$$

which together with formula (14) finishes the proof of this lemma. □

3.5. Approximating F_a^u for $a \in \{n, n+1\}$, $u \in \{x,y\}$ and Large n

Lemma 6. *If (m,n,x,y) is a positive integer solution of equation (3) with $n \geq 3$, $x \neq y$ and $2x \geq y$, then the estimates*

$$F_a^u = \frac{\alpha^{au}}{5^{u/2}}\left(1+\zeta_{a,u}\right), \qquad \textit{where} \qquad |\zeta_{a,u}| < \frac{2}{\alpha^n}, \tag{17}$$

hold for $a \in \{n, n+1\}$ and $u \in \{x,y\}$.

Proof. The proof is based on inequality (iv) of Lemma 2, which in the particular case $y \le 2x$ implies

$$y(n+1) > (m-2)x \ge \frac{(m-2)y}{2}, \qquad \text{so} \qquad n > \frac{m-2}{2} - 1 = \frac{m}{2} - 2.$$

Now for $a \in \{n, n+1\}$ and $u \in \{x, y\}$, we have, by the Binet formula (4),

$$F_a^u = \frac{\alpha^{au}}{5^{u/2}} \left(1 - \frac{(-1)^a}{\alpha^{2a}}\right)^u.$$

Observe that, by Lemma 4,

$$\frac{u}{\alpha^{2a}} \le \frac{y}{\alpha^{2n}} \le \frac{2 \times 10^{13} m^2 \log m}{\alpha^{2n}} \le \frac{1}{\alpha^n},$$

where the last inequality is implied by

$$\alpha^n \ge \alpha^{m/2-2} \ge 2 \times 10^{13} m^2 \log m,$$

which holds for all $m \ge 182$. The conclusion of the lemma follows as in the proof of Lemma 5. □

3.6. A Small Linear Form in α and $\sqrt{5}$

Lemma 7. *If* (m, n, x, y) *is a positive integer solution to equation* (3) *with* $n \ge 3$ *and* $x \ne y$ *such that inequalities* (17) *hold, then putting* $\lambda = \min\{n, (m-n-1)y\}$, *we have*

$$\left|1 - \alpha^{ay-mx} 5^{(x-y)/2}\right| < \frac{13}{\alpha^\lambda} \qquad \textit{for some} \qquad a \in \{n, n+1\}. \tag{18}$$

Proof. By Lemma 5, we have

$$\left|F_m^x - \frac{\alpha^{mx}}{5^{x/2}}\right| < \frac{2}{\alpha^m} \left(\frac{\alpha^{mx}}{5^{x/2}}\right).$$

Since $m > n \ge 3$ and $\alpha^m > \alpha^n = 2\alpha + 1 = 2 + \sqrt{5} > 4$, it follows that $2/\alpha^m < 1/2$, therefore the above estimate implies that

$$\left|F_m^x - \frac{\alpha^{mx}}{5^{x/2}}\right| < \frac{1}{2}\left(\frac{\alpha^{mx}}{5^{x/2}}\right), \qquad \text{so} \qquad \frac{1}{2}\left(\frac{\alpha^{mx}}{5^{x/2}}\right) < F_m^x < \frac{3}{2}\left(\frac{\alpha^{mx}}{5^{x/2}}\right). \tag{19}$$

In particular,

$$\left|F_m^x - \frac{\alpha^{mx}}{5^{x/2}}\right| < \left(\frac{4}{\alpha^m}\right) F_m^x.$$

Since we are assuming that estimates (17) hold, we get, by a similar argument, that the estimates

$$\left|F_a^u - \frac{\alpha^{au}}{5^{u/2}}\right| < \left(\frac{4}{\alpha^n}\right) F_a^u \qquad \text{hold with} \qquad a \in \{n, n+1\},\ u \in \{x, y\}.$$

Thus, in the case of the left equation (3), we get

$$\begin{aligned}\left|\frac{\alpha^{mx}}{5^{x/2}}-\frac{\alpha^{nx}}{5^{x/2}}-\frac{\alpha^{(n+1)y}}{5^{y/2}}\right| &\le \left|\frac{\alpha^{mx}}{5^{x/2}}-F_m^x\right|+\left|\frac{\alpha^{nx}}{5^{x/2}}-F_n^x\right|+\left|\frac{\alpha^{(n+1)y}}{5^{y/2}}-F_{n+1}^y\right|\\ &\le \left(\frac{4}{\alpha^m}\right)F_m^x+\left(\frac{4}{\alpha^n}\right)(F_n^x+F_{n+1}^y)\\ &< \left(\frac{8}{\alpha^n}\right)F_m^x<\left(\frac{12}{\alpha^n}\right)\left(\frac{\alpha^{mx}}{5^{x/2}}\right),\end{aligned}$$

so that

$$\left|1-\alpha^{(n+1)y-mx}5^{(x-y)/2}\right|<\frac{1}{\alpha^{(m-n)x}}+\frac{12}{\alpha^n}\le\frac{13}{\alpha^{\min\{n,(m-n)x\}}}. \tag{20}$$

In the case of the right equation (3), a similar argument gives

$$\left|\frac{\alpha^{mx}}{5^{x/2}}-\frac{\alpha^{ny}}{5^{y/2}}-\frac{\alpha^{(n+1)x}}{5^{x/2}}\right|<\left(\frac{12}{\alpha^n}\right)\left(\frac{\alpha^{mx}}{5^{x/2}}\right),$$

leading to the similar looking inequality as (20), namely

$$\left|1-\alpha^{ny-mx}5^{(x-y)/2}\right|<\frac{13}{\alpha^{\min\{n,(m-n-1)x\}}}, \tag{21}$$

which together with (20) finishes the proof of this lemma. □

3.7. An Upper Bound for λ

Lemma 8. *If (m,n,x,y) is a positive integer solution of equation (3) with $n\ge 3$ and $x\ne y$ such that inequality (18) holds, then*

$$\lambda<5\times 10^{10}\log y. \tag{22}$$

Proof. We put

$$\Lambda=1-\alpha^{ay-mx}5^{(x-y)/2},\qquad \text{where}\qquad a\in\{n,n+1\}$$

is the expression appearing in the left hand side of the inequality (18) from Lemma 7. Since α and 5 are multiplicatively independent, and $x\ne y$, it follows that $\Lambda\ne 0$. Lemmas 6 and 7 show that

$$\log|\Lambda|\le\log 13-\lambda\log\alpha. \tag{23}$$

We next find a lower bound on $\log|\Lambda|$. For this, we use Theorem 2 with the choices $t=2$, $\mathbb{K}=\mathbb{Q}(\sqrt{5})$, $\gamma_1=\alpha$, $\gamma_2=\sqrt{5}$, $b_1=ay-mx$ and $b_2=x-y$. We have $D=2$. Since $b_2<0$ and

$$|\alpha^{b_1}5^{(x-y)/2}-1|<\frac{13}{\alpha^\lambda}\le\frac{13}{\alpha}<\alpha^5-1,$$

we have

$$\alpha^{b_1}5^{(x-y)/2} < \alpha^5, \quad \text{so} \quad b_1 \le (y-x)\frac{\log\sqrt{5}}{\log\alpha} + 5 < 2(y-x) + 5 \le 2y + 3.$$

Hence, we can take $B = 2y + 3$. We can also take $A_1 = \log\alpha$ and $A_2 = \log 5$. Theorem 2 now tells us that

$$\log|\Lambda| \ge -1.4 \times 30^5 \times 2^{4.5} \times 2^2(1 + \log 2)(1 + \log(2y + 3))(\log\alpha)(\log 5). \tag{24}$$

Putting together inequalities (23) and (24), we get

$$\lambda\log\alpha - \log 13 < 1.4 \times 30^5 \times 2^{4.5} \times 2^2(1 + \log 2)(1 + \log(2y + 3))(\log\alpha)(\log 5),$$

or

$$\lambda < \frac{\log 13}{\log\alpha} + 1.4 \times 30^5 \times 2^{4.5} \times 2^2(1 + \log 2)(1 + \log(2y + 3))(\log 5).$$

Since $1 + \log(2y+3) \le 1 + \log(4y) \le 5\log y$ for all $y \ge 2$, the above inequality gives

$$\lambda < \frac{\log 13}{\log\alpha} + 1.4 \times 30^5 \times 2^{4.5} \times 2^2(1 + \log 2)(\log 5)(5\log y) < 5 \times 10^{10}\log y, \tag{25}$$

which finishes the proof of this lemma. □

3.8. The Case When $x < y \le 2x$

Lemma 9. *Equation* (3) *has no positive integer solution* (m, n, x, y) *with* $n \ge 3$ *and* $x < y \le 2x$.

Proof. We exploit the conclusion of Lemma 8. We use the fact that

$$\lambda = \min\{(m - n - 1)x, n\} \ge \min\{y/2, m/2 - 2\}. \tag{26}$$

If the minimum on the right above is $y/2$, then inequality (25) gives

$$y < 10^{11}\log y \qquad \text{giving} \qquad y < 2 \times 10^{11} \times \log(10^{11}) < 6 \times 10^{12},$$

so

$$B < 1.5 \times 10^{13}.$$

If on the other hand the minimum in (26) is $m/2-2$, we then get, using also Lemma 2, that

$$\begin{aligned} m/2 - 2 &< 5 \times 10^{10}\log(2 \times 10^{13}m^2\log m) \\ &< 5 \times 10^{10}(\log(2 \times 10^{13}) + 3\log m) \\ &< 5 \times 10^{10} \times (23\log m), \end{aligned}$$

where we used the fact that $\log(2\times 10^{13})+3\log m<30+3\log m<23\log m$ for $m\geq 5$ (in fact, $m\geq 1000$, so a slightly better inequality holds at this step). Hence,

$$m/2-2<105\times 10^{10}\log m \qquad \text{giving} \qquad m<3\times 10^{12}\log m.$$

This last inequality leads to

$$m<2\times 3\times 10^{12}\log(3\times 10^{12})<2\times 10^{14},$$

so that

$$B=2y+3\leq 3+2\times 2\times 10^{13}m^2\log m<10^{44}.$$

Suppose now that $\lambda>10$. Then $13/\alpha^{\lambda}<1/2$, and so inequality (18) implies by a standard argument

$$|(ay-mx)\log\alpha-(y-x)\log\sqrt{5}|<\frac{26}{\alpha^{\lambda}},$$

or

$$\left|\frac{ay-mx}{y-x}-\frac{\log\sqrt{5}}{\log\alpha}\right|<\frac{26}{(\log\alpha)(y-x)\alpha^{\lambda}}<\frac{55}{(y-x)\alpha^{\lambda}}. \tag{27}$$

Let $[a_0,a_1,\ldots,a_{99}]=p_{99}/q_{99}$ be the 99th convergent of $\eta=(\log\sqrt{5})/\log\alpha$. The maximal a_i for $i=0,\ldots,99$ is $a_{20}=29$. Furthermore, we also have $q_{99}>10^{48}>B$. Hence,

$$\left|\frac{ay-mx}{y-x}-\frac{\log\sqrt{5}}{\log\alpha}\right|>\frac{1}{(29+2)(y-x)^2}=\frac{1}{31(y-x)^2}. \tag{28}$$

Thus, we get, from inequalities (27) and (28),

$$\frac{1}{31(y-x)^2}<\frac{54}{(y-x)\alpha^{\lambda}}$$

giving

$$\alpha^{\lambda}\leq 54\times 31\times(y-x)<2000\times B<2\times 10^{47},$$

leading to

$$\lambda\leq\frac{\log(2\times 10^{47})}{\log\alpha}<227.$$

If $\lambda=n$, we then get that $m/2-2\leq n<227$, so $m<458$, contradicting the fact that $m\geq 1000$. If $\lambda=(m-n-1)x$, then $(m-n-1)x<227$. In particular, $x<227$ and $(m-n)x\leq 2(m-n-1)x<454$. Further, inequality (27) shows that

$$\frac{ay-mx}{y-x}<\frac{\log\sqrt{5}}{\log\alpha}+\frac{55}{(y-x)\alpha}<2+\frac{34}{y-x},$$

so

$$ay-mx<2(y-x)+34.$$

If $a = n$, then

$$ay - mx = ny - mx = n(y-x) - (m-n)x < 2(y-x) + 34,$$

so

$$n \leq \frac{(m-n)x}{y-x} + 2 + \frac{34}{y-x} < 454 + 2 + 34 = 490,$$

therefore $m \leq (m-n)+n < 454+490 = 944$, contradicting the fact that $m \geq 1000$. If $a = n+1$, then

$$ay - mx = (n+1)y - mx = (n+1)(y-x) - (m-n-1)x < 2(y-x) + 34,$$

so

$$n+1 < \frac{(m-n-1)x}{y-x} + 2 + \frac{34}{y-x} < 227 + 2 + 34 = 263,$$

so $m \leq (m-n) + n < 454 + 262 = 716$, contradicting the fact that $m \geq 1000$. □

3.9. A Small Linear Form in Three Logarithms

From now on, we assume that $y > 2x$.

Lemma 10. *Any positive integer solution (m, n, x, y) of equation (3) with $n \geq 3$ and $x \neq y$ satisfies*

$$\left|F_a^y \alpha^{-mx} 5^{x/2} - 1\right| < \frac{4}{\alpha^\mu} \quad \textit{for some} \quad a \in \{n, n+1\}, \tag{29}$$

where $\mu = \min\{m, (m-n-2)x\}$.

Proof. Suppose that we work with the left equation (3). Then, by Lemma 5, we have

$$F_n^x + F_{n+1}^y = F_m^x = \frac{\alpha^{mx}}{5^{x/2}} (1 + \zeta_{m,x}),$$

so

$$\left|F_{n+1}^y \alpha^{-mx} 5^{x/2} - 1\right| \leq |\zeta_{m,x}| + \frac{F_n^x}{\alpha^{mx}/5^{x/2}}. \tag{30}$$

Estimate (19) gives

$$\frac{F_n^x}{\alpha^{mx}/5^{x/2}} < \frac{3}{2}\left(\frac{F_n^x}{F_m^x}\right) < \frac{2}{\alpha^{(m-n-1)x}},$$

where we used inequalities (5) to say that $F_n < \alpha^{n-1}$ and $F_m > \alpha^{m-2}$. Since $|\zeta_{m,x}| < 2/\alpha^m$ by Lemma 5, we get, from inequality (30), that

$$\left|F_{n+1}^y \alpha^{-mx} 5^{x/2} - 1\right| < \frac{2}{\alpha^m} + \frac{2}{\alpha^{(m-n-1)x}} \leq \frac{4}{\alpha^{\min\{m,(m-n-1)x\}}}, \tag{31}$$

A similar argument applies to the right equation (3). In that case, we get

$$F_n^y + F_{n+1}^x = F_m^x = \frac{\alpha^{mx}}{5^{x/2}}(1+\zeta_{m,x}),$$

therefore

$$\begin{aligned}\left|F_n^y \alpha^{-mx} 5^{x/2} - 1\right| &< |\zeta_{m,x}| + \frac{F_{n+1}^x}{\alpha^{mx}/5^{x/2}} < \frac{2}{\alpha^m} + \frac{3}{2}\left(\frac{F_{n+1}^x}{F_m^x}\right) \\ &< \frac{2}{\alpha^m} + \frac{2}{\alpha^{(m-n-2)x}} \le \frac{4}{\alpha^{\min\{m,(m-n-2)x\}}}, \end{aligned} \tag{32}$$

which together with inequality (31) completes the proof of this lemma. □

Remark. Lemma 2 (iv) shows that

$$(m-1)x > (n-2)y > 2(n-2)x, \quad \text{so} \quad m-1 > 2n-4 \quad \text{so} \quad m \ge 2n-2.$$

In particular, $m-n-2 \ge (m-6)/2$. This will be useful later.

Lemma 11. *Any positive solution (m,n,x,y) of equation (3) with $n \ge 3$ and $x \ne y$ satisfies:*

(i) $n > 10^{-14} m/\log m$;

(ii) $n \ge 1000$.

Proof. We put $\Lambda = F_a^y \alpha^{-mx} 5^{x/2} - 1$ for the form that appears in the left hand side of inequality (29). Since $mx > 0$ and no power of α of positive integer exponent can be a rational number, it follows that $\Lambda \ne 0$. Inequality (29) shows that

$$\log|\Lambda| < \log 4 - \mu \log \alpha. \tag{33}$$

We find a lower bound on $\log|\Lambda|$. We use Theorem 2 with the choices of parameters $t=3$, $\gamma_1 = F_a$, $\gamma_2 = \alpha$, $\gamma_3 = \sqrt{5}$, $b_1 = y$, $b_2 = -mx$, $b_3 = -x$. We have $\mathbb{K} = \mathbb{Q}(\sqrt{5})$ for which $D=2$. We take $A_1 = 2n\log\alpha$, $A_2 = \log\alpha$, and $A_3 = \log 5$. We take $B = my$. We then have

$$\log|\Lambda| > -1.4\times 30^6 \times 3^{4.5} \times 2^2 \times (1+\log 2)(1+\log(my))(2n\log\alpha)(\log\alpha)(\log 5). \tag{34}$$

Comparing estimates (33) and (34), we get

$$\mu < \frac{4}{\log\alpha} + 1.4\times 30^6 \times 3^{4.5} \times 2^2 \times (1+\log 2) \times 2 \times (\log\alpha) \times (\log 5) n (2\log(my)).$$

giving

$$\mu < 4\times 10^{12} n \log(my).$$

With Lemmas 3 and 10 and the remark following Lemma 10, we get

$$\begin{aligned} m-6 &< 8\times 10^{12} n \log(2\times 10^{13} m^4) = 8\times 10^{12} n(\log(2\times 10^{13}) + 4\log m) \\ &< 8\times 10^{12} n \times (9\log m), \end{aligned}$$

where we used the fact that $\log(2\times 10^{13}) + 4\log m < 31 + 4\log m < 9\log m$ for $m \geq 1000$. Thus, we get

$$m < 6 + 8\times 9\times 10^{12} n\log m < 10^{14} n\log m, \tag{35}$$

which leads to (i). For (ii), assuming that $n < 1000$, we get, by inequality (35), that

$$m < 10^{17}\log m \quad \text{therefore} \quad m < 2\times 10^{17}\log(10^{17}) < 10^{19}.$$

By (ii) of Lemma 3, we have

$$B = my < 10^{13} m^2 \log m < 10^{53}.$$

Clearly, since $\mu \geq m/2 - 3 > 10$, it follows that $4/\alpha^{\mu} < 1/2$. A standard argument implies that inequality (29) leads to

$$|y\log F_a - mx\log\alpha + x\log\sqrt{5}| < \frac{8}{\alpha^{\mu}}, \tag{36}$$

where $a \leq n+1 \leq 1000$ and $\max\{y, mx, x\} \leq B < 10^{53}$. However, the minimum of the expression appearing in the left–hand side of inequality (36) even over all the indices $n < 3000$ and coefficients at most 5×10^{65} in absolute value was bounded from below using LLL in Section 6 of [4]. The lower bound there was $100/1.5^{750}$. Hence, we get that

$$\frac{100}{1.5^{750}} < \frac{8}{\alpha^{\mu}}, \qquad \text{therefore} \qquad \mu < 750\left(\frac{\log 1.5}{\log\alpha}\right) - \frac{\log 12.5}{\log\alpha} < 630.$$

Since in fact $\mu = \min\{m, (m-n-1)x\} \geq \min\{m, (m-6)x/2\}$ and $m \geq 1000$, the only possibility is when $\mu = (m-n-2)x$ and $x = 1$. If $y \geq 3$, then, Lemma 2 (iv) shows that

$$m - 1 > (n-2)y \geq 3n - 6 \quad \text{so} \quad m \geq 3n - 4 \quad \text{so} \quad (m-n-2) \geq 2(m-5)/3,$$

implying that $\mu = (m-n-2)x \geq 2(m-5)/3 > 663$, a contradiction with $\mu < 630$. Hence, $y = 2$. Let us see that this is impossible. Suppose that we work with the left equation (3). Then

$$F_m = F_n + F_{n+1}^2 < F_n^2 + F_{n+1}^2 = F_{2n+1}$$

so $F_m < F_{2n+1}$, therefore $m < 2n$. The case $m = 2n$ is not convenient because F_n and F_{n+1} are coprime, so $m \leq 2n - 1$, which is impossible because then

$$F_m \leq F_{2n-1} = F_{n-1}^2 + F_n^2 < (F_n + F_{n-1})^2 = F_{n+1}^2 < F_{n+1}^2 + F_n = F_m.$$

Suppose now that we work with the right equation (3). Then

$$F_m = F_n^2 + F_{n+1} < F_n^2 + F_{n-1}^2 = F_{2n-1} \qquad \text{for} \qquad n > 10.$$

The inequality $n > 10$ holds because $m \geq 1000$, and the last inequality above is implied by $F_{n+1} < F_{n-1}^2$, which holds because $F_{n+1} < 2F_n < 4F_{n-1} < F_{n-1}^2$ for $n > 10$. Hence, $m < 2n-1$. The case $m \leq 2n-3$ leads to a contradiction since then

$$F_m \leq F_{2n-3} = F_{n-1}^2 + F_{n-2}^2 < (F_{n-1} + F_{n-2})^2 = F_n^2 < F_n^2 + F_{n+1} = F_m.$$

Finally, the case $m = 2n-2$, gives

$$F_{2n-2} = F_n^2 + F_{n+1} = F_n(F_n + 1) + F_{n-1}.$$

Since $F_{n-1} \mid F_{2n-2}$ and F_{n-1} is coprime to F_n, we get that F_{n-1} is a divisor of $F_n + 1 = (F_{n-2} + 1) + F_{n-1}$, so F_{n-1} divides $F_{n-2} + 1$, which in turn implies that $F_{n-2} + 1 \geq F_{n-1} = F_{n-2} + F_{n-3}$, or $1 \geq F_{n-3}$, which is false for $n > 10$. □

Lemma 12. *Estimates* (17) *hold.*

Proof. As in the proofs of Lemma 5 and 6, it is enough, in light of the Binet formula (4), to show that the inequality

$$y < \alpha^n \tag{37}$$

holds. By Lemma 3 (ii) and Lemma 11 (i), it suffices that the inequality

$$\log(10^{13} m \log m) < 10^{-14} (\log \alpha) m / \log m$$

holds. The above inequality holds for $m > 10^{18}$. On the other hand, if $m \leq 10^{18}$, then again by Lemma 3 (ii) and Lemma 11 (ii), we have

$$y < 10^{13} m \log m < 10^{13}(10^{18}) \log(10^{18}) < 10^{33} < \alpha^{1000} \leq \alpha^n.$$

This finishes the proof of this lemma. □

Lemma 13. *Equation* (3) *has no positive integer solution* (m, n, x, y) *with* $n \geq 3$ *and* $x \neq y$.

Proof. By Lemmas 6 and 12, inequalities (18) hold. Recall $\lambda = \min\{n, (m-n-1)x\}$. Inequality (22) is

$$\lambda < 5 \times 10^{10} \log y.$$

By the remark following Lemma 10, $m - 1 > 2n - 4$, so $(m - n - 1) \geq n - 4$. Hence, for us, $\lambda \geq n - 4$. By Lemma 11 (i) and 3 (ii), we get

$$10^{-14} m / \log m - 4 < n - 4 \leq \lambda < 5 \times 10^{10} \log y < 5 \times 10^{10} \log(10^{13} m \log m),$$

giving $m < 10^{30}$, so $y < 10^{13} m \log m < 10^{45}$. We thus get inequality (27), which we recall here under the form

$$\left| \frac{ay - mx}{y - x} - \frac{\log \sqrt{5}}{\log \alpha} \right| < \frac{55}{(y - x)\alpha^{996}}.$$

The calculation with the 99th convergent of $\log \sqrt{5}/\log \alpha$ from the proof of Lemma 10 shows that the left hand side of the above inequality is at least $1/(31(y - x)^2)$. So, we get $\alpha^{996} < 55 \times 31(y - x) < 55 \times 31 \times 10^{45} < 10^{50}$, which is absurd. This finishes the proof of the lemma and of the theorem. □

Acknowledgements. We thank the referee for useful suggestions. F. L. worked on this project during a visit to the University of French Polynesia in Faa'a, Tahiti in May 2012. He thanks the people of the GAATI team for their hospitality. He also thanks Sofitel Hotel Marara Beach Bora–Bora and Moorea Pearl & Spa for excellent working conditions.

References

[1] Y. Bugeaud, M. Mignotte and S. Siksek, "Classical and modular approaches to exponential Diophantine equations I. Fibonacci and Lucas perfect powers", *Annals of Mathematics* **163** (2006), 969–1018.

[2] Y. Bugeaud, F. Luca, M. Mignotte and S. Siksek, "Fibonacci numbers at most one away from a perfect power", *Elem. Math.* **63** (2008), 65–75.

[3] R. D. Carmichael, "On the numerical factors of the arithmetic forms $\alpha^n \pm \beta^n$", *Ann. Math. (2)* **15** (1913), 30–70.

[4] S. Díaz Alvarado, A. Dujella and F. Luca, "On a conjecture regarding balancing with powers of Fibonacci numbers", *INTEGERS* **12A** (2012), A2.

[5] A. Dujella and A. Pethő, "A generalization of a theorem of Baker and Davenport", *Quart. J. Math. Oxford Ser. (2)* **49** (1998), no. 195, 291–306.

[6] N. Hirata Kohno and F. Luca, "On the Diophantine equation $F_n^x + F_{n+1}^x = F_m^y$", *Rocky Mtn. J. Math.*, to appear.

[7] F. Luca and R. Oyono, "On the sum of powers of two consecutive Fibonacci numbers", *Proc. Japan Acad. Ser. A.* **87** (2011), 45–50.

[8] E. M. Matveev, "An explicit lower bound for a homogeneous rational linear form in logarithms of algebraic numbers. II", *Izv. Ross. Akad. Nauk Ser. Mat.* **64** (2000), 125–180; English transl. in *Izv. Math.* **64** (2000), 1217–1269.

[9] T. Miyazaki, "On exponential diophantine equations concerning Fibonacci polynomials", *Preprint*, 2012.

[10] K. Yu: "p-adic logarithmic forms and group varieties II, *Acta Arith.* **89** (1999), 337–378.

A NOTE ON THE MULTIPLICATIVE STRUCTURE OF AN ADDITIVELY SHIFTED PRODUCT SET $AA+1$

Steven Senger
Department of Mathematics, University of Delaware, Newark, Deleware
senger@math.udel.edu

Received: 7/19/12, Revised: 3/1/13, Accepted: 4/28/13, Published: 6/14/13

Abstract

We consider the multiplicative structure of sets of the form $AA+1$, where A is a large, finite set of real numbers. In particular, we show that the additively shifted product set, $AA+1$, must have a large part outside of any proper non-degenerate generalized geometric progression of comparable length. We prove an analogous result in finite fields as well.

1. Introduction

There are many problems in additive combinatorics which seek to differentiate between additive and multiplicative structure. By additive (resp. multiplicative) structure in a set, we refer to some arrangement or information that is largely undisturbed by addition (resp. multiplication). A prime example is the sums and products problem. Let A be a large, finite set of n natural numbers. Define the sum set of A to be

$$A+A=\{a+b : a \in A, b \in A\}.$$

Define the product set of A to be

$$AA=\{ab : a \in A, b \in A\}.$$

Let $|\cdot|$ denote the size of a set. The sums and products problem conjectures that,

$$\max\{|A+A|, |AA|\} \geq cn^{2-\epsilon},$$

for any $\epsilon > 0$, and some constant, $c = c(\epsilon) > 0$, which is independent of n. In [1], Elekes made progress with an elegant proof based on the celebrated Szemerédi-Trotter point-line incidence theorem, from [9]. Many of the results in this area have seen largely geometric proofs, including the current record in [8]. Therein, Solymosi proved the remarkable result that either the set of sums or the set of products must have more than about $n^{\frac{4}{3}}$ elements. See the book by Nathanson, [7], or the book by Tao and Vu, [10], for more.

One indication of multiplicative structure is how the size of a finite set A compares to the size of AA. If AA is not much bigger that A, then there must be some multiplicative structure in the set A. The construction of a product set will build in some multiplicative structure. This can often be estimated using tools such as multiplicative energy and the Plünnecke-Rusza inequalities. Again, see [7] and [10] for more details. Our main focus is to show that the multiplicative structure inherent in the product set of a large, finite set of numbers cannot be maximal after an additive shift. This will be made precise in the statement of the main theorem. First, however, we need to introduce some definitions and notation.

In what follows, we use the following asymptotic notation. If two positive quantities, X and Y, vary with respect to some parameter, n, we say $X \lesssim Y$ if $X \leq CY$, for some constant, $C > 0$, which does not depend on n. We write $X \approx Y$ when $X \lesssim Y$ and $Y \lesssim X$. In what follows, the implied constant may depend on some other parameters (such as δ or ϵ below), but it will be independent of the size of our set A.

If A is a set of numbers, then define its *additive shift* to be

$$A + 1 = \{a + 1 : a \in A\}.$$

Similarly, a *scaling* of A by some number s will be

$$sA = \{sa : a \in A\}.$$

We note that multiplicative behavior of additive shifts have been studied in relation to product sets by Garaev and Shen in [2], and Jones and Roche-Newton in [6]. However, they consider sets of the form $A(A + 1)$, which exhibit behavior which is quite different from that of sets of the form $AA + 1$, considered here.

Let $r_0, r_1, \ldots, r_d$ be real numbers called *generators*, and let $l_1, \ldots, l_d$ be positive integers greater than 2. We define the *d-dimensional generalized arithmetic progression*

$$\begin{aligned} R &= R(r_0, r_1, \ldots, r_d, l_1, \ldots, l_d) \\ &= \{r_0 + x_1 r_1 + \cdots + x_d r_d : x_j \in \mathbb{Z}, 0 \leq x_j < l_j, j = 1, \ldots, d\}. \end{aligned}$$

The related notion of a *d-dimensional generalized geometric progression* is defined as

$$G = G(g_0, R) = \{g_0^r : r \in R\},$$

where g_0 is some positive real number and R is some d-dimensional arithmetic progression. Either type of progression is said to be *proper* if every choice of $(x_1, x_2, \ldots, x_d)$ yields a distinct element of the set. Furthermore, we will call either type of generalized progression *degenerate* if d grows asymptotically with the size of the progression. That is to say, if the number of generators is not like a constant compared to the length of the progression, it is degenerate.

Such progressions exhibit maximality in arithmetic, or, respectively, geometric structure. We clarify this with the following elementary proposition.

Proposition 1.1. *If R is a non-degenerate generalized arithmetic progression, we have that*

$$|R+R| \approx |R|.$$

Also, if G is a non-degenerate generalized geometric progression, we have that

$$|GG| \approx |G|.$$

We now state the main result.

Theorem 1.2. *Let $A \subset \mathbb{R}$, be a large, finite set of numbers. Let G be any proper non-degenerate generalized geometric progression with $|G| \approx |AA|$. We have that for any $\delta > 0$*

$$|(AA+1) \setminus G| \gtrsim |A|^{1-\delta}.$$

The proof of Theorem 1.2 works for some slightly more general progressions. Remark 2.3 provides some insight into how to modify the proof to slightly relax the assumptions of properness and non-degeneracy. Two direct corollaries follow.

Corollary 1.3. *Let $A \subset \mathbb{R}$, be a large, finite set of numbers. Let G be any proper non-degenerate generalized geometric progression with $|G| \approx |AA|$. We have that*

$$(AA+1) \not\subset G.$$

Notice that for any non-degenerate geometric progression, $H \subset \mathbb{R}$, there exists a set $H' \subset \mathbb{R}$ such that $|H'H'| \approx |H|$ and $H \subset H'H'$. If we apply Theorem 1.2 with $A = H'$, we get the following corollary.

Corollary 1.4. *Let G and H be any two large, finite, proper non-degenerate generalized geometric progressions with $|G| \approx |H|$. We have that*

$$(H+1) \not\subset G.$$

We suspect that the additive shift disrupts multiplicative structure even more than Theorem 1.2 indicates, as suggested by the following conjecture.

Conjecture 1.5. *Let $A \subset \mathbb{R}$, be a large, finite set of numbers. If*

$$|(AA+1) \cap BC| \approx |AA|, \text{ and } |BC| \approx |AA|,$$

where B and C are also finite sets of numbers, then $\min\{|B|, |C|\} \lesssim 1$.

The next result is of a similar type, but in the setting of finite fields.

Theorem 1.6. *Let $A \subset \mathbb{F}_q$ such that the following two conditions hold:*

1. *There exists a real number $\epsilon > 0$ such that $|A||AA| \gtrsim q^{\frac{3}{2}+\epsilon}$.*
2. *There exists a real number $\delta > 0$ such that $|AA| \lesssim q^{1-\delta}$.*

Let G be any proper non-degenerate generalized geometric progression with $|G| \approx |AA|$. We have $|(AA+1) \setminus G| \gtrsim q^{\delta}$.

One should note that some condition like (2), the bound on $|AA|$, is necessary to avoid triviality. However, it is reasonable to expect that one can improve this result in either of two ways. First, condition (1), the bound on $|A||AA|$, could probably be lowered, and second, the bound on $|(AA+1) \setminus G|$ could probably be raised. As before, the proof will still work for some more general types of progressions. Again, see the Remark 2.3.

2. Proof of Theorem 1.2

The basic outline of the proof is to start with a given large, finite subset of $\mathbb{R}$. Then, with this set and any appropriate generalized geometric progression, we construct two large, finite sets of points in $\mathbb{R}^2$. We will then apply a recent result by Iosevich, Roche-Newton, and Rudnev regarding the set of dot products determined by our point sets, from [5]. The underlying arithmetic of the dot product set will allow us to compare the elements of the shifted product set to the elements of the progression.

The key ingredient to their proof is inspired by recent developments in the study of the Erdős distance problem. The classical Erdős distance problem asks for the minimum number of distinct distances which can be determined by any large, finite set of n points. The conjecture in the plane was that any such set must determine at least $n^{1-\epsilon}$ distinct distances, for any $\epsilon > 0$. Guth and Katz proved this in [3] with a blend of cell-decomposition and algebraic geometry, applied to an incidence problem in three dimensions. Shortly thereafter, Iosevich, Roche-Newton, and Rudnev, used similar techniques to prove a related result on the number of distinct dot products determined by such point sets in the plane. Specifically, they proved the following theorem.

Theorem 2.1. *Consider any large finite point sets $E, F \subset \mathbb{R}^2$ of n points each, neither of which is contained in a single line. Let $\Pi(E,F)$ denote the set of dot products $\Pi(E,F) = \{x \cdot y : x \in E, y \in F\}$. Then, for any $\epsilon > 0$, the number of distinct dot products is bounded below by $|\Pi(E,F)| \gtrsim n^{1-\epsilon}$.*

We now prove Theorem 1.2.

Proof. Fix any large finite set $A \subset \mathbb{R}$, and a real number $\delta > 0$. Let $G = G(g_0, R)$ be any non-degenerate generalized geometric progression with $|G| \approx |AA|$. Consider $R = R(r_0, r_1, \ldots, r_d, l_1, \ldots, l_d)$, the d-dimensional arithmetic progression defining the exponents of g_0 which make up G. Since G is non-degenerate, R must also be non-degenerate. We will define g_1 to be the "first element" of G, namely, $g_1 := g_0^{r_0}$. In what follows, we need to work with the normalized progression, G' which will be defined as

$$G' := \frac{G}{g_1} = \left\{ \frac{g}{g_1} : g \in G \right\}.$$

Notice that $|G| = |G'|$. Now, define the set B to be

$$B := \{g \in G' : gg \in G'\} = \{g \in G' : g = g_0^{x_1 r_1 + \cdots + x_d r_d}, x_j < l_j, x_j \in 2\mathbb{N}, j = 1, \ldots, d\}.$$

Claim 2.2. $|B| \approx |G|$.

Since $B \subset G'$, it is clear that $|B| \lesssim |G|$. Now we need only to show that $|B| \gtrsim |G|$. Notice that for an element to be in B, its square must be in G', hence the evenness condition on the x_j in the definition of B. So, we can count the number of elements in B by counting the number of elements of G' whose corresponding x_j are all even. By definition of the l_j, we get

$$|B| \geq \prod_{j=1}^{d} \left\lfloor \frac{l_j}{2} \right\rfloor \geq \prod_{j=1}^{d} \frac{l_j}{3} \geq \frac{|G|}{3^d}, \tag{2.1}$$

as $|G|$ is equal to the product of the l_j, and the claim is proved.

By Claim 2.2, $|B| \approx |G|$, so we have that $|BB| \approx |AA|$. Now, we construct E and F, finite subsets of $\mathbb{R}^2$,

$$E := \left\{(g_1 b, g_1 ba) \in \mathbb{R}^2 : b \in B, a \in A\right\}, \text{and } F := \left\{(b, ba) \in \mathbb{R}^2 : b \in B, a \in A\right\}.$$

These sets will have size $|E| = |F| = |A||B| \approx |A||G|$.

Remark 2.3. Note that the size estimate in Claim 2.2 could be satisfied by some slightly more general types of progressions. Specifically, as long as Claim 2.2 holds for progression G, we can get the desired result.

We will consider $\Pi(E, F)$, the set of distinct dot products determined by pairs in $E \times F$. Notice that

$$\begin{aligned}
\Pi(E, F) &= \{(g_1 b, g_1 ba) \cdot (b', b'a') : a, a' \in A, b, b' \in B\} \\
&= \{g_1 bb'(aa' + 1) : a, a' \in A, b, b' \in B\} \\
&= g_1 BB(AA + 1) \subset G(AA + 1).
\end{aligned}$$

By construction, $|\Pi(E, F)| = |g_1 BB(AA + 1)| \leq |G(AA + 1)|$.

Set $\epsilon = \delta/3$. Since $\epsilon > 0$, Theorem 2.1 gives us that

$$|\Pi(E,F)| \gtrsim |E|^{1-\epsilon} \gtrsim (|A||B|)^{1-\epsilon}.$$

Comparing upper and lower bounds on $|\Pi(E,F)|$ gives us

$$|G(AA+1)| \gtrsim (|A||B|)^{1-\epsilon}. \tag{2.2}$$

Our aim is to get a lower bound on the exceptional set $C := (AA+1) \setminus G$. From (2), we get

$$(|A||B|)^{1-\epsilon} \lesssim |G(AA+1)| = |G((G \cap (AA+1)) \cup C)| = |G(G \cap (AA+1)) \cup GC|.$$

Notice that the first term in the above union is a subset of GG, and therefore has size $\lesssim |G|$, by Proposition 1.1. So we can conclude that

$$|GC| \gtrsim (|A||B|)^{1-\epsilon} \approx (|A||G|)^{1-\epsilon},$$

which, by simple counting, gives us that

$$|C| \gtrsim \frac{(|A||G|)^{1-\epsilon}}{|G|} = |A|^{1-\epsilon}|G|^{-\epsilon}. \tag{2.3}$$

Again, by a simple counting argument, we see that $|A|^2 \gtrsim |AA|$. Since $|G| \approx |AA|$, we can rewrite (3) as

$$|C| \gtrsim |A|^{1-\epsilon}|G|^{-\epsilon} \gtrsim |A|^{1-\epsilon}|A|^{-2\epsilon} \gtrsim |A|^{1-\delta},$$

where the last line follows by definition of ϵ. □

3. Proof of Theorem 1.6

We follow a similar program to the proof of Theorem 1.2. Therefore, some details are omitted. The dot product set estimate in the finite field setting is in a slightly different form. It is due to Hart, Iosevich, Koh, and Rudnev. The statement of the theorem in [4] is for one set, but the proof works, with obvious modifications, for two sets. We consider the special case that both sets have the same size. For this section, define $\Pi(E,F)$ as before, except for subsets of $\mathbb{F}_q^2$ instead of $\mathbb{R}^2$. Also, let $\mathbb{F}_q^*$ denote the multiplicative group of $\mathbb{F}_q$.

Theorem 3.1. *Let $E, F \subset \mathbb{F}_q^d$ such that $|E| = |F| > q^{\frac{d+1}{2}}$. Then*

$$\mathbb{F}_q^* \subset \Pi(E,F) = \{x \cdot y : x \in E, y \in F\}.$$

Notice that the size condition in Theorem 3.1 is given with constant 1. This is why we include the ϵ in the size condition of Theorem 1.6, although a slightly more general statement is also true. Also, notice that we only use the case that $d = 2$.

Proof. Let $A \subset \mathbb{F}_q$ be given, and suppose that it satisfies the two size conditions:

1. There exists a real number $\epsilon > 0$ such that $|A||AA| \gtrsim q^{\frac{3}{2}+\epsilon}$.
2. There exists a real number $\delta > 0$ such that $|AA| \lesssim q^{1-\delta}$.

Now, let $G = G(g_0, R)$ be any non-degenerate generalized geometric progression with $|G| \approx |AA|$. Again, let

$$g_1 := g_0^{r_0}, G' := \frac{G}{g_1} = \left\{ \frac{g}{g_1} : g \in G \right\}, \text{ and } B := \{g \in G' : gg \in G'\}.$$

Claim 2.2 still holds in this context, so, as before, we have $|B| \approx |BB| \approx |AA|$. Now, we construct E and F, finite subsets of $\mathbb{F}_q^2$,

$$E := \left\{(g_1 b, g_1 ba) \in \mathbb{F}_q^2 : b \in B, a \in A\right\}, \text{and } F := \left\{(b, ba) \in \mathbb{F}_q^2 : b \in B, a \in A\right\}.$$

As in the proof of Theorem 1.2, the set of dot products determined by pairs in $E \times F$ will be

$$\Pi(E, F) = g_1 BB(AA + 1).$$

We also know that $|E| = |F| = |A||B| \approx |A||G|$. So, by the first size condition satisfied by A, and the fact that $|G| \approx |AA|$, we see that $|E| \gtrsim q^{\frac{3}{2}+\epsilon}$. Since E is large enough to satisfy the hypotheses of Theorem 3.1, we are guaranteed that $|\Pi(E, F)| \geq q - 1$. Specifically, using the proof of Theorem 1.2 as a guide, we get

$$q - 1 \leq |\Pi(E, F)| = |g_1 BB(AA + 1)| \leq |G(AA + 1)|. \tag{3.1}$$

We again seek a lower bound on the exceptional set. Define $C \subset \mathbb{F}_q$ to be $(AA + 1) \setminus G$. By (3), we get

$$q - 1 \leq |G(AA + 1)| = |G((G \cap (AA + 1)) \cup C)| \leq |G(G \cap (AA + 1)) \cup GC|$$

Again, the first term in the union will have size $\lesssim |G| \approx |AA|$. The second size condition satisfied by A tells us that $|AA| \lesssim q^{1-\delta}$, so the second term dominates. This gives us that $|GC| \approx q$, which, by simple counting and the fact that $|G| \approx |AA|$, yields

$$|C| \gtrsim \frac{q}{|AA|} = q^{\delta},$$

as claimed. □

noindent**Acknowledgment** I would like to thank the anonymous referee for the careful analysis and useful comments, which improved the final presentation.

References

[1] Gy. Elekes, *On the number of sums and products,* Acta Arith. **81** (1997), 365–367.

[2] M. Z. Garaev and C.-Y. Shen, *On the size of the set* $A(A+1)$, Mathematische Z. **265** (2010), 125–132.

[3] L. Guth and N. H. Katz, *On the Erdős distinct distance problem in the plane,* preprint, arXiv:1011:4015, (2010).

[4] D. Hart, A. Iosevich, D. Koh, and M. Rudnev, *Averages over hyperplanes, sum-product theory in finite fields, and the Erdős-Falconer distance conjecture,* Trans. Amer. Math. Soc. **363** (2011), 3255–3275.

[5] A. Iosevich, O. Roche-Newton, and M. Rudnev, *On an application of the Guth-Katz theorem,* Math. Res. Lett. **18** (2011), no. 4, 691-697.

[6] T. Jones and O. Roche-Newton, *Improved bounds on the set* $A(A+1)$, J. Combin. Theory Ser. A **120** (2013), 515–526

[7] M. Nathanson, *Additive Number Theory: Inverse Problems and the Geometry of Sumsets,* Grad. Texts in Math., Vol. 165, Springer-Verlag, 1996.

[8] J. Solymosi, *Bounding the multiplicative energy by the sumset,* Adv. Math., Volume 222, Issue 2, 1 October 2009, 402–408.

[9] E. Szemerédi and W. T. Trotter, Jr., *Extremal problems in discrete geometry,* Combinatorica **3** (1983), no. 3-4, 381–392.

[10] T. Tao and V. Vu, *Additive Combinatorics.* Cambridge Stud. Adv. Math., 2006.

REPRESENTATIONS OF SQUARES BY CERTAIN SEPTENARY QUADRATIC FORMS

Shaun Cooper
Institute of Natural and Mathematical Sciences, Massey University, Auckland, New Zealand
s.cooper@massey.ac.nz

Heung Yeung Lam
Institute of Natural and Mathematical Sciences, Massey University, Auckland, New Zealand
h.y.lam@massey.ac.nz

Dongxi Ye
Institute of Natural and Mathematical Sciences, Massey University, Auckland, New Zealand
lawrencefrommath@gmail.com

Received: 11/14/12, Accepted: 5/1/13, Published: 6/14/13

Abstract

For any positive integer n and for certain fixed positive integers a_1, $a_2, \ldots, a_7$, we study the number of solutions in integers of

$$a_1x_1^2 + a_2x_2^2 + a_3x_3^2 + a_4x_4^2 + a_5x_5^2 + a_6x_6^2 + a_7x_7^2 = n^2.$$

When $a_1 = a_2 = \cdots = a_7 = 1$, this reduces to the classical formula for the number of representations of a square as a sum of seven squares. A further eighteen analogous results will be given.

1. Introduction

Let n be a positive integer and let its prime factorization be given by $n = \prod_p p^{\lambda_p}$. Let $r_k(n)$ denote the number of solutions in integers of $x_1^2 + x_2^2 + \cdots + x_k^2 = n$.

Three classical results are given by Hurwitz (see [6] and [5]) and Sandham (see [9]), respectively:

$$r_3(n^2) = 6\prod_{p\geq 3}\left(\frac{p^{\lambda_p+1}-1}{p-1} - \left(\frac{-1}{p}\right)\frac{p^{\lambda_p}-1}{p-1}\right), \tag{1.1}$$

$$r_5(n^2) = 10\left(\frac{2^{3\lambda_2+3}-1}{2^3-1}\right)\prod_{p\geq 3}\left(\frac{p^{3\lambda_p+3}-1}{p^3-1} - p\,\frac{p^{3\lambda_p}-1}{p^3-1}\right) \tag{1.2}$$

and

$$r_7(n^2) = 14\left(\frac{5\times 2^{5\lambda_2+3}-9}{2^5-1}\right)\prod_{p\geq 3}\left(\frac{p^{5\lambda_p+5}-1}{p^5-1} - p^2\left(\frac{-1}{p}\right)\frac{p^{5\lambda_p}-1}{p^5-1}\right), \tag{1.3}$$

where the values of the Legendre symbol are given by

$$\left(\frac{-1}{p}\right) = \begin{cases} 1 & \text{if } p \equiv 1 \pmod 4, \\ -1 & \text{if } p \equiv -1 \pmod 4). \end{cases}$$

In recent work [3], the number of solutions in integers of

$$n^2 = x_1^2 + bx_2^2 + cx_3^2 \tag{1.4}$$

was investigated for certain values of b and c. When $b = c = 1$, the number of solutions of (1.4) is given by (1.1). The number of solutions of (1.4) in the case $b = 1$, $c = 2$ is given by

Theorem 1.1. [3, Theorem 1.2] *The number of* $(x_1, x_2, x_3) \in \mathbf{Z}^3$ *such that*

$$n^2 = x_1^2 + x_2^2 + 2x_3^2$$

is given by

$$4\,b(\lambda_2)\prod_{p\geq 3}\left[\frac{p^{\lambda_p+1}-1}{p-1} - \left(\frac{-8}{p}\right)\frac{p^{\lambda_p}-1}{p-1}\right], \tag{1.5}$$

where

$$b(\lambda_2) = \begin{cases} 1 & \text{if } \lambda_2 = 0, \\ 3 & \text{if } \lambda_2 \geq 1 \end{cases}$$

and the values of the Legendre symbol are given by

$$\left(\frac{-8}{p}\right) = \begin{cases} 1 & \text{if } p \equiv 1 \text{ or } 3 \pmod 8, \\ -1 & \text{if } p \equiv 5 \text{ or } 7 \pmod 8). \end{cases}$$

Further three-variable analogues of Theorem 1.1 were given in [3] and five-variable analogues were analyzed in [4]. In this work we study seven-variable analogues of Theorem 1.1. That is, we investigate the number of solutions in integers of

$$a_1x_1^2 + a_2x_2^2 + a_3x_3^2 + a_4x_4^2 + a_5x_5^2 + a_6x_6^2 + a_7x_7^2 = n^2$$

for certain fixed positive integers a_1, a_2, a_3, a_4, a_5, a_6 and a_7. The number of solutions in the case $a_1 = a_2 = \cdots = a_7 = 1$ is given by Sandham's identity (1.3).

This work is organized as follows. In Section 2 we define some notation and list all the results by grouping them into three theorems. The results in the first theorem are different from the others and they are treated in Section 3. Proofs of results in the second theorem are given in Section 4. Finally, results in the third theorem can be deduced from the results in the second theorem. A proof of one of them is given as an illustration in Section 5.

2. Notation and Results

Let n be a positive integer and let its prime factorization be $n = 2^{\lambda_2}\prod_{p\geq 3} p^{\lambda_p}$. Let $m = \prod_{p\geq 3} p^{\lambda_p}$ so that $n = 2^{\lambda_2}m$ and define $s(n)$ and $t(n)$ by

$$s(n) = \prod_{p\geq 3}\left(\frac{p^{5\lambda_p+5}-1}{p^5-1} - p^2\left(\frac{-8}{p}\right)\frac{p^{5\lambda_p}-1}{p^5-1}\right) \tag{2.1}$$

and

$$t(n) = \prod_{p\geq 3}\left(\frac{p^{5\lambda_p+5}-1}{p^5-1} - p^2\left(\frac{-1}{p}\right)\frac{p^{5\lambda_p}-1}{p^5-1}\right). \tag{2.2}$$

Let $b(n)$ be defined by

$$\sum_{n=1}^{\infty} b(n)q^n = q\prod_{i=1}^{\infty}\left(1-q^{2i}\right)^{12}. \tag{2.3}$$

Then let $h(n)$ and $k(n)$ be defined by

$$h(n) = b(m)\prod_{p\geq 3}\left(1 - p^2\left(\frac{-8}{p}\right)\frac{b(m/p)}{b(m)}\right) \tag{2.4}$$

and

$$k(n) = b(m)\prod_{p\geq 3}\left(1 - p^2\left(\frac{-1}{p}\right)\frac{b(m/p)}{b(m)}\right). \tag{2.5}$$

We note that $s(n)$, $t(n)$, $h(n)$ and $k(n)$ do not depend on λ_2, thus $s(n) = s(m)$, $t(n) = t(m)$, $h(n) = h(m)$ and $k(n) = k(m)$. The theta functions $\varphi(q)$ and $\psi(q)$ are defined for $|q| < 1$ by

$$\varphi(q) = \sum_{j=-\infty}^{\infty} q^{j^2} \quad\text{and}\quad \psi(q) = \sum_{j=0}^{\infty} q^{j(j+1)/2}$$

and for any positive integer k we define φ_k and ψ_k by

$$\varphi_k = \varphi(q^k) \qquad \text{and} \qquad \psi_k = \psi(q^k).$$

In addition, for positive integers a_1, a_2, a_3, a_4, a_5, a_6 and a_7 and for any non-negative integer n let $r_{(a_1,a_2,a_3,a_4,a_5,a_6,a_7)}(n)$ denote the number of solutions in integers of

$$a_1x_1^2 + a_2x_2^2 + a_3x_3^2 + a_4x_4^2 + a_5x_5^2 + a_6x_6^2 + a_7x_7^2 = n.$$

We note that Sandham's result (1.3) is equivalent to

$$r_{(1,1,1,1,1,1,1)}(n^2) = 14\left(\frac{5\times 2^{5\lambda_2+3}-9}{2^5-1}\right)t(n). \tag{2.6}$$

Analogously, we have the following results:

Theorem 2.1.

$$r_{(1,1,1,1,1,1,2)}(n^2) = 12\left|\frac{2^{5\lambda_2+5}-63}{2^5-1}\right|s(n). \tag{2.7}$$

$$r_{(1,1,1,1,2,2,2)}(n^2) = \begin{cases} 6s(n) + 2h(n) & \text{if } n \text{ is odd,} \\ \frac{198\times 2^{5\lambda_2}-756}{2^5-1}s(n) & \text{if } n \text{ is even.} \end{cases} \tag{2.8}$$

$$r_{(1,1,2,2,2,2,2)}(n^2) = \begin{cases} 3s(n) + h(n) & \text{if } n \text{ is odd,} \\ \frac{105\times 2^{5\lambda_2}-756}{2^5-1}s(n) & \text{if } n \text{ is even.} \end{cases} \tag{2.9}$$

Theorem 2.2.

$$r_{(1,2,2,2,2,2,2)}(n^2) = 2\left|\frac{2^{5\lambda_2+5}-63}{2^5-1}\right|t(n). \tag{2.10}$$

$$r_{(1,1,1,1,1,2,2)}(n^2) = \begin{cases} 8t(n) + 2k(n) & \text{if } n \text{ is odd,} \\ \frac{250\times 2^{5\lambda_2}-126}{2^5-1}t(n) & \text{if } n \text{ is even.} \end{cases} \tag{2.11}$$

Theorem 2.3.

$$r_{(1,1,1,2,2,2,2)}(n^2) = \begin{cases} 4t(n) + 2k(n) & \text{if } n \text{ is odd,} \\ \frac{126\times 2^{5\lambda_2}-126}{2^5-1}t(n) & \text{if } n \text{ is even.} \end{cases} \tag{2.12}$$

$$r_{(1,1,1,1,1,1,4)}(n^2) = \begin{cases} 6t(n) + 6k(n) & \text{if } n \text{ is odd,} \\ \frac{250\times 2^{5\lambda_2} - 126}{2^5-1} t(n) & \text{if } n \text{ is even.} \end{cases} \tag{2.13}$$

$$r_{(1,1,1,1,1,4,4)}(n^2) = \begin{cases} 3t(n) + 7k(n) & \text{if } n \text{ is odd,} \\ \frac{95\times 2^{5\lambda_2} - 126}{2^5-1} t(n) & \text{if } n \text{ is even.} \end{cases} \tag{2.14}$$

$$r_{(1,1,1,1,4,4,4)}(n^2) = \begin{cases} 2t(n) + 6k(n) & \text{if } n \text{ is odd,} \\ \frac{33\times 2^{5\lambda_2} - 126}{2^5-1} t(n) & \text{if } n \text{ is even.} \end{cases} \tag{2.15}$$

$$r_{(1,1,1,4,4,4,4)}(n^2) = \begin{cases} \frac{3}{2}t(n) + \frac{9}{2}k(n) & \text{if } n \text{ is odd,} \\ \frac{35\times 2^{5\lambda_2-1} - 126}{2^5-1} t(n) & \text{if } n \text{ is even.} \end{cases} \tag{2.16}$$

$$r_{(1,1,4,4,4,4,4)}(n^2) = \begin{cases} t(n) + 3k(n) & \text{if } n \text{ is odd,} \\ \frac{35\times 2^{5\lambda_2-1} - 126}{2^5-1} t(n) & \text{if } n \text{ is even.} \end{cases} \tag{2.17}$$

$$r_{(1,4,4,4,4,4,4)}(n^2) = \begin{cases} \frac{1}{2}t(n) + \frac{3}{2}k(n) & \text{if } n \text{ is odd,} \\ \frac{35\times 2^{5\lambda_2-1} - 126}{2^5-1} t(n) & \text{if } n \text{ is even.} \end{cases} \tag{2.18}$$

$$r_{(1,1,1,1,2,2,4)}(n^2) = \begin{cases} 4t(n) + 4k(n) & \text{if } n \text{ is odd,} \\ \frac{126\times 2^{5\lambda_2} - 126}{2^5-1} t(n) & \text{if } n \text{ is even.} \end{cases} \tag{2.19}$$

$$r_{(1,1,1,2,2,4,4)}(n^2) = \begin{cases} 2t(n) + 4k(n) & \text{if } n \text{ is odd,} \\ \frac{64\times 2^{5\lambda_2} - 126}{2^5-1} t(n) & \text{if } n \text{ is even.} \end{cases} \tag{2.20}$$

$$r_{(1,1,2,2,4,4,4)}(n^2) = \begin{cases} t(n) + 3k(n) & \text{if } n \text{ is odd,} \\ \frac{33\times 2^{5\lambda_2}-126}{2^5-1} t(n) & \text{if } n \text{ is even.} \end{cases} \tag{2.21}$$

$$r_{(1,1,2,2,2,2,4)}(n^2) = \begin{cases} 2t(n) + 2k(n) & \text{if } n \text{ is odd,} \\ \frac{64\times 2^{5\lambda_2}-126}{2^5-1} t(n) & \text{if } n \text{ is even.} \end{cases} \tag{2.22}$$

$$r_{(1,2,2,2,2,4,4)}(n^2) = \begin{cases} t(n) + k(n) & \text{if } n \text{ is odd,} \\ \frac{33\times 2^{5\lambda_2}-126}{2^5-1} t(n) & \text{if } n \text{ is even.} \end{cases} \tag{2.23}$$

$$r_{(1,2,2,4,4,4,4)}(n^2) = \begin{cases} \frac{1}{2}t(n) + \frac{3}{2}k(n) & \text{if } n \text{ is odd,} \\ \frac{35\times 2^{5\lambda_2-1}-126}{2^5-1} t(n) & \text{if } n \text{ is even.} \end{cases} \tag{2.24}$$

All three theorems give values of $r_{(a_1,\ldots,a_7)}(n^2)$. Theorems 2.2 and 2.3 account for all instances for which $1 = a_1 \leq a_2 \leq \cdots \leq a_7 \leq 4$ with $a_1, \ldots, a_7 \in \{1, 2, 4\}$ and for which the product $a_1 a_2 \cdots a_7$ is an even power of 2. Theorem 2.1 accounts for those instances for which $1 = a_1 \leq a_2 \leq \cdots \leq a_7 = 2$ with $a_1, \ldots, a_7 \in \{1, 2\}$ and for which the product $a_1 a_2 \cdots a_7$ is an odd power of 2. Some further comments about other values of $a_1, \ldots, a_7$ are given in the concluding remarks at the end of the paper.

3. Proof of Theorem 2.1

In this section we will outline the proof of Theorem 2.1.

Lemma 3.1. *Fix an odd integer m. For any nonnegative integer k, let*

$$u_1(k) = r_{(1,1,1,1,1,1,2)}(2^{2k}m^2),$$
$$u_2(k) = r_{(1,1,1,1,2,2,2)}(2^{2k}m^2)$$

and

$$u_3(k) = r_{(1,1,2,2,2,2,2)}(2^{2k}m^2).$$

Then

$$u_i(k+3) = 33u_i(k+2) - 32u_i(k+1) \quad \textit{for } i{=}1, 2 \textit{ or } 3.$$

Moreover

$$u_1(1) = 31u_1(0), \qquad u_1(2) = 1055u_1(0)$$
$$u_2(1) = 15u_1(0), \qquad u_2(2) = 543u_1(0)$$

and

$$u_3(1) = 7u_1(0), \qquad u_3(2) = 287u_1(0).$$

Hence, on solving the recurrence relation, we have: for $k \geq 1$

$$r_{(1,1,1,1,1,1,2)}(2^{2k}m^2) = \left|\frac{2^{5k+5} - 63}{2^5 - 1}\right| r_{(1,1,1,1,1,1,2)}(m^2),$$
$$r_{(1,1,1,1,2,2,2)}(2^{2k}m^2) = \frac{33 \times 2^{5k} - 126}{(2^5 - 1) \times 2} r_{(1,1,1,1,1,1,2)}(m^2)$$

and

$$r_{(1,1,2,2,2,2,2)}(2^{2k}m^2) = \frac{35 \times 2^{5k} - 252}{(2^5 - 1) \times 4} r_{(1,1,1,1,1,1,2)}(m^2).$$

Proof. These can all be deduced by the methods in [3, Section 4]. □

It remains to determine the values of $r_{(1,1,1,1,1,1,2)}(m^2)$, $r_{(1,1,1,1,2,2,2)}(m^2)$ and $r_{(1,1,2,2,2,2,2)}(m^2)$ in the case that m is odd.

Proposition 3.2. *Let* m *be a positive odd number and let its prime factorization be given by*

$$m = \prod_{p \geq 3} p^{\lambda_p}.$$

Let $s(m)$ *and* $h(m)$ *be defined by* (2.1) *and* (2.4). *Then*

$$r_{(1,1,1,1,1,1,2)}(m^2) = 12s(m), \tag{3.1}$$
$$r_{(1,1,1,1,2,2,2)}(m^2) = 6s(m) + 2h(m) \tag{3.2}$$

and

$$r_{(1,1,2,2,2,2,2)}(m^2) = 3s(m) + h(m). \tag{3.3}$$

We may note that Lemma 3.1 and Proposition 3.2 immediately imply (2.7)–(2.9) in Theorem 2.1.

To prove Proposition 3.2, we will need:

Lemma 3.3. *Let f_1, f_2 and f_3 be defined by*

$$f_1(q) = \frac{1}{8}\varphi_1^5\varphi_2 - \frac{1}{8}\varphi_1\varphi_2^5,$$
$$f_2(q) = \frac{1}{2}\varphi_1^5\varphi_2 - 2\varphi_1\varphi_2^5$$

and

$$f_3(q) = -\frac{1}{4}\varphi_1^5\varphi_2 + \frac{3}{4}\varphi_1^3\varphi_2^3 - \frac{1}{2}\varphi_1\varphi_2^5.$$

And let their series expansions be given by

$$f_1(q) = \sum_{n=0}^{\infty} a_1(n)q^n, \quad f_2(q) = \sum_{n=0}^{\infty} a_2(n)q^n \tag{3.4}$$

and

$$f_3(q) = \sum_{n=0}^{\infty} a_3(n)q^n. \tag{3.5}$$

Then for any nonnegative integer n and any prime p we have

$$a_j(pn) = a_j(p)a_j(n) - \chi(p)a_j(n/p) \qquad \text{for } j \in \{1, 2, 3\} \tag{3.6}$$

where χ is the completely multiplicative function defined on the positive integers by

$$\chi(r) = r^2\left(\frac{-8}{r}\right) \tag{3.7}$$

and $\left(\frac{-8}{r}\right)$ is the Kronecker symbol, and $a_j(x)$ is defined to be zero if x is not an integer.

Proof. The results for j=1 or 2 follow from work of Alaca et al. [1, pp. 291–292]. The result for j=3 was given by Martin [8, pp. 4828–4833]. □

We may note that

$$\varphi^5(q)\varphi(q^2) = \frac{32}{3}f_1(q) - \frac{2}{3}f_2(q), \tag{3.8}$$
$$\varphi^3(q)\varphi^3(q^2) = \frac{16}{3}f_1(q) - \frac{2}{3}f_2(q) + \frac{4}{3}f_3(q) \tag{3.9}$$

and

$$\varphi(q)\varphi^5(q^2) = \frac{8}{3}f_1(q) - \frac{2}{3}f_2(q). \tag{3.10}$$

Let $A_j(n)$ be defined by

$$\sum_{n=0}^{\infty} A_j(n)q^n = \left(\sum_{n=0}^{\infty} a_j(n)q^n\right)^2 \qquad \text{for } j \in \{1,2,3\}. \tag{3.11}$$

The next result is due to Hurwitz.

Lemma 3.4. *Suppose that $a(n)$ is a function, defined for all non-negative integers n, that satisfies the property*

$$a(pn) = a(p)a(n) - \chi(p)a\left(\frac{n}{p}\right)$$

for all primes p, where χ is a completely multiplicative function. Then the coefficient of q^{n^2} in

$$\left(\sum_{j=-\infty}^{\infty} q^{j^2}\right) \times \left(\sum_{k=0}^{\infty} a(k)q^k\right)$$

is equal to

$$\sum_{r=1}^{\infty} A\left(\frac{2n}{r}\right)\chi(r)\mu(r)$$

where μ is the Möbius function, $A(n)$ is defined by

$$\sum_{n=0}^{\infty} A(n)q^n = \left(\sum_{k=0}^{\infty} a(k)q^k\right)^2$$

and $A(x)$ is defined to be 0 if x is not a non-negative integer.

Proof. See the work of Sandham [9, Section 2]. □

Before starting the next lemma, let us define $[q^k]f(q)$ to be the coefficient of q^k in the Taylor expansion of $f(q)$.

Lemma 3.5. *Let m be a positive odd number and let its prime factorization be given by*

$$m = \prod_{p \geq 3} p^{\lambda_p}.$$

Let $c_1(m)$, $c_2(m)$ and $c_3(m)$ be the coefficients of q^{2m} in

$$\frac{32}{3}f_1^2(q) - \frac{2}{3}f_2^2(q), \qquad \frac{16}{3}f_1^2(q) - \frac{2}{3}f_2^2(q) + \frac{4}{3}f_3^2(q) \quad \text{and} \quad \frac{8}{3}f_1^2(q) - \frac{2}{3}f_2^2(q),$$

respectively. Let $b(m)$ be defined by (2.3). Then

$$c_1(m) = \frac{1}{22}[q^{2m}]\varphi^{12}(q), \tag{3.12}$$

$$c_2(m) = \frac{1}{44}[q^{2m}]\varphi^{12}(q) + 2b(m) \tag{3.13}$$

and

$$c_3(m) = \frac{1}{88}[q^{2m}]\varphi^{12}(q) + b(m). \tag{3.14}$$

And thus

$$c_1(m) = 12\sum_{d|m} d^5 = 12\prod_{p\geq 3} \frac{p^{5\lambda_p+5}-1}{p^5-1}, \tag{3.15}$$

$$c_2(m) = 6\sum_{d|m} d^5 + 2b(m) = 6\prod_{p\geq 3} \frac{p^{5\lambda_p+5}-1}{p^5-1} + 2b(m) \tag{3.16}$$

and

$$c_3(m) = 3\sum_{d|m} d^5 + b(m) = 3\prod_{p\geq 3} \frac{p^{5\lambda_p+5}-1}{p^5-1} + b(m). \tag{3.17}$$

Proof. (3.12)–(3.14) may be deduced by the methods in [3, Section 4], and (3.15)–(3.17) follows from the result of $[q^{2m}]\varphi^{12}(q)$ given in [7]. □

We are now ready for

Proof of Proposition 3.2. We will deal with (3.1) first. The proofs of (3.2) and (3.3) will be similar.

By (3.4) and (3.8),

$$\begin{aligned} r_{(1,1,1,1,1,1,2)}(m^2) &= [q^{m^2}]\left(\varphi(q)\left(\frac{32}{3}f_1(q) - \frac{2}{3}f_2(q)\right)\right) \\ &= \frac{32}{3}[q^{m^2}]\left(\varphi(q)\sum_{j=0}^{\infty} a_1(j)q^j\right) - \frac{2}{3}[q^{m^2}]\left(\varphi(q)\sum_{j=0}^{\infty} a_2(j)q^j\right). \end{aligned}$$

By Lemmas 3.3 and 3.4 and (3.11) this is equivalent to

$$r_{(1,1,1,1,1,1,2)}(m^2) = \frac{32}{3}\sum_{r=1}^{\infty} A_1\left(\frac{2m}{r}\right)\chi(r)\mu(r) - \frac{2}{3}\sum_{r=1}^{\infty} A_2\left(\frac{2m}{r}\right)\chi(r)\mu(r)$$

where $\chi(r)$ is the completely multiplicative function defined by (3.7). Since $\chi(r) = 0$ if r is even, the last sum in the above is over odd values of r only. Moreover, since m is odd, we may apply Lemma 3.5 to deduce that

$$r_{(1,1,1,1,1,1,2)}(m^2) = \sum_{r=1}^{\infty} c_1(m/r)\chi(r)\mu(r)$$

$$
\begin{aligned}
&= c_1(m)\sum_{r|m}\frac{c_1(m/r)}{c_1(m)}\chi(r)\mu(r)\\
&= c_1(m)\prod_{p\geq 3}\left(1-\chi(p)\frac{c_1(m/p)}{c_1(m)}\right)\\
&= \left(12\prod_{p\geq 3}\frac{p^{5\lambda_p+5}-1}{p^5-1}\right)\times\left(\prod_{p\geq 3}\left(1-p^2\left(\frac{-8}{p}\right)\frac{p^{5\lambda_p}-1}{p^{5\lambda_p+5}-1}\right)\right)\\
&= 12\prod_{p\geq 3}\left(\frac{p^{5\lambda_p+5}-1}{p^5-1}-p^2\left(\frac{-8}{p}\right)\frac{p^{5\lambda_p}-1}{p^5-1}\right)\\
&= 12s(m).
\end{aligned}
$$

Similarly, we can deduce:

$$
\begin{aligned}
r_{(1,1,1,1,2,2,2)}(m^2) &= \sum_{r=1}^{\infty}c_2(m/r)\chi(r)\mu(r)\\
&= \sum_{r=1}^{\infty}\left(\frac{1}{44}[q^{2m/r}]\varphi(q)^{12}+2b(m/r)\right)\chi(r)\mu(r)\\
&= \left(6\prod_{p\geq 3}\frac{p^{5\lambda_p+5}-1}{p^5-1}\right)\times\left(\prod_{p\geq 3}\left(1-p^2\left(\frac{-8}{p}\right)\frac{p^{5\lambda_p}-1}{p^{5\lambda_p+5}-1}\right)\right)\\
&\quad + 2b(m)\prod_{p\geq 3}\left(1-p^2\left(\frac{-8}{p}\right)\frac{b(m/p)}{b(m)}\right)\\
&= 6\prod_{p\geq 3}\left(\frac{p^{5\lambda_p+5}-1}{p^5-1}-p^2\left(\frac{-8}{p}\right)\frac{p^{5\lambda_p}-1}{p^5-1}\right)\\
&\quad + 2b(m)\prod_{p\geq 3}\left(1-p^2\left(\frac{-1}{p}\right)\frac{b(m/p)}{b(m)}\right)\\
&= 6s(m)+2h(m)
\end{aligned}
$$

and

$$
\begin{aligned}
r_{(1,1,2,2,2,2,2)}(m^2) &= \sum_{r=1}^{\infty}c_3(m/r)\chi(r)\mu(r)\\
&= \sum_{r=1}^{\infty}\left(\frac{1}{88}[q^{2m/r}]\varphi(q)^{12}+b(m/r)\right)\chi(r)\mu(r)\\
&= \left(3\prod_{p\geq 3}\frac{p^{5\lambda_p+5}-1}{p^5-1}\right)\times\left(\prod_{p\geq 3}\left(1-p^2\left(\frac{-8}{p}\right)\frac{p^{5\lambda_p}-1}{p^{5\lambda_p+5}-1}\right)\right)
\end{aligned}
$$

$$
\begin{aligned}
&+ b(m) \prod_{p\geq 3} \left(1 - p^2 \left(\frac{-8}{p}\right) \frac{b(m/p)}{b(m)}\right) \\
&= 3 \prod_{p\geq 3} \left(\frac{p^{5\lambda_p+5}-1}{p^5-1} - p^2 \left(\frac{-8}{p}\right) \frac{p^{5\lambda_p}-1}{p^5-1}\right) \\
&\quad + b(m) \prod_{p\geq 3} \left(1 - p^2 \left(\frac{-8}{p}\right) \frac{b(m/p)}{b(m)}\right) \\
&= 3s(m) + h(m).
\end{aligned}
$$

□

4. Proof of Theorem 2.2

In this section, we will outline proofs of results in Theorem 2.2. The proof of (2.10) depends on:

Lemma 4.1. *Fix an odd integer m. For any nonnegative integer k let*

$$u(k) = r_{(1,2,2,2,2,2,2)}(2^{2k}m^2).$$

Then

$$u(k+3) = 33u(k+2) - 32u(k+1),$$

$$u(1) = 31u(0), \qquad u(2) = 1055u(0)$$

and

$$u(0) = \frac{1}{7} r_{(1,1,1,1,1,1,1)}(m^2).$$

Hence, on solving the recurrence relation, we have

$$r_{(1,2,2,2,2,2,2)}(2^{2k}m^2) = \frac{1}{7}\left|\frac{2^{5k+5}-63}{2^5-1}\right| r_{(1,1,1,1,1,1,1)}(m^2).$$

Proof. These may be deduced by the methods in [3, Section 4]. □

We are now ready for:

Proof of (2.10). By Lemma 4.1 and Sandham's result (1.3), we can immediately deduce:

$$
\begin{aligned}
r_{(1,2,2,2,2,2,2)}(n^2) &= 2\left|\frac{2^{5\lambda_2+5}-63}{2^5-1}\right| \prod_{p\geq 3} \left(\frac{p^{5\lambda_p+5}-1}{p^5-1} - p^2 \left(\frac{-1}{p}\right) \frac{p^{5\lambda_p}-1}{p^5-1}\right) \\
&= 2\left|\frac{2^{5\lambda_2+5}-63}{2^5-1}\right| t(n).
\end{aligned}
$$

This proves (2.10). □

Now we will outline the proof of (2.11). This will be achieved in two steps according to whether n is even or odd. We begin with the case when n is even.

Lemma 4.2. *Fix an odd integer m. For any nonnegative integer k let*

$$u(k) = r_{(1,1,1,1,1,2,2)}(2^{2k}m^2).$$

Then

$$u(k+3) = 33u(k+2) - 32u(k+1),$$
$$u(1) = \frac{127}{7} r_{(1,1,1,1,1,1,1)}(m^2) \quad \text{and} \quad u(2) = \frac{4127}{7} r_{(1,1,1,1,1,1,1)}(m^2).$$

Hence, on solving the recurrence relation, we have: for $k \geq 1$

$$r_{(1,1,1,1,1,2,2)}(2^{2k}m^2) = \frac{125 \times 2^{5k} - 63}{(2^5-1) \times 7} r_{(1,1,1,1,1,1,1)}(m^2)$$

and thus

$$r_{(1,1,1,1,1,2,2)}(n^2) = \frac{250 \times 2^{5\lambda_2} - 126}{2^5 - 1} t(n) \tag{4.1}$$

where n is even.

Proof. These may be deduced by the methods in [3, Section 4], and (4.1) follows from (2.6). □

It remains to deal with the case when n is odd. We will need two lemmas.

Lemma 4.3. *Let g_1, g_2 and g_3 be defined by*

$$g_1(q) = q\varphi^2(q)\psi^4(q^2), \; g_2(q) = -\frac{1}{4}\varphi^2(q)\varphi^4(-q), \; \text{and} \; g_3(q) = q\psi^2(q^4)\varphi^4(-q^2);$$

and let their series expansions be given by

$$g_1(q) = \sum_{n=0}^{\infty} a_1(n)q^n, \quad g_2(q) = \sum_{n=0}^{\infty} a_2(n)q^n \tag{4.2}$$

and

$$g_3(q) = \sum_{n=0}^{\infty} a_3(n)q^n. \tag{4.3}$$

Then for any nonnegative integer n and any prime p we have

$$a_j(pn) = a_j(p)a_j(n) - \chi(p)a_j(n/p) \qquad \text{for } j \in \{1,2,3\}$$

where χ is the completely multiplicative function defined on the positive integers by

$$\chi(r) = \begin{cases} r^2\left(\frac{-1}{r}\right) & \text{if } r \text{ is odd} \\ 0 & \text{if } r \text{ is even} \end{cases} \tag{4.4}$$

and $\left(\frac{-1}{r}\right)$ is the Kronecker symbol, and $a_j(x)$ is defined to be zero if x is not an integer.

Proof. This follows from [2, Theorem 2.4]. □

We note that

$$\varphi^4(q)\varphi^2(q^2) = 8g_1(q) - 4g_2(q) + 4g_3(q), \tag{4.5}$$

and we let $A_j(n)$ be defined by

$$\sum_{n=0}^{\infty} A_j(n)q^n = \left(\sum_{n=0}^{\infty} a_j(n)q^n\right)^2 \qquad \text{for } j \in \{1,2,3\}. \tag{4.6}$$

Lemma 4.4. *Let m be an positive odd number and $c(m)$ be the coefficient of q^{2m} in $8g_1^2(q) - 4g_2^2(q) + 4g_3^2(q)$. Let $b(m)$ be defined by (2.3). Then*

$$c(m) = \frac{1}{33}[q^{2m}]\varphi^{12}(q) + 2b(m). \tag{4.7}$$

And thus

$$c(m) = 8\sum_{d|m} d^5 + 2b(m) = 8\prod_{p\geq 3}\frac{p^{5\lambda_p+5}-1}{p^5-1} + 2b(m). \tag{4.8}$$

Proof. Equation (4.7) may be deduced by the methods in [3, Section 4], and (4.8) follows from the result of $[q^{2m}]\varphi^{12}(q)$ given in [7]. □

Proposition 4.5. *Let m be a positive odd number and let its prime factorization be given by $m = \prod_{p\geq 3} p^{\lambda_p}$. Let $t(m)$ and $k(m)$ be defined by (2.2) and (2.5) respectively. Then $r_{(1,1,1,1,1,2,2)}(m^2) = 8t(m) + 2k(m)$.*

Proof. By (4.2), (4.3) and (4.5),

$$\begin{aligned} r_{(1,1,1,1,1,2,2)}(m^2) &= [q^{m^2}]\left(\varphi(q)\left(8g_1(q) - 4g_2(q) + 4g_3(q)\right)\right) \\ &= 8[q^{m^2}]\left(\varphi(q)\sum_{j=0}^{\infty} a_1(j)q^j\right) - 4[q^{m^2}]\left(\varphi(q)\sum_{j=0}^{\infty} a_2(j)q^j\right) \\ &\quad + 4[q^{m^2}]\left(\varphi(q)\sum_{j=0}^{\infty} a_3(j)q^j\right). \end{aligned}$$

By Lemmas 3.4 and 4.3, and (4.6), this is equivalent to

$$\begin{aligned} r_{(1,1,1,1,1,2,2)}(m^2) &= 8\sum_{r=1}^{\infty} A_1\left(\frac{2m}{r}\right)\chi(r)\mu(r) - 4\sum_{r=1}^{\infty} A_2\left(\frac{2m}{r}\right)\chi(r)\mu(r) \\ &\quad + 4\sum_{r=1}^{\infty} A_3\left(\frac{2m}{r}\right)\chi(r)\mu(r), \end{aligned}$$

where $\chi(r)$ is the completely multiplicative function defined by (4.4). Since $\chi(r)=0$ if r is even, the last sum in the above is over odd values of r only.

Moreover, since m is odd, we may apply Lemma 4.4 to deduce that

$$\begin{aligned}
r_{(1,1,1,1,1,2,2)}(m^2) &= \sum_{r=1}^{\infty} c(m/r)\chi(r)\mu(r)\\
&= \sum_{r=1}^{\infty}\left(\frac{1}{33}[q^{2m/r}]\varphi(q)^{12} + 2b(m/r)\right)\chi(r)\mu(r)\\
&= \left(8\prod_{p\geq 3}\frac{p^{5\lambda_p+5}-1}{p^5-1}\right)\times\left(\prod_{p\geq 3}\left(1-p^2\left(\frac{-1}{p}\right)\frac{p^{5\lambda_p}-1}{p^{5\lambda_p+5}-1}\right)\right)\\
&\quad + 2b(m)\prod_{p\geq 3}\left(1-p^2\left(\frac{-1}{p}\right)\frac{b(m/p)}{b(m)}\right)\\
&= 8\prod_{p\geq 3}\left(\frac{p^{5\lambda_p+5}-1}{p^5-1} - p^2\left(\frac{-1}{p}\right)\frac{p^{5\lambda_p}-1}{p^5-1}\right)\\
&\quad + 2b(m)\prod_{p\geq 3}\left(1-p^2\left(\frac{-1}{p}\right)\frac{b(m/p)}{b(m)}\right)\\
&= 8t(m) + 2k(m).
\end{aligned}$$

□

5. Proof of Theorem 2.3

In this section, we will give the proof of (2.12) and regard it as an illustration. Proofs of the remaining results are all similar: for n is even, the value can be deduced from the value of $r_{(1,1,1,1,1,1,1,1)}(n^2)$ and for n is odd, the value can be deduced from Theorem 2.2.

Lemma 5.1. *Fix an odd integer m. For any nonnegative integer k let*

$$u(k) = r_{(1,1,1,2,2,2,2)}(2^{2k}m^2).$$

Then $u(k+3) = 33u(k+2) - 32u(k+1)$,

$$u(1) = 9r_{(1,1,1,1,1,1,1,1)}(m^2) \quad \textit{and} \quad u(2) = 297r_{(1,1,1,1,1,1,1,1)}(m^2).$$

Hence, on solving the recurrence relation, we have: for $k \geq 1$

$$r_{(1,1,1,2,2,2,2)}(2^{2k}m^2) = \frac{9\times 2^{5k}-9}{2^5-1}r_{(1,1,1,1,1,1,1,1)}(m^2)$$

and thus

$$r_{(1,1,1,2,2,2,2)}(n^2) = \frac{126 \times 2^{5\lambda_2} - 126}{2^5 - 1} t(n) \tag{5.1}$$

where n is even.

Proof. These may be deduced by the methods in [3, Section 4], and (5.1) follows from (2.6). □

It remains to deal with the case when n is odd.

Lemma 5.2. *Let n be a positive integer and let its prime factorization be given by*

$$n = 2^{\lambda_2} m \quad \textit{where} \quad m = \prod_{p \geq 3} p^{\lambda_p}.$$

Let $t(m)$ and $k(m)$ be defined by (2.2) and (2.5). Then $t(m) = \frac{1}{2} r_{(1,2,2,2,2,2,2)}(m^2)$ and $k(m) = \frac{1}{2} r_{(1,1,1,1,1,2,2)}(m^2) - 2r_{(1,2,2,2,2,2,2)}(m^2)$.

Proof. This follows from Lemma 4.1 and Proposition 4.5. □

Lemma 5.3. *Let n be a positive integer and let its prime factorization be given by*

$$n = 2^{\lambda_2} m \quad \textit{where} \quad m = \prod_{p \geq 3} p^{\lambda_p}.$$

Then

$$r_{(1,1,1,2,2,2,2)}(m^2) = r_{(1,1,1,1,1,2,2)}(m^2) - 2r_{(1,2,2,2,2,2,2)}(m^2). \tag{5.2}$$

Hence,

$$r_{(1,1,1,2,2,2,2)}(m^2) = 4t(m) + 2k(m). \tag{5.3}$$

Proof. (5.2) can be deduced by the method in [3, Section 4], and (5.3) immediately follows from Lemma 5.2 and (5.2). □

By Lemma 5.1 and Lemma 5.3, we immediately deduce (2.12).

Finally, let $\mathbf{a} = (a_1, a_2, a_3, a_4, a_5, a_6, a_7)$. Let n be a positive integer and let its prime factorization be given by

$$n = 2^{\lambda_2} m \quad \text{where} \quad m = \prod_{p \geq 3} p^{\lambda_p}.$$

Then similar to (2.12), we can deduce that

$$r_{\mathbf{a}}(n^2) = \begin{cases} d_{\mathbf{a}} r_{(1,1,1,1,1,2,2)}(m^2) + e_{\mathbf{a}} r_{(1,2,2,2,2,2,2)}(m^2) & \text{if } n \text{ is odd,} \\ c_{\mathbf{a}}(\lambda_2) r_{(1,1,1,1,1,1,1)}(m^2) & \text{if } n \text{ is even.} \end{cases}$$

Hence,

$$r_{\mathbf{a}}(n^2) = \begin{cases} (8d_{\mathbf{a}} + 2e_{\mathbf{a}})t(n) + 2d_{\mathbf{a}}k(n) & \text{if } n \text{ is odd,} \\ 14c_{\mathbf{a}}(\lambda_2)t(n) & \text{if } n \text{ is even.} \end{cases}$$

for the values of $\mathbf{a}$, $c_{\mathbf{a}}(\lambda_2)$, $d_{\mathbf{a}}$ and $e_{\mathbf{a}}$ given in the Table 1, below.

$\mathbf{a}$	$c_{\mathbf{a}}$	$d_{\mathbf{a}}$	$e_{\mathbf{a}}$
$(1,1,1,1,1,1,4)$	$\frac{125\times 2^{5\lambda_2}-63}{(2^5-1)\times 7}$	3	-9
$(1,1,1,1,1,4,4)$	$\frac{95\times 2^{5\lambda_2-1}-63}{(2^5-1)\times 7}$	$\frac{7}{2}$	$-\frac{25}{2}$
$(1,1,1,1,4,4,4)$	$\frac{33\times 2^{5\lambda_2-1}-63}{(2^5-1)\times 7}$	3	-11
$(1,1,1,4,4,4,4)$	$\frac{5\times 2^{5\lambda_2-2}-9}{2^5-1}$	$\frac{9}{4}$	$-\frac{33}{4}$
$(1,1,4,4,4,4,4)$	$\frac{5\times 2^{5\lambda_2-2}-9}{2^5-1}$	$\frac{3}{2}$	$-\frac{11}{2}$
$(1,4,4,4,4,4,4)$	$\frac{5\times 2^{5\lambda_2-2}-9}{2^5-1}$	$\frac{3}{4}$	$-\frac{11}{4}$
$(1,1,1,1,2,2,4)$	$\frac{9\times 2^{5\lambda_2}-9}{2^5-1}$	2	-4
$(1,1,1,2,2,4,4)$	$\frac{32\times 2^{5\lambda_2}-63}{(2^5-1)\times 7}$	2	-7
$(1,1,2,2,4,4,4)$	$\frac{33\times 2^{5\lambda_2-1}-63}{(2^5-1)\times 7}$	$\frac{3}{2}$	$-\frac{11}{2}$
$(1,1,2,2,2,2,4)$	$\frac{32\times 2^{5\lambda_2}-63}{(2^5-1)\times 7}$	1	-3
$(1,2,2,2,2,4,4)$	$\frac{33\times 2^{5\lambda_2-1}-63}{(2^5-1)\times 7}$	$\frac{1}{2}$	$-\frac{3}{2}$
$(1,2,2,4,4,4,4)$	$\frac{5\times 2^{5\lambda_2-2}-9}{2^5-1}$	$\frac{3}{4}$	$-\frac{11}{4}$

Table 1: Data for (2.13)–(2.24)

6. Concluding Remarks

It is natural to ask if Theorem 2.1 can be extended by allowing some of the a_j to be equal to 4, that is, to consider the case $1 = a_1 \le a_2 \le \cdots \le a_7 = 4$ for which the product $a_1 a_2 \cdots a_7$ is an odd power of 2. For example, consider the case $r_{(1,1,1,1,1,2,4)}(n^2)$. The methods in [3, Section 4] can be used to find a formula in the case that n is even. For odd values of n, it would be necessary to study the sextenary form $r_{(1,1,1,1,2,4)}(n)$ and be able to express $\varphi^4(q)\varphi(q^2)\varphi(q^4)$ as a linear combination of functions whose coefficients satisfy (3.6). Such a formula is not known. This could be the subject of further investigation.

Acknowledgment. We sincerely thank the anonymous referee for his/her valuable comments that improve the quality of this paper. The third author is highly grateful to the first author for his kind guidance and encouragement.

References

[1] A. Alaca, Ş. Alaca, F. Uygul and K. S. Williams, *Representations by sextenary quadratic forms whose coefficients are 1, 2 and 4,* Acta Arith. **141** (2010), 289–309.

[2] A. Alaca, Ş. Alaca and K. S. Williams, *Liouville's sextenary quadratic forms* $x^2 + y^2 + z^2 + t^2 + 2u^2 + 2v^2$, $x^2 + y^2 + 2z^2 + 2t^2 + 2u^2 + 2v^2$ *and* $x^2 + 2y^2 + 2z^2 + 2t^2 + 2u^2 + 4v^2$, Far East J. Math. Sci. **30** (2008), 547–556.

[3] S. Cooper and H. Y. Lam, *On the Diophantine equation* $n^2 = x^2 + by^2 + cz^2$, J. Number Theory **133** (2013), 719–737.

[4] S. Cooper, H. Y. Lam and D. Ye, *Representations of squares by certain quinary quadratic forms,* Acta Arith. **157** (2013), 147–168.

[5] A. Hurwitz, *Sur la décomposition des nombres en cinq carrés,* Paris C. R. Acad. Sci. **98** (1884), 504–507.

[6] A. Hurwitz, L'Intermédiaire des Mathematiciens, **14** (1907), 107.

[7] J. Liouville, *Nombre des représentations du double d'un entier impair sous la forme d'une somme de douze carrés,* J. Math. Pures et Appl. (2) **5** (1860), 143–146.

[8] Y. Martin, *Multiplicative* η*-quotients,* Trans. Amer. Math. Soc. **348** (1996), 4825–4856.

[9] H. F. Sandham, *A square as the sum of 7 squares,* Quart. J. Math. (2) **4** (1953), 230–236.

CATALAN NUMBERS MODULO A PRIME POWER

Yong-Gao Chen[1]
School of Mathematical Sciences and Institute of Mathematics, Nanjing Normal University, Nanjing, P. R. CHINA
ygchen@njnu.edu.cn

Wen Jiang
School of Mathematical Sciences and Institute of Mathematics, Nanjing Normal University, Nanjing, P. R. CHINA

Received: 9/6/12, Accepted: 5/1/13, Published: 6/14/13

Abstract

Let $C_n = (2n)!/((n+1)!n!)$ be the n-th Catalan number. It is proved that for any odd prime p and integers a, k with $0 \le a < p$ and $k > 0$, if $0 \le a < (p+1)/2$, then the Catalan numbers $C_{p^1-a}, \dots, C_{p^k-a}$ are all distinct modulo p^k, and the sequence $(C_{p^n-a})_{n\ge1}$ modulo p^k is constant from $n = k$ on; if $(p+1)/2 \le a < p$, then the Catalan numbers $C_{p^1-a}, \dots, C_{p^{k+1}-a}$ are all distinct modulo p^k, and the sequence $(C_{p^n-a})_{n\ge1}$ modulo p^k is constant from $n = k+1$ on. The similar conclusion is proved for $p = 2$ recently by Lin.

1. Introduction

Let $C_n = (2n)!/((n+1)!n!)$ be the n-th Catalan number. In 2011, Lin [4] proved a conjecture of Liu and Yeh by showing that for all $k \ge 2$, the Catalan numbers $C_{2^1-1}, \dots, C_{2^{k-1}-1}$ are all distinct modulo 2^k, and the sequence $(C_{2^n-1})_{n\ge1}$ modulo 2^k is constant from $n = k-1$ on. For $k = 2, 3$, this is proved by Eu, Liu and Yeh [2]. In this paper, the following result is proved.

Theorem 1. *Let p be an odd prime and a, k be two integers with $0 \le a < p$ and $k > 0$. Then*

(i) for $0 \le a < \frac{1}{2}(p+1)$, the Catalan numbers $C_{p^1-a}, \dots, C_{p^k-a}$ are all distinct modulo p^k, and the sequence $(C_{p^n-a})_{n\ge1}$ modulo p^k is constant from $n = k$ on;

(ii) for $\frac{1}{2}(p+1) \le a < p$, the Catalan numbers $C_{p^1-a}, \dots, C_{p^{k+1}-a}$ are all distinct modulo p^k, and the sequence $(C_{p^n-a})_{n\ge1}$ modulo p^k is constant from $n = k+1$ on.

[1]This work was supported by the National Natural Science Foundation of China, Grant No 11071121.

2. Proof of the Theorem

We begin with the following lemmas.

Lemma 1. ([1]) *For any odd prime p and any positive integer k, we have*

$$p \nmid C_{p^k-1}.$$

Lemma 2. ([3, Theorem 129]) *If p is an odd prime and k is a positive integer, then*

$$\prod_{\substack{0<d<p^k \\ (d,p)=1}} d \equiv -1 \pmod{p^k}.$$

Lemma 3. *Let p be an odd prime, and a, i be integers with $0 \le a < p$ and $i > 0$. Then*

(i) for $0 \le a < \frac{1}{2}(p+1)$, we have

$$C_{p^{i+1}-a} \equiv C_{p^i-a} \pmod{p^i}, \quad C_{p^{i+1}-a} \not\equiv C_{p^i-a} \pmod{p^{i+1}};$$

(ii) for $\frac{1}{2}(p+1) \le a < p$, we have

$$C_{p^{i+1}-a} \equiv C_{p^i-a} \pmod{p^{i-1}}, \quad C_{p^{i+1}-a} \not\equiv C_{p^i-a} \pmod{p^i}.$$

Proof. First we deal with the case $a = 1$.

Define $\tau_p(n) = n/p^\alpha$ for $p^\alpha \mid n$ and $p^{\alpha+1} \nmid n$. By Lemma 1, we have

$$2(2p^{i+1}-1)C_{p^{i+1}-1} = \frac{2\cdot(2p^{i+1}-1)!}{p^{i+1}!(p^{i+1}-1)!} = \frac{\tau_p(2\cdot(2p^{i+1}-1)!)}{\tau_p(p^{i+1}!(p^{i+1}-1)!)} = \frac{\tau_p((2p^{i+1})!)}{(\tau_p(p^{i+1}!))^2}. \quad (1)$$

Similarly, we have

$$2(2p^i-1)C_{p^i-1} = \frac{\tau_p((2p^i)!)}{(\tau_p(p^i!))^2}. \quad (2)$$

Since

$$p^{i+1}! = \prod_{\substack{0<d<p^{i+1} \\ (d,p)=1}} d \cdot \prod_{v=1}^{p^i} vp,$$

by Lemma 2, we have

$$\tau_p(p^{i+1}!) \equiv -\tau_p(p^i!) \pmod{p^{i+1}}. \quad (3)$$

Similarly, we have

$$\tau_p((2p^{i+1})!) \equiv \tau_p((2p^i)!) \pmod{p^{i+1}}. \quad (4)$$

By (1), (2), (3) and (4), we have

$$2(2p^{i+1}-1)C_{p^{i+1}-1} \equiv 2(2p^i-1)C_{p^i-1} \pmod{p^{i+1}}.$$

That is,

$$C_{p^{i+1}-1} \equiv (1-2p^i)C_{p^i-1} \pmod{p^{i+1}}.$$

Now Lemma 3 for $a=1$ follows immediately from Lemma 1 and the above congruence.

If $a=0$, then

$$C_{p^i-a} = C_{p^i} = \frac{(2p^i)!}{p^i!(p^i+1)!} = \frac{(2p^i)(2p^i-1)}{p^i(p^i+1)}C_{p^i-1} = \frac{2(2p^i-1)}{p^i+1}C_{p^i-1}.$$

Thus, by Lemma 1 we have $p \nmid C_{p^i}$. Hence

$$(p^i+1)C_{p^i} = \frac{\tau_p((2p^i)!)}{(\tau_p(p^i!))^2}. \tag{5}$$

Similarly, we have

$$(p^{i+1}+1)C_{p^{i+1}} = \frac{\tau_p((2p^{i+1})!)}{(\tau_p(p^{i+1}!))^2}. \tag{6}$$

By (3)-(6) we have

$$(p^i+1)C_{p^i} \equiv (p^{i+1}+1)C_{p^{i+1}} \pmod{p^{i+1}}.$$

Now Lemma 3 for $a=0$ follows immediately.

Now we assume that $2 \le a < p$. Then

$$\begin{aligned} C_{p^i-a} &= \frac{(2p^i-2a)!}{(p^i-a)!(p^i-a+1)!} \\ &= \frac{(p^i-a+1)\cdots(p^i-1)(p^i-a+2)\cdots p^i}{(2p^i-2a+1)\cdots(2p^i-2)} \cdot \frac{(2p^i-2)!}{(p^i-1)!p^i!} \\ &= \frac{(p^i-a+1)\cdots(p^i-1)(p^i-a+2)\cdots p^i}{(2p^i-2a+1)\cdots(2p^i-2)} C_{p^i-1}. \end{aligned} \tag{7}$$

By Lemma 1 we have $p \nmid C_{p^i-1}$ for $i \ge 1$. If $2 \le a < \frac{1}{2}(p+1)$, then, by (7), we have

$$p^i \mid C_{p^i-a}, \quad p^{i+1} \nmid C_{p^i-a}.$$

Similarly, we have

$$p^{i+1} \mid C_{p^{i+1}-a}, \quad p^{i+2} \nmid C_{p^{i+1}-a}.$$

Hence, if $2 \le a < \frac{1}{2}(p+1)$, then

$$C_{p^{i+1}-a} \equiv C_{p^i-a} \pmod{p^i}, \quad C_{p^{i+1}-a} \not\equiv C_{p^i-a} \pmod{p^{i+1}}.$$

If $\frac{1}{2}(p+1) \leq a < p$, then, by (7), we have

$$p^{i-1} \mid C_{p^i-a}, \quad p^i \nmid C_{p^i-a}.$$

Similarly, we have

$$p^i \mid C_{p^{i+1}-a}, \quad p^{i+1} \nmid C_{p^{i+1}-a}.$$

Hence, if $\frac{1}{2}(p+1) \leq a < p$, then

$$C_{p^{i+1}-a} \equiv C_{p^i-a} \pmod{p^{i-1}}, \quad C_{p^{i+1}-a} \not\equiv C_{p^i-a} \pmod{p^i}.$$

This completes the proof of Lemma 3. □

Proof of Theorem 1. We prove (i). Case (ii) is similar. Assume that $0 \leq a < \frac{1}{2}(p+1)$. For any $u \geq v$, by Lemma 3 (i) and $p^v \mid p^u$, we have

$$C_{p^{u+1}-a} \equiv C_{p^u-a} \pmod{p^v}. \tag{8}$$

For $1 \leq i < j \leq k$, by (8) and Lemma 3 (i) we have

$$C_{p^j-a} \equiv C_{p^{j-1}-a} \equiv \cdots \equiv C_{p^{i+1}-a} \not\equiv C_{p^i-a} \pmod{p^{i+1}}. \tag{9}$$

Since $p^{i+1} \mid p^k$, it follows from (9) that

$$C_{p^j-a} \not\equiv C_{p^i-a} \pmod{p^k}.$$

For $n > k$, by (8) we have

$$C_{p^n-a} \equiv C_{p^{n-1}-a} \equiv \cdots \equiv C_{p^k-a} \pmod{p^k}.$$

This completes the proof of Theorem 1. □

References

[1] R. Alter and K. Kubota, *Prime and prime power divisibility of Catalan numbers,* J. Combin. Theory Ser. A 15 (1973), 243-256.

[2] S.-P. Eu, S.-C. Liu and Y.-N. Yeh, *Catalan and Motzkin numbers modulo 4 and 8,* European J. Combin. 29 (2008), 1449-1466.

[3] G. H. Hardy and E. M. Wright, An Introduction to the Theory of Numbers, 5th ed., Clarendon Press, Oxford Univ. Press, New York, 1979.

[4] H.-Y. Lin, *Odd Catalan numbers modulo* 2^k, Integers 12 (2012), no. 2, 161-165.

GONI: PRIMES REPRESENTED BY BINARY QUADRATIC FORMS

Pete L. Clark
Department of Mathematics, University of Georgia, Athens, Georgia
pete@math.uga.edu

Jacob Hicks
Department of Mathematics, University of Georgia, Athens, Georgia
jhicks@math.uga.edu

Hans Parshall
Department of Mathematics, University of Georgia, Athens, Georgia
hans@math.uga.edu

Katherine Thompson
Department of Mathematics, University of Georgia, Athens, Georgia
thompson@math.uga.edu

Received: 8/4/12, Revised: 2/28/13, Accepted: 5/2/13, Published: 6/14/13

Abstract

We list 2779 regular primitive positive definite integral binary quadratic forms, and show that, conditional on the Generalized Riemann Hypothesis, this is the complete list of *regular*, positive definite binary integral quadratic forms (up to $\mathrm{SL}_2(\mathbb{Z})$-equivalence). For each of these 2779 forms we determine the primes that they represent by elementary combinatorial methods, avoiding Gauss's genus theory. The key intermediate result is a *Small Multiple Theorem* for representations of primes by integral binary forms.

1. Introduction

An **imaginary quadratic discriminant** is a negative integer Δ which is 0 or 1 modulo 4. For a given imaginary quadratic disciminant Δ, let $C(\Delta)$ be the set of $\mathrm{SL}_2(\mathbb{Z})$-equivalence classes of primitive positive definite integral binary quadratic forms of discriminant Δ. Then $C(\Delta)$ is a finite set [3, Thm. 2.13] which, when endowed with Gauss's composition law, becomes a finite abelian group, the **class group of discriminant** Δ [3, Thm. 3.9].

Thus a form q of discriminant Δ determines an element $[q] \in C(\Delta)$. A quadratic form q is **ambiguous** if $[q]^2 = 1$. For a $q = \langle A, B, C \rangle$, the form $\overline{q} = \langle A, -B, C \rangle$ represents the inverse of $[q]$ in $C(\Delta)$ [3, Thm. 3.9]. Note that q and $\overline{q}$ are $\mathrm{GL}_2(\mathbb{Z})$-equivalent: $\overline{q}(x, y) = q(x, -y)$, so q and $\overline{q}$ represent the same integers.

A discriminant Δ is **idoneal** if every $q \in C(\Delta)$ is ambiguous; this holds if and only if $C(\Delta) \cong (\mathbb{Z}/2\mathbb{Z})^r$ for some $r \in \mathbb{N}$. A quadratic form is **idoneal** if its discriminant is idoneal. A discriminant Δ is **bi-idoneal** if $C(\Delta) \cong (\mathbb{Z}/4\mathbb{Z}) \oplus (\mathbb{Z}/2\mathbb{Z})^r$ for some $r \in \mathbb{N}$. A quadratic form q is **bi-idoneal** if Δ is bi-idoneal and q is *not* ambiguous.

A **full congruence class of primes** is the set of all primes $p \nmid 2\Delta$ with $p \equiv n \pmod{N}$ for fixed coprime positive integers n and N. We say q is **regular** if the set of primes $p \nmid 2\Delta$ represented by q is a union of full congruence classes.

Recall Fermat's Two Squares Theorem: an odd prime p is of the form $x^2 + y^2$ if and only if $p \equiv 1 \pmod 4$. In our terminology then the form $q(x, y) = x^2 + y^2$ is regular. Indeed, much classical work on quadratic forms can be phrased as showing that certain specific binary quadratic forms represent full congruence classes of primes, or are regular. Among primitive, positive definite, integral binary quadratic forms, how many are regular? How many represent full congruence classes of primes? Remarkably, this problem has recently been solved (conditionally on GRH) but the answer does not appear explicitly in the literature. Here it is:

Theorem 1. *Let q be a primitive, positive definite integral binary quadratic form.*

a) The following are equivalent:
(i) q is regular.
(ii) q represents a full congruence class of primes.
(iii) q is either idoneal or bi-idoneal.

b) There are at least 425 *and at most* 432 *imaginary quadratic discriminants which are either idoneal or bi-idoneal. These* 425 *known discriminants give rise to precisely* 2779 $\mathrm{SL}_2(\mathbb{Z})$*-equivalence classes of regular forms: see Table 1.*

c) The list of idoneal and bi-idoneal discriminants of part b) is complete among all imaginary quadratic discriminants Δ with $|\Delta| \leq 80604484$. Assuming the Riemann Hypothesis for Dedekind zeta functions of imaginary quadratic fields, there are precisely 425 *imaginary discriminants which are idoneal or bi-idoneal.*

For these 2779 regular forms, it is natural to ask for explicit congruence conditions, as in Fermat's Two Squares Theorem. The following result accomplishes this.

Theorem 2. *Let $q = \langle A, B, C \rangle$ be one of the* 2779 *primitive, positive definite integral binary quadratic forms in Table 1, and let $\Delta = B^2 - 4AC$ be the discriminant of q. For a prime $p \nmid 2\Delta$, the following are equivalent:*

a) The form q integrally represents p: there are $x, y \in \mathbb{Z}$ with $q(x, y) = p$.

b) All of the following conditions hold:
(i) $\left(\frac{\Delta}{p}\right) = 1$.

(ii) For each odd prime $m \mid \Delta$, if $m \nmid A$, then $(\frac{p}{m}) = (\frac{A}{m})$, and if $m \nmid C$, then $(\frac{p}{m}) = (\frac{C}{m})$.

(iii) If $16 \mid \Delta$ and $2 \nmid A$, then $p \equiv A \pmod 4$. If $16 \mid \Delta$ and $2 \nmid C$, then $p \equiv C \pmod 4$.

(iv) If $32 \mid \Delta$ and $2 \nmid A$, then $p \equiv A \pmod 8$. If $32 \mid \Delta$ and $2 \nmid C$, then $p \equiv C \pmod 8$.

We will prove Theorem 1 by deducing it from Gauss's genus theory together with results of Meyer, Weinberger, Louboutin, Kaplan-Williams and Voight. We do this mostly for completeness and perspective. Our main goal is quite different: we will give a new proof of Theorem 2 using none of Gauss's genus theory but instead using elementary ideas from the **Geometry of Numbers**. Our methods build on the classical proof of the Two Squares Theorem via Minkowski's Convex Body Theorem and its recent generalization to the 65 principal idoneal forms $x^2 + Dy^2$ of T. Hagedorn [5], although we find it simpler to use (sharp) bounds on minima of binary quadratic forms going back to Lagrange and Legendre.

We may compare the two methods as follows: let q be a binary form of discriminant Δ, and let $p \nmid 2\Delta$ be a prime. To analyze the question of whether q represents p, genus theory begins with the observation that $(\frac{\Delta}{p}) = 1$ if and only if some $q' \in C(\Delta)$ represents p and attempts to rule out the representation of p by all forms $q' \neq q$. Our method begins with a *small multiple theorem*: if $(\frac{\Delta}{p}) = 1$, then q represents some multiple kp of p with k bounded in terms of Δ and via a combination of *elimination* and *reduction* attempts to show that we may take $k = 1$. Our method is more computational – at present it is more a technique than a theory – and the reasons for its success in all 2779 cases are rather mysterious! However, our technique can be used in settings where genus theory does not apply: in [2] and [8] some of us use these ideas to establish universality of most (but not all) of the 112 positive definite quaternary universal forms of square discriminant. In [1], the first author extends the method to a technique for proving representation theorems for quadratic forms in $2d$ variables over a normed Dedekind domain.

2. Proof of Theorem 1

2.1. Part a)

(i) $\Longrightarrow$ (ii): By [3, Thm. 9.12], q represents infinitely many prime numbers. Having established this, the implication is immediate.

(ii) $\Longrightarrow$ (iii): Suppose that there are coprime integers n and N such that for all primes p, if $p \nmid 2\Delta$ and $p \equiv n \pmod N$, then q represents p. By [9, Thm. 2], if q is ambiguous then Δ is idoneal hence so is q; whereas if q is not ambiguous then Δ is bi-idoneal and hence – since q is not ambiguous – so is q.

(iii) $\implies$ (i): Let $G(\Delta) = C(\Delta)/C(\Delta)^2$, and let $r : C(\Delta) \to G(\Delta)$ be the quotient map. The fibers of r are called **genera**; they are cosets of $C(\Delta)^2$, the **principal genus**. Let $c = \#C(\Delta)/\#G(\Delta)$. Thus Δ is idoneal if and only if $c = 1$ and bi-idoneal if and only if $c = 2$. For $q \in C(\Delta)$, we define $g(q)$ to be the set of $n \in (\mathbb{Z}/\Delta\mathbb{Z})^\times$ which are represented by q. We will need the following tenets of genus theory:
- For all $q, q' \in C(\Delta)$, $g(q) = g(q') \iff r(q) = r(q')$ [3, pp. 53-54].
- If $q \in C(\Delta)^2$, then $g(q)$ is a subgroup, H, of $(\mathbb{Z}/\Delta\mathbb{Z})^\times$ [3, Lem. 2.24, Thm. 3.15].
- For all $q \in C(\Delta)$, $g(q)$ is a coset of H in $(\mathbb{Z}/\Delta\mathbb{Z})^\times$ [3, Lem. 2.24, Thm. 3.15].
- Let n be a positive integer which is relatively prime to 2Δ. Then there is $q \in C(\Delta)$ representing n if and only if $\left(\frac{\Delta}{n}\right) = 1$ [3, Thm. 2.16].

In particular, let $p \nmid 2\Delta$ be an odd prime. Then if $\left(\frac{\Delta}{p}\right) = -1$, no $q \in C(\Delta)$ represents p, whereas if $\left(\frac{\Delta}{p}\right) = 1$, then some $q \in C(\Delta)$ represents p, and if $q, q' \in C(\Delta)$ both represent p, then $r(q) = r(q')$.

Suppose Δ is idoneal, let $q \in C(\Delta)$, and let $p \nmid 2\Delta$ be a prime. If q represents p then $p \in g(q)$; conversely, if $p \in g(q)$ then $\left(\frac{\Delta}{p}\right) = 1$, so some $q' \in C(\Delta)$ represents p and any such q' must lie in $g(q)$. But since Δ is idoneal, $c = 1$, and q is the only form in $r(q)$. Thus q represents p if and only if $p \in g(q)$, so q is regular.

Suppose Δ is bi-idoneal, let $q \in C(\Delta)$ be a nonambiguous form, and let $p \nmid 2\Delta$ be a prime. As above, if q represents p then $p \in g(q)$; conversely, if $p \in g(q)$ then some $q' \in r(q)$ repesents p. But since $c = 2$, $r(q) = \{[q], [\overline{q}]\} = \{[q], [q]^{-1}\}$, and q and $\overline{q}$ represent the same primes. Thus q represents p if and only if $p \in g(q)$, so q is regular.

Remark. The implication (ii) $\implies$ (iii) for *fundamental* discriminants was first proven by Kusaba [10], using methods of class field theory. In [9] the general case is proved, and their proof uses Gauss's genus theory together with a theorem of Meyer [12]: if n and N are coprime positive integers, $p \equiv n \pmod{N}$, $p \nmid 2\Delta$ is a prime number and $q \in C(\Delta)$ represents p, then q represents infinitely many prime numbers $q \equiv n \pmod{N}$. See [6] for a proof of Meyer's theorem and a second proof of (ii) $\implies$ (iii), both using class field theory.

2.2. Part b)

That the total number of idoneal and bi-idoneal discriminants lies between 425 and 432 is [14, Thm. 8.2].

2.3. Part c)

This is [14, Prop. 5.1] and [14, Thm. 8.6]. The latter result builds on work of Weinberger [15] and Louboutin [11].

3. A Small Multiple Theorem

Let $q = \langle A, B, C\rangle$ be a real binary quadratic form with discriminant $\Delta \neq 0$. Recall:
• If $\Delta > 0$, then q is **indefinite**: it assumes both positive and negative values.
• If $\Delta < 0$ and $A, C > 0$, then q is **positive definite**: it assumes only positive values except at $(x, y) = (0, 0)$.
• If $\Delta < 0$ and $A, C < 0$, then q is **negative definite**: it assumes only negative values except at $(x, y) = (0, 0)$. Since q is negative definite if and only if $-q$ is positive definite, negative definite forms do not require separate consideration.

Theorem 3. *Let $q = \langle A, B, C\rangle$ be a binary form over $\mathbb{R}$ with discriminant Δ.*
a) If $\Delta < 0$, there are integers x and y, not both zero, such that $|q(x, y)| \leq \sqrt{\frac{|\Delta|}{3}}$.
b) If $\Delta > 0$, there are integers x and y, not both zero, such that $|q(x, y)| \leq \sqrt{\frac{\Delta}{5}}$.

Proof. The core of the proof is the following "reduction lemma": if x_0, y_0 are coprime integers with $q(x_0, y_0) = M \neq 0$, then there are $b, c \in \mathbb{R}$ such that q is $\mathrm{SL}_2(\mathbb{Z})$-equivalent to $Mx^2 + bxy + cy^2$ with $-|M| < b \leq |M|$. For the details, see e.g. [7, Thm. 453, Thm. 454]. □

A **lattice** $\Lambda \subset \mathbb{R}^N$ is the set of all $\mathbb{Z}$-linear combinations of an $\mathbb{R}$-basis $\mathfrak{b} = \{v_1, \ldots, v_N\}$ for $\mathbb{R}^N$. If $M_{\mathfrak{b}} \in M_N(\mathbb{R})$ is the matrix with columns $v_1, \ldots, v_N$, then $\Lambda = M_{\mathfrak{b}}\mathbb{Z}^N$.

Proposition 1. *Let $q = \langle A, B, C\rangle$ be an integral form of discriminant Δ. Let p be an odd prime with $(\frac{\Delta}{p}) = 1$. Then there is an index p sublattice $\Lambda_p \subset \mathbb{Z}^2$ such that for all $(x, y) \in \Lambda_p$, $q(x, y) \equiv 0 \pmod{p}$.*

Proof. If $p \mid A$, take $M_p = \begin{bmatrix} 1 & 0 \\ 0 & p \end{bmatrix}$ and $\Lambda_p = M_p\mathbb{Z}^2$. If $p \nmid A$, by the quadratic formula in $\mathbb{Z}/p\mathbb{Z}$, there is $r \in \mathbb{Z}$ with $Ar^2 + Br + C \equiv 0 \pmod{p}$; set $M_p = \begin{bmatrix} p & r \\ 0 & 1 \end{bmatrix}$ and $\Lambda_p = M_p\mathbb{Z}^2$. In either case, $q(x, y) \equiv 0 \pmod{p}$ for all $(x, y) \in \Lambda_p$. □

Theorem 4. *Let $q = \langle A, B, C\rangle$ be an integral form of discriminant Δ. Let p be an odd prime with $(\frac{\Delta}{p}) = 1$.*
a) If q is positive definite, there are $x, y, k \in \mathbb{Z}$ with $q(x, y) = kp$ and $1 \leq k \leq \sqrt{\frac{|\Delta|}{3}}$.
b) If q is indefinite, there are $x, y, k \in \mathbb{Z}$ with $q(x, y) = kp$ and $1 \leq |k| \leq \sqrt{\frac{\Delta}{5}}$.

Proof. By Proposition 1, there is an index p sublattice $\Lambda_p = M_p\mathbb{Z}^2 \subset \mathbb{Z}^2$ with $q(x, y) \equiv 0 \pmod{p}$ for all $(x, y) \in \Lambda_p$. Thus the quadratic form $q'(x, y) = q(M_p(x, y))$ has discriminant $(\det M_p)^2\Delta = p^2\Delta$ and is such that $q'(x, y) \equiv 0 \pmod{p}$ for all $(x, y) \in \mathbb{Z}^2$. Apply Theorem 3 to q': if q is positive definite, there are integers x and y, not both zero, such that $|q(M_p(x,y)| = |q'(x,y)| \leq \left(\sqrt{\frac{|\Delta|}{3}}\right)p$. Thus $q(x,y) = kp$ with $1 \leq |k| \leq \sqrt{\frac{|\Delta|}{3}}$; since q is positive definite, $k > 0$. If $\Delta > 0$, there are integers x and y, not both zero, such that $|q(M_p(x,y)| = |q'(x,y)| \leq \left(\sqrt{\frac{\Delta}{5}}\right)p$, so $q(x,y) = kp$ with $1 \leq |k| \leq \sqrt{\frac{\Delta}{5}}$. □

Remark. Taking $q = \langle 1,1,1\rangle$ (resp. $\langle 1,1,-1\rangle$) shows that the bound in Theorem 4a) (resp. Theorem 4b) is sharp.

Remark. Let $q = \langle A,B,C\rangle$ be positive definite with $|\Delta| < 12$. Then $\sqrt{\frac{|\Delta|}{3}} < 2$, and Theorem 4 takes the form: every odd prime p with $\left(\frac{\Delta}{p}\right) = 1$ is $\mathbb{Z}$-represented by q. It is easy to see that these are the only odd primes $p \nmid 2\Delta$ which are represented by q (c.f. Proposition 2), so this proves Theorem 2 for these forms, namely for $\langle 1,1,1\rangle$, $\langle 1,0,1\rangle$, $\langle 1,1,2\rangle$, $\langle 1,0,2\rangle$, and $\langle 1,1,3\rangle$.

4. 2779 Regular Forms

In this section we will use Theorem 4 to prove Theorem 2.

Henceforth "forms" are primitive, positive definite integral binary quadratic forms.

4.1. Necessity

Proposition 2. *Let $q = \langle A,B,C\rangle$ be a form with discriminant Δ. Let p be an odd prime not dividing Δ. Suppose there exist $x,y \in \mathbb{Z}$ with $q(x,y) = p$. Then p satisfies conditions (i) - (iv) from Theorem 2.*

Proof. Via the discriminant-preserving transformation $\langle A,B,C\rangle \mapsto \langle C,B,A\rangle$ we may assume in $m \nmid A$ in part (ii) and $2 \nmid A$ in parts (iii) and (iv); otherwise, q would not be primitive.
(i) If both x and y were divisible by p, this would imply $p^2 \mid q(x,y) = p$, a contradiction. If $p \nmid y$, then we have $A(xy^{-1})^2 + B(xy^{-1}) + C \equiv 0 \pmod{p}$. Let $r \in \mathbb{Z}$ with $r \equiv xy^{-1} \pmod{p}$. Then

$$(2Ar+B)^2 = 4A(Ar^2+Br+C) + B^2 - 4AC \equiv \Delta \pmod{p}$$

As $p \nmid \Delta$, we conclude $(\frac{\Delta}{p}) = 1$. The case $p \nmid x$ follows similarly.
(ii) Let m be an odd prime such that $m \mid \Delta$ and $m \nmid A$. Via a change of variables we can diagonalize q over $\mathbb{Z}/m\mathbb{Z}$ as $\langle A, 0, C - B^2(4A)^{-1}\rangle$, so there are $w, z \in \mathbb{Z}$ with

$$p = q(x,y) \equiv Aw^2 + (C - B^2(4A)^{-1})z^2 \pmod{m}.$$

Multiplying by $4A$ gives $4Ap \equiv 4A^2w^2 \pmod{m}$. Hence $p \equiv Aw^2 \pmod{m}$. It follows that $(\frac{p}{m}) = (\frac{A}{m})$.
(iii) Suppose $2 \nmid A$ and $\Delta \equiv 0 \pmod{16}$. We have $B^2 \equiv 4AC \pmod{16}$, so $B = 2B_0$ for some $B_0 \in \mathbb{Z}$. Then $4(B_0^2 - AC) \equiv 0 \pmod{16}$, so $B_0^2 - AC \equiv 0 \pmod{4}$.
Case 1: B_0 is odd. Then $A \equiv C \equiv \pm 1 \pmod{4}$. Now, $Ax^2 + 2B_0xy + Cy^2 = p$, so $x^2 + y^2 \equiv p \equiv 1 \pmod{2}$, and $x \not\equiv y \pmod{2}$. If $y \equiv 0 \pmod{2}$, $p \equiv A \pmod{4}$ as claimed. Similarly if $x \equiv 0 \pmod{2}$, $p \equiv C \pmod{4}$. But since $A \equiv C \pmod{4}$, $p \equiv A \pmod{4}$ as claimed.
Case 2: B_0 is even. Then $AC \equiv 0 \pmod{4}$. As $2 \nmid A$, $C \equiv 0 \pmod{4}$. Hence, $Ax^2 \equiv p \pmod{4}$, and so $p \equiv A \pmod{4}$ as claimed.
(iv) Suppose $2 \nmid A$ and $\Delta \equiv 0 \pmod{32}$. Put $B = 2B_0$, so $B_0^2 - AC \equiv 0 \pmod{8}$.
Case 1: B_0 is odd, Then $A \equiv C \pmod{2}$ and in fact $A \equiv C \pmod{8}$. Thus $x^2 + y^2 \equiv p \equiv 1 \pmod{2}$, so $x \not\equiv y \pmod{2}$. If $y \equiv 0 \pmod{2}$, set $y = 2y_0$. Then $Ax^2 + 4y_0(B_0x + Cy_0) = p$. If y_0 is even, then $Ax^2 \equiv A \equiv p \pmod{8}$. If instead y_0 is odd, then since B_0, x, and C are odd, $B_0x + Cy_0$ is even and $Ax^2 \equiv A \equiv p \pmod{8}$. Similarly if $x \equiv 0 \pmod{2}$, then $p \equiv C \equiv A \pmod{8}$.
Case 2: B_0 is even. Put $B_0 = 2B_1$ and $C = 4C_0$, so $B_1^2 \equiv AC_0 \pmod{2}$ and

$$p = Ax^2 + Bxy + Cy^2 = Ax^2 + 4y(B_1x + C_0y).$$

Thus x is odd and $x^2 \equiv 1 \pmod{8}$. If y is even, then $p \equiv Ax^2 \equiv A \pmod{8}$. If y is odd then either $B_1 \equiv C_0 \equiv 0 \pmod{2}$ so $p \equiv Ax^2 \equiv A \pmod{8}$ or $B_1 \equiv C_0 \equiv 1 \pmod{2}$, so $B_1x + C_0y$ is even and once again $p \equiv Ax^2 \equiv A \pmod{8}$. □

4.2. Sufficiency

Our proof that (b) implies (a) in Theorem 2 is handled individually for each of the 2779 forms. For each form, we apply a three step process. First, we use Theorem 4 to demonstrate that our form represents a small multiple of a prime. In the second step, we *eliminate* certain multiples from consideration. In the final step, we *reduce* the remaining multiples to find a representation of p.

Example. Consider $q = \langle 3, 3, 5 \rangle$ with $\Delta = -51$. Let p be an odd prime not dividing Δ that satisfies conditions (i) - (iv) of Theorem 2.

`Step 1.` From condition (i) of Theorem 2, $(\frac{\Delta}{p}) = 1$. Apply Theorem 4: there are $x, y, k \in \mathbb{Z}$ with $q(x, y) = kp$ and $1 \le k \le \sqrt{\frac{51}{3}} = 4.123\ldots$.

`Step 2 (Elimination).` We will show that the cases $k = 2$ and $k = 3$ cannot occur.

- Suppose $q(x, y) = 2p$. Then x and y are both even, so $q(x, y) = 2p \equiv 0 \pmod{4}$, contradicting the fact that p is odd.
- Suppose $q(x, y) = 3p$. Then $q(x, y) \equiv 5y^2 \equiv 0 \pmod{3}$, so $3 \mid y$. Hence,

$q(x,y) \equiv 3x^2 \equiv 3p \pmod 9$, so $\left(\frac{p}{3}\right) = 1$. As $3 \mid \Delta$, from condition (ii) of Theorem 2, $\left(\frac{p}{3}\right) = \left(\frac{5}{3}\right) = -1$: contradiction.

`Step 3 (Reduction).` Note that we cannot hope to eliminate the possibility of $k = 4$: indeed, we want to show that there are $x, y \in \mathbb{Z}$ such that $q(x,y) = p$, and then necessarily $q(2x, 2y) = 4p$. (A similar argument will be needed for any value of k which is a perfect square). We must instead argue that a representation of $4p$ by q implies a representation of p by q. In this case, this is easy: suppose $q(x,y) = 4p$. Then as above x and y are both even, so $q(\frac{x}{2}, \frac{y}{2}) = p$.

In Lemmas 1 and 2, we collect a number of congruence restrictions that apply assuming a form q represents kp. In particular, for our 2779 forms, we use Lemma 1 in the elimination step and Lemma 2 in the reduction step.

Lemma 1 (Elimination). *Let $q = \langle A, B, C\rangle$ be a form of discriminant Δ. Let $p \nmid 2\Delta$ be a prime. Suppose there are $x, y, k \in \mathbb{Z}$, $k \geq 1$, with $q(x,y) = kp$.*

a) Let $a \in \mathbb{Z}$, $a > 1$. Suppose $2^{a+2} \mid \Delta$ and $2^a \mid B$. If $p \equiv A \pmod{2^a}$, then k is a square modulo 2^a.

b) If k is even, A, C are odd, $B \equiv 0 \pmod 4$ and $A + C \not\equiv 2 \pmod 4$, then $4 \mid k$.

c) Let m be an odd prime dividing Δ. If $\left(\frac{p}{m}\right) = \left(\frac{A}{m}\right)$, then k is a square modulo m.

d) Let m be an odd prime dividing k. If $\left(\frac{\Delta}{m}\right) = -1$ or $m^2 \mid \Delta$, then $m^2 \mid k$.

e) Let m be an odd prime dividing $\gcd(\Delta, k)$ such that $m^2 \nmid k$. If $\left(\frac{p}{m}\right) = \left(\frac{A}{m}\right)$ then $\left(\frac{k/m}{m}\right) = \left(\frac{-\Delta/m}{m}\right)$.

Proof. a) Since $\Delta \equiv B^2 \equiv 0 \pmod{2^{a+2}}$, and A is odd, $2^a \mid C$. Then $kp \equiv Ax^2 \equiv px^2 \pmod{2^a}$, and since p is odd, this implies $k \equiv x^2 \pmod{2^a}$.

b) We have $q(x,y) \equiv Ax^2 + Cy^2 \equiv A(x^2 - y^2) \equiv kp \pmod 4$. Since k is even, $x \equiv y \pmod 2$ and thus $kp \equiv A(x^2 - y^2) \equiv 0 \pmod 4$. Since p is odd, $4 \mid k$.

c) Via a change of variables we can diagonalize q over $\mathbb{Z}/m\mathbb{Z}$ as $\langle A, 0, C - B^2(4A)^{-1}\rangle$, so there are $w, z \in \mathbb{Z}$ with $kp = q(x,y) \equiv Aw^2 + (C - B^2(4A)^{-1})z^2 \pmod m$. Thus, $4Akp \equiv 4A^2w^2 \pmod m$, implying $kp \equiv Aw^2 \pmod m$. As $\left(\frac{p}{m}\right) = \left(\frac{A}{m}\right) \neq 0$, k is a square modulo m.

d) Suppose first that $\left(\frac{\Delta}{m}\right) = -1$. We have $q(x,y) \equiv 0 \pmod m$. If $m \nmid y$, then $q(xy^{-1}, 1) \equiv 0 \pmod m$, so Δ is a square modulo m: contradiction. So $m \mid y$. Then $Ax^2 \equiv 0 \pmod m$, and $m \nmid A$, since otherwise $\Delta \equiv B^2 \pmod m$. Hence $m \mid x$. Then $m^2 \mid q(x,y) = kp$, and since $\left(\frac{\Delta}{p}\right) = 1$, we have $p \neq m$ and $m^2 \mid k$.

Next suppose $m^2 \mid \Delta$. If $m \mid \gcd(A, C)$, since $m \mid \Delta$ we would also have $m \mid B$, contradicting the primitivity of q. We may assume without loss of generality that $m \nmid A$. As $B^2 - 4AC \equiv 0 \pmod m$, $C \equiv B^2(4A)^{-1} \pmod m$. Hence,

$Ax^2 + Bxy + B^2(4A)^{-1}y^2 \equiv 0 \pmod{m}$, so by multiplying through by $4A$,

$$4A^2x^2 + 4ABxy + B^2y^2 \equiv (2Ax + By)^2 \equiv 0 \pmod{m}.$$

Since m is prime, $2Ax+By \equiv 0 \pmod{m}$, so $4A^2x^2+4ABxy+B^2y^2 \equiv 0 \pmod{m^2}$. As $B^2 - 4AC \equiv 0 \pmod{m^2}$, we have $B^2(4A)^{-1} \equiv C \pmod{m^2}$. Then

$$4Akp \equiv 4A^2x^2 + 4ABxy + B^2y^2 \equiv 0 \pmod{m^2}.$$

Since $p \nmid \Delta$, $m \neq p$. Then m does not divide $4Ap$, so $m^2 \mid k$.

e) Since $m \mid \Delta$ and $p \nmid \Delta$, $m \neq p$. We may write $\Delta = m\Delta_0$ and $k = mk_0$ with $\Delta_0, k_0 \in \mathbb{Z}$ and $m \nmid k_0$. Then $Ax^2 + Bxy + Cy^2 \equiv mk_0p \pmod{m^2}$. As in part d),

$$Ax^2 + Bxy + (B^2(4A)^{-1})y^2 \equiv 0 \pmod{m^2}.$$

Subtracting gives $(C - B^2(4A^{-1}))y^2 \equiv mk_0p \pmod{m^2}$. Since $\gcd(m, k_0p) = 1$, it follows that $m \nmid y$. Multiplying through by $4A$, we get

$$-m\Delta_0y^2 \equiv (4AC - B^2)y^2 \equiv 4Amk_0p \pmod{m^2}.$$

Then $(4Ak_0p + \Delta_0y^2)m \equiv 0 \pmod{m^2}$, so $4Apk_0 \equiv -\Delta_0y^2 \pmod{m}$. It follows that $(\frac{-\Delta_0}{m}) = (\frac{-\Delta_0 y^2}{m}) = (\frac{4Apk_0}{m}) \equiv (\frac{A}{m})(\frac{p}{m})(\frac{k_0}{m}) = (\frac{k_0}{m})$. □

Lemma 2 (Reduction). *Let $q = \langle A, B, C\rangle$ have discriminant Δ. Let p be an odd prime not dividing Δ. Suppose there exist $x, y, k \in \mathbb{Z}$ with $q(x, y) = kp$ and $k \geq 1$.*

a) Let $a \in \mathbb{Z}$ with $a \geq 1$. If $p \equiv A \pmod{2^a}$, then $q(x, y) \equiv Ak \pmod{2^ak}$.

b) Let $a \in \mathbb{Z}$ with $a \geq 0$, and let $m \mid \Delta$ be an odd prime. If $m^{2a} \mid k$, $m^{2a+1} \nmid k$, and $(\frac{p}{m}) = (\frac{A}{m})$, then we have $(\frac{q(x,y)/m^{2a}}{m}) = (\frac{Ak/m^{2a}}{m})$.

Proof. For a), write $p = 2^a\ell + A$. Then $q(x, y) \equiv k(2^a\ell + A) \equiv Ak \pmod{2^ak}$. For b), write $k = m^{2a}k_0$. Then $(\frac{q(x,y)/m^{2a}}{m}) = (\frac{k_0p}{m}) = (\frac{Ak_0}{m})$. □

4.3. Proof of Theorem 2

(a) $\Longrightarrow$ (b): This is Proposition 2.

(b) $\Longrightarrow$ (a): Let $q = \langle A, B, C\rangle$ be one of the 2779 regular forms, and let $p \nmid 2\Delta$ be a prime satisfying conditions (i) - (iv) from Theorem 2.

`Step 1.` Using condition (i), Theorem 4 implies there exist $x, y, k \in \mathbb{Z}$ such that $q(x, y) = kp$ with $1 \leq k \leq \sqrt{\frac{|\Delta|}{3}}$.

`Step 2 (Elimination).` For each $k \in \{2, \dots, \lfloor\sqrt{\frac{|\Delta|}{3}}\rfloor\}$, assume $q(x, y) = kp$. If k does not satisfy the conditions imposed on it by Lemma 1, we have a contradiction.

We similarly have a contradiction if k does not satisfy the conditions imposed on it by applying Lemma 1 to the equivalent forms $q(y,x) = \langle C, B, A\rangle$ and $q(x+y, x+2y) = \langle A+B+C, 2A+3B+4C, A+2B+4C\rangle$ representing kp. We eliminate these k from consideration.

`Step 3 (Reduction).` For each $k \in \{2, \ldots, \lfloor\sqrt{\frac{|\Delta|}{3}}\rfloor\}$ that was not eliminated in Step 2, assume $q(x,y) = kp$. Using a computer, we have verified that this assumption leads to a representation of p by q in every case. Our algorithm is as follows. First, we construct the finite set of matrices

$$\mathcal{M} = \left\{ \begin{bmatrix} a & b \\ c & d \end{bmatrix} \in M_2(\mathbb{Z}) \;\middle|\; a \geq 0,\ q(a,c) = kA \text{ and } q(b,d) = kC \right\}$$

by enumerating the representations of kA and kC by q. Given $M = \begin{bmatrix} a & b \\ c & d \end{bmatrix} \in \mathcal{M}$,

$$q(M(x,y)) = kAx^2 + (2abA + (ad+bc)B + 2cdC)xy + kCy^2.$$

In particular, $q(M(x,y)) = kq(x,y)$ whenever $2abA + (ad+bc)B + 2cdC = kB$. By iterating over $\mathcal{M}$ and checking this condition, we verify that there exists some $M \in \mathcal{M}$ such that $q(M(x,y)) = kq(x,y)$. Fixing such an M, we further check whether for each $(x,y) \in \mathbb{Z}^2$ with $q(x,y) \equiv 0 \pmod{k}$ that also satisfies the congruence restrictions imposed by Lemma 2, the pair $(x_0, y_0) = M(x,y)$ satisfies $x_0 \equiv y_0 \equiv 0 \pmod{k}$. It suffices to check this condition modulo $k\Delta$ by an exhaustive search. In every case we've considered, this search successfully produces such an $M \in \mathcal{M}$. Once such an M has been found, we can set $x_0 = kw$ and $y_0 = kz$. Then $q(M(x,y)) = q(kw, kz) = k^2p$, so $q(w,z) = p$. Therefore, we've shown that q represents p.

Example. Consider $q = \langle 2, 1, 7\rangle$ with $\Delta = -55$. Let p be an odd prime not dividing Δ that satisfies conditions (i) - (iv) of Theorem 2.

`Step 1.` From condition (i) of Theorem 2, $(\frac{\Delta}{p}) = 1$. Thus, applying Theorem 4 yields $x, y, k \in \mathbb{Z}$ with $q(x,y) = kp$ and $1 \leq k \leq \sqrt{\frac{55}{3}} = 4.28\ldots$.

`Step 2 (Elimination).` By Lemma 1(c), k is a square modulo 5. As $(\frac{2}{5}) = (\frac{3}{5}) = -1$, $k \in \{1, 4\}$.

`Step 3 (Reduction).` Suppose $q(x,y) = 4p$. One might try to argue, as in the example in Section 4.2, that both x and y are even. However, this need not be the case: e.g. q represents 7 and $q(3,1) = 4 \cdot 7$. Applying the algorithm described above we obtain

$$\mathcal{M} = \left\{ \begin{bmatrix} 1 & -3 \\ -1 & -1 \end{bmatrix}, \begin{bmatrix} 1 & -1 \\ -1 & 2 \end{bmatrix}, \begin{bmatrix} 1 & 0 \\ -1 & -2 \end{bmatrix}, \begin{bmatrix} 1 & 0 \\ -1 & 2 \end{bmatrix}, \begin{bmatrix} 1 & 1 \\ -1 & -2 \end{bmatrix}, \begin{bmatrix} 1 & 3 \\ -1 & 1 \end{bmatrix}, \right.$$
$$\left. \begin{bmatrix} 2 & -3 \\ 0 & -1 \end{bmatrix}, \begin{bmatrix} 2 & -1 \\ 0 & 2 \end{bmatrix}, \begin{bmatrix} 2 & 0 \\ 0 & -2 \end{bmatrix}, \begin{bmatrix} 2 & 0 \\ 0 & 2 \end{bmatrix}, \begin{bmatrix} 2 & 1 \\ 0 & -2 \end{bmatrix}, \begin{bmatrix} 2 & 3 \\ 0 & 1 \end{bmatrix} \right\}.$$

Set $M = \begin{bmatrix} 1 & -3 \\ -1 & -1 \end{bmatrix}$. Set $(x_0, y_0) = M(x, y) = (x - 3y, -x - y)$ and note $q(x_0, y_0) = 4q(x, y) = 16p$. If we knew $x_0 \equiv y_0 \equiv 0 \pmod 4$, then we could divide through by 4 to obtain an integer representation of p. Certainly we need only consider $(x, y) \in \mathbb{Z}^2$ with $q(x, y) \equiv 0 \pmod 4$. Further, since we're assuming $(\frac{p}{5}) = (\frac{2}{5}) = -1$ and $(\frac{p}{11}) = (\frac{2}{11}) = -1$, condition (ii) of Theorem 2 implies we need only consider $(x, y) \in \mathbb{Z}^2$ with $(\frac{q(x,y)}{5}) = (\frac{4p}{5}) = -1$ and $(\frac{q(x,y)}{11}) = (\frac{4p}{11}) = -1$. By an exhaustive search modulo 220, we verify the only such $(x, y) \in \mathbb{Z}^2$ yield $x_0 \equiv y_0 \equiv 0 \pmod 4$. Setting $x_0 = 4w$ and $y_0 = 4z$, we have $q(x_0, y_0) = 32w^2 + 16wz + 224z^2 = 16p$. Dividing through by 16, we see $q(w, z) = 2w^2 + wz + 7z^2 = p$. Therefore, we've shown that q represents p.

Acknowledgements. This work was done in the context of a VIGRE Research Group at the University of Georgia during the 2011-2012 academic year. The group was led by the first author, with participants the other three authors together with Christopher Drupieski (postdoc), Brian Bonsignore, Harrison Chapman, Lauren Huckaba, David Krumm, Allan Lacy Mora, Nham Ngo, Alex Rice, James Stankewicz, Lee Troupe, Nathan Walters (doctoral students) and Jun Zhang (master's student).

References

[1] P.L. Clark, *Geometry of numbers explained*, in preparation.

[2] P.L. Clark, J. Hicks, K. Thompson and N. Walters, *GoNII: Universal quaternary quadratic forms.* Integers 12 (2012), A50, 16 pp.

[3] D.A. Cox, *Primes of the Form $x^2 + ny^2$*, John Wiley & Sons Inc., 1989.

[4] C.F. Gauss, *Disquisitiones Arithmeticae (English Edition)*, trans. A.A. Clarke, Springer-Verlag, 1986.

[5] T.R. Hagedorn, *Primes of the Form $x^2 + ny^2$ and the Geometry of (Convenient) Numbers*, preprint.

[6] F. Halter-Koch, *Representation of prime powers in arithmetical progressions by binary quadratic forms.* Les XXIIèmes Journées Arithmetiques (Lille, 2001). J. Théor. Nombres Bordeaux 15 (2003), no. 1, 141-149.

[7] G.H. Hardy and E.M. Wright, *An Introduction to the Theory of Numbers*. Sixth edition. Revised by D. R. Heath-Brown and J. H. Silverman. Oxford, 2008.

[8] J. Hicks and K. Thompson, *GoNIII: More universal quaternary quadratic forms*, in preparation.

[9] P. Kaplan and K.S. Williams, *Representation of primes in arithmetic progression by binary quadratic forms*. J. Number Theory (1993), 61-67.

[10] T. Kusaba, *Remarque sur la distribution des nombres premiers*. C. R. Acad. Sci. Paris Sér. A-B 265 (1967), A405-A407.

[11] S. Louboutin, *Minorations (sous l'hypothèse de Riemann généralisée) des nombres de classes des corps quadratiques imaginaires. Application*. C. R. Acad. Sci. Paris Sér. I Math. 310 (1990), no. 12, 795-800.

[12] A. Meyer, *Über einen satz von Dirichlet*. J. Reine Angew. Math. 103 (1888), 98–117.

[13] W. A. Stein et al., *Sage Mathematics Software (Version 4.7.1)*, The Sage Development Team, 2011, `http://www.sagemath.org`

[14] J. Voight, *Quadratic forms that represent almost the same primes*. Math. Comp. 76 (2007), 1589-1617.

[15] P.J. Weinberger, *Exponents of the class groups of complex quadratic fields*. Acta Arith. 22 (1973), 117-124.

Appendix

In Table 1, we list the reduced representative for each of the 2779 $SL_2(\mathbb{Z})$ equivalence classes of regular forms. The discriminants were calculated by Voight in [14]. We redid this calculation, and in so doing found a minor error of tabulation which Voight confirmed. The forms were generated using Sage.

Table 1: Representatives for 2779 $SL_2(\mathbb{Z})$-equivalence Classes of Regular Forms

$\|\Delta\|$	$\langle A,B,C\rangle$	$\|\Delta\|$	$\langle A,B,C\rangle$	$\|\Delta\|$	$\langle A,B,C\rangle$	$\|\Delta\|$	$\langle A,B,C\rangle$
3	$\langle 1,1,1\rangle$	4	$\langle 1,0,1\rangle$	7	$\langle 1,1,2\rangle$	8	$\langle 1,0,2\rangle$
11	$\langle 1,1,3\rangle$	12	$\langle 1,0,3\rangle$	15	$\langle 1,1,4\rangle$	15	$\langle 2,1,2\rangle$
16	$\langle 1,0,4\rangle$	19	$\langle 1,1,5\rangle$	20	$\langle 1,0,5\rangle$	20	$\langle 2,2,3\rangle$
24	$\langle 1,0,6\rangle$	24	$\langle 2,0,3\rangle$	27	$\langle 1,1,7\rangle$	28	$\langle 1,0,7\rangle$
32	$\langle 1,0,8\rangle$	32	$\langle 3,2,3\rangle$	35	$\langle 1,1,9\rangle$	35	$\langle 3,1,3\rangle$
36	$\langle 1,0,9\rangle$	36	$\langle 2,2,5\rangle$	39	$\langle 2,\pm 1,5\rangle$	40	$\langle 1,0,10\rangle$
40	$\langle 2,0,5\rangle$	43	$\langle 1,1,11\rangle$	48	$\langle 1,0,12\rangle$	48	$\langle 3,0,4\rangle$
51	$\langle 1,1,13\rangle$	51	$\langle 3,3,5\rangle$	52	$\langle 1,0,13\rangle$	52	$\langle 2,2,7\rangle$
55	$\langle 2,\pm 1,7\rangle$	56	$\langle 3,\pm 2,5\rangle$	60	$\langle 1,0,15\rangle$	60	$\langle 3,0,5\rangle$
63	$\langle 2,\pm 1,8\rangle$	64	$\langle 1,0,16\rangle$	64	$\langle 4,4,5\rangle$	67	$\langle 1,1,17\rangle$
68	$\langle 3,\pm 2,6\rangle$	72	$\langle 1,0,18\rangle$	72	$\langle 2,0,9\rangle$	75	$\langle 1,1,19\rangle$
75	$\langle 3,3,7\rangle$	80	$\langle 3,\pm 2,7\rangle$	84	$\langle 1,0,21\rangle$	84	$\langle 2,2,11\rangle$
84	$\langle 3,0,7\rangle$	84	$\langle 5,4,5\rangle$	88	$\langle 1,0,22\rangle$	88	$\langle 2,0,11\rangle$
91	$\langle 1,1,23\rangle$	91	$\langle 5,3,5\rangle$	96	$\langle 1,0,24\rangle$	96	$\langle 3,0,8\rangle$
96	$\langle 4,4,7\rangle$	96	$\langle 5,2,5\rangle$	99	$\langle 1,1,25\rangle$	99	$\langle 5,1,5\rangle$
100	$\langle 1,0,25\rangle$	100	$\langle 2,2,13\rangle$	112	$\langle 1,0,28\rangle$	112	$\langle 4,0,7\rangle$
115	$\langle 1,1,29\rangle$	115	$\langle 5,5,7\rangle$	120	$\langle 1,0,30\rangle$	120	$\langle 2,0,15\rangle$
120	$\langle 3,0,10\rangle$	120	$\langle 5,0,6\rangle$	123	$\langle 1,1,31\rangle$	123	$\langle 3,3,11\rangle$
128	$\langle 3,\pm 2,11\rangle$	132	$\langle 1,0,33\rangle$	132	$\langle 2,2,17\rangle$	132	$\langle 3,0,11\rangle$
132	$\langle 6,6,7\rangle$	136	$\langle 5,\pm 2,7\rangle$	144	$\langle 5,\pm 4,8\rangle$	147	$\langle 1,1,37\rangle$

Table 1: Representatives for 2779 $SL_2(\mathbb{Z})$-equivalence Classes of Regular Forms

$\lvert\Delta\rvert$	$\langle A, B, C\rangle$	$\lvert\Delta\rvert$	$\langle A, B, C\rangle$	$\lvert\Delta\rvert$	$\langle A, B, C\rangle$	$\lvert\Delta\rvert$	$\langle A, B, C\rangle$
147	$\langle 3, 3, 13\rangle$	148	$\langle 1, 0, 37\rangle$	148	$\langle 2, 2, 19\rangle$	155	$\langle 3, \pm1, 13\rangle$
156	$\langle 5, \pm2, 8\rangle$	160	$\langle 1, 0, 40\rangle$	160	$\langle 4, 4, 11\rangle$	160	$\langle 5, 0, 8\rangle$
160	$\langle 7, 6, 7\rangle$	163	$\langle 1, 1, 41\rangle$	168	$\langle 1, 0, 42\rangle$	168	$\langle 2, 0, 21\rangle$
168	$\langle 3, 0, 14\rangle$	168	$\langle 6, 0, 7\rangle$	171	$\langle 5, \pm3, 9\rangle$	180	$\langle 1, 0, 45\rangle$
180	$\langle 2, 2, 23\rangle$	180	$\langle 5, 0, 9\rangle$	180	$\langle 7, 4, 7\rangle$	184	$\langle 5, \pm4, 10\rangle$
187	$\langle 1, 1, 47\rangle$	187	$\langle 7, 3, 7\rangle$	192	$\langle 1, 0, 48\rangle$	192	$\langle 3, 0, 16\rangle$
192	$\langle 4, 4, 13\rangle$	192	$\langle 7, 2, 7\rangle$	195	$\langle 1, 1, 49\rangle$	195	$\langle 3, 3, 17\rangle$
195	$\langle 5, 5, 11\rangle$	195	$\langle 7, 1, 7\rangle$	196	$\langle 5, \pm2, 10\rangle$	203	$\langle 3, \pm1, 17\rangle$
208	$\langle 7, \pm4, 8\rangle$	219	$\langle 5, \pm1, 11\rangle$	220	$\langle 7, \pm2, 8\rangle$	224	$\langle 3, \pm2, 19\rangle$
224	$\langle 5, \pm4, 12\rangle$	228	$\langle 1, 0, 57\rangle$	228	$\langle 2, 2, 29\rangle$	228	$\langle 3, 0, 19\rangle$
228	$\langle 6, 6, 11\rangle$	232	$\langle 1, 0, 58\rangle$	232	$\langle 2, 0, 29\rangle$	235	$\langle 1, 1, 59\rangle$
235	$\langle 5, 5, 13\rangle$	240	$\langle 1, 0, 60\rangle$	240	$\langle 3, 0, 20\rangle$	240	$\langle 4, 0, 15\rangle$
240	$\langle 5, 0, 12\rangle$	252	$\langle 8, \pm6, 9\rangle$	256	$\langle 5, \pm2, 13\rangle$	259	$\langle 5, \pm1, 13\rangle$
260	$\langle 3, \pm2, 22\rangle$	260	$\langle 6, \pm2, 11\rangle$	264	$\langle 5, \pm4, 14\rangle$	264	$\langle 7, \pm4, 10\rangle$
267	$\langle 1, 1, 67\rangle$	267	$\langle 3, 3, 23\rangle$	275	$\langle 3, \pm1, 23\rangle$	276	$\langle 5, \pm2, 14\rangle$
276	$\langle 7, \pm2, 10\rangle$	280	$\langle 1, 0, 70\rangle$	280	$\langle 2, 0, 35\rangle$	280	$\langle 5, 0, 14\rangle$
280	$\langle 7, 0, 10\rangle$	288	$\langle 1, 0, 72\rangle$	288	$\langle 4, 4, 19\rangle$	288	$\langle 8, 0, 9\rangle$
288	$\langle 8, 8, 11\rangle$	291	$\langle 5, \pm3, 15\rangle$	292	$\langle 7, \pm4, 11\rangle$	308	$\langle 3, \pm2, 26\rangle$
308	$\langle 6, \pm2, 13\rangle$	312	$\langle 1, 0, 78\rangle$	312	$\langle 2, 0, 39\rangle$	312	$\langle 3, 0, 26\rangle$
312	$\langle 6, 0, 13\rangle$	315	$\langle 1, 1, 79\rangle$	315	$\langle 5, 5, 17\rangle$	315	$\langle 7, 7, 13\rangle$
315	$\langle 9, 9, 11\rangle$	320	$\langle 3, \pm2, 27\rangle$	320	$\langle 7, \pm4, 12\rangle$	323	$\langle 3, \pm1, 27\rangle$
328	$\langle 7, \pm6, 13\rangle$	336	$\langle 5, \pm2, 17\rangle$	336	$\langle 8, \pm4, 11\rangle$	340	$\langle 1, 0, 85\rangle$
340	$\langle 2, 2, 43\rangle$	340	$\langle 5, 0, 17\rangle$	340	$\langle 10, 10, 11\rangle$	352	$\langle 1, 0, 88\rangle$
352	$\langle 4, 4, 23\rangle$	352	$\langle 8, 0, 11\rangle$	352	$\langle 8, 8, 13\rangle$	355	$\langle 7, \pm3, 13\rangle$
360	$\langle 7, \pm2, 13\rangle$	360	$\langle 9, \pm6, 11\rangle$	363	$\langle 7, \pm1, 13\rangle$	372	$\langle 1, 0, 93\rangle$
372	$\langle 2, 2, 47\rangle$	372	$\langle 3, 0, 31\rangle$	372	$\langle 6, 6, 17\rangle$	384	$\langle 5, \pm4, 20\rangle$
384	$\langle 7, \pm6, 15\rangle$	387	$\langle 9, \pm3, 11\rangle$	388	$\langle 7, \pm2, 14\rangle$	400	$\langle 8, \pm4, 13\rangle$
403	$\langle 1, 1, 101\rangle$	403	$\langle 11, 9, 11\rangle$	408	$\langle 1, 0, 102\rangle$	408	$\langle 2, 0, 51\rangle$
408	$\langle 3, 0, 34\rangle$	408	$\langle 6, 0, 17\rangle$	420	$\langle 1, 0, 105\rangle$	420	$\langle 2, 2, 53\rangle$
420	$\langle 3, 0, 35\rangle$	420	$\langle 5, 0, 21\rangle$	420	$\langle 6, 6, 19\rangle$	420	$\langle 7, 0, 15\rangle$
420	$\langle 10, 10, 13\rangle$	420	$\langle 11, 8, 11\rangle$	427	$\langle 1, 1, 107\rangle$	427	$\langle 7, 7, 17\rangle$
435	$\langle 1, 1, 109\rangle$	435	$\langle 3, 3, 37\rangle$	435	$\langle 5, 5, 23\rangle$	435	$\langle 11, 7, 11\rangle$
448	$\langle 1, 0, 112\rangle$	448	$\langle 4, 4, 29\rangle$	448	$\langle 7, 0, 16\rangle$	448	$\langle 11, 6, 11\rangle$
456	$\langle 5, \pm2, 23\rangle$	456	$\langle 10, \pm8, 13\rangle$	468	$\langle 7, \pm6, 18\rangle$	468	$\langle 9, \pm6, 14\rangle$
475	$\langle 7, \pm1, 17\rangle$	480	$\langle 1, 0, 120\rangle$	480	$\langle 3, 0, 40\rangle$	480	$\langle 4, 4, 31\rangle$
480	$\langle 5, 0, 24\rangle$	480	$\langle 8, 0, 15\rangle$	480	$\langle 8, 8, 17\rangle$	480	$\langle 11, 2, 11\rangle$
480	$\langle 12, 12, 13\rangle$	483	$\langle 1, 1, 121\rangle$	483	$\langle 3, 3, 41\rangle$	483	$\langle 7, 7, 19\rangle$
483	$\langle 11, 1, 11\rangle$	504	$\langle 5, \pm4, 26\rangle$	504	$\langle 10, \pm4, 13\rangle$	507	$\langle 7, \pm5, 19\rangle$
520	$\langle 1, 0, 130\rangle$	520	$\langle 2, 0, 65\rangle$	520	$\langle 5, 0, 26\rangle$	520	$\langle 10, 0, 13\rangle$
528	$\langle 7, \pm2, 19\rangle$	528	$\langle 8, \pm4, 17\rangle$	532	$\langle 1, 0, 133\rangle$	532	$\langle 2, 2, 67\rangle$
532	$\langle 7, 0, 19\rangle$	532	$\langle 13, 12, 13\rangle$	544	$\langle 5, \pm4, 28\rangle$	544	$\langle 7, \pm4, 20\rangle$
552	$\langle 7, \pm6, 21\rangle$	552	$\langle 11, \pm8, 14\rangle$	555	$\langle 1, 1, 139\rangle$	555	$\langle 3, 3, 47\rangle$
555	$\langle 5, 5, 29\rangle$	555	$\langle 13, 11, 13\rangle$	564	$\langle 5, \pm4, 29\rangle$	564	$\langle 10, \pm6, 15\rangle$
568	$\langle 11, \pm2, 13\rangle$	576	$\langle 5, \pm2, 29\rangle$	576	$\langle 9, \pm6, 17\rangle$	580	$\langle 7, \pm6, 22\rangle$
580	$\langle 11, \pm6, 14\rangle$	592	$\langle 8, \pm4, 19\rangle$	595	$\langle 1, 1, 149\rangle$	595	$\langle 5, 5, 31\rangle$
595	$\langle 7, 7, 23\rangle$	595	$\langle 13, 9, 13\rangle$	600	$\langle 7, \pm4, 22\rangle$	600	$\langle 11, \pm4, 14\rangle$
603	$\langle 9, \pm3, 17\rangle$	612	$\langle 7, \pm2, 22\rangle$	612	$\langle 11, \pm2, 14\rangle$	616	$\langle 5, \pm2, 31\rangle$
616	$\langle 10, \pm8, 17\rangle$	624	$\langle 5, \pm4, 32\rangle$	624	$\langle 11, \pm6, 15\rangle$	627	$\langle 1, 1, 157\rangle$
627	$\langle 3, 3, 53\rangle$	627	$\langle 11, 11, 17\rangle$	627	$\langle 13, 7, 13\rangle$	640	$\langle 7, \pm2, 23\rangle$
640	$\langle 11, \pm8, 16\rangle$	651	$\langle 5, \pm3, 33\rangle$	651	$\langle 11, \pm3, 15\rangle$	660	$\langle 1, 0, 165\rangle$
660	$\langle 2, 2, 83\rangle$	660	$\langle 3, 0, 55\rangle$	660	$\langle 5, 0, 33\rangle$	660	$\langle 6, 6, 29\rangle$
660	$\langle 10, 10, 19\rangle$	660	$\langle 11, 0, 15\rangle$	660	$\langle 13, 4, 13\rangle$	667	$\langle 11, \pm9, 17\rangle$
672	$\langle 1, 0, 168\rangle$	672	$\langle 3, 0, 56\rangle$	672	$\langle 4, 4, 43\rangle$	672	$\langle 7, 0, 24\rangle$
672	$\langle 8, 0, 21\rangle$	672	$\langle 8, 8, 23\rangle$	672	$\langle 12, 12, 17\rangle$	672	$\langle 13, 2, 13\rangle$
708	$\langle 1, 0, 177\rangle$	708	$\langle 2, 2, 89\rangle$	708	$\langle 3, 0, 59\rangle$	708	$\langle 6, 6, 31\rangle$
715	$\langle 1, 1, 179\rangle$	715	$\langle 5, 5, 37\rangle$	715	$\langle 11, 11, 19\rangle$	715	$\langle 13, 13, 17\rangle$
720	$\langle 7, \pm6, 27\rangle$	720	$\langle 8, \pm4, 23\rangle$	723	$\langle 11, \pm5, 17\rangle$	736	$\langle 5, \pm2, 37\rangle$
736	$\langle 11, \pm10, 19\rangle$	760	$\langle 1, 0, 190\rangle$	760	$\langle 2, 0, 95\rangle$	760	$\langle 5, 0, 38\rangle$
760	$\langle 10, 0, 19\rangle$	763	$\langle 13, \pm11, 17\rangle$	768	$\langle 7, \pm4, 28\rangle$	768	$\langle 13, \pm8, 16\rangle$
772	$\langle 11, \pm8, 19\rangle$	792	$\langle 9, \pm6, 23\rangle$	792	$\langle 13, \pm12, 18\rangle$	795	$\langle 1, 1, 199\rangle$
795	$\langle 3, 3, 67\rangle$	795	$\langle 5, 5, 41\rangle$	795	$\langle 15, 15, 17\rangle$	819	$\langle 5, \pm1, 41\rangle$
819	$\langle 9, \pm3, 23\rangle$	820	$\langle 11, \pm4, 19\rangle$	820	$\langle 13, \pm8, 17\rangle$	832	$\langle 7, \pm6, 31\rangle$
832	$\langle 11, \pm2, 19\rangle$	840	$\langle 1, 0, 210\rangle$	840	$\langle 2, 0, 105\rangle$	840	$\langle 3, 0, 70\rangle$
840	$\langle 5, 0, 42\rangle$	840	$\langle 6, 0, 35\rangle$	840	$\langle 7, 0, 30\rangle$	840	$\langle 10, 0, 21\rangle$

Table 1: Representatives for 2779 $SL_2(\mathbb{Z})$-equivalence Classes of Regular Forms

$\|\Delta\|$	$\langle A,B,C\rangle$	$\|\Delta\|$	$\langle A,B,C\rangle$	$\|\Delta\|$	$\langle A,B,C\rangle$	$\|\Delta\|$	$\langle A,B,C\rangle$
840	$\langle 14,0,15\rangle$	852	$\langle 7,\pm4,31\rangle$	852	$\langle 14,\pm10,17\rangle$	868	$\langle 11,\pm10,22\rangle$
868	$\langle 13,\pm4,17\rangle$	880	$\langle 7,\pm4,32\rangle$	880	$\langle 13,\pm2,17\rangle$	900	$\langle 9,\pm6,26\rangle$
900	$\langle 13,\pm6,18\rangle$	912	$\langle 8,\pm4,29\rangle$	912	$\langle 11,\pm10,23\rangle$	915	$\langle 7,\pm3,33\rangle$
915	$\langle 11,\pm3,21\rangle$	928	$\langle 1,0,232\rangle$	928	$\langle 4,4,59\rangle$	928	$\langle 8,0,29\rangle$
928	$\langle 8,8,31\rangle$	952	$\langle 11,\pm4,22\rangle$	952	$\langle 13,\pm6,19\rangle$	955	$\langle 7,\pm5,35\rangle$
960	$\langle 1,0,240\rangle$	960	$\langle 3,0,80\rangle$	960	$\langle 4,4,61\rangle$	960	$\langle 5,0,48\rangle$
960	$\langle 12,12,23\rangle$	960	$\langle 15,0,16\rangle$	960	$\langle 16,16,19\rangle$	960	$\langle 17,14,17\rangle$
987	$\langle 11,\pm5,23\rangle$	987	$\langle 13,\pm1,19\rangle$	1003	$\langle 11,\pm3,23\rangle$	1008	$\langle 9,\pm6,29\rangle$
1008	$\langle 11,\pm2,23\rangle$	1012	$\langle 1,0,253\rangle$	1012	$\langle 2,2,127\rangle$	1012	$\langle 11,0,23\rangle$
1012	$\langle 17,12,17\rangle$	1027	$\langle 7,\pm3,37\rangle$	1032	$\langle 7,\pm2,37\rangle$	1032	$\langle 14,\pm12,21\rangle$
1035	$\langle 7,\pm1,37\rangle$	1035	$\langle 9,\pm3,29\rangle$	1056	$\langle 5,\pm2,53\rangle$	1056	$\langle 7,\pm6,39\rangle$
1056	$\langle 13,\pm6,21\rangle$	1056	$\langle 15,\pm12,20\rangle$	1060	$\langle 7,\pm2,38\rangle$	1060	$\langle 14,\pm2,19\rangle$
1092	$\langle 1,0,273\rangle$	1092	$\langle 2,2,137\rangle$	1092	$\langle 3,0,91\rangle$	1092	$\langle 6,6,47\rangle$
1092	$\langle 7,0,39\rangle$	1092	$\langle 13,0,21\rangle$	1092	$\langle 14,14,23\rangle$	1092	$\langle 17,8,17\rangle$
1120	$\langle 1,0,280\rangle$	1120	$\langle 4,4,71\rangle$	1120	$\langle 5,0,56\rangle$	1120	$\langle 7,0,40\rangle$
1120	$\langle 8,0,35\rangle$	1120	$\langle 8,8,37\rangle$	1120	$\langle 17,6,17\rangle$	1120	$\langle 19,18,19\rangle$
1128	$\langle 11,\pm4,26\rangle$	1128	$\langle 13,\pm4,22\rangle$	1131	$\langle 5,\pm3,57\rangle$	1131	$\langle 15,\pm3,19\rangle$
1140	$\langle 7,\pm6,42\rangle$	1140	$\langle 11,\pm2,26\rangle$	1140	$\langle 13,\pm2,22\rangle$	1140	$\langle 14,\pm6,21\rangle$
1152	$\langle 11,\pm6,27\rangle$	1152	$\langle 16,\pm8,19\rangle$	1155	$\langle 1,1,289\rangle$	1155	$\langle 3,3,97\rangle$
1155	$\langle 5,5,59\rangle$	1155	$\langle 7,7,43\rangle$	1155	$\langle 11,11,29\rangle$	1155	$\langle 15,15,23\rangle$
1155	$\langle 17,1,17\rangle$	1155	$\langle 19,17,19\rangle$	1204	$\langle 5,\pm4,61\rangle$	1204	$\langle 10,\pm6,31\rangle$
1227	$\langle 11,\pm7,29\rangle$	1240	$\langle 11,\pm6,29\rangle$	1240	$\langle 17,\pm16,22\rangle$	1243	$\langle 17,\pm7,19\rangle$
1248	$\langle 1,0,312\rangle$	1248	$\langle 3,0,104\rangle$	1248	$\langle 4,4,79\rangle$	1248	$\langle 8,0,39\rangle$
1248	$\langle 8,8,41\rangle$	1248	$\langle 12,12,29\rangle$	1248	$\langle 13,0,24\rangle$	1248	$\langle 19,14,19\rangle$
1275	$\langle 11,\pm1,29\rangle$	1275	$\langle 13,\pm5,25\rangle$	1288	$\langle 13,\pm8,26\rangle$	1288	$\langle 17,\pm2,19\rangle$
1312	$\langle 7,\pm2,47\rangle$	1312	$\langle 13,\pm12,28\rangle$	1320	$\langle 1,0,330\rangle$	1320	$\langle 2,0,165\rangle$
1320	$\langle 3,0,110\rangle$	1320	$\langle 5,0,66\rangle$	1320	$\langle 6,0,55\rangle$	1320	$\langle 10,0,33\rangle$
1320	$\langle 11,0,30\rangle$	1320	$\langle 15,0,22\rangle$	1332	$\langle 9,\pm6,38\rangle$	1332	$\langle 18,\pm6,19\rangle$
1344	$\langle 5,\pm4,68\rangle$	1344	$\langle 11,\pm8,32\rangle$	1344	$\langle 15,\pm6,23\rangle$	1344	$\langle 17,\pm4,20\rangle$
1360	$\langle 8,\pm4,43\rangle$	1360	$\langle 11,\pm2,31\rangle$	1380	$\langle 1,0,345\rangle$	1380	$\langle 2,2,173\rangle$
1380	$\langle 3,0,115\rangle$	1380	$\langle 5,0,69\rangle$	1380	$\langle 6,6,59\rangle$	1380	$\langle 10,10,37\rangle$
1380	$\langle 15,0,23\rangle$	1380	$\langle 19,8,19\rangle$	1387	$\langle 13,\pm11,29\rangle$	1395	$\langle 13,\pm3,27\rangle$
1395	$\langle 17,\pm13,23\rangle$	1408	$\langle 13,\pm10,29\rangle$	1408	$\langle 16,\pm8,23\rangle$	1411	$\langle 5,\pm3,71\rangle$
1428	$\langle 1,0,357\rangle$	1428	$\langle 2,2,179\rangle$	1428	$\langle 3,0,119\rangle$	1428	$\langle 6,6,61\rangle$
1428	$\langle 7,0,51\rangle$	1428	$\langle 14,14,29\rangle$	1428	$\langle 17,0,21\rangle$	1428	$\langle 19,4,19\rangle$
1435	$\langle 1,1,359\rangle$	1435	$\langle 5,5,73\rangle$	1435	$\langle 7,7,53\rangle$	1435	$\langle 19,3,19\rangle$
1440	$\langle 7,\pm4,52\rangle$	1440	$\langle 9,\pm6,41\rangle$	1440	$\langle 11,\pm10,35\rangle$	1440	$\langle 13,\pm4,28\rangle$
1443	$\langle 11,\pm3,33\rangle$	1443	$\langle 17,\pm11,23\rangle$	1467	$\langle 9,\pm3,41\rangle$	1488	$\langle 8,\pm4,47\rangle$
1488	$\langle 17,\pm12,24\rangle$	1507	$\langle 13,\pm1,29\rangle$	1540	$\langle 1,0,385\rangle$	1540	$\langle 2,2,193\rangle$
1540	$\langle 5,0,77\rangle$	1540	$\langle 7,0,55\rangle$	1540	$\langle 10,10,41\rangle$	1540	$\langle 11,0,35\rangle$
1540	$\langle 14,14,31\rangle$	1540	$\langle 22,22,23\rangle$	1555	$\langle 17,\pm3,23\rangle$	1560	$\langle 7,\pm6,57\rangle$
1560	$\langle 14,\pm8,29\rangle$	1560	$\langle 17,\pm2,23\rangle$	1560	$\langle 19,\pm6,21\rangle$	1600	$\langle 13,\pm8,32\rangle$
1600	$\langle 17,\pm10,25\rangle$	1632	$\langle 1,0,408\rangle$	1632	$\langle 3,0,136\rangle$	1632	$\langle 4,4,103\rangle$
1632	$\langle 8,0,51\rangle$	1632	$\langle 8,8,53\rangle$	1632	$\langle 12,12,37\rangle$	1632	$\langle 17,0,24\rangle$
1632	$\langle 23,22,23\rangle$	1635	$\langle 11,\pm9,39\rangle$	1635	$\langle 13,\pm9,33\rangle$	1659	$\langle 5,\pm1,83\rangle$
1659	$\langle 15,\pm9,29\rangle$	1672	$\langle 7,\pm6,61\rangle$	1672	$\langle 14,\pm8,31\rangle$	1680	$\langle 8,\pm4,53\rangle$
1680	$\langle 11,\pm6,39\rangle$	1680	$\langle 13,\pm6,33\rangle$	1680	$\langle 19,\pm12,24\rangle$	1683	$\langle 7,\pm5,61\rangle$
1683	$\langle 9,\pm3,47\rangle$	1716	$\langle 5,\pm2,86\rangle$	1716	$\langle 10,\pm2,43\rangle$	1716	$\langle 15,\pm12,31\rangle$
1716	$\langle 17,\pm16,29\rangle$	1752	$\langle 13,\pm4,34\rangle$	1752	$\langle 17,\pm4,26\rangle$	1768	$\langle 11,\pm6,41\rangle$
1768	$\langle 22,\pm16,23\rangle$	1771	$\langle 5,\pm3,89\rangle$	1771	$\langle 13,\pm7,35\rangle$	1780	$\langle 13,\pm12,37\rangle$
1780	$\langle 19,\pm14,26\rangle$	1792	$\langle 11,\pm10,43\rangle$	1792	$\langle 16,\pm8,29\rangle$	1824	$\langle 5,\pm4,92\rangle$
1824	$\langle 13,\pm10,37\rangle$	1824	$\langle 15,\pm6,31\rangle$	1824	$\langle 20,\pm4,23\rangle$	1827	$\langle 17,\pm3,27\rangle$
1827	$\langle 19,\pm15,27\rangle$	1848	$\langle 1,0,462\rangle$	1848	$\langle 2,0,231\rangle$	1848	$\langle 3,0,154\rangle$
1848	$\langle 6,0,77\rangle$	1848	$\langle 7,0,66\rangle$	1848	$\langle 11,0,42\rangle$	1848	$\langle 14,0,33\rangle$
1848	$\langle 21,0,22\rangle$	1860	$\langle 7,\pm4,67\rangle$	1860	$\langle 13,\pm8,37\rangle$	1860	$\langle 14,\pm10,35\rangle$
1860	$\langle 21,\pm18,26\rangle$	1920	$\langle 11,\pm4,44\rangle$	1920	$\langle 13,\pm2,37\rangle$	1920	$\langle 16,\pm8,31\rangle$
1920	$\langle 17,\pm16,32\rangle$	1947	$\langle 13,\pm9,39\rangle$	1947	$\langle 17,\pm5,29\rangle$	1992	$\langle 13,\pm6,39\rangle$
1992	$\langle 23,\pm20,26\rangle$	1995	$\langle 1,1,499\rangle$	1995	$\langle 3,3,167\rangle$	1995	$\langle 5,5,101\rangle$
1995	$\langle 7,7,73\rangle$	1995	$\langle 15,15,37\rangle$	1995	$\langle 19,19,31\rangle$	1995	$\langle 21,21,29\rangle$
1995	$\langle 23,11,23\rangle$	2016	$\langle 5,\pm2,101\rangle$	2016	$\langle 13,\pm8,40\rangle$	2016	$\langle 19,\pm6,27\rangle$
2016	$\langle 20,\pm12,27\rangle$	2020	$\langle 11,\pm2,46\rangle$	2020	$\langle 22,\pm2,23\rangle$	2035	$\langle 7,\pm3,73\rangle$
2035	$\langle 19,\pm13,29\rangle$	2040	$\langle 7,\pm2,73\rangle$	2040	$\langle 13,\pm12,42\rangle$	2040	$\langle 14,\pm12,39\rangle$
2040	$\langle 21,\pm12,26\rangle$	2067	$\langle 11,\pm1,47\rangle$	2067	$\langle 19,\pm17,31\rangle$	2080	$\langle 1,0,520\rangle$
2080	$\langle 4,4,131\rangle$	2080	$\langle 5,0,104\rangle$	2080	$\langle 8,0,65\rangle$	2080	$\langle 8,8,67\rangle$

Table 1: Representatives for 2779 $SL_2(\mathbb{Z})$-equivalence Classes of Regular Forms

$\lvert\Delta\rvert$	$\langle A, B, C\rangle$	$\lvert\Delta\rvert$	$\langle A, B, C\rangle$	$\lvert\Delta\rvert$	$\langle A, B, C\rangle$	$\lvert\Delta\rvert$	$\langle A, B, C\rangle$
2080	$\langle 13, 0, 40\rangle$	2080	$\langle 20, 20, 31\rangle$	2080	$\langle 23, 6, 23\rangle$	2088	$\langle 9, \pm 6, 59\rangle$
2088	$\langle 18, \pm 12, 31\rangle$	2100	$\langle 11, \pm 10, 50\rangle$	2100	$\langle 17, \pm 12, 33\rangle$	2100	$\langle 19, \pm 16, 31\rangle$
2100	$\langle 22, \pm 10, 25\rangle$	2112	$\langle 7, \pm 4, 76\rangle$	2112	$\langle 17, \pm 8, 32\rangle$	2112	$\langle 19, \pm 4, 28\rangle$
2112	$\langle 21, \pm 18, 29\rangle$	2115	$\langle 9, \pm 3, 59\rangle$	2115	$\langle 13, \pm 11, 43\rangle$	2128	$\langle 8, \pm 4, 67\rangle$
2128	$\langle 13, \pm 2, 41\rangle$	2139	$\langle 5, \pm 1, 107\rangle$	2139	$\langle 15, \pm 9, 37\rangle$	2163	$\langle 11, \pm 9, 51\rangle$
2163	$\langle 17, \pm 9, 33\rangle$	2208	$\langle 7, \pm 2, 79\rangle$	2208	$\langle 11, \pm 6, 51\rangle$	2208	$\langle 17, \pm 6, 33\rangle$
2208	$\langle 21, \pm 12, 28\rangle$	2212	$\langle 17, \pm 10, 34\rangle$	2212	$\langle 19, \pm 12, 31\rangle$	2244	$\langle 5, \pm 4, 113\rangle$
2244	$\langle 10, \pm 6, 57\rangle$	2244	$\langle 15, \pm 6, 38\rangle$	2244	$\langle 19, \pm 6, 30\rangle$	2272	$\langle 11, \pm 4, 52\rangle$
2272	$\langle 13, \pm 4, 44\rangle$	2275	$\langle 19, \pm 9, 31\rangle$	2275	$\langle 23, \pm 5, 25\rangle$	2280	$\langle 7, \pm 4, 82\rangle$
2280	$\langle 14, \pm 4, 41\rangle$	2280	$\langle 17, \pm 10, 35\rangle$	2280	$\langle 21, \pm 18, 31\rangle$	2340	$\langle 11, \pm 6, 54\rangle$
2340	$\langle 19, \pm 4, 31\rangle$	2340	$\langle 22, \pm 6, 27\rangle$	2340	$\langle 23, \pm 12, 27\rangle$	2368	$\langle 19, \pm 8, 32\rangle$
2368	$\langle 23, \pm 22, 31\rangle$	2392	$\langle 7, \pm 4, 86\rangle$	2392	$\langle 14, \pm 4, 43\rangle$	2400	$\langle 7, \pm 6, 87\rangle$
2400	$\langle 11, \pm 8, 56\rangle$	2400	$\langle 21, \pm 6, 29\rangle$	2400	$\langle 25, \pm 20, 28\rangle$	2436	$\langle 5, \pm 2, 122\rangle$
2436	$\langle 10, \pm 2, 61\rangle$	2436	$\langle 15, \pm 12, 43\rangle$	2436	$\langle 23, \pm 18, 30\rangle$	2451	$\langle 5, \pm 3, 123\rangle$
2451	$\langle 15, \pm 3, 41\rangle$	2464	$\langle 5, \pm 4, 124\rangle$	2464	$\langle 17, \pm 16, 40\rangle$	2464	$\langle 19, \pm 14, 35\rangle$
2464	$\langle 20, \pm 4, 31\rangle$	2475	$\langle 23, \pm 3, 27\rangle$	2475	$\langle 25, \pm 15, 27\rangle$	2496	$\langle 5, \pm 2, 125\rangle$
2496	$\langle 11, \pm 10, 59\rangle$	2496	$\langle 15, \pm 12, 44\rangle$	2496	$\langle 20, \pm 12, 33\rangle$	2520	$\langle 9, \pm 6, 71\rangle$
2520	$\langle 17, \pm 8, 38\rangle$	2520	$\langle 18, \pm 12, 37\rangle$	2520	$\langle 19, \pm 8, 34\rangle$	2580	$\langle 11, \pm 4, 59\rangle$
2580	$\langle 17, \pm 2, 38\rangle$	2580	$\langle 19, \pm 2, 34\rangle$	2580	$\langle 22, \pm 18, 33\rangle$	2632	$\langle 19, \pm 16, 38\rangle$
2632	$\langle 23, \pm 6, 29\rangle$	2640	$\langle 8, \pm 4, 83\rangle$	2640	$\langle 13, \pm 8, 52\rangle$	2640	$\langle 19, \pm 18, 39\rangle$
2640	$\langle 24, \pm 12, 29\rangle$	2667	$\langle 17, \pm 11, 41\rangle$	2667	$\langle 23, \pm 1, 29\rangle$	2688	$\langle 13, \pm 4, 52\rangle$
2688	$\langle 16, \pm 8, 43\rangle$	2688	$\langle 17, \pm 10, 41\rangle$	2688	$\langle 23, \pm 16, 32\rangle$	2715	$\langle 7, \pm 1, 97\rangle$
2715	$\langle 21, \pm 15, 35\rangle$	2755	$\langle 13, \pm 1, 53\rangle$	2755	$\langle 17, \pm 13, 43\rangle$	2760	$\langle 11, \pm 10, 65\rangle$
2760	$\langle 13, \pm 10, 55\rangle$	2760	$\langle 22, \pm 12, 33\rangle$	2760	$\langle 26, \pm 16, 29\rangle$	2772	$\langle 13, \pm 6, 54\rangle$
2772	$\langle 17, \pm 4, 41\rangle$	2772	$\langle 26, \pm 6, 27\rangle$	2772	$\langle 27, \pm 24, 31\rangle$	2788	$\langle 19, \pm 10, 38\rangle$
2788	$\langle 23, \pm 8, 31\rangle$	2832	$\langle 8, \pm 4, 89\rangle$	2832	$\langle 24, \pm 12, 31\rangle$	2880	$\langle 7, \pm 2, 103\rangle$
2880	$\langle 23, \pm 8, 32\rangle$	2880	$\langle 27, \pm 24, 32\rangle$	2880	$\langle 27, \pm 12, 28\rangle$	2907	$\langle 27, \pm 21, 31\rangle$
2907	$\langle 27, \pm 15, 29\rangle$	2968	$\langle 13, \pm 10, 59\rangle$	2968	$\langle 26, \pm 16, 31\rangle$	3003	$\langle 1, 1, 751\rangle$
3003	$\langle 3, 3, 251\rangle$	3003	$\langle 7, 7, 109\rangle$	3003	$\langle 11, 11, 71\rangle$	3003	$\langle 13, 13, 61\rangle$
3003	$\langle 21, 21, 41\rangle$	3003	$\langle 29, 19, 29\rangle$	3003	$\langle 31, 29, 31\rangle$	3040	$\langle 1, 0, 760\rangle$
3040	$\langle 4, 4, 191\rangle$	3040	$\langle 5, 0, 152\rangle$	3040	$\langle 8, 0, 95\rangle$	3040	$\langle 8, 8, 97\rangle$
3040	$\langle 19, 0, 40\rangle$	3040	$\langle 20, 20, 43\rangle$	3040	$\langle 29, 18, 29\rangle$	3060	$\langle 9, \pm 6, 86\rangle$
3060	$\langle 11, \pm 8, 71\rangle$	3060	$\langle 18, \pm 6, 43\rangle$	3060	$\langle 22, \pm 14, 37\rangle$	3108	$\langle 11, \pm 4, 71\rangle$
3108	$\langle 13, \pm 8, 61\rangle$	3108	$\langle 22, \pm 18, 39\rangle$	3108	$\langle 26, \pm 18, 33\rangle$	3168	$\langle 9, \pm 6, 89\rangle$
3168	$\langle 13, \pm 2, 61\rangle$	3168	$\langle 19, \pm 10, 43\rangle$	3168	$\langle 23, \pm 12, 36\rangle$	3172	$\langle 19, \pm 18, 46\rangle$
3172	$\langle 23, \pm 18, 38\rangle$	3192	$\langle 11, \pm 8, 74\rangle$	3192	$\langle 17, \pm 2, 47\rangle$	3192	$\langle 22, \pm 8, 37\rangle$
3192	$\langle 31, \pm 30, 33\rangle$	3220	$\langle 11, \pm 6, 74\rangle$	3220	$\langle 13, \pm 2, 62\rangle$	3220	$\langle 22, \pm 6, 37\rangle$
3220	$\langle 26, \pm 2, 31\rangle$	3243	$\langle 17, \pm 15, 51\rangle$	3243	$\langle 19, \pm 5, 43\rangle$	3315	$\langle 1, 1, 829\rangle$
3315	$\langle 3, 3, 277\rangle$	3315	$\langle 5, 5, 167\rangle$	3315	$\langle 13, 13, 67\rangle$	3315	$\langle 15, 15, 59\rangle$
3315	$\langle 17, 17, 53\rangle$	3315	$\langle 29, 7, 29\rangle$	3315	$\langle 31, 23, 31\rangle$	3355	$\langle 13, \pm 5, 65\rangle$
3355	$\langle 23, \pm 7, 37\rangle$	3360	$\langle 1, 0, 840\rangle$	3360	$\langle 3, 0, 280\rangle$	3360	$\langle 4, 4, 211\rangle$
3360	$\langle 5, 0, 168\rangle$	3360	$\langle 7, 0, 120\rangle$	3360	$\langle 8, 0, 105\rangle$	3360	$\langle 8, 8, 107\rangle$
3360	$\langle 12, 12, 73\rangle$	3360	$\langle 15, 0, 56\rangle$	3360	$\langle 20, 20, 47\rangle$	3360	$\langle 21, 0, 40\rangle$
3360	$\langle 24, 0, 35\rangle$	3360	$\langle 24, 24, 41\rangle$	3360	$\langle 28, 28, 37\rangle$	3360	$\langle 29, 2, 29\rangle$
3360	$\langle 31, 22, 31\rangle$	3432	$\langle 17, \pm 6, 51\rangle$	3432	$\langle 19, \pm 8, 46\rangle$	3432	$\langle 23, \pm 8, 38\rangle$
3432	$\langle 31, \pm 28, 34\rangle$	3480	$\langle 13, \pm 2, 67\rangle$	3480	$\langle 19, \pm 4, 46\rangle$	3480	$\langle 23, \pm 4, 38\rangle$
3480	$\langle 26, \pm 24, 39\rangle$	3507	$\langle 13, \pm 9, 69\rangle$	3507	$\langle 23, \pm 9, 39\rangle$	3520	$\langle 7, \pm 6, 127\rangle$
3520	$\langle 13, \pm 4, 68\rangle$	3520	$\langle 17, \pm 4, 52\rangle$	3520	$\langle 28, \pm 20, 35\rangle$	3588	$\langle 11, \pm 8, 83\rangle$
3588	$\langle 17, \pm 4, 53\rangle$	3588	$\langle 22, \pm 14, 43\rangle$	3588	$\langle 33, \pm 30, 34\rangle$	3627	$\langle 9, \pm 3, 101\rangle$
3627	$\langle 11, \pm 5, 83\rangle$	3640	$\langle 11, \pm 10, 85\rangle$	3640	$\langle 17, \pm 10, 55\rangle$	3640	$\langle 22, \pm 12, 43\rangle$
3640	$\langle 31, \pm 24, 34\rangle$	3648	$\langle 11, \pm 2, 83\rangle$	3648	$\langle 23, \pm 20, 44\rangle$	3648	$\langle 29, \pm 8, 32\rangle$
3648	$\langle 32, \pm 24, 33\rangle$	3712	$\langle 16, \pm 8, 59\rangle$	3712	$\langle 31, \pm 16, 32\rangle$	3795	$\langle 13, \pm 1, 73\rangle$
3795	$\langle 17, \pm 9, 57\rangle$	3795	$\langle 19, \pm 9, 51\rangle$	3795	$\langle 29, \pm 27, 39\rangle$	3808	$\langle 11, \pm 8, 88\rangle$
3808	$\langle 13, \pm 12, 76\rangle$	3808	$\langle 19, \pm 12, 52\rangle$	3808	$\langle 29, \pm 22, 37\rangle$	3828	$\langle 7, \pm 6, 138\rangle$
3828	$\langle 14, \pm 6, 69\rangle$	3828	$\langle 21, \pm 6, 46\rangle$	3828	$\langle 23, \pm 6, 42\rangle$	3840	$\langle 16, \pm 8, 61\rangle$
3840	$\langle 17, \pm 6, 57\rangle$	3840	$\langle 19, \pm 6, 51\rangle$	3840	$\langle 23, \pm 22, 47\rangle$	3843	$\langle 9, \pm 3, 107\rangle$
3843	$\langle 17, \pm 13, 59\rangle$	4020	$\langle 13, \pm 6, 78\rangle$	4020	$\langle 17, \pm 14, 62\rangle$	4020	$\langle 26, \pm 6, 39\rangle$
4020	$\langle 31, \pm 14, 34\rangle$	4032	$\langle 9, \pm 6, 113\rangle$	4032	$\langle 11, \pm 4, 92\rangle$	4032	$\langle 23, \pm 4, 44\rangle$
4032	$\langle 29, \pm 12, 36\rangle$	4048	$\langle 8, \pm 4, 127\rangle$	4048	$\langle 17, \pm 10, 61\rangle$	4123	$\langle 17, \pm 5, 61\rangle$
4123	$\langle 29, \pm 13, 37\rangle$	4128	$\langle 7, \pm 4, 148\rangle$	4128	$\langle 21, \pm 18, 53\rangle$	4128	$\langle 23, \pm 14, 47\rangle$
4128	$\langle 28, \pm 4, 37\rangle$	4180	$\langle 17, \pm 6, 62\rangle$	4180	$\langle 23, \pm 12, 47\rangle$	4180	$\langle 29, \pm 24, 41\rangle$
4180	$\langle 31, \pm 6, 34\rangle$	4260	$\langle 13, \pm 2, 82\rangle$	4260	$\langle 23, \pm 8, 47\rangle$	4260	$\langle 26, \pm 2, 41\rangle$
4260	$\langle 31, \pm 24, 39\rangle$	4323	$\langle 19, \pm 3, 57\rangle$	4323	$\langle 23, \pm 1, 47\rangle$	4368	$\langle 8, \pm 4, 137\rangle$
4368	$\langle 17, \pm 16, 68\rangle$	4368	$\langle 23, \pm 18, 51\rangle$	4368	$\langle 24, \pm 12, 47\rangle$	4420	$\langle 7, \pm 2, 158\rangle$

Table 1: Representatives for 2779 $SL_2(\mathbb{Z})$-equivalence Classes of Regular Forms

$\lvert\Delta\rvert$	$\langle A,B,C\rangle$	$\lvert\Delta\rvert$	$\langle A,B,C\rangle$	$\lvert\Delta\rvert$	$\langle A,B,C\rangle$	$\lvert\Delta\rvert$	$\langle A,B,C\rangle$
4420	$\langle 14,\pm2,79\rangle$	4420	$\langle 19,\pm8,59\rangle$	4420	$\langle 35,\pm30,38\rangle$	4440	$\langle 11,\pm2,101\rangle$
4440	$\langle 19,\pm14,61\rangle$	4440	$\langle 22,\pm20,55\rangle$	4440	$\langle 33,\pm24,38\rangle$	4452	$\langle 11,\pm6,102\rangle$
4452	$\langle 17,\pm6,66\rangle$	4452	$\langle 22,\pm6,51\rangle$	4452	$\langle 33,\pm6,34\rangle$	4480	$\langle 16,\pm8,71\rangle$
4480	$\langle 17,\pm12,68\rangle$	4480	$\langle 19,\pm2,59\rangle$	4480	$\langle 32,\pm16,37\rangle$	4488	$\langle 13,\pm6,87\rangle$
4488	$\langle 26,\pm20,47\rangle$	4488	$\langle 29,\pm6,39\rangle$	4488	$\langle 31,\pm10,37\rangle$	4512	$\langle 11,\pm8,104\rangle$
4512	$\langle 13,\pm8,88\rangle$	4512	$\langle 31,\pm18,39\rangle$	4512	$\langle 33,\pm30,41\rangle$	4515	$\langle 13,\pm3,87\rangle$
4515	$\langle 19,\pm11,61\rangle$	4515	$\langle 23,\pm19,53\rangle$	4515	$\langle 29,\pm3,39\rangle$	4680	$\langle 9,\pm6,131\rangle$
4680	$\langle 18,\pm12,67\rangle$	4680	$\langle 23,\pm14,53\rangle$	4680	$\langle 31,\pm30,45\rangle$	4740	$\langle 11,\pm10,110\rangle$
4740	$\langle 22,\pm10,55\rangle$	4740	$\langle 29,\pm4,41\rangle$	4740	$\langle 33,\pm12,37\rangle$	4788	$\langle 9,\pm6,134\rangle$
4788	$\langle 13,\pm10,94\rangle$	4788	$\langle 18,\pm6,67\rangle$	4788	$\langle 26,\pm10,47\rangle$	4960	$\langle 11,\pm10,115\rangle$
4960	$\langle 17,\pm2,73\rangle$	4960	$\langle 23,\pm10,55\rangle$	4960	$\langle 29,\pm12,44\rangle$	4992	$\langle 16,\pm8,79\rangle$
4992	$\langle 19,\pm10,67\rangle$	4992	$\langle 29,\pm24,48\rangle$	4992	$\langle 32,\pm16,41\rangle$	5083	$\langle 19,\pm3,67\rangle$
5083	$\langle 31,\pm1,41\rangle$	5115	$\langle 7,\pm3,183\rangle$	5115	$\langle 17,\pm11,77\rangle$	5115	$\langle 21,\pm3,61\rangle$
5115	$\langle 35,\pm25,41\rangle$	5152	$\langle 13,\pm10,101\rangle$	5152	$\langle 17,\pm4,76\rangle$	5152	$\langle 19,\pm4,68\rangle$
5152	$\langle 31,\pm26,47\rangle$	5160	$\langle 13,\pm12,102\rangle$	5160	$\langle 17,\pm12,78\rangle$	5160	$\langle 26,\pm12,51\rangle$
5160	$\langle 34,\pm12,39\rangle$	5187	$\langle 11,\pm7,119\rangle$	5187	$\langle 17,\pm7,77\rangle$	5187	$\langle 29,\pm27,51\rangle$
5187	$\langle 33,\pm15,41\rangle$	5208	$\langle 19,\pm6,69\rangle$	5208	$\langle 23,\pm6,57\rangle$	5208	$\langle 37,\pm34,43\rangle$
5208	$\langle 38,\pm32,41\rangle$	5280	$\langle 1,0,1320\rangle$	5280	$\langle 3,0,440\rangle$	5280	$\langle 4,4,331\rangle$
5280	$\langle 5,0,264\rangle$	5280	$\langle 8,0,165\rangle$	5280	$\langle 8,8,167\rangle$	5280	$\langle 11,0,120\rangle$
5280	$\langle 12,12,113\rangle$	5280	$\langle 15,0,88\rangle$	5280	$\langle 20,20,71\rangle$	5280	$\langle 24,0,55\rangle$
5280	$\langle 24,24,61\rangle$	5280	$\langle 33,0,40\rangle$	5280	$\langle 37,14,37\rangle$	5280	$\langle 40,40,43\rangle$
5280	$\langle 41,38,41\rangle$	5355	$\langle 9,\pm3,149\rangle$	5355	$\langle 13,\pm1,103\rangle$	5355	$\langle 23,\pm21,63\rangle$
5355	$\langle 31,\pm15,45\rangle$	5412	$\langle 13,\pm10,106\rangle$	5412	$\langle 23,\pm4,59\rangle$	5412	$\langle 26,\pm10,53\rangle$
5412	$\langle 39,\pm36,43\rangle$	5440	$\langle 11,\pm4,124\rangle$	5440	$\langle 31,\pm4,44\rangle$	5440	$\langle 32,\pm24,47\rangle$
5440	$\langle 32,\pm8,43\rangle$	5460	$\langle 1,0,1365\rangle$	5460	$\langle 2,2,683\rangle$	5460	$\langle 3,0,455\rangle$
5460	$\langle 5,0,273\rangle$	5460	$\langle 6,6,229\rangle$	5460	$\langle 7,0,195\rangle$	5460	$\langle 10,10,139\rangle$
5460	$\langle 13,0,105\rangle$	5460	$\langle 14,14,101\rangle$	5460	$\langle 15,0,91\rangle$	5460	$\langle 21,0,65\rangle$
5460	$\langle 26,26,59\rangle$	5460	$\langle 30,30,53\rangle$	5460	$\langle 35,0,39\rangle$	5460	$\langle 37,4,37\rangle$
5460	$\langle 42,42,43\rangle$	5467	$\langle 19,\pm9,73\rangle$	5467	$\langle 31,\pm19,47\rangle$	5520	$\langle 8,\pm4,173\rangle$
5520	$\langle 19,\pm16,76\rangle$	5520	$\langle 24,\pm12,59\rangle$	5520	$\langle 37,\pm20,40\rangle$	5712	$\langle 8,\pm4,179\rangle$
5712	$\langle 19,\pm8,76\rangle$	5712	$\langle 24,\pm12,61\rangle$	5712	$\langle 29,\pm28,56\rangle$	5952	$\langle 17,\pm10,89\rangle$
5952	$\langle 29,\pm14,53\rangle$	5952	$\langle 32,\pm24,51\rangle$	5952	$\langle 32,\pm8,47\rangle$	6160	$\langle 8,\pm4,193\rangle$
6160	$\langle 23,\pm2,67\rangle$	6160	$\langle 31,\pm28,56\rangle$	6160	$\langle 40,\pm20,41\rangle$	6195	$\langle 11,\pm3,141\rangle$
6195	$\langle 31,\pm25,55\rangle$	6195	$\langle 33,\pm3,47\rangle$	6195	$\langle 37,\pm13,43\rangle$	6240	$\langle 7,\pm2,223\rangle$
6240	$\langle 17,\pm4,92\rangle$	6240	$\langle 19,\pm12,84\rangle$	6240	$\langle 21,\pm12,76\rangle$	6240	$\langle 23,\pm4,68\rangle$
6240	$\langle 28,\pm12,57\rangle$	6240	$\langle 29,\pm16,56\rangle$	6240	$\langle 35,\pm30,51\rangle$	6307	$\langle 19,\pm1,83\rangle$
6307	$\langle 23,\pm15,71\rangle$	6420	$\langle 11,\pm2,146\rangle$	6420	$\langle 22,\pm2,73\rangle$	6420	$\langle 31,\pm20,55\rangle$
6420	$\langle 33,\pm24,53\rangle$	6435	$\langle 9,\pm3,179\rangle$	6435	$\langle 17,\pm5,95\rangle$	6435	$\langle 19,\pm5,85\rangle$
6435	$\langle 37,\pm15,45\rangle$	6528	$\langle 16,\pm8,103\rangle$	6528	$\langle 23,\pm2,71\rangle$	6528	$\langle 32,\pm16,53\rangle$
6528	$\langle 37,\pm24,48\rangle$	6580	$\langle 11,\pm8,151\rangle$	6580	$\langle 17,\pm4,97\rangle$	6580	$\langle 22,\pm14,77\rangle$
6580	$\langle 34,\pm30,55\rangle$	6612	$\langle 17,\pm16,101\rangle$	6612	$\langle 23,\pm14,74\rangle$	6612	$\langle 34,\pm18,51\rangle$
6612	$\langle 37,\pm14,46\rangle$	6688	$\langle 7,\pm2,239\rangle$	6688	$\langle 28,\pm12,61\rangle$	6688	$\langle 31,\pm16,56\rangle$
6688	$\langle 37,\pm34,53\rangle$	6708	$\langle 23,\pm10,74\rangle$	6708	$\langle 29,\pm22,62\rangle$	6708	$\langle 31,\pm22,58\rangle$
6708	$\langle 37,\pm10,46\rangle$	6720	$\langle 11,\pm10,155\rangle$	6720	$\langle 13,\pm12,132\rangle$	6720	$\langle 19,\pm14,91\rangle$
6720	$\langle 31,\pm10,55\rangle$	6720	$\langle 32,\pm24,57\rangle$	6720	$\langle 32,\pm8,53\rangle$	6720	$\langle 33,\pm12,52\rangle$
6720	$\langle 39,\pm12,44\rangle$	6820	$\langle 19,\pm18,94\rangle$	6820	$\langle 29,\pm16,61\rangle$	6820	$\langle 37,\pm32,53\rangle$
6820	$\langle 38,\pm18,47\rangle$	6840	$\langle 9,\pm6,191\rangle$	6840	$\langle 18,\pm12,97\rangle$	6840	$\langle 29,\pm2,59\rangle$
6840	$\langle 43,\pm30,45\rangle$	7008	$\langle 13,\pm8,136\rangle$	7008	$\langle 17,\pm8,104\rangle$	7008	$\langle 39,\pm18,47\rangle$
7008	$\langle 43,\pm42,51\rangle$	7035	$\langle 11,\pm7,161\rangle$	7035	$\langle 23,\pm7,77\rangle$	7035	$\langle 31,\pm23,61\rangle$
7035	$\langle 33,\pm15,55\rangle$	7072	$\langle 11,\pm10,163\rangle$	7072	$\langle 23,\pm14,79\rangle$	7072	$\langle 29,\pm2,61\rangle$
7072	$\langle 41,\pm12,44\rangle$	7140	$\langle 13,\pm6,138\rangle$	7140	$\langle 19,\pm2,94\rangle$	7140	$\langle 23,\pm6,78\rangle$
7140	$\langle 26,\pm6,69\rangle$	7140	$\langle 29,\pm20,65\rangle$	7140	$\langle 37,\pm36,57\rangle$	7140	$\langle 38,\pm2,47\rangle$
7140	$\langle 39,\pm6,46\rangle$	7315	$\langle 13,\pm11,143\rangle$	7315	$\langle 29,\pm15,65\rangle$	7315	$\langle 31,\pm1,59\rangle$
7315	$\langle 37,\pm23,53\rangle$	7392	$\langle 1,0,1848\rangle$	7392	$\langle 3,0,616\rangle$	7392	$\langle 4,4,463\rangle$
7392	$\langle 7,0,264\rangle$	7392	$\langle 8,0,231\rangle$	7392	$\langle 8,8,233\rangle$	7392	$\langle 11,0,168\rangle$
7392	$\langle 12,12,157\rangle$	7392	$\langle 21,0,88\rangle$	7392	$\langle 24,0,77\rangle$	7392	$\langle 24,24,83\rangle$
7392	$\langle 28,28,73\rangle$	7392	$\langle 33,0,56\rangle$	7392	$\langle 43,2,43\rangle$	7392	$\langle 44,44,53\rangle$
7392	$\langle 47,38,47\rangle$	7395	$\langle 7,\pm5,265\rangle$	7395	$\langle 21,\pm9,89\rangle$	7395	$\langle 31,\pm13,61\rangle$
7395	$\langle 35,\pm5,53\rangle$	7480	$\langle 19,\pm14,101\rangle$	7480	$\langle 23,\pm8,82\rangle$	7480	$\langle 38,\pm24,53\rangle$
7480	$\langle 41,\pm8,46\rangle$	7540	$\langle 17,\pm12,113\rangle$	7540	$\langle 23,\pm2,82\rangle$	7540	$\langle 34,\pm22,59\rangle$
7540	$\langle 41,\pm2,46\rangle$	7755	$\langle 7,\pm1,277\rangle$	7755	$\langle 19,\pm15,105\rangle$	7755	$\langle 21,\pm15,95\rangle$
7755	$\langle 35,\pm15,57\rangle$	7968	$\langle 13,\pm12,156\rangle$	7968	$\langle 23,\pm6,87\rangle$	7968	$\langle 29,\pm6,69\rangle$
7968	$\langle 39,\pm12,52\rangle$	7995	$\langle 19,\pm17,109\rangle$	7995	$\langle 23,\pm3,87\rangle$	7995	$\langle 29,\pm3,69\rangle$
7995	$\langle 37,\pm21,57\rangle$	8008	$\langle 17,\pm4,118\rangle$	8008	$\langle 29,\pm24,74\rangle$	8008	$\langle 34,\pm4,59\rangle$

Table 1: Representatives for 2779 $\mathrm{SL}_2(\mathbb{Z})$-equivalence Classes of Regular Forms

$\lvert\Delta\rvert$	$\langle A, B, C\rangle$	$\lvert\Delta\rvert$	$\langle A, B, C\rangle$	$\lvert\Delta\rvert$	$\langle A, B, C\rangle$	$\lvert\Delta\rvert$	$\langle A, B, C\rangle$
8008	$\langle 37, \pm 24, 58\rangle$	8052	$\langle 19, \pm 2, 106\rangle$	8052	$\langle 31, \pm 16, 67\rangle$	8052	$\langle 38, \pm 2, 53\rangle$
8052	$\langle 41, \pm 36, 57\rangle$	8160	$\langle 7, \pm 4, 292\rangle$	8160	$\langle 13, \pm 2, 157\rangle$	8160	$\langle 21, \pm 18, 101\rangle$
8160	$\langle 28, \pm 4, 73\rangle$	8160	$\langle 35, \pm 10, 59\rangle$	8160	$\langle 39, \pm 24, 56\rangle$	8160	$\langle 41, \pm 32, 56\rangle$
8160	$\langle 43, \pm 28, 52\rangle$	8320	$\langle 16, \pm 8, 131\rangle$	8320	$\langle 23, \pm 12, 92\rangle$	8320	$\langle 31, \pm 22, 71\rangle$
8320	$\langle 32, \pm 16, 67\rangle$	8352	$\langle 9, \pm 6, 233\rangle$	8352	$\langle 31, \pm 24, 72\rangle$	8352	$\langle 36, \pm 12, 59\rangle$
8352	$\langle 37, \pm 26, 61\rangle$	8512	$\langle 13, \pm 4, 164\rangle$	8512	$\langle 32, \pm 24, 71\rangle$	8512	$\langle 32, \pm 8, 67\rangle$
8512	$\langle 41, \pm 4, 52\rangle$	8547	$\langle 17, \pm 15, 129\rangle$	8547	$\langle 23, \pm 3, 93\rangle$	8547	$\langle 31, \pm 3, 69\rangle$
8547	$\langle 43, \pm 15, 51\rangle$	8580	$\langle 7, \pm 4, 307\rangle$	8580	$\langle 14, \pm 10, 155\rangle$	8580	$\langle 21, \pm 18, 106\rangle$
8580	$\langle 29, \pm 2, 74\rangle$	8580	$\langle 31, \pm 10, 70\rangle$	8580	$\langle 35, \pm 10, 62\rangle$	8580	$\langle 37, \pm 2, 58\rangle$
8580	$\langle 42, \pm 18, 53\rangle$	8680	$\langle 13, \pm 2, 167\rangle$	8680	$\langle 26, \pm 24, 89\rangle$	8680	$\langle 29, \pm 22, 79\rangle$
8680	$\langle 43, \pm 36, 58\rangle$	8715	$\langle 19, \pm 5, 115\rangle$	8715	$\langle 23, \pm 5, 95\rangle$	8715	$\langle 41, \pm 31, 59\rangle$
8715	$\langle 43, \pm 33, 57\rangle$	8835	$\langle 11, \pm 3, 201\rangle$	8835	$\langle 33, \pm 3, 67\rangle$	8835	$\langle 41, \pm 29, 59\rangle$
8835	$\langle 43, \pm 25, 55\rangle$	8932	$\langle 13, \pm 8, 173\rangle$	8932	$\langle 19, \pm 6, 118\rangle$	8932	$\langle 26, \pm 18, 89\rangle$
8932	$\langle 38, \pm 6, 59\rangle$	9108	$\langle 9, \pm 6, 254\rangle$	9108	$\langle 17, \pm 2, 134\rangle$	9108	$\langle 18, \pm 6, 127\rangle$
9108	$\langle 34, \pm 2, 67\rangle$	9120	$\langle 7, \pm 6, 327\rangle$	9120	$\langle 17, \pm 14, 137\rangle$	9120	$\langle 21, \pm 6, 109\rangle$
9120	$\langle 28, \pm 20, 85\rangle$	9120	$\langle 31, \pm 26, 79\rangle$	9120	$\langle 35, \pm 20, 68\rangle$	9120	$\langle 41, \pm 8, 56\rangle$
9120	$\langle 51, \pm 48, 56\rangle$	9240	$\langle 13, \pm 4, 178\rangle$	9240	$\langle 17, \pm 12, 138\rangle$	9240	$\langle 23, \pm 12, 102\rangle$
9240	$\langle 26, \pm 4, 89\rangle$	9240	$\langle 34, \pm 12, 69\rangle$	9240	$\langle 37, \pm 26, 67\rangle$	9240	$\langle 39, \pm 30, 65\rangle$
9240	$\langle 46, \pm 12, 51\rangle$	9568	$\langle 7, \pm 6, 343\rangle$	9568	$\langle 28, \pm 20, 89\rangle$	9568	$\langle 43, \pm 8, 56\rangle$
9568	$\langle 53, \pm 48, 56\rangle$	9867	$\langle 29, \pm 15, 87\rangle$	9867	$\langle 37, \pm 7, 67\rangle$	9867	$\langle 43, \pm 25, 61\rangle$
9867	$\langle 47, \pm 35, 59\rangle$	10080	$\langle 9, \pm 6, 281\rangle$	10080	$\langle 17, \pm 16, 152\rangle$	10080	$\langle 19, \pm 16, 136\rangle$
10080	$\langle 36, \pm 12, 71\rangle$	10080	$\langle 37, \pm 24, 72\rangle$	10080	$\langle 43, \pm 38, 67\rangle$	10080	$\langle 45, \pm 30, 61\rangle$
10080	$\langle 47, \pm 42, 63\rangle$	10528	$\langle 19, \pm 6, 139\rangle$	10528	$\langle 23, \pm 12, 116\rangle$	10528	$\langle 29, \pm 12, 92\rangle$
10528	$\langle 41, \pm 38, 73\rangle$	10560	$\langle 13, \pm 10, 205\rangle$	10560	$\langle 19, \pm 2, 139\rangle$	10560	$\langle 29, \pm 24, 96\rangle$
10560	$\langle 32, \pm 24, 87\rangle$	10560	$\langle 32, \pm 8, 83\rangle$	10560	$\langle 39, \pm 36, 76\rangle$	10560	$\langle 41, \pm 10, 65\rangle$
10560	$\langle 52, \pm 36, 57\rangle$	10920	$\langle 11, \pm 6, 249\rangle$	10920	$\langle 19, \pm 10, 145\rangle$	10920	$\langle 22, \pm 16, 127\rangle$
10920	$\langle 29, \pm 10, 95\rangle$	10920	$\langle 33, \pm 6, 83\rangle$	10920	$\langle 38, \pm 28, 77\rangle$	10920	$\langle 55, \pm 50, 61\rangle$
10920	$\langle 57, \pm 48, 58\rangle$	10948	$\langle 37, \pm 2, 74\rangle$	10948	$\langle 41, \pm 32, 73\rangle$	10948	$\langle 43, \pm 24, 67\rangle$
10948	$\langle 47, \pm 12, 59\rangle$	11040	$\langle 11, \pm 2, 251\rangle$	11040	$\langle 13, \pm 6, 213\rangle$	11040	$\langle 29, \pm 26, 101\rangle$
11040	$\langle 33, \pm 24, 88\rangle$	11040	$\langle 39, \pm 6, 71\rangle$	11040	$\langle 43, \pm 22, 67\rangle$	11040	$\langle 44, \pm 20, 65\rangle$
11040	$\langle 52, \pm 20, 55\rangle$	11067	$\langle 13, \pm 3, 213\rangle$	11067	$\langle 37, \pm 25, 79\rangle$	11067	$\langle 39, \pm 3, 71\rangle$
11067	$\langle 47, \pm 5, 59\rangle$	11328	$\langle 31, \pm 24, 96\rangle$	11328	$\langle 32, \pm 24, 93\rangle$	11328	$\langle 32, \pm 8, 89\rangle$
11328	$\langle 43, \pm 14, 67\rangle$	11715	$\langle 17, \pm 7, 173\rangle$	11715	$\langle 29, \pm 1, 101\rangle$	11715	$\langle 43, \pm 29, 73\rangle$
11715	$\langle 51, \pm 27, 61\rangle$	11872	$\langle 13, \pm 6, 229\rangle$	11872	$\langle 31, \pm 30, 103\rangle$	11872	$\langle 41, \pm 10, 73\rangle$
11872	$\langle 52, \pm 20, 59\rangle$	12160	$\langle 16, \pm 8, 191\rangle$	12160	$\langle 29, \pm 22, 109\rangle$	12160	$\langle 32, \pm 16, 97\rangle$
12160	$\langle 43, \pm 40, 80\rangle$	12180	$\langle 13, \pm 12, 237\rangle$	12180	$\langle 17, \pm 14, 182\rangle$	12180	$\langle 26, \pm 14, 119\rangle$
12180	$\langle 34, \pm 14, 91\rangle$	12180	$\langle 37, \pm 20, 85\rangle$	12180	$\langle 39, \pm 12, 79\rangle$	12180	$\langle 51, \pm 48, 71\rangle$
12180	$\langle 53, \pm 40, 65\rangle$	12768	$\langle 11, \pm 6, 291\rangle$	12768	$\langle 17, \pm 4, 188\rangle$	12768	$\langle 31, \pm 2, 103\rangle$
12768	$\langle 33, \pm 6, 97\rangle$	12768	$\langle 37, \pm 16, 88\rangle$	12768	$\langle 44, \pm 28, 77\rangle$	12768	$\langle 47, \pm 4, 68\rangle$
12768	$\langle 51, \pm 30, 67\rangle$	12915	$\langle 9, \pm 3, 359\rangle$	12915	$\langle 19, \pm 9, 171\rangle$	12915	$\langle 45, \pm 15, 73\rangle$
12915	$\langle 53, \pm 21, 63\rangle$	13195	$\langle 11, \pm 7, 301\rangle$	13195	$\langle 43, \pm 7, 77\rangle$	13195	$\langle 47, \pm 23, 73\rangle$
13195	$\langle 55, \pm 15, 61\rangle$	13440	$\langle 16, \pm 8, 211\rangle$	13440	$\langle 29, \pm 4, 116\rangle$	13440	$\langle 31, \pm 18, 111\rangle$
13440	$\langle 32, \pm 16, 107\rangle$	13440	$\langle 37, \pm 18, 93\rangle$	13440	$\langle 41, \pm 34, 89\rangle$	13440	$\langle 47, \pm 40, 80\rangle$
13440	$\langle 48, \pm 24, 73\rangle$	13728	$\langle 17, \pm 12, 204\rangle$	13728	$\langle 19, \pm 16, 184\rangle$	13728	$\langle 23, \pm 16, 152\rangle$
13728	$\langle 31, \pm 6, 111\rangle$	13728	$\langle 37, \pm 6, 93\rangle$	13728	$\langle 51, \pm 12, 68\rangle$	13728	$\langle 53, \pm 30, 69\rangle$
13728	$\langle 57, \pm 54, 73\rangle$	13860	$\langle 9, \pm 6, 386\rangle$	13860	$\langle 18, \pm 6, 193\rangle$	13860	$\langle 23, \pm 20, 155\rangle$
13860	$\langle 31, \pm 20, 115\rangle$	13860	$\langle 41, \pm 30, 90\rangle$	13860	$\langle 45, \pm 30, 82\rangle$	13860	$\langle 46, \pm 26, 79\rangle$
13860	$\langle 62, \pm 42, 63\rangle$	13920	$\langle 13, \pm 4, 268\rangle$	13920	$\langle 19, \pm 8, 184\rangle$	13920	$\langle 23, \pm 8, 152\rangle$
13920	$\langle 39, \pm 30, 95\rangle$	13920	$\langle 41, \pm 26, 89\rangle$	13920	$\langle 52, \pm 4, 67\rangle$	13920	$\langle 57, \pm 30, 65\rangle$
13920	$\langle 61, \pm 54, 69\rangle$	14280	$\langle 11, \pm 8, 326\rangle$	14280	$\langle 22, \pm 8, 163\rangle$	14280	$\langle 23, \pm 16, 158\rangle$
14280	$\langle 33, \pm 30, 115\rangle$	14280	$\langle 46, \pm 16, 79\rangle$	14280	$\langle 47, \pm 14, 77\rangle$	14280	$\langle 55, \pm 30, 69\rangle$
14280	$\langle 59, \pm 36, 66\rangle$	14560	$\langle 11, \pm 2, 331\rangle$	14560	$\langle 17, \pm 14, 217\rangle$	14560	$\langle 31, \pm 14, 119\rangle$
14560	$\langle 41, \pm 6, 89\rangle$	14560	$\langle 43, \pm 24, 88\rangle$	14560	$\langle 44, \pm 20, 85\rangle$	14560	$\langle 53, \pm 42, 77\rangle$
14560	$\langle 55, \pm 20, 68\rangle$	14763	$\langle 23, \pm 7, 161\rangle$	14763	$\langle 47, \pm 29, 83\rangle$	14763	$\langle 53, \pm 17, 71\rangle$
14763	$\langle 59, \pm 39, 69\rangle$	14820	$\langle 17, \pm 2, 218\rangle$	14820	$\langle 29, \pm 12, 129\rangle$	14820	$\langle 34, \pm 2, 109\rangle$
14820	$\langle 43, \pm 12, 87\rangle$	14820	$\langle 47, \pm 28, 83\rangle$	14820	$\langle 51, \pm 36, 79\rangle$	14820	$\langle 58, \pm 46, 73\rangle$
14820	$\langle 59, \pm 44, 71\rangle$	16192	$\langle 17, \pm 14, 241\rangle$	16192	$\langle 32, \pm 24, 131\rangle$	16192	$\langle 32, \pm 8, 127\rangle$
16192	$\langle 61, \pm 20, 68\rangle$	16555	$\langle 29, \pm 27, 149\rangle$	16555	$\langle 37, \pm 13, 113\rangle$	16555	$\langle 41, \pm 3, 101\rangle$
16555	$\langle 47, \pm 41, 97\rangle$	17220	$\langle 17, \pm 16, 257\rangle$	17220	$\langle 29, \pm 8, 149\rangle$	17220	$\langle 31, \pm 4, 139\rangle$
17220	$\langle 34, \pm 18, 129\rangle$	17220	$\langle 43, \pm 18, 102\rangle$	17220	$\langle 51, \pm 18, 86\rangle$	17220	$\langle 58, \pm 50, 85\rangle$
17220	$\langle 62, \pm 58, 83\rangle$	17472	$\langle 17, \pm 2, 257\rangle$	17472	$\langle 23, \pm 10, 191\rangle$	17472	$\langle 32, \pm 24, 141\rangle$
17472	$\langle 32, \pm 8, 137\rangle$	17472	$\langle 47, \pm 24, 96\rangle$	17472	$\langle 51, \pm 36, 92\rangle$	17472	$\langle 59, \pm 46, 83\rangle$
17472	$\langle 68, \pm 36, 69\rangle$	17760	$\langle 11, \pm 4, 404\rangle$	17760	$\langle 19, \pm 10, 235\rangle$	17760	$\langle 33, \pm 18, 137\rangle$
17760	$\langle 44, \pm 4, 101\rangle$	17760	$\langle 47, \pm 10, 95\rangle$	17760	$\langle 55, \pm 40, 88\rangle$	17760	$\langle 57, \pm 48, 88\rangle$

Table 1: Representatives for 2779 $SL_2(\mathbb{Z})$-equivalence Classes of Regular Forms

$\lvert\Delta\rvert$	$\langle A, B, C\rangle$	$\lvert\Delta\rvert$	$\langle A, B, C\rangle$	$\lvert\Delta\rvert$	$\langle A, B, C\rangle$	$\lvert\Delta\rvert$	$\langle A, B, C\rangle$
17760	$\langle 61, \pm 28, 76\rangle$	17952	$\langle 13, \pm 12, 348\rangle$	17952	$\langle 29, \pm 12, 156\rangle$	17952	$\langle 31, \pm 20, 148\rangle$
17952	$\langle 37, \pm 20, 124\rangle$	17952	$\langle 39, \pm 12, 116\rangle$	17952	$\langle 47, \pm 40, 104\rangle$	17952	$\langle 52, \pm 12, 87\rangle$
17952	$\langle 53, \pm 42, 93\rangle$	18720	$\langle 9, \pm 6, 521\rangle$	18720	$\langle 23, \pm 18, 207\rangle$	18720	$\langle 31, \pm 2, 151\rangle$
18720	$\langle 36, \pm 12, 131\rangle$	18720	$\langle 45, \pm 30, 109\rangle$	18720	$\langle 53, \pm 28, 92\rangle$	18720	$\langle 67, \pm 24, 72\rangle$
18720	$\langle 72, \pm 48, 73\rangle$	19320	$\langle 17, \pm 14, 287\rangle$	19320	$\langle 29, \pm 20, 170\rangle$	19320	$\langle 34, \pm 20, 145\rangle$
19320	$\langle 41, \pm 14, 119\rangle$	19320	$\langle 51, \pm 48, 106\rangle$	19320	$\langle 53, \pm 48, 102\rangle$	19320	$\langle 58, \pm 20, 85\rangle$
19320	$\langle 73, \pm 68, 82\rangle$	19380	$\langle 13, \pm 4, 373\rangle$	19380	$\langle 23, \pm 20, 215\rangle$	19380	$\langle 26, \pm 22, 191\rangle$
19380	$\langle 39, \pm 30, 130\rangle$	19380	$\langle 43, \pm 20, 115\rangle$	19380	$\langle 46, \pm 26, 109\rangle$	19380	$\langle 65, \pm 30, 78\rangle$
19380	$\langle 69, \pm 66, 86\rangle$	19635	$\langle 19, \pm 7, 259\rangle$	19635	$\langle 31, \pm 9, 159\rangle$	19635	$\langle 37, \pm 7, 133\rangle$
19635	$\langle 41, \pm 39, 129\rangle$	19635	$\langle 43, \pm 39, 123\rangle$	19635	$\langle 53, \pm 9, 93\rangle$	19635	$\langle 57, \pm 45, 95\rangle$
19635	$\langle 59, \pm 37, 89\rangle$	20020	$\langle 19, \pm 14, 266\rangle$	20020	$\langle 23, \pm 6, 218\rangle$	20020	$\langle 37, \pm 16, 137\rangle$
20020	$\langle 38, \pm 14, 133\rangle$	20020	$\langle 46, \pm 6, 109\rangle$	20020	$\langle 47, \pm 40, 115\rangle$	20020	$\langle 61, \pm 54, 94\rangle$
20020	$\langle 74, \pm 58, 79\rangle$	20640	$\langle 13, \pm 2, 397\rangle$	20640	$\langle 17, \pm 10, 305\rangle$	20640	$\langle 39, \pm 24, 136\rangle$
20640	$\langle 51, \pm 24, 104\rangle$	20640	$\langle 52, \pm 28, 103\rangle$	20640	$\langle 61, \pm 10, 85\rangle$	20640	$\langle 65, \pm 50, 89\rangle$
20640	$\langle 68, \pm 44, 83\rangle$	20832	$\langle 19, \pm 12, 276\rangle$	20832	$\langle 23, \pm 12, 228\rangle$	20832	$\langle 37, \pm 6, 141\rangle$
20832	$\langle 41, \pm 18, 129\rangle$	20832	$\langle 43, \pm 18, 123\rangle$	20832	$\langle 47, \pm 6, 111\rangle$	20832	$\langle 57, \pm 12, 92\rangle$
20832	$\langle 69, \pm 12, 76\rangle$	21120	$\langle 16, \pm 8, 331\rangle$	21120	$\langle 32, \pm 16, 167\rangle$	21120	$\langle 37, \pm 28, 148\rangle$
21120	$\langle 41, \pm 6, 129\rangle$	21120	$\langle 43, \pm 6, 123\rangle$	21120	$\langle 48, \pm 24, 113\rangle$	21120	$\langle 61, \pm 48, 96\rangle$
21120	$\langle 71, \pm 40, 80\rangle$	21840	$\langle 8, \pm 4, 683\rangle$	21840	$\langle 24, \pm 12, 229\rangle$	21840	$\langle 37, \pm 8, 148\rangle$
21840	$\langle 40, \pm 20, 139\rangle$	21840	$\langle 43, \pm 2, 127\rangle$	21840	$\langle 53, \pm 46, 113\rangle$	21840	$\langle 56, \pm 28, 101\rangle$
21840	$\langle 59, \pm 52, 104\rangle$	22080	$\langle 19, \pm 6, 291\rangle$	22080	$\langle 32, \pm 24, 177\rangle$	22080	$\langle 32, \pm 8, 173\rangle$
22080	$\langle 37, \pm 34, 157\rangle$	22080	$\langle 57, \pm 6, 97\rangle$	22080	$\langle 59, \pm 24, 96\rangle$	22080	$\langle 71, \pm 70, 95\rangle$
22080	$\langle 76, \pm 44, 79\rangle$	22848	$\langle 19, \pm 16, 304\rangle$	22848	$\langle 29, \pm 2, 197\rangle$	22848	$\langle 32, \pm 24, 183\rangle$
22848	$\langle 32, \pm 8, 179\rangle$	22848	$\langle 57, \pm 54, 113\rangle$	22848	$\langle 61, \pm 24, 96\rangle$	22848	$\langle 73, \pm 72, 96\rangle$
22848	$\langle 76, \pm 60, 87\rangle$	24640	$\langle 23, \pm 4, 268\rangle$	24640	$\langle 31, \pm 6, 199\rangle$	24640	$\langle 32, \pm 24, 197\rangle$
24640	$\langle 32, \pm 8, 193\rangle$	24640	$\langle 41, \pm 40, 160\rangle$	24640	$\langle 59, \pm 50, 115\rangle$	24640	$\langle 61, \pm 2, 101\rangle$
24640	$\langle 67, \pm 4, 92\rangle$	27360	$\langle 9, \pm 6, 761\rangle$	27360	$\langle 29, \pm 4, 236\rangle$	27360	$\langle 36, \pm 12, 191\rangle$
27360	$\langle 43, \pm 26, 163\rangle$	27360	$\langle 45, \pm 30, 157\rangle$	27360	$\langle 59, \pm 4, 116\rangle$	27360	$\langle 72, \pm 48, 103\rangle$
27360	$\langle 72, \pm 24, 97\rangle$	29568	$\langle 16, \pm 8, 463\rangle$	29568	$\langle 32, \pm 16, 233\rangle$	29568	$\langle 43, \pm 4, 172\rangle$
29568	$\langle 47, \pm 18, 159\rangle$	29568	$\langle 48, \pm 24, 157\rangle$	29568	$\langle 53, \pm 18, 141\rangle$	29568	$\langle 73, \pm 56, 112\rangle$
29568	$\langle 83, \pm 48, 96\rangle$	29920	$\langle 19, \pm 10, 395\rangle$	29920	$\langle 23, \pm 16, 328\rangle$	29920	$\langle 41, \pm 16, 184\rangle$
29920	$\langle 53, \pm 48, 152\rangle$	29920	$\langle 67, \pm 30, 115\rangle$	29920	$\langle 76, \pm 28, 101\rangle$	29920	$\langle 79, \pm 10, 95\rangle$
29920	$\langle 92, \pm 76, 97\rangle$	31395	$\langle 17, \pm 15, 465\rangle$	31395	$\langle 31, \pm 15, 255\rangle$	31395	$\langle 43, \pm 9, 183\rangle$
31395	$\langle 47, \pm 1, 167\rangle$	31395	$\langle 51, \pm 15, 155\rangle$	31395	$\langle 61, \pm 9, 129\rangle$	31395	$\langle 71, \pm 49, 119\rangle$
31395	$\langle 85, \pm 15, 93\rangle$	32032	$\langle 17, \pm 8, 472\rangle$	32032	$\langle 29, \pm 10, 277\rangle$	32032	$\langle 37, \pm 26, 221\rangle$
32032	$\langle 59, \pm 8, 136\rangle$	32032	$\langle 68, \pm 60, 131\rangle$	32032	$\langle 71, \pm 42, 119\rangle$	32032	$\langle 79, \pm 68, 116\rangle$
32032	$\langle 89, \pm 50, 97\rangle$	33915	$\langle 11, \pm 3, 771\rangle$	33915	$\langle 33, \pm 3, 257\rangle$	33915	$\langle 41, \pm 19, 209\rangle$
33915	$\langle 55, \pm 25, 157\rangle$	33915	$\langle 61, \pm 1, 139\rangle$	33915	$\langle 67, \pm 11, 127\rangle$	33915	$\langle 77, \pm 63, 123\rangle$
33915	$\langle 79, \pm 23, 109\rangle$	34720	$\langle 13, \pm 4, 668\rangle$	34720	$\langle 29, \pm 14, 301\rangle$	34720	$\langle 43, \pm 14, 203\rangle$
34720	$\langle 52, \pm 4, 167\rangle$	34720	$\langle 65, \pm 30, 137\rangle$	34720	$\langle 79, \pm 44, 116\rangle$	34720	$\langle 89, \pm 48, 104\rangle$
34720	$\langle 91, \pm 56, 104\rangle$	36960	$\langle 13, \pm 8, 712\rangle$	36960	$\langle 17, \pm 10, 545\rangle$	36960	$\langle 23, \pm 22, 407\rangle$
36960	$\langle 37, \pm 22, 253\rangle$	36960	$\langle 39, \pm 18, 239\rangle$	36960	$\langle 51, \pm 24, 184\rangle$	36960	$\langle 52, \pm 44, 187\rangle$
36960	$\langle 65, \pm 60, 156\rangle$	36960	$\langle 67, \pm 52, 148\rangle$	36960	$\langle 68, \pm 44, 143\rangle$	36960	$\langle 69, \pm 24, 136\rangle$
36960	$\langle 85, \pm 10, 109\rangle$	36960	$\langle 89, \pm 8, 104\rangle$	36960	$\langle 91, \pm 70, 115\rangle$	36960	$\langle 92, \pm 68, 113\rangle$
36960	$\langle 104, \pm 96, 111\rangle$	40755	$\langle 23, \pm 1, 443\rangle$	40755	$\langle 31, \pm 17, 331\rangle$	40755	$\langle 41, \pm 9, 249\rangle$
40755	$\langle 43, \pm 3, 237\rangle$	40755	$\langle 69, \pm 45, 155\rangle$	40755	$\langle 79, \pm 3, 129\rangle$	40755	$\langle 83, \pm 9, 123\rangle$
40755	$\langle 93, \pm 45, 115\rangle$	43680	$\langle 11, \pm 10, 995\rangle$	43680	$\langle 19, \pm 18, 579\rangle$	43680	$\langle 29, \pm 20, 380\rangle$
43680	$\langle 33, \pm 12, 332\rangle$	43680	$\langle 44, \pm 12, 249\rangle$	43680	$\langle 55, \pm 10, 199\rangle$	43680	$\langle 57, \pm 18, 193\rangle$
43680	$\langle 61, \pm 22, 181\rangle$	43680	$\langle 67, \pm 2, 163\rangle$	43680	$\langle 76, \pm 20, 145\rangle$	43680	$\langle 77, \pm 56, 152\rangle$
43680	$\langle 83, \pm 12, 132\rangle$	43680	$\langle 87, \pm 78, 143\rangle$	43680	$\langle 88, \pm 56, 133\rangle$	43680	$\langle 88, \pm 32, 127\rangle$
43680	$\langle 95, \pm 20, 116\rangle$	57120	$\langle 11, \pm 6, 1299\rangle$	57120	$\langle 23, \pm 14, 623\rangle$	57120	$\langle 33, \pm 6, 433\rangle$
57120	$\langle 44, \pm 28, 329\rangle$	57120	$\langle 47, \pm 28, 308\rangle$	57120	$\langle 55, \pm 50, 271\rangle$	57120	$\langle 59, \pm 46, 251\rangle$
57120	$\langle 69, \pm 60, 220\rangle$	57120	$\langle 77, \pm 28, 188\rangle$	57120	$\langle 79, \pm 32, 184\rangle$	57120	$\langle 88, \pm 72, 177\rangle$
57120	$\langle 88, \pm 16, 163\rangle$	57120	$\langle 89, \pm 14, 161\rangle$	57120	$\langle 92, \pm 60, 165\rangle$	57120	$\langle 109, \pm 66, 141\rangle$
57120	$\langle 115, \pm 60, 132\rangle$	77280	$\langle 17, \pm 6, 1137\rangle$	77280	$\langle 29, \pm 18, 669\rangle$	77280	$\langle 41, \pm 28, 476\rangle$
77280	$\langle 51, \pm 6, 379\rangle$	77280	$\langle 53, \pm 10, 365\rangle$	77280	$\langle 68, \pm 28, 287\rangle$	77280	$\langle 73, \pm 10, 265\rangle$
77280	$\langle 85, \pm 40, 232\rangle$	77280	$\langle 87, \pm 18, 223\rangle$	77280	$\langle 107, \pm 98, 203\rangle$	77280	$\langle 109, \pm 108, 204\rangle$
77280	$\langle 116, \pm 76, 179\rangle$	77280	$\langle 119, \pm 28, 164\rangle$	77280	$\langle 123, \pm 54, 163\rangle$	77280	$\langle 136, \pm 96, 159\rangle$
77280	$\langle 136, \pm 40, 145\rangle$	87360	$\langle 32, \pm 24, 687\rangle$	87360	$\langle 32, \pm 8, 683\rangle$	87360	$\langle 37, \pm 16, 592\rangle$
87360	$\langle 43, \pm 4, 508\rangle$	87360	$\langle 53, \pm 14, 413\rangle$	87360	$\langle 59, \pm 14, 371\rangle$	87360	$\langle 96, \pm 72, 241\rangle$
87360	$\langle 96, \pm 24, 229\rangle$	87360	$\langle 101, \pm 56, 224\rangle$	87360	$\langle 111, \pm 90, 215\rangle$	87360	$\langle 113, \pm 92, 212\rangle$
87360	$\langle 127, \pm 4, 172\rangle$	87360	$\langle 129, \pm 90, 185\rangle$	87360	$\langle 139, \pm 40, 160\rangle$	87360	$\langle 148, \pm 132, 177\rangle$
87360	$\langle 159, \pm 120, 160\rangle$						

A COMBINATORIAL PROOF OF TWO EQUIVALENT IDENTITIES BY FREE 2-MOTZKIN PATHS

Alina F. Y. Zhao[1]
School of Mathematical Sciences and Institute of Mathematics, Nanjing Normal University, Nanjing, P.R. China
alinazhao@njnu.edu.cn

Received: 9/23/12, Accepted: 5/4/13, Published: 6/14/13

Abstract
We present a combinatorial proof of the equivalence between a formula of MacMahon and a formula of Gould using free 2-Motzkin paths, and give a generalization for a formula asked by Gould as well.

1. Introduction

More than a century ago, MacMahon [4] derived a well-known formula as follows:

$$\sum_{k=0}^{n}\binom{n}{k}^3=\sum_{0\le k\le \frac{n}{2}}\binom{n}{2k}\binom{2k}{k}\binom{n+k}{k}2^{n-2k}, \tag{1}$$

and further obtained that

$$\sum_{k=0}^{n}\binom{n}{k}^3 x^k y^{n-k}=\sum_{0\le k\le \frac{n}{2}}\binom{n}{2k}\binom{2k}{k}\binom{n+k}{k}x^k y^k (x+y)^{n-2k}. \tag{2}$$

Recently, Gould [2] found that

$$\sum_{k=0}^{n}\binom{n}{k}^3=\sum_{0\le k\le \frac{n}{2}}\binom{n}{2k}\binom{2k}{k}\binom{2n-2k}{n-k} \tag{3}$$

by using the Carlitz's formula [1]

$$\sum_{k=0}^{n}\binom{n}{k}^3=[x^n](1-x^2)^n P_n\left(\frac{1+x}{1-x}\right), \tag{4}$$

[1]This work was supported by the National Science Foundation of China (#11226301).

where $[x^n]f(x)$ means the coefficient of x^n in the series expansion of $f(x)$, and $P_n(x)$ is the Legendre polynomial defined by

$$\sum_{n=0}^{\infty} P_n(x)t^n = \frac{1}{\sqrt{1-2xt+t^2}}.$$

In deriving the equivalence between the right-hand sides of formulas (1) and (3), Gould [2] explored the identity

$$\sum_{k\le j\le \frac{n}{2}} \binom{n}{2j}\binom{2j}{j}\binom{j}{k}2^{n-2j} = \binom{n}{k}\binom{2n-2k}{n}, \tag{5}$$

and asked for an extension analogous to (2) of MacMahon.

In this note, we aim to prove the equivalence of the formulas (1) and (3) combinatorially in terms of free 2-Motzkin paths and meanwhile give a generalization of the identity (5).

2. Free 2-Motzkin paths

We first introduce some definitions about a special kind of lattice path. A free 2-Motzkin path of length n is a lattice path starting at $(0,0)$, ending at $(n,0)$, with possible up steps $(1,1)$, down steps $(1,-1)$ and level steps $(1,0)$, where the level steps can be either of two kinds: straight or wavy. Let $\mathcal{M}(n)$ be the set of free 2-Motzkin paths of length n. More generally, let $\mathcal{M}(n-k,-k)$ be the set of lattice paths starting at $(0,0)$, ending at $(n-k,-k)$, with the same possible steps as free 2-Motzkin paths. It is easy to see that the set of free 2-Moztkin paths is just the set $\mathcal{M}(n,0)$.

Proposition 2.1. *The number of free 2-Motzkin paths in $\mathcal{M}(n)$ with a total of k up and straight level steps equals $\binom{n}{k}^2$.*

Proof. For a free 2-Motzkin path in $\mathcal{M}(n)$ with a total of k up and straight level steps, if there are i $(0 \le i \le k)$ up steps, then there are $k-i$ straight level steps. We can generate such a path by first choosing i positions from the n positions to place the i up steps and another i positions from the $n-i$ positions that are left to place the i down steps in $\binom{n}{i}\binom{n-i}{i}$ ways; then choosing $k-i$ positions from the remaining $n-2i$ positions for the straight level steps, and putting the wavy level steps on the positions that are left. Thus we have $\binom{n}{i}\binom{n-i}{i}\binom{n-2i}{k-i}$ choices to generate such a path with i up steps and $k-i$ straight level steps. Since

$$\binom{n}{i}\binom{n-i}{i}\binom{n-2i}{k-i} = \binom{n}{k}\binom{k}{i}\binom{n-k}{n-k-i},$$

summing up all possible i and using Vandermonde's convolution formula [3]

$$\sum_{i=0}^{k}\binom{k}{i}\binom{n-k}{n-k-i}=\binom{n}{k},$$

the proposition is followed. □

Proposition 2.2. *The number of paths in* $\mathcal{M}(n-k,-k)$ $(k\geq 0)$ *equals* $\binom{2n-2k}{n}$.

Proof. If there are i $(k\leq i\leq n-k)$ down steps in a path of $\mathcal{M}(n-k,-k)$, then there are $i-k$ up steps and $n-2i$ level steps. Such a path can be constructed as follows: first we could arrange those up and down steps in $\binom{2i-k}{i}$ ways, and then the remaining $n-2i$ level steps can be repeatedly put in the $2i-k+1$ positions in between the $2i-k$ determined steps in $\binom{2i-k+1+n-2i-1}{n-2i}=\binom{n-k}{n-2i}$ ways. Since a level step can be straight or wavy, the number of such paths with i down steps is $\binom{2i-k}{i}\binom{n-k}{n-2i}2^{n-2i}$.

Because of $\binom{2i-k}{i}\binom{n-k}{n-2i}2^{n-2i}=\binom{n-k}{i}\binom{n-k-i}{n-2i}2^{n-2i}$, it remains to show that

$$\sum_{i=k}^{n-k}\binom{n-k}{i}\binom{n-k-i}{n-2i}2^{n-2i}=\binom{2n-2k}{n}.$$

Evaluating in two different ways in the expansion on t of the expression $(1+t)^{2n-2k}$, we have

$$(1+t)^{2n-2k}=\sum_{n\geq 0}\binom{2n-2k}{n}t^{n};$$

on the other hand,

$$(1+t)^{2n-2k}=(1+2t+t^2)^{n-k}=\sum_{i\geq 0}\binom{n-k}{i}t^{2i}(1+2t)^{n-k-i}$$

$$=\sum_{i\geq 0}\binom{n-k}{i}t^{2i}\sum_{j\geq 0}\binom{n-k-i}{j}2^{j}t^{j}=\sum_{n\geq 0}\sum_{i\geq 0}\binom{n-k}{i}\binom{n-k-i}{n-2i}2^{n-2i}t^{n}.$$

Comparison of the coefficient of t^n in the above leads to the desired result. □

3. The Combinatorial Proof

A colored free 2-Motzkin path is a free 2-Motzkin path with some steps (up, down or level) being colored, say yellow. Define $\mathcal{CM}(n)$ be the set of colored free 2-Motzkin paths of length n in which the number of yellow steps equals the sum of the number of up steps and straight level steps. If there are k up and straight level steps in a path of $\mathcal{CM}(n)$, then we have $\binom{n}{k}$ choices to color the yellow steps. From Proposition 2.1, we get

$$|\mathcal{CM}(n)| = \sum_{k=0}^{n} \binom{n}{k}^3. \tag{6}$$

The proof of formula (3) is based on a different enumeration order for the counting of the set $\mathcal{CM}(n)$.

Theorem 3.1. *We have*

$$|\mathcal{CM}(n)| = \sum_{0 \le k \le \frac{n}{2}} \binom{n}{2k}\binom{2k}{k}\binom{2n-2k}{n-k}. \tag{7}$$

Proof. Suppose that there are k up steps and i straight level steps in a colored free 2-Motzkin path of $\mathcal{CM}(n)$. We can construct such a path by first choosing k positions to place the k up steps, and then choosing k down steps from the remaining $n-k$ steps. After this we choose i positions for the straight level steps and the wavy level steps are placed on the positions that are left. Finally, we pick $i+k$ steps from the n steps to color them yellow. Thus the total number of such paths is

$$\binom{n}{k}\binom{n-k}{k}\binom{n-2k}{i}\binom{n}{k+i}.$$

Summing up all the possible i and k, and using the formula

$$\sum_{0 \le i \le n-2k} \binom{n-2k}{i}\binom{n}{k+i} = \binom{2n-2k}{n-k},$$

we obtain that the number of colored free 2-Motzkin paths of length n equals

$$\sum_{0 \le k \le \frac{n}{2}} \binom{n}{k}\binom{n-k}{k} \sum_{0 \le i \le n-2k} \binom{n-2k}{i}\binom{n}{k+i} = \sum_{0 \le k \le \frac{n}{2}} \binom{n}{k}\binom{n-k}{k}\binom{2n-2k}{n-k}.$$

The proof is completed by noticing that $\binom{n}{k}\binom{n-k}{k} = \binom{n}{2k}\binom{2k}{k}$. □

For formula (1), we introduce another combinatorial structure which is motivated by observing that the number of the elements in the set $\mathcal{CM}(n)$ can also be evaluated as follows:

Proposition 3.2. *We have*

$$|\mathcal{CM}(n)| = \sum_{0 \le k \le \frac{n}{2}} \binom{n}{k}^2 \binom{2n-2k}{n}. \tag{8}$$

Proof. If there are k up steps and i straight level steps in a colored free 2-Motzkin path in $\mathcal{CM}(n)$, then we could construct such a path by first choosing k positions for the steps with yellow color. After this, we pick k positions to place the k up steps, k other positions to place the k down steps, and i positions for the straight level steps. Finally, we choose another i steps to color them yellow. Since we have colored k steps initially, the choice of the remaining i yellow steps equals $\binom{n-k}{i}/\frac{(k+i)!}{k!i!}$. Thus the total number of such colored paths is

$$\binom{n}{k}\binom{n}{k}\binom{n-k}{k}\binom{n-2k}{i}\binom{n-k}{i}\Big/\frac{(k+i)!}{k!i!}.$$

Summing up all the possible i, we have

$$\sum_{0\le i\le n-2k}\binom{n-k}{k}\binom{n-2k}{i}\binom{n-k}{i}\Big/\frac{(k+i)!}{k!i!}$$
$$=\sum_{0\le i\le n-2k}\binom{n-k}{k+i}\binom{n-k}{n-k-i}=\binom{2n-2k}{n}.$$

Thus, the number of colored free 2-Motzkin paths of length n is given by

$$\sum_{0\le k\le \frac{n}{2}}\binom{n}{k}^2\binom{2n-2k}{n},$$

which completes the proof. □

The above proposition can be regarded as a combinatorial explanation for the identity

$$\binom{n}{2k}\binom{2k}{k}\binom{2n-2k}{n-k}=\binom{n}{k}^2\binom{2n-2k}{n}.$$

Now we are in the position to introduce the desired structure in the process of the proof of (1). A double colored free 2-Motzkin path is a free 2-Motzkin path such that some steps (up, down or level) can be colored by two colors, say yellow and green. Define $\mathcal{DM}(n)$ be the set of double colored free 2-Motzkin paths of length n such that the up steps can be colored by yellow or green and other steps can be colored only by green, moreover the number of yellow steps equals the number of green steps.

Theorem 3.3. *We have*

$$|\mathcal{DM}(n)|=\sum_{0\le k\le \frac{n}{2}}\binom{n}{k}^2\binom{2n-2k}{n}.$$

Proof. Suppose that there are k yellow up steps in a path of $\mathcal{DM}(n)$. We can construct a double colored free 2-Motzkin path by first choosing k positions for the up steps with yellow color, and the remaining $n-k$ steps can be regarded as a free 2-Motzkin path in $\mathcal{M}(n-k,-k)$. From Proposition 2.2, we see that the number of such paths is $\binom{2n-2k}{n}$. Finally it suffices to choose k steps to color them green. Therefore, we produce a double colored free 2-Motzkin path with k up steps. Summing up all the possible k, we derive that the total number of paths in $\mathcal{DM}(n)$ is

$$\sum_{0\le k\le \frac{n}{2}} \binom{n}{k}\binom{2n-2k}{n}\binom{n}{k},$$

and this completes the proof. □

From Proposition 3.2 and Theorem 3.3, we see that $|\mathcal{CM}(n)| = |\mathcal{DM}(n)|$. The equivalence between formulas (3) and (1) is established by showing the following:

Theorem 3.4. *We have*

$$|\mathcal{DM}(n)| = \sum_{0\le k\le \frac{n}{2}} \binom{n}{2k}\binom{2k}{k}\binom{n+k}{k}2^{n-2k}.$$

Proof. Suppose that there are k up steps in a path of $\mathcal{DM}(n)$, and such a path can be constructed by first establishing a free 2-Motzkin path with k up steps, and then coloring its steps. If there are i yellow up steps, then the total number of colorings is

$$\sum_{0\le i\le k} \binom{k}{i}\binom{n}{i} = \binom{n+k}{n},$$

and the number of free 2-Motzkin path of length n with k up steps equals

$$\binom{n}{k}\binom{n-k}{k}2^{n-2k}.$$

This theorem follows by summing up all possible k. □

In order to give a generalization of the formula (5), we will use the weighted path as an underlying set. A weighted colored free 2-Motzkin path is a path in which every step is endowed with a weight. The weight of a path is the product of the weights of the steps and the weight of a set of paths means the sum of the weights of the paths in it. For our purpose, we define the weight of the paths in $\mathcal{CM}(n)$ as follows: the up steps and straight level steps are given the weight x, whereas the down steps and wavy level steps are given the weight y. We present an extension of the formula (5) as follows.

Theorem 3.5. *The following holds:*

$$\sum_{k\le j\le \frac{n}{2}} \binom{n}{2j}\binom{2j}{j}\binom{j}{k} x^j y^j (x+y)^{n-2j} = \sum_{j=k}^{n-k} \binom{n}{k}\binom{n-k}{j}\binom{n-k}{n-j} x^j y^{n-j}. \quad (9)$$

Proof. By our weight assignment, the left-hand side of the formula (9) is the sum of the weighted colored free 2-Motzkin paths of length n with k yellow up steps.

Meanwhile, we can generate such a path by the following procedure: i) we choose k positions to put the k yellow up steps in $\binom{n}{k}$ ways; ii) if there are j down steps in such a colored free 2-Motzkin paths, then we have $j-k$ uncolored up steps, and we choose j positions from the remaining $n-k$ positions to place the down steps; iii) we pick $j-k$ positions from the $n-k-j$ positions that are left to place the uncolored up steps, and iv) the remaining $n-2j$ positions can be put by a straight level step with weight x or a wavy level step with weight y. Thus we have

$$\sum_{j=k}^{n-k} \binom{n}{k}\binom{n-k}{j}\binom{n-k-j}{j-k} x^j y^j (x+y)^{n-2j}$$

weighted paths of this kind. By expanding the expression $\left[1+(x+y)t+xyt^2\right]^{n-k}$, we see that

$$\begin{aligned}\left[1+(x+y)t+xyt^2\right]^{n-k} &= \sum_{j=0}^{n-k} \binom{n-k}{j} x^j y^j t^{2j} \left[1+(x+y)t\right]^{n-k-j} \\ &= \sum_{j=0}^{n-k} \binom{n-k}{j} x^j y^j t^{2j} \sum_{i=0}^{n-k-j} \binom{n-k-j}{i} (x+y)^i t^i.\end{aligned}$$

Also, we have

$$\begin{aligned}\left[1+(x+y)t+xyt^2\right]^{n-k} &= (1+xt)^{n-k}(1+yt)^{n-k} \\ &= \sum_{j=0}^{n-k} \binom{n-k}{j} x^j t^j \sum_{i=0}^{n-k} \binom{n-k}{i} y^i t^i.\end{aligned}$$

Equating the coefficient of t^n in the above two expansions, we have

$$\sum_{k\le j\le \frac{n}{2}} \binom{n-k}{j}\binom{n-k}{n-j} x^j y^{n-j} = \sum_{j=k}^{n-k} \binom{n-k}{j}\binom{n-k-j}{n-2j} x^j y^j (x+y)^{n-2j}.$$

This implies that weight of such kind of paths also equals

$$\binom{n}{k}\sum_{j=k}^{n-k} \binom{n-k}{j}\binom{n-k}{n-j} x^j y^{n-j},$$

and identity (9) follows. □

We see that the identity (9) is reduced to the identity (5) by setting $x = 1, y = 1$, and using the convolution

$$\sum_{j=k}^{n-k} \binom{n-k}{j} \binom{n-k}{n-j} = \binom{2n-2k}{n}.$$

Acknowledgements. The author would like to thank the referee for many helpful comments on a previous version of this paper.

References

[1] L. Carlitz, Problem 352, Math. Mag., 32 (1958), 47–48.

[2] H.W. Gould, Sums of powers of binomial coefficients via Legendre polynomials, Part 2, Ars Combin., 86 (2008), 161–173.

[3] R.L. Graham, D.E. Knuth and O. Patashnik, Concrete Mathematics, 2nd edition, Addison-Wesley, 1994.

[4] Major P.A. MacMahon, The sum of powers of the binomial coefficients, Q. J. Math., 33 (1902), 274–288.

IMPROVING THE CHEN AND CHEN RESULT FOR ODD PERFECT NUMBERS

Kevin A. Broughan
Department of Mathematics, University of Waikato, Hamilton, New Zealand
kab@waikato.ac.nz

Daniel Delbourgo
Department of Mathematics, University of Waikato, Hamilton, New Zealand
delbourgo@waikato.ac.nz

Qizhi Zhou
Department of Mathematics, University of Waikato, Hamilton, New Zealand
qz49@waikato.ac.nz

Received: 9/18/12, Revised: 2/1/13, Accepted: 5/2/13, Published: 6/14/13

Abstract

If q^α is the Euler factor of an odd perfect number N, then we prove that its so-called index $\sigma(N/q^\alpha)/q^\alpha \geq 3^2 \times 5 \times 7 = 315$. It follows that for any odd perfect number, the ratio of the non-Euler part to the Euler part is greater than $3^2 \times 5 \times 7/2$.

1. Introduction

The main motivation for studying the structure of an odd perfect number is ultimately to establish that such a number cannot exist. It is known that any odd perfect number N must have at least 9 distinct prime factors [10], be larger than 10^{1500} [12], have a squarefree core which is less than $2N^{\frac{17}{26}}$ [9], and every prime divisor is less than $(3N)^{\frac{1}{3}}$ [1]. These results represent recent progress on what must be one of the oldest current problems in mathematics.

Following Dris [5], in this paper we define the index m of a prime power dividing N. Using a lower bound for the index one can derive an upper bound, in terms of N, for the Euler factor of N. Dris found the bound $m \geq 3$; then Dris and Luca [6] improved this to $m \geq 6$. In [4] a list of forms in terms of products of prime powers, which includes the results of Dris and Dris-Luca, is derived. We improve the method of [4], obtaining an expanded list of prime power products which cannot occur as the value of an index. This enables us to conclude, in the case of the Euler factor, that $m \geq 315$; for any other prime, if the Euler factor divides N to a power at least 2 then $m \geq 630$, and if the Euler factor divides N to the power 1 then $m \geq 210$.

Notations: $\Omega(n)$ is the total number of prime divisors of n counted with multiplicity, $\omega(n)$ the number of distinct prime divisors of n, $\omega_0(n)$ is the number of distinct odd prime divisors of n, $\sigma(n)$ the sum of the divisors of n, $d(n)$ the number of divisors of n, $\log_2 n$ the logarithm to base 2, (a, b) the greatest common divisor, $p^e \| n$ means p^e divides n but p^{e+1} does not, $\nu_p(n)$ the highest power of p which divides n, and $\operatorname{ord}_p a$ is the smallest power of a which is congruent to 1 modulo p. The symbol $\square$, when not being used to denote the end of a proof, represents the square of an integer.

Let N denote an odd perfect number, and q a prime divisor with $q^\alpha \| N$ say. We write the standard factorization of N as

$$N = q^\alpha \times \prod_{i=1}^{k} p_i^{\lambda_i} \times \prod_{j=k+1}^{s} p_j^{\lambda_j}$$

where for $1 \leq i \leq k$ we have

$$\sigma\left(p_i^{\lambda_i}\right) = m_i q^{\beta_i}, \ \beta_i \geq 0, \ (m_i, q) = 1, \ m_i > 1. \tag{1}$$

These prime numbers p_i are called primes of **type 1**. For $k+1 \leq j \leq s$

$$\sigma\left(p_j^{\lambda_j}\right) = q^{\beta_j}, \ \beta_j > 0 \tag{2}$$

and the p_j are called primes of **type 2**.

One defines the **index** or **perfect number index at prime** q to be the integer

$$m := \frac{\sigma(N/q^\alpha)}{q^\alpha}; \tag{3}$$

in particular $m = m_1 \cdots m_k$.

In fact $4 \nmid m$, $q \nmid m$, and if an odd prime p satisfies $p^e \mid m$ then $p^e \mid N$. Furthermore if q is the Euler prime, then m is odd and each m corresponding to any other prime is even. Lastly we have the fundamental equation

$$m \times \sigma(q^\alpha) = 2 \times \prod_{i=1}^{k} p_i^{\lambda_i} \times \prod_{j=k+1}^{s} p_j^{\lambda_j} = \frac{2N}{q^\alpha}. \tag{4}$$

2. Preliminary Results

First we state the theorem of Chen and Chen [4].

Theorem 1 *If N is an odd perfect number with a prime power $q^\alpha \| N$, then the index $m := \sigma(N/q^\alpha)/q^a$ is not equal to any of the six forms*

$$\{p_1,\ p_1^2,\ p_1^3,\ p_1^4,\ p_1p_2,\ p_1^2p_2\}$$

where p_1 and p_2 are any distinct primes.

The following lemma comes from [6]. Here we give an alternative proof.

Lemma 2 *If for some j with $k+1 \le j \le s$ (so p_j is a prime of type 2) and for some γ with $2 \le \gamma \le \lambda_j$ we have $p_j^\gamma \mid (q^{\alpha+1}-1)/(q-1)$, then $p_j^{\gamma-1} \mid \alpha+1$.*

Proof. Because $p_j(1+p_j+\cdots+p_j^{\lambda_j-1}) = q^{\beta_j}-1$ one deduces $p_j^1 \| q^{\beta_j}-1$, in which case $p_j^1 \| q^{\operatorname{ord}_{p_j}(q)}-1$. However

$$2 \le \gamma \le \nu_{p_j}\left(\frac{q^{\alpha+1}-1}{q-1}\right) = \nu_{p_j}\left(\frac{q^{\operatorname{ord}_{p_j}(q)}-1}{q-1}\right) + \nu_{p_j}\left(\frac{\alpha+1}{\operatorname{ord}_{p_j}(q)}\right). \tag{5}$$

If $\operatorname{ord}_{p_j}(q) = 1$ then $\gamma \le \nu_{p_j}(\alpha+1)$ and $p_j^\gamma \mid \alpha+1$, whereas if $\operatorname{ord}_{p_j}(q) > 1$ one has $\gamma \le 1+\nu_{p_j}(\alpha+1)$ and therefore $p_j^{\gamma-1} \mid \alpha+1$. □

Lemma 3 (Ljunggren, see [7]) *The only integer solutions $(x,\ n,\ y)$ with $|x| > 1$, $n > 2$, $y > 0$ to the equation $(x^n-1)/(x-1) = y^2$ are $(7,\ 4,\ 20)$ and $(3,\ 5,\ 11)$, i.e. $(7^4-1)/(7-1) = 20^2$ and $(3^5-1)/(3-1) = 11^2$.*

Lemma 4 *[7] The only solutions in non-zero integers with $n > 1$ to the equation $y^n = x^2+x+1$ are $n = 3$, $y = 7$ and $x = 18$ or $x = -19$.*

The following well known result [2, 3, 13] guarantees the existence of primitive prime divisors for expressions of the form $a^n - 1$ with fixed $a > 1$.

Lemma 5 *Let a and n be integers greater than 1. Then there exists a prime $p \mid a^n-1$ which does not divide any of a^m-1 for each $m \in \{2,\dots,n-1\}$, except possibly in the two cases $n = 2$ and $a = 2^\beta-1$ for some $\beta \ge 2$, or $n = 6$ and $a = 2$. Such a prime is called a* **primitive prime factor**.

We complete this set of preliminary results by filling in the missing case from the proof of the fundamental lemma [4, Lemma 2.4].

Lemma 6 *Let N be an odd perfect number. Then $d(\alpha+1) \le \omega(N)$ whenever a prime power $q^\alpha \| N$.*

Proof. Let $n_1,\ n_2,\dots,\ n_w$ denote all the distinct positive divisors of $\alpha+1$ which are greater than 1.

If $2 \mid \alpha+1$ then α is odd, and thus $q \equiv \alpha \equiv 1 \bmod 4$. Therefore q cannot be of the form $2^\beta - 1$ and must be odd. By Lemma 5 there exists a primitive prime factor $q_i \mid q^{n_i} - 1$; since $2 \mid q^1 - 1$ the q_i are all odd, and as they are primitive, one finds $q_i \nmid q^1 - 1$ also. Hence

$$q_i \mid \frac{q^{n_i}-1}{q-1} \mid \frac{q^{\alpha+1}-1}{q-1}$$

so that $q_{n_1} \cdots q_{n_w} \mid (q^{\alpha+1}-1)/(q-1)$. But $m \times \sigma(q^\alpha) = 2N/q^\alpha$ thus, including the divisor 1 and recalling $2 \mid \sigma(q^\alpha)$, one obtains the inequalities

$$d(\alpha+1) = w+1 \le \omega\big(\sigma(q^\alpha)\big) \le \omega\big(m\sigma(q^\alpha)\big) = \omega\left(\frac{2N}{q^\alpha}\right) = \omega(N).$$

Alternatively if $2 \nmid \alpha+1$ then α is even so, again by Lemma 5, we obtain distinct odd primes q_{n_i} with

$$q_{n_1} \cdots q_{n_w} \mid \frac{q^{\alpha+1}-1}{q-1}.$$

Because in this case $2 \mid m$ and $2 \nmid \sigma(q^\alpha)$, we deduce that

$$d(\alpha+1) = 1+w \le 1+\omega\big(\sigma(q^\alpha)\big) \le \omega\big(m\sigma(q^\alpha)\big) = \omega\left(\frac{2N}{q^a}\right) = \omega(N)$$

which completes the proof of the lemma. □

3. The Proof

We now amend the proof of Theorem 1.1 of [4].

Lemma 7 *Let N be an odd perfect number, and m the index at some prime divisor of N. Then*

$$\Omega(m) + \omega_0(m) \ge \omega(N) - \log_2 \sqrt{\omega(N)} - \eta$$

where $\eta = 1$ if m is odd, $\eta = \frac{1}{2}$ if m is even and the Euler prime divides N to a power greater than 1, and $\eta = \frac{3}{2}$ if m is even and the Euler prime divides N exactly to the power 1.

Proof. Whenever $(m, p_{k+1} \cdots p_s) = p_{k+1} \cdots p_s$, one has an inequality

$$s - k \le \omega_0(m) = t$$

and it follows that

$$k + t \ge s = \omega(N) - 1.$$

Because $k \leq \Omega(m)$, $t = \omega_0(m)$ and $\omega(N) \geq 9$, we quickly deduce

$$\Omega(m) + \omega_0(m) \geq k + t \geq \omega(N) - 2 \geq \omega(N) - \log_2 \sqrt{\omega(N)} - 0.42.$$

The non-trivial case occurs when $(m, p_{k+1} \cdots p_s) \neq p_{k+1} \cdots p_s$. By suitably reordering the p_i, we can always write for some l with $k \leq l < s$:

$$\frac{p_{k+1} \cdots p_s}{(m, p_{k+1} \cdots p_s)} = p_{l+1} \cdots p_s. \tag{6}$$

Applying [4] Equation (2.2) and (6), we see that

$$p_{l+1}^{\lambda_{l+1}} \cdots p_s^{\lambda_s} \mid \sigma(q^\alpha).$$

Moreover using [4] Equation (2.1) and [4] Lemma 2.3,

$$p_i^{\lambda_i - 1} \mid \alpha + 1, \quad l + 1 \leq i \leq s$$

hence

$$p_{l+1}^{\lambda_{l+1}-1} \cdots p_s^{\lambda_s - 1} \mid \alpha + 1.$$

Now for $k + 1 \leq i \leq s$ one knows $\sigma(p_i^{\lambda_i}) = q^{\beta_i}$, and q is odd so we must have λ_i even. It follows for $l + 1 \leq i \leq s$ each $\lambda_i \geq 2$, thus $p_{l+1} \cdots p_s \mid \alpha + 1$. Note also that $l < s$ in which case $s - l \geq 1$.

If $s - l = 1$ then because $\omega(N) \geq 9$,

$$\Omega(m) + \omega_0(m) \geq k + t \geq l = s - 1 \geq \omega(N) - 2 \geq \omega(N) - \log_2 \sqrt{\omega(N)} - 0.42$$

as in the previous case.

If $s - l \geq 2$ then we **claim** at most one of the $\lambda_i = 2$ and the remainder have $\lambda_i \geq 4$. To see this, consider the equations

$$p_i^2 + p_i + 1 = q^{\beta_i}.$$

If $\beta_i > 1$ then, by Lemma 4, the only solution is $\beta_i = 3$, $q = 7$ and $p_i = 18$ which is not prime, so the solution cannot occur in this context. Hence $\beta_i = 1$ and the form of the equation is $q = x^2 + x + 1$. But this, for given q, has at most one positive integer solution, therefore at most one prime solution p_i.

By renumbering the p_i if necessary, when $s - l \geq 2$ we can write

$$p_{l+1}^3 p_{l+2}^3 \cdots p_{s-1}^3 p_s \mid \alpha + 1.$$

`Case 1.` Suppose that the index m is odd. Then q is the Euler prime, and consequently $2 \mid \alpha + 1$. Hence

$$2p_{l+1}^3 p_{l+2}^3 \cdots p_{s-1}^3 p_s \mid \alpha + 1,$$

and thus, by Lemma 6, we have

$$2^{2s-2l} \le d(\alpha+1) \le \omega(N),$$

or in other words $s-l \le \log_2 \sqrt{\omega(N)}$, which implies

$$l \ge \omega(N) - \log_2 \sqrt{\omega(N)} - 1.$$

As $\omega_0(m) = t$ then by Equation (6) we have $l-k \le t$, so $l \le \Omega(m)+\omega_0(m)$. Lastly because $\omega(N) \ge 9$,

$$6.41 \le \omega(N) - \log_2 \sqrt{\omega(N)} - 1 \le l \le \Omega(m) + \omega_0(m).$$

Case 2. Here we assume the Euler prime divides N to a power at least 2. Let m be even. Now $m = m_1 \cdots m_k$ and $2 \| m$ so, for a unique i, one knows that $2 \mid m_i$. We **claim** that $2 \neq m_i$. If not, then

$$\sigma(p_i^{\lambda_i}) = 2q^{\beta_i}$$

whence p_i is the Euler prime and $\lambda_i + 1$ is even; we can write

$$\frac{p_i^{\lambda_i+1}-1}{2(p_i-1)} = \left(\frac{p_i^{\frac{\lambda_i+1}{2}}-1}{p_i-1}\right) \times \left(\frac{p_i^{\frac{\lambda_i+1}{2}}+1}{2}\right) = q^{\beta_i}$$

but this cannot hold since the two factors in the middle term are coprime and greater than 1, thus $2 \neq m_i$.

It follows that $k \le \Omega(m) - 1$. In this scenario with $s - l \ge 2$, we also know

$$p_{l+1}^3 p_{l+2}^3 \cdots p_{s-1}^3 p_s \mid \alpha + 1$$

thus

$$2^{2s-2l-1} \le d(\alpha+1) \le \omega(N),$$

which in turn implies

$$l \ge s - \frac{1}{2} - \log_2 \sqrt{\omega(N)} = \omega(N) - \log_2 \sqrt{\omega(N)} - \frac{3}{2}.$$

It follows from the discussion that

$$\omega(N) - \log_2 \sqrt{\omega(N)} - \frac{3}{2} \le l \le k+t \le \Omega(m) - 1 + \omega_0(m)$$

and therefore

$$6.91 \le \omega(N) - \log_2 \sqrt{\omega(N)} - \frac{1}{2} \le \Omega(m) + \omega_0(m).$$

Case 3. We shall now assume the Euler prime divides N exactly to the power 1 and that m is even. Here we have only the weaker inequality $k \le \Omega(m)$, and using an identical argument to Case 2:

$$5.91 \le \omega(N) - \log_2 \sqrt{\omega(N)} - \frac{3}{2} \le \Omega(m) + \omega_0(m). \qquad \square$$

Lemma 8 *If the index m is a square, then $\alpha = 1$.*

Proof. If $m = \square$ then necessarily q is the Euler prime. We must have $\sigma(q^\alpha) = 2\square$ and α is odd. Assuming $\alpha > 1$ then

$$\frac{1}{2}\left(\frac{q^{\alpha+1}-1}{q-1}\right) = \left(\frac{q^{(\alpha+1)/2}-1}{q-1}\right) \times \left(\frac{q^{(\alpha+1)/2}+1}{2}\right) = \square$$

and the two factors in the penultimate term are coprime, in which case

$$\frac{q^{(\alpha+1)/2}-1}{q-1} = \square.$$

By Lemma 3 one has $(\alpha+1)/2 \leq 2$, and as $\alpha \equiv 1 \bmod 4$ we deduce $\alpha = 1$, thereby yielding a contradiction. $\square$

Lemma 9 *If the index m is odd, then it cannot be the sixth power of a prime.*

Proof. Firstly the index being odd means it corresponds to the Euler prime. Assume $m = p^6 = \square$. By Lemma 8, we have $\alpha = 1$. If $p = p_I$ is of type 1 then $\sigma(p_I^{\lambda_I}) = p_I^\theta q^{\beta_I}$ for some $\theta > 0$, which is false. Hence p_I will be type 2. If any other prime p_j were also of type 2, then due to the equality $\sigma(q^\alpha) = \frac{2N}{q^\alpha p_I^6}$ we would have $p_j^2 \mid \sigma(q^\alpha)$ and also $q^{\beta_j} = \frac{p_j^{\lambda_j+1}-1}{p_j-1}$; however from Lemma 2 there is a divisibility $p_j \bigm| \alpha+1 = 2$, which is clearly false as $p_j \geq 3$.

Consequently there exists exactly one type 2 prime, p_I. Note that $\lambda_I \geq 6$. If $\lambda_I \neq 6$ we would have λ_I even and greater than 6, implying $p_I^2 \mid \sigma(q^\alpha)$ and by Lemma 2, $p_I \mid \alpha+1$ which is false. Hence $\lambda_I = 6$ and we can write $\sigma(q^\alpha) = 2p_1^{\lambda_1} \cdots p_k^{\lambda_k}$. But $m = p^6 = m_1 \cdots m_k$ has at most 6 factors, in which case $k \leq 6$; therefore $9 \leq \omega(N) = k+2 \leq 8$ a clear contradiction, completing the proof that $m \neq p^6$. $\square$

Applying Lemmas 7 and 9, we have shown

Theorem 10 *If N is an odd perfect number and the odd prime $q^\alpha \| N$ then the index $\sigma(N/q^\alpha)/q^\alpha$ is either odd when q is the Euler prime, or even but not divisible by 4 when q is not the Euler prime.*
(i) If q is the Euler prime, it cannot take any of the 11 forms $\{p, p^2, p^3, p^4, p^5, p^6, p_1p_2,\ p_1^2p_2,\ p_1^3p_2,\ p_1^2p_2^2,\ p_1p_2p_3\}$ where p is any odd prime and p_1, p_2, p_3 are any distinct odd primes.

(ii) If q is not the Euler prime and the Euler prime divides N to a power greater than 1, it cannot take any of the 7 forms $\{2,\ 2p,\ 2p^2,\ 2p^3,\ 2p^4,\ 2p_1p_2,\ 2p_1^2p_2\}$.

(iii) If q is not the Euler prime and the Euler prime divides N to the power 1, it cannot take any of the 5 forms $\{2,\ 2p,\ 2p^2,\ 2p^3,\ 2p_1p_2\}$.

Therefore the smallest possible value of the index m *is, respectively:*

$$3^2 \times 5 \times 7 = 315 \quad \textit{in case (i)},$$

$$2 \times 3^2 \times 5 \times 7 = 630 \quad \textit{in case (ii)},$$

$$\textit{and} \quad 2 \times 3 \times 5 \times 7 = 210 \quad \textit{in case (iii)}.$$

Corollary 11 *It follows directly that for any odd perfect number, the ratio of the non-Euler part to the Euler part is greater than* $315/2$.

Acknowledgements. The first version of this paper was based on [6]. We are most grateful to a referee who alerted us to the recent work of Chen and Chen [4], and pointed out some errors in an earlier version.

References

[1] P. Acquaah and S. Konyagin, *On prime factors of odd perfect numbers*, Int. J. Number Theory **8** (2012), 1537–1540.

[2] A. S. Bang, *Taltheoretiske undersogelser* Tidsskrift for Mat. **5** (1886), 70–80, 130–137.

[3] G. D. Birkhoff and H. S. Vandiver, *On the integral divisors of* $a^n - b^n$, Ann. of Math. (2) **5** (1904), 173–180.

[4] F.-J. Chen and Y.-G. Chen, *On odd perfect numbers*, Bull. Aust. Math. Soc. **86** (2012), 510–514.

[5] J. A. B. Dris, *Solving the odd perfect number problem: some old and new approaches*, M. Sc. Thesis, De La Salle University, Philippines, 2008.

[6] J. A. B. Dris and F. Luca, *A note on odd perfect numbers*, preprint.

[7] D. Estes, R. Guralnick, M. Schacher and E. Straus, *Equations in prime powers*, Pacific J. Math. **118** (1985), 359–367.

[8] R. Heath-Brown, *Odd perfect numbers*, Math. Proc. Cambridge Phil. Soc. **115** (1994), 191–196.

[9] F. Luca and C. Pomerance, *On the radical of a perfect number*, New York J. Math. **16** (2010), 23–30.

[10] P. P. Nielsen, *An upper bound for odd perfect numbers*, Integers, **3** (2003), #A14.

[11] P. P. Nielsen, *Odd perfect numbers have at least nine prime factors*, Math. Comp. **76** (2007), 2109–2126.

[12] P. Ochem and M. Rao, *Odd perfect numbers are greater than* 10^{1500}, Math. Comp. **81** (2012), 1869–1877.

[13] M. Roitman, *On Zsigmondy primes*, Proc. Amer. Math. Soc. **125** (1997), 1913–1919.

FINITE SUMS THAT INVOLVE RECIPROCALS OF PRODUCTS OF GENERALIZED FIBONACCI NUMBERS

R. S. Melham[1]
School of Mathematical Sciences, University of Technology, Sydney, Australia
ray.melham@uts.edu.au

Received: 12/4/12, Accepted: 4/30/13, Published: 6/24/13

Abstract

In this paper we find closed forms for certain finite sums. In each case the denominator of the summand consists of products of generalized Fibonacci numbers. Furthermore, we express each closed form in terms of *rational numbers*.

1. Introduction

The Fibonacci and Lucas numbers are defined, respectively, for all integers n, by

$$F_n = F_{n-1} + F_{n-2}, \quad F_0 = 0, \quad F_1 = 1,$$

$$L_n = L_{n-1} + L_{n-2}, \quad L_0 = 2, \quad L_1 = 1.$$

Define, for all integers n, the sequences $\{U_n\}$ and $\{V_n\}$ by

$$U_n = pU_{n-1} + U_{n-2}, \quad U_0 = 0, \quad U_1 = 1, \tag{1}$$

$$V_n = pV_{n-1} + V_{n-2}, \quad V_0 = 2, \quad V_1 = p, \tag{2}$$

in which p is a positive integer. Then $\{U_n\}$ and $\{V_n\}$ are integer sequences that generalize the Fibonacci and Lucas numbers, respectively. Throughout this paper p is taken to be a positive integer. Let $\Delta = p^2 + 4$. Then with the use of standard difference techniques it can be shown that the closed forms (the Binet forms) for U_n and V_n are

$$U_n = \frac{\alpha^n - \beta^n}{\alpha - \beta}, \quad V_n = \alpha^n + \beta^n, \tag{3}$$

where $\alpha = \left(p + \sqrt{\Delta}\right)/2$, and $\beta = \left(p - \sqrt{\Delta}\right)/2$.

Next we define, for all integers n, the sequence $\{W_n\}$ by

$$W_n = pW_{n-1} + W_{n-2}, \quad W_0 = a, \quad W_1 = b, \tag{4}$$

[1] I dedicate this paper to my mother Maria. She continues to be a fountain of love and support.

where $a \geq 0$ and $b \geq 0$ are integers with $(a, b) \neq (0, 0)$. These conditions on a and b, together with the fact that p is assumed to be a positive integer, ensure that each of the reciprocal sums in this paper is well defined. It can be shown that

$$W_n = \frac{A\alpha^n - B\beta^n}{\alpha - \beta}, \tag{5}$$

where $A = b - a\beta$ and $B = b - a\alpha$. Associated with $\{W_n\}$ is the constant $e_W = AB = b^2 - pab - a^2$, which occurs in the sequel. Accordingly, $e_F = 1$ and $e_L = -5$.

An identity linking the Fibonacci and Lucas numbers is $L_n = F_{n-1} + F_{n+1}$, and a similar identity links the sequences $\{U_n\}$ and $\{V_n\}$. Motivated by this we define a companion sequence $\{\overline{W}_n\}$ of $\{W_n\}$ by $\overline{W}_n = W_{n-1} + W_{n+1}$. With the use of (5) we see that

$$\overline{W}_n = A\alpha^n + B\beta^n. \tag{6}$$

The sequences $\{W_n\}$ and $\{\overline{W}_n\}$ generalize the sequences $\{U_n\}$ and $\{V_n\}$, respectively. Furthermore $\overline{U}_n = V_n$, and $\overline{V}_n = \Delta U_n$, so that $\overline{F}_n = L_n$, and $\overline{L}_n = 5F_n$.

Let $k \geq 1$, $m \geq 0$, and $n \geq 2$ be integers. In this paper (in Section 3) we give a closed expression, in terms of *rational numbers*, for each of the following sums:

$$S(k,m,n) = \sum_{i=1}^{n-1} \frac{(-1)^{ki}}{W_{ki+m}W_{k(i+1)+m}}, \tag{7}$$

$$T_1(k,m,n) = \sum_{i=1}^{n-1} \frac{(-1)^{ki}U_{k(i+1)+m}}{W_{ki+m}W_{k(i+1)+m}W_{k(i+2)+m}}, \tag{8}$$

$$T_2(k,m,n) = \sum_{i=1}^{n-1} \frac{(-1)^{ki}V_{k(i+1)+m}}{W_{ki+m}W_{k(i+1)+m}W_{k(i+2)+m}}, \tag{9}$$

$$T_3(k,m,n) = \sum_{i=1}^{n-1} \frac{(-1)^{ki}\overline{W}_{k(i+1)+m}}{W_{ki+m}W_{k(i+1)+m}W_{k(i+2)+m}}, \tag{10}$$

$$X(k,m,n) = \sum_{i=1}^{n-1} \frac{1}{W_{ki+m}W_{k(i+1)+m}W_{k(i+2)+m}W_{k(i+3)+m}}. \tag{11}$$

In Section 2 we present some background and motivation for our study. In Section 3 we present our main results, and in Section 4 we present a detailed proof of one of our results. The method of proof that we outline can be used to prove all the results in this paper. It is reasonable to surmise that there are sums, analogous to (7)-(11), with longer products in their denominators, for which closed forms can be found. In Sections 5, 6, and 7 we select some of (7)-(11) and show that this is the case.

2. Background and Motivation

The broad question of summation of reciprocals that involve Fibonacci or generalized Fibonacci numbers has a long history. In this short paper it is not our intention to give this history. Instead, in the next few paragraphs, we give a brief commentary on work that we have cited on reciprocal sums that involve the summands in (7)-(11). After this, we indicate the motivation for the present paper.

André-Jeannin [1] considered the summand in (7) for the particular cases $W_n = U_n$ and $W_n = V_n$. Taking $m = 0$ and k an odd integer, he expressed the infinite sums in terms of the Lambert Series

$$L(x) = \sum_{i=1}^{\infty} \frac{x^i}{1-x^i}, \quad |x| < 1.$$

Inspired by the work of André-Jeannin [1], the authors in [5] considered the analogues of U_n and V_n for the recurrence $W_n = pW_{n-1} - W_{n-2}$, and obtained analogues of André-Jeannin's results for these sequences. The authors first obtained two finite sums in terms of the irrational roots of $x^2 - px + 1 = 0$. They then took the appropriate limits to obtain the corresponding infinite sums. Interestingly, these infinite sums did not involve the Lambert Series, but were expressed in terms of the irrational roots of $x^2 - px + 1 = 0$.

Filipponi [3] considered the summand in (7) for the particular cases $W_n = U_n$ and $W_n = V_n$. Taking $m = 0$ and k an even integer, he expressed the infinite sums in terms of α and β.

André-Jeannin [2] considered the more general sequence of integers defined by $W_n = pW_{n-1} - qW_{n-2}$, in which the initial values W_0 and W_1 are integers, and p and q are integers with $pq \neq 0$. For this sequence he studied the sums

$$\sum_{i=1}^{\infty} \frac{1}{W_{ki+m}W_{k(i+i_0)+m}} \quad \text{and} \quad \sum_{i=1}^{\infty} \frac{q^{ki+m}}{W_{ki+m}W_{k(i+i_0)+m}}, \tag{12}$$

for integers $k \geq 1$, $m \geq 0$, and $i_0 \geq 1$. In the course of his analysis, André-Jeannin expressed any finite sums in terms of one of the roots of $x^2 - px + q = 0$. Furthermore, in (12), he evaluated only the sum on the right in terms of rational numbers.

In [6] we studied the infinite sums

$$\sum_{i=1}^{\infty} \frac{\overline{W}_{k(i+m)}}{W_{ki}W_{k(i+m)}W_{k(i+2m)}}, \tag{13}$$

and

$$\sum_{i=1}^{\infty} \frac{(-1)^i}{W_{ki}W_{k(i+m)}W_{k(i+2m)}W_{k(i+3m)}}, \tag{14}$$

in which $k \geq 1$ and $m \geq 1$ are odd integers.

In [7] we studied the infinite sums

$$\sum_{i=1}^{\infty} \frac{\overline{W}_{k(i+m)}}{W_{ki}W_{k(i+m)}W_{k(i+2m)}}, \tag{15}$$

and

$$\sum_{i=1}^{\infty} \frac{1}{W_{ki}W_{k(i+m)}W_{k(i+2m)}W_{k(i+3m)}}, \tag{16}$$

in which $k \geq 1$ and $m \geq 1$ are integers with k even. For (13)-(16) it was only in the case of (15) that we managed to express the infinite sum as a finite sum of rational numbers.

Reflecting upon the results in the foregoing paragraphs, we realized that sums (finite or infinite) involving summands that are similar to those in (7)-(11) are not usually evaluated in terms of rational numbers. It was this realization that prompted us to embark upon the investigation that led to the present paper. We list two finite sums (see [8]) that we recently evaluated in terms of rational numbers.

$$\sum_{i=1}^{n-1} \frac{1}{F_{ki+m}F_{k(i+1)+m}} = \frac{F_{k(n-1)}}{F_kF_{k+m}F_{kn+m}}, \; n > 1, \tag{17}$$

where $k > 0$ is an even integer, and $m > 0$ is any integer.

$$\sum_{i=1}^{n-1} \frac{1}{L_{ki+m}L_{k(i+1)+m}} = \frac{F_{k(n-1)}}{F_kL_{k+m}L_{kn+m}}, \; n > 1, \tag{18}$$

where $k \neq 0$ is an even integer, and m is any integer. In the present paper our evaluation of $S_1(k,m,n)$ generalizes both (17) and (18).

3. The Main Results

We now state our main results. As stated in Section 1, in (19)-(23) $k \geq 1$, $m \geq 0$, and $n \geq 2$ are assumed to be integers. We have

$$U_kW_{k+m}S(k,m,n) = \frac{(-1)^kU_{k(n-1)}}{W_{kn+m}}, \tag{19}$$

$$\begin{aligned} e_WU_{2k}W_{k+m}T_1(k,m,n) &= -\frac{1}{U_k}\left(\frac{W_{-k}U_{k(n-1)}}{W_{kn+m}} + \frac{(-1)^{k+1}W_kU_{kn}}{W_{k(n+1)+m}}\right) \\ &\quad + \frac{(-1)^{k+1}W_k}{W_{2k+m}}, \end{aligned} \tag{20}$$

$$e_W U_{2k} W_{k+m} T_2(k,m,n) = \frac{1}{U_k}\left(\frac{\overline{W}_{-k} U_{k(n-1)}}{W_{kn+m}} + \frac{(-1)^{k+1}\overline{W}_k U_{kn}}{W_{k(n+1)+m}}\right) + \frac{(-1)^k \overline{W}_k}{W_{2k+m}}, \tag{21}$$

$$U_k W_{k+m} T_3(k,m,n) = \frac{(-1)^k}{W_{2k+m}} - \frac{(-1)^{kn} W_{k+m}}{W_{kn+m} W_{k(n+1)+m}}, \tag{22}$$

$$e_W V_k U_{3k} W_{k+m} X(k,m,n) = \frac{(-1)^m \overline{W}_{4k+m}}{W_{2k+m} W_{3k+m}} + \frac{(-1)^{k+m}}{U_k^2}\left(\frac{U_{k(n-1)}}{W_{kn+m}} + \frac{(-1)^{k+1} V_{2k} U_{kn}}{W_{k(n+1)+m}} + \frac{U_{k(n+1)}}{W_{k(n+2)+m}}\right). \tag{23}$$

To illustrate, we present two examples of (23). Let $k = 1$ and $m = 0$. Then for $W_n = F_n$ (23) becomes

$$\sum_{i=1}^{n-1} \frac{1}{F_i F_{i+1} F_{i+2} F_{i+3}} = \frac{7}{4} - \frac{1}{2}\left(\frac{F_{n-1}}{F_n} + \frac{3F_n}{F_{n+1}} + \frac{F_{n+1}}{F_{n+2}}\right). \tag{24}$$

Let $k = 2$ and $m = 0$. Then for $W_n = L_n$ (23) becomes

$$\sum_{i=1}^{n-1} \frac{1}{L_{2i} L_{2(i+1)} L_{2(i+2)} L_{2(i+3)}} = -\frac{1}{432} - \frac{1}{360}\left(\frac{F_{2(n-1)}}{L_{2n}} - \frac{7F_{2n}}{L_{2(n+1)}} + \frac{F_{2(n+1)}}{L_{2(n+2)}}\right). \tag{25}$$

4. The Method of Proof

In this section, to illustrate our method of proof, we prove (23) in detail. We require the following four identities, which can be proved with the use of the closed forms:

$$U_{kn} W_{kn+m} - U_{k(n-1)} W_{k(n+1)+m} = (-1)^{k(n+1)} U_k W_{k+m}, \tag{26}$$

$$U_{4k} W_{2k+m} - (-1)^k U_k W_{3k+m} - U_k^2 \overline{W}_{4k+m} = (-1)^k U_{3k} W_{k+m}, \tag{27}$$

$$W_{2k+m} W_{3k+m} - W_{k+m} W_{4k+m} = (-1)^{k+m} e_W U_k^2 V_k, \tag{28}$$

and

$$W_{kn+m} W_{k(n+1)+m} - V_{2k} W_{kn+m} W_{k(n+3)+m} + W_{k(n+2)+m} W_{k(n+3)+m} = (-1)^{kn+m} e_W U_k V_k U_{3k}. \tag{29}$$

For $n \geq 2$ denote the right side of (23) by $R(k,m,n)$. Then

$$\begin{aligned} R(k,m,n+1) - R(k,m,n) &= \frac{(-1)^{k+m}}{U_k^2}\left(\frac{U_{kn}}{W_{k(n+1)+m}} - \frac{U_{k(n-1)}}{W_{kn+m}}\right) \\ &\quad + \frac{(-1)^{m+1}V_{2k}}{U_k^2}\left(\frac{U_{k(n+1)}}{W_{k(n+2)+m}} - \frac{U_{kn}}{W_{k(n+1)+m}}\right) \\ &\quad + \frac{(-1)^{k+m}}{U_k^2}\left(\frac{U_{k(n+2)}}{W_{k(n+3)+m}} - \frac{U_{k(n+1)}}{W_{k(n+2)+m}}\right). \end{aligned}$$

With the use of (26) the right side becomes

$$\begin{aligned} &\frac{(-1)^{kn+m}W_{k+m}}{U_k}\left(\frac{1}{W_{kn+m}W_{k(n+1)+m}} - \frac{V_{2k}}{W_{k(n+1)+m}W_{k(n+2)+m}}\right. \\ &\qquad\qquad \left. + \frac{1}{W_{k(n+2)+m}W_{k(n+3)+m}}\right), \end{aligned} \tag{30}$$

and with the use of (29) we see that (30) simplifies to

$$\frac{e_W V_k U_{3k} W_{k+m}}{W_{kn+m}W_{k(n+1)+m}W_{k(n+2)+m}W_{k(n+3)+m}}. \tag{31}$$

Thus

$$\begin{aligned} &R(k,m,n+1) - R(k,m,n) \\ &= e_W V_k U_{3k} W_{k+m}\left(X(k,m,n+1) - X(k,m,n)\right). \end{aligned} \tag{32}$$

Next,

$$\begin{aligned} R(k,m,2) &= \frac{(-1)^{k+m}}{U_k^2}\left(\frac{U_k}{W_{2k+m}} + \frac{(-1)^{k+1}U_{4k}}{W_{3k+m}} + \frac{U_{3k}}{W_{4k+m}}\right) \\ &\quad + \frac{(-1)^m \overline{W}_{4k+m}}{W_{2k+m}W_{3k+m}}. \end{aligned} \tag{33}$$

Expressing the right side of (33) as a fraction with denominator $U_k^2 W_{2k+m}W_{3k+m}W_{4k+m}$, the numerator is

$$\begin{aligned} &(-1)^{m+1}W_{4k+m}\left(U_{4k}W_{2k+m} - (-1)^k U_k W_{3k+m} - U_k^2 \overline{W}_{4k+m}\right) \\ &+ (-1)^{k+m}U_{3k}W_{2k+m}W_{3k+m}. \end{aligned} \tag{34}$$

Then, with the use of (27) and (28), we see that

$$\begin{aligned} R(k,m,2) &= \frac{(-1)^{k+m}U_{3k}\left(W_{2k+m}W_{3k+m} - W_{k+m}W_{4k+m}\right)}{U_k^2 W_{2k+m}W_{3k+m}W_{4k+m}} \\ &= \frac{e_W U_k^2 V_k U_{3k}}{U_k^2 W_{2k+m}W_{3k+m}W_{4k+m}} \\ &= \frac{e_W V_k U_{3k}}{W_{2k+m}W_{3k+m}W_{4k+m}} \\ &= e_W V_k U_{3k} W_{k+m} X(k,m,2). \end{aligned} \tag{35}$$

Taken together, (32) and (35) show that (23) is true.

We remark that all the results in this paper can be proved in a similar manner. Above we made use of identities (26)-(29) to assist in the proof. An alternative method is to simply use brute force. Specifically, to prove that two quantities are equal we substitute the closed forms of the sequences in question and expand with the use of a computer algebra system. All of the results in this paper can be proved with this method. With this method any occurrence of e_W is replaced by AB.

5. Sums that Belong to the Same Family as $T_3(k,m,n)$

Throughout the remainder of this paper $k \geq 1$, $m \geq 0$, and $n \geq 2$ are assumed to be integers. Let us write (22) as

$$U_k W_{k+m} W_{2k+m} T_3(k,m,n) = (-1)^k - \frac{(-1)^{kn} W_{k+m} W_{2k+m}}{W_{kn+m} W_{k(n+1)+m}}. \tag{36}$$

Now define the following sum, where the denominator of the summand consists of seven factors.

$$T_7(k,m,n) = \sum_{i=1}^{n-1} \frac{(-1)^{ki} \overline{W}_{k(i+3)+m}}{W_{ki+m} W_{k(i+1)+m} \cdots W_{k(i+6)+m}}. \tag{37}$$

In (37) we have modified our notation to make it more suggestive. Specifically, the number 7 in the subscript of T_7 denotes seven factors in the denominator of the summand. We continue this convention in what follows.

Next, define the following sum, where the denominator of the summand consists of eleven factors.

$$T_{11}(k,m,n) = \sum_{i=1}^{n-1} \frac{(-1)^{ki} \overline{W}_{k(i+5)+m}}{W_{ki+m} W_{k(i+1)+m} \cdots W_{k(i+10)+m}}. \tag{38}$$

Finally, define the following sum, where the denominator of the summand consists of fifteen factors.

$$T_{15}(k,m,n) = \sum_{i=1}^{n-1} \frac{(-1)^{ki} \overline{W}_{k(i+7)+m}}{W_{ki+m} W_{k(i+1)+m} \cdots W_{k(i+14)+m}}. \tag{39}$$

In each of (37)-(39) the analogy with (10) is clear. Furthermore, the pattern in (36) can be extended to yield

$$U_{3k} W_{k+m} \cdots W_{6k+m} T_7(k,m,n) = (-1)^k - \frac{(-1)^{kn} W_{k+m} \cdots W_{6k+m}}{W_{kn+m} \cdots W_{k(n+5)+m}}, \tag{40}$$

$$U_{5k}W_{k+m}\cdots W_{10k+m}T_{11}\left(k,m,n\right)=(-1)^{k}-\frac{(-1)^{kn}W_{k+m}\cdots W_{10k+m}}{W_{kn+m}\cdots W_{k(n+9)+m}}, \tag{41}$$

and

$$U_{7k}W_{k+m}\cdots W_{14k+m}T_{15}\left(k,m,n\right)=(-1)^{k}-\frac{(-1)^{kn}W_{k+m}\cdots W_{14k+m}}{W_{kn+m}\cdots W_{k(n+13)+m}}. \tag{42}$$

The lists of formulas (37)-(39) and (40)-(42) have clearly defined patterns and can easily be extended by the reader.

Are there similar sums, for which closed forms exist, that have $5, 9, 13, \ldots$ factors in the denominator of the summand? We have discovered such sums. The first few representatives of these sums are

$$t_5\left(k,m,n\right) = \sum_{i=1}^{n-1}\frac{\overline{W}_{k(i+2)+m}}{W_{ki+m}W_{k(i+1)+m}\cdots W_{k(i+4)+m}}, \tag{43}$$

$$t_9\left(k,m,n\right) = \sum_{i=1}^{n-1}\frac{\overline{W}_{k(i+4)+m}}{W_{ki+m}W_{k(i+1)+m}\cdots W_{k(i+8)+m}}, \tag{44}$$

$$t_{13}\left(k,m,n\right) = \sum_{i=1}^{n-1}\frac{\overline{W}_{k(i+6)+m}}{W_{ki+m}W_{k(i+1)+m}\cdots W_{k(i+12)+m}}. \tag{45}$$

We have found that

$$U_{2k}W_{k+m}\cdots W_{4k+m}t_5\left(k,m,n\right) = 1-\frac{W_{k+m}\cdots W_{4k+m}}{W_{kn+m}\cdots W_{k(n+3)+m}}, \tag{46}$$

$$U_{4k}W_{k+m}\cdots W_{8k+m}t_9\left(k,m,n\right) = 1-\frac{W_{k+m}\cdots W_{8k+m}}{W_{kn+m}\cdots W_{k(n+7)+m}}, \tag{47}$$

$$U_{6k}W_{k+m}\cdots W_{12k+m}t_{13}\left(k,m,n\right) = 1-\frac{W_{k+m}\cdots W_{12k+m}}{W_{kn+m}\cdots W_{k(n+11)+m}}. \tag{48}$$

We have not been able to find closed forms for those counterparts to (43)-(45) where each summand is multiplied by $(-1)^{ki}$. The sums (43)-(45) together with their respective closed forms (46)-(48) have clearly defined patterns and can easily be extended by the reader.

Let $k=2$ and $m=0$. Then for $W_n=F_n$ (40) becomes

$$31933440\sum_{i=1}^{n-1}\frac{L_{2(i+3)}}{F_{2i}F_{2(i+1)}\cdots F_{2(i+6)}}=1-\frac{3991680}{F_{2n}F_{2(n+1)}\cdots F_{2(n+5)}}. \tag{49}$$

Let $k=1$ and $m=0$. Then for $W_n=L_n$ (40) becomes

$$33264\sum_{i=1}^{n-1}\frac{5(-1)^{i}F_{i+3}}{L_iL_{i+1}\cdots L_{i+6}}=-1-\frac{16632(-1)^{n}}{L_nL_{n+1}\cdots L_{n+5}}. \tag{50}$$

6. Sums that Belong to the Same Family as $S(k,m,n)$

Define the following sum, where the denominator of the summand consists of six factors.

$$S_6(k,m,n) = \sum_{i=1}^{n-1} \frac{(-1)^{ki}}{W_{ki+m}W_{k(i+1)+m}\cdots W_{k(i+5)+m}}. \tag{51}$$

We have managed to find a closed form for (51). To present this closed form succinctly we define three quantities $c_i = c_i(k)$ as follows:

$$\begin{aligned} c_0 &= 1, \\ c_1 &= (-1)^{k+1}V_kV_{3k}, \\ c_2 &= \frac{(-1)^k\left(V_{6k}+V_{2k}+2(-1)^k\right)}{2}. \end{aligned}$$

Then

$$e_W^2 U_k \cdots U_{5k}\left(S_6(k,m,n) - S_6(k,m,2)\right) = \tag{52}$$
$$U_{k(n-2)}\sum_{i=0}^{2} c_i\left(\frac{1}{W_{(2+i)k+m}W_{(n+i)k+m}} + \frac{1}{W_{(6-i)k+m}W_{(n+4-i)k+m}}\right).$$

Indeed, numerical evidence suggests that a similar closed form exists for the analogous sum $S_{10}(k,m,n)$ with ten factors in the denominator of the summand, and for the analogous sum $S_{14}(k,m,n)$ with fourteen factors in the denominator of the summand. Numerical evidence also suggests that similar closed forms exist for analogous sums that have $18, 22, 26, \ldots$ factors in the denominator of the summand.

Let $k=2$ and $m=0$. Then for $W_n = F_n$ (52) becomes

$$\begin{aligned} 27720\sum_{i=1}^{n-1}\frac{1}{F_{2i}F_{2(i+1)}\cdots F_{2(i+5)}} - \frac{1}{144} &= F_{2(n-2)}\left(\frac{1}{3F_{2n}} - \frac{27}{4F_{2(n+1)}}\right. \\ &\quad + \frac{331}{21F_{2(n+2)}} - \frac{54}{55F_{2(n+3)}} \\ &\quad \left. + \frac{1}{144F_{2(n+4)}}\right). \end{aligned} \tag{53}$$

Let $k=1$ and $m=0$. Then for $W_n = L_n$ (52) becomes

$$\begin{aligned} 750\sum_{i=1}^{n-1}\frac{(-1)^i}{L_iL_{i+1}\cdots L_{i+5}} + \frac{125}{2772} &= F_{n-2}\left(\frac{1}{3L_n} + \frac{1}{L_{n+1}} - \frac{19}{7L_{n+2}}\right. \\ &\quad \left. + \frac{4}{11L_{n+3}} + \frac{1}{18L_{n+4}}\right). \end{aligned} \tag{54}$$

7. Sums that Belong to the Same Family as $X(k,m,n)$

Define the following sum, where the denominator of the summand consists of eight factors.

$$X_8(k,m,n) = \sum_{i=1}^{n-1} \frac{1}{W_{ki+m}W_{k(i+1)+m}\cdots W_{k(i+7)+m}}. \tag{55}$$

We have managed to find a closed form for (55). In order to present this closed form we define four quantities $f_i = f_i(k)$ as follows:

$$\begin{aligned} f_0 &= 1, \\ f_1 &= -\frac{U_{3k}V_{4k}}{U_k}, \\ f_2 &= \frac{(-1)^k U_{3k}\left(V_{8k} + (-1)^k V_{2k} + 1\right)}{U_k}, \\ f_3 &= \frac{(-1)^{k+1} V_{4k}\left(V_{8k} + V_{4k} + 2(-1)^k V_{2k} + 2\right)}{2}. \end{aligned}$$

Then

$$e_W^3 U_k \cdots U_{7k}\left(X_8(k,m,n) - X_8(k,m,2)\right) = \tag{56}$$
$$(-1)^m U_{k(n-2)} \sum_{i=0}^{3} f_i \left(\frac{1}{W_{(2+i)k+m}W_{(n+i)k+m}} + \frac{1}{W_{(8-i)k+m}W_{(n+6-i)k+m}}\right).$$

Numerical evidence suggests that similar closed forms exist for analogous sums that have $12, 16, 20, \ldots$ factors in the denominator of the summand.

Let $k=1$ and $m=0$. Then for $W_n = F_n$ (56) becomes

$$\begin{aligned} 3120\sum_{i=1}^{n-1} \frac{1}{F_iF_{i+1}\cdots F_{i+7}} - \frac{1}{21} &= F_{n-2}\left(\frac{1}{F_n} - \frac{7}{F_{n+1}} - \frac{30}{F_{n+2}} + \frac{70}{F_{n+3}}\right. \\ &\quad \left. - \frac{45}{4F_{n+4}} - \frac{14}{13F_{n+5}} + \frac{1}{21F_{n+6}}\right). \end{aligned} \tag{57}$$

8. Concluding Comments

We have discovered closed forms for variants of (8)-(10) that we do not present here. For instance, we have discovered a closed form for

$$\sum_{i=1}^{n-1} \frac{(-1)^{ki} U_{ki+m}}{W_{ki+m}W_{k(i+1)+m}W_{k(i+2)+m}}. \tag{58}$$

Furthermore, in the spirit of Sections 5-7, we have discovered lengthier analogues of these variants. The possibilities seem endless.

Acknowledgment of priority: Just prior to submitting this article I was made aware of Theorem 1 in [4]. In fact our result (19) is a consequence of Theorem 1 in [4], as is our main result in [8]. I thank Zhi-Wei Sun for alerting me to [4].

References

[1] R. André-Jeannin, Lambert series and the summation of reciprocals in certain Fibonacci-Lucas-type sequences, *Fibonacci Quart.* **28** (1990), 223–226.

[2] R. André-Jeannin, Summation of reciprocals in certain second-order recurring sequences, *Fibonacci Quart.* **35** (1997), 68–74.

[3] P. Filipponi, A note on two theorems of Melham and Shannon, *Fibonacci Quart.* **36** (1998), 66–67.

[4] Hong Hu, Zhi-Wei Sun & Jian-Xin Liu, Reciprocal sums of second-order recurrent sequences, *Fibonacci Quart.* **39** (2001), 214–220.

[5] R. S. Melham & A. G. Shannon, On reciprocal sums of Chebyshev related sequences, *Fibonacci Quart.* **33** (1995), 194–202.

[6] R. S. Melham, Summation of reciprocals which involve products of terms from generalized Fibonacci sequences, *Fibonacci Quart.* **38** (2000), 294–298.

[7] R. S. Melham, Summation of reciprocals which involve products of terms from generalized Fibonacci sequences-Part II, *Fibonacci Quart.* **39** (2001), 264–267.

[8] R. S. Melham, On finite sums of Good and Shar that involve reciprocals of Fibonacci numbers, *Integers* **12** (2012), #A61.

COMPOSITIONS CONSTRAINED BY GRAPH LAPLACIAN MINORS

Benjamin Braun[1]
Department of Mathematics, University of Kentucky, Lexington, Kentucky
benjamin.braun@uky.edu

Robert Davis
Department of Mathematics, University of Kentucky, Lexington, Kentucky
davis.robert@uky.edu

Jessica Doering
Department of Mathematics, University of Kentucky, Lexington, Kentucky
jessica.mckim@uky.edu

Ashley Harrison
Department of Mathematics, University of Kentucky, Lexington, Kentucky
ashley.harrison0525@uky.edu

Jenna Noll
Department of Mathematics, University of Utah, Salt Lake City, Utah
noll@math.utah.edu

Clifford Taylor
Department of Mathematics, University of Kentucky, Lexington, Kentucky
clifford.taylor@uky.edu

Received: 2/23/12, Revised: 11/29/12, Accepted: 5/29/13, Published: 6/24/13

Abstract

Motivated by examples of symmetrically constrained compositions, super convex partitions, and super convex compositions, we initiate the study of partitions and compositions constrained by graph Laplacian minors. We provide a complete description of the multivariate generating functions for such compositions in the case of trees. We answer a question due to Corteel, Savage, and Wilf regarding super convex compositions, which we describe as compositions constrained by Laplacian minors for cycles; we extend this solution to the study of compositions constrained by Laplacian minors of leafed cycles. Connections are established and conjectured between compositions constrained by Laplacian minors of leafed cycles of prime length and algebraic/combinatorial properties of reflexive simplices.

[1]All the authors were partially supported by the NSF through award DMS-0758321 of the first author.

1. Introduction

A *partition* of a positive integer n is a sequence $\lambda := (\lambda_1, \lambda_2, \ldots, \lambda_k)$ of non-negative integers satisfying $\lambda_k \leq \lambda_{k-1} \leq \cdots \leq \lambda_1$ such that $\sum_i \lambda_i = n$. We call each λ_i a *part* of λ and say that λ has k *parts.* A *(weak) composition* of n is a sequence $\lambda := (\lambda_1, \lambda_2, \ldots, \lambda_k)$ of non-negative integers satisfying $\sum_i \lambda_i = n$. Beginning with [1] and through eleven subsequent papers, George Andrews and various coauthors successfully revived Percy MacMahon's technique of *Partition Analysis* (PA) from obscurity as a tool for computing generating functions for partitions and compositions. PA is particularly well-suited to the study of partitions and compositions constrained by a linear system of inequalities, i.e., of the form $A\lambda \geq b$ for some integral matrix A and integral vector b; this is equivalent to the study of integer points in a rational polyhedral cone. Motivated in part by this renewed interest in PA, there has been growing interest in the application of polyhedral-geometric techniques to the study of integer partitions and compositions.

For example, in [21], Igor Pak considers several partition theory problems from the point of view of lattice-point enumeration, leading to the establishment of unique bijective correspondences. In [4], Matthias Beck, Ira Gessel, Sunyoung Lee, and Carla Savage consider the general case of determining integer-point transforms of symmetrically constrained families of compositions; their techniques include both unimodular decompositions of cones and the use of permutation statistics. In [11], Katie Bright and Carla Savage produce a bijection between certain "boxes of width t" of lecture hall partitions and integer points in the cube $[0,t]^d$; the methods for defining this bijection are motivated via geometry and again involve permutation statistics. In [3], the first author, Matthias Beck, and Nguyen Le give geometric proofs of several of the results in the PA papers by George Andrews et al; these proofs often produce multivariate generating functions in cases where PA produced only univariate identities. Finally, motivated by [4] and [11], the first author and Matthias Beck use geometric techniques to establish new multivariate Euler-Mahonian distributions in [10].

This paper continues this line of investigation by initiating the explicit study of partitions and compositions constrained by graph Laplacian minors. Algebraic graph theory is a rich source of interesting constraint matrices, the first being the signed incidence matrix of a graph, defined as follows. Let $G = (\{0, 1, \ldots, n\}, E)$ be a simple graph with $n + 1$ vertices and $|E| = m$ edges.

Definition 1.1. For any fixed edge orientation ε, the *signed vertex-edge incidence matrix* of G with respect to ε is the $(n + 1) \times m$ matrix ∂, with rows indexed by vertices of G and columns indexed by edges of G, whose (v, e)-th entry is 1 if v is the positive end of e, -1 if v is the negative end of e, and 0 otherwise.

The study of compositions with constraint matrices given by ∂^T for various graphs was recently advanced in [13] by using five guidelines from MacMahon's partition analysis along with the Omega software package. Additionally, such compositions often fall within the realm of the theory of P-partitions [24, Chapter 4]. Our goal is to extend this line of investigation by using as constraint matrices various minors of graph Laplacians, defined as follows.

Definition 1.2. The *vertex Laplacian* of G is $L := D - A$, where D is the $(n+1) \times (n+1)$ matrix indexed by the vertices of G such that the matrix entry $d_{i,j}$ is the degree of vertex i if $i = j$ and 0 otherwise and A is the 0/1-valued adjacency matrix of G indexed in the same way as D. Given a graph G, the *i-th Laplacian minor* of G is L_i, the Laplacian matrix L with row i and column i deleted.

The connection between the incidence matrix of a graph and its Laplacian is the well-known equation $L = \partial\partial^T$ [8, Chapter 6]. Kirchoff's Matrix-Tree Theorem [8] asserts that given a graph G on $\{0, \ldots, n\}$ and given any i between 0 and n, $\det(L_i)$ is equal to the number of labeled spanning trees of G. This establishes a strong connection between Laplacian minors and enumerative properties of graphs.

There are several examples of partitions and compositions occurring in the literature on partition and composition enumeration that are constrained by graph Laplacian minors. The following three such examples provide further motivation for the study of these objects; note that these examples were not originally introduced in the context of graph Laplacian minors.

Example 1.3. Let K_{n+1} be the complete graph on $\{0, 1, \ldots, n\}$. The compositions determined by $L_0\lambda \geq 0$ are one example of the class of *symmetrically constrained compositions.* A study of multivariate generating functions for symmetrically constrained compositions was given by Beck, Gessel, Lee, and Savage in [4].

Example 1.4. Let P_n be the path graph with vertices $\{0, 1, \ldots, n-1\}$. The partitions determined by $L_0\lambda \geq 0$ are named *super convex partitions* by Corteel, Savage, and Wilf in Example 3 of [12]; they were first studied using techniques from MacMahon's Partition Analysis by Andrews in [1], where they are named *partitions with mixed difference conditions.* The multivariate generating function for such partitions is described in [12].

Example 1.5. Let C_n be the cycle on vertices $\{0, 1, \ldots, n-1\}$. The compositions determined by $L_0\lambda \geq 0$ were introduced by Corteel, Savage, and Wilf in the concluding remarks of [12], where they are named *super convex compositions.* The determination of the multivariate generating function for super convex compositions was given as an open problem at the end of [12].

We now introduce our primary object of study.

Definition 1.6. Given a graph G with Laplacian L and a vertex i of G, the *compositions constrained by the i-th Laplacian minor of* G are those satisfying $L_i\lambda \geq 0$.

Our paper is structured as follows. In Section 2, we review techniques from integer point enumeration in polyhedral cones that will be useful for producing generating function identities. In Section 3, we investigate compositions constrained by Laplacian minors for trees. Theorem 3.1 gives a combinatorial interpretation for the entries of the matrix inverse for tree Laplacian minors, leading to Corollary 3.2, the main result of the section.

In Section 4, we provide a solution to the open problem of Corteel, Savage, and Wilf discussed in Example 1.5. However, as shown in the section, our solution is less clear than one might want. Thus, motivated by the underlying graphical structure of our compositions, in Section 5 we examine compositions constrained by Laplacian minors for leafed cycles, i.e., for cycles with an additional vertex forming a leaf appended. These compositions are closely related to the super convex compositions in Example 1.5. The addition of a leaf to the cycle causes a change in the structure of the resulting compositions, leading to Theorem 5.1. We also consider in Conjecture 5.5 a possible combinatorial interpretation for the compositions constrained by leafed cycles, and we prove special cases of this conjecture.

In Section 6, we consider generating functions for certain subfamilies of compositions constrained by Laplacian minors for leafed cycles of prime length. Surprisingly, these generating functions are Ehrhart series for a family of reflexive simplices, as shown in Theorem 6.3. After introducing Ehrhart series and reflexive polytopes and proving Theorem 6.3, we indicate some possible connections between our enumerative problem and the algebraic/combinatorial structure of these simplices.

We close in Section 7 with a conjecture regarding compositions constrained by Laplacian minors for leafed cycles of length 2^k.

2. Cones and Generating Functions

The problem of enumerating integer compositions and partitions satisfying a system of linear constraints $A\lambda \geq b$ is identical to that of enumerating integer points in the polytope or polyhedral cone defined by these constraints. For further background related to integer point enumeration in polyhedra, see [6].

Given a square matrix A with $\det(A) \neq 0$, the solution set

$$\mathcal{C} = \{x \in \mathbb{R}^n : Ax \geq 0\}$$

is called an *n-dimensional simplicial cone*; the reason for the use of "simplicial" comes from $\mathcal{C}$ being an n-dimensional cone with n bounding hyperplanes, also called *facets*. The following lemma, whose proof is straightforward, indicates that simplicial cones can be described either by specifying these n facets or by specifying the n ray generators for the cone.

Lemma 2.1. *Let A be an invertible matrix in $\mathbb{R}^{n\times n}$. Then*

$$\{x \in \mathbb{R}^n : Ax \geq 0\} = \{A^{-1}y : y \geq 0\}$$

where each inequality is understood componentwise.

We encode the integer points in a cone $\mathcal{C}$ using the *integer point transform* of $\mathcal{C}$ given by the multivariate generating function

$$\sigma_{\mathcal{C}}(z_1, z_2, \ldots, z_n) := \sum_{m \in \mathcal{C}\cap\mathbb{Z}^n} z_1^{m_1} z_2^{m_2} \cdots z_n^{m_n} .$$

This function encodes the integer points in $\mathcal{C}$ as a formal sum of monomials. Throughout this paper, we will typically need to specialize $\sigma_{\mathcal{C}}$ in one of two ways. If we set $z_1 = z_2 = \cdots = z_n = q$, we obtain

$$\sigma_{\mathcal{C}}(q, q, \ldots, q) = \sigma_{\mathcal{C}}(q) = \sum_{m \in \mathcal{C}\cap\mathbb{Z}^n} q^{|m|} ,$$

where $|m| = m_1 + m_2 + \cdots + m_n$. Setting $z_1 = q$ and $z_2 = z_3 = \cdots = z_n = 1$, we obtain

$$\sigma_{\mathcal{C}}(q, 1, \ldots, 1) = \sigma_{\mathcal{C}}(q) = \sum_{m \in \mathcal{C}\cap\mathbb{Z}^n} q^{m_1} .$$

When A is an integer matrix having $\det(A) = 1$, the ray generating matrix A^{-1} for the n-dimensional simplicial cone $\mathcal{C} = \{x \in \mathbb{R}^n : Ax \geq 0\}$ has integer entries; hence, every integer point in $\mathcal{C}$ can be expressed as $A^{-1}y$ for a unique integer vector $y \geq 0$. In this case, the integer point transform can be simplified as a product of geometric series as follows.

Lemma 2.2. *Let $A = (a_{i,j})$ be an integer matrix with $\det(A) = 1$ and inverse matrix $A^{-1} = (b_{i,j})$. Then*

$$\sigma_{\mathcal{C}}(z_1, z_2, \ldots, z_n) := \frac{1}{\prod_{j=1}^{n}(1 - z_1^{b_{1,j}} z_2^{b_{2,j}} \cdots z_n^{b_{n,j}})}$$

More generally, suppose that $\det(A) \neq 1$. In this case, $\sigma_{\mathcal{C}}(z_1, z_2, \ldots, z_n)$ is still rational, but A^{-1} need not have integer entries. As a result, representations of integer points in $\mathcal{C}$ are not necessarily obtained by using integral combinations of the minimal integral vectors on the generating rays of $\mathcal{C}$. We can work around this difficulty in the following manner.

Let $v_1, v_2, \ldots, v_n$ be the (integral) ray generators of $\mathcal{C}$. Define the *(half-open) fundamental parallelepiped* (FPP) to be

$$\Pi = \{\lambda_1 v_1 + \lambda_2 v_2 + \cdots + \lambda_n v_n \ : 0 \leq \lambda_1, \lambda_2, \ldots, \lambda_n < 1\} .$$

The cone $\mathcal{C}$ can be tiled by integer translates of Π, but there may be integer points contained in Π that can only be obtained using strictly rational scaling factors of the ray generators of $\mathcal{C}$. This leads to our main tool for computing generating functions.

Lemma 2.3. *Using the preceding notation, if $w_1, w_2, \ldots, w_m$ are the integer points contained in Π, then the integer point transform of $\mathcal{C}$ is given by*

$$\sigma_{\mathcal{C}}(z_1, z_2, \ldots, z_n) := \frac{\sum_{i=1}^{m} z_1^{w_{i,1}} z_2^{w_{i,2}} \cdots z_n^{w_{i,n}}}{\prod_{j=1}^{n}(1 - z_1^{v_{1,j}} z_2^{v_{2,j}} \cdots z_n^{v_{n,j}})}. \tag{1}$$

3. Trees

We begin our study of compositions constrained by graph Laplacian minors with the special case of trees. Since a tree T has a unique spanning tree, namely itself, the Matrix-Tree Theorem implies that $\det(L_i) = 1$ for any vertex i. Thus, compositions constrained by tree Laplacian minors will have a generating function of the form

$$\sigma_{\mathcal{C}}(z_1, z_2, \ldots, z_n) = \frac{1}{\prod_{j=1}^{n}(1 - z_1^{b_{1,j}} z_2^{b_{2,j}} \cdots z_n^{b_{n,j}})}.$$

Since the $b_{k,j}$'s arise as the entries in the inverse matrix for L_i, we are interested in calculating this inverse.

Observe that without loss of generality we may consider only those vertices i which are leaves of T. This is due to the fact that taking the minor at a non-leaf vertex produces a matrix L_i which is a block matrix. Each individual block consists of a Laplacian minor from a tree formed by re-attaching vertex i to a connected component of $T \setminus \{i\}$, then taking the minor of the resulting tree at i. Thus, the generating function for compositions constrained by L_i is the product of those for the compositions constrained by the blocks of L_i.

3.1. Inverses of Laplacian Minors for Trees

To determine the denominator of $\sigma_{\mathcal{C}}$ for an arbitrary tree T minored at a leaf, we first find a combinatorial interpretation for the columns of the inverse of the Laplacian minor for T. Throughout this section, we assume that T is a tree on the vertex set $\{0, \ldots, n\}$ with an arbitrary orientation of the edges of T. The following theorem is a special case of Corollary 4.3 in [26]; we include a simple proof for the sake of completeness.

Theorem 3.1. *Consider a tree T and its corresponding Laplacian matrix L. Minor at a leaf n of T and let $L_n^{-1} = (l_{i,j})$. Then entry $l_{i,j}$ is the distance from n to the path connecting the vertices i and j. Equivalently, $l_{i,j}$ is the length of the intersection of the path from n to i with the path from n to j.*

Theorem 3.1 has the following implication on the level of generating functions. Let $\sigma_{\mathcal{C}}(q)$ denote the specialized generating function $\sigma_{\mathcal{C}}(q, q, \ldots, q)$.

Corollary 3.2. *The specialized generating function for a tree T minored at leaf n is given by*

$$\sigma_C(q) = \frac{1}{\prod_{i=0}^{n-1}(1-q^{b_i})},$$

where b_i is the sum of the distances between n and the paths connecting vertex i with the other vertices of T.

For example, let G have vertex set $\{1,2,\ldots,7\}$ and edges $\{1,2\}$, $\{2,3\}$, $\{3,4\}$, $\{4,5\}$, $\{3,6\}$, and $\{6,7\}$. Thus, G has three branches of length two emanating from 3. If $n=7$, then $b_1 = 14 = 1+2+2+2+3+4$, where for example the distance from 7 to the path connecting 1 to 4 contributes a summand of two, since the path from 1 to 4 is $(1,2,3,4)$ and the path from 7 to this path is $(7,6,3)$.

To prove Theorem 3.1, we return to the fact that $L = \partial\partial^T$ and proceed combinatorially. We say the *n-th subminor of ∂* (resp. of ∂^T) imposed by a vertex n of a tree is the square matrix in which the n-th row, (resp. column) is deleted. We denote the n-th subminor of ∂ as ∂_n, with the understanding that ∂_n^T is $(\partial_n)^T$. It is easy to see that since $L = \partial\partial^T$, we have $L_n = \partial_n\partial_n^T$; thus, $L_n^{-1} = (\partial_n^T)^{-1}\partial_n^{-1}$.

Proposition 3.3. *Let $G_n = (g_{e,j})$, $0 \le j \le n-1$, where $g_{e,j}$ is determined as follows:*

- *1 if edge e is on the path between j and n and e points away from n;*
- *-1 if edge e is on the path between j and n and e points towards n;*
- *0 if edge e is not on the path between j and n.*

Then $G_n\partial_n = I$, hence $\partial_n^{-1} = G_n$.

Proof. Set $G_n\partial_n = M = (m_{i,j})$ and first consider $m_{i,i}$. Entries of row i of ∂_n are non-zero for each edge, i.e., column, adjacent to i. The path connecting vertices i and n overlaps with exactly one of the edges adjacent to i, which is reflected in G_n. By construction, the entries in ∂_n and G_n corresponding to this vertex/edge pairing have the same sign. So $m_{i,i} = 1$ for all i.

Now consider $m_{i,j}$ with $i \ne j$. The path from j to n overlaps with either no edges or exactly two edges connected to i. If there is no overlap, then the product of row i in G_n and column j in ∂_n is zero. If there is overlap, suppose both edges point either toward or away from n. Then the entries in G_n have the same sign, but those in ∂_n have different signs, thus summing to zero. A similar canceling occurs if the edges point in different directions. So, $m_{i,j} = 0$ if $i \ne j$, implying $M = I$. Therefore, $G_n = \partial_n^{-1}$. □

Proof of Theorem 3.1. Since $L_n^{-1} = (\partial_n^T)^{-1}\partial_n^{-1}$, we may multiply columns i and j in ∂_n^{-1} component-wise to find $l_{i,j}$. Consider an arbitrary edge of the graph. If the edge belongs to only one of the paths from i to n or from j to n, then there will be a zero corresponding to that edge in either column i or j, contributing nothing to entry $l_{i,j}$. Otherwise we obtain a 1, since the sign of a row for any edge is constant. Thus, summing the component-wise product will record the length of the path from n to the path between i and j. □

Theorem 3.1 provides some insight into the form of the generating function for compositions constrained by an arbitrary tree. However, determining the lengths of paths between vertices in trees is not an easy exercise in general. In the next subsection, we analyze a special case.

3.2. k-ary Trees

We next provide an explicit formula for a specialized generating function for compositions constrained by Laplacian minors of k-ary trees. Recall that a k-ary tree is a tree formed from a root node by adding k leaves to the root, then adding k leaves to each of these vertices, etc., stopping after a finite number of iterations of this process. To study the compositions constrained by a Laplacian minor of a k-ary tree T, we add a leaf labeled 0 to the root of T and minor at this new vertex. We define the *level* of a vertex as its distance to vertex 0 and a *subtree* of a vertex as the tree which includes the vertex and all of its children (we assume all edges are oriented away from vertex 0). Recall that $[n]_k := 1 + k + k^2 + k^3 + \cdots + k^{n-1}$.

Theorem 3.4. *For a k-ary tree with n levels,*

$$\sigma_C(q) = \frac{1}{\prod_{j=1}^{n} \left(1 - q^{j[n-j+1]_k + \sum_{i=1}^{j-1}(j-i)k^{n-(j-i)}}\right)^{k^{j-1}}}.$$

Proof. Suppose a k-ary tree T is given and consider a column of L_0^{-1} corresponding to a vertex $v \in T$. This column sum will be the sum of the distances between 0 and all distinct paths in the tree with v as an endpoint. By the symmetry of k-ary trees, we choose an arbitrary vertex v at level $j \in [n]$, since the corresponding column sum will be the same as any other vertex on that level.

We must show that $j[n-j+1]_k + \sum_{i=1}^{j-1}(j-i)(k^{n-(j-i)})$ is the column sum corresponding to any vertex at level j. For any vertex c that is a descendent of v (and including v), the distance between 0 and the path from v to c will be j. There are

$$1 + k + k^2 + \cdots + k^{n-j} = \frac{k^{n-j+1} - 1}{k - 1} = [n-j+1]_k$$

such vertices and each of these vertices contributes j to the column sum, yielding our first summand above.

Consider the path between 0 and v, $P = \{v = w_0, w_1, w_2, \ldots, w_{j-1}, 0\}$. For each $i \in [j-1]$, define S_i to be the set of vertices which are in the subtree with w_i as the root but not in the subtree with w_{i-1} as the root; note that the vertex w_i is on level $j-i$ of T. There are $k^0 + k^1 + \cdots + k^{n-(j-i)}$ vertices in the subtree with w_i as a root, and $k^0 + k^1 + \cdots + k^{n-(j-i)-1}$ vertices in the subtree of w_{i-1}, yielding

$$k^0 + k^1 + \cdots + k^{n-(j-i)} - (k^0 + k^1 + \cdots + k^{n-(j-i)-1}) = k^{n-(j-i)}$$

vertices in S_i. Since the path connecting v and s passes through w_i, which is the closest point on this path to 0, the shortest distance between 0 and the path connecting v and any vertex $s \in S_i$ is $j-i$. Hence, the column sum corresponding to an element in S_i will be $(j-i)(k^{n-(j-i)})$, yielding $\sum_{i=1}^{j-1}(j-i)(k^{n-(j-i)})$ as the total column sum for a vertex not a descendent of v, which is our second desired summand. The exponents in the denominator of our generating function follow after noting that there are k^{j-1} vertices on the j-th level. □

When $k = 2$, the formulas for our exponents simplify nicely, as follows.

Corollary 3.5. *For a binary tree with n levels,*

$$\sigma_{\mathcal{C}}(q) = \frac{1}{\prod_{j=1}^{n}\left(1 - q^{2^{n-j+1}(2^j-1)-j}\right)^{2^{j-1}}}$$

We omit the proof of Corollary 3.5, as it results from a tedious calculation where the key step is the use of $\sum_{i=1}^{j-1}(j-i)2^i = \sum_{i=1}^{j-1}(2 + 2^2 + \cdots + 2^i)$.

4. n-Cycles

In this section we consider the open problem due to Corteel, Savage, and Wilf given in Example 1.5. Recall that the Laplacian minor for a general n-cycle on vertices $1, 2, \ldots, n$, labelled cyclically and minored at n, has the following form when rows (resp. columns) are indexed $1, 2, \ldots, n-1$ from top to bottom (resp. left to right).

$$L_{n,cyc} := \begin{pmatrix} 2 & -1 & 0 & 0 & \cdots & 0 & 0 & 0 \\ -1 & 2 & -1 & 0 & \cdots & 0 & 0 & 0 \\ 0 & -1 & 2 & -1 & \cdots & 0 & 0 & 0 \\ \vdots & \vdots & \vdots & \vdots & \ddots & \vdots & \vdots & \vdots \\ 0 & 0 & 0 & 0 & \cdots & -1 & 2 & -1 \\ 0 & 0 & 0 & 0 & \cdots & 0 & -1 & 2 \end{pmatrix}.$$

Define

$$\mathcal{K}_n := \{\lambda \in \mathbb{R}^{n-1} : L_{n,cyc}\lambda \geq 0\}$$

and

$$\sigma_{\mathcal{K}_n}(z_1,\ldots,z_{n-1}) := \sum_{\lambda\in\mathcal{K}_n\cap\mathbb{Z}^{n-1}} z_1^{\lambda_1}\cdots z_{n-1}^{\lambda_{n-1}}.$$

Our goal is to study the integer point transform for $\mathcal{K}_n$ by enumerating the integer points in a fundamental parallelepiped for $\mathcal{K}_n$ and applying Lemma 2.3; note that a fundamental parallelepiped is determined by a choice of integral ray generators, which we must first identify. By Lemma 2.1, the columns of $L_{n,cyc}^{-1}$ provide (non-integral) ray generators of $\mathcal{K}_n$, with entries given as follows.

Proposition 4.1. *For an n-cycle and $1 \le i,j \le n-1$, $L_{n,cyc}^{-1} = (b_{i,j})$ where*

$$b_{i,j} = \begin{cases} \frac{i(n-j)}{n} & \text{if } i \le j \\ \frac{j(n-i)}{n} & \text{if } i > j \end{cases}.$$

Proof. Define the matrix $B = (b_{i,j})$ using the b-values given above. Let $B\cdot L_{n,cyc} = (a_{i,j})$; since $L_{n,cyc}$ and B are symmetric matrices, it follows that $a_{i,j} = a_{j,i}$. Entry $a_{i,j}$ is given by $2b_{i,j} - b_{i,x} - b_{i,y}$ where x and y are the neighbors of j in the cycle, where we drop the b-terms for x or y equal to n. It is straightforward to show that $BL_{n,cyc} = I$ through a case-by-case analysis.

As a representative example, for $i < j$ we have

$$\begin{aligned} a_{i,j} &= 2b_{i,j} - b_{i,j-1} - b_{n,j+1} \\ &= 2\left(\frac{i(n-j)}{n}\right) - \left(\frac{i(n-j+1)}{n}\right) - \left(\frac{i(n-j-1)}{n}\right) \\ &= 0. \end{aligned}$$

□

We next wish to classify the integer points in the fundamental parallelepiped Π_n, as defined prior to Lemma 2.3, for the ray generators of $\mathcal{K}_n$ given by the columns of $nL_{n,cyc}^{-1}$. Observe first that every integer point $\lambda \in \mathcal{K}_n$, hence every integer point in $\Pi_n \subset \mathcal{K}_n$, can be expressed in the form $\lambda = L_{n,cyc}^{-1}c = nL_{n,cyc}^{-1}(\frac{c}{n})$ for an integer vector c; this follows from the fact that $L_{n,cyc}\lambda = c$ is an integer vector for all such λ. By definition, every integer point in Π_n is a linear combination of the columns of $nL_{n,cyc}^{-1}$ with coefficients in $[0,1)$. Thus, if $\lambda \in \Pi_n$, it follows that

$$\lambda = \left(nL_{n,cyc}^{-1}\right)\left(\frac{c}{n}\right)$$

where the entries in c are necessarily taken from $\{0,1,2,\ldots,n-1\}$.

Our next two propositions identify and use an underlying group-theoretic structure for the columns of $nL_{n,cyc}^{-1}$ to classify the integer points in Π_n.

Proposition 4.2. *Let $M := nL_{n,cyc}^{-1} \pmod{n}$, i.e., reduce all entries in $nL_{n,cyc}^{-1}$ modulo n. Then*

$$M = \begin{pmatrix} | & | & \cdots & | & | \\ v_1 & v_2 & \cdots & v_{n-2} & v_{n-1} \\ | & | & \cdots & | & | \end{pmatrix}$$

where $v_1 = (n-1, n-2, \ldots, 2, 1)^T \in \mathbb{Z}_n^{n-1}$ and $v_k = kv_1 \in \mathbb{Z}_n^{n-1}$.

Proof. Let $M = (m_{i,j})$. Since $m_{i,j}$ is equal to either $i(n-j)$ or $j(n-i)$, it follows that $m_{i,j} \equiv -ij \bmod n$. Consider column v_1. We find

$$\begin{aligned} v_1 &= (-1 \cdot 1, -2 \cdot 1, \ldots, -(n-1) \cdot 1)^T \bmod n, \\ &= (n-1, n-2, \ldots, 1)^T \in \mathbb{Z}_n^{n-1}. \end{aligned}$$

Evaluating an arbitrary column v_k, $1 \leq k \leq n-1$, we have

$$\begin{aligned} v_k &= (-1 \cdot k, -2 \cdot k, \ldots, -(n-1) \cdot k)^T \bmod n, \\ &= k \cdot v_1 \in \mathbb{Z}_n^{n-1}. \end{aligned}$$

□

Proposition 4.3. *If the point $x = (x_1, x_2, \ldots, x_{n-1}) \in \Pi_n$ is of the form $L_{n,cyc}^{-1}c$ for some vector c with entries taken from $\{0, 1, 2, \ldots, n-1\}$, then x is an integer point if and only if x_1 is an integer.*

Proof. If x is an integer vector, then x_1 is certainly an integer. For the converse, suppose that x_1 is an integer. By the symmetry of $nL_{n,cyc}^{-1}$, v_1 is the first row of $nL_{n,cyc}^{-1}$ (mod$n$), where v_1 is as defined in Proposition 4.2. Since x_1 is an integer, it must be that $v_1 \cdot c \equiv \sum_{j=1}^{n} v_{1,j}c_j \equiv 0 \bmod n$. Since $v_k = kv_1$,

$$\sum_{j=1}^{n} v_{k,j}c_j \equiv \sum_{j=1}^{n} kv_{1,j}c_j \equiv k\sum_{j=1}^{n} v_{1,j}c_j \equiv 0 \bmod n.$$

Therefore, x_k is also an integer. □

The integer points in Π_n are therefore parametrized by the set of solutions to the system

$$\sum_{j=1}^{n-1}(n-j)c_j \equiv 0 \pmod{n}$$

where $c_j \in \{0, 1, 2, \ldots, n-1\}$ (which has also appeared in the study of "Hermite reciprocity" given in [15]). Equivalently, these points are parametrized by the set of partitions of multiples of n into positive parts not exceeding $n-1$, with no more than $n-1$ of each part. Our next theorem results from combining the preceding discussion with Lemma 2.3.

Theorem 4.4. *Let S_n denote the set of solutions to the system $\sum_{j=1}^{n-1}(n-j)c_j \equiv 0 \pmod{n}$ where $c_j \in \{0,1,2,\ldots,n-1\}$. Let $l_{i,j}$ denote the i,j-th entry of $nL_{n,cyc}^{-1}$. Then*

$$\sigma_{\mathcal{K}_n}(z_1,\ldots,z_n) = \frac{\sum_{c\in S_n}\left(z_1^{\sum_{j=1}^n l_{1,j}c_j} z_2^{\sum_{j=1}^n l_{2,j}c_j} \cdots z_n^{\sum_{j=1}^n l_{n,j}c_j}\right)}{\prod_{j=1}^n (1 - z_1^{l_{1,j}} z_2^{l_{2,j}} \cdots z_n^{l_{n,j}})}.$$

While Theorem 4.4 and its specializations resolve the question of Corteel, Savage, and Wilf, this resolution is not as clean as one might like. In the next section, we show how a slight modification of the graph underlying this problem leads to much more simply stated and elegant, but still closely related, univariate generating function identities.

5. Leafed n-Cycles

In this section we examine the case of compositions constrained by n-cycles augmented with a leaf; we will refer to such a graph as a *leafed n-cycle*. Throughout this section, we label the leaf vertex as n and label the vertices of the n-cycle cyclically as $0, 1, \ldots, n-1$, where n is adjacent to 0 as depicted in Figure 1.

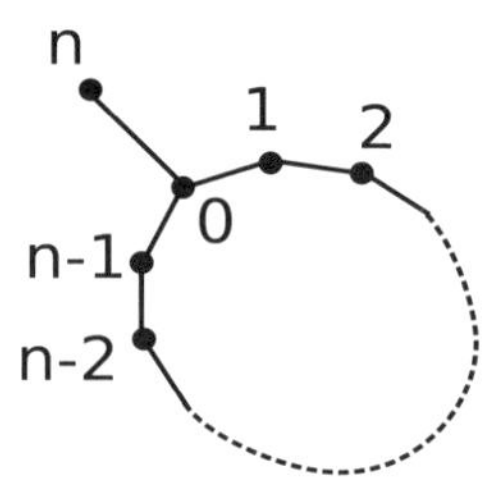

Figure 1: A leafed n-cycle

For a leafed n-cycle, we denote by L_n the minor of the Laplacian taken at vertex n and observe that it has the form

$$L_n = \begin{pmatrix} 3 & -1 & 0 & 0 & 0 & \cdots & 0 & 0 & -1 \\ -1 & 2 & -1 & 0 & 0 & \cdots & 0 & 0 & 0 \\ 0 & -1 & 2 & -1 & 0 & \cdots & 0 & 0 & 0 \\ 0 & 0 & -1 & 2 & -1 & \cdots & 0 & 0 & 0 \\ \vdots & \vdots & \vdots & \vdots & \vdots & \ddots & \vdots & \vdots & \vdots \\ 0 & 0 & 0 & 0 & 0 & \cdots & -1 & 2 & -1 \\ -1 & 0 & 0 & 0 & 0 & \cdots & 0 & -1 & 2 \end{pmatrix}$$

where the columns (resp. rows) are labeled from top to bottom (resp. left to right) by $0, 1, 2, \ldots, n-1$. Define

$$\mathcal{C}_n := \{\lambda \in \mathbb{R}^n : L_n \lambda \geq 0\} \quad \text{and} \quad \sigma_{\mathcal{C}_n}(q) := \sum_{\lambda \in \mathcal{C}_n \cap \mathbb{Z}^n} q^{\lambda_0} .$$

Note that we will focus on the first coordinate of the integer points in $\mathcal{C}_n$.

5.1. A Generating Function Identity

Our main result regarding leafed cycles is the following identity.

Theorem 5.1. *Let S_n denote the set of solutions to the system*

$$0 \cdot c_0 + \sum_{j=1}^{n-1} (n-j) c_j \equiv 0 \,(\mathrm{mod}\, n) ,$$

where $c_j \in \{0, 1, 2, \ldots, n-1\}$. For $c \in S_n$, define $\phi(c) := \sum_{j=0}^{n-1} c_j$. Then

$$\sigma_{\mathcal{C}_n}(q) = \frac{\sum_{c \in S_n} q^{\phi(c)}}{(1-q^n)^n} .$$

The sum $\sum_{c \in S_n} q^{\phi(c)}$ is the ordinary generating function for the "number of parts" statistic on the set of all partitions of multiples of n into parts not exceeding $n-1$, with no more than $n-1$ of each part, where we may include 0 as a part. The similarity between Theorem 5.1 and Theorem 4.4 is immediate. However, by switching from n-cycles to leafed n-cycles, a more elegant univariate identity is obtained than those obtained by straightforward specializations of Theorem 4.4. We remark that our proof of Theorem 5.1 can easily be extended to produce a multivariate statement analogous to Theorem 4.4; we omit this, since it is the specialization above that we find most interesting. As in the previous section, our goal is to prove Theorem 5.1 through a careful study of the integer points in a fundamental parallelepiped for $\mathcal{C}_n$ and an application of Lemma 2.3.

We first explicitly describe the entries in L_n^{-1}, which are closely related to the entries of $L_{n,cyc}^{-1}$.

Proposition 5.2. *For a leafed n-cycle and $0 \leq i, j \leq n-1$, $L_n^{-1} = (b_{i,j})$ where*

$$b_{i,j} = \begin{cases} \frac{i(n-j)}{n} + 1 & \text{if } i \leq j \\ \frac{j(n-i)}{n} + 1 & \text{if } i > j \end{cases} .$$

Proof. This is a straightforward adaptation of the proof of Proposition 4.1. □

Every entry in the top row of L_n^{-1} is equal to 1; it is this property of the inverse that allows us to effectively study the generating function recording only the first entry of each composition in $\mathcal{C}_n$.

As in the previous section, we next wish to classify the integer points in the fundamental parallelepiped Π_n, which now denotes the fundamental parallelepiped for $\mathcal{C}_n$, for the ray generators of $\mathcal{C}_n$ given by the columns of nL_n^{-1}. As in the situation for $\mathcal{K}_n$, every integer point $\lambda \in \mathcal{C}_n$, hence every integer point in $\Pi_n \subset \mathcal{C}_n$, can be expressed in the form $\lambda = L_n^{-1}c = nL_n^{-1}(\frac{c}{n})$ for an integer vector c. Thus, for the same reasons as given in the last section, if $\lambda \in \Pi_n$, then $\lambda = \left(nL_n^{-1}\right)\left(\frac{c}{n}\right)$ where the entries in c are necessarily taken from $\{0, 1, 2, \ldots, n-1\}$.

The next two propositions are analogues of Propositions 4.2 and 4.3.

Proposition 5.3. *Let the matrix $M = nL_n^{-1} (\mathrm{mod}\ n)$. Then*

$$M = \begin{pmatrix} | & | & | & \cdots & | & | \\ v_0 & v_1 & v_2 & \cdots & v_{n-2} & v_{n-1} \\ | & | & | & \cdots & | & | \end{pmatrix}$$

where $v_1 = (0, n-1, n-2, \ldots, 2, 1)^T \in \mathbb{Z}_n^n$ and $v_k = kv_1 \in \mathbb{Z}_n^n$.

Proof. This proof is a straightforward extension of that for Proposition 4.2. □

Proposition 5.4. *If the point $x = (x_0, x_1, \ldots, x_{n-1}) \in \Pi_n$ is of the form $L_n^{-1}c$ for some vector c with entries taken from $\{0, 1, 2, \ldots, n-1\}$, then x is an integer point if and only if x_1 is an integer.*

Proof. The proof is almost identical to that for Proposition 4.3. □

Proof of Theorem 5.1. By Proposition 5.3 and 5.4, the integer points in Π_n are parametrized by the set of solutions to the system

$$0 \cdot c_0 + \sum_{j=1}^{n-1} (n-j)c_j \equiv 0 \,(\mathrm{mod}\, n)\,,$$

where $c_j \in \{0, 1, 2, \ldots, n-1\}$. Equivalently, these points are parametrized by the set of partitions of multiples of n into non-negative parts not exceeding $n-1$, with no more than $n-1$ of each part.

The key observation needed to apply Lemma 2.3 and complete our proof is that $\lambda_0 = \sum_i c_i$ for any point λ such that $L_n\lambda = c \in \Pi_n$. This is a consequence of Proposition 5.2, which shows that the first row (and column) of L_n^{-1} are vectors of all ones. □

5.2. Cyclically Distinct Compositions

We define two compositions, each having n parts, to be *cyclically distinct* if one cannot be obtained from the other via a cyclic shift of the parts. For example, $(0, 3, 1)$ is not cyclically distinct from $(1, 0, 3)$ or $(3, 1, 0)$, since the latter two arise as cyclic shifts of the entires of the first, while $(0,3,1)$ and $(1,3,0)$ are cyclically distinct. This subsection is devoted to the following conjecture, which provides a combinatorial interpretation for the coefficients of $\sigma_{\mathcal{C}_n}(q)$.

Conjecture 5.5. *The number of integer points $\lambda \in \mathcal{C}_n$ with $\lambda_0 = m$ is equal to the number of cyclically distinct classes of compositions of m into n parts.*

Conjecture 5.5 is based on experimental evidence for small values of n; further, we used LattE [18] to check the conjecture for all prime values of n up to $n = 17$. An obvious way to attack Conjecture 5.5 is to attempt to prove that for any $\lambda \in \mathcal{C}_n$ with $\lambda_0 = m$, we have that $\lambda = L_n^{-1}c$ for an integer vector c whose entries form a weak composition of m that is cyclically distinct from all other such c vectors forming elements of $\mathcal{C}_n$. Unfortunately, this does not work out, as the following example demonstrates.

Example 5.6. The weak compositions of 3 into 3 parts are given by the following families F_1, F_2, F_3, and F_4; each family is the orbit of a single composition under the cyclic shift operation.

$$\begin{array}{llll} F_1: & (3,0,0) & (0,3,0) & (0,0,3) \\ F_2: & (2,1,0) & (0,2,1) & (1,0,2) \\ F_3: & (1,2,0) & (0,1,2) & (2,0,1) \\ F_4: & (1,1,1) & & \end{array}$$

The four compositions which satisfy $L_3^{-1}c \in \mathcal{C}_3$ are the elements of F_1 and F_4, which correspond to the compositions $(3,3,3)$, $(3,5,4)$, $(3,4,5)$, and $(3,4,4)$ in $\mathcal{C}_3$. Thus, while this shows that Conjecture 5.5 holds in this case, it is not a consequence of the c-vectors being themselves cyclically distinct.

While this obvious approach to Conjecture 5.5 fails in certain cases, it is successful in others, as seen in the following theorem.

Theorem 5.7. *Let n be prime and fix m such that $\gcd(m,n) = 1$. The number of integer points in $\mathcal{C}_n$ with first coordinate m is equal to the number of cyclically distinct classes of compositions of m into n parts.*

Proof. We must show that for every composition c of m with n parts, exactly one of the cyclic shifts of c will produce an integer point when L_n^{-1} is applied. Begin with a composition $c = (c_0, c_1, \ldots, c_{n-1})$ of m such that $L_n^{-1}c \in \mathcal{C}_n \cap \mathbb{Z}^n$. Then for all j,

$$\sum_{i=0}^{n-1} c_i v_{j,i} \equiv 0 \pmod{n}$$

where $v_j = (v_{j,0}, v_{j,1}, \ldots, v_{j,n-1})$ is the jth row of $nL_n^{-1} \pmod{n}$. Now consider another cyclic ordering c' of c where we shift the position of each entry by k, so

$c'_k = (c_k, c_{k+1}, \ldots, c_{k+n-1})$ with the indices taken mod n. Then the jth coordinate of $L_n^{-1}c'$ is given by

$$\sum_{i=0}^{n-1} c_{i+k}v_{j,i} = \sum_{i=0}^{n-1} c_i v_{j,i-k}.$$

By properties of v_j given in Proposition 5.3, shifting the ith coordinate of v_j by k is equivalent to subtracting the row v_k from the row v_j component-wise. So,

$$\sum_{i=0}^{n-1} c_i v_{j,i-k} \equiv \sum_{i=0}^{n-1} c_i(v_{j,i} - v_{j,k}) \equiv \sum_{i=0}^{n-1} c_i v_{j,i} - \sum_{i=0}^{n-1} c_i v_{j,k} \equiv \; 0 - v_{j,k}\sum_{i=0}^{n-1} c_i \; .$$

Since $\sum_{i=0}^{n-1} c_i \not\equiv 0 \bmod n$ by assumption, and since $v_{j,k}\sum_{i=0}^{n-1} c_i \not\equiv 0 \bmod n$ for $j, k \not\equiv 0$ due to n being prime, we have that

$$\sum_{i=0}^{n-1} c_i v_{j,i-k} \not\equiv 0\,.$$

Thus, the non-trivial cyclic shifts of c do not correspond to integer points in $\mathcal{C}_n$. □

It is straightforward to show that for the case where n is prime and $n|m$, for any composition c of m into n parts that satisfies $L_n^{-1}c \in \mathcal{C}_n \cap \mathbb{Z}^n$ all the cyclic shifts of c also satisfy this condition. Thus, the remaining challenge regarding Conjecture 5.5 in the case where n is prime is to show that this union of orbits of compositions under the cycle action contains as many compositions as there are orbits of the cycle action. While we have not been successful in proving this, the study of these compositions reveals a beautiful underlying geometric structure for our cone, which we discuss in the following section.

6. Reflexive Polytopes and Leafed Cycles of Prime Length

For leafed cycles of prime length, the remaining cases of Conjecture 5.5 turn out to be equivalent to the problem of enumerating points in integer dilates of an integer polytope, which is the subject of Ehrhart theory. The polytope in question is a simplex, and we will prove that it has the additional structure of being reflexive, a property we now introduce.

6.1. Reflexive Polytopes

An n-dimensional *polytope* P in $\mathbb{R}^n$ may equivalently [25, Chapter 1] be described as the convex hull of at least $n+1$ points in $\mathbb{R}^n$ or as the intersection of at least $n+1$ linear halfspaces in $\mathbb{R}^n$. When $P = \operatorname{conv}(V)$ and V is the minimal set such that this holds, then we say the points in V are the *vertices* of P. If $P = \text{conv}(V)$ and $|V| = n+1$, then we say P is a *simplex*. We say P is *integral* if the vertices of P are contained in $\mathbb{Z}^n$.

For $P \subset \mathbb{R}^n$ an integral polytope of dimension n, and for $t \in \mathbb{Z}_{>0}$, set $tP := \{tp : p \in P\}$ and $L_P(t) := |\mathbb{Z}^n \cap tP|$, i.e., the number of integer points in tP is $L_P(t)$. Define the *Ehrhart series for* P to be $\text{Ehr}_P(x) := 1 + \sum_{t\geq 1} L_P(t)x^t$. A fundamental theorem due to E. Ehrhart [14] states that for an n-dimensional integral polytope P in $\mathbb{R}^n$, there exist complex values h_j^* so that

$$\text{Ehr}_P(x) = \frac{\sum_{j=0}^n h_j^* x^j}{(1-x)^{n+1}} \quad \text{and} \quad \sum_j h_j^* \neq 0\,.$$

A stronger result, originally due to R. Stanley in [23], is that the h_j^*'s are actually non-negative integers; Stanley's proof of this used commutative algebra, though several combinatorial and geometric proofs have since appeared. We call the coefficient vector $h_P^* := (h_0^*, \dots, h_d^*)$ in the numerator of the rational generating function for $\text{Ehr}_P(x)$ the *h-star vector* for P. The volume of P can be recovered as $(\sum_j h_j^*)/n!$.

Obtaining a general understanding of the structure of h^*-vectors of integral polytopes is currently of great interest. Recent activity, e.g., [2, 5, 7, 9, 17, 16, 20, 22], has focused on the class of reflexive polytopes, which we now introduce as they will be needed in Section 6. Given an n-dimensional polytope P, the *polar* or *dual* polytope to P is

$$P^\Delta = \{x \in \mathbb{R}^n : x \cdot p \leq 1 \text{ for all } p \in P\}\,.$$

Let P° denote the topological interior of P.

Definition 6.1. An n-dimensional polytope P is *reflexive* if $0 \in P^\circ$ and both P and P^Δ are integral.

Reflexive polytopes have many rich properties, as seen in the following lemma.

Lemma 6.2. [2, 17] *P is reflexive if and only if P is an integer polytope with $0 \in P^\circ$ that satisfies one of the following (equivalent) conditions:*

1. *P^Δ is integral.*

2. *$L_{P^\circ}(t+1) = L_P(t)$ for all $t \in \mathbb{Z}_{\geq 0}$, i.e., all lattice points in $\mathbb{R}^n$ sit on the boundary of some non-negative integral dilate of P.*

3. *$h_i^* = h_{n-i}^*$ for all i, where h_i^* is the ith coefficient in the numerator of the Ehrhart series for P.*

Reflexive polytopes are simultaneously a very large class of integral polytopes and a very small one, in the following sense. Due to a theorem of Lagarias and Ziegler [19], there are only finitely many reflexive polytopes (up to unimodular equivalence) in each dimension. On the other hand, Haase and Melnikov [16] proved that every integral polytope is a face of some reflexive polytope. Because of this "large yet small" tension, combined with their importance in other areas of mathematics as noted before, whenever a reflexive polytope arises unexpectedly it is a surprise, an indication that something interesting is happening.

6.2. Reflexive Slices of Laplacian Minor Cones

Our main result in this subsection is Theorem 6.3, which asserts that for prime n values, reflexive polytopes arise as slices of the cone constrained by Laplacian minors for leafed n-cycles. Note that when n is not prime, this construction does not always yield reflexive simplices; this has been verified experimentally with LattE [18] and Lemma 6.2. However, exactly when the construction described below produces reflexive polytopes is at this time not clear.

Let p be an odd prime and let lCP_p be the simplex obtained by intersecting the cone constrained by a leafed p-cycle with the hyperplane $\lambda_1 = p$ in $\mathbb{R}^p$.

Theorem 6.3. *lCP_p is reflexive (after translation by an integral vector).*

Proof. By elementary results about polar polytopes found in [25, Chapter 2], P is a reflexive polytope if and only if P is integral, contains the origin in its interior, and has a half-space description of the form $\{x \in \mathbb{R}^d : Ax \geq -\mathbf{1}\}$ where A is an integral matrix and $\mathbf{1}$ denotes the vector of all ones.

We first observe that lCP_p is clearly an integer polytope, as its vertices are the columns of pL_p^{-1}, which are integral. A halfspace description of lCP_p is given by $\lambda_0 = p$ and

$$\begin{bmatrix} 3 & -1 & 0 & \cdots & -1 \\ -1 & 2 & -1 & \cdots & 0 \\ 0 & -1 & 2 & \cdots & 0 \\ \vdots & \vdots & \vdots & \ddots & \vdots \\ -1 & 0 & 0 & \cdots & 2 \end{bmatrix} \begin{bmatrix} \lambda_0 \\ \lambda_1 \\ \lambda_2 \\ \vdots \\ \lambda_{p-1} \end{bmatrix} \geq \begin{bmatrix} 0 \\ 0 \\ 0 \\ \vdots \\ 0 \end{bmatrix}.$$

Let w_i be the ith column of L_p^{-1}, and observe that $L_p(\sum_i w_i) = L_p(\sum_i L_p^{-1}(e_i)) = \mathbf{1}$. Thus, $\sum_i w_i$ is an interior integer point of our simplex lCP_p; we want to change coordinates so that $\sum_i w_i$ is translated to the origin. We therefore consider solutions to our halfspace description of the form $\lambda + \sum_i w_i$. It follows that the description of our translated lCP_p does not use a λ_0-coordinate (since $\lambda_0 + p = p$ implies $\lambda_0 = 0$), and is given by

$$\begin{bmatrix} -1 & 0 & \cdots & -1 \\ 2 & -1 & \cdots & 0 \\ -1 & 2 & \cdots & 0 \\ \vdots & \vdots & \ddots & \vdots \\ 0 & 0 & \cdots & 2 \end{bmatrix} \begin{bmatrix} \lambda_1 \\ \lambda_2 \\ \vdots \\ \lambda_{p-1} \end{bmatrix} \geq \begin{bmatrix} -1 \\ -1 \\ \vdots \\ -1 \end{bmatrix}.$$

Since lCP_p is an integral simplex, and $\sum_i w_i$ is an integer vector by which we translated lCP_p, our translated lCP_p is still integral. Thus lCP_p is reflexive, and our proof is complete. □

While the proof above is elementary, it ties the study of compositions constrained by graph Laplacian minors into an interesting circle of questions regarding reflexive polytopes. The only remaining case of Conjecture 5.5 is that the number of integer points in $m \cdot lCP_p$ is equal to the number of compositions of mp with p parts, up to cyclic equivalence, which is now asking for a combinatorial interpretation for the Ehrhart series of a reflexive simplex. Because of reflexivity, the generating function for this series yields a rational function with a symmetric numerator, which yields a functional relation on the Ehrhart polynomial for this simplex.

The study of reflexive polytopes goes hand in hand with the study of several other interesting classes of polytopes; one such example are normal polytopes, where an integer polytope P is *normal* if every integer point in the m-th dilate of P is a sum of exactly m integer points in P. In our attempts to prove the prime case of Conjecture 5.5, we noticed that our techniques (though unsuccessful at providing a proof) provided evidence suggesting that lCP_p is normal. Normality is implied by the presence of a unimodular triangulation for an integral polytope; while we are not yet certain of the existence of such a triangulation for lCP_p or of the normality of lCP_p, it would not surprise us if such a triangulation can be found for all prime p.

Our final remark regarding lCP_p, reflexivity, and normality, regards the unimodality of the h^* vector of lCP_p. It is a major open question (see [22] and the references therein) whether all normal reflexive polytopes have unimodal h^*-vectors. For $p \leq 7$, we found using LattE [18] that the h^*-vector for lCP_p is unimodal. We again suspect that this holds in general.

7. Leafed Cycles of Length a Power of Two

We conclude our paper by considering leafed cycles for non-prime values of n; specifically, we study when $n = 2^k$ for some k. Using LattE [18], one observes that the cone constrained by a leafed 8-cycle does not have a reflexive 8-th slice (we suspect this to be the case for all non-prime values of n). Nevertheless, the integer point transform of this cone exhibits some interesting "near-symmetry" in the numerator for small powers of 2. Based on experimental evidence, we offer the following conjecture.

Conjecture 7.1. *Let $\mathcal{C}_{2^k}$ be the cone constrained by the Laplacian minor of a leafed n-cycle where $n = 2^k$ for some integer $k \geq 2$. Then the generating function has the form*

$$\sigma_{\mathcal{C}_{2^k}}(q, 1, 1, \ldots, 1) = \frac{f(q)}{\left(1 - q^{2^k}\right) \prod_{i=0}^{k-1} \left(1 - q^{2^{k-(i+1)}}\right)^{2^i}}$$

where $f(q)$ has the following property. Let $(a_0, \ldots, a_j)$ denote the coefficient list of $f(q)$. If we append a 0 to the end of this coefficient list, then take the difference between the appended coefficient list and its reverse, we obtain the coefficient list of the polynomial

$$\sum_{i=0}^{n-2} \binom{n-2}{i} (-1)^i q^{ni}.$$

The near-symmetry of the numerator polynomial indicates that there should be some interesting structure to the finite group obtained by quotienting the semigroup of integer points in $\mathcal{C}_{2^k}$ by the semigroup generated by a specific choice of integral ray generators for the cone.

Acknowledgements. This paper is the result of a graduate/undergraduate research experience in Summer 2011. The authors thank the anonymous referee for their helpful comments. The first author thanks Matthias Beck, Matthias Koeppe, Peter Paule, Carla Savage, and Zafeirakis Zafeirakopoulos for many interesting discussions during the American Institute of Mathematics SQuaREs on Polyhedral Geometry and Partition Theory.

References

[1] George E. Andrews. MacMahon's partition analysis. I. The lecture hall partition theorem. In *Mathematical essays in honor of Gian-Carlo Rota (Cambridge, MA, 1996)*, volume 161 of *Progr. Math.*, pages 1–22. Birkhäuser Boston, Boston, MA, 1998.

[2] Victor V. Batyrev. Dual polyhedra and mirror symmetry for Calabi-Yau hypersurfaces in toric varieties. *J. Algebraic Geom.*, 3(3):493–535, 1994.

[3] Matthias Beck, Benjamin Braun, and Nguyen Le. Mahonian partition identities via polyhedral geometry. In *From Fourier analysis and number theory to radon transforms and geometry*, volume 28 of *Dev. Math.*, pages 41–54. Springer, New York, 2013.

[4] Matthias Beck, Ira Gessel, Sunyoung Lee, and Carla Savage. Symmetrically constrained compositions. *Ramanujan J.*, 23:355–369, 2010.

[5] Matthias Beck and Serkan Hoşten. Cyclotomic polytopes and growth series of cyclotomic lattices. *Math. Res. Let.*, 13(4):607–622, 2006.

[6] Matthias Beck and Sinai Robins. *Computing the continuous discretely.* Undergraduate Texts in Mathematics. Springer, New York, 2007.

[7] Christian Bey, Martin Henk, and Jörg M. Wills. Notes on the roots of Ehrhart polynomials. *Discrete Comput. Geom.*, 38(1):81–98, 2007.

[8] Norman Biggs. *Algebraic graph theory.* Cambridge Mathematical Library. Cambridge University Press, Cambridge, second edition, 1993.

[9] Benjamin Braun. An Ehrhart series formula for reflexive polytopes. *Electron. J. Combin.*, 13(1):Note 15, 5 pp. (electronic), 2006.

[10] Benjamin Braun and Matthias Beck. Euler-Mahonian permutation statistics via polyhedral geometry. Preprint.

[11] Katie L. Bright and Carla D. Savage. The geometry of lecture hall partitions and quadratic permutation statistics. In *22nd International Conference on Formal Power Series and Algebraic Combinatorics (FPSAC 2010)*, Discrete Math. Theor. Comput. Sci. Proc., AN, pages 569–580. Assoc. Discrete Math. Theor. Comput. Sci., Nancy, 2010.

[12] Sylvie Corteel, Carla D. Savage, and Herbert S. Wilf. A note on partitions and compositions defined by inequalities. *Integers*, 5(1):A24, 11 pp. (electronic), 2005.

[13] J. William Davis, Erwin D'Souza, Sunyoung Lee, and Carla D. Savage. Enumeration of integer solutions to linear inequalities defined by digraphs. In *Integer points in polyhedra—geometry, number theory, representation theory, algebra, optimization, statistics*, volume 452 of *Contemp. Math.*, pages 79–91. Amer. Math. Soc., Providence, RI, 2008.

[14] Eugène Ehrhart. Sur les polyèdres rationnels homothétiques à n dimensions. *C. R. Acad. Sci. Paris*, 254:616–618, 1962.

[15] A. Elashvili, M. Jibladze, and D. Pataraia. Combinatorics of necklaces and "Hermite reciprocity". *J. Algebraic Combin.*, 10(2):173–188, 1999.

[16] Christian Haase and Ilarion V. Melnikov. The reflexive dimension of a lattice polytope. *Ann. Comb.*, 10(2):211–217, 2006.

[17] Takayuki Hibi. Dual polytopes of rational convex polytopes. *Combinatorica*, 12(2):237–240, 1992.

[18] Matthias Köppe. Latte macchiato, version 1.2-mk-0.6. Available from URL http://www.math.ucdavis.edu/%7emkoeppe/latte/, 2009.

[19] Jeffrey C. Lagarias and Günter M. Ziegler. Bounds for lattice polytopes containing a fixed number of interior points in a sublattice. *Canad. J. Math.*, 43(5):1022–1035, 1991.

[20] Mircea Mustaţă and Sam Payne. Ehrhart polynomials and stringy Betti numbers. *Math. Ann.*, 333(4):787–795, 2005.

[21] Igor Pak. Partition identities and geometric bijections. *Proc. Amer. Math. Soc.*, 132(12):3457–3462 (electronic), 2004.

[22] Sam Payne. Ehrhart series and lattice triangulations. *Discrete Comput. Geom.*, 40(3):365–376, 2008.

[23] Richard P. Stanley. Decompositions of rational convex polytopes. *Ann. Discrete Math.*, 6:333–342, 1980. Combinatorial mathematics, optimal designs and their applications (Proc. Sympos. Combin. Math. and Optimal Design, Colorado State Univ., Fort Collins, Colo., 1978).

[24] Richard P. Stanley. *Enumerative combinatorics. Vol. 1*, volume 49 of *Cambridge Studies in Advanced Mathematics*. Cambridge University Press, Cambridge, 1997. With a foreword by Gian-Carlo Rota, Corrected reprint of the 1986 original.

[25] Günter M. Ziegler. *Lectures on polytopes*, volume 152 of *Graduate Texts in Mathematics*. Springer-Verlag, New York, 1995.

[26] Georg Zimmermann. A minimax-condition for the characteristic center of a tree. *Linear and Multilinear Algebra*, 45(2-3):161–187, 1998.

EDGE GROWTH IN GRAPH SQUARES

Michael Goff
Department of Mathematics, Vanderbilt University, Nashville, Tennessee
michael.goff@vanderbilt.edu

Received: 3/4/12, Revised: 5/13/13, Accepted: 6/7/13, Published: 6/24/13

Abstract

We resolve a conjecture of Hegarty regarding the number of edges in the square of a regular graph. If G is a connected d-regular graph with n vertices, the graph square of G is not complete, and G is not a member of two narrow families of graphs, then the square of G has at least $(2 - o_d(1))n$ more edges than G.

1. Introduction

In this paper, we consider the following problem. Let G be a d-regular graph, and let G^2 be the graph with the same vertex set as G and an edge uv whenever u and v are within distance 2 in G. Then find a lower bound on the number of edges of G^2, or $e(G^2)$. With the assumptions that G is connected and that G^2 is not a complete graph, this question was posed by Hegarty [1, Conjecture 1.8].

In his work, Hegarty discussed general graph powers. Let G^k be the graph with an edge uv whenever u and v are within distance k in G. Several authors have considered lower bounds on $e(G^k)$. Hegarty found that $e(G^3) \geq (1+c)dn/2$ if G is a d-regular graph with diameter at least three, with $c = 0.087$. Pokrovskiy [5] found a value of $c = 1/6$, and DeVos and Thomassé [3] improved the value of c to $3/4$ and provided examples to demonstrate that no higher value of c is possible. The latter authors also weakened the d-regular condition to a minimum degree of d.

DeVos, McDonald, and Scheide [2] considered higher powers of G. They found that if G has a minimum degree of $d \geq 2$ and G has at least $\frac{8}{3}d$ vertices, then G^4 has an average degree of at least $\frac{7}{3}d$. Examples demonstrate that neither the $8/3$ nor the $7/3$ constants may be improved. They also found that when the diameter is at least $3k+3$, then the average degree of G^{3k+2} is at least $(2k+1)(d+1) - k(k+1)(d+1)^2/n - 1$, and examples show that this cannot be improved.

Inspiration for Hegarty's work comes from the Cauchy-Davenport theorem, which states that if A is a subset of $\mathbb{Z}_p$ for a prime p, and kA denotes the set of sums of collections of k elements of A, then $|kA| \geq \min(p, k|A| - (k-1))$. Kneser [4] generalized the Cauchy-Davenport theorem to an abelian group H. Now suppose that $A = -A$ and that A contains the identity element. The connection to graph theory comes through the Cayley graph. The Cayley graph $G(H, A)$ has vertex set H and an edge g_1g_2 whenever $g_1 - g_2 \in A$. Then $(G(H, A))^k = G(H, kA)$, and the growth in kA is equivalent to the growth in the vertex degrees of $(G(H, A))^k$.

For large values of d and $k > 2$, $e(G^k)$ exceeds $e(G)$ by at least a constant factor, so long as G^k is not a complete graph and G is d-regular and connected. This is not true for $k = 2$, as examples demonstrate. In that case, Hegarty conjectures the following [1, Conjecture 1.8].

Conjecture 1.1. Let G be a connected d-regular graph with n vertices such that G^2 is not a complete graph. Then $e(G^2) - e(G) \geq (2 - o_d(1))n$.

This conjecture is not true as stated, and we shall see some counterexamples below, but the counterexamples are confined to narrow families of graphs known as snake graphs and peanut graphs. The modifcation of Conjecture 1.1 that $e(G^2) - e(G) \geq (3/2 - o_d(1))n$ is true. However, our main theorem is as follows.

Theorem 1.2. *Let G be a connected d-regular graph with n vertices such that G^2 is not a complete graph and $d > 6$. Also suppose that G is not a snake graph or a peanut graph. Then*

$$e(G^2) - e(G) > 2n\left(1 - \frac{2}{d+1} - \frac{3}{d-3}\right).$$

Our approach is as follows. We define basic terms in Section 2. We rephase the problem by counting ordered pairs of vertices (u, v) such that u and v are at distance 2. In Section 3, we divide G into what we call regions, and into superregions in Section 4, and for each superregion R, we associate a collection of pairs of vertices S_R such that $|S_R| > 4|R|\left(1 - \frac{2}{d+1} - \frac{3}{d-3}\right)$. A particularly important type of superregion is the class of tails, which we discuss in Section 5. It is necessary to show that the superregions are a partition of the vertices of G, which we do in Section 6. In Section 7, we discuss the snake graph and peanut graph in detail. We complete the proof in Section 8 by showing that $S_R \cap S_{R'} = \emptyset$ for distinct superregions S_R and $S_{R'}$.

2. Definitions

This section contains basic definitions that are used for the rest of the paper.

Let G be a graph without multiple edges or loops. $V(G)$ denotes the vertex set of G, and $e(G)$ is the edge set of G. If X is a subset of vertices of G, then $G[X]$ is the *induced subgraph* on X, or the maximal subgraph of G with vertex set X.

The *distance* between two vertices u and v, denoted by $d(u,v)$, is the number of edges in a shortest path between u and v. Thus $d(u,u) = 0$, $d(u,v) = 1$ if there is an edge uv, and so on. For each $i \geq 1$ and vertex v, let $N_i(v)$ be the set of vertices that are distance i from v. We also say that $N(v) := N_1(v)$, and $\deg_i(v) := |N_i(v)|$. Also, $N_2'(v)$ is the set of vertices $u \in N_2(v)$ such that $u \in N(w)$ for some $w \in N_3(v)$. A d-regular graph is a graph such that every vertex v satisfies $\deg(v) := \deg_1(v) = d$.

Let the *graph power* G^k be the graph with $V(G^k) = V(G)$ and an edge uv whenever $d(u,v) \leq k$ in G.

A *low degree vertex* v is a vertex v satisfying $\deg_2(v) \leq 3$. Note that if v is a low degree vertex and $N_2'(v) = \emptyset$, then G contains at most $d+4$ vertices and thus G^2 is complete when G is d-regular with $d > 6$.

Lemma 2.1. *Let $v \in V(G)$ and let $u \in N_2'(v)$. Then* $\deg_2(u) \geq d - \deg_2(v) + 1$. *In particular, if v is low degree, then* $\deg_2(u) \geq d - 2$.

Proof. $G - N_2'(v)$ is disconnected, with one component of $G - N_2'(v)$ being $G_v := \{v\} \cup N(v) \cup (N_2(v) - N_2'(v))$ and another component G' containing a vertex $x' \in N(u)$ that is distance 3 from v. Since $|N_2'(v)| \leq \deg_2(v)$, x' has degree at least $d - \deg_2(v)$ in G', and let $N'(x')$ be the sets of neighbors of x' in G'. Also, choose $x \in G_v \cap N(u)$. Since $N(x) \subset \{v\} \cup N(v) \cup N_2(v)$, the sets $N(x) \cup \{x\}$ and $N'(x') \cup \{x'\}$ are disjoint, and thus $|N(x) \cup N'(x') \cup \{x, x'\}| \geq 2d - \deg_2(v) + 2$. Every vertex of $N(x) \cup N'(x') \cup \{x, x'\}$ is within distance 2 of u, and since at most $d+1$ of them are within distance 1 of u, we have that $\deg_2(u) \geq d - \deg_2(v) + 1$. □

For the remainder of this paper, we assume that G is a connected d-regular graph such that $d > 6$ and G^2 is not complete.

3. Regions

In this section, we discuss regions, a key tool in the proof of our main theorem. Let X be the set of vertices $x \in G$ that satisfy $\deg_2(x) < 4$. Define an equivalence relation $\sim$ on X by saying that $u \sim v$ if there exists a sequence of vertices $(u = v_0, v_1, \ldots, v_t = v)$ such that for $0 \leq i \leq t-1$, $v_i \in X$ and $d(v_i, v_{i+1}) \leq 2$.

Definition 3.1. Let W be the union of an equivalence class $X' \subset X$ under $\sim$ and all neighbors of vertices in X'. Then $G[W]$ is a *region.*

Note that some vertices might not be contained in any region. We prove some basic properties of regions.

Lemma 3.2. *Let X be the set of vertices x of G that satisfy* $\deg_2(x) < 4$, *and let R be a region that contains $v \in X$. Then*
1) R contains at least $d+1$ vertices.

2) $R \subset \{v\} \cup N(v) \cup N_2(v)$.
3) Let $t := \min\{\deg_2(x)|x \in R\}$. *Then* R *contains at most* $d + t + 1$ *vertices.*
4) R *contains at most* $d + 4$ *vertices.*
5) R *is disjoint from any other region* R'.

Proof. Part 1 follows from the fact that, by definition, R contains v and all neighbors of v.

Note that $G - N_2'(v)$ is disconnected, and the component of $G - N_2'(v)$ that contains v is $G_v := \{v\} \cup N(v) \cup (N_2(v) - N_2'(v))$. Let G' be a different component, if there is one. By Lemma 2.1 and $d > 6$, no vertex in $N_2'(v)$ is in X. Consider $u \in N_2'(v)$. If some $w \in N(u) \cap G_v$ is in X, then no $w' \in N(u) \cap G'$ is also in X. To see this, observe that $N_2(w)$ contains $N(u) - G_v$, and so $|N(u) - G_v| < 4$ and $|N(u) \cap G_v| > d - 4$. Since $N_2(w')$ contains $N(u) \cap G_v$, $\deg_2(w') > d - 4 \geq 3$ and $w' \notin X$.

Suppose by way of contradiction that there exists $v' \in R \cap X$ outside of G_v. By definition of a region, there is a sequence of vertices $\{v = v_0, v_1, \ldots, v_k = v'\}$ such that each $v_i \in X$ and $d(v_i, v_{i+1}) \leq 2$ for all $0 \leq i \leq k - 1$. Let v_j be the first vertex in the sequence that is not in G_v. Since $G - N_2'(v)$ is disconnected and no vertex in $N_2'(v)$ is also in X, $d(v_{j-1}, v_j) = 2$, and there exists a vertex $u \in N_2'(v)$ adjacent to both v_{j-1} and v_j. This contradicts the previous paragraph. Part 2 follows. There exists $v \in R \cap X$ with $\deg_2(v) = t$, and Part 3 follows. Part 4 follows by $t < 4$.

Now consider $x \in R \cap R'$ for some region R'. By definition of a region, there exist vertices $v \in R \cap X$ and $v' \in R' \cap X$ such that $d(v, x) \leq 1$ and $d(v', x) \leq 1$. Then $d(v, v') \leq 2$ and thus $R = R'$. This proves Part 5. □

We define several classes of regions. Let R be a region with a vertex v with $\deg_2(v) = 1$ and $N_2(v) = \{u\}$. Let G_v be the component of $G - u$ that contains v. Then G_v contains $d+1$ vertices, namely $\{v\} \cup N(v)$. In G_v, all neighbors of u have degree $d-1$, and all other vertices have degree d. Hence the complement of G_v is a matching on the neighbors of u. Let $t := |G_v \cap N(u)|$, and consider $w \in G_v \cap N(u)$. Since u contains $d - t$ neighbors outside of G_v, $\deg_2(w) = d - t + 1$, and w is low degree if and only if $t \geq d - 2$. Then R is either G_v or $G_v \cup \{u\}$, and the latter holds if and only if $t \geq d-2$. If $t \neq d-1$, then we say that R is an *A region*, and if $t = d - 1$, we say that R is a *B region*. Since a B region contains the complement of a matching on $d - 1$ vertices, a B region can exist only when d is odd.

Next suppose that R is not an A or B region, but R contains a vertex v with $\deg_2(v) = 2$. If R contains a vertex v' with $\deg_2(v') = 2$ and $|N_2'(v')| = 1$, then we say that R is a *C region*. Otherwise, we say that R is a *D region*.

Now suppose that R is a region such that $\deg_2(v) = 3$ for all low degree vertices v in R. Let k be the minimum size of $N_2'(v)$ for low degree vertices in R. In the respective cases that $k = 1, 2$, or 3, we say that R is an *E region*, an *F region*, or a *G region*.

4. Superregions

We now define *superregions.* We show that the superregions of G are a partition of $V(G)$ in Section 6. Before we specify the superregions, we first define sets of vertices associated with G and the superregions.

Let $\mathcal{U}$ be the set of vertices u in G that satisfy $\deg_2(u) \geq d-2$. By Lemma 2.1, if v is a low degree vertex, then $N_2'(v) \subset \mathcal{U}$. In defining superregions, we will also designate special sets $\mathcal{W}$ and $\mathcal{N}$ such that if R is a superregion, then $|R \cap (\mathcal{W} \cup \mathcal{N} - \mathcal{U})| \leq \frac{2}{d+1}|R|$. Let $\mathcal{V} := V(G) - \mathcal{U} - \mathcal{W} - \mathcal{N}$.

Lemma 4.1. *Theorem 1.2 holds for G if $|\mathcal{U}| \geq \frac{3}{d-3}n$.*

Proof. If $v \notin \mathcal{U}$, then $\deg_2(v) \geq 1$. We then have that

$$\sum_{v \in V(G)} \deg_2(v) \geq \frac{3}{d-3}n(d-2) + \frac{d-6}{d-3}n(1) \geq 4n,$$

which implies Theorem 1.2. □

We therefore assume that

$$|\mathcal{U}| < |V(G)|\frac{3}{d-3},$$

and then, since superregions partition $V(G)$,

$$|\mathcal{V}| = |V(G)| - |\mathcal{W}| - |\mathcal{N}| - |\mathcal{U}| > |V(G)|\left(1 - \frac{2}{d+1} - \frac{3}{d-3}\right).$$

For every superregion R, we associate a collection S_R of at least $4|R \cap \mathcal{V}|$ ordered pairs of vertices of the form (x, y) such that $d(x, y) = 2$. Since the superregions partition $V(G)$, we have that $\sum_R |S_R| \geq 4|\mathcal{V}|$. This proves Theorem 1.2 as long as the S_R are disjoint, a matter that is partially addressed below and addressed more fully in Section 8. By the following constructions, S_R may be partitioned into subsets S_{Ri}, for $1 \leq i \leq 4$, as follows. For each pair $(x, y) \in S_R$,

- if x and y are both in R, then $(x, y) \in S_{R1}$;
- if $x \in \mathcal{V} \cap R$ and $y \notin R$, then $(x, y) \in S_{R2}$;
- if $y \in \mathcal{V} \cap R$, y is low degree, and $x \in N_2'(y) - R$, then $(x, y) \in S_{R3}$;
- if $x \in \mathcal{W} \cap R$ and $y \notin R$, then $(x, y) \in S_{R4}$.

Lemma 4.2. *Suppose that $(x, y) \in S_R \cap S_{R'}$ for distinct superregions R and R'. Then either $(x, y) \in S_{R4} \cap S_{R'3}$ or $(x, y) \in S_{R3} \cap S_{R'4}$.*

Proof. By definition of the S_{Ri}, one of x or y is in R and the other is in R'. Without loss of generality, assume that $x \in R$ and $y \in R'$. Then $(x, y) \in S_{R'3}$. Then $x \notin \mathcal{V}$, and so $(x, y) \in S_{R4}$. □

As we specify the superregions and the sets S_R over the rest of this section and Section 5, observe that the following holds by construction.

Lemma 4.3. *Let v be a low degree vertex of a superregion R. Then there are at most $4 - \deg_2(v)$ vertices u such that $(u, v) \in S_{R3}$.*

4.1. Single Vertex Superregions

If v is a vertex that is not contained in any region, then say that $\{v\}$ is a superregion R. Since v is not in a region, $\deg_2(v) \geq 4$. If $v \notin \mathcal{V}$, then $S_R = \emptyset$. Otherwise, let $S_R = S_{R2}$ be $\{(v, u) : u \in N_2(v)\}$.

4.2. D, E, F, G Regions

If R is a region that is not contained in any other superregion, then R is a superregion. If R is a D, E, F, or G region, for all $v \in R \cap \mathcal{V}$, $\deg_2(v) + |N_2'(v)| \geq 4$. Choose $A_v \subseteq N_2'(v)$ so that $A_v = N_2'(v)$ if $\deg_2(v) = 2$, $|A_v| = 1$ if $\deg_2(v) = 3$, and otherwise $A_v = \emptyset$. Set

$$S_v := \{(v, a)\}_{a \in N_2(v)} \cup \{(a, v)\}_{a \in A_v}.$$

Then set $S_R := \cup_{v \in R \cap \mathcal{V}} S_v$. The pairs (v, a) are either in S_{R1} or S_{R2}, and the pairs (a, v) are either in S_{R1} or S_{R3}. Since $A_v \cap \mathcal{V} = \emptyset$, the S_v are disjoint, and thus $|S_R| \geq 4|R \cap \mathcal{V}|$. The sets A_v are not chosen arbitrarily, but the choice will be made strategically in order to insure that $S_R \cap S_{R'} = \emptyset$ for all superregions $R' \neq R$. This matter is discussed more fully below.

4.3. A Regions

Next, suppose that R is an A region and a superregion. Let V be the set of vertices $v \in R$ with $\deg_2(v) = 1$, let u be the unique vertex in $N_2(v)$ for each vertex in $v \in V$, and let X be the set of remaining vertices of R. Choose distinct $w_1, w_2 \in X$ and set $R \cap \mathcal{W} = \{w_1, w_2\}$. By definition of an A region, $|V| \geq 3$. Note that $\deg_2(x) = |V|$ for $x \in X$. Let S_R be the the union of the following sets of pairs:

- if $|V| < d - 2$ (so that $X \cap \mathcal{U} = \emptyset$), all $(|X| - 2)(|V|)$ pairs of the form (x, y) for $x \in X - \mathcal{W}$ and $y \in N_2(x)$ (these pairs are either in S_{R1} or S_{R2});
- $(|X| - 2) \max(0, 4 - |V|)$ pairs of the form (y, x) for $x \in X - \mathcal{W}$ and $y \in N_2(x) - R$ (call this set S'_R);
- all $|V|$ pairs (v, u) for $v \in V$ (these pairs are either in S_{R1} or S_{R2});

- all $|V|$ pairs (u, v) for $v \in V$ (these pairs are either in S_{R1} or S_{R3});
- and all $2|V|$ pairs of the form (w, y) for $w \in \{w_1, w_2\}$ and $y \in N_2(w)$ (these pairs are in S_{R4} if $y \notin R$ and in S_{R1} otherwise).

Then $|S_R| \geq 4(|V| + |X| - 2) \geq 4|R \cap \mathcal{V}|$.

Observe that $S'_R \subseteq S_{R3}$. To see this, if $(|X| - 2)\max(0, 4 - |V|) \neq 0$, which happens only if $|V| = 3$, then for $x \in X - \mathcal{W}$, $\deg_2(x) = 3$. Let $N_2(x) = \{x', y, y'\}$ for $x' \in R$ and $y, y' \notin R$. The only possible neighbors of y in $N_1(x) \cup N_2(x)$ are u and y', and thus y is adjacent to some vertex that is distance 3 from x. The same is true for y'. Thus $\{y, y'\} \subset N'_2(x)$, showing that $(y, x), (y', x) \in S_{R3}$.

4.4. C Regions

Suppose that R is a C region. Let v be a vertex in R with $\deg_2(v) = 2$ and $N'_2(v) = \{u\}$. Let G_v be the component of $G - \{u\}$ that contains v and $V := N_2(u) \cap G_v$. Choose a vertex w in $R \cap N(u)$ and set $R \cap \mathcal{W} = \{w\}$, and let X be the set of remaining vertices in $R \cap N(u)$. We prove several lemmas on the structure of C regions.

Lemma 4.4. *With all quantities as above, $V(R) \subseteq V \cup X \cup \{w, u\}$.*

Proof. By Lemma 3.2, $V(R) \subset G_v \cup \{u\}$. It suffices to show that if $v' \in R$, then $d(v', u) \leq 2$. Since $G_v = \{v\} \cup N(v) \cup N_2(v) - \{u\}$, $|G_v| = d + 2$. Suppose by way of contradiction that $d(v', u) \geq 3$. Then $\{v'\} \cup N(v') \cup N_2(v') \subseteq G_v$, and thus $\deg_2(v') \leq 1$, a contradiction to the definition of a C region. □

Lemma 4.5. *Let all quantities be as above. Then $G_v = R$.*

Proof. First we show that $G_v \subset R$. As in the proof of Lemma 4.4, G_v has $d + 2$ vertices. Since $N(v) \subset G_v$, there exists one vertex $z \in G_v$ that is not adjacent to v. By definition of a region, all vertices of $G_v - \{z\}$ are in R. Each $v' \in V$ is in R since v' is within distance 2 of v and is low degree. Since u must have at least one neighbor outside of G_v, u has at most $d - 1$ neighbors in G_v. Some vertex $z' \in \{z\} \cup N(z)$ is not in $\{u\} \cup N(u)$ and thus $z' \in V$, and therefore $z \in R$ by definition of a region.

Finally, we must show that $u \notin R$. Note that $|V| = 3$ is impossible, since then G_v would contain $d - 1$ vertices of degree $d - 1$ and 3 vertices of degree d, and the sum of the degrees of all vertices would be odd. Thus $|V| \geq 4$, and since u has $d + 2 - |V|$ neighbors in G_v, u has $|V| - 2 \geq 2$ neighbors outside of G_v. If $x \in R \cap N(u)$, then $N_2(x)$ contains 2 vertices in G_v and $|V| - 2 \geq 2$ outside of G_v, and thus x is not low degree. By construction of a region, $u \notin R$. □

Lemma 4.6. *With all quantities as above, $|R \cap N(u)| \geq 2$.*

Proof. Suppose by way of contradiction that $R \cap N(u) = \{w\}$. By Lemma 4.5, $R = \{v\} \cup N(v) \cup N_2(v) - N_2'(v)$. Thus some vertex x of R is not adjacent to w, and $d(x, u) \geq 3$. This contradicts Lemma 4.4. □

Let y_1 and y_2 be two distinct neighbors of u outside of R, which exist as in the proof of Lemma 4.5. Also observe that $u \in \mathcal{U}$. Let S_R be the union of the following sets:

- the $|V| + 2|X|$ pairs of the form (x, y) for $x \in V \cup X$ and $y \in R \cap N_2(x)$. These pairs are each in S_{R1}, and they exist since every vertex in V and X have, respectively 1 and 2 non-neighbors in G_v. Furthermore, $(G_v)^2$ is complete since in G_v, every vertex has degree at least $d-1$ and there are $d+2$ vertices in total;
- the $|V|$ pairs (v', u) for $v' \in V$ (these pairs are in S_{R2});
- the $|V|$ pairs (u, v') for $v' \in V$ (these pairs are in S_{R3});
- the $|V|$ pairs of the form (w, y) for $y \in N_2(w)$ (these pairs are either in S_{R1} or S_{R4});
- and if $|V| < d-2$, the $2|X|$ pairs $(x, y_1), (x, y_2)$ for each $x \in X$ (call this set of pairs S_R').

As in the proof of Lemma 4.5, $\deg_2(x) = |V|$ for all $x \in R \cap N(u)$, and the condition that $|V| < d-2$ is equivalent to $x \notin \mathcal{U}$ for each $x \in R \cap N(u)$, and so $S_R' \subset S_{R2}$. We have that $|S_R| \geq 4|R \cap \mathcal{V}|$.

B regions are discussed in the context of tails in the next section.

5. Tails and Superregions

In this section we define several types of superregions that are based on a tail subgraph. Every superregion described in this section contains a B region, and thus these superregions exist only when d is odd.

5.1. Tails

The following construction, a tail, is a subgraph of all superregions defined in this section. A tail itself is a superregion unless it is contained in a larger superregion.

Definition 5.1. Let R_1 be a B region, and for $2 \leq i \leq k$, let R_i be a clique on $d+1$ vertices with the edge $v_i v_i'$ removed. Let v be the degree $d-1$ vertex of R_1. Suppose that there are edges vv_2 and $v_i' v_{i+1}$ for $2 \leq i \leq k-1$. Then $G[R_1 \cup \ldots \cup R_k]$ is a *tail*. See Figure 1.

The R_i are the *segments* of T. We may have $k = 1$, in which case the tail is a B region. An *improper tail* is contained in a larger tail, and otherwise a tail is *proper*. If T is a tail, let u_T be the unique vertex that is adjacent to a vertex in T but not itself in T, and let w_T be the unique vertex in T that is adjacent to u_T.

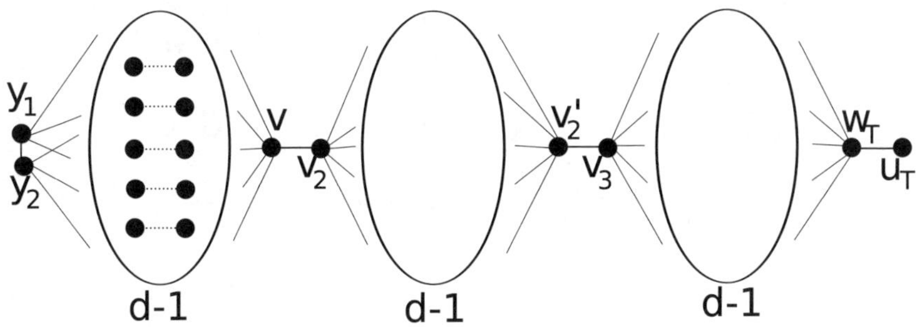

Figure 1: Tail. Note that u_T is not part of the tail.

Let T be a tail. Let $\mathcal{N} \cap T$ be $\{y_1, y_2\} = R_1 \cap N_2(v)$. Associate the following sets of pairs of vertices with T:

- S_{T1} is the set of $(4k-3)(d-1)$ pairs of vertices (x, y) and (y, x) such that $x, y \in T$, $d(x, y) = 2$, and $\deg_2(x) = 2$;
- S_{T2} (resp. S_{T3}) is the set of $d - 1$ pairs of vertices of the form (x, u_T) (resp. (u_T, x)) such that $x \in T$ and $d(x, u_T) = 2$.

In the event that T is a superregion, let Y be the set of $d - 1$ neighbors of u_T that are not in T and set $T \cap \mathcal{W} = \{w_T\}$. Then say that

- S_{T4} is the set of $d - 1$ pairs (w_T, y) for $y \in Y$.

Then $T \cap \mathcal{V}$ contains $d-1$ vertices in each segment of T, and $|S_T| = |S_{T1}| + |S_{T2}| + |S_{T3}| + |S_{T4}| = 4|T \cap \mathcal{V}|$.

We now define a snake graph, our first exception to Conjecture 1.1. See Figure 2 for an illustration.

Definition 5.2. A *snake graph* G is the union of tails $T \cup T'$ with $u_T = w_{T'}$.

The two superregions of G above are $G - R_1$ and $G - R_1'$, where R_1 and R_1' are the two B regions. These superregions intersect if either T or T' contain more than 1 segment.

We prove an important lemma on tails.

Lemma 5.3. *Let T and T' be tails with nonempty intersection. Then either G is a snake graph or one of T or T' is contained in the other.*

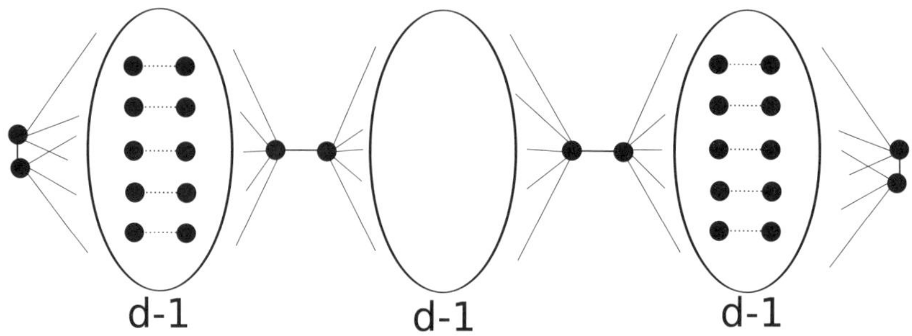

Figure 2: Snake graph

Proof. Let the segments of T and T' be $R_1, R_2, \ldots, R_k$ and $R'_1, \ldots, R'_j$ respectively, and without loss of generality, $k \geq j$. Also suppose that R_1 and R'_1 are B regions, and that R_i and R_{i+1} are joined by an edge for $1 \leq i \leq k-1$, as are R'_i and R'_{i+1} for $1 \leq i \leq j-1$. Since a tail is a union of regions, $T \cap T'$ is also a union of regions.

First suppose that $R_1 = R'_1$, and we show by induction that $T' \subset T$. By construction of a B region, there is only one vertex outside of R_1 that is adjacent to a vertex in R_1, and thus R_2 and R'_2 intersect and are thus equal. Now suppose that $R_{i-1} = R'_{i-1}$ for some $3 \leq i \leq j$. By construction, there are two vertices outside of R_{i-1} that are adjacent to a vertex in R_{i-1}, one of which is in $R_{i-2} = R'_{i-2}$. Thus the other must be in both R_i and R'_i, and we conclude that $R_i = R'_i$. Thus if $R_1 = R'_1$, then $T' \subseteq T$.

Now consider the case that $R_1 \neq R'_1$. Choose k' and j' so that $R_{k'} = R'_{j'}$ and the sum $j' + k'$ is minimized. Let T_1 be $R_1 \cup \ldots \cup R_{k'-1}$, and let T_2 be $R'_1 \cup \ldots \cup R'_{j'-1}$. Let v and v' be the two vertices of $R_{k'} = R'_{j'}$ that are adjacent to vertices outside of $R_{k'}$. Since T_1 and T_2 are disjoint, $w_{T_1} \neq w_{T_2}$. Then either v or v' is adjacent to w_{T_1}; assume without loss of generality an edge vw_{T_1}. Either vw_{T_2} or $v'w_{T_2}$ is an edge, and since v is adjacent to only one vertex outside of R_k, there is an edge $v'w_{T_2}$. There can be no other vertices or edges in G, as the above establishes a d-regular graph. Thus G is a snake graph. □

5.2. Multitails

Let $\{T_1, \ldots, T_m\}$ be a maximal collection of tails such that $u_{T_1} = \cdots = u_{T_m}$. If $m \geq 2$, then $R = T_1 \cup \cdots \cup T_m$ is an *m-multitail*, and multitails are superregions. Let S_R, each S_{Ri} for $1 \leq i \leq 4$, $R \cap \mathcal{N}$, and $R \cap \mathcal{W}$ be the unions of the corresponding sets over the T_j. Then $|S_R| = 4|R \cap \mathcal{V}|$.

5.3. α-Tails

We define our next superregion R, an *α-tail*, as follows. See Figure 3 for an illustration.

Definition 5.4. Let T be a tail with k segments. Let H' be a subgraph of G consisting of the following vertices:

- u_T,
- a vertex z,
- a set X with $|X| = d - 2$,
- a subset $X' \subset X$ of even cardinality less than $d - 3$, and
- vertices y_1 and y_2.

Let the edge set of H' be as follows:

- $u_T z$,
- $u_T x$ for all $x \in X$,
- a complete subgraph on X with a matching on X' removed,
- xy_1 and xy_2 for all $x \in X$,
- $y_1 y_2$,
- zy_1 and zy_2, and
- xz for all $x \in X'$.

All vertices $x \in X - X'$ are low degree and are all in the same region H. Then we say that $T \cup H$ is an α-tail.

Either $H = H'$ or $H = H' - \{z\}$; this follows from Lemma 3.2 and the observations that $N(x) = H' - \{z\}$ for $x \in X - X'$, and $N_2(x) = \{w_T, z\}$, whereas $w_T \notin H$ since $w_T \in T$. Let $R := T \cup H$. We call H the *head region* of R. Note that $G - \{z\}$ is disconnected with $R - \{z\}$ a component. Also note that if we were to allow $|X'| = d - 3$, then z would have d neighbors in R, and G would be a snake graph.

Take $R \cap \mathcal{N}$ to be the two vertices $v_1, v_2 \in T$ that satisfy $\deg_2(v_1) = \deg(v_2) = 1$, together with y_1 and y_2. We take $R \cap \mathcal{W} = \{u_T\}$. Note that $z \in \mathcal{U}$. Also, $X - X' \subset \mathcal{V}$, whereas X' might or might not be a subset of $\mathcal{V}$. Then $|R \cap \mathcal{V}| = k(d-1) + |X \cap \mathcal{V}|$. Let Z be the set of vertices adjacent to z and not in R; $|Z| = d - |X'| - 3 \geq 2$ by $|X'| < d - 3$, $|X'|$ even, and d odd. Let b_1 and b_2 be two distinct vertices of Z.

Now we let S_R be the following sets of pairs:

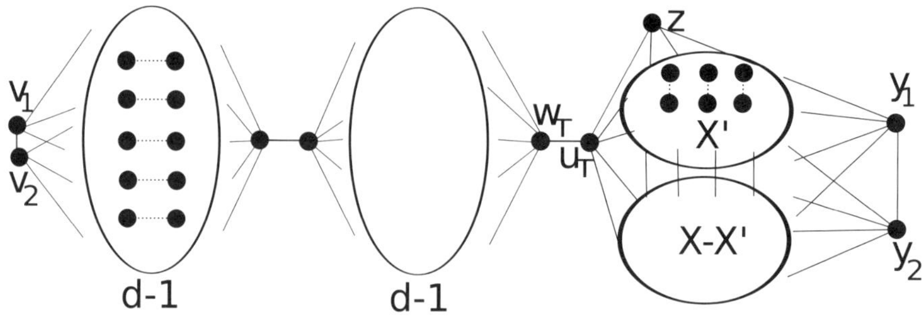

Figure 3: α-tail

- the $(4k-1)(d-1)$ elements of $S_{T1} \cup S_{T2} \cup S_{T3}$ (these are in S_{R1});
- the 2 pairs (u_T, y_1) and (u_T, y_2) (these are in S_{R1});
- the $|X - X'| - 1$ pairs (u_T, s) for all $s \in Z$ (these are in S_{R4});
- the $|X'|$ pairs (x, x') for $x, x' \in X'$ (these are in S_{R1});
- the $2|X - X'|$ pairs (x, w_T), (w_T, x) for $x \in X - X'$ (these are in S_{R1});
- the $|X - X'|$ pairs (x, z) for $x \in X - X'$ (these are in S_{R1} if $z \in R$ and otherwise in S_{R2});
- the $|X - X'|$ pairs (z, x) for $x \in X - X'$ (these are in S_{R1} if $z \in R$ and otherwise in S_{R3});
- the $2|X' \cap \mathcal{V}|$ pairs (x, w_T), (w_T, x) for $x \in X' \cap \mathcal{V}$ (these are in S_{R1});
- and the $2|X' \cap \mathcal{V}|$ pairs (x, b_1), (x, b_2) for $x \in X' \cap \mathcal{V}$ (these are in S_{R2}).

Then $|S_R| = 4|\mathcal{V} \cap R|$.

5.4. β-Tails

Our final superregion is a *β-tail*, defined as follows. See Figure 4 for an illustration.

Definition 5.5. Let T be a tail with k segments, and let H' be a subgraph of G with the following vertices:

- X with $|X| = d - 1$;
- $X' \subset X$ such that $|X'|$ is even and not equal to 0 or $d - 1$;
- a vertex w;

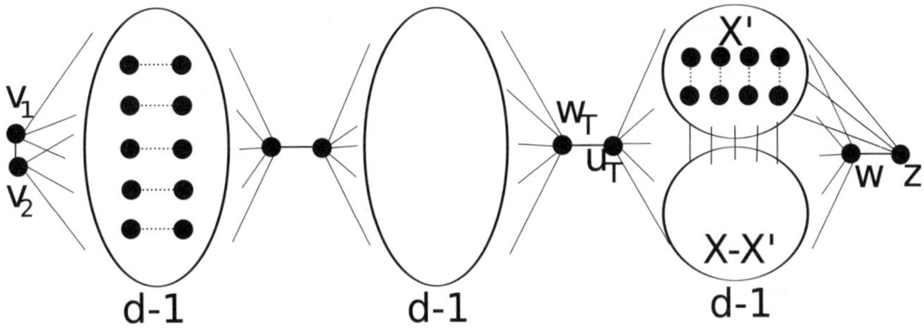

Figure 4: β-tail

- a vertex z;
- and u_T.

Suppose that H' consists of the following edges:

- a complete graph on X with a matching on X' removed;
- all edges $u_T x$ for $x \in X$;
- all edges wx for $x \in X$;
- all edges zx for $x \in X'$;
- and wz.

Each $x \in X - X'$ is low degree and is contained in a common region H. Then $T \cup H$ is a β-tail.

For $x \in X - X'$, $N_2(x) = N_2'(x) = \{z, w_T\}$. Since $w_T \in T$, $w_T \notin H$. Also, since d is odd, $|X'| \leq d - 3$ and z has $|X - X'| \geq 2$ neighbors outside of R. Therefore, no vertex of X' is low degree, and w is low degree if and only if $|X'| = d - 3$. We conclude that $H = H'$ exactly when $|X'| = d - 3$, and otherwise $H = H' - \{z\}$. We say that H is the *head region* of R. Note that $G - \{z\}$ is disconnected with $R - \{z\}$ a component. Also note that if we were to allow $|X'| = d - 1$, then R would be a snake graph, whereas if $|X'| = 0$, then R would be an ordinary tail.

We take $R \cap \mathcal{W} = \{w\}$ and $R \cap \mathcal{N}$ to be the two vertices v_1 and v_2 of T such that $\deg_2(v_1) = \deg_2(v_2) = 1$. Also, $X - X' \subset \mathcal{V}$, whereas X' might or might not be a subset of $\mathcal{V}$. Note that $|R \cap \mathcal{V}| = k(d-1) + |X \cap \mathcal{V}|$. Also, z has $|X - X'|$ neighbors outside of R, and since $|X - X'| \geq 2$, consider distinct $b_1, b_2 \in N(z) - R$. Then we define S_R as the union of the following sets:

- the $(4k-1)(d-1)$ elements of $S_{T1} \cup S_{T2} \cup S_{T3}$ (these are in S_{R1});

- the $|X - X'|$ pairs (w, s) for $s \in N(z) - R$ (these are in S_{R4});
- the $|X'|$ pairs (x, x') for $x, x' \in X'$ and xx' not an edge (these are in S_{R1});
- the $2|X|$ pairs $(x, w_T), (w_T, x)$ for $x \in X$ (these are in S_{R1});
- the $|X - X'|$ pairs (x, z) for each $x \in X - X'$ (these are in S_{R1} if $z \in H$ and otherwise in S_{R2});
- the $|X - X'|$ pairs (z, x) for each $x \in X - X'$ (these are in S_{R1} if $z \in H$ and otherwise in S_{R3});
- the $2|X' \cap \mathcal{V}|$ pairs (x, b_1), (x, b_2) for $x \in X' \cap \mathcal{V}$ (these are in S_{R2}).

Then $|S_R| = 4k(d-1) + 4|X - X'| + 2|X'| + 2|X' \cap \mathcal{V}| \geq 4|\mathcal{V} \cap R|$.

5.5. Identifying α- and β-Tails

In this section we prove an important lemma on the structure of α- and β-tails.

Lemma 5.6. *A region H is the head region of at most one α- or β-tail.*

Proof. We show that an α- or β-tail $R = T \cup H$ containing H is uniquely determined, given H. By construction, H contains a set of vertices M, $|M| \geq 2$, such that each $m \in M$ is adjacent to some vertex outside of H. First consider the case that $M = \{a, b\}$. Then, by construction, one of a or b, say a, is adjacent to exactly one vertex outside of H, and b is adjacent to at least 2 vertices outside of H. Then a must be u_T, and T and R are uniquely determined since w_T is the only vertex outside of H adjacent to a and by Lemma 5.3.

Now suppose $|M| \geq 3$. By construction, R is an α-tail, and there exists a unique vertex $w \notin H$ such that w is adjacent to exactly one vertex in H. Thus $w = w_T$, and by Lemma 5.3, T and R are uniquely determined. □

6. Superregions as a Partition

In this section, we show that the superregions of G partition $V(G)$ unless G is a snake graph.

Theorem 6.1. *Suppose that G is not a snake graph. Then $V(G)$ is partitioned by the superregions of G.*

Proof. First, every vertex is contained in a superregion, since singleton sets are superregions if not contained in any larger superregion. We need only to show that $R \cap R' = \emptyset$ if R and R' are distinct superregions. If either R or R', say R, is a single vertex, then either $R \subset R'$ or $R \cap R' = \emptyset$. The former is impossible by definition. Now we assume that R and R' each consist of multiple vertices.

By construction, all superregions with multiple vertices are unions of regions. Therefore, if R is both a region and a superregion, then either $R \subset R'$ or $R \cap R' = \emptyset$. Again, the former is impossible by definition, and now we assume that R and R' each consist of multiple regions. By construction, both R and R' contain tails.

Next, suppose that R and R' are both either a tail or a multitail, and suppose that $R \cap R' \neq \emptyset$. By Lemma 5.3, $R \cap R'$ is a collection of proper tails, and let T be a proper tail in $R \cap R'$. By $T \subset R \cap R'$, for all proper tails $T' \subset R \cup R'$, $u_T = u_{T'}$. Thus $R \cup R' = R = R'$.

Now let R be an α- or β-tail with proper tail T and head H, and let R' be either a tail or multitail. Write $R' = T_1' \cup \cdots \cup T_k'$ as a union of disjoint tails. The head region of an α- or β-tail is not isomorphic to any segment of a tail. Thus, if $R \cap R' \neq \emptyset$, then $H \cap R' = \emptyset$ and $T \cap R' \neq \emptyset$. By Lemma 5.3, $T = T_i$ for some $1 \leq i \leq k$. Furthermore, $H \cup T_j'$ is also an α- or β-tail for each $1 \leq j \leq k$ by $u_{T_1} = u_T = u_{T_j}$, which implies that $k = 1$ by Lemma 5.6. Then $T = R'$ and R' is not a superregion, a contradiction.

Finally, if R and R' are both α- or β-tails such that $R \cap R' \neq \emptyset$, then we show that $R = R'$. Let T and T' be the respective tails of R and R', and let H and H' be the respective head regions. By Lemma 5.3, if $T \cap T' \neq \emptyset$, then $T = T'$. Since only one vertex outside of T is adjacent to T, then $H = H'$ and thus $R = R'$. Next, we have that $H \cap T' = H' \cap T = \emptyset$ by the fact that H and H' are not isomorphic to any segment of a tail. Finally, if $H = H'$, then $R = R'$ by Lemma 5.6.

This establishes that superregions partition $V(G)$. □

7. Exceptions

There are two families of graphs that are exceptions to Conjecture 1.1. In this section we discuss these exceptions in more detail.

7.1. Snake Graphs

A snake graph G consisting of $k \geq 2$ regions, as described above, has $n = k(d+1)+2$ vertices. Since a snake graph contains a B region, a snake graph exists only if d is odd. A snake graph is determined, to isomorphism, by d and k.

By construction, we may calculate that

$$\sum_{v \in V(G)} \deg_2(v) = (4k-2)(d-1) + 8 = \frac{(4k-2)(d-1)+8}{k(d+1)+2} n.$$

For large d, this quantity is approximately $(4 - 2/k)n$.

7.2. Peanut Graphs

A *peanut graph* G is defined as follows. Partition $V(G)$ into sets R_1 and R_2 with $d+1$ and $d+2$ vertices respectively. The only edge in the complement of $G[R_1]$ is w_1w_2. The only edges in the complement of $G[R_2]$ are uv_1, uv_2, uv_3 and a matching on the remaining vertices. The only edges between R_1 and R_2 are uw_1 and uw_2. Note that R_1 and R_2 are both A regions. A peanut graph exists only when d is even, due to the matching in the complement of R_2, and is determined up to isomorphism by d. See Figure 5 for an illustration.

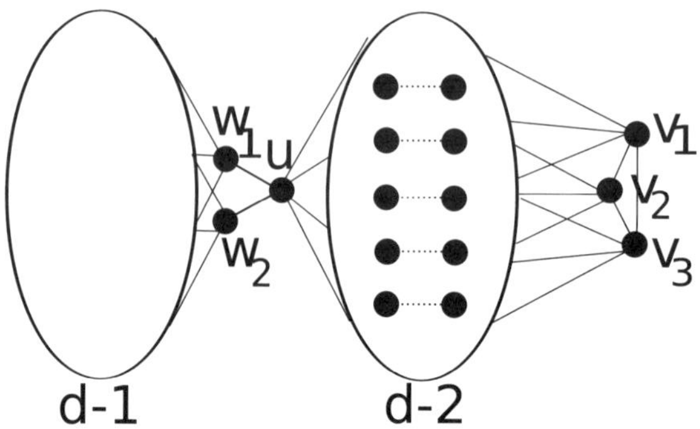

Figure 5: Peanut graph

One may check that $n = 2d+3$ and

$$\sum_{v\in V(G)} \deg_2(v) = 7d-4 = \frac{7d-4}{2d+3}n.$$

For large d, this quantity is approximately $(7/2)n$.

8. Proof of Theorem 1.2

In this section, we conclude the proof of Theorem 1.2. We need to show that the S_R are disjoint over all superregions R. By Lemma 4.2, we need only to consider $(x,y) \in S_{R4}$ for a superregion R and show that, for each superregion R', that $(x,y) \notin S_{R'3}$, assuming that $S_{R'}$ is properly chosen. Now suppose that $(x,y) \in S_{R'3}$. Note that $x \in R\cap\mathcal{W}$, and y is a low degree vertex of R' with $x \in N_2'(y)$. We consider several cases on R.

8.1. *R* is a Single Vertex, or a D, E, F, or G Region

This case is trivial, since by construction, $\mathcal{W} \cap R = \emptyset$ and $S_{R4} = \emptyset$.

8.2. *R* is an A Region

Let $X = R \cap N_2(y)$. Since the complement of $G[X]$ is a matching, $|X|$ is even, and since y is low degree, $|X| = 2$. We must have $\deg_2(y) = 3$; otherwise, let u be the one vertex adjacent to both x and y, and let $Y := N(y) - \{u\}$ so that $|Y| = d - 1$. No vertex of Y may have a neighbor outside of $Y \cup \{y, u\}$ if $\deg_2(y) = 2$, and thus for $y' \in Y$, $N(y') = Y - \{y'\} \cup \{u, y\}$. But then $X \cup Y \subset N(u)$ is a contradiction to $\deg(u) = d$.

Let z be the unique element of $N_2(y) - X$. By construction, since y is a low-degree vertex with $|N_2(y) - R'| \geq 2$ and $S_{R'3}$ contains a pair of the form $(-, y)$, R' must be a A, D, E, F, or G region.

Consider the case that $z \in N_2(y) - N_2'(y)$. Then $V(G) = V(R) \cup Y \cup \{u, y, z\}$. Counting degrees on $G[Y \cup \{u, y, z\}]$, which has $d + 2$ vertices, we see that u has degree $d - 2$ and all other vertices have degree d. Let a, b, c be the three vertices of $G[Y \cup \{u, y, z\}]$ that are not adjacent to u. Then by degree considerations, the complement of $G[Y \cup \{u, y, z\}]$ contains edges ua, ub, uc and a matching on all other vertices. Thus G is a peanut graph. If R' is an A region, then $z \in N_2(y) - N_2'(y)$ and G is a peanut graph. Then suppose that R' is a D, E, F, or G region.

In $S_{R'}$, we may replace (x, y) with (z, y), unless, by Lemma 4.2, $(z, y) \in S_{R''4}$ for some superregion R''. Then $z \in \mathcal{W}$, and since z is the only vertex in $R'' \cap N_2(y)$, by the choice of $\mathcal{W}$ for the various classes of superregions, R'' must be a tail with $z = w_{R''}$. It must be that $u_{R''} \neq u$; otherwise, since $X \subset N(u)$, at most $d - 2$ vertices of $Y = N(y) - \{u\}$ are adjacent to u, and so let $y' \in Y$ be not adjacent to u. Then $N(y') \subseteq Y - \{y'\} \cup \{y\}$ by $\deg_2(y) = 3$, but then y' has degree at most $d-1$, a contradiction. Now, $H := G[\{y\} \cup N(y)]$ has $d+1$ vertices. By $\deg_2(y) = 3$, all vertices of H have degree d in H, except that u has degree $d - 2$ and $u_{R''}$ has degree $d - 1$. This would imply an odd degree sum on H, which is impossible.

8.3. *R* is an *m*-Multitail

First suppose that $m \geq 3$. Then y is distance 2 from each vertex in $R \cap \mathcal{W}$, and since y is a low degree vertex, $m = 3$. Let u be the vertex adjacent to both x and y, and let $Y := N(y) - \{u\}$. Since u is adjacent to y and 3 vertices in R, u has at most $d-4$ neighbors in Y. Choose $y' \in Y - N(u)$. Since $\deg_2(y) = 3$, $N(y') \subset Y - \{y'\} \cup \{y\}$. Then y' has degree at most $d - 1$, a contradiction. We conclude that $m = 2$.

With $m = 2$, we may perform the same analysis as in Section 8.2 and conclude that R' is the complement of a graph with $d+2$ vertices containing edges ua, ub, uc and a matching on all other vertices. This implies that d is even, and the existence of a tail implies that d is odd, and so R cannot be a multitail.

8.4. *R* is a C Region

Let u be the unique vertex that is adjacent to both x and y. By Lemma 4.6, R contains a vertex $x' \neq x$ in $N(u)$, and $\{x, x'\} \subseteq N_2(y)$. Suppose, by way of contradiction, that $\deg_2(y) = 2$. Let $Y := N(y) - \{u\}$. For each $y' \in Y$, $N(y') \subset Y - \{y'\} \cup \{u, y\}$, and since $\deg_2(y') = d$, then $N(y') = Y - \{y'\} \cup \{u, y\}$. Then $Y \cup \{x, x'\} \subset N(u)$, a contradiction to $\deg_2(u) = d$. We conclude that $\deg_2(y) = 3$.

Note that $x' \notin \mathcal{W}$. By Lemma 4.3 and the fact that $\deg_2(y) = 3$, if $(\tilde{x}, y) \in S_{R3}$, then $\tilde{x} = x$. Then in $S_{R'3}$, we may replace (x, y) by (x', y), and since $x' \notin \mathcal{W}$, $(x, y) \notin S_{R''4}$ for any superregion R''.

8.5. *R* is an *α*-Tail or a *β*-Tail

Let u be the unique vertex that is adjacent to both x and y. By construction, there are at least three vertices x, x', x'' in $R \cap N_2(y)$. Let $Y := N(y) - \{u\}$. Since y is low degree, for each $y' \in Y$, $N(y') \subseteq Y - \{y'\} \cup \{u, y\}$. Since $\deg(y') = d$, $N(y') = Y - \{y'\} \cup \{u, y\}$. Then $Y \cup \{x, x', x''\} \subset N(u)$, a contradiction to $\deg_2(u) = d$.

8.6. *R* is a Tail

For the final case, that R is a tail, we consider several cases on R'. Let u be the one vertex that is adjacent to both x and y.

8.6.1. *R′* is an A Region or a Tail

If R' is an A region, then let $Y := N(u) \cap R'$. By definition of an A region, $|Y| < d-1$. Also $|Y|$ must be even, and d must be odd by the existence of a tail R, and so $|Y| \leq d - 3$. Then $N(u)$ consists of at least 3 vertices outside of R'. Also, $N(u) - R' \subset N_2(y)$, and $N_2(y)$ consists of at least one vertex in R', contradicting the definition of a low degree vertex. Thus R' is not an A region. If R' is a tail, then G is a snake graph.

8.6.2. *R′* is a C Region

By construction, $\deg_2(y) = 2$ and $|N_2'(y)| = 1$. Let y' be the one vertex in $N_2(y) - N_2'(y)$. Since u has $d - 2$ neighbors outside of R, excluding y, and y has $d - 1$ neighbors excluding u, y has a neighbor z that is not adjacent to u. Then $N(z) \subseteq N(y) \cup \{y, y'\} - \{z, u\}$, and since $\deg(z) = d$, $N(z) = N(y) \cup \{y, y'\} - \{z, u\}$. The only vertices in the component of $G - \{u\}$ that contains y are $N(y) - \{u\} \cup \{y, y'\}$, and thus $N_2(z) = \{u\}$, which implies that R' is an A or B region, a contradiction.

8.6.3. $\deg_2(y) = 2$ and $|N'_2(y)| = 2$

Let $N'_2(y) = \{x, z\}$. We consider two cases: if z and u are neighbors, and if they are not neighbors.

If z and u are neighbors, then u has a set Y of $d-3$ neighbors outside of $\{x, y, z\}$, and y has $d-1$ neighbors, excluding u. Let w and w' be distinct vertices in $N(y) - N(u) - \{u\}$. Every vertex in Y is within distance 2 of y, and since $N_2(y) \cap Y = \emptyset$, $Y \subset N(y)$. Thus $Y \subset R'$. Also, w and w' are adjacent to each vertex in $Y \cup \{y, z\}$ as well as each other, since neither is adjacent to u, x, or any vertex of distance 3 or more from y. Similarly, every vertex in Y can only have neighbors among $Y \cup \{y, w, w', z, u\}$, a set of size $d+2$, and so $y' \in Y$ is adjacent to all but possibly one other vertex in Y. The complement of $G[Y]$ has no edges except a (possibly empty) matching, and let Y' be the set of such vertices that are in such a matching. Every vertex in Y' is adjacent to z. If $y' \in Y - Y'$, then y' is not adjacent to z since y' has been established to be adjacent to every other vertex in $Y \cup \{y, w, w', z, u\}$ besides itself. We conclude that $R \cup R'$ is an α-tail, a contradiction to the assumption that R is a superregion.

Now, if u and z are not neighbors, let $Y := N(y) \cap N(u)$. Since $\{x, y\} \subseteq N(u) - N(y)$, $|Y| \leq d-2$. Since $(N(u) - N(y) - \{y\}) \sqcup \{z\} \subseteq N_2(y)$, in fact $N(u) - N(y) = \{x, y\}$ and $|Y| = d-2$. Choose u' so that $N(y) = Y \cup \{u, u'\}$. By $\deg_2(y) = 2$, u' can have no neighbors outside of $Y \cup \{y, z\}$, and so $N(u') = Y \cup \{y, z\}$. For $y' \in Y$, the only possible neighbors of y' are in $Y \cup \{u, y, u', z\}$, since y' is not adjacent to x or any vertex of distance 3 or more from y. Of the vertices of $Y \cup \{u, y, u', z\} - \{y'\}$, y' is adjacent to all but 1. Let $Y' := Y \cap N(z)$. Each $y' \in Y'$ is adjacent to u, u', y, z and thus all but 1 other vertex of $Y - \{y'\}$. If $y' \in Y - Y'$, then y' is adjacent to every vertex of $Y - \{y'\}$, and so the complement of $G[Y]$ is a matching on Y'. If $Y' \neq \emptyset$, then $R \cup R'$ is a β-tail, a contradiction to the assumption that R is a superregion, and if $Y = \emptyset$, then $R \cup R'$ is a tail, also a contradiction.

8.6.4. $\deg_2(y) = 3$ and $|N'_2(y)| = 1$

Let G_y be the component of $G - x$ that contains y. Then G_y has $d+3$ vertices, namely y, all neighbors of y, and $N_2(y) - \{x\}$. Each vertex in G_y has degree d, except that u has degree $d-1$. This is impossible since the degree sum would be odd.

8.6.5. $\deg_2(y) = 3$ and $|N'_2(y)| = 2$

Let $N'_2(y) = \{x, z\}$. By construction, R' is not a tail, multitail, or α- or β-tail. Assume that R' is a D, E, F, or G region. By Lemma 4.3, in $S_{R'}$, we may replace (x, y) with (z, y), unless $(z, x) \in S_{R''4}$ for some superregion R''. In this case, since $z \in \mathcal{W}$, and R'' has no vertex besides z in $N_2(y)$, R'' is a tail. Let u' be the unique that is adjacent to both y and z. Then $u \neq u'$, since otherwise R is a multitail. Let G_y be the component of $G - \{x, z\}$ that contains y. Then G_y has $d + 2$ vertices, namely y, $N(y)$, and the one vertex of $N_2(y) - N_2'(y)$. In G_y, all vertices have degree d except for u and u', which each have degree $d - 1$. This requires d to be even, so that sum of the degrees of all vertices in G_y is even. However, the existence of a tail R requires d to be odd. We conclude that $(z, x) \notin S_{R''4}$ as desired.

8.6.6. $\deg_2(y) = 3$ and $|N_2'(y)| = 3$

As above, we may assume that R' is a D, E, F, or G region. Let $N_2(y) = N_2'(y) = \{x, z, z'\}$. By Lemma 4.3, if $z \notin \mathcal{W}$, then in $S_{R'}$, we may replace (x, y) with (z, y). Likewise, if $z' \notin \mathcal{W}$, then in $S_{R'3}$, we may replace (x, y) with (z', y). Now suppose that both z and z' are in $\mathcal{W}$. If z and z' are in the same region R'', then R'' is either an A region or a multitail. In that case, let u' be the unique vertex adjacent to each of y, z, z'. Then in $G[\{y\} \cup N(y)]$, all vertices have degree d except for u and u'. If $u = u'$, then $\deg(u) = d - 3$ in $G[\{y\} \cup N(y)]$, while if $u \neq u'$, then u and u' have degrees $d - 1$ and $d - 2$ respectively in $G[\{y\} \cup N(y)]$. Both of these cases are impossible since the degree sum would be odd.

Now suppose that x, z, and z' are all in different regions. Since the regions that contain x, z, z' respectively each have exactly one vertex in $N_2(y)$, they must all be tails. Let u, u_1, u_2 be the vertices adjacent to y and respectively x, z, z'. Since each of x, z, z' are contained in tails and not multitails, u, u_1, u_2 are distinct. Then G_y, the induced subgraph consisting of y and its neighbors, has $d + 1$ vertices, and all vertices have degree d except for u, u_1, u_2, which each have degree $d - 1$. This is also impossible, since the sum of the degrees would be odd.

This enumerates all cases.

References

[1] P. Hegarty, *A Cauchy-Davenport type result for arbitrary regular graphs*, Integers **11** A19 (2011), 8 pp.

[2] M. DeVos, J. McDonald, D. Scheide, *Average degree in graph powers*, J. Graph Theory **72(1)**, (2013), 7-18.

[3] M. DeVos, S. Thomassé, *Edge growth in graph cubes*, arXiv: 1009.0343 (2010).

[4] M. Kneser, *Abschaetzung der asymptotischen Dichte von Summenmengen*, Math. Z. **58** (1953), 459-484.

[5] A. Pokrovskiy, *Growth of graph powers*, Electron. J. Combin. **18(1)**, P88 (2011), 7 pp.

NEW BOUNDS AND COMPUTATIONS ON PRIME-INDEXED PRIMES

Jonathan Bayless
Department of Mathematics, Husson University, Bangor, Maine
baylessj@husson.edu

Dominic Klyve
Dept. of Mathematics, Central Washington University, Ellensburg, Washington
klyved@cwu.edu

Tomás Oliveira e Silva
Electronics, Telecommunications, and Informatics Department, Universidade de Aveiro, Aveiro, Portugal
tos@ua.pt

Received: 5/25/12, Revised: 5/15/13, Accepted: 5/16/13, Published: 7/10/13

Abstract

In a 2009 article, Barnett and Broughan considered the set of prime-index primes. If the prime numbers are listed in increasing order (2, 3, 5, 7, 11, 13, 17, ...), then the prime-index primes are those which occur in a prime-numbered position in the list (3, 5, 11, 17, ...). Barnett and Broughan established a prime-indexed prime number theorem analogous to the standard prime number theorem and gave an asymptotic for the size of the n-th prime-indexed prime.

We give explicit upper and lower bounds for $\pi^2(x)$, the number of prime-indexed primes up to x, as well as upper and lower bounds on the n-th prime-indexed prime, all improvements on the bounds from 2009. We also prove analogous results for higher iterates of the sequence of primes. We present empirical results on large gaps between prime-index primes, the sum of inverses of the prime-index primes, and an analog of Goldbach's conjecture for prime-index primes.

1. Introduction

Many of the classes of primes typically studied by number theorists concern properties of primes dictated by the positive integers. Twin primes, for example, are consecutive primes with an absolute difference of 2. It is interesting, however, to consider a sequence of primes whose members are determined by the primes themselves.

In this work, we consider the set of prime-indexed-primes (or PIPs), which are prime numbers whose index in the increasing list of all primes is itself prime. In particular we have:

Definition 1.1. Let P be the sequence of primes written in increasing order $\{p_i\}_{i\geq 1}$. The sequence of *prime-indexed-primes*, or PIPs, is the subsequence of P where the index i is itself prime. Specifically, the sequence of prime-indexed primes is given by $\{q_i\}$ where $q_i = p_{p_i}$ for all $i \geq 1$.

Prime-indexed-primes seem to have been first considered in 1965 by Jordan [1], who also considered primes indexed by other arithmetic sequences. They were again studied in 1975 by Dressler and Parker [2], who showed that every positive integer greater than 96 is representable by the sum of distinct PIPs. Later, Sándor [3] built on Jordan's work of considering the general question of primes indexed by an arithmetic sequence by studying reciprocal sums of such primes and limit points of the difference of two consecutive primes. Some of these results can be found in [4, p. 248-249].

The direct inspiration of this work is the recent work of Broughan and Barnett [5], who have demonstrated many properties of PIPs, giving bounds on the n-th PIP, a PIP counting function (analogous to the prime number theorem), and some results on small gaps between consecutive PIPs.

We follow Broughan and Barnett by explaining PIPs via example. In the following list of primes up to 109, all those with prime index (the PIPs) are in bold type: 2, **3**, **5**, 7, **11**, 13, **17**, 19, 23, 29, **31**, 37, **41**, 43, 47, 53, **59**, 61, **67**, 71, 73, 79, **83**, 89, 97, 101, 103, 107, **109**.

We note also the equivalent definition of PIPs, which may be of some use. If we let, as usual, $\pi(x)$ be the number of primes not greater than x, then an integer p is a PIP if both p and $\pi(p)$ are prime.

2. Preliminary Lemmas

In this section we state a few lemmas which will be helpful in the proofs of the theorems in the remaining sections of this paper. The proofs consist of technical details, and contain no insight of direct relevance to the rest of the paper, so we defer their proofs to Section 10. First, we give a pair of bounds comparing combinations of n and $\log n$ to $\log\log n$.

Lemma 2.1. *For $n \geq 3$,*

$$\log\left(\log\left(n\log n\right)\right) < \log\log n + \frac{\log\log n}{\log n}.$$

Lemma 2.2. *For $n \geq 3$,*

$$\log\log\left(n\log\left(n\log n\right)\right) < \left(1 + \frac{1}{\log n} + \frac{1}{\log^2 n}\right)\log\log n.$$

We will also need two lemmas bounding the reciprocal of $\log \pi(x)$, one for explicit bounds and one for asymptotic bounds.

Lemma 2.3. *For all $x \geq 33$,*

$$\frac{1}{\log x}\left(1 + \frac{\log\log x}{\log x}\right) < \frac{1}{\log \pi(x)} < \frac{1}{\log x}\left(1 + \frac{\log\log x}{\log x} + \frac{(\log\log x)^2}{\log^2 x} + \cdots\right).$$

Lemma 2.4. *For all $k \geq 1$, we have*

$$\frac{1}{\log^k \pi(x)} = \frac{1}{\log^k x}\left(1 + \frac{k\log\log x}{\log x} + O_k\left(\frac{(\log\log x)^2}{\log^2 x}\right)\right)$$

and

$$\frac{\log\log \pi(x)}{\log \pi(x)} = \frac{\log\log x}{\log x} + O\left(\frac{(\log\log x)^2}{\log^2 x}\right).$$

3. Bounds on PIPs

Broughan and Barnett were the first to put upper and lower bounds on q_n, the n-th PIP. In particular, they show

$$\begin{aligned} q_n &< n(\log n + 2\,\mathrm{loglog}\, n)(\log n + \mathrm{loglog}\, n) - n\log n + O(n\,\mathrm{loglog}\, n), \\ q_n &> n(\log n + 2\,\mathrm{loglog}\, n)(\log n + \mathrm{loglog}\, n) - 3n\log n + O(n\log\log n). \end{aligned}$$

By using some theorems of Dusart [6], we show that these bounds can both be improved and made explicit.

Theorem 3.1. *For all $n \geq 3$, we have*

$$q_n > n\left(\log n + \log\log n - 1\right)\left(\log n + 2\log\log n - 1\right),$$

and for all $n \geq 71$, we have

$$q_n < n\left(\log n + \log\log n - \frac{1}{2}\right)\left(\log n + 2\log\log n + \frac{3\log\log n}{\log n} - \frac{1}{2}\right).$$

Proof. We begin with the lower bound. Using the result of Dusart [6] that

$$p_n > n\left(\log n + \log\log n - 1\right)$$

for all $n \geq 2$, we have for all $n \geq 3$ that

$$q_n > n(\log n + \log\log n - 1)\Big(\log(n(\log n + \log\log n - 1)) + \log\log(n(\log n + \log\log n - 1)) - 1\Big).$$

Now, using the fact that

$$\log(n\log n + n\log\log n - n) > \log n + \log\log n,$$

we may bound

$$\begin{aligned} q_n &> n(\log n + \log\log n - 1)(\log n + \log\log n + \log(\log n + \log\log n) - 1) \\ &> n(\log n + \log\log n - 1)(\log n + 2\log\log n - 1). \end{aligned}$$

Turning to the upper bound, we use a result of Rosser and Schoenfeld [7], that:

$$p_n < n\left(\log n + \log\log n - \frac{1}{2}\right)$$

for $n \geq 20$. To simplify the notation here, let

$$N = n\left(\log n + \log\log n - \frac{1}{2}\right). \tag{3.1}$$

This gives that for $n \geq 71$,

$$q_n < N\left(\log N + \log\log N - \frac{1}{2}\right). \tag{3.2}$$

Now, Lemma 2.1 gives that

$$\log N < \log n + \log\log n + \frac{\log\log n}{\log n}.$$

Also, by Lemma 2.2,

$$\log\log N < \left(1 + \frac{1}{\log n} + \frac{1}{\log^2 n}\right)\log\log n < \log\log n + \frac{2\log\log n}{\log n}.$$

Putting these into (2), we can bound

$$q_n < n\left(\log n + \log\log n - \frac{1}{2}\right)\left(\log n + 2\log\log n + \frac{3\log\log n}{\log n} - \frac{1}{2}\right),$$

which is the bound in the theorem. □

In fact, using an upper bound on p_n due to Robin [8], namely

$$p_n < n(\log n + \log\log n - 0.9385) \text{ for } n \geq 7022,$$

we are able to prove the somewhat stronger upper bound

$$q_n < n(\log n + \log\log n - 0.9385)\left(\log n + 2\log\log n + \frac{2\log\log n}{\log n} - 0.9191\right)$$

for all $n \geq 70919$.

4. Bounds for the Number of Prime-Indexed-Primes

Define $\pi^2(x)$ to be the number of PIPs not greater than x. Note that it follows immediately from the definition that $\pi^2(x) = \pi(\pi(x))$. Broughan and Barnett show that

$$\pi^2(x) \sim \frac{x}{\log^2 x},$$

and also give the slightly more sophisticated

$$\pi^2(x) \sim \frac{x}{\log^2 x} + O\left(\frac{x\log\log x}{\log^3 x}\right).$$

With respect to the classical prime number theorem, a study of the error term has driven much research in number theory. Among the known results [6, page 16] is

$$\pi(x) = \frac{x}{\log x}\left(1 + \frac{1}{\log x} + \frac{2}{\log^2 x} + O\left(\frac{1}{\log^3 x}\right)\right). \tag{4.1}$$

We also have a number of known explicit bounds on $\pi(x)$, including [6, page 2]

$$\frac{x}{\log x}\left(1 + \frac{1}{\log x}\right) \leq \pi(x) \leq \frac{x}{\log x}\left(1 + \frac{1.2762}{\log x}\right), \tag{4.2}$$

where the lower bound holds for $x \geq 599$ and the upper bound holds for $x > 1$. Based on this work, we are able to prove the following theorem.

Theorem 4.1. *For all $x \geq 3$, we have*

$$\pi^2(x) < \frac{x}{\log^2 x}\left(1 + \frac{1.5}{\log x}\right)^2\left(1 + \frac{\log\log x}{\log x} + \frac{1.5(\log\log x)^2}{\log^2 x}\right),$$

while, for all $x \geq 179$,

$$\pi^2(x) > \frac{x}{\log^2 x}\left(1 + \frac{1}{\log x}\right)^2\left(1 + \frac{\log\log x}{\log x}\right).$$

Proof. We begin with the upper bound. First, note that

$$\sum_{i=2}^{\infty}\left(\frac{\log\log x}{\log x}\right)^i = \frac{\left(\frac{\log\log x}{\log x}\right)^2}{1-\left(\frac{\log\log x}{\log x}\right)} < 1.5\left(\frac{\log\log x}{\log x}\right)^2 \tag{4.3}$$

for $x \geq 179$. We may bound $\pi^2(x)$ with bounds on $\pi(\pi(x))$, using Lemma 2.3 and (5) to see that

$$\begin{aligned}\pi^2(x) &\leq \frac{\pi(x)}{\log \pi(x)}\left(1+\frac{1.2762}{\log \pi(x)}\right)\\ &< \frac{x}{\log^2 x}\left(1+\frac{1.2762}{\log x}\right)\left(1+\frac{1.2762}{\log \pi(x)}\right)\left(1+\frac{\log\log x}{\log x}+\frac{1.5(\log\log x)^2}{\log^2 x}\right).\end{aligned}$$

Now, to complete the upper bound's proof, we need only show

$$\left(1+\frac{1.2762}{\log x}\right)\left(1+\frac{1.2762}{\log \pi(x)}\right) < \left(1+\frac{1.5}{\log x}\right)^2. \tag{4.4}$$

By Lemma 2.3 and (5),

$$\frac{1.2762}{\log \pi(x)} \leq \frac{1.2762}{\log x}\left(1+\frac{\log\log x}{\log x}+\frac{1.5(\log\log x)^2}{\log^2 x}\right) < \frac{1.7238}{\log x},$$

for $x \geq 3030$. Then,

$$\left(1+\frac{1.2762}{\log x}\right)\left(1+\frac{1.7238}{\log x}\right) \leq 1+\frac{3}{\log x}+\frac{2.22}{\log^2 x} < \left(1+\frac{1.5}{\log x}\right)^2,$$

which establishes (6) and thus the upper bound in the theorem for $x \geq 3030$. Finally, a computer check verifies the upper bound for $3 \leq x \leq 3030$.

For the lower bound, we can use (4) to see that

$$\begin{aligned}\pi^2(x) &\geq \frac{\pi(x)}{\log \pi(x)}\left(1+\frac{1}{\log \pi(x)}\right)\\ &\geq \frac{\pi(x)}{\log \pi(x)}\left(1+\frac{1}{\log x}\right)\\ &\geq \frac{x}{\log x}\left(1+\frac{1}{\log x}\right)^2\cdot\frac{1}{\log \pi(x)}\\ &\geq \frac{x}{\log^2 x}\left(1+\frac{1}{\log x}\right)^2\left(1+\frac{\log\log x}{\log x}\right)\end{aligned}$$

by Lemma 2.3, for all $x \geq 4397$. A computer check shows that the bound also holds for all $179 \leq x \leq 4397$, completing the proof of the theorem. □

n	$\pi(10^n)$	$\pi^2(10^n)$	$\pi^3(10^n)$
1	4	2	1
2	25	9	4
3	168	39	12
4	1229	201	46
5	9592	1184	194
6	78498	7702	977
7	664579	53911	5492
8	5761455	397557	33666
9	50847534	3048955	220068
10	455052511	24106415	1513371
11	4118054813	195296943	10833076
12	37607912018	1613846646	80104927
13	346065536839	13556756261	608455060
14	3204941750802	115465507935	4726881850
15	29844570422669	995112599484	37431015268
16	279238341033925	8663956207026	301327263751
17	2623557157654233	76105984161825	2460711566651
18	24739954287740860	673776962356604	20348625806080
19	234057667276344607	6006525919368810	170149286304116
20	2220819602560918840	53878729390812464	1436870802519360
21	21127269486018731928	485986685605473234	12241980697771924
22	201467286689315906290	4405654516157364292	105136072207222852
23	1925320391606803968923	40121204955640303216	909475787902559408
24	18435599767349200867866	366893555203205479291	7919305232077304848

Table 1: Values of $\pi^2(x)$ and of $\pi^3(x)$ for small powers of 10

Small improvements can be made to the bounds in Theorem 4.1, at the expense of the relative simplicity in the statement. However, the lower bound is much closer to the truth. Using a stronger version of (4) and a finer version of Lemma 2.3 it is possible to prove the following theorem.

Theorem 4.2. *We have*

$$\pi^2(x) = \frac{x}{\log^2 x}\left(1 + \frac{\log\log x}{\log x} + \frac{2}{\log x}\right) + O\left(\frac{x(\log\log x)^2}{\log^4 x}\right).$$

Table 4 gives the values of $\pi(x)$ and $\pi^2(x)$ for powers of 10 up to 10^{24}. The table also gives values of $\pi^3(x)$, the number of PIPs of prime index up to x (see a definition and further generalization in Section 7). The values of $\pi(x)$ up to 10^{23}, as well as all values of $\pi^2(x)$ and of $\pi^3(x)$, were computed using an implementation of the Lagarias-Miller-Odlyzko algorithm [9] described in [10]. The value of $\pi\left(10^{24}\right)$ was computed in 2010 by Buethe, Franke, Jost, and Kleinjung [11] using a conditional (on the Riemann hypothesis) analytic method, and latter confirmed in 2012 by Platt [12] using an unconditional analytic method.

5. Twin PIPs

Twin primes are pairs of primes which are as close together as possible – namely, they are consecutive odd numbers which are both prime. We will define *twin prime-indexed primes*, or *twin PIPs*, with a similar motivation, that is, they are PIPs which are as close together as possible. Since two consecutive indices cannot be prime, there must always be at least one prime between consecutive PIPs (except for the trivial case $q_2 = 3, q_3 = 5$). Futhermore, we know that p, $p+2$, and $p+4$ cannot be simulatenously prime, since 3 always divides one of them, and thus the smallest distance between consecutive PIPs is 6. To this end, we make the following definition.

Definition 5.1. Let p and q be PIPs. We say that they are *twin PIPs* if $p-q=6$.

The twin prime conjecture states that there are infinitely many twin primes, and Broughan and Barnett conjecture that there are infinitely many consecutive PIPs with a difference of 6 (and indeed they conjecture that all gaps of even size at least 6 appear infinitely often). However, the Twin Prime Conjecture has a strong form, which states that where $\pi_2(x)$ is the number of twin primes not greater than x,

$$\pi_2(x) \sim 2C_{\text{twin}} \int_0^x \frac{dt}{\log^2 t} \sim 2C_{\text{twin}} \frac{x}{\log^2 x},$$

where

$$C_{\text{twin}} = \prod_{p \geq 3} \frac{p(p-2)}{(p-1)^2} \approx 0.6601618158$$

is the twin prime constant. One of the reasons that number theorists have faith in the twin prime conjecture is that the strong form of the conjecture seems to be very accurate. Experimentally we see that the data conform to this conjecture remarkably closely. The twin prime conjecture predicts that the number of primes up to 4×10^{18} is about

$$2C_{\text{twin}} \int_0^{4\times 10^{18}} \frac{dx}{\log^2 x} \approx 3023463139207178.4.$$

The third author has calculated this value precisely [13], and found

$$\pi_2(4 \times 10^{18}) = 3023463123235320,$$

which impressively agree in the first eight digits.

The basic idea behind the strong form of the twin prime conjecture is that the events "p is prime" and "$p+2$ is prime" are not independent events (details can be found in, say, [14, pp. 14–16]). Similar reasoning applies to twin PIPs. In particular, we consider the following question: if q is a PIP, what is the probability that $q+6$ is also a PIP? We must have either that $q, q+2, q+6$ are all prime, or that $q, q+4, q+6$ are all prime. But it is not enough for $q+6$ to be prime; it must also be a PIP. Combining these ideas, we find the following

Theorem 5.2. *If a prime q with index n is the first of a pair of twin PIPs, then one of two cases hold. Either: The triple $(q,\ q+2,\ q+6)$ are all prime, or the triple $(q,\ q+4,\ q+6)$ are all prime. Furthermore, the index of q and $q+6$ must each be prime.*

From this theorem we can construct a heuristic bound on the density of twin PIPs.

Conjecture 5.3. The number of twin PIPs up to x, $\pi_2^2(x)$, is asymptotically

$$\prod_{p>3} \frac{p^3(p-2)(p-3)}{(p-1)^5} \cdot \int_2^x \frac{dt}{\log^3 t(\log t - \log\log t)^2}.$$

Argument. Hardy and Little established conjectures on the density of prime constellations a century ago [15], and though unproven they are widely accepted, and enjoy considerable empirical support. Their conjectured density of either triple given in Theorem 5.2 is asymptotically

$$P_x(q, q+2, q+6) \sim \prod_{p>3} \frac{p^2(p-3)}{(p-1)^3} \cdot \int_2^x \frac{dt}{\log^3 t}.$$

We expect, as we have throughout this paper, that we can treat the primality of the index of a prime q as independent of q. Let n be the index of q. Then $n+2$ is the index of the prime $q+6$. The probability of both n and $n+2$ being prime is heuristically

$$\prod_{p>2} \frac{p(p-2)}{(p-1)^2} \cdot \frac{1}{\log^2 n} = \prod_{p>2} \frac{p(p-2)}{(p-1)^2} \cdot \frac{1}{\log^2(q/\log q)}.$$

If we simply multiply our earlier heuristic by the probability of this additional restriction, we find an expected density of twin PIPs to be

$$\begin{aligned} L(x) &= \prod_{p>3} \left(\frac{p^2(p-3)}{(p-1)^3}\right)\left(\frac{p(p-2)}{(p-1)^2}\right) \int_2^x \frac{dt}{\log^3 t} \frac{1}{\log^2(t/\log t)} \\ &= \prod_{p>3} \frac{p^3(p-2)(p-3)}{(p-1)^5} \int_2^x \frac{dt}{\log^3 t(\log t - \mathrm{loglog}\, t)^2}. \end{aligned}$$

Evaluating the product gives

$$L(x) \approx 7.5476417 \int_2^x \frac{dt}{\log^3 t(\log t - \mathrm{loglog}\, t)^2}. \qquad \square$$

This heuristic seems to describe the distribution of twin PIPs quite well. The following table gives the predicted number and actual number of twin PIPs up to various powers of 10, together with the absolute and relative error at each stage.

i	$\pi_2^2(10^i)$	$L(10^i)$	$L(10^i) - \pi_2^2(10^i)$	$(L(10^i) - \pi_2^2(10^i))/\sqrt{\pi_2^2(10^i)}$
1	1	19.5	18.5	18.5348
2	2	22.7	20.71	14.6437
3	3	24.6	21.6	12.4549
4	3	27.7	24.6	14.2297
5	7	35.8	28.8	10.8928
6	32	64.4	32.4	5.7360
7	149	184.5	35.5	2.9117
8	733	756.8	23.8	0.8774
9	3783	3754.2	-28.8	-0.4676
10	20498	20650.7	152.7	1.0664
11	119901	121621.9	1720.9	4.9700
12	750092	754446.3	4354.23	5.0276
13	4864965	4880705.2	15740.2	7.1363
14	32618201	32699568.8	81367.8	14.2470
15	225217352	225689240.9	471888.9	31.4441

Table 2: Actual counts, predicted values, absolute and relative errors for $\pi_2^2(x)$ at small powers of 10.

6. The Sum of the Reciprocals of the PIPs

The asymptotic density of the PIPs is $O(x/\log^2 x)$, from which it follows that the sum of the reciprocals of the PIPs converges (as noted first in [1]). Reciprocal sums have some interest in themselves; bounding Brun's constant, the sum of the reciprocals of the twin primes, has been a goal of many mathematicians since at least 1974 [16, 17, 18, 19, 20]. However, the accuracy of bounds on reciprocal sums also measures in an important way how much we understand a particular class of numbers. Bounding a reciprocal sum well requires two things: first, a computationally determined bound on small integers from the class; and second, good explicit bounds on the density of large integers from the class.

By this measure, twin primes are understood much better than, say, amicable numbers. Let B be the sum of the reciprocals of the twin primes. Then B has been shown to satisfy [14]

$$1.83 < B < 2.347,$$

and in fact [20]

$$1.83 < B < 2.15,$$

assuming the Extended Riemann Hypothesis. By contrast, let P be the Pomerance Constant – the sum of the reciprocals of the amicable numbers. Then the best known bounds on P [21] are the fairly weak

$$0.01198 < P < 6.56 \times 10^8.$$

Comparing the size of these intervals shows that, in a measurable way, current mathematical knowledge about twin primes is better than that of amicable numbers. With this is mind, we are interested in determining the accuracy to which we can bound the sum of the reciprocals of the PIPs.

By generating all PIPs up to 10^{15} using the obvious expedient of employing two segmented Erathosthenes sieves [22], one to check the primality of each odd n in that interval and another to check the primality of $\pi(n)$, and by accumulating the sum of the inverses of the PIPs using a 192-bits fractional part, it we find that

$$\sum_{q\leq 10^{15}} \frac{1}{q} \approx 1.01243131879802898253. \tag{6.1}$$

When u and v are not PIPs, the sum

$$T(u,v) = \sum_{u<q<v} \frac{1}{q} = \int_u^v \frac{d\pi\big(\pi(x)\big)}{x}$$

can be reasonably well approximated by replacing $\pi(x)$ by $\mathrm{li}(x) = \int_0^x \frac{dt}{\log t}$. This yields

$$T(u,v) \approx \hat{T}(u,v) = \int_u^v \frac{dx}{x\log x \log \mathrm{li}(x)} = \int_{\log u}^{\log v} \frac{dx}{x \log \mathrm{li}(e^x)}.$$

Due to potential arithmetic overflow problems when x is large, in this last integral $\log\mathrm{li}(e^x)$ should be evaluated by replacing it by its asymptotic expansion

$$\mathrm{li}(e^x) \approx x - \log x + \log\left(\sum_{k=0}^{N} \frac{k!}{x^k}\right);$$

$N = \lfloor x \rfloor$ delivers an approximation with relative error close to $\sqrt{2\pi x}\,e^{-x}$. This is not enough to evaluate $\hat{T}(10^{15},\infty)$ directly with an absolute error smaller that 5×10^{-21}, so $\hat{T}(10^{15},10^{100})$ can be numerically integrated without using the asymptotic expansion, and then $\hat{T}(10^{100},\infty)$ can be numerically integrated using the asymptotic expansion. Both Mathematica and pari-gp agree that

$$\hat{T}(10^{15},\infty) = 0.03077020549198786752.$$

It follows that

$$\sum_{q} \frac{1}{q} \approx 1.04320152429001685005. \tag{6.2}$$

Comparing $T(k\times 10^{14},(k+1)\times 10^{14})$ with $\hat{T}(k\times 10^{14},(k+1)\times 10^{14})$ for $k = 1,\ldots,10$ suggests that the absolute value of the relative error of the latter is, with high probability, smaller in 10^{-6}. The error of our estimate of $\sum_q \frac{1}{q}$ is therefore expected to be of order 10^{-8}.

Another approach to finding the sum of the reciprocals of the PIPs is to use the explicit upper and lower bounds on $\pi^2(x)$ from Theorem 4.1, our calculations to 10^{15}, and partial summation to bound the sum. Let us label the upper bound on $\pi^2(x)$ as $\pi_u^2(x)$, and the lower bound as $\pi_l^2(x)$. Then we have

$$\sum_q \frac{1}{q} < \sum_{q \le 10^{15}} \frac{1}{q} - \frac{\pi^2\left(10^{15}\right)}{10^{15}} + \int_{10^{15}}^{\infty} \frac{\pi_u^2(t)}{t^2} dt,$$

and

$$\sum_q \frac{1}{q} > \sum_{q \le 10^{15}} \frac{1}{q} - \frac{\pi^2\left(10^{15}\right)}{10^{15}} + \int_{10^{15}}^{\infty} \frac{\pi_l^2(t)}{t^2} dt.$$

In fact, these functions $\pi_u^2(x)$ and $\pi_l^2(x)$ do a fairly good job bounding $\pi^2(x)$ past 10^{15}, as determined by the difference in the integrals above. Numerical calculation with Mathematica gives the following:

$$\int_{10^{15}}^{\infty} \frac{\pi_l^2(t)}{t^2} dt \approx 0.0315569; \qquad \int_{10^{15}}^{\infty} \frac{\pi_u^2(t)}{t^2} dt \approx 0.0322135.$$

Combining these values with the calculation in (7), we can show that the sum of the reciprocals of the PIPs satisfies

$$1.04299 < \sum_q \frac{1}{q} < 1.04365,$$

in good agreement with (8).

7. The Generalized Prime Number Theorem

It is natural to consider a further generalization of prime-index primes. If the set of primes is listed in order, the subsequence of prime-index primes could be called 2-primes. Similarly, if the set of 2-primes is listed in increasing order, we may call the subsequence with prime index 3-primes. Let a k-prime be a member of the k-th iteration of this process. One may ask for the analogous results on the n-th k-prime and the number of k-primes up to x.

As noted in Broughan and Barnett [5], it is not hard to establish an analog to the Prime Number Theorem. Namely, defining $\pi^k(x)$ as the number of k-primes less than or equal to x, it is easy to show that

$$\pi^k(x) = \frac{x}{\log^k x} + O\left(\frac{x \log\log x}{\log^k x}\right).$$

In fact, it can be shown that $\pi^k(x) \sim \mathrm{Li}^k(x)$ as $x \to \infty$. The proof of this statement is not difficult, but is also not very enlightening. The theorem we prove here is slightly weaker, but shows more of the shape of the three main terms in this asymptotic.

Theorem 7.1. *For all* $k \geq 1$,

$$\pi^k(x) = \frac{x}{\log^k x}\left(1 + \frac{(k-1)\log\log x}{\log x} + \frac{k}{\log x}\right) + O_k\left(\frac{x(\log\log x)^2}{\log^{k+2} x}\right).$$

Proof. The proof proceeds by induction on k. The case $k = 1$ is given in (3), so we assume the statement holds up to k. Then,

$$\begin{aligned}\pi^{k+1}(x) &= \pi^k(\pi(x)) \\ &= \frac{\pi(x)}{\log^k \pi(x)}\left(1 + \frac{(k-1)\log\log \pi(x)}{\log \pi(x)} + \frac{k}{\log \pi(x)}\right) \\ &\qquad + O_k\left(\frac{\pi(x)(\log\log \pi(x))^2}{\log^{k+2} \pi(x)}\right)\end{aligned}$$

by the induction hypothesis. Now, Lemma 2.4 gives that

$$\frac{1}{\log^k \pi(x)}\left(1 + \frac{(k-1)\log\log \pi(x)}{\log \pi(x)} + \frac{k}{\log \pi(x)}\right)$$

is equivalent to

$$\frac{1}{\log^k x}\left(1 + \frac{k\log\log x}{\log x} + \frac{k}{\log x} + O_k\left(\frac{(\log\log x)^2}{\log^2 x}\right)\right).$$

Putting this together with (3) gives that

$$\begin{aligned}\pi^{k+1}(x) &= \frac{\pi(x)}{\log^k x}\left(1 + \frac{k\log\log x}{\log x} + \frac{k}{\log x} + O_k\left(\frac{(\log\log x)^2}{\log^2 x}\right)\right) \\ &= \frac{x}{\log^{k+1} x}\left(1 + \frac{k\log\log x}{\log x} + \frac{k+1}{\log x}\right) + O_k\left(\frac{x(\log\log x)^2}{\log^2 x}\right),\end{aligned}$$

completing the theorem's proof. □

Using Theorem 4.1 as the base case and Lemma 2.3 for the inductive step, we can prove the following theorem giving explicit bounds on $\pi^k(x)$.

Theorem 7.2. *For all* $k \geq 2$, *there exists a computable* $x_0(k)$ *such that*

$$\pi^k(x) < \frac{x}{\log^k x}\left(1 + \frac{1.5}{\log x}\right)^k \left(1 + \frac{1.5\log\log x}{\log x}\right)^{k-1}$$

and

$$\pi^k(x) > \frac{x}{\log^k x}\left(1 + \frac{1}{\log x}\right)^k \left(1 + \frac{\log\log x}{\log x}\right)^{k-1}$$

for all $x \geq x_0(k)$.

Note that from the proof of Lemma 2.3 and Theorem 4.1 we may choose any $x_0(k)$ satisfying

$$\pi^k\left(x_0(k)\right) \geq 13,$$

as $\pi^2(179) = 13$.

It is also not difficult to adapt an argument from [5] to prove a proposition on $\pi^k(x)$ for all $k \geq 1$.

Proposition 7.3. *The following inequalities are true for every integer $n > 1$ and $k \geq 1$ and for all sufficiently large real numbers x, y:*

(a) $\pi^k(nx) < n\pi^k(x)$,

(b) $\pi^k(x+y) \leq \pi^k(x) + 2^k\pi^k(y)$, *and*

(c) $\pi^k(x+y) - \pi^k(x) \ll_k \frac{y}{\log^k y}$.

Proof. Each of the statements in this theorem have been shown for $k = 2$ in [5] and for $k = 1$ elsewhere ((a) in [23], (b) and (c) in [24]). We assume these base cases and follow the argument in [5] to complete the induction for each statement.

(a) From Panaitopol [23], we have $\pi(nx) < n\pi(x)$ for sufficiently large x. The induction hypothesis, followed by an application of Panaitopol's result gives

$$\pi^{k+1}(nx) < \pi\left(n\pi^k(x)\right) < n\pi^{k+1}(x)$$

for sufficiently large x.

(b) Using Montgomery and Vaughan's [24] bound $\pi(x+y) \leq \pi(x)+2\pi(y)$ together with the induction hypothesis, we have

$$\pi^{k+1}(x+y) \leq \pi\left(\pi^k(x) + 2^k\pi(y)\right) \leq \pi^{k+1}(x) + 2^{k+1}\pi(y)$$

for sufficiently large x and y.

(c) This follows from part (b) and Theorem 7.1. □

Note that these bounds are certainly not the best possible. Inequality (b) in particular seems rather weak. Proving a stronger general theorem, however, seems difficult.

8. Gaps Between PIPs

Our consideration in Section 5 of twin PIPs, or consecutive PIPs with difference 6, is just a special case of a more general question about gaps between consecutive PIPs. In this section we consider some computational data on other gap sizes. Let

$$q(h) = \min_{q_{i+1}-q_i=h} q_i$$

be the first occurrence of a gap of size h between PIPs (or infinity if no such gap exists), let

$$Q(x;h) = \sum_{\substack{q_i \le x \\ q_{i+1}-q_i=h}} 1$$

be the number of gaps of h between PIPs up to x, and let

$$F(h) = \prod_{\substack{p>2 \\ p|h}} \frac{p-1}{p-2}$$

be the corresponding Hardy-Littlewood correction factor. As expected due to the prime k-tuples conjecture, it was found that the graph of $Q(10^{15};h)$ exhibited a rapid "oscillation" (cf., for example, [25]), which disappeared in a graph of $Q(10^{15};h)/F(h)$. Contrary to what happens with the graphs of smoothed counts of prime gaps (i.e., counts divided by $F(h)$), the graphs of smoothed counts of PIP gaps first increase, then attain a maximum at an absissa which grows with the count limit x, and only then start to decrease exponentially (this behavior can be observed in figure 1 of [5]).

Table 8 presents the record gaps (also known as maximal gaps [26]) that were observed up to 10^{15}. Since a large gap between primes very likely corresponds to a large gap between PIPs (having the large prime gap between their indices), the first ten occurrences of each prime gap up to 4×10^{18}, obtained as a colateral result of the third author's extensive verification of the Goldbach conjecture [?], were used to locate large gaps between PIPs. This was done as follows:

1. given an index i (the first prime of a large prime gap), an approximation $\hat{p}_i$ of p_i was found by solving $|\hat{\pi}(\hat{p}_i) - i| < 10$, where $\hat{\pi}(x)$ is the Riemann's formula for $\pi(x)$, truncated to the first one million complex conjugate zeros on the critical line, and with lower order terms replaced by simpler asymptotic approximations;

2. using the algorithm described in [10], $\pi(\hat{p}_i)$ was computed (this was by far the most time consuming step);

3. using a segmented sieve and using $\hat{p}_i$ as a starting point, going backwards if necessary, p_i, which is by construction a PIP, was located;

4. since the gap between indices was known *a priori*, the next PIP was also located, and the difference between the two was computed.

The maximal gap candidates above 10^{15} that resulted from this effort are presented in Table 8; below 10^{15}, the results of Table 8 were reproduced exactly. Based on the data from these two tables, it appears that $h/\log^3 q(h)$ is bounded.

h	$q(h)$	h	$q(h)$	h	$q(h)$
2	3	1380	3733111	7524	49868272577
6	5	1500	5188297	7648	57757941919
14	17	1766	5336477	7884	60381716303
18	41	1784	7244099	8994	93033321509
26	83	1800	9026123	9208	104673577891
30	127	1852	12818959	9328	215587773169
36	241	1998	21330371	10254	271208089553
48	283	2200	21459517	10702	521584527307
92	617	2280	24931771	12388	655650146791
112	1297	2628	32637571	13436	1139727488171
114	1913	2992	79689091	13826	3565879112657
122	2099	3000	182395973	13898	5144378650811
150	3761	3770	315619631	14570	8549998218191
168	5869	4406	390002363	15102	8724860034481
190	9103	4506	2199880757	15218	12118597117331
348	10909	4872	2515605941	16006	13479163888087
372	46751	4938	3443579963	16814	31885486594523
384	104827	5214	3994122787	17010	36971628663863
458	114089	5256	4043156627	18312	40798355884309
474	152953	5844	6111419117	19680	60418125851197
498	177791	5974	8440859467	21820	81040555147807
642	219931	6486	9568037147	22804	229922915352703
738	293123	6864	21472440259	24658	452388122520163
1028	368153	7098	29861568733	25694	647593721749763
1244	2101553	7446	35005449181	26148	804920613659501

Table 3: Record gaps between PIPs up to 10^{15}

h	$q(h)$	h	$q(h)$
27324	1451492253702853	50932	1797828789776991187
27462	3031506479624729	51282	3367200144283080467
31184	3149270807374079	51496	5303163766511877793
33348	7759035095377103	54766	5948139313109849407
34428	19843157450989771	55438	8686480782592200319
34902	44370362884634417	56964	13131568510506112637
35560	48210577082615809	57744	14471372274538980343
35964	58458364312779077	60646	15209204300586561877
36276	63536060873650711	62244	18108618970703357989
45390	63775464504542041	65278	35376288156449516509
46910	770359508644782761	67136	63526302908206766003
46948	1186416917758809991	67236	146174033905511020897
47838	1263062213472998429	67356	170912819272488312527

Table 4: Potential record gaps between PIPs after 10^{15}

9. A Goldbach-Like Conjecture for PIPs

Let $R(n)$ be the number of pairs $(q, n-q)$ such that both q and $n-q$ are PIPs. Just like for the classical Goldbach conjecture [27], the identity

$$\sum_{n=1}^{L} R(n)\, x^n = \Big(\sum\nolimits_{q \le L} x^q\Big)^2 \bmod x^{L+1},$$

coupled with a fast polynomial multiplication algorithm based on the Fast Fourier Transform, makes it possible to compute $R(n)$ for all $n \le L$ using only $O(L^{1+\epsilon})$ time and space. For L a positive even integer, let

$$R_{\text{lower}}(x; L) = \min_{x \le 2n \le L} R(2n) \quad \text{and} \quad R_{\text{upper}}(x) = \max_{n \le x} R(n).$$

For n even, these two non-decreasing functions are useful lower and upper bounds of the value of $R(n)$. Figure 9 shows how these two functions behave (their points of increase up to $L = 10^9$ were computed with the help of a simple matlab script). Based on our empirical data, the following conjecture is almost certainly true.

Conjecture 9.1. All even integers larger than 80612 can be expressed as the sum of two prime-indexed primes.

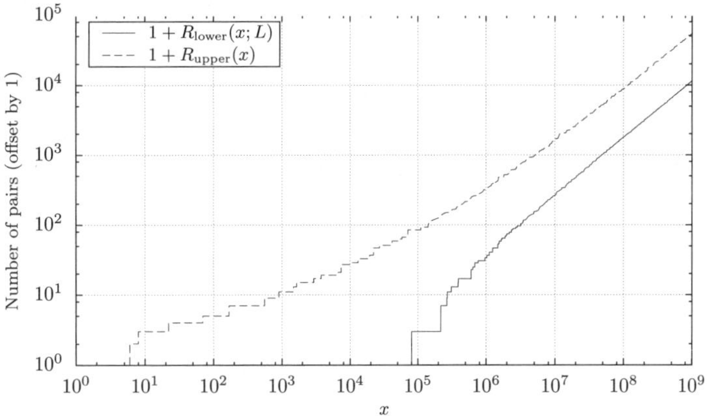

FIGURE 1. Lower and upper bounds of the number of ways, offset by one, of expressing an even number by an ordered sum of two PIPs.

10. Proofs of Lemmas

Proof of Lemma 2.1. First, note that $n \log n = n^{1+\frac{\log\log n}{\log n}}$. From this,

$$\log\left(\log\left(n\log n\right)\right) = \log\left(\log n^{1+\frac{\log\log n}{\log n}}\right) = \log\left(1+\frac{\log\log n}{\log n}\right) + \log\log n.$$

Since $\log(1+x) < x$ for $x > 0$, we have $\log\left(1+\frac{\log\log n}{\log n}\right) < \frac{\log\log n}{\log n}$, which completes the lemma's proof. □

Proof of Lemma 2.2. First, note that

$$n\log\left(n\log n\right) = n\log\left(n^{1+\frac{\log\log n}{\log n}}\right) = \left(n\log n\right)\left(1+\frac{\log\log n}{\log n}\right).$$

Now,

$$1+\frac{\log\log n}{\log n} = n^{\frac{\log\left(1+\frac{\log\log n}{\log n}\right)}{\log n}} < n^{\frac{\log\log n}{\log^2 n}}$$

because $\log(1+x) < x$ for $x > 0$. Thus,

$$\begin{aligned}
\log\log\left(n\log\left(n\log n\right)\right) &= \log\log\left(\left(n\log n\right)\left(1+\frac{\log\log n}{\log n}\right)\right)\\
&< \log\log n^{1+\frac{\log\log n}{\log n}+\frac{\log\log n}{\log^2 n}}\\
&= \log\log n + \log\left(1+\frac{\log\log n}{\log n}+\frac{\log\log n}{\log^2 n}\right)\\
&< \log\log n + \frac{\log\log n}{\log n}+\frac{\log\log n}{\log^2 n},
\end{aligned}$$

where the last inequality comes from again using $\log(1+x) < x$. This proves the lemma. □

Proof of Lemma 2.3. We begin with the upper bound. From [7], we know that $\pi(x) \geq \frac{x}{\log x}$ for all $x \geq 17$. Thus, in this range,

$$\log\pi(x) \geq \left(1-\frac{\log\log x}{\log x}\right)\log x,$$

and so

$$\frac{1}{\log\pi(x)} < \frac{1}{\log x}\left(\frac{1}{1-\frac{\log\log x}{\log x}}\right). \tag{10.1}$$

Writing the second factor as a geometric series proves the bound.

Considering the lower bound, we use the upper bound on $\pi(x)$ for $x > 1$ in (4) to see that

$$\begin{aligned}\log \pi(x) \le \log\left(\frac{x}{\log x}\left(1+\frac{1.2762}{\log x}\right)\right) &= \log\frac{x}{\log x} + \log\left(1+\frac{1.2762}{\log x}\right)\\ &= \left(1-\frac{\log\log x}{\log x}\right)\log x + \log\left(1+\frac{1.2762}{\log x}\right)\\ &< \left(1-\frac{\log\log x}{\log x}\right)\log x + \frac{1.2762}{\log x},\end{aligned}$$

where the last inequality uses $\log(1+x) < x$ for $x > 0$.

Taking the reciprocal of this inequality, we have

$$\frac{1}{\log \pi(x)} \ge \frac{1}{\log x}\left(\frac{1}{1-\frac{\log\log x}{\log x}+\frac{1.2762}{\log^2 x}}\right).$$

Thus, to establish the lemma, we need to bound

$$\frac{1}{1-\frac{\log\log x}{\log x}+\frac{1.2762}{\log^2 x}} \ge 1+\frac{\log\log x}{\log x}. \tag{10.2}$$

This is indeed the case, as we may rewrite the fraction as the sum of a geometric series. That is, we may write

$$\frac{1}{1-\frac{\log\log x}{\log x}+\frac{1.2762}{\log^2 x}} = 1+\sum_{k=1}^{\infty}\left(\frac{\log\log x}{\log x}-\frac{1.2762}{\log^2 x}\right)^k.$$

Now, for $x \ge 33$,

$$\sum_{k=2}^{\infty}\left(\frac{\log\log x}{\log x}-\frac{1.2762}{\log^2 x}\right)^k = \frac{\left(\frac{\log\log x}{\log x}-\frac{1.2762}{\log^2 x}\right)^2}{1-\frac{\log\log x}{\log x}+\frac{1.2762}{\log^2 x}} > \frac{1.2762}{\log^2 x},$$

establishing (10). This completes the lemma's proof. □

Proof of Lemma 2.4. Using Lemma 2.3, together with (3), we have

$$\begin{aligned}\frac{1}{\log^k \pi(x)} &= \left(\frac{1}{\log x}\left(1+\frac{\log\log x}{\log x}+O\left(\frac{(\log\log x)^2}{\log^2 x}\right)\right)\right)^k\\ &= \frac{1}{\log^k x}\left(1+\frac{k\log\log x}{\log x}+O_k\left(\frac{(\log\log x)^2}{\log^2 x}\right)\right),\end{aligned} \tag{10.3}$$

which establishes the first half of the lemma.

We know that

$$\pi(x) = \frac{x}{\log x}\left(1+\frac{1}{\log x}+O\left(\frac{1}{\log^2 x}\right)\right),$$

and so the same argument used in the proof of Lemma 2.2 gives that

$$\begin{aligned}\log\log\pi(x) &= \log\log x + \log\left(1 - \frac{\log\log x}{\log x} + O\left(\frac{(\log\log x)^2}{\log^2 x}\right)\right)\\ &= \log\log x + O\left(\frac{\log\log x}{\log x}\right).\end{aligned}$$

Using this and (11), we see

$$\begin{aligned}\frac{\log\log\pi(x)}{\log\pi(x)} &= \left(\log\log x + O\left(\frac{\log\log x}{\log x}\right)\right)\cdot\left(\frac{1}{\log x}\left(1 + O\left(\frac{\log\log x}{\log x}\right)\right)\right)\\ &= \frac{\log\log x}{\log x} + O\left(\frac{(\log\log x)^2}{\log^2 x}\right),\end{aligned}$$

completing the proof of the lemma. □

References

[1] J.H. Jordan, *On sums of inverses of primes*, Math. Magazine **38** (1965), no. 5, 259–262.

[2] R.E. Dressler and S. T. Parker, *Primes with a prime subscript*, J. Assoc. Comput. Mach. **22** (1975), 380–381.

[3] J. Sándor, *On certain sequences and series with applications in prime number theory*, Gaz. Mat. Met. Inf **6** (1985), no. 1-2, 38–48.

[4] D.S. Mitrinovic, J. Sándor, and B. Crstici, *Handbook of Number Theory, Mathematics and its Applications 351*, Kluwer Academic Pub., 1996

[5] K.A. Broughan and A.R. Barnett, *On the subsequence of primes having prime subscripts*, J. Integer Seq. **12** (2009), no. 2, Article 09.2.3, 10pp.

[6] P. Dusart, *Autour de la Fonction qui Compte le Nombre de Nombres Premiers*, Ph.D. thesis, Université de Limoges, Limoges, France, 1998.

[7] J.B. Rosser and L. Schoenfeld, *Sharper bounds for the Chebyshev functions $\theta(x)$ and $\psi(x)$*, Math. Comp. **29** (1975), no. 129, 243–269.

[8] G. Robin, *Estimation de la fonction de Tchebychef θ sur le kieme nombre premier et grandes valeurs de la fonction ω (n) nombre de diviseurs premiers de n*, Acta Arith **42** (1983), no. 4, 367–389.

[9] J.C. Lagarias, V.S. Miller, and A.M. Odlyzko, *Computing $\pi(x)$: The Meissel-Lehmer method*, Math. Comp. **44** (1985), no. 170, 537–560.

[10] T. Oliveira e Silva, *Computing $\pi(x)$: the combinatorial method*, Revista do DETUA **4** (2006), no. 6, 759–768.

[11] J. Buethe, J. Franke, A. Jost, and T. Kleinjung, 2010, personal comunication.

[12] D. Platt, 2012, personal comunication.

[13] T. Oliveira e Silva, S. Herzog, and S. Pardi, *Empirical verification of the even Goldbach conjecture, and computation of prime gaps, up to $4 \cdot 10^{18}$*, Math. Comp., to appear.

[14] R. Crandall and C. Pomerance, *Prime Numbers: A Computational Perspective*, Second ed., Springer, New York, 2005.

[15] G.H. Hardy and J.E. Littlewood, *Some problems of Partitio numerorum; III: On the expression of a number as a sum of primes*, Acta Mathematica **44** (1923), no. 1, 1–70.

[16] D. Shanks and J.W. Wrench Jr, *Brun's constant*, Math. Comp. (1974), 293–299.

[17] R. P. Brent, *Irregularities in the distribution of primes and twin primes*, Math. Comp. **29** (1975), no. 129, 43–56.

[18] T. R. Nicely, *Enumeration to* 10^{14} *of the twin primes and Brun's constant*, Virginia J. Science **46** (1995), 195–204.

[19] T. R. Nicely, *A new error analysis for Brun's constant*, Virginia J. Science **52** (2001), no. 1, 45–56.

[20] D. Klyve, *Explicit bounds on twin primes and Brun's constant*, ProQuest LLC, Ann Arbor, MI, 2007, Ph.D. Thesis, Dartmouth College.

[21] J. Bayless and D. Klyve, *On the sum of reciprocals of amicable numbers*, Integers **11A** (2011), #A5.

[22] C. Bays and R. Hudson, *The segmented sieve of Eratosthenes and primes in arithmetic progressions to* 10^{12}, Nordisk Tidskrift for Informationsbehandling (BIT) **17** (1977), no. 2, 121–127.

[23] L. Panaitopol, *Eine eigenschaft der funktion über die verteilung der primzahlen*, Bull. Math. Soc. Sci. Math. RS Roumanie **23** (1979), 189–194.

[24] H.L. Montgomery and R.C. Vaughan, *The large sieve*, Mathematika **20** (1973), no. 02, 119–134.

[25] A. Odlyzko, M. Rubinstein, and M. Wolf, *Jumping champions*, Experimental Math. **8** (1999), no. 2, 107–118.

[26] D. Shanks, *On maximal gaps between successive primes*, Math. Comp. **18** (1964), no. 88, 646–651.

[27] J. Richstein, *Computing the number of Goldbach partitions up to* 5×10^8, Algorithmic Number Theory: ANTS-IV Proceedings (Wieb Bosma, ed.), Lecture Notes in Computer Science, vol. 1838, Springer, Berlin / New York, 2000, 475–490.

GAP DISTRIBUTION OF FAREY FRACTIONS UNDER SOME DIVISIBILITY CONSTRAINTS

Florin P. Boca[1]
Department of Mathematics, University of Illinois, Urbana, Illinois
fboca@illinois.edu

Byron Heersink
Department of Mathematics, University of Illinois, Urbana, Illinois
heersin2@illinois.edu

Paul Spiegelhalter
Department of Mathematics, University of Illinois, Urbana, Illinois
spiegel3@illinois.edu

Received: 1/2/13, Revised: 4/10/13, Accepted: 6/9/13, Published: 7/10/13

Abstract

For a given positive integer ℓ, we show the existence of the limiting gap distribution measure for the sets of Farey fractions $\frac{a}{q}$ of order Q with $\ell \nmid a$, and respectively with $(q, \ell) = 1$, as $Q \to \infty$.

1. Introduction

The set $\mathcal{F}_Q$ of Farey fractions of order Q consists of those rational numbers $\frac{a}{q} \in (0, 1]$ with $(a, q) = 1$ and $q \leqslant Q$. The spacing statistics of the increasing sequence $(\mathcal{F}_Q)$ of finite subsets of $(0, 1]$ have been investigated by several authors [9, 1, 7]. Recently Badziahin and Haynes considered a problem related to the distribution of gaps in the subset $\mathcal{F}_{Q,d}$ of $\mathcal{F}_Q$ of those fractions $\frac{a}{q}$ with $(q, d) = 1$, where d is a fixed positive integer and $Q \to \infty$. They proved [2] that, for each $k \in \mathbb{N}$, the number $N_{Q,d}(k)$ of pairs $\left(\frac{a}{q}, \frac{a'}{q'}\right)$ of consecutive elements in $\mathcal{F}_{Q,d}$ with $a'q - aq' = k$ satisfies the asymptotic formula

$$N_{Q,d}(k) = c(d, k)Q^2 + O_{d,k}(Q \log Q) \qquad (Q \to \infty), \tag{1.1}$$

for some positive constant $c(d, k)$ that can be expressed using the measure of certain cylinders associated with the area-preserving transformation introduced by Cobeli,

[1]Member of the Institute of Mathematics "Simion Stoilow" of the Romanian Academy, 21 Calea Griviţei, 010702 Bucharest, Romania

Zaharescu, and the first author in [4]. The pair correlation function of $(\mathcal{F}_{Q,d})$ was studied and shown to exist by Xiong and Zaharescu [11], even in the more general situation where $d = d_Q$ is no longer constant but increases according to the rules $d_{Q_1} \mid d_{Q_2}$ as $Q_1 < Q_2$ and $d_Q \ll Q^{\log\log Q/4}$.

This paper is concerned with the gap distribution of the sequence of sets $(\mathcal{F}_{Q,d})$, and respectively of $(\widetilde{\mathcal{F}}_{Q,\ell})$, the sequence of sets $\widetilde{\mathcal{F}}_{Q,\ell}$ of Farey fractions $\gamma = \frac{a}{q} \in \mathcal{F}_Q$ with $\ell \nmid a$. Our peculiar interest in $\widetilde{\mathcal{F}}_{Q,\ell}$ arises from the problem studied in [5], concerning the distribution of the free path associated to the linear flow through $(0,0)$ in $\mathbb{R}^2$ in the small scatterer limit, in the case of circular scatterers of radius $\varepsilon > 0$ placed at the points $(m,n) \in \mathbb{Z}^2$ with $\ell \nmid (m-n)$. When $\ell = 3$ this corresponds, after suitable normalization, to the situation of scatterers distributed at the vertices of a honeycomb tessellation, and the linear flow passing through the center of one of the hexagons. When $\ell = 2$ the scatterers are placed at the vertices of a square lattice and the linear flow passes through the center of one the squares. Arithmetic properties of the number ℓ are shown to be explicitly reflected by the gap distribution of the elements of $(\widetilde{\mathcal{F}}_{Q,\ell})$. The symmetry $x \mapsto 1-x$ shows that for the purpose of studying the gap distribution of these fractions on $[0,1]$ one can replace the condition $\ell \nmid (m-n)$ by the more esthetic one $\ell \nmid n$.

The *gap distribution* (or *nearest neighbor distribution*) of a numerical sequence, or more generally of a sequence of finite subsets of $[0,1)$, measures the distribution of lengths of gaps between the elements of the sequence. Let $A = \{x_0 \leqslant x_1 \leqslant \ldots \leqslant x_N\}$ be a finite list of numbers in $[0,1)$, not all equal, scaled to $\tilde{x}_j = \frac{Nx_j}{x_N - x_0}$ with mean spacing $\frac{\tilde{x}_N - \tilde{x}_0}{N} = 1$. The *gap distribution measure* of A is the finitely supported probability measure on $[0,\infty)$ defined by

$$\nu_A(-\infty,\xi] = \nu_A[0,\xi] := \frac{1}{N}\#\{j \in [1,N] : \tilde{x}_j - \tilde{x}_{j-1} \leqslant \xi\}, \qquad \xi \geqslant 0.$$

If it exists, the weak limit $\nu = \nu_{\mathcal{A}}$ of the sequence (ν_{A_n}) of probability measures associated with an increasing sequence $\mathcal{A} = (A_n)$ of finite lists of numbers in $[0,1)$, is called the *limiting gap measure* of $\mathcal{A}$.

It is elementary (see, e.g., Lemma 1 below) that

$$\#\widetilde{\mathcal{F}}_{Q,\ell} = \widetilde{K}_\ell Q^2 + O_\ell(Q\log Q), \qquad \#\mathcal{F}_{Q,d} = K_d Q^2 + O_d(Q\log Q), \tag{1.2}$$

where

$$\widetilde{K}_\ell = \frac{1}{2\zeta(2)} - \frac{C(\ell)}{2\ell}, \quad K_d = \frac{C(d)}{2}, \quad \text{with} \quad C(\ell) = \frac{1}{\zeta(2)} \prod_{\substack{p\in\mathcal{P} \\ p\mid\ell}} \left(1 + \frac{1}{p}\right)^{-1}.$$

We prove the following result:

Theorem 1. *Given positive integers ℓ and d, the limiting gap measures $\widetilde{\nu}_\ell$ of $(\widetilde{\mathcal{F}}_{Q,\ell})$, and respectively ν_d of $(\mathcal{F}_{Q,d})$, exist. Their densities are continuous on $[0,\infty)$ and real analytic on each component of $(0,\infty) \setminus \mathbb{N}\widetilde{K}_\ell$, and respectively of $(0,\infty) \setminus \mathbb{N}K_d$.*

The existence of $\widetilde{\nu}_\ell$ is proved in Section 2 and the limiting gap distribution is explicitly computed in (2.9) using tools from [4], [8] and [5]. The result on ν_d is proved in Section 4. When d is a prime power, an explicit computation can be done as for $\tilde{\nu}_\ell$. In general the repartition function of ν_d depends on the measure of some cylinders associated with the transformation T from (2.7), and on the length of strings of consecutive elements in $\mathcal{F}_Q$ with at least one denominator relatively prime with d.

The upper bound $4d^3$ for $L(d) = \min\{L : \forall i, \forall Q, \exists j \in [0, L], (q_{i+j}, d) = 1\}$ was found in [2], where $q_i, \dots, q_{i+L}$ denote the denominators of a string $\gamma_i < \cdots < \gamma_{i+L}$ of consecutive elements in $\mathcal{F}_Q$. Although we expect this bound to be considerably smaller, we could only improve it in a limited number of situations. In Section 3 we lower it to $4\omega(d)^3$ for integers d with the property that the smallest prime divisor of d is $\geqslant \omega(d)$, where $\omega(d)$ denotes as usual the number of distinct prime factors of d. The bound $L(d) = 1$ is trivial when d is a prime power. Employing properties of the transformation T^2 we show that $L(d) \leqslant 5$ when d is the product of two prime powers, which is sharp. Finding better bounds on $L(d)$ when $\omega(d) \geqslant 3$ appears to be an interesting problem in combinatorial number theory.

2. The Gap Distribution of $\widetilde{\mathcal{F}}_{Q,\ell}$

Let $\mathcal{F}_Q^{(\ell)} = \mathcal{F}_Q \setminus \widetilde{\mathcal{F}}_{Q,\ell}$ denote the set of Farey fractions $\gamma = \frac{a}{q} \in \mathcal{F}_Q$ with $\ell \mid a$, and let $N_Q^{(\ell)}$ denote the cardinality of $\mathcal{F}_Q^{(\ell)}$. Consider also:

$$\begin{aligned} \mathcal{G}_Q(\xi) &:= \left\{ (\gamma, \gamma') : \gamma,\, \gamma' \text{ consecutive in } \mathcal{F}_Q, 0 < \gamma' - \gamma \leqslant \frac{\xi}{Q^2} \right\}, \\ \mathcal{G}_Q^{(\ell)}(\xi) &:= \left\{ (\gamma, \gamma') : \gamma,\, \gamma' \text{ consecutive in } \widetilde{\mathcal{F}}_{Q,\ell}, 0 < \gamma' - \gamma \leqslant \frac{\xi}{Q^2} \right\}, \\ N_Q(\xi) &: = \#\mathcal{G}_Q(\xi), \quad N_Q^{(\ell)}(\xi) := \#\mathcal{G}_Q^{(\ell)}(\xi). \end{aligned}$$

Lemma 1. *$N_Q^{(\ell)} = \dfrac{C(\ell)}{2\ell} Q^2 + O_\ell(Q \log Q)$ as $Q \to \infty$.*

Proof. It is clear that

$$N_Q^{(\ell)} = \#\mathcal{F}_Q^{(\ell)} = \sum_{\substack{q=1 \\ (\ell,q)=1}}^{Q} \sum_{\substack{a=1 \\ (a,q)=1 \\ \ell \mid a}}^{q} 1.$$

Letting $k = \frac{a}{\ell}$ and noting that, whenever $(\ell, q) = 1$, we have $(k\ell, q) = 1$ if and only

$(k,q)=1$, the sum above becomes

$$\sum_{\substack{q=1\\(\ell,q)=1}}^{Q}\ \sum_{\substack{k=1\\(k,q)=1}}^{[q/\ell]} 1.$$

Standard Möbius summation, cf. (A.1) and (A.2), and $\sum_{q=1}^{Q}\sigma_0(q)=O(Q\log Q)$, where $\sigma_0(q)=\sum_{d|q}1$, yield

$$\sum_{\substack{q=1\\(\ell,q)=1}}^{Q}\ \sum_{\substack{k=1\\(k,q)=1}}^{[q/\ell]} 1=\sum_{\substack{q=1\\(\ell,q)=1}}^{Q}\left(\frac{\varphi(q)}{q}\cdot\frac{q}{\ell}+O(\sigma_0(q))\right)=\frac{C(\ell)}{2\ell}Q^2+O_\ell(Q\log Q),$$

concluding the proof. □

This also establishes the first equality in (1.2) because

$$\#\widetilde{\mathcal{F}}_{Q,\ell}=\#\mathcal{F}_Q-\#\mathcal{F}_Q^{(\ell)}\sim\left(\frac{1}{2\zeta(2)}-\frac{C(\ell)}{2\ell}\right)Q^2.$$

Letting $\xi>0$ and $Q,\ell\in\mathbb{N}$ with $\ell\geqslant 2$, we set out to asymptotically estimate the number $N_Q^{(\ell)}(\xi)$ as $Q\to\infty$. Now if $\gamma=\frac{a}{q}$ and $\gamma'=\frac{a'}{q'}$ are consecutive elements in $\mathcal{F}_Q$ and $\gamma'\in\mathcal{F}_Q^{(\ell)}$, then $1=a'q-aq'\equiv -aq'\pmod{\ell}$, which implies that $(a,\ell)=1$, and thus $\gamma\notin\mathcal{F}_Q^{(\ell)}$. Similarly, if $\gamma\in\mathcal{F}_Q^{(\ell)}$, then $\gamma'\notin\mathcal{F}_Q^{(\ell)}$; and so no two consecutive elements of $\mathcal{F}_Q$ belong simultaneously to $\mathcal{F}_Q^{(\ell)}$. This means that if $\gamma<\gamma'$ are consecutive elements in $\widetilde{\mathcal{F}}_{Q,\ell}$, then two cases can occur:
Case 1. γ and γ' are consecutive elements in $\mathcal{F}_Q$ and $\gamma,\gamma'\notin\mathcal{F}_Q^{(\ell)}$. In this case the number of gaps in consecutive fractions of length $\leqslant\frac{\xi}{Q^2}$ is equal to $\mathcal{N}_1(Q,\xi)=N_Q(\xi)-M_1(Q,\xi)-M_2(Q,\xi)$, where $M_1(Q,\xi)$ is the number of pairs $(\gamma,\gamma')\in\mathcal{G}_Q(\xi)$ with $\gamma'\in\mathcal{F}_Q^{(\ell)}$, and $M_2(Q,\xi)$ is the number of pairs $(\gamma,\gamma')\in\mathcal{G}_Q(\xi)$ with $\gamma\in\mathcal{F}_Q^{(\ell)}$.

The number $N_Q(\xi)$ is estimated employing the well-known fact that $\gamma<\gamma'$ are consecutive elements in $\mathcal{F}_Q$ if and only if $q,q'\in\{1,\ldots,Q\}$, $q+q'>Q$, and $a'q-aq'=1$. Furthermore, $\frac{a'}{q'}-\frac{a}{q}=\frac{1}{qq'}$, and so $\frac{a'}{q'}-\frac{a}{q}\leqslant\frac{\xi}{Q^2}$ if and only if $qq'\geqslant\frac{Q^2}{\xi}$. This establishes the equality

$$\begin{aligned}N_Q(\xi)&=\#\left\{(q,q')\in\mathbb{N}^2: q,q'\leqslant Q, q+q'>Q, (q,q')=1, qq'\geqslant\frac{Q^2}{\xi}\right\}\\&=\sum_{q'=1}^{Q}\ \sum_{\substack{q\in I_Q(q')\\(q,q')=1}}1,\end{aligned}\tag{2.1}$$

where $I_Q(q')=Q\cdot\big[\eta_Q(q'),1\big]$ and $\eta_Q(q')=\max\left\{1-\frac{q'-1}{Q},\frac{Q}{\xi q'}\right\}$.

Standard Möbius summation provides

$$\begin{aligned}N_Q(\xi) &= \sum_{q'=1}^{Q}\left(\frac{\varphi(q')}{q'}|I_Q(q')| + O(\sigma_0(q'))\right) = \sum_{q'=1}^{Q}\frac{\varphi(q')}{q'}|I_Q(q')| + O(Q\log Q)\\ &= \frac{A(\xi)}{\zeta(2)}Q^2 + O(Q\log Q),\end{aligned}$$

where

$$\begin{aligned}A(\xi) &= \left|\left\{(x,y)\in(0,1]^2 : x+y>1, xy\geqslant\frac{1}{\xi}\right\}\right|\\ &= \begin{cases}0 & \text{if } 0<\xi\leqslant 1\\ 1-\frac{\log\xi+1}{\xi} & \text{if } 1\leqslant\xi\leqslant 4\\ 1-\frac{1}{\xi}-\frac{1}{2}\sqrt{1-\frac{4}{\xi}}+\frac{2}{\xi}\log\left(\frac{1+\sqrt{1-4/\xi}}{2}\right) & \text{if } \xi\geqslant 4.\end{cases}\end{aligned} \tag{2.2}$$

Next, we estimate $M_1(Q,\xi)$. Clearly $M_1(Q,\xi)=0$ if $\xi\in(0,1]$, and so assume $\xi>1$. If $\frac{a'}{q'}\in\mathcal{F}_Q^{(\ell)}$, then $(a',q')=1$ and $\ell\mid a'$. Since $(a',q')=1$, we have $(\ell,q')=1$. Therefore, we have to count all pairs of integers $(q,q')\in(0,Q]^2$ with $q+q'>Q$, $(q,q')=1$, $qq'\geqslant\frac{Q^2}{\xi}$, in which $(\ell,q')=1$, and there is an $a'\in\{1,\dots,q'\}$ such that $a'q\equiv 1 \pmod{q'}$ and $\ell\mid a'$. As a result, after also letting $k=\frac{a'}{\ell}$, $\ell\bar{\ell}\equiv 1 \pmod{q'}$, $M_1(Q,\xi)$ can be expressed as

$$M_1(Q,\xi) = \sum_{\substack{q'=1\\(\ell,q')=1}}^{Q}\ \sum_{\substack{q\in I_Q(q')\\(q,q')=1}}\ \sum_{\substack{a'=1\\a'q\equiv 1 \ (\mathrm{mod}\ q')\\ \ell\mid a'}}^{q'} 1 = \sum_{\substack{q'=1\\(\ell,q')=1}}^{Q}\ \sum_{\substack{q\in I_Q(q')\\(q,q')=1}}\ \sum_{\substack{k\in(0,q'/\ell]\\kq\equiv\bar{\ell}\ (\mathrm{mod}\ q')}} 1. \tag{2.3}$$

Now by (2.3) and (A.4), for any $\delta>0$,

$$\begin{aligned}M_1(Q,\xi) &= \sum_{\substack{q'=1\\(\ell,q')=1}}^{Q}\left(\frac{\varphi(q')}{q'^2}\iint_{I_Q(q')\times[0,q'/\ell]} dx\,dy + O_\delta\big(q'^{1/2+\delta}\big)\right)\\ &= \frac{1}{\ell}\sum_{\substack{q'=1\\(\ell,q')=1}}^{Q}\frac{\varphi(q')}{q'}|I_Q(q')| + O_{\ell,\delta}\big(Q^{3/2+\delta}\big).\end{aligned}$$

Then using (A.2), we have

$$\begin{aligned}\frac{1}{\ell}\sum_{\substack{q'=1\\(\ell,q')=1}}^{Q}\frac{\varphi(q')}{q'}|I_Q(q')| &= \frac{C(\ell)}{\ell}\int_0^Q |I_Q(q')|\,dq' + O_\ell(Q\log Q)\\ &= \frac{C(\ell)}{\ell}A(\xi)Q^2 + O_\ell(Q\log Q).\end{aligned}$$

This proves $M_1(Q,\xi) \sim \frac{C(\ell)}{\ell} A(\xi) Q^2$ if $\xi > 1$. The formula for $M_2(Q,\xi)$ is analogous and we infer

$$\begin{aligned} \mathcal{N}_1(Q,\xi) &= N_Q(\xi) - M_1(Q,\xi) - M_2(Q,\xi) \\ &= \left(\frac{1}{\zeta(2)} - \frac{2C(\ell)}{\ell}\right) A(\xi) Q^2 + O_{\ell,\delta}\big(Q^{3/2+\delta}\big). \end{aligned} \tag{2.4}$$

Case 2. There is exactly one fraction in $\mathcal{F}_Q$ between γ and γ' that belongs to $\mathcal{F}_Q^{(\ell)}$. It is more convenient to change γ' to γ'', so we shall consider triples $\gamma < \gamma' < \gamma''$ of elements in $\mathcal{F}_Q$ with $\gamma' \in \mathcal{F}_Q^{(\ell)}$ and with $\gamma'' - \gamma \leqslant \frac{\xi}{Q^2}$. The equalities

$$\frac{a''+a}{a'} = \frac{q''+q}{q'} = K \quad \text{and} \quad \gamma'' - \gamma = \frac{K}{qq''}, \tag{2.5}$$

involving the number

$$K = \nu_2(\gamma) = \left[\frac{Q+q}{q'}\right],$$

called the *index* of the Farey fraction $\gamma = \frac{a}{q} \in \mathcal{F}_Q$, will be useful here. In particular, the inequality $\gamma'' - \gamma \leqslant \frac{\xi}{Q^2}$ enforces $K \leqslant \xi$. Consider the set $J_{Q,K,\xi}(q')$ of elements $q \in (Q-q', Q] \cap \big[Kq' - Q, (K+1)q' - Q\big)$ that satisfy $\frac{K}{q(Kq'-q)} \leqslant \frac{\xi}{Q^2}$. This set is either empty, an interval, or the union of two intervals. The number $\mathcal{N}_2(Q,\xi)$ of gaps of consecutive elements in $\widetilde{\mathcal{F}}_{Q,\ell}$ of length $\leqslant \frac{\xi}{Q^2}$ that arise in this case can now be expressed, with k and $\overline{\ell}$ as in (2.3), as

$$\begin{aligned} \mathcal{N}_2(Q,\xi) &= \sum_{1\leqslant K\leqslant \xi} \sum_{q'\leqslant Q} \sum_{\substack{q\in J_{Q,K,\xi}(q') \\ (q,q')=1}} \sum_{\substack{a'=1 \\ a'q\equiv 1 \pmod{q'} \\ \ell | a'}}^{q'} 1 \\ &= \sum_{1\leqslant K\leqslant \xi} \sum_{\substack{q'\leqslant Q \\ (\ell,q')=1}} \sum_{\substack{q\in J_{Q,K,\xi}(q') \\ k\in(0,q'/\ell] \\ kq\equiv \overline{\ell} \pmod{q'}}} 1. \end{aligned} \tag{2.6}$$

We will employ elementary properties of the area preserving invertible transformation $T : \mathcal{T} \to \mathcal{T}$ defined [4] by

$$T(x,y) = \big(y, \kappa(x,y)y - x\big), \qquad (x,y) \in \mathcal{T}, \qquad \text{where} \tag{2.7}$$

$$\mathcal{T} = \{(x,y) \in (0,1]^2 : x+y > 1\} \quad \text{and} \quad \kappa(x,y) = \left[\frac{1+x}{y}\right].$$

An important connection with Farey fractions is given by the equality

$$T\left(\frac{q_i}{Q}, \frac{q_{i+1}}{Q}\right) = \left(\frac{q_{i+1}}{Q}, \frac{q_{i+2}}{Q}\right). \tag{2.8}$$

For each $K \in \mathbb{N}$ consider the subset $\mathcal{T}_K = \{(x,y) \in \mathcal{T} : \kappa(x,y) = K\}$ of $\mathcal{T}$, described by the inequalities $0 < x, y \leqslant 1$, $x + y > 1$, and $Ky - 1 \leqslant x < (K+1)y - 1$.

Denote $V_{Q,K,\xi}(q') = |J_{Q,K,\xi}(q')|$, so $V_{Q,K,\xi}(Qu) = QW_{K,\xi}(u)$, where

$$W_{K,\xi}(u) = \big|\{v : (v,u) \in \mathcal{T}_K\} \cap \{v : K \leqslant \xi v(Ku - v)\}\big|.$$

Similar arguments as in the proof of (2.4) lead to

$$\begin{aligned}\mathcal{N}_2(Q,\xi) &= \frac{C(\ell)}{\ell}\, Q^2 \sum_{K \leqslant \xi} \int_0^1 W_{K,\xi}(u)\, du + O_{\ell,\delta,\xi}(Q^{3/2+\delta}) \\ &= \frac{C(\ell)}{\ell} Q^2 \sum_{K \leqslant \xi} A_K(\xi) + O_{\ell,\delta,\xi}(Q^{3/2+\delta}),\end{aligned}$$

uniformly in ξ on compact subsets of $[0,\infty)$, where

$$A_K(\xi) = \operatorname{Area}\big(\Omega_K(\xi)\big), \qquad \Omega_K(\xi) = \left\{(v,u) \in \mathcal{T}_K : u \geqslant f_{K,\xi}(v) := \frac{v}{K} + \frac{1}{\xi v}\right\}.$$

Summarizing, we have shown

$$N_Q^{(\ell)}(\xi) = G_\ell(\xi) Q^2 + O_{\ell,\xi,\delta}(Q^{3/2+\delta}) \qquad (\text{as } Q \to \infty),$$

where

$$G_\ell(\xi) = \left(\frac{1}{\zeta(2)} - \frac{2C(\ell)}{\ell}\right) A(\xi) + \frac{C(\ell)}{\ell} \sum_{K \leqslant \xi} A_K(\xi). \tag{2.9}$$

Taking also into account Lemma 1 we conclude that the gap limiting measure of $(\widetilde{\mathcal{F}}_{Q,\ell})$ exists and its distribution function is given by

$$\widetilde{F}_\ell(\xi) = \int_0^\xi d\tilde{\nu}_\ell = \frac{1}{\widetilde{K}_\ell} G_\ell\left(\frac{\xi}{\widetilde{K}_\ell}\right).$$

2.1. Explicit Expressions of $A_K(\xi)$

2.1.1. $K = 1$

$\mathcal{T}_1$ is the triangle with vertices $(0,1)$, $(1,1)$, and $\left(\frac{1}{3}, \frac{2}{3}\right)$. When $\xi \leqslant 4$ we have $f_{1,\xi}(v) \geqslant 1$ for every $v > 0$, so $A_1(\xi) = 0$. When $\xi > 4$ we have

$$A_1(\xi) = \int_{u_1}^{u_2} \left(1 - \max\left\{f_{1,\xi}(v), 1 - v, \frac{v+1}{2}\right\}\right) dv,$$

where $u_{1,2} = \frac{1}{2}\left(1 \pm \sqrt{1 - \frac{4}{\xi}}\right)$, $0 < u_1 < u_2 < 1$, are the solutions of $f_{1,\xi}(v) = 1$. When $4 < \xi \leqslant 8$ we have $f_{1,\xi}(v) \geqslant \max\left\{1 - v, \frac{1+v}{2}\right\}$, so $A_1(\xi)$ is the area of the region defined by $v \in [u_1, u_2]$ and $u \in [f_{1,\xi}(v), 1]$. When $\xi \geqslant 8$ let $v_{1,2} =$

$\frac{1}{4}\left(1\pm\sqrt{1-\frac{8}{\xi}}\right)$, $v_1 < v_2$, denote the solutions of $f_{1,\xi}(v) = 1-v$ and by $w_{1,2} := 2v_{1,2}$ the solutions of $f_{1,\xi}(w) = \frac{w+1}{2}$. If $8 \leqslant \xi \leqslant 9$, then $0 < u_1 < v_1 \leqslant v_2 \leqslant \frac{1}{3} \leqslant w_1 \leqslant w_2 < u_2 < 1$. In this case $A_1(\xi)$ is the area of the region described by $v \in [u_1, v_1] \cup [v_2, w_1] \cup [w_2, u_2]$ and $u \in [f_{1,\xi}(v), 1]$, $v \in [v_1, v_2]$ and $u \in [1-v, 1]$, or $v \in [w_1, w_2]$ and $u \in \left[\frac{1+v}{2}, 1\right]$. Finally, if $\xi > 9$, then $0 < u_1 < v_1 < w_1 < \frac{1}{3} < v_2 < w_2 < u_2 < 1$, and $A_1(\xi)$ is the area of the region described by $v \in [u_1, v_1] \cup [w_2, u_2]$ and $u \in [f_{1,\xi}(v), 1]$, or $v \in \left[v_1, \frac{1}{3}\right]$ and $u \in [1-v, 1]$, or $v \in \left[\frac{1}{3}, w_2\right]$ and $u \in \left[\frac{1+v}{2}, 1\right]$. A plain calculation gives

$$A_1(\xi) = \begin{cases} 0 & \text{if } 0 < \xi \leqslant 4 \\ \frac{1}{2}\sqrt{1-\frac{4}{\xi}} - \frac{1}{\xi}\ln\left(\frac{u_2}{u_1}\right) & \text{if } 4 \leqslant \xi \leqslant 8 \\ \frac{1}{2}\sqrt{1-\frac{4}{\xi}} - \frac{1}{\xi}\ln\left(\frac{u_2}{u_1}\right) - \frac{1}{2}\sqrt{1-\frac{8}{\xi}} + \frac{2}{\xi}\ln\left(\frac{v_2}{v_1}\right) & \text{if } 8 \leqslant \xi \leqslant 9 \\ \frac{1}{2}\sqrt{1-\frac{4}{\xi}} - \frac{1}{\xi}\ln\left(\frac{u_2}{u_1}\right) - \frac{1}{4}\sqrt{1-\frac{8}{\xi}} - \frac{1}{12} + \frac{1}{\xi}\ln\left(\frac{2v_2}{v_1}\right) & \text{if } \xi \geqslant 9. \end{cases}$$

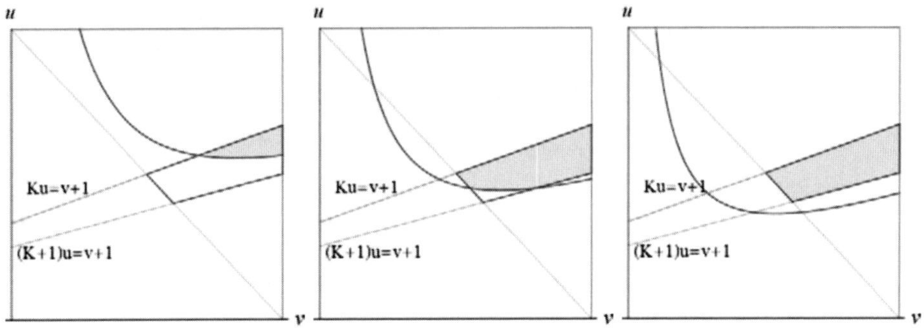

Figure 1: The intersection between the quadrilateral $\mathcal{T}_K$ and the curve $u = f_{K,\xi}(v)$ when $K < \xi < \frac{K(K+1)}{K-1}$, $\frac{K(K+1)}{K-1} \leqslant \xi < \frac{(K+2)^2}{K}$, and respectively $\xi \geqslant \frac{(K+2)^2}{K}$

2.1.2. $K \geqslant 2$

Note that $f_{K,\xi}(1) = f_{K,\xi}\left(\frac{K}{\xi}\right) = \frac{1}{K} + \frac{1}{\xi}$. The situation is described by Figure 1. The solution of $f_{K,\xi}(v) = \frac{v+1}{K}$ is $v = \frac{K}{\xi}$, so the curve $u = f_{K,\xi}(v)$ intersects the upper edge of $\mathcal{T}_K$ if and only if $K < \xi < \frac{K(K+1)}{K-1}$, in which case it does not intersect the two lower edges of $\mathcal{T}_K$ and

$$A_K(\xi) = \int_{K/\xi}^{1} \left(\frac{v+1}{K} - f_{K,\xi}(v)\right) dv = \int_{K/\xi}^{1} \left(\frac{1}{K} - \frac{1}{\xi v}\right) dv.$$

The solution of $f_{K,\xi}\left(\frac{K}{K+2}\right) > \frac{2}{K+2}$ is $\xi < \frac{(K+2)^2}{K}$. This shows that when $\frac{K(K+1)}{K-1} \leqslant \xi < \frac{(K+2)^2}{K}$ the graph of $u = f_{K,\xi}(u)$ intersects the segment $u = 1-v$, $v \in$

$\left[\frac{K-1}{K+1}, \frac{K}{K+2}\right]$, exactly when $v = v_K = \frac{K}{2(K+1)}\left(1+\sqrt{1-\frac{4}{\xi}(1+\frac{1}{K})}\right)$, and the segment $u = \frac{v+1}{K+1}$, $v \in \left[\frac{K}{K+2}, 1\right]$, exactly at $v = w_K = \frac{K}{2}\left(1-\sqrt{1-\frac{4}{\xi}(1+\frac{1}{K})}\right)$, so in this case

$$A_K(\xi) = \operatorname{Area}(\mathcal{T}_K) - \int_{v_K}^{w_K} f_{K,\xi}(v)\,dv + \int_{v_K}^{K/(K+2)} (1-v)\,dv + \int_{K/(K+2)}^{w_K} \frac{v+1}{K+1}\,dv.$$

Finally, when $\xi > \frac{(K+2)^2}{K}$, the graph of $u = f_{K,\xi}(v)$ does not intersect any of the edges of $\mathcal{T}_K$ and

$$A_K(\xi) = \operatorname{Area}(\mathcal{T}_K).$$

In summary, a quick calculation leads to

$$A_K(\xi) = \begin{cases} 0 & \text{if } 0 \leqslant \xi \leqslant K \\ \frac{1}{K} - \frac{1}{\xi} - \frac{1}{\xi}\ln\left(\frac{\xi}{K}\right) & \text{if } K \leqslant \xi \leqslant \frac{K(K+1)}{K-1} \\ \frac{K^3+8}{2K(K+1)(K+2)} - \frac{1}{\xi}\ln\left(\frac{w_K}{v_K}\right) - \frac{v_K}{2} + \frac{w_K}{2(K+1)} & \text{if } \frac{K(K+1)}{K-1} \leqslant \xi \leqslant \frac{(K+2)^2}{K} \\ \frac{4}{K(K+1)(K+2)} & \text{if } \xi \geqslant \frac{(K+2)^2}{K}. \end{cases}$$

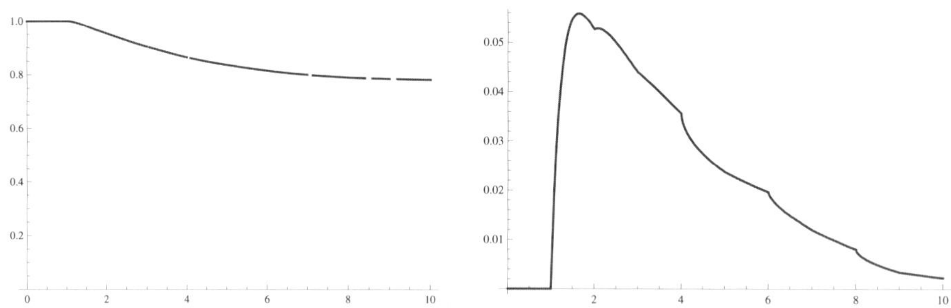

Figure 2: The repartition function $1 - G_3(\xi)$ and the density $-G_3'(\xi)$

3. Consecutive Elements in $\mathcal{F}_Q$ with Denominator Relatively Prime to d

In this section we comment on the first two steps in the proof of (1.1) from [2].

3.1. Upper Bounds on the Number of Consecutive Farey Fractions Whose Denominators Are Not Relatively Prime to d

One of the key steps in the proof of (1.1) in [2] is to show that for any Q and any d, any string of consecutive elements in $\mathcal{F}_Q$ of length $4d^3$ contains at least one element whose denominator is coprime with d. Next we provide two arguments which show that the upper bound $L(d)$ should actually be much smaller than $4d^3$.

Lemma 2. *If* $\omega(d) \leqslant \min\{p \in \mathcal{P} : p \mid d\}$, *then* $L(d) \leqslant 4\omega(d)^3$.

Proof. We first revisit the proof of the first part of Step (i) in the proof of Theorem 1 in [2] (pp. 210–211). Suppose Q and $i_1 < i_2$ are chosen such that, for every $j \in [i_1, i_2]$,

$$\max\{q_{i_1}, q_{i_2}\} \leqslant q_j \quad \text{and} \quad (q_j, d) > 1.$$

Then $(q_{i_1}, q_{i_2}) = 1$ and

$$\{q_j : i_1 < j < i_2\} \subset \{mq_{i_1} + nq_{i_2} : m, n \in \mathbb{N}, (m,n) = 1, mq_{i_1} + nq_{i_2} \leqslant Q\}. \quad (3.1)$$

Let $d_1 = p_1^{\alpha_1} \cdots p_\omega^{\alpha_\omega}$, with $p_1 < \cdots < p_\omega$ primes, be the largest divisor of d which is coprime to q_{i_1}. Then $\omega < \omega(d) \leqslant \min\{p \in \mathcal{P} : p \mid d\} \leqslant p_1$. Fix some integer L with $\omega + 1 \leqslant L \leqslant p_1$. We claim that there exists $m_1 \in \mathbb{N}$, $m_1 \leqslant L$ such that $(m_1 q_{i_1} + q_{i_2}, d_1) = 1$. If not, then $(\ell q_{i_1} + q_{i_2}, d_1) > 1$ for all $\ell \in \{1, \ldots, L\}$. Since $L > \omega$, the Pigeonhole Principle shows that there exist $i_0 \in \{1, \ldots, \omega\}$ and $1 \leqslant \ell < \ell' \leqslant L$ such that $p_{i_0} \mid (\ell q_{i_1} + q_{i_2})$ and $p_{i_0} \mid (\ell' q_{i_1} + q_{i_2})$, and so $p_{i_0} \mid (\ell' - \ell) q_{i_1}$. But $(p_{i_0}, q_{i_1}) = 1$, hence $L > \ell' - \ell \geqslant p_{i_0} \geqslant p_1$, which contradicts $L \leqslant p_1$.

So if $(m_1 q_{i_1} + q_{i_2}, d) > 1$, then there exists p prime with $p \mid q_{i_1}$ and $p \mid (m_1 q_{i_1} + q_{i_2})$, thus contradicting $(q_{i_1}, q_{i_2}) = 1$. Hence $(m_1 q_{i_1} + q_{i_2}, d) = 1$, which in turn yields $Q \leqslant m_1 q_{i_1} + q_{i_2} \leqslant L q_{i_1} + q_{i_2}$. In a similar way one has $Q \leqslant q_{i_1} + L q_{i_2}$, thus (3.1) leads to

$$\{q_j : i_1 < j < i_2\} \subset \{mq_{i_1} + nq_{i_2} : 1 \leqslant m, n \leqslant L\},$$

and in particular $i_2 - i_1 \leqslant L^2$.

The second part of the proof proceeds ad litteram as in the proof of Step (i) [2, pp. 211–212] replacing d there by L. □

When d is the product of two prime powers the bound above can be lowered. In this case we show that $L(d) \leqslant 5$, which is sharp for $d = 6$ because $\frac{1}{4} < \frac{1}{3} < \frac{1}{2} < \frac{2}{3} < \frac{3}{4}$ are consecutive in $\mathcal{F}_4$. Our proof employs elementary properties of the transformation T from (2.7). In particular (2.8) and the following inclusions will be useful in the proof of Lemma 3:

$$T\mathcal{T}_k \subseteq \mathcal{T}_1 \ \text{ if } k \geqslant 5, \qquad T(\mathcal{T}_3 \cup \mathcal{T}_4) \subseteq \mathcal{T}_1 \cup \mathcal{T}_2,$$
$$T\mathcal{T}_2 \subseteq \mathcal{T}_1 \cup \mathcal{T}_2 \cup \mathcal{T}_3 \cup \mathcal{T}_4, \qquad T(T\mathcal{T}_3 \cap \mathcal{T}_2) \subseteq \mathcal{T}_1 \cup \mathcal{T}_2.$$

Lemma 3. *If* $d = p^\alpha q^\beta$, *then for each* $i \in \{0, \ldots, \#\mathcal{F}_Q - 5\}$ *there exists* $j \in \{0, \ldots, 5\}$ *such that* $(q_{i+j}, d) = 1$, *and so* $L(d) \leqslant 5$.

Proof. We have $q_{i+2} = Kq_{i+1} - q_i$, $q_{i+3} = K' q_{i+2} - q_{i+1}$, $q_{i+4} = K'' q_{i+3} - q_{i+2}$, $q_{i+5} = K''' q_{i+4} - q_{i+3}$, where $K = \kappa\left(\frac{q_i}{Q}, \frac{q_{i+1}}{Q}\right)$, $K' = \kappa\left(\frac{q_{i+1}}{Q}, \frac{q_{i+2}}{Q}\right)$, $K'' = \kappa\left(\frac{q_{i+2}}{Q}, \frac{q_{i+3}}{Q}\right)$, and $K''' = \kappa\left(\frac{q_{i+3}}{Q}, \frac{q_{i+4}}{Q}\right)$. Suppose that $(q_i, d), \ldots, (q_{i+5}, d) > 1$. Then either $p \mid (q_i, q_{i+2}, q_{i+4})$ and $q \mid (q_{i+1}, q_{i+3}, q_{i+5})$, or vice versa.

Without loss of generality we can work in the first case. The equality $q_{i+2}+q_i = Kq_{i+1}$ and $p \nmid q_{i+1}$ yield $p \mid K$. Similarly we have $q \mid K'$. Assume first that $K \geqslant 5$. Since $\left(\frac{q_i}{Q}, \frac{q_{i+1}}{Q}\right) \in \mathcal{T}_K$ and $T\mathcal{T}_K \subseteq \mathcal{T}_1$ we must have $K' = 1$, which contradicts $q \geqslant 2$. In particular $p \geqslant 5$ cannot occur.

When $p = 3$ and $K = 3$, from $T\mathcal{T}_3 \subseteq \mathcal{T}_1 \cup \mathcal{T}_2$ it follows that $K' \in \{1, 2\}$. Since $q \mid K'$, we infer $q = 2$. The region $T\mathcal{T}_3 \cap \mathcal{T}_2$ is the quadrilateral with vertices at $\left(\frac{1}{2}, \frac{1}{2}\right)$, $\left(\frac{2}{5}, \frac{3}{5}\right)$, $\left(\frac{3}{5}, \frac{4}{5}\right)$, and $\left(\frac{3}{7}, \frac{5}{7}\right)$, being further mapped by T into a subset of $\mathcal{T}_1 \cup \mathcal{T}_2$ whence $K'' \in \{1, 2\}$. Again $K'' = 1$ leads to an immediate contradiction, while $K'' = 2$ yields $q_{i+2}+q_{i+4} = 2q_{i+3}$, showing that $p = 2$, another contradiction.

When $p = 2$ and $K < 5$, we have $K \in \{2, 4\}$. Assume first $K = 2$. As $T\mathcal{T}_2 \subseteq \mathcal{T}_1 \cup \mathcal{T}_2 \cup \mathcal{T}_3 \cup \mathcal{T}_4$ and $K' \neq 1$ it remains that $K' \in \{2, 3, 4\}$. Since $q \geqslant 3$ divides K', we infer $q = 3$. Furthermore, $T\mathcal{T}_3 \subseteq \mathcal{T}_1 \cup \mathcal{T}_2$ and $K'' \neq 1$ yield $K'' = 2$. Employing again $T(T\mathcal{T}_3 \cap \mathcal{T}_2) \subseteq \mathcal{T}_1 \cup \mathcal{T}_2$, we infer $K''' = 2$, and so $q_{i+3}+q_{i+5} = 2q_{i+4}$. This is again a contradiction, because 3 divides $q_{i+3}+q_{i+5}$ and cannot divide $2q_{i+4}$. Finally, assume $K = 4$, so $K' \in \{1, 2\}$, which is not possible because $q \geqslant 3$ divides K'. □

Note that if (p_n) is the sequence of primes, then none of the denominators of the fractions in $\mathcal{F}_{p_n} \setminus \{1\}$ are relatively prime to $\prod_{i=1}^{n} p_i$. This gives the lower bound $\#\mathcal{F}_{p_n} - 1$ on the size of the largest string of consecutive fractions in $\mathcal{F}_Q \setminus \mathcal{F}_{Q,d}$ for some $Q, d \in \mathbb{N}$ with $\omega(d) = n$. Since $p_n \sim n \log n$ as $n \to \infty$ and $\#\mathcal{F}_Q \sim \frac{3}{\pi^2} Q^2$ as $Q \to \infty$, there exists $A > 0$ such that $\#\mathcal{F}_{p_n} - 1 \geqslant A(n \log n)^2$. Thus any upper bound on $L(d)$ involving only $\omega(d)$ must be greater than $A(\omega(d) \log \omega(d))^2$.

3.2. The Index and the Continuant

The second step in the proof of (1.1) in [2] relies on [2, Lemma 1], which is actually exactly Remark 2.6 in [6] (see also [4, Lemma 5]), and on a result relating the ℓ-index of a Farey fraction and the continuant of regular continued fractions. The ℓ-index of $\gamma_i = \frac{a_i}{q_i} \in \mathcal{F}_Q$ is the positive integer $\nu_\ell(\gamma_i) = a_{i+\ell-1}q_{i-1} - a_{i-1}q_{i+\ell-1}$ where $\frac{a_{i+k}}{q_{i+k}}$ denotes the kth successor of γ_i in $\mathcal{F}_Q$. The (regular continued fraction) *continuants* are defined as usual by $K_0(\cdot) = 1$, $K_1(x_1) = 1$, and

$$K_\ell(x_1, \ldots, x_\ell) = x_\ell K_{\ell-1}(x_1, \ldots, x_{\ell-1}) + K_{\ell-2}(x_1, \ldots, x_{\ell-2}) \quad \text{if } \ell \geqslant 2.$$

In [10] the identity

$$\nu_\ell(\gamma_i) = \epsilon_\ell K_{\ell-1}\big(-\nu_2(\gamma_i), \nu_2(\gamma_{i+1}), \ldots, (-1)^{\ell-1}\nu_2(\gamma_{i+\ell-2})\big) \tag{3.2}$$

was proved, with $\epsilon_\ell = 1$ if $\ell \in \{0, 1\} \pmod 4$ and $\epsilon_\ell = -1$ if $\ell \in \{2, 3\} \pmod 4$.

We give a very short proof of (3.2). We define the *Farey continuants* K_ℓ^F by $K_0^F(\cdot) = 1$, $K_1^F(x_1) = x_1$, and

$$K_\ell^F(x_1, \ldots, x_\ell) = x_\ell K_{\ell-1}^F(x_1, \ldots, x_{\ell-1}) - K_{\ell-2}^F(x_1, \ldots, x_{\ell-2}) \quad \text{if } \ell \geqslant 2.$$

The defining equalities for K_ℓ and K_ℓ^F plainly yield, for all $\ell \geqslant 2$,

$$\begin{pmatrix} x_1 & 1 \\ 1 & 0 \end{pmatrix} \cdots \begin{pmatrix} x_\ell & 1 \\ 1 & 0 \end{pmatrix} = \begin{pmatrix} K_\ell(x_1,\dots,x_\ell) & K_{\ell-1}(x_1,\dots,x_{\ell-1}) \\ K_{\ell-1}(x_2,\dots,x_\ell) & K_{\ell-2}(x_2,\dots,x_{\ell-1}) \end{pmatrix}, \tag{3.3}$$

$$\begin{pmatrix} x_1 & 1 \\ -1 & 0 \end{pmatrix} \cdots \begin{pmatrix} x_\ell & 1 \\ -1 & 0 \end{pmatrix} = \begin{pmatrix} K_\ell^F(x_1,\dots,x_\ell) & K_{\ell-1}^F(x_1,\dots,x_{\ell-1}) \\ -K_{\ell-1}^F(x_2,\dots,x_\ell) & -K_{\ell-2}^F(x_2,\dots,x_{\ell-1}) \end{pmatrix}. \tag{3.4}$$

From (3.4) and the definition of $\nu_\ell(\gamma_i)$ we now infer

$$\nu_\ell(\gamma_i) = K_{\ell-1}^F\big(\nu_2(\gamma_i), \nu_2(\gamma_{i+1}), \dots, \nu_2(\gamma_{i+\ell-2})\big). \tag{3.5}$$

The equality (3.2) follows immediately from (3.3), (3.4), (3.5) and

$$\begin{pmatrix} x & 1 \\ -1 & 0 \end{pmatrix} \begin{pmatrix} y & 1 \\ -1 & 0 \end{pmatrix} = -\begin{pmatrix} -x & 1 \\ 1 & 0 \end{pmatrix} \begin{pmatrix} y & 1 \\ 1 & 0 \end{pmatrix}.$$

4. The Gap Distribution of $\mathcal{F}_{Q,d}$

Letting $d \in \mathbb{N}$ and $\xi > 0$, we wish to asymptotically estimate the number of pairs of consecutive elements $\gamma < \gamma'$ in $\mathcal{F}_{Q,d}$ with $\gamma' - \gamma \leqslant \frac{\xi}{Q^2}$ as $Q \to \infty$. It is plain that

$$\#\mathcal{F}_{Q,d} = \sum_{\substack{q=1 \\ (q,d)=1}}^{Q} \varphi(q) = C(d) \int_0^Q q\, dq + O_d(Q \log Q) = \frac{C(d)}{2} Q^2 + O_d(Q \log Q),$$

showing the second equality in (1.2). Denote $N_Q = \#\mathcal{F}_Q$ and $\gamma_j = \frac{a_j}{q_j}$, so the number of pairs of fractions we wish to estimate is

$$\begin{aligned} N_d(Q,\xi) &= \sum_{\ell=1}^{L(d)} \# \left\{ i \in [1, N_Q] : \begin{array}{c} \frac{\nu_\ell(\gamma_i)}{q_{i-1}q_{i+\ell-1}} \leqslant \frac{\xi}{Q^2}, \quad (q_{i-1}, d) = (q_{i+\ell-1}, d) = 1 \\ (q_i, d) > 1, \dots, (q_{i+\ell-2}, d) > 1 \end{array} \right\} \\ &= \sum_{\ell=1}^{L(d)} \sum_{k=1}^{[\xi]} \# \left\{ i \in [1, N_Q] : \begin{array}{c} \frac{k}{q_{i-1}q_{i+\ell-1}} \leqslant \frac{\xi}{Q^2}, \quad \nu_\ell(\gamma_i) = k \\ (q_{i-1}, d) = (q_{i+\ell-1}, d) = 1 \\ (q_i, d) > 1, \dots, (q_{i+\ell-2}, d) > 1 \end{array} \right\}. \end{aligned}$$

It is shown in [4, 5] that given $i \in [1, N_Q]$ and $k, \ell \in \mathbb{N}$ with $\ell \geqslant 2$, if $\nu_\ell(\gamma_i) = k$, then the $(\ell-1)$-tuple $\big(\nu_2(\gamma_i), \dots, \nu_2(\gamma_{i+\ell-2})\big)$ can take on $n(k,\ell)$ values, where $n(k,\ell) \in \mathbb{N} \cup \{0\}$ depends only on k and ℓ and not on i or Q; and in [10], it is proven that $\nu_\ell(\gamma_i)$ can be determined if $\big(\nu_2(\gamma_i), \dots, \nu_2(\gamma_{i+\ell-2})\big)$ is known (cf. identity (3.2) above). Therefore, letting $\{x(k,\ell,m)\}_{m=1}^{n(k,\ell)}$ be the $(\ell-1)$-tuples for which $\nu_\ell(\gamma_i) = k$ whenever $x(k,\ell,m) = \big(\nu_2(\gamma_i), \dots, \nu_2(\gamma_{i+\ell-2})\big)$ for some $m \in \{1, \dots, n(k,\ell)\}$, we have

$$N_d(Q,\xi) = \#\left\{ i \in [1,N_Q] : (q_{i-1},d) = (q_i,d) = 1, \quad q_{i-1}q_i \geqslant \frac{Q^2}{\xi} \right\}$$
$$+ \sum_{\ell=2}^{L(d)} \sum_{k=1}^{[\xi]} \sum_{m=1}^{n(k,\ell)} \#\left\{ i \in [1,N_Q] : \begin{array}{c} q_{i-1}q_{i+\ell-1} \geqslant \frac{kQ^2}{\xi},\ (q_{i-1},d) = (q_{i+\ell-1},d) = 1 \\ (q_i,d) > 1, \ldots, (q_{i+\ell-2},d) > 1 \\ x(k,\ell,m) = (\nu_2(\gamma_i), \ldots, \nu_2(\gamma_{i+\ell-2})) \end{array} \right\}.$$

Since $q_{j+1} = \nu_2(\gamma_j)q_j - q_{j-1}$ for $j \in [1, N_Q - 1]$, the residue classes of the denominators $q_{i-1}, \ldots, q_{i+\ell-1}$ can be determined once the residue classes of q_{i-1} and q_i, and the $(\ell-1)$-tuple $\big(\nu_2(\gamma_i), \ldots, \nu_2(\gamma_{i+\ell-2})\big)$ are known. Thus, there is a subset $\mathcal{A}_{k,\ell,m} \subseteq \{1, \ldots, d\}^2$ such that when $\big(\nu_2(\gamma_i), \ldots, \nu_2(\gamma_{i+\ell-2})\big) = x(k,\ell,m)$, we have $(q_{i-1},d) = (q_{i+\ell-1},d) = 1$ and $(q_{i+j-1},d) > 1$ for $1 \leqslant j < \ell$ if and only if $(q_{i-1},q_i) \pmod{d} \in \mathcal{A}_{k,\ell,m}$. (Note clearly that $(a,d) = 1$ for $(a,b) \in \mathcal{A}_{k,\ell,m}$.) Furthermore, if we let $x(k,\ell,m) = \big(x_1(k,\ell,m), \ldots, x_{\ell-1}(k,\ell,m)\big)$ and denote $\mathbb{Z}^2_{\text{vis}} = \{(a,b) \in \mathbb{Z}^2 : (a,b) = 1\}$, it is clear that $\big(\nu_2(\gamma_i), \ldots, \nu_2(\gamma_{i+\ell-2})\big) = x(k,\ell,m)$ if and only if

$$(q_{i-1},q_i) \in Q \cdot (\mathcal{T}_{x_1(k,\ell,m)} \cap T^{-1}\mathcal{T}_{x_2(k,\ell,m)} \cap \cdots \cap T^{-(\ell-2)}\mathcal{T}_{x_{\ell-1}(k,\ell,m)}) \cap \mathbb{Z}^2_{\text{vis}}.$$

Now if we let $\pi_1, \pi_2 : \mathbb{R}^2 \to \mathbb{R}$ be the canonical projections, then

$$\frac{q_{i-1}q_{i+\ell-1}}{Q^2} = \pi_1\left(\frac{q_{i-1}}{Q}, \frac{q_i}{Q}\right) \cdot (\pi_2 \circ T^{\ell-1})\left(\frac{q_{i-1}}{Q}, \frac{q_i}{Q}\right),$$

and so

$$q_{i-1}q_{i+\ell-1} \geqslant \frac{kQ^2}{\xi} \quad \Longleftrightarrow \quad (q_{i-1},q_i) \in Qg_\ell^{-1}\left[\frac{k}{\xi}, \infty\right),$$

where $g_\ell = \pi_1 \cdot (\pi_2 \circ T^{\ell-1})$. Now set $g_1(x,y) = xy$ and

$$\Omega_{k,\ell,m}(\xi) = \mathcal{T}_{x_1(k,\ell,m)} \cap T^{-1}\mathcal{T}_{x_2(k,\ell,m)} \cap \cdots \cap T^{-(\ell-2)}\mathcal{T}_{x_{\ell-1}(k,\ell,m)} \cap g_\ell^{-1}\left[\frac{k}{\xi}, \infty\right),$$
$$\Omega_1(\xi) = \mathcal{T} \cap g_1^{-1}\left[\frac{1}{\xi}, \infty\right), \qquad \mathcal{A}_1 = \{(a,b) : a,b \in [1,d], (a,d) = (b,d) = 1\}.$$

We then have

$$N_d(Q,\xi) = \sum_{(a,b)\in\mathcal{A}_1} \#Q\Omega_1(\xi) \cap \big((a,b) + d\mathbb{Z}^2\big) \cap \mathbb{Z}^2_{\text{vis}}$$
$$+ \sum_{\ell=2}^{L(d)} \sum_{k=1}^{[\xi]} \sum_{m=1}^{n(k,\ell)} \sum_{(a,b)\in\mathcal{A}_{k,\ell,m}} \#Q\Omega_{k,\ell,m}(\xi) \cap \big((a,b) + d\mathbb{Z}^2\big) \cap \mathbb{Z}^2_{\text{vis}},$$

where we have used the fact that if $(a,b) \in Q\mathcal{T} \cap \mathbb{Z}^2_{\text{vis}}$, then there is an i such that $a = q_{i-1}$ and $b = q_i$. One can prove in a similar manner to [2, Lemma 2] that for all bounded $\Omega \subseteq \mathbb{R}^2$ whose boundary can be covered by the images of finitely many Lipschitz functions from $[0,1]$ to $\mathbb{R}^2$, and for all $\mathcal{A} \subseteq \{1,\ldots,d\}^2$ in which $(a,d)=1$ for all $(a,b) \in \mathcal{A}$, we have

$$\sum_{(a,b)\in\mathcal{A}} \#Q\Omega \cap \big((a,b)+d\mathbb{Z}^2\big) \cap \mathbb{Z}^2_{\text{vis}} = \frac{\text{Area}(\Omega)\#\mathcal{A}}{\zeta(2)d^2} \prod_{\substack{p\in\mathcal{P}\\ p|d}} \left(1-\frac{1}{p^2}\right)^{-1} Q^2 + O_d(Q\log Q)$$

as $Q \to \infty$. It is easily seen that the boundaries of $\Omega_1(\xi)$ and $\Omega_{k,\ell,m}(\xi)$ can be covered by finitely many Lipschitz functions from $[0,1]$ to $\mathbb{R}^2$, and so we have

$$N_d(\xi,Q) = C_d(\xi)Q^2 + O_d(Q\log Q),$$

where

$$C_d(\xi) = \frac{1}{\zeta(2)d^2} \prod_{\substack{p\in\mathcal{P}\\ p|d}} \left(1-\frac{1}{p^2}\right)^{-1} \cdot \left(\varphi(d)^2 \,\text{Area}(\Omega_1(\xi)) + \sum_{\ell=2}^{L(d)} \sum_{k=1}^{[\xi]} \sum_{m=1}^{n(k,\ell)} \text{Area}(\Omega_{k,\ell,m}(\xi))\#\mathcal{A}_{k,\ell,m}\right),$$

noting that $\#\mathcal{A}_1 = \varphi(d)^2$.

The gap limiting measure of $(\mathcal{F}_{Q,d})_Q$ exists with distribution function given by

$$F_d(\xi) = \int_0^\xi d\nu_d = \frac{1}{K_d} C_d\left(\frac{\xi}{K_d}\right).$$

When d is a prime power this can be expressed more explicitly as in (2.9).

Acknowledgements. The second author acknowledges support from Department of Education Grant P200A090062, "University of Illinois GAANN Mathematics Fellowship Project." The third author acknowledges support from National Science Foundation grant DMS 08-38434 "EMSW21-MCTP: Research Experience for Graduate Students." We are grateful to the referee for careful reading and pertinent remarks.

References

[1] V. Augustin, F. P. Boca, C. Cobeli, A. Zaharescu, *The h-spacing distribution between Farey points*, Math. Proc. Cambridge Philos. Soc. **131** (2001), 23–38.

[2] D. A. Badziahin, A. K. Haynes, *A note on Farey fractions with denominators in arithmetic progressions*, Acta Arith. **147** (2011), 205–215.

[3] F. P. Boca, C. Cobeli, and A. Zaharescu, *Distribution of lattice points visible from the origin*, Comm. Math. Phys. 213 (2000), 433-470.

[4] F. P. Boca, C. Cobeli, A. Zaharescu, *A conjecture of R. R. Hall on Farey arcs*, J. Reine Angew. Math. **535** (2001), 207–236.

[5] F. P. Boca, R. N. Gologan, *On the distribution of the free path length of the linear flow in a honeycomb*, Ann. Inst. Fourier (Grenoble) **59** (2009), 1043–1075.

[6] F. P. Boca, R. N. Gologan, A. Zaharescu, *On the index of Farey sequences*, Q. J. Math. **53** (2002), 377–391.

[7] F. P. Boca, A. Zaharescu, *The correlations of Farey fractions*, J. Lond. Math. Soc. (2) **72** (2005), 25–39.

[8] F. P. Boca, A. Zaharescu, *On the correlation of directions in the Euclidean space*, Trans. Amer. Math. Soc. **358** (2006), 1797–1825.

[9] R. R. Hall, *A note on Farey series*, J. Lond. Math. Soc. (2) **2** (1970), 139–148.

[10] A. K. Haynes, *Numerators of differences of nonconsecutive Farey fractions*, Int. J. Number Theory **6** (2010), 655–666.

[11] M. Xiong, A. Zaharescu, *Correlation of fractions with divisibility constraints*, Math. Nachr. **284** (2011), 393–407.

A. Appendix

For the convenience of the reader we collect in this appendix the asymptotic formulas used in this paper.

Assuming that f is a C^1 function on the interval of integration in (A.1)-(A.3) and that I, J are intervals and $f \in C^1(I \times J)$ in (A.4), we have

$$\sum_{\substack{a<k\leqslant b \\ (k,q)=1}} f(k) = \frac{\varphi(q)}{q} \int_a^b f(x)\,dx + O\Big(\sigma_0(q)\big(\|f\|_\infty + T_a^b f\big)\Big). \tag{A.1}$$

$$\sum_{\substack{1\leqslant k\leqslant N \\ (k,q)=1}} \frac{\varphi(k)}{k} f(k) = C(\ell) \int_0^N f(x)\,dx + O_\ell\Big(\big(\|f\|_\infty + T_0^N f\big)\log N\Big). \tag{A.2}$$

$$\sum_{1\leqslant k\leqslant N} \frac{\varphi(\ell k)}{k} f(k) = \ell C(\ell) \int_0^N f(x)\,dx + O_{\ell,\delta}\Big(\big(\|f\|_\infty + T_0^N f\big)N^\delta\Big). \tag{A.3}$$

$$\begin{aligned}\sum_{\substack{a\in I, b\in J \\ ab\equiv h \pmod q \\ (b,q)=1}} f(a,b) = &\frac{\varphi(q)}{q^2} \iint_{I\times J} f(x,y)\,dxdy + O_\delta\big(T^2\|f\|_\infty q^{1/2+\delta}(h,q)^{1/2}\big) \\ &+ O_\delta\Big(T\|\nabla f\|_\infty q^{3/2+\delta}(h,q)^{1/2} + \frac{1}{T}\|\nabla f\|_\infty |I|\cdot|J|\Big).\end{aligned} \tag{A.4}$$

Proofs can be found for instance in [3, Lemma 2.2], [5, Lemmas 2.1 and 2.2], and respectively in [7, Proposition A4].

ON FINITE SUMS AND INTEGRAL REPRESENTATIONS

Anthony Sofo
Victoria University, Melbourne City, Australia.
anthony.sofo@vu.edu.au

N. Batir
Department of Mathematics, Faculty of Sciences and Arts, Nevşehır University, Nevşehır, Turkey
nbatir@hotmail.com

Received: 7/26/12, Revised: 4/29/13, Accepted: 7/6/13, Published: 8/12/13

Abstract

In this paper we give an alternative proof for a reciprocal binomial coefficient sum given by Sun: $\sum_{k=1}^{n} \frac{1}{k^2\binom{k+n}{k}}$. We then generalize the result to $\sum_{k=1}^{n} \frac{1}{k^m\binom{k+n}{n}}$, for m, n positive integers, and give its integral representation.

1. Introduction and Preliminaries

In the paper [15], Sun states and proves the following new lemma.

Lemma 1. *For each $n = 1, 2, 3, ...,$ we have:*

$$\sum_{k=1}^{n} \frac{1}{k^2 \binom{k+n}{k}} = 3\sum_{k=1}^{n} \frac{1}{k^2 \binom{2k}{k}} - \sum_{k=1}^{n} \frac{1}{k^2}. \tag{1}$$

Sun's proof relies on a known identity and induction. The induction proof is perfectly valid; however, it does not shed light on how and if (1) may be generalized. In this paper we give an integral representation for the left-hand side of (1) and then generalize the result in various directions. For the sake of completeness we first highlight the induction proof as given in [15].

Proof. Observe that, for $n \in \mathbb{N}$ (where $\mathbb{N} := \{1,2,3...\}$),

$$\sum_{k=1}^{n}\left(\frac{1}{k^2\binom{k+n}{k}}-\frac{1}{k^2\binom{k+n+1}{k}}\right)=\sum_{k=1}^{n}\frac{\binom{k+n}{k-1}}{k^2\binom{k+n}{k}\binom{k+n+1}{k}}$$

$$=\frac{1}{(n+1)(n+2)}\sum_{k=0}^{n-1}\frac{1}{\binom{k+n+2}{n+2}}.$$

There is an identity [4] that states: for b an integer greater than 1 :

$$\sum_{k=0}^{m}\frac{1}{\binom{x+k}{b}}=\frac{b}{b-1}\left(\frac{1}{\binom{x-1}{b-1}}-\frac{1}{\binom{x+m}{b-1}}\right);$$

and for $b=1$,

$$\sum_{k=0}^{m}\frac{1}{x+k} = H_{m+x}-H_{x-1}.$$

Now

$$\sum_{k=1}^{n}\left(\frac{1}{k^2\binom{k+n}{k}}-\frac{1}{k^2\binom{k+n+1}{k}}\right) = \frac{1}{(n+1)^2}\left(1-\frac{1}{\binom{2n+1}{n+1}}\right)$$

and hence

$$\sum_{k=1}^{n+1}\frac{1}{k^2\binom{k+n+1}{k}}-\sum_{k=1}^{n}\frac{1}{k^2\binom{k+n}{k}} = \frac{3}{(n+1)^2\binom{2n+2}{n+1}}-\frac{1}{(n+1)^2}.$$

Therefore (1) follows by induction. □

In this paper we shall give a different proof of (1), then generalize the result, and also give integral representations of the specified sums. These results will in turn allow for stronger results on the congruences of similar sums. First we give some definitions which will be useful throughout this paper.

The generalized harmonic numbers in power α are defined as:

$$H_n^{(\alpha)}=\sum_{r=1}^{n}\frac{1}{r^\alpha},$$

and the nth harmonic number for $\alpha = 1$ is:

$$H_n^{(1)} = H_n = \int_0^1 \frac{1-t^n}{1-t} dt = \sum_{r=1}^{n} \frac{1}{r} = \gamma + \psi(n+1),$$

where γ denotes the Euler-Mascheroni constant, defined by

$$\gamma = \lim_{n \to \infty} \left(\sum_{r=1}^{n} \frac{1}{r} - \log(n) \right) = -\psi(1) \approx 0.5772156649.......$$

Let $\mathbb{C}$ be the set of complex numbers, $\mathbb{R}$ the set of real numbers and $\mathbb{N}$ the set of natural numbers. Then for $w \in \mathbb{C} \backslash \mathbb{Z}_0^-$, $(w)_n$ is Pochhammer's symbol defined by

$$(w)_n = \begin{cases} \frac{\Gamma(w+n)}{\Gamma(n)} = w(w+1)....(w+n-1), & \text{for } n \in \mathbb{N} \\ 1, & \text{for } n = 0 \end{cases}, \tag{2}$$

where $\mathbb{Z}_0^-$ denotes the set of non-positive integers, and the Gamma and beta functions are defined respectively as

$$\Gamma(z) = \int_0^\infty w^{z-1} e^{-w} dw, \text{ for } \operatorname{Re}(z) > 0,$$

and

$$B(s,z) = B(z,s) = \int_0^1 w^{s-1} (1-w)^{z-1} dw = \frac{\Gamma(s)\Gamma(z)}{\Gamma(s+z)}$$

for $\operatorname{Re}(s) > 0$ and $\operatorname{Re}(z) > 0$. The binomial coefficient is defined as

$$\binom{z}{w} = \frac{\Gamma(z+1)}{\Gamma(w+1)\Gamma(z-w+1)}$$

for z and w non-negative integers, where $\Gamma(x)$ is the Gamma function. The polygamma functions $\psi^{(k)}(z)$, $k \in \mathbb{N}$ are defined by:

$$\begin{aligned} \psi^{(k)}(z) &:= \frac{d^{k+1}}{dz^{k+1}} \log \Gamma(z) = \frac{d^k}{dz^k} \left(\frac{\Gamma'(z)}{\Gamma(z)} \right) \\ &= -\int_0^1 \frac{[\log(t)]^k \ t^{z-1}}{1-t} dt, \ k \in \mathbb{N}, \end{aligned} \tag{3}$$

and $\psi^{(0)}(z) = \psi(z)$ denotes the psi, or digamma function, defined by:

$$\psi(z) = \frac{d}{dz} \log \Gamma(z) = \frac{\Gamma'(z)}{\Gamma(z)}.$$

We also recall the relation for $m = 1, 2, 3, \ldots,$

$$H_{z-1}^{(m+1)} = \zeta(m+1) + \frac{(-1)^m}{m!}\psi^{(m)}(z). \tag{4}$$

There are many results of the type (1) for finite, infinite and alternating sums. For example see [1], [2], [3], [5], [6], [7], [8], [9], [10], [11], [16], [18] and references therein.

2. Generalizations

The next lemma deals with an alternative proof of (1) from which more generalized identities can be ascertained.

Lemma 2. *Let n be a positive integer. Then*

$$\sum_{k=1}^{n} \frac{1}{k^2 \binom{k+n}{n}} = \int_0^1 \int_0^1 \frac{(1-x)^n (1-(xy)^n)}{1-xy} dydx \tag{5}$$

$$= -(H_n)^2 + \sum_{k=1}^{n} \frac{(-1)^{k+1} \binom{n}{k} H_{n+k}}{k}. \tag{6}$$

Proof.

$$\sum_{k=1}^{n} \frac{1}{k^2 \binom{k+n}{n}} = \sum_{k=1}^{n} \frac{n!}{k^2 \prod_{j=1}^{n}(k+j)} = \sum_{k=1}^{n} \frac{n!}{k^2 (k+1)_{n+1}}$$

$$= \sum_{k=1}^{n} \frac{n!}{k^2} \sum_{j=1}^{n} \frac{A_j}{k+j},$$

where

$$A_j = (-1)^{j+1} \frac{j}{n!} \binom{n}{j}$$

which can be easily shown by multiplying the previous equation by $k+i$ and then taking the limit $k \to (-i)$.

Now

$$\sum_{k=1}^{n} \frac{1}{k^2 \binom{k+n}{n}} = \sum_{j=1}^{n} (-1)^{j+1} j \binom{n}{j} \sum_{k=1}^{n} \frac{1}{k^2 (k+j)}$$

$$= \sum_{j=1}^{n} (-1)^{j+1} j \binom{n}{j} \sum_{k=1}^{n} \left[\frac{1}{k^2 j} - \frac{1}{kj^2} + \frac{1}{j^2 (k+j)} \right]$$

$$= \sum_{j=1}^{n} (-1)^{j+1} j \binom{n}{j} \left[\frac{H_n^{(2)}}{j} - \frac{H_n}{j^2} + \frac{H_{n+j}}{j^2} - \frac{H_j}{j^2} \right]$$

$$= H_n^{(2)} - H_n H_n - H_n^{(2)} + \sum_{j=1}^{n} \frac{(-1)^{j+1} \binom{n}{j} H_{n+j}}{j},$$

which is the result (6). The identity (6) is an alternate representation to (1) and shows that:

$$\sum_{k=1}^{n} \frac{(-1)^{k+1} \binom{n}{k} H_{n+k}}{k} = 3 \sum_{k=1}^{n} \frac{1}{k^2 \binom{2k}{k}} + (H_n)^2 - H_n^{(2)}.$$

From (6), (1) and some manipulations we can ascertain

$$-(H_n)^2 + \sum_{k=1}^{n} \frac{(-1)^{k+1} \binom{n}{k} H_{n+k}}{k} = 3 \sum_{k=1}^{n} \frac{1}{k^2 \binom{2k}{k}} - H_n^{(2)};$$

$$\sum_{k=1}^{n} \left(\frac{(-1)^{k+1} \binom{n}{k} H_{n+k}}{k} - \frac{3}{k^2 \binom{2k}{k}} \right) = (H_n)^2 - H_n^{(2)} = \frac{2}{n!} \left[\begin{matrix} n+1 \\ 3 \end{matrix} \right]$$

$$= 2 \sum_{k=1}^{n} (-1)^k \binom{n}{k} \left(\frac{H_k}{k} - \frac{1}{k^2} \right) \qquad (7)$$

where $\left[\begin{matrix} \cdot \\ \cdot \end{matrix} \right]$ are the unsigned Stirling numbers of the first kind (see [17]). Also from (7), after some algebraic manipulations, we recover the alternating sum identity [14], given by:

$$\sum_{k=1}^{n} (-1)^{k+1} \binom{n}{k} \frac{H_k}{k} = H_n^{(2)}.$$

The integral representation (5) is obtained as follows. Let $m \in \mathbb{R}\backslash\{-1,-2,-3,..\}$ and expand,

$$\sum_{k=1}^{n} \frac{1}{k^2 \binom{k+n}{n} \binom{k+m}{m}} = \sum_{k=1}^{n} \frac{\Gamma(k)\Gamma(n+1)\Gamma(k)\Gamma(m+1)}{\Gamma(k+n+1)\Gamma(k+m+1)}$$

$$= \sum_{k=1}^{n} B(k,n+1)\ B(k,m+1) = \int_0^1\int_0^1 \frac{(1-x)^n(1-y)^m}{xy} \sum_{k=1}^{n} (xy)^k\, dydx$$

$$= \int_0^1\int_0^1 (1-x)^n(1-y)^m \left(\frac{1-(xy)^n}{1-xy}\right) dydx,$$

and putting $m=0$ we arrive at (5). □

Remark 3. As a matter of interest, in hypergeometric form (5) is written as:

$$\sum_{k=1}^{n} \frac{1}{k^2 \binom{k+n}{n}} = \zeta(2) - H_n^{(2)} - \frac{{}_3F_2\left[\begin{array}{c} 1, n+1, n+1 \\ 2+n, 2+2n \end{array}\middle| 1\right]}{(n+1)^2 \binom{2n+1}{n}}$$

from which we see:

$$\frac{{}_3F_2\left[\begin{array}{c} 1, n+1, n+1 \\ 2+n, 2+2n \end{array}\middle| 1\right]}{(n+1)^2 \binom{2n+1}{n}} = \zeta(2) - 3\sum_{k=1}^{n} \frac{1}{k^2 \binom{2k}{k}}$$

and

$$\frac{1}{(n+1)}\ {}_3F_2\left[\begin{array}{c} 1,1,1 \\ 2,2+n \end{array}\middle| 1\right] = \zeta(2) - H_n^{(2)} = \sum_{k=n+1}^{\infty} \frac{1}{k^2},$$

where ${}_pF_q\left[\begin{array}{c} \cdot,\cdot,\cdot,\ldots.. \\ \cdot,\cdot,\cdot,\ldots.. \end{array}\middle| 1\right]$ is the hypergeometric function. Also from (5), we obtain the very useful integral identity:

$$\sum_{k=1}^{n} \frac{H_k}{k^2} = -\int_0^1\int_0^1 \frac{\ln(1-x)(1-(xy)^n)}{1-xy} dydx.$$

To the authors knowledge no identity, in terms of special numbers, exists for $\sum_{k=1}^{n} \frac{H_k}{k^2}$, but we have the known identities:

$$\sum_{k=1}^{n} \left(\frac{H_k}{k^2} + \frac{H_k^{(2)}}{k} \right) = H_n H_n^{(2)} + H_n^{(3)} \text{ and } \sum_{k=1}^{n} \frac{H_k^{(2)}}{k^2} = \frac{H_n^{(4)} + \left(H_n^{(2)} \right)^2}{2}.$$

The next lemma deals with the derivatives of binomial coefficients.

Lemma 4. *Let q and p be positive integers. Also let $Q(p,x) = \binom{p+x}{x}^{-1}$ be an analytic function in x and*

$$Q^{(q)}(p,x) = \frac{d^q}{dx^q} \left(\binom{p+x}{x}^{-1} \right).$$

Then,

$$(-1)^{q-1} q! \sum_{r=0}^{p} (-1)^r \binom{p}{r} \frac{r}{(r+x)^{q+1}} = Q^{(q)}(p,x)$$

$$= \frac{(-1)^q q!\, p}{(x+1)^{q+1}} \;{}_{q+2}F_{q+1} \left[\begin{matrix} \overbrace{x+1, \ldots\ldots, x+1}^{(q+1)-terms}, 1-p \\ \underbrace{x+2, \ldots\ldots, x+2}_{(q+1)-terms} \end{matrix} \middle| 1 \right].$$

For $x = 0$,

$$\begin{aligned} Q^{(q)}(p,0) &= (-1)^{q-1} q! \sum_{r=1}^{p} (-1)^r \binom{p}{r} \frac{1}{r^q} \\ &= (-1)^q q!\, p \;{}_{q+2}F_{q+1} \left[\begin{matrix} \overbrace{1, \ldots\ldots, 1}^{(q+1)-terms}, 1-p \\ \underbrace{2, \ldots\ldots, 2}_{(q+1)-terms} \end{matrix} \middle| 1 \right]. \end{aligned}$$

Proof. A proof and the following examples have been given in [12]. □

Now we list some particular cases of Lemma 4:

$$Q^{(1)}(p,x) = -\binom{p+x}{x}^{-1}\sum_{r=1}^{p}\frac{1}{r+x},$$

$$\begin{aligned} Q^{(2)}(p,x) &= \binom{p+x}{x}^{-1}\left[\left(\sum_{r=1}^{p}\frac{1}{r+x}\right)^2+\sum_{r=1}^{p}\frac{1}{(r+x)^2}\right] \\ &= \binom{p+x}{x}^{-1}\left[\sum_{r=1}^{p}\sum_{s=1}^{r}\frac{2}{(r+x)(s+x)}\right], \end{aligned}$$

$$Q^{(3)}(p,x) = -\binom{p+x}{x}^{-1}\left[\begin{array}{c}\left(\sum_{r=1}^{p}\frac{1}{r+x}\right)^3+2\sum_{r=1}^{p}\frac{1}{(r+x)^3} \\ +3\sum_{r=1}^{p}\frac{1}{(r+x)^2}\sum_{r=1}^{p}\frac{1}{r+x}\end{array}\right]$$

and

$$Q^{(4)}(p,x) = \binom{p+x}{x}^{-1}\left[\begin{array}{c} 6\sum_{r=1}^{p}\frac{1}{(r+x)^2}\left(\sum_{r=1}^{p}\frac{1}{r+x}\right)^2 \\ +8\sum_{r=1}^{p}\frac{1}{(r+x)^3}\sum_{r=1}^{p}\frac{1}{r+x}+3\left(\sum_{r=1}^{p}\frac{1}{(r+x)^2}\right)^2 \\ +\left(\sum_{r=1}^{p}\frac{1}{r+x}\right)^4+6\sum_{r=1}^{p}\frac{1}{(r+x)^4}\end{array}\right].$$

In the special case when $x=0$ we may write:

$$Q^{(1)}(p,0) = -H_p^{(1)}, \tag{8}$$

$$Q^{(2)}(p,0) = \left(H_p^{(1)}\right)^2 + H_p^{(2)}, \tag{9}$$

$$Q^{(3)}(p,0) = -\left(H_p^{(1)}\right)^3 - 3H_p^{(1)}H_p^{(2)} - 2H_p^{(3)} \tag{10}$$

and

$$Q^{(4)}(p,0) = \left(H_p^{(1)}\right)^4 + 6\left(H_p^{(1)}\right)^2 H_p^{(2)} + 8H_p^{(1)}H_p^{(3)} + 3\left(H_p^{(2)}\right)^2 + 6H_p^{(4)}. \tag{11}$$

We can generalize (5) as follows.

Theorem 5. *Let m and n be positive integers. Then:*

$$\sum_{k=1}^{n}\frac{1}{k^m\binom{k+n}{n}} = \underbrace{\int_0^1\cdots\int_0^1}_{m-fold}\frac{(1-x_1)^n\left[1-\left(\prod_{j=1}^{m}x_j\right)^n\right]}{1-\prod_{j=1}^{m}x_j}dx_j \tag{12}$$

$$= \sum_{k=1}^{n} \frac{(-1)^{m+k+1}\binom{n}{k}(H_{n+k}-H_k)}{k^{m-1}} + \sum_{s=1}^{m} \frac{H_n^{(s)}\, Q^{(m-s)}(n,0)}{(m-s)!} \tag{13}$$

where $Q^{(0)}(n,0) = 1$. *For* $m \geq 3$,

$$\sum_{k=1}^{n} \frac{1}{k^m \binom{k+n}{n}} = \frac{1}{n+1}\ {}_{m+1}F_m\left[\begin{array}{c} \overbrace{1,\ldots\ldots,1}^{(m+1)-terms} \\ \underbrace{2,\ldots\ldots,2}_{(m-1)-terms}\ ,2+n \end{array}\middle|\, 1\right] \tag{14}$$

$$-\frac{1}{(n+1)^m\binom{2n+1}{n}}\ {}_{m+1}F_m\left[\begin{array}{c} 1,\overbrace{1+n,\ldots\ldots,1+n}^{(m)-terms} \\ \underbrace{2+n,\ldots\ldots,2+n}_{(m-1)-terms},2+2n \end{array}\middle|\, 1\right].$$

Proof. For m and n positive integers we have:

$$\sum_{k=1}^{n} \frac{1}{k^m \binom{k+n}{n}} = \sum_{k=1}^{n} \frac{n!}{k^m \prod\limits_{r=1}^{n}(k+r)} = \sum_{k=1}^{n} \frac{n!}{k^m (k+1)_{n+1}} = \sum_{k=1}^{n} \frac{n!}{k^m} \sum_{r=1}^{n} \frac{A_r}{k+r},$$

where A_r is given in Lemma 2. Now

$$\sum_{k=1}^{n} \frac{1}{k^m \binom{k+n}{n}} = \sum_{r=1}^{n} (-1)^{r+1}\, r \binom{n}{r} \sum_{k=1}^{n} \frac{1}{k^m (k+r)}$$

$$= \sum_{r=1}^{n} (-1)^{r+1}\, r \binom{n}{r} \sum_{k=1}^{n} \left(\frac{(-1)^m}{r^m (k+r)} + \sum_{s=1}^{m} \frac{(-1)^{m-s}}{r^{m+1-s} k^s} \right)$$

$$= \sum_{r=1}^{n} \frac{(-1)^{m+r+1}\binom{n}{r}(H_{n+r}-H_r)}{r^{m-1}} + \sum_{s=1}^{m}\sum_{r=1}^{n} (-1)^{r+1}\, r \binom{n}{r} \sum_{k=1}^{n} \frac{(-1)^{m-s}}{r^{m+1-s} k^s}$$

$$= \sum_{r=1}^{n} \frac{(-1)^{m+r+1}\binom{n}{r}(H_{n+r}-H_r)}{r^{m-1}} + \sum_{s=1}^{m} (-1)^{m-s} H_n^{(s)} \sum_{r=1}^{n} (-1)^{r+1} \binom{n}{r} \frac{1}{r^{m-s}},$$

where the last sum is defined in Lemma 4, and hence equals

$$\sum_{r=1}^{n} \frac{(-1)^{m+r+1} \binom{n}{r} (H_{n+r} - H_r)}{r^{m-1}} + \sum_{s=1}^{m} \frac{H_n^{(s)} Q^{(m-s)}(n,0)}{(m-s)!},$$

arriving at the result (13). The integral (12) is obtained in the same way as was done in Lemma 2. □

Some illustrative cases follow.

Example 6. For $m = 0$ and 1, we have:

$$\sum_{k=1}^{n} \frac{1}{\binom{k+n}{n}} = \frac{1}{n-1}\left(1 - \frac{n+1}{\binom{2n}{n}}\right), \text{ for } n > 1;$$

$$\sum_{k=1}^{n} \frac{1}{k\binom{k+n}{n}} = \frac{1}{n}\left(1 - \frac{1}{\binom{2n}{n}}\right).$$

For $m = 2$, we recover the result (5), and for $m = 4$,

$$\sum_{k=1}^{n} \frac{1}{k^4\binom{k+n}{n}} = \sum_{k=1}^{n} \frac{(-1)^{k+1}\binom{n}{k}(H_{n+k} - H_k)}{k^3} - \left(\frac{1}{6}H_n^4 + \frac{4}{3}H_n H_n^{(3)} - H_n^{(4)} - \frac{1}{2}\left(H_n^{(2)}\right)^2\right)$$

$$= \frac{1}{n+1}\, {}_5F_4\left[\begin{matrix} 1,1,1,1,1 \\ 2,2,2,2+n \end{matrix}\middle| 1\right] - \frac{1}{(n+1)^4\binom{2n+1}{n}}\, {}_5F_4\left[\begin{matrix} 1,1+n,1+n,1+n,1+n \\ 2+n,2+n,2+n,2+2n \end{matrix}\middle| 1\right],$$

and

$$\sum_{k=1}^{n} \frac{1}{k^3\binom{k+2}{2}} = \frac{n(13n+27)}{8(n+1)(n+2)} - \frac{3}{2}H_n^{(2)} + H_n^{(3)}.$$

Remark 7. Some infinite versions of (12) have been evaluated (see [13], for example):

$$\sum_{k\geq 1} \frac{1}{k^4 \binom{k+n}{n}} = \zeta(4) - H_n \zeta(3) + \frac{\zeta(2)}{2} \left[(H_n)^2 + H_n^{(2)} \right] + \sum_{r=1}^{n} \frac{(-1)^r}{r^3} \binom{n}{r} H_r,$$

$$\sum_{k\geq 1} \frac{(-1)^k}{k^4 \binom{k+n}{n}} = -\frac{7}{8}\zeta(4) + \frac{3}{4} H_n \zeta(3) - \frac{\zeta(2)}{4} \left[(H_n)^2 + H_n^{(2)} \right] + \sum_{r=1}^{n} \frac{(-1)^r}{r^3} \binom{n}{r} \left(H_{\frac{r}{2}} - H_r \right)$$

and

$$\sum_{k\geq 1} \frac{1}{k^2 \binom{k+n}{n}} = \zeta(2) - H_n^{(2)}.$$

Acknowledgement. The authors wish to thank the anonymous referee for many valuable comments which have improved the presentation of the paper.

References

[1] H. Belbachir, M. Rahmani and B. Sury. *Alternating sums of the reciprocals of binomial coefficients.* J. Integer Seq. **15** (2012), Article 12.2.8.

[2] J. Choi. *Certain summation formulas involving harmonic numbers and generalized harmonic numbers.* Appl. Math. Comput. **218** (2011), 734-740.

[3] J. Choi and H. M. Srivastava. *Some summation formulas involving harmonic numbers and generalized harmonic numbers.* Math. Comp. Modelling. **54** (2011), 2220-2234.

[4] H. W. Gould. *Combinatorial Identities.* Morgantown Printing and Binding Co., New York, 1972.

[5] R. L. Graham, D. E. Knuth, and O. Patashnik. *Concrete Mathematics.* Addison-Wesley, Massachusetts, 2nd Ed., 1994.

[6] H. Liu and W. Wang. *Harmonic number identities via hypergeometric series and Bell polynomials.* Int. Trans. Sp. Functions. **23** (2012), 49-68.

[7] A. Sofo. *Computational techniques for the summation of series.* Kluwer Academic/Plenum Publishers, 2003.

[8] A. Sofo. *Summation formula involving harmonic numbers.* Analysis Mathematica. **37** (2011), 51-64.

[9] A. Sofo. *Harmonic sums and integral representations.* J. Appl. Analysis. **16** (2010), 265-277.

[10] A. Sofo. *Integral forms of sums associated with harmonic numbers.* Appl. Math. Comput. **207** (2009), 365-372.

[11] A. Sofo. *Sums of derivatives of binomial coefficients.* Advances Appl. Math. **42** (2009), 123-134.

[12] A. Sofo. *Reciprocal power sums.* Integers. **12** (2012), #A39.

[13] A. Sofo. *Triple integral identities and Zeta functions.* Appl. Anal. Disc. Math. **4** (2010), 347-360.

[14] A. Sofo. *New classes of harmonic number identities.* J. Integer Seq. **15** (2012), Article 12.7.4.

[15] Z. W. Sun. *Super congruences and Euler numbers.* Science China Math. **54** (2011), 2509-2535

[16] W. Y. G. Wa. *Identities involving powers and inverse of binomial coefficients.* J. Math. Research & Exposition. **31** (2011), 1021-1029.

[17] W. Wang. *Riordan arrays and harmonic number identities.* Comp. Math. Appl. **60** (2010), 1494-1509.

[18] Z. Zhang and H. Song. *A generalization of an identity involving the inverse of binomial coefficients.* Tamkang J. Math. **39** (2008), 219-226.

FINE-WILF GRAPHS AND THE GENERALIZED FINE-WILF THEOREM

Stuart A. Rankin
Department of Mathematics, Western University, London, Ontario, Canada
srankin@uwo.ca

Received: 7/3/12, Revised: 5/30/13, Accepted: 6/21/13, Published: 8/12/13

Abstract

The Fine-Wilf theorem was generalized to finite sequences with three periods by M. G. Castelli, F. Mignosi, and A. Restivo. They introduced a function f from the set of all ordered triples of nonnegative integers to the set of positive integers which was critical to their analysis, and they introduced the graphs that we shall refer to as Fine-Wilf graphs. The work of Castelli et al. was generalized by R. Tijdeman and L. Zamboni, who introduced a function fw from the set of all sequences of nonnegative integers to the set of positive integers that was essential to their analysis. In this paper, we obtain an alternative formulation of f and fw, and we use this formulation to establish important properties of f and fw, obtaining in particular new upper and lower bounds for each. We also carry out an investigation of Fine-Wilf graphs for arbitrary finite sequences, showing how they are related to f and fw.

1. Introduction

A finite sequence $w = (a_1, a_2, \ldots, a_n)$ is said to have period $r \geq 1$, or to be r-periodic, if for every positive integer i for which $i + r \leq n$, $a_i = a_{i+r}$. In 1962, R. C. Lyndon and M. P. Schützenberger [4] established that for any integers $r, s \geq 1$, if w is both r-periodic and s-periodic and $|w| \geq r + s$, then w is $\gcd(r, s)$-periodic. Shortly thereafter (1965), N. J. Fine and H. S. Wilf [2] proved that for any integers $r, s \geq 1$, if $\{a_i\}$ is an infinite sequence of period r and $\{b_i\}$ is an infinite sequence of period s such that $a_i = b_i$ for all i with $1 \leq i \leq r + s - \gcd(r, s)$, then $a_i = b_i$ for all i. This is equivalent to the following result, which is commonly referred to as the Fine-Wilf theorem: for any integers $r, s \geq 1$, if w is a finite sequence that is both r-periodic and s-periodic, and $|w| \geq r + s - \gcd(r, s)$, then w is $\gcd(r, s)$-periodic. It was also shown in [2] that this bound is best possible, in the sense that for any positive integers r and s (with neither a divisor of the other), there exists a sequence w of length $r + s - \gcd(r, s) - 1$ that is both r-periodic and s-periodic, but not $\gcd(r, s)$-periodic. It is known that for such r and s, there is a unique (up to relabelling) two-symbol sequence of length $r + s - \gcd(r, s) - 1$ that is both r-periodic and s-periodic, but not $\gcd(r, s)$-periodic. For example, for $r = 2$ and $s = 3$, the sequence is $(0, 1, 0)$.

Nearly thirty-five years later (1999), the Fine-Wilf theorem was generalized to finite sequences with three periods by M. G. Castelli, F. Mignosi, and A. Restivo [1]. They introduced a function f from the set of all ordered triples of positive integers to the set $\mathbb{Z}^+$ of positive integers with the property that if w is a finite sequence with periods p_1, p_2, and p_3, and $|w| \geq f(p)$, where $p = (p_1, p_2, p_3)$, then w is $\gcd(p)$-periodic as well. They further established a condition on p under which the bound $f(p)$ is best possible. The sequences p that met this condition were precisely those for which the unique (up to relabelling) finite sequence of greatest length and with the greatest possible number of distinct entries that had periodicity p_1, p_2, and p_3, but not $\gcd(p_1, p_2, p_3)$ had exactly three distinct entries. In support of their work, they introduced the difference graphs that we shall refer to as Fine-Wilf graphs $G_p(n)$, where $p = (p_1, p_2, p_3)$ and $p_1 < p_2 < p_3$ and n are positive integers and $G_p(n)$ denotes the graph with vertex set $\{1, 2, \ldots, n\}$ and edge set

$$\{\{i, j\} \mid |i - j| \in \{p_1, p_2, p_3\}\}.$$

The work of Castelli et al. was followed immediately (2000) by work of J. Justin [3], who extended the definition of the function f to all finite sequences of positive integers, with analogous results.

A broader generalization of the work of Castelli et al. was then given by R. Tijdeman and L. Zamboni [5] (2003). They introduced a function, which we shall denote as fw, from the set of all finite sequences of positive integers to $\mathbb{Z}^+$, and they proved that for a sequence $p = (p_1, p_2, \ldots, p_n)$, a finite sequence w with periods p_i, $i = 1, 2, \ldots, n$ and length at least $fw(p)$ must be $\gcd(p)$-periodic as well, and that there exists a sequence w of length $fw(p) - 1$ that is p_i-periodic for all i, but not $\gcd(p)$-periodic.

In this paper, we establish new properties of the functions f and fw. In particular, we introduce new upper and lower bounds for f. We also begin an investigation of Fine-Wilf graphs for arbitrary $p_1, p_2, \ldots, p_n$, with a view to understanding how the graph depends on the values $p_1, p_2, \ldots, p_n$, and how the properties of fw are reflected in the graphs.

2. Generalization of the Fine-Wilf Theorem

Let $\mathscr{F}$ denote the set of all strictly increasing finite sequences with entries from $\mathbb{Z}^+$. For $p \in \mathscr{F}$, let $\gcd(p)$ denote the greatest common divisor of the entries in p, let $|p|$ denote the length of p, and for i such that $1 \leq i \leq |p|$, let p_i denote the ith entry of p and let $p|_i$ denote the truncated sequence $(p_1, p_2, \ldots, p_i)$. In particular, for $p \in \mathscr{F}$

with $|p| > 1$, we shall let $p^- = p|_{|p|-1}$. Finally, let $\max(p) = p_{|p|}$ and $\min(p) = p_1$.

Definition 2.1. *For $p \in \mathscr{F}$, let $p' = p$ if $|p| = 1$, otherwise let p' denote the element of $\mathscr{F}$ whose entries form the set $\{\min(p), p_2 - \min(p), \ldots, \max(p) - \min(p)\}$. Moreover, define $p^{(i)} \in \mathscr{F}$ for $i \geq 0$ as follows: $p^{(0)} = p$, and for $k \geq 0$, $p^{(k+1)} = (p^{(k)})'$. Finally, let $ht(p)$, the height of p, be the least nonnegative integer m for which $|p^{(m)}| = 1$, and $wt(p)$, the weight of p, be given by $wt(p) = \sum_{i=1}^{|p|} p_i$.*

Note that $|p'| = |p| - 1$ if $p_i = 2\min(p)$ for some i; otherwise, $|p'| = |p|$. As well, for any $p \in \mathscr{F}$, $\gcd(p) = \gcd(p')$, and this is the single entry in $p^{(ht(p))}$.

Definition 2.2. *Define $f : \mathscr{F} \to \mathbb{Z}^+$ by $f(p) = \sum_{i=0}^{ht(p)} \min(p^{(i)})$ for $p \in \mathscr{F}$. Moreover, the column of sequences whose i^{th} row is $p^{(i)}$, $0 \leq i \leq ht(p)$, shall be called the* tableau *for the calculation of $f(p)$.*

For example, the tableaux for the calculation of $f(p)$ for $p = (4, 7)$ and $p = (4, 7, 9)$ are shown below. In each case, $ht(p) = 4$, and $f(p) = 10$.

$p^{(0)}$	4,7	4,7,9
$p^{(1)}$	3,4	3,4,5
$p^{(2)}$	1,3	1,2,3
$p^{(3)}$	1,2	1,2
$p^{(4)}$	1	1

Note that for any $p \in \mathscr{F}$ with $|p| > 1$, $f(p) = \min(p) + f(p')$.

Lemma 2.3. *For any $p \in \mathscr{F}$, $f(p) \geq \max(p)$; and if $|p| > 1$, then $f(p) \geq 2\min(p)$.*

Proof. We use induction on $ht(p)$. If $p \in \mathscr{F}$ has $ht(p) = 0$, then $|p| = 1$ and $\max(p) = \min(p) = f(p)$. Suppose now that $p \in \mathscr{F}$ has $ht(p) > 0$, and the result holds for lower sequences. Note that $|p| > 1$ since $ht(p) > 0$. Thus $\min(p) < \max(p)$ and $f(p) = \min(p) + f(p')$. Since $ht(p') = ht(p) - 1$, the inductive hypothesis applies to p' and we obtain $f(p') \geq \max(p')$. If $\max(p) - \min(p) > \min(p)$, then $\max(p') = \max(p) - \min(p)$ and then $f(p) = \min(p) + f(p')) \geq \min(p) + \max(p) - \min(p) = \max(p) > 2\min(p)$. Otherwise, $\max(p) - \min(p) \leq \min(p)$, so $\max(p') = \min(p)$ and thus $f(p) = \min(p) + f(p') \geq \min(p) + \max(p') = 2\min(p) \geq \max(p)$. The result follows now by induction. □

Definition 2.4. *Define $fw : \mathscr{F} \to \mathbb{Z}^+$ as follows. For $p \in \mathscr{F}$,*

$$fw(p) = \begin{cases} fw(p^-) & \text{if } |p| > 1,\ \gcd(p^-) = \gcd(p),\ \text{and } \max(p) \geq f(p^-) \\ f(p) & \text{otherwise.} \end{cases}$$

We shall show later (see Proposition 3.12) that if $p \in \mathscr{F}$ satisfies $|p| > 1$ and $\gcd(p^-) = \gcd(p)$, then $fw(p) \leq fw(p^-)$.

The following lemma can be easily proven by induction on $\max(p)$.

Lemma 2.5. *Let $c \in \mathbb{Z}^+$, and $p \in \mathscr{F}$, say $p = (p_1, p_2, \ldots, p_n)$. Let $cp = (cp_1, cp_2, \ldots,$
$cp_n)$. Then $cp \in \mathscr{F}$ and $f(cp) = c\,f(p)$ and $fw(cp) = c\,fw(p)$.*

Proposition 2.6. *Let $p \in \mathscr{F}$. If $min(p) = \gcd(p)$, then $fw(p) = \min(p)$ and $f(p) = \max(p)$, while if $\min(p) \neq \gcd(p)$, then $fw(p) \geq 2\min(p)$.*

Proof. First, we prove by induction on $ht(p)$ that if $\min(p) = \gcd(p)$, then $f(p) = \max(p)$. The base case of $ht(p) = 0$ is immediate, so suppose that $ht(p) > 1$ with $\min(p) = \gcd(p)$, and the assertion holds for all elements of $\mathscr{F}$ of lower height. Then $f(p) = \min(p) + f(p')$ and $ht(p') < ht(p)$. Since every element of p is a multiple of $\gcd(p) = \min(p)$, it follows that $\min(p') = \min(p) = \gcd(p) = \gcd(p')$, and so we may apply the induction hypothesis to p' to obtain that $f(p') = \max(p') = \max(p) - \min(p)$. Thus $f(p) = \max(p)$, as required.

Next, we prove by induction on $|p|$ that if $\min(p) = \gcd(p)$, then $fw(p) = \gcd(p)$, while if $\min(p) > \gcd(p)$, then $fw(p) \geq 2\min(p)$. The base case $|p| = 1$ is immediate, so suppose that $p \in \mathscr{F}$ has $|p| > 1$, and the assertion holds for all shorter sequences. Consider first the case when $\min(p) = \gcd(p)$. Then $\gcd(p) = \gcd(p^-)$ and so $\min(p^-) = \min(p) = \gcd(p^-)$. Since $|p^-| < |p|$, we may apply the induction hypothesis to p^- to obtain that $fw(p^-) = \gcd(p^-)$. Since $\gcd(p) = \gcd(p^-)$, and by the preceding part, $f(p^-) = \max(p^-) < \max(p)$, the definition of fw yields $fw(p) = fw(p^-) = \gcd(p)$. Now suppose that $\min(p) > \gcd(p)$. If $fw(p) = f(p)$, the result follows from Lemma 2.3, so we may suppose that $fw(p) \neq f(p)$. Then by definition of fw, $|p| > 1$ and $\gcd(p^-) = \gcd(p)$, and $fw(p) = fw(p^-)$. We thereby obtain that $\min(p^-) = \min(p) > \gcd(p) = \gcd(p^-)$, and so upon application of the induction hypothesis to p^-, we obtain $fw(p) = fw(p^-) \geq 2\min(p^-) = 2\min(p)$. The result follows now by induction. □

The next result gives an important lower bound for $f(p)$, and this result can be viewed as one generalization of the Fine-Wilf theorem. We will later obtain an upper bound (see Proposition 2.14, also Proposition 4.11) for $f(p)$ and the combination of these upper and lower bounds, when applied in the case of $|p| = 2$, yields the classical Fine-Wilf theorem.

Theorem 2.7. *Let $p \in \mathscr{F}$. If $|p| > 1$, then $f(p) \geq \dfrac{-\gcd(p) + wt(p)}{|p| - 1}$, and if equality holds and i is such that $p_i = 2\min(p)$, then $i = |p|$ and $f(p) = 2\min(p) = \max(p)$.*

Proof. The proof is by induction on $ht(p)$. If $p \in \mathscr{F}$ has $ht(p) = 0$, then $|p| = 1$, and the assertion is vacuously true. Suppose now that $p \in \mathscr{F}$ has $ht(p) > 0$, so $|p| > 1$, and that the assertion holds for all elements of $\mathscr{F}$ of lower height. In particular, the assertion holds for p', since $ht(p') = ht(p) - 1$.

Case 1: $|p'| = |p| > 1$. Then

$$wt(p') = \min(p) + \sum_{i=2}^{|p|}(p_i - \min(p)) = wt(p) - (|p|-1)\min(p),$$

and by the induction hypothesis and the facts $|p'| = |p|$ and $\gcd(p') = \gcd(p)$, we have $f(p) = \min(p) + f(p') \geq \min(p) + \frac{-\gcd(p')+wt(p')}{|p'|-1} = \frac{-\gcd(p)+wt(p)}{|p|-1}$. In this case, there is no value of i such that $p_i = 2\min(p)$.
Case 2: $|p'| = |p| - 1$. Then there exists i with $1 \leq i \leq |p|$ such that $p_i = 2\min(p)$, and $wt(p') = \sum_{j=2}^{|p|}(p_j - \min(p)) = wt(p) - |p|\min(p)$.

If $|p'| = 1$, then $p = (\min(p), 2\min(p))$ and the result holds. Suppose that $|p'| \geq 2$. Then $f(p') \geq \frac{-\gcd(p')+wt(p')}{|p'|-1} = \frac{-\gcd(p)+wt(p)-|p|\min(p)}{|p|-2}$ by the induction hypothesis applied to p', and so $f(p) \geq \min(p) + \frac{-\gcd(p)+wt(p)-|p|\min(p)}{|p|-2} = \frac{-2\min(p)-\gcd(p)+wt(p)}{|p|-2}$.

Consider first the possibility that $\frac{-2\min(p)-\gcd(p)+wt(p)}{|p|-2} \geq \frac{-\gcd(p)+wt(p)}{|p|-1}$, in which case $f(p) \geq \frac{-\gcd(p)+wt(p)}{|p|-1}$, as required. If as well, $f(p) = \frac{-\gcd(p)+wt(p)}{|p|-1}$, then $\frac{-2\min(p)-\gcd(p)+wt(p)}{|p|-2} = \frac{-\gcd(p)+wt(p)}{|p|-1}$, and thus $-\gcd(p)+wt(p) = (|p|-1)2\min(p)$, yielding $f(p) = 2\min(p)$. Then by Lemma 2.3, $2\min(p) = f(p) \geq \max(p) \geq p_i = 2\min(p)$ and thus $p_i = \max(p)$.

Now suppose that $\frac{-2\min(p)-\gcd(p)+wt(p)}{|p|-2} < \frac{-\gcd(p)+wt(p)}{|p|-1}$. Then $-\gcd(p)+wt(p)$ is less than $(|p|-1)2\min(p)$ and so $\frac{-\gcd(p)+wt(p)}{|p|-1} < 2\min(p) = p_i \leq \max(p) \leq f(p)$. This completes the proof of the inductive step, and so the result follows. □

Proposition 2.8. *For any $p \in \mathscr{F}$, the following hold.*

1. *If $|p| > 1$, $\gcd(p) = \gcd(p^-)$, and $\max(p) \geq f(p^-)$, then $f(p) = \max(p)$.*

2. *$f(p) \geq fw(p)$.*

Proof. We prove (1) by induction on $ht(p)$, with trivial base case of $ht(p) = 0$. Suppose now that $p \in \mathscr{F}$ has $ht(p) > 0$ with $\gcd(p) = \gcd(p^-)$ and $\max(p) \geq f(p^-)$, and that the result holds for all sequences of lower height. If $|p| = 2$, then $\min(p) = \gcd(p)$, and so by Proposition 2.6, $f(p) = \max(p)$. Suppose that $|p| > 2$. Then by Lemma 2.3, we have $f(p^-) \geq 2\min(p)$, so $\max(p) \geq 2\min(p)$ and thus $\max(p') = \max(p) - \min(p)$. Suppose first that $\max(p) > 2\min(p)$. Then we have $(p')^- = (p^-)'$, so $\gcd((p')^-) = \gcd((p^-)') = \gcd(p^-) = \gcd(p) = \gcd(p')$ and $\max(p') = \max(p) - \min(p) \geq f(p^-) - \min(p) = f((p^-)') = f((p')^-)$. Since $ht(p') < ht(p)$, we may apply the induction hypothesis to p' to obtain that $f(p') = \max(p') = \max(p) - \min(p)$, and thus $f(p) = \max(p)$ in this case. Otherwise, $\max(p) = 2\min(p)$ and then $p' = (p^-)'$ and from $\max(p) \geq f(p^-) \geq 2\min(p)$ it follows that $f(p^-) = 2\min(p)$. We have $f(p') = f((p^-)') = f(p^-) - \min(p^-) =$

$2\min(p) - \min(p) = \min(p)$ and so $f(p) = \min(p) + f(p') = 2\min(p) = \max(p)$, as required. Statement (1) of the theorem follows now by induction.

Next, we prove (2) by induction on $|p|$, with obvious base case $|p| = 1$. Suppose now that $p \in \mathscr{F}$ has $|p| > 1$, and that the result holds for all shorter sequences. Suppose that $f(p) \neq fw(p)$, so $\gcd(p) = \gcd(p^-)$ and $\max(p) \geq f(p^-)$. Then by (1) and the induction hypothesis, we have $f(p) = \max(p) \geq f(p^-) \geq fw(p^-) = fw(p)$. The result follows now by induction. $\square$

Definition 2.9. *$p \in \mathscr{F}$ is said to be* trim *if either $|p| = 1$ or else $|p| > 1$ and either $\gcd(p) \neq \gcd(p^-)$ or $\gcd(p) = \gcd(p^-)$ but $\max(p) < f(p^-)$. For any $p \in \mathscr{F}$, there exists i with $1 \leq i \leq |p|$ such that $p|_i$ is trim. The trimmed form of p, denoted by p^t, is $p|_i$, where i is maximal with respect to the property $p|_i$ is trim.*

Note that p is not trim if and only if $|p| > 1$, $\gcd(p) = \gcd(p^-)$, and $\max(p) \geq f(p^-)$. We remark that even if p is trim, there may exist i with $1 < i < |p|$ such that $p|_i$ is not trim. We shall examine such situations in the final section.

By Definition 2.4, we have

$$fw(p) = \begin{cases} fw(p^-) & \text{if } p \text{ is not trim} \\ f(p) & \text{if } p \text{ is trim.} \end{cases}$$

Corollary 2.10. *Let $p \in \mathscr{F}$. If p is not trim, then $f(p^{(i)}) = \max(p^{(i)})$ for every i, $0 \leq i \leq ht(p)$.*

Proof. Since p is not trim, we have $|p| > 1$, $\gcd(p) = \gcd(p^-)$, and $\max(p) \geq f(p^-)$, so we may apply Proposition 2.8 to obtain $f(p) = \max(p)$. We now prove by induction on $ht(p)$ that if $f(p) = \max(p)$, then $f(p^{(i)}) = \max(p^{(i)})$ for every i, $0 \leq i \leq ht(p)$. The base case of $ht(p) = 0$ is true by definition, so suppose that $p \in \mathscr{F}$ has $ht(p) > 0$ and $f(p) = \max(p)$, and that the result holds for all sequences of lower height. By Lemma 2.3, $f(p) \geq 2\min(p)$ and so we have $\max(p) = f(p) \geq 2\min(p)$. Hence $\max(p) - \min(p) \geq \min(p)$ and so $\max(p') = \max(p) - \min(p)$. Now, $f(p') = f(p) - \min(p) = \max(p) - \min(p) = \max(p')$, and since $ht(p') < ht(p)$, we may apply the inductive hypothesis to p' to obtain that $f((p')^{(i)}) = \max((p')^{(i)})$ for every i with $0 \leq i \leq ht(p')$, and so $f(p^{(i)}) = \max(p^{(i)})$ for every i with $0 \leq i \leq ht(p)$. The result follows now by induction. $\square$

If $p \in \mathscr{F}$ is not trim, then $|p| > 1$, $p^t = (p^-)^t$, and $fw(p) = fw(p^-)$. We shall say that p^- is obtained by trimming p. Evidently, for any $p \in \mathscr{F}$, we may iteratively apply the trimming operation to obtain p^t, and it follows that $fw(p) = fw(p^t)$.

Proposition 2.11. *Let $p \in \mathscr{F}$. Then $fw(p) = fw(p^t) = f(p^t)$, and $\gcd(p) = \gcd(p^t)$. Furthermore, if $|p^t| > 1$, then $\min(p) > \gcd(p^t)$.*

Proof. If p is trim, then by definition of fw, $fw(p) = f(p)$. Otherwise, $fw(p) = fw(p^-)$, and so the first assertion follows by induction on $|p|$. Next, note that $\min(p) = \min(p^t) \geq \gcd(p^-)$. Suppose that $|p^t| > 1$. If $\min(p^t) = \gcd(p^t)$, then by Proposition 2.6, $fw(p^t) = \min(p^t) < \max(p^t) = f(p^t)$, which is not the case, and so $\min(p) > \gcd(p^t)$. That $\gcd(p) = \gcd(p^t)$ is immediate from the definition of p^t. □

Corollary 2.12. *Let $p \in \mathscr{F}$. If p is not trim, then $p_{|p^t|+1} \geq fw(p)$.*

Proof. Suppose that p is not trim. Then by Definition 2.9, $p_{|p^t|+1} \geq fw(p^t)$, and by Proposition 2.11, $fw(p^t) = fw(p)$. □

Proposition 2.13. *Let $p \in \mathscr{F}$. If p is trim and $|p| > 1$, then $f(p) > \max(p)$. In addition, if p' is not trim, then $f(p) = 2\min(p)$.*

Proof. The proof is by induction on $ht(p)$, with vacuous base case. Suppose that $p \in \mathscr{F}$ is trim with $ht(p) > 1$, and the implication holds for every sequence of lower height. Since $ht(p) > 1$, we have $|p| > 1$, and so $f(p) = \min(p) + f(p')$. Consider first the situation when p' is trim. If $|p'| = 1$, then $\max(p) - \min(p) = \min(p)$ and $p = (\min(p), 2\min(p))$, which is not possible since p is trim. Thus $|p'| > 1$, and then by the inductive hypothesis, we have $f(p') > \max(p') \geq \max(p) - \min(p)$, and so $f(p) > \max(p)$.

Now suppose that p' is not trim, in which case $|p'| > 1$, $\gcd(p') = \gcd((p')^-)$, and $\max(p') \geq f((p')^-)$, and by Corollary 2.10, $f(p') = \max(p')$. Thus $f(p) = \min(p) + \max(p')$. If $\max(p) < 2\min(p)$, then $\max(p') = \min(p)$, and so $f(p) = 2\min(p) > \max(p)$, as required. The proof of the inductive step will be complete once we prove that $\max(p) \geq 2\min(p)$ is not possible. Indeed, suppose that $\max(p) \geq 2\min(p)$. Then $\max(p') = \max(p) - \min(p)$. If $\max(p) - \min(p) = \min(p)$, then $p' = (p^-)'$, while if $\max(p) - \min(p) > \min(p)$, then $(p')^- = (p^-)'$. In the first case, we have $\gcd(p^-) = \gcd((p^-)') = \gcd(p') = \gcd(p)$, while in the second case, we have $\gcd(p^-) = \gcd((p^-)') = \gcd((p')^-) = \gcd(p') = \gcd(p)$. In either case therefore, we have $\gcd(p^-) = \gcd(p)$. Since p is trim, this implies that $|p| > 2$ and $\max(p) < f(p^-) = \min(p^-) + f((p^-)') = \min(p) + f((p^-)')$. Now, $(p^-)' = p'$ or $(p')^-$, so either $f((p^-)') = f(p') = \max(p')$, or else $f((p^-)') = f((p')^-) \leq \max(p')$. Thus $\max(p) < \min(p) + \max(p') = \max(p)$, which is impossible. □

The following result gives an upper bound for f that is reminiscent of the Fine-Wilf theorem. Later (see Proposition 4.11), we shall establish a generalization of this which for p trim with $|p| \geq 3$ offers a slightly improved upper bound for fw.

Proposition 2.14. *For $p \in \mathscr{F}$, $f(p) \leq \min(p) + \max(p) - \gcd(p)$.*

Proof. The proof is by induction on $ht(p)$, with immediate base case. Suppose now that $p \in \mathscr{F}$ has $ht(p) > 0$, and the result holds for all lower elements of $\mathscr{F}$.

Since $ht(p) > 0$, the induction hypothesis applies to p', and since $ht(p) > 0$ implies that $|p| > 1$, we have $f(p) = \min(p) + f(p') \le \min(p) + \min(p') + \max(p') - \gcd(p') = \min(p) + \min(p') + \max(p') - \gcd(p)$. It therefore suffices to prove that $\min(p') + \max(p') \le \max(p)$. We consider three cases. The first occurs when $\min(p) \le p_2 - \min(p)$, in which case $\min(p') = \min(p)$ and $\max(p') = \max(p) - \min(p)$, so $\min(p') + \max(p') = \max(p)$. Next, suppose that $p_2 - \min(p) < \min(p) \le \max(p) - \min(p)$. Then $\min(p') = p_2 - \min(p)$ and $\max(p') = \max(p) - \min(p)$, so $\min(p') + \max(p') = \max(p) + p_2 - 2\min(p) < \max(p)$. Finally, suppose that $\max(p) - \min(p) < \min(p)$, so that $\min(p') = p_2 - \min(p)$ and $\max(p') = \min(p)$ and thus $\min(p') + \max(p') = p_2 \le \max(p)$. This completes the proof of the inductive step. □

Theorem 2.15 (Fine-Wilf). *Let $p \in \mathscr{F}$ with $|p| = 2$. Then $f(p) = \min(p) + \max(p) - \gcd(p)$. Furthermore, $fw(p) = \min(p) + \max(p) - \gcd(p)$ if p is trim, otherwise $fw(p) = \min(p) < f(p)$.*

Proof. By Theorem 2.7, $f(p) \ge \min(p) + \max(p) - \gcd(p)$, while by Proposition 2.14, $f(p) \le \min(p) + \max(p) - \gcd(p)$. If p is trim, then $fw(p) = f(p)$, while if p is not trim, then $fw(p) = fw(p^-) = \min(p)$, while $f(p) = \max(p)$. □

3. The Fine-Wilf Graphs $G_p(k)$

Definition 3.1. *Let $p \in \mathscr{F}$. For any $k \in \mathbb{Z}^+$, $G_p(k)$ denotes the simple graph with vertex set $\{1, \ldots, k\}$ and edge set $\{\{i,j\} \mid |i-j| = p_t$ for some $t = 1, 2, \ldots, |p|\}$. The values $k = fw(p)$ and $k = fw(p) - 1$ feature prominently in the development of the theory, and we shall let $G_p = G_p(fw(p))$ and $G_p^- = G_p(fw(p) - 1)$.*

Proposition 3.2. *Let $p \in \mathscr{F}$. Then for any $k \le fw(p)$, $G_p(k) = G_{p^t}(k)$.*

Proof. We need only consider p not trim. In such a case, by Corollary 2.12, $fw(p^t) \le p_{|p^t|+1}$. Thus for any i, j, k such that $1 \le i < j \le k \le fw(p) = fw(p^t)$, $j - i \le fw(p^t) - 1 < p_{|p^t|+1}$. It follows that if $\{i,j\}$ is an edge of $G_p(k)$, then it is an edge $G_{p^t}(k)$ (the converse is obvious) and so $G_p(k) = G_{p^t}(k)$. □

Our first goal in this section is to establish that for $p \in \mathscr{F}$, the graph G_p has exactly $d = \gcd(p)$ connected components, each isomorphic to $G_{p/d}$. For example, for $p = (6, 8, 10)$, $fw(p) = 12$, $\gcd(p) = 2$, and G_p has two connected components, each isomorphic to the connected graph $G_{(3,4,5)}$, where by Lemma 2.5, $fw((3,4,5)) = fw((6,8,10))/2 = 6$.

```
4 — 1 — 6 — 3          7 — 1 — 11 — 5    8 — 2 — 12 — 6
    |   |                  |   |              |    |
    5 — 2                  9 — 3             10 — 4
```

$G_{(3,4,5)}$ $\qquad\qquad$ $G_{(6,8,10)}$

The second major objective of the section is to establish that $\kappa(G_p^-) > d$.

Proposition 3.3. *Let $p \in \mathscr{F}$ with $|p| > 1$, and let $k \geq 1$. Then for any i and j with $1 \leq i < j \leq k$, i and j belong to the same connected component of $G_{p'}(k)$ if and only if $i + \min(p)$ and $j + \min(p)$ belong to the same connected component of $G_p(k + \min(p))$.*

Proof. Let i and j be such that $1 \leq i < j \leq k$. For the first implication, it suffices to show that if $\{i, j\}$ is an edge in $G_{p'}(k)$, then $i+\min(p)$ and $j+\min(p)$ are connected in $G_p(k + \min(p))$. Suppose that $\{i, j\}$ is an edge in $G_{p'}(k)$. If $j - i = \min(p)$, then $\{j, i\}$, $\{i, i + \min(p)\}$ and $\{j, j + \min(p)\}$ are edges in $G_p(k + \min(p))$, and so $i + \min(p)$ and $j + \min(p)$ are connected in $G_p(k + \min(p))$. Otherwise, $j - i = p_r - \min(p)$ for some r, and then $\{j + \min(p), i\}$ and $\{i, i + \min(p)\}$ are edges in $G_p(k + \min(p))$, so $i + \min(p)$ and $j + \min(p)$ are connected in $G_p(\min(p) + k)$.

We prove the converse by induction on path length. Our hypothesis is that if $i + \min(p)$ and $j + \min(p)$ are connected by a path of length n in $G_p(k + \min(p))$, then i and j are connected in $G_{p'}(k)$. We consider first the case $n = 1$; that is, $\{i+\min(p), j+\min(p)\}$ is an edge in $G_p(k+\min(p))$. Then $(j+\min(p))-(i+\min(p)) = p_r$ for some r, and so $j-i = p_r \geq \min(p)$. But then $(j-\min(p))-i = p_r-\min(p) \geq 0$. Thus either $j - \min(p) = i$ or $\{j - \min(p), i\}$ is an edge in $G_{p'}(k)$. In either case, since $\{j, j-\min(p)\}$ is an edge in $G_{p'}(k)$, i and j are connected in $G_{p'}(k)$. Suppose now that $n \geq 1$ is an integer and the hypothesis holds for all smaller integers. We consider i and j with $1 \leq i < j \leq k$ such that $i + \min(p)$ and $j + \min(p)$ are joined by a path of length $n+1$ in $G_p(k+\min(p))$, say $i+\min(p), i_1, \ldots, i_{n+1} = j+\min(p)$.
Case 1: $i_1 > \min(p)$. Since $i_1 \leq k+\min(p)$, we have $i' = i_1 - \min(p) \leq k$. But then $1 \leq i', j \leq k$ and $i' + \min(p) = i_1$ and $j + \min(p)$ are connected in $G_p(k + \min(p))$ by a path of length n, so by hypothesis, i' and j are connected in $G_{p'}(k)$. We prove now that i and i' are connected in $G_{p'}(k)$. We have $|i + \min(p) - i_1| = p_r$ for some r. If $i \geq i_1 - \min(p)$, then $i = i_1 + p_r - \min(p) \geq i_1 > \min(p)$, which means that $(i - \min(p)) - (i_1 - \min(p)) = p_r - \min(p)$, and $i_1 - \min(p) \leq k$, so $\{i_1 - \min(p), i - \min(p)\}$ and $\{i - \min(p), i\}$ are edges in $G_{p'}(k)$, whence i and $i' = i_1 - \min(p)$ are connected in $G_{p'}(k)$. Otherwise, $i < i_1 - \min(p)$, and so $i_1 - \min(p) - i = p_r \geq \min(p)$. Then $i_1 - 2\min(p) - i = p_r - \min(p)$, and so $\{i_1 - \min(p), i_1 - 2\min(p)\}$ is an edge in $G_{p'}(k)$ and either $i_1 - 2\min(p) = i$ or else $\{i_1 - 2\min(p), i\}$ is an edge in $G_{p'}(k)$. Thus i and i' are connected in $G_{p'}(k)$, and so i and j are connected in $G_{p'}(k)$.
Case 2: $i_1 \leq \min(p)$. Then $i + \min(p) - i_1 = p_r$ for some r, and so $i - i_1 = p_r - \min(p) \geq 0$. Thus $k \geq i \geq i_1$. As well, we have $|i_2 - i_1| = p_t \geq \min(p)$ for some t, which means that $i_2 - i_1 = p_t$ since $i_1 - i_2 < \min(p)$. Thus $(i_2 - \min(p)) - i_1 = p_t - \min(p)$, which means that $\{i_2 - \min(p), i_1\}$ is an edge in $G_{p'}(k)$. By hypothesis, since $(i_2 - \min(p)) + \min(p) = i_2, \ldots, i_{n+1} = j + \min(p)$ is a path of length $n - 1$ in $G_p(k + \min(p))$, $i_2 - \min(p)$ and j are connected in $G_{p'}(k)$. Since $\{i_2 - \min(p), i_1\}$

and $\{i, i_1\}$ are edges in $G_{p'}(k)$ (possibly $i = i_1$ or $i_2 - \min(p) = i_1$), i and j are connected in $G_{p'}(k)$.

This completes the proof of the inductive step, and so the result follows. □

Definition 3.4. *Let $p \in \mathscr{F}$ with $|p| > 1$. For any $k \geq 1$, define the function $\alpha_p : G_{p'}(k) \to G_p(\min(p) + k)$ by $\alpha_p(i) = \min(p) + i$.*

We remark that in general, α_p is not a graph homomorphism. However, the preceding proposition establishes that if C is a connected component of $G_{p'}(k)$, then $\alpha_p(C)$ is contained in a component of $G_p(\min(p) + k)$, and more generally, α_p induces an injective map from the set of components of $G_{p'}(k)$ into the set of components of $G_p(k + \min(p))$.

Our next objective is to show that for any $p \in \mathscr{F}$ with $\gcd(p) = 1$, G_p is connected. For this, the following lemma will be useful.

Lemma 3.5. *Let $p \in \mathscr{F}$ and $k \in \mathbb{Z}^+$ be such that $\min(p) \leq k$. If the interval $\{1, 2, \ldots, \min(p)\}$ is contained within a component of $G_p(k)$, then $G_p(k)$ is connected.*

Proof. Every $i > \min(p)$ is connected to $j \leq \min(p)$ by a path of length q, where $i = q\min(p) + j$ and $0 < j \leq \min(p)$. □

Proposition 3.6. *Let $p \in \mathscr{F}$ and $k \geq \mathit{fw}(p)$. If $\gcd(p) = 1$, then $G_p(k)$ is connected.*

Proof. The proof is by induction on $\max(p)$. The base case occurs when $\max(p) = 1$, in which case $p = (1)$ and $G_p(k)$ is the chain graph on k vertices. Suppose now that $\max(p) > 1$ and $\gcd(p) = 1$, and the result holds for all sequences with smaller maximum entry. Note that $\gcd(p) = 1$ and $\max(p) > 1$ imply that $|p| > 1$.

We consider first the case when p is not trim. Then $1 = \gcd(p) = \gcd(p^-)$ and $\mathit{fw}(p^-) = \mathit{fw}(p) \leq f(p) = \max(p)$, so $G_p = G_p(\mathit{fw}(p^-)) = G_{p^-}$. As well, $\max(p^-) < \max(p)$, so by hypothesis, $G_p = G_{p^-}$ is connected. Suppose now that $k \geq \mathit{fw}(p)$. Then G_p is a connected subgraph of $G_p(k)$, and since $\mathit{fw}(p) \geq \min(p)$, Lemma 3.5 implies that $G_p(k)$ is connected.

Now suppose that p is trim. Since $|p| > 1$, we have $\mathit{fw}(p) = f(p) = \min(p) + f(p')$, so $G_p = G_p(\min(p) + f(p'))$. Now, since $1 = \gcd(p) = \gcd(p')$ and $\max(p') < \max(p)$, we may apply the induction hypothesis to p' to obtain that $G_{p'}(f(p'))$ is connected since $f(p') \geq \mathit{fw}(p')$. Thus $\{i + \min(p) \mid 1 \leq i \leq f(p')\}$, the image of $G_{p'}(f(p'))$ under α_p, is contained within a connected component of $G_p(f(p))$. Since $f(p') \geq \min(p)$, it follows that $\{1, 2, \ldots, \min(p)\}$ is contained within a connected component of $G_p(f(p))$, and thus by Lemma 3.5, G_p is connected. But then for any $k \geq \mathit{fw}(p)$, G_p is a connected subgraph of $G_p(k)$ and thus by Lemma 3.5, $G_p(k)$ is connected. □

Lemma 3.7. *Let $p \in \mathscr{F}$, and $i, j, k \in \mathbb{Z}^+$. If i and j belong to the same connected component of $G_p(k)$, then* $\gcd(p)$ *divides $i - j$. Moreover, if $k \geq$* $\gcd(p)$, *then* $\kappa(G_p(k)) \geq \gcd(p)$.

Proof. It suffices to observe that if i and j are adjacent in $G_p(k)$, then $|i - j| = p_r$ for some r with $1 \leq r \leq |p|$. Thus if $1 \leq i < j \leq \gcd(p)$, it follows that i and j cannot be in the same connected component of $G_p(k)$. □

Proposition 3.8. *Let $p \in \mathscr{F}$, and let $d =$* $\gcd(p)$. *Then for each $i = 1, 2, \ldots, d$, the map $\gamma_i : G_{p/d} \to G_p$ defined by $\gamma_i(j) = i + (j-1)d$ for $1 \leq j \leq fw(p/d)$ is a graph isomorphism from $G_{p/d}$ onto the subgraph $\gamma_i(G_{p/d})$ of G_p. Moreover, G_p has exactly d components, whose vertex sets are the images of γ_i, $i = 1, 2, \ldots, d$; that is, the congruence classes of the interval $\{1, 2, \ldots, fw(p)\}$ modulo d.*

Proof. It is immediate from Lemma 3.7 that each component of G_p is contained in the image of γ_i for some i with $1 \leq i \leq d$, and by Proposition 3.6, $G_{p/d}$ is connected. Let j, k be vertices of $G_{p/d}$. Since $|\gamma_i(j) - \gamma_i(k)| = |(j-1)d - (k-1)d| = |j-k|d$, it follows that $|j - k| = p_t/d$ if and only if $|\gamma_i(j) - \gamma_i(k)| = p_t$. Thus γ_i is a graph isomorphism from $G_{p/d}$ onto the subgraph $\gamma_i(G_{p/d})$ of G_p. □

Corollary 3.9. *Let $p \in \mathscr{F}$ and let $k \geq fw(p)$. Then $G_p(k)$ has exactly* $\gcd(p)$ *components, the congruence classes of the interval $\{1, 2, \ldots, k\}$ modulo* $\gcd(p)$.

Proof. G_p is a subgraph of $G_p(k)$, and by Proposition 3.8, $\kappa(G_p) = \gcd(p)$. Moreover, by Lemma 3.7, $\kappa(G_p(k)) \geq \gcd(p)$, and since each vertex i of $G_p(k)$ for which $fw(p) < i \leq k$ is connected to a vertex in the subgraph G_p, the result follows. □

Proposition 3.10. *Let $p \in \mathscr{F}$. If* $\gcd(p) < \min(p) \leq k < fw(p)$, *then* $\kappa(G_p(k)) >$ $\gcd(p)$.

Proof. Suppose that $\gcd(p) < \min(p)$. Then by Proposition 2.6, $\min(p) < fw(p)$. Since for any k with $\min(p) \leq k < fw(p)$, $\kappa(G_p(k)) \geq \kappa(G_p(k+1))$ (since $k \geq \min(p)$, in $G_p(k)$, any $i > \min(p)$ is connected to either $\min(p)$ if i is a multiple of $\min(p)$, or some $r < \min(p)$ if i is not a multiple of p), it suffices to prove that $\kappa(G_p^-) > \gcd(p)$. The proof will be by induction on $\max(p)$, and the result is vacuously true for $\max(p) = 1$. Suppose now that $p \in \mathscr{F}$ has $\max(p) > 1$ and the result holds for all elements of $\mathscr{F}$ with smaller maximum entry. Suppose further that $\gcd(p) < \min(p)$. We first consider the case when p is not trim. Since $\max(p) > \max(p^-)$, we may apply the induction hypothesis to p^-. Since p is not trim, we have $\gcd(p) = \gcd(p^-)$, and $fw(p) = fw(p^-)$, so $\min(p^-) = \min(p) > \gcd(p) = \gcd(p^-)$. Thus $\kappa(G_{p^-}^-) > \gcd(p^-) = \gcd(p)$. By Corollary 2.10, $\max(p) = f(p) > f(p) - 1 \geq fw(p) - 1 = fw(p^-) - 1$, and so $G_{p^-}^- = G_p^-$. Thus $\kappa(G_p^-) > \gcd(p)$.

Suppose now that p is trim. There are two possibilities, p' trim or not. Suppose first that p' is trim. By Proposition 2.11, either $\min(p') > \gcd(p')$ or $|p'| = 1$. If $|p'| = 1$, then $\gcd(p') = \min(p') = \min(p) > \gcd(p) = \gcd(p')$, which is not possible. Thus $\min(p') > \gcd(p')$. Now since $\max(p') < \max(p)$, we may apply the induction hypothesis to p' to obtain that $G^-_{p'} = G_{p'}(\mathit{fw}(p')-1)$ has more than $\gcd(p') = \gcd(p)$ components. By Proposition 3.3, this implies that $G_p(\min(p) + \mathit{fw}(p') - 1)) = G_p(\mathit{fw}(p) - 1) = G^-_p$ has more than $\gcd(p)$ components, as required.

Now consider the case when p is trim but p' is not trim. By Proposition 2.13, $f(p) = 2\min(p)$. Since p is trim, we have $\mathit{fw}(p) = f(p) = 2\min(p)$, and so it follows that $G^-_p = G_p(\mathit{fw}(p) - 1) = G_p(2\min(p) - 1)$ has $\{\min(p)\}$ as a component. By Proposition 3.3, the map $\alpha_p : G_{p'}(\min(p) - 1) = G_{p'}(\mathit{fw}(p) - \min(p) - 1) \to G^-_p$ induces an injective map on components. Since $\min(p)$ is not in the image of α_p, $\kappa(G^-_p) \geq 1 + \kappa(G_{p'}(\min(p) - 1))$. We have $\min(p) = f(p') \geq \mathit{fw}(p')$. By Corollary 3.9, if $\mathit{fw}(p') < \min(p)$, then $\kappa(G_{p'}(\min(p) - 1)) = \gcd(p') = \gcd(p)$, which then implies that $\kappa(G^-_p) > \gcd(p)$. Suppose now that $\mathit{fw}(p') = \min(p)$. Then $G_{p'}(\min(p) - 1) = G^-_{p'}$. If $\min(p') = \gcd(p')$, then by Proposition 2.6, $\mathit{fw}(p') = \min(p')$ and so $\min(p) = \mathit{fw}(p') = \min(p') = \gcd(p') = \gcd(p)$, which is not the case. Thus $\min(p') > \gcd(p')$. Again by Proposition 2.6, $\mathit{fw}(p') \geq 2\min(p') > \min(p')$, and so we may apply the induction hypothesis to p' to obtain that $G^-_{p'} = G_{p'}(\mathit{fw}(p') - 1)$ has more than $\gcd(p') = \gcd(p)$ components. By Proposition 3.3, this implies that $\kappa(G_p(\min(p) + \mathit{fw}(p') - 1)) > \gcd(p)$. Since $\mathit{fw}(p') = \min(p)$, $\min(p) + \mathit{fw}(p') - 1 = 2\min(p) - 1 = \mathit{fw}(p) - 1$, and so $G_p(\min(p) + \mathit{fw}(p') - 1) = G^-_p$. Thus $\kappa(G^-_p) > \gcd(p)$ in this case was well, and this completes the proof of the inductive step. □

Corollary 3.11. *For $p \in \mathscr{F}$, $\mathit{fw}(p) = \min\{k \mid k \geq \min(p),\ \kappa(G_p(k)) = \gcd(p)\}$.*

Proof. If $\min(p) > \gcd(p)$, the result follows from Proposition 3.10 and Corollary 3.9. Suppose that $\min(p) = \gcd(p)$. Then $\mathit{fw}(p) = \gcd(p)$, and for $k < \gcd(p) = \min(p)$, $G_p(k)$ is a null graph, so has k components, while for $k \geq \gcd(p) = \mathit{fw}(p)$, Corollary 3.9 asserts that $\kappa(G_p(k)) = \gcd(p)$. □

Proposition 3.12. *If $p \in \mathscr{F}$ has $|p| > 1$ and $\gcd(p^-) = \gcd(p)$, then $\mathit{fw}(p^-) \geq \mathit{fw}(p)$.*

Proof. By Lemma 2.5, we need only consider $p \in \mathscr{F}$ for which $\gcd(p) = 1$. Note that by Proposition 2.6, the result holds if $\min(p) = 1$. Suppose that $p \in \mathscr{F}$ is such that $|p| > 1$, $\gcd(p^-) = \gcd(p) = 1 < \min(p)$, and, contrary to our claim, $\mathit{fw}(p^-) < \mathit{fw}(p)$. Then G_{p^-} is a subgraph of G^-_p, and by Corollary 3.9, G_{p^-} is connected. Thus $\{1, 2, \ldots, \min(p^-)\}$ is contained within a component of G^-_p, and since $\min(p^-) = \min(p)$, G^-_p is connected by Lemma 3.5. However, by Corollary 3.11, G^-_p is not connected, and so we have obtained a contradiction. □

We remark that if $\gcd(p^-) = \gcd(p)$ and $\max(p) \geq fw(p^-)$, then p is not trim and so $fw(p) = fw(p^-)$. However, when $\gcd(p^-) = \gcd(p)$ and $\max(p) < fw(p^-)$, it is possible that we may actually have $fw(p^-) > fw(p)$. The lexically first such example is $p = (5, 7, 8)$, where $fw(p) = 10$, while $fw((5, 7)) = 11$.

It might be tempting to believe that fw grows monotonically with respect to the product order on sequences of a given length and greatest common divisor 1, and, as the Fine-Wilf theorem tells us, this is indeed the case for sequences of length 2. However, this observation does not hold even for sequences of length 3. For example, $fw((7, 9, 11)) = 15$, while $fw((7, 9, 13)) = 14$.

Our next observation relates $\kappa(G_p(f(p)-1))$ to the tableau for the computation of $f(p)$. Let $p \in \mathscr{F}$ with $\gcd(p) = 1$ and $|p| > 1$, and consider the tableau for the computation of $f(p)$. Let $m = ht(p)$. Then $p^{(m)} = (1)$, and $p^{(m-1)} = (1, 2)$. For each i with $0 \leq i \leq m$, we shall call $p^{(i)}$ a *jump* if $f(p^{(i)}) = 2\min(p^{(i)})$, and in the tableau for the computation of $f(p)$, we shall prefix each jump with a plus sign (+). Furthermore, let $J(p)$ denote the number of jumps in the tableau for the calculation of $f(p)$. For example, $p = (6, 10, 13)$ has tableau

	6,10,13
+	4,6,7
+	2,3,4
+	1,2
	1

and so $J(p) = 3$. We observe that $p^{(m)}$ is never a jump, while $p^{(m-1)}$ is always a jump. For each $i = 0, \ldots, m$, let $G^i = G_{p^{(m-i)}}(f(p^{(m-i)}) - 1)$, so that G^0 is the null graph on a single vertex, and $G^m = G_p(f(p) - 1)$. Now for each $i = 1, \ldots, m$, let $\alpha_i : G^{i-1} \to G^i$ denote $\alpha_{p^{(m-(i-1))}}$, so that for a vertex j, $\alpha_i(j) = \min(p^{(m-i)}) + j$. By Proposition 3.3, for each i, α_i induces an injective map from the set of components of G^{i-1} into the set of components of G^i, and G^0 has a single component. Moreover, the image of α_i is the set $\{\min(p^{(m-i)}) + 1, \ldots, \min(p^{(m-i)}) + f(p^{(m-i+1)}) - 1\}$, which is equal to $\{\min(p^{(m-i)}) + 1, \ldots, f(p^{(m-i)}) - 1\}$. If $f(p^{(m-i)}) > 2\min(p^{(m-i)})$, then for each $k \in \{1, 2, \ldots, \min(p^{(m-i)})\}$, $k + \min(p^{(m-i)}) \leq 2\min(p^{(m-i)}) \leq f(p^{(m-i)}) - 1$, and thus $\{k, k + \min(p^{(m-i)})\}$ is an edge in G^i joining k to a vertex in the image of α_i. Consequently, $\kappa(G^i) = \kappa(G^{i-1})$. On the other hand, if $f(p^{(m-i)}) = 2\min(p^{(m-i)})$, then $\min(p^{(m-i)})$ has degree 0 in G^i, so $\{\min(p^{(m-i)})\}$ is a component of G^i that is not contained in the image of α_i. For any k with $1 \leq k < \min(p^{(m-i)})$, $\{k, k + \min(p^{(m-i)})\}$ is an edge in G^i joining k to a vertex in the image of α_i, and so $\kappa(G^i) = 1 + \kappa(G^{i-1})$. This proves the following result.

Proposition 3.13. *If $p \in \mathscr{F}$ has $|p| > 1$ and $\gcd(p) = 1$, then $\kappa(G_p(f(p) - 1)) = J(p)$, the number of jumps in the tableau for the computation of $f(p)$.*

Note that for p trim, $fw(p) = f(p)$ and so $G_p^- = G_p(f(p) - 1)$, and the finite sequence a of length $fw(p) - 1$ formed by labelling the components of G_p^-, then setting a_i equal to the label of the component containing vertex i of G_p^-, is the unique sequence (up to labelling) of length $fw(p) - 1$ with the greatest number of distinct entries that has periods the entries of p, but not $\gcd(p)$. By Proposition 3.13, the number of distinct entries in a is equal to $J(p)$. We observe as well that a can be calculated from the tableau for the calculation of $f(p)$. Let $m = ht(p)$. We begin at row $p^{(m-1)}$ with sequence 0. Then at stage $p^{(i)}$, shift the preceding word $\min(p^{(i)})$ spaces to the right. If $p^{(i)}$ is not a jump, then the preceding sequence has length at least $\min(p^{(i)})$ and we fill in the first $\min(p^{(i)})$ locations of the new sequence with the first $\min(p^{(i)})$ entries in the preceding sequence, while if $p^{(i)}$ is a jump, then the preceding sequence has length $\min(p^{(i)}) - 1$, and we fill in the first $\min(p^{(i)}) - 1$ spaces with the entries of the preceding sequence and then introduce a new symbol for the vertex at position $\min(p(i))$.

4. Reduction

In this final section, we show that the reduction concept first introduced with the trimming operation can be developed further.

Definition 4.1. *For $p \in \mathscr{F}$ and j such that $2 \leq j \leq |p|$ and $\gcd(p|_j) = \gcd(p|_{j-1})$, we shall say that p_j is* type I redundant *in p if p_j is a multiple of p_i for some i with $1 \leq i < j$, or* type II redundant *in p if $p_j \geq f(p|_{j-1})$. If p_j is either type I or type II redundant in p, we shall say that p_j is* redundant *in p.*

Definition 4.2. *For $p \in \mathscr{F}$ and j such that $1 \leq j \leq |p|$, let $p - p_j$ denote the element of $\mathscr{F}$ that is formed by deleting p_j from p.*

Proposition 4.3. *Let $p \in \mathscr{F}$ and p_j be type I redundant in p. Then*

1. *For every $k \in \mathbb{Z}^+$, the vertex sets of the components of $G_p(k)$ are identical to those of $G_{p-p_j}(k)$.*

2. *$fw(p) = fw(p - p_j)$.*

Proof. Let i be such that $1 \leq i < j$ and $p_j = tp_i$ for some integer t. Then every edge of $G_p(k)$ that is determined by p_j has its endpoints joined by a path of length t with edges determined by p_i, so the edges of $G_p(k)$ that are determined by p_j can be deleted (thereby forming $G_{p-p_j}(k)$) with no change in component vertex sets. Thus the vertex sets of the components of $G_p(k)$ are identical to those of $G_{p-p_j}(k)$. By Corollary 3.11, since $\gcd(p) = \gcd(p - p_j)$, $fw(p) = \min\{k \mid \kappa(G_p(k)) = \gcd(p)\} = \min\{k \mid \kappa(G_{p-p_j}(k)) = \gcd(p - p_j)\} = fw(p - p_j)$. □

One might suspect that if p_j is type I redundant in p, then p_j is type II redundant as well, but this is not necessarily the case. For example, if $p = (5, 13, 15)$, then

$f(p) = 17 = f(p^-) > max(p)$, while $\gcd(p) = 1 = \gcd(p^-)$, so p is trim; that is, $\max(p)$ is type I redundant but not type II redundant.

Lemma 4.4. *If $p \in \mathscr{F}$ and j is such that $1 < j < |p|$ and p_j is type II redundant in p, then $f(p) = f(p - p_j)$.*

Proof. Let $q = p|_j$. Since p_j is type II redundant in p, q is not trim. By Lemma 2.10 and Lemma 2.3, $\max(q^{(i)}) = f(q^{(i)}) \geq 2\min(q^{(i)})$ for every i, $0 \leq i < ht(q)$, and so for each $i = 0, 1, \ldots, ht(q) - 1$, $\max(q^{(i)}) - \min(q^{(i)}) \geq \min(q^{(i)})$. Thus in the formation of the tableau for the calculation of $f(q)$, if $\min(q^{(i)})$ needed to be inserted in the formation of $q^{(i+1)}$, it would be inserted prior to the last entry of $q^{(i)}$. Let $n = ht(q)$, and let $d = \gcd(q)$. Then $q^{(n)} = (d)$, and $q^{(n-1)} = (d, 2d)$. Since $p_j < p_{j+1}$, there exists $e \in \mathbb{Z}^+$ such that the tableau for the calculation of $f(p)$ looks like

$$\begin{array}{l} p \\ \vdots \\ p^{(n-1)} = d, 2d, e, \ldots \\ p^{(n)} = d, e - d, \ldots \\ \vdots \end{array}$$

where the entry $2d$ in $p^{(n-1)}$ is derived from p_j in p, while the entry e is derived from p_{j+1}. Thus for each $i \geq n$, $p^{(i)} = (p - p_j)^{(i)}$, while for each i with $1 \leq i \leq n$, $\min(p^{(i)}) = \min((p - p_j)^{(i)})$. It follows that $ht(p) = ht(p - p_j)$ and $f(p) = \sum_{i=1}^{ht(p)} \min(p^{(i)}) = \sum_{i=1}^{ht(p-p_j)} \min((p - p_j)^{(i)}) = f(p - p_j)$. □

We note that the requirement $j < |p|$ in the preceding lemma is essential, as the example discussed prior to Lemma 3.5 illustrates.

Proposition 4.5. *If $p \in \mathscr{F}$ and p_j is redundant in p, then $fw(p) = fw(p - p_j)$.*

Proof. If p_j is type I redundant, then the result follows from Proposition 4.3, so we may assume that p_j is type II redundant and thus $\gcd(p|_j) = \gcd(p|_{j-1})$ and $p_j \geq f(p|_{j-1})$. If $j \geq |p^t|$, then by Lemma 4.4, $f(p|_i) = f(p|_i - p_j)$ for all $i > j$, and thus $p^t = (p - p_j)^t$, in which case by Proposition 2.11, we have $fw(p) = fw(p^t) = fw((p - p_j)^t) = fw(p - p_j)$. Thus we may further assume that p is trim (and thus $j < |p|$).

Suppose that $fw(p - p_j) > fw(p)$. Then by Proposition 3.10, $G_{p-p_j}(fw(p))$ is not connected, and by Proposition 2.13, $p_j < \max(p) < f(p) = fw(p)$. Thus $fw(p|_{j-1}) \leq f(p|_{j-1}) \leq p_j < fw(p)$, and so $G_{p|_{j-1}}(fw(p|_{j-1}))$ is a connected subgraph of $G_{p-p_j}(fw(p))$. Since $fw(p|_{j-1}) \geq \min(p)$, the interval $\{1, 2, \ldots, \min(p)\}$ is contained within a component of $G_{p-p_j}(fw(p))$, and so by Lemma 3.5, $G_{p-p_j}(fw(p))$ is connected. This contradiction implies that $fw(p-p_j) \leq fw(p)$. Suppose now that $fw(p - p_j) < fw(p)$. Then $G_{p-p_j}(fw(p - p_j))$ is a connected subgraph of G_p^-. By

Proposition 2.6 applied to $p - p_j$, $fw(p - p_j) \geq 2\min(p) > \min(p)$ and so the interval $\{1, 2, \ldots, \min(p)\}$ is contained within a component of G_p^-. But then by Lemma 3.5, G_p^- is connected, which is not the case. This last contradiction implies that $fw(p) = fw(p - p_j)$, as required. □

Definition 4.6. *For $p \in \mathscr{F}$, let p^r denote the sequence obtained from p by deleting all type I redundant entries of p. Then p^r shall be called the* reduced form *of p, and is said to be obtained by* reducing p. *If $p = p^r$, we say that p is* reduced.

The next result is an immediate consequence of Proposition 4.5, since it is clear that we may form p^r by removing type I redundant entries one after the other in any order (that is, no new type I redundant entry can be formed nor can an existing type I redundant entry be made non-redundant by the deletion of an existing type I redundant entry).

Corollary 4.7. *For $p \in \mathscr{F}$, $fw(p) = fw(p^r)$.*

Proposition 4.8. *Let $p \in \mathscr{F}$ and p_j be type II redundant in p, where $1 < j < |p|$. Then for any $i \neq j$ with $1 \leq i \leq |p|$, p_i is type II redundant in p if and only if p_i is type II redundant in $p - p_j$.*

Proof. We have $\gcd(p|_j) = \gcd(p|_{j-1})$ and $p_j \geq f(p|_{j-1})$. The result is obvious for $i < j$, so we consider the case when $i > j$. Suppose first of all that p_i is type II redundant in p. Then $\gcd(p|_i) = \gcd(p|_{i-1})$, and $p_i \geq f(p|_{i-1})$. Note that the index of p_i in $p - p_j$ is $i - 1$. Since $i > j$, we have $\gcd((p - p_j)|_{i-1}) = \gcd(p|_i - p_j) = \gcd(p|_i) = \gcd(p|_{i-1}) = \gcd(p|_{i-1} - p_j) = \gcd((p - p_j)|_{i-2})$. If $i = j+1$, then $(p-p_j)_{i-1} = p_i > p_j \geq f(p|_{j-1}) = f((p-p_j)|_{j-1}) = f((p-p_j)|_{i-2})$, while if $i > j + 1$, then we have $(p - p_j)_{i-1} = p_i \geq f(p|_{i-1})$ and by Lemma 4.4 (since $j < i - 1 = |p|_{i-1}|$), $f(p|_{i-1}) = f(p|_{i-1} - p_j) = f((p - p_j)|_{i-2})$, so $(p - p_j)_{i-1} \geq f((p - p_j)|_{i-2})$. Thus in either case, p_i is a type II redundant entry of $p - p_j$.

Conversely, suppose that p_i is type II redundant in $p-p_j$. Then $p_i = (p-p_j)_{i-1} \geq f((p - p_j)|_{i-2})$ and $\gcd((p - p_j)|_{i-1}) = \gcd((p - p_j)|_{i-2})$. We have $\gcd(p|_i) = \gcd(p|_i - p_j) = \gcd((p-p_j)|_{i-1}) = \gcd((p-p_j)|_{i-2}) = \gcd(p|_{i-1} - p_j) = \gcd(p|_{i-1})$. As well, if $i > j + 1$, then by Proposition 4.4, we have $f(p|_{i-1}) = f(p|_{i-1} - p_j)$, and so $p_i \geq f((p - p_j)|_{i-2}) = f(p|_{i-1} - p_j) = f(p|_{i-1})$, while if $i = j + 1$, then by Proposition 2.8 applied to $p|_j$, $p_i > p_j = f(p|_j) = f(p|_{i-1})$. In either case, p_i is a type II redundant entry of p. □

Definition 4.9. *For $p \in \mathscr{F}$, let $\hat{p}$ denote the totally reduced form of p; namely the sequence obtained from p^r by deleting all type II redundant entries of p^r, and let $r(p) = |p| - |\hat{p}|$. If $p = \hat{p}$, we say that p is totally reduced.*

Note that $\hat{p}$ contains no redundant entries and is thus totally reduced. As well, note that as a result of Proposition 4.8, we may form $\hat{p}$ by deleting the type II redundant elements of p^r one by one, in any order whatsoever, and so induction on $r(p)$ proves the next result.

Proposition 4.10. *Let $p \in \mathscr{F}$. Then $fw(p) = fw(\hat{p})$.*

We are now in a position to give an upper bound for $fw(p)$ that is an improvement over that given in Proposition 2.14 (provided that $r(p) < |p| - 1$, its maximum possible value).

Theorem 4.11. *For $p \in \mathscr{F}$, $fw(p) \leq \min(p) + \max(p) - \gcd(p)(|p| - 1 - r(p))$.*

Proof. Let $d = \gcd(p)$. By Lemma 2.5, $r(p) = r(p/d)$, $d\, fw(p/d) = fw(p)$, while by definition of $p/d, \min(p) = d \min(p/d)$, $\max(p) = d \max(p/d)$ and $|p| = |p/d|$. It suffices therefore to prove the result for $p \in \mathscr{F}$ with $\gcd(p) = 1$, and this we shall do by induction on $\max(p)$. If $\gcd(p) = 1$ and $\max(p) = 1$, then $p = (1)$ and the result holds.

Suppose now that $\gcd(p) = 1$, $\max(p) > 1$, and the result holds for all elements of $\mathscr{F}$ with greatest common divisor 1 and smaller maximum entry. Since $|p| = 1$ would imply that $1 = \gcd(p) = \max(p) > 1$, it follows that $|p| > 1$. Consider $\hat{p}$, the totally reduced form of p. Since $\min(\hat{p}) = \min(p)$ and $\max(\hat{p}) \leq \max(p)$, it follows that if we are able to prove that $fw(\hat{p}) \leq \min(\hat{p}) + \max(\hat{p}) - (|\hat{p}| - 1)$, then by Proposition 4.10, $fw(p) = fw(\hat{p}) \leq \min(\hat{p}) + \max(\hat{p}) - (|\hat{p}| - 1) \leq \min(p) + \max(p) - (|p| - r(p) - 1)$, as required. Thus we may assume that p is totally reduced, and we are to prove that $fw(p) \leq \min(p) + \max(p) - (|p| - 1)$. Since p is totally reduced, it is in particular trim, and so $fw(p) = f(p) = \min(p) + f(p')$. If p' is not trim, then by Proposition 2.13, $f(p) = 2\min(p)$, and since $\max(p) - \min(p) \geq |p| - 1$, $fw(p) = f(p) = 2\min(p) \leq \min(p) + \max(p) - (|p| - 1)$, as required. Thus we may assume that p' is trim, so $fw(p') = f(p')$. Furthermore, by Proposition 4.3, the fact that p is totally reduced means in particular that no entry of p is a multiple of $\min(p)$. Thus $\min(p) \neq p_j - \min(p)$ for every j with $2 \leq j \leq |p|$, and so $|p'| = |p|$. Apply the induction hypothesis to p' to obtain $fw(p) = \min(p) + f(p') = \min(p) + fw(p') \leq \min(p) + \min(p') + \max(p') - (|p'| - 1 - r(p')) = \min(p) + \min(p') + \max(p') - (|p| - 1 - r(p'))$. It will suffice to prove that $\min(p') + \max(p') + r(p') \leq \max(p)$. Let us first treat the case when p' is totally reduced; that is, $r(p') = 0$. There are three subcases to consider. If $\min(p) \leq p_2 - \min(p)$, then $\min(p') = \min(p)$ and $\max(p') = \max(p) - \min(p)$, so $\min(p') + \max(p') + r(p') = \max(p)$. Suppose now that $p_2 - \min(p) < \min(p) \leq \max(p) - \min(p)$, so that $\min(p') = p_2 - \min(p)$ and $\max(p') = \max(p) - \min(p)$. Then $\min(p') + \max(p') + r(p') = p_2 - \min(p) + \max(p) - \min(p) = \max(p) + p_2 - 2\min(p)$. But from $p_2 - \min(p) < \min(p)$, we have $p_2 - 2\min(p) < 0$ and so $\max(p) + p_2 - 2\min(p) < \max(p)$. Finally, suppose that $\max(p) - \min(p) < \min(p)$, so that $\min(p') = p_2 - \min(p)$ and $\max(p') = \min(p)$, which implies that $\min(p')+\max(p')+r(p') = p_2-\min(p)+\min(p) = p_2 \leq \max(p)$, as required.

We now treat the case when p' is not totally reduced, so that $r(p') > 0$. Let j be such that $p_j - \min(p)$ is redundant in p'. Since p is totally reduced and therefore reduced, $\min(p) \neq p_i - \min(p)$ for every i, and so in particular, $\min(p) \neq p_j - \min(p)$. Consider first the possibility that $\min(p) < p_j - \min(p)$. Then $(p|_j)' = p'|_j$ has maximum entry $p_j - \min(p)$ and is not trim, so by Proposition 2.13, $f(p|_j) = 2\min(p)$. As p is totally reduced, $p|_j$ is trim, and since $j > 1$, Proposition 2.13 implies that $f(p|_j) > \max(p|_j) = p_j$. Thus $2\min(p) > p_j$, contradicting our assumption that $p_j - \min(p) > \min(p)$. Hence $p_j - \min(p) < \min(p)$ (which implies that $j > 2$ since $p_2 - \min(p) = \min(p')$), and so we have established that if j is any index such that $p_j - \min(p)$ is redundant in p', then $p_2 < p_j < 2\min(p)$. Thus $r(p') \leq |\{j \mid p_2 < p_j < 2\min(p)\}| + 1$, where we have added 1 to acknowledge that $\min(p)$ might be redundant in p'. Thus $0 < r(p') \leq (2\min(p) - p_2 - 1) + 1 = 2\min(p) - p_2$, and so $p_2 < 2\min(p)$, which means that $p_2 - \min(p) = \min(p')$. There are two subcases to consider.

Subcase 1: $\min(p) < \max(p) - \min(p)$. Then $\max(p') = \max(p) - \min(p)$, and so $\min(p')+\max(p')+r(p') \leq p_2-\min(p)+\max(p)-\min(p)+2\min(p)-p_2 = \max(p)$.

Subcase 2: $\min(p) > \max(p) - \min(p)$. Then $\max(p') = \min(p)$, and $\max(p) < 2\min(p)$. Since p' is trim, $\max(p')$ is not type II redundant in p' and thus in this case, we have $r(p') \leq |\{j \mid p_2 < p_j < 2\min(p)\}| = |\{j \mid p_2 < p_j \leq \max(p)\}| = |p| - 2$ and so $\min(p')+\max(p')+r(p') \leq p_2-\min(p)+\min(p)+|p|-2 = p_2+|p|-2 \leq \max(p)$. This completes the proof of the inductive step. □

Corollary 4.12. *Let $p \in \mathscr{F}$ be totally reduced. Then $fw(p) \leq \min(p) + \max(p) - \gcd(p)(|p| - 1)$.*

References

[1] M. G. Castelli, F. Mignosi, A. Restivo, Fine and Wilf's theorem for three periods and a generalization of Sturmian words, Theor. Comput. Sci. 218(1999), 83-94.

[2] N. J. Fine and H. S. Wilf, Uniqueness theorems for periodic functions, Proc. Amer. Math. Soc., 16(1965) 109–114.

[3] J. Justin, On a paper by Castelli, Mignosi, Restivo, Theor. Inform. Appl., 34(2000), 373–377.

[4] R. C. Lyndon and M. P. Schützenberger, The equation $a^M = b^N c^P$ in a free group, Michigan Math. J., 9(1962), 289–298.

[5] R. Tijdeman and L. Zamboni, Fine and Wilf words for any periods, Indag. Math., 14(2003), 135–147.

ON THE LEAST SIGNIFICANT 2-ADIC AND TERNARY DIGITS OF CERTAIN STIRLING NUMBERS

Tamás Lengyel
Mathematics Department, Occidental College, Los Angeles, California
lengyel@oxy.edu

Received: 1/4/13, Accepted: 7/1/13, Published: 8/12/13

Abstract

Our main goal is to effectively calculate the p-ary digits of certain Stirling numbers of the second kind. We base our study on an observation regarding these numbers: as m increases, more and more p-adic digits match in $S(i(p-1)p^m, k)$ with integer $i \geq 1$.

1. Introduction

Let n and k be positive integers, p be a prime, $d_p(k)$ and $\nu_p(k)$ denote the sum of digits in the base p representation of k and the highest power of p dividing k, i.e., the p-adic order of k, respectively. For the rational n/k we set $\nu_p(n/k) = \nu_p(n) - \nu_p(k)$. In 1808, Legendre showed

Lemma 1. ([2]) *For any positive integer k, we have $\nu_p(k!) = (k - d_p(k))/(p-1)$.*

We define the 2-free part of $k!$ (or unit factor of $k!$ with respect to 2), b_k, as

$$k! = 2^{k-d_2(k)} b_k,$$

or more explicitly,

$$b_k = \prod_{\substack{3 \leq p \leq k \\ p \text{ prime}}} p^{\frac{k-d_p(k)}{p-1}}.$$

In general, b_k is the p-free part of $k!$ (or unit factor of $k!$ with respect to p), i.e., $k! = p^{\frac{k-d_p(k)}{p-1}} b_k$ with

$$b_k = \prod_{\substack{2 \leq p' \leq k \\ p' \neq p \text{ and prime}}} p'^{\frac{k-d'_p(k)}{p'-1}}.$$

We have the identity (cf. [1]) for the Stirling numbers of the second kind

$$S(n,k) = \frac{1}{k!} \sum_{j=0}^{k} \binom{k}{j} (-1)^j (k-j)^n.$$

Our main goal is to effectively calculate the p-ary digits of certain Stirling numbers of the second kind. For example, if $k=2$ then $S(m,2)=2^{m-1}-1, m\geq 2$; thus, the binary representation consists of all ones. We try to find similar properties for other values of k. We base our study on an observation (cf. [6]) regarding these numbers: as m increases, more and more p-adic digits match in $S(i(p-1)p^m,k)$ with integer $i\geq 1$.

We claim the main results (cf. Theorems 2, 4, and 5) in Section 2, and illustrate and prove them in Sections 3-5. We discuss the case with $p=2$ in Sections 3 and 4 and derive additional results (cf. Lemmas 8 and 9). A general approach is presented in Section 4. Options and limitations (cf. Theorems 12-18 based on [4] and [6]) for other primes are discussed in Section 5. Two examples are provided to demonstrate the cases of 2-adic and ternary digits.

2. Main Results

First, we deal with the binary digits and obtain

Theorem 2. *With the above introduced notation,*

$$S(2^m i,k)\equiv\frac{1}{k!}\sum_{\substack{j=0\\ k-j\ odd}}^{k}\binom{k}{j}(-1)^j(k-j)^{2^m i}$$

$$\equiv 2^{d_2(k)-1}(-1)^{k-1}b_k^{2^m-1} \bmod 2^{m+2-k+d_2(k)} \tag{2.1}$$

for $m+2\geq k-d_2(k)$, $m\geq 2$, and $i\geq 1$.

Remark 3. Recall that (2.1) implies that $\nu_2(S(2^m i,k))=d_2(k)-1$ if $d_2(k)-1< m+2-k+d_2(k)$, i.e., $m\geq k-2$, cf. [5] and [7] for the generalized version.

We make the calculation more explicit in Theorem 4 and generalize it for $p=3$ in Theorem 5, and in Theorems 12 and 17, in general.

We set $u_k\equiv b_k\equiv b_k^{-1} \bmod 4$ to be the least positive residue of the 2-free part b_k of $k!$ modulo 4 which is the same as that of its inverse modulo 4,

$$c_k=\begin{cases}-1, & \text{if } u_k=3,\\ +1, & \text{if } u_k=1,\end{cases}$$

and

$$a_k=\begin{cases}\lceil\frac{b_k}{4}\rceil, & \text{if } u_k=3,\\ \lceil\frac{b_k}{4}\rceil-1, & \text{if } u_k=1,\end{cases} \tag{2.2}$$

which yields that $b_k=4a_k+c_k$. We end up with the following theorem that gives $S(2^m i,k)$ explicitly, modulo a high power of two, and in terms of k, m, and r ($r\geq 0$ integer).

Theorem 4. *With the above introduced notation, for $k \geq 3$ we have*

$$S(2^m i, k) \equiv 2^{d_2(k)-1}(-1)^{k-1} c_k \sum_{j=0}^{r} (-4c_k a_k)^j \bmod 2^{e(m,k,r)} \tag{2.3}$$

with $e(m,k,r) = \min\{m+2-k+d_2(k), (r+1)(2+\nu_2(a_k)) + d_2(k) - 1\}$.

With $p = 3$, we set $u_k \equiv b_k \equiv b_k^{-1} \bmod p$ to be the least positive residue of the p-free part b_k of $k!$ modulo p which is the same that of its inverse modulo p,

$$c_k = \begin{cases} -1, & \text{if } u_k = p-1, \\ +1, & \text{if } u_k = 1, \end{cases}$$

and

$$a_k = \begin{cases} \lceil \frac{b_k}{p} \rceil, & \text{if } u_k = p-1, \\ \lceil \frac{b_k}{p} \rceil - 1, & \text{if } u_k = 1, \end{cases}$$

which yields that $b_k = p \cdot a_k + c_k$. We get that

Theorem 5. *For $p = 3$ and $k \equiv 2$ or $4 \pmod 6$, we have*

$$S(i(p-1)p^m, k) \equiv p^{\frac{d_p(k)}{p-1}-1}(-1)^{\frac{kp}{p-1}} c_k \sum_{j=0}^{r} (-pc_k a_k)^j \bmod p^{e(m,k,r)} \tag{2.4}$$

where

$$\begin{aligned} e(m,k,r) = \min\{&m+1-\frac{k-d_p(k)}{p-1}, m+1+\nu_p(a_k)+\frac{d_p(k)}{p-1}-1, \\ &(r+1)(1+\nu_p(a_k)) + \frac{d_p(k)}{p-1} - 1\}. \end{aligned}$$

3. Proof of Theorem 2

We need a well-known theorem and two lemmas.

Theorem 6. (Kummer, 1852) *The power of a prime p that divides the binomial coefficient $\binom{n}{k}$ is given by the number of carries when we add k and $n-k$ in base p.*

The first lemma is an improvement of the Fermat–Euler Theorem which claims only that $t^{2^{m+1}} \equiv 1 \bmod 2^{m+2}$ for $p = 2, m \geq 0$, and $t \geq 1$ odd.

Lemma 7. (Lemma 3 in [3]) *For any integer $m \geq 1$ and any odd integer t,*

$$t^{2^m} \equiv 1 \bmod 2^{m+2}.$$

This lemma can be proven by induction on m and further generalized to higher 2-power moduli (cf. [3]). The following lemma is an improvement of the well-known congruence $\binom{p^t-1}{j} \equiv (-1)^j \bmod p, 0 \le j \le p^t - 1$ for prime p and $t \ge 1$ integer.

Lemma 8. *If p is a prime, $(a,p) = 1$, $t \ge 1$, and $1 \le j \le p^t - 1$, then*

$$\nu_p\left(\binom{ap^t}{j}\right) = t - \nu_p(j) \tag{3.1}$$

and

$$\binom{ap^t - 1}{j} \equiv (-1)^j \bmod p^{t-\lfloor \log_p j \rfloor}.$$

Proof. Clearly, identity (3.1) is true by Theorem 6. Using the fact that $\binom{ap^t-1}{0} = 1$ and

$$\binom{ap^t}{j} = \binom{ap^t - 1}{j-1} + \binom{ap^t - 1}{j},$$

it implies that

$$\binom{ap^t - 1}{j} \equiv (-1)^j \bmod p^{t-\lfloor \log_p j \rfloor}$$

by step-by-step increasing j from $j = 1$ on. □

Proof of Theorem 2. The proof relies on the fact that terms with $k - j$ even will not contribute to the congruence since $2^m i \ge m + 2$ as $m \ge 2$, and on Lemma 7, since

$$\frac{1}{k!} \sum_{\substack{j=0 \\ k-j \text{ odd}}}^{k} \binom{k}{j} (-1)^j (k-j)^{2^m i} \equiv \frac{1}{k!} (-1)^{k-1} \sum_{\substack{j=0 \\ k-j \text{ odd}}}^{k} \binom{k}{j}$$

$$\equiv \frac{(-1)^{k-1} 2^{k-1}}{2^{k-d_2(k)} b_k} \bmod 2^{m+2-k+d_2(k)}.$$

Note that since b_k is odd, $b_k^{-1} \equiv b_k^{2^m-1} \bmod 2^{m+2}$ by Lemma 7. □

We note that it is easy to see that

$$S(n,5) = \frac{1}{24}(5^{n-1} - 4^n + 2 \cdot 3^n - 2^{n+1} + 1) \tag{3.2}$$

holds which yields

$$S(2^m i, 5) \equiv 2 \cdot 15^{2^m - 1} \bmod 2^{m-1} \tag{3.3}$$

for $i, m \ge 1$. Indeed, we have

$$3^{-1} \equiv 3^{2^m - 1} \equiv 3^{2^m i - 1} \bmod 2^{m+2},$$

$$5^{-1} \equiv 5^{2^m - 1} \equiv 5^{2^m i - 1} \bmod 2^{m+2}$$

and

$$S(2^m i,5) \equiv \frac{1}{8\cdot 3}\frac{1}{5}(5^{2^m i}+10\cdot 3^{2^m i}+5) \equiv \frac{1}{15}\frac{1}{8}16 \equiv 2\cdot 15^{2^m-1} \bmod 2^{m-1}$$

by identity (3.2) and Lemma 7, if $m \geq 1$ and $i \geq 1$, with direct calculations and without using Theorem 2. Moreover, we get

Lemma 9. *For any integer $r \geq 0$ and $i, m \geq 1$, we have*

$$S(2^m i,5) \equiv -2\sum_{j=0}^{r} 2^{4j} \bmod 2^{\min\{m-1,4r+5\}}. \tag{3.4}$$

Proof of Lemma 9. In fact, the statement holds if $2^m i < 5$. Otherwise, we rewrite

$$\begin{aligned}15^{2^m-1} &= (4^2-1)^{2^m-1} = -1+\binom{2^m-1}{1}4^2-\binom{2^m-1}{2}4^4+\cdots\\ &\equiv -\sum_{j=0}^{r}2^{4j} \bmod 2^{\min\{m+4,4r+4\}}\end{aligned}$$

by Lemma 8, which already implies (3.4) by (3.3) since $m+2-k+d_2(k)=m-1$. □

The congruence (3.4) guarantees that the binary representation of $S(2^m i,5)$ ends in (0111)*011110 if m is large enough. (With d being any finite word formed over the alphabet $\{0,1\}$, $(d)^*$ denotes any finite number $t, t \geq 0$, of copies of the "word" d.) If $r=0$ and $m \geq 6$ then we have

$$S(2^m i,5) \equiv 30 \bmod 32.$$

If $r \geq (m-6)/4$ then the congruence (3.4) turns into

$$S(2^m i,5) \equiv -2\sum_{j=0}^{r} 2^{4j} \bmod 2^{m-1},$$

and the terms beyond $j = \lceil (m-6)/4 \rceil$ effectively do not contribute to the sum.

4. 2-adic Digits: A General Approach for Effective Calculation and the Proof of Theorem 4

If $k=5$ then we get $d_2(5)=2, b_5=15$ and $S(2^m i,5)$ satisfies congruence (3.3) by Theorem 2. For larger values of k, we use (4.1) below since we do not need the exact value of b_k. In fact, to effectively calculate $S(2^m i,k)$ modulo a large 2-power, it suffices to use b_k modulo that 2-power. It can be calculated by the congruence

$$b_k = \frac{k!}{p^{\sum_{j\geq 1}\lfloor \frac{k}{p^j}\rfloor}} \equiv \delta^{\sum_{j\geq q}\lfloor \frac{k}{p^j}\rfloor}\prod_{j\geq 0}(K_j!)_p \bmod p^q, \tag{4.1}$$

with $\delta = \delta(p^q) = -1$ except if $p = 2, q \geq 3$ when $\delta = 1$, K_j is the least positive residue of $\lfloor k/p^j \rfloor \pmod{p^q}, 0 \leq j \leq d$, if $p^d \leq k < p^{d+1}$, and

$$(K!)_p = \frac{K!}{p^{\lfloor \frac{K}{p} \rfloor} \lfloor \frac{K}{p} \rfloor !}$$

is the product of those positive integers not exceeding K that are not divisible by p; cf. [2, Proposition 1, p8]. With $p = 2$, we have $\delta = 1$ if q is large enough. This implies that

$$b_k \equiv \prod_{j \geq 0} (K_j!)_2 \bmod 2^q.$$

Now we can gain a more in-depth look at the binary digits of $S(2^m i, k)$ by evaluating the right-hand side of (2.1) more effectively via Theorem 4.

Proof of Theorem 4. In a similar fashion to the case with $k = 5$ and depending upon $u_k \pmod 4$, we rewrite

$$\begin{aligned} b_k^{2^m-1} &= (4a_k + c_k)^{2^m-1} = c_k + \binom{2^m-1}{1}(4a_k)c_k^2 + \binom{2^m-1}{2}(4a_k)^2(c_k)^3 + \cdots \\ &\equiv c_k \sum_{j=0}^{r} (-4a_k c_k)^j \bmod 2^{\min\{m+2+\nu_2(a_k),(r+1)(2+\nu_2(a_k))\}} \end{aligned}$$

by Lemma 8, which already implies (2.3) by Theorem 2 since $\min\{m + 2 - k + d_2(k), m + 2 + \nu_2(a_k) + d_2(k) - 1, (r+1)(2+\nu_2(a_k)) + d_2(k) - 1\} = \min\{m + 2 - k + d_2(k), (r+1)(2+\nu_2(a_k)) + d_2(k) - 1\}$. □

Example 10. For $k = 3, 4, 5$, and 7, we get $b_3 = b_4 = 3, b_5 = 15, b_7 = 315$, $u_k = 3$, $c_k = -1$, $a_3 = a_4 = 1, a_5 = 4$, and $a_7 = 79$, which yield that

$$S(2^m i, 3) \equiv -2 \sum_{j=0}^{r} 4^j \bmod 2^{\min\{m+1, 2(r+1)+1\}},$$

$$S(2^m i, 4) \equiv \sum_{j=0}^{r} 4^j \bmod 2^{\min\{m-1, 2(r+1)\}},$$

$$S(2^m i, 5) \equiv -2 \sum_{j=0}^{r} 16^j \bmod 2^{\min\{m-1, 4(r+1)+1\}},$$

in agreement with (3.4), and

$$S(2^m i, 7) \equiv -2^2 \sum_{j=0}^{r} 316^j \bmod 2^{\min\{m-2, 2(r+1)+2\}}.$$

On the other hand, if $k = 6$ then $b_6 = 45$, $u_6 = 1$, $c_6 = 1$, $a_6 = 11$, and

$$S(2^m i, 6) \equiv -2^2 \sum_{j=0}^{r} (-44)^j \bmod 2^{\min\{m-2, 2(r+1)+1\}}.$$

Remark 11. Note that the "best use" of the congruence (2.3) comes with values of a_k that are powers of two, e.g., if $k = 3, 4, 5$, etc. It will be interesting to see the general solution to this problem, i.e., find all k so that a_k, which is derived from the 2-free part b_k of $k!$ by (2.2), is a power of two. Indeed, beyond the small cases, we look for any $k \geq 4$, for which $k!$ is the difference or sum of two powers of two (depending on the sign of c_k), or equivalently, whose binary representation is of the form $1(0)^*1(0)^*0$ or $1(1)^*(0)^*0$. This follows by the identity $k! = 2^{k-d_2(k)} b_k = 2^{k-d_2(k)}(4a_k + c_k)$. (Of course, for $k \geq 2$, we get an even $k!$ so it must end with a binary zero.)

5. Other primes

As m increases, more and more p-adic digits match in $S(i(p-1)p^m, k)$. However, to effectively calculate these matching digits we need another approach. We rely on papers [4] and [6]. We need the following combination of Lemma 5 and Theorem 3 of [4]. This helps in generalizing Theorem 4 for odd primes if k is divisible by $p-1$.

Theorem 12. ([4]) *For any odd prime p, integer t, $n = i(p-1)p^m$, $1 \leq k \leq n$, and $m > \frac{k}{p-1} - 2$, we have*

$$(-1)^{k+1} k! S(n,k) \equiv \sum_{p|i} \binom{k}{i} (-1)^i \bmod p^{m+1} \tag{5.1}$$

and

$$\sum_{i \equiv t \bmod p} \binom{k}{i} (-1)^i \equiv \begin{cases} (-1)^{\frac{k}{p-1}-1} p^{\frac{k}{p-1}-1} \bmod p^{\frac{k}{p-1}}, & \text{if } k \text{ is divisible by } p-1, \\ 0 \bmod p^{\lfloor \frac{k}{p-1} \rfloor}, & \text{otherwise.} \end{cases} \tag{5.2}$$

Therefore, if k is divisible by $p-1$ then

$$S(n,k) \equiv p^{\frac{d_p(k)}{p-1}-1} (-1)^{\frac{kp}{p-1}} b_k^{-1} \bmod p^{\min\{m+1-\frac{k-d_p(k)}{p-1}, \frac{d_p(k)}{p-1}\}}$$

where b_k is the p-free part of $k!$ as defined in the introduction and by the Fermat–Euler Theorem

$$b_k^{-1} \equiv b_k^{(p-1)p^m - 1} \bmod p^{m+1}.$$

Remark 13. Note that the p-adic order of $S(i(p-1)p^m,k)$ does not depend on i and m. This does not exclude the possibility that by increasing m we can get more insight into the base p representation of $S(i(p-1)p^m,k)$. Indeed, if $p=2$ then (2.1) provides us with the right tool since $\sum_{2|i}\binom{k}{i}(-1)^i = 2^{k-1}$, and it leads to Theorem 4. However, in general, increasing m does not help in getting more p-ary digits in a computationally effective way, for (5.2) cannot be significantly improved; although, according to Theorem 17, we get more and more matching digits in $S(i(p-1)p^m,k)$ and $S(i(p-1)p^{m+1},k)$ (starting with the least significant bit). We can avoid the use of (5.2) if a closed form exists for $\sum_{p|i}\binom{k}{i}(-1)^i$ in (5.1), at least for some k, e.g., if $p=3$ or 5.

In fact, for example, if k is even and $3 \nmid k$, we get that $\sum_{3|i}\binom{k}{i}(-1)^i = (-1)^{k/2+1}3^{k/2-1}$. Theorem 5 provides us with a tool to calculate the ternary digits of $S(i(p-1)p^m,k)$ if $k \equiv 2$ or $4 \pmod 6$. Its proof is a straightforward generalization of that of Theorem 4. We demonstrate its use in the next example.

Example 14. If $p=3$ then $u_k \equiv b_k \equiv b_k^{-1} \bmod 3$ is the least positive residue of the 3-free part b_k of $k!$ modulo 3 which is the same as that of its inverse modulo 3. For instance, if $k=4$ we get then $b_4=8$, $u_k=2$, $c_4=-1$ and

$$a_4 = \left\lceil \frac{b_4}{3} \right\rceil = 3,$$

which yields that $b_4 = 9-1$. We obtain that

$$S(2i\cdot 3^m,4) \equiv -\sum_{j=0}^{r} 3^{2j} \bmod 3^{e(m,4,r)} \tag{5.3}$$

with

$$\begin{aligned} e(m,4,r) &= \min\{m+1-\frac{4-d_3(4)}{2}, m+1+\nu_3(3)+\frac{d_3(4)}{2}-1, \\ &\quad (r+1)(1+\nu_3(3))+\frac{d_3(4)}{2}-1\} \\ &= \min\{m, 2(r+1)\}. \end{aligned}$$

This implies that $S(2i\cdot 3^m,4)$ ends in $(12)^*122$ in base 3.

Remark 15. Since $k! = 3^{\frac{k-d_3(k)}{2}} b_k = 3^{\frac{k-d_3(k)}{2}}(3a_k+c_k)$ we get the "best use" of Theorem 5 when a_k is a power of three, i.e., when $k!$ is the difference or sum of two powers of three. For example, in Example 14, $4! = 24 = 3^3-3$ leads to (5.3).

Remark 16. In a similar fashion to the case with $p=3$, if $p=5$ then we can use the fact that $\sum_{5|i}\binom{k}{i}(-1)^i$ can be expressed explicitly in terms of Fibonacci or Lucas numbers, with a formula depending on k modulo 20 (cf. [4]).

The idea of getting more p-ary digits of $S(i(p-1)p^m, k)$ by increasing m is well supported and the rate of increase is made effective by the following theorem which is based on Theorems 11 and 14 of [6]. This theorem can be used in getting the digits successively although not in a direct fashion as in (2.3), (2.4), and (5.3).

Theorem 17. *Let $p \geq 2$ be a prime, $c, n, k \in \mathbb{N}$ with $1 \leq k \leq p^n$ and $(c,p) = 1$, and u be a nonnegative integer, then*

$$\nu_p(S(cp^{n+1} + u, k) - S(cp^n + u, k)) \geq n - \lceil \log_p k \rceil + 2.$$

It was also conjectured in Conjecture 2 in [6] that for $n, k \in \mathbb{N}$, $3 \leq k \leq 2^n$, and $c \geq 1$ odd integer, we have

$$\nu_2(S(c2^{n+1}, k) - S(c2^n, k)) = n + 1 - f(k)$$

for some function $f(k)$ which is independent of n (for any sufficiently large n). In fact, for small values of k, numerical experimentation suggests that

$$f(k) = 1 + \lceil \log_2 k \rceil - d_2(k) - \gamma(k),$$

with $\gamma(4) = 2$ and otherwise it is zero except if k is a power of two or one less, in which cases $\gamma(k) = 1$. This would imply that $f(k) \geq 0$, cf. [6].

In connection with Theorem 12, we note that if k is divisible by $p-1$ then k/p is not an odd integer. On the other hand, if k/p is an odd integer then we observe a behavior which is somewhat different from that of Theorem 12.

Theorem 18. (Theorem 2 in [4]) *For any odd prime p, if k/p is an odd integer then $\nu_p(k!S(i(p-1)p^m, k)) > m$.*

Acknowledgment The author wishes to thank Gregory P. Tollisen and the referee for helpful comments that improved the presentation of the paper.

References

[1] L. Comtet, Advanced combinatorics, D. Reidel Publishing Co., Dordrecht, 1974.

[2] A. Granville, Arithmetic properties of binomial coefficients. I. Binomial coefficients modulo prime powers, in Organic mathematics (Burnaby, BC, 1995), volume 20 of CMS Conf. Proc., 253–276, Amer. Math. Soc., Providence, RI, 1997. Electronic version (a dynamic survey): http://www.dms.umontreal.ca/~andrew/Binomial/.

[3] C. D. Bennett and E. Mosteig, Congruence classes of 2-adic valuations of Stirling numbers of the second kind, arXiv:1204.6361v1, April 2012.

[4] I. M. Gessel and T. Lengyel, On the order of Stirling numbers and alternating binomial coecient sums, Fibonacci Quart. **39** (2001), 444–454.

[5] T. Lengyel, On the divisibility by 2 of the Stirling numbers of the second kind, Fibonacci Quart. **32** (1994), 194–201.

[6] T. Lengyel, On the 2-adic order of Stirling numbers of the second kind and their differences, 21st International Conference on Power Series and Algebraic Combinatorics (FPSAC 2009), Hagenberg, Austria, Discrete Math. Theor. Comput. Sci. Proceedings **AK**, 561–572, 2009. Downloadable from: `http://www.dmtcs.org/dmtcs-ojs/index.php/proceedings/article/view/dmAK0147/`.

[7] T. Lengyel, Alternative proofs on the 2-adic order of Stirling numbers of the second kind, Integers **10** (2010), 453–463.

ON THE DENSITY OF HAPPY NUMBERS

J. Gilmer
Department of Mathematics, Rutgers, Piscataway, New Jersey
jmgilmer@math.rutgers.edu

Received: 1/4/13, Accepted: 7/1/13, Published: 8/12/13

Abstract

The happy function $H : \mathbb{N} \to \mathbb{N}$ sends a positive integer to the sum of the squares of its digits. A number x is said to be happy if the sequence $\{H^n(x)\}_{n=1}^{\infty}$ eventually reaches 1 (here $H^n(x)$ denotes the nth iteration of H on x). A basic open question regarding happy numbers is what bounds on the density can be proved. This paper uses probabilistic methods to reduce this problem to experimentally finding suitably large intervals containing a high (or low) density of happy numbers as a subset. Specifically, we show that $\bar{d} > .18577$ and $\underline{d} < .1138$, where $\bar{d}$ and $\underline{d}$ denote the upper and lower density of happy numbers respectively. We also prove that the asymptotic density does not exist for several generalizations of happy numbers.

1. Introduction

It is well known that if you iterate the process of sending a positive integer to the sums of the squares of its digits, you eventually arrive at either 1 or the cycle

$$4 \to 16 \to 37 \to 58 \to 89 \to 145 \to 42 \to 20 \to 4.$$

If we change the map, instead sending an integer to the sum of the cubes of its digits, then there are 9 different possible cycles (see Section 5.2.1). Many generalizations of these kinds of maps have been studied. For instance, [3] considered the map which sends an integer n to the sum of the eth power of its base-b digits. In this paper, we study a more general class of functions.

Definition 1.1. Let $b > 1$ be an integer, and let h be a sequence of b non-negative integers such that $h(0) = 0$ and $h(1) = 1$. Define $H : \mathbb{Z}^+ \to \mathbb{Z}^+$ to be the following function: for $n \in \mathbb{Z}^+$, with base-b representation $n = \sum_{i=0}^{k} a_i b^i$, $H(n) := \sum_{i=0}^{k} h(a_i)$. We say H is the b-happy function with digit sequence h.

As a special case, the b-happy function with digit sequence $\{0, 1, 2^e, \ldots, (b-1)^e\}$ is called the (e, b)-function.

Definition 1.2. Let H be any b-happy function and let $C \subseteq \mathbb{N}$. We say $n \in \mathbb{N}$ is type-C if there exists k such that $H^k(n) \in C$.

For the (e, b)-function we refer to type-$\{1\}$ integers as (e, b)-happy.

Fix a b-happy function H and let $\alpha := \max_{i=0,\ldots,b-1} \big(H(i)\big)$. If n is a d-digit integer in base-b, then $H(n) \leq \alpha d$. If d^* is the smallest $d \in \mathbb{N}$ such that $\alpha d < b^{d-1}$, then for all n with $d \geq d^*$ digits, $H(n) \leq \alpha d < b^{d-1} \leq n$. This implies the following

Fact 1.3. For all $n \in \mathbb{N}$, there exists an integer i such that $H^i(n) < b^{d^*} - 1$.

Moreover, to find all possible cycles for a b-happy function, it suffices to perform a computer search on the trajectories of the integers in the interval $[0, b^{d^*} - 1]$.

Richard Guy asks a number of questions regarding $(2, 10)$-happy numbers and their generalizations, including the existence (or not) of arbitrarily long sequences of consecutive happy numbers and whether or not the asymptotic density exists [4, problem E34]. To date, there have been a number of papers in the literature addressing the former question ([1],[3],[6]). This paper addresses the latter question. Informally, our main result says that if the asymptotic density exists, then the density function must quickly approach this limit.

Theorem 1.4. *Fix a b-happy function H. Let I be a sufficiently large interval and let $S \subseteq I$ be a set of type-C integers. If $\frac{|S|}{|I|} = d$, then the upper density of type-C integers is at least $d\,(1 - o(1))$.*

Note as a corollary we can get an upper bound on the lower density by taking C to be the union of all cycles except the one in which we are interested. In Sections 3 and 4 we will define explicitly what constitutes a sufficiently large interval and provide an expression for the $o(1)$ term. Using Theorem 1.4, one can prove the asymptotic density of (e, b)-happy numbers (or more generally type-$\{C\}$ numbers) does not exist by finding two large intervals I_1, I_2 for which the density in I_1 is large and in I_2 is small. In the case of $(2, 10)$-happy numbers, taking $I_1 = [10^{403}, 10^{404} - 1]$ and $I_2 = [10^{2367}, 10^{2368} - 1]$, we show that $\bar{d} \geq .185773(1 - 10^{-49})$ and $\underline{d} \leq .11379(1 + 10^{-100})$ respectively.

We also show that the asymptotic density does not exist for 8 of the cycles for the $(3, 10)$-function (see Section 5). It should be noted that our methods only give one sided bounds. In an earlier version of this manuscript, we asked if $\bar{d} < 1$ for $(2, 10)$-happy numbers. Recently, [5] has announced a proof of this. Specifically, he proves that $.1962 < \bar{d} < .38$, and $0.002937 < \underline{d} < .1217$.

2. Preliminaries

Unless otherwise noted, we regard an interval $I = [a, b]$ as the integer interval $\{n \in \mathbb{Z}^+ : a \leq n \leq b\}$ where, in general, $a, b \in \mathbb{R}$. We denote $|I|$ to be the cardinality of this set. We also denote $[n]$ as the set $\{0, 1, \ldots, n\}$. Throughout this section let H denote an arbitrary b-happy function.

Definition 2.1. Let I be an interval and Y the random variable uniformly distributed amongst the set of integers in I. Then we say the random variable Y is induced by the interval I.

Definition 2.2. The type-C density of an integer interval I is defined to be the quantity

$$d_C(I) := \frac{|\{n \in I : n \text{ is type-}C\}|}{|I|}.$$

Observation 2.3. If Y is the random variable induced by an interval I, then

$$d_C(I) = \mathbb{P}\big(H(Y) \text{ is type-C}\big).$$

Usually, we take C to be one of the cycles arising from a b-happy function H. However, if we wish to upper bound the lower density of type-C integers, then we study the density of type-C' integers, where C' is the union of all cycles except C.

2.1. The Random Variable $H(Y_m)$

Consider the random variable Y_m induced by the interval $[0, b^m - 1]$, i.e., Y_m is a random m-digit number. If X_i is the random variable corresponding to the coefficient of b^i in the base-b expansion of Y_m, then

$$H(Y_m) = \sum_{i=0}^{m-1} H(X_i). \tag{1}$$

In this paper, we will be interested in the mean and variance of $H(Y_1)$ (i.e., the image of a random digit) which we refer to as the *digit mean* (μ) and *digit variance* (σ^2) of H. The random variables $H(X_i)$ in (1) are all independent and identically distributed (i.i.d.), thus,

$$\mathbf{E}[H(Y_m)] = \mu m \qquad \mathbf{Var}[H(Y_m)] = \sigma^2 m. \tag{2}$$

The random variable $H(Y_m)$ is equivalent to rolling m times a b-sided die with faces $0, 1, H(2), \ldots, H(b-1)$ and taking the sum. Since it is a sum of m i.i.d. random variables, it approaches a normal distribution as m gets large. Also, the distribution of $H(Y_m)$ is concentrated near the mean. This implies the following key insight:

Observation 2.4. The density of happy numbers amongst m-digit integers depends almost entirely on the distribution of happy numbers near μm.

2.2. Computing Densities

Let $P_{m,i} := \mathbb{P}\big(H(Y_m) = i\big)$. Then

$$P_{m,i} = \frac{\left|\{(a_1, a_2, \ldots, a_m) : a_k \in H([b-1]) \text{ and } \sum_{k=1}^{m} a_k = i\}\right|}{b^m}.$$

For fixed m, the sequence $\{P_{m,i}\}_{i=1}^{\infty}$ has generating function

$$f_m(x) = \sum_{i=0}^{\infty} P_{m,i} x^i = \left(\frac{1 + x + x^{H(2)} + \cdots + x^{H(b-1)}}{b}\right)^m. \tag{3}$$

This implies the following recurrence relation with initial conditions $P_{0,0} = 1$, and $P_{0,i} = 0$ for $i \in \mathbb{Z} - \{0\}$.

$$P_{m,i} = \frac{P_{m-1,i} + P_{m-1,i-1} + P_{m-1,i-H(2)} + \cdots + P_{m-1,i-H(b-1)}}{b}. \tag{4}$$

To see this, write $f_m(x) = \left(\frac{1+x+x^{H(2)}+\cdots+x^{H(b-1)}}{b}\right)^{m-1} \left(\frac{1+x+x^{H(2)}+\cdots+x^{H(b-1)}}{b}\right)$ and consider the coefficient of x^i.

If $\alpha = \max_{i=0,\ldots,b-1} \big(H(i)\big)$, then $H(Y_m) \subseteq [0, m\alpha]$. In particular, $P_{m,i} = 0$ if $i > m\alpha$. Using this fact combined with (4), we can implement the following simple algorithm for quickly calculating the type-C density of the interval $[0, b^m - 1]$.

1. First, using the recurrence (4), calculate $P_{m,i}$ for $i = 0, \ldots, m\alpha$.

2. Using brute force, find the type-C integers in the interval $[0, m\alpha]$.

3. Output $\sum_{\substack{i \in [0,m\alpha] \\ i \text{ type-C}}} P_{m,i}$.

Using this algorithm, calculating the density for large n becomes computationally feasible. Figure 1 graphs the density of $(2, 10)$-happy numbers $< 10^n$ for n up to 8000.

The peak near 10^{400} and valley near 10^{2350} will be used to imply the bounds obtained in this paper.

2.3. A Local Limit Law

The random variable $H(Y_m)$ approaches a normal distribution as m becomes large. The following theorem[1], presented in [2, p. 593], gives a bound.

[1] We quote a simpler version, with a minor typo corrected

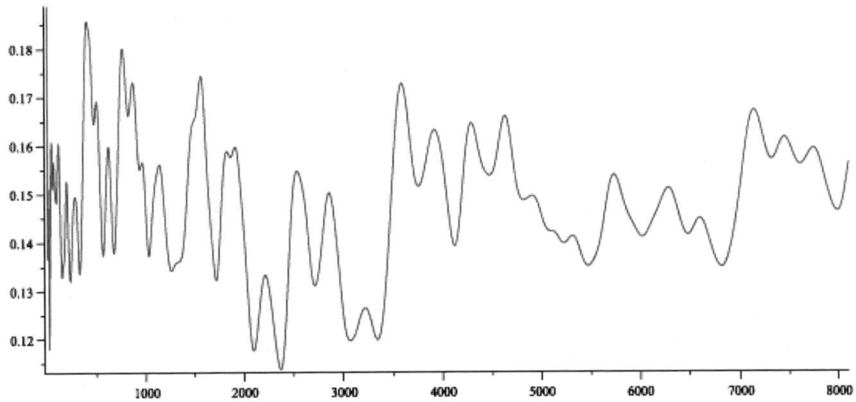

Figure 1: Relative Density of $(2, 10)$-Happy Numbers $< 10^n$

Theorem 2.5. (Local limit law for sums) *Let $X_1, \ldots, X_n$ be i.i.d. integer-valued variables with probability generating function (PGF) $B(z)$, mean μ, and variance σ^2, where it is assumed that the X_i are supported on $\mathbb{Z}^+$. Assume that $B(z)$ is analytic in some disc that contains the unit disc in its interior and that $B(z)$ is aperiodic with $B(0) \neq 0$. Then the sum,*

$$S_n := X_1 + X_2 + \cdots + X_n$$

satisfies a local limit law of the Gaussian type: for t in any finite interval, one has

$$\mathbb{P}(S_n = \lfloor \mu n + t\sigma\sqrt{n} \rfloor) = \frac{e^{-t^2/2}}{\sqrt{2\pi n}\sigma}\left(1 + O(n^{-1/2})\right).$$

Here aperiodic means that the $\gcd\{j : b_j > 0, j > 0\} = 1$, where $B(z) = \sum_{j=0}^{\infty} b_j z^j$ (or more informally, the digit sequence for H cannot all be divisible by some integer larger than 1). In our case, the PGF of the $H(X_i)$ is the polynomial

$$p(x) = \frac{x^{H(0)} + x^{H(1)} + \cdots + x^{H(b-1)}}{b}.$$

It is important in our definition of b-happy functions that we assume that $H(0) = 0$, and $H(1) = 1$. This guarantees that $p(x)$ is aperiodic and in particular that the above theorem applies for the sum $H(Y_m)$. As a consequence, for a fixed interval $[-T, T]$, if $i = \mu m + t\sigma\sqrt{m}$ for some $t \in [-T, T]$, then

$$P_{m,i} = \frac{e^{-t^2/2}}{\sqrt{2\pi m}\sigma}\left(1 + O\left(m^{-1/2}\right)\right).$$

The above error term, $O\left(m^{-1/2}\right)$, will prove to a technical difficulty which will be discussed later.

2.4. Overview of the Proof

The following heuristic will provide the general motivation for the proofs. Recall that the random variable Y_m is concentrated near its mean μm.

Observation 2.6. Suppose I is a large interval with type-C density d. Consider the choices of m such that the mean of $H(Y_m)$ is in the interval I; then for some choices of m we likely have

$$\mathbb{P}\big(H(Y_m) \text{ is type-}C\big) \geq d.$$

The key idea to turn this heuristic into a proof is to average over all reasonable choices of m in order to imply there is an m with the desired property.

We will use Theorem 2.5 to show that, for small k, $H(Y_m)$ and $H(Y_{m+k})$ have essentially the same distribution only shifted by a factor of μk. Thus, as k varies, the distributions $H(Y_{m+k})$ should uniformly cover the interval I. It is crucial here to use the fact that $H(Y_m)$ is locally normal, otherwise the proof will fail. For example, suppose all the happy numbers in I are odd. In this case, if $H(Y_m)$ is not locally normal and instead is supported on the even numbers for all m, then every shift $H(Y_{m+k})$ will miss all of the happy numbers in I.

Unfortunately, the fuzzy term in the local limit law prevents us from obtaining explicit bounds on the error (and any explicit bounds seem unsatisfactory for our purposes). Section 3 adds a necessary step, which is to construct an interval within $[b^{n-1}, b^n]$ with high type-C density for n arbitrarily large. The trick is to consider intervals of the form $I_k := [1^k 0^{n-k}, 1^k 0^m (b-1)^{n-k-m}]$ (here 1^k denotes k consecutive 1's). This solves the issue where $H(Y_m)$ and $H(Y_{m+k})$ are not exact shifts of each other, as the distributions induced by the I_k *are* exact shifts under the image of H. These distributions will uniformly cover the base interval I with much better provable bounds. The main result is presented in Section 4, the proof uses the local limit law with the result from Section 3.

3. Constructing Intervals

Throughout this section, if Y is a random variable and k is an integer, let $\tau_k(Y)$ denote the random variable $Y + k$.

Definition 3.1. We say an integer interval I is n-strict if $I \subseteq [b^{n-1}, b^n - 1]$ and $|I| = b^{3n/4}$.

The primary goal of this section is to construct n-strict intervals of high type-C density for arbitrarily large n.

Our choice of the definition of n-strict is only for the purpose of simplifying calculations, there is nothing special about the value $\frac{3}{4}$. In fact, any ratio $> \frac{1}{2}$ would work. Note if 4 does not divide n, then no n-strict intervals exist.

For the entirety of this section we will make the following assumptions:

- H is a b-happy function with digit mean μ and digit variance σ^2.
- We wish to lower bound the upper density of type-C integers for some $C \subset \mathbb{N}$.
- We have found, by computer search, an appropriate starting interval I_1, which is n_1-strict and has suitably large type-C density $d_C(I_1)$.

The results in this section apply only if this n_1 is sufficiently large, so we state here exactly how large n_1 must be so one knows where to look for the interval I_1. In particular, we say an integer n satisfies the bounds (**B**) if

B1: $4\left(1+3\mu+\sqrt{2}\sigma b^{5n/8}\right) \le b^{n-1}$,

B2: $\sqrt{3\mu b}\sigma \le b^{3n/8}$,

B3: $4\mu\left(3\mu+1+b^{3n/4}+2\sigma\mu^{-1/2}b^{5n/8}\right) \le b^{n-1}$.

Generally, n need not be too large to satisfy these bounds. For example, if H is a $(2,10)$-happy function, assuming $n > 13$ is enough to guarantee that it satisfies bounds (**B**). This is well within the scope of the average computer as it is possible to compute the density of type-C integers in $[0, b^n - 1]$ for n up to (and beyond) 1000 using the algorithm in Section 2. These bounds are necessary in the proof of Theorem 3.5.

Our first goal is to use an arbitrary n-strict interval I to construct a second interval, I_2, which is n_2-strict for some n_2 much larger than n and contains a similar density of type-C integers as I. The next lemma will be a helpful tool.

Lemma 3.2. *Let $I := [i_1, i_2]$, $J := [j_1, j_2]$ be integer intervals. Let $S \subseteq I$ and Y be an integer-valued random variable whose support is in J. For $k \in \mathbb{Z}$, denote the random variable $Y + k$ as $\tau_k(Y)$. Then there exists an integer $k \in [i_1 - j_2, i_2 - j_1]$ such that $\mathbb{P}\big(\tau_k(Y) \in S\big) \ge \frac{|S|}{|I|+|J|-1}$.*

Proof. The idea of the proof is that by averaging over all appropriate k, the distributions of $\tau_k(Y)$ should uniformly cover I. More formally, let $k_1 := i_1 - j_2$, $k_2 := i_2 - j_1$, and let K be the set of integers in the interval $[k_1, k_2]$. Note that $|K| = |I| + |J| - 1$. Pick k uniformly at random from K and consider the random variable $Z := \mathbb{P}\big(\tau_k(Y) \in S\big)$. Then

$$\mathbf{E}[Z] = \frac{1}{|K|}\sum_{k=k_1}^{k_2} \mathbb{P}\big(\tau_k(Y) \in S\big)$$

$$= \frac{1}{|I|+|J|-1}\sum_{k=k_1}^{k_2}\sum_{i\in S} \mathbb{P}\big(\tau_k(Y) = i\big)$$

$$= \frac{1}{|I|+|J|-1} \sum_{i\in S} \sum_{k=k_1}^{k_2} \mathbb{P}(\tau_k(Y)=i). \tag{5}$$

Note that $\mathbb{P}\big(\tau_k(Y)=i\big) = \mathbb{P}\big(Y=i-k\big)$ and for $i \in S \subseteq I$ we have,

$$J \subseteq [i-k_2, i-k_1].$$

Thus, for all $i \in S$,

$$\sum_{k=k_1}^{k_2} \mathbb{P}\big(\tau_k(Y)=i\big) = \mathbb{P}(Y \in [i-k_2, i-k_1]) = 1.$$

Therefore,

$$(5) = \frac{|S|}{|I|+|J|-1}.$$

So there exists k such that $\mathbb{P}\big(\tau_k(Y) \in S\big) \geq \mathbf{E}[Z] = \frac{|S|}{|I|+|J|-1}$.

□

Using Lemma 3.2, we will not lose much density assuming that $|I|$ is much larger than $|J|$. However, if Y_m is induced by the interval $[0, b^m - 1]$, then the random variable $H(Y_m)$ will be supported on a set J that is much too large. As a result, it will be more useful to consider a smaller interval where the bulk of the distribution lies.

Lemma 3.3. *Let Y be an integer-valued random variable with mean μ_Y and variance σ_Y^2, and let $\lambda > 0$. Let $S \subseteq [i_1, i_2] = I$ be a set of integers where $|S|/|I| = d$. Then there exists an integer $k \in [i_1 - (\mu_Y + \sigma_Y\lambda), i_2 - (\mu_Y - \sigma_Y\lambda)]$ such that*

$$\mathbb{P}\big(\tau_k(Y) \in S\big) \geq \left(1 - \frac{1}{\lambda^2}\right)\left(\frac{d}{1+\frac{2\sigma_Y\lambda}{|I|}}\right).$$

Proof. By Chebyshev's Inequality[2] we have $\mathbb{P}(|Y-\mu_Y| < \sigma_Y\lambda) > 1 - \frac{1}{\lambda^2}$. Let Y' be Y conditioned on being in the interval $J := [\mu_Y - \sigma_Y\lambda, \mu_Y + \sigma_Y\lambda]$. Note that

$$|J| \leq 2\sigma_Y\lambda + 1.$$

Then, by Lemma 3.2, there exists $k \in [i_1 - (\mu_Y + \sigma_Y\lambda), i_2 - (\mu_Y - \sigma_Y\lambda)]$ such that $\mathbb{P}\big(\tau_k(Y') \in S\big) \geq \frac{|S|}{|I|+|J|-1} = \frac{d}{1+\frac{2\sigma_Y\lambda}{|I|}}$. Therefore, we have

$$\mathbb{P}\big(\tau_k(Y) \in S\big) \geq \mathbb{P}(Y \in J)\mathbb{P}\big(\tau_k(Y') \in S\big) \geq \left(1 - \frac{1}{\lambda^2}\right)\left(\frac{d}{1+\frac{2\sigma_Y\lambda}{|I|}}\right).$$

□

[2] We certainly could do better than Chebyshev's Inequality here. However, the bounds it gives will suit our purposes fine.

It is possible to construct sets of intervals which, under the image of H, act as shifts of each other. For example, in base-10 (recall $H(0) = 0, H(1) = 1$) if the random variable X_1 is induced by $[1100, 1199]$ and X_2 is induced by $[0, 99]$, then $H(X_1) = H(X_2) + 2$.

We will now further expand on the example above. Let $n \in \mathbb{N}$ be divisible by 4. Let $B_0 := [0, b^{3n/4} - 1]$ and, for $k = 1, \ldots, \frac{n}{4}$, consider the interval

$$B_k := [b^{n-1} + b^{n-2} + \cdots + b^{n-k}, b^{n-1} + b^{n-2} + \cdots + b^{n-k} + b^{3n/4} - 1].$$

Then the intervals B_k will all be n-strict (with exception of B_0), and a random integer $x \in B_k$ will have the following base-b expansion:

$$x = \underbrace{11\ldots1}_{\text{k digits}}\ \underbrace{00\ldots0}_{\frac{n}{4}-k \text{ digits}}\ \underbrace{X_iX_{i-1}\ldots X_1}_{\frac{3n}{4} \text{ digits}}.$$

That is, x will have its first k digits equal to 1, the next $\frac{n}{4} - k$ digits equal to 0, and the remaining $\frac{3n}{4}$ digits will be i.i.d. random variables X_i taking values uniformly in the set $\{0, 1, \ldots, b-1\}$.

Let Y_0 be the random variable induced by B_0, and Y_k be induced by B_k. Then

$$H(Y_k) = H(Y_0) + k = \tau_k(H(Y_0)).$$

Recall $H(Y_0)$ has mean $\frac{3n}{4}\mu$, and variance $\frac{3n}{4}\sigma^2$. Consider an interval $I = [i_1, i_2]$ containing a set of type-C integers S, and let $\lambda > 0$. By Lemma 3.3, there exists $k' \in \left[i_1 - \left(\frac{3n}{4}\mu + \sqrt{\frac{3n}{4}}\lambda\sigma\right), i_2 - \left(\frac{3n}{4}\mu - \sqrt{\frac{3n}{4}}\lambda\sigma\right)\right]$ such that

$$\mathbb{P}\Big(\tau_{k'}(H(Y_0)) \in S\Big) \geq \left(1 - \frac{1}{\lambda^2}\right)\left(\frac{d_C(I)}{1 + \frac{\sqrt{3n}\lambda\sigma}{|I|}}\right). \tag{6}$$

Thus, if $I \subseteq \left[1 + \frac{3}{4}n\mu + \lambda\sigma\sqrt{\frac{3}{4}n}, \frac{1}{4}n + \frac{3}{4}n\mu - \lambda\sigma\sqrt{\frac{3}{4}n}\right]$, then $1 \leq k' \leq \frac{n}{4}$. Setting $k := k'$ produces the interval B_k, which will be n-strict with

$$d_C(B_k) \geq \left(1 - \frac{1}{\lambda^2}\right)\left(\frac{d}{1 + \frac{\sqrt{3n}\lambda\sigma}{|I|}}\right).$$

In fact, we have proven the following

Theorem 3.4. *Let $n \in \mathbb{N}$ be divisible by 4 and let $C \subset \mathbb{N}$. For $\lambda > 0$, define $J_{n,\lambda} := \left[1 + \frac{3}{4}n\mu + \lambda\sigma\sqrt{\frac{3}{4}n}, \frac{1}{4}n + \frac{3}{4}n\mu - \lambda\sigma\sqrt{\frac{3}{4}n}\right]$. Fix an interval $I \subseteq J_{n,\lambda}$. Then there exists an n-strict interval, I_2, such that $d_C(I_2) \geq d_C(I)\left(1 - \frac{1}{\lambda^2}\right)\left(\frac{1}{1+\frac{\sqrt{3n}\sigma\lambda}{|I|}}\right)$.*

The goal for the rest of the section is to use the previous theorem iteratively to construct a sequence of intervals $\{I_i\}_{i=1}^{\infty}$, each with high type-C density, such that each I_i is n_i-strict and the sequence $\{n_i\}_{i=1}^{\infty}$ grows quickly. One technical issue to worry about is that $d_C(I_{i+1}) < d_C(I_i)$ for all i. How much smaller $d_C(I_{i+1})$ is depends on how large we choose λ_i to be in each step. We wish to choose λ_i as large as possible, but choose λ_i too large and two bad things can happen: First, I_i will not be contained in J_{n_{i+1},λ_i} for any choice of n_{i+1}. Second, $\frac{\sqrt{3n}\sigma\lambda_i}{|I|}$ will not be small. We are helped by the fact that the sequence $\{n_i\}_{i=1}^{\infty}$ will grow super exponentially (in fact, $n_{i+1} = \Omega(b^{n_i})$). Choosing $\lambda_i = b^{n_i/8}$ in each step will work well; however, we will need the initial n_1 to be sufficiently large. The next theorem gives precise calculations. The proof follows from a number of routine calculations and estimations, some of which we have left for the appendix.

Theorem 3.5. *Suppose I is n-strict, where n satisfies bounds (**B**). Then there exists $n_2 \geq \frac{b^{n-1}}{\mu}$, and an n_2-strict interval I_2 such that*

$$d_C(I_2) \geq d_C(I)\left(1 - b^{-n/4}\right)\left(1 - \frac{2\sigma}{\sqrt{\mu}}b^{-n/8}\right).$$

Proof. As before, let $J_{m,\lambda} := \left[1 + \frac{3}{4}m\mu + \lambda\sigma\sqrt{\frac{3}{4}m}, \frac{1}{4}m + \frac{3}{4}m\mu - \lambda\sigma\sqrt{\frac{3}{4}m}\right]$. We assumed that I is n-strict, so $|I| = b^{3n/4}$. Write I as $[a, a + b^{3n/4} - 1]$. Setting $\lambda := b^{n/8}$, we attempt to find an n_2 divisible by 4 such that $I \subseteq J_{n_2,\lambda}$. It would be prudent to consider $f(m) := 1 + \frac{3}{4}m\mu + \lambda\sigma\sqrt{\frac{3}{4}m}$, which is the left endpoint of $J_{m,\lambda}$. We first find an integer n_2 such that $f(n_2) \leq a$ and $a - f(n_2)$ is small. By Lemma 6.2 in the Appendix, assuming n satisfies bounds (**B**), it follows that there exists n_2 such that:

- $4 \mid n_2$,
- $\frac{b^{n-1}}{\mu} \leq n_2 \leq \frac{4}{3\mu}b^n$,
- $0 \leq a - f(n_2) \leq 3\mu + 1$.

We now check that $I \subseteq J_{n_2,\lambda}$ in order to invoke Theorem 3.4. We already have that the left endpoint $f(n_2) \leq a$. It remains to check the right endpoints of I and J_{n_2}. We need to show that

$$a - 1 + b^{3n/4} \leq \frac{n_2}{4} + \frac{3n_2}{4}\mu - \lambda\sigma\sqrt{\frac{3n_2}{4}}. \tag{7}$$

The above is equivalent to

$$a - \left(\frac{3n_2}{4}\mu + \lambda\sigma\sqrt{\frac{3n_2}{4}} + 1\right) + b^{3n/4} \leq \frac{n_2}{4} - 2\lambda\sigma\sqrt{\frac{3n_2}{4}}.$$

Simplifying, the above follows from showing that

$$a - f(n_2) + b^{3n/4} + 2\lambda\sigma\sqrt{\frac{3n_2}{4}} \leq \frac{n_2}{4}.$$

Now let

$$\text{LHS} := a - f(n_2) + b^{3n/4} + 2\lambda\sigma\sqrt{\frac{3n_2}{4}}.$$

Then

$$\text{LHS} \leq 3\mu + 1 + b^{3n/4} + \lambda\sigma\sqrt{3n_2}.$$

Using the facts that $\lambda = b^{n/8}$ and $n_2 \leq \frac{4b^n}{3\mu}$, we get that

$$\text{LHS} \leq 3\mu + 1 + b^{3n/4} + 2\frac{\sigma}{\sqrt{\mu}}b^{5n/8}.$$

Now consider RHS $:= \frac{n_2}{4}$. By the assumptions on n_2, we have

$$\text{RHS} \geq \frac{b^n}{4b\mu}.$$

So (7) follows from showing that

$$3\mu + 1 + b^{3n/4} + 2\frac{\sigma}{\sqrt{\mu}}b^{5n/8} \leq \frac{b^n}{4b\mu}.$$

The above is exactly the bound (**B3**). Therefore, $I \subseteq J_{n_2,\lambda}$. Thus, by applying Theorem 3.4 with $\lambda = b^{n/8}$, there exists an n_2-strict interval I_2 such that

$$d_C(I_2) \geq d_C(I)\Big(1 - \frac{1}{b^{n/4}}\Big)\Big(\frac{1}{1 + \frac{\sqrt{3n_2}\sigma b^{n/8}}{b^{3n/4}}}\Big).$$

Since $n_2 \leq \frac{4}{3\mu}b^n$, it follows that

$$\frac{1}{1 + \frac{\sqrt{3n_2}\sigma b^{n/8}}{b^{3n/4}}} \geq \frac{1}{1 + \frac{2\sigma}{\sqrt{\mu}}b^{-n/8}} \geq 1 - \frac{2\sigma}{\sqrt{\mu}}b^{-n/8}.$$

Thus, we conclude that $d_C(I_2) \geq d_C(I)\left(1 - b^{-n/4}\right)\left(1 - \frac{2\sigma}{\sqrt{\mu}}b^{-n/8}\right)$. □

Apply the previous theorem to our starting n_1-strict interval I_1 to get an n_2-strict interval I_2. Since $n_2 > n_1$, we can apply Theorem 3.5 again on I_2. Continuing in this manner produces a sequence of integers $\{n_i\}_{i=1}^{\infty}$ and n_i-strict intervals $\{I_i\}_{i=1}^{\infty}$ such that, for all i:

- $n_{i+1} \geq \frac{b^{n_i-1}}{\mu}$,
- $d_C(I_{i+1}) \geq d_C(I_i)\left(1 - b^{-n_i/4}\right)\left(1 - \frac{2\sigma}{\sqrt{\mu}}b^{-n_i/8}\right)$.

The second condition implies that, for all i,

$$d_C(I_i) \geq d_C(I_1)\prod_{i=1}^{\infty}\left(\left(1 - b^{-n_i/4}\right)\left(1 - \frac{2\sigma}{\sqrt{\mu}}b^{-n_i/8}\right)\right). \tag{8}$$

The following fact will help simplify the above expression. For positive real numbers x and α, if $x \geq 2\alpha > 0$, then

$$1 - \alpha x^{-1} \geq \frac{1}{1 + 2\alpha x^{-1}} \geq e^{-2\alpha x^{-1}}.$$

Therefore, (8) implies that

$$d_C(I_i) \geq d_C(I_1) \cdot \exp\left(\sum_{i=1}^{\infty} -2b^{-n_i/4} - \frac{4\sigma}{\sqrt{\mu}} b^{-n_i/8}\right).$$

For all $i \in \mathbb{N}$, it holds that $n_i \geq i n_1$ (it may happen that $n_2 < 2n_1$ if μ is very large, but assuming the bounds (**B**) this will not be the case). The sum in the previous inequality is the sum of two geometric series, one with ratio $r = b^{-n_1/4}$ and first term $a = -2b^{-n_1/4}$. The second has $r = b^{-n_1/8}$ and $a = \frac{-4\sigma}{\sqrt{\mu}} b^{-n_1/8}$. Recall that an infinite geometric series with $|r| < 1$, and first term a sums to

$$\frac{a}{1-r}.$$

Therefore, the first series sums to $\frac{-2b^{-n_1/4}}{1-b^{-n_1/4}}$, the second sums to $\frac{-4\sigma b^{-n_1/8}}{\sqrt{\mu}(1-b^{-n_1/8})}$. After simplifying we conclude that, for all i,

$$d_C(I_i) \geq d_C(I_1) \cdot \exp\left(\frac{-2}{b^{n_1/4} - 1} + \frac{-4\sigma}{\sqrt{\mu}(b^{n_1/8} - 1)}\right).$$

Thus, we have proven the following

Theorem 3.6. *Assume there exists n_1 satisfying the bounds (**B**) and an n_1-strict interval I_1. Then, for all $N \in \mathbb{N}$, there exists $n > N$ and an n-strict interval I such that*

$$d_C(I) \geq d_C(I_1) \cdot exp\left(\frac{2}{1 - b^{n_1/4}} + \frac{4\sigma}{\sqrt{\mu}(1 - b^{n_1/8})}\right).$$

4. Main Result

As in the previous section we continue to assume that H is a b-happy function with digit mean μ and digit variance σ^2. Also, we assume that we have experimentally found a suitable starting n_1-strict interval, I_1, with large type-C density for some $C \subset \mathbb{N}$. As in Section 2, for positive integers m, let Y_m denote the random variable induced by the interval $[0, b^m - 1]$.

In this section we give a proof of the following

Theorem 4.1. *Suppose I_1 is n_1-strict, where n_1 satisfies bounds (**B**). Let $\bar{d}$ denote the upper density of the set of type-C integers. Then*

$$\bar{d} \geq d_C(I_1) \cdot exp\left(\frac{2}{1-b^{n_1/4}} + \frac{4\sigma}{\sqrt{\mu}(1-b^{n_1/8})}\right).$$

The digit mean and digit variance for the case $(e, b) = (2, 10)$ are 28.5 and 721.05 respectively. In this case, if $n > 13$, then it satisfies bounds (**B**). After performing a computer search we find that the density of happy numbers in the interval $[10^{403}, 10^{404} - 1]$ is at least .185773; thus, there exists a 404-strict interval containing at least this density of happy numbers as a subset. Consider

$$\delta(n) := \left(\frac{2}{1-b^{n/4}} + \frac{4\sigma}{\sqrt{\mu}(1-b^{n/8})}\right).$$

Plugging in the value for n, we find that $e^{\delta(404)} > 1-10^{-49}$. Thus, by Theorem 4.1, the upper density of type-$\{1\}$ integers is at least .1857729. For the lower density, the type-$\{1\}$ density of $[10^{2367}, 10^{2368}-1]$ is at most .11379. This implies that there is a 2368-strict interval with type-$\{4\}$ density at least $1-.11379$. We can then apply the main result to conclude that the upper density of type-$\{4\}$ integers is at least $1 - .1138$. This gives the following

Corollary 4.2. *Let $\underline{d}$ and $\bar{d}$ be the lower and upper density of $(2, 10)$-happy numbers respectively. Then $\underline{d} < .1138$ and $\bar{d} > .18577$.*

The proof of Theorem 4.1 is somewhat technical despite having a rather intuitive motivation. For the sake of clarity we first give a sketch of how to use Theorems 3.6 and 2.5 in order to prove a lower bound on the upper density of type-C numbers.

Given our starting interval I_1, apply Theorem 3.6 to construct an n-strict interval I, where

$$d_C(I) \geq (1 - o(1))d_C(I_1).$$

Do this with n large enough as to make all the following error estimations arbitrarily small. Pick m_1 such that μm_1 (i.e., the mean of $H(Y_{m_1})$) lands in the interval I. Since $I \subseteq [b^{n-1}, b^n]$, we have that $m_1 = \Theta(b^n)$. This implies that the standard deviation of $H(Y_{m_1})$ is roughly $b^{n/2}$. This will be much less than $|I| = b^{3n/4}$ and thus the bulk of the distribution of $H(Y_{m_1})$ will lie in the interval I.

Next, use Lemma 3.3 with a large λ to find an integer k for which

$$\mathbb{P}\left(\tau_k\big(H(Y_{m_1})\big) \text{ is type-C}\right) \geq (1 - o(1))d_C(I_1).$$

Note that this k will be smaller than $|I| = b^{3n/4}$ and that the mean of $\tau_k\big(H(Y_{m_1})\big)$ is equal to $\mu m_1 + k$. Clearly, there exists an integer m_2 such that

$$|\mu m_2 - (\mu m_1 + k)| \leq \mu.$$

Consider the random variable $H(Y_{m_2})$. The means of $H(Y_{m_2})$ and $\tau_k\left(H(Y_{m_1})\right)$ are almost equal. Since k is much smaller relative to m_1 and m_2, the variance of these two distributions will be close. Furthermore, the distributions of $H(Y_{m_2})$ and $\tau_k\big(H(Y_{m_1})\big)$ are asymptotically locally normal, so we may apply the local limit law to conclude that the distributions are point-wise close near the means. Thus,

$$\mathbb{P}\big(H(Y_{m_2}) \text{ is type-C}\big) \geq (1 - o(1))d_C(I_1).$$

This implies that the interval $[0, b^{m_2} - 1]$ has type-C density at least $d_C(I_1)\,(1 - o(1))$. Note that in the above analysis, we may take n (and therefore m_2) to be arbitrarily large. This lower bounds the upper density of type-C integers by $d_C(I_1)(1 - o(1))$. In fact, the only contribution to the error term is from the application of Theorem 3.6 (the rest of the error tends to 0 as n tends to infinity).

4.1. Some Lemmas

We will now begin to prove the main result. We have broken some of the pieces down for 3 lemmas. The proofs primarily consist of calculations and we leave them for after the proof of the main result. Note that Lemma 4.5 (part 1) is the only place where the local limit law is used.

Lemma 4.3. *There exists a sufficiently large N such that, if $n > N$ and I is an n-strict interval, then there exists $m \in \mathbb{N}$ with the property that*

$$[\mu m - \sigma m^{5/8}, \mu m + \sigma m^{5/8}] \subseteq I.$$

Lemma 4.4. *Let $\epsilon > 0$ be given (assume as well that $\epsilon \leq 1$). Let $\lambda := \sqrt{\frac{6}{\epsilon}}$. Then there exists a sufficiently large N such that, if n, m_1, and I satisfy:*

- $n > N$,
- $m_1 \in [\frac{b^{n-1}}{\mu}, \frac{b^n}{\mu}]$,
- *I is n-strict,*

then the following hold:

1. $\lambda \leq {m_1}^{1/8}$,
2. $\left|1 - \frac{1}{1+\frac{2\lambda\sigma\sqrt{m_1}}{b^{3n/4}}}\right| \leq \frac{\epsilon}{6}$.

Lemma 4.5. *Let $\epsilon > 0$ be given (assume as well that $\epsilon \leq 1$). Let $T := \frac{2\sqrt{6}}{\sqrt{\epsilon}}$, and $\lambda := \frac{\sqrt{6}}{\sqrt{\epsilon}}$. Then there exists a sufficiently large N such that, if n, m_1, m_2, k, and I satisfy:*

- $n > N$,
- $m_1 \in [\frac{b^{n-1}}{\mu}, \frac{b^n}{\mu}], m_2 \in [\frac{b^{n-2}}{\mu}, \frac{b^{n+1}}{\mu}]$,
- $|k| \leq b^{3n/4}$,
- $|\mu m_1 + k - \mu m_2| \leq \mu$,
- *I is n-strict,*

then the following hold:

1. *For* $i \in \{1,2\}$, $\max_{|t|\leq T} \left| 1 - \frac{\mathbb{P}(H(Y_{m_i}) = \lfloor \mu m_i + t\sigma\sqrt{m_i} \rfloor)}{\frac{e^{t^2/2}}{\sqrt{2\pi m_i}\sigma}} \right| \leq \frac{\epsilon}{6}$.
2. $\left|1 - \sqrt{\frac{m_1}{m_2}}\right| \leq \frac{\epsilon}{6}$.
3. *For any real numbers* t_1 *and* t_2*, where* $t_1 \in [\frac{-T}{2}, \frac{T}{2}]$ *and*

$$\mu m_1 + k + t_1\sigma\sqrt{m_1} = \mu m_2 + t_2\sigma\sqrt{m_2},$$

it holds that $t_2 \in [-T, T]$ *and* $|1 - e^{{t_1}^2 - {t_2}^2/2}| \leq \frac{\epsilon}{6}$.

4.2. Proof of Theorem 4.1

Proof. In order to lower bound the upper density of type-C integers, it suffices to show that, for all $\epsilon > 0$ and $N_1 \in \mathbb{N}$, there exists $m > N_1$ such that

$$d_C([0, b^m - 1]) \geq d_C(I_1) \cdot \exp\left(\frac{2}{1 - b^{n/4}} + \frac{4\sigma}{\sqrt{\mu}(1 - b^{n/8})}\right)(1 - \epsilon).$$

Let ϵ and N_1 be arbitrary (with $\epsilon \leq 1$). Set $T := \frac{2\sqrt{6}}{\sqrt{\epsilon}}$. Also, in anticipation of applying Lemma 3.3, set $\lambda := \frac{\sqrt{6}}{\sqrt{\epsilon}}$.

First, pick $N > N_1$ large enough to apply Lemmas 4.3, 4.4, and 4.5. By Theorem 3.6, there exists an n-strict interval I, where $n > N$ and

$$d_C(I) \geq d_C(I_1) \cdot \exp\left(\frac{2}{1 - b^{n_1/4}} + \frac{4\sigma}{\sqrt{\mu}(1 - b^{n_1/8})}\right). \tag{9}$$

For $m \in \mathbb{N}$, let

$$J_m := [\mu m - \sigma m^{5/8}, \mu m + \sigma m^{5/8}].$$

Recall that $\mathbf{E}[H(Y_m)] = \mu m$ and $\mathbf{Var}[H(Y_m)] = \sigma^2 m$. Hence, J_m is where the bulk of the distribution of $H(Y_m)$ lands. Pick m_1 such that $J_{m_1} \subseteq I$ (the existence of such m_1 is guaranteed by Lemma 4.3). Note that $m_1 \in [\frac{b^{n-1}}{\mu}, \frac{b^n}{\mu}]$ since I is n-strict.

Let S be the set of type-C integers in I. Apply Lemma 3.3 on the random variable $H(Y_{m_1})$ to find an integer k such that

$$\mathbb{P}\Big(\tau_k\big(H(Y_{m_1})\big) \in S\Big) \geq d_C(I)\left(1 - \frac{1}{\lambda^2}\right)\left(\frac{1}{1 + \frac{2\lambda\sigma\sqrt{m_1}}{|I|}}\right). \tag{10}$$

Since $J_{m_1} \subseteq I$ and $|I| = b^{3n/4}$, it follows that $k \leq b^{3n/4}$.

Let $\tau_k(J_{m_1}) := [a+k, b+k]$, where $J_{m_1} = [a,b]$. Let S' be the set of type-C integers in interval $\tau_k(J_{m_1})$. Recall the proof of Lemma 3.3. In particular, we applied Lemma 3.2 after ignoring the tails of the distribution of $H(Y_{m_1})$ outside of $\lambda\sigma\sqrt{m_1}$ from the mean. Since $\lambda \leq {m_1}^{1/8}$ (by Lemma 4.4, part 1), we may replace (10) by the stronger conclusion that

$$\sum_{i \in S'} \mathbb{P}\Big(\tau_k\big(H(Y_{m_1})\big) = i\Big) \geq d_C(I)\Big(1 - \frac{1}{\lambda^2}\Big)\Big(\frac{1}{1 + \frac{2\lambda\sigma\sqrt{m_1}}{|I|}}\Big).$$

Using the assumption that $\lambda = \sqrt{\frac{6}{\epsilon}}$ and part 2 of Lemma 4.4, we simplify the above as

$$\sum_{i \in S'} \mathbb{P}\Big(\tau_k\big(H(Y_{m_1})\big) = i\Big) \geq d_C(I)\left(1 - \frac{\epsilon}{6}\right)^2. \tag{11}$$

Now pick $m_2 \in \mathbb{N}$ such that $|m_1\mu + k - m_2\mu| \leq \mu$. Since $|k| \leq b^{3n/4}$, it follows that $m_2 \in [\frac{b^{n-2}}{\mu}, \frac{b^{n+1}}{\mu}]$. In particular m_1, m_2, n, k, and I now all satisfy the conditions of Lemma 4.5. It remains to show that near the mean of $\tau_k\big(H(Y_{m_1})\big)$, the distributions of $\tau_k\big(H(Y_{m_1})\big)$ and $H(Y_{m_2})$ are similar. This will imply that the interval $[0, b^{m_2} - 1]$ contains a large density of type-C integers. Making this precise, we prove the following

Claim 1. For integers $i \in \tau_k\big(J_{m_1}\big)$,

$$\frac{\mathbb{P}\big(H(Y_{m_2}) = i\big)}{\mathbb{P}\Big(\tau_k\big(H(Y_{m_1})\big) = i\Big)} \geq \left(1 - \frac{\epsilon}{6}\right)^4.$$

Proof. Let $i \in \tau_k(J_{m_1})$ be fixed and pick t_1, t_2 such that

$$i = \mu m_1 + k + t_1\sigma\sqrt{m_1} = \mu m_2 + t_2\sigma\sqrt{m_2}.$$

It is important now that we had chosen $\lambda = \frac{T}{2}$, this implies that $|t_2| \leq T$ (see Lemma 4.5 part 3). We can use the local limit law to estimate the distributions of $\tau_k\big(H(Y_{m_1})\big)$ and $H(Y_{m_2})$. By Lemma 4.5 part 1,

$$\mathbb{P}\big(H(Y_{m_2}) = i\big) = \mathbb{P}\big(H(Y_{m_2}) = \mu m_2 + t_2\sigma\sqrt{m_2}\big) \geq \frac{e^{-{t_2}^2/2}}{2\pi\sigma\sqrt{m_2}}\left(1 - \frac{\epsilon}{6}\right)$$

and

$$\mathbb{P}\Big(\tau_k\big(H(Y_{m_1})\big) = i\Big) = \mathbb{P}\big(H(Y_{m_1}) = \mu m_1 + t_1\sigma\sqrt{m_1}\big) \leq \frac{e^{-{t_1}^2/2}}{2\pi\sigma\sqrt{m_1}}\left(1+\frac{\epsilon}{6}\right).$$

Hence,

$$\frac{\mathbb{P}\big(H(Y_{m_2}) = i\big)}{\mathbb{P}\Big(\tau_k\big(H(Y_{m_1})\big) = i\Big)} \geq \exp\left(({t_1}^2 - {t_2}^2)/2\right)\frac{\sqrt{m_1}}{\sqrt{m_2}}\frac{(1-\frac{\epsilon}{6})}{(1+\frac{\epsilon}{6})}.$$

The above, by Lemma 4.5 parts 2 and 3, is at least $\left(1-\frac{\epsilon}{6}\right)^4$. □

Putting it all together, we have shown that

$$\begin{aligned}
d_C([0, b^{m_2}-1]) &\geq \sum_{i\in S'} \mathbb{P}\big(H(Y_{m_2}) = i\big) \\
&= \sum_{i\in S'} \frac{\mathbb{P}\big(H(Y_{m_2}) = i\big)}{\mathbb{P}\Big(\tau_k\big(H(Y_{m_1})\big) = i\Big)}\mathbb{P}\Big(\tau_k\big(H(Y_{m_1})\big) = i\Big). \\
&\geq \left(1-\frac{\epsilon}{6}\right)^4 \sum_{i\in S'} \mathbb{P}\Big(\tau_k\big(H(Y_{m_1})\big) = i\Big) && \text{(Claim 1)} \\
&\geq d_C(I)\left(1-\frac{\epsilon}{6}\right)^6 && \text{(Equation 11)} \\
&\geq d_C(I)(1-\epsilon).
\end{aligned}$$

To conclude the proof, equation (9) implies that

$$d_C([0, b^{m_2}-1]) \geq d_C(I_1)\cdot\exp\left(\frac{2}{1-b^{n_1/4}} + \frac{4\sigma}{\sqrt{\mu}(1-b^{n_1/8})}\right)(1-\epsilon).$$

□

We conclude this section with the proofs of the lemmas used in the previous theorem.

Proof of Lemma 4.3

Proof. For $m \in \mathbb{N}$, let $J_m := [\mu m - \sigma m^{5/8}, \mu m + \sigma m^{5/8}]$. If I is an n-strict interval, then $I \subseteq [b^{n-1}, b^n - 1]$. Note that $\mu m \in I$ implies that $m = O(b^n)$. This in turn shows that

$$|J_m| = O(b^{5n/8}) \ll |I| = b^{3n/4}.$$

Comparing the growth rates of $|J_m|$ and $|I|$ it is clear that we can pick N_1 large enough such that $n > N_1$ implies that there exists m with $J_m \subseteq I$. □

Proof of Lemma 4.4

Proof. We find N_1, N_2 for the two parts respectively and then choose $N = \max(N_1, N_2)$.

1. λ is a fixed constant here and it is assumed that $m_1 \geq \frac{b^{n-1}}{\mu}$, so the result is trivial (this gives N_1).

2. For $x > 0$, to show that $\left|1 - \frac{1}{1+x}\right| \leq \frac{\epsilon}{6}$, it is equivalent to prove that

$$\left(1 - \frac{\epsilon}{6}\right)(1+x) \leq 1 \leq \left(1 + \frac{\epsilon}{6}\right)(1+x).$$

The above follows if $x \leq \frac{\epsilon}{6}$. Thus, the result will follow by finding N large enough such that $\frac{2\sigma\lambda\sqrt{m_1}}{b^{3n/4}} \leq \frac{\epsilon}{6}$. Using the assumption that $m_1 \leq \frac{b^n}{\mu}$, we get

$$\frac{2\sigma\lambda\sqrt{m_1}}{b^{3n/4}} \leq \frac{2\sigma\lambda}{\sqrt{\mu}b^{n/4}}.$$

This is equivalent to

$$\frac{12\sigma\lambda}{\sqrt{\mu}\epsilon} \leq b^{n/4}.$$

Hence, picking $N_2 \geq 4\log_b(\frac{12\sigma\lambda}{\sqrt{\mu}\epsilon})$ suffices.

□

Proof of Lemma 4.5

Proof. We first find N_1, N_2, and N_3 for the three parts respectively, and then define $N := \max(N_1, N_2, N_3)$.

1. For each m, we have

$$H(Y_m) = \sum_{i=1}^{m} H(X_i),$$

where each X_i is uniform in the set $\{0, 1, \cdots, b-1\}$. Recall that it is assumed that $H(0) = 0$ and $H(1) = 1$. In particular, the random variables $H(X_i)$ satisfy the aperiodic condition required by Theorem 2.5. Thus, the result follows from applying Theorem 2.5 to the sum $\sum_{i=1}^{m} H(X_i)$ with finite interval $[-T, T]$. Fix M large enough such that $m > M$ implies that the $O(m^{-1/2})$ term in Theorem 2.5 is less than $\frac{\epsilon}{6}$. By assumption, we have that both m_1 and m_2 are larger than $\frac{b^{n-2}}{\mu}$. Hence, setting $N_1 = \log_b(\mu M) + 2$ suffices.

2. Ignoring the square root, it suffices to show that

$$\left|1 - \frac{m_1}{m_2}\right| \leq \frac{\epsilon}{6}. \tag{12}$$

By assumption

$$|\mu m_1 + k - \mu m_2| \leq \mu.$$

Dividing through by μm_2, it follows that

$$\left|1 - \frac{m_1}{m_2} - \frac{k}{\mu m_2}\right| \leq \frac{1}{m_2}.$$

This implies that

$$\frac{k}{\mu m_2} - \frac{1}{m_2} \leq 1 - \frac{m_1}{m_2} \leq \frac{1}{m_2} + \frac{k}{\mu m_2}.$$

Thus, (12) follows from showing that $\frac{1}{m_2} + \frac{k}{\mu m_2} \leq \frac{\epsilon}{6}$. Using the assumption that $m_2 \geq \frac{b^{n-2}}{\mu}$ and $|k| \leq b^{3n/4}$, it follows that

$$\frac{1}{m_2} + \frac{k}{\mu m_2} \leq \mu b^{-(n-2)} + b^{2-(n/4)}.$$

Therefore, picking $N_2 := \max(\log_b(\frac{12\mu}{\epsilon}) + 2, 4\log_b(\frac{12}{\epsilon}) + 2)$ suffices.

3. We first find N' such that $n > N'$ implies that $t_2 \in [-T, T]$. We start with the assumption that

$$\mu m_1 + k + t_1\sigma\sqrt{m_1} = \mu m_2 + t_2\sigma\sqrt{m_2}.$$

Using the facts that $|\mu m_1 + k - \mu m_2| \leq \mu$ and $|t_1| \leq \frac{T}{2}$, this implies that

$$|t_2| \leq \frac{\mu}{\sigma\sqrt{m_2}} + \frac{T\sqrt{m_1}}{2\sqrt{m_2}}.$$

We assumed that $m_2 \geq \frac{b^{n-2}}{\mu}$. Also, in part (2) we showed that there exists N_2 such that $n > N_2$ implies that $\frac{\sqrt{m_1}}{\sqrt{m_2}} \leq \left(1 + \frac{\epsilon}{6}\right) \leq \frac{7}{6}$. Hence, if we take $N' > N_2$, it follows that

$$|t_2| \leq \frac{\mu^2}{\sigma b^{n-2/2}} + \frac{7T}{12}.$$

Pick $N' > N_2$ large enough such that $n > N'$ implies that $\frac{\mu^2}{\sigma b^{n-2/2}} \leq \frac{5T}{12}$. This will take care of the size of t_2.

Now we must show that there exists N'' large enough such that $n > N''$ implies that

$$\left|1 - e^{({t_1}^2 - {t_2}^2)/2}\right| \leq \frac{\epsilon}{6}.$$

For a real number x, if we wish to show that $|1 - e^x| \leq \frac{\epsilon}{6}$, it is equivalent to prove that

$$\ln\left(1 - \frac{\epsilon}{6}\right) \leq x \leq \ln\left(1 + \frac{\epsilon}{6}\right).$$

Set $\epsilon^* := \min\left(\ln\left(1 + \frac{\epsilon}{6}\right), \left|\ln\left(1 - \frac{\epsilon}{6}\right)\right|\right)$. We find N'' such that $n > N''$ implies that

$$\left|\frac{{t_2}^2 - {t_1}^2}{2}\right| \leq \epsilon^*.$$

It was assumed that

$$\mu m_1 + k + t_1\sigma\sqrt{m_1} = \mu m_2 + t_2\sigma\sqrt{m_2}.$$

Equivalently

$$\mu m_1 + k - \mu m_2 = t_2\sigma\sqrt{m_2} - t_1\sigma\sqrt{m_1}.$$

Applying the assumption that the left hand side is at most μ and dividing both sides by $\sigma\sqrt{m_2}$, we get

$$\left|t_2 - t_1\sqrt{\frac{m_1}{m_2}}\right| \leq \frac{\mu}{\sqrt{m_2}\sigma}.$$

Rearranging, this gives

$$\left|t_2 - t_1 + t_1(1 - \sqrt{\frac{m_1}{m_2}})\right| \leq \frac{\mu}{\sqrt{m_2}\sigma}.$$

This implies that

$$|t_2 - t_1| \leq \frac{\mu}{\sqrt{m_2}\sigma} + \left|t_1(1 - \sqrt{\frac{m_1}{m_2}})\right|.$$

We assumed that $m_2 \geq \frac{b^{n-2}}{\mu}$ and $|t_1| \leq T$. By part (2), if we chose $N'' > N_2$, then

$$\left|1 - \sqrt{\frac{m_1}{m_2}}\right| \leq \mu b^{-(n-2)} + b^{-(n-2)/4}.$$

Putting this together, it follows that

$$\left|\frac{{t_2}^2 - {t_1}^2}{2}\right| = \left|\left(\frac{t_2 + t_1}{2}\right)(t_2 - t_1)\right| \leq T\Big(\frac{\mu^{3/2}b^{-(n-2)/2}}{\sigma} + T(\mu b^{-(n-2)} + b^{-(n-2)/4})\Big).$$

Now, since T, μ, σ, and b are all constants, it follows that the right hand side tends to 0 as n goes to infinity. Therefore, there exists N'' such that $n > N''$ implies that the right hand side is at most ϵ^*. Finally, set $N_3 := \max(N', N'')$. □

5. Experimental Data

The data[3] presented in this section is the result of short computer searches, so the bounds surely can be improved with more computing time. Floating point approximation with conservative rounding was used.

[3]Data generated by fellow graduate student, Patrick Devlin.

5.1. Finding an Appropriate n-Strict Interval

If n is divisible by 4 and the interval $[b^{n-1}, b^n - 1]$ has type-C density d, then there exists an n-strict interval with type-C density at least d to which we may apply Theorem 4.1. The type-C density of $[b^{n-1}, b^n - 1]$ for various n can be quickly calculated by first computing the densities of intervals of the form $[0, b^n - 1]$; the algorithm which does this was discussed in Section 2. After an appropriate n-strict interval is found, we check to see that n satisfies bounds (**B**), compute the error term, and find the desired bound. Our results show that in almost all cases, the asymptotic density of type-C numbers does not exist.

5.2. Explanation of Results

The following information is given in tables (in the order of column in which they appear):

1. The cycle C in which type-C densities are being computed.
2. The lower bound on the upper density (UD) implied by Theorem 4.1.
3. The upper bound on the lower density (LD) implied by Theorem 4.1.
4. The n such that the interval $[b^{n-1}, b^n - 1]$ is used to find the bound (denoted as UD n or LD n).
5. The $\delta(n) = \left(\frac{2}{1-b^{n/4}} + \frac{4\sigma}{\sqrt{\mu}(1-b^{n/8})}\right)$ part of the error term for Theorem 4.1 (we only present an upper bound on $|\delta(n)|$, the true number is always negative). In all cases the error is small enough not to affect the bounds as we only give precision of about 5 or 6 decimal places.

5.2.1. Cubing the Digits in Base-10

In this case, if $n > 16$, then it satisfies bounds (**B**). Table 1 shows the results for the cycles when studying the $(3, 10)$-happy function. There are 9 possible cycles. Figure 2 graphs the density of type-$\{1\}$ integers less than 10^n. It is easy to prove, in this case, that $3 \mid n$ if an only if n is type-$\{153\}$.

5.2.2. A More General Function

In order to emphasize the generality of Theorem 4.1, we consider the function in base-7 with digit sequence $[0, 1, 7, 4, 17, 9, 13]$. There are only two cycles for this function, both are fixed points. Written in base-10 the cycles are $\{1\}$ and $\{20\}$. Figure 3 graphs the relative density of type-$\{1\}$ numbers. Table 2 shows the bounds derived. As there are only two cycles, we focus on the cycle $\{1\}$. In this case, if $n > 12$, then it satisfies bounds (**B**).

Cycle	UD	LD	UD n	LD n	UD $\delta(n)$	LD $\delta(n)$
{1}	$> .028219$	$< .0049761$	10^{864}	10^{132}	$< 10^{-106}$	$< 10^{-14}$
{55,250,133}	$> .06029$	$< .0447701$	10^{208}	10^{964}	$< 10^{-24}$	$< 10^{-118}$
{136,244}	$> .024909$	$< .006398$	10^{204}	10^{420}	$< 10^{-23}$	$< 10^{-51}$
{153}	$= \frac{1}{3}$	$= \frac{1}{3}$	N/A	N/A	N/A	N/A
{160,217,352}	$> .050917$	$< .03184$	10^{160}	10^{456}	$< 10^{-18}$	$< 10^{-56}$
{370}	$> .19905$	$< .16065$	10^{276}	10^{560}	$< 10^{-32}$	$< 10^{-68}$
{371}	$> .30189$	$< .288001$	10^{836}	10^{420}	$< 10^{-102}$	$< 10^{-50}$
{407}	$> .04532$	$< .0314401$	10^{420}	10^{836}	$< 10^{-50}$	$< 10^{-103}$
{919,1459}	$> .04425$	$< .01843$	10^{916}	10^{120}	$< 10^{-112}$	$< 10^{-13}$

Table 1: Bounds for the cycles appearing for the $(3, 10)$-function

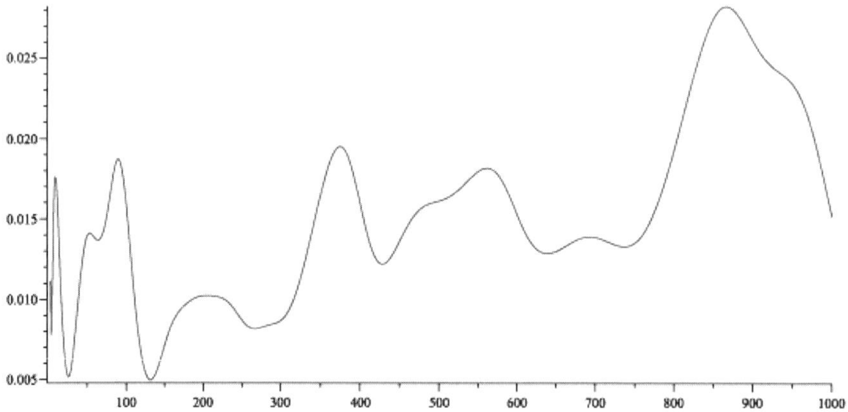

Figure 2: Density of type-{1} integers in the interval $[0, 10^n - 1]$ for the $(3, 10)$-function

6. Appendix

Lemma 6.1. *Fix $a > 0$. Assume that $f : \mathbb{R}^+ \to \mathbb{R}^+$ has continuous first and second derivatives such that, for all $x \in \mathbb{R}^+$, $f'(x) > 0$ and $f''(x) < 0$. Also, assume that $\lim_{x\to\infty} f(x) = \infty$. Furthermore, suppose we have $x^* \in \mathbb{R}^+$ such that $f(x^* + 1) \leq a$. Then there exists $n \in \mathbb{N}$ such that $n \geq x^*$ and $0 \leq a - f(n) \leq f'(x^*)$.*

Proof. This follows from a first order Taylor approximation of the function f. Let x^* such that $f(x^*+1) \leq a$ be given. Set $n := \sup\{m \in \mathbb{N} | f(m) \leq a\}$. Since f is strictly

Cycle	UD	LD	UD n	LD n	UD $\delta(n)$	LD $\delta(n)$
$\{1\}$	$> .9858$	$< .94222$	7^{176}	7^{384}	$< 10^{-17}$	$< 10^{-40}$

Table 2: Bounds for the cycles appearing for the function with digit sequence $[0, 1, 7, 4, 17, 9, 13]$

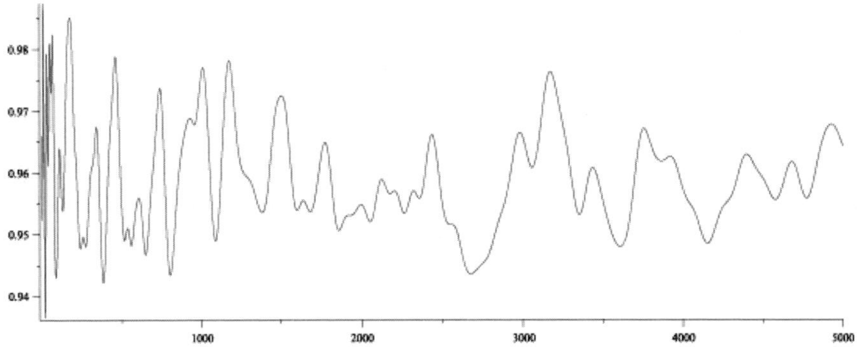

Figure 3: Density of type-$\{1\}$ integers in the interval $[0, 7^n - 1]$ for digit sequence $[0, 1, 7, 4, 17, 9, 13]$

increasing and unbounded, this n exists. Note that $f(n) \le a$ and $f(n+1) > a$. It also follows that $n \ge x^*$, as otherwise $\lceil x^* \rceil$ would be the supremum. By the concavity of f, we have

$$f(n+1) - f(n) \le f'(n) \le f'(x^*).$$

However, $f(n+1) > a$, so we conclude that $0 \le a - f(n) \le f'(x^*)$. □

Lemma 6.2. *Let n be a positive integer, $\lambda = b^{n/8}$, and $a \in [b^{n-1}, b^n]$. Let μ and σ be the digit mean and variance of some b-happy function H. Also, assume that n satisfies bounds (**B**). Let $f(n) := 1 + \frac{3}{4}\mu n + \lambda\sigma\sqrt{\frac{3}{4}n}$. Then there exists an integer n_2 such that:*

- $\frac{b^{n-1}}{\mu} \le n_2 \le \frac{4}{3\mu} b^n$,
- $4 \mid n_2$,
- $0 \le a - f(n_2) \le 3\mu + 1$.

Proof. Since we require that $4 \mid n_2$, we apply Lemma 6.1 on the function

$$g(m) = f(4m) = 1 + 3\mu m + \lambda\sigma\sqrt{3m}.$$

Let $x^* := \frac{b^{n-1}}{4\mu}$. We first check that $g(x^*+1) \leq a$. By assumption, $a \geq b^{n-1}$. Therefore, we need to show that

$$1 + 3\mu\Big(\frac{b^{n-1}}{4\mu} + 1\Big) + b^{n/8}\sigma\sqrt{3\Big(\frac{b^{n-1}}{4\mu} + 1\Big)} \leq b^{n-1}.$$

Simplifying the above, it suffices to show that

$$1 + 3\mu + b^{5n/8}\sigma\sqrt{\frac{3}{4b\mu} + 3b^{-n}} \leq \frac{b^{n-1}}{4}. \tag{13}$$

To keep the results of this paper as general as possible, we only assumed that $\mu \geq \frac{1}{b}$ (this would correspond to the quite uninteresting b-happy function H which maps all digits to 0 except for the digit 1). Also it is clear that $b^n \geq 3$, and therefore

$$\frac{3}{4b\mu} + 3b^{-n} \leq \frac{3}{4} + 1 \leq 2.$$

Plugging this in and rearranging, we see that equation (13) follows if

$$4(1 + 3\mu + \sqrt{2}\sigma b^{5n/8}) \leq b^{n-1}.$$

This is exactly the bound (**B1**) and is true by assumption. Therefore, by Lemma 6.1, there exists $m \in \mathbb{N}$ such that

$$0 \leq a - g(m) \leq g'(x^*).$$

Also,

$$g'(x^*) = 3\mu + \frac{\sqrt{3}\sigma b^{n/8}}{2\sqrt{\frac{b^{n-1}}{4\mu}}} = 3\mu + \sqrt{3\mu b}\sigma b^{-3n/8}.$$

Again, by the assumption (**B2**) on n, the previous statement is bounded above by $3\mu + 1$. Set $n_2 := 4m$. Then $4 \mid n_2, n_2 \geq \frac{b^{n-1}}{\mu}$, and $0 \leq a - f(n_2) \leq 3\mu + 1$. Finally, we note that $f(\frac{4}{3\mu}b^n) > a$ and, since f is strictly increasing, we conclude that $n_2 \leq \frac{4}{3\mu}b^n$. □

Acknowledgements The author would like to thank Dr. Zeilberger, Dr. Saks, and Dr. Kopparty for their advice and support. He would also like to thank his fellow graduate students Simao Herdade and Kellen Myers for their help in editing, and fellow graduate student Patrick Devlin for generating the necessary data.

References

[1] E. El-Sedy and S. Siksek, On happy numbers. Rocky Mountain J. Math. **30** (2000), 565-570.

[2] P. Flajolet and R. Sedgewick, Analytic Combinatorics, Cambridge University Press, 2009.

[3] H.G. Grundman, E.A. Teeple, Sequences of generalized happy numbers with small bases, J. Integer Seq. (2007), Article 07.1.8.

[4] R.K. Guy, Unsolved Problems in Number Theory, Springer-Verlag, New York, 2nd Edition, 2004.

[5] D. Moews, Personal communication, November 2011.

[6] Hao Pan, On consecutive happy numbers, J. Number Theory **128 (6)** (2008), 1646-1654.

A REMARK ON RELATIVELY PRIME SETS

Prapanpong Pongsriiam
Department of Mathematics, Faculty of Science, Silpakorn University, Nakornpathom, Thailand
prapanpong@gmail.com

Received: 1/7/13, Revised: 6/12/13, Accepted: 7/4/13, Published: 8/12/13

Abstract

Four functions counting the number of subsets of $\{1, 2, \ldots, n\}$ having particular properties are defined by Nathanson and generalized by many authors. They derive explicit formulas for all four functions. In this paper, we point out that we need to compute only one of them as the others will follow as a consequence. Moreover, our method is simpler and leads to more general results than those in the literature.

1. Introduction

There are a number of articles concerning relatively prime subsets and Euler phi function for sets. Most of them show the calculation of explicit formulas for certain functions. Their main tools are the Möbius inversion formula and the inclusion-exclusion principle. In this paper, we give simpler and shorter calculations which lead to the results extending those in the literature. To be precise, we cover the results of Nathanson [12], Nathanson and Orosz [13], El Bachraoui [5], [6], [7], [8], [9], El Bachraoui and Salim [10], Ayad and Kihel [2], [3], and Shonhiwa [14], [15]. We show how to apply our method to obtain all results mentioned above and their generalization. Now, let us introduce the following notations and definitions which will be used throughout this paper.

Unless stated otherwise, we let a, b, k, m, n be positive integers, $\gcd(a, b)$ the greatest common divisor of a and b, $[a, b] = \{a, a+1, \ldots, b\}$, A, X finite subsets of positive integers, $|A|$ the cardinality of the set A, $\gcd(A)$ the greatest common divisor of the elements of A, $\gcd(A, n)$ means $\gcd(A \cup \{n\})$, $\lfloor x \rfloor$ the greatest integer less than or equal to x, and μ the Möbius function.

A nonempty finite subset A of positive integers is said to be relatively prime if $\gcd(A) = 1$ and is said to be relatively prime to n if $\gcd(A, n) = 1$. The function counting the number of relatively prime subsets of $\{1, 2, \ldots, n\}$ and other related functions are defined by Nathanson [12] and generalized by many authors. We summarize them in the following definition.

Definition 1. Let X be a nonempty finite subset of positive integers. Define $f(X)$ to be the number of relatively prime subsets of X, $f_k(X)$ the number of such subsets with cardinality k, $\Phi(X, n)$ the number of subsets A of X which is relatively prime to n and $\Phi_k(X, n)$ the number of such subsets A with $|A| = k$.

Nathanson [12] first considered the case $X = [1, n]$. Using the Möbius inversion formula for functions of several variables, El Bachraoui [5], [6], Nathanson and Orosz [13] generalized the formulas to the case $X = [m, n]$ and $X = [1, m]$ (see details in the table). Then Ayad and Kihel [2], [3] generalized all results mentioned above to the case of an arithmetic progression $X = \{a, a+b, \ldots, a+(m-1)b\}$ by using the inclusion-exclusion principle. In another direction, El Bachraoui and Salim [10] obtained the formulas for the case $X = [\ell_1, m_1] \cup [\ell_2, m_2] \cup \cdots \cup [\ell_k, m_k]$. In addition, Shonhiwa [14], [15] and Toth [17] gave results where various constraints are assumed. We summarize the development in the table below.

Authors	Formulas for $f(X)$, $f_k(X)$, $\Phi(X, n)$, $\Phi_k(X, n)$
Nathanson [12]	$X = [1, n]$
El Bachraoui [5] Nathanson and Orosz [13]	$X = [m, n]$
El Bachraoui [6]	$X = [1, m]$
Ayad and Kihel [2], [3]	$X = \{a, a+b, a+(n-1)b\}$, $X = [\ell, m]$, $X = \{a, a+b, a+(m-1)b\}$
El Bachraoui [7], [9] El Bachraoui and Salim [10]	$X = [1, m_1] \cup [\ell_2, m_2]$ $X = [\ell_1, m_1] \cup [\ell_2, m_2]$ $X = [\ell_1, m_1] \cup [\ell_2, m_2] \cup \cdots \cup [\ell_k, m_k]$
Shonhiwa [14], [15]	$X = [1, n]$ with various contraints
Ayad and Kihel [4], El Bachraoui [8], [9] Tang [16], Toth [17]	Different direction such as congruence properties, divisor sum types, combinatorial identities

In this paper, we give shorter and simpler calculations for these formulas. In Section 3, we show that we need only to derive the formula for $\Phi_k(X, n)$ as the others will follow as a consequence. This will cover the results in [2], [3], [5], [6], [12], and [13]. In Section 4, we extend the formulas obtained by Ayad and Kihel [2], [3], by El Bachraoui [7], [8], [9], and by El Bachraoui and Salim [10]. In Section 5, we show how our method can be used to obtain Shonhiwa's results [14], [15] in a simpler and shorter way. We conclude this paper by giving a possible research related to the work of Ayad and Kihel [4], Tang [16], and Toth [17].

2. Lemmas

In this section, we give a formula for the number of terms in an arithmetic progression which are divisible by a fixed positive integer.

Lemma 2. *For integers $d, m \geq 1$ and nonzero integers a and b, let $A = \{a, a+b, \ldots, a+(m-1)b\}$ be an m-arithmetic progression, $A_d = \{x \in A : d \mid x\}$, and $k = \gcd(d, b)$. Then*

(i) If $k \nmid a$, then $|A_d| = 0$.

(ii) If $k \mid a$, then $|A_d| = \left\lfloor \frac{mk}{d} \right\rfloor + \varepsilon_d$ where

$$\varepsilon_d = \begin{cases} 1, & \text{if } d \nmid mk \text{ and } -\frac{a}{k}\left(\frac{b}{k}\right)^{-1} \bmod \frac{d}{k} \in \left\{0, 1, \ldots, m-1-\left\lfloor \frac{(m-1)k}{d} \right\rfloor \frac{d}{k}\right\} \\ & \text{where } \left(\frac{b}{k}\right)^{-1} \text{ is the multiplication inverse of } \frac{b}{k} \text{ modulo } \frac{d}{k}. \\ 0, & \text{otherwise} \end{cases}$$

Proof. From the definition of A and A_d, we see that $|A_d|$ is equal to the number of $x \in \{0, 1, 2, \ldots, m-1\}$ such that $a + xb \equiv 0 (\operatorname{mod} d)$. So we consider the congruence

$$bx \equiv -a (\operatorname{mod} d) \tag{1}$$

If k does not divide a, then there is no x satisfying (1) and thus $|A_d| = 0$. This proves (i). Next, we assume that $k \mid a$. Then (1) becomes

$$\frac{b}{k} x \equiv -\frac{a}{k} \left(\operatorname{mod} \frac{d}{k}\right) \tag{2}$$

Since $k = (d, b)$, $\left(\frac{d}{k}, \frac{b}{k}\right) = 1$. So (2) has a unique solution mod $\frac{d}{k}$ which is

$$x \equiv -\frac{a}{k} \left(\frac{b}{k}\right)^{-1} \left(\operatorname{mod} \frac{d}{k}\right) \tag{3}$$

where $\left(\frac{b}{k}\right)^{-1}$ is the multiplication inverse of $\frac{b}{k}$ modulo $\frac{d}{k}$. So we want to count the number of elements in the set $\{0, 1, 2, \ldots, m-1\}$ which satisfy (3). Each of the following sets contain a unique element satisfying (3)

$$\left\{0, 1, \ldots, \frac{d}{k} - 1\right\}, \left\{\frac{d}{k}, \frac{d}{k} + 1, \ldots, \frac{2d}{k} - 1\right\}, \ldots, \left\{\left(\left\lfloor \frac{mk}{d} \right\rfloor - 1\right) \frac{d}{k}, \ldots, \left\lfloor \frac{mk}{d} \right\rfloor \frac{d}{k} - 1\right\}.$$

There are $\left\lfloor \frac{mk}{d} \right\rfloor$ sets. This implies that $|A_d| = \left\lfloor \frac{mk}{d} \right\rfloor + \varepsilon_d$ where $\varepsilon_d = 1$ if $\left\lfloor \frac{mk}{d} \right\rfloor \frac{d}{k} - 1 < m - 1$ and the set $\left\{\left\lfloor \frac{mk}{d} \right\rfloor \frac{d}{k}, \left\lfloor \frac{mk}{d} \right\rfloor \frac{d}{k} + 1, \ldots, m-1\right\}$ contains an element satisfying (3), otherwise $\varepsilon_d = 0$.

It is easy to see that $\left\lfloor\frac{mk}{d}\right\rfloor\frac{d}{k}-1<m-1$ if and only if $d\nmid mk$, we also see that $\left\lfloor\frac{mk}{d}\right\rfloor\frac{d}{k}$, $\left\lfloor\frac{mk}{d}\right\rfloor\frac{d}{k}+1,\ldots,m-1$ are congruent to $0,1,2,\ldots,m-1-\left\lfloor\frac{(m-1)k}{d}\right\rfloor\frac{d}{k}$ modulo $\frac{d}{k}$, respectively. Hence

$$\varepsilon_d=\begin{cases}1, & \text{if } d\nmid mk \text{ and } -\frac{a}{k}\left(\frac{b}{k}\right)^{-1}\bmod\frac{d}{k}\in\left\{0,1,\ldots,m-1-\left\lfloor\frac{(m-1)k}{d}\right\rfloor\frac{d}{k}\right\}\\ 0, & \text{otherwise.}\end{cases}$$

This completes the proof. □

If we consider the case $\gcd(a,b)=1$, we obtain a lemma of Ayad and Kihel as a corollary. We record it in the next lemma.

Lemma 3. *[3] For an integer $d,m\geq 1$, and for nonzero integers a and b with $\gcd(a,b)=1$, let $X=\{a,a+b,\ldots,a+(m-1)b\}$ be an m-arithmetic progression, $A_d=\{x\in X\ :\ d\mid x\}$, and $k=\gcd(d,b)$. Then*

i) If $k\neq 1$, then $|A_d|=0$.

ii) If $k=1$, then $|A_d|=\left\lfloor\frac{m}{d}\right\rfloor+\varepsilon_d$ where

$$\varepsilon_d=\begin{cases}1 & \text{if } d\nmid m \text{ and } \left(-ab^{-1}\right)\bmod d\in\{0,\ldots,m-\left\lfloor\frac{m}{d}\right\rfloor d-1\},\\ & \text{where } b^{-1} \text{ is the multiplication inverse of } b \text{ modulo } d.\\ 0 & \text{otherwise.}\end{cases}$$

Proof. Since $\gcd(a,b)=1$, we see that $k=1$ if and only if $k\mid a$. So if $k\neq 1$, then $|A_d|=0$ and if $k=1$, then $|A_d|=\left\lfloor\frac{m}{d}\right\rfloor+\varepsilon_d$, by Lemma 2. Notice also that $\left\lfloor\frac{m}{d}\right\rfloor=\left\lfloor\frac{m}{d}-\frac{1}{d}\right\rfloor$ if and only if $d\nmid m$. So the conditions determining ε_d in Lemma 2 and Lemma 3 are the same. This completes the proof. □

The next lemma will be used throughout this paper.

Lemma 4. $$\sum_{d\mid n}\mu(d)=\begin{cases}1 & \text{if } n=1\\ 0 & \text{if } n>1.\end{cases}$$

Proof. This is a well-known result. For the proof see, for example, ([1], p.25). □

3. Only One Formula is Enough

In this section, we give a simple proof of the formula for $\Phi_k^{(a,b)}(m,n)$ and show that the formulas for $\Phi^{(a,b)}(m,n)$, $f_k^{(a,b)}(m)$, and $f^{(a,b)}(m)$ can be obtained as a consequence. In the notation used in [2], [3], $f^{(a,b)}(m)$, $f_k^{(a,b)}(m)$, $\Phi^{(a,b)}(m,n)$ and $\Phi_k^{(a,b)}(m,n)$ are $f(X)$, $f_k(X)$, $\Phi(X)$, and $\Phi_k(X)$, respectively, where $X=\{a,a+b,\ldots,a+(m-1)b\}$. The following are the results obtained by Ayad and Kihel in [2] and [3].

Theorem 5. *[3] For all positive integers m, a and b, let $f^{(a,b)}(m)$ denote the number of relatively prime subsets of $\{a, a+b, \ldots, a+(m-1)b\}$ and $f_k^{(a,b)}(m)$ denote of the number of relatively prime subsets of $\{a, a+b, \ldots, a+(m-1)b\}$ of cardinality k. Suppose that* $\gcd(a,b) = 1$. *Then*

$$f^{(a,b)}(m) = \sum_{\substack{d=1 \\ \gcd(d,b)=1}}^{a+(m-1)b} \mu(d)\left(2^{\lfloor m/d \rfloor + \varepsilon_d} - 1\right) \quad \textit{and}$$

$$f_k^{(a,b)}(m) = \sum_{\substack{d=1 \\ \gcd(d,b)=1}}^{a+(m-1)b} \mu(d)\binom{\lfloor m/d \rfloor + \varepsilon_d}{k}$$

where ε_d is the function defined in Lemma 3. If $\gcd(a,b) \neq 1$, *then* $f^{(a,b)}(m) = f_k^{(a,b)}(m) = 0$.

Theorem 6. *[2] For positive integers m, a, and b, let $\Phi^{(a,b)}(m,n)$ be the number of nonempty subsets of $\{a+, a+b, \ldots, a+(m-1)b\}$ which are relatively prime to n and let $\Phi_k^{(a,b)}(m,n)$ be the number of such subsets of cardinality k. Suppose that* $\gcd(a,b) = 1$. *Then*

$$\Phi^{(a,b)}(m,n) = \sum_{\substack{d|n \\ \gcd(b,d)=1}} \mu(d)\left(2^{\lfloor \frac{m}{d} \rfloor + \varepsilon_d} - 1\right) \quad \textit{and}$$

$$\Phi_k^{(a,b)}(m,n) = \sum_{\substack{d|n \\ \gcd(b,d)=1}} \mu(d)\binom{\lfloor \frac{m}{d} \rfloor + \varepsilon_d}{k}$$

where ε_d is the function defined in Lemma 3

Corollary 7. *([2], [3]) The formulas for $f(m,k)$, $f_k(m,\ell)$, $\Phi(m,\ell)$, $\Phi_k(m,\ell)$ $\Phi([1,m],n)$, and $\Phi_k([1,m],n)$ obtained in [12], [5], [13], and [6] are consequences of Theorem 5 and Theorem 6.*

Now we will give a proof of the formula for $\Phi_k^{(a,b)}(m,n)$ and show that the other formulas follow as a consequence.

Proof. Let $X = \{a, a+b, \ldots, a+(m-1)b\}$ and for each d, we let $A_d = \{x \in X : d \mid x\}$. Then the definition of $\Phi_k^{(a,b)}(m,n)$ can be written as

$$\Phi_k^{(a,b)}(m,n) = \sum_{\substack{\emptyset \neq A \subseteq X \\ |A|=k \\ \gcd(A,n)=1}} 1.$$

Now we capture the condition $\gcd(A, n) = 1$ by Lemma 4 and write

$$\Phi_k^{(a,b)}(m,n) = \sum_{\substack{\emptyset \neq A \subseteq X \\ |A|=k}} \sum_{d \mid \gcd(A,n)} \mu(d) = \sum_{\substack{\emptyset \neq A \subseteq X \\ |A|=k}} \sum_{\substack{d \mid \gcd(A) \\ d \mid n}} \mu(d).$$

Changing the order of summation, the above sum becomes

$$\sum_{d \mid n} \mu(d) \sum_{\substack{\emptyset \neq A \subseteq X \\ |A|=k \\ d \mid \gcd(A)}} 1.$$

Now, $d \mid \gcd(A)$ if and only if d divides all elements of A. So the condition $\emptyset \neq A \subseteq X$ and $d \mid \gcd(A)$ is equivalent to $\emptyset \neq A \subseteq A_d$. Hence the above sum is equal to

$$\sum_{d \mid n} \mu(d) \sum_{\substack{\emptyset \neq A \subseteq A_d \\ |A|=k}} 1.$$

By Lemma 3, $|A_d| = 0$ if $\gcd(d, b) \neq 1$. So for nonzero contribution, we can restrict our attention to the case $\gcd(d, b) = 1$. Therefore the above sum is equal to

$$\sum_{\substack{d \mid n \\ \gcd(d,b)=1}} \mu(d) \sum_{\substack{\emptyset \neq A \subseteq A_d \\ |A|=k}} 1 = \sum_{\substack{d \mid n \\ \gcd(d,b)=1}} \mu(d) \binom{|A_d|}{k}. \tag{4}$$

Applying Lemma 3 again, we substitute $|A_d| = \left\lfloor \frac{m}{d} \right\rfloor + \varepsilon_d$ to get

$$\Phi_k^{(a,b)}(m,n) = \sum_{\substack{d \mid n \\ \gcd(d,b)=1}} \mu(d) \binom{\lfloor m/d \rfloor + \varepsilon_d}{k}.$$

To obtain the formula of $\Phi^{(a,b)}(m, n)$, we use the well-known identity

$$\sum_{k=1}^{n} \binom{n}{k} = 2^n - 1.$$

This can also be written as $\displaystyle\sum_{k=1}^{\infty} \binom{n}{k} = 2^n - 1$ since $\binom{n}{k} = 0$ when $k > n$. Now by the definition of $\Phi^{(a,b)}(m, n)$, we have

$$
\begin{aligned}
\Phi^{(a,b)}(m,n) &= \sum_{k=1}^{m} \Phi_k^{(a,b)}(m,n) = \sum_{k=1}^{m} \sum_{\substack{d \mid n \\ \gcd(d,b)=1}} \mu(d) \binom{|A_d|}{k} \\
&= \sum_{\substack{d \mid n \\ \gcd(d,b)=1}} \mu(d) \sum_{k=1}^{m} \binom{|A_d|}{k} \\
&= \sum_{\substack{d \mid n \\ \gcd(d,b)=1}} \mu(d) \left(2^{|A_d|} - 1\right) \\
&= \sum_{\substack{d \mid n \\ \gcd(d,b)=1}} \mu(d) \left(2^{\lfloor m/d \rfloor + \varepsilon_d} - 1\right), \quad \text{as required.}
\end{aligned}
$$

Next, we put $n = (a + (m-1)b)!$ in the formula of $\Phi_k^{(a,b)}(m,n)$. For a nonempty subset A of $\{a, a+b, \ldots, a+(m-1)b\}$, we have $\gcd(A,n) = 1$ if and only if $\gcd(A) = 1$. Therefore by the definition of $\Phi_k^{(a,b)}(m,n)$ and $f_k^{(a,b)}(m)$, we have

$$
\Phi_k^{(a,b)}(m,n) = f_k^{(a,b)}(m). \tag{5}
$$

On the other hand, we have from (4) that

$$
\Phi_k^{(a,b)}(m,n) = \sum_{\substack{d \mid n \\ \gcd(d,b)=1}} \mu(d) \binom{|A_d|}{k}.
$$

Notice that $d = 1, 2, \ldots, a + (m-1)b$ are divisors of n and if $d > a + (m-1)b$, then d is larger than all elements of A and thus $|A_d| = 0$. Therefore the above sum is

$$
\Phi_k^{(a,b)}(m,n) = \sum_{\substack{d=1 \\ \gcd(d,b)=1}}^{a+(m-1)b} \mu(d) \binom{|A_d|}{k} \tag{6}
$$

From (5), (6) and Lemma 3, we obtain

$$
f_k^{(a,b)}(m) = \sum_{\substack{d=1 \\ \gcd(d,b)=1}}^{a+(m-1)b} \mu(d) \binom{|A_d|}{k} = \sum_{\substack{d=1 \\ \gcd(d,b)=1}}^{a+(m-1)b} \mu(d) \binom{\lfloor m/d \rfloor + \varepsilon_d}{k}.
$$

Similar to the proof of $\Phi^{(a,b)}(m,n)$, we sum $f_k^{(a,b)}(m)$ over all k to get $f^{(a,b)}(m)$. This completes the proof. □

Remark 8. As noted by Ayad and Kihel in [3] that we can easily deduced from Theorem 6 the results for the case when a and b are integers not necessary positive, or the case $((a,b),n) \neq 1$ or $(a,b) \neq 1$ but $((a,b),n) = 1$. For the details, see Remark 11 and Remark 12 in [3]. Combining this with Corollary 7, we see that we cover the results given by Ayad and Kihel ([2], [3]), El Bachraoui ([5], [6]), Nathanson ([12]) and Nathanson and Orosz ([13]).

4. Extending the Formulas to Finite Union of Arithmetic Progressions

In this section, we will give formulas for $f(X), f_k(X), \Phi(X,n), \Phi_k(X,n)$ when

$$\begin{aligned} X &= [a_1, b_1] \cup [a_2, b_2] \cup \ldots \cup [a_\ell, b_\ell] \quad \text{or} \\ X &= \{a_1, a_1 + b_1, \ldots, a_1 + (m_1 - 1)b_1\} \cup \{a_2, a_2 + b_2, \ldots, a_2 + (m_2 - 1)b_2\} \\ &\quad \cup \ldots \cup \{a_\ell, a_\ell + b_\ell, \ldots, a_\ell + (m_\ell - 1)b_\ell\}. \end{aligned}$$

Considering our method carefully, we see that it can be applied in any situation where the number of elements divisible by a fixed positive integer can be calculated. We illustrate this idea explicitly below.
Let X be a nonempty finite subset of integers and for each d, let $X_d = \{x \in X : d \mid x\}$. By applying Lemma 4 and changing the order of summation, we have

$$\begin{aligned} \Phi_k(X,n) = \sum_{\substack{\emptyset \neq A \subseteq X \\ |A|=k \\ \gcd(A,n)=1}} 1 = \sum_{\substack{\emptyset \neq A \subseteq X \\ |A|=k}} \sum_{d \mid \gcd(A,n)} \mu(d) &= \sum_{d \mid n} \mu(d) \sum_{\substack{\emptyset \neq A \subseteq X \\ |A|=k \\ d \mid \gcd(A)}} 1 \\ &= \sum_{d \mid n} \mu(d) \sum_{\substack{\emptyset \neq A \subseteq X_d \\ |A|=k}} 1 \\ &= \sum_{d \mid n} \mu(d) \binom{|X_d|}{k}. \end{aligned} \tag{7}$$

Summing over all k, we see that

$$\Phi(X,n) = \sum_{k=1}^{|X|} \Phi_k(X,n) = \sum_{d \mid n} \mu(d) \sum_{k=1}^{|X|} \binom{|X_d|}{k} = \sum_{d \mid n} \mu(d) \left(2^{|X_d|} - 1\right). \tag{8}$$

Again, applying Lemma 4 and changing the order of summation, we have

$$f_k(X) = \sum_{\substack{\emptyset \neq A \subseteq X \\ |A|=k}} \sum_{d \mid \gcd(A)} \mu(d) = \sum_{d=1}^{\max X} \mu(d) \sum_{\substack{\emptyset \neq A \subseteq X \\ |A|=k \\ d \mid \gcd(A)}} 1$$

$$= \sum_{d=1}^{\max X} \mu(d) \sum_{\substack{\emptyset \neq A \subseteq X_d \\ |A|=k}} 1 = \sum_{d=1}^{\max X} \mu(d) \binom{|X_d|}{k}. \tag{9}$$

Summing $f_k(X)$ over all k, we obtain

$$f(X) = \sum_{d=1}^{\max X} \mu(d) \left(2^{|X_d|} - 1\right) \tag{10}$$

Remark 9. 1) If $n > 1$, by Lemma 4, the formula in (8) can be reduced to

$$\Phi(X, n) = \sum_{d \mid n} \mu(d) 2^{|X_d|}.$$

2) From (7), (8), (9), and (10), we see that explicit formulas for $\Phi_k(X,n), \Phi(X,n)$, $f_k(X)$ and $f(X)$ can be obtained whenever we can compute $|X_d|$ for all d.

With equations (7), (8), (9), and (10), we obtain the following theorem.

Theorem 10. *Let $X = [a_1, b_1] \cup [a_2, b_2] \cup \ldots \cup [a_\ell, b_\ell]$ where $a_1 \leq b_1 < a_2 \leq b_2 < \ldots < a_\ell \leq b_\ell$. Assume that $[a_i, b_i] \cap [a_j, b_j] = \emptyset$ for $i \neq j$. Then*

(i) $\Phi_k(X, n) = \sum_{d \mid n} \mu(d) \binom{\sum_{i=1}^{\ell} \lfloor \frac{b_i}{d} \rfloor - \lfloor \frac{a_i - 1}{d} \rfloor}{k}$

(ii) $\Phi(X, n) = \sum_{d \mid n} \mu(d) \left(2^{\sum_{i=1}^{\ell} \lfloor \frac{b_i}{d} \rfloor - \lfloor \frac{a_i - 1}{d} \rfloor} - 1\right)$

(iii) $f_k(X) = \sum_{d=1}^{b_\ell} \mu(d) \binom{\sum_{i=1}^{\ell} \lfloor \frac{b_i}{d} \rfloor - \lfloor \frac{a_i - 1}{d} \rfloor}{k}$

(iv) $f(X) = \sum_{d=1}^{b_\ell} \mu(d) \left(2^{\sum_{i=1}^{\ell} \lfloor \frac{b_i}{d} \rfloor - \lfloor \frac{a_i - 1}{d} \rfloor} - 1\right).$

Proof. The number of integers $x \in [1, n]$ divisible by d is equal to $\lfloor \frac{n}{d} \rfloor$. So the number of integers $x \in [a, b]$ such that $d \mid x$ is equal to $\lfloor \frac{b}{d} \rfloor - \lfloor \frac{a-1}{d} \rfloor$. This implies that

$$|X_d| = \sum_{i=1}^{\ell} \left\lfloor \frac{b_i}{d} \right\rfloor - \left\lfloor \frac{a_i - 1}{d} \right\rfloor.$$

Substituting this in (7), (8), (9), and (10), we obtain the desired result. □

Note that the formulas in Theorem 10 are also obtained by El Bachraoui [8], [9] in a different form but his proof does not seem to be applicable in more general situations such as [14], [15]. However, our method still works well in this case (see section 5).

Theorem 11. *Let* $X = \bigcup_{i=1}^{\ell} I_i$ *where* $I_i = \{a_i, a_i + b_i, \ldots, a_i + (m_i - 1)b_i\}$ *be an* m_i*-arithmetic progression. Assume that* $(a_i, b_i) = 1$ *for all* i *and* $I_i \cap I_j = \emptyset$ *for all* $i \neq j$*. Then*

(i) $\Phi_k(X, n) = \sum_{d|n} \mu(d) \binom{\sum_{i=1}^{\ell} |I_{id}|}{k}$

(ii) $\Phi(X, n) = \sum_{d|n} \mu(d) \left(2^{\sum_{i=1}^{\ell} |I_{id}|} - 1\right)$

(iii) $f_k(X) = \sum_{d=1}^{\max X} \mu(d) \binom{\sum_{i=1}^{\ell} |I_{id}|}{k}$

(iv) $f(X) = \sum_{d=1}^{\max X} \mu(d) \left(2^{\sum_{i=1}^{\ell} |I_{id}|} - 1\right)$

where for each i *and* d, $|I_{id}| = 0$ *if* $(d, b_i) \neq 1$ *and if* $(d, b_i) = 1$, *then* $|I_{id}| = \left\lfloor \frac{m_i}{d} \right\rfloor + \varepsilon_{id}$ *where*

$$\varepsilon_{id} = \begin{cases} 1 & \text{if } d \nmid m_i \text{ and } -a_i b_i^{-1} \bmod d \in \{0, \ldots, m_i - 1 - \left\lfloor \frac{m_i}{d} \right\rfloor d\}, \\ & \text{where } b_i^{-1} \text{ is the multiplication inverse of } b_i \text{ modulo } d. \\ 0 & \text{otherwise.} \end{cases}$$

Proof. We have $|X_d| = \sum_{i=1}^{\ell} |I_{id}|$ and $|I_{id}|$ can be obtained by Lemma 3. This completes the proof. □

Remark 12. 1) If $(a_i, b_i) > 1$ for some i, we can apply Lemma 2 to obtain the corresponding result to Theorem 11.
2) Theorem 11 extends Ayad and Kihel's results [2], [3] to the case of finite union of arithmetic progressions. Replacing $\ell = 1$ and $X = \{a, a + b, \ldots, a + (m - 1)b\}$, we obtain their result in [2] and [3].

5. Cover Shonhiwa's Theorems

Shonhiwa considers the case $X = [1, n]$ with various constraints. He [14], [15] uses the Möbius inversion formula, the inclusion-exclusion principle, generating functions, and standard formulas in enumerative combinatorics. In this section, we illustrate again how our method can be used to obtain Shonhiwa's results in a faster and simpler way. So let us recall his theorems in [14], [15].

Theorem 13. *([14], [15]) Let*

$$\begin{aligned}
(i)\ S_k^m(n) &= \sum_{\substack{1\le a_1,a_2,\ldots,a_k\le n\\(a_1,a_2,\ldots,a_k,m)=1}} 1; \quad \forall n\ge k\ge 1, m\ge 1\\
(ii)\ G_k(n) &= \sum_{\substack{1\le a_1,a_2,\ldots,a_k\le n\\(a_1,a_2,\ldots,a_k)=1}} 1; \quad \forall n\ge k\ge 1\\
(iii)\ L_k^m(n) &= \sum_{\substack{1\le a_1\le a_2\le\ldots\le a_k\le n\\(a_1,a_2,\ldots,a_k,m)=1}} 1; \quad \forall n\ge k\ge 1, m\ge 1\\
(iv)\ H_k(n) &= \sum_{\substack{1\le a_1\le a_2\le\ldots\le a_k\le n\\(a_1,a_2,\ldots,a_k)=1}} 1; \quad \forall n\ge k\ge 1\\
(v)\ T_k^m(n) &= \sum_{\substack{1\le a_1< a_2<\ldots< a_k\le n\\(a_1,a_2,\ldots,a_k,m)=1}} 1; \quad \forall n\ge k\ge 1, m\ge 1.
\end{aligned}$$

Then

$$\begin{aligned}
S_k^m(n) &= \sum_{d\mid m}\mu(d)\left\lfloor\frac{n}{d}\right\rfloor^k,\\
G_k(n) &= \sum_{d\le n}\mu(d)\left\lfloor\frac{n}{d}\right\rfloor^k,\\
L_k^m(n) &= \sum_{d\mid m}\mu(d)\binom{\lfloor\frac{n}{d}\rfloor+k-1}{k}\\
H_k(n) &= \sum_{d\le n}\mu(d)\binom{\lfloor\frac{n}{d}\rfloor+k-1}{k}\\
T_k^m(n) &= \sum_{d\mid m}\mu(d)\binom{\lfloor\frac{n}{d}\rfloor}{k}.
\end{aligned}$$

Proof. Throughout the proof, we let

$$\begin{aligned}
&A=\{1,2,\ldots,n\} \quad \text{and}\\
&A_d=\{a\in A\ :\ d\mid a\}=\{d,2d,3d,\ldots,\left\lfloor\frac{n}{d}\right\rfloor d\}.
\end{aligned}$$

For (i), we apply Lemma 4 and change the order of summation to obtain

$$\begin{aligned} S_k^m(n) &= \sum_{1\le a_1,a_2,\ldots,a_k\le n} \sum_{d|(a_1,a_2,\ldots,a_k,m)} \mu(d) \\ &= \sum_{d|m} \mu(d) \sum_{\substack{1\le a_1,a_2,\ldots,a_k\le n \\ d|a_1,d|a_2,\ldots,d|a_k}} 1. \end{aligned}$$

The condition $1 \le a \le n$ and $d \mid a$ is the same as $a \in A_d$. Therefore the number of choices for a is $|A_d| = \lfloor \frac{n}{d} \rfloor$. So we have

$$S_k^m(n) = \sum_{d|m} \mu(d) \sum_{a_1,a_2,\ldots,a_k\in A_d} 1 = \sum_{d|m} \mu(d)|A_d|^k = \sum_{d|m} \mu(d) \left\lfloor \frac{n}{d} \right\rfloor^k.$$

For (ii) we put $m = n!$ in $S_k^m(n)$ and argue as in the proof of $\Phi^{(a,b)}(m,n)$ in the previous section. We see that

$$G_k(n) = S_k^m(n) = \sum_{d|m} \mu(d) \left\lfloor \frac{n}{d} \right\rfloor^k = \sum_{d|n!} \mu(d) \left\lfloor \frac{n}{d} \right\rfloor^k = \sum_{d=1}^{n} \mu(d) \left\lfloor \frac{n}{d} \right\rfloor^k.$$

Before giving the proof of (iii), let us recall an elementary formula in enumeration. The number of ways to select k objects from n different objects with repetition allowed is equal to ([11], p.47)

$$\binom{k+n-1}{k}. \tag{11}$$

Now similar to (i), we apply Lemma 4 and change the order of summation to obtain

$$L_k^m(n) = \sum_{d|m} \mu(d) \sum_{\substack{1\le a_1\le a_2\le\ldots\le a_k\le n \\ d|a_1,d|a_2,\ldots,d|a_k}} 1 = \sum_{d|m} \mu(d) \sum_{\substack{a_1,a_2,\ldots,a_k\in A_d \\ a_1\le a_2\le\ldots\le a_k}} 1.$$

The inner sum is equal to the number of ways to select k objects from $|A_d| = \lfloor \frac{n}{d} \rfloor$ different objects with repetition allowed. So it is equal to $\binom{k+\lfloor \frac{n}{d} \rfloor-1}{k}$ by (11). Thus

$$L_k^m(n) = \sum_{d|m} \mu(d) \binom{k+\lfloor \frac{n}{d} \rfloor - 1}{k}. \tag{12}$$

For (iv), we put $m = n!$ in (12) and argue as before to get $H_k(n)$.
Similar to (i) and (iii), we have

$$T_k^m(n) = \sum_{d|m} \mu(d) \sum_{\substack{1\le a_1< a_2<\ldots< a_k\le n \\ d|a_1,d|a_2,\ldots,d|a_k}} 1 = \sum_{d|m} \mu(d) \sum_{\substack{a_1,a_2,\ldots,a_k\in A_d \\ a_1< a_2<\ldots< a_k}} 1.$$

The inner sum is the same as k-combinations of $|A_d| = \lfloor \frac{n}{d} \rfloor$ distinct objects, which equals $\binom{\lfloor \frac{n}{d} \rfloor}{k}$. This gives (v). Hence the proof is complete. □

6. Possible Research Questions

There are some interesting problems that can be investigated further. For example, combinatorial identities given by Bachraoui and Salim in [8], divisor-type functions by Toth [17], arithmetic properties of the sequence $f(n) = f([1,n])$, and $\Phi(n) = \Phi([1,n])$ by Ayad and Kihel [4] and Min Tang [16] can be developed more. The possible research questions are the following:

(i) Is $f(n)$ a cube, a fourth power, or a perfect power for some $n \geq 2$? (Ayad and Kihel [4] proved that $f(n)$ is never a square for $n \geq 2$.) We conjecture that $f(n)$ is never a perfect power for any $n \geq 2$.

(ii) What are congruence relations for $f(n), f_k(n), \Phi(n)$, and $\Phi_k(n)$? (Ayad and Kihel [4] showed that $\Phi(n) \equiv 0 (\bmod 3)$ for $n \geq 3$.)

(iii) Does the sequence $f(n)$ contain infinitely many primes?

(iv) What are the solutions of $f(n) = \Phi(m)$? We conjecture that all solutions are $(m,n) = (1,1)$ or $(2,2)$.

(v) What are the properties of the divisor sums $D(n) = \sum_{d|n} f(d)$? Currently, we obtain a combinatorial and a congruence property of $D(n)$ but there are some questions that we do not have a complete answer. Looking at the values of $D(n)$ and $\Phi(n)$, we see that $D(n) > \Phi(n)$ for all even integers $2 \leq n \leq 30$ and $D(n) < \Phi(n)$ for all odd integers $5 \leq n \leq 30$. This pattern might hold for all larger values of n or at least for infinitely many n but we do not have a proof yet.

Acknowledgement The author's research is financially supported by Faculty of Science, Silpakorn University, Thailand, contract number RGP 2555-07. The author also wishes to thank the anonymous referee and the editor (Professor Bruce Landman) for careful reading and pointing out some typographical and grammatical errors.

References

[1] T.M. Apostol, Introduction to Analytic Number Theory, Springer, Berline, 2010.

[2] M. Ayad and O. Kihel, On the number of subsets relatively prime to an integer, J. Integer Seq. 11 (2008), no. 5, Article 08.5.5, 4 pp.

[3] M. Ayad and O. Kihel, On relatively prime sets, Integers 9 (2009), A28, 343-352.

[4] M. Ayad and O. Kihel, The number of relatively prime subsets of $\{1, 2, \ldots, N\}$, Integers 9 (2009), A14, 163-166.

[5] M. El Bachraoui, The number of relatively prime subsets and phi functions for $\{m, m+1, \ldots, n\}$, Integers 7 (2007), A43, 8 pp.

[6] M. El Bachraoui, On the number of subsets of $[1, M]$ relatively prime to N and asymptotic estimates, Integers 8 (2008), A41, 5 pp.

[7] M. El Bachraoui, On relatively prime subsets and supersets, Integers 10 (2010), A47, 565-574.

[8] M. El Bachraoui, Combinatorial identities via phi functions and relatively prime subsets, preprint arxiv1002.3254vi.

[9] M. El Bachraoui, On Relatively Prime Subsets, Combinatorial identities, and Diophantine Equations, J. Integers Seq. 15 (2012), Article 12.3.6, 9pp.

[10] M. El Bachraoui, M. Salim Combinatorial identities involving the Möbius function, J. Integer Seq. 13 (2010), no. 8, Article 10.8.6, 7 pp.

[11] C. Chuan-Chong, K. Khee-Meng, Principles and Techniques in Combinatorics, World Scientific, River Edge, New Jersey, 1992.

[12] M.B. Nathanson, Affine invariants, relatively prime sets, and a phi function for subsets of $\{1, 2, \ldots, n\}$, Integers 7 (2007), A1, 7 pp.

[13] M.B. Nathanson and B. Orosz, Asymptotic estimates for phi functions for subsets of $\{m+1, m+2, \ldots, n\}$, Integers 7 (2007), A54, 5 pp.

[14] T. Shonhiwa, On relatively prime sets counting functions, Integers 10 (2010), A39, 465 - 476.

[15] T. Shonhiwa, A generalization of the Euler and Jordan totient functions. Fibonacci Quart. 37 (1999), no. 1, 67-76.

[16] M.Tang, Relatively prime sets and a phi function for subsets of $\{1, 2, \ldots, n\}$, J. Integers Seq. 13 (2010), Article 10.7.6, 5pp.

[17] L. Toth, On the number of certain relatively prime subsets of $\{1, 2, \ldots, n\}$, Integers 10 (2010), A35, 407-421.

INTERSECTING RATIONAL BEATTY SEQUENCES

Jamie Simpson
Dept. of Mathematics and Statistics, Curtin University, Perth, WA, Australia
Jamie.Simpson@curtin.edu.au

Received: 3/5/12, Revised: 3/24/13, Accepted: 7/6/13, Published: 8/12/13

Abstract

A rational Beatty sequence has the form $\{\lfloor pi/q+b\rfloor : i \in \mathbb{Z}\}$ where $p > q > 0$ and $\gcd(p,q) = 1$. We call p/q the *modulus* of the sequence and b the *offset*. Morikawa gave a condition on the moduli of two Beatty sequences such that they would be disjoint for a suitable choice of offsets. Holzman and Fraenkel showed that the sequence formed by the intersection of two Beatty sequences with moduli p_1/q_1 and p_2/q_2, $q_2 \leq q_1$, could have as many as q_2+3 distinct consecutive differences. In this note we show that if the moduli satisfy the Morikawa condition but the sequences do intersect then the consecutive differences take on at most three different values.

1. Introduction

A Beatty sequence has the form $\{\lfloor i\alpha + \beta\rfloor : i \in \mathbb{Z}\}$. We call α the *modulus* and β the *offset* of the sequence. The sequences were named for Samuel Beatty [1] who asked for a proof that two Beatty sequences, with offsets equal to 0 and moduli α_1 and α_2, partition the positive integers if both moduli are irrational and $1/\alpha_1 + 1/\alpha_2 = 1$. A proof of this pleasing result appeared in [2]. A Beatty sequence is *rational* or *irrational* according to whether its modulus is rational or irrational. Covering properties of irrational Beatty sequences are now well understood. See, for instance, [8] and its bibliography. This is not so for coverings by collections of rational Beatty sequences which are the subject of this paper. We write $S(p/q, b)$, where $\gcd(p,q) = 1$, for the Beatty Sequence $\{\lfloor pi/q+b\rfloor : i \in \mathbb{Z}\}$. We will assume throughout that b here is an integer – this involves no loss of generality by a result in [9]. The following famous conjecture is due to Aviezri Fraenkel [4].

Conjecture 1. If the collection of Beatty sequences $\{S(p_i/q_i, b_i) : i = 1, \ldots, t\}$ partitions the integers with $t > 2$ then $\{p_1/q_1, \ldots, p_t/q_t\} = \{(2^t-1)/2^{t-i} : 1 \leq i \leq t\}$.

This conjecture has generated a considerable literature. The strongest result to date is by Bark and Varjú [3] who showed that any counterexample must have $t > 7$. See also the surveys [8], and Section F14 of [6].

A Beatty sequence may be regarded as an approximation to an arithmetic progression in that its consecutive differences take two values ($\lfloor \alpha \rfloor$ and $\lceil \alpha \rceil$) rather than one. The intersection properties of arithmetic progressions are given by the Chinese Remainder Theorem which we give, in a long-winded way, here.

Theorem 2. (Chinese Remainder Theorem) *Let a_1, a_2, b_1, b_2 be integers with a_1 and a_2 positive.*
(a) There exist integers b_1 and b_2 such that $S(a_1, b_1)$ and $S(a_2, b_2)$ are disjoint if and only if $\gcd(a_1, a_2) > 1$.
(b) If $\gcd(a_1, a_2) = 1$, *then the intersection of $S(a_1, b_1)$ and $S(a_2, b_2)$ is an arithmetic progression with common difference $a_1 a_2$.*
(c) If $\gcd(a_1, a_2) > 1$, *and $S(a_1, b_1)$ and $S(a_2, b_2)$ do intersect, then their intersection is an arithmetic progression with common difference* $\mathrm{lcm}(a_1 a_2)$.

The situation for Beatty sequences is more complicated. Instead of part (a) we have the following result of Ryozu Morikawa [7], [10].

Theorem 3. (Japanese Remainder Theorem) *With $p = (p_1, p_2)$, $q = (q_1, q_2)$, $q_1 = u_1 q$ and $q_2 = u_2 q$, there exist numbers b_1 and b_2 such that $S(p_1/q_1, b_1)$ and $S(p_2/q_2, b_2)$ are disjoint if and only if there exist positive integers x and y such that*

$$xu_1 + yu_2 = p - 2u_1u_2(q - 1). \tag{1}$$

When this is so we say that p_1/q_1 and p_2/q_2 satisfy the *Morikawa condition.*

Definition 4. If $a_1, \ldots, a_n$ is an increasing sequence of integers then we say that the differences $a_{i+1} - a_i$ are the *gap sizes* of the sequence. If S is a set of residues modulo p, whose members have been reduced modulo p to integers in the interval $[0, p-1]$, and labeled $g_1 \leq g_2 \leq \cdots \leq g_n$, then the set of *gap sizes* of S is $\{g_{i+1} - g_i : i = 1, \ldots, n-1\} \cup \{p + g_1 - g_n\}$.

Instead of part (b) of Theorem 2 we have the following result of Fraenkel and Holzman [5].

Theorem 5. *If $S(p_1/q_1, b_1)$ and $S(p_2/q_2, b_2)$ are Beatty sequences whose moduli do not satisfy the Morikawa condition, and $q_1 \geq q_2 \geq 2$, then the intersection of the two sequences has at most $q_2 + 3$ distinct gap sizes.*

The bound here is best possible. In this paper we obtain an analogy of part (c) of Theorem 2 by giving a precise description of the intersection of two Beatty sequences whose moduli satisfy the Morikawa condition. In particular, it follows that in this case the intersection has at most three gap sizes.

2. Results

Notation 6. Throughout this section we will use the following notation. We will be considering Beatty sequences $S(p_1/q_1, b_1)$ and $S(p_2/q_2, b_2)$. We assume, without loss of generality, that $q_1 \leq q_2$. We put $p = \gcd(p_1, p_2)$, $p_1 = mp$, and $p_2 = np$. This implies

$$\gcd(m, q_1) = \gcd(n, q_2) = 1.$$

We set $\overline{q}_1$ and $\overline{q}_2$ to be the least non-negative residues satisfying $q_1\overline{q}_1 \equiv -1 \pmod{p}$ and $q_2\overline{q}_2 \equiv -1 \pmod{p}$ respectively. Similarly, $\overline{q}_m$ and $\overline{q}_n$ are the least non-negative residues satisfying $q_1\overline{q}_m \equiv -1 \pmod{m}$ and $q_2\overline{q}_n \equiv -1 \pmod{n}$, respectively. We set $k_1 = (q_1\overline{q}_1 + 1)/p$ and $k_2 = (q_2\overline{q}_2 + 1)/p$.

The argument proceeds in three steps. In Theorem 11 we obtain an expression for the intersection of $S(p/q_1, b_1)$ and $S(p/q_2, b_2)$. This is used in Theorem 15 to obtain an expression for the intersection of $S(pm/q_1, b_1)$ and $S(p/q_2, b_2)$, and that result is used in Theorem 16 to obtain an expression for the intersection of $S(pm/q_1, b_1)$ and $S(pn/q_2, b_2)$.

Definition 7. Let b, n, p, d be positive integers with $n \leq p$, $\gcd(p, d) = 1$ and $S = \{id + b \mod p : i = 0, \ldots, n-1\}$. Reduce each member of S to an integer in $[0, p-1]$ and label them $g_1, \ldots, g_n$, such that $g_1 \leq g_2 \leq \cdots \leq g_n$. We say that this sequence is a *modular arithmetic progression modulo* p with *additive difference* d.

The following is easily derived from the usual Three Gap Theorem, see [11].

Theorem 8 (Three Gap Theorem). *The set of gap sizes of a modular arithmetic progression has cardinality at most* 3*, and if the cardinality equals* 3 *then the largest member of the set equals the sum of the other two.*

The following corollary follows immediately from the preceding theorem and Definition 7.

Corollary 9. *If* $g_1, \ldots, g_n$ *is a modular arithmetic progression modulo* p*, then the set of gap sizes in the doubly infinite increasing sequence with range* $\{g_i + jp : 1 \leq i \leq n, j \in \mathbb{Z}\}$ *has cardinality at most* 3*, and if the cardinality equals* 3 *then the largest member of the set equals the sum of the other two.*

The Beatty sequence $S = S(p/q, b)$ has period p in the sense that $a \in S$ if and only if $a + p \in S$, and so is characterised by a set of residues modulo p. The following is Theorem 3 of [10].

Theorem 10. *The sequence* $S(p/q, b)$ *with* $\gcd(p, q) = 1$ *coincides with the set of integers congruent modulo* p *to a member of* $\{i\overline{q} + b : 0 \leq i \leq q-1\}$*, where* $q\overline{q} \equiv -1$ $\pmod{p}$.

Thus the set of residues in a Beatty sequence forms a modular arithmetic progression, and the Beatty sequence itself fulfils the conditions of Corollary 9. In fact the Beatty sequence has at most two gap sizes. These are $\lfloor p/q \rfloor$ and $\lceil p/q \rceil$ (which are equal when $q = 1$).

Theorem 11. *Let p, q_1 and q_2 be positive integers with $\gcd(p, q_1) = \gcd(p, q_2) = 1$ such that p/q_1 and p/q_2 satisfy the Morikawa condition. If b_1, b_2 are integers such that $S(p/q_1, b_1)$ and $S(p/q_2, b_2)$ intersect then the intersection is the set of residues*

$$\{a\bar{q}_1 + iG_1\bar{q}_1 + b_1 : 0 \leq i \leq t-1\} \tag{2}$$

modulo p for some positive integer t where G_1 is the smallest gap size in $\{-q_1 i\bar{q}_2 - q_1 b_2 \pmod{p} : i = 0, \dots, q_2 - 1\}$ if $t > 2$, and the second or third smallest gap size if $t = 2$, and a is a non-negative integer satisfying

$$a + G_1(t-1) < q_1. \tag{3}$$

The ideas of the following proof are illustrated in the accompanying figure.

Proof. Without loss of generality suppose $b_1 = 0$. Let B_1 be the set of residues modulo p in $S(p/q_1, b_1)$ and B_2 be the set in $S(p/q_2, b_2)$. Theorem 10 implies that $B_1 \equiv \{i\bar{q}_1 : i = 0, \dots, q_1 - 1\} \pmod{p}$ and $B_2 \equiv \{i\bar{q}_2 + b_2 : i = 0 \dots q_2 - 1\} \pmod{p}$. Let

$$\begin{aligned} B_1^* &\equiv \{-q_1 i\bar{q}_1 : 0 \leq i \leq q_1 - 1\} \pmod{p} \\ &\equiv \{0, \dots, q_1 - 1\} \pmod{p}, \end{aligned}$$

and

$$B_2^* \equiv \{-q_1 i\bar{q}_2 - q_1 b_2 \bmod p : 0 \leq i \leq q_2 - 1\} \pmod{p}.$$

Clearly

$$-q_1(B_1 \cap B_2) \equiv B_1^* \cap B_2^* \pmod{p} \tag{4}$$

so $|B_1 \cap B_2| = |B_1^* \cap B_2^*|$. If $|B_1 \cap B_2| = 1$ then we have nothing to prove, and if $|B_1 \cap B_2| = 2$ then a simpler version of the proof applies (but note the comments at the end of the proof), so we assume $|B_1 \cap B_2| \geq 3$. Consider the set of gaps in the modular arithmetic progression B_2^*. By Theorem 8 there are at most 3 gap sizes. We will assume there are 3 (if there are less, then a simpler version of the proof applies) and that the gaps are $G_1 < G_2 < G_3$. Since the moduli of the Beatty sequences satisfy the Morikawa condition there will be some value for b_2 which makes $|B_1 \cap B_2|$ empty, and therefore $B_1^* \cap B_2^* = \{0, \dots, q_1 - 1\} \cap B_2^*$ empty, which implies $G_3 > q_1$. Now consider a different value of b_2 for which the sequences do intersect (note that this doesn't change the gap sizes of B_2^*) and let the intersection be the sequence

$$0 \leq a_1 < a_2 < \cdots < a_t \leq q_1 - 1 \tag{5}$$

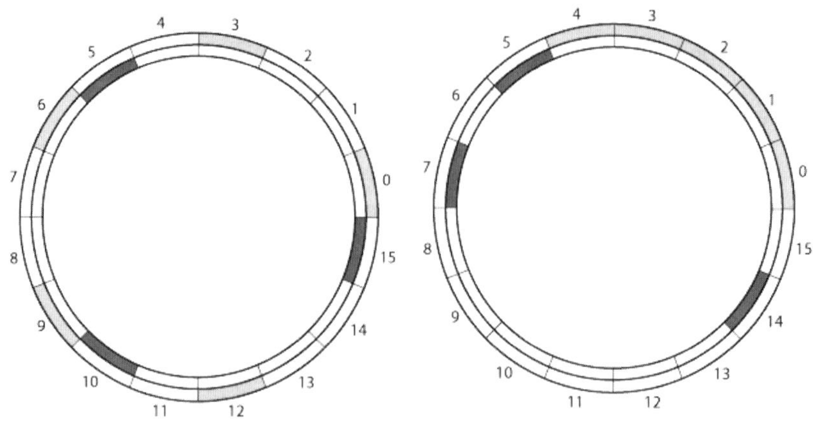

Figure 1: The diagram on the left shows the set of residues modulo 16 of $B_1 = S(16/5, 0)$ in the outer ring and $B_2 = S(16/3, 5)$ in the inner ring. The diagram on the right shows $B_1^* \equiv -5B_1 \pmod{16}$ and $B_2^* \equiv -5B_2 \pmod{16}$. The Beatty sequences do not intersect. Making a suitable change to the offset of B_1 will cause the outer rings to rotate so that they do.

where, by the assumption above, $t \geq 3$. We claim that the only gap size in this sequence is G_1. Clearly no gap can equal G_3 since

$$G_3 > q_1 > a_t - a_1. \tag{6}$$

Suppose G_2 occurs in the sequence. Then, since $t \geq 3$, there is an adjacent gap of size at least G_1. This implies $a_t - a_1 \geq G_1 + G_2$, but $G_1 + G_2 = G_3$ by Theorem 8 and we get a contradiction as in (6). Thus all gaps equal G_1 and

$$B_1^* \cap B_2^* = \{a + iG_1 : 0 \leq i \leq t-1\}$$

for some integer a. By (5) we have

$$a + (t-1)G_1 \leq q_1 - 1. \tag{7}$$

Therefore from (4)

$$\begin{aligned} B_1 \cap B_2 &\equiv \overline{q}_1\{a + iG_1 : 0 \leq i \leq t-1\} \pmod{p} \\ &\equiv \{a\overline{q}_1 + iG_1\overline{q}_1, 0 \leq i \leq t-1\} \pmod{p} \end{aligned}$$

which is (2). If there are only two elements in the intersection we cannot conclude that $a_2 - a_1 < G_2$. This observation leads to the anomalous case in the theorem. $\square$

This completes the analysis of the case when two Beatty sequences with the same numerator in their moduli intersect. Before progressing to the more general case we prove the following theorem and its corollary.

Theorem 12. *Let m and t be positive integers, and a_1, a_2, b_1 and b_2 be integers in the interval $[0, t-1]$. If*

$$\{a_1 i + b_1 : i = 0, \ldots, t-1\} \equiv \{a_2 i + b_2 : i = 0, \ldots, t-1\} \pmod{m} \tag{8}$$

and

$$(t+1)\gcd(a_1, a_2, m) < m \tag{9}$$

then either $a_1 \equiv a_2 \pmod{m}$ and $b_1 \equiv b_2 \pmod{m}$, or $a_1 \equiv -a_2 \pmod{m}$ and $b_2 \equiv b_1 + a_1(t-1) \pmod{m}$.

Proof. Let $S = \{a_1 i + b_1 : i = 0, \ldots, t-1\}$. We first show that neither $a_1 t + b_1$ nor $a_1(t+1) + b_1$ is congruent modulo m to a member of S. Suppose otherwise. If $a_1 t + b_1$ is congruent modulo m to a member of S, then

$$\begin{aligned} & a_1 t + b_1 \equiv a_1 i + b_1 \pmod{m} \text{ for some } i \in [0, t-1] \\ \Rightarrow\ & a_1(t-i) \equiv 0 \pmod{m}, \end{aligned}$$

which implies $i = 0$ else the members of S would not be distinct. Hence m divides $a_1 t$ but m does not divide $a_1 i$ for any $i \in [1, t-1]$. Hence t divides m and m/t divides a_1. Thus $a_1 = Am/t$ for some integer A where $\gcd(A, m) = 1$ and

$$S \equiv \{(Am/t)i + b_1 : i = 0, \ldots, t-1\} \pmod{m}.$$

Clearly

$$\{(Am/t)i + b_1 : i = 0, \ldots, t-1\} \equiv \{(m/t)i + b_1 : i = 0, \ldots, t-1\} \pmod{m}$$

so $S \equiv \{mi/t + b_1 : i = 0, \ldots, t-1\} \pmod{m}$. It follows that m/t divides a_2. In fact we have

$$m \mid \gcd(a_1, m)t \text{ and } m \mid \gcd(a_2, m)t,$$

and thus m divides $\gcd(a_1, a_2, m)t$ which implies $t\gcd(a_1, a_2, m) \geq m$, contradicting (9). We conclude $a_1 t + b_1$ is not congruent modulo m to any member of S.

Now suppose $a_1(t+1) + b_1$ is congruent modulo m to a member of S. As above this leads to

$$a_1(t+1-i) \equiv 0 \pmod{m}$$

for some i in $[0, t-1]$. In order for the members of S to be distinct this implies $i = 0$ or $i = 1$. If $i = 1$ we get $a_1 t \equiv 0 \pmod{m}$ which is impossible as in the previous case. Using similar reasoning to the previous case we see that if $i = 0$ which implies that m divides $a_1(t+1)$, $t+1$ divides m, and

$$S \cup \{a_1 t + b_1\} \equiv \{mi/(t+1) + b_1 : i = 0, \ldots, t\} \pmod{m}.$$

Then $m/(t+1)$ divides a_2 which leads to

$$m \mid \gcd(a_1, a_2, m)(t+1),$$

implying that $(t+1)\gcd(a_1, a_2, m) \geq m$, again contradicting (9). We conclude that $a_1(t+1) + b_1$ is not congruent modulo m to any member of S.

By similar reasoning we conclude that neither $a_2 t + b_2$ nor $a_2(t+1) + b_2$ is congruent modulo m to a member of S.

Now consider the set $S' = \{a_2 i + b_2 : i = 1, \ldots, t\}$ modulo m. That is, S' is formed by adding a_2 to each member of S. Note that $|S \cap S'| = t - 1$ since $a_2(t+1) + b_2$ is not congruent modulo m to any member of S. Hence

$$S \cap S' = \{a_1 i + b_1 : i = 0, \ldots, t-1, i \neq j\}, \tag{10}$$

for some $j \in [0, t-1]$. We will show that j equals 0 or $t-1$. Suppose, for the sake of contradiction, that $0 < j < t-1$. Since $j > 0$, $a_1(j-1) + b_1$ belongs to $S \cap S'$. Then,

$$a_1(j-1) + b_1 \equiv a_1(t-1) + b_1 + a_2 \pmod{m}, \tag{11}$$

since if $t-1$ were replaced by k with $0 \leq k < t-1$, then $a_1 j + b_1$ would be congruent modulo m to $a_1(k+1) + b_1 + a_2$ and so belong to S'. From (11) we therefore get

$$a_2 \equiv (j-t)a_1 \pmod{m}.$$

Now from our assumption that $j < t-1$ we have $a_1(j+1)+b_1$ congruent to a member of S, and by (11) congruent to $a_1(t+1) + b_1 + a_2$ modulo m. So $a_1(t+1) + b_1 + a_2$ is congruent modulo m to a member of S', which implies that $a_1(t+1) + b_1$ is congruent to a member of S which we showed earlier to be impossible. We conclude that $j = 0$ or $j = t-1$.

If $j = 0$ then (10) gives

$$S' \cap S \equiv \{a_1 i + b_1 : i = 1, \ldots, t-1\} \pmod{m},$$

so that no member is congruent modulo m to b_1, and $a_1 + b_1 \equiv a_1 k + b_1 + a_2$ for some k in $\{0, \ldots, t-1\}$. We must have $k = 0$, else $S' \cap S$ would include an element congruent to $a_1(k-1) + b_1 + a_2 \equiv b_1 \pmod{m}$. Hence we get $a_1 \equiv a_2 \pmod{m}$, and from (8) we see that $b_1 = b_2$.

Similarly, if $j = t-1$ then (10) gives $a_1(t-2) + b_1 \equiv a_1 k + b_1 + a_2 \pmod{m}$ for some k in $\{0, \ldots, t-1\}$, and this k must equal $t-1$ else $a_1(t-1) + b_1$ would be congruent modulo m to a member of S' and we get $a_1 \equiv -a_2 \pmod{m}$. In this case (8) then gives

$$\begin{aligned} & \{a_1 i + b_1 : 0 \leq i \leq t-1\} \\ \equiv\ & \{a_1 i + b_2 : 0 \leq i \leq t-1\} \pmod{m} \\ \equiv\ & \{a_1(t-1-i) + b_2 - a_1(t-1) : 0 \leq i \leq t-1\} \pmod{m} \\ \equiv\ & \{a_1 j + b_2 - a_1(t-1) : 0 \leq j \leq t-1\} \pmod{m}, \end{aligned}$$

which implies

$$b_1 \equiv b_2 - a_1(t-1) \pmod{m},$$

as required. □

Corollary 13. *Using Notation 6 we have either*

$$a_2\overline{q}_2 + b_2 \equiv a\overline{q}_1 + b_1 \pmod p$$

or

$$a_2\overline{q}_2 + b_2 \equiv \overline{q}_1(a + (q_2 - 1)G_1) + b_1 \pmod p.$$

Proof. Let H be the smallest gap size in $\{-q_2 i\overline{q}_1 - q_2 b_1 \mod p : 0 \le i \le q_1 - 1\}$. By swapping the roles of $S(p/q_1, b_1)$ and $S(p/q_2, b_2)$ in Theorem 11 we can rewrite (2) as follows. The set of residues modulo p in the intersection of $S(p/q_1, b_1)$ and $S(p/q_2, b_2)$ is

$$\{a_2\overline{q}_2 + iH\overline{q}_2 + b_2 : 0 \le i \le t-1\},$$

where a_2 satisfies $0 \le a_2 + H(t-1) \le q_2$. Note that we don't need a different t value as the size of the intersection doesn't change. We thus have, using the notation of the theorem,

$$\{a\overline{q}_1 + iG_1\overline{q}_1 + b_1 : 0 \le i \le t-1\} \equiv \{a_2\overline{q}_2 + iH\overline{q}_2 + b_2 : 0 \le i \le t-1\} \tag{12}$$

modulo p. We will show that either $H\overline{q}_2 \equiv G_1\overline{q}_1$ or $H\overline{q}_2 \equiv -G_1\overline{q}_1$ modulo p. This is immediate if $t = 1$ or $t = 2$, so we assume $t \ge 3$. In Notation 6 we assumed that $q_1 \le q_2$. Since the moduli of our Beatty sequences satisfy the Morikawa condition (so the sequences would be disjoint for suitable offsets) we must have $q_1 + q_2 \le p$, and so $q_1 \le p/2$. Then from (7) $(t-1)G_1 < q_1 \le p/2$, so that, for $t \ge 3$,

$$G_1(t+1) < p. \tag{13}$$

We now apply the theorem with (12) in the role of (8). Since $\overline{q}_1$ and $\overline{q}_2$ are relatively prime to p,

$$\gcd(G_1\overline{q}_1, H\overline{q}_2, p) = \gcd(G_1, H, p) \le G_1.$$

So with (13) we have $\gcd(G_1\overline{q}_1, H\overline{q}_2, p)(t+1) < p$, which plays the role of (9). We conclude that either $H\overline{q}_2 \equiv G_1\overline{q}_1$ and

$$a_2\overline{q}_2 + b_2 \equiv a\overline{q}_1 + b_1 \pmod p,$$

or $H\overline{q}_2 \equiv -G_1\overline{q}_1$ modulo p and

$$\begin{aligned} a_2\overline{q}_2 + b_2 &\equiv a_1\overline{q}_1 + b_1 + G_1\overline{q}_1(t-1) \pmod p \\ &\equiv \overline{q}_1(a_1 + G_1(t-1)) + b_1 \pmod p, \end{aligned}$$

as required. □

We now analyse the intersection $S(p_1/q_1, b_1) \cap S(p/q_2, b_2)$.

Lemma 14. *The set of residues modulo pm in $S(pm/q_1, b)$ is*

$$\{i(\overline{q}_1 + p\overline{q}_m k_1) + b_1 : 0 \leq j \leq q_1 - 1\}.$$

Proof. By Theorem 10 the set of residues modulo pm in $S(pm/q_1, b_1)$ is $\{i\overline{Q} + b_1 : 0 \leq i \leq q_1 - 1\}$, where $\overline{Q}$ is the least non-negative residue modulo pm satisfying $q_1\overline{Q} \equiv -1 \pmod{pm}$. Using Notation 6 we then have $\overline{Q} \equiv \overline{q}_1 \pmod{p}$ so that $\overline{Q} = \overline{q}_1 + lp$ for some integer l. Then

$$\begin{aligned}\overline{Q}q_1 &= (\overline{q}_1 + lp)q_1 \\ &= -1 + k_1 p + lpq_1.\end{aligned}$$

But $\overline{Q}q_1 \equiv -1 \pmod{pm}$ so $k_1 + lq_1 \equiv 0 \pmod{m}$, which implies that $l \equiv \overline{q}_m k_1 \pmod{m}$, and the result follows. □

Theorem 15. *We use Notation 6, recalling that $p_1 = pm$. If p_1/q_1 and p/q_2 satisfy the Morikawa condition then $S(p_1/q_1, b_1) \cap S(p/q_2, b_2)$ equals*

$$\{(a + iG_1)(\overline{q}_1 + p\overline{q}_m k_1) + b_1 + \mu mp : 0 \leq i \leq t - 1, \mu \in \mathbb{Z}\},$$

where a, G_1 and t have the same meaning as in Theorem 11.

Proof. We write S_1, S_2, and S_m for $S(p/q_1, b_1)$, $S(p/q_2, b_2)$, and $S(pm/q_1, b_1)$ respectively. Since $S_m \subseteq S_1$ we have

$$S_m \cap S_2 = (S_1 \cap S_2) \cap S_m.$$

By Theorem 11

$$S_1 \cap S_2 \equiv \{a\overline{q}_1 + iG_1\overline{q}_1 + b_1 : 0 \leq i \leq t - 1\} \pmod{p}$$

where a and t are positive integers satisfying $0 < a + G_1(t-1) \leq q_1$. By Lemma 14

$$S_m \equiv \{j(\overline{q}_1 + p\overline{q}_m k_1) + b_1 : 0 \leq j \leq q_1 - 1\} \pmod{pm}.$$

Suppose $x \in S_2 \cap S_m$. Since $x \in S_1 \cap S_2$ we have

$$x = a\overline{q}_1 + i_1 G_1 \overline{q}_1 + b_1 + lp,$$

for some i_1 in $\{0, \ldots, t-1\}$ and $l \in \mathbb{Z}$. Then, since $x \in S_m$,

$$a\overline{q}_1 + i_1 G_1\overline{q}_1 + b_1 + lp = j_1(\overline{q}_1 + p\overline{q}_m k_1) + b_1 + \mu pm \tag{14}$$

for some $\mu \in \mathbb{Z}$ and j_1 in $\{0, \ldots, q_1\}$. It follows that

$$(a + i_1 G_1)\overline{q}_1 \equiv j_1 \overline{q}_1 \pmod{p}.$$

Since $0 < a + G_1(t-1) \leq q_1 < p$ and $\gcd(\overline{q}_1, p) = 1$, we have $a + i_1 G_1 = j_1$. Then (14) gives

$$l = (a + i_1 G_1)\overline{q}_m k_1 + \mu m.$$

The implications here can be reversed, so that $S_m \cap S_2$ equals

$$\{(a + iG_1)(\overline{q}_1 + p\overline{q}_m k_1) + b_1 + \mu mp : 0 \leq i \leq t-1, \mu \in \mathbb{Z}\},$$

as required. □

Now we obtain our main result.

Theorem 16. *We use Notation 6 recalling that $p_1 = pm$ and $p_2 = pn$. If p_1/q_1 and p_2/q_2 satisfy the Morikawa condition then the intersection of the Beatty sequences $S_m = S(pm/q_1, b_1)$ and $S_n = S(pn/q_2, b_2)$ is the modular arithmetic progression given below, where a, G_1 and t have the same meaning as in Theorem 11:*

$$\{a(\overline{q}_1 + pk_1\overline{q}_m) + b_1 + \mu_0 mp + i(G_1(\overline{q}_1 + pk_1\overline{q}_m) + \mu_1 mp) : 0 \leq i \leq t-1\}$$

modulo mnp, and one of the following cases holds.
Case 1 *In this case μ_0 is the least non-negative residue satisfying*

$$\mu_0 m \equiv \frac{a_2\overline{q}_2 + b_2 - a_1\overline{q}_1 - b_1}{p} + a_2 k_2 \overline{q}_n - a_1 k_1 \overline{q}_m \pmod{n},$$

and μ_1 is the least non-negative residue satisfying

$$\mu_1 m \equiv \frac{G_2\overline{q}_2 - G_1\overline{q}_1}{p} + G_2 k_2 \overline{q}_n - G_1 k_1 \overline{q}_m \pmod{n}.$$

Case 2 *In this case μ_0 is the least non-negative residue satisfying*

$$\mu_0 m \equiv \frac{a_1\overline{q}_1 + b_1 - a_2\overline{q}_2 - b_2}{p} + G_2(\overline{q}_2 + pk_2\overline{q}_n) - a_1 k_1 \overline{q}_m + a_2 k_2 \overline{q}_n \pmod{n},$$

and μ_1 is the least non-negative residue satisfying

$$\mu_1 m \equiv -\frac{G_1\overline{q}_1 + G_2\overline{q}_2}{p} - G_1 k_1 \overline{q}_m - G_2 k_2 \overline{q}_n \pmod{n}.$$

Proof. By Theorem 15 $S_m \cap S_2$ equals

$$\{(a_1 + iG_1)(\overline{q}_1 + pk_1\overline{q}_m) + b_1 + \mu mp : 0 \leq i \leq t-1, \mu \in \mathbb{Z}\}, \tag{15}$$

and $S_n \cap S_1$ equals

$$\{(a_2 + iG_2)(\overline{q}_2 + pk_2\overline{q}_n) + b_2 + \nu np : 0 \leq i \leq t-1, \nu \in \mathbb{Z}\}. \tag{16}$$

Since $S_m \subseteq S_1$ and $S_n \subseteq S_2$ we can obtain $S_m \cap S_n$ by evaluating $(S_m \cap S_2) \cap (S_n \cap S_1)$. Suppose $x \in S_m \cap S_n$. Then there exist integers $i_1, i_2 \in \{1, \ldots, t\}$ and $\mu, \nu \in \mathbb{Z}$ such that

$$\begin{aligned} x &= i_1 G_1(\overline{q}_1 + pk_1\overline{q}_m) + a_1(\overline{q}_1 + pk_1\overline{q}_m) + b_1 + \mu mp \\ &= i_2 G_2(\overline{q}_2 + pk_2\overline{q}_n) + a_2(\overline{q}_2 + pk_2\overline{q}_n) + b_2 + \nu np. \end{aligned} \tag{17}$$

Considering this modulo p we get

$$i_1 G_1 \overline{q}_1 + a_1\overline{q}_1 + b_1 \equiv i_2 G_2 \overline{q}_2 + a_2 \overline{q}_2 + b_2 \pmod{p}. \tag{18}$$

This is the congruence considered in Corollary 13. We therefore have either $i_2 = i_1$ or $i_2 = p - i_1$. We suppose the first of these holds and return to the other case at the end of the proof. Then (17) gives

$$\begin{aligned} & i_1(G_1(\overline{q}_1 + pk_1\overline{q}_m) - G_2(\overline{q}_2 + pk_2\overline{q}_n)) + a_1(\overline{q}_1 + pk_1\overline{q}_m) \\ & - a_2(\overline{q}_2 + pk_2\overline{q}_n) + b_1 - b_2 = \nu np - \mu mp \end{aligned} \tag{19}$$

for some integers μ and ν. Using Corollary 13 again we may divide through by p getting

$$\begin{aligned} & i_1\left(\frac{G_1\overline{q}_1 - G_2\overline{q}_2}{p} + G_1 k_1 \overline{q}_m - G_2 k_2 \overline{q}_n\right) + \frac{a_1\overline{q}_1 + b_1 - a_2\overline{q}_2 - b_2}{p} \\ & + a_1 k_1 \overline{q}_m - a_2 k_2 \overline{q}_n = \nu n - \mu m. \end{aligned}$$

Since m and n are relatively prime, we can find μ and ν satisfying this for any i_1. Consider the case $i_1 = 0$, and let μ_0 and ν_0 be the unique values of $\mu \in \{0, \ldots, n-1\}$ and $\nu \in \{0, \ldots, m-1\}$ satisfying,

$$\nu_0 n - \mu_0 m \equiv \frac{a_1\overline{q}_1 + b_1 - a_2\overline{q}_2 - b_2}{p} + a_1 k_1 \overline{q}_m - a_2 k_2 \overline{q}_n \pmod{mn}, \tag{20}$$

and let μ_1 and ν_1 be the unique values of $\mu \in \{0, \ldots, n-1\}$ and $\nu \in \{0, \ldots, m-1\}$ satisfying,

$$\nu_1 n - \mu_1 m \equiv \frac{G_1\overline{q}_1 - G_2\overline{q}_2}{p} + G_1 k_1 \overline{q}_m - G_2 k_2 \overline{q}_n \pmod{mn}. \tag{21}$$

We see that (19) will be satisfied for any $i_1 \in \{0, \ldots, t-1\}$ if we set

$$\begin{aligned} \mu &= \mu_0 + i_1\mu_1 + sn \\ \nu &= \nu_0 + i_1\nu_1 + sm, \end{aligned}$$

where s is any integer. Substituting in (15) and replacing i_1 with i we get,

$$\begin{aligned} S_m \cap S_n &= \{(a_1 + iG_1)(\overline{q}_1 + pk_1\overline{q}_m) + b_1 + (\mu_0 + i\mu_1 + sn)mp \\ &\qquad : 0 \le i \le t-1, s \in \mathbb{Z}\} \\ &= \{a_1(\overline{q}_1 + pk_1\overline{q}_m) + b_1 + \mu_0 mp + i(G_1(\overline{q}_1 + pk_1\overline{q}_m) + \mu_1 mp) \\ &\qquad + smnp : 0 \le i \le t-1, s \in \mathbb{Z}\}, \end{aligned}$$

as required.

If we put $i_2 = p - i_1$ instead of $i_2 = i_1$ as a consequence of (18) then (17) gives the values of μ_0 and μ_1 in Case 2.

□

Corollary 17. *If p_1/q_1 and p_2/q_2 satisfy the Morikawa condition then the intersection of $S_1 = S(pm/q_1, b_1)$ and $S_2 = S(pn/q_2, b_2)$ contains at most 3 distinct gap sizes.*

Proof. Immediate from Theorem 16 and Corollary 9. □

We end with an example. Consider the pair of Beatty Sequences $S(737/10, 0)$, and $S(469/15, 2)$. This gives $p = 67$, $m = 11$, $n = 7$, $q_1 = 10$, $q_2 = 15$, $b_1 = 0$, $b_2 = 2$, $\overline{q}_1 = 20$, $\overline{q}_2 = 58$, $\overline{q}_m = 1$, $\overline{q}_n = 6$, $k_1 = 3$, $k_2 = 13$, $G_1 = 2$, $G_2 = 3$, and $t = 4$. By Theorem 11 the intersection of $S(67/10, 0)$ and $S(67/15, 2)$ is $\{20a_1 + 40i : 0 \leq i \leq t-1\} \pmod{67}$. With $a_1 = 3$ this gives the intersection $\{6, 33, 46, 6\}$. We similarly have $a_2 = 1$. Now consider Theorem 16. We have μ_0 being the least non-negative residue satisfying

$$\begin{aligned} \mu_0 m &\equiv \frac{a_2\overline{q}_2 + b_2 - a_1\overline{q}_1 - b_1}{p} + a_2k_2\overline{q}_n - a_1k_1\overline{q}_m \pmod{n} \\ \Longrightarrow\ 11\mu_0 &\equiv 69 \pmod{7} \\ \Longrightarrow\ \mu_0 &\equiv 5 \pmod{7}, \end{aligned}$$

and μ_1 is the least non-negative residue satisfying

$$\begin{aligned} \mu_1 m &\equiv \frac{G_2\overline{q}_2 - G_1\overline{q}_1}{p} + G_2k_2\overline{q}_n - G_1k_1\overline{q}_m \pmod{n} \\ \Longrightarrow\ 11\mu_1 &\equiv 230 \pmod{7} \\ \Longrightarrow\ \mu_1 &\equiv 5 \pmod{7}. \end{aligned}$$

Then $S_m \cap S_n$ is

$$\begin{aligned} &\{a_1(\overline{q}_1 + pk_1\overline{q}_m) + b_1 + \mu_0 mp + i(G_1(\overline{q}_1 + pk_1\overline{q}_m) \\ &\quad + \mu_1 mp) : 0 \leq i \leq t-1\} \pmod{pmn} \\ &\equiv \{4384 + 4127i : 0 \leq i \leq 3\} \pmod{5159} \\ &\equiv \{1252, 2284, 3316, 4384\} \pmod{5159}. \end{aligned}$$

Acknowledgements: Thanks to Julie Caddy, Aviezri Fraenkel, Greg Gamble and Amy Glen for advice and assistance, and to the referee for very careful reading.

References

[1] Samuel Beatty, Problem 3173, *Amer. Math. Monthly* **33** (1926),159.

[2] S. Beatty, A. Ostrowski, J. Hyslop, A. C. Aitken, Solutions to Problem 3173, *Amer. Math. Monthly* **34** (1927), 159-160.

[3] J. Bark, P.P. Varjú, Partitioning the positive integers to seven Beatty sequences, *Indag. Math.* **14** (2003), 149-161.

[4] A.S. Fraenkel, Complementing and exactly covering sequences, *J. Comb. Th. Series A* **14** (1973), 8-20.

[5] A.S. Fraenkel, R. Holzman, Gap problems for integer part and fractional part sequences, *J. Number Theory* **50** (1995), 66-86.

[6] Richard K. Guy, Unsolved Problems in Number Theory, Springer, Berlin, 2004.

[7] R. Morikawa, Disjoint sequences generated by the bracket function, *Bull. Fac. Liberal Arts, Nagasaki Univ.* **26** (1985), 1-13.

[8] Š. Porubský and J. Schönheim, Covering systems of Paul Erdős: Past, present and future, in: Paul Erdős and his Mathematics (edited by G. Halász, L. Lovász, M. Simonvits, S.T. Sós), Bolyai Soc. Math. Studies 11, Budapest, 2002, Vol. I, 581-627.

[9] R.J. Simpson, Disjoint covering systems of rational Beatty sequences, *Discrete Math.* **92** (1991), 361-369.

[10] R.J. Simpson, Disjoint Beatty sequences, *Integers* **4** (2004), A12, 10pp. (electronic).

[11] Tony Van Ravenstein, The Three Gap Theorem (Steinhaus Conjecture), *J. Austral. Math. Soc. Series A* 45 (1988), 360-370.

NEW PROOFS FOR THE P, Q-ANALOGUE OF CHU-VANDERMONDE'S IDENTITY

George Teşeleanu
Department of Mathematics, Bucharest University, Romania
george.teseleanu@yahoo.com

Received: 10/3/12, Revised: 5/4/13, Accepted: 7/7/13, Published: 9/5/13

Abstract

In this paper we provide five new proofs for the p,q-analogue of Chu-Vandermonde's identity. The presented proofs can be considered as generalizations of the q-version and $q = 1$-version.

1. Introduction

The binomial coefficients are an important tool in combinatorics and they are denoted by $\binom{n}{k}$. They naturally occur as the coefficients in the binomial expansion of $(x+y)^n$; that is,

$$(x+y)^n = \sum_{k=0}^{n} \binom{n}{k} x^k y^{n-k}.$$

If we set $y = 1$ the above relation is called the horizontal generating function for the binomial coefficients. Their explicit value is given by

$$\binom{n}{k} = \frac{n!}{k!(n-k)!},$$

if $0 \leq k \leq n$ and zero otherwise. Some of the most important relations satisfied by the binomial coefficients are the triangular recurrence relation or Pascal's identity:

$$\binom{n+1}{k} = \binom{n}{k} + \binom{n}{k-1},$$

and Chu-Vandermonde's identity:

$$\binom{m+n}{k} = \sum_{j=0}^{k} \binom{m}{k-j} \binom{n}{j}. \tag{1}$$

The relations for the binomial coefficient and other identities can be found for example in [4].

The q-binomial coefficients are a generalization of the binomial coefficients introduced by C.F. Gauss. Choosing particular values for the integer parameters, they evaluate to polynomials in $\mathbb{Z}[q]$. One of their natural occurrences is as a counting tool for the number of k-dimensional subspaces of an n-dimensional vector space over a finite field of order q. Their explicit values are given by

$$\begin{bmatrix} n \\ k \end{bmatrix}_q = \frac{[n]_q!}{[k]_q![n-k]_q!},$$

where we use the notation $[k]_q = 1 + q + q^2 + \ldots + q^{k-1}$ and where $[k]_q!$ denotes the q-factorial which is given by $[k]_q! = [k]_q[k-1]_q \ldots [2]_q[1]_q$. It can easily be verified that, when $q = 1$, the q-binomial coefficients turn into the usual binomial coefficients. Unlike the usual binomial coefficients, the q-binomial coefficients have a limited array of identities. The most basic ones are the triangular recurrence relations:

$$\begin{aligned} \begin{bmatrix} n+1 \\ k \end{bmatrix}_q &= q^k \begin{bmatrix} n \\ k \end{bmatrix}_q + \begin{bmatrix} n \\ k-1 \end{bmatrix}_q, \\ \begin{bmatrix} n+1 \\ k \end{bmatrix}_q &= \begin{bmatrix} n \\ k \end{bmatrix}_q + q^{n-k+1} \begin{bmatrix} n \\ k-1 \end{bmatrix}_q, \end{aligned}$$

and the q-analogues of Chu-Vandermonde's identity:

$$\begin{bmatrix} m+n \\ k \end{bmatrix}_q = \sum_{j=0}^{k} q^{j(m-k+j)} \begin{bmatrix} m \\ k-j \end{bmatrix}_q \begin{bmatrix} n \\ j \end{bmatrix}_q, \tag{2}$$

$$\begin{bmatrix} m+n \\ k \end{bmatrix}_q = \sum_{j=0}^{k} q^{(k-j)(n-j)} \begin{bmatrix} m \\ k-j \end{bmatrix}_q \begin{bmatrix} n \\ j \end{bmatrix}_q. \tag{3}$$

We also have a generalization of the horizontal generating function that is satisfied by the q-binomial coefficient:

$$\begin{aligned} \prod_{i=0}^{n-1} (q^i + x) &= \sum_{k=0}^{n} q^{(n-k-1)(n-k)/2} \begin{bmatrix} n \\ k \end{bmatrix}_q x^k, \\ \prod_{i=0}^{n-1} (1 + q^i x) &= \sum_{k=0}^{n} q^{k(k-1)/2} \begin{bmatrix} n \\ k \end{bmatrix}_q x^k. \end{aligned}$$

The relations for the q-binomial coefficient and other identities can be found for example in [3] or [5].

The p,q-binomial coefficients are a generalization of the q-binomial coefficient introduced by Robert B. Corcino in [1]. Choosing particular values for the integer parameters they evaluate to polynomials in $\mathbb{Z}[p,q]$. Their explicit value is given by

$$\begin{bmatrix} n \\ k \end{bmatrix}_{p,q} = \frac{[n]_{p,q}!}{[k]_{p,q}![n-k]_{p,q}!},$$

where $[k]_{p,q} = p^{k-1} + p^{k-2}q + p^{k-3}q^2 + \ldots + q^{k-1}$, and $[k]_{p,q}!$ denotes the p,q-factorial which is given by $[k]_{p,q}! = [k]_{p,q}[k-1]_{p,q}\cdots[2]_{p,q}[1]_{p,q}$. It can easily be verified that, when $p = 1$, the p,q-binomial coefficients turn into the q-binomial coefficients. It can be noticed that the p,q-binomial coefficients is symmetrical, i.e. $\begin{bmatrix} n \\ k \end{bmatrix}_{p,q} = \begin{bmatrix} n \\ k \end{bmatrix}_{q,p}$. Because they are symmetrical, through the rest of the paper, we will write and prove only one variant of a relation. Some of the identities satisfied by the p,q-binomial coefficients are the triangular recurrence relation:

$$\begin{bmatrix} n+1 \\ k \end{bmatrix}_{p,q} = q^k \begin{bmatrix} n \\ k \end{bmatrix}_{p,q} + p^{n-k+1} \begin{bmatrix} n \\ k-1 \end{bmatrix}_{p,q}, \tag{4}$$

Chu-Vandermonde's identity:

$$\begin{bmatrix} m+n \\ k \end{bmatrix}_{p,q} = \sum_{j=0}^{k} p^{(k-j)(n-j)} q^{j(m-k+j)} \begin{bmatrix} m \\ k-j \end{bmatrix}_{p,q} \begin{bmatrix} n \\ j \end{bmatrix}_{p,q}, \tag{5}$$

and the horizontal generating function:

$$\prod_{i=0}^{n-1} (q^i + p^i x) = \sum_{k=0}^{n} q^{(n-k-1)(n-k)/2} p^{k(k-1)/2} \begin{bmatrix} n \\ k \end{bmatrix}_{p,q} x^k. \tag{6}$$

We may also see that the following basic identity occurs:

$$\begin{bmatrix} n \\ k \end{bmatrix}_{p,q} = \begin{bmatrix} n \\ n-k \end{bmatrix}_{p,q}. \tag{7}$$

These relations and more identities for the p,q-binomial coefficients can be found in [1] and [7]. Also in [7] there is a combinatorial interpretation of the p,q-binomial coefficients.

The aim of this paper is to provide new proofs for Chu-Vandermonde's identity (5). The identity is already proven in [7] as Identity 3.6 and in [2] as Proposition 4. In the first proof we homogenize the q-variant of the Chu-Vandermonde relation (2). In the second and third proof we generalize some proofs for the classical Chu-Vandermonde relation (1); specifically, in the second proof we make repeated use of Pascal's identity (4), an idea found for the classical relation (1) in [6] as Property 10.7, and for the third proof we use a generalization of the generating horizontal relation (6), an idea for (1) found in [4, Chapter 5.4]. In the fourth proof we use a generalization of quantum calculus ideas found in [5] as Propositions IV.2.2 and IV.2.3. In the last proof we use induction to prove (1) directly.

Through the rest of the paper we will use the following conventions: $\begin{bmatrix} n \\ 0 \end{bmatrix}_{p,q} = \begin{bmatrix} n \\ n \end{bmatrix}_{p,q} = 1$; $\begin{bmatrix} n \\ j \end{bmatrix}_{p,q} = 0$, for $j > n$ or $j < 0$. We will also use the notations LHS and RHS for the left-hand side, respectively right-hand side of an equality.

2. Chu-Vandermonde's Identity

Theorem 2.1. (Chu-Vandermonde identity) *The p,q-binomial coefficients satisfy the identity:*

$$\begin{bmatrix} m+n \\ k \end{bmatrix}_{p,q} = \sum_{j=0}^{k} p^{(k-j)(n-j)} q^{j(m-k+j)} \begin{bmatrix} m \\ k-j \end{bmatrix}_{p,q} \begin{bmatrix} n \\ j \end{bmatrix}_{p,q}.$$

Proof 1. For this proof of (5) we will work in the rational function field $\mathbb{Q}(p,q)$. We will use the Chu-Vandermonde identity (2) for r-binomial coefficients with the parameter $r = \frac{p}{q}$ to prove our identity. Due to our choice of parameter we have the following relation:

$$[k]_r = \frac{r^k - 1}{r - 1} = \frac{\left(\frac{p}{q}\right)^k - 1}{\frac{p}{q} - 1} = q^{-(k-1)} \cdot \frac{p^k - q^k}{p - q} = q^{-(k-1)} \cdot [k]_{p,q}. \tag{8}$$

Expanding the Chu-Vandermonde's identity (2) for r-binomial coefficients,

$$\begin{bmatrix} m+n \\ k \end{bmatrix}_{r} = \sum_{j=0}^{k} r^{j(m-k+j)} \begin{bmatrix} m \\ k-j \end{bmatrix}_{r} \begin{bmatrix} n \\ j \end{bmatrix}_{r},$$

we obtain:

$$\frac{[m+n]_r!}{[k]_r![m+n-k]_r!} = \sum_{j=0}^{k} r^{j(m-k+j)} \frac{[m]_r!}{[k-j]_r![m-k+j]_r!} \frac{[n]_r!}{[j]_r![n-j]_r!}. \tag{9}$$

Now making use of (8), relation (9) becomes:

$$\begin{aligned} LHS &= \frac{q^{1+2+\ldots+(k-1)} \cdot q^{1+2+\ldots+(m+n-k-1)}}{q^{1+2+\ldots+(m+n-1)}} \frac{[m+n]_{p,q}!}{[k]_{p,q}![m+n-k]_{p,q}!} \\ &= q^{[k(k-1)+(m+n-k)(m+n-k-1)-(m+n)(m+n-1)]/2} \frac{[m+n]_{p,q}!}{[k]_{p,q}![m+n-k]_{p,q}!}, \end{aligned}$$

$$
\begin{aligned}
RHS &= \sum_{j=0}^{k} \frac{q^{1+2+\ldots+(k-j-1)} \cdot q^{1+2+\ldots+(m-k+j-1)}}{q^{1+2+\ldots+(m-1)}} \frac{[m]_{p,q}!}{[k-j]_{p,q}![m-k+j]_{p,q}!} \\
&\quad \frac{q^{1+2+\ldots+(j-1)} \cdot q^{1+2+\ldots+(n-j-1)}}{q^{1+2+\ldots+(n-1)}} \frac{[n]_{p,q}!}{[j]_{p,q}![n-j]_{p,q}!} \left(\frac{p}{q}\right)^{j(m-k+j)} \\
&= \sum_{j=0}^{k} p^{j(m-k+j)} \cdot q^{A} \frac{[m]_{p,q}!}{[k-j]_{p,q}![m-k+j]_{p,q}!} \frac{[n]_{p,q}!}{[j]_{p,q}![n-j]_{p,q}!},
\end{aligned}
$$

where $A = [(k-j)(k-j-1) + (m-k+j)(m-k+j-1) - m(m-1) + j(j-1) + (n-j)(n-j-1) - n(n-1)]/2 - j(m-k+j)$. By multiplying the LHS and RHS with $q^{-[k(k-1)+(m+n-k)(m+n-k-1)-(m+n)(m+n-1)]/2}$, we obtain identity (5). □

Proof 2. In this proof we make repeated use of Pascal's identity (4). So, we have:

$$
\begin{aligned}
\begin{bmatrix} m+n \\ k \end{bmatrix}_{p,q} &= q^{k} \begin{bmatrix} m+n-1 \\ k \end{bmatrix}_{p,q} + p^{n+m-k} \begin{bmatrix} m+n-1 \\ k-1 \end{bmatrix}_{p,q} \\
&= q^{k} \left\{ q^{k} \begin{bmatrix} m+n-2 \\ k \end{bmatrix}_{p,q} + p^{n+m-1-k} \begin{bmatrix} m+n-2 \\ k-1 \end{bmatrix}_{p,q} \right\} \\
&\quad + p^{n+m-k} \left\{ q^{k-1} \begin{bmatrix} m+n-2 \\ k-1 \end{bmatrix}_{p,q} + p^{n+m-k} \begin{bmatrix} m+n-2 \\ k-2 \end{bmatrix}_{p,q} \right\} \\
&= q^{2k} \begin{bmatrix} m+n-2 \\ k \end{bmatrix}_{p,q} + p^{n+m-k-1} q^{k-1} [2]_{p,q} \begin{bmatrix} m+n-2 \\ k-1 \end{bmatrix}_{p,q} \\
&\quad + p^{2(n+m-k)} \begin{bmatrix} m+n-2 \\ k-2 \end{bmatrix}_{p,q} \\
&= q^{3k} \begin{bmatrix} m+n-3 \\ k \end{bmatrix}_{p,q} + p^{n+m-k-2} q^{2(k-1)} [3]_{p,q} \begin{bmatrix} m+n-3 \\ k-1 \end{bmatrix}_{p,q} \\
&\quad + p^{2(n+m-k-1)} q^{k-2} [3]_{p,q} \begin{bmatrix} m+n-3 \\ k-2 \end{bmatrix}_{p,q} + p^{3(n+m-k)} \begin{bmatrix} m+n-3 \\ k-3 \end{bmatrix}_{p,q} \\
&= q^{4k} \begin{bmatrix} m+n-4 \\ k \end{bmatrix}_{p,q} + p^{n+m-k-3} q^{3(k-1)} [4]_{p,q} \begin{bmatrix} m+n-4 \\ k-1 \end{bmatrix}_{p,q} \\
&\quad + p^{2(n+m-k-2)} q^{2(k-2)} \frac{[3]_{p,q} \cdot [4]_{p,q}}{[2]_{p,q}} \begin{bmatrix} m+n-4 \\ k-2 \end{bmatrix}_{p,q} \\
&\quad + p^{3(n+m-k-1)} q^{k-3} [4]_{p,q} \begin{bmatrix} m+n-4 \\ k-3 \end{bmatrix}_{p,q} + p^{4(n+m-k)} \begin{bmatrix} m+n-4 \\ k-4 \end{bmatrix}_{p,q} \\
&= \ldots \\
&= q^{mk} \begin{bmatrix} n \\ k \end{bmatrix}_{p,q} \begin{bmatrix} m \\ 0 \end{bmatrix}_{p,q} + p^{n-k-1} q^{(m-1)(k-1)} \begin{bmatrix} n \\ k-1 \end{bmatrix}_{p,q} \begin{bmatrix} m \\ 1 \end{bmatrix}_{p,q}
\end{aligned}
$$

$$+\ldots+p^{(k-1)(n-1)}q^{m-k+1}\begin{bmatrix}n\\1\end{bmatrix}_{p,q}\begin{bmatrix}m\\k-1\end{bmatrix}_{p,q}+p^{kn}\begin{bmatrix}n\\0\end{bmatrix}_{p,q}\begin{bmatrix}m\\k\end{bmatrix}_{p,q}.$$

Taking the first and last term of the equalities we obtain identity (5). □

For the next proof of Theorem 2.1 we utilize the following proposition.

Proposition 2.1. *The p,q-binomial coefficients satisfy the following identity:*

$$\prod_{i=0}^{n-1}(q^{a+i}+p^i x) = \sum_{k=0}^{n} q^{(n-k-1+2a)(n-k)/2}p^{k(k-1)/2}\begin{bmatrix}n\\k\end{bmatrix}_{p,q} x^k, \tag{10}$$

for any nonnegative integer a.

Remark. When $a=0$, relation (10) becomes relation (6) that can be found in [1] as Theorem 3 and in [7] as Identity 3.3.

Proof. We will make use of induction and Pascal's identity (4) to prove this identity. When $n=1$ we have $q^a+x=q^a\begin{bmatrix}1\\0\end{bmatrix}_{p,q}+\begin{bmatrix}1\\1\end{bmatrix}_{p,q}x$, which is true. When $n=2$ we have $(q^a+x)(q^{a+1}+px)=q^{2a+1}+q^a(p+q)x+px^2=q^{2a+1}\begin{bmatrix}2\\0\end{bmatrix}_{p,q}+q^a\begin{bmatrix}2\\1\end{bmatrix}_{p,q}x+p\begin{bmatrix}2\\2\end{bmatrix}_{p,q}x^2$, which is true. Now we assume that the identity holds for n and we prove the identity for $n+1$. So we have:

$$\begin{aligned}\prod_{i=0}^{n}(q^{a+i}+p^i x) &= \prod_{i=0}^{n-1}(q^{a+i}+p^i x)\cdot(q^{a+n}+p^n x)\\
&= \left[\sum_{k=0}^{n} q^{(n-k-1+2a)(n-k)/2}p^{k(k-1)/2}\begin{bmatrix}n\\k\end{bmatrix}_{p,q} x^k\right](q^{a+n}+p^n x)\\
&= \sum_{k=0}^{n} q^{((n-k-1+2a)(n-k)/2)+a+n}p^{k(k-1)/2}\begin{bmatrix}n\\k\end{bmatrix}_{p,q} x^k\\
&+ \sum_{k=0}^{n} q^{(n-k-1+2a)(n-k)/2}p^{(k(k-1)/2)+n}\begin{bmatrix}n\\k\end{bmatrix}_{p,q} x^{k+1}.\end{aligned}$$

We observe that the coefficients $q^{(n+2a)(n+1)/2}$ of x^0 and $p^{n(n+1)/2}$ of x^{n+1}, correspond with the ones in the formula (10). Now let us check the other coefficients. Let us take a j, $1\le j\le n$, and calculate the coefficient of x^j:

$$\begin{aligned}
C &= q^{((n-j-1+2a)(n-j)/2)+a+n} p^{j(j-1)/2} \begin{bmatrix} n \\ j \end{bmatrix}_{p,q} \\
&\quad + q^{(n-j+2a)(n-j+1)/2} p^{((j-1)(j-2)/2)+n} \begin{bmatrix} n \\ j-1 \end{bmatrix}_{p,q} \\
&= q^{(n-j+2a)(n-j+1)/2} p^{(j(j-1)/2)} \left(q^j \begin{bmatrix} n \\ j \end{bmatrix}_{p,q} + p^{n-(j-1)} \begin{bmatrix} n \\ j-1 \end{bmatrix}_{p,q} \right) \\
&= q^{(n-j+2a)(n-j+1)/2} p^{(j(j-1)/2)} \begin{bmatrix} n+1 \\ j \end{bmatrix}_{p,q},
\end{aligned}$$

where for the last equality we used Pascal's identity (4). So we have proved identity (10). □

Proof 3. In this proof we make use of Proposition 2.1 for $a = 0$ and $a = n$. So, we have:

$$\prod_{k=0}^{n+m-1} (q^k + p^k x) = \prod_{i=0}^{n-1} (q^i + p^i x) \cdot \prod_{j=0}^{m-1} .(q^{n+j} + p^j (p^n x)).$$

Now using Proposition 2.1 we have:

$$\begin{aligned}
LHS &= \sum_{k=0}^{n+m} q^{(n+m-k-1)(n+m-k)/2} p^{k(k-1)/2} \begin{bmatrix} n+m \\ k \end{bmatrix}_{p,q} x^k, \\
RHS &= \Big(\sum_{i=0}^{n} q^{(n-i-1)(n-i)/2} p^{i(i-1)/2} \begin{bmatrix} n \\ i \end{bmatrix}_{p,q} x^i \Big) \\
&\quad \cdot \Big(\sum_{j=0}^{m} q^{(m-j-1+2n)(m-j)/2} p^{[j(j-1)/2]+nj} \begin{bmatrix} m \\ j \end{bmatrix}_{p,q} x^j \Big) \\
&= \sum_{i,j=0}^{m} q^B p^{[i(i-1)+j(j-1)+2nj]/2} \begin{bmatrix} n \\ i \end{bmatrix}_{p,q} \begin{bmatrix} m \\ j \end{bmatrix}_{p,q} x^{i+j},
\end{aligned}$$

where $B = [(n-i-1)(n-i) + (m-j-1+2n)(m-j)]/2$. Now we identify the coefficient of x^k, $0 \le k \le n+m$, on both sides and we obtain:

$$q^{(n+m-k-1)(n+m-k)/2} p^{k(k-1)/2} \begin{bmatrix} n+m \\ k \end{bmatrix}_{p,q} = \sum_{i=0, j=k-i}^{n} q^V p^W \begin{bmatrix} n \\ i \end{bmatrix}_{p,q} \begin{bmatrix} m \\ k-i \end{bmatrix}_{p,q}, \quad (11)$$

where $V = [(n-i-1)(n-i) + (m-k+i-1+2n)(m-k+i)]/2$ and $W = [i(i-1) + (k-i)(k-i-1) + 2n(k-i)]/2$. Now by multiplying relation (11) by $q^{-(n+m-k-1)(n+m-k)/2} p^{-k(k-1)/2}$, we obtain identity (5). □

For the fourth proof of Theorem 2.1. we need Proposition 2.2. To prepare the stage, the following remark is crucial.

Remark. For Proposition 2.2 and Proof 4 we will work in the quotient algebra $R = k\{X,Y\}/I_{p,q}$, where k is a ring, $k\{X,Y\}$ is the free algebra of two indeterminates and $I_{p,q}$ is the two-sided ideal of the free algebra generated by $qYX - pXY$, $p,q \in k$. Clearly $qyx = pxy$, where x and y are the equivalence classes of X and Y in R.

Proposition 2.2. *The p,q-binomial coefficients satisfy the following identity:*

$$q^{n(n-1)/2}(x+y)^n = \sum_{j=0}^{n} q^{[j(j-1)+(n-j)(n-j-1)]/2} \begin{bmatrix} n \\ n-j \end{bmatrix}_{p,q} x^{n-j}y^j. \quad (12)$$

Proof. We will prove the above relation using induction on n and Pascal's identity (4). For $n = 2$ we check that

$$\begin{aligned} q(x+y)^2 &= qx^2 + qxy + qyx + qy^2 = q(x^2+y^2) + (q+p)xy \\ &= q\begin{bmatrix} 2 \\ 0 \end{bmatrix}_{p,q} x^2 + \begin{bmatrix} 2 \\ 1 \end{bmatrix}_{p,q} xy + q\begin{bmatrix} 2 \\ 2 \end{bmatrix}_{p,q} y^2. \end{aligned}$$

Now we assume that the identity holds for n and we prove the identity for $n+1$. So we have:

$$\begin{aligned} q^{n(n+1)/2}(x+y)^{n+1} &= q^n(x+y)q^{n(n-1)/2}(x+y)^n \\ &= q^n(x+y)\Bigg(q^{n(n-1)/2}(x^n+y^n) \\ &\quad + \sum_{j=1}^{n-1} q^{[j(j-1)+(n-j)(n-j-1)]/2} \begin{bmatrix} n \\ n-j \end{bmatrix}_{p,q} x^{n-j}y^j\Bigg) \\ &= q^{n(n+1)/2}(x^n+y^n) + q^{n(n+1)/2}xy^n + q^{n(n+1)/2}yx^n \\ &\quad + \sum_{j=1}^{n-1} q^{[j(j-1)+(n-j)(n-j-1)+2n]/2} \begin{bmatrix} n \\ n-j \end{bmatrix}_{p,q} x^{n-j+1}y^j \\ &\quad + \sum_{j=1}^{n-1} q^{[j(j-1)+(n-j)(n-j-1)+2n]/2} \begin{bmatrix} n \\ n-j \end{bmatrix}_{p,q} yx^{n-j}y^j. \end{aligned}$$

Now we check the coefficient of xy^n:

$$\begin{aligned} &q^{n(n+1)/2}xy^n + q^{[(n-1)(n-2)+2n]/2}yxy^{n-1}\begin{bmatrix} n \\ 1 \end{bmatrix}_{p,q} \\ &= \left(q^{n(n+1)/2} + pq^{[(n-1)(n-2)+2(n-1)]/2}\begin{bmatrix} n \\ 1 \end{bmatrix}_{p,q}\right)xy^n \end{aligned}$$

$$= q^{n(n-1)/2}\{q^n + p(q^{n-1} + q^{n-2}p + \ldots + p^{n-1})\}xy^n$$
$$= q^{n(n-1)/2}xy^n \begin{bmatrix} n+1 \\ 1 \end{bmatrix}_{p,q},$$

and we see that it is the right coefficient. Similarly for $x^n y$.

We group the terms with similar powers and we see that the coefficient for $x^{n-j}y^{j+1}$, where $1 \le j \le n-2$, is:

$$q^{[j(j+1)+(n-j-1)(n-j-2)+2n]/2}x^{n-j}y^{j+1}\begin{bmatrix} n \\ n-j-1 \end{bmatrix}_{p,q}$$
$$+q^{[j(j-1)+(n-j)(n-j-1)+2n]/2}yx^{n-j}y^{j}\begin{bmatrix} n \\ n-j \end{bmatrix}_{p,q}$$
$$= q^{[j(j+1)+(n-j)(n-j-1)]/2}\left(q^{j+1}\begin{bmatrix} n \\ n-j-1 \end{bmatrix}_{p,q} + p^{n-j}\begin{bmatrix} n \\ n-j \end{bmatrix}_{p,q}\right)x^{n-j}y^{j+1}$$
$$= q^{[j(j+1)+(n-j)(n-j-1)]/2}x^{n-j}y^{j+1}\begin{bmatrix} n+1 \\ n-j \end{bmatrix}_{p,q},$$

where for the last equality we used Pascal's identity (4). We observe that for this j we end up with the right coefficient and this ends our proof. □

Proof 4. For this proof we use Proposition 2.2. In the ring R we have:

$$q^{(n+m)(n+m-1)/2}(x+y)^{n+m} = q^{nm} \cdot q^{n(n-1)/2}(x+y)^n \cdot q^{m(m-1)/2}(x+y)^m.$$

We expand both sides using Proposition 2.2.,

$$LHS = \sum_{i=0}^{n+m} q^{[i(i-1)+(n+m-i)(n+m-i-1)]/2}x^{n+m-i}y^i \begin{bmatrix} n+m \\ n+m-i \end{bmatrix}_{p,q},$$
$$RHS = q^{nm}\left(\sum_{j=0}^{n} q^{[j(j-1)+(n-j)(n-j-1)]/2}x^{n-j}y^j \begin{bmatrix} n \\ n-j \end{bmatrix}_{p,q}\right)$$
$$\cdot\left(\sum_{k=0}^{m} q^{[k(k-1)+(m-k)(m-k-1)]/2}x^{m-k}y^k \begin{bmatrix} m \\ m-k \end{bmatrix}_{p,q}\right)$$
$$= \sum_{j=0}^{n+m}\sum_{k=0}^{n+m} q^Z \begin{bmatrix} n \\ n-j \end{bmatrix}_{p,q}\begin{bmatrix} m \\ m-k \end{bmatrix}_{p,q} x^{n-j}y^jx^{m-k}y^k$$
$$= \sum_{j=0}^{n+m}\sum_{k=0}^{n+m} q^T p^{j(m-k)} \begin{bmatrix} n \\ n-j \end{bmatrix}_{p,q}\begin{bmatrix} m \\ m-k \end{bmatrix}_{p,q} x^{(n-j)+(m-k)}y^{j+k},$$

where $Z = [j(j-1)+(n-j)(n-j-1)+k(k-1)+(m-k)(m-k-1)]/2$ and $T = Z - j(m-k)$. For a given i, the coefficient of $x^{n+m-i}y^i$ in LHS is:

$$C1 = q^{[i(i-1)+(n+m-i)(n+m-i-1)]/2}\begin{bmatrix} n+m \\ n+m-i \end{bmatrix}_{p,q},$$

and in RHS is:

$$C2 = \sum_{j=0,k=i-j}^{n} q^{\alpha}p^{j(m-i+j)}\begin{bmatrix} n \\ n-j \end{bmatrix}_{p,q}\begin{bmatrix} m \\ m-i+j \end{bmatrix}_{p,q},$$

where $2\alpha = j(j-1)+(n-j)(n-j-1)+(i-j)(i-j-1)+(m-i-j)(m-i-j-1)-j(m-i-j)$. Now reducing $C1$ and $C2$ with $q^{[i(i-1)+(n+m-i)(n+m-i-1)]/2}$ we obtain:

$$\begin{bmatrix} n+m \\ n+m-i \end{bmatrix}_{p,q} = \sum_{j=0}^{n} q^{(i-j)(n-j)}p^{j(m-i+j)}\begin{bmatrix} n \\ n-j \end{bmatrix}_{p,q}\begin{bmatrix} m \\ m-i+j \end{bmatrix}_{p,q}.$$

Using identity (7), the conclusion follows. □

Proof 5. In our last proof we make use of induction on n and Pascal's identity (4). When $n = 1$, we have:

$$\begin{aligned}\begin{bmatrix} m+1 \\ k \end{bmatrix}_{p,q} &= p^k\begin{bmatrix} m \\ k \end{bmatrix}_{p,q}\begin{bmatrix} 1 \\ 0 \end{bmatrix}_{p,q} + q^{m-k+1}\begin{bmatrix} m \\ k-1 \end{bmatrix}_{p,q}\begin{bmatrix} 1 \\ 1 \end{bmatrix}_{p,q} \\ &= p^k\begin{bmatrix} m \\ k \end{bmatrix}_{p,q} + q^{m-k+1}\begin{bmatrix} m \\ k-1 \end{bmatrix}_{p,q},\end{aligned}$$

which is exactly Pascal's identity (4). We assume that the identity holds for n and we prove it for $n+1$. Alternatively, using (4) and the induction statement, we have:

$$\begin{aligned}\begin{bmatrix} m+n+1 \\ k \end{bmatrix}_{p,q} &= q^k\begin{bmatrix} m+n \\ k \end{bmatrix}_{p,q} + p^{m+n+1-k}\begin{bmatrix} m+n \\ k-1 \end{bmatrix}_{p,q} \\ &= q^k\sum_{j=0}^{k} p^{(k-j)(n-j)}q^{j(m-k+j)}\begin{bmatrix} m \\ k-j \end{bmatrix}_{p,q}\begin{bmatrix} n \\ j \end{bmatrix}_{p,q} \\ &\quad + p^{m+n+1-k}\sum_{j=0}^{k-1} p^{(k-1-j)(n-j)}q^{j(m-k+1+j)}\begin{bmatrix} m \\ k-1-j \end{bmatrix}_{p,q}\begin{bmatrix} n \\ j \end{bmatrix}_{p,q} \\ &= \sum_{j=0}^{k-1} p^{(k-j)(n-j)}q^{j(m-k+1+j)}\begin{bmatrix} n \\ j \end{bmatrix}_{p,q}\left\{q^{k-j}\begin{bmatrix} m \\ k-j \end{bmatrix}_{p,q}\right.\end{aligned}$$

$$+p^{m-k+j+1}\begin{bmatrix} m \\ k-j-1 \end{bmatrix}_{p,q} \Bigg\} + q^{k(m+1)} \begin{bmatrix} m \\ 0 \end{bmatrix}_{p,q} \begin{bmatrix} n \\ k \end{bmatrix}_{p,q}$$

$$= \sum_{j=0}^{k-1} p^{(k-j)(n-j)} q^{j(m-k+1+j)} \begin{bmatrix} n \\ j \end{bmatrix}_{p,q} \begin{bmatrix} m+1 \\ k-j \end{bmatrix}_{p,q}$$

$$+q^{k(m+1)} \begin{bmatrix} m+1 \\ 0 \end{bmatrix}_{p,q} \begin{bmatrix} n \\ k \end{bmatrix}_{p,q}$$

$$= \sum_{j=0}^{k} p^{(k-j)(n-j)} q^{j(m+1-k+j)} \begin{bmatrix} m+1 \\ k-j \end{bmatrix}_{p,q} \begin{bmatrix} n \\ j \end{bmatrix}_{p,q},$$

so we end up with identity (5). □

References

[1] Roberto B. Corcino, *On p, q-binomial coefficients*, Integers 8, A29, 2008.

[2] M. Dziemiańczuk, *First remark on a ξ-analogue of the Stirling numbers*, Integers 11, A9, 2011.

[3] George Gasper and Mizan Rahman, *Basic Hypergeometric Series*, Cambridge University Press, 2004.

[4] Ronald L. Graham, Donald E. Knuth, Oren Patashnik, *Concrete Mathematics*, Addison-Wesley, Upper Saddle River, New Jersey, 1989.

[5] Christian Kassel, *Quantum Groups*, Springer, New York, 1995.

[6] Dragoş-Radu Popescu, *Combinatorică şi Teoria Grafurilor*, Societatea de Ştiinţe Matematice din România, 2005.

[7] Mark Shattuck, *On some relations satisfied by the p, q-binomial coefficient*, Šiauliai Math. Semin. 6 (14), 2011, 69-84.

REMARKS ON TINY ZERO-SUM SEQUENCES

Yushuang Fan
Center for Combinatorics, LPMC-TJKLC, Nankai University, Tianjin, P.R. China
fys850820@163.com

Weidong Gao
Center for Combinatorics, LPMC-TJKLC, Nankai University, Tianjin, P.R. China
wdgao1963@yahoo.com.cn

Jiangtao Peng
College of Science, Civil Aviation University of China, Tianjin, P.R.China
jtpeng1982@yahoo.com.cn

Linlin Wang
Center for Combinatorics, LPMC-TJKLC, Nankai University, Tianjin, P.R. China
wanglinlin_1986@yahoo.cn

Qinghai Zhong
Center for Combinatorics, LPMC-TJKLC, Nankai University, Tianjin, P.R. China
zhongqinghai@yahoo.com.cn

Received: 12/10/12, Revised: 4/30/13, Accepted: 7/7/13, Published: 9/5/13

Abstract

Let G be an additive finite abelian group with exponent $\exp(G)$. Let $S = g_1 \cdot \ldots \cdot g_l$ be a sequence over G and $\mathsf{k}(S) = \mathrm{ord}(g_1)^{-1} + \cdots + \mathrm{ord}(g_l)^{-1}$ be its cross number. Let $\eta(G)$ (resp. $\mathsf{t}(G)$) be the smallest integer t such that every sequence of t elements (repetition allowed) from G contains a non-empty zero-sum subsequence T of length $|T| \le \exp(G)$ (resp. $\mathsf{k}(T) \le 1$). It is easy to see that $\mathsf{t}(G) \ge \eta(G)$ for all finite abelian groups G, and a previous result showed that for every positive integer $r \ge 4$, there exist finite abelian groups of rank r such that $\mathsf{t}(G) > \eta(G)$. In this paper we provide the first example of groups G of rank three with $\mathsf{t}(G) > \eta(G)$. We also prove that $\mathsf{t}(G) = \eta(G)$ for $G = C_2 \oplus C_{2p}$ where p is a prime.

1. Introduction

Let G be an additively written finite abelian group with $\exp(G)$ its exponent. A sequence $S = g_1 \cdot \ldots \cdot g_l$ over G is said to be a zero-sum sequence, if $\sum_{i=1}^{l} g_i = 0$. S is called a minimal zero-sum sequence, if it contains no proper zero-sum subsequence. The cross number $\mathsf{k}(S)$ of a sequence S is defined by

$$\mathsf{k}(S) = \sum_{i=1}^{l} \frac{1}{\mathrm{ord}(g_i)}.$$

The cross number is an important concept in factorization theory. For recent work on the cross number we refer to ([10], [12], [13]).

By $\mathsf{t}(G)$ we denote the smallest integer $t \in \mathbb{N}$ such that every sequence S over G of length $|S| \geq t$ contains a non-empty zero-sum subsequence $S'|S$ with $\mathsf{k}(S') \leq 1$. Such a subsequence will be called a *tiny zero-sum subsequence.*

The study of $\mathsf{t}(G)$ goes back to the late 1980s, Lemke and Kleitman [17] proved that $\mathsf{t}(C_n) = n$, which confirmed a conjecture by Erdős and Lemke, where C_n denotes the cyclic group of n elements.

In the general case, Kleitman and Lemke [17] conjectured that $\mathsf{t}(G) \leq |G|$ holds for every finite abelian group G. This conjecture was confirmed by Geroldinger [9] in 1993. A different proof was found by Elledge and Hurlbert [4] in 2005 using graph pebbling. For more work on applications of graph pebbling to zero-sum problems we refer to ([2], [3], [15], [16]).

Quite recently, Girard [14] proved that, by using a result of Alon and Dubiner [1], for finite abelian groups of fixed rank, $\mathsf{t}(G)$ grows linearly in the exponent of G, which gives the correct order of magnitude.

Let $\eta(G)$ denote the smallest integer $t \in \mathbb{N}$ such that every sequence S over G of length $|S| \geq t$ contains a non-empty zero-sum subsequence $S'|S$ with $|S'| \leq \exp(G)$. Such a subsequence is called a *short zero-sum subsequence.* For more information on $\eta(G)$ we refer to [5] and [6].

Since $\mathsf{k}(T) \leq 1$ implies $|T| \leq \exp(G)$, we know that $\eta(G) \leq \mathsf{t}(G)$ always holds. The constant $\eta(G)$ is one of many classical invariants in so-called zero-sum theory. For zero-sum theory and its application, the interested reader is referred to [7] and [11].

Girard [14] noticed that if $\mathsf{t}(G) = \eta(G)$ for some finite abelian group G, then $\eta(H) \leq \eta(G)$ for any subgroup H of G, and then he deduced that for any positive integer $r \geq 4$, there is a finite abelian group of rank r such that $\mathsf{t}(G) > \eta(G)$. Girard [14] also proved that $\mathsf{t}(C_{p^\alpha}^2) = \eta(C_{p^\alpha}^2) = 3p^\alpha - 2$ for any prime p and conjectured that $\mathsf{t}(G) = \eta(G)$ for all finite abelian groups of rank two.

Conjecture 1.1. For all positive integers m, n with $m|n$, we have

$$\mathsf{t}(C_m \oplus C_n) = \eta(C_m \oplus C_n) = 2m + n - 2.$$

Conjecture 1.1 is wide open. For the case that G has rank three, he asked the following question.

Question 1.2. ([14], page 1856) Does $\mathsf{t}(G) = \eta(G)$ hold for all finite abelian groups G of rank three?

In this paper, we offer a negative answer to this question by showing

Theorem 1.3. *Let $n > 1$ be a positive integer, and let $G = C_2 \oplus C_2 \oplus C_{2n}$. Then* $\mathsf{t}(G) > \eta(G) = 2n + 4$.

We also prove the following.

Theorem 1.4. *Let p be a prime, and let $G = C_2 \oplus C_{2p}$. Then, $\mathsf{t}(G) = \eta(G)$.*

2. Notations and Preliminaries

Let $\mathbb{P}$ denote the set of prime numbers, $\mathbb{N}$ denote the set of positive integers, and $\mathbb{N}_0 = \mathbb{N} \cup \{0\}$. For any two integers $a, b \in \mathbb{N}_0$, we set $[a, b] = \{x \in \mathbb{N}_0 : a \leq x \leq b\}$. Throughout this paper, all abelian groups will be written additively, and for $n, r \in \mathbb{N}$, we denote by C_n the cyclic group of order n, and denote by C_n^r the direct sum of r copies of C_n.

Let $\mathcal{F}(G)$ be the free abelian monoid, multiplicatively written, with basis G. The elements of $\mathcal{F}(G)$ are called sequences over G. We write sequences $S \in \mathcal{F}(G)$ in the form

$$S = \prod_{g \in G} g^{v_g(S)}, \text{ with } v_g(S) \in \mathbb{N}_0 \text{ for all } g \in G.$$

We call $v_g(G)$ the multiplicity of g in S, and we say that S contains g if $v_g(S) > 0$. The unit element $1 \in \mathcal{F}(G)$ is called the empty sequence. A sequence S_1 is called a subsequence of S if $S_1 \mid S$ in $\mathcal{F}(G)$. For a subset A of G we denote $S_A = \Pi_{g \in A} g^{v_g(S)}$. If a sequence $S \in \mathcal{F}(G)$ is written in the form $S = g_1 \cdot \ldots \cdot g_l$, we tacitly assume that $l \in \mathbb{N}_0$ and $g_1, \ldots, g_l \in G$.

For a sequence

$$S = g_1 \cdot \ldots \cdot g_l = \prod_{g \in G} g^{v_g(S)} \in \mathcal{F}(G),$$

we call

- $|S| = l = \sum_{g \in G} v_g(G) \in \mathbb{N}_0$ the *length* of S,
- $\text{supp}(S) = \{g \in G | v_g(S) > 0\} \subset G$ the *support* of S,
- $\sigma(S) = \sum_{i=1}^{l} g_i = \sum_{g \in G} v_g(S) g \in G$ the *sum* of S,

The sequence S is called *zero-sumfree* if it contains no nonempty zero-sum subsequence.

We denote by $\mathcal{A}(G) \subset \mathcal{F}(G)$ the set of all minimal zero-sum sequences over G. Every map of abelian groups $\varphi : G \to H$ extends to a homomorphism $\varphi : \mathcal{F}(G) \to$

$\mathcal{F}(H)$ where $\varphi(S) = \varphi(g_1) \cdot \ldots \cdot \varphi(g_l)$. If φ is a homomorphism then $\varphi(S)$ is a zero-sum sequence if and only if $\sigma(S) \in \ker(\varphi)$.

We shall use the following invariants on zero-sum sequences.

Definition 2.1. Let $n, t \in \mathbb{N}$ and $\exp(G) = n$. We denote by

- $\mathsf{D}(G)$ the smallest integer $t \in \mathbb{N}$ such that every sequence $S \in \mathcal{F}(G)$ of length $|S| \geq t$ contains a non-empty zero-sum subsequence. The invariant $\mathsf{D}(G)$ is called the *Davenport constant* of G.
- $\mathsf{s}(G)$ the smallest integer $t \in \mathbb{N}$ such that every sequence $S \in \mathcal{F}(G)$ of length $|S| \geq t$ contains a zero-sum subsequence of length $\exp(G)$.

Throughout this paper, let p always denote an odd prime.

Lemma 2.2. *[11, Theorem 5.4.5] Let $n > 1$ be a positive integer, and let $S \in \mathcal{F}(C_n)$ be a sequence of length $n-1$. If S is zero-sumfree then $S = g^{n-1}$ for some generating element $g \in C_n$.*

Lemma 2.3. *Let $n > 1$ be a positive integer, and let $S \in \mathcal{F}(C_n)$ be a sequence of length $2n-1$. If S contains no two disjoint nonempty zero-sum subsequences then $S = g^{2n-1}$ for some generating element $g \in C_n$.*

Proof. Let T be an arbitrary subsequence of S of length $|T| = n-1$. Then, $|ST^{-1}| = n = \mathsf{D}(C_n)$. Therefore, ST^{-1} contains a zero-sum subsequence. It follows from the hypothesis of the lemma that T is zero-sumfree. Hence, $T = g^{n-1}$ for some generating element $g \in C_n$ by Lemma 2.2. Now the result follows from the arbitrariness of the choice of T. □

Lemma 2.4. *Let $n > 1$ be a positive integer, $G = C_2 \oplus C_2 \oplus C_{2n}$, and let (e_1, e_2, e_3) be a basis of G with $\mathrm{ord}(e_1) = \mathrm{ord}(e_2) = 2$ and $\mathrm{ord}(e_3) = 2n$. Then, the sequence $S = e_3^{2n-1} e_1 e_2 (e_1 + e_3)(e_1 + e_2 + e_3)(e_1 + e_2)$ contains no tiny zero-sum subsequence and therefore $\mathsf{t}(G) > 2n + 4$.*

Proof. Let $W = e_3^{2n-1} e_1 e_2 (e_1 + e_3)(e_1 + e_2 + e_3)$. It is easy to see that W contains no short zero-sum subsequence, and the sequence $W_1 = e_1 e_2 (e_1 + e_2)$ is the only short zero-sum subsequence of S. But $\mathsf{k}(W_1) = \frac{3}{2} > 1$. Therefore, S contains no tiny zero-sum subsequence. Hence, $\mathsf{t}(G) > |S| = 2n + 4$. □

3. Proof of the Main Results

Proof of Theorem 1.3. By Lemma 2.4, it suffices to prove that $\eta(C_2 \oplus C_2 \oplus C_{2n}) = 2n+4$. Let (e_1, e_2, e_3) be a basis of G with $\mathrm{ord}(e_1) = \mathrm{ord}(e_2) = 2$ and $\mathrm{ord}(e_3) = 2n$. Let $W = e_3^{2n-1} e_1 e_2 (e_1 + e_3)(e_1 + e_2 + e_3)$. Clearly, W contains no short zero-sum subsequence. Therefore, $\eta(G) \geq 1 + |W| = 2n + 4$.

So, it remains to prove $\eta(G) \leq 2n+4$. Let $S \in \mathcal{F}(G)$ be a sequence of length $2n+4$, we need to show that S contains a short zero-sum subsequence. Assume to the contrary that S contains no short zero-sum subsequence.

Let $\varphi : G \to C_2^3$ be the homomorphism with $\ker(\varphi) = C_n$. Since $\eta(C_2^3) = 8$ and $|S| = 2n+4$, S has a decomposition $S = T_1 \cdots T_{n-1}T$ with $\sigma(T_i) \in \ker(\varphi) \setminus \{0\}$ and $|T_i| \leq 2$ for each $i \in [1, n-1]$. It follows that $|T| \geq |S| - 2(n-1) \geq 6$.

Since S contains no short zero-sum subsequence and $\mathsf{D}(\ker(\varphi)) = \mathsf{D}(C_n) = n$, the sequence $\sigma(T_1) \cdot \ldots \cdot \sigma(T_{n-1})$ is zero-sumfree over C_n and $\varphi(T)$ contains no short zero-sum subsequence over C_2^3. It follows that $|\operatorname{supp}(\varphi(T))| = |\varphi(T)| = |T|$. Recall that $|T| \geq 6$. Let T' be a subsequence of T of length $|T'| = 6$. So, $\varphi(T')$ is a subset of C_2^3. Suppose that $(C_2^3 \setminus \{0\}) \setminus \varphi(T') = \{\alpha\}$. Let x_1, x_2, x_3 be a basis of C_2^3. Then, $C_2^3 = \{0, x_1, x_2, x_3, x_1+x_2, x_1+x_3, x_2+x_3, x_1+x_2+x_3\}$. Let ψ be an automorphism over C_2^3 with $\psi(\alpha) = x_1 + x_2 + x_3$. Then, $\varphi(T') = \{\alpha_1, \alpha_2, \alpha_3, \alpha_4, \alpha_5, \alpha_6\}$ with $\alpha_1 = \psi^{-1}(x_1), \alpha_2 = \psi^{-1}(x_2), \alpha_3 = \psi^{-1}(x_3), \alpha_4 = \psi^{-1}(x_1 + x_2), \alpha_5 = \psi^{-1}(x_1 + x_3)$ and $\alpha_6 = \psi^{-1}(x_2 + x_3)$. It is easy to check that the following subsequences of $\varphi(T')$ are all zero-sum.

$$\alpha_1\alpha_2\alpha_4, \alpha_1\alpha_3\alpha_5, \alpha_2\alpha_3\alpha_6, \alpha_4\alpha_5\alpha_6, \alpha_1\alpha_2\alpha_5\alpha_6, \alpha_1\alpha_3\alpha_4\alpha_6, \alpha_2\alpha_3\alpha_4\alpha_5. \tag{1}$$

Let $T' = g_1g_2g_3g_4g_5g_6$ with $\varphi(g_i) = \alpha_i$ for every $i \in [1, 6]$. From (1) we know that each of the following subsequences of T' has sum in $\ker(\varphi)$ and each is of length in $[3, 4]$:

$$g_1g_2g_4, g_1g_3g_5, g_2g_3g_6, g_4g_5g_6, g_1g_2g_5g_6, g_1g_3g_4g_6, g_2g_3g_4g_5. \tag{2}$$

Let T_n be any sequence listed in (2). Then, $\sigma(T_1) \cdot \ldots \cdot \sigma(T_{n-1}) \cdot \sigma(T_n)$ is a sequence over $\ker(\varphi) = C_n$ of length n. If there is a subset $I \subset [1, n]$ such that $1 \leq |I| \leq n-1$ and such that $\sum_{i \in I} \sigma(T_i) = 0$, then $\prod_{i \in I} T_i$ is a zero-sum subsequence of S of length

$$\Big|\prod_{i \in I} T_i\Big| \leq 2(|I| - 1) + 4 \leq 2(n-2) + 4 = 2n,$$

a contradiction. Therefore, every subsequence of $\sigma(T_1) \cdot \ldots \cdot \sigma(T_{n-1}) \cdot \sigma(T_n)$ of length $n-1$ is zero-sumfree. Therefore, $\sigma(T_1) = \sigma(T_2) = \cdots = \sigma(T_n)$ by Lemma 2.2. This proves that every sequence listed in (2) has sum $\sigma(T_1)$. Therefore,

$$\begin{aligned} g_1 + g_2 + g_4 &= g_1 + g_3 + g_5 = g_2 + g_3 + g_6 = g_4 + g_5 + g_6 \\ &= g_1 + g_2 + g_5 + g_6 = g_1 + g_3 + g_4 + g_6 = g_2 + g_3 + g_4 + g_5. \end{aligned}$$

From $g_1+g_2+g_4 = g_1+g_2+g_5+g_6$ we get that $g_4 = g_5+g_6$. Similarly, we obtain that $g_5 = g_4+g_6$ and $g_6 = g_4+g_5$. Therefore, $g_4+g_5+g_6 = (g_5+g_6)+(g_4+g_6)+(g_4+g_5)$ and $g_4 + g_5 + g_6 = 0$ follows. Hence, $g_4g_5g_6$ is a short zero-sum subsequence of S, a contradiction. This proves that $\eta(C_2 \oplus C_2 \oplus C_{2n}) = 2n + 4$. □

Proof of Theorem 1.4. As mentioned in the introduction we always have $\mathsf{t}(G) \geq \eta(G)$. From a result in [11, Theorem 5.8.3] we know that $\mathsf{t}(G) \geq \eta(G) = 2p+2$. So, it remains to prove that $\mathsf{t}(G) \le 2p+2$. Let $S \in \mathcal{F}(G)$ be of length $|S| = 2p+2$. We need to show that S contains a tiny zero-sum subsequence. Assume to the contrary that S contains no tiny zero-sum subsequence.

For every integer d, let S_d denote the subsequence of S consisting of all terms of S of order d. Then $S = S_2 S_p S_{2p}$ and

$$|S_2| + |S_p| + |S_{2p}| = 2p + 2. \tag{3}$$

Let $\varphi : G \to C_2^2$ be the homomorphism with $\ker(\varphi) = C_p$ and $\psi : G \to C_p$ be the homomorphism with $\ker(\psi) = C_2^2$. For any element $g \mid S_2$ we have $g \in \ker(\psi)$ and since $\eta(C_2^2) = 4$ we deduce that

$$|S_2| \le 3. \tag{4}$$

Similarly, as $\eta(C_p) = p$ we obtain that $|S_p| \le p - 1$ and $|S_{2p}| \ge p$ follows.

Since $\eta(\varphi(G)) = \eta(C_2^2) = 4$, S_{2p} has a decomposition

$$S_{2p} = T_1 \cdots T_m T$$

with $|T_i| = 2, \sigma(T_i) \in \ker(\varphi) = C_p$ for every $i \in [1, m]$ and $|T| \le 3$.

If there is a short zero-sum subsequence of $\sigma(T_1) \cdot \ldots \cdot \sigma(T_m) \cdot S_p$, i.e., there is a subset $I \subset [1, m]$ and a subsequence $T_0 | S_p$ such that $T_0 \prod_{i \in I} \sigma(T_i)$ is a short zero-sum sequence over C_p, then $T_0 \prod_{i \in I} T_i$ is a zero-sum subsequence of S with $\mathsf{k}(T_0 \prod_{i \in I} T_i) = \mathsf{k}(T_0) + \sum_{i \in I} \mathsf{k}(T_i) = \frac{|T_0|}{p} + \frac{|I|}{p} \le 1$, a contradiction. Therefore, $\sigma(T_1) \cdot \ldots \cdot \sigma(T_m) \cdot S_p$ contains no short zero-sum subsequence over $\ker(\varphi) = C_p$. It follows that

$$m + |S_p| \le \eta(C_p) - 1 = p - 1. \tag{5}$$

From $|T| \le 3$ and $|T_i| = 2$ we derive that $m \ge \frac{|S_{2p}| - 3}{2}$. This together with (5) gives that

$$|S_{2p}| + 2|S_p| \le 2p + 1. \tag{6}$$

By (3), (4), and (6) we obtain that

$$2p + 2 - 3 + |S_p| \le |S| - |S_2| + |S_p| = |S_{2p}| + 2|S_p| \le 2p + 1$$

and so $|S_p| \le 2$. Hence, $|S_{2p}| \ge 2p - 3 \ge \eta(C_p)$. Therefore, there exists a subsequence $R \mid S_{2p}$ such that $\sigma(R) \in \ker(\psi)$ and $|R| \le p$. It follows that $\mathsf{k}(R) = \frac{|R|}{2p} \le \frac{1}{2}$. If $|S_2| = 3$ then the sequence $\sigma(R) \cdot S_2 \in \mathcal{F}(C_2^2)$ is of length 4, and it follows from $\eta(C_2^2) = 4$ that the sequence $\sigma(R) \cdot S_2$ contains a short zero-sum subsequence W over C_2^2. By the contradiction hypothesis we must have that W is of the form $\sigma(R) g$ where g is a term of S. So, $R \cdot g$ is a zero-sum subsequence of S with $\mathsf{k}(W) = \mathsf{k}(R) + \mathsf{k}(g) \le 1$, a contradiction. Therefore, $|S_2| \le 2$. Similarly to above, by (3) and (6) we deduce that $|S_p| \le 1$ and $|S_{2p}| \ge 2p - 1$.

We show next that $|S_2| \le 1$. Assume to the contrary that $|S_2| = 2$. We assert that $\psi(S_{2p})$ contains no two disjoint short zero-sum subsequences over $\psi(G) =$

C_p. Otherwise, there exist two disjoint subsequences W_1, W_2 of S_{2p} such that $\sigma(W_1), \sigma(W_2) \in \ker(\psi) = C_2^2$ and $\mathsf{k}(W_1) \le \frac{1}{2}$, $\mathsf{k}(W_2) \le \frac{1}{2}$. Now the sequence $\sigma(W_1) \cdot \sigma(W_2) \cdot S_2 \in \mathcal{F}(C_2^2)$ is of length 4. Similarly to the case that $|S_2| = 3$ we can find a tiny zero-sum subsequence of $W_1W_2S_2$, a contradiction. It follows from $\eta(\psi(G)) = \eta(C_p) = p$ and $|S_{2p}| \ge 2p-1$ that every subsequence of $\psi(S_{2p})$ of length $p-1$ is zero-sumfree. Therefore, $\psi(S_{2p}) = \beta^{|S_{2p}|}$ for some $\beta \in \psi(G) = C_p$ by Lemma 2.3. Let W' be any subsequence of S_{2p} of length p. Then, $\sigma(W') \in \ker(\psi) = C_2^2$. Let $C_2^2 \setminus \{0, \operatorname{supp}(S_2)\} = \{y\}$. Since S contains no tiny zero-sum subsequence, similarly to above we infer that $\sigma(W') = y$. By the arbitrariness of the choice of W' we obtain that $S_{2p} = g^{|S_{2p}|}$. Now m in equation (5) can be chosen satisfying $m \ge \frac{|S_{2p}|-1}{2}$ and therefore the equation (6) can be improved to $|S_{2p}|+2|S_p| \le 2p-1$. But the above inequality is impossible as $|S_{2p}| + |S_p| = 2p + 2 - |S_2| = 2p$. This proves that $|S_2| \le 1$. It follows from equation (6) and (3) that

$$|S_2| = 1, |S_p| = 0 \text{ and } |S_{2p}| = 2p + 1.$$

By (5) we have that $p-1 = \frac{2p+1-3}{2} = \frac{|\varphi(S_{2p})|-3}{2} \le m \le p-1$. Therefore, $m = p-1$. Since S contains no tiny zero-sum subsequence, the sequence $\sigma(T_1) \cdot \ldots \cdot \sigma(T_{p-1})$ is a zero-sumfree sequence over $\ker(\varphi) = C_p$. It follows from Lemma 2.2 that

$$\sigma(T_1) = \cdots = \sigma(T_{p-1}) = h$$

for some $h \in \ker(\varphi) = C_p$.

Let $S(T_1 \cdots T_{p-1})^{-1} = g_0g_1g_2g_3$ with $S_2 = g_0$. Since S contains no tiny zero-sum subsequence, it follows from $m = p-1$ and $\eta(C_p) = p$ that $\varphi(g_1g_2g_3)$ contains no short zero-sum subsequence over C_2^2. Therefore, $\varphi(g_1)$, $\varphi(g_2)$ and $\varphi(g_3)$ are distinct in $C_2^2 \setminus \{0\}$. Without loss of generality we assume that $\varphi(g_0) = \varphi(g_1) = \varphi(g_2)+\varphi(g_3)$. Let $U_1 = g_0g_1$, $U_2 = g_0g_2g_3$, $U_3 = g_1g_2g_3$. Then $\sigma(U_i) \in \ker(\varphi)$ for every $i \in [1,3]$. So, for every $i \in [1,3]$, the sequence $\sigma(T_1) \cdot \ldots \cdot \sigma(T_{p-1}) \cdot \sigma(U_i) = h^{p-1}\sigma(U_i)$ contains a zero-sum subsequence V_i over $\ker(\varphi)$, i.e., there exists a subset $J_i \subseteq [1, p-1]$ such that $V_i = (\Pi_{j \in J_i} \sigma(T_j)) \cdot \sigma(U_i)$ for each $i \in [1,3]$. Let $X_i = \Pi_{j \in J_i} T_j \cdot U_i$ for each $i \in [1,3]$. Then, X_1, X_2 and X_3 are zero-sum subsequences of S. Let $t_i = |J_i|$ for each $i \in [1,3]$. Then,

$$V_1 = h^{t_1}(g_0 + g_1),\ V_2 = h^{t_2}(g_0 + g_2 + g_3),\ V_3 = h^{t_3}(g_1 + g_2 + g_3).$$

Since $\mathsf{k}(X_i) > 1$ for every $i \in [1,3]$, by a straightforward computation we obtain that $\frac{p+1}{2} \le t_1 \le p-1$, $\frac{p-1}{2} \le t_2 \le p-1$ and $t_3 = p-1$. Therefore,

$$p + 1 \le t_1 + t_2 + 1 \le 2p - 1. \tag{7}$$

From V_i is zero-sum over $\ker(\varphi) = C_p$ we infer that

$$t_1h + g_0 + g_1 = t_2h + g_0 + g_2 + g_3 = (p-1)h + g_1 + g_2 + g_3 = 0.$$

Therefore,

$$(t_1h + g_0 + g_1) + (t_2h + g_0 + g_2 + g_3) - ((p-1)h + g_1 + g_2 + g_3) = 0.$$

This together with $2g_0 = 0$ gives that $(t_1 + t_2 + 1)h = 0 \in \ker(\varphi) = C_p$. Therefore, $t_1 + t_2 + 1 \equiv 0 \pmod p$, a contradiction to (7). This completes the proof. □

Acknowledgments. This work was supported by the National Key Basic Research Program of China (Grant No. 2013CB834204), the PCSIRT Project of the Ministry of Science and Technology, and the National Science Foundation of China. The authors would like to thank the referee for his/her very useful suggestions.

References

[1] N. Alon and M. Dubiner, *A lattice point problem and additive number theroy*, Combinatorica 15 (1995), 301-309.

[2] F. Chung, *Pebbling in hypercubers*, SIAM J. Discrete Math. 2 (1989), 467-472.

[3] T. Denley, *On a result of Lemke and Kleitman*, Combin. Probab. Comput. 6 (1997), 39-43.

[4] S. Elledge and G.H. Hurlbert, *An application of graph pebbling to zero-sum sequences in abelian groups*, Integers 5 (2005), #A17.

[5] P. van Emde Boas and D. Kruyswijk, *A combinatorial problem on finite abelian groups III*, ZW 1969-008, Math. Centrum, Amsterdam.

[6] Y.S. Fan, W.D. Gao, L.L. Wang, and Q.H. Zhong, *Two zero-sum invariants on finite abelian groups,* European J. Combinatorics, to appear.

[7] W.D. Gao and A. Geroldinger, *Zero-sum problems in finite abelian groups: a survey*, Expo. Math. 24 (2006), 337-369.

[8] W. Gao, Q.H. Hou, W.A. Schmid, and R. Thangadurai, *On short zero-sum subsequences II*, Integers 7 (2007), #A21.

[9] A. Geroldinger, *On a conjecture of Lemke and Kleitman*, J. Number Theory 44 (1993), 60-65.

[10] A. Geroldinger and D. Grynkiewicz, *On the structure of minimal zero-sum sequences with maximal cross number*, J. Combinatorics and Number Theory 1 (2009), 9-26.

[11] A. Geroldinger and F. Halter-Koch, *Non-Unique Factorizations. Algebraic, Combinatorial and Analytic Theory*, Pure Appl. Math. 278, Chapman & Hall/CRC, 2006.

[12] A. Geroldinger and R. Schneider, *The cross number of finite abelian groups II*, European J. Combinatorics 15 (1994), 399-405.

[13] B. Girard, *A new upper bound for the cross number of finite abelian groups*, Israel J. Math 172 (2009), 253-278.

[14] B. Girard, *On a combinatorial problem of Erdős, Kleitman and Lemke*, Advances Math. 231 (2012), 1843-1857.

[15] G. Hurlbert, *A survey of graph pebbling*, in: Proceedings of the Thirtieth Southeastern International Conference on Combinatorics, in: Graph Theory and Computing, vol. 139, 1999, pp. 41-64.

[16] G. Hurlbert, *Recent progress in graph pebbling*, Graph Theory Notes N. Y. 49 (2005), 25-37.

[17] P. Lemke and D. Kleitman, *An addition theorem on the integers modulo n*, J. Number Theory 31 (1989), 335-345.

A NOTE ON IRREGULARITIES OF DISTRIBUTION

R. L. Graham

Received: 11/25/12, Accepted: 7/23/13, Published: 9/5/13

Abstract

Let $d \geq 0$ be a fixed integer. Suppose $X = (x_1, x_2, x_3, \ldots)$ is a sequence in $[0, 1)$ with the property that for every $n \leq N$, each of the intervals $[\frac{k-1}{n}, \frac{k}{n})$, $1 \leq k \leq n$, contains *at least one* x_i with $1 \leq i \leq n + d$. We show that this implies that $N = O(d^3)$. This is a generalization of a question raised by Steinhaus some 50 years ago.

1. Introduction

In [4], H. Steinhaus [4] raised the following question. Does there exist for every positive integer N, a sequence of real numbers $x_1, x_2, \ldots, x_N$ in $[0, 1)$ such that every $n \in \{1, 2, \ldots, N\}$ and every $k \in \{1, 2, \ldots, n\}$, we have

$$\frac{k-1}{n} \leq x_i < \frac{k}{n}$$

for some $i \in \{1, 2, \ldots, n\}$? If not, what is the largest possible value of N for which this can hold?

It was first shown by M. Warmus [5] (see also [4]) that the largest possible value of N is 17. In fact, he showed there are 768 essentially different solutions for $N = 17$. For example, one of these solutions is given by choosing the x_i to satisfy the following inequalities:

$$\frac{4}{7} \leq x_1 < \frac{7}{12}, \quad \frac{2}{7} \leq x_2 < \frac{5}{17}, \quad \frac{16}{17} \leq x_3 < 1, \quad \frac{1}{14} \leq x_4 < \frac{1}{13},$$
$$\frac{8}{11} \leq x_5 < \frac{11}{15}, \quad \frac{5}{11} \leq x_6 < \frac{6}{13}, \quad \frac{1}{7} \leq x_7 < \frac{2}{13}, \quad \frac{14}{17} \leq x_8 < \frac{5}{6},$$
$$\frac{3}{8} \leq x_9 < \frac{5}{13}, \quad \frac{11}{17} \leq x_{10} < \frac{2}{3}, \quad \frac{3}{14} \leq x_{11} < \frac{3}{13}, \quad \frac{15}{17} \leq x_{12} < \frac{11}{12},$$
$$\frac{1}{12} \leq x_{13} < \frac{9}{17}, \quad 0 \leq x_{14} < \frac{1}{17}, \quad \frac{13}{17} \leq x_{15} < \frac{4}{5}, \quad \frac{5}{16} \leq x_{16} < \frac{6}{17},$$
$$\frac{10}{17} \leq x_{17} < \frac{11}{17}.$$

A generalization of Steinhaus' question was treated in 1970 in a paper of E. R. Berlekamp and the author [1]. Here, for each integer $d \geq 0$, we define $s(d)$ to be the largest integer such that there is a sequence $x_1, x_2, \ldots, x_{s(d)}$ with $x_i \in [0,1)$ so that for all $n \leq s(d) - d$ and all $k \in \{1, 2, \ldots, n\}$, there is *at least one* $i \in \{1, 2, \ldots, n+d\}$ such that $\frac{k-1}{n} \leq x_i < \frac{k}{n}$. The result of Warmus shows that $s(0) = 17$.

In addition to giving a different proof that $s(0) = 17$, the authors of [1] also presented an argument asserting that $s(d) < 4^{(d+2)^2}$ for general d. There are several things to be said about the claim in this paper.

First, the form of this bound already follows (as pointed out in [1]) from a fundamental inequality of Roth [3]. In particular, he proves the following. Let $X = (x_1, x_2, \ldots)$ be a sequence of points in the interval $[0,1)$, $I \subseteq [0,1)$ a subinterval, $|I|$ its length and $X_n(I)$ the number of $x_m \in I$ with $m \leq n$. Then

$$\sup_{I \subseteq [0,1)} | X_n(I) - n|I| | > c\sqrt{\log n} \tag{1}$$

for some suitable absolute constant $c > 0$. This easily implies that $s(d) < \exp(c'd^2)$ for an appropriate absolute constant c'.

The second point to make is that it was recently pointed out by David and Moshe Newman [2] that the proof given in [1] is incomplete. The purpose of this note is to fill this gap by a different argument which also yields a much stronger upper bound on $s(d)$.

2. The Main Result

Theorem 1 *There is an absolute constant c such that for all* $d \geq 1$, $s(d) < cd^3$.

Proof. (sketch) Fix $d \geq 1$ and let $X = (x_1, x_2, \ldots)$ be a sequence of points in $[0,1)$. Let $I_{k,n}$ denote the subinterval $[\frac{k-1}{n}, \frac{k}{n})$. Assume that for $1 \leq n \leq N$, and $1 \leq k \leq n$, each subinterval $I_{k,n}$ contains at least one x_i with $1 \leq i \leq n+d$. Our goal is to show that $N = N(d) = O(d^3)$. We first want to transform our problem as follows. Namely, at time n, let us expand the time line by a factor of n. Thus, the interval $I_{k,n}$ becomes the (unit) interval $I_k = [k-1, k)$ and each point x_i gets replaced by nx_i. In this form, our condition now is that for $1 \leq k \leq n$, each of the intervals I_k contains at least one of the points nx_i for some i satisfying $1 \leq i \leq n+d$. Furthermore, this must hold for all n satisfying $1 \leq n \leq N$.

With this interpretation, we can think of each point x_i as starting from the origin at time 0, and then, as time progresses, continuously moving to the right along the positive time axis with a velocity x_i. Let us use the notation $x_i(t) = tx_i$ to denote the position of x_i at time t. Thus, $x_i(0) = 0$ for all i. If we stop the clock at some integer time n and see where the first $n+d$ points $x_i(n)$ are, the hypothesis implies that each of the intervals $I_k = [k-1, k)$, $1 \le k \le n$, has at least one of these points in it.

Our proof sketch proceeds as follows. We first choose some (large) time t_0 and identify a (large) set of disjoint pairs of points $(y_i(t_0), z_i(t_0))$, with $y_i(t_0) \in I_{4i}$ and $z_i(t_0) \in I_{4i+2}$. Thus, the differences $g_i(t_0) = z_i(t_0) - y_i(t_0)$ are bounded between 1 and 3. We will next select $d+1$ of these, say $g_{i_1}(t_0), g_{i_2}(t_0), \ldots, g_{i_{d+1}}(t_0)$, with $g_{i_j}(t_0) = z_{i_j}(t_0) - y_{i_j}(t_0)$ which all are close to each other in size, i.e., so that all $|g_{i_j}(t_0) - g_{i_k}(t_0)| < \alpha$ for some small $\alpha = \alpha(d)$. We will then let time advance from t_0 to some later time $t_1 > t_0$. This will have the effect of (linearly) expanding the $g_{i_j}(t)$ by a factor of roughly $\frac{t_1}{t_0}$ (since $y_{i_j}(t_0) = t_0 y_{i_j}$ goes to $y_{i_j}(t_1) = t_1 y_{i_j}$, etc). We will choose t_1 so that all the sizes of the (expanded) differences $g_{i_j}(t_1) = z_{i_j}(t_1) - y_{i_j}(t_1)$ now lie between 3 and $3+\delta$ for a small $\delta = \delta_i(d)$. (Strictly speaking, the differences should be bounded away from 3, but this does not affect the basic argument). Note that we can always choose $t_1 \le 3t_0$. The next (and final) step is to start letting the time variable advance, keeping track of where the terms of each pair $y_{i_j}(t)$ and $z_{i_j}(t)$ are as the time variable assumes successive integer values.

Suppose at some time t a particular pair $(y_{i_j}(t), z_{i_j}(t))$ has $y_{i_j}(t)$ just barely less than some integer m. That is, at time t, $y_{i_j}(t) \in I_m = [m-1, m)$ but at time $t+1$, $y_{i_j}(t+1) \in I_{m+1} = [m, m+1)$. However, at time t, we still have $z_{i_j}(t) \in I_{m+4}$ (since $g_{i_j}(t) = z_{i_j}(t) - y_{i_j}(t)$ is just barely greater than 3). By the hypothesis on the original sequence X, the intervals I_{m+1}, I_{m+2} and I_{m+3} must each also have points $x_m(t)$ in them at time t. However, when we advance the time to $t+1$ and $y_{i_j}(t+1)$ now moves out of I_m and into I_{m+1}, its mate $z_{i_j}(t+1)$ is still in I_{m+4} (since $g_{i_j}(t+1)$ is still only slightly larger than 3). Consequently, at time $t+1$, the 4 intervals $I_{m+1}, I_{m+2}, I_{m+3}$ and I_{m+4} now have **5** points in them! Furthermore, the situation persists until the time advances to the point where $z_{i_j}(t)$ leaves I_{m+4}, which is most of the time that $y_{i_j}(t)$ is still moving across I_{m+1} . We will call these times at which there are 5 $x_j's$ in 4 consecutive $I_r's$ "good" times for the pair (y_{i_j}, z_{i_j}). While we don't know exactly when these various transitions occur for different pairs, it is enough to show that for a suitable finishing time t_2, the *fraction* of the number of time steps that are "good" for each pair exceeds the fraction $\frac{d}{d+1}$ of the total number of time steps rom t_1 to t_2. From this it will follow that there must be some common time which is simultaneously good for all $d+1$ pairs. This is a contradiction since at this time, we would have $d+1$ disjoint sets of 4 consecutive intervals I_k which each have 5 points x_i in them, forcing some other interval I_l to have no point x_i in it, contradicting our initial hypothesis.

To make this all work, it is simply a matter of choosing the various parameters carefully (actually, one doesn't have to be too careful). For example, we need to make sure that that the values y_i (and z_i) are not too small (since it takes roughly $\frac{1}{y_i}$ time steps for $y_i(t)$ to go from m to $m+1$). We would like this minimum numbers of steps to cross a unit interval to be relatively large.

It turns out that with a modest amount of computation, the following choice of parameters can be shown to work. Namely, start by taking $t = 5000d^3$ (for example). We will choose our y_i and z_i from the intervals I_{4r} to I_{8r+2} where $r = 20d^2$. This will guarantee that we can find $d+1$ differences $g_{i_j}(t) = |z_{i_j}(t) - y_{i_j}(t)|$ which differ in size from each other by at most $\frac{1}{10d}$. We then expand these (by letting time increase to $t_1 \leq 3t_0 \leq 15000d^3$) so that the sizes of all the enlarged $g_{i_j}(t_0)$ satisfy $3 < g_{i_j}(t_0) < 3 + \frac{1}{9d}$. Finally, we compute the fraction of time steps which are "good" time for each difference $z_{i_j}(t) - y_{i_j}(t)$ from times t_1 to $t_2 = t_1 + 1000d^2$. This turns out to be at least the fraction $\frac{d}{d+1}$ of the total number of times steps during time period from t_1 to t_2. This in turn implies that there is some time t during this period which is good for *all* $d+1$ of the differences $z_{i_j}(t) - y_{i_j}(t)$, which is a contradiction. This implies that the initial value of N must be less than $t_1 + 1000d^2 < 16000d^3$ and consequently $s(d) < 16000d^3$ (this bound is actually rather loose). Checking the details is not difficult (although messy) and we leave this to the interested reader. This completes our proof sketch.

3. Some Remarks

It is not clear what the truth for the upper bound of $s(d)$ really is. Could it actually be *linear* in d? Or even sublinear? It would be interesting to know what the correct growth rate for s(d) is. While it is known that $s(1) \geq 23$, we have no nontrivial estimates for lower bounds for $s(d)$. It would appear that some new ideas may be needed to attack this problem.

References

[1] E. R. Berlekamp and R. L. Graham, Irregularities in the Distributions of Finite Sequences, *J. Number Th.* **2**, (1970), 152–161.

[2] David and Moshe Newman (personal communication).

[3] K. F. Roth, On irregularities of distribution, *Mathematika* **1** (1954), 73–79.

[4] H. Steinhaus, One Hundred Problems in Elementary Mathematics, *Basic Books*, New York, (1964).

[5] M. Warmus, A Supplementary Note on the Irregularities of Distributions, *J. Number Th.* **8**, (1976), 260–263.

GENERALIZING DELANNOY NUMBERS VIA COUNTING WEIGHTED LATTICE PATHS

M. Dziemiańczuk
Institute of Informatics, University of Gdańsk, Poland
mdziemia@inf.ug.edu.pl

Received: 11/12/12, Revised: 5/15/13, Accepted: 7/10/13, Published: 9/26/13

Abstract

The aim of this paper is to introduce a generalization of Delannoy numbers. The standard Delannoy numbers count lattice paths from $(0,0)$ to (n,k) consisting of horizontal $(1,0)$, vertical $(0,1)$, and diagonal $(1,1)$ steps called segments. We assign weights to the segments of the lattice paths, and we sum weights of all lattice paths from any (a,b) to (n,k). Generating functions for the generalized Delannoy numbers and weighted central Delannoy numbers are delivered and some consequences of them are drawn. Several identities, summations, and convolution-like formulas for the general case are given. We show how our approach may be used in counting lattice paths with certain restrictions. Namely, we remove certain classes of segments from the complete lattice and then we ask about the number of paths in such modified lattices. In the last section we show how these numbers generalize, among others, binomial, q-binomial coefficients, ordinary Stirling numbers of both kinds, and p,q-Stirling numbers. Applying the results of generalized Delannoy numbers we deliver a q-Vandermonde identity and a generalized Carlitz-like formula for p,q-binomial coefficients.

1. Introduction

Let $L(n,k)$ denote the family of lattice paths from $(0,0)$ to (n,k), where $n,k \geq 0$, consisting of vertical $(0,1)$, horizontal $(1,0)$, and diagonal $(1,1)$ steps (segments). An example of a lattice path from $L(4,2)$ is given in Fig. 1. Let $D(n,k)$ denote the size of $L(n,k)$. The numbers $D(n,k)$ satisfy

$$D(n,k) = D(n,k-1) + D(n-1,k) + D(n-1,k-1)$$

with initial values $D(n,0) = D(0,k) = D(0,0) = 1$.

From the recurrence relation one can obtain the generating function

$$\sum_{n,k\geq 0} D(n,k)x^k y^n = \frac{1}{1-x-y-xy}. \tag{1.1}$$

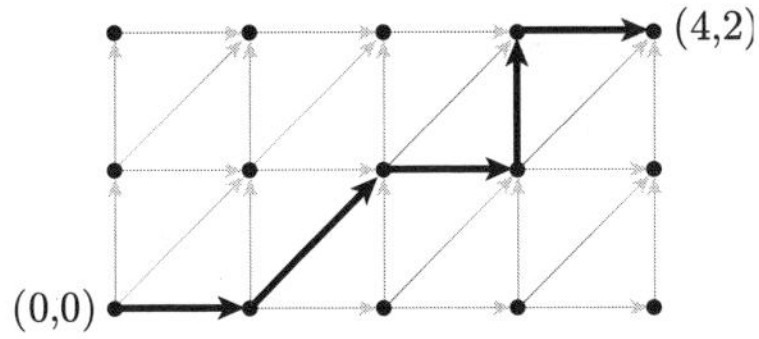

Figure 1: A lattice path from the set $L(4,2)$.

Recall that $D(n,k)$ are called *Delannoy numbers* from the name of Henri Delannoy [8] and $D(n) = D(n,n)$ *central Delannoy numbers.* These numbers are denoted in OEIS [23] by A152250 and A001850, respectively. See Comtet [6, p. 81] and Stanley [24, p. 185]. Delannoy numbers appear also in Stanton and Cowan [25]. We refer the reader to Banderier and Schwer [3]. From the generating function it follows that

$$D(n,k) = \sum_{i=0}^{k} \binom{n}{i} \binom{n+k-i}{n}. \tag{1.2}$$

There are several ways of generalizing Delannoy numbers in the literature. Wagner [28] defines some statistics on lattice paths. Autebert and Schwer [2] extend this 2-dimensional case to the d-dimensional space $\mathbb{Z}^d$, where $d > 2$. See also Caughman, Haithcock and Veerman [5]. The main idea of our approach is to assign weights to the segments of a lattice and to sum weights of paths. This idea has been previously used by Fray and Roselle [13], Hetyei [17] and Loehr and Savage [21, Sec. 3.2].

Let $\mathfrak{L}$ denote the 4-tuple $\langle \mathbf{v}, \mathbf{h}, \mathbf{d}, \zeta \rangle$ of three sequences of complex numbers $\mathbf{v} = (v_0, v_1, \ldots)$, $\mathbf{h} = (h_0, h_1, \ldots)$, $\mathbf{d} = (d_0, d_1, \ldots)$, and a complex number ζ. A *step* is a pair of two lattice points. We consider only three types of steps: horizontal $((a,b),(a+1,b))$, vertical $((a,b),(a,b+1))$, and diagonal $((a,b),(a+1,b+1))$. We represent a path $\pi \in L(n,k)$ by the sequence $(\pi_1, \pi_2, \ldots, \pi_s)$ of steps π_i from the origin $(0,0)$ to (n,k). By the *weight* of a step $\pi_i = ((a,b),(c,e))$ we mean the value of $w^{\mathfrak{L}}(\pi_i)$ defined as follows

$$w^{\mathfrak{L}}(\pi_i) = \begin{cases} v_c, & \text{if } \pi_i \text{ is a vertical step;} \\ h_c\zeta^e, & \text{if } \pi_i \text{ is a horizontal step;} \\ d_c\zeta^{e-1}, & \text{if } \pi_i \text{ is a diagonal step.} \end{cases}$$

Incoming and outcoming steps of a lattice point with corresponding weights are given in Fig. 2. We define the *weight* of a path π, denoted by $w^{\mathfrak{L}}(\pi)$, to be $w^{\mathfrak{L}}(\pi_1) \cdots w^{\mathfrak{L}}(\pi_s)$. By the (n,k) $\mathfrak{L}$-*Delannoy number*, expressed in angular brackets, we mean the sum

$$\left\langle {n \atop k} \right\rangle_{\mathfrak{L}} = \sum_{\pi \in L(n,k)} w^{\mathfrak{L}}(\pi). \tag{1.3}$$

The $\mathfrak{L}$-Delannoy numbers generalize a wide class of combinatorial numbers,

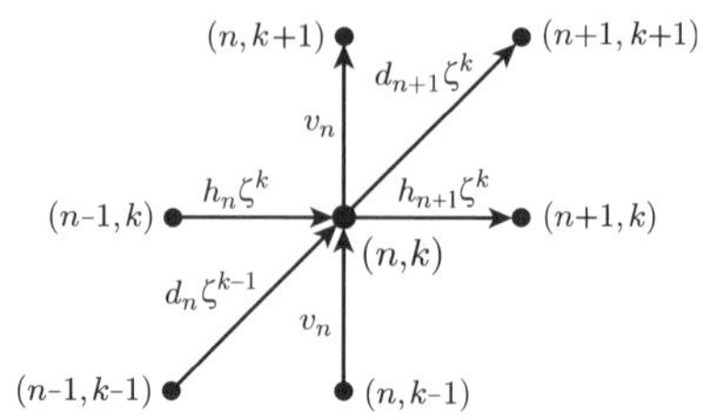

Figure 2: Weights of incoming and outcoming steps.

including binomial, q-binomial coefficients, Stirling numbers of both kinds, p,q-Stirling numbers [27], ζ-analogues [12] (see Section 8). For $\mathfrak{B} = \langle \mathbf{1}, \mathbf{1}, \mathbf{1}, 1\rangle$, where $\mathbf{1} = (1, 1, \ldots)$, the $\mathfrak{B}$-Delannoy numbers reduce to the ordinary Delannoy numbers, i.e.,

$$\left\langle {n \atop k} \right\rangle_{\mathfrak{B}} = D(n, k). \tag{1.4}$$

In Section 3 we show that the generating function for the general case of the $\mathfrak{L}$-Delannoy numbers is

$$\sum_{k\geq 0} \left\langle {n \atop k} \right\rangle_{\mathfrak{L}} z^k = \frac{1}{(1 - v_0\zeta^n z)} \prod_{i=1}^{n} \frac{h_i + d_i\zeta^{n-i}z}{1 - v_i\zeta^{n-i}z}.$$

In Section 4 we consider central weighted Delannoy numbers studied by Fray and Roselle [13], and Hetyei [17]. Namely, let $\mathfrak{W} = \langle \mathbf{v}^c, \mathbf{h}^c, \mathbf{d}^c, 1\rangle$, where $\mathbf{v}^c = (v, v, \ldots), \mathbf{h}^c = (h, h, \ldots), \mathbf{d}^c = (d, d, \ldots)$ are constant sequences of $v, h, d \in \mathbb{C}$. We show that

$$\sum_{n\geq 0} \left\langle {n \atop n} \right\rangle_{\mathfrak{W}} z^n = \frac{1}{\sqrt{d^2 z^2 - z(2d + 4hv) + 1}}.$$

We define two reduced cases of the generalized Delannoy numbers. Namely, for a given 4-tuple $\mathfrak{L} = \langle \mathbf{v}, \mathbf{h}, \mathbf{d}, \zeta\rangle$ we set $\mathfrak{L}_1 = \langle \mathbf{0}, \mathbf{h}, \mathbf{d}, \zeta\rangle$ and $\mathfrak{L}_2 = \langle \mathbf{v}, \mathbf{1}, \mathbf{0}, \zeta\rangle$, where $\mathbf{0} = (0, 0, \ldots)$. By the *$\mathfrak{L}$-Delannoy numbers of the first kind* (expressed in square brackets) we mean $\mathfrak{L}_1$-Delannoy numbers, and by the *$\mathfrak{L}$-Delannoy numbers of the second kind* (expressed in curly brackets) we mean $\mathfrak{L}_2$-Delannoy numbers, i.e.,

$$\left[{n \atop k} \right]_{\mathfrak{L}} \equiv \left\langle {n \atop k} \right\rangle_{\mathfrak{L}_1}, \qquad \left\{ {n \atop k} \right\}_{\mathfrak{L}} \equiv \left\langle {n \atop k} \right\rangle_{\mathfrak{L}_2}.$$

It is shown in Section 3 that

$$\sum_{k\geq 0} \left[{n \atop k} \right]_{\mathfrak{L}} z^k = \prod_{i=1}^{n} \left(h_i + d_i\zeta^{n-i}z\right),$$

$$\sum_{k\geq 0} \left\{ {n \atop k} \right\}_{\mathfrak{L}} z^k = \prod_{i=0}^{n} \frac{1}{(1 - v_i\zeta^{n-i}z)},$$

It turns out that (1.2) has a natural generalization for $\mathfrak{L}$-Delannoy numbers of the first and second kind, that is,

$$\left\langle {n \atop k} \right\rangle_{\mathfrak{L}} = \sum_{i=0}^{k} \left[{n \atop i} \right]_{\mathfrak{L}} \left\{ {n \atop k-i} \right\}_{\mathfrak{L}},$$

which follows from the generating function. However, a combinatorial proof is given in Section 2. The generalization of the Carlitz formula is delivered in Section 3 (Theorem 10) for the $\mathfrak{L}$-Delannoy numbers of the first and second kind. That is,

$$\left[{n \atop k} \right]_{\mathfrak{H}} = \sum_{j=0}^{k} \binom{n-k+j}{j} (\alpha_2 \zeta^n)^j \left[{n \atop k-j} \right]_{\mathfrak{H}^*},$$

$$\left\{ {n \atop k} \right\}_{\mathfrak{H}} = \sum_{j=0}^{k} \binom{n+k}{j} (\alpha_1 \zeta^n)^j \left\{ {n \atop k-j} \right\}_{\mathfrak{H}^*},$$

where $\mathfrak{H} = \langle \mathbf{v}, \mathbf{1}, \mathbf{d}, \zeta \rangle$ and $\mathfrak{H}^* = \langle \mathbf{v}^*, \mathbf{1}, \mathbf{d}^*, \zeta \rangle$ such that $v_i = v_i^* + \alpha_1 \zeta^i$ and $d_i = d_i^* + \alpha_2 \zeta^i$, where $\alpha_1, \alpha_2 \in \mathbb{C}$.

In Section 5 we study the case where the sequences $\mathbf{v}, \mathbf{h}$, and $\mathbf{d}$ take values over the set $\{0, 1\}$ which allows to count Delannoy paths with certain restrictions. Namely, paths whose steps have weight equal to zero are omitted in the sum. Therefore, by setting certain elements of $\mathbf{v}, \mathbf{h}$, and $\mathbf{d}$ to zero we remove corresponding steps of the lattices. Let $H = H(\sigma, \tau, \eta)$ be a family of steps (see Section 5), we denote by $D(n, k; H)$ the number of Delannoy paths from $(0, 0)$ to (n, k) which consist of steps from H. We show that

$$D(n,k;H) = \sum_{j=0}^{k} \binom{n-|\sigma|-|\tau|}{j-|\sigma|} \binom{n-|\eta|+k-j}{n-|\eta|}, \tag{1.5}$$

where $\sigma, \tau \subseteq \{1, 2, \ldots, n\}$ such that $\tau \cap \sigma = \emptyset$, and $\eta \subseteq \{0, 1, \ldots, n\}$.

In Section 6 we generalize previous results to the case where weighted Delannoy paths begin at any lattice point (m, l). Let x, y be any two lattice points, and let us denote by $L[x, y]$ the set of all Delannoy paths from x to y. Let

$$\left\langle {n \atop k} \right\rangle_{\mathfrak{L}}^{(m,l)} = \sum_{\pi \in L[x,y]} w^{\mathfrak{L}}(\pi),$$

where $x = (m, l)$ and $y = (m+n, l+k)$. Although, this generalization does not give essentially new results but provides a convenient notation for some new convolution-like formulae given in Section 6.

In Section 7 we show the connection between these generalized Delannoy numbers and partitions of numbers. We show, for instance, that for certain 4-tuples $\mathfrak{P}(H)$, the $\mathfrak{P}(H)$-Delannoy numbers of the first and second kind are generating functions of certain sequences counting partitions of numbers, i.e.,

$$\begin{bmatrix} m \\ k \end{bmatrix}_{\mathfrak{P}(H)} = \sum_{n\geq 0} |\mathcal{D}(H;n,k)|\zeta^n,$$

$$\begin{Bmatrix} m \\ k \end{Bmatrix}_{\mathfrak{P}(H)} = \sum_{n\geq 0} |\mathcal{R}(H;n,k)|\zeta^n,$$

where $\mathcal{D}(H;n,k)$, and $\mathcal{R}(H;n,k)$ are the families of partitions of n into k distinct parts from the set H without and with repetition of parts allowed, respectively.

In Section 8 we show how $\mathfrak{L}$-Delannoy numbers generalize binomial and q-binomial coefficients, ordinary Stirling numbers of both kinds, and p,q-Stirling numbers.

2. Counting Lattice Paths with Diagonals

For $n,k \geq 0$, the $\mathfrak{L}$-Delannoy numbers satisfy the following recurrence relation

$$\left\langle {n \atop k} \right\rangle_{\mathfrak{L}} = v_n \left\langle {n \atop k-1} \right\rangle_{\mathfrak{L}} + h_n\zeta^k \left\langle {n-1 \atop k} \right\rangle_{\mathfrak{L}} + d_n\zeta^{k-1} \left\langle {n-1 \atop k-1} \right\rangle_{\mathfrak{L}} \tag{2.1}$$

with initial value $\left\langle {0 \atop 0} \right\rangle_{\mathfrak{L}} = 1$. We assume that $\left\langle {n \atop k} \right\rangle_{\mathfrak{L}} = 0$ for $n < 0$ or $k < 0$. We denote by $L_1(n,k)$ the set of all lattice paths from $(0,0)$ to (n,k) consisting of only horizontal and diagonal steps, and by $L_2(n,k)$ the set of lattice paths from $(0,0)$ to (n,k) consisting of vertical and horizontal steps (see Fig. 3). With this notation, we have

$$\begin{bmatrix} n \\ k \end{bmatrix}_{\mathfrak{L}} = \sum_{\pi \in L_1(n,k)} w^{\mathfrak{L}_1}(\pi), \qquad \begin{Bmatrix} n \\ k \end{Bmatrix}_{\mathfrak{L}} = \sum_{\pi \in L_2(n,k)} w^{\mathfrak{L}_2}(\pi).$$

Let $D_1(n,k)$ and $D_2(n,k)$ denote the cardinalities of the sets $L_1(n,k)$ and $L_2(n,k)$, respectively. It is well-known that

$$D_1(n,k) = \binom{n}{k}, \qquad D_2(n,k) = \binom{n+k}{k}.$$

The $\mathfrak{L}$-Delannoy numbers of the first and second kind satisfy the following recurrence relations

$$\begin{bmatrix} n \\ k \end{bmatrix}_{\mathfrak{L}} = h_n\zeta^k \begin{bmatrix} n-1 \\ k \end{bmatrix}_{\mathfrak{L}} + d_n\zeta^{k-1} \begin{bmatrix} n-1 \\ k-1 \end{bmatrix}_{\mathfrak{L}}, \tag{2.2a}$$

$$\begin{Bmatrix} n \\ k \end{Bmatrix}_{\mathfrak{L}} = v_n \begin{Bmatrix} n \\ k-1 \end{Bmatrix}_{\mathfrak{L}} + \zeta^k \begin{Bmatrix} n-1 \\ k \end{Bmatrix}_{\mathfrak{L}}, \tag{2.2b}$$

with initial values $\left[{0 \atop 0} \right]_{\mathfrak{L}} = \left\{ {0 \atop 0} \right\}_{\mathfrak{L}} = 1$.

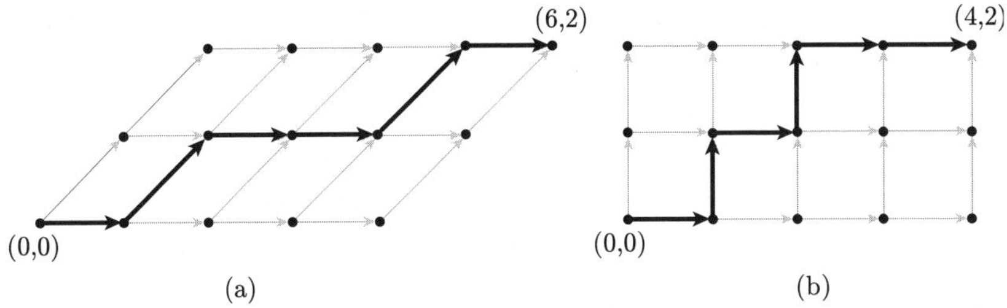

Figure 3: Two lattice paths: (a) from $L_1(6,2)$, and (b) from $L_2(4,2)$.

Let us denote by $L(n,k,j)$ the set of those paths from $L(n,k)$ which contain exactly j diagonal steps. Observe that any path contains at most $\min(n,k)$ diagonal steps, and thus

$$D(n,k) = \sum_{j=0}^{\min(n,k)} |L(n,k,j)|.$$

One shows that

$$|L(n,k,j)| = \binom{n}{j}\binom{n+k-j}{k-j},$$

which immediately implies (1.2). In fact, we show a generalization of this result for the $\mathfrak{L}$-Delannoy numbers.

Theorem 1. *For $n, k \geq 0$, we have*

$$\left\langle {n \atop k} \right\rangle_{\mathfrak{L}} = \sum_{i=0}^{k} \left[{n \atop i} \right]_{\mathfrak{L}} \left\{ {n \atop k-i} \right\}_{\mathfrak{L}}. \tag{2.3}$$

Proof. We divide the proof into three lemmas. Let us consider the left-hand side of (2.3). By Lemma 2 and Lemma 4, we show that

$$\begin{aligned}
\left\langle {n \atop k} \right\rangle_{\mathfrak{L}} &= \sum_{i=0}^{\min(n,k)} \sum_{\pi \in L(n,k,i)} w^{\mathfrak{L}}(\pi) \\
&= \sum_{i=0}^{\min(n,k)} \sum_{\substack{\alpha \in L_1(n,i) \\ \beta \in L_2(n,k-i)}} w^{\mathfrak{L}_1}(\alpha) w^{\mathfrak{L}_2}(\beta) \\
&= \sum_{i=0}^{\min(n,k)} \left[{n \atop i} \right]_{\mathfrak{L}} \left\{ {n \atop k-i} \right\}_{\mathfrak{L}}.
\end{aligned}$$

□

Note that the ith summand of the right-hand side of (2.3) is the sum of weights of these Delannoy paths which consist of exactly i diagonal steps.

Let $\sigma = ((a,b),(c,d))$ and $\phi = ((e,f),(g,h))$ be two lattice segments. We write $\sigma \sim \phi$ if $a = e$ and $c = g$. We set $\mathtt{x}(\sigma) = c$, and $\mathtt{y}(\sigma) = b$. We define two operations on segments: $\mathtt{up}(\sigma) = ((a,b+1),(c,d+1))$, and if σ is horizontal, then $\mathtt{dg}(\sigma) = ((a,b),(c,d+1))$, and $\mathtt{dg}(\sigma) = \sigma$ otherwise.

Lemma 2. *There is a bijection $f : L_1(n,j) \times L_2(n,k-j) \to L(n,k,j)$.*

Proof. The proof consists in the construction of the bijection $f : L_1(n,j) \times L_2(n,k-j) \to L(n,k,j)$.

($\to$) Let us take two paths $\alpha \in L_1(n,j)$ and $\beta \in L_2(n,k-j)$. On one hand, the path α contains j diagonal and $n-j$ horizontal segments, respectively. On the other hand, the path β contains n horizontal and $k-j$ vertical steps, respectively. Let $\alpha = (\alpha_1, \alpha_2, \ldots, \alpha_n)$ and $\beta = (\beta_1, \beta_2, \ldots, \beta_{n+k-j})$. Now, we describe the construction of the path $\pi = f(\alpha,\beta) \in L(n,k,j)$ in $j+1$ consecutive steps. Every path from $L(n,k,j)$ consists of exactly $n+k-j$ segments: j diagonal, $n-j$ horizontal and $k-j$ vertical, respectively. Let $\pi = (\pi_1, \pi_2, \ldots, \pi_{n+k-j})$.

Step 0. We set $\pi := \beta$. Step i for $i = 1, 2, \ldots, j$. Suppose that the ith diagonal segment of α is α_l (counting from the left-hand side). There is only one segment π_r of π such that $\alpha_l \sim \pi_r$. We apply the operation $\mathtt{dg}$ on π_r, i.e., $\pi_r := \mathtt{dg}(\pi_r)$. Next, for all next segments $\pi_{r+1}, \ldots, \pi_{n+k-j}$ of π, we apply the operation $\mathtt{up}$, i.e., for all π_m such that $m = r+1, r+2, \ldots, n+k-j$, we set $\pi_m := \mathtt{up}(\pi_m)$. It is easy to check that after these $j+1$ steps the path π is one of the set $L(n,k,j)$.

($\leftarrow$) Simply executing f in reverse order gives f^{-1}. $\square$

Two paths $\alpha \in L_1(3,2)$, $\beta \in L_2(3,2)$ and corresponding $f(\alpha,\beta) \in L(3,4)$ are given in Fig. 4.

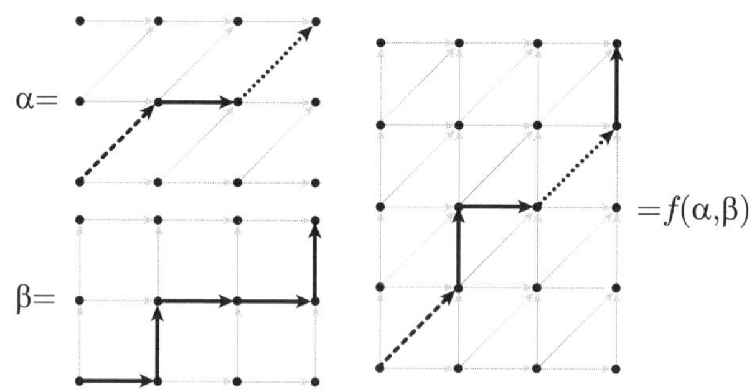

Figure 4: Illustration of the bijection f from Lemma 2.

Lemma 3. *Let $\pi \in L(n,k,j)$ and $\alpha \in L_1(n,j)$, $\beta \in L_2(n,k-j)$ such that $\pi = f(\alpha,\beta)$. For $i = 1,2,\dots,n$, if α_i is the ith segment of α, and β_l, π_l are segments of β and π, respectively, such that $\alpha_i \sim \beta_l \sim \pi_l$, then*

$$\mathtt{y}(\pi_l) = \mathtt{y}(\alpha_i) + \mathtt{y}(\beta_l).$$

Proof. Consider the construction of f given in the proof of Lemma 2. From the 0th step we see that $\mathtt{y}(\pi_l) = \mathtt{y}(\beta_l)$. In the next j consecutive steps we raise up, by the operation $\mathtt{up}$, certain segments of π such that the first segment of π begins at $(0,0)$ and the last one ends at (n,k).

Observe that the value of $\mathtt{y}(\alpha_i)$ determines the number of diagonal segments $\alpha_b \in \alpha$ for which we have $\mathtt{x}(\alpha_b) < \mathtt{x}(\alpha_i)$. Thus the value of $\mathtt{y}$ for π_l is increased from $\mathtt{y}(\beta_l)$ by the number $\mathtt{y}(\alpha_i)$ which is equivalent to the number of $\mathtt{up}$ operations applied to π_l. □

Lemma 4. *Let $\pi \in L(n,k,j)$ and $\alpha \in L_1(n,j)$, $\beta \in L_2(n,k-j)$ such that $\pi = f(\alpha,\beta)$ for $1 \le j \le \min(n,k)$. We have*

$$w^{\mathfrak{L}}(\pi) = w^{\mathfrak{L}_1}(\alpha) w^{\mathfrak{L}_2}(\beta). \tag{2.4}$$

Proof. Let $\pi^v = (\pi_1^v, \dots, \pi_{k-j}^v)$, $\pi^h = (\pi_1^h, \dots, \pi_{n-j}^h)$, and $\pi^d = (\pi_1^d, \dots, \pi_j^d)$ be the sequences of all vertical, horizontal, and diagonal segments of π, respectively, in the ascending order of $\mathtt{x}$. In the same manner we define the sets α^h, α^d, and β^h, β^v. With this notation, we have

$$w^{\mathfrak{L}}(\pi) = \prod_{i=1}^{k-j} w^{\mathfrak{L}}(\pi_i^v) \prod_{i=1}^{n-j} w^{\mathfrak{L}}(\pi_i^h) \prod_{i=1}^{j} w^{\mathfrak{L}}(\pi_i^d).$$

We rewrite (2.4) as

$$w^{\mathfrak{L}}(\pi) = \left(\prod_{i=1}^{n-j} w^{\mathfrak{L}_1}(\alpha_i^h) \prod_{i=1}^{j} w^{\mathfrak{L}_1}(\alpha_i^d) \right) \left(\prod_{i=1}^{n} w^{\mathfrak{L}_2}(\beta_i^h) \prod_{i=1}^{k-j} w^{\mathfrak{L}_2}(\beta_i^v) \right).$$

The proof is divided into three parts.

(a) First, we show that $\prod_{i=1}^{k-j} w^{\mathfrak{L}}(\pi_i^v) = \prod_{i=1}^{k-j} w^{\mathfrak{L}_2}(\beta_i^v)$. To see this, observe that the weight function $w^{\mathfrak{L}}$ on vertical segments depends only on the sequence $\{v_n\}_{n\ge 0}$. For $i = 1,2,\dots,k-j$, denote $x_i = \mathtt{x}(\pi_i^v)$. By the definition of $w^{\mathfrak{L}}$, we see that

$$\prod_{i=1}^{k-j} w^{\mathfrak{L}}(\pi_i^v) = v_{x_1} v_{x_2} \cdots v_{x_{k-j}}.$$

Let us observe that the corresponding vertical segments β^v of the path β and π^v of π share a common value of $\mathtt{x}$, i.e., $\mathtt{x}(\pi_i^v) = \mathtt{x}(\beta_i^v)$ for $i = 1,2,\dots,k-j$. Indeed, operations $\mathtt{up}$ and $\mathtt{dg}$ do not change the first coordinate of the segments' points.

Let us divide $\beta^h = (\beta_1^h, \ldots, \beta_n^h)$ into two sequences $B^d = (b_1^d, \ldots, b_j^d)$ and $B^h = (b_1^h, \ldots, b_{n-j}^h)$ as follows. For $i = 1, 2, \ldots, n$, let $\alpha_l \in \alpha$, be a segment such that $\alpha_l \sim \beta_i^h$, if α_l is diagonal then $\beta_i^h \in B^d$, and $\beta_i^h \in B^h$ otherwise.

(b) Next, we show $\prod_{i=1}^{n-j} w^{\mathfrak{L}}(\pi_i^h) = \prod_{i=1}^{n-j} w^{\mathfrak{L}_1}(\alpha_i^h) w^{\mathfrak{L}_2}(b_i^h)$. For $i = 1, 2, \ldots, n-j$, denote $x_i = \mathtt{x}(\pi_i^h)$ and $y_i = \mathtt{y}(\pi_i^h)$. By definition, we see that

$$\prod_{i=1}^{n-j} w^{\mathfrak{L}}(\pi_i^h) = h_{x_1} \cdots h_{x_{n-j}} \zeta^{y_1 + \cdots + y_{n-j}}. \tag{2.5}$$

Let us denote $y_i' = \mathtt{y}(\alpha_i^h)$, $y_i'' = \mathtt{y}(b_i^h)$, and observe that for $i = 1, \ldots, n-j$, we have $x_i = \mathtt{x}(\pi_i^h) = \mathtt{x}(\alpha_i^h) = \mathtt{x}(b_i^h)$. Thus by the definitions of $w^{\mathfrak{L}_1}$ and $w^{\mathfrak{L}_2}$, we have

$$\begin{aligned}\prod_{i=1}^{n-j} w^{\mathfrak{L}_1}(\alpha_i^h) w^{\mathfrak{L}_2}(b_i^h) &= \left(h_{x_1} \cdots h_{x_{n-j}} \zeta^{y_1' + \cdots y_{n-j}'}\right)\left(\zeta^{y_1'' + \cdots + y_{n-j}''}\right) \\ &= h_{x_1} \cdots h_{x_{n-j}} \zeta^{y_1' + y_1'' + \cdots + y_{n-j}' + y_{n-j}''}.\end{aligned}$$

Now, observe that $\pi_i^h \sim \alpha_i^h \sim b_i^h$ for $i = 1, \ldots, n-j$. Thus we apply Lemma 3 to the values of $\mathtt{y}$ for these segments to get $y_i = y_i' + y_i''$, for $i = 1, 2, \ldots, n-j$. Thus

$$\prod_{i=1}^{n-j} w^{\mathfrak{L}_1}(\alpha_i^h) w^{\mathfrak{L}_2}(b_i^h) = h_{x_1} \cdots h_{x_{n-j}} \zeta^{y_1 + \cdots + y_{n-j}},$$

which is exactly (2.5).

(c) In the same manner we can show the equivalent formula for diagonal segments, i.e., $\prod_{i=1}^{j} w^{\mathfrak{L}}(\pi_i^d) = \prod_{i=1}^{j} w^{\mathfrak{L}_1}(\alpha_i^d) w^{\mathfrak{L}_2}(b_i^d)$. If we set $x_i = \mathtt{x}(\pi_i^d)$ and $y_i = \mathtt{y}(\pi_i^d)$, then the left-hand side of the above is equal to

$$\prod_{i=1}^{j} w^{\mathfrak{L}}(\pi_i^d) = d_{x_1} \cdots d_{x_j} \zeta^{y_1 + \cdots + y_j}.$$

The rest of the proof of (c) goes much the same way as in (b).

What is left is to observe that the product of the left-hand sides of (a), (b), and (c) is equal to

$$w^{\mathfrak{L}}(\pi) = \prod_{i=1}^{k-j} w^{\mathfrak{L}}(\pi_i^v) \prod_{i=1}^{n-j} w^{\mathfrak{L}}(\pi_i^h) \prod_{i=1}^{j} w^{\mathfrak{L}}(\pi_i^d),$$

and the product of the right-hand sides of (a), (b), and (c) is equal to

$$w^{\mathfrak{L}_1}(\alpha) w^{\mathfrak{L}_2}(\beta) = \prod_{i=1}^{k-j} w^{\mathfrak{L}_2}(\beta_i^v) \prod_{i=1}^{n-j} w^{\mathfrak{L}_1}(\alpha_i^h) w^{\mathfrak{L}_2}(b_i^h) \prod_{i=1}^{j} w^{\mathfrak{L}_1}(\alpha_i^d) w^{\mathfrak{L}_2}(b_i^d),$$

which is equivalent to $w^{\mathfrak{L}}(\pi) = w^{\mathfrak{L}_1}(\alpha) w^{\mathfrak{L}_2}(\beta)$. □

Proposition 5 (Vertical summation). *For $n, k \in \mathbb{N}$, we have*

$$\left\langle {n \atop k} \right\rangle_{\mathfrak{L}} = (v_n h_n + d_n) \sum_{i=0}^{k-1} \left\langle {n-1 \atop i} \right\rangle_{\mathfrak{L}} v_n^{k-i-1} \zeta^i + h_n \zeta^k \left\langle {n-1 \atop k} \right\rangle_{\mathfrak{L}}. \tag{2.6}$$

Proof. Let us consider the left-hand side of the equation, and observe that the family $L(n,k)$ can be partitioned into classes $A_0, \ldots, A_k$, such that A_i contains paths which have the diagonal segment $\sigma_i^d = ((n-1,i),(n,i+1))$ or horizontal segment $\sigma_i^h = ((n-1,i),(n,i))$.

(a) Let us consider a path π from the class A_k. Note that the last segment of π is horizontal $\sigma_k^h = ((n-1,k),(n,k))$. Thus the weight of π is equal to the product of its segments' weights from $(0,0)$ to $(n-1,k)$ times the weight of the segment σ_k^h. Therefore, by definition, we have

$$\sum_{\pi \in A_k} w^{\mathfrak{L}}(\pi) = h_n \zeta^k \left\langle {n-1 \atop k} \right\rangle_{\mathfrak{L}}.$$

(b) For $i = 0, 1, \ldots, k-1$, let π be a path of A_i which contains σ_i^d. Let $\pi = (\pi_1, \ldots, \pi_l, \sigma_i^d, \pi_{l+2}, \ldots, \pi_s)$. The weight of π is equal to the product

$$w^{\mathfrak{L}}(\pi) = w^{\mathfrak{L}}(\pi_1) \cdots w^{\mathfrak{L}}(\pi_l) w^{\mathfrak{L}}(\sigma_i^d) w^{\mathfrak{L}}(\pi_{l+2}) \cdots w^{\mathfrak{L}}(\pi_s).$$

All the remaining segments $\pi_{l+2}, \ldots, \pi_s$ are vertical, and their number equals $(k-i-1)$. Thus, by definition, we have

$$w^{\mathfrak{L}}(\pi) = \left\langle {n-1 \atop i} \right\rangle_{\mathfrak{L}} w^{\mathfrak{L}}(\sigma_i^d) v_n^{k-i-1}.$$

(c) In the same manner we compute the weight of a path π from the class A_i which contains the horizontal segment σ_i^h, where $i = 0, 1, \ldots, k-1$. We need to observe that the number of vertical segments in this case is equal to $(k-i)$, and thus

$$w^{\mathfrak{L}}(\pi) = \left\langle {n-1 \atop i} \right\rangle_{\mathfrak{L}} w^{\mathfrak{L}}(\sigma_i^h) v_n^{k-i}.$$

The rest of the proof is straightforward. □

Proposition 6 (Horizontal summation). *For $n, k \in \mathbb{N}$, we have*

$$\left\langle {n \atop k} \right\rangle_{\mathfrak{L}} = \sum_{i=0}^{n-1} H_{i+2,n} \left\langle {i \atop k-1} \right\rangle_{\mathfrak{L}} \zeta^{k(n-i)-1} (d_{i+1} + v_i h_{i+1} \zeta) + v_n \left\langle {n \atop k-1} \right\rangle_{\mathfrak{L}}, \tag{2.7}$$

where $H_{i,n} = h_i h_{i+1} \cdots h_n$, and $H_{i,n} = 1$ for $i > n$.

Proof. We only outline the proof. We partition the set of paths $L(n,k)$ into $A_0, A_1, \dots, A_n$ defined as follows. Take a path π from $L(n,k)$. If π contains the vertical segment $\sigma_i^v = ((i,k-1),(i,k))$ or the diagonal segment $\sigma_i^d = ((i,k-1),(i+1,k))$, then $\pi \in A_i$ for $i = 0,1,\dots,n$. Note that the class A_n contains paths whose last segment is vertical σ_k^v, and $w^{\mathfrak{L}}(\sigma_k^v) = v_n$. The union of $A_0,\dots,A_n$ is the whole $L(n,k)$. Therefore,

$$\left\langle {n \atop k} \right\rangle_{\mathfrak{L}} = \sum_{i=0}^{n-1} \sum_{\pi \in A_i} w^{\mathfrak{L}}(\pi) + \sum_{\pi \in A_n} w^{\mathfrak{L}}(\pi).$$

The rest of the proof can be handled in much the same way as the proof of Proposition 5. □

3. Generating Functions

We denote by $\mathcal{P}_k(n)$ the family of subsets $X \subseteq \{1,2,\dots,n\}$ such that $|X| = k$. If $\pi \in \mathcal{P}_k(n)$ then we denote by $\overline{\pi}$ the set $\{1,2,\dots,n\} \setminus \pi$. We denote by $\mathcal{P}(n)$ the union $\mathcal{P}_0(n) \cup \mathcal{P}_1(n) \cup \dots \cup \mathcal{P}_n(n)$. For $n \geq 0$, let

$$F_n(z) = \sum_{k \geq 0} \left\langle {n \atop k} \right\rangle_{\mathfrak{L}} z^k. \tag{3.1}$$

Theorem 7. *$F_0(z) = 1/(1 - v_0 z)$, and for $n = 1,2,\dots$, we have*

$$F_n(z) = \frac{1}{(1 - v_0 \zeta^n z)} \prod_{i=1}^{n} \frac{h_i + d_i \zeta^{n-i} z}{1 - v_i \zeta^{n-i} z}. \tag{3.2}$$

Proof. Using the recurrence relation we see that $F_n(z) = (h_n + d_n z)/(1 - v_n z) F_{n-1}(\zeta z)$. Solving this recurrence with initial value $F_0(z) = 1/(1 - v_0 z)$ we obtain the formula. □

The corresponding generating functions for the $\mathfrak{L}$-Delannoy numbers of the first and second kind are denoted by $C_n(z)$ and $S_n(z)$, respectively. We have $C_0(z) = 1$,

$$C_n(z) = \sum_{k \geq 0} \left[{n \atop k} \right]_{\mathfrak{L}} z^k = \prod_{i=1}^{n} \left(h_i + d_i \zeta^{n-i} z \right), \qquad \text{for } n \geq 1, \tag{3.3a}$$

$$S_n(z) = \sum_{k \geq 0} \left\{ {n \atop k} \right\}_{\mathfrak{L}} z^k = \prod_{i=0}^{n} \frac{1}{(1 - v_i \zeta^{n-i} z)}, \qquad \text{for } n \geq 0. \tag{3.3b}$$

Note that there is another, very simple proof of Theorem 1 using the generating functions. Indeed, for every $n \geq 0$, we have $F_n(z) = C_n(z) S_n(z)$.

Proposition 8 (Symmetric-like functions). *For $n, k \geq 0$, we have*

$$\begin{bmatrix} n \\ k \end{bmatrix}_{\mathfrak{L}} = \sum_{\pi \in \mathcal{P}_k(n)} \prod_{i=1}^{k} d_{\pi_i} \zeta^{n-\pi_i} \prod_{i=1}^{n-k} h_{\overline{\pi}_i}, \tag{3.4a}$$

$$\begin{Bmatrix} n \\ k \end{Bmatrix}_{\mathfrak{L}} = \sum_{0 \leq \sigma_1 \leq \sigma_2 \leq \cdots \leq \sigma_k \leq n} \prod_{i=1}^{k} v_{\sigma_i} \zeta^{n-\sigma_i}, \tag{3.4b}$$

where $\pi = \{\pi_1, \ldots, \pi_k\}$, and $\overline{\pi} = \{\overline{\pi}_1, \ldots, \overline{\pi}_{n-k}\}$.

Proof. From (3.3a) and (3.3b), we obtain (3.4a) and (3.4b), respectively. □

Corollary 9. *For $n, k \geq 0$, we have*

$$\left\langle \begin{matrix} n \\ k \end{matrix} \right\rangle_{\mathfrak{L}} = \sum_{j=0}^{k} \sum_{\substack{1 \leq \pi_1 < \cdots < \pi_j \leq n \\ 0 \leq \sigma_1 \leq \cdots \leq \sigma_{k-j} \leq n}} \zeta^{kn-S} \prod_{i=1}^{j} d_{\pi_i} \prod_{i=1}^{n-j} h_{\overline{\pi}_i} \prod_{i=1}^{k-j} v_{\sigma_i}, \tag{3.5}$$

where $S = \pi_1 + \cdots + \pi_j + \sigma_1 + \cdots + \sigma_{k-j}$.

Proof. Combining Proposition 8 with (2.3) we obtain the formula. □

Theorem 10 (Generalization of the Carlitz formula). *Let $\mathfrak{H} = \langle \mathbf{v}, \mathbf{1}, \mathbf{d}, \zeta \rangle$ and $\mathfrak{H}^* = \langle \mathbf{v}^*, \mathbf{1}, \mathbf{d}^*, \zeta \rangle$ such that $v_i = v_i^* + \alpha_1 \zeta^i$ and $d_i = d_i^* + \alpha_2 \zeta^i$, where $\alpha_1, \alpha_2 \in \mathbb{C}$. Then*

$$\begin{bmatrix} n \\ k \end{bmatrix}_{\mathfrak{H}} = \sum_{j=0}^{k} \binom{n-k+j}{j} (\alpha_2 \zeta^n)^j \begin{bmatrix} n \\ k-j \end{bmatrix}_{\mathfrak{H}^*}, \tag{3.6a}$$

$$\begin{Bmatrix} n \\ k \end{Bmatrix}_{\mathfrak{H}} = \sum_{j=0}^{k} \binom{n+k}{j} (\alpha_1 \zeta^n)^j \begin{Bmatrix} n \\ k-j \end{Bmatrix}_{\mathfrak{H}^*}. \tag{3.6b}$$

Proof. (a) Substituting $d_i = d_i^* + \alpha_2 \zeta^i$ and $h_i = 1$ to (3.4a) we get

$$\begin{bmatrix} n \\ k \end{bmatrix}_{\mathfrak{H}} = \sum_{1 \leq \pi_1 < \cdots < \pi_k \leq n} \prod_{i=1}^{k} (d_{\pi_i}^* \zeta^{n-\pi_i} + \alpha_2 \zeta^n). \tag{3.7}$$

Let us consider a power series of the product $\prod_{i=1}^{k} (d_{\pi_i}^* \zeta^{n-\pi_i} + \alpha_2 \zeta^n)$ in the parameter $\alpha_2 \zeta^n$. That is,

$$\sum_{j=0}^{k} \left(\sum_{1 \leq s_1 < \cdots < s_{k-j} \leq k} \prod_{i=1}^{k-j} d_{\gamma_i}^* \zeta^{n-\gamma_i} \right) (\alpha_2 \zeta^n)^j,$$

where $\gamma_i = \pi_{s_i}$ for $i = 1, 2, \ldots, k-j$. What is left is to combine the above with (3.7), and prove that

$$\sum_{\substack{1\leq\pi_1<\cdots<\pi_k\leq n\\ 1\leq s_1<\cdots<s_{k-j}\leq k}} \prod_{i=1}^{k-j} d^*_{\gamma_i}\zeta^{n-\gamma_i} = \binom{n-k+j}{j} \sum_{1\leq\nu_1<\cdots<\nu_{k-j}\leq n} \prod_{i=1}^{k-j} d^*_{\nu_i}\zeta^{n-\nu_i}.$$

Indeed, $k-j$ variables $\gamma_i = \pi_{s_i}$ take values over the set $\{1,\ldots,n\}$, and the remaining j variables π_l of $\pi_1,\ldots,\pi_k$ we can choose from the set $[n]\backslash\{\gamma_1,\ldots,\gamma_{k-j}\}$ in $\binom{n-(k-j)}{j}$ ways. By definition, the right-hand side of the above is the value of the $\mathfrak{H}^*$-Delannoy number of the first kind.

(b) Substituting $v_i = v_i^* + \alpha_1\zeta^i$ to (3.4b) we obtain

$$\begin{aligned}
\left\{ n \atop k \right\}_{\mathfrak{L}} &= \sum_{0\leq\sigma_1\leq\cdots\leq\sigma_k\leq n} \prod_{i=1}^{k} (v^*_{\sigma_i}\zeta^{n-\sigma_i} + \alpha_1\zeta^n)\\
&= \sum_{0\leq\sigma_1\leq\cdots\leq\sigma_k\leq n} \sum_{j=0}^{k} \left(\sum_{1\leq s_1<\cdots<s_{j-k}\leq k} \prod_{i=1}^{k-j} v^*_{\sigma_{s_i}}\zeta^{n-\sigma_{s_i}} \right) (a_2\zeta^n)^j\\
&= \sum_{j=0}^{k} \left(\sum_{\substack{0\leq\sigma_1\leq\cdots\leq\sigma_k\leq n\\ 1\leq s_1<\cdots<s_{j-k}\leq k}} \prod_{i=1}^{k-j} v^*_{\sigma_{s_i}}\zeta^{n-\sigma_{s_i}} \right) (a_2\zeta^n)^j.
\end{aligned}$$

What is left is to show

$$\sum_{\substack{0\leq\sigma_1\leq\cdots\leq\sigma_k\leq n\\ 1\leq s_1<\cdots<s_{k-j}\leq k}} \prod_{i=1}^{k-j} v^*_{\sigma_{s_i}}\zeta^{n-\sigma_{s_i}} = \binom{n+k}{j} \sum_{0\leq\nu_1\leq\cdots\leq\nu_{k-j}\leq n} \prod_{i=1}^{k-j} v^*_{\nu_i}\zeta^{n-\nu_i}.$$

First, observe that $(k-j)$ variables $\nu_1 = \sigma_{s_1}, \ldots, \nu_{k-j} = \sigma_{s_{k-j}}$ take values over the set $\{0, 1, \ldots, n\}$ with repetition allowed. The remaining j variables σ_l variables lie between these ν_i and depend on values $\nu_1, \nu_2 - \nu_1, \ldots, \nu_{k-j} - \nu_{k-j-1}, n - \nu_{k-j}$. Denote by $A = A(n, k, j, \nu_1, \ldots, \nu_{k-j})$ the number of ways to select these remaining ν_l, where $l \neq s_i$ for $i = 1, 2, \ldots, k-j$. We have

$$A = \sum_{\substack{i_1+\cdots+i_{k-j+1}=j\\ i_1,\ldots,i_{k-j+1}\geq 0}} \binom{\nu_1+i_1}{i_1}\binom{\nu_2-\nu_1+i_2}{i_2}\cdots\binom{n-\nu_{k-j}+i_{k-j+1}}{i_{k-j+1}}.$$

Using standard methods of generating functions [29] we obtain

$$A = [z^j]\frac{1}{(1-z)^{\nu_1+1}}\frac{1}{(1-z)^{\nu_2-\nu_1+1}}\cdots\frac{1}{(1-z)^{\nu_{k-j}-\nu_{k-j-1}+1}}\frac{1}{(1-z)^{n-\nu_{k-j}+1}}.$$

Since there are $k-j+1$ terms we have

$$A=[z^j]\frac{1}{(1-z)^{n+k-j+1}}=\binom{n+k}{j}.$$

which completes the proof. □

Note that the generalization of Carlitz formulae for the generalized Stirling numbers delivered by Médicis and Leroux [22, Th.2.2] is a special case of the above with $\zeta=1$ and $\alpha_2=w_0$ (see Section 8.6 for more details).

Corollary 11. *Let $\mathfrak{I}=\langle \mathbf{v},\mathbf{1},\mathbf{v}^*,\zeta\rangle$, where $\mathbf{v}=(v_0,v_1,\ldots)$ and $\mathbf{v}^*=(0,v_0,v_1,\ldots)$. Then for all $n,k\in\mathbb{N}$, we have*

$$\sum_{i=0}^{k}(-1)^i\begin{bmatrix}n\\ i\end{bmatrix}_{\mathfrak{I}}\begin{Bmatrix}n-1\\ k-i\end{Bmatrix}_{\mathfrak{I}}=0, \tag{3.8a}$$

$$\sum_{i=0}^{k}(-1)^{k-i}\begin{bmatrix}n\\ k-i\end{bmatrix}_{\mathfrak{I}}\begin{Bmatrix}n-1\\ i\end{Bmatrix}_{\mathfrak{I}}=0. \tag{3.8b}$$

Proof. We need to observe that for every integer $n\geq 1$, the generating functions $C_n(z)$ and $S_n(z)$ of the $\mathfrak{I}$-Delannoy numbers of the first and second kind satisfy $C_n(-z)S_{n-1}(z)\equiv 1$. □

Proposition 12. *Let $n,k\geq 0$. If for every $0\leq i,j\leq n$ such that $i\neq j$ we have $v_i\neq v_j\zeta^{i-j}$, then*

$$\left\langle {n \atop k} \right\rangle_{\mathfrak{L}}=\sum_{i=0}^{n}\frac{\prod_{j=1}^{n}(h_jv_i+d_j\zeta^{i-j})}{\prod_{j=0,\, j\neq i}^{n}(v_i-v_j\zeta^{i-j})}\left(v_i\zeta^{n-i}\right)^k. \tag{3.9}$$

Proof. Using the partial fraction decomposition of the generating function (3.2) we obtain $F_n(z)=\sum_{i=0}^{n}W_i/(1-v_i\zeta^{n-i}z)$. To find the coefficients $W_0,W_1,\ldots,W_n$ we multiply both sides by $\prod_{i=0}^{n}(1-v_i\zeta^{n-i}z)$. For $s=0,1,\ldots,n$, we evaluate it with $z=(v_s\zeta^{n-s})^{-1}$ to obtain

$$\prod_{j=1}^{n}(h_j+\frac{d_j}{v_s}\zeta^{s-j})=W_s\prod_{\substack{j=0\\ j\neq s}}^{n}(1-\frac{v_j}{v_s}\zeta^{s-j}).$$

The rest of the proof is straightforward. □

Theorem 13. *Let $n,k\geq 0$, if $h_i\neq 0$ for $i\geq 1$, then*

$$\left\langle {n \atop k+1} \right\rangle_{\mathfrak{L}}=\frac{1}{k+1}\sum_{i=0}^{k}\left\langle {n \atop i} \right\rangle_{\mathfrak{L}}\sigma_n(k-i), \tag{3.10}$$

where

$$\sigma_n(i) = \sum_{j=0}^{n} \left(v_j \zeta^{n-j}\right)^{i+1} + (-1)^i \sum_{j=1}^{n} \left(\frac{d_j}{h_j} \zeta^{n-j}\right)^{i+1}. \tag{3.11}$$

Proof. Let us consider the following equality $\partial/\partial z \log(F_n(z)) = 1/F_n(z)\partial/\partial z F_n(z)$. We insert the power series expansion of $F_n(z)$ and $\partial/\partial z F_n(z)$ to get

$$\frac{\partial}{\partial z} \log \left\{ \frac{1}{1 - v_0 \zeta^n z} \prod_{i=1}^{n} \frac{h_i + d_i \zeta^{n-i} z}{1 - v_i \zeta^{n-i} z} \right\} F_n(z) = \sum_{k \geq 0} (k+1) \left\langle {n \atop k+1} \right\rangle_{\mathfrak{L}} z^k.$$

Using the Cauchy product of the series in the left-hand side and equating coefficients of like powers of z we obtain the formula. □

Applying the above to the $\mathfrak{B}$-Delannoy numbers we obtain a recurrence relation for $D(n,k)$, that is,

$$D(n,k) = \frac{1}{k} \sum_{i=0}^{k-1} D(n,i) \left(1 + n + (-1)^{k-i-1} n\right).$$

4. Weighted Central Delannoy Numbers

Let $\mathfrak{W} = \langle \mathbf{v}^c, \mathbf{h}^c, \mathbf{d}^c, 1 \rangle$, where $\mathbf{v}^c = (v, v, \ldots), \mathbf{h}^c = (h, h, \ldots), \mathbf{d}^c = (d, d, \ldots)$ are constant sequences of $v, h, d \in \mathbb{C}$. The generalized Delannoy numbers reduce to the so-called unrestricted weighted lattice paths studied, among others, by Fray and Roselle [13]. They showed, for instance, that

$$\sum_{n \geq 0} \left\langle {n \atop k} \right\rangle_{\mathfrak{W}} z^n = \frac{(v + dz)^k}{(1 - hz)^{k+1}} = \sum_{n \geq 0} \sum_{r=0}^{n} h^{n-r} v^{k-r} d^r \binom{k}{r} \binom{n+k-r}{k} z^n.$$

The central weighted Delannoy numbers were studied by Hetyei [17] who showed that the total weight of all Delannoy paths may be expressed by substitution into a shifted Jacobi polynomial with the appropriate parameters. Namely,

$$\left\langle {n+\beta \atop n} \right\rangle_{\mathfrak{W}} = h^{\beta} (-d)^n \widetilde{P}_n^{(0,\beta)} \left(-\frac{hv}{d}\right),$$

where $\widetilde{P}_n^{(0,\beta)}(x)$ is a shifted Legendre polynomial defined with the help of nth Jacobi polynomial $P_n^{(0,\beta)}$ of type $(0,\beta)$, i.e., $\widetilde{P}_n^{(\alpha,\beta)}(x) = P_n^{(\alpha,\beta)}(2x-1)$. We refer the reader also to Sulanke [26].

Let us define auxiliary numbers. We denote by $S(n)$ the sum of weights of all Delannoy paths from the origin to (n,n) that do not go above the line $y = x$. For

$v = h = d = 1$ we obtain the sequence of the so-called *large Schroeder numbers* [9]. The sequence

$$\{S(n)\}_{n\geq 0} = (1, 2, 6, 21, 77, 286, 1066, 3977, 14841, \dots)$$

is denoted by A006318 in OEIS [23].

Proposition 14. *$S(0) = 1$, and for $n = 1, 2, \dots,$ we have*

$$S(n) = d\,S(n-1) + h\,v \sum_{i=0}^{n-1} S(i)S(n-1-i). \tag{4.1}$$

Proof. We partition the family of paths from $(0,0)$ to (n,n) that do not go above the line $y = x$ into two classes A and B, such that the last step of paths from A is diagonal, and the last one from B is vertical. The weighted sum of paths in A is $S(n-1)$. To calculate the size of B we partition B into pairwise disjoint classes $B_0, B_1, \dots, B_{n-1}$ such that B_i contains these paths of B such that the last horizontal step $a_l = ((l,l),(l+1,l)$ of π is a_i, for $0 \leq l \leq n-1$. It is clear that the size of B_i is $S(i)S(n-1-i)$. □

Let $G(z) = \sum_{n\geq 0} S(n)z^n$. From (4.1), the function $G(z)$ satisfies the following functional equation

$$G(z) = 1 + d\,zG(z) + h\,v\,zG(z)^2, \tag{4.2}$$

which yields

$$G(z) = \frac{1 - d\,z - \sqrt{d^2\,z^2 - z(2\,d + 4\,h\,v) + 1}}{2\,h\,v\,z}. \tag{4.3}$$

We take the negative square root due to the initial condition $[z^0]G(z) = 1$.

Theorem 15. *For $t \geq 1$ we have $[z^0]G(z)^t = 1$, and for $n = 1, 2, \dots,$ we have*

$$[z^n]G(z)^t = \sum_{l=0}^{t} \sum_{k=0}^{n} \sum_{i=0}^{n-l-2k} \sum_{j=0}^{k+l} \frac{l}{n} \binom{t}{l} \binom{n}{k} \binom{n-k}{l+k} \binom{n-l-2k}{i} \binom{k+l}{j} \cdot d^{i+j} 2^{n-l-2k-i} (hv)^{n-i-j}. \tag{4.4}$$

Proof. The function $G(z)$ satisfies the functional equation (4.2). Letting $u = G(z) - 1$, we obtain $u = z\phi(u)$, where $\phi(u) = d + h\,v + u(d + 2\,h\,v) + h\,v\,u^2$. From the Lagrange Inversion Formula [24, Th. 5.4.2] we obtain

$$\begin{aligned}
[z^n]u(z)^t &= \frac{1}{n}[u^{n-1}]\left\{tu^{t-1}\phi(u)^n\right\} \\
&= \frac{t}{n}[u^{n-t}]\left\{(d + h\,v + u(d + 2\,h\,v) + h\,v\,u^2)^n\right\} \\
&= \frac{t}{n} \sum_{k=0}^{n} \sum_{i=0}^{n-t-2k} \sum_{j=0}^{k+t} \binom{n}{k} \binom{n-k}{n-t-2k} \binom{n-t-2k}{i} \binom{k+t}{j} \\
&\quad \cdot d^{i+j} 2^{n-t-2k-i} (hv)^{n-i-j}.
\end{aligned}$$

Finally, to get the series expansion of $G(z)^t = (u+1)^t$ we use the binomial theorem,

$$[z^n]G(z)^t = \sum_{l=0}^{t} \binom{t}{l} [z^n]u(z)^l$$

and the formula follows. □

Proposition 16. *For $n \geq 1$, we have*

$$\left\langle {n \atop n} \right\rangle_{\mathfrak{W}} = \sum_{k=1}^{n} \sum_{\substack{i_1+\cdots+i_k=n \\ i_1,\ldots,i_k \geq 1}} \prod_{j=1}^{k} \left(2\,h\,v\,S(i_j-1) + d\,\delta_{i_j,1}\right), \tag{4.5}$$

where $\delta_{a,b} = 1$ when $a = b$, and zero otherwise.

Proof. Let us partition $L(n,n)$ into n subclasses $A_1, A_2, \ldots, A_n$ such that paths from A_k contains exactly k lattice points from the set $\{(1,1), \ldots, (n,n)\}$. We call (j,j) the jth *diagonal point.* Let us consider the subclass A_k. These k diagonal points (j,j) can be determined by the solution of the equation $i_1 + \cdots + i_k = n$, where $i_1, \ldots, i_k \geq 1$. Indeed, the first diagonal point is (i_1, i_1), the second one is $(i_1 + i_2, i_1 + i_2)$, and so on. For the solution $i_1, \ldots, i_k$, the number of paths that contains corresponding k diagonal points is the product of the numbers $g_1, \ldots, g_k$ of weighted Delannoy paths π from $(j-1)$th diagonal point to the jth diagonal point, for $j = 1, 2, \ldots, k$, and π does not contain any other diagonal point between these fixed ones. For the convenience, we assume that the 0th diagonal point is $(0,0)$. To calculate g_j we observe that for every such path π we have three cases: (a) π does not go above the line $y = x$, (b) π does not go below the line $y = x$, and (c) π is a single diagonal step. Simple verification shows that $g_j = 2\,v\,h\,S(i_j-1) + d\,\delta_{i_j,1}$. □

Theorem 17. *We have*

$$F(z) = \sum_{n \geq 0} \left\langle {n \atop n} \right\rangle_{\mathfrak{W}} z^n = \frac{1}{\sqrt{d^2\,z^2 - z(2\,d + 4\,h\,v) + 1}}. \tag{4.6}$$

Proof. Let us denote by $g(z) = \sum_{n \geq 1}(2\,h\,v\,S(n-1) + d\,\delta_{n,1})z^n$ which implies $g(z) = h\,v\,2G(z)z + d\,z$. Using the generating function (4.3) of the numbers $S(n)$ we get explicit form of $g(z) = 1 - \sqrt{d^2 z^2 - z(2\,d + 4\,h\,v) + 1)}$. On the other hand, by (4.5) we obtain $[z^n]F(z) = [z^n]\sum_{k=1}^{n} g(z)^k = [z^n]\sum_{k \geq 1} g(z)^k$. The last equality is due to $g(0) = 0$. The value of the central $\mathfrak{W}$-Delannoy number is one for $n = 0$, thus we can change the summation range from $k \geq 1$ to $k \geq 0$. Therefore, $F(z) = 1/(1 - g(z))$ which simplifying gives the formula. □

Corollary 18. *For every $n \geq 0$, we have*

$$\left\langle {n \atop n} \right\rangle_{\mathfrak{W}} = \sum_{k=0}^{n} \sum_{s=0}^{k} \binom{k}{s} h^s v^s 2^s d^{k-s} [z^{n-k}] G(z)^s.$$

Proof. By Theorem 17 we see that

$$\left\langle {n \atop n} \right\rangle_{\mathfrak{W}} = \sum_{k=0}^{n} [z^n](h\,v\,2\,G(z)z + d\,z)^k = \sum_{k=0}^{n}\sum_{s=0}^{k} \binom{k}{s} h^s v^s 2^s d^{k-s} [z^{n-k}] G(z)^s.$$

We use the binomial theorem and the explicit formula for $[z^n]G(z)^s$ from Theorem 15. □

5. Counting Lattice Paths with Restrictions

Let us define three classes of segments: vertical $S^v(k)$, horizontal $S^h(k)$ and diagonal $S^d(k)$, as follows

$$\begin{aligned} S^v(k) &= \{((k,i),(k,i+1)) : i = 0,1,\ldots\}, && \text{for } k \geq 0,\\ S^h(k) &= \{((k-1,i),(k,i)) : i = 0,1,\ldots\}, && \text{for } k \geq 1,\\ S^d(k) &= \{((k-1,i),(k,i+1)) : i = 0,1,\ldots\}, && \text{for } k \geq 1. \end{aligned}$$

Let $S^v = \bigcup_{k\geq 0} S^v(k)$, we denote S^h and S^d analogously. With this notation, the family $\mathcal{S}$ of all lattice segments (vertical, horizontal, and diagonal), is $S^v \cup S^h \cup S^d$. Let $H \subseteq \mathcal{S}$, we denote by $L(n,k;H)$ the family of all lattice paths from $(0,0)$ to (n,k) consisting of steps from H. Let $D(n,k;H) = |L(n,k;H)|$, and let

$$G_n(z;H) = \sum_{k\geq 0} D(n,k;H)z^k. \tag{5.1}$$

For the case $H = \mathcal{S}$ we have $L(n,k) = L(n,k;\mathcal{S})$ and $D(n,k) = D(n,k;\mathcal{S})$, and thus $G_n(z;\mathcal{S}) = (1+z)^n/(1-z)^{n+1}$. We are interested in finding the numbers $D(n,k;H)$ for sets H which can be obtained from the whole $\mathcal{S}$ by removing certain classes of segments S^v, S^h, and S^d.

Proposition 19. *If $r \geq 1$, then for every $k \geq 0$ and $n \geq r$, we have*

$$S^h(r) \cap H = \emptyset \quad \text{and} \quad S^d(r) \cap H = \emptyset \quad \text{implies} \quad L(n,k;H) = \emptyset. \tag{5.2}$$

Fix $n \geq 1$. Let H' be the set of lattice segments obtained from $\mathcal{S}$ by removing r horizontal classes of segments $S^h(i_1),\ldots,S^h(i_r)$, where $0 \leq r \leq n$. It is clear that we can remove at most $(n-r)$ diagonal classes of segments $S^d(j_1),\ldots,S^d(j_{n-r})$ from the set H' to obtain a set H'' such that $L(n,k;H'') \neq \emptyset$. Indeed, the sets of indices $I = \{i_1,\ldots,i_r\}$ and $J = \{j_1,\ldots,j_{n-r}\}$ can not share a common element, i.e., $I \cap J = \emptyset$.

Let $[n] = \{1,2,\ldots,n\}$ and $[n]_0 = \{0,1,\ldots,n-1\}$. Let A be a set of non-negative integers. From now on we denote by $\mathcal{P}(A)$ the set of all subsets of A. Let $\sigma \in \mathcal{P}([n])$, we denote by $S^v[\sigma]$ the union

$$S^v[\sigma] = \bigcup_{\sigma_i \in \sigma} S^v(\sigma_i).$$

We denote $S^h[\sigma]$ and $S^d[\sigma]$ similarly.

Theorem 20. *For a fixed $n \geq 0$, let $H = \mathcal{S} \setminus (S^v[\eta] \cup S^h[\sigma] \cup S^d[\tau])$, where $\tau \in \mathcal{P}([n])$, $\sigma \in \mathcal{P}([n])$, and $\eta \in \mathcal{P}([n+1]_0)$ such that $\tau \cap \sigma = \emptyset$. Then*

$$G_n(z; H) = \frac{z^{|\sigma|}(1+z)^{n-|\sigma|-|\tau|}}{(1-z)^{n+1-|\eta|}}, \tag{5.3}$$

$$D(n,k;H) = \sum_{j=0}^{k} \binom{n-|\sigma|-|\tau|}{j-|\sigma|}\binom{n-|\eta|+k-j}{n-|\eta|}. \tag{5.4}$$

Proof. Let $\tau = \{\tau_1, \ldots, \tau_a\}$, $\sigma = \{\sigma_1, \ldots, \sigma_b\}$, and $\eta = \{\eta_1, \ldots, \eta_c\}$. Let us consider the generating function (3.2) for the $\mathfrak{R}$-Delannoy numbers, where the 4-tuple $\mathfrak{R} = \langle \mathbf{v}, \mathbf{h}, \mathbf{d}, \zeta \rangle$ is defined as follows. We set $\zeta = 1$, $h_0 = 0$, $d_0 = 0$, and

(a) for $i \in [n+1]_0$, if $i \in \eta$ then $v_i = 0$, and $v_i = 1$ otherwise, (vertical);
(b) for $i \in [n]$, if $i \in \sigma$ then $h_i = 0$, and $h_i = 1$ otherwise, (horizontal);
(c) for $i \in [n]$, if $i \in \tau$ then $d_i = 0$, and $d_i = 1$ otherwise, (diagonal).

On one hand, $v_i = 0$ ($d_i = 0$ and $h_i = 0$, respectively) means that the weight $w^{\mathfrak{R}}(\pi)$ of $\pi \in L(n,k)$ consisting of a segment from $S^v(i)$ ($S^d(i)$ and $S^h(i)$, respectively) is equal to zero. On the other hand, for every π consisting of segments from H the weight function $w^{\mathfrak{R}}(\pi)$ is equal to one. Thus the value of the corresponding $\mathfrak{R}$-Delannoy number is equal to the number of paths from $(0,0)$ to (n,k) whose segments lie in H, i.e.,

$$\left\langle {n \atop k} \right\rangle_{\mathfrak{R}} = \sum_{\pi \in L(n,k)} w^{\mathfrak{R}}(\pi) = \sum_{\pi \in L(n,k;H)} 1 = D(n,k;H). \tag{5.5}$$

□

Example 21. Let $H = \mathcal{S} \setminus (S^d(1) \cup S^d(n))$. Then

$$G_n(z;H) = \frac{(1+z)^{n-2}}{(1-z)^{n+1}}, \quad D(n,k;H) = \sum_{j=0}^{k} \binom{n-2}{j}\binom{n+k-j}{n}.$$

See Fig. 5 (a) for $n = 4$.

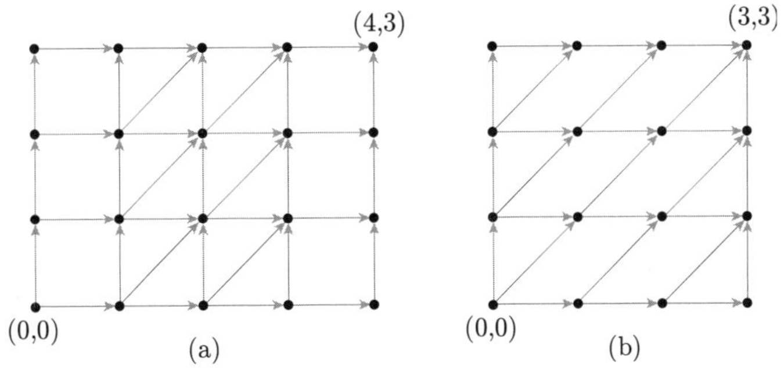

Figure 5: Lattices with removed segments.

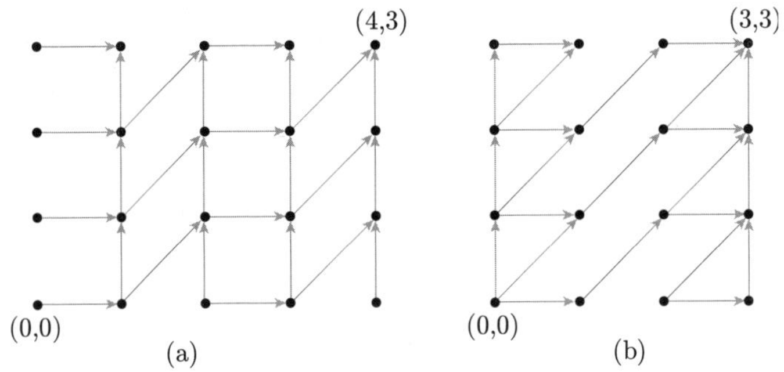

Figure 6: Lattices with removed segments.

Example 22. Let $H = \mathcal{S} \setminus (S^v(1) \cup S^v(2) \cup \cdots \cup S^v(n-1))$. Then

$$G_n(z;H) = \frac{(1+z)^n}{(1-z)^2}, \quad D(n,k;H) = \sum_{j=0}^{k} \binom{n}{j}(k-j+1).$$

See Fig. 5 (b) for $n = 3$.

Example 23. Let $H = (S^v \setminus S^v(0)) \cup B$, where $B = S^h(1) \cup S^d(2) \cup S^h(3) \cup S^d(4) \cup \cdots$. Then

$$G_n(z;H) = \frac{z^{\lfloor n/2 \rfloor}}{(1-z)^n}, \quad D(n,k;H) = \sum_{j=0}^{k} \binom{0}{j - \lfloor n/2 \rfloor}\binom{n-1+k-j}{n-1}.$$

See Fig. 6 (a) for $n = 4$.

Example 24. Let $H = S^d \cup S^v(0) \cup S^v(n) \cup B$, where $B = S^h(1) \cup S^h(3) \cup S^h(5) \cup \cdots$. Then

$$G_n(z;H) = \frac{z^{\lfloor n/2 \rfloor}(1+z)^{n-\lfloor n/2 \rfloor}}{(1-z)^2}, \quad D(n,k;H) = \sum_{j=0}^{k} \binom{n - \lfloor n/2 \rfloor}{n-j}(k-j+1).$$

See Fig. 6 (b) for $n = 3$.

6. A General Case of Generalized Delannoy Numbers

In previous sections we have been working under the assumption that lattice paths begin at $(0,0)$. To study the general case let x, y be any two lattice points, and let us denote by $L[x,y]$ the set of all Delannoy paths from x to y. Let

$$\left\langle {n \atop k} \right\rangle_{\mathfrak{L}}^{(m,l)} = \sum_{\pi \in L[x,y]} w^{\mathfrak{L}}(\pi), \tag{6.1}$$

where $x = (m,l)$ and $y = (m+n, l+k)$. Denote by $D^{(m,l)}(n,k)$ the cardinality of the family $L[(m,l),(m+n,l+k)]$. It is clear that $D^{(m,l)}(n,k) = D(n,k)$. For all $n,k,m,l \geq 0$, we have

$$\begin{aligned} \left\langle {n \atop k} \right\rangle_{\mathfrak{L}}^{(m,l)} &= v_{m+n} \left\langle {n \atop k-1} \right\rangle_{\mathfrak{L}}^{(m,l)} + h_{m+n}\, \zeta^{l+k} \left\langle {n-1 \atop k} \right\rangle_{\mathfrak{L}}^{(m,l)} \\ &\quad + d_{m+n}\, \zeta^{l+k-1} \left\langle {n-1 \atop k-1} \right\rangle_{\mathfrak{L}}^{(m,l)}. \end{aligned} \tag{6.2}$$

Likewise,

$$\begin{aligned} \left\langle {n \atop k} \right\rangle_{\mathfrak{L}}^{(m,l)} &= v_{m} \left\langle {n \atop k-1} \right\rangle_{\mathfrak{L}}^{(m,l+1)} + h_{m+1}\, \zeta^{l} \left\langle {n-1 \atop k} \right\rangle_{\mathfrak{L}}^{(m+1,l)} \\ &\quad + d_{m+1}\, \zeta^{l} \left\langle {n-1 \atop k-1} \right\rangle_{\mathfrak{L}}^{(m+1,l+1)} \end{aligned} \tag{6.3}$$

with $\left\langle {0 \atop 0} \right\rangle_{\mathfrak{L}}^{(m,l)} = 1$. For $n < 0$ or $k < 0$, we assume that $\left\langle {n \atop k} \right\rangle_{\mathfrak{L}}^{(m,l)} = 0$. See Fig. 7.

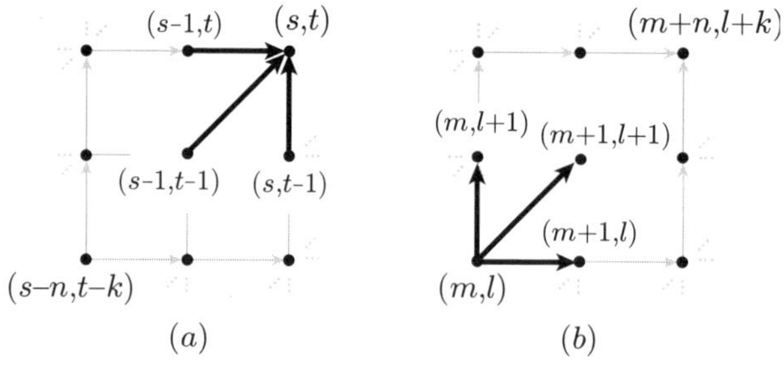

Figure 7: Recurrence relations for the general case.

Theorem 25. *For every $n \geq 0$, let*

$$F_n^{(m,l)}(z) = \sum_{k\geq 0} \left\langle {n \atop k} \right\rangle_{\mathfrak{L}}^{(m,l)} z^k. \tag{6.4}$$

Then for every $n \geq 1$, we have

$$F_n^{(m,l)}(z) = \frac{\zeta^{nl}}{(1 - v_m\zeta^n z)} \prod_{i=1}^{n} \frac{h_{m+i} + d_{m+i}\zeta^{n-i}z}{1 - v_{m+i}\zeta^{n-i}z}, \tag{6.5}$$

and $F_0^{(m,l)}(z) = 1/(1 - v_m z)$.

Proof. Applying (6.2) to (6.4) we obtain

$$F_n^{(m,l)}(z) = \frac{(h_{m+n}\zeta^l + d_{m+n}\zeta^l z)}{1 - v_{m+n}z} F_{n-1}^{(m,l)}(\zeta z) \tag{6.6}$$

with $F_0^{(m,l)}(z) = 1/(1 - v_m z)$. Solving this recurrence relation yields (6.5). □

In a similar way we define generalized $\mathfrak{L}$-Delannoy numbers of the first and the second kind, respectively. For a given 4-tuple $\mathfrak{L} = \langle \mathbf{v}, \mathbf{h}, \mathbf{d}, \zeta \rangle$ we set $\mathfrak{L}_1 = \langle \mathbf{0}, \mathbf{h}, \mathbf{d}, \zeta \rangle$ and $\mathfrak{L}_2 = \langle \mathbf{v}, \mathbf{1}, \mathbf{0}, \zeta \rangle$. We have

$$\begin{bmatrix} n \\ k \end{bmatrix}_{\mathfrak{L}}^{(m,l)} = \left\langle {n \atop k} \right\rangle_{\mathfrak{L}_1}^{(m,l)}, \qquad \left\{ {n \atop k} \right\}_{\mathfrak{L}}^{(m,l)} = \left\langle {n \atop k} \right\rangle_{\mathfrak{L}_2}^{(m,l)}.$$

Let

$$C_n^{(m,l)}(z) = \sum_{k\geq 0} \begin{bmatrix} n \\ k \end{bmatrix}_{\mathfrak{L}}^{(m,l)} z^k, \qquad S_n^{(m,l)}(z) = \sum_{k\geq 0} \left\{ {n \atop k} \right\}_{\mathfrak{L}}^{(m,l)} z^k, \tag{6.7}$$

then $C_0(z) = 1$, and

$$C_n^{(m,l)}(z) = \zeta^{nl} \prod_{i=1}^{n} \left(h_{m+i} + d_{m+i}\zeta^{n-i}z\right), \qquad \text{for } n \geq 1, \tag{6.8a}$$

$$S_n^{(m,l)}(z) = \zeta^{nl} \prod_{i=0}^{n} \frac{1}{(1 - v_{m+i}\zeta^{n-i}z)}, \qquad \text{for } n \geq 0. \tag{6.8b}$$

Proposition 26 (Symmetric-like functions). *For all $n, k, m, l \geq 0$, we have*

$$\begin{bmatrix} n \\ k \end{bmatrix}_{\mathfrak{L}}^{(m,l)} = \zeta^{nl} \sum_{\pi\in\mathcal{P}_k(n)} \prod_{i=1}^{k} d_{m+\pi_i}\zeta^{n-\pi_i} \prod_{i=1}^{n-k} h_{m+\overline{\pi}_i}, \tag{6.9a}$$

$$\left\{ {n \atop k} \right\}_{\mathfrak{L}}^{(m,l)} = \zeta^{nl} \sum_{0\leq j_1\leq j_2\leq\cdots\leq j_k\leq n} \prod_{s=1}^{k} v_{m+j_s}\zeta^{n-j_s}, \tag{6.9b}$$

where $\pi = \{\pi_1, \ldots, \pi_k\}$ and $\overline{\pi} = \{\overline{\pi}_1, \ldots, \overline{\pi}_{n-k}\}$.

Proposition 27 (Convolution-like formulae). *For all $n, m, k \geq 1$, we have*

$$\begin{bmatrix} n+m \\ k \end{bmatrix}_{\mathfrak{L}} = \sum_{j=0}^{k} \begin{bmatrix} n \\ j \end{bmatrix}_{\mathfrak{L}} \begin{bmatrix} m \\ k-j \end{bmatrix}_{\mathfrak{L}}^{(n,j)}, \tag{6.10a}$$

$$\begin{Bmatrix} n+m \\ k+m \end{Bmatrix}_{\mathfrak{L}} = \sum_{j=0}^{m} \begin{Bmatrix} j \\ m-j \end{Bmatrix}_{\mathfrak{L}} \begin{Bmatrix} n+m-j \\ k+j \end{Bmatrix}_{\mathfrak{L}}^{(j,m-j)}. \tag{6.10b}$$

Proof. (a) To show the first equation let us consider the set $L_1(n+m,k)$ of lattice paths with diagonal and horizontal segments, and observe that every such path π can be "cut" at the point (n,j) into two paths π_1 and π_2, where $0 \leq j \leq k$. The weight of π is the product of weights of π_1 and π_2. See Fig. 8 (a).

(b) In the second equality we consider the set $L_2(n+m, k+m)$ of lattice paths with horizontal and vertical segments. As in (a) we can show that $\pi \in L_2(n+m, k+m)$ can be "cut" at the point $(j, m-j)$, where $0 \leq j \leq m$. See Fig. 8 (b). □

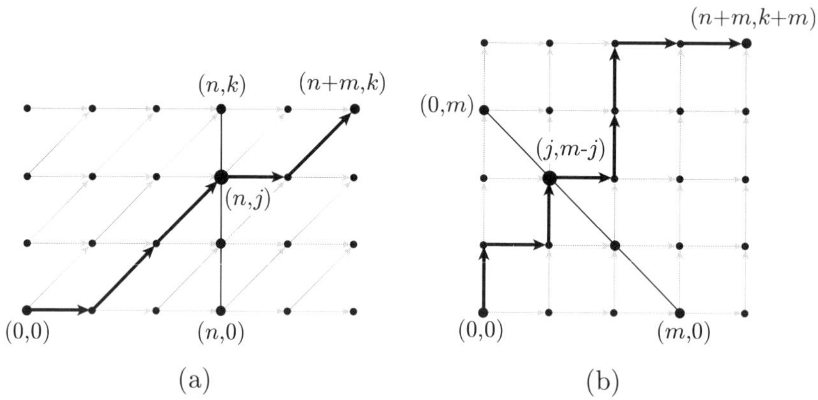

Figure 8: Illustration of the convolution-like formulae.

Let $\pi \in L[x,y]$, and let α be a segment (vertical, horizontal or diagonal) of the lattice. We write $\alpha \in \pi$ if the path π contains the segment α. A set of segments $\lambda = \{\alpha_1, \ldots, \alpha_s\}$ is called the *layer* of the set $L[x,y]$ if for every $\pi \in L[x,y]$ there is exactly one segment $\alpha_i \in \lambda$ such that $\alpha_i \in \pi$. Denote by $\Lambda[x,y]$ the family of all layers of the set $L[x,y]$. If $x = (0,0)$, then we write $\Lambda[y]$ for short.

Example 28. Let $\lambda = \{\alpha^v, \alpha^h, \alpha^d\}$, where $\alpha^v = ((0,0),(0,1))$, $\alpha^h = ((0,0),(1,0))$, and $\alpha^d = ((0,0),(1,1))$. The set λ is a layer of $L(n,k)$. Indeed, every path from $L(n,k)$ contains exactly one segment from λ.

Lemma 29. *For $n, k \geq 0$ and $\lambda \in \Lambda[(n,k)]$, we have*

$$\left\langle \begin{matrix} n \\ k \end{matrix} \right\rangle_{\mathfrak{L}} = \sum_{\alpha \in \lambda} \left\langle \begin{matrix} m \\ l \end{matrix} \right\rangle_{\mathfrak{L}} w^{\mathfrak{L}}(\alpha) \left\langle \begin{matrix} n-m' \\ k-l' \end{matrix} \right\rangle_{\mathfrak{L}}^{(m',l')}, \tag{6.11}$$

where $\alpha = ((m, l), (m', l'))$.

Proof. Let $\lambda = \{\alpha_1, \alpha_2, \ldots, \alpha_s\}$. By definition, the set $\lambda \in \Lambda[(n, k)]$ determines a partition of the family $L(n, k)$ into pairwise-disjoint classes of paths $A_1, \ldots, A_s$ such that $\pi \in A_i$ if and only if $\alpha_i \in \pi$, for $i = 1, 2, \ldots, s$. Therefore, we have

$$\sum_{\pi \in L[(n,k)]} w^{\mathfrak{L}}(\pi) = \sum_{\alpha \in \lambda} \sum_{\pi \in L[(m,l)]} w^{\mathfrak{L}}(\pi) \cdot w^{\mathfrak{L}}(\alpha) \cdot \sum_{\eta \in L[(m',l'),(n,k)]} w^{\mathfrak{L}}(\eta),$$

where $\alpha = ((m, l), (m', l'))$. □

Corollary 30. *For all* $n, k \geq 0$ *and* $\lambda \in \Lambda[(n, k)]$, *we have*

$$D(n,k) = \sum_{((m,l),(m',l')) \in \lambda} D(m,l) D(n-m', k-l'). \tag{6.12}$$

Proposition 31 (Vertical summation). *For* $m, n, k \geq 1$, *we have*

$$\left\langle {m+n \atop k} \right\rangle_{\mathfrak{L}} = h_{m+1} \sum_{i=0}^{k} \zeta^i \left\langle {m \atop i} \right\rangle_{\mathfrak{L}} \left\langle {n-1 \atop k-i} \right\rangle_{\mathfrak{L}}^{(m+1,i)} + d_{m+1} \sum_{i=0}^{k-1} \zeta^i \left\langle {m \atop i} \right\rangle_{\mathfrak{L}} \left\langle {n-1 \atop k-i-1} \right\rangle_{\mathfrak{L}}^{(m+1,i+1)}. \tag{6.13}$$

Proof. Let us define two classes of segments:

- $H = \{(x, y) : x = (m, i), y = (m + 1, i), i = 0, 1, \ldots, k\}$ (horizontal);
- $S = \{(x, y) : x = (m, i), y = (m + 1, i + 1), i = 0, 1, \ldots, k - 1\}$ (diagonal).

The union $H \cup S$ is a layer of the set $L(n, k)$. Applying Lemma 29 and simplifying the formula we obtain our claim. See Fig. 9. □

The convolution-like formula for the $\mathfrak{L}$-Delannoy numbers of the second kind takes a form

$$\left\{ {m+n \atop k} \right\}_{\mathfrak{L}} = \sum_{i=0}^{k} \left\{ {m \atop i} \right\}_{\mathfrak{L}} \zeta^i \left\{ {n-1 \atop k-i} \right\}_{\mathfrak{L}}^{(m+1,i)}.$$

Proposition 32 (Horizontal summation). *For all* $n, k, m \geq 1$, *we have*

$$\left\langle {n \atop k+m} \right\rangle_{\mathfrak{L}} = \sum_{i=0}^{n} v_i \left\langle {i \atop m} \right\rangle_{\mathfrak{L}} \left\langle {n-i \atop k-1} \right\rangle_{\mathfrak{L}}^{(i,m+1)} + \zeta^m \sum_{i=0}^{n-1} d_{i+1} \left\langle {i \atop m} \right\rangle_{\mathfrak{L}} \left\langle {n-i-1 \atop k-1} \right\rangle_{\mathfrak{L}}^{(i+1,m+1)}. \tag{6.14}$$

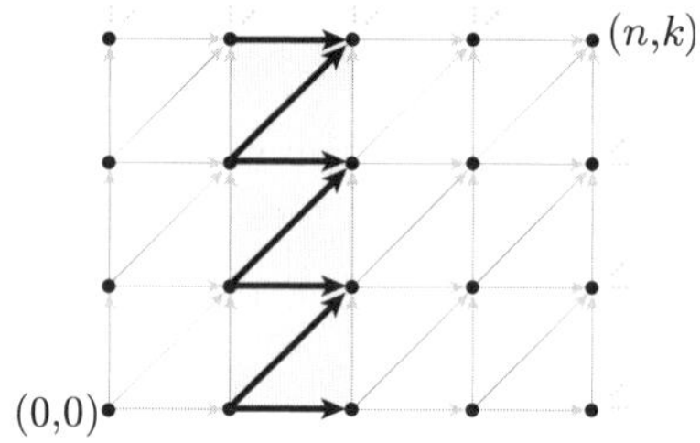

Figure 9: A layer of $L[(n,k)]$ used in vertical summation.

Proof. Let us define two classes of segments:

- $V = \{(x,y) : x = (i,m), y = (i,m+1), i = 0,1,\ldots,n\}$,
- $D = \{(x,y) : x = (i,m), y = (i+1,m+1), i = 0,1,\ldots,n-1\}$.

The union $V \cup D$ is a layer of the set $L(n,k)$. See Fig. 10. □

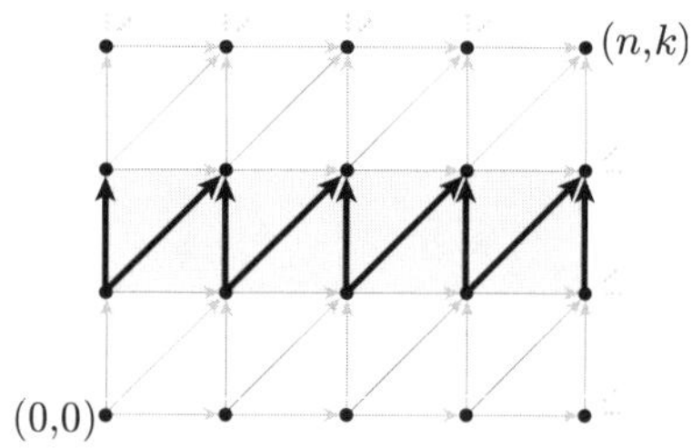

Figure 10: A layer of $L[(n,k)]$ used in horizontal summation.

Proposition 33 (Diagonal summation). *For all $n,k,m \geq 1$, we have*

$$\left\langle {n+m \atop k+m} \right\rangle_{\mathfrak{L}} = \sum_{i=0}^{m} r(i) \left\langle {i \atop m-i} \right\rangle_{\mathfrak{L}} + \sum_{i=0}^{m-1} p(i) \left\langle {i \atop m-1-i} \right\rangle_{\mathfrak{L}}, \tag{6.15}$$

where

$$\begin{aligned} r(i) &= v_i \left\langle {n+m-i \atop k+i-1} \right\rangle_{\mathfrak{L}}^{(i,m-i+1)} \\ &\quad + h_{i+1}\zeta^{m-i} \left\langle {n+m-i-1 \atop k+i} \right\rangle_{\mathfrak{L}}^{(i+1,m-i)} \\ &\quad + d_{i+1}\zeta^{m-i} \left\langle {n+m-i-1 \atop k+i-1} \right\rangle_{\mathfrak{L}}^{(i+1,m-i+1)}, \end{aligned} \tag{6.16}$$

$$p(i) = d_{i+1}\zeta^{m-1-i} \left\langle {n+m-i-1 \atop k+i} \right\rangle_{\mathfrak{L}}^{(i+1,m-i)}. \tag{6.17}$$

Proof. In this case we have four classes of segments that together form a layer of the set $L(n,k)$ (see Fig. 11). Namely,

- $V = \{(x,y) : x = (i, m-i), y = (i, m-i+1), i = 0, 1, \ldots, m\}$,
- $H = \{(x,y) : x = (i, m-i), y = (i+1, m-i), i = 0, 1, \ldots, m\}$,
- $D = \{(x,y) : x = (i, m-i), y = (i+1, m-i+1), i = 0, 1 \ldots, m\}$,
- $D' = \{(x,y) : x = (i, m-1-i), y = (i+1, m-i), i = 0, 1 \ldots, m-1\}$.

□

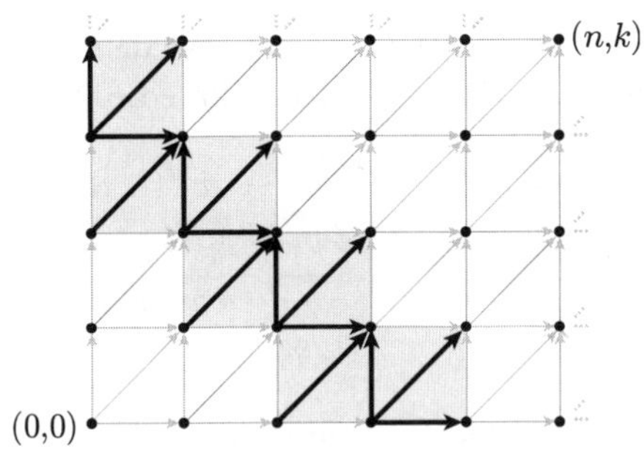

Figure 11: A layer of $L[(n,k)]$ used in diagonal summation.

The above three propositions and Corollary 30 applied to the $\mathfrak{B}$-numbers give the following recurrence relations for the standard Delannoy numbers $D(n,k)$.

Corollary 34. *For $n, k, m \geq 1$, we have*

$$D(m+n,k) = D(m,k) + \sum_{i=0}^{k-1} D(m,i)(D(n-1,k-i) + D(n-1,k-i-1)),$$

$$D(n,k+m) = D(n,m) + \sum_{i=0}^{n-1} D(i,m)(D(n-i,k-1) + D(n-i-1,k-1)),$$

$$D(n+m,k+m) = \sum_{i=0}^{m} D(i,m-i)r(i) + \sum_{i=0}^{m-1} D(i,m-1-i)D(n+m-i-1,k+i),$$

where

$$r(i) = (D(n+m-i,k+i-1) + D(n+m-i-1,k+i) + D(n+m-i-1,k+i-1)).$$

7. Connection with Partitions of Numbers

In this section we deal with the 4-tuple $\mathfrak{P} = \langle \mathbf{v}, \mathbf{1}, \mathbf{d}, \zeta \rangle$. We follow the notation of Andrews [1]. By the *partition* λ of a non-negative integer n we mean a finite non-increasing sequence of non-negative integers $(\lambda_1, \lambda_2, \ldots, \lambda_r)$ such that $\lambda_1 + \cdots + \lambda_r = n$. Each $\lambda_i > 0$ is called a *part* of λ.

We define two functions on partitions. Namely, let

$$w^d(\lambda, m) = \prod_{i=1}^{r} d_{m-\lambda_i}, \qquad w^v(\lambda, m) = \prod_{i=1}^{r} v_{m-\lambda_i}.$$

For all $n, k \geq 1$, we denote by $\mathcal{D}(m, n, k)$ the set of all partitions of the number n into k distinct parts from the set $\{0, 1, \ldots, m\}$. We denote by $\mathcal{R}(m, n, k)$ the set of such partitions with repetition of their parts allowed. Then we set

$$\alpha(m, n, k) = \sum_{\lambda \in \mathcal{D}(m,n,k)} w^d(\lambda, m), \tag{7.1}$$

$$\beta(m, n, k) = \sum_{\lambda \in \mathcal{R}(m,n,k)} w^v(\lambda, m). \tag{7.2}$$

Let us consider the generating function of the $\mathfrak{L}$-Delannoy numbers of the first kind $C_m(z)$ and the second kind $S_m(z)$, respectively. Observe that

$$\prod_{i=0}^{m-1} \left(1 + d_{m-i}\zeta^i z\right) = \sum_{k=0}^{m-1} \sum_{n \geq 0} \alpha(m-1, n, k)\zeta^n z^k,$$

$$\prod_{i=0}^{m} \frac{1}{(1 - v_{m-i}\zeta^i z)} = \sum_{k \geq 0} \sum_{n \geq 0} \beta(m, n, k)\zeta^n z^k.$$

Therefore, the $\mathfrak{P}$-Delannoy numbers of the first and second kind, respectively, are generating functions of α and β, in the sense that

$$\begin{bmatrix} m \\ k \end{bmatrix}_{\mathfrak{P}} = \sum_{n \geq 0} \alpha(m-1, n, k)\zeta^n, \qquad \begin{Bmatrix} m \\ k \end{Bmatrix}_{\mathfrak{P}} = \sum_{n \geq 0} \beta(m, n, k)\zeta^n.$$

Thus

$$\left\langle \begin{matrix} m \\ k \end{matrix} \right\rangle_{\mathfrak{P}} = \sum_{j=0}^{k} \begin{bmatrix} m \\ j \end{bmatrix}_{\mathfrak{P}} \begin{Bmatrix} m \\ k-j \end{Bmatrix}_{\mathfrak{P}} = \sum_{n \geq 0} \gamma(m, n, k)\zeta^n,$$

where

$$\gamma(m, n, k) = \sum_{j=0}^{k} \sum_{l=0}^{n} \alpha(m-1, l, j)\beta(m, n-l, k-j).$$

But what do the numbers $\gamma(m,n,k)$ really count in terms of partitions? From Corollary 9 we have

$$\left\langle {m \atop k} \right\rangle_{\mathfrak{P}} = \sum_{j=0}^{k} \sum_{\substack{1\le\pi_1<\cdots<\pi_j\le m \\ 0\le\sigma_1\le\cdots\le\sigma_{k-j}\le m}} d_{\pi_1}\cdots d_{\pi_j} v_{\sigma_1}\cdots v_{\sigma_{k-j}} \zeta^{km-S} \tag{7.3}$$

$$= \sum_{n\ge 0} \gamma(m,n,k)\zeta^n, \tag{7.4}$$

where $S = \pi_1 + \cdots + \pi_j + \sigma_1 + \cdots + \sigma_{k-j}$, and

$$\gamma(m,n,k) = \sum_{I} \prod_{i=1}^{j} d_{m-\pi_i} \prod_{i=1}^{k-j} v_{m-\sigma_i},$$

where the summation range I is given by

$$I = \begin{cases} 0 \le \pi_1 < \pi_2 < \cdots < \pi_j \le m-1; \\ 0 \le \sigma_1 \le \sigma_2 \le \cdots \le \sigma_{k-j} \le m; \\ \pi_1 + \cdots + \pi_j + \sigma_1 + \cdots + \sigma_{k-j} = n. \end{cases}$$

Let H be a finite set of non-negative integers. We denote by $\mathcal{D}(H;n,k)$ the family of partitions of n into k distinct parts from the set H. We define $\mathcal{R}(H;n,k)$ similarly (repetition of parts is allowed). It is well-known that

$$\sum_{n,k\ge 0} |\mathcal{D}(H;n,k)| z^k y^n = \prod_{i\in H} (1 + y^i z), \tag{7.5a}$$

$$\sum_{n,k\ge 0} |\mathcal{R}(H;n,k)| z^k y^n = \prod_{i\in H} \frac{1}{(1 - y^i z)}. \tag{7.5b}$$

Fix $m \ge 1$, and for a given subset $H \subseteq \{0, 1, \ldots, m-1\}$, let $\mathfrak{P}(H)$ be the 4-tuple $\langle \mathbf{a}, \mathbf{1}, \mathbf{a}, \zeta \rangle$ such that $\mathbf{a} = \{0, a_1, \ldots, a_m, 0, 0 \ldots\}$ is defined as follows. For $i = 0, 1, \ldots, m-1$, if $i \in H$ then $a_{m-i} = 1$, and $a_{m-i} = 0$ otherwise.

Theorem 35. *We have*

$$\begin{bmatrix} m \\ k \end{bmatrix}_{\mathfrak{P}(H)} = \sum_{n\ge 0} |\mathcal{D}(H;n,k)| \zeta^n, \tag{7.6a}$$

$$\begin{Bmatrix} m \\ k \end{Bmatrix}_{\mathfrak{P}(H)} = \sum_{n\ge 0} |\mathcal{R}(H;n,k)| \zeta^n. \tag{7.6b}$$

Proof. (a) Due to the definition of the sequence $\mathbf{a}$ of the 4-tuple $\mathfrak{P}(H) = \langle \mathbf{a}, \mathbf{1}, \mathbf{a}, \zeta \rangle$, we have

$$\sum_{k\ge 0} \begin{bmatrix} m \\ k \end{bmatrix}_{\mathfrak{P}(H)} z^k = \prod_{i=1}^{m} \left(1 + a_i \zeta^{m-i} z\right)$$

$$= \prod_{i\in H} \left(1 + \zeta^i z\right) = \sum_{k\ge 0} \sum_{\pi} \zeta^{\pi_1+\cdots+\pi_k}\, z^k,$$

where the summation range π is over all subsets $A \subseteq H$ such that $|A| = k$. Comparing coefficients of z^k we obtain (7.5a). (b) In a similar way we show the second equality, i.e.,

$$\sum_{k\geq 0} \left\{ {m \atop k} \right\}_{\mathfrak{P}(H)} z^k = \prod_{i=1}^{m} \frac{1}{(1-a_i\zeta^{m-i}z)} = \prod_{i\in H} \frac{1}{(1-\zeta^i z)}.$$

□

8. Examples and Remarks

In this section we present some numbers of general interest which are special cases of $\mathfrak{L}$-Delannoy numbers.

8.1. The Binomial Coefficients

Setting $\mathfrak{B} = \langle \mathbf{1}, \mathbf{1}, \mathbf{1}, 1 \rangle$ gives

$$\begin{bmatrix} n \\ k \end{bmatrix}_{\mathfrak{B}} = \binom{n}{k}, \qquad \left\{ {n \atop k} \right\}_{\mathfrak{B}} = \binom{n+k}{k}, \qquad \left\langle {n \atop k} \right\rangle_{\mathfrak{B}} = D(n,k).$$

8.2. The p, q-Binomial Coefficients

The p, q-binomial coefficient [4, 7, 11] is defined as

$$\binom{n}{k}_{p,q} = \frac{(q^n - p^n)(q^{n-1} - p^{n-1}) \cdots (q^{n-k+1} - p^{n-k+1})}{(q^k - p^k)(q^{k-1} - p^{k-1}) \cdots (q - p)}.$$

One shows that

$$\prod_{i=1}^{n} (1 + q^{i-1}p^{n-i}\, z) = \sum_{k\geq 0} q^{\binom{k}{2}} p^{\binom{k}{2}} \binom{n}{k}_{p,q} z^k,$$

$$\prod_{i=0}^{n} \frac{1}{(1 - q^i p^{n-i}\, z)} = \sum_{k\geq 0} \binom{n+k}{k}_{p,q} z^k.$$

Setting $\mathfrak{P} = \langle \mathbf{q}, \mathbf{1}, \mathbf{q}^*, p \rangle$, where $\mathbf{q} = (1, q, q^2, \ldots)$ and $\mathbf{q}^* = (1, 1, q, q^2, \ldots)$, we obtain

$$\begin{bmatrix} n \\ k \end{bmatrix}_{\mathfrak{P}} = p^{\binom{k}{2}} q^{\binom{k}{2}} \binom{n}{k}_{p,q}, \qquad \left\{ {n \atop k} \right\}_{\mathfrak{P}} = \binom{n+k}{k}_{p,q}.$$

Setting $p = (1+\sqrt{5})/2$ and $q = (1-\sqrt{5})/2$ we obtain the Fibonomial coefficients [10, 15, 16, 18, 19, 20] defined as

$$\binom{n}{k}_{Fib} = \frac{F_n F_{n-1} \cdots F_{n-k+1}}{F_k F_{k-1} \cdots F_1},$$

where F_n is the n-th Fibonacci number. Thus we have

$$\begin{bmatrix} n \\ k \end{bmatrix}_{\mathfrak{P}} = (-1)^{\binom{k}{2}} \binom{n}{k}_{Fib}, \qquad \begin{Bmatrix} n \\ k \end{Bmatrix}_{\mathfrak{P}} = \binom{n+k}{k}_{Fib}.$$

For the general case of the $\mathfrak{P}$-Delannoy numbers we have

$$\begin{bmatrix} n \\ k \end{bmatrix}_{\mathfrak{P}}^{(m,l)} = p^{n\,l} q^{k\,m} \begin{bmatrix} n \\ k \end{bmatrix}_{\mathfrak{P}}, \quad \begin{Bmatrix} n \\ k \end{Bmatrix}_{\mathfrak{P}}^{(m,l)} = p^{n\,l} q^{k\,m} \begin{Bmatrix} n \\ k \end{Bmatrix}_{\mathfrak{P}}.$$

Due to the above correspondence, all the results from Section 6 take simple forms. For instance, Proposition 27 yields

$$\begin{bmatrix} n+m \\ k \end{bmatrix}_{\mathfrak{P}} = \sum_{j=0}^{k} \begin{bmatrix} n \\ j \end{bmatrix}_{\mathfrak{P}} p^{m\,j} q^{(k-j)n} \begin{bmatrix} m \\ k-j \end{bmatrix}_{\mathfrak{P}},$$

$$\begin{Bmatrix} n+m \\ k+m \end{Bmatrix}_{\mathfrak{P}} = \sum_{j=0}^{m} \begin{Bmatrix} j \\ m-j \end{Bmatrix}_{\mathfrak{P}} p^{(n+m-j)(m-j)} q^{(k+j)j} \begin{Bmatrix} n+m-j \\ k+j \end{Bmatrix}_{\mathfrak{P}}.$$

Corcino [7] asked about the generalization of q-Vandermonde identity for the p,q-binomials. From the above we answer to his question.

Corollary 36 (**p,q-Vandermonde like identity**). *For $n,m,k \geq 1$, we have*

$$\binom{n+m}{k}_{p,q} = \sum_{j=0}^{k} p^{(m-k+j)\,j} q^{(k-j)(n-j)} \binom{n}{j}_{p,q} \binom{m}{k-j}_{p,q}, \tag{8.1}$$

$$\binom{n+2m+k}{k+m}_{p,q} = \sum_{j=0}^{m} \binom{m}{m-j}_{p,q} p^{(n+m-j)(m-j)} q^{(k+j)j} \binom{n+m+k}{k+j}_{p,q}.$$

8.3. The q-Binomial Coefficients

For $p=1$, we obtain the q-binomial coefficients

$$\begin{bmatrix} n \\ k \end{bmatrix}_{\mathfrak{Q}} = q^{\binom{k}{2}} \binom{n}{k}_q, \quad \begin{Bmatrix} n \\ k \end{Bmatrix}_{\mathfrak{Q}} = \binom{n+k}{k}_q.$$

Recall that $\binom{n}{k}_q$ is the number of k-dimensional subspaces of the n-dimensional vector space over the finite field of order q. From Theorem 1 we have

$$\left\langle \begin{matrix} n \\ k \end{matrix} \right\rangle_{\mathfrak{Q}} = \sum_{i=0}^{k} q^{\binom{i}{2}} \binom{n}{i}_q \binom{n+k-i}{k-i}_q.$$

There is natural question about an analogous combinatorial interpretation of the above in terms of counting certain subspaces. The convolution (8.1) reduces to the q-Vandermonde identity [14].

8.4. The Stirling Numbers

Let us define the ordinary Stirling numbers of the first and second kind as follows:

$$\prod_{i=1}^{n}(1+iz)=\sum_{k=0}^{n}\begin{bmatrix} n+1 \\ n+1-k \end{bmatrix} z^k, \quad \prod_{i=1}^{n}\frac{1}{(1-iz)}=\sum_{k=0}^{n}\left\{\begin{matrix} n+k \\ n \end{matrix}\right\} z^k.$$

Setting $\mathfrak{S}=\langle \mathbf{n}, \mathbf{1}, \mathbf{n}, 1\rangle$, where $\mathbf{n}=(0,1,2,3,\ldots)$ we obtain

$$\begin{bmatrix} n \\ k \end{bmatrix}_{\mathfrak{S}}=\begin{bmatrix} n+1 \\ n+1-k \end{bmatrix}, \qquad \left\{\begin{matrix} n \\ k \end{matrix}\right\}_{\mathfrak{S}}=\left\{\begin{matrix} n+k \\ n \end{matrix}\right\}.$$

From Theorem 1 we have

$$\left\langle \begin{matrix} n \\ k \end{matrix} \right\rangle_{\mathfrak{S}}=\sum_{i=0}^{k}\begin{bmatrix} n+1 \\ n+1-i \end{bmatrix}\left\{\begin{matrix} n+k-i \\ n \end{matrix}\right\}$$

and as in previous example, we may ask about its interpretation.

8.5. The p, q-Stirling Numbers

Wachs and White [27] define the p, q-Stirling numbers by the following recursion

$$S_{p,q}(n,k)=[k]_{p,q}S_{p,q}(n-1,k)+p^{k-1}S_{p,q}(n-1,k-1)$$

with $S_{p,q}(0,0)=1$, where $[k]_{p,q}=\sum_{i=1}^{k}p^{k-i}q^{i-1}$ and $[0]_{p,q}=0$. The generating function is

$$\sum_{k\geq 0}S_{p,q}(n+k,n)z^k=p^{\binom{n}{2}}\prod_{i=0}^{n}\frac{1}{(1-z[i]_{p,q})}.$$

Setting $\mathfrak{S}=\langle \mathbf{i}, \mathbf{1}, \mathbf{i}, 1\rangle$, where $\mathbf{i}=([0]_{p,q},[1]_{p,q},\ [2]_{p,q}\ldots)$ we obtain

$$p^{\binom{n}{2}}\left\{\begin{matrix} n \\ k \end{matrix}\right\}_{\mathfrak{S}}=S_{p,q}(n+k,n).$$

8.6. The $\mathfrak{A}$-Stirling Numbers

Let $\mathbf{w}=(w_0,w_1,\ldots)$. In the notation of Médicis and Leroux [22] we define $\mathfrak{A}$-Stirling numbers of the first kind $c^{\mathfrak{A}}$ and the second kind $S^{\mathfrak{A}}$ as follows

$$\sum_{k\geq 0}c^{\mathfrak{A}}(n,k)z^k=(1+w_0z)(1+w_1z)\cdots(1+w_{n-1}z),$$

$$\sum_{k\geq 0}S^{\mathfrak{A}}(n+k,n)z^k=\frac{1}{(1+w_0z)}\frac{1}{(1+w_1z)}\cdots\frac{1}{(1+w_nz)}.$$

Setting $\mathfrak{A} = \langle \mathbf{w}, \mathbf{1}, \mathbf{w}^*, 1\rangle$, where $\mathbf{w}^* = (1, w_0, w_1, \ldots)$ we obtain

$$\begin{bmatrix} n \\ k \end{bmatrix}_{\mathfrak{A}} = c^{\mathfrak{A}}(n,k), \qquad \begin{Bmatrix} n \\ k \end{Bmatrix}_{\mathfrak{A}} = S^{\mathfrak{A}}(n+k,n).$$

The generalization of Carlitz formulae for the $\mathfrak{A}$-Stirling numbers delivered by Médicis and Leroux [22, Th.2.2] is a special case of Theorem 10 for $\zeta = 1$ and $\alpha_2 = w_0$, that is,

$$S^{\mathfrak{A}}(n,k) = \begin{Bmatrix} k \\ n-k \end{Bmatrix}_{\mathfrak{H}} = \sum_{j=k}^{n} \binom{n}{j} w_0^{n-j} \begin{Bmatrix} k \\ j-k \end{Bmatrix}_{\mathfrak{H}^*}. \tag{8.2}$$

8.7. The ζ-Analogues of the Stirling Numbers

Let $w_1, w_2, \ldots$ be a sequence of complex numbers. For $n \geq 1$, let $\mathbf{w}_n(\zeta) = (w_1\zeta^{n-1}, w_2\zeta^{n-2}, \ldots, w_n)$, and recall that ζ-analogues [12] of the Stirling numbers are defined as

$$\hat{C}_k^n(\hat{\mathbf{w}}_n) = \sum_{1 \leq i_1 < i_2 < \cdots < i_k \leq n} \hat{w}_{i_1}\hat{w}_{i_2}\cdots\hat{w}_{i_k},$$
$$\hat{S}_k^n(\hat{\mathbf{w}}_n) = \sum_{1 \leq i_1 \leq i_2 \leq \cdots \leq i_k \leq n} \hat{w}_{i_1}\hat{w}_{i_2}\cdots\hat{w}_{i_k},$$

where $\hat{w}_i = w_i\zeta^{n-i}$. If we set $\mathfrak{Z} = \langle \mathbf{w}, \mathbf{1}, \mathbf{w}, \zeta\rangle$, where $\mathbf{w}$ is the sequence $(0, w_1, w_2, \ldots)$, then

$$\begin{bmatrix} n \\ k \end{bmatrix}_{\mathfrak{Z}} = \hat{C}_k^n(\mathbf{w}_n(\zeta)), \qquad \begin{Bmatrix} n \\ k \end{Bmatrix}_{\mathfrak{Z}} = \hat{S}_k^n(\mathbf{w}_n(\zeta)).$$

Acknowledgements I would like to thank the referee for very careful reading of the manuscript and for bringing some papers about Delannoy numbers to my attention.

References

[1] George E. Andrews. *The Theory of Partitions*. Encycl. Math. Appl. Vol. 2. Addison-Wesley, Reading, 1976.

[2] J. M. Autebert and S. R. Schwer. On generalized Delannoy paths. *SIAM J. Discrete Math*, 16(2), 2003.

[3] C. Banderier and S. Schwer. Why Delannoy numbers. *J. Statist. Plann. Inference*, 135(1):40–54, 2005.

[4] R. D. Carmichael. On the numerical factors of the arithmetic forms $\alpha^n \pm \beta^n$. *Ann. of Math. Stud.*, 15:30–70, 1913.

[5] John S. Caughman, Clifford R. Haithcock, and J. J. P. Veerman. A note on lattice chains and Delannoy numbers. *Discrete Math.*, 308:2623–2628, 2008.

[6] Louis Comtet. *Advanced Combinatorics: The Art of Finite and Infinite Expansions.* D. Reidel Publishing Company, Boston, U.S.A., 1974.

[7] Roberto B. Corcino. On p, q-binomial coefficients. *Integers*, 8(A29):1–16, 2008.

[8] Henri Delannoy. Employ d'échiquier pour la résolution de certains problèmes de probabilités. *Assoc. Franc. Bordeaux*, 24:70–90, 1895.

[9] Emeric Deutsch. A bijective proof of the equation linking the Schröeder numbers, large and small. *Discrete Math.*, 241:235–240, 2001.

[10] M. Dziemiańczuk. A combinatorial interpretation of the Lucas-nomial coefficients in terms of tiling of rectangular boxes. *To appear in: J. Combin. Math. Combin. Comput.*

[11] M. Dziemiańczuk. Generalization of Fibonomial coefficients. ArXiv:0908.3248, 2009.

[12] M. Dziemiańczuk. First remark on a ζ-analogue of the Stirling numbers. *Integers*, 11(A9), January 2011.

[13] R.D. Fray and D.P. Roselle. Weighted lattice paths. *Pacific J. Math.*, 37(1):85–96, 1971.

[14] G. Gasper and M. Rahman. *Basic Hypergeometric Series.* Encyclopedia of Mathematics and its Applications. Cambridge University Press, Cambridge, second edition, 2004.

[15] Ira M. Gessel and X. G. Viennot. Determinant paths and plane partitions. preprint http://citeseer.nj.nec.com/gessel89determinants.html, 1992.

[16] H. W. Gould. The bracket function and Fontené-Ward generalized binomial coefficients with application to Fibonomial coefficients. *Fibonacci Quartely*, 7(55):23–40, 1969.

[17] Gábor Hetyei. Shifted Jacobi polynomials and Delannoy numbers. arXiv 0909.5512, December 2009.

[18] V. E. Hoggatt. Fibonacci numbers and generalized binomial coefficients. *Fibonacci Quartely*, 5:383–400, 1967.

[19] A. Krzysztof Kwaśniewski. Comments on combinatorial interpretation of Fibonomial coefficients - an e-mail style letter. *Bull. Inst. Combin. Appl.*, 42:10–11, 2004.

[20] A. Krzysztof Kwaśniewski. On cobweb posets and their combinatorially admissible sequences. *Adv. Stud. Contemp. Math.*, 18(1):17–32, 2009.

[21] Nicholas A. Loehr and Carla D. Savage. Generalizing the combinatorics of binomial coefficients via l-nomials. *Integers*, 10:531–558, 2010.

[22] A. De Médicis and P. Leroux. Generalized Stirling numbers, convolution formulae and p, q-analogues. *Canad. J. Math.*, 47:474–499, 1995.

[23] N. J. A. Sloane. The on-line encyclopedia of integer sequences. published electronically at `http://www.research.att.com/~njas/sequences/`.

[24] Richard P. Stanley. *Enumerative Combinatorics*, volume II. Cambridge University Press, 1999.

[25] R. G. Stanton and D. D. Cowan. Note on a square functional equation. *SIAM Review*, 12:277–279, 1970.

[26] Robert A. Sulanke. Objects counted by the central Delannoy numbers. *J. Integer Seq.*, 6, 2003. Article 03.1.5.

[27] M. Wachs and D. White. p, q-Stirling numbers and set partition statistics. *J. Combin. Theory Ser. A*, 56:27–46, 1991.

[28] Carl G. Wagner. The Carlitz lattice path polynomials. *Discrete Math.*, 222:291–298, 2000.

[29] Herbert S. Wilf. *Generatingfunctionology.* Academic Press, 1994.

INTEGERS: 13 (2013) #A55

ON THE LEAST PRIMITIVE ROOT EXPRESSIBLE AS A SUM OF TWO SQUARES

Christopher Ambrose[1]
Mathematisches Institut, Georg-August Universität Göttingen, Göttingen, Deutschland
ambrose@uni-math.gwdg.de

Received: 2/11/13, Accepted: 7/18/13, Published: 9/26/13

Abstract

For a positive integer n, a λ-root modulo n is an integer q coprime to n which has maximal order in $(\mathbb{Z}/n\mathbb{Z})^*$. We establish upper bounds for $s^*(n)$, the least λ-root modulo n which is expressible as a sum of two squares, in particular proving that for $\varepsilon > 0$, and n large enough there always exists a λ-root q modulo n in the range $1 \le q \le n^{\frac{1}{2}+\varepsilon}$ such that q is a sum of two squares.

1. Introduction and Statement

According to a folklore conjecture, all but finitely many primes p admit a positive *prime* primitive root q modulo p which is smaller than p, a problem that has not been resolved yet. Writing $g^*(p)$ for the least prime primitive root q, Martin [3] showed $g^*(p) \ll (\log p)^{B(\varepsilon)}$ for all but $O(Y^\varepsilon)$ primes $p \le Y$, and Shoup [6] proved $g^*(p) \ll (\log p)^6$ for all p, conditionally on GRH (his result is stated for primitive roots, but it holds for prime primitive roots, too). Further information on this topic, and related problems is provided in [5].

In this paper we are interested in unconditional results valid for all primes, and therefore weaken the condition that the primitive root be a prime. Instead, as a natural variation, we ask for the smallest primitive root expressible as a sum of two squares. From a sieve-theoretic point of view this may be regarded as half way towards prime primitive roots. As it amounts to no extra effort, we treat the case of arbitrary moduli n. Of course, $(\mathbb{Z}/n\mathbb{Z})^*$ may not have a primitive root in general. Therefore we define a λ-root modulo n to be an integer coprime to n of maximal order in $(\mathbb{Z}/n\mathbb{Z})^*$. We write $s^*(n)$ for the smallest λ-root modulo n expressible as a sum of two squares, and prove the following unconditional result.

[1]The author wants to thank the Volkswagen Foundation for their support and Valentin Blomer for fruitful discussions and comments on earlier versions of this work.

Theorem 1. *For a positive integer n let n_c denote the largest odd cube-free divisor of n. Then, for any $\varepsilon > 0$ we have*

$$s^*(n) \ll_\varepsilon n_c^{\frac{1}{2}+\varepsilon}.$$

The proof of Theorem 1 uses ideas of Martin [4] who treated the case of almost-primitive roots. It is based on a semi-linear lower bound sieve and Burgess' bound for short character sums. The semi-linear sieve is applicable because primes represented by $x^2 + y^2$ have density 1/2 in the set of all primes. We thus note that, as a generalization of Theorem 1, our method also works if one considers λ-roots represented by any binary quadratic form of class number 1.

The necessary notations and results from sieve theory will be provided in Section 2. In Section 3 we develop a sieve setting appropriate to our problem and note some preliminary results which are needed in the sequel. Finally, in Section 4 we proceed with the proof of Theorem 1.

Throughout this paper, $\mathbb{N}$ shall denote the set of positive integers. For any $n \in \mathbb{N}$ we denote by $\varphi(n)$ the order, and by $\lambda(n)$ the exponent of $(\mathbb{Z}/n\mathbb{Z})^*$. By $\mathrm{rad}(n)$, and $\omega(n)$ we mean the largest squarefree divisor, and the number of distinct prime divisors of n, respectively.

2. The Semi-Linear Sieve

As usual in sieve theory we denote by $\mathcal{A}$ a set of positive integers, and for any $d \in \mathbb{N}$ we let $\mathcal{A}_d := \{a \in \mathcal{A} \mid a \equiv 0 \ (d)\}$. For a set $\mathcal{P}$ of rational primes, and a positive parameter z we let $\mathcal{P}(z)$ be the product of all primes $p \in \mathcal{P}$ smaller than z. The predominant goal in sieve theory consist of estimating the quantity

$$\mathcal{S}(\mathcal{A}, z) := \sharp \{a \in \mathcal{A} \mid (a, \mathcal{P}(z)) = 1\}.$$

To this end one assumes that for any $d \in \mathbb{N}$ one has an asymptotic formula

$$\sharp \mathcal{A}_d = g(d)X + r_d \tag{1}$$

for the size of $\mathcal{A}_d$. Here the so called density function $g(d)$ is a multiplicative function, and X is a real parameter which serves as an approximation to the size of $\mathcal{A}$. The remainder terms r_d are intended to be rather small (at least on average over d). In a probabilistic sense one may then expect that X multiplied with

$$V(z) := \prod_{\substack{p \in \mathcal{P} \\ p < z}} (1 - g(p))$$

yields a good approximation for $\mathcal{S}(\mathcal{A}, z)$. This is indeed true under appropriate assumptions, and we state the following result which is a special case of a beta sieve of sieve dimension $\kappa = \frac{1}{2}$ (cf. [2, p.207,275]). Since we are merely interested in lower bounds for $\mathcal{S}(\mathcal{A}, z)$ we omit the upper bound version.

Theorem 2. *Assume that the function $g(d)$ in (1) satisfies*

$$\prod_{\substack{w \le p < z \\ p \in \mathcal{P}}} (1 - g(p))^{-1} \le \left(\frac{\log z}{\log w}\right)^{\frac{1}{2}} \left(1 + \frac{L}{\log w}\right) \tag{2}$$

for all $2 \le w \le z$, with some constant $L \ge 1$. Then, for $s \ge 1$ we have

$$\mathcal{S}(\mathcal{A}, z) \ge XV(z)\left\{f(s) + O\left((\log D)^{-\frac{1}{6}}\right)\right\} + O\Bigg(\sum_{\substack{d \mid \mathcal{P}(z) \\ d < D}} |r_d|\Bigg), \tag{3}$$

where $s = \log D / \log z$, and the first implied constant depends on L. For $1 \le s \le 3$ the function $f(s)$ is given by

$$f(s) = \sqrt{\frac{e^{\gamma}}{\pi s}} \log\left(1 + 2(s-1) + 2\sqrt{s(s-1)}\right)$$

with γ denoting the Euler-Mascheroni constant.

3. Description of the Method and Preliminary Results

We now translate our problem into an appropriate sieve setting in order to make Theorem 2 applicable. From now on, let n be a fixed positive integer, $x > 0$ a real parameter, and set

$$\mathcal{A} := \{1 \le a < x \mid a \text{ is a } \lambda\text{-root modulo } n,\ a \equiv 1\ (4)\}.$$

Furthermore $\mathcal{P}$ will be the set of primes $p \equiv 3$ (4). Then we aim to bound

$$\mathcal{S}(\mathcal{A}, z) := \sharp\{a \in \mathcal{A} \mid (a, \mathcal{P}(z)) = 1\}$$

from below for a suitable choice of the parameters z and x. Indeed, if $\mathcal{S}(\mathcal{A}, z) \ge 1$ for $z > \sqrt{x}$, there exists a λ-root modulo n less than x which is expressible as a sum of two squares, since an odd number in $\mathcal{A}$ must have an even number of prime divisors in $\mathcal{P}$.

Before we can apply Theorem 2 to our problem, it is essential to derive an asymptotic formula for $\sharp\,\mathcal{A}_d$ as in (1). Therefore we let $\gamma(k)$ denote the characteristic function of λ-roots modulo n, i.e. $\gamma(k) = 1$, if k is a λ-root modulo n, and $\gamma(k) = 0$, otherwise. Since $\gamma(k)$ is periodic with period n, and with support inside the set of integers coprime to n, it admits a unique expression as a linear combination of Dirichlet characters modulo n. This linear combination has been determined in Lemma 4 and 5 of [4] which we summarize in the following lemma.

Lemma 3. *Let G be the subgroup of Dirichlet characters modulo n given by*

$$G=\left\{\chi^{\frac{\lambda(n)}{\mathrm{rad}(\lambda(n))}}\mid\chi\pmod n\right\}.$$

For every prime p dividing $\varphi(n)$, let $m(p)$ denote the number of independent characters of order p in G. For every character χ modulo n let $\sigma(\chi)$ denote its order. Then, for any integer k we have

$$\gamma(k)=\sum_{\chi(n)}c_\chi\chi(k),$$

where the coefficients c_χ are given by

$$c_\chi=\begin{cases}\prod_{p\mid\sigma(\chi)}\left(\frac{-1}{p^{m(p)}}\right)\prod_{\substack{p\mid\varphi(n)\\ p\nmid\sigma(\chi)}}\left(1-\frac{1}{p^{m(p)}}\right), & \text{if } \chi\in G,\\ 0, & \text{otherwise,}\end{cases}$$

and satisfy the equation

$$\sum_{\chi(n)}|c_\chi|=2^{\omega(\varphi(n))}c_0,$$

where $c_0:=c_{\chi_0}$, and χ_0 is the principal character modulo n.

The error term in the asymptotic formula for $\sharp\mathcal{A}_d$, which we proceed to prove in the subsequent section, involves short sums of consecutive values of characters $\chi\in G$. To this end we state the following result (cf. Lemma 7 in [4]) which yields appropriate estimates for such sums, and is based on a more general result of Burgess (cf. [1]).

Lemma 4. *For every $G\ni\chi\neq\chi_0$, $M,N\geq 1$, and $0<\eta<1$ we have*

$$\sum_{M<k\leq M+N}\chi(k)\ll N\left(\frac{n_c^{1/4+\eta}}{N}\right)^\eta.$$

4. Proof of Theorem 1

We begin with the deduction of an asymptotic formula for $\sharp\mathcal{A}_d$.

Lemma 5. *For a positive integer $d\leq x$ we have*

$$\sharp\mathcal{A}_d=g(d)X+r_d, \tag{4}$$

where $X:=\frac{c_0x\varphi(2n)}{4n}$ with c_0 as in Lemma 3, and $g(d)$ is a multiplicative function given by

$$g(d)=\begin{cases}0, & \text{if } (d,2n)>1,\\ \frac{1}{d}, & \text{otherwise.}\end{cases}$$

For any $0<\eta\le 1$, $\varepsilon>0$, *and* $D\le x$ *the remainder terms* r_d *satisfy*

$$\sum_{\substack{d\mid\mathcal{P}(z)\\ d<D}} |r_d| \ll_{\varepsilon,\eta} X n_c^{\varepsilon}\left(\frac{n_c^{1/4+\eta}}{x}\right)^{\eta} D^{\eta}.$$

Proof. If $(d,n)>1$ or $2\mid d$ we clearly have $\mathcal{A}_d=\varnothing$. For d odd and $(d,n)=1$ we deduce by Lemma 3

$$\begin{aligned}
\sharp\mathcal{A}_d &= \sum_{\substack{k\le x\\ k\equiv 1\ (4)\\ k\equiv 0\ (d)}} \gamma(k)\\
&= \sum_{\substack{k\le x/d\\ kd\equiv 1\ (4)}} \sum_{\chi\in G} c_\chi \chi(kd)\\
&= c_0\chi_0(d)\sum_{\substack{k\le x/d\\ k\equiv d\ (4)}} \chi_0(k) + \sum_{\substack{\chi\in G\\ \chi\ne\chi_0}} c_\chi\chi(d)\sum_{\substack{k\le x/d\\ k\equiv d\ (4)}}\chi(k). \qquad (5)
\end{aligned}$$

By Möbius inversion the first sum in (5) equals

$$\begin{aligned}
\sum_{\substack{k\le x/d\\ k\equiv d\ (4)\\ (k,n)=1}} 1 &= \sum_{\substack{k\le x/d\\ k\equiv d\ (4)}}\sum_{\substack{f\mid k\\ f\mid n}} \mu(f)\\
&= \sum_{f\mid n}\mu(f)\sum_{\substack{lf\le x/d\\ lf\equiv d\ (4)}} 1\\
&= \frac{x}{4d}\sum_{\substack{f\mid n\\ 2\nmid f}}\frac{\mu(f)}{f} + O\left(2^{\omega(n)}\right)\\
&= \frac{x}{4d}\frac{\varphi(2n)}{n} + O\left(2^{\omega(n)}\right). \qquad (6)
\end{aligned}$$

The second sum in (5) can be estimated using Lemma 3 and 4. If $d_0\equiv d\ (4)$ with $d_0\in\{1,3\}$, and m is an integer satisfying $4m\equiv 1\ (n)$, we obtain

$$\begin{aligned}
\sum_{\substack{\chi\in G\\ \chi\ne\chi_0}} c_\chi\chi(d)\sum_{\substack{k\le x/d\\ k\equiv d\ (4)}}\chi(k) &\ll \sum_{\substack{\chi\in G\\ \chi\ne\chi_0}} |c_\chi|\left|\chi(m)\sum_{0\le l\le \frac{x}{4d}-\frac{d_0}{4}}\chi(4l+d_0)\right|\\
&\ll \sum_{\substack{\chi\in G\\ \chi\ne\chi_0}} |c_\chi|\left|\sum_{0\le l\le \frac{x}{4d}-\frac{d_0}{4}}\chi(l+md_0)\right|\\
&\ll c_0 2^{\omega(\varphi(n))}\frac{x}{d}\left(\frac{d}{x}n_c^{1/4+\eta}\right)^{\eta}.
\end{aligned}$$

By (5) and (6) the remainder term r_d is therefore

$$\ll c_0 2^{\omega(n)} + c_0 2^{\omega(\varphi(n))} \frac{x}{d} \left(\frac{d}{x} n_c^{1/4+\eta} \right)^{\eta} \ll_{\varepsilon} \frac{1}{d} \frac{c_0 x \varphi(2n)}{4n} n_c^{\varepsilon} \left(\frac{d}{x} n_c^{1/4+\eta} \right)^{\eta},$$

since $d \leq x$ and $\eta \leq 1$. Using the definition of X, we finally deduce

$$\sum_{\substack{d \mid \mathcal{P}(z) \\ d<D}} |r_d| \leq \sum_{d<D} |r_d| \ll_{\varepsilon} X n_c^{\varepsilon} \left(\frac{n_c^{1/4+\eta}}{x} \right)^{\eta} \sum_{d<D} d^{\eta-1} \ll_{\eta} X n_c^{\varepsilon} \left(\frac{n_c^{1/4+\eta}}{x} \right)^{\eta} D^{\eta}.$$

□

Using Theorem 2, and the preceding lemma we may now prove the following modified version of Theorem 1.

Theorem 6. *For a positive integer n, and x a positive real parameter, let $\mathcal{S}(\mathcal{A})$ denote the number of odd λ-roots q modulo n in the range $0 < q < x$, such that q is expressible as a sum of two squares, and set $X := \frac{c_0 x \varphi(2n)}{4n}$. Then, for any $0 < \eta < \frac{1}{2}$ there exists $x_0(\eta) \geq 1$ such that*

$$\mathcal{S}(\mathcal{A}) \gg_{\eta} \frac{X}{\sqrt{\log x}}$$

holds, whenever $x > \max\{x_0(\eta), n_c^{\frac{1}{2}+\frac{5\eta}{1-2\eta}}\}$.

Proof. Inserting the multiplicative function $g(d)$ from Lemma 5 into the definition of $V(z)$ one can easily verify that (2) is satisfied, and $V(z) \sim \frac{c_n}{\sqrt{\log z}}$ holds with some constant $c_n \geq 1$ depending on n (cf. [2, p.277f]). Hence Theorem 2 is applicable. If $z > \sqrt{x}$, we clearly have

$$\begin{aligned} \mathcal{S}(\mathcal{A}) &\geq \mathcal{S}(\mathcal{A}, z) \\ &\gg \frac{X}{\sqrt{\log x}} \left\{ f(s) + O\left((\log D)^{-\frac{1}{6}} \right) \right\} + O_{\eta,\varepsilon} \left(X n_c^{\varepsilon} \left(\frac{n_c^{1/4+\eta}}{x} \right)^{\eta} D^{\eta} \right) \end{aligned}$$

by Theorem 2 and Lemma 5. Now we define $D := \frac{x^{1-\eta}}{n_c^{2\eta+1/4}}$, and set $\varepsilon = \eta^2$. With these choices the above error term becomes $o(X/\sqrt{\log x})$. If we choose x according to the condition

$$x > n_c^{\frac{1}{2}+\frac{5\eta}{1-2\eta}},$$

we obtain $D > x^{1/2}$, and hence $f(s) > 0$. This completes the proof. □

Finally, Theorem 1 is a simple consequence of the previous theorem. Indeed, for any $0 < \eta < \frac{1}{2}$ Theorem 6 implies that $\mathcal{S}(\mathcal{A})$ is positive, whenever

$$x > x_0(\eta) n_c^{\frac{1}{2}+\frac{5\eta}{1-2\eta}}$$

is satisfied. Since $\mathcal{S}(\mathcal{A})$ is a counting function, it must therefore be ≥ 1, and Theorem 1 follows.

References

[1] D. A. Burgess, *On character sums and L-series. II*, Proc. Lond. Math. Soc. (3) **13** (1963), 524–536.

[2] J. Friedlander and H. Iwaniec, *Opera de cribro*, vol. 57 of Amer. Math. Soc. Colloq. Publ., Amer. Math. Soc., Providence, RI, 2010.

[3] G. Martin, *The least prime primitive root and the shifted sieve*, Acta Arith. **80** (1997), 277–288.

[4] G. Martin, *Uniform bounds for the least almost-prime primitive root*, Mathematika **45** (1998), 191–207.

[5] P. Moree, *Artin's primitive root conjecture—a survey*, Integers **12** (2012), 1305–1416.

[6] V. Shoup, *Searching for primitive roots in finite fields*, Math. Comp. **58** (1992), 369–380.

STABILIZATION OF COEFFICIENTS FOR PARTITION POLYNOMIALS

Robert P. Boyer
Department of Mathematics, Drexel University, Philadelphia, Pennsylvania
boyerrp@drexel.edu

William J. Keith[1]
Dept. of Mathematics, Michigan Technological University, Houghton, Michigan
william.keith@gmail.com

Received: 11/20/11, Revised: 5/1/13, Accepted: 7/24/13, Published: 9/26/13

Abstract
We find that a wide variety of families of partition statistics stabilize in a fashion similar to $p_k(n)$, the number of partitions of n with k parts, which satisfies $p_k(n) = p_{k+1}(n+1)$, $k \geq n/2$. We bound the regions of stabilization, discuss variants on the phenomenon, and give the limiting sequence in many cases as the coefficients of a single-variable generating function. Examples include many statistics that have an Euler product form, partitions with prescribed subsums, and plane overpartitions.

1. Introduction

Consider a set of combinatorial objects $\{\alpha\}$ with statistics $wt(\alpha)$ and $t(\alpha)$, thinking of $wt(\alpha)$ as the primary descriptor. Let $G(z,q)$ be its two-variable generating function, that is, if $p(n,k)$ is the number of objects α with $wt(\alpha) = n$ and $t(\alpha) = k$,

$$G(z,q) = \sum_{n,k \in \mathbb{N} \bigcup \{0\}} p(n,k) q^n z^k .$$

An important example occurs when the generating function $G(z,q)$ has the Euler product form

$$G(z,q) = \prod_{i \geq 1} \frac{1}{(1 - zq^i)^{a_i}} \tag{1}$$

with $a_i \in \mathbb{Z}^+ \bigcup \{0\}$. We let $F_n(z)$ denote the q^n coefficient of $G(z,q)$ which is a

[1]Work on this paper began with the second author at Drexel and concluded at CELC, Universidade Lisboa, Portugal

polynomial in z, so

$$G(z,q)=\sum_n F_n(z)q^n, \quad F_n(z)=\sum_k p(n,k)z^k .$$

For $a_i=1$ this is the generating function for partitions of weight $wt(\alpha)=n$ with number of parts $t(\alpha)=k$. It is a well-known fact in partition theory that $p(n,k)$ for k nearly equal to n has a value independent of n: $p(n,n-b)$ is the number of partitions of b for $b\le \frac{n}{2}$. A similar result holds for plane partitions of n indexed by their trace [1, p. 199] or [7, Corollary 5.3].

This phenomenon, which we call *stabilization,* is widespread in generating functions of combinatorial interest, even those of greater complexity. The purpose of this paper is to describe this behavior in more general cases, and consider some illustrative examples and variations. We found the polynomial framework to be well-suited to these problems rather than a direct approach. The arguments should be adaptable to a wide variety of cases.

2. Basic Infinite Product Generating Functions

Let $G(z,q)$ be an infinite product generating function of the form

$$G(z,q)=\sum_{n=1}^{\infty} F_n(z)q^n=\prod_{j=1}^{\infty}\frac{1}{(1-z^{b(j)}q^{c(j)})^{a_j}}, \tag{2}$$

where we assume that the number of j for which $c(j)=t$ for any t is finite, so that the series converges. We find that if the c grow sufficiently faster than the b, the upper ends of the $F_n(z)$ stabilize, to the coefficients of a single-variable generating function which we can give.

Let $\mathcal{F}$ denote the set of all nonnegative integer sequences with finite support. For $\mathbf{e}=(e_1,e_2,\dots)\in\mathcal{F}$ set

$$\mu(\mathbf{e})=(c(1)^{e_1}c(2)^{e_2}\cdots), \quad \nu(\mathbf{e})=(b(1)^{e_1}b(2)^{e_2}\cdots)$$

to denote the partitions with parts $c(j)$ (resp. $b(j)$) appearing e_j times. A direct expansion of the generating functions yields an explicit form for the polynomial $F_n(z)$:

$$F_n(z) \;=\; \sum_{k=0}^{n} z^k \sum_{\substack{\mu(\mathbf{e})\vdash n\\ \nu(\mathbf{e})\vdash k}} \prod_{i\ge 1}\binom{a_i+e_i-1}{e_i}. \tag{3}$$

We will compare the coefficients of $F_n(z)$ with those of the expansion of

$$\prod_{i=1}^{\infty}\frac{1}{(1-z^{b(i)})^{a_i}}=\sum_{k=0}^{\infty} z^k \sum_{\nu(\mathbf{e})\vdash k}\prod_{i\ge 1}\binom{a_i+e_i-1}{e_i}.$$

Theorem 1. *Suppose that the exponents satisfy* $a_1 = 1 = b(1) = c(1)$. *If there exists a positive integer* $m \geq 2$ *such that*

$$m \cdot b(j) \leq c(j), \quad j \geq 2, \tag{4}$$

then
(a) for $k > n/m$, $\left[z^k\right] F_n(z) = \left[z^{k+1}\right] F_{n+1}(z)$.
(b) If $c(j) - b(j) > 0$ *for all* $j > 1$ *and the set of* j *for which* $c(j) - b(j)$ *takes a given value is finite for any fixed difference, then for* $\ell \leq \lfloor n/m \rfloor$,

$$\left[z^{n-\ell}\right] F_n(z) = \left[z^{\ell}\right] \prod_{j \geq 2} \frac{1}{(1 - z^{c(j)-b(j)})^{a_j}}.$$

Proof. (a) From the explicit form (3) of the polynomial $F_n(z)$, we can expand it as

$$\begin{aligned} F_n(z) &= \sum_{k=0}^{n} z^k \sum_{\substack{\mu(\mathbf{e}) \vdash n \\ \nu(\mathbf{e}) \vdash k}} \prod_{i \geq 1} \binom{a_i + e_i - 1}{e_i} \\ &= \sum_{k=0}^{n} \sum_{e_1=0}^{k} z^{b(1)e_1} \sum_{\substack{\mu^-(\mathbf{e}) \vdash n - c(1)e_1 \\ \nu^-(\mathbf{e}) \vdash k - b(1)e_1}} z^{k-b(1)e_1} \prod_{i \geq 2} \binom{a_i + e_i - 1}{e_i}. \end{aligned}$$

Now if the integer sequence $\mathbf{e}$ gives a contribution to $[z^{k+1}]F_{n+1}(z)$ and $e_1 > 0$, then we define $\mathbf{e}'$ as the integer sequence all of whose terms agree with $\mathbf{e}$ except for $j = 1$ where we set $e_1' = e_1 - 1$. In this way, we obtain all the possible terms contributing to $[z^k]F_n(z)$. Conversely, any term for $[z^k]F_n(z)$ gives a contribution to $[z^{k+1}]F_{n+1}(z)$ by simply adding 1 to its first component.

The result reduces to showing that any contribution to $[z^{k+1}]F_{n+1}(z)$ indexed by $\mathbf{e}$ must have $e_1 > 0$.

We introduce the notation for the modified partitions

$$\mu^-(\mathbf{e}) = (c(2)^{e_2} c(3)^{e_3} \cdots), \quad \nu^-(\mathbf{e}) = (b(2)^{e_2} b(3)^{e_3} \cdots).$$

Now assume that the partition with $\mu(\mathbf{e}) \vdash n$ and $\nu(\mathbf{e}) \vdash k$ gives a contribution to $[z^{k+1}]F_{n+1}(z)$ but $e_1 = 0$. So $\mu^-(\mathbf{e}) \vdash n$ and $\nu^-(\mathbf{e}) \vdash k$, but if $c(j) \geq m \cdot b(j)$ for all $j \geq 2$, then $|\mu^-(\mathbf{e})| \geq m|\nu^-(\mathbf{e})|$. Hence if $k > \frac{n}{m}$, we find that $|\mu^-(\mathbf{e})| > n$, a contradiction. Thus all terms in both expansions are the same, and so the coefficients are equal. This proves part (a).

For part (b), we begin by forming a new partition as follows. First subtract $b(j)$ from each $c(j)$ and consider the partition $\lambda(\mathbf{e}) = ((c(2) - b(2))^{e_2} \ldots)$. This removes exactly the amount $|\nu^-(\mathbf{e})|$ from $|\mu^-(\mathbf{e})|$, so

$$\begin{aligned} F_n(z) &= \sum_{k=0}^{n}\sum_{e_1=0}^{k} z^k \sum_{\substack{\lambda(\mathbf{e})\vdash n-k \\ \nu^-(\mathbf{e})\vdash k-b(1)e_1}} \prod_{i\geq 2}\binom{a_i+e_i-1}{e_i} \\ &= \sum_{k=0}^{n} z^k \sum_{\substack{\lambda(\mathbf{e})\vdash n-k \\ |\nu^-(\mathbf{e})|\leq k}} \prod_{i\geq 2}\binom{a_i+e_i-1}{e_i}. \end{aligned}$$

By hypothesis, the parts of λ satisfy $c(j)-b(j)\geq (m-1)b(j)$. If $k=n-\ell>\lfloor n/m\rfloor$, then $\ell<\lfloor n\frac{m-1}{m}\rfloor$. If $\lambda(\mathbf{e})\vdash n-k=\ell$, then $\nu^-(\mathbf{e})\leq \frac{\ell}{m-1}\leq k$. Thus the sum runs over all $\mathbf{e}$ for which $\lambda(\mathbf{e})\vdash \ell$ with parts $c(j)-b(j)$. But this is exactly the coefficient of z^ℓ in the expansion claimed:

$$\left[z^{n-\ell}\right]F_n(z)=\left[z^\ell\right]\prod_{j\geq 2}\frac{1}{\left(1-z^{c(j)-b(j)}\right)^{a_j}}.$$

□

By the technique of proof, we have a slight improvement in a special case of hand enumerators of prefabs ([2], [8, page 92]).

Corollary 2. *Suppose* $a_1=1$. *If* $F_n(z)=[q^n]\prod_{j=1}^{\infty}(1-zq^j)^{-a_j}$, *then for* $k\geq n/2$, $\left[z^k\right]F_n(z)=\left[z^{k+1}\right]F_{n+1}(z)$.

By the proof, we find that the generating function for the stabilized coefficients gives an upper bound outside the range of stability.

Corollary 3. *With the hypotheses of Theorem 1,*

$$\left[z^{n-\ell}\right]F_n(z)\leq\left[z^\ell\right]\prod_{j\geq 2}\frac{1}{\left(1-z^{c(j)-b(j)}\right)^{a_j}},\quad 0\leq\ell\leq n.$$

Variants of stabilization exist in several guises. If the b grow faster than the bound of the previous theorem, we find that the smaller end of the polynomials stabilize instead of the larger (and b and c staying within a given ratio range will permit both phenomena). We also note that it is possible for the larger coefficients of a sequence of polynomials to stabilize in periods, i.e., the coefficients match those of a polynomial every 2 or more steps further along.

Theorem 4. *(a) Let* $\{b(j)\}$ *and* $\{c(j)\}$ *be two strictly increasing sequences of integers, positive except that* $b(1)=0$. *Let* $F_n(z)=[q^n]\prod_{j=1}^{\infty}(1-z^{b(j)}q^{c(j)})^{-a_j}$ *with* $a_1=1$. *If there exists a positive integer* m *such that for all* $j\geq 2$

$$(m+1)b(j)>c(j),$$

then

$$[z^k]F_n(z) = [z^k]F_{n+c(1)}(z), \quad 0 \le k \le n/(m+1).$$

(b) Let $a_1 = 0, a_2 = 1$. *If* $F_n(z) = [q^n]\prod_{j=2}^{\infty}(1 - zq^j)^{-a_j}$, *then* $\deg(F_n) = \lfloor n/2 \rfloor$ *and*

$$[z^k]F_n(z) = [z^{k+1}]F_{n+2}(z), \quad k \ge n/3.$$

Proof. Let $\mathbf{e}$ be a sequence of non-negative integers with finite support. By (3), for the coefficients $[z^k]F_n(z)$ and $[z^k]F_{n+c(1)}(z)$, we need to consider the following two sets of finitely supported nonnegative integer sequences:

$$\begin{aligned} S_n &= \{\mathbf{e} \in \mathcal{F} : |\mu(\mathbf{e})| = n,\, |\nu(\mathbf{e})| = k\}, \\ S_{n+c(1)} &= \{\mathbf{f} \in \mathcal{F} : |\mu(\mathbf{f})| = n + c(1),\, |\nu(\mathbf{f})| = k\}. \end{aligned}$$

We construct a bijection between these two sets. Given $\mathbf{e} \in S_n$, we take the corresponding $\mathbf{f}$ with $f_i = e_i$ for $2 \le i$ since $b(1) = 0$. Next write out $|\mu(\mathbf{e})| = n$ and $|\mu(\mathbf{f})| = n + c(1)$:

$$n = \sum_{j \ge 1} c(j)e_j, \quad n + c(1) = c(1)f_1 + \sum_{j \ge 2} c(j)e_j.$$

The term f_1 uniquely determines a preimage e_1 provided $f_1 > 0$. But f_1 must be positive for $k \le n/(m+1)$ since

$$\sum_{j \ge 2} c(j)f_j = \sum_{j \ge 2} c(j)e_j < (m+1)\sum_{j \ge 2} b(j)e_j = (m+1)k \le n.$$

Finally, the coefficients themselves agree; that is, $[z^k]F_n(z) = [z^k]F_{n+c(1)}(z)$:

$$\prod\left\{\binom{a_i + e_i - 1}{e_i} : \mathbf{e} \in S_n\right\} = \prod\left\{\binom{a_i + f_i - 1}{f_i} : \mathbf{f} \in S_{n+c(1)}\right\}$$

since the above binomial coefficients are all equal for $i \ge 2$ and when $i = 1$ they both reduce to 1 since $a_1 = 1$.

For part (b), let $S_n = \{\mathbf{e} \in \mathcal{F} : \mu(\mathbf{e}) \vdash n, \nu(\mathbf{e}) \vdash k\}$ while $S_{n+2} = \{\mathbf{f} \in \mathcal{F} : \mu(\mathbf{f}) \vdash n+2, \nu(\mathbf{f}) \vdash k+1\}$. We can construct a bijection between these two sets as in part (a) provided $f_2 > 0$ when $k \ge n/3$. Assume that $f_2 = 0$ is possible. Then $\sum f_j = k + 1$ while $\sum_{j \ge 3} jf_j = n + 2$. On the other hand, $3(k+1) \ge \sum_{j \ge 3} jf_j$ which yields a contradiction. □

3. Partitions With Prescribed Subsums

Fix a positive integer m and integer i so $1 \le i \le m$. Canfield-Savage-Wilf [4, Section 3] introduced the generating function

$$G_{m,i}(z, q) = \prod_{j=1}^{\infty}\prod_{b=1}^{i-1}\frac{1}{1 - z^{j-1}q^{(j-1)m+b}}\prod_{b=i}^{m}\frac{1}{1 - z^j q^{(j-1)m+b}}$$

to describe partitions with prescribed subsums. They let $\Lambda_{m,i}(n,k)$ be the number of partitions $\lambda = (\lambda_1, \cdots, \lambda_n)$ of n such that the sum of those parts λ_j whose indices j are congruent to i modulo m is k; that is,

$$\sum_{j:j\equiv i \pmod m} \lambda_j = k.$$

Then they found

$$\sum_{n,k\geq 0} \Lambda_{m,i}(n,k) z^k q^n = G_{m,i}(z,q).$$

We begin by recovering a result for $\Lambda_{2,2}(n,k)$ in [4, Theorem 1] and [9] by reformulating it in terms of the generating function $G_{2,2}(z,q)$ and stabilization of polynomial coefficients.

Proposition 5. *Let $m \geq 2$ and $1 \leq b < m$. Let $G(z,q) = \prod_{j=1}^{\infty}(1-z^{j-1}q^{(j-1)m+b})^{-1}$ and $A_n(z) = [q^n]G(z,q)$. Then for $0 \leq k \leq n/(m+1)$ we have*

1. $[z^k]A_n(z) = [z^k]A_{n+b}(z)$,
2. *if $n - mk$ is not divisible by b, then $[z^k]A_n(z) = 0$,*
3. *if $n - mk$ is divisible by b and $bk \leq n/(m+1)$, then $[z^k]A_n(z) = p(k)$.*

Proof. The first part is a direct consequence of the last theorem.

For part (2), in Theorem 4 let $c(j) = m(j-1) + b$ and $b(j) = j - 1$. When all $a_j = 1$, $[z^j]F_n(z)$ is the number of all $\mathbf{e} \in \mathcal{F}$ such that $\mu(\mathbf{e}) \vdash n$ and $\nu(\mathbf{e}) \vdash k$. Consider

$$\begin{aligned} n &= |\mu(\mathbf{e})| = \sum_{j\geq 1} e_j[m(j-1)+b] \\ &= m\sum_{j\geq 1} e_j(j-1) + b\sum_{j\geq 1} e_j \\ &= mk + b\sum_{j\geq 1} e_j. \end{aligned} \tag{5}$$

Hence $n - mk$ must be divisible by b for any nonzero choice of $\mathbf{e}$.

For part (3), assume $0 \leq bk \leq n/(m+1)$ and that $n - mk$ is divisible by b. Let $\mathbf{e}$ be any solution to $\nu(\mathbf{e}) = k$. Note that this does not give a constraint for the choice of e_1. On the other hand, the choice of e_1 by (5) must be

$$e_1 = (n-mk)/b - \sum_{j\geq 2} e_j, \quad e_1 \geq 0$$

to yield $\mu(\mathbf{e}) = n$. Hence, $0 \leq [z^k]F_n(z) \leq p(k)$. Finally, the inequality $0 \leq bk \leq n/(m+1)$ shows that $e_1 \geq 0$ always holds. Hence, $[z^k]F_n(z) = p(k)$. □

Theorem 6. *Let* $G_A(z,q) = \prod_{k=1}^{\infty}(1 - z^k q^{2k-1})^{-1}$ *and* $G_B(z,q) = \prod_{k=1}^{\infty}(1 - z^k q^{2k})^{-1}$, *so* $G_{2,2}(z,q) = G_A(z,q)G_B(z,q)$. *Then the coefficients of the polynomials* $F_n(z) = [q^n]G_{2,2}(z,q)$ *satisfy*

$$[z^k]F_n(z) = [z^k]F_{n+1}(z) = \sum_{\ell=0}^{k} p(\ell)p(k-\ell), \quad 0 \le k \le n/3,$$

where $p(\ell)$ *is the number of partitions of* ℓ, *as usual.*

Proof. Let $A_n(z) = [q^n]G_A(z,q)$ and $B_n(z) = [q^n]G_B(z,q)$. The generating function $G_B(z,q)$ has the explicit expansion

$$G_B(z,q) = \sum_{j=0}^{\infty} p(j) z^j q^{2j}$$

so $B_{2k+1}(z) = 0$ while $B_{2k}(z) = p(k)z^k$. We also know by Proposition 5 that

$$[z^j]A_n(z) = p(j), \quad 0 \le j \le n/3.$$

Next we have that

$$F_n(z) = \sum_{\ell=0}^{n/2} A_{n-2\ell}(z)B_{2\ell}(z) = \sum_{\ell=0}^{n/2} p(\ell)z^\ell A_{n-2\ell}(z)$$

Examine the coefficient $[z^k]F_n(z)$ for $0 \le k \le n/3$:

$$[z^k]F_n(z) = [z^k]\sum_{\ell=0}^{n/2} p(\ell)z^\ell A_{n-2\ell}(z) = \sum_{\ell=0}^{n/2} p(\ell)\,[z^{k-\ell}]A_{n-2\ell}(z)$$

Since $k-\ell \le (n-2\ell)/3$ for $0 \le \ell \le k$ and $k \le n/3$, we find $[z^{k-\ell}]A_{n-2\ell}(z) = p(k-\ell)$. We conclude that

$$[z^k]F_n(z) = \sum_{\ell=0}^{k} p(\ell)\,p(k-\ell).$$

□

In order to investigate more general cases for the polynomials $F_n(z) = [q^n]G_{m,i}(z,q)$ it is convenient to rewrite the generating function as

$$G_{m,i}(z,q) = \left(\prod_{b=1}^{i-1} \frac{1}{1-q^b}\right) \prod_{a=1}^{\infty} \prod_{d=0}^{m-1} \frac{1}{1 - z^a q^{i+(a-1)m+d}}. \tag{6}$$

We wish to find a useful form for the coefficient $[z^j]F_n(z)$. As usual, we have

$$\begin{aligned}[q^n]G_{m,i}(z,q) &= \sum_{s=0}^{n}[q^{n-s}]\left(\prod_{b=1}^{i-1}\frac{1}{1-q^b}\right)[q^s]\prod_{a=1}^{\infty}\prod_{d=0}^{m-1}\frac{1}{1-z^aq^{i+(a-1)m+d}}\\ &= \sum_{s=0}^{n}[q^{n-s}]\left(\prod_{b=1}^{i-1}\sum_{f_b=0}^{\infty}(q^b)^{f_b}\right)\\ &\times\ [q^s]\prod_{a=1}^{\infty}\prod_{d=0}^{m-1}\sum_{f_{(a,d)}=0}^{\infty}\left(z^aq^{i+(a-1)m+d}\right)^{f_{(a,d)}}\end{aligned}$$

A typical term of the above expansion is indexed by a pair of partitions (ρ, μ) where $\rho \vdash n-s$ where each part of ρ is $< i$ while $\mu \vdash s$ where each part of μ is $\geq i$. We write the parts of μ as $\mu_{(a,d)}$ with multiplicity $e_{\mu(a,d)}$ where $a \geq 1$ and $0 \leq d < m$. By construction, we find that

$$s = \sum_{a,d} e_{\mu(a,d)}\mu_{(a,d)}.$$

while, of course, no part of ρ has a contribution to z.

Each part $\mu_{(a,d)}$ of μ with multiplicity $e_{\mu(a,d)}$ contributes the power of z

$$(z^a)^{e_{\mu(a,d)}} = z^{ae_{\mu(a,d)}},$$

so the total contribution from μ is

$$\sum_{a,d} ae_{\mu(a,d)}.$$

We prove a few lemmas in preparation for the next theorem.

Lemma 7. *We have*

$$[z^j]F_n(z) = \sum_{s=ji}^{n} p(n-s,<i)\,\#\{\mu \vdash s : \textit{all parts of } \mu \textit{ are } \geq i, \sum_{a,d} ae_{\mu(a,d)} = j\} \quad (7)$$

where $p(n-s,<i)$ denotes the number of all partitions of $n-s$ all of whose parts are strictly less than i.

Proof. This expression for $[z^j]F_n(z)$ follows directly from the generating function except for the range of summation. Let μ be a partition of s. Then we have

$$s = \sum_{a,d}[i+m(a-1)+d]e_{\mu(a,d)} \geq \sum_{a,d}[i+i(a-1)]e_{\mu(a,d)} = i\sum_{a,d} ae_{\mu(a,d)} = ij.$$

□

Let r denote the number of parts of μ while t denotes the number of parts with $a \geq 2$ so $r-t$ gives the number of parts with $a=1$. Note that

$$r = \sum_{a,d} e_{\mu(a,d)}.$$

Lemma 8. *Given the partition μ, we have the bound $j - r \geq t$, and can rewrite s as*

$$s = ir + m(j-r) + \sum_{a,d} de_{\mu(a,d)}.$$

Proof. Since $\mu \vdash s$, we find that

$$\begin{aligned} s = \sum_{a,d} e_{\mu(a,d)}\left[i+(a-1)m+d\right] &= i\sum_{a,d} e_{\mu(a,d)} + m\sum_{a,d}(a-1)e_{\mu(a,d)} + \sum_{a,d} de_{\mu(a,d)} \\ &= ir + m(j-r) + \sum_{a,d} de_{\mu(a,d)}. \end{aligned}$$

For the bound, consider

$$j - r = \sum_{a,d}(a-1)e_{\mu(a,d)} \geq \sum_{a\geq 2,d} e_{\mu(a,d)} \geq t.$$

□

Lemma 9. *Let μ be a partition of s all of whose parts are $\geq i$. If $m \geq i+2$ and $j > s/(i+1)$, then μ must have at least one part of size i.*

Proof. We now assume that $m \geq i+2$ and that the part i does not appear in μ. In particular, for any part of μ of the form $i+d$, with $a=1$, d must be strictly positive. Then we have a refinement of the above bounds:

$$\begin{aligned} s &\geq ir + m(j-r) + \sum_{a,d} de_{\mu(a,d)} \geq ir + (i+2)(j-r) + \sum_{d} de_{1,d} \\ &\geq ir + (i+2)(j-r) + (r-t) \\ &\geq ir + (i+1)(j-r) + t + (r-t) \\ &= (i+1)[r + (j-r)] = (i+1)j. \end{aligned}$$

Hence the partition μ must have i as a part; otherwise, $(i+1)j > s$ which contradicts our assumption. □

With these preparations, we will show that

Theorem 10. *Let $m \geq 1$ and $1 \leq i \leq m$. Set $F_n(z) = [q^n]G_{m,i}(z,q)$ where $G_{m,i}(z,q)$ is given by (6). If $m > i+1$ and $j > \frac{n}{i+1}$, then*

$$[z^j]F_n(z) = [z^{j-1}]F_{n-i}(z).$$

Proof. By (7), we need to show that

$$\sum_{s=ji}^{n} p(n-s,<i)\,\#\{\mu\vdash s : \text{all parts of } \mu \text{ are } \geq i, \sum_{a,d} ae_{\mu(a,d)} = j\}$$

$$= \sum_{s'=(j-1)i}^{n-i} p(n-s'-i,<i)$$

$$\times \quad \#\{\nu\vdash s' : \text{all parts of } \nu \text{ are } \geq i, \sum_{a,d} ae_{\nu(a,d)} = j-1\}$$

These two coefficients are equal provided we construct a bijection T between

$$U_s = \{\mu\vdash s : \text{all parts of } \mu \text{ are } \geq i, \sum_{a,d} ae_{\mu(a,d)} = j\}$$

$$V_{s'} = \{\nu\vdash s' : \text{all parts of } \nu \text{ are } \geq i, \sum_{a,d} ae_{\nu(a,d)} = j-1\}.$$

when $s' = s - i$. Let $\mu \in U_s$. By Lemma 7, the partition μ must have a part equal to i. Let $\nu = T(\mu)$ be the partition of $s-i$ obtained by deleting one part from μ of size i. It is easy to verify that $\nu \in V_{s'}$. The inverse of T is simply adding a part of size i to $\nu \in V_{s'}$. □

In other words, Theorem 10 shows that if $m > i+1$ and $j > n/(i+1)$, then the subsums satisfy $\Lambda_{m,i}(n,j) = \Lambda_{m,i}(n-i,j-1)$.

4. Laurent Type Polynomials

A more general case consists of generating functions that involve z raised to different powers; ultimately, we might treat the case of the generating function

$$G(z,q) = \prod_{(i,j)\in\mathbb{Z}^2\setminus(0,0)} \left(1 - z^i q^j\right)^{a_{ij}}.$$

An important example comes from the generating function for the crank statistic for partitions. Let

$$C(z,q) = \prod_{k\geq 1} \frac{1-q^k}{(1-zq^k)(1-z^{-1}q^k)} = \sum_{n=0}^{\infty} M_n(z) q^n$$

where $M_n(z)$ is a symmetric Laurent polynomial. From the definition of the crank, the coefficient of z^{n-k} in $M_n(z)$, for $k \leq \frac{n}{2}$, equals the number of partitions of k that include no 1s. In particular, the coefficients of $M_n(z)$ stabilize in the ranges for powers $n-k$ and $-n+k$ for $0 \leq k \leq \lfloor n/2 \rfloor$. It is suggested in [3] that the zeros for the crank polynomial converge to the unit circle. Another example is the generating function

$$\prod_{k=1}^{\infty} \frac{1}{(1-zq^k)^k} \frac{1}{(1-z^{-1}q^k)^k} \frac{1}{(1-q^k)^{2k}},$$

which comes from the Donaldson-Thomas Theory in algebraic geometry and whose asymptotics were studied in [6].

Lemma 11. *Let $\{A_n(z)\}_{n=0}^{\infty}$ be a sequence of polynomials, such that the degree of $A_n(z)$ is n, whose coefficients satisfy*

$$[z^{n-k}]A_n(z) = [z^{n+1-k}]A_{n+1}(z), \quad 0 \leq k \leq n/m$$

for some integer $m \geq 2$. Let $\{B_n(z)\}_{n=0}^{\infty}$ be another sequence of polynomials. Then the coefficients of the polynomial sequence $\{F_n(z)\}_{n=0}^{\infty}$ where

$$F_n(z) = \sum_{\ell=0}^{n} A_\ell(z) B_{n-\ell}(z^{-1})$$

also satisfy

$$[z^{n-k}]F_n(z) = [z^{n+1-k}]F_{n+1}(z), \quad 0 \leq k \leq n/m.$$

Proof. Let $0 \leq k \leq \lfloor n/m \rfloor$. Consider $[z^{n-k}]F_n(z)$. We have

$$\begin{aligned} [z^{n-k}]F_n(z) &= [z^{n-k}] \sum_{\ell=0}^{n} A_\ell(z)\, B_{n-\ell}(z^{-1}) \\ &= \sum_{\ell=0}^{n} \sum_{a=0}^{k} [z^{n-k+a}]A_\ell(z)\, [z^{-a}]B_{n-\ell}(z^{-1}) \\ &= \sum_{\ell=n-k}^{n} \sum_{a=0}^{k} [z^{n-k+a}]A_\ell(z)\, [z^{-a}]B_{n-\ell}(z^{-1}) \end{aligned} \tag{8}$$

where in the last step we note that $[z^{n-k+a}]A_\ell(z) = 0$ if $\ell < n-k$.

For $[z^{n+1-k}]F_{n+1}(z)$, we reindex ℓ by 1 and note that $[z^{n+1-k}]A_0(z) = 0$ for all n, giving us the expression for $[z^{n+1-k}]F_{n+1}(z)$:

$$\sum_{\ell=n-k}^{n} \sum_{a=0}^{k} [z^{n+1-k+a}]A_{\ell+1}(z)\, [z^{-a}]B_{n-\ell}(z^{-1}) \tag{9}$$

By assumption, we know that

$$[z^{n-k+a}]A_\ell(z) = [z^{n+1-k+a}]A_{\ell+1}(z), \quad n-k \leq \ell \leq n$$

since $0 \leq k \leq \lfloor n/m \rfloor$. The coefficients of $[z^{-a}]B_{n-\ell}(z^{-1})$ are the same, and consequently the two sums (8) and (9) are equal. □

Lemma 12. *Given the generating function* $G(z,q) = \prod_{i\geq 1}(1-zq^i)^{-a_i}$ *where* $a_1 = 1$, *let* $A_n(z) = [q^n]G(z,q)$. *Let* $Q(q) = \sum_{j=0}^{\infty} c_j q^j$. *Let* $F_n(z) = [q^n]G(z,q)Q(q)$. *Then the tail coefficients of* $F_n(z)$ *stabilize; that is,*

$$[z^k]F_n(z) = [z^{k+1}]F_{n+1}(z), \quad k \geq n/2.$$

Proof. By construction, the polynomials $F_n(z)$ have the form

$$F_n(z) = \sum_{\ell=0}^{n} c_{n-\ell} A_\ell(z).$$

Let $k \geq n/2$. Then the coefficients for $[z^k]F_n(z)$ and $[z^{k+1}]F_{n+1}(z)$ are given by

$$[z^k]F_n(z) = \sum_{\ell=k}^{n} c_{n-\ell}[z^k]A_\ell(z), \quad [z^{k+1}]F_{n+1}(z) = \sum_{j=k+1}^{n+1} c_{n+1-j}[z^{k+1}]A_j(z).$$

As a consequence of Corollary 1, $c_{n-\ell}[z^k]A_\ell(z) = c_{n+1-j}[z^{k+1}]A_j(z)$ for $j = \ell+1$ and $k \leq \ell \leq n$. □

Theorem 13. *If* $a_1 = b_1 = 1$, *and*

$$G(z,q) = \sum_{n=0}^{\infty} F_n(z)q^n = \prod_{i\geq 1}(1-zq^i)^{-a_i}(1-z^{-1}q^i)^{-b_i}(1\pm q^i)^{c_i},$$

then the coefficients of the Laurent polynomials $F_n(z)$ *satisfy*

$$[z^{n-k}]F_n(z) = [z^{n+1-k}]F_{n+1}(z), \quad [z^{-(n-k)}]F_n(z) = [z^{-(n+1-k)}]F_{n+1}(z),$$

for $0 \leq k \leq n/2$.

5. Plane Overpartition Stabilization

A plane partition is an array of positive integers, conventionally justified to the upper left corner of the fourth quadrant, which are weakly descending left in rows and down in columns. A plane overpartition is a plane partition whose entries may be overlined or not according to certain rules [5]: in each row, the last occurrence of an integer may be overlined (or not) and in every column, all but the first occurrence of an integer are overlined, while the first occurrence may or may not be overlined. In [5, Proposition 4], the generating function for the weighted plane overpartitions is found to be

$$\sum_{\Pi \text{ is a plane overpartition}} z^{o(\Pi)} q^{|\Pi|} = \prod_{n=1}^{\infty} \frac{(1+zq^n)^n}{(1-q^n)^{\lceil n/2 \rceil}(1-z^2q^n)^{\lfloor n/2 \rfloor}}$$

where $o(\Pi)$ is the number of overlined parts of the plane overpartition Π.

Theorem 14. *Let $G(z,q)$ be the generating function for the polynomials $F_n(z)$:*

$$G(z,q) = \prod_{n=1}^{\infty} \frac{(1+zq^n)^n}{(1-q^n)^{\lceil n/2 \rceil}(1-z^2q^n)^{\lfloor n/2 \rfloor}} = \sum_{n=0}^{\infty} F_n(z)q^n.$$

Then the coefficients of the polynomials $F_n(z)$ satisfy the stabilization condition

$$[z^{k+1}]F_{n+1}(z) = [z^k]F_n(z),$$

for $k \geq 2n/3$.

Proof. Let $\{A_n(z)\}$ be the polynomial sequence with generating function $G_A(z,q)$ where

$$G_A(z,q) = \prod_{n=2}^{\infty} \frac{1}{(1-z^2q^n)^{\lfloor n/2 \rfloor}} = \sum_{n=0}^{\infty} A_n(z)q^n$$

and $\{B_n(z)\}$ with generating function $G_B(z,q)$:

$$G_B(z,q) = \prod_{n=2}^{\infty} (1+zq^n)^n = \sum_{n=0}^{\infty} B_n(z)q^n.$$

Easily, we have that $\deg(A_n) = 2\lfloor n/2 \rfloor$. By Theorem 4, replacing z by z^2, we also find

$$[z^k]A_n = [z^{k+2}]A_{n+2}, \quad k \geq 2n/3.$$

The degree of the polynomial $B_N(z)$ is the largest number of parts in a possible partition of N drawn from a multiset of two 2s, three 3s, etc. For $N \geq 21 = 2+2+3+3+3+4+4$, the average size of part is at least 3 and so

$$\deg(B_N) \leq \frac{1}{3}N\,.$$

For smaller N, direct calculation shows that $\deg(B_N) \leq \frac{2}{3}N$.

It is more convenient to work with the intermediate polynomials

$$Q_n(z) = [q^n]G_A(z,q)G_B(z,q) = \sum_{\ell=0}^{n} A_{n-\ell}(z)\, B_\ell(z).$$

We have the summation formula for $[z^k]Q_n(z)$:

$$\begin{aligned} [z^k]Q_n(z) &= \sum_{\ell=0}^{n}\sum_{a=0}^{k} [z^{k-a}]A_{n-\ell}(z)\cdot[z^a]B_\ell(z) \\ &= \sum_{\ell \in I_{k,n}} \sum_{a=0}^{\deg(B_\ell)} [z^{k-a}]A_{n-\ell}(z)\cdot[z^a]B_\ell(z) \end{aligned}$$

where $I_{k,n}$ is the set of indices

$$I_{k,n} = \{\ell : \deg(B_\ell) + \deg(A_{n-\ell}) \geq k,\ 0 \leq \ell \leq n\}.$$

We have $I_{k+2,n+2} = I_{k,n}$ since for any B_ℓ the matching $A_{n-\ell}$ for sums of k map to the matching $A_{n+2-\ell}$ for sums of $k+2$.

We next show that

$$\sum_{\ell \in I_{k,n}} \sum_{a=0}^{\deg(B_\ell)} [z^{k-a}]A_{n-\ell}(z) \cdot [z^a]B_\ell(z) = \sum_{\ell \in I_{k+2,n+2}} \sum_{a=0}^{\deg(B_\ell)} [z^{k+2-a}]A_{n+2-\ell}(z) \cdot [z^a]B_\ell(z);$$

To do this, we need that all the terms $[z^{k-a}]A_{n-\ell}$ fall into the stable range of indices. Now to get into the stable range, we need

$$k - a \geq \frac{2}{3}(n - \ell)$$

where $0 \leq a \leq \deg B_\ell$ and $\ell \in I_{k,n}$. We consider the stronger condition

$$\frac{2}{3}n - \deg B_\ell \ \geq\ \frac{2}{3}(n - \ell),$$

which reduces to

$$\deg(B_\ell) \leq \frac{2}{3}\ell, \quad \ell \in I_{k,n}$$

which does indeed hold.

Our next step is to define

$$P_n(z) = (1 + zq)Q_n(z).$$

(Leaving out this factor earlier simplified the degree analysis, since without it we had the single exceptional case of $\deg(B_1) = 1$.) We observe that

$$\begin{aligned} [z^k]P_n(z) &= [z^k]Q_n(z) + [z^{k-1}]Q_{n-1}(z), \\ [z^{k+1}]P_{n+1}(z) &= [z^{k+1}]Q_{n+1}(z) + [z^k]Q_n(z). \end{aligned}$$

Hence, we have

$$[z^k]P_n(z) = [z^{k+1}]P_{n+1}(z), \quad k \geq 2n/3.$$

To finish the proof, we define a polynomial family $C_n(q)$ by

$$\prod_{n=1}^{\infty} \frac{1}{(1 - q^n)^{\lceil n/2 \rceil}} = \sum_{n=0}^{\infty} C_n(q).$$

Then the polynomials $F_n(z)$ are given by

$$F_n(z) = [q^n]C(q) \sum_{\ell=0}^{\infty} P_\ell(z) q^\ell.$$

By Lemma 12, we see that this construction maintains the stability of the coefficients of the polynomial family $\{P_\ell(z)\}$. □

Corollary 15. *Let* $\overline{pp}_k(n)$ *be the number of plane overpartitions of* n *with* k *overlined parts. If* $k \geq 2n/3$, *then*

$$\overline{pp}_k(n) = \overline{pp}_{k+1}(n+1).$$

Acknowledgement We would like to thank the anonymous referee for a careful readthrough: noting typos, improving exposition, and suggesting occasional strengthening of theorems for which we had been hesitant to extend our reach. This article is substantially improved from his or her efforts.

References

[1] George E. Andrews, "The Theory of Partitions," Reading, MA: Addison-Wesley Press, 1976.

[2] Edward A. Bender and Jay R. Goodman, *Enumerative uses of generating functions*, Indiana Univ. Math. J. **20** (1971) 753-765.

[3] Robert P. Boyer and William M. Y. Goh, *Polynomials Associated with Partitions: Their Asymptotics and Zeros*, Contemp. Math. **471** (2008), 33-45.

[4] E. Rodney Canfield, Carla D. Savage, Herbert S. Wilf, *Regularly spaced subsums of integer partitions*, Acta Arith. **115** (2004) 205-216.

[5] Sylvie Corteel, Cyrille Savelief, and Mirjana Vuletí, *Plane overpartitions and cylindric partitions*, J. Comb. Theory, Ser. A **118** (2011) 1239-1269.

[6] Daniel Panario, Bruce Richmond, Benjamin Young, *Bivariate asymptotics for striped plane partitions,* 2010 Proceedings of the Seventh Workshop on Analytic Algorithmics and Combinatorics (ANALCO), 18-26, 2010.

[7] Richard P. Stanley, *The conjugate trace and trace of a plane partition*, J. Comb. Theory, Ser. A (1973) **14** 53-65.

[8] Herbert S. Wilf, "generatingfunctionology," Second Edition, Academic Press, 1994.

[9] Yuri Yakubovich, *Integer partitions with fixed subsums*, Electron. J. Combin. **12** (2005), #A7

ON SETS WITH MORE RESTRICTED SUMS THAN DIFFERENCES

David Penman
Department of Mathematical Sciences, University of Essex, Wivenhoe Park, Colchester, United Kingdom
dbpenman@essex.ac.uk

Matthew Wells
Department of Mathematical Sciences, University of Essex, Wivenhoe Park, Colchester, United Kingdom
mwells@essex.ac.uk

Received: 8/2/12, Revised: 4/3/13, Accepted: 7/27/13, Published: 9/26/13

Abstract

Given a finite set A of integers, we define its restricted sumset $A\hat{+}A$ to be the set of sums of two *distinct* elements of A - a subset of the sumset $A+A$ - and its difference set $A-A$ to be the set of differences of two elements of A. We say A is a restricted-sum-dominant set if $|A\hat{+}A| > |A-A|$. Though intuition suggests that such sets should be rare, we present various constructions of such sets and prove that a positive proportion of subsets of $\{0,1,\dots n-1\}$ are restricted-sum-dominant sets. As a by-product, we improve on the previous record for the maximum value of $\ln(|A+A|)/\ln(|A-A|)$, and give some related discussion.

1. Introduction

Let A be a finite set of integers. We define its *sumset* $A+A$ to be $\{a+b : a,b \in A\}$, its *difference set* $A-A$ to be $\{a-b : a,b \in A\}$ and its *restricted sumset* $A\hat{+}A$ to be $\{a+b : a \neq b, a,b \in A\}$. It is a natural intuition that, since addition is commutative but subtraction is not, that 'often' we should have $|A+A| \leq |A-A|$. However it has been known for some time that this is not always the case: for example, the set $C = \{0,2,3,4,7,11,12,14\}$, which is attributed to Conway, has $|C+C| = 26$, but $|C-C| = 25$. In this paper, sets with this property are called sum-dominant: in some other literature, they are described as MSTD (for 'more sums than differences') sets, see, e.g., Nathanson [6]. It is now known by work of Martin and O'Bryant [5] that sum-dominant sets are less rare than they might initially appear: they prove that, for $n \geq 15$, the proportion of subsets of $\{0,1,2\dots n-1\}$

which are sum-dominant is at least 2×10^{-7}. The constant was sharpened, and the existence of a limit shown, by Zhao [11].

In this paper we investigate what might appear to be an even more demanding condition on a set, namely what we will call the restricted-sum-dominant property.

Definition 1. A set A of integers is said to be *restricted-sum-dominant* if $|A \hat{+} A| > |A - A|$.

There are examples of this. For example, we find the set from Hegarty [3]

$$A_{15} = \{0, 1, 2, 4, 5, 9, 12, 13, 17, 20, 21, 22, 24, 25, 29, 32, 33, 37, 40, 41, 42, 44, 45\}$$

has $|A_{15} \hat{+} A_{15}| = 86$ whilst $|A_{15} - A_{15}| = 83$.

Clearly any restricted-sum-dominant set is sum-dominant. The converse is false as Conway's set is sum-dominant but not restricted-sum-dominant ($|C \hat{+} C| = 21$).

Note that the property of being restricted-sum-dominant is preserved when we apply a bijection of the form $x \to ax + b$ with $a, b \in \mathbb{Z}$, $a \neq 0$. It therefore suffices to consider sets $A \subset \mathbb{Z}$ with $\min(A) = 0$ and $\gcd(A) = 1$. We shall refer to such sets as being *normalised.*

The organisation of this paper is as follows. In Section 2 we exhibit several sequences of restricted-sum-dominant sets, addressing some natural questions about the relative sizes of the restricted sumset and difference sets. In Section 3, we show that a strictly positive proportion of subsets of $\{0, 1, 2, \dots n-1\}$ are restricted-sum-dominant sets. In Section 4 we obtain a new record high value of each of

$$f(A) = \frac{\ln(|A + A|)}{\ln(|A - A|)} \text{ and } g(A) = \frac{\ln(|A + A|/|A|)}{\ln(|A - A|/|A|)}$$

and give some related discussion. Finally, in Section 5 we improve somewhat the bounds on the order of the smallest restricted-sum-dominant set.

We shall, slightly unusually, use the notation $[a, b]$, when $a < b$ are integers, to denote $\{a, a + 1, \dots b\}$.

We are grateful to the referee for suggestions which have non-trivially improved the organisation and exposition of this paper, especially in Section 5.

2. Explicit Sequences of Restricted-Sum-Dominant Sets

Our first sequence of restricted-sum-dominant sets arose by considering the set $B = \{0, 1, 2, 4, 5, 9, 12, 13, 17, 20, 21, 25, 28, 30, 32, 33\}$ which appears in [7] and [9] as a set of integers with $|B \hat{+} B| > |(B - B) \setminus \{0\}|$). We then noted that replacing 33 with 29 gives a 16-element restricted-sum-dominant set (which will be T_3' below). To get the subsequent terms of the sequence, we used (here and elsewhere in the paper) the idea from [9], Conjecture 6, that repetition of certain so-called interior blocks when the set is written in order as a sequence of differences can increase the size of the sumset more than the difference set: see [9] for details.

Theorem 2. *For every integer $j \geq 1$ we define*

$$T'_j = \{0,2\} \cup \{1,9,\dots,1+8j\} \cup \{4,12,\dots,4+8j\} \cup \{5,13,\dots,5+8j\} \cup \{6+8j, 8(j+1)\}.$$

Then

$$T'_j \hat{+} T'_j = [1, 6+8(2j+1)] \setminus \{8, 8(2j+1)\},$$
$$T'_j + T'_j = [0, 8(2j+2)] \setminus \{7+8(2j+1)\} \text{ and}$$
$$T'_j - T'_j = [-8(j+1), 8(j+1)] \setminus \{\pm 6, \dots \pm (6+8(j-1))\}.$$

Proof. We deal first with the restricted sumset. Since $0 \in T'_j$, $T'_j \setminus \{0\} \subseteq T'_j \hat{+} T'_j$, giving all elements congruent to 1,4 or 5 mod 8 less than $8(j+1)$. Also

$$8(j+1)\hat{+}\{1,9,\dots,1+8j\} = \{1+8(j+1),\dots,1+8(2j+1)\}$$
$$8(j+1)\hat{+}\{4,12,\dots,4+8j\} = \{4+8(j+1),\dots,4+8(2j+1)\}$$
$$8(j+1)\hat{+}\{5,13,\dots,5+8j\} = \{5+8(j+1),\dots,5+8(2j+1)\}$$

so $T'_j \hat{+} T'_j$ contains all the elements congruent modulo 8 to 1,4 or 5 stated. For integers congruent to 2 modulo 8 the restricted sumset contains 0+2 and

$$\{1,9,\dots,1+8j\}\hat{+}\{1,9,\dots,1+8j\} = \{10,18,\dots,2+8(2j-1)\}$$

gives most of the rest: the two missing elements are $(4+8j)+(6+8j) = 2+8(2j+1)$ and $4+8(j-1)+6+8j = 2+8(2j)$.

For integers congruent to 3 modulo 8, note that

$$\{1,9,\dots,1+8j\}\hat{+}(2) = \{3,11,\dots,3+8j\}$$

and

$$(6+8j)\hat{+}\{5,13,\dots 5+8j\} = \{3+8(j+1),\dots 3+8(2j+1)\}.$$

For integers congruent to 6 modulo 8,

$$\{1,9,\dots,1+8j\}\hat{+}\{5,13,\dots,5+8j\} = \{6,14,\dots 6+8(2j)\}$$

and $(6+8j)+8(j+1) = 6+8(2j+1) \in T'_j \hat{+} T'_j$ also. The elements congruent to 7 modulo 8 are obtained from

$$(2)+\{5,13,\dots,5+8j\} = \{7,15,\dots,7+8j\}$$

and

$$(6+8j)+\{1,9,\dots,1+8j\} = \{7+8j,\dots,7+8(2j)\}$$

in $T'_j\hat{+}T'_j$. Finally, the required multiples of 8 are obtained from

$$\{4, 12, \ldots, 4 + 8j\}\hat{+}\{4, 12, \ldots, 4 + 8j\} = \{16, 24, \ldots, 8(2j)\}.$$

Finally we note that the alleged omitted elements $0, 8$ and $8(2j + 1)$ are not in $T'_j\hat{+}T'_j$. The claim for 0 is clear, the only way to get 8 is as $4 + 4$ which is not a restricted sum, for $8(2j+1)$ the large elements of T'_j are $5+8j, 6+8j, 8(j+1) \in T'_j$ but $3 + 8j, 2 + 8j, 8j \notin T'_j$ so it could only be obtained as $(4 + 8j) + (4 + 8j)$ which is not a restricted sum.

Next we address the sumset $T'_j + T'_j$. All we need do here is note that $0 = 0+0$, $8 = 4+4$, $7+8(2j+1)$ is still not attained and that $8(2j+2) = 8(j+1)+8(j+1)$.

We finally deal with $T'_j - T'_j$. Given that $d \in T_j - T_j \iff -d \in T_j - T_j$ it suffices to consider the positive differences. Firstly we show that $\{6, \ldots, 6 + 8(j - 1)\} \notin T'_j - T'_j$. Given that T'_j has the form

$$T'_j = \{0, 1 + 8x, 2, 4 + 8y, 5 + 8z, 6 + 8j, 8(j + 1)\}$$

(where $0 \leq x, y, z, \leq j$), considering the difference set $T'_j - T'_j$ we see that the only difference of the form $6 + 8t$ (where t is a non-negative integer) is $6 + 8j$, as stated. To confirm $T'_j - T'_j$ does contain the other elements in the interval specified, note that, as $0 \in T'_j$, $T'_j \subseteq T'_j - T'_j$. The other elements are obtained as follows:

$$\begin{aligned}
\{1, 9, \ldots, 1 + 8j\} - (1) &= \{0, 8, \ldots, 8j\} \\
\{4, 12, \ldots, 4 + 8j\} - 1 &= \{3, 11, \ldots, 3 + 8j\} \\
\{4, 12, \ldots, 4 + 8j\} - 2 &= \{2, 10, \ldots, 2 + 8j\} \\
\{12, 20, \ldots, 4 + 8j\} - (5) &= \{7, 15, \ldots, 7 + 8(j - 1)\} \\
8(j + 1) - (1) &= 7 + 8j.
\end{aligned}$$

Thus all the elements of the right-hand side are in $T'_j - T'_j$ as required. □

Corollary 3. *For every integer $j \geq 1$ the set $T'_j \subset \mathbb{Z}$ has*

$$|T'_j| = 3j + 7,\ |T'_j\hat{+}T'_j| = 16j + 12, |T'_j + T'_j| = 16j + 16 \quad \textit{and} \quad |T'_j - T'_j| = 14j + 17.$$

Therefore

$$|T'_j\hat{+}T'_j| - |T'_j - T'_j| = 2j - 5, \quad |T'_j + T'_j| - |T'_j - T'_j| = 2j - 1$$

and T'_j is an restricted-sum-dominant set for every integer $j \geq 3$.

T'_3 of order 16 is one of the two smallest restricted-sum-dominant sets we have.

The set T'_j has a superset $T_j = T'_j \cup 1 + 8(j + 1)$, which is also restricted-sum-dominant for $j \geq 3$:

Theorem 4. *For every integer $j \geq 1$ define*

$$T_j = \{0,2\} \cup \{1,9,\ldots,1+8(j+1)\} \cup \{4,12,\ldots,4+8j\} \\ \cup \{5,13,\ldots,5+8j\} \cup \{6+8j, 8(j+1)\}.$$

Then

$$T_j \hat{+} T_j = [1, 1+8(2j+2)] \backslash \{8, 8(2j+1), 8(2j+2)\},$$
$$T_j + T_j = [0, 2+8(2j+2)] \text{ and}$$
$$T_j - T_j = [-(1+8(j+1)), 1+8(j+1)] \backslash \{\pm 6, \ldots \pm (6+8(j-1))\}.$$

Proof. Firstly since $T_j \supset T'_j$ we have $T_j \hat{+} T_j \supset [1, 6+8(2j+1)] \setminus \{8, 8(2j+1)\}$. With $1+8(j+1) \in T_j$ we now also have that

$$8(j+1) + (1+8(j+1)) = 1+8(2j+2) \quad \text{and}$$
$$(6+8j) + (1+8(j+1)) = 7+8(2j+1)$$

are in $T_j \hat{+} T_j$ as well. Furthermore

$$(1+8(j+1)) + (1+8(j+1)) = 2+8(2j+2) \in T_j + T_j.$$

This completes the claims for the sumset and restricted sumset, noting that clearly 8 and $8(2j+2)$ are not in $T_j \hat{+} T_j$ and checking that $8(2j+1) \notin T_j \hat{+} T_j$.

As regards the difference set, with $0 \leq x \leq j+1$ the positive differences resulting from the introduction of the new element have the form

$$(1+8(j+1)) - \{0, 2, 1+8x, 4+8y, 5+8z, 6+8j, 8(j+1)\}$$
$$= \{1+8(j+1), 8j+7, 8(j-x+1), 8(j-y)+5, 8(j-z)+4, 3, 1, 0\}.$$

This shows that $T_j - T_j = T'_j - T'_j \cup \pm(1+8(j+1))$ and the result follows. □

Corollary 5. *For every integer $j \geq 1$ the set $T_j \subset \mathbb{Z}$ has*

$$|T_j| = 3j+8,\ |T_j \hat{+} T_j| = 16j+14, |T_j + T_j| = 16j+19 \quad \text{and} \quad |T_j - T_j| = 14j+19.$$

Therefore

$$|T_j \hat{+} T_j| - |T_j - T_j| = 2j-5, \quad |T_j + T_j| - |T_j - T_j| = 2j$$

and T_j is an restricted-sum-dominant set for every integer $j \geq 3$.

In [5], Martin and O'Bryant construct, for all integers x, subsets S of $[0, 17|x|]$ with $|S+S| - |S-S| = x$. Corollary 3 shows that for each positive odd integer x there is $T'_j \subset \mathbb{Z}$ with $|T'_j + T'_j| - |T'_j - T'_j| = x$, and Corollary 5 shows each positive even integer can be expressed as the difference of the cardinalities of the sumset and the difference set of some $T_j \subset \mathbb{Z}$.

Recall that the *diameter* of a finite set A of integers is $\max(A) - \min(A)$. There is some interest in finding sets of integers of small diameter with prescribed relationships between the order of the sumset (or restricted sumset) and the difference set: see, e.g., [5] Theorem 4 where sets S_x of diameter at most $17|x|$ are constructed with $|S_x + S_x| - |S_x - S_x|$ equal to x. Our sets T_j' and T_j have respective diameters $8j+8$ and $8j+9$, which is smaller than the sets S_x in [5] for $j \geq 3$.

Further Corollary 5 makes it clear that the difference between the size of the restricted sumset and the difference set can be any odd positive integer. We will get any even difference for $|A\hat{+}A| - |A - A|$ in our next construction. This was motivated by the sum-dominant (but not restricted-sum-dominant) set called $A_{13} = \{0, 1, 2, 4, 7, 8, 12, 14, 15, 18, 19, 20\}$ in Hegarty [3]. We exhibit, addressing his remark about the desirability of generalising A_{13}, two infinite sequences of (eventually) restricted-sum dominant sets derived from A_{13} (which shall be our R_1).

Theorem 6. *For each integer $j \geq 1$ define $R_j \subset \mathbb{Z}$ to be the set*

$$\begin{aligned} R_j =& \{1,4\} \cup \{0, 12, \ldots, 12j\} \cup \{2, 14, \ldots, 2+12j\} \\ & \cup \{7, 19, \ldots, 7+12j\} \cup \{8, 20, \ldots, 8+12j\} \cup \{3+12j, 6+12j\}. \end{aligned}$$

For each integer $j \geq 2$ we have

$$\begin{aligned} R_j\hat{+}R_j &= [1, 3+12(2j+1)] \setminus \{\{17, \ldots, 5+12(j-1)\} \cup \{12(2j), 12(2j+1)\}\}, \\ R_j + R_j &= [0, 4+12(2j+1)] \setminus \{17, \ldots, 5+12(j-1)\} \quad \textit{and} \\ R_j - R_j &= [-(8+12j), 8+12j] \setminus \{\pm 9, \ldots, \pm(9+12(j-1))\}. \end{aligned}$$

Proof. We first verify the claim for the restricted sumset. For multiples of 12,

$$\{0, 12, \ldots, 12j\}\hat{+}\{0, 12, \ldots, 12j\} = \{12, 24, \ldots, 12(2j-1)\}.$$

The elements congruent to 1 modulo 12 are given by

$$(1) + \{0, 12, \ldots, 12j\} = \{1, 13, \ldots, 1+12j\}.$$

and

$$(6+12j) + \{7, 19, \ldots, 7+12j\} = \{1+12(j+1), \ldots, 1+12(2j+1)\}.$$

For those congruent to 2 modulo 12

$$\{0, 12, \ldots, 12j\}\hat{+}\{2, 14, \ldots, 2+12j\} = \{2, 14, \ldots, 2+12(2j)\}$$

and also $(6+12j) + (8+12j) = 2 + 12(2j+1) \in R_j\hat{+}R_j$. For 3 modulo 12 clearly $3 = 1 + 2 \in R_j\hat{+}R_j$ and the rest follow from

$$\{7, 19, \ldots, 7+12j\}\hat{+}\{8, 20, \ldots, 8+12j\} = \{15, 27, \ldots, 3+12(2j+1)\}.$$

For elements congruent to 4 modulo 12, we clearly have that 4 and 16 are in $R_j\hat{+}R_j$ as well as

$$\{8,20,\ldots,8+12j\}\hat{+}\{8,20,\ldots,8+12j\}=\{28,40,\ldots,4+12(2j)\}.$$

The elements congruent to 6 modulo 12 in $R_j\hat{+}R_j$ can be obtained as the union of

$$(4)\hat{+}\{2,14,\ldots,2+12j\}=\{6,18,\ldots,6+12j\}$$

and

$$(6+12j)+\{0,12,\ldots,12j\}.$$

The elements congruent to 7 (respectively 8) modulo 12 are obtained from

$$\{0,12,\ldots,12j\}\hat{+}\{7,19,\ldots,7+12j\}=\{7,19,\ldots,7+12(2j)\}.$$

and

$$\{0,12,\ldots,12j\}\hat{+}\{8,20,\ldots,8+12j\}=\{8,20,\ldots,8+12(2j)\}.$$

For 9 (respectively 10) modulo 12 use

$$\{2,14,\ldots,2+12j\}\hat{+}\{7,19,\ldots,7+12j\}=\{9,21,\ldots,9+12(2j)\}$$

respectively

$$\{2,14,\ldots,2+12j\}\hat{+}\{8,20,\ldots,8+12j\}=\{10,22,\ldots,10+12(2j)\}.$$

Finally the elements congruent to 11 modulo 12 are obtained from

$$(4)+\{7,19,\ldots,7+12j\}=\{11,23,\ldots,11+12j\}$$

and

$$(3+12j)+\{8,20,\ldots,8+12j\}=\{11+12j,\ldots,11+12(2j)\}.$$

To see that the restricted sumset does not contain any of $\{17,\ldots,5+12(j-1)\}$, note that none of the sumsets of the progressions with common difference 12 give elements which are congruent to 5 modulo 12 and neither can translates of the progressions by 1 or 4). The remaining elements congruent to 5 modulo 12 are obtained as clearly $5\in R_j\hat{+}R_j$, and also

$$(3+12j)+\{2,14,\ldots,2+12j\}=\{5+12j,\ldots,5+12(2j)\}\subseteq R_j\hat{+}R_j.$$

Finally, to see that $R_j\hat{+}R_j$ does not contain $12(2j)$ or $12(2j+1)$, note that it is impossible to obtain $12(2j)$ as a sum of distinct elements of R_j since the only elements of R_j greater than $12j$ are $S=\{2+12j,3+12j,6+12j,7+12j,8+12j\}$ but none of the numbers in $2(12j)-S$ (namely $10+12(j-1),9+12(j-1)$,

$6+12(j-1), 5+12(j-1), 4+12(j-1))$ are in R_j. Further as $12(j+1) \notin R_j$ $12(2j+1)$ is excluded from $R_j \hat{+} R_j$. This completes the argument for $R_j \hat{+} R_j$.

However, we do have that $12j+12j = 12(2j) \in R_j+R_j$ and $(6+12j)+(6+12j) = 12(2j+1) \in R_j + R_j$, so both these missing elements get into $R_j + R_j$. Since we readily see that none of the numbers congruent to 7 mod 12 ruled out of $R_j \hat{+} R_j$ are in $R_j + R_j$ either, the sumset is as stated.

To confirm the claim for the difference set as before we consider the positive differences. Writing R_j as

$$\{1, 4, 12w, 2+12x, 7+12y, 8+12z, 3+12j, 6+12j\}$$

the remainders which occur in $R_j - R_j$ are exactly the set $[0, 11] \setminus \{9\}$. On the other hand, to see that $R_j - R_j$ contains all the claimed differences, note that as $0 \in R_j$ we have $R_j \subset R_j - R_j$. Also the right hand sides of

$$\begin{aligned}
\{0, 12, \ldots, 12j\} - (1) &= \{-1, 11, \ldots, 11+12(j-1)\} \\
\{2, 14, \ldots, 2+12j\} - (1) &= \{1, 13, \ldots, 1+12j\} \\
\{7, 19, \ldots, 7+12j\} - (4) &= \{3, 15, \ldots, 3+12j\} \\
\{8, 20, \ldots, 8+12j\} - (4) &= \{4, 16, \ldots, 4+12j\} \\
\{7, 19, \ldots, 7+12j\} - (2) &= \{5, 17, \ldots, 5+12j\} \\
\{7, 19, \ldots, 7+12j\} - (1) &= \{6, 18, \ldots, 6+12j\} \\
\{2, 14, \ldots, 2+12j\} - (4) &= \{-2, 10, \ldots, 10+12(j-1)\}.
\end{aligned}$$

are in the difference set which completes the claim. □

Corollary 7. *For every integer $j \geq 2$ the set $R_j \subset \mathbb{Z}$ has*

$$|R_j| = 4j+8,\ |R_j \hat{+} R_j| = 23j+14,\ |R_j+R_j| = 23j+18 \quad \textit{and} \quad |R_j - R_j| = 22j+17.$$

Therefore

$$|R_j \hat{+} R_j| - |R_j - R_j| = j-3, \quad |R_j + R_j| - |R_j - R_j| = j+1$$

and R_j is an restricted-sum-dominant set for every integer $j \geq 4$.

This indeed confirms that any positive integer can be obtained as $|R_j \hat{+} R_j| - |R_j - R_j|$.

Our fourth sequence of sets, the M_js, also has R_1 (Hegarty's A_{13}) as its first member, but this time we focus not on prescribing $|M_j \hat{+} M_j| - |M_j - M_j|$ but instead on getting a reduced diameter $9 + 11j$ rather than the diameter $8 + 12j$ of R_j. (We were first led to this family by considering Marica's sum-dominant set [4] $M = \{1, 2, 3, 5, 8, 9, 13, 15, 16\}$, normalising it and trying to expand it to a restricted-sum-dominant set).

Theorem 8. *For $j \geq 1$ define*

$$\begin{aligned} M_j =& \{0,2\} \cup \{1,12,\ldots,1+11j\} \cup \{4,15,\ldots,4+11j\} \\ & \cup \{7,18,\ldots,7+11j\} \cup \{8,19,\ldots,8+11j\} \cup \{3+11j, 9+11j\} \end{aligned}$$

We then have that

$$\begin{aligned} M_j \hat{+} M_j &= [1, 6+11(2j+1)] \setminus \{3+11(2j+1)\}, \\ M_j + M_j &= [0, 7+11(2j+1)] \text{ and} \\ M_j - M_j &= [-(9+11j), 9+11j] \setminus \{\pm 9, \ldots, \pm(9+11(j-1))\}. \end{aligned}$$

Proof. Firstly we show that $M_j \hat{+} M_j$ consists of

$$\bigcup_{a=1,2,4,5,6} \{a, a+11, \ldots, a+11(2j+1)\}$$

and

$$\bigcup_{a=3,7,8,9,10,11} \{a, a+11, \ldots, a+11(2j)\}$$

and then show that the sumset contains the additional elements claimed. In the case where $a = 1$ we have

$$\{4,15,\ldots,4+11j\} \hat{+} \{8,19,\ldots,8+11j\} = \{12,23,\ldots,12+11(2j) = 1+11(2j+1)\}$$

and $0+1 \in M_j \hat{+} M_j$ also. For the case $a = 2$

$$\{1,12,\ldots,1+11j\} \hat{+} \{1,12,\ldots,1+11j\} = \{13,24,\ldots,2+11(2j-1)\}$$

and $0+2$, $(4+11(j-1))+(9+11j) = 2+11(2j)$, $(4+11j)+(9+11j) = 2+11(2j+1)$ are also in $M_j \hat{+} M_j$.

For the case $a = 4$,

$$\{7,18,\ldots,7+11j\} \hat{+} \{8,19,\ldots,8+11j\} = \{15,26,\ldots,15+11(2j) = 4+11(2j+1)\}$$

and $0+4 \in M_j \hat{+} M_j$.

For the case $a = 5$,

$$\{8,19,\ldots,8+11j\} \hat{+} \{8,19,\ldots,8+11j\} = \{27,\ldots,16+11(2j-1) = 5+11(2j)\}$$

and also $5 = 1+4$, $16 = 12+4$ and $(7+11j)+(9+11j) = 5+11(2j+1)$.

For the case $a = 6$

$$\begin{aligned} (2) + \{4,15,\ldots,4+11j\} &= \{6,17,\ldots,6+11j\} \\ (9+11j) + \{8,19,\ldots,8+11j\} &= \{6+11(j+1),\ldots,6+11(2j+1)\}. \end{aligned}$$

For the case $a = 3$

$$\{7, 18, \ldots, 7 + 11j\}\hat{+}\{7, 18, \ldots, 7 + 11j\} = \{25, 36, \ldots, 3 + 11(2j)\}$$

and $3 = 1 + 2, 14 = 2 + 12$ are in $M_j\hat{+}M_j$.

For the case $a = 7$

$$(0) + \{7, 18, \ldots, 7 + 11j\} = \{7, 18, \ldots, 7 + 11j\}$$
$$(3 + 11j) + \{4, 15, \ldots, 4 + 11j\} = \{7 + 11j, \ldots, 7 + 11(2j)\}.$$

For the case $a = 8$

$$\{1, 12, \ldots, 1 + 11j\}\hat{+}\{7, 18, \ldots, 7 + 11j\} = \{8, 19, \ldots, 8 + 11(2j)\}.$$

For the case $a = 9$

$$\{1, 12, \ldots, 1 + 11j\}\hat{+}\{8, 19, \ldots, 8 + 11j\} = \{9, 20, \ldots, 9 + 11(2j)\}.$$

For $a = 10$

$$(2)\hat{+}\{8, 19, \ldots, 8 + 11j\} = \{10, 21, \ldots, 10 + 11j\}$$
$$(3 + 11j)\hat{+}\{7, 18, \ldots, 7 + 11j\} = \{10 + 11j, \ldots, 10 + 11(2j)\}.$$

For $a = 11$

$$\{4, 15, \ldots, 4 + 11j\}\hat{+}\{7, 18, \ldots, 7 + 11j\} = \{11, 22, \ldots, 11 + 11(2j)\}.$$

To see that $3 + 11(2j + 1) \notin M\hat{+}M$, if it did not we would have a sum of the form $(a + 11j) + (c + 11j) = 14 + 22j$ from elements of M_j with $a + c = 14$, however, since a and c are distinct elements of $\{1, 3, 4, 7, 8, 9\}$ this is impossible and hence $3 + 11(2j + 1) \notin M_j\hat{+}M_j$. This confirms the claim for the restricted sumset. Furthermore for each $m \in M_j$ the sumset contains $0, 2(7 + 11j) = 3 + 11(2j + 1)$ and $2(9 + 11j) = 7 + 11(2j + 1)$ which completes the claim for the sumset.

For the difference set to see that $\{\pm 9, \ldots, \pm(9 + 11(j - 1))\} \notin M_j - M_j$ let

$$M_j = \{0, 2, 1 + 11w, 4 + 11x, 7 + 11y, 8 + 11z, 3 + 11j, 9 + 11j\},$$

where $0 \leq w, x, y, z \leq j$. It suffices to consider just the positive differences. Calculation of $M_j - M_j$ reveals that the only positive difference congruent to 9 modulo 11 is $(9 + 11j) - 0$, which is outside the range claimed.

To see that $M_j - M_j$ contains the remaining elements in the interval, firstly note that as $0 \in M_j$ we have $M_j - M_j \supset M_j$. Furthermore $M_j - M_j$ also contains the right-hand sides of the following:

$$\begin{aligned}
\{1, 12, \ldots, 1+11j\} - (1) &= \{0, 11, \ldots, 11j\} \\
\{4, 15, \ldots, 4+11j\} - (1) &= \{3, 14, \ldots, 3+11j\} \\
\{7, 18, \ldots, 7+11j\} - (1) &= \{6, 17, \ldots, 6+11j\} \\
\{1, 12, \ldots, 1+11j\} - (2) &= \{-1, 10, 21, \ldots, 10+11(j-1)\} \\
\{4, 15, \ldots, 4+11j\} - (2) &= \{2, 13, \ldots, 2+11j\} \\
\{7, 18, \ldots, 7+11j\} - (2) &= \{5, 16, \ldots, 5+11j\} \\
9+11j-0 &= 9+11j.
\end{aligned}$$

This completes the claim of the theorem. □

Corollary 9. *For every integer $j \geq 1$ the set $M_j \subset \mathbb{Z}$ has*

$$|M_j| = 4j+8,\ |M_j \hat{+} M_j| = 22j+16,\ |M_j+M_j| = 22j+19 \quad \textit{and} \quad |M_j-M_j| = 20j+19.$$

Hence

$$|M_j \hat{+} M_j| - |M_j - M_j| = 2j-3, \quad |M_j + M_j| - |M_j - M_j| = 2j$$

and M_j is an restricted-sum-dominant set for every $j \geq 2$.

Note that the set M_2 has slightly smaller diameter 31 than the other 16-element restricted-sum-dominant set T_3'.

Martin and O'Bryant refer to sets with $|A+A| = |A-A|$ as *sum-difference balanced.* Similarly we can consider sets with $|A \hat{+} A| = |A-A|$ as *restricted-sum-difference balanced.* The results above show such sets exist (e.g., R_3). The smallest such set we have found has order 14: it is is

$$M' = \{0, 1, 2, 4, 7, 8, 12, 14, 15, 19, 22, 25, 26, 27\},$$

so $|M' \hat{+} M'| = |[1, 53] \setminus \{43, 50\}| = 51$ and $|M' - M'| = |[-27, 27] \setminus \{\pm 9, \pm 16\}| = 51$. We show that by taking the union of translates of M' by non-negative integer multiples of its maximum element one can obtain arbitrarily large restricted-sum-difference balanced sets.

Lemma 10. *Let $k \geq 2$ and $A_0 = A = \{0 = a_1 < a_2 < \cdots < a_k = m\} \subset \mathbb{Z}$ and $A_i = A \cup (A+m) \cup \cdots \cup (A+im)$. Then*

$$\begin{aligned}
|A_i \hat{+} A_i| - |A_{i-1} \hat{+} A_{i-1}| &= c_1 \quad \forall\, i \geq 2, \\
|A_i + A_i| - |A_{i-1} + A_{i-1}| &= c_1 \quad \forall\, i \geq 1
\end{aligned}$$

and

$$|A_i - A_i| - |A_{i-1} - A_{i-1}| = c_2 \quad \forall\, i \geq 1.$$

where c_1 and c_2 are positive constants.

Proof. We first note

$$|A_i \hat{+} A_i| - |A_{i-1} \hat{+} A_{i-1}| = |(A_i \hat{+} A_i) \setminus (A_{i-1} \hat{+} A_{i-1})|$$

and show that the right-hand side is a constant by showing that the set of new elements introduced on each iteration is a translate of the set of new elements introduced on the previous iteration. We have

$$A_i \hat{+} A_i = \cup_{r,s=0}^{i} ((A + rm) \hat{+} (A + sm)).$$

If $|r - s| \geq 2$, it is clear that $A + rm$ and $A + sm$ are disjoint so their restricted sum is just their sum. If $i - 1 \geq r = s \geq 1$, then $(A + rm) \hat{+} (A + rm) = (A + (r-1)m) + (A + (r+1)m)$. The only case needing a little thought is $|r - s| = 1$: without loss of generality, $r = s + 1$. Then

$$(A + (s+1)m) \hat{+} (A + sm) = \{a + b + (2s+1)m : \ a + m \neq b\}$$

the only way we can have $a + m = b$ is if $a = 0, b = m$, but in this case

$$(0 + (s+1)m) + (m + sm) = (m + (s+1)m) \hat{+} (0 + sm)$$

We deduce that, for all $i \geq 2$

$$A_i \hat{+} A_i = (A \hat{+} A) \cup (A + (A + m)) \cup \cdots \cup (A + A + (2i-1)m) \cup (A \hat{+} A + 2im).$$

Similarly

$$A_{i-1} \hat{+} A_{i-1} = (A \hat{+} A) \cup (A + A + m) \cup \cdots \cup (A \hat{+} A + (2i-2)m).$$

Now some elements of $(A + A + (2i-2)m) \setminus (A \hat{+} A + (2i-2)m)$ may be in $A + A + (2i-3)m$ and thus in $A_{i-1} \hat{+} A_{i-1}$. (Translates of $A + A$ by less than $(2i-3)m$ need not be considered). We have

$$\begin{aligned}(A_i \hat{+} A_i) \setminus (A_{i-1} \hat{+} A_{i-1}) = ((A + A + (2i-2)m) \cup (A + A + (2i-1)m) \cup \\ (A \hat{+} A + 2im)) \setminus ((A + A + (2i-3)m) \cup (A \hat{+} A + (2i-2)m)).\end{aligned} \qquad (1)$$

Likewise

$$\begin{aligned}(A_{i+1} \hat{+} A_{i+1}) \setminus (A_i \hat{+} A_i) = ((A + A + 2im) \cup (A + A + (2i+1)m) \cup \\ (A \hat{+} A + (2i+2)m)) \setminus ((A + A + (2i-1)m) \cup (A \hat{+} A + (2i)m)).\end{aligned} \qquad (2)$$

The right-hand side of (2) is a translation of the right-hand side of (1) by $2m$. (To see this, note it is easy to check for sets of integers that if $C_i + 2m = C_{i+1}$ and $D_i + 2m = D_{i+1}$, then $(C_i \setminus D_i) + 2m = (C_{i+1} \setminus D_{i+1})$: apply this with the obvious choices of C_i and D_i). Thus

$$(A_{i+1} \hat{+} A_{i+1}) \setminus (A_i \hat{+} A_i) = ((A_i \hat{+} A_i) \setminus (A_{i-1} \hat{+} A_{i-1})) + 2m.$$

Since translation by a constant leaves the cardinality of the set difference unaltered it follows that

$$|(A_{i+1}\hat{+}A_{i+1}) \setminus (A_i\hat{+}A_i)| = |(A_i\hat{+}A_i) \setminus (A_{i-1}\hat{+}A_{i-1})|$$

as required.

To see that

$$|A_i + A_i| - |A_{i-1} + A_{i-1}| = |A_i\hat{+}A_i| - |A_{i-1}\hat{+}A_{i-1}| \tag{3}$$

for all $i \geq 1$ we show that the number of additional elements $A_i + A_i$ contains is constant. All the elements of

$$(A + A) \setminus (A\hat{+}A)$$

except for $2m$, which is in $A_i\hat{+}A_i$ for $i \geq 1$ due to $0 + 2m$, are excluded from $A_i\hat{+}A_i$ for all $i \geq 1$. Similarly the elements of

$$((A + A) \setminus (A\hat{+}A)) + 2im$$

except for $2im$ are excluded from $A_i\hat{+}A_i$. This means that for all $i \geq 1$

$$|A_i + A_i| - |A_i\hat{+}A_i| = 2(|(A + A) \setminus (A\hat{+}A)| - 1).$$

In other words the difference between the cardinalities of the sumset and the restricted sumset is a constant for all $i \geq 1$ and (3) holds.

To verify the claim for the difference set, write

$$A_i - A_i = \cup_{j=-i}^{i}(A - A + jm).$$

Thus we have

$$(A_i - A_i) \setminus (A_{i-1} - A_{i-1}) = (A - A - im) \cup (A - A + im) \setminus \bigcup_{j=-(i-1)}^{i-1} (A - A - jm).$$

But the only sets in $\cup_{j=-(i-1)}^{i-1}(A - A - jm)$ which could intersect $(A - A - im)$ or $(A - A + im)$ are for $j = (i-1)$, $j = (i-2)$ (which will intersect $A - A - im$ in precisely the one element $(1-i)m$), $j = -(i-2)$ (which will intersect it in precisely the one element $(i-1)m$) and $j = -(i-1)$. Thus for all $i \geq 1$

$$\begin{aligned}(A_i - A_i) \setminus (A_{i-1} - A_{i-1}) =&((A - (A + im)) \setminus (A - (A + (i-1)m))) \\ &\cup ((A - A + im) \setminus (A - A + (i-1)m)).\end{aligned}$$

Similarly

$$\begin{aligned}(A_{i+1} - A_{i+1}) \setminus (A_i - A_i) =&((A - (A + (i+1)m)) \setminus (A - (A + im))) \\ &\cup ((A - A + (i+1)m) \setminus (A - A + im)).\end{aligned}$$

The sets $(A-(A+(i+1)m))\setminus(A-(A+im))$ and $(A-A+(i+1)m)\setminus(A-A+im)$ are disjoint for all $i\geq 1$. Also $(A-(A+(i+1)m))\setminus(A-(A+im))$ is a translation of $(A-(A+im))\setminus(A-(A+(i-1)m))$ by $-m$ and $(A-A+(i+1)m)\setminus(A-A+im)$ is a translation of $(A-A+im)\setminus(A-A+(i-1)m)$ by m. These translations leave the cardinalities of the sets unchanged, therefore

$$|(A_{i+1}-A_{i+1})\setminus(A_i-A_i)|=|(A_i-A_i)\setminus(A_{i-1}-A_{i-1})|$$

and the overall result follows. □

Setting $M_1'=M'\cup(M'+27)$ we easily check

$$|M_1'\hat{+}M_1'|=|[1,107]\setminus\{97,104\}|=|[-54,54]\setminus\{\pm 36,\pm 43\}|=|M_1'-M_1'|$$

and $M_2'=M'\cup(M'+27)\cup(M'+54)$ gives

$$|M_2'\hat{+}M_2'|=|[1,161]\setminus\{151,158\}|=|[-81,81]\setminus\{\pm 63,\pm 70\}|=|M_2'-M_2'|.$$

It follows from Lemma 10 that

Corollary 11. *There exist arbitrarily large restricted-sum-difference balanced subsets of* $\mathbb{Z}$.

Our final sequence of restricted-sum-dominant sets is constructed with a view to obtaining high values of $f(A)$ as defined in the introduction. Again, this set is a modification of one in [9], which describes $Q_j\setminus\{1+4(4j+7)\}$ for $j=1,2,3$ as sets giving large sumset relative to the difference set. Including $1+4(4j+7)$ increases the sumset but does not change the difference set.

Theorem 12. *Let*

$$\begin{aligned}Q_j =&\{0,2,4,12\}\cup\{1,5,\ldots,1+4(4j+8)\}\cup\{24,40,\ldots,8+16j\}\\ &\cup\{4+16(j+1),12+16(j+1),14+16(j+1),16(j+2)\}\end{aligned}$$

for an integer $j\geq 1$. *Then*

$$\begin{aligned}Q_j\hat{+}Q_j =&[1,1+4(8j+16)]\\ &\setminus\{8,20,32,48,4(8j+4),4(8j+8),4(8j+11),4(8j+14),4(8j+16)\}\end{aligned}$$

for $j\geq 2$, *whilst*

$$Q_j+Q_j=[0,2+4(8j+16)]\setminus\{20,32,4(8j+8),4(8j+11)\}$$

for $j\geq 1$ *and*

$$\begin{aligned}Q_j-Q_j =&[-(1+4(4j+8)),1+4(4j+8)]\setminus\pm\{\{6\},\{14,\ldots,14+16j\},\\ &\{18,\ldots,2+16j\},\{26,\ldots,10+16j\},6+16(j+1)\}\end{aligned}$$

for $j\geq 1$.

Proof. To verify these claims, consider elements of Q_j in terms of the union of

$$Q_{\text{odd}} = \{1, 5, \ldots, 1 + 4(4j + 8)\}$$

and

$$\begin{aligned} Q_{\text{even}} =& \{0, 2, 4, 12\} \cup \{24, \ldots, 8 + 16j\} \\ & \cup \{4 + 16(j + 1), 12 + 16(j + 1), 14 + 16(j + 1), 16(j + 2)\}. \end{aligned}$$

Firstly $Q_j \hat{+} Q_j$ contains all the odd numbers in the interval since we have

$$\begin{aligned} (0)\hat{+}\{1, 5, \ldots, 1 + 4(4j + 8)\} =& \{1, 5, \ldots, 1 + 4(4j + 8)\} \\ 16(j + 2)\hat{+}\{1, 5, \ldots, 1 + 4(4j + 8)\} =& \{1 + 4(4j + 8), 5 + 4(4j + 8), \\ & \ldots, 1 + 4(8j + 16)\} \\ (2)\hat{+}\{1, 5, \ldots, 1 + 4(4j + 8)\} =& \{3, 7, \ldots, 3 + 4(4j + 8)\} \\ 14 + 16(j + 1)\hat{+}\{1, 5, \ldots, 1 + 4(4j + 8)\} =& \{3 + 4(4j + 7), 7 + 4(4j + 7), \\ & \ldots, 3 + 4(8j + 15)\}. \end{aligned}$$

The union of the right hand sides of the above is indeed

$$\{1, 3, \ldots, 3 + 4(8j + 15), 1 + 4(8j + 16)\} = \{1, 3, \ldots, 1 + 2(4(4j + 8))\}.$$

To see that the sumset contains all the even elements claimed, note first that $Q_{\text{odd}} \hat{+} Q_{\text{odd}}$ gives the following elements congruent to 2 mod 4:

$$Q_{\text{odd}} \hat{+} Q_{\text{odd}} = \{6, 10, \ldots, 2 + 4(8j + 15)\} \subseteq Q_j \hat{+} Q_j.$$

Clearly $0 + 2$ is also in $Q_j \hat{+} Q_j$, however whilst $\max(Q_j + Q_j) = 2 + 4(8j + 16)$ this is not in the restricted sumset. As regards the multiples of four, clearly none of these can be obtained from $Q_{\text{odd}} \hat{+} Q_{\text{odd}}$ or $Q_{\text{odd}} \hat{+} Q_{\text{even}}$. To confirm the elements we claim to be excluded cannot be present note that Q_{even} is symmetric w.r.t. 16(j+2): $Q_{even} = 16(j + 2) - Q_{even}$. Hence $Q_{even} \hat{+} Q_{even} = 16(2j + 4) - (Q_{even} \hat{+} Q_{even})$ and $Q_{even} + Q_{even} = 16(2j + 4) - (Q_{even} + Q_{even})$. The restricted sumset of the elements of Q_{even} less than or equal to 32 is

$$\{0, 2, 4, 12, 24\} \hat{+} \{0, 2, 4, 12, 24\} = \{2, 4, 6, 12, 14, 16, 24, 26, 28, 36\}.$$

Thus $0, 8, 20, 32$ and 48 are excluded from $Q_j \hat{+} Q_j$. Whilst $Q_j + Q_j$ contains $0, 8$ and 48 as the doubles of $0, 4$ and 24 respectively, it is easy to check that neither 20 nor 32 are in $Q_j + Q_j$. By symmetry

$$16(2j+4) - \{0, 8, 20, 32, 48\} = \{4(8j+4), 4(8j+8), 4(8j+11), 4(8j+14), 4(8j+16)\}$$

which has empty intersection with $Q_j \hat{+} Q_j$.

It remains to show that all other (relevant) multiples of 4 are in the (restricted) sumset; we consider the cases 0,4,8 and 12 modulo 16 separately. We have the following multiples of 16 in $Q_j \hat{+} Q_j$:

$$\{24, 40, \ldots, 16j + 8\} \hat{+} \{24, 40, \ldots, 16j + 8\} = \{64, 80, \ldots, 16(2j)\}$$
$$(4 + 16(j + 1)) \hat{+} (12 + 16(j + 1)) = 4(8j + 12) = 16(2j + 3).$$

Furthermore $Q_j + Q_j$ contains 48 and $16(2j + 1) = 2(16j + 8)$ and also $16(j + 2) + 16(j + 2) = 4(8j + 16) = 16(2j + 4)$. We already saw $16(2j + 2) = 4(8j + 8)$ is not in $Q_j + Q_j$.

We obtain those congruent to 4 modulo 16 from

$$(12) \hat{+} \{24, 40, \ldots, 16j + 8\} = \{36, 52, \ldots, 4 + 16(j + 1)\}$$
$$(4) \hat{+} (16(j + 2)) = 4 + 16(j + 2)$$
$$(12 + 16(j + 1)) \hat{+} \{24, \ldots, 8 + 16j\} = \{4 + 16(j + 3), \ldots, 4 + 16(2j + 2)\}$$
$$(4 + 16(j + 1)) \hat{+} (16(j + 2)) = 4 + 16(2j + 3).$$

The elements congruent to 8 modulo 16 are given by

$$(0) \hat{+} \{24, 40, \ldots, 8 + 16j\} = \{24, 40, \ldots, 8 + 16j\}$$
$$(4) \hat{+} (4 + 16(j + 1)) = 8 + 16(j + 1)$$
$$(12) \hat{+} (12 + 16(j + 1)) = 8 + 16(j + 2)$$
$$(16(j + 2)) \hat{+} \{24, 40, \ldots, 8 + 16j\} = \{8 + 16(j + 3), \ldots, 8 + 16(2j + 2)\}.$$

Also $(12 + 16(j + 1)) + (12 + 16(j + 1)) = 8 + 16(2j + 3) \in Q_j + Q_j$. Finally the elements congruent to 12 modulo 16 follow from

$$(4) \hat{+} \{24, \ldots, 8 + 16j\} = \{28, \ldots, 12 + 16j\}$$
$$(0) \hat{+} (12 + 16(j + 1)) = 12 + 16(j + 1)$$
$$(4 + 16(j + 1)) \hat{+} \{24, \ldots, 8 + 16j\} = \{12 + 16(j + 2), \ldots, 12 + 16(2j + 1)\}$$
$$(12 + 16(j + 1)) \hat{+} (16(j + 2)) = 12 + 16(2j + 3).$$

We now deal with the difference set. Again, it suffices to consider the non-negative differences. Since all the differences which we claim are excluded are even we need only consider differences of pairs of elements of Q_j of the same parity and therefore divide into cases accordingly. The non-negative elements of $Q_{\text{odd}} - Q_{\text{odd}}$ are

$$\{0, 4, \ldots, 4(4j + 8)\}.$$

The even elements of Q_j have the form

$$Q_{\text{even}} = \{0, 2, 4, 12, 8 + 16x, 4 + 16(j + 1), 12 + 16(j + 1), 14 + 16(j + 1), 16(j + 2)\}$$

where $x \in \mathbb{Z}$ with $1 \leq x \leq j$. The positive differences of the elements of Q_{even} are

$$\{2, 4, 8, 10, 12, 12 + 16(x-1), 4 + 16x, 6 + 16x, 8 + 16x,\\ 12 + 16(j-x), 4 + 16(j-x+1), 6 + 16(j-x+1), 8 + 16(j-x+1),\\ 8 + 16j, 16(j+1), 2 + 16(j+1), 4 + 16(j+1), 8 + 16(j+1),\\ 10 + 16(j+1), 12 + 16(j+1), 14 + 16(j+1), 16(j+2)\}.$$

Thus none of the differences in $Q_j - Q_j$ have the form which we claim is excluded. To confirm the presence of the remaining differences we have that all the differences congruent to 1 modulo 4 are present since

$$\{1, 5, \ldots, 1 + 4(4j+8)\} - \{0\} = \{1, 5, \ldots, 1 + 4(4j+8)\} \subseteq Q_j - Q_j.$$

The elements congruent to 3 modulo 4 follow from

$$\{1, 5, \ldots, 1 + 4(4j+8)\} - \{2\} = \{-1, 3, \ldots, 3 + 4(4j+7)\} \subseteq Q_j - Q_j.$$

The multiples of 4 are obtained from

$$\{1, 5, \ldots, 1 + 4(4j+8)\} - \{1\} = \{0, 4, \ldots, 4(4j+8)\}.$$

For elements congruent to 2 mod 4, the only elements congruent to 2 mod 16 we are claiming to get are 2 and $2 + 16(j+1)$; 2 is clearly in, and $2 + 16(j+1) = 14 + 16(j+1) - 12$.

The elements congruent to 6 modulo 16 can be obtained from

$$\{24, 40, \ldots, 8 + 16j\} - \{2\} = \{22, 38, \ldots, 6 + 16j\}.$$

The only elements congruent to 10 mod 16 we are claiming are $10 + 16(j+1) = 12 + 16(j+1) - 2$ and $10 = 12 - 2$. Finally, the only element congruent to 14 mod 16 we claim is present is $14 + 16(j+1) \in Q_j$. □

Corollary 13. *For the set Q_j defined above we have*

$$|Q_j| = 5j + 17, |Q_j \hat{+} Q_j| = 32j + 56 \text{ for } j \geq 2, |Q_j + Q_j| = 32j + 63 \text{ for } j \geq 1,\\ |Q_j - Q_j| = 26j + 61 \text{ for } j \geq 1$$

(and $|Q_1 \hat{+} Q_1| = 90$). Thus Q_j is an restricted-sum-dominant set for all $j \geq 1$.

3. The Proportion of Restricted-Sum-Dominant Sets Is Strictly Positive

Martin and O'Bryant prove that for $n \geq 15$ the number of sum-dominant subsets of $[0, n-1]$ is at least $(2 \times 10^{-7})2^n$ (see Theorem 1 of [5]). Their result has been improved by Zhao [11] who shows that the proportion of sum-dominant sets tends to a limit and that that limit is at least 4.28×10^{-4}. In this section we will show that the proportion of subsets of $\{0,1,2,\dots n-1\}$ which are restricted-sum-dominant is bounded below by a much weaker constant. It may well be that Zhao's techniques, or others, can be modified to improve the result but at least a substantial piece of computation would appear to be required and our concern at present is simply to show that a positive proportion of sets are restricted-sum-dominant sets. Note that the fact that a positive proportion of sets have more differences than restricted sums is an immediate consequence of Theorem 14 in [5]. Many lemmas etc. in what follows are very slight modifications of corresponding results in [5] and we merely present these proofs without further comment. However the construction of the two 'fringe sets' U and L is notably more involved.

Lemma 14. *Let n,ℓ and u be integers such that $n\geq \ell+u$. Fix $L\subseteq[0,\ell-1]$ and $U\subseteq[n-u,n-1]$. Suppose R is a uniformly randomly selected subset of $[\ell,n-u-1]$ (where each element is chosen with probability 1/2) and set $A=L\cup R\cup U$. Then for every integer k satisfying $2\ell-1\leq k\leq n-u-1$, we have*

$$\mathbb{P}(k\notin A\hat{+}A)=\begin{cases}\left(\frac{1}{2}\right)^{|L|}\left(\frac{3}{4}\right)^{(k+1)/2-\ell}, & \text{if } k \text{ is odd,}\\ \left(\frac{1}{2}\right)^{|L|}\left(\frac{3}{4}\right)^{k/2-\ell}, & \text{if } k \text{ is even.}\end{cases}$$

Proof. Define an indicator variable

$$X_j=\begin{cases}1, & \text{if } j\in A,\\ 0, & \text{otherwise.}\end{cases}$$

Since $A=L\cup R\cup U$ the X_j are independent random variables for $\ell\leq j\leq n-u-1$, each taking values 0 or 1 equiprobably. For $0\leq j\leq \ell-1$ and $n-u\leq j\leq n-1$ the values of X_j are dictated by the choices of L and U.

Now, $k\notin A\hat{+}A$ if and only if $X_jX_{k-j}=0$ for all $0\leq j\leq k/2-1$. ($j=k/2$ would not give a *restricted* sum). The random variables X_jX_{k-j} for $0\leq j\leq k/2$ are independent of each other. Hence

$$\mathbb{P}(k\notin A\hat{+}A)=\Pi_{0\leq j\leq k/2-1}\mathbb{P}(X_jX_{k-j}=0).$$

When k is odd we have

$$\begin{aligned}\mathbb{P}(k\notin A\hat{+}A)&=\prod_{j=0}^{\ell-1}\mathbb{P}(X_jX_{k-j}=0)\prod_{j=\ell}^{(k-1)/2}\mathbb{P}(X_jX_{k-j}=0)\\&=\prod_{j\in L}\mathbb{P}(X_{k-j}=0)\prod_{j=\ell}^{(k-1)/2}\mathbb{P}(X_j=0\text{ or }X_{k-j}=0)\\&=\left(\frac{1}{2}\right)^{|L|}\left(\frac{3}{4}\right)^{(k+1)/2-\ell}.\end{aligned}$$

When k is even

$$\mathbb{P}(k \notin A\hat{+}A) = \prod_{j=0}^{\ell-1} \mathbb{P}(X_j X_{k-j} = 0) \prod_{j=\ell}^{k/2-1} \mathbb{P}(X_j X_{k-j} = 0)$$
$$= \prod_{j\in L} \mathbb{P}(X_{k-j} = 0) \prod_{j=\ell}^{k/2-1} \mathbb{P}(X_j = 0 \text{ or } X_{k-j} = 0) = \left(\frac{1}{2}\right)^{|L|} \left(\frac{3}{4}\right)^{k/2-\ell}.$$

□

Lemma 15. *Let n, ℓ, u, L, U, R and A be defined as in Lemma 14. Then for every integer k satisfying $n + \ell - 1 \le k \le 2n - 2u - 1$, we have*

$$\mathbb{P}(k \notin A\hat{+}A) = \begin{cases} \left(\frac{1}{2}\right)^{|U|} \left(\frac{3}{4}\right)^{n-(k+1)/2-u}, & \text{if } k \text{ is odd,} \\ \left(\frac{1}{2}\right)^{|U|} \left(\frac{3}{4}\right)^{n-1-k/2-u}, & \text{if } k \text{ is even.} \end{cases}$$

Proof. This is similar to the previous lemma, but we consider different intervals for the summands. For k odd, we have

$$\mathbb{P}(k \notin A\hat{+}A) = \prod_{j=(k+1)/2}^{n-u-1} \mathbb{P}(X_j X_{k-j} = 0) \prod_{j=n-u}^{n-1} \mathbb{P}(X_j X_{k-j} = 0)$$
$$= \prod_{j=(k+1)/2}^{n-u-1} \mathbb{P}(X_j = 0 \text{ or } X_{k-j} = 0) \prod_{j\in U} \mathbb{P}(X_{k-j} = 0)$$
$$= \left(\frac{3}{4}\right)^{n-(k+1)/2-u} \left(\frac{1}{2}\right)^{|U|}.$$

For k even, as $k = k/2 + k/2$ is forbidden,

$$\mathbb{P}(k \notin A\hat{+}A) = \prod_{j=k/2+1}^{n-u-1} \mathbb{P}(X_j X_{k-j} = 0) \prod_{j=n-u}^{n-1} \mathbb{P}(X_j X_{k-j} = 0)$$
$$= \prod_{j=k/2+1}^{n-u-1} \mathbb{P}(X_j = 0 \text{ or } X_{k-j} = 0) \prod_{j\in U} \mathbb{P}(X_{k-j} = 0)$$
$$= \left(\frac{3}{4}\right)^{n-1-k/2-u} \left(\frac{1}{2}\right)^{|U|}.$$

□

Proposition 16. *Let n, ℓ and u be integers such that $n \ge \ell + u$. Fix $L \subseteq [0, \ell - 1]$ and $U \subseteq [n - u, n - 1]$. Suppose R is a uniformly randomly selected subset of $[\ell, n - u - 1]$ (where each element is chosen, independently of all other elements,* *with probability 1/2) and set* $A = L \cup R \cup U$. *Then for every integer* k *satisfying* $2\ell - 1 \leq n - u - 1$,

$$\mathbb{P}([2\ell-1, n-u-1] \cup [n+\ell-1, 2n-2u-1] \subseteq A\hat{+}A) > 1 - 8(2^{-|L|} + 2^{-|U|}).$$

Proof. We crudely estimate

$$\begin{aligned}&\mathbb{P}([2\ell-1, n-u-1] \cup [n+\ell-1, 2n-2u-1] \not\subseteq A\hat{+}A)\\ &\leq \sum_{k=2\ell-1}^{n-u-1} \mathbb{P}(k \notin A\hat{+}A) + \sum_{k=n+\ell-1}^{2n-2u-1} \mathbb{P}(k \notin A\hat{+}A).\end{aligned}$$

The left summation of the line above can be bounded using Lemma 14:

$$\begin{aligned}\sum_{k=2\ell-1}^{n-u-1} \mathbb{P}(k \notin A\hat{+}A) &< \sum_{\substack{k\geq 2\ell-1\\ k \text{ odd}}} \left(\frac{1}{2}\right)^{|L|} \left(\frac{3}{4}\right)^{(k+1)/2-\ell} + \sum_{\substack{k\geq 2\ell-1\\ k \text{ even}}} \left(\frac{1}{2}\right)^{|L|} \left(\frac{3}{4}\right)^{k/2-\ell}\\ &= \left(\frac{1}{2}\right)^{|L|} \sum_{m=0}^{\infty} \left(\frac{3}{4}\right)^{m} + \left(\frac{1}{2}\right)^{|L|} \sum_{m=0}^{\infty} \left(\frac{3}{4}\right)^{m} = 8\left(\frac{1}{2}\right)^{|L|}.\end{aligned}$$

The summation on the right can be bounded similarly, using Lemma 15, to give

$$\sum_{k=n+\ell-1}^{2n-2u-1} \mathbb{P}(k \notin A\hat{+}A) < 8\left(\frac{1}{2}\right)^{|U|}.$$

Thus $\mathbb{P}([2\ell, n-u-1] \cup [n+\ell-1, 2n-2u-1] \subseteq A\hat{+}A)$ is bounded above by $8((1/2)^{|L|} + (1/2)^{|U|})$, which is equivalent to the claim of Proposition 16. □

We now come to the main result. Whilst the respective lower and upper fringes $U = \{0, 2, 3, 7, 8, 9, 10\}$ and $L = \{n-11, n-10, n-9, n-8, n-6, n-3, n-2, n-1\}$ used by Martin and O'Bryant are sufficient for the sum-dominant case these fall some way short of what is required for a restricted-sum-dominant result. However we can again use Spohn's idea of repeating interior blocks. After a few iterations we get the new fringes, which we shall henceforth refer to as L and U, to fit with the earlier lemmas. Thus from now on

$$\begin{aligned}L &= \{0, 2, 3, 7, 9, 10, 14, 16, 17, 21, 23, 24, 28, 30, 31, 35,\\ &\qquad 37, 38, 42, 44, 45, 49, 51, 52, 56, 57, 58, 59, 60\},\\ U &= n - \{59, 58, 57, 55, 52, 51, 50, 48, 45, 44, 43, 41, 38, 37, 36, 34, 31,\\ &\qquad 30, 29, 27, 24, 23, 22, 20, 17, 16, 15, 13, 10, 9, 8, 6, 3, 2, 1\}.\end{aligned}$$

Theorem 17. *For* $n \geq 120$, *the number of restricted-sum-dominant subsets of* $[0, n-1]$ *is at least* $(7.52 \times 10^{-37})2^n$.

Proof. With L and U as just defined, one can check that

$$U - L = [n-119, n-1] \setminus \{n-7, n-14, n-21, n-28, \\ n-35, n-42, n-49, n-56\}.$$

Now since $n-7, n-14, n-21, n-28, n-35, n-42, n-49, n-56 \notin U-L$ it follows that $\pm(n-7), \pm(n-14), \pm(n-21), \pm(n-28), \pm(n-35), \pm(n-42), \pm(n-49), \pm(n-56) \notin A - A \subseteq [-(n-1), n-1]$. With eight pairs of differences excluded from $A - A$ we have $|A - A| \leq 2n - 17$. On the other hand one can check

$$\begin{aligned} L\hat{+}L &= [0, 120] \setminus \{0, 1, 4, 6, 8, 15, 22, 29, 36, 43, 50, 120\} \\ U\hat{+}L = U + L &= [n-59, n+59] \\ U\hat{+}U &= [2n-118, 2n-2] \setminus \{2n-118, 2n-6, 2n-2\}. \end{aligned}$$

Hence for $120 \leq n \leq 178$ we have that $A\hat{+}A$ contains

$$[0, 2n-2] \setminus \{0, 1, 4, 6, 8, 15, 22, 29, 36, 43, 50, 120, 2n-118, 2n-6, 2n-2\}$$

so that $|A\hat{+}A| \geq 2n - 16$. There are $n - 120$ numbers between 61 and $n - 60$ inclusive. Therefore the number of such A is 2^{n-120}.

For $n \geq 178$ applying Proposition 16 with $\ell = 61$ and $u = 59$ implies that when A is chosen uniformly randomly from all such sets, the probability that $A\hat{+}A$ contains $[61, n-60] \cup [n+60, 2n-119]$ is at least

$$1 - 8(2^{-|L|} + 2^{-|U|}) = 1 - 8(2^{-29} + 2^{-35}) = \frac{4294967231}{4294967296}.$$

That is, there are at least $2^{n-120}\frac{4294967231}{4294967296} > (7.52 \times 10^{-37})2^n$ such sets A with

$$A\hat{+}A = [0, 2n-2] \setminus \{0, 1, 4, 6, 8, 15, 22, 29, 36, 43, 50, 120, 2n-118, 2n-6, 2n-2\},$$

whilst at the same time eight pairs of differences are excluded from $A - A$. Thus all such sets A are restricted-sum-dominant sets. □

Martin and O'Bryant's Lemma 7 and Theorem 16 for a subset S of an arithmetic progression of length n can also be adapted to give the following result.

Theorem 18. *Given a subset S of an arithmetic progression P of length n for every positive integer n, we have*

$$\sum_{S \subseteq P} |S\hat{+}S| = 2^n(2n-15) + \begin{cases} 26 \cdot 3^{(n-1)/2}, & \text{if } n \text{ is odd,} \\ 15 \cdot 3^{n/2}, & \text{if } n \text{ is even.} \end{cases} \tag{4}$$

Thus $\frac{1}{2^n}\sum_{S\subseteq P}|S\hat{+}S| \sim 2n - 15$. This combined with Martin and O'Bryant's Theorem 3, that $\frac{1}{2^n}\sum_{S\subseteq P}|S - S| \sim 2n - 7$ gives that on average the difference set has eight elements more than the restricted sumset. Details will appear in [10].

4. How Much Larger Can the Sumset Be?

As in Section 4 of [3] we consider this question in terms of $f(A) = \ln|A+A|/\ln|A-A|$ (and the analogous quantity $\hat{f}(A) = \ln|A\hat{+}A|/\ln|A-A|$). It is known – see, e.g., [1] – that $\frac{3}{4} \leq f(A) \leq \frac{4}{3}$. The reason for considering the ratio of logarithms rather than (say) the ratio is explained in [3] in terms of the *base expansion method.* Some authors, e.g., Granville in [2], prefer to use $g(A) = \ln(|A+A|/|A|)/\ln(|A-A|/|A|)$ for which the analogous bounds are $1/2 \leq g(A) \leq 2$.

Hegarty's set A_{15} is easily checked to have $f(A_{15}) = 1.0208\ldots$, which is often quoted as the largest known value of $f(A)$. In fact, the set X (our T_2) which Hegarty uses to write $A_{15} = X \cup (X+20)$ already does fractionally better:

Lemma 19. *Let* $X = \{0,1,2,4,5,9,12,13,17,20,21,22,24,25\}$*. Then* $X+X = [0,50]$ *but* $X-X = [-25,25]\backslash\{\pm 6, \pm 14\}$*. Thus* $f(X) = \ln(51)/\ln(47) \simeq 1.0212$.

Proof. This is just a short calculation. □

We do better than either of these using the sets Q_j at the end of Section 2.

Theorem 20. *There is a set* A *of integers for which*

$$f(A) = \frac{\ln(|A+A|)}{\ln(|A-A|)} \simeq 1.030597781\ldots$$

and another set B *of integers for which*

$$\hat{f}(B) = \frac{\ln(|B\hat{+}B|)}{\ln(|B-B|)} \simeq 1.028377107\ldots$$

Proof. Take $A = Q_{10}$ for the first claim and $A = Q_{19}$ for the second claim. □

It is easy to check that neither any other Q_j, nor any of the T_j, T'_j, M_j or R_j give better results than the two Q_js listed above.

The function g has a slightly different behaviour, as it is monotone increasing as j increases in our sequences. The result here is

Theorem 21. *Given* $\epsilon > 0$*, there is a set* C *of integers for which*

$$g(C) = \frac{\ln(|C+C|/|C|)}{\ln(|C-C|/|C|)} > \frac{\ln(32/5)}{\ln(26/5)} - \epsilon \simeq 1.125944426$$

Proof. Take Q_j for j sufficiently large. □

(For comparison, $g(A_{15}) \simeq 1.0717$).

The corresponding suprema are $\ln(16/3)/\ln(14/3) \simeq 1.0867$ for both $(g(T'_j))$ and $(g(T_j))$, $\ln(23/4)/\ln(11/2) \simeq 1.0261$ for $(g(R_j))$ and $\ln(11/2)/\ln(5) \simeq 1.0592$ for $(g(M_j))$. None of these do as well as the supremum for the (Q_j).

Note also that because the sumsets and restricted sumsets in each of our families T'_j, T_j, M_j, R_j and Q_j only differ in order by a constant, the function

$$\hat{g}(A) = \frac{\ln(|A\hat{+}A|/|A|)}{\ln(|A-A|/|A|)}$$

will give similar insights to g.

5. The Smallest Order of a Restricted-Sum-Dominant Set

We noted above that we have two restricted-sum-dominant sets of order 16, namely T'_3 and M_2: we know of no smaller examples. In this section we reduce the range in which the smallest restricted-sum-dominant set can be.

Hegarty ([3], Theorem 1) proves that no 7-element subset of the integers is sum-dominant, and that up to linear transformations Conway's set is the unique 8-element sum-dominant subset of $\mathbb{Z}$. As Conway's set is not a restricted-sum-dominant set there is no 8-element restricted-sum-dominant set of integers.

Further Hegarty finds all 9-element sum-dominant sets A of integers with the additional property that for some $x \in A+A$ there are at least four ordered pairs $(a, a') \in A \times A$ with $a+a' = x$. There are, up to linear transformations, 9 such sets, listed in [3] as A_2 and A_4 through to A_{11}. It is easy to check that none of these nine sets is restricted-sum-dominant.

Thus, the only possible 9-element restricted-sum-dominant sets of integers have the property that for every $x \in A+A$ there are fewer than four ordered pairs (a, a') such that $x = a+a'$. This condition implies that there is no solution of $x+y = u+v$ with x, y, u, v all distinct, so such a set is a weak Sidon set in the sense of Ruzsa [8].

Defining $\delta(n)$ for $n \in A - A$ to be the number of ordered pairs (x, y) such that $x - y = n$, it is shown in the proof of Theorem 4.7 in [8] that for a weak Sidon set, $\delta(n) \leq 2$ whenever $n \neq 0$ and at most $2|A|$ elements n have $\delta(n) = 2$.

Thus, noting 0 has $|A| = 9$ representations and putting $m = |A - A|$,

$$81 \leq 9 + (2 \times 9) \times 2 + (m - 19) \Rightarrow m \geq 55$$

so if such a set were to be sum-dominant its sumset would have to have order at least 56. But of course $|A + A| \leq 9 \times 10/2 = 45$, and we have proven

Theorem 22. *All sum-dominant sets of integers of order 9 are linear transformations of one of Hegarty's nine sets A_2 and A_4 to A_{11}. None of these is restricted-sum-dominant, so there is no restricted-sum-dominant set of order 9.*

We thus know that the smallest restricted-sum-dominant set of integers has order between 10 and 16. It appears a non-trivial computational challenge to find the order of the smallest restricted-sum-dominant set.

References

[1] Gregory A. Freiman and V.P. Pigarev, *The relation between the invariants r and t* (Russian), Kalinin. Gos. Univ. Moscow (1973), 172-174.

[2] Andrew Granville, *An Introduction to Additive Combinatorics.* Lecture notes, available at http://www.dms.umontreal.ca/∼ andrew/PDF/ProcAddPap.pdf

[3] Peter V. Hegarty, *Some explicit constructions of sets with more sums than differences*, Acta Arith. **130** (2007), no.1, 61-77. Available online at http://arxiv.org/abs/math/0611582

[4] John Marica, *On A Conjecture of Conway*, Canad. Math. Bull. **12** (1969), 233-234. Available online at http://www.math.ca/cmb/v12/cmb1969v12.0233-0234.pdf

[5] Greg Martin and Kevin O'Bryant, *Many sets have more sums than differences*, in: Additive Combinatorics, 287-305, CRM Proc. Lecture Notes, 43, Amer. Math. Soc., Providence, RI, 2007. Available online at http://arxiv.org/abs/math/0608131

[6] Melvyn B. Nathanson *Sets with more sums than differences*, Integers **7** (2007), #a05. Available online at http://www.emis.de/journals/INTEGERS/papers/h5/h5.pdf

[7] Sheila Oates McDonald and Anne Penfold Street, *On Conway's conjecture for integer sets*, Bull. Austral. Math. Soc. **8** (1973), 355-358. Available online at http://dx.doi.org/10.1017/S0004972700042647

[8] Imre Z. Ruzsa, *Solving a linear equation in a set of integers I,* Acta Arith. **65** (1993) no 3, 259-282. Available online at http://matwbn.icm.edu.pl/ksiazki/aa/aa65/aa6537.pdf

[9] William G. Spohn Jr. *On Conway's conjecture for integer sets*, Canad. Math. Bull. **14** (3), 1971 461-462. Available online at http://cms.math.ca/10.4153/CMB-1971-085-4

[10] Matthew Wells, Ph.D. Thesis (in preparation), University of Essex.

[11] Yufei Zhao, *Sets Characterized by missing sums and differences*, J. Number Theory **131** (2011) 2107-2134. Available online at http://yufeizhao.com/papers/sum-dominant_density.pdf

ON AN INCOMPLETE ARGUMENT OF ERDŐS ON THE IRRATIONALITY OF LAMBERT SERIES

Joseph Vandehey
University of Illinois at Urbana-Champaign, Urbana, Illinois
vandehe2@illinois.edu

Received: 12/18/12, Accepted: 7/28/13, Published: 9/26/13

Abstract

We show that the Lambert series $f(x) = \sum d(n)x^n$ is irrational at $x = 1/b$ for negative integers $b < -1$ using an elementary proof that finishes an incomplete proof of Erdős.

1. Introduction

Chowla [5] conjectured that the functions

$$f(x) = \sum_{n=1}^{\infty} \frac{x^n}{1-x^n} \qquad \text{and} \qquad g(x) = \sum_{n=1}^{\infty} \frac{x^n}{1-x^n}(-1)^{n+1}$$

are irrational at all rational values of x satisfying $|x| < 1$. For such x the above functions may be rewritten as

$$f(x) = \sum_{n=1}^{\infty} d(n)x^n \qquad \text{and} \qquad g(x) = \frac{1}{4}\sum_{n=1}^{\infty} r(n)x^n,$$

where $d(n)$ is the number of divisors of n and $r(n)$ is the number of representations of n as a sum of two squares.

Erdős [6] proved that for any integer $b > 1$, the value $f(1/b)$ is irrational. He did so by showing that $f(1/b)$ written in base b contains arbitrary long strings of 0's without terminating on 0's completely. If we take $b < -1$ to be a negative integer, then Erdős' methods show by the same method that $f(1/b)$ in base $|b|$ contains arbitrary long strings of 0's; however, Erdős claims without proof that showing it will not terminate on 0's can be done using similar methods. It is not clear what method Erdős intended, and in later papers (including his review of similar irrationality results [7]) Erdős only refers to proving the case of positive b.

Since then, several proofs have been offered for the irrationality of the $b < -1$ case and far more general theorems besides. Borwein [3] is credited for proving the first major generalizations of these results, using Padeé approximants to show that all the numbers

$$\sum_{n=1}^{\infty} \frac{1}{q^n - r}, \qquad q \in \mathbb{N}_{\geq 2}, \quad r \in \mathbb{Q} \setminus \{0\}, \quad r \neq -q^m \text{ for any } m \geq 1$$

are irrational. Bundschuh and Väänänen [4], using a result of Bézivin [2] on the linear independence over $\mathbb{Q}$ of the infinite product

$$E_q(z) = \prod_{n=1}^{\infty} (1 + zq^{-j})$$

and its derivative, provide a quick second proof of Borwein's theorem. Other results can often be found in the literature under the term of the q-analogue of the logarithm or, simply, the q-logarithm.

However, all of these results are proved using entirely different techniques than those Erdős used and leave open the question of whether his method could have finished the proof.

Erdős' method can be extended to the following stronger result with a virtually identical proof.

Theorem 1.1. *Let $b > 1$ be a positive integer and $\mathcal{A}$ be any finite set of non-negative integers. Then for any sequence $\{a_n\}_{n=1}^{\infty}$ taking values in $\mathcal{A}$ such that the sequence does not end on repeated 0's, we have that*

$$\sum_{n=1}^{\infty} d(n) \frac{a_n}{b^n}$$

is irrational.

Theorem 1.1 has the following curious corollary. Let $a_n(x)$ be the nth base b digit of a number x in $(0, 1)$. (If x has two base b expansions, then we chose the one which does not end on repeated 0's.) Then the map

$$x = \sum_{n=1}^{\infty} \frac{a_n(x)}{b^n} \longmapsto \sum_{n=1}^{\infty} d(n) \frac{a_n(x)}{b^n}$$

has its image in $\mathbb{R} \backslash \mathbb{Q}$ and is also continuous at all x that do not have a representation as a finite base b expansion.

We could replace the condition that a_n be in the finite set $\mathcal{A}$ with a restriction that $0 \leq a_n \leq \phi(n)$ for some sufficiently slowly growing integer-valued function ϕ. It would be interesting to know what the fastest growing ϕ for which the Theorem 1.1 holds would be.

In this paper, we will prove the following extension of Theorem 1.1.

Theorem 1.2. *Let $b > 1$ be a positive integer and $\mathcal{A}$ be any finite set of integers* that does not contain 0. *Then for any sequence of $\{a_n\}_{n=1}^{\infty}$ taking values in $\mathcal{A}$, we have that*

$$\sum_{n=1}^{\infty} d(n)\frac{a_n}{b^n}$$

is irrational.

The new ingredient to extend Erdős' method is finding arbitrarily long strings of zeros *that are known to be preceded by a non-zero number*, and to find these strings arbitrarily far into the base $|b|$ expansion.

In particular, by taking $a_n = (-1)^n$, this proves that $f(1/b)$ is irrational for negative integers $b < 1$ as well, completing Erdős' proof.

Theorem 1.2 has parallels with a result of Tachiya [8], which showed that if $\{a_n\}$ is a sequence of rational numbers of period 2 (not identically zero), then the function

$$f(q) = \sum_{n=1}^{\infty} a_n \frac{1}{1-q^n}$$

is irrational for integers q with $|q| > 1$.

2. Proof of Theorem 1.2

We will require a result mentioned by Alford, Granville, and Pomerance [1, p. 705]. The function $\pi(N; d, a)$ equals the number of primes up to N that are congruent to a modulo d.

Proposition 2.1. *Let $0 < \delta < 5/12$. Then there exist positive integers N_0 and $\overline{\mathcal{D}}$ dependent only on δ, such that the bound*

$$\pi(N; d, a) \geq \frac{N}{2\varphi(d)\log N}$$

holds for all $N > N_0$; for all moduli d with $1 \leq d \leq N^{\delta}$, except, possibly for those d that are multiples of some element in $\mathcal{D}(N)$, a set of at most $\overline{\mathcal{D}}$ different integers that all exceed $\log N$; *and for all integers a that are relatively prime to d.*

We begin our proof much as Erdős did his. Let $b \geq 2$ be a fixed positive integer, let $\mathcal{A}$ be a finite set of integers that does not contain 0, and let N be a large positive integer that is allowed to vary. Define k in terms of N by

$$k = k(N) := \lfloor (\log N)^{1/10} \rfloor.$$

Let j_0 be a fixed integer, independent of N, so that $2\max_{a\in\mathcal{A}} |a|/b^{j_0} < 1$.

Let $0 < \delta < 5/12$ be some sufficiently small fixed constant, and let $N_0 = N_0(\delta)$ and $\overline{\mathcal{D}} = \overline{\mathcal{D}}(\delta)$ be the corresponding constants from Proposition 2.1. Let $N_1 > N_0$ be large enough so that for any $N > N_1$, the interval $((\log N)^2, 2(\log N)^2)$ cotains at least $u + \overline{\mathcal{D}}$ primes, where $u = u(N) = k(k+1)/2$. In addition, for such $N > N_1$, let $\mathcal{D}(N)$ be the set of exceptional moduli from Proposition 2.1. Since we assume that δ is constant, $|\mathcal{D}(N)| \leq \overline{\mathcal{D}}$ is bounded.

For each D in $\mathcal{D}(N)$, let $\tilde{p}_D$ denote the smallest prime strictly greater than $(\log N)^2$ that divides D, if such a prime exists, and then let $p_1 < p_2 < \cdots < p_u$ be the smallest u primes strictly greater than $(\log N)^2$ that are not equal to $\tilde{p}_D$ for any $D \in \mathcal{D}(N)$; by assumption on N, we have that each such p_i is less than $2(\log N)^2$. Finally, let

$$A := \prod_{i=1}^{j_0(j_0+1)/2} p_i^b \prod_{i=(j_0+1)(j_0+2)/2+1}^{u} p_i^b,$$

so that, in particular, A is not a multiple of any D in $\mathcal{D}(N)$; moreover, provided N is sufficiently large, we have

$$A < (2(\log N)^2)^{bk(k+1)/2} \leq N^\delta.$$

By the Chinese remainder theorem, there exists an integer r, with $0 \leq r \leq A-1$, such that

$$r + j \equiv \prod_{i=j(j+1)/2+1}^{(j+1)(j+2)/2} p_i^{b-1} \pmod{\prod_{i=j(j+1)/2+1}^{(j+1)(j+2)/2} p_i^b}, \qquad 0 \leq j \leq k-1, j \neq j_0.$$

(The exception $j \neq j_0$ marks the key difference between this proof and Erdős'.) Since all the p_i's are bounded below by $(\log N)^2$, we have that r necessarily tends to infinity as N does, although possibly much slower.

With this value of r, for any integer of the form

$$r + mA, \qquad 0 \leq m < \lfloor N/A \rfloor,$$

we have that

$$d(r + mA + j) \equiv 0 \pmod{b^{j+1}}, \qquad 0 \leq j < k,\ j \neq j_0 \tag{1}$$

by the multiplicitivity of $d(\cdot)$. Moreover, $r + j_0$ is relatively prime to A, since each p dividing A also divides some $r + j$, with $0 \leq j < k$, $j \neq j_0$; the largest j can be is $k \leq (\log N)^{1/10}$, but all primes dividing A are at least $(\log N)^2$.

We can also apply Proposition 2.1 to see that

$$\pi(N, A, r + j_0) \geq \frac{N}{2\varphi(A)\log N}. \tag{2}$$

Erdős in [6] also proved the following result, which we give here without reproof. (While our construction of A and r are different from Erdős', they satisfy all the requirements for Erdős' proof technique to still hold.)

Lemma 2.2. *With A, r, b, and k all as above, the number of $m < \lfloor N/A \rfloor$ such that*

$$\sum_{n>r+k+mA} d(n)\frac{1}{b^n} > \frac{1}{b^{r+k/2+mA}}$$

is less than

$$\frac{10cN(\log N)^2}{A2^{k/4}}$$

for some constant c independent of all variables.

Regardless of how large c is, we have, for sufficently large N, that

$$\frac{N}{2\varphi(A)\log N} \geq \frac{10cN(\log N)^2}{A2^{k/4}}.$$

Therefore, by combining Lemma 2.2 with (1) and (2), we see that for sufficiently large N there exists some $m_0 < \lfloor N/A \rfloor$ such that

$$b^{j+1} | d(r+m_0A+j), \qquad 0 \leq j < k, j \neq j_0, \tag{3}$$

$$r + m_0A + j_0 \text{ is prime,} \tag{4}$$

and

$$\sum_{n>r+k+m_0A} \left| d(n)\frac{a_n}{b^n} \right| \leq \frac{\max_{a\in\mathcal{A}} |a|}{b^{r+k/2+mA}}. \tag{5}$$

Now consider a particular sequence (a_n) with each $a_n \in \mathcal{A}$ together with the sum

$$\sum_{n=1}^{\infty} d(n)\frac{a_n}{b^n}.$$

By (3), the partial sum

$$\sum_{\substack{n\leq r+k+m_0A \\ n\neq r+j_0+m_0A}} d(n)\frac{a_n}{b^n},$$

when written in base b, has its last non-zero digit in the $(r-1+m_0A)$th place or earlier.[1] In addition, by (5), the partial sum

$$\sum_{n>r+k+m_0A} d(n)\frac{a_n}{b^n}$$

[1]Here we switch back to the convention that finite expansions are assumed to end on an infinite string of zeros.

when written in base b has its first non-zero digit in the

$$(r + k/2 + m_0 A - \lceil \log_b \max_{a \in \mathcal{A}} |a| \rceil)\text{th}$$

place or later. The number

$$d(r + j_0 + m_0 A) \frac{a_{r+j_0+m_0 A}}{b^{r+j_0+m_0 A}} = \frac{2a_{r+j_0+m_0 A}}{b^{r+j_0+m_0 A}}$$

when written in base b has its non-zero digits only between the $(r + m_0 A)$th and $(r + j_0 + m_0 A)$th place, and it has at least one such non-zero digit. Thus the full sum has a string of at least $k/2 + O(1)$ zeroes immediately preceded by a non-zero digit starting somewhere between the $(r + m_0 A)$th and $(r + j_0 + m_0 A)$th place.

So as N increases to infinity, we can find arbitrarily long strings of 0's (this corresponds to k increasing to infinity) immediately preceded by a non-zero digit, and we find these strings arbitrarily far out in the expansion (since r also tends to infinity). The base b digits cannot therefore be periodic and hence the sum is irrational. This completes the proof.

Acknowledgements The author acknowledges support from National Science Foundation grant DMS 08-38434 "EMSW21-MCTP: Research Experience for Graduate Students." The author would also like to thank Paul Pollack and Paul Spiegelhalter for their assistance.

References

[1] W. Alford, A. Granville, and C. Pomerance. There are infinitely many Carmichael numbers, *Annals of Math.*, 140 (1994), 703–722.

[2] J.-P. Bézivin. Indépendance linéaire des valeurs des solutions transcendantes de certaines équations fonctionnelles, *Manuscripta Math*, 61 (1988), no. 1, 103–129.

[3] P. Borwein. On the irrationality of $\sum(1/(q^n + r))$, *J. Number Theory*, 37 (1991), no. 3, 253–259.

[4] P. Bundschuh and K. Väänänen. Arithmetical investigations of a certain infinite product, *Compos. Math.* 91 (1994), no. 2, 175–199

[5] S. Chowla. On series of the Lambert type which assume irrational values for rational values of the argument, *Proc. Nat. Inst. of Sciences of India*, 13 (1947).

[6] P. Erdős. On arithmetical properties of Lambert series, *J. Indian Math. Soc.*, 12 (1948), 63–66.

[7] P. Erdős. On the irrationality of certain series: problems and results, *New Advances in Transcendence Theory*, 102–109, Cambridge Univ. Press, Cambridge, 1988.

[8] Y. Tachiya. Irrationality of certain Lambert series, *Tokyo J. Math.* 27 (2004), 75–85.

GENERALIZATION OF UNIVERSAL PARTITION AND BIPARTITION THEOREMS

Hacène Belbachir[1]
USTHB, Faculty of Mathematics, RECITS Laboratory, Algiers, Algeria
hbelbachir@usthb.dz, hacenebelbachir@gmail.com

Miloud Mihoubi[2]
USTHB, Faculty of Mathematics, RECITS Laboratory, Algiers, Algeria.
mmihoubi@usthb.dz, miloudmihoubi@gmail.com

Received: 5/24/12, Revised: 5/4/13, Accepted: 8/10/13, Published: 9/26/13

Abstract

Let $A = (a_{i,j})$, $i = 1, 2, \ldots$, $j = 0, 1, 2, \ldots$, be an infinite matrix with elements $a_{i,j} = 0$ or 1; $p(n, k; A)$ the number of partitions of n into k parts whose number y_i of parts which are equal to i belongs to the set $Y_i = \{j : a_{i,j} = 1\}$, $i = 1, 2, \ldots$. The universal theorem on partitions states that

$$\sum_{n=0}^{\infty}\sum_{k=0}^{n} p(n, k; A)\, u^k t^n = \prod_{i=1}^{\infty}\left(\sum_{j=0}^{\infty} a_{i,j} u^j t^{ij}\right).$$

In this paper, we present a generalization of this result. We show that this generalization remains true when $a_{i,j}$ are indeterminate. We also take into account the bi-partite and multi-partite situations.

1. Introduction

Let $A = (a_{i,j})$, $i = 1, 2, \ldots$, $j = 0, 1, 2, \ldots$ be an infinite matrix with elements $a_{i,j} = 0$ or 1; $p(n, k; A)$ the number of partitions of n into k parts whose number y_i of parts which are equal to i belongs to the set $Y_i = \{j : a_{i,j} = 1\}$, $i = 1, 2, \ldots$. The universal theorem on partitions states that

$$\sum_{n=0}^{\infty}\left(\sum_{k=0}^{n} p(n, k; A)\, u^k\right) t^n = \prod_{i=1}^{\infty}\left(\sum_{j=0}^{\infty} a_{i,j} u^j t^{ij}\right); \qquad (1)$$

[1]The research is partially supported by the LITIS Laboratory of Rouen University and the PNR project 8/u160/664.

[2]This research is supported by the PNR project 8/u160/3172.

see for instance [2] and [3].

In Section 2, we will provide an extension of the above identity and show that it remains true when $a_{i,j}$ are indeterminate. In Section 3, we will present an equivalent version in terms of complete Bell polynomials when $a_{i,0} = 1$, $i \geq 1$.

Similarly, a partition of an ordered pair $(m,n) \neq (0,0)$, of nonnegative integers, is a non-ordered collection of nonnegative integers $(x_i, y_i) \neq (0,0)$, $i = 1,2,\ldots$, whose sum equals (m,n). Given a partition of (m,n), let $k_{i,j}$ be the number of parts which are equal to (i,j), $i = 0,1,2,\ldots,m$, $j = 0,1,2,\ldots,n$, $(i,j) \neq (0,0)$, such that

$$\sum_{i=0}^{m} i \sum_{j=0}^{n} k_{i,j} = m, \quad \sum_{j=0}^{n} j \sum_{i=0}^{m} k_{i,j} = n. \tag{2}$$

For a partition of (m,n) into k parts, we add

$$\sum_{i=0}^{m}\sum_{j=0}^{n} k_{i,j} = k. \tag{3}$$

Let $p(m,n)$ be the number of partitions of the bi-partite number (m,n) with $p(0,0) = 1$ and $p(m,n,k)$ be the number of partitions of (m,n) into k parts with $p(0,0,0) = 1$. The universal bipartition theorem states that

$$F(t,u,w) = \sum_{m,n\geq 0} \left(\sum_{k=0}^{m+n} p(m,n,k) w^k \right) t^m u^n = \prod_{\substack{j=0}}^{\infty} \prod_{\substack{i=0 \\ (i,j)\neq(0,0)}}^{\infty} \left(1 - wt^i u^j\right)^{-1}; \tag{4}$$

see [2, p. 403, pb. 24]. A generalization of identity (4) is dealt with in Section 4. Section 5 is devoted to the concept of multipartition.

2. Generalized Universal Partition Theorem

Theorem 1. *Let* $X = (x_{i,j})$, $i = 1,2,\ldots$, $j = 0,1,2,\ldots$, *be an infinite matrix of indeterminates;* $\pi(n,k)$ *the set of all nonnegative integer solutions of*

$$k_1 + k_2 + \cdots + k_n = k \quad \text{and} \quad k_1 + 2k_2 + \cdots + nk_n = n;$$

and $\pi(n) = \bigcup_{k=1}^{n} \pi(n,k)$ *be the set of all nonnegative integer solutions of* $k_1 + 2k_2 + \cdots + nk_n = n$. *For every solution* $k_1, k_2, \ldots, k_n$, *we set*

$$p(n,k;X) := \sum_{\pi(n,k)} x_{1,k_1} x_{2,k_2} \cdots x_{n,k_n}.$$

Then

$$G(t,u;X) := \sum_{n=0}^{\infty} \left(\sum_{k=0}^{n} p(n,k;X) u^k \right) t^n = \prod_{i=1}^{\infty} \left(\sum_{j=0}^{\infty} x_{i,j} u^j t^{ij} \right).$$

Proof. We have

$$G(t,u;X)=\sum_{n=0}^{\infty}\sum_{k=0}^{n}\left(\sum_{\pi(n,k)}x_{1,k_1}x_{2,k_2}\cdots x_{n,k_n}u^{k_1+\cdots+k_n}t^{k_1+2k_2+\cdots+nk_n}\right).$$

Since these sums apply for all $k=0,1,\ldots,n$ and $n=0,1,\ldots$, it follows that

$$\begin{aligned}G(t,u;X)&=\sum_{n=0}^{\infty}\sum_{k=0}^{n}\left(\sum_{\pi(n,k)}\left(x_{1,k_1}(ut)^{k_1}\right)\left(x_{2,k_2}\left(ut^2\right)^{k_2}\right)\cdots\left(x_{n,k_n}(ut^n)^{k_n}\right)\right)\\&=\sum_{n=0}^{\infty}\left(\sum_{\pi(n)}\left(x_{1,k_1}(ut)^{k_1}\right)\left(x_{2,k_2}\left(ut^2\right)^{k_2}\right)\cdots\left(x_{n,k_n}(ut^n)^{k_n}\right)\right)\\&=\left(\sum_{k_1=0}^{\infty}x_{1,k_1}(ut)^{k_1}\right)\left(\sum_{k_2=0}^{\infty}x_{2,k_2}\left(ut^2\right)^{k_2}\right)\cdots\\&=\prod_{i=1}^{\infty}\left(\sum_{j=0}^{\infty}x_{i,j}\left(ut^i\right)^j\right),\end{aligned}$$

which is the required expression. □

For $x_{i,j}=a_{i,j}$ with $i=1,2,\ldots,\ j=0,1,2,\ldots$, in Theorem 1, we obtain the universal theorem on partitions. For $x_{i,j}=\frac{a_{i,j}}{j!i^j}$ with $i=1,2,\ldots,\ j=0,1,2,\ldots$, in Theorem 1, we obtain:

Corollary 2. *Let $A=(a_{i,j})$, $i=1,2,\ldots,\ j=0,1,2,\ldots$, be an infinite matrix with elements $a_{i,j}=0$ or 1 and $c(n,k;A)$ the number of permutations of a finite set W_n, of n elements, that are decomposed into k cycles such that the number of cycles of length i belongs to the set $Y_i=\{j:a_{i,j}=1\}$. Then*

$$\sum_{n=0}^{\infty}\sum_{k=0}^{n}c(n,k;A)\,u^k\frac{t^n}{n!}=\prod_{i=1}^{\infty}\left(\sum_{j=0}^{\infty}a_{i,j}\frac{u^j}{j!}\left(\frac{t^i}{i}\right)^j\right).$$

For $x_{i,j}=\frac{1}{j!}\left(\frac{z_i}{i!}\right)^j a_{i,j},\ z_i\in\mathbb{C},\ i=1,2,\ldots,\ j=0,1,2,\ldots$, in Theorem 1, we obtain a remarkable identity according to the partial Bell polynomials:

Corollary 3. *Let $A=(a_{i,j})$, $i=1,2,\ldots,\ j=0,1,2,\ldots$, be an infinite matrix with elements $a_{i,j}=0$ or 1 and*

$$B_{n,k;A}(z_1,z_2,\ldots,z_n):=\sum_{\pi(n,k)}\frac{n!}{k_1!\cdots k_n!}\left(\frac{z_1}{1!}\right)^{k_1}\cdots\left(\frac{z_n}{n!}\right)^{k_n}a_{1,k_1}\cdots a_{n,k_n}.$$

Then we obtain

$$\sum_{n=0}^{\infty}\sum_{k=0}^{n}B_{n,k;A}\left(z_1,z_2,\ldots,z_n\right)u^k\frac{t^n}{n!}=\prod_{i=1}^{\infty}\left(\sum_{j=0}^{\infty}\frac{a_{i,j}}{j!}\left(uz_i\frac{t^i}{i!}\right)^j\right).$$

Remark 4. For $a_{i,j}=1$, $i=1,2,\ldots$ and $j=0,1,\ldots$, Corollary 3 gives

$$\sum_{n=0}^{\infty}\sum_{k=0}^{n}B_{n,k}\left(z_1,z_2,\ldots,z_n\right)u^k\frac{t^n}{n!}=\exp\left(u\sum_{i=1}^{\infty}z_i\frac{t^i}{i!}\right),$$

which is the definition of the partial Bell polynomials $B_{n,k}\left(z_1,\ldots,z_n\right)$. See [1, 3, 4].

3. Connection With the Complete Bell Polynomials

Recall that the complete Bell polynomials $A_n\left(x_1,x_2,\ldots\right)$ are defined by

$$\sum_{n=k}^{\infty}A_n\left(x_1,x_2,\ldots\right)\frac{t^n}{n!}=\exp\left(\sum_{m=1}^{\infty}x_m\frac{t^m}{m!}\right).$$

See [1, 3, 4].

In this section, we provide another formulation of Theorem 1 according to the complete Bell polynomials. We determine the generating functions of the sequences $\left(p\left(n,k;X\right)\right)_n$ and $\left(p\left(n,k;X\right)\right)_k$, where $X=\left(x_{i,j}\right)$, $i=1,2,\ldots$, $j=0,1,2,\ldots$, is an infinite matrix with indeterminates $x_{i,j}$ such that $x_{i,0}=1$ for every $i\geq 1$.

Theorem 5. *Let q,u be indeterminate. Then, for $n\geq 1$, we have*

$$\sum_{j=n}^{\infty}p\left(j,n;X\right)q^j = \frac{1}{n!}A_n\left(\rho_1\left(q;X\right),\rho_2\left(q;X\right),\ldots,\rho_n\left(q;X\right)\right), \tag{5}$$

$$\sum_{j=0}^{n}p\left(n,j;X\right)u^j = \frac{1}{n!}A_n\left(\sigma_1\left(u;X\right),\sigma_2\left(u;X\right),\ldots,\sigma_n\left(u;X\right)\right), \tag{6}$$

where

$$\rho_n\left(q;X\right):=\sum_{i=1}^{\infty}b_n\left(i\right)q^{ni} \quad \text{and} \quad \sigma_n\left(u;X\right):=n!\sum_{k|n}b_k\binom{n}{k}\frac{u^k}{k!},$$

with

$$b_n\left(i\right)=\sum_{k=1}^{n}\left(-1\right)^{k-1}\left(k-1\right)!B_{n,k}\left(1!x_{i,1},2!x_{i,2},\ldots,j!x_{i,j},\ldots\right).$$

Proof. From Theorem 1, we get

$$G(q,u;X) := \sum_{n=0}^{\infty}\sum_{k=0}^{n} p(n,k;X)\,u^k q^n = \exp\left(\sum_{i=1}^{\infty} \ln\left(1+\sum_{j=1}^{\infty}(j!x_{i,j})\frac{(uq^i)^j}{j!}\right)\right).$$

Using the following known expansion (see [2, Theorem 11.17])

$$\ln\left(1+\sum_{k=1}^{\infty} g_k \frac{q^k}{k!}\right) = \sum_{n=1}^{\infty} c_n \frac{q^n}{n!},$$

with $c_n = \sum_{k=1}^{n}(-1)^{k-1}(k-1)!B_{n,k}(g_1,g_2,\ldots)$, we obtain

$$\begin{aligned} G(q,u;X) = \exp\left(\sum_{i=1}^{\infty}\sum_{k=1}^{\infty} b_k(i)\frac{u^k}{k!}q^{ki}\right) &= \exp\left(\sum_{k=1}^{\infty}\frac{u^k}{k!}\sum_{i=1}^{\infty} b_k(i)\,q^{ki}\right) \\ &= \exp\left(\sum_{k=1}^{\infty}\rho_k(q;X)\frac{u^k}{k!}\right) \\ &= 1+\sum_{k=1}^{\infty} A_k(\rho_1(q;X),\ldots,\rho_k(q;X))\frac{u^k}{k!}. \end{aligned}$$

On the other hand, we have

$$G(q,u;X) = \sum_{n=0}^{\infty}\sum_{k=0}^{n} p(n,k;X)\,u^k q^n = \sum_{k=0}^{\infty}\left(\sum_{n=k}^{\infty} p(n,k;X)\,q^n\right)u^k.$$

The first identity follows from identification, where as the second identity follows from the expansion

$$\begin{aligned} G(q,u;X) &= \sum_{n=0}^{\infty}\left(\sum_{k=0}^{n} p(n,k;X)\,u^k\right)q^n \\ &= \exp\left(\sum_{i=1}^{\infty}\ln\left(1+\sum_{j=1}^{\infty}(j!x_{i,j})\frac{(uq^i)^j}{j!}\right)\right) \\ &= \exp\left(\sum_{i=1}^{\infty}\sum_{k=1}^{\infty} b_k(i)\,u^k\frac{q^{ik}}{k!}\right) \\ &= \exp\left(\sum_{n=1}^{\infty} q^n \sum_{k|n} b_k\left(\frac{n}{k}\right)\frac{u^k}{k!}\right) \\ &= \exp\left(\sum_{n=1}^{\infty}\sigma_n(u;X)\frac{q^n}{n!}\right) \\ &= 1+\sum^{\infty} A \ldots \, q^n \end{aligned}$$

Corollary 6. *Let α and q be two indeterminates. We then have*

$$A_n\left(\frac{1}{1-q}-\alpha,\ldots,(n-1)!\left(\frac{1}{1-q^n}-\alpha^n\right)\right)=n!\left(\frac{1}{1-q^n}-\alpha\right)\prod_{i=1}^{n-1}\left(1-q^i\right)^{-1}$$

and

$$A_n\left((1-\alpha)\,u,\ldots,(n-1)!\left(\sum_{k|n}ku^{n/k}-(\alpha u)^n\right)\right)=n!\left(p_n(u)-\alpha u p_{n-1}(u)\right),$$

where $p_n(u):=\sum_{j=0}^{n}p(n,j)\,u^j$.

Proof. We put in identity (5) $x_{1,0}=1,\ x_{1,j}=q^{-j}(1-\alpha)$ for $j\geq 1$ and $x_{i,j}=q^{-j}$ for $i\geq 2,\ j\geq 0$, and use identity $B_{n,k}(1!,2!,3!,\ldots)=\frac{(n-1)!}{(k-1)!}\binom{n}{k}$ (Lah numbers). We obtain

$$\begin{aligned}
b_n(1)&=(n-1)!\,(1-\alpha^n)\,q^{-n},\ b_n(i)=(n-1)!q^{-n},\quad i\geq 2,\\
\rho_n(q;X)&=\sum_{i=1}^{\infty}b_n(i)\,q^{ni}=(n-1)!\left(\frac{1}{1-q^n}-\alpha^n\right),\\
p(n,k;X)&=q^{-k}\sum_{\pi(n,k),\ k_1=0}1+q^{-k}\sum_{\pi(n,k),\ k_1\geq 1}(1-\alpha)\\
&=q^{-k}\sum_{\pi(n-k,k)}1+(1-\alpha)\,q^{-k}\sum_{\pi(n-1,k-1)}1\\
&=q^{-k}\left[p(n-k,k)+(1-\alpha)\,p(n-1,k-1)\right],
\end{aligned}$$

where $p(n,k)$ is the number of partitions of n into k parts, which satisfy

$$p(n,k)=p(n-k,k)+p(n-1,k-1).$$

Thus, we obtain

$$\begin{aligned}
\sum_{j=n}^{\infty}p(j,n;X)\,q^j&=q^{-n}\left(\sum_{j=n}^{\infty}p(j,n)\,q^j-\alpha\sum_{j=n}^{\infty}p(j-1,n-1)\,q^j\right)\\
&=\left(\frac{1}{1-q^n}-\alpha\right)\prod_{i=1}^{n-1}\left(1-q^i\right)^{-1},
\end{aligned}$$

which gives the first identity.
For the second identity, we take $x_{1,0}=1,\ x_{1,j}=1-\alpha$ for $j\geq 1$ and $x_{i,j}=1$ for

$i \geq 2$, $j \geq 0$, in relation (6) to get

$$b_n(1) = -(n-1)!(\alpha^n - 1), \; b_n(i) = (n-1)!, \quad i \geq 2,$$
$$\sigma_n(u; X) = n! \sum_{k|n} b_k \binom{n}{k} \frac{u^k}{k!} = -(n-1)!(\alpha u)^n + (n-1)! \sum_{k|n} k u^{n/k},$$
$$p(n, k; X) = p(n, k) - \alpha p(n-1, k-1),$$

thus

$$\sum_{j=0}^{n} p(n, j; X) u^j = p_n(u) - \alpha u p_{n-1}(u),$$

which provides the second identity. □

4. Generalized Universal Bipartition Theorem

In this section, we provide a generalization of identity (4) and deduce some known identities. Let us start with the following example: how do we partition $(2,3)$ into different parts? Let $p(2,3,k)$ be the number of partitions of the bi-partite number $(2,3)$ into k parts, $k = 1, \ldots, 5$ and $p(2,3)$ be the total number of partitions of $(2,3)$. We have

$(2,3)$

– $p(2,3,1) = 1$

$(0,1) + (2,2) = (0,2) + (2,1) = (0,3) + (2,0) = (1,0) + (1,3) = (1,1) + (1,2)$

– $p(2,3,2) = 5$

$(0,1) + (0,1) + (2,1) = (0,1) + (0,2) + (2,0) = (0,1) + (1,1) + (1,1) =$
$(0,1) + (1,0) + (1,2) = (0,3) + (1,0) + (1,0)$

– $p(2,3,3) = 5$

$(0,1) + (0,1) + (0,1) + (2,0) = (0,1) + (0,1) + (1,0) + (1,1) =$
$(0,1) + (1,0) + (1,0) + (0,2)$

– $p(2,3,4) = 3$

$(0,1) + (0,1) + (0,1) + (1,0) + (1,0)$

– $p(2,3,5) = 1$

$$p(2,3) = \sum_{k=1}^{5} p(2,3,k) = 1 + 5 + 5 + 3 + 1 = 15.$$

Theorem 7. *Let* $X = (x_{i,j,s})$, $i, j, s = 0, 1, 2, \ldots,$ *be a sequence of indeterminates with* $x_{0,0,s} = 0$, $\Pi(m, n, k)$ *the set of all nonnegative integers* $k_{i,j}$ *satisfying (2) and (3) and* $\Pi(m, n) := \bigcup_{k=1}^{n+m} \Pi(m, n, k)$ *the set of all nonnegative integers satisfying (2). For every partition of the bi-partite number* (m, n) *into* k *parts, we set*

$$\mathbf{p}(m, n, k; X) := \sum_{\Pi(m,n,k)} \prod_{i=0}^{m} \prod_{j=0}^{n} x_{i,j,k_{i,j}}.$$

Then we have

$$F(t,u,\omega;X) := \sum_{m,n\geq 0}\left(\sum_{k=0}^{m+n}\mathbf{p}(m,n,k;X)\,\omega^k\right)t^m u^n = \prod_{i+j\geq 1}\left(\sum_{s=0}^{\infty}x_{i,j,s}\left(\omega t^i u^j\right)^s\right).$$

Proof. We have

$$\begin{aligned}F(t,u,\omega;X) &= \sum_{m,n\geq 0}\left(\sum_{k=0}^{m+n}\mathbf{p}(m,n,k;X)\,\omega^k\right)t^m u^n\\ &= \sum_{m,n\geq 0}\sum_{k=0}^{m+n}\left(\sum_{\Pi(m,n,k)} t^{\sum\limits_{i=0}^{m} i\sum\limits_{j=0}^{n}k_{i,j}}\, u^{\sum\limits_{j=0}^{n} j\sum\limits_{i=0}^{m}k_{i,j}}\, \omega^{\sum\limits_{i=0}^{m}\sum\limits_{j=0}^{n}k_{i,j}}\prod_{i=0}^{m}\prod_{j=0}^{n}x_{i,j,k_{i,j}}\right)\\ &= \sum_{m,n\geq 0}\sum_{\Pi(m,n)}\left(\prod_{i=0}^{m}\prod_{j=0}^{n}x_{i,j,k_{i,j}}\left(\omega t^i u^j\right)^{k_{i,j}}\right)\\ &= \prod_{i+j\geq 1}\left(\sum_{k_{i,j}=0}^{\infty}x_{i,j,k_{i,j}}\left(\omega t^i u^j\right)^{k_{i,j}}\right).\end{aligned}$$

□

Corollary 8. *Let* $A=(a_{i,j,s})$, $i,j,s=0,1,2,\ldots$, *with* $a_{i,j,s}=0$ *or* 1 *for* $(i,j)\neq(0,0)$ *and let* $\mathbf{p}(m,n,k;A)$ *be the number of partitions of* (m,n) *into* k *parts whose number* $y_{i,j}$ *of parts which are equal to* (i,j) *belongs to the set* $Y_{i,j}=\{s: a_{i,j,s}=1\}$, $i,j=0,1,2,\ldots$, $(i,j)\neq(0,0)$. *Then*

$$F(t,u,\omega;A) := \sum_{m,n\geq 0}\left(\sum_{k=0}^{m+n}\mathbf{p}(m,n,k;A)\,\omega^k\right)t^m u^n = \prod_{i+j\geq 1}\left(\sum_{s=0}^{\infty}a_{i,j,s}\left(\omega t^i u^j\right)^s\right).$$

For $x_{i,j,s}=1$ for all $i,j,s=0,1,2,\ldots$, $(i,j)\neq(0,0)$, Theorem 7 becomes:

Corollary 9. *Let* $\mathbf{p}(m,n)$ *be the number of partitions of the bi-partite number* (m,n) *with* $\mathbf{p}(0,0)=1$ *and* $\mathbf{p}(m,n,k)$ *the number of partitions of* (m,n) *into* k *parts, with* $\mathbf{p}(0,0,0)=1$. *Then*

$$\sum_{m,n\geq 0}\left(\sum_{k=0}^{m+n}\mathbf{p}(m,n,k)\,\omega^k\right)t^m u^n = \prod_{i+j\geq 1}\left(1-\omega t^i u^j\right)^{-1}.$$

Remark 10. Let $(y_{i,j})$, $i,j=0,1,\ldots$, be a sequence of indeterminates and let $x_{i,j,s}=\frac{1}{s!}\left(\frac{y_{i,j}}{i!j!}\right)^s$, $i,j=0,1,2,\ldots$, we have

$$\mathbf{p}(m,n,k;X) = \sum_{\Pi(m,n,k)}\prod_{i=0}^{m}\prod_{j=0}^{n}\frac{1}{k_{i,j}!}\left(\frac{y_{i,j}}{i!j!}\right)^{k_{i,j}} = \frac{A_{m,n,k}}{m!n!},$$

where

$$A_{m,n,k} := A_{m,n,k}\left(y_{0,1}, y_{1,0}, \ldots, y_{m,n}\right),$$

and

$$F\left(t,u,\omega;X\right) = \sum_{m,n\geq 0}\left(\sum_{k=0}^{m+n} A_{m,n,k}\omega^{k}\right)\frac{t^{m}}{m!}\frac{u^{n}}{n!}.$$

From Theorem 7, we obtain

$$F\left(t,u,\omega;X\right) = \prod_{i+j\geq 1}\left(\sum_{k_{i,j}\geq 0}\frac{1}{k_{i,j}!}\left(\omega y_{i,j}\frac{t^{i}}{i!}\frac{u^{j}}{j!}\right)^{k_{i,j}}\right) = \exp\left(\omega\left(\sum_{i+j\geq 1} y_{i,j}\frac{t^{i}}{i!}\frac{u^{j}}{j!}\right)\right).$$

From the two expressions of $F\left(t,u;\omega,X\right)$, we retrieve the exponential partial bi-partitional polynomials:

$$\sum_{m,n\geq 0}\left(\sum_{k=0}^{m+n} A_{m,n,k}\left(y_{0,1}, y_{1,0}, \ldots, y_{m,n}\right)\omega^{k}\right)\frac{t^{m}}{m!}\frac{u^{n}}{n!} = \exp\left(\omega\left(\sum_{i+j\geq 1} y_{i,j}\frac{t^{i}}{i!}\frac{u^{j}}{j!}\right)\right);$$

see [2, pp: 454–457].

5. Universal Multipartition Theorem

More generally, a multipartition of order r of $\mathbf{n} = (n_1, \ldots, n_r)$, different from $\mathbf{0} = (0, \ldots, 0)$, of nonnegative integers, is a non-ordered collection of nonnegative integers $\left(x_i^{(1)}, \ldots, x_i^{(r)}\right)$, $i = 1, 2, \ldots$, whose sum equals $\mathbf{n}$. In a partition of an r-partite number $\mathbf{n}$, let $k_{\mathbf{i}} := k_{i_1,\ldots,i_r}$ be the number of ordered r numbers that are equal to $\mathbf{i} = (i_1, \ldots, i_r) \in \{0, 1, 2, \ldots, n_1\} \times \cdots \times \{0, 1, 2, \ldots, n_r\}$, $(i_1, \ldots, i_r) \neq \mathbf{0}$, such that

$$\sum_{i_1=0}^{n_1}\cdots\sum_{i_r=0}^{n_r} i_j k_{i_1,\ldots,i_r} = n_j, \quad j = 1, \ldots, r. \tag{7}$$

For the partition of $\mathbf{n}$ into k parts, we add

$$\sum_{i_1=0}^{n_1}\cdots\sum_{i_r=0}^{n_r} k_{i_1,\ldots,i_r} = k. \tag{8}$$

Let $\mathbf{p}(\mathbf{n})$ be the number of partitions of the r-partite $\mathbf{n}$ with $\mathbf{p}(\mathbf{0}) = 1$ and $\mathbf{p}(\mathbf{n}, k)$ the number of partitions of the r-partite number $\mathbf{n}$ into k parts, with $\mathbf{p}(\mathbf{0}, 0) = 1$.

Theorem 11. *Let* $X = (x_{\mathbf{i},s})$, $\mathbf{i} = (i_1, \ldots, i_r) \in \mathbb{N}^r$, $\mathbf{i} \neq \mathbf{0}$, $s = 0, 1, 2, \ldots$, *be a sequence of indeterminates with* $r+1$ *indices, with* $x_{\mathbf{0},s} = 0$, $\Pi(\mathbf{n}, k)$ *the set of all nonnegative integers* $k_{i_1,\ldots,i_r}$ *satisfying (7) and (8) and* $\Pi(\mathbf{n}) := \bigcup_{k=1}^{n_1+\cdots+n_r} \Pi(\mathbf{n}, k)$

the set of all nonnegative integers solutions of (7). For every partition of $\mathbf{n}$ *into* k *parts, we set*

$$\mathbf{p}(\mathbf{n};k,X) := \sum_{\Pi(\mathbf{n},k)} \prod_{\mathbf{i}=\mathbf{0}}^{\mathbf{n}} x_{\mathbf{i},k_{\mathbf{i}}}.$$

Then

$$F(\mathbf{t},\omega;X) = \sum_{\mathbf{n}\geq\mathbf{0}} \left(\sum_{k=0}^{\mathbf{n}\cdot\mathbf{1}} p(\mathbf{n},k;X)\,\omega^k \right) \mathbf{t}^{\mathbf{n}} = \prod_{\mathbf{i}\cdot\mathbf{1}\geq 1} \left(\sum_{s=0}^{\infty} x_{\mathbf{i},s} \left(\omega \mathbf{t}^{\mathbf{i}}\right)^s \right).$$

where $\mathbf{t}^{\mathbf{n}} := t_1^{n_1} \dots t_r^{n_r}$, $\mathbf{n}\cdot\mathbf{1} :=n_1+\cdots+n_r$, $\mathbf{n}\geq\mathbf{0} \Leftrightarrow n_1\geq 0,\dots,n_r\geq 0$.

Proof. We have the following

$$\begin{aligned} F(\mathbf{t},\omega;X) &= \sum_{\mathbf{n}\geq\mathbf{0}} \left(\sum_{k=0}^{\mathbf{n}\cdot\mathbf{1}} p(\mathbf{n},k;X)\,\omega^k \right) \mathbf{t}^{\mathbf{n}} \\ &= \sum_{\mathbf{n}\geq\mathbf{0}} \sum_{k=0}^{\mathbf{n}\cdot\mathbf{1}} \left(\sum_{\Pi(\mathbf{n},k)} \left(\prod_{\mathbf{i}=\mathbf{0}}^{\mathbf{n}} x_{\mathbf{i},k_{\mathbf{i}}} \right) \omega^{\sum_{i_j\leq n_j} k_{\mathbf{i}}} \prod_{j=1}^{r} t_j^{\sum_{i_j\leq n_j} i_j k_{\mathbf{i}}} \right) \\ &= \sum_{\mathbf{n}\geq\mathbf{0}} \sum_{\Pi(\mathbf{n})} \left(\prod_{\mathbf{i}=\mathbf{0}}^{\mathbf{n}} x_{\mathbf{i},k_{\mathbf{i}}} \left(\omega\mathbf{t}^{\mathbf{i}}\right)^{k_{\mathbf{i}}} \right), \end{aligned}$$

and exploiting $x_{\mathbf{0},0}=0$, the last expression becomes

$$\prod_{\mathbf{i}\geq\mathbf{0}} \left(\sum_{k_{\mathbf{i}}\geq 0} x_{\mathbf{i},k_{\mathbf{i}}} \left(\omega\mathbf{t}^{\mathbf{i}}\right)^{k_{\mathbf{i}}} \right) = \prod_{\mathbf{i}\cdot\mathbf{1}\geq 1} \left(\sum_{k_{\mathbf{i}}\geq 0} x_{\mathbf{i},k_{\mathbf{i}}} \left(\omega\mathbf{t}^{\mathbf{i}}\right)^{k_{\mathbf{i}}} \right).$$

□

For $x_{\mathbf{i},s}=a_{\mathbf{i},s}\in\{0,1\}$ for all $\mathbf{i}\in\mathbb{N}^r$, $\mathbf{i}\neq\mathbf{0}$, $s=0,1,2,\dots$, we obtain:

Corollary 12. *Let* $A=(a_{\mathbf{i},s})$, $\mathbf{i}\in\mathbb{N}^r, \mathbf{i}\neq\mathbf{0}$, $s=0,1,2,\dots$, *with* $a_{\mathbf{i},s}=0$ *or* 1 *and* $\mathbf{p}(\mathbf{n},k;A)$ *be the number of partitions of* $\mathbf{n}$ *into* k *parts whose number* $y_{\mathbf{i}}$ *of parts which are equal to* $\mathbf{i}$ *belongs to the set* $Y_{\mathbf{i}}=\{s: a_{\mathbf{i},s}=1\}$, $\mathbf{i}\in\mathbb{N}^r$, $\mathbf{i}\neq\mathbf{0}$. *Then*

$$F(t,u,\omega;A) := \sum_{\mathbf{n}\geq\mathbf{0}} \left(\sum_{k=0}^{\mathbf{n}\cdot\mathbf{1}} \mathbf{p}(\mathbf{n},k;A)\,\omega^k \right) \mathbf{t}^{\mathbf{n}} = \prod_{\mathbf{i}\cdot\mathbf{1}\geq 1} \left(\sum_{s\geq 0} a_{\mathbf{i},s} \left(\omega\mathbf{t}^{\mathbf{i}}\right)^s \right).$$

For $x_{\mathbf{i},s}=1$ for all $\mathbf{i}\in\mathbb{N}^r$, $\mathbf{i}\neq\mathbf{0}$, $s=0,1,2,\dots$, we obtain:

Corollary 13. *Let* $\mathbf{p}(\mathbf{n})$ *be the number of partitions of the* r*-partite number* $\mathbf{n}$ *with* $\mathbf{p}(\mathbf{0}) = 1$ *and* $\mathbf{p}(\mathbf{n}, k)$ *the number of partitions of the* r*-bipartite number* $\mathbf{n}$ *into* k *parts, with* $\mathbf{p}(\mathbf{0}, 0) = 1$. *Then*

$$\sum_{\mathbf{n}\geq\mathbf{0}} \left(\sum_{k=0}^{\mathbf{n}\cdot\mathbf{1}} p(\mathbf{n}, k)\, \omega^k \right) \mathbf{t}^{\mathbf{n}} = \prod_{\mathbf{i}\cdot\mathbf{1}\geq 1} \left(1 - \omega \mathbf{t}^{\mathbf{i}}\right)^{-1}.$$

Consequently

$$\sum_{\mathbf{n}\geq\mathbf{0}} p(\mathbf{n})\, \mathbf{t}^{\mathbf{n}} = \prod_{\mathbf{i}\cdot\mathbf{1}\geq 1} \left(1 - \mathbf{t}^{\mathbf{i}}\right)^{-1}.$$

Remark 14. If we take $t_1 = \cdots = t_r = t$, we obtain

$$\begin{aligned}\prod_{i\geq 1} \left(1 - \omega t^i\right)^{-\binom{i+r-1}{r-1}} &= \sum_{n\geq 0}\sum_{k=0}^{n} \left(\sum_{n_1+\cdots+n_r=n} \mathbf{p}(\mathbf{n}, k) \right) \omega^k t^n \\ &= \sum_{n\geq 0} \left(\sum_{n_1+\cdots+n_r=n} \mathbf{p}_{\mathbf{n}}(\omega) \right) t^n,\end{aligned}$$

and more generally, for nonnegative integers $a_1, \ldots, a_r$ and $t_1 = t^{a_1}, \ldots, t_r = t^{a_r}$, we obtain

$$\prod_{i\geq 1} \left(1 - \omega t^i\right)^{-f(i,r)} = \sum_{n\geq 0} \left(\sum_{a_1n_1+\cdots+a_rn_r=n} \mathbf{p}_{\mathbf{n}}(\omega) \right) t^n,$$

where

$$\mathbf{p}_{\mathbf{n}}(\omega) = \sum_{k=0}^{a_1n_1+\cdots+a_rn_r} \mathbf{p}(\mathbf{n}, k)\, \omega^k,$$

and where $f(n, r)$ is the number of solutions of the integer equation

$$a_1n_1 + a_2n_2 + \cdots + a_rn_r = n.$$

Acknowledgments The authors wish to express their gratitude to the referee for his/her valuable advice and comments which helped to greatly in improving the quality of the paper.

References

[1] E. T. Bell, Exponential polynomials. *Ann. Math.* **35** (1934), 258-277.

[2] Charalambos A. Charalambides, *Enumerative Combinatorics.* Chapman & Hall/CRC, A CRC Press Company, Boca Raton, London, New York, Washington, D.C. (2001).

[3] L. Comtet, *Advanced Combinatorics.* D. Reidel Publishing Company, Dordrecht-Holland / Boston-U.S.A, (1974).

[4] M. Mihoubi, Bell polynomials and binomial type sequences. *Discrete Math.*, 308 (2008), 2450–2459.

SOME WEIGHTED SUMS OF POWERS OF FIBONACCI POLYNOMIALS

Claudio Pita Ruiz
Department of Mathematics, Universidad Panamericana, Mexico City, Mexico
cpita@up.edu.mx

Received: 10/8/12, Revised: 4/6/13, Accepted: 8/10/13, Published: 9/26/13

Abstract

We obtain closed formulas for some weighted sums of powers of bivariate Fibonacci and Lucas polynomials of the form $\sum_{n=0}^{q} \lambda^n(x,y) F_{tsn+l}^k(x,y)$ and $\sum_{n=0}^{q} \lambda^n(x,y) L_{tsn+l}^k(x,y)$, in the cases $k = 1,2,3,4$, and some specific values of the parameters t and l. We express these sums as linear combinations of the Fibopolynomials $\binom{q+m}{tk}_{F_s(x,y)}$, $m = 1,2,\ldots,tk$.

1. Introduction

The sequence of generalized bivariate Fibonacci polynomials $G_n(x,y)$ is defined by the second-order recurrence $G_{n+2}(x,y) = xG_{n+1}(x,y) + yG_n(x,y)$, with arbitrary initial conditions $G_0(x,y)$ and $G_1(x,y)$. When $G_0(x,y) = 0$ and $G_1(x,y) = 1$ we have the bivariate Fibonacci polynomials $F_n(x,y)$, and when $G_0(x,y) = 2$ and $G_1(x,y) = x$ we have the bivariate Lucas polynomials $L_n(x,y)$. (What we will use about these polynomials is contained in reference [2].) The corresponding extensions to negative indices are given by $F_{-n}(x,y) = -(-y)^{-n} F_n(x,y)$ and $L_{-n}(x,y) = (-y)^{-n} L_n(x,y)$, $n \in \mathbb{N}$, respectively. In the case $y = 1$, we have the Fibonacci and Lucas polynomials (in the variable x), $F_n(x,1)$ and $L_n(x,1)$, denoted simply as $F_n(x)$ and $L_n(x)$. By setting $x = 1$ in these polynomials, we obtain the numerical sequences $F_n(1)$ and $L_n(1)$, denoted as F_n and L_n, corresponding to the Fibonacci sequence $F_n = (0,1,1,2,3,5,\ldots)$ and the Lucas sequence $L_n = (2,1,3,4,7,11,\ldots)$.

In a recent work [7] we showed that the sequence $G_{tsn+\mu}^k(x,y)$, where $t,s,k \in \mathbb{N}$ and $\mu \in \mathbb{Z}$ are given parameters, can be written as a linear combination of the (s-)Fibopolynomials $\binom{n+tk-i}{tk}_{F_s(x,y)}$, $i = 0,1,\ldots,tk$, according to

$$G_{tsn+\mu}^k(x,y) = (-1)^{s+1} \sum_{i=0}^{tk} \sum_{j=0}^{i} (-1)^{\frac{(sj+2(s+1))(j+1)}{2}} \binom{tk+1}{j}_{F_s(x,y)} \times G_{ts(i-j)+\mu}^k(x,y)\, y^{\frac{sj(j-1)}{2}} \binom{n+tk-i}{tk}_{F_s(x,y)}. \quad (1)$$

Recall that for integers $n, p \geq 0$, we have $\binom{n}{0}_{F_s(x,y)} = \binom{n}{n}_{F_s(x,y)} = 1$ and

$$\binom{n}{p}_{F_s(x,y)} = \frac{F_{sn}(x,y)\,F_{s(n-1)}(x,y)\cdots F_{s(n-p+1)}(x,y)}{F_s(x,y)\,F_{2s}(x,y)\cdots F_{ps}(x,y)}, \qquad 0<p<n.$$

If n or p are negative, or $p>n$, we have $\binom{n}{p}_{F_s(x,y)} = 0$. It is known that $\binom{n}{p}_{F_s(x,y)}$ are indeed polynomials in x and y. When $x=y=1$ we have the s-Fibonomials $\binom{n}{p}_{F_s}$ (see [6]), and when $x=y=s=1$, we have the (usual) Fibonomials $\binom{n}{p}_F$, introduced and studied by Hoggatt [3] in 1967.

If $\lambda(x,y)$ is a (non-zero) given real function of the real variables x and y, we see from (1) that

$$\begin{aligned}
&\sum_{n=0}^{q}\lambda^n(x,y)\,G^k_{tsn+\mu}(x,y)\\
&= (-1)^{s+1}\sum_{i=0}^{tk}\sum_{j=0}^{i}(-1)^{\frac{(sj+2(s+1))(j+1)}{2}}\binom{tk+1}{j}_{F_s(x,y)}\\
&\quad\times G^k_{ts(i-j)+\mu}(x,y)\,y^{\frac{sj(j-1)}{2}}\sum_{n=0}^{q}\lambda^n(x,y)\binom{n+tk-i}{tk}_{F_s(x,y)}.
\end{aligned} \tag{2}$$

Expression (2) can be written as

$$\begin{aligned}
&\sum_{n=0}^{q}\lambda^n(x,y)\,G^k_{tsn+\mu}(x,y)\\
&= (-1)^{s+1}\sum_{m=1}^{tk}\sum_{i=0}^{tk-m}\sum_{j=0}^{i}(-1)^{\frac{(sj+2(s+1))(j+1)}{2}}\binom{tk+1}{j}_{F_s(x,y)}\\
&\quad\times G^k_{ts(i-j)+\mu}(x,y)\,y^{\frac{sj(j-1)}{2}}\lambda^{i-tk+q+m}(x,y)\binom{q+m}{tk}_{F_s(x,y)}\\
&\quad+(-1)^{s+1}\lambda^{-tk}(x,y)\sum_{i=0}^{tk}\sum_{j=0}^{i}(-1)^{\frac{(sj+2(s+1))(j+1)}{2}}\binom{tk+1}{j}_{F_s(x,y)}\\
&\quad\times G^k_{ts(i-j)+\mu}(x,y)\,y^{\frac{sj(j-1)}{2}}\lambda^{i}(x,y)\sum_{n=0}^{q}\lambda^n(x,y)\binom{n}{tk}_{F_s(x,y)}.
\end{aligned} \tag{3}$$

This simple observation derived from (1) allows us to obtain closed formulas, in terms of Fibopolynomials, for the weighted sums $\sum_{n=0}^{q}\lambda^n(x,y)\,G^k_{tsn+\mu}(x,y)$, as the following proposition states (with a straightforward proof using (3)).

Proposition 1. *If $z=\lambda(x,y)$ is a root of*

$$\sum_{i=0}^{tk}\sum_{j=0}^{i}(-1)^{\frac{(sj+2(s+1))(j+1)}{2}}\binom{tk+1}{j}_{F_s(x,y)}G^k_{ts(i-j)+\mu}(x,y)\,y^{\frac{sj(j-1)}{2}}z^i=0, \tag{4}$$

then the weighted sum of k-th powers $\sum_{n=0}^{q} \lambda^n(x,y) G_{tsn+\mu}^k(x,y)$ *can be expressed as a linear combination of the s-Fibopolynomials* $\binom{q+m}{tk}_{F_s(x,y)}$, $m = 1, 2, \ldots, tk$, *according to*

$$\begin{aligned}
&\sum_{n=0}^{q} \lambda^{n-q}(x,y) G_{tsn+\mu}^k(x,y) \\
&\quad = (-1)^{s+1} \sum_{m=1}^{tk} \sum_{i=0}^{tk-m} \sum_{j=0}^{i} (-1)^{\frac{(sj+2(s+1))(j+1)}{2}} \binom{tk+1}{j}_{F_s(x,y)} \\
&\qquad \times G_{ts(i-j)+\mu}^k(x,y)\, y^{\frac{sj(j-1)}{2}} \lambda^{i-tk+m}(x,y) \binom{q+m}{tk}_{F_s(x,y)}.
\end{aligned} \tag{5}$$

Moreover, suppose expression (5) is valid for some weight function $\lambda(x,y)$. *Then* $z = \lambda(x,y)$ *is a root of (4).*

Let us consider the simplest case $t = k = 1$. Equation (4) is in this case

$$\begin{aligned}
0 &= \sum_{i=0}^{1} \sum_{j=0}^{i} (-1)^{\frac{(sj+2(s+1))(j+1)}{2}} \binom{2}{j}_{F_s(x,y)} G_{s(i-j)+\mu}(x,y)\, y^{\frac{sj(j-1)}{2}} z^i \\
&= G_\mu(x,y) + (G_{s+\mu}(x,y) - L_s(x,y) G_\mu(x,y))\, z.
\end{aligned} \tag{6}$$

If $G_{s+\mu}(x,y) - L_s(x,y) G_\mu(x,y) \neq 0$, we have from (6) the weight

$$\lambda(x,y) = \frac{G_\mu(x,y)}{L_s(x,y) G_\mu(x,y) - G_{s+\mu}(x,y)}. \tag{7}$$

Expression (5) for the corresponding weighted sum is in this case

$$\sum_{n=0}^{q} \lambda^n(x,y) G_{sn+\mu}(x,y) = G_\mu(x,y) \lambda^q(x,y) \binom{q+1}{1}_{F_s(x,y)}. \tag{8}$$

In the Fibonacci case we have, from (7) and (8), that if $\mu \neq s$ then

$$\sum_{n=0}^{q} \left(\frac{F_\mu(x,y)}{(-y)^s F_{\mu-s}(x,y)} \right)^{n-q} F_{sn+\mu}(x,y) = \frac{F_\mu(x,y)}{F_s(x,y)} F_{s(q+1)}(x,y). \tag{9}$$

Similarly, from (7) and (8), we have in the Lucas case that

$$\sum_{n=0}^{q} \left(\frac{L_\mu(x,y)}{(-y)^s L_{\mu-s}(x,y)} \right)^{n-q} L_{sn+\mu}(x,y) = \frac{L_\mu(x,y)}{F_s(x,y)} F_{s(q+1)}(x,y). \tag{10}$$

Thus, if $\mu \neq 0, s$ we can combine (9) and (10) together as

$$\begin{aligned}
&\frac{1}{F_\mu(x,y)}\sum_{n=0}^{q}\left(\frac{F_\mu(x,y)}{(-y)^s F_{\mu-s}(x,y)}\right)^{n-q} F_{sn+\mu}(x,y)\\
=\;&\frac{1}{L_\mu(x,y)}\sum_{n=0}^{q}\left(\frac{L_\mu(x,y)}{(-y)^s L_{\mu-s}(x,y)}\right)^{n-q} L_{sn+\mu}(x,y)\\
=\;&\frac{1}{F_s(x,y)}F_{s(q+1)}(x,y). \qquad (11)
\end{aligned}$$

We call attention to the fact that the term $\frac{1}{F_s(x,y)}F_{s(q+1)}(x,y)$ in (11) *does not depend on the parameter* μ. That is, *in (11) we have infinitely many weighted sums of* $q+1$ *bivariate Fibonacci and Lucas polynomials that are equal to* $\frac{1}{F_s(x,y)}F_{s(q+1)}(x,y)$.

We can have a slight generalization of (11) if we substitute x by $L_r(x,y)$ and y by $(-1)^{r+1}y^r$, where $r \in \mathbb{Z}$. By using the equations

$$\begin{aligned}
F_r(x,y)\,F_n\left(L_r(x,y),(-1)^{r+1}y^r\right) &= F_{rn}(x,y),\\
L_n\left(L_r(x,y),(-1)^{r+1}y^r\right) &= L_{rn}(x,y),
\end{aligned}$$

we get for $r \neq 0$ (and $\mu \neq 0, s$ in the Fibonacci sums)

$$\begin{aligned}
&\frac{1}{F_{r\mu}(x,y)}\sum_{n=0}^{q}\left(\frac{F_{r\mu}(x,y)}{(-y)^{rs} F_{r(\mu-s)}(x,y)}\right)^{n-q} F_{r(sn+\mu)}(x,y)\\
=\;&\frac{1}{L_{r\mu}(x,y)}\sum_{n=0}^{q}\left(\frac{L_{r\mu}(x,y)}{(-y)^{rs} L_{r(\mu-s)}(x,y)}\right)^{n-q} L_{r(sn+\mu)}(x,y)\\
=\;&\frac{1}{F_{rs}(x,y)}F_{rs(q+1)}(x,y). \qquad (12)
\end{aligned}$$

Some examples from (12) (corresponding to some specific values of μ) are the following:

- With $s = 1$ we have

$$\begin{aligned}
&\frac{1}{F_{-2r}(x,y)}\sum_{n=0}^{q}\left(\frac{F_{2r}(x,y)}{F_{3r}(x,y)}\right)^{n-q} F_{r(n-2)}(x,y) \qquad (13)\\
=\;&\frac{1}{F_{-r}(x,y)}\sum_{n=0}^{q}\left(\frac{1}{L_r(x,y)}\right)^{n-q} F_{r(n-1)}(x,y)\\
=\;&\frac{1}{F_{2r}(x,y)}\sum_{n=0}^{q}\left(\frac{L_r(x,y)}{(-y)^r}\right)^{n-q} F_{r(n+2)}(x,y)\\
=\;&\frac{1}{F_{3r}(x,y)}\sum_{n=0}^{q}\left(\frac{F_{3r}(x,y)}{(-y)^r F_{2r}(x,y)}\right)^{n-q} F_{r(n+3)}(x,y)
\end{aligned}$$

$$= \frac{1}{L_{-2r}(x,y)} \sum_{n=0}^{q} \left(\frac{L_{2r}(x,y)}{L_{3r}(x,y)} \right)^{n-q} L_{r(n-2)}(x,y)$$

$$= \frac{1}{2} \sum_{n=0}^{q} \left(\frac{2}{L_r(x,y)} \right)^{n-q} L_{rn}(x,y) = \frac{1}{L_r(x,y)} \sum_{n=0}^{q} \left(\frac{L_r(x,y)}{2(-y)^r} \right)^{n-q} L_{r(n+1)}(x,y)$$

$$= \frac{1}{L_{-r}(x,y)} \sum_{n=0}^{q} \left(\frac{L_r(x,y)}{L_{2r}(x,y)} \right)^{n-q} L_{r(n-1)}(x,y) = \frac{F_{r(q+1)}(x,y)}{F_r(x,y)}.$$

- With $s = 2$ we have

$$\frac{1}{F_{-3r}(x,y)} \sum_{n=0}^{q} \left(\frac{F_{3r}(x,y)}{F_{5r}(x,y)} \right)^{n-q} F_{r(2n-3)}(x,y) \tag{14}$$

$$= \frac{1}{F_{-r}(x,y)} \sum_{n=0}^{q} \left(\frac{F_r(x,y)}{F_{3r}(x,y)} \right)^{n-q} F_{r(2n-1)}(x,y)$$

$$= \frac{1}{F_{3r}(x,y)} \sum_{n=0}^{q} \left(\frac{F_{3r}(x,y)}{y^{2r} F_r(x,y)} \right)^{n-q} F_{r(2n+3)}(x,y)$$

$$= \frac{1}{L_{-3r}(x,y)} \sum_{n=0}^{q} \left(\frac{L_{3r}(x,y)}{L_{5r}(x,y)} \right)^{n-q} L_{r(2n-3)}(x,y)$$

$$= \frac{1}{L_{-r}(x,y)} \sum_{n=0}^{q} \left(\frac{L_r(x,y)}{L_{3r}(x,y)} \right)^{n-q} L_{r(2n-1)}(x,y)$$

$$= \frac{(-1)^{rq}}{L_r(x,y)} \sum_{n=0}^{q} \frac{(-1)^{rn} L_{r(2n+1)}(x,y)}{y^{r(n-q)}} = \frac{(-1)^{(r+1)q}}{F_r(x,y)} \sum_{n=0}^{q} \frac{(-1)^{(r+1)n} F_{r(2n+1)}(x,y)}{y^{r(n-q)}}$$

$$= \frac{1}{L_{3r}(x,y)} \sum_{n=0}^{q} \left(\frac{L_{3r}(x,y)}{y^{2r} L_r(x,y)} \right)^{n-q} L_{r(2n+3)}(x,y) = \frac{F_{2r(q+1)}(x,y)}{F_{2r}(x,y)}.$$

- With $s = 3$ we have

$$\frac{1}{F_r(x,y)} \sum_{n=0}^{q} \left(\frac{-1}{(-y)^r L_r(x,y)} \right)^{n-q} F_{r(3n+1)}(x,y) \tag{15}$$

$$= \frac{1}{F_{2r}(x,y)} \sum_{n=0}^{q} \left(-\frac{L_r(x,y)}{y^{2r}} \right)^{n-q} F_{r(3n+2)}(x,y)$$

$$= \frac{1}{F_{4r}(x,y)} \sum_{n=0}^{q} \left(\frac{L_r(x,y)\, L_{2r}(x,y)}{(-y)^{3r}} \right)^{n-q} F_{r(3n+4)}(x,y)$$

$$= -\frac{(-y)^r}{F_r(x,y)} \sum_{n=0}^{q} \left(\frac{1}{L_r(x,y)\, L_{2r}(x,y)} \right)^{n-q} F_{r(3n-1)}(x,y)$$

$$= \frac{1}{L_{-r}(x,y)}\sum_{n=0}^{q}\left(\frac{L_r(x,y)}{L_{4r}(x,y)}\right)^{n-q}L_{r(3n-1)}(x,y)$$
$$= \frac{1}{L_r(x,y)}\sum_{n=0}^{q}\left(\frac{L_r(x,y)}{(-y)^r L_{2r}(x,y)}\right)^{n-q}L_{r(3n+1)}(x,y)$$
$$= \frac{1}{L_{2r}(x,y)}\sum_{n=0}^{q}\left(\frac{L_{2r}(x,y)}{y^{2r}L_r(x,y)}\right)^{n-q}L_{r(3n+2)}(x,y)$$
$$= \frac{1}{L_{4r}(x,y)}\sum_{n=0}^{q}\left(\frac{L_{4r}(x,y)}{(-y)^{3r}L_r(x,y)}\right)^{n-q}L_{r(3n+4)}(x,y) = \frac{F_{3r(q+1)}(x,y)}{F_{3r}(x,y)}.$$

In particular we have the following numerical identities (from (13), (14) and (15) with $x = y = r = 1$)

$$\begin{aligned} F_{q+1} &= -\sum_{n=0}^{q} 2^{q-n}F_{n-2} = \sum_{n=0}^{q} F_{n-1} = \sum_{n=0}^{q}(-1)^{n+q}F_{n+2} \\ &= \frac{1}{2}\sum_{n=0}^{q}(-2)^{n-q}F_{n+3} = \frac{1}{3}\sum_{n=0}^{q}\left(\frac{3}{4}\right)^{n-q}L_{n-2} = \sum_{n=0}^{q}\frac{-1}{3^{n-q}}L_{n-1} \\ &= \sum_{n=0}^{q} 2^{n-q-1}L_n = \sum_{n=0}^{q}(-2)^{q-n}L_{n+1}. \end{aligned} \tag{16}$$

$$\begin{aligned} F_{2(q+1)} &= \frac{1}{2}\sum_{n=0}^{q}\left(\frac{2}{5}\right)^{n-q}F_{2n-3} = \sum_{n=0}^{q} 2^{q-n}F_{2n-1} = \sum_{n=0}^{q} F_{2n+1} \\ &= \sum_{n=0}^{q} 2^{n-q-1}F_{2n+3} = -\frac{1}{4}\sum_{n=0}^{q}\left(\frac{4}{11}\right)^{n-q}L_{2n-3} = -\sum_{n=0}^{q} 4^{q-n}L_{2n-1} \\ &= \sum_{n=0}^{q}(-1)^{n+q}L_{2n+1} = \sum_{n=0}^{q} 4^{n-q-1}L_{2n+3}. \end{aligned} \tag{17}$$

$$\begin{aligned} \frac{1}{2}F_{3(q+1)} &= \sum_{n=0}^{q} F_{3n+1} = \sum_{n=0}^{q}(-1)^{n-q}F_{3n+2} = \frac{1}{3}\sum_{n=0}^{q}(-3)^{n-q}F_{3n+4} \\ &= \sum_{n=0}^{q}\frac{1}{3^{n-q}}F_{3n-1} = -\sum_{n=0}^{q}\frac{1}{7^{n-q}}L_{3n-1} = \sum_{n=0}^{q}\left(-\frac{1}{3}\right)^{n-q}L_{3n+1} \\ &= \sum_{n=0}^{q} 3^{n-q-1}L_{3n+2} = \frac{1}{7}\sum_{n=0}^{q}(-7)^{n-q}L_{3n+4}. \end{aligned} \tag{18}$$

Of course, some of the numerical results shown in (16), (17) and (18) are well-known identities. For example, in [1]: (i) formula $\sum_{n=0}^{q} 2^{n-q-1}L_n = F_{q+1}$ of (16) is identity 236, (ii) formula $\sum_{n=0}^{q} F_{2n+1} = F_{2(q+1)}$ of (17) is essentially identity 2, and (iii) formula $\sum_{n=0}^{q} F_{3n+1} = \frac{1}{2} F_{3(q+1)}$ in (18) is essentially identity 24 (corrected).

In the remaining sections of this work we consider only the case $y = 1$ of Proposition 1. The corresponding weight function $\lambda(x, 1)$ will be denoted as $\lambda(x)$.

2. Weighted Sums of Cubes

In this section we will obtain expressions for some weighted sums of cubes of Fibonacci polynomials. More specifically, we will consider the Fibonacci case of Proposition 1 when $k = 3$, $\mu = 0$ and $t = 1$.

In this case, equation (4) is (after some simplifications)

$$F_s^3(x)\, z \left(z^2 + 2L_s(x)\, z + (-1)^s\right) = 0. \tag{19}$$

Thus, we have two weights, namely

$$\lambda_1(x) = -L_s(x) + \sqrt{L_s^2(x) - (-1)^s}\ ,\ \lambda_2(x) = -L_s(x) - \sqrt{L_s^2(x) - (-1)^s}. \tag{20}$$

The corresponding weighted sum (5) can be written as

$$\sum_{n=0}^{q} \lambda^{n-q}(x)\, F_{sn}^3(x) = F_s^3(x)\left((-1)^{s+1}\lambda(x)\binom{q+1}{3}_{F_s(x)} + \binom{q+2}{3}_{F_s(x)}\right), \tag{21}$$

where $\lambda(x)$ is any of the weights (20).

In the case $x = 1$, we have in particular the following numerical identities (obtained by setting $s = 1$ and $s = 2$ in (21) with the corresponding weights (20))

$$\sum_{n=0}^{q} \left(-1 \pm \sqrt{2}\right)^{n-q} F_n^3 = \frac{F_{q+1}F_q}{2}\left(\left(-1 \pm \sqrt{2}\right) F_{q-1} + F_{q+2}\right). \tag{22}$$

$$\sum_{n=0}^{q} \left(-3 \pm 2\sqrt{2}\right)^{n-q} F_{2n}^3 = \frac{F_{2(q+1)}F_{2q}}{24}\left(-\left(-3 \pm 2\sqrt{2}\right) F_{2(q-1)} + F_{2(q+2)}\right). \tag{23}$$

We will show now some different versions of (21), involving Chebyshev polynomials of the first kind $T_n(x)$, or of the second kind $U_n(x)$. The results are the following.

Proposition 2. *(a) If s is even, we have the following weighted sums of cubes of Fibonacci polynomials, valid for any integer $l \geq 0$*

$$\sum_{n=1}^{q} T_{n+l}(-L_s(x))\, F_{sn}^3(x) \tag{24}$$

$$= F_s^3(x)\left(-T_{q+l+1}(-L_s(x))\binom{q+1}{3}_{F_s(x)} + T_{q+l}(-L_s(x))\binom{q+2}{3}_{F_s(x)}\right).$$

$$\sum_{n=1}^{q} U_{n+l-1}\left(-L_s\left(x\right)\right) F_{sn}^3\left(x\right) \tag{25}$$
$$= F_s^3\left(x\right)\left(-U_{q+l}\left(-L_s\left(x\right)\right)\binom{q+1}{3}_{F_s(x)} + U_{q+l-1}\left(-L_s\left(x\right)\right)\binom{q+2}{3}_{F_s(x)}\right).$$

(b) If s is odd, we have the following weighted sums of cubes of Fibonacci polynomials (where $i^2=-1$), valid for any integer $l \geq 0$

$$\sum_{n=1}^{q} \left(-i\right)^n T_{n+l}\left(-iL_s\left(x\right)\right) F_{sn}^3\left(x\right) \tag{26}$$
$$= F_s^3\left(x\right)\left(\left(-i\right)^{q+1} T_{q+l+1}\left(-iL_s\left(x\right)\right)\binom{q+1}{3}_{F_s(x)} + \left(-i\right)^q T_{q+l}\left(-iL_s\left(x\right)\right)\binom{q+2}{3}_{F_s(x)}\right).$$

$$\sum_{n=1}^{q} \left(-i\right)^{n-1} U_{n+l-1}\left(-iL_s\left(x\right)\right) F_{sn}^3\left(x\right) \tag{27}$$
$$= F_s^3\left(x\right)\left(\left(-i\right)^q U_{q+l}\left(-iL_s\left(x\right)\right)\binom{q+1}{3}_{F_s(x)} + \left(-i\right)^{q-1} U_{q+l-1}\left(-iL_s\left(x\right)\right)\binom{q+2}{3}_{F_s(x)}\right).$$

Proof. Recall that Chebyshev polynomials of the first kind $T_n\left(x\right)$ can be calculated as

$$T_n\left(x\right) = \frac{1}{2}\left(\left(x+\sqrt{x^2-1}\right)^n + \left(x-\sqrt{x^2-1}\right)^n\right), \tag{28}$$

and that Chebyshev polynomials of the second kind $U_n\left(x\right)$ can be calculated as

$$U_n\left(x\right) = \frac{1}{2\sqrt{x^2-1}}\left(\left(x+\sqrt{x^2-1}\right)^{n+1} - \left(x-\sqrt{x^2-1}\right)^{n+1}\right). \tag{29}$$

We can write (21) as

$$\sum_{n=0}^{q} \lambda^{n+l}\left(x\right) F_{sn}^3\left(x\right) \tag{30}$$
$$= F_s^3\left(x\right)\left(\left(-1\right)^{s+1}\lambda^{q+l+1}\left(x\right)\binom{q+1}{3}_{F_s(x)} + \lambda^{q+l}\left(x\right)\binom{q+2}{3}_{F_s(x)}\right),$$

where l is a non-negative integer. If s is even, the weights (20) are

$$\lambda_1(x) = -L_s(x) + \sqrt{L_s^2(x) - 1} \quad , \quad \lambda_2(x) = -L_s(x) - \sqrt{L_s^2(x) - 1}. \tag{31}$$

Thus, (24) follows from (28), (30) and (31). Similarly, (25) follows from (29), (30) and (31).

On the other hand, if s is odd the weights (20) are $\lambda_1(x) = -L_s(x) + \sqrt{L_s^2(x) + 1}$, $\lambda_2(x) = -L_s(x) - \sqrt{L_s^2(x) + 1}$. These weights can be written as

$$\begin{aligned} \lambda_1(x) &= -i\left(-iL_s(x) + \sqrt{(-iL_s(x))^2 - 1}\right), \\ \lambda_2(x) &= -i\left(-iL_s(x) - \sqrt{(-iL_s(x))^2 - 1}\right). \end{aligned} \tag{32}$$

From (28) and (32) we see that

$$(-i)^n T_n(-iL_s(x)) = \frac{1}{2}\left(\lambda_1^n(x) + \lambda_2^n(x)\right), \tag{33}$$

and from (29) and (32) we see that

$$(-i)^{n+1} U_n(-iL_s(x)) = \frac{1}{2i\sqrt{L_s^2(x) + 1}}\left(\lambda_1^{n+1}(x) - \lambda_2^{n+1}(x)\right). \tag{34}$$

Thus, (26) follows from (30), (32) and (33). Similarly, (27) follows from (30), (32) and (34). □

We note that the integer parameter $l \geq 0$ gives us, in (24) and (25), infinitely many weighted sums of cubes of Fibonacci polynomials for each even s, and in (26) and (27) gives us infinitely many weighted sums of cubes of Fibonacci polynomials for each odd s.

We show some numerical examples from formulas (24) to (27). We set $x = 1$ and $q = 5$ in them.

If $s = 2$, the sequence $(T_n(-L_2))_{n=1}^{\infty}$ involved in (24) is

$$(T_n(-L_2))_{n=1}^{\infty} = (-3, 17, -99, 577, -3363, 19601, -114243, 665857, \ldots).$$

(See [5, A001541] for the unsigned version.) Thus, we have the following weighted sums of cubes of Fibonacci numbers (corresponding to $l = 0, 1$)

$$\begin{aligned} -3F_2^3 + 17F_4^3 - 99F_6^3 + 577F_8^3 - 3363F_{10}^3 &= \frac{F_{10}F_{12}}{L_2F_6}(-19601F_8 - 3363F_{14}), \\ 17F_2^3 - 99F_4^3 + 577F_6^3 - 3363F_8^3 + 19601F_{10}^3 &= \frac{F_{10}F_{12}}{L_2F_6}(114243F_8 + 19601F_{14}). \end{aligned}$$

Similarly, the sequence $(U_n(-L_2))_{n=1}^{\infty}$ involved in (25) is

$$(U_n(-L_2))_{n=1}^{\infty} = (-6, 35, -204, 1189, -6930, 40391, -235416, 1372105, \ldots).$$

(See [5, A001109] for the unsigned version.) Thus, we have (for $l = 1, 2$)

$$-6F_2^3 + 35F_4^3 - 204F_6^3 + 1189F_8^3 - 6930F_{10}^3 = \frac{F_{10}F_{12}}{L_2F_6}\left(-40391F_8 - 6930F_{14}\right),$$

$$35F_2^3 - 204F_4^3 + 1189F_6^3 - 6930F_8^3 + 40391F_{10}^3 = \frac{F_{10}F_{12}}{L_2F_6}\left(235416F_8 + 40391F_{14}\right).$$

If $s = 1$, the sequence $\left((-i)^n T_n(-iL_1)\right)_{n=1}^{\infty}$ involved in (26) is

$$\left((-i)^n T_n(-iL_1)\right)_{n=1}^{\infty} = (-1, 3, -7, 17, -41, 99, -239, 577, -1393, 3363, -8119, \ldots).$$

(See [5, A001333] for the unsigned version.) Then, we have the weighted sums (for $l = 0, 1$)

$$\begin{aligned} -F_1^3 + 3F_2^3 - 7F_3^3 + 17F_4^3 - 41F_5^3 &= \frac{F_6F_5}{2}\left(99F_4 - 41F_7\right), \\ 3F_1^3 - 7F_2^3 + 17F_3^3 - 41F_4^3 + 99F_5^3 &= \frac{F_6F_5}{2}\left(-239F_4 + 99F_7\right). \end{aligned}$$

Similarly, if $s = 3$, the sequence $\left((-i)^n U_n(-iL_3)\right)_{n=1}^{\infty}$ involved in (27) is

$$\begin{aligned} &\left((-i)^n U_n(-iL_3)\right)_{n=1}^{\infty} \\ &\qquad = (-8, 65, -528, 4289, -34840, 283009, -2298912, 18674305, \ldots). \end{aligned}$$

(See [5, A041025] for the unsigned version.) Then, we have the weighted sums (for $l = 1, 2$)

$$\begin{aligned} &-8F_3^3 + 65F_6^3 - 528F_9^3 + 4289F_{12}^3 - 34840F_{15}^3 \\ &\qquad = \frac{F_3^2F_{18}F_{15}}{F_6F_9}\left(283009F_{12} - 34840F_{21}\right), \end{aligned}$$

$$\begin{aligned} &65F_3^3 - 528F_6^3 + 4289F_9^3 - 34840F_{12}^3 + 283009F_{15}^3 \\ &\qquad = \frac{F_3^2F_{18}F_{15}}{F_6F_9}\left(-2298912F_{12} + 283009F_{21}\right). \end{aligned}$$

3. Other Weighted Sums

In this section we will obtain expressions for some other weighted sums of Fibonacci and Lucas polynomials. More specifically, we will set $\mu = 0$ in Proposition 1 and consider the following cases: (i) $k = t = 2$ and $G = F$ or $G = L$; (ii) $k = 4$, $t = 1$, $G = F$.

Let us begin with the case (i). When $G = F$, equation (4) is

$$\begin{aligned} 0 &= \sum_{i=0}^{4}\sum_{j=0}^{i}(-1)^{\frac{(sj+2(s+1))(j+1)}{2}}\binom{5}{j}_{F_s(x)} F_{2s(i-j)}^2(x)\,z^i \\ &= (-1)^{s+1}F_{2s}^2(x,y)\,z\,(z+1)\left(z^2-(-1)^s L_{2s}(x)\,z+1\right), \end{aligned} \tag{35}$$

and with $G = L$ is

$$\begin{aligned} 0 &= \sum_{i=0}^{4}\sum_{j=0}^{i}(-1)^{\frac{(sj+2(s+1))(j+1)}{2}}\binom{5}{j}_{F_s(x)} L_{2s(i-j)}^2(x)\,z^i \\ &= (-1)^{s+1}\left(L_{2s}^2(x)\,z^2-(3L_{4s}(x)+2)\,z+4\right)\left(z^2-(-1)^s L_{2s}(x)\,z+1\right). \end{aligned} \tag{36}$$

Observe that

$$z^2-(-1)^s L_{2s}(x)\,z+1 = \left(z-(-1)^s\alpha^{2s}(x)\right)\left(z-(-1)^s\beta^{2s}(x)\right),$$

where $\alpha(x) = \frac{1}{2}\left(x+\sqrt{x^2+4}\right)$ and $\beta(x) = \frac{1}{2}\left(x-\sqrt{x^2+4}\right)$.

Since $z^2-(-1)^s L_{2s}(x)\,z+1$ is a factor of the right-hand sides of (35) and (36), we have for both, the Fibonacci and the Lucas cases, the following weights

$$\lambda_1(x) = (-1)^s\alpha^{2s}(x) \quad , \quad \lambda_2(x) = (-1)^s\beta^{2s}(x). \tag{37}$$

In the Fibonacci case we have in addition the weight $\lambda(x) = -1$ (from the factor $z+1$ of the right-hand side of (35)). In the Lucas case we have in addition the following weights (from the factor $L_{2s}^2(x)\,z^2-(3L_{4s}(x)+2)\,z+4$ of the right-hand side of (36))

$$\begin{aligned} \lambda_3(x) &= \frac{2}{L_{2s}(x)}\left(\frac{3L_{4s}(x)+2}{4L_{2s}(x)}+\sqrt{\left(\frac{3L_{4s}(x)+2}{4L_{2s}(x)}\right)^2-1}\right), \\ \lambda_4(x) &= \frac{2}{L_{2s}(x)}\left(\frac{3L_{4s}(x)+2}{4L_{2s}(x)}-\sqrt{\left(\frac{3L_{4s}(x)+2}{4L_{2s}(x)}\right)^2-1}\right). \end{aligned} \tag{38}$$

In the Fibonacci case the corresponding sum (5) is

$$\begin{aligned} \frac{1}{F_{2s}^2(x)}\sum_{n=0}^{q}\lambda^{n-q}(x)\,F_{2sn}^2(x) &= -\lambda(x)\binom{q+1}{4}_{F_s(x)}+\binom{q+3}{4}_{F_s(x)} \\ &\quad +\left(\lambda^{-1}(x)+(-1)^{s+1}\frac{L_{3s}(x)}{L_s(x)}\right)\binom{q+2}{4}_{F_s(x)}, \end{aligned} \tag{39}$$

(where $\lambda(x)$ is any of the weights (37) or $\lambda(x) = -1$). In the Lucas case the sum (5) is

$$\sum_{n=0}^{q} \lambda^{n-q}(x) L_{2sn}^2(x) = -L_{2s}^2(x)\lambda(x)\binom{q+1}{4}_{F_s(x)} + 4\binom{q+4}{4}_{F_s(x)} \tag{40}$$
$$+\left(\left(-1\right)^s L_s^3(x) L_{3s}(x)\lambda(x) - L_{2s}^2(x)\lambda^2(x)\right)\binom{q+2}{4}_{F_s(x)}$$
$$+\left(4\lambda^{-1}(x) - L_s^2(x)\left(3L_{2s}(x) - 2(-1)^s\right)\right)\binom{q+3}{4}_{F_s(x)},$$

(where $\lambda(x)$ is any of the weights (37) or (38)).

For the weight $\lambda(x) = -1$ of the Fibonacci case, we have from (39) the following alternating sum of squares of Fibonacci polynomials

$$\sum_{n=0}^{q} (-1)^{n+q} F_{2sn}^2(x) \tag{41}$$
$$= F_{2s}^2(x)\left(\binom{q+1}{4}_{F_s(x)} + (-1)^{s+1} L_{2s}(x)\binom{q+2}{4}_{F_s(x)} + \binom{q+3}{4}_{F_s(x)}\right).$$

When $x = s = 1$, the weights (37) are $-\frac{3\pm\sqrt{5}}{2}$. In this case we have the following numerical formulas for weighted sums of squares of Fibonacci and Lucas numbers in terms of Fibonomials:

$$\sum_{n=0}^{q} \left(-\tfrac{3\pm\sqrt{5}}{2}\right)^{n-q} F_{2n}^2 = \tfrac{3\pm\sqrt{5}}{2}\binom{q+1}{4}_F + \tfrac{5\pm\sqrt{5}}{2}\binom{q+2}{4}_F + \binom{q+3}{4}_F,$$
$$\sum_{n=0}^{q} \left(-\tfrac{3\pm\sqrt{5}}{2}\right)^{n-q} L_{2n}^2 = \tfrac{9\left(3\pm\sqrt{5}\right)}{2}\binom{q+1}{4}_F - \tfrac{51\pm 23\sqrt{5}}{2}\binom{q+2}{4}_F$$
$$+\left(-17\pm 2\sqrt{5}\right)\binom{q+3}{4}_F + 4\binom{q+4}{4}_F.$$

By using the weights (37) in (39) and (40), we can obtain expressions for weighted sums of squares of Fibonacci and Lucas polynomials, in which the weight functions are in turn certain Fibonacci or Lucas polynomials. This is the content of the following proposition.

Proposition 3. *For $l \in \mathbb{Z}$ we have the following weighted sums of squares of Fibonacci and Lucas polynomials*

(a)

$$\frac{1}{F_{2s}^2(x)} \sum_{n=0}^{q} (-1)^{s(n+q+1)+1} F_{2s(n+l)}(x) F_{2sn}^2(x) \tag{42}$$
$$= F_{2s(q+l+1)}(x) \binom{q+1}{4}_{F_s(x)} + F_s(x) L_{s(2q+2l+1)}(x) \binom{q+2}{4}_{F_s(x)}$$
$$+ (-1)^{s+1} F_{2s(q+l)}(x) \binom{q+3}{4}_{F_s(x)}.$$

(b)

$$\frac{1}{F_{2s}^2(x)} \sum_{n=0}^{q} (-1)^{s(n+q+1)+1} L_{2s(n+l)}(x) F_{2sn}^2(x) \tag{43}$$
$$= L_{2s(q+l+1)}(x) \binom{q+1}{4}_{F_s(x)} + \left(x^2+4\right) F_s(x) F_{s(2q+2l+1)}(x) \binom{q+2}{4}_{F_s(x)}$$
$$+ (-1)^{s+1} L_{2s(q+l)}(x) \binom{q+3}{4}_{F_s(x)}.$$

(c)

$$\sum_{n=0}^{q} (-1)^{s(n+q)} F_{2s(n+l)}(x) L_{2sn}^2(x) \tag{44}$$
$$= L_{2s}^2(x) (-1)^{s+1} F_{2s(q+l+1)}(x) \binom{q+1}{4}_{F_s(x)} + 4F_{2s(q+l)}(x) \binom{q+4}{4}_{F_s(x)}$$
$$+ \left(L_s^3(x) L_{3s}(x) F_{2s(q+l+1)}(x) - L_{2s}^2(x) F_{2s(q+l+2)}(x)\right) \binom{q+2}{4}_{F_s(x)}$$
$$+ \left(4(-1)^s F_{2s(q+l-1)}(x) - L_s^2(x) \left(3L_{2s}(x) - 2(-1)^s\right) F_{2s(q+l)}(x)\right) \binom{q+3}{4}_{F_s(x)}.$$

(d)

$$\sum_{n=0}^{q} (-1)^{s(n+q)} L_{2s(n+l)}(x) L_{2sn}^2(x) \tag{45}$$
$$= (-1)^{s+1} L_{2s}^2(x) L_{2s(q+l+1)}(x) \binom{q+1}{4}_{F_s(x)} + 4L_{2s(q+l)}(x) \binom{q+4}{4}_{F_s(x)}$$
$$+ \left(L_s^3(x) L_{3s}(x) L_{2s(q+l+1)}(x) - L_{2s}^2(x) L_{2s(q+l+2)}(x)\right) \binom{q+2}{4}_{F_s(x)}$$
$$+ \left(4(-1)^s L_{2s(q+l-1)}(x) - L_s^2(x) \left(3L_{2s}(x) - 2(-1)^s\right) L_{2s(q+l)}(x)\right) \binom{q+3}{4}_{F_s(x)}.$$

Proof. First write the sum (39) as

$$\begin{aligned}\frac{1}{F_{2s}^{2}(x)}\sum_{n=0}^{q}\lambda^{n+l}(x)F_{2sn}^{2}(x) &= -\lambda^{q+l+1}(x)\binom{q+1}{4}_{F_s(x)}+\lambda^{q+l}(x)\binom{q+3}{4}_{F_s(x)} \\ &+\left(\lambda^{q+l-1}(x)+(-1)^{s+1}\frac{L_{3s}(x)}{L_s(x)}\lambda^{q+l}(x)\right)\binom{q+2}{4}_{F_s(x)},\end{aligned} \tag{46}$$

where $l\in\mathbb{Z}$. Substitute the weights (37) in (46), take the difference of the resulting expressions, multiply both sides of this difference by $(x^2+4)^{-\frac{1}{2}}$, and use the Binet's formula $F_r(x)=\frac{1}{\sqrt{x^2+4}}(\alpha^r(x)-\beta^r(x))$, to obtain that

$$\begin{aligned}\frac{1}{F_{2s}^{2}(x)}\sum_{n=0}^{q}(-1)^{s(n+q+1)+1}F_{2s(n+l)}(x)F_{2sn}^{2}(x) &= F_{2s(q+l+1)}(x)\binom{q+1}{4}_{F_s(x)}+(-1)^{s+1}F_{2s(q+l)}(x)\binom{q+3}{4}_{F_s(x)} \\ &+\left(\frac{L_{3s}(x)}{L_s(x)}F_{2s(q+l)}(x)-F_{2s(q+l+1)}(x)\right)\binom{q+2}{4}_{F_s(x)}.\end{aligned}$$

Finally, use the identity

$$\frac{L_{3s}(x)}{L_s(x)}F_{2s(q+l)}(x)-F_{2s(q+l-1)}(x)=F_s(x)L_{s(2q+2l+1)}(x),$$

to obtain (42). Similarly, if we substitute the weights (37) in (46), then take the sum of the resulting expressions, then use the Binet's formula $L_r(x)=\alpha^r(x)+\beta^r(x)$, and then use the identity

$$\frac{L_{3s}(x)}{L_s(x)}L_{2s(q+l)}(x)-L_{2s(q+l-1)}(x)=(x^2+4)F_s(x)F_{s(2q+2l+1)}(x),$$

we obtain (43).

In a similar fashion, if we begin now with (40), written as

$$\begin{aligned}\sum_{n=0}^{q}\lambda^{n+l}(x)L_{2sn}^{2}(x) &= -L_{2s}^{2}(x)\lambda^{q+l+1}(x)\binom{q+1}{4}_{F_s(x)}+4\lambda^{q+l}(x)\binom{q+4}{4}_{F_s(x)} \\ &+\left((-1)^s L_s^3(x)L_{3s}(x)\lambda^{q+l+1}(x)-L_{2s}^2(x)\lambda^{q+l+2}(x)\right)\binom{q+2}{4}_{F_s(x)} \\ &+\left(4\lambda^{q+l-1}(x)-L_s^2(x)\left(3L_{2s}(x)-2(-1)^s\right)\lambda^{q+l}(x)\right)\binom{q+3}{4}_{F_s(x)},\end{aligned} \tag{47}$$

we see that (44) and (45) are obtained by using the weights (37) in (47), together with Binet's formulas. □

Observe that the case $l = 0$ of (42) and (45) gives us sums of cubes of Fibonacci and Lucas polynomials. More precisely, if s is even we have the following (unweighted) sums of cubes

$$\frac{1}{F_{2s}^2(x)}\sum_{n=0}^{q} F_{2sn}^3(x) \tag{48}$$

$$= -F_{2s(q+1)}(x)\binom{q+1}{4}_{F_s(x)} - F_s(x)L_{s(2q+1)}(x)\binom{q+2}{4}_{F_s(x)} + F_{2sq}(x)\binom{q+3}{4}_{F_s(x)}.$$

$$\sum_{n=0}^{q} L_{2sn}^3(x) = -L_{2s}^2(x)L_{2s(q+1)}(x)\binom{q+1}{4}_{F_s(x)} + 4L_{2sq}(x)\binom{q+4}{4}_{F_s(x)} \tag{49}$$

$$+\left(L_s^3(x)L_{3s}(x)L_{2s(q+1)}(x) - L_{2s}^2(x)L_{2s(q+2)}(x)\right)\binom{q+2}{4}_{F_s(x)}$$

$$+\left(4L_{2s(q-1)}(x) - L_s(x)\left(3L_{3s}(x)+L_s(x)\right)L_{2sq}(x)\right)\binom{q+3}{4}_{F_s(x)}.$$

If s is odd, we have the following alternating sums of cubes

$$\frac{1}{F_{2s}^2(x)}\sum_{n=0}^{q} (-1)^{n+q} F_{2sn}^3(x) \tag{50}$$

$$= F_{2s(q+1)}(x)\binom{q+1}{4}_{F_s(x)} + F_s(x)L_{s(2q+1)}(x)\binom{q+2}{4}_{F_s(x)} + F_{2sq}(x)\binom{q+3}{4}_{F_s(x)}.$$

$$\sum_{n=0}^{q} (-1)^{n+q} L_{2sn}^3(x) \tag{51}$$

$$= L_{2s}^2(x)L_{2s(q+1)}(x)\binom{q+1}{4}_{F_s(x)} + 4L_{2sq}(x)\binom{q+4}{4}_{F_s(x)}$$

$$+\left(L_s^3(x)L_{3s}(x)L_{2s(q+1)}(x) - L_{2s}^2(x)L_{2s(q+2)}(x)\right)\binom{q+2}{4}_{F_s(x)}$$

$$-\left(4L_{2s(q-1)}(x) + L_s(x)\left(3L_{3s}(x)-L_s(x)\right)L_{2sq}(x)\right)\binom{q+3}{4}_{F_s(x)}.$$

For the remaining weights (38) of the Lucas case, we have in particular the following numerical identity (case $s = x = 1$)

$$\sum_{n=0}^{q} \left(\tfrac{23\pm\sqrt{385}}{18}\right)^{n-q} L_{2n}^2 = -\tfrac{23\pm\sqrt{385}}{2}\binom{q+1}{4}_F - \tfrac{61\pm3\sqrt{385}}{2}\binom{q+2}{4}_F - \tfrac{-1\pm\sqrt{385}}{2}\binom{q+3}{4}_F + 4\binom{q+4}{4}_F.$$

In fact, we can proceed in a similar fashion of the proof of Proposition 2 (using now (38) and (40)) to obtain that, if

$$\Delta_s(x) = \frac{3L_{4s}(x)+2}{4L_{2s}(x)},$$

we have the following weighted sum of squares of Lucas polynomials, valid for integers $l \geq 0$,

$$\sum_{n=0}^{q}\left(\frac{2}{L_{2s}(x)}\right)^{n-q} T_{n+l}(\Delta_s(x)) L_{2sn}^2(x) \tag{52}$$
$$= -2L_{2s}(x) T_{q+l+1}(\Delta_s(x)) \binom{q+1}{4}_{F_s(x)}$$
$$+ \left((-1)^s \frac{2L_s^3(x) L_{3s}(x)}{L_{2s}(x)} T_{q+l+1}(\Delta_s(x)) - 4T_{q+l+2}(\Delta_s(x))\right)\binom{q+2}{4}_{F_s(x)}$$
$$+ \left(2L_{2s}(x) T_{q+l-1}(\Delta_s(x)) - L_s^2(x)(3L_{2s}(x) - 2(-1)^s) T_{q+l}(\Delta_s(x))\right)\binom{q+3}{4}_{F_s(x)}$$
$$+ 4T_{q+l}(\Delta_s(x))\binom{q+4}{4}_{F_s(x)}.$$

For example, if $x = 1$ and $s = 2$, the sequence $\left(T_n\left(\frac{143}{28}\right)\right)_{n=0}^{\infty}$ involved in (52) is

$$\left(T_n\left(\tfrac{143}{28}\right)\right)_{n=0}^{\infty} = \left(1, \tfrac{143}{28}, \tfrac{20057}{392}, \tfrac{2840123}{5488}, \tfrac{402206417}{76832}, \tfrac{56958853523}{1075648}, \tfrac{8066283596057}{15059072}, \tfrac{1142314618945643}{210827008}, \ldots\right).$$

For $q = 4$ and $l = 0, 1$, we have the weighted sums

$$\left(\tfrac{2}{7}\right)^{-4} L_0^2 + \left(\tfrac{2}{7}\right)^{-3}\tfrac{143}{28} L_4^2 + \left(\tfrac{2}{7}\right)^{-2}\tfrac{20057}{392} L_8^2 + \left(\tfrac{2}{7}\right)^{-1}\tfrac{2840123}{5488} L_{12}^2 + \tfrac{402206417}{76832} L_{16}^2$$
$$= -\tfrac{56958853523}{76832}\binom{5}{1}_{F_2} + \tfrac{400320800329}{76832}\binom{6}{2}_{F_2} - \tfrac{68220633199}{76832}\binom{7}{3}_{F_2} + \tfrac{402206417}{19208}\binom{8}{4}_{F_2},$$

and

$$\left(\tfrac{2}{7}\right)^{-4}\tfrac{143}{28} L_0^2 + \left(\tfrac{2}{7}\right)^{-3}\tfrac{20057}{392} L_4^2 + \left(\tfrac{2}{7}\right)^{-2}\tfrac{2840123}{5488} L_8^2 + \left(\tfrac{2}{7}\right)^{-1}\tfrac{402206417}{76832} L_{12}^2 + \tfrac{56958853523}{1075648} L_{16}^2$$
$$= -\tfrac{8066283596057}{1075648}\binom{5}{1}_{F_2} + \tfrac{56691820586491}{1075648}\binom{6}{2}_{F_2} - \tfrac{9661131494701}{1075648}\binom{7}{3}_{F_2} + \tfrac{56958853523}{268912}\binom{8}{4}_{F_2}.$$

Finally, let us consider the case (ii) mentioned at the beginning of this section (Fibonacci case of Proposition 1, with $\mu = 0$, $k = 4$ and $t = 1$). In this case, equation (4) is

$$\begin{aligned} 0 &= \sum_{i=0}^{4}\sum_{j=0}^{i}(-1)^{\frac{(sj+2(s+1))(j+1)}{2}}\binom{5}{j}_{F_s(x)} F^4_{s(i-j)}(x)\,z^i \\ &= (-1)^{s+1}F^4_s(x)\,z\,(z+1)\left(z^2+(-1)^s\left(3L_{2s}(x)+4(-1)^s\right)z+1\right). \end{aligned} \tag{53}$$

Thus we have the weights $\lambda_1(x) = -1$ and

$$\begin{aligned} \lambda_2(x) &= -\frac{3L_{2s}(x)+4(-1)^s}{2(-1)^s}+\sqrt{\left(\frac{3L_{2s}(x)+4(-1)^s}{2}\right)^2-1}\,, \\ \lambda_3(x) &= -\frac{3L_{2s}(x)+4(-1)^s}{2(-1)^s}-\sqrt{\left(\frac{3L_{2s}(x)+4(-1)^s}{2}\right)^2-1}\,. \end{aligned} \tag{54}$$

The corresponding weighted sum (5) (of fourth powers of Fibonacci polynomials) is in this case

$$\begin{aligned} &\frac{1}{F^4_s(x)}\sum_{n=0}^{q}\lambda^{n-q}(x)F^4_{sn}(x) \\ &\quad= -\lambda(x)\binom{q+1}{4}_{F_s(x)}+\binom{q+3}{4}_{F_s(x)} \\ &\quad\quad -\left(\left(3(-1)^s L_{2s}(x)+5\right)\lambda(x)+\lambda^2(x)\right)\binom{q+2}{4}_{F_s(x)}. \end{aligned} \tag{55}$$

With the weight $\lambda_1(x) = -1$ we obtain from (55) the following alternating sum of fourth powers of Fibonacci polynomials

$$\begin{aligned} &\sum_{n=0}^{q}(-1)^n F^4_{sn}(x) \\ &=(-1)^q F^4_s(x)\left(\binom{q+1}{4}_{F_s(x)}+\left(3(-1)^s L_{2s}(x)+4\right)\binom{q+2}{4}_{F_s(x)}+\binom{q+3}{4}_{F_s(x)}\right). \end{aligned} \tag{56}$$

(The case $x = 1$ of (56), after some transformations, is contained in [4].) By setting $s = 1$ in the weights (54), we have in particular the numerical identity from (55)

$$\sum_{n=0}^{q}\left(\tfrac{5\pm\sqrt{21}}{2}\right)^{n-q}F^4_n = -\tfrac{5\pm\sqrt{21}}{2}\tbinom{q+1}{4}_F - \tfrac{3\pm\sqrt{21}}{2}\tbinom{q+2}{4}_F + \tbinom{q+3}{4}_F. \tag{57}$$

However, following the ideas of the proof of Proposition 2, we can see from (54) and (55) that for any integer $l \geq 0$ we have

$$\frac{1}{F_s^4(x)} \sum_{n=1}^{q} T_{n+l}(\Theta_s(x)) F_{sn}^4(x) \tag{58}$$
$$= -T_{q+l+1}(\Theta_s(x)) \binom{q+1}{4}_{F_s(x)} + T_{q+l}(\Theta_s(x)) \binom{q+3}{4}_{F_s(x)}$$
$$- \left(\left(3(-1)^s L_{2s}(x) + 5 \right) T_{q+l+1}(\Theta_s(x)) + T_{q+l+2}(\Theta_s(x)) \right) \binom{q+2}{4}_{F_s(x)},$$

where

$$\Theta_s(x) = -\frac{3L_{2s}(x) + 4(-1)^s}{2(-1)^s}.$$

For example, if $x = 1$ and $s = 2$, we have $\Theta_2(1) = -\frac{25}{2}$, and the sequence $\left(T_n\left(-\frac{25}{2}\right)\right)_{n=1}^{\infty}$ involved in (58) is

$$\left(T_n\left(-\tfrac{25}{2}\right)\right)_{n=1}^{\infty}$$
$$= \left(-\tfrac{25}{2}, \tfrac{623}{2}, -7775, \tfrac{388127}{2}, -\tfrac{9687625}{2}, 120901249, -\tfrac{6035374825}{2}, \tfrac{150642568127}{2}, \ldots\right).$$

(See [5, A090733] for $2T_n\left(\frac{25}{2}\right)$.) With $q = 5$ and $l = 0, 1$ we have the weighted sums

$$-\tfrac{23}{2}F_2^4 + \tfrac{623}{2}F_4^4 - 7775F_6^4 + \tfrac{388127}{2}F_8^4 - \tfrac{9687625}{2}F_{10}^4$$
$$= -120901249\binom{6}{2}_{F_2} - \tfrac{251490123}{2}\binom{7}{3}_{F_2} - \tfrac{9687625}{2}\binom{8}{4}_{F_2},$$

$$\tfrac{623}{2}F_2^4 - 7775F_4^4 + \tfrac{388127}{2}F_6^4 - \tfrac{9687625}{2}F_8^4 + 120901249F_{10}^4$$
$$= \tfrac{6035374825}{2}\binom{6}{2}_{F_2} + \tfrac{6277177323}{2}\binom{7}{3}_{F_2} + 120901249\binom{8}{4}_{F_2}.$$

Acknowledgements I thank the referee for a careful reading of the first version of this work, and all comments and suggestions that certainly contributed to present a more readable version of the article.

References

[1] A. T. Benjamin and J. J. Quinn, *Proofs that Really Count. The Art of Combinatorial Proof,* The Dolciani Mathematical Expositions **27**. The Mathematical Association of America. 2003.

[2] M. Catalani, Some formulae for bivariate Fibonacci and Lucas polynomials, arXiv:math/0406323v1

[3] V. E. Hoggatt, Jr. Fibonacci numbers and generalized binomial coefficients, *Fibonacci Quart.* **5** (1967), 383–400.

[4] R. S. Melham, Alternating sums of fourth powers of Fibonacci and Lucas numbers, *Fibonacci Quart.* **38** (2000), 254–259.

[5] OEIS Foundation Inc. (2011), The On-Line Encyclopedia of Integer Sequences, http://oeis.org.

[6] C. Pita, On s-Fibonomials, *J. Integer Seq.* **14** (2011). Article 11.3.7.

[7] C. Pita, On bivariate s-Fibopolynomials, arXiv:1203.6055v1

CONGRUENCES OF CONCAVE COMPOSITION FUNCTIONS

Keenan Monks
Harvard University, Cambridge, Massachusetts
monks@college.harvard.edu

Lynnelle Ye
Stanford University, Stanford, California
lynnelle@stanford.edu

Received: 9/1/12, Revised: 7/24/13, Accepted: 8/17/13, Published: 9/26/13

Abstract

Concave compositions are ordered partitions whose parts are decreasing towards a central part. We study the distribution modulo a of the number of concave compositions. Let $c(n)$ be the number of concave compositions of n having even length. It is easy to see that $c(n)$ is even for all $n \geq 1$. Refining this fact, we prove that

$$\#\{n < X : c(n) \equiv 0 \pmod 4\} \gg \sqrt{X}$$

and also that for every $a > 2$ and at least two distinct values of $r \in \{0, 1, \ldots, a-1\}$,

$$\#\{n < X : c(n) \equiv r \,(\mathrm{mod}\; a)\} > \frac{\log_2 \log_3 X}{a}.$$

We obtain similar results for concave compositions of odd length.

1. Introduction and Statement of Results

In their 1967 paper, Parkin and Shanks [5] conjectured that the partition function $p(n)$ takes on even and odd values with equal likelihood. Very little is known about the distribution of the parity of $p(n)$. Recently, Ahlgren [1] (see also the work of Berndt, Yee, and Zaharescu [3] as well as the works referenced therein) proved that the number of integers with an even number of partitions less than X is on the order of $\sqrt{X}$. This improved on work of Mirsky [4] showing that about $\log \log X$ numbers less than X have partition values in some nonzero residue class modulo any integer a, for the special case $a = 2$.

A combinatorial object similar to the set of partitions is the set of concave compositions — ordered partitions whose summands decrease towards a center summand.

We break up these compositions into the following three types, as defined by Andrews in [2].

Concave compositions of even length are ordered partitions of the form $a_1 + a_2 + \cdots + a_m + b_1 + b_2 + \cdots + b_m$ where

$$a_1 > a_2 > \cdots > a_m = b_m < b_{m-1} < \cdots < b_1$$

and $a_m \geq 0$. We denote the number of concave compositions of even length of an integer n by $\mathrm{ce}(n)$.

Concave compositions of odd length of type 1 are ordered partitions of the form $a_1 + a_2 + \cdots + a_{m+1} + b_1 + b_2 + \cdots + b_m$ where

$$a_1 > a_2 > \cdots > a_{m+1} < b_m < b_{m-1} < \cdots < b_1$$

and $a_{m+1} \geq 0$. We denote the number of concave compositions of odd length of type 1 of n by $\mathrm{co}_1(n)$.

Finally, *concave compositions of odd length of type* 2 are ordered partitions of the form $a_1 + a_2 + \cdots + a_{m+1} + b_1 + b_2 + \cdots + b_m$ where

$$a_1 > a_2 > \cdots > a_{m+1} \leq b_m < b_{m-1} < \cdots < b_1$$

and $a_{m+1} \geq 0$. We denote the number of concave compositions of odd length of type 2 of n by $\mathrm{co}_2(n)$. Note that all concave compositions of odd length of type 1 are also of type 2.

It is natural to consider the distribution of these functions modulo a. To this end, we define

$$E_f(r, a; X) = \#\{n < X : f(n) \equiv r \pmod{a}\}.$$

Since the function $\mathrm{co}_1(n)$ is odd exactly when n is a triangular number (see the remark following Lemma 2.1), we define the function $\mathrm{co}_1'(n)$ by subtracting 1 from $\mathrm{co}_1(n)$ if n is triangular and keeping it the same otherwise. Then we have the following theorem.

Theorem 1.1. *The following are true.*
(i) There exists an explicit constant $c > 0$ such that for sufficiently large X we have

$$E_{\mathrm{ce}}(0, 4; X) > c\sqrt{X}.$$

(ii) There exists an explicit constant $c > 0$ and $0 < \alpha < 1$ such that for sufficiently large X we have

$$E_{\mathrm{co}_1'}(0, 4; X) > \frac{X \log^\alpha X - cX}{(\sqrt{24X + 1} + 1) \log^\alpha X}.$$

(iii) There exists an explicit constant $c > 0$ such that for sufficiently large X we have

$$E_{\mathrm{co}_2}(0, 2; X) > c\sqrt{X}.$$

If we consider the more general case of a modulus a, we get a result similar to that of Mirsky [4].

Theorem 1.2. *For every $a > 2$ and at least two distinct values of r among $\{0, 1, \ldots, a-1\}$, we have*

$$E_{\text{ce}}(r, a; X) > \frac{\log_2 \log_3 X}{a}, \tag{1.1}$$

$$E_{\text{co}_1}(r, a; X) > \frac{\log_2 \log_3 X}{a}, \text{ and} \tag{1.2}$$

$$E_{\text{co}_2}(r, a; X) > \frac{\log_2 \log_3 X}{a} \tag{1.3}$$

for X sufficiently large. In the cases co_1 and co_2 this also applies when $a = 2$.

Although the above bound is the best we can prove, we expect the true distributions to be much more balanced.

Conjecture. For any modulus $a \geq 2$, we have

$$E_{\text{co}_1}(r, a; X) \sim \frac{X}{a}, \ a \text{ odd}$$

$$E_{\text{co}_1}(r, a; X) \sim \frac{2X}{a}, \ a \text{ even}, \ r \text{ even}$$

$$E_{\text{co}_1}(r, a; X) \sim \frac{2\sqrt{2X}}{a}, \ a \text{ even}, \ r \text{ odd}.$$

For ce, we expect similar asymptotics to hold, but without the last case. For co_2, we expect uniformity across residue classes for any a.

2. Proofs

2.1. Generating Functions

When faced with combinatorial objects such as concave compositions, it is natural to consider the generating functions for each object. Andrews found q-series expansions for each type of concave composition in Theorems 1-3 of [2], which we restate here.

Lemma 2.1. *Define the generating functions $\mathcal{CE}(q) = \sum_{n=0}^{\infty} \text{ce}(n) q^n$, $\mathcal{CO}_1(q) = \sum_{n=0}^{\infty} \text{co}_1(n) q^n$, and $\mathcal{CO}_2(q) = \sum_{n=0}^{\infty} \text{co}_2(n) q^n$. Then we have*

$$\mathcal{CE}(q) = \frac{1 + \sum_{n=1}^{\infty} \left(-q^{\frac{3n^2-n}{2}} + q^{\frac{3n^2+n}{2}}\right)}{1 + \sum_{n=1}^{\infty} (-1)^n \left(q^{\frac{3n^2-n}{2}} + q^{\frac{3n^2+n}{2}}\right)} = 1 + 2q^2 + 2q^3 + 4q^4 + 4q^5 + \cdots,$$

$$\mathcal{CO}_1(q) = \frac{1 + \sum_{n=1}^{\infty} \left(-q^{6n^2-2n} + q^{6n^2+2n}\right)}{1 + \sum_{n=1}^{\infty} (-1)^n \left(q^{\frac{3n^2-n}{2}} + q^{\frac{3n^2+n}{2}}\right)} = 1 + q + 2q^2 + 3q^3 + 4q^4 + 6q^5 + \cdots,$$

$$\mathcal{CO}_2(q) = \frac{1 + \sum_{n=1}^{\infty} \left(q^{6n^2-8n+3} - q^{6n^2-4n+1}\right)}{1 + \sum_{n=1}^{\infty} (-1)^n \left(q^{\frac{3n^2-n}{2}} + q^{\frac{3n^2+n}{2}}\right)} = 1+2q+3q^2+4q^3+7q^4+10q^5+\cdots.$$

Remark. It is easy to show that $\mathcal{CO}_1(q) \equiv \sum_{n=0}^{\infty} q^{n(n+1)/2} \pmod 2$, as follows. We have by Lemma 2.1 that

$$\mathcal{CO}_1(q) = \frac{1 - \sum_{n=1}^{\infty} \left(q^{4n(3n-1)/2} - q^{4n(3n+1)/2}\right)}{1 + \sum_{n=1}^{\infty} (-1)^n \left(q^{n(3n-1)/2} + q^{n(3n+1)/2}\right)} \equiv (q)_{\infty}^3 \pmod 2$$

where $(q)_{\infty} = \prod_{n=1}^{\infty} (1-q^n)$. By Lemma 12 of [2], we have that

$$\sum_{n=0}^{\infty} (-1)^n q^{n(n+1)/2} = (q)_{\infty} \sum_{n=0}^{\infty} \frac{q^{n(n+1)}}{(q)_n^2} \equiv (q)_{\infty} \left(\sum_{n=0}^{\infty} \frac{q^{n(n+1)/2}}{(q)_n} \right)^2 \equiv (q)_{\infty}^3 \pmod 2$$

since the partitions into distinct parts are conjugate to the partitions into $1, 2, \ldots, n$ missing nothing. The desired congruence follows.

Motivated by this congruence, we define $\mathcal{CO}_1'(q) = \sum_{n=0}^{\infty} \mathrm{co}_1'(n) q^n = \mathcal{CO}_1(q) - \sum_{n=0}^{\infty} q^{n(n+1)/2}$, so that all of the coefficients of $\mathcal{CO}_1'(q)$ are even. This will ease our study of the series modulo 4. With these, we can now proceed to the proofs of our main theorems.

2.2. Proofs of Theorems 1.1 and 1.2

Combining the expansions given by Andrews [2] with a generalization of Ahlgren's argument in [1], we prove our main theorem.

Proof of Theorem 1.1. (i) First, we show that the coefficients $\mathrm{ce}(n)$ are all even for $n > 1$. There is a natural pairing between non-palindromic concave compositions of even length given by mirroring the sequence. There is furthermore a natural pairing between palindromic compositions given by inserting or removing a pair of zeroes at the center of the sequence. Alternatively, Lemma 2.1 makes it clear that $\mathcal{CE}(q) \equiv 1 \pmod 2$.

Thus we have that $\mathrm{ce}(n)$ is congruent to either 0 or 2 mod 4. From Lemma 2.1, we have the following:

$$\left(1 + \sum_{n=1}^{\infty} (-1)^n \left(q^{\frac{3n^2-n}{2}} + q^{\frac{3n^2+n}{2}}\right)\right) \left(\sum_{n=0}^{\infty} \mathrm{ce}(n) q^n\right) = 1 + \sum_{n=1}^{\infty} \left(-q^{\frac{3n^2-n}{2}} + q^{\frac{3n^2+n}{2}}\right).$$

We will denote the right-hand side by $\sum_{n=0}^{\infty} a(n)q^n$. Thus, since $-2 \equiv 2 \pmod 4$, we have that

$$a(n) \equiv \mathrm{ce}(n)+\mathrm{ce}(n-1)+\mathrm{ce}(n-2)+\mathrm{ce}(n-5)+\cdots+\mathrm{ce}\left(n-\frac{3k^2 \pm k}{2}\right)+\cdots \pmod 4. \tag{2.1}$$

If $a(n)$ is divisible by 4 and there are an odd number of summands, we can conclude that one of the summands is also divisible by 4. It is easy to see that there will be an odd number of summands exactly when $\frac{3k^2+k}{2} < n < \frac{3(k+1)^2-(k+1)}{2}$. As X tends to infinity, it is easy to show that the number of such $n < X$ tends to $\frac{2}{3}X$ from below very quickly, since $\frac{3(k+1)^2-(k+1)}{2}$ is almost exactly two-thirds of the way from $\frac{3k^2+k}{2}$ to $\frac{3(k+1)^2+(k+1)}{2}$.

Thus for approximately $\frac{2}{3}X$ values of n, one of the terms $\mathrm{ce}(i)$ must be congruent to 0 modulo 4. These terms may be overcounted by the number of decompositions of the form (2.1) in which they appear. This is bounded above by twice the number of pentagonal numbers less than X, which is $\frac{\sqrt{24X+1}+1}{3}$. Thus we can conclude that we have, for some small constant ϵ,

$$E_{\mathrm{ce}}(0,4;X) > \frac{(2-\epsilon)\,X}{\sqrt{24X+1}+1},$$

as desired.

(ii) We first write out the expansion of $\mathcal{CO}_1'$ as in (i):

$$\left(1+\sum_{n=1}^{\infty}(-1)^n\left(q^{\frac{3n^2-n}{2}}+q^{\frac{3n^2+n}{2}}\right)\right)\left(\sum_{n=0}^{\infty}\mathrm{co}_1'(n)q^n\right)$$
$$=1+\sum_{n=1}^{\infty}\left(-q^{6n^2-2n}+q^{6n^2+2n}\right)-\sum_{j,k=1}^{\infty}\left(q^{\frac{j^2-j+3k^2-k}{2}}+q^{\frac{j^2-j+3k^2+k}{2}}\right).$$

Again writing the right-hand side as $\sum_{n=0}^{\infty} a(n)q^n$, we notice that $a(n)=0$ whenever n is not expressible as $6k^2 \pm 2k$ or as the sum of a triangular and a pentagonal number. To obtain a bound on how many such terms there are, we first notice that there are at most $\frac{\sqrt{6X+1}-1}{3}$ values of $n < X$ expressible as $6k^2 \pm 2k$. To find how many numbers less than X are expressible as the sum of a triangular and a pentagonal number, we use the result that if $Q(x,y)$ is a positive definite binary quadratic form,

$$\#\{n < X : n = Q(x,y) \text{ for some } x,y \in \mathbb{Z}\} \asymp \frac{X}{\log^{\alpha} X}$$

for some $0 < \alpha < 1$. For a presentation of a similar result, see Section 2 in [6].

Thus we have that there are at least $X\left(\frac{\log^\alpha X - c}{\log^\alpha X}\right)$ numbers $n < X$ such that $a(n) = 0$. Then by an argument analogous to that in (i), we can conclude that

$$E_{\mathrm{co}_1'}(0,4;X) > \frac{X\log^\alpha X - cX}{(\sqrt{24X+1}+1)\log^\alpha X}.$$

(iii) This proof is analogous to the proof of (i), the only difference being that we need to exclude the values of n for which $a(n)$ is nonzero. Thus the bound we get in this case is, for some small constant ϵ,

$$E_{\mathrm{co}_2}(0,2;X) > \frac{(2-\epsilon)X - 4 - \sqrt{6X-2}}{\sqrt{24X+1}+1},$$

implying the desired result. □

We adapt the strategy of Mirsky in [4] to prove Theorem 1.2.

Proof of Theorem 1.2. Let $E^*_{\mathrm{ce}}(r,a;X) = \#\{n \le X : \mathrm{ce}(n) \not\equiv r \pmod{a}\}$, and define $E^*_{\mathrm{co}_1}(r,a;X)$ and $E^*_{\mathrm{co}_2}(r,a;X)$ similarly. Fix r; then, in the case of $\mathcal{CE}$, we claim that $E^*_{\mathrm{ce}}(r,a;X) > \log_2\log_3 X - C$ for some constant C.

Recall that $(q)_\infty = \prod_{n=1}^{\infty}(1-q^n)$. Then we have by Lemma 2.1, considering the coefficient of $q^{\ell(3\ell+1)/2+2}$ in $(q)_\infty\mathcal{CE}(q)$, that

$$\mathrm{ce}(\ell(3\ell+1)/2+2) + \sum_{k=1}^{\ell}(-1)^k\,\mathrm{ce}\left(\frac{\ell(3\ell+1)}{2}+2-\frac{k(3k-1)}{2}\right)$$
$$+\sum_{k=1}^{\ell-1}(-1)^k\,\mathrm{ce}\left(\frac{\ell(3\ell+1)}{2}+2-\frac{k(3k+1)}{2}\right) + (-1)^\ell\,\mathrm{ce}(2) = 0$$

for all $\ell \ge 1$. For $\ell = 2m-1$ this becomes

$$\mathrm{ce}((2m-1)(3m-1)+2) + \sum_{k=1}^{2m-1}(-1)^k\,\mathrm{ce}((2m-1)(3m-1)+2-k(3k-1)/2) \tag{2.2}$$

$$+\sum_{k=1}^{2m-2}(-1)^k\,\mathrm{ce}((2m-1)(3m-1)+2-k(3k+1)/2) = 2. \tag{2.3}$$

We claim that some element of $\mathrm{ce}(2m+1), \mathrm{ce}(2m+2), \ldots, \mathrm{ce}((2m-1)(3m-1)+2)$ is not congruent to $r \pmod{a}$. Suppose otherwise; then the above equation gives

$$0 \equiv r + \sum_{k=1}^{2m-1}(-1)^k r + \sum_{k=1}^{2m-2}(-1)^k r \equiv 2 \pmod{a},$$

which is a contradiction. We would like to construct a sequence m_j so that the terms appearing in Equation 2.2 for m_j and m_i do not overlap for $i \neq j$, hence giving a value of ce not congruent to $r \pmod{a}$ for each j. Since the lowest-indexed term in Equation 2.2 for m_j is $2m_j+1$ and the highest is $(2m_j-1)(3m_j-1)+2$, it suffices to set $2m_j+1 > (2m_{j-1}-1)(3m_{j-1}-1)+2$, or $2m_j > 6m_{j-1}^2-5m_{j-1}+2$. Hence we can choose $m_1 = 1$, $m_j = 3m_{j-1}^2$, so that $m_j = 3^{2^{j-1}-1}$. This gives $E^*_{\text{ce}}(r, a; 3^{2^{j-1}-1}) \geq j$ for all j. Setting $j = \lfloor \log_2 \log_3 X \rfloor$ gives $E^*_{\text{ce}}(r, a; X) \geq \log_2 \log_3 X - C$, as desired.

Since $E^*_{\text{ce}}(0, a; X) = \sum_{r' \neq 0} E_{\text{ce}}(r', a; X)$, we can write

$$\sum_{r' \neq 0} E_{\text{ce}}(r', a; X) > \log_2 \log_3 X - C,$$

from which there is some r_1 so that

$$E_{\text{ce}}(r_1, a; X) > \frac{1}{a-1} \log_2 \log_3 X - C > \frac{1}{a} \log_2 \log_3 X$$

for X sufficiently large. Then we also have $\sum_{r' \neq r_1} E_{\text{ce}}(r', a; X) = E^*_{\text{ce}}(r_1, a; X) > \log_2 \log_3 X - C$, giving some $r_2 \neq r_1$ so that $E_{\text{ce}}(r_2, a; X) > \frac{1}{a} \log_2 \log_3 X$. This gives us two residues, r_1 and r_2, with the desired bound in the $\mathcal{CE}$ case.

For $\mathcal{CO}_1$ and $\mathcal{CO}_2$, the same argument applies almost verbatim, with the remark that we use the coefficient of $q^{(2m-1)(3m-1)}$ in $(q)_\infty \mathcal{CO}_1(q)$, $(q)_\infty \mathcal{CO}_2(q)$ respectively, rather than $q^{(2m-1)(3m-1)+2}$. In both cases we can guarantee that this coefficient must equal 0. In the case $\mathcal{CO}_1$ the difference between $(2m-1)(3m-1)$ and the nearest exponent of a nonzero coefficient is $3m-1$; in the case $\mathcal{CO}_2$ it is m. Noting that $\text{co}_1(0) = \text{co}_2(0) = 1 \neq 0$, the rest of the argument follows. □

References

[1] S. Ahlgren, *Distribution of the Parity of the Partition Function in Arithmetic Progressions*, Indag. Math. N. S. 10 (2), (1999), 172-181.

[2] G. Andrews, *Concave Compositions*, Electronic J. of Comb. 18 (2011), P6.

[3] B. C. Berndt, A. J. Yee, A. Zaharescu, *On the Parity of Partition Functions*, Int. J. Math. 14 (2003), 437-459.

[4] L. Mirsky, *The Distribution of Values of the Partition Function in Residue Classes*, J. of Math. Analysis and Applications 93 (1983), 593-598.

[5] T. R. Parkin and D. Shanks, *On the distribution of parity in the partition function*, Math. Comp. 21 (1967), 466-480.

[6] J.-P. Serre, *Divisibilitie de certaines fonctions arithmetiques*, L'Ens. Math. 22 (1976) 227-260.

DISJUNCTIVE RADO NUMBERS FOR $ax_1 + x_2 = x_3$ AND $bx_1 + x_2 = x_3$

Liz Lane-Harvard
Department of Mathematics, Colorado State University, Fort Collins, Colorado
lane@math.colostate.edu

Daniel Schaal
Dept. of Mathematics, South Dakota State University, Brookings, South Dakota
daniel.schaal@sdstate.edu

Received: 5/15/12, Revised: 5/13/13, Accepted: 8/3/13, Published: 9/26/13

Abstract

If S is a set of linear equations, the 2-color *disjunctive Rado number* for the set S is the least integer n, provided it exists, such that every coloring of the set $\{1, 2, ..., n\}$ with two colors admits a monochromatic solution to at least one equation in the set S. If no such integer n exists, then the 2-color disjunctive Rado number for the set S is infinite. In this paper, the disjunctive Rado number for the set consisting of the two equations $ax_1 + x_2 = x_3$ and $bx_1 + x_2 = x_3$ is determined for all integers a and b, where $1 \leq a < b$.

1. Introduction

Let $\mathbb{N}$ represent the set of natural numbers and let $[a, b]$ denote the set $\{n \in \mathbb{N} : a \leq n \leq b\}$. For every $t \in \mathbb{N}$, a function $\Delta : [1, n] \to [0, t-1]$ is referred to as a t-coloring of the set $[1, n]$, and the set $[0, t-1]$ is referred to as the set of colors. Given a t-coloring Δ and a system L of linear equations or inequalities in m variables, a solution $(x_1, x_2, ..., x_m)$ to the system L is *monochromatic* if and only if

$$\Delta(x_1) = \Delta(x_2) = ... = \Delta(x_m).$$

In 1916, I. Schur [31] proved that for every $t \geq 2$, there exists a least integer $n = S(t)$ such that for every t-coloring of the set $[1, n]$, there exists a monochromatic solution to $x_1 + x_2 = x_3$.

The integers $S(t)$ are called *Schur numbers.* It is known that $S(2) = 5$, $S(3) = 14$, and $S(4) = 45$, but no other Schur numbers are known [33]. In 1933, R. Rado generalized the concept of Schur numbers to arbitrary systems of linear equations. Rado found necessary and sufficient conditions to determine if an arbitrary system of linear equations admits a monochromatic solution under every t-coloring of the natural numbers [16, 17, 18]. For a given system L of linear equations, the least integer n, provided that it exists, such that for every t-coloring of the set $[1,n]$ there exists a monochromatic solution to L is called the t-color *Rado number* (or t-color *generalized Schur number*) for the system L. If such an integer n does not exist, then the t-color Rado number for the system L is infinite. In recent years, the exact Rado numbers for several families of equations and inequalities have been found [4, 6, 7, 8, 11, 12, 15, 23, 24, 25, 27, 28, 32].

We will need the following notation to proceed. For all integers $m \geq 3$, let $L(m)$ represent the equation

$$L(m)\colon\ x_1 + x_2 + \ldots + x_{m-1} = x_m,$$

and let $r(L(m))$ represent the 2-color Rado number for this equation. In [1], it was determined that for every $m \geq 3$, the 2-color Rado number for the equation $L(m)$ is

$$r(L(m)) = m^2 - m - 1.$$

It is routine to verify that for every $m \geq 3$, it is impossible to 2-color the values in the set

$$\{1, m-1, m, m^2-2m+1, m^2-m-1\}$$

and avoid a monochromatic solution to $L(m)$, showing that $r(L(m)) \leq m^2 - m - 1$.

Several variations of $L(m)$ have recently been considered. One such variation will require the following notation. For all integers $a \geq 1$, let $S(a)$ represent the equation

$$S(a)\colon\ ax_1 + x_2 = x_3,$$

and let $r(S(a))$ represent the 2-color Rado number for this equation. For every integer $a \geq 1$, if (α, β, γ) is a solution to $S(a)$, then the $(a+2)$-tuple $(\alpha, \alpha, \ldots, \alpha, \beta, \gamma)$ is a solution to $L(a+2)$. So, every 2-coloring that admits a monochromatic solution to $S(a)$ also admits a monochromatic solution to $L(a+2)$. It follows that for every integer $a \geq 1$,

$$r(S(a)) \geq r(L(a+2)).$$

In [10], it was shown that for every integer $a \geq 1$,

$$r(S(a)) = a^2 + 3a + 1 = (a+2)^2 - (a+2) - 1 = r(L(a+2)).$$

It is also routine to verify that for every $a \geq 1$, it is impossible to 2-color the values in the set

$$\{1, a+1, a+2, a^2+2a+1, a^2+3a+1\}$$

and avoid a monochromatic solution to $S(a)$, showing that $r(S(a)) \leq a^2 + 3a + 1$. It should be noted that this result determines the 2-color Rado numbers for a more general family of equations. Let positive integers a and l be given and let $T(a)$ represent any equation of the form

$$T(a)\colon\ a_1x_1 + a_2x_2 + \ldots + a_lx_l + x_{l+1} = x_{l+2}$$

where $a_i \in \mathbb{N}$ for $i \in [1, l]$ and $\sum_{i=1}^{l} a_i = a$. Let $r(T(a))$ represent the 2-color Rado number for this equation. For reasons similar to the argument above, every 2-coloring that admits a monochromatic solution to $S(a)$ also admits a monochromatic solution to $T(a)$, and every 2-coloring that admits a monochromatic solution to $T(a)$ also admits a monochromatic solution to $L(a+2)$. Hence,

$$r(S(a)) \geq r(T(a)) \geq r(L(a+2)) \quad \text{and} \quad r(T(a)) = a^2 + 3a + 1.$$

Recently, several variations of the classical Rado numbers have been developed [2, 3, 5, 14, 19, 20, 21, 26, 29, 30]. Specifically, the concept of t-color disjunctive Rado numbers has been introduced [9, 13, 22]. Given a set S of linear equations, the least integer n, provided that it exists, such that for every t-coloring of the set $[1, n]$ there exists a monochromatic solution to at least one equation in S is called the t-color *disjunctive Rado number* for the set S. If such an integer n does not exist, then the t-color disjunctive Rado number for the set S is infinite. Given a set of linear equations, it is clear that the t-color disjunctive Rado number for this set is less than or equal to the t-color Rado number for each equation in the set.

Specifically, the 2-color disjunctive Rado numbers for the set $S = \{L(m), L(n)\}$ were recently considered. Let $r_d(L(m), L(n))$ represent the 2-color disjunctive Rado number for this set. In [13], it was determined that

$$r_d(L(m), L(n)) = \begin{cases} m^2 - m - 1 & \text{if } n = m + 1 \\ m^2 - 2m + 1 & \text{if } m + 2 \leq n \leq m^2 - 2m + 2 \\ n - 1 & \text{if } m^2 - 2m + 3 \leq n \leq m^2 - m - 1 \\ m^2 - m - 1 & \text{if } n \geq m^2 - m \end{cases}$$

for all integers m and n such that $3 \leq m < n$. In this paper we consider the 2-color disjunctive Rado numbers for the set $S = \{S(a), S(b)\}$. Let $r_d(S(a), S(b))$ represent the 2-color disjunctive Rado number for this set. Just as it was easy to see that $r(S(a)) \geq r(L(a+2))$, it is also easy to see that for all integers a and b where $1 \leq a < b$,

$$r_d(S(a), S(b)) \geq r_d(L(a+2), L(b+2)).$$

One may wonder if equality also holds in this situation. The answer is sometimes, but not always. For ease of comparison, we will express the formula for $r_d(L(m), L(n))$ as $r_d(L(a+2), L(b+2))$. For all integers a and b where $1 \leq a < b$,

$$r_d(L(a+2),L(a+2))=\begin{cases} a^2+3a+1 & \text{if } b=a+1\\ a^2+2a+1 & \text{if } a+2\le b\le a^2+2a\\ b+1 & \text{if } a^2+2a+1\le b\le a^2+3a-1\\ a^2+3a+1 & \text{if } b\ge a^2+3a\end{cases}$$

The main result of this paper is the following theorem. We will need to introduce the value r, which is the remainder when a^2+2a+1 is divided by b.

Theorem 1. *For all integers a and b where $1\le a<b$,*
$r_d(S(a),S(b))=r_d(L(a+2),L(b+2))+$

$$\begin{cases} 1 & \text{if } a+3\le b\le \frac{a^2+3a-4}{2} \text{ and } r=0\\ a & \text{if } a+3\le b\le \frac{a^2+3a-4}{2} \text{ and } r\ge a+1 \text{ and } r\le b-a+1\\ b-r+1 & \text{if } a+3\le b\le \frac{a^2+3a-4}{2} \text{ and } r\ge a+1 \text{ and } r\ge b-a+2\\ a-2 & \text{if } b=\frac{a^2+3a-2}{2}\\ a & \text{if } \frac{a^2+3a}{2}\le b\le a^2+a\\ 0 & \text{otherwise}\end{cases}$$

where $a^2+2a+1=qb+r$ and $0\le r\le b-1$.

It should be noted that this theorem also gives upper bounds for the 2-color disjunctive Rado numbers for a more general family of sets of equations. Let positive integers a and b be given where $a<b$, and let equations $T(a)$ and $T(b)$ of the form described above be given. Let $r_d(T(a),T(b))$ represent the 2-color disjunctive Rado number for the set $\{T(a),T(b)\}$. If (α,β,γ) is a solution to $\{S(a),S(b)\}$, then (α,β,γ) is a solution to $S(a)$ or $S(b)$. For reasons similar to the argument above, we have that

$$r_d(S(a),S(b))\ge r_d(T(a),T(b))\ge r_d(L(a+2),L(b+2)).$$

This gives upper and lower bounds for $r_d(T(a),T(b))$ for all values of a and b, with the difference between the bounds the difference between $r_d(S(a),S(b))$ and $r_d(L(a+2),L(b+2))$ given in Theorem 1. For values of a and b where the difference is zero, the inequality gives the exact values of $r_d(T(a),T(b))$.

II. Main Result

Before proving Theorem 1, we will first state and prove two lemmas. For the remainder of this paper we will assume, without loss of generality, that every 2-coloring Δ colors the first integer 0. We will need the following definition.

Definition 1. For all positive integers a and n, a 2-coloring $\Delta : [1,n] \to [0,1]$ is said to be *a-good* if Δ avoids a monochromatic solution to the equation $S(a)$.

Lemma 2. *For every positive integer a, every a-good coloring $\Delta : [1, a^2+2a+1] \to [0,1]$ satisfies the property*

$$\Delta(x) = \begin{cases} 0 & \text{if } 1 \le x \le a \\ 1 & \text{if } a+1 \le x \le a^2+a+1 \\ 0 & \text{if } x = a^2+2a+1. \end{cases} \tag{1}$$

Proof. Let a positive integer a be given and let an a-good coloring $\Delta : [1, a^2 + 2a+1] \to [0,1]$ be given. If $\Delta(a+1) = 0$, then $(1, 1, a+1)$ is a solution to $S(a)$ monochromatic in color 0. So, we are able to conclude that $\Delta(a+1) = 1$.

If $\Delta(a^2+2a+1) = 1$, then $(a+1, a+1, a^2+2a+1)$ is a solution to $S(a)$ monochromatic in color 1, so we have $\Delta(a^2+2a+1) = 0$.

If $\Delta(a^2+a+1) = 0$, then $(1, a^2+a+1, a^2+2a+1)$ is a solution to $S(a)$ monochromatic in color 0, so we have

$\Delta(a^2+2a+1) = 1$.

We will now prove the following claim.

Claim. *For every $k \in [1, a-1]$, we have $\Delta(a+1-k) = 0$.*

We will prove this via induction on k. First assume that $k = 1$. If $\Delta(a+1-1) = \Delta(a) = 1$, then $(a, a+1, a^2+a+1)$ would be a solution to $S(a)$ monochromatic is color 1, so we have $\Delta(a) = 0$. Now, let $k \in [1, a-2]$ and assume that $\Delta(a+1-k) = 0$. If $\Delta(a^2+a(1-k)+1) = 0$, then $(a+1-k, 1, a^2+a(1-k)+1)$ is a solution to $S(a)$ monochromatic in color 0, so we have

$$\Delta(a^2+a(1-k)+1) = 1.$$

Now, if $\Delta(a-k) = 1$, then $(a-k, a+1, a^2+a(1-k)+1)$ is a solution to $S(a)$ monochromatic is color 1, so we have $\Delta(a-k) = 0$ and the proof of the claim is complete.

To complete the proof of Lemma 2 it remains to be shown that $\Delta(x) = 1$ for $x \in [a+2, a^2+a]$. Let $x \in [a+2, a^2+a]$ be given, and note that there exists unique integers $\alpha, \beta \in [1, a]$ such that $x = a\alpha + \beta$. If $\Delta(x) = 0$, then (α, β, x) is a solution to $S(a)$ monochromatic in color 0, so $\Delta(x) = 1$ and the proof of Lemma 2 is complete. □

It can be verified that property (1) characterizes all a-good 2-colorings on the set $[1, a^2+2a+1]$. In fact, we can arbitrarily color all the numbers in the set $[a^2+a+2, a^2+2a]$ and the colorings will be a-good, as long as property (1) is satisfied. As we will see in Lemma 3, the number of a-good colorings decreases as the length of the interval being colored increases until there is a unique a-good 2-coloring of the set $[1, a^2+3a]$.

Lemma 3. *For every positive integer a and every integer $p \in [1, a-1]$, every a-good coloring $\Delta : [1, a^2 + 2a + 1 + p] \to [0, 1]$ satisfies the property*

$$\Delta(x) = \begin{cases} 0 & \text{if } 1 \le x \le a \\ 1 & \text{if } a+1 \le x \le a^2 + a + 1 + p \\ 0 & \text{if } a^2 + 2a + 1 \le x \le a^2 + 2a + 1 + p. \end{cases} \tag{2}$$

Proof. Let $a \in \mathbb{N}$ and $p \in [1, a-1]$ be given and let an a-good coloring $\Delta : [1, a^2 + 2a + 1 + p] \to [0, 1]$ be given. Now, the coloring Δ restricted to the domain $[1, a^2 + 2a + 1 + p]$ will also be a-good, and hence, satisfies property (1). So, we will only need to show that

$$\Delta(a^2 + a + 1 + q) = 1 \text{ and } \Delta(a^2 + 2a + 1 + q) = 0 \text{ for } q \in [1, p].$$

Let $q \in [1, p]$ be given. If $\Delta(a^2 + 2a + 1 + q) = 1$, then

$$(a + 1, a + 1 + q, a^2 + 2a + 1 + q)$$

is a solution to $S(a)$ monochromatic is color 1, so we have

$$\Delta(a^2 + 2a + 1 + q) = 0.$$

If $\Delta(a^2 + a + 1 + q) = 0$, then

$$(1, a^2 + a + 1 + q, a^2 + 2a + 1 + q)$$

is a solution to $S(a)$ monochromatic is color 0, so we have

$$\Delta(a^2 + a + 1 + q) = 1.$$

The proof of Lemma 3 is complete. □

Before proving Theorem 1, we first restate the theorem in a more expanded form.

Theorem 4. *For all integers a and b where $1 \le a < b$, $r_d(S(a), S(b)) =$*

$$\begin{cases} a^2 + 3a + 1 & \text{if } b = a + 1 \\ a^2 + 2a + 1 & \text{if } b = a + 2 \\ a^2 + 2a + 2 & \text{if } a + 3 \le b \le \frac{a^2+3a-4}{2} \text{ and } r = 0 \\ a^2 + 2a + 1 & \text{if } a + 3 \le b \le \frac{a^2+3a-4}{2} \text{ and } 1 \le r \le a \\ a^2 + 3a + 1 & \text{if } a + 3 \le b \le \frac{a^2+3a-4}{2} \text{ and } r \ge a + 1 \text{ and } r \le b - a + 1 \\ a^2 + 2a + b - r + 2 & \text{if } a + 3 \le b \le \frac{a^2+3a-4}{2} \text{ and } r \ge a + 1 \text{ and } r \ge b - a + 2 \\ a^2 + 3a - 1 & \text{if } b = \frac{a^2+3a-2}{2} \\ a^2 + 3a + 1 & \text{if } \frac{a^2+3a}{2} \le b \le a^2 + a \\ a^2 + 2a + 1 & \text{if } a^2 + a + 1 \le b \le a^2 + 2a \\ b + 1 & \text{if } a^2 + 2a + 1 \le b \le a^2 + 3a - 1 \\ a^2 + 3a + 1 & \text{if } a^2 + 3a \le b \end{cases}$$

where $a^2+2a+1=qb+r$ *and* $0\le r\le b-1$.

Proof. Let integers a and b be given where $1\le a<b$. Let r and q be the unique integers such that $a^2+2a+1=qb+r$ and $0\le r\le b-1$. Note that when $a+3\le b\le \frac{a^2+3a-4}{2}$, it follows that $q\in[2,a-1]$. Recall from the introduction that

$$r_d(S(a),S(b))\le r(S(a))=a^2+3a+1 \tag{3}$$

and $r_d(S(a),S(b))\ge r_d(L(a+2),L(b+2))=$

$$\begin{cases} a^2+3a+1 & \text{if } b=a+1\\ a^2+2a+1 & \text{if } a+2\le b\le a^2+2a\\ b+1 & \text{if } a^2+2a+1\le b\le a^2+3a-1\\ a^2+3a+1 & \text{if } b\ge a^2+3a. \end{cases} \tag{4}$$

We will use eleven cases based on the values of b and r. Note that when $a\in[1,2]$, some cases have no values of b that satisfy the assumption. Also note that $r_d(S(1),S(2))$, $r_d(S(1),S(3))$, and $r_d(S(2),S(4))$ are each in two cases, but the values given in the two cases agree. For all the cases where the lower bound is not given by (4), the lower bound will be established by the coloring $\widehat{\Delta}:[1,a^2+3a]\to[0,1]$ defined by

$$\widehat{\Delta}(x)=\begin{cases} 0 & \text{if } 1\le x\le a\\ 1 & \text{if } a+1\le x\le a^2+2a\\ 0 & \text{if } a^2+2a+1\le x\le a^2+3a \end{cases}$$

or a truncation of this coloring. It is easy to check that the coloring $\widehat{\Delta}$ is a-good, so every truncation of $\widehat{\Delta}$ is also a-good. In fact, from Lemma 2, we see that $\widehat{\Delta}$ is the unique a-good 2-coloring of the set $[1,a^2+3a]$, and if we are considering only colorings where 1 is colored 0 then $\widehat{\Delta}$ is the unique coloring. For every $\theta\in[a^2+2a+1,a^2+3a]$, let $\widehat{\Delta}\lceil[1,\theta]:[1,\theta]\to[0,1]$ represent the coloring $\widehat{\Delta}$ restricted to the set $[1,\theta]$. To establish the lower bounds we will need to show that the coloring $\widehat{\Delta}\lceil[1,\theta]$ is b-good for some values of b and θ. That is, we will need to show that if the triple (x_1,x_2,x_3) is monochromatic, then (x_1,x_2,x_3) is not a solution to $S(b)$. Let $b\ge a+1$ and $\theta\in[a^2+2a,a^2+3a]$ be given. If the triple (x_1,x_2,x_3) is monochromatic in color 1, then

$$bx_1+x_2\ge b(a+1)+(a+1)=ba+b+a+1>a^2+2a\ge x_3,$$

so (x_1,x_2,x_3) is not a solution to $S(b)$. If the triple (x_1,x_2,x_3) is monochromatic in color 0 and $x_i\in[1,a]$ for $i\in[1,3]$, then

$$bx_1+x_2\ge b(1)+1>a\ge x_3,$$

so (x_1, x_2, x_3) is not a solution to $S(b)$. If the triple (x_1, x_2, x_3) is monochromatic in color 0 and either $x_1 \in [a^2 + 2a + 1, \theta]$ or $x_2 \in [a^2 + 2a + 1, \theta]$, then

$$bx_1 + x_2 \geq b(1) + (a^2 + 2a + 1) > a^2 + 3a + 1 > x_3,$$

so (x_1, x_2, x_3) is not a solution to $S(b)$. Hence, to show the coloring $\widehat{\Delta}\restriction[1, \theta]$ is both a-good and b-good, it only remains to be demonstrated that if the triple (x_1, x_2, x_3) is monochromatic in color 0 and $x_i \in [1, a]$ for $i \in [1, 2]$ and $x_3 \in [a^2 + 2a + 1, \theta]$, then (x_1, x_2, x_3) is not a solution to $S(b)$.

`Case 1:` Assume that $b = a + 1$. The upper bound follows from (3) and the lower bound follows from (4).

`Case 2:` Assume that $b = a + 2$. The lower bound follows from (4). For the upper bound, let a coloring $\Delta : [1, a^2 + 2a + 1] \to [0, 1]$ be given. We must show that Δ is either not a-good or not b-good. Since $b = a + 2$, the triple $(a, 1, a^2 + 2a + 1)$ is a solution to $S(b)$. If Δ is a-good, then from Lemma 2, this triple is monochromatic in color 0, so Δ is not b-good.

`Case 3:` Assume that $a+3 \leq b \leq \frac{a^2+3a-4}{2}$ and $r = 0$. It follows that $a^2+2a+1 = qb$. For the lower bound, we will show that the coloring $\widehat{\Delta}\restriction[1, a^2 + 2a + 1]$ is both a-good and b-good. If $x_1 \in [q, q - 1]$, $x_2 \in [1, a]$, and $x_3 = a^2 + 2a + 1$, then

$$bx_1 + x_2 \leq b(q - 1) + a = a^2 + 2a + 1 - b + a < a^2 + 2a + 1 = x_3.$$

If $x_1 \in [q, a]$, $x_2 \in [1, a]$, and $x_3 = a^2 + 2a + 1$, then

$$bx_1 + x_2 \geq bq + 1 = a^2 + 2a + 2 > x_3.$$

From this and the argument above, the lower bound is complete. For the upper bound, let a coloring $\Delta : [1, a^2 + 2a + 2] \to [0, 1]$ be given. We must show that Δ is either not a-good or not b-good. Since $a^2+2a+1 = qb$, the triple $(a, 1, a^2+2a+2)$ is a solution to $S(b)$. If Δ is a-good, then from Lemma 2, this triple is monochromatic in color 0, so Δ is not b-good.

`Case 4:` Assume that $a+3 \leq b \leq \frac{a^2+3a-4}{2}$ and $1 \leq r \leq a$. The lower bound follows from (4). For the upper bound, let a coloring $\Delta : [1, a^2 + 2a + 1] \to [0, 1]$ be given. We must show that Δ is either not a-good or not b-good. Since $a^2+2a+1 = qb+r$, the triple $(q, r, a^2 + 2a + 1)$ is a solution to $S(b)$. If Δ is a-good, then from Lemma 2 and the fact that $r \leq a$, this triple is monochromatic in color 0. Thus, Δ is not b-good.

`Case 5:` Assume that $a + 3 \leq b \leq \frac{a^2+3a-4}{2}$ and $r \geq a + 1$ and $r \leq b - a + 1$. The upper bound follows from (3). For the lower bound, we will show that the coloring $\widehat{\Delta}$ is both a-good and b-good. If $x_1 \in [1, q]$, $x_2 \in [1, a]$, and $x_3 \in [a^2+2a+1, a^2+3a]$, then, since $r \geq a + 1$,

$$bx_1 + x_2 \leq bq + a = a^2 + 2a + 1 - r + a < a^2 + 2a + 1 \leq x_3.$$

If $x_1 \in [q+1,a]$, $x_2 \in [1,a]$, and $x_3 \in [a^2+2a+1, a^2+3a]$, then since $r \leq b-a+1$,

$$bx_1 + x_2 \geq b(q+1)+1 \geq bq+r+a = a^2+3a+1 \geq x_3.$$

From this and the argument above, the lower bound is complete.

Case 6: Assume that $a+3 \leq b \leq \frac{a^2+3a-4}{2}$ and $r \geq a+1$ and $r \geq b-a+2$. For the lower bound we will show that the coloring $\widehat{\Delta}\upharpoonright[1, a^2+2a+b-r+1]$ is both a-good and b-good. Note that since $r \geq b-a+2$, it follows that

$$a^2+2a+b-r+1 \in [a^2+2a+1, a^2+3a+1].$$

If $x_1 \in [1,q]$, $x_2 \in [1,a]$, and $x_3 \in [a^2+2a+1, a^2+2a+b-r+1]$, then since $r \geq a+1$,

$$bx_1 + x_2 \leq bq+a = a^2+2a+1-r+a < a^2+2a+1 \leq x_3.$$

If $x_1 \in [q+1,a]$, $x_2 \in [1,a]$, and $x_3 \in [a^2+2a+1, a^2+2a+b-r+1]$, then

$$bx_1 + x_2 \geq b(q+1)+1 = bq+r+b-r+1 = a^2+2a+b-r+2 > x_3.$$

From this and the argument above, the lower bound is complete. For the upper bound let a coloring $\Delta : [1, a^2+2a+b-r+2] \to [0,1]$ be given. We must show that Δ is either not a-good or not b-good. Since $qb+r = a^2+2a+1$, the triple $(q+1, 1, a^2+2a+b-r+2)$ is a solution to $S(b)$. If Δ is a-good, then from Lemma 2, this triple is monochromatic in color 0, so Δ is not b-good.

Case 7: Assume $b = \frac{a^2+3a-2}{2}$. For the lower bound we will show that the coloring $\widehat{\Delta}\upharpoonright[1, a^2+3a-2]$ is both a-good and b-good. If $x_1 = 1$, $x_2 \in [1,a]$, and $x_3 \in [a^2+2a+1, a^2+3a-2]$, then since $b = \frac{a^2+3a-2}{2}$,

$$bx_1 + x_2 \leq b+a = \frac{a^2+5a-2}{2} < a^2+2a+1 \leq x_3.$$

If $x_1 \in [2,a]$, $x_2 \in [1,a]$, and $x_3 \in [a^2+2a+1, a^2+3a-2]$, then since $b = \frac{a^2+3a-2}{2}$,

$$bx_1 + x_2 \geq 2b+1 = a^2+3a-1 > a^2+3a-2 \geq x_3.$$

From this and the argument above, the lower bound is complete. For the upper bound, let a coloring $\Delta : [1, a^2+3a-1] \to [0,1]$ be given. We must show that Δ is either not a-good or not b-good. Since $b = \frac{a^2+3a-2}{2}$, the triple $(2, 1, a^2+3a-1)$ is a solution to $S(b)$. If Δ is a-good, then from Lemma 2 this triple is monochromatic in color 0, so Δ is not b-good.

Case 8: Assume that $\frac{a^2+3a}{2} \leq b \leq a^2+a$. The upper bound follows from (3). For the lower bound we will show that the coloring $\widehat{\Delta}$ is both a-good and b-good. If $x_1 = 1$, $x_2 \in [1,a]$, and $x_3 \in [a^2+2a+1, a^2+3a]$, then since $b \leq a^2+a$,

$$bx_1 + x_2 \leq b+a \leq a^2+2a < a^2+2a+1 \leq x_3.$$

If $x_1 \in [2, a]$, $x_2 \in [1, a]$, and $x_3 \in [a^2 + 2a + 1, a^2 + 3a]$, then since $b \geq \frac{a^2+3a}{2}$,

$$bx_1 + x_2 \geq 2b + 1 \geq a^2 + 3a + 1 > a^2 + 3a \geq x_3.$$

From this and the argument above, the lower bound is complete.

Case 9: Assume that $a^2 + a + 1 \leq b \leq a^2 + 2a$. The lower bound follows from (4). For the upper bound, let a coloring $\Delta : [1, a^2 + 2a + 1] \to [0, 1]$ be given. We must show that Δ is either not a-good or not b-good. Let $b = a^2 + a + k$ for $k \in [1, a]$, so the triple $(1, a + 1 - k, a^2 + 2a + 1)$ is a solution to $S(b)$. If Δ is a-good, then from Lemma 2, this triple is monochromatic in color 0, so Δ is not b-good.

Case 10: Assume that $a^2+2a+1 \leq b \leq a^2+3a-1$. The lower bound follows from (4). For the upper bound let a coloring $\Delta : [1, b + 1] \to [0, 1]$ be given. We must show that Δ is either not a-good or not b-good. The triple $(1, 1, b+1)$ is a solution to $S(b)$. Note that $b + 1 = a^2 + 2a + 1 + p$ where $p \in [1, a - 1]$, so if Δ is a-good, then from Lemma 3, this triple is monochromatic in color 0, so Δ is not b-good.

Case 11: Assume that $a^2 + 3a \leq b$. The upper bound follows from (3) and the lower bound follows from (4).

Since we have shown upper and lower bounds for all cases, the proof is complete. □

References

[1] A. Beutelspacher, W. Brestovansky, Generalized Schur numbers, *Lecture Notes Math.* 969, Springer-Verlag, Berlin, 1982, 30-38.

[2] A. Bialostocki, G. Bialostocki, and D. Schaal, A zero-sum theorem, *J. Combin. Theorey Ser.* A 101 (2003), 147-152.

[3] A. Bialostocki, R. Sabar, and D. Schaal, On a Zero-sum Generalization of a Variation of Schur's Equation, *Graphs Comb.* 24 (2008), no. 6, 511-518.

[4] A. Bialostocki, D. Schaal, On a Variation of Schur Numbers, *Graphs Comb.* 16 (2000), 139-147.

[5] D. Flint, B. Lowery, D. Schaal, Distinct Rado Numbers for $x_1 + x_2 + c = x_3$, to appear in *Ars Combin.*

[6] H. Harborth, S. Maasberg, Rado numbers for $a(x + y) = bz$, *J. Combin. Theory Ser.* A 80, num. 2 (1997) 356-363.

[7] H.Harborth, S. Maasberg, All two-color Rado numbers for $a(x + y) = bz$, *Discrete Math.* 197/198 (1999), 397-407.

[8] B. Hopkins, D. Schaal, Rado Numbers for $\sum_{i=1}^{m-1} a_i x_i = x_m$, *Adv. in Appl. Math.* 35 (2005), no. 4, 433-441.

[9] B. Johnson, D. Schaal, Disjunctive Rado Numbers, *J. Combin. Theory Ser. A.* 112 (2005), 263-276.

[10] S. Jones, D. Schaal, Some 2-color Rado numbers, *Congr. Numer.* 152 (2001), 197-199.

[11] S. Jones, D. Schaal, A class of two-color Rado numbers, *Discrete Math.* 289 (2004) no. 1-3, 63-69.

[12] W. Kosek, D. Schaal, Rado Numbers for the equation $\sum_{i=1}^{m-1} x_i + c = x_m$ for negative values of c, *Adv. in Appl. Math.* 27 (2001), 805-815.

[13] W. Kosek, D. Schaal, A Note on Disjunctive Rado Numbers, *Adv. in Appl. Math.* 31 (2003), iss. 2, 433-439.

[14] W. Kosek, A. Robertson, D. Sabo, D. Schaal, Multiplicity of Monochromatic Solutions to $x+y<z$, *J. Combin. Theory Ser. A.* 117 (2010), no. 8, 1127-1135.

[15] B. Martinelli, D. Schaal, On Generalized Schur Numbers for $x_1+x_2+c=kx_3$, *Ars Combin.* 85 (2007), 33-42.

[16] R. Rado, Verallgemeinerung eines Satzes von van der Waerden mit Anwendungen auf ein Problem der Zahlentheorie, *Sonderausg. Sitzungsber. Preuss. Akad. Wiss. Phys.- Math. Klasse* 17 (1933), 1-10.

[17] R. Rado, Studien zur Kombinatorik, *Math. Z.*, 36 (1933), 242-280.

[18] R. Rado, Note on Combinatorial Analysis, *Proc. London Math.* Soc. 48 (1936), 122-160.

[19] A. Robertson, D. Schaal, Off-Diagonal Generalized Schur Numbers, *Adv. in Appl. Math.* 26 (2001), 252-257.

[20] A. Robertson, Difference Ramsey Numbers and Issai Numbers, *Adv. in Appl. Math.* 25 (2000), 153-162.

[21] A. Robertson, D. Zeilberger, A 2-Coloring of $[1,N]$ Can Have $\frac{N^2}{22}+O(N)$ Monochromatic Schur Triples, But Not Less!, *Electron. J. Combin.* 5 (1998), R19.

[22] D. Sabo, D. Schaal, and J. Tokaz, Disjunctive Rado numbers for $x_1+x_2+c=x_3$, *Integers* 7 (2007), A29, 5pp.

[23] D. Schaal, A Family of 3-Color Rado Numbers, *Congr. Numer.* 111 (1995), 150-160.

[24] D. Schaal, On Generalized Schur Numbers, *Congr. Numer.* 98 (1993), 178-187.

[25] D. Schaal, Upper and lower bounds for some Rado numbers for inequalities, *Util. Math.* 53 (1998), 63-69.

[26] D. Schaal, H. Snevily, A Multiplicity problem related to Schur Numbers, *Integers* 8 (2008), A26, 7 pp.

[27] D. Schaal, D. Vestal, Rado numbers for $x_1+x_2+...+x_{m-1}=2x_m$, *Congr. Numer.* vol. 191 (2008), 105-116.

[28] D. Schaal, S. Wise, Rado numbers for some inequalities and an arbitrary number of colors, *Congr. Numer.* 121 (1996), 147-153.

[29] D. Schaal, M. Zinter, Real number Rado numbers for $\sum_{i=1}^{m-1} x_i + c = x_m$, *Congr. Numer.* 202 (2010), 113-118.

[30] D. Schaal, M. Zinter, Continuous Rado numbers for the equation $a_1x_1 + a_2x_2 + \cdots + a_{m-1}x_{m-1} + c = x_m$, *Congr. Numer.* 207 (2011), 97-104.

[31] I. Schur, Uber die Kongruenz $x^m + y^m \equiv z^m (mod p)$, *Jahresber. Deutsch. Math. Verein.* 25 (1916), 114-117.

[32] K. Rendall Truman, D. Schaal, Three-color Rado numbers for $x_1+x_2+c=x_3$ for negative values of c, *Congr. Numer.* 183 (2006), 5-10.

[33] W. Wallis, A. Street, J. Wallis, Combinatorics: Room Squares, Sum-free Sets, Hadamard Matrices, *Lecture Notes in Math.* 292, Springer-Verlag, Berlin and New York, 1972.

HOW TO WRITE A PERMUTATION AS A PRODUCT OF INVOLUTIONS (AND WHY YOU MIGHT CARE)

T. Kyle Petersen[1]
Department Mathematical Sciences, DePaul University, Chicago, IL 60614, USA
tkpeters@math.depaul.edu

Bridget Eileen Tenner[2]
Department Mathematical Sciences, DePaul University, Chicago, IL 60614, USA
bridget@math.depaul.edu

Received: 9/10/12, Revised: 3/23/13, Accepted: 8/18/13, Published: 9/26/13

Abstract

It is well-known that any permutation can be written as a product of two involutions. We provide an explicit formula for the number of ways to do so, depending only on the cycle type of the permutation. In many cases, these numbers are sums of absolute values of irreducible characters of the symmetric group evaluated at the same permutation, although apart from the case where all cycles are the same size, we have no good explanation for why this should be so.

1. Introduction

The authors were interested in finding a combinatorial model counted by the following sequence of integers:

$$1, 4, 9, 27, 61, 185, 469, \ldots, \tag{1}$$

And upon putting these terms into The On-Line Encyclopedia of Integer Sequences, we received a match in sequence A164342 [3]. That entry was a stub however, simply

[1]Research partially supported by an NSA Young Investigator grant
[2]Research partially supported by a DePaul University Faculty Summer Research Grant

defined as the row sums of the following table, given in entry A164341:

$n\backslash m$	1	2	3	$\cdots$											
1	1														
2	2	2													
3	3	2	4												
4	4	3	6	4	10										
5	5	4	6	6	6	8	26								
6	6	5	8	12	8	6	20	12	12	20	76				
7	7	6	10	12	10	8	12	18	16	12	20	30	24	52	232

(2)

But this was fantastic news, for our sequence could be refined in exactly the same fashion!

We had generated the terms in (1) by computing the sums of the absolute values of the irreducible characters of the symmetric group S_n, more precisely:

$$\sum_{\lambda,\mu\vdash n} |\chi_\mu^\lambda|,$$

and the entries in array (2) by fixing μ and considering

$$\sum_{\lambda\vdash n} |\chi_\mu^\lambda|.$$

On the other hand, A164341 said entry (n,m) "counts the decompositions into involutions of a permutation that has a cycle structure given by the mth partition of n."

Could this mean there might be a way to compute irreducible characters of S_n by counting involution products in the right way? Well, no. Our refinement of the sequence in (1) and the array in A164341 agreed for $n \le 7$, but diverged at $n = 8$. Here is the row for $n = 8$ in our table, sitting atop the row for $n = 8$ in A164341.

$\sum_{\lambda\vdash 8} \lvert\chi_\mu^\lambda\rvert$	8	7	12	15	20	12	10	12	**20**	24	20	**12**	24	18	76
A164341	8	7	12	15	20	12	10	12	**24**	24	20	**16**	24	18	76

$\cdots$	**36**	24	**36**	78	**52**	**148**	764
	40	24	**40**	78	**60**	**152**	764

(The differences are highlighted in boldface.) And here are the next few terms in sequence (1), with the next few terms of A164342 below.

n	8	9	10	11	12	13
$\sum_{\lambda,\mu\vdash n} \lvert\chi_\mu^\lambda\rvert$	1428	4292	14456	50040	186525	724023
A164342	1456	4368	14720	50800	190149	735451
(overcount)	28	76	264	760	3624	11428

In this note, we study the ways in which a permutation can be expressed as a product of two involutions. This puts entries A164341 and A164342 of [3] on solid footing, as there seem to be no references to this question in the literature. Because the motivation for this problem is the study of irreducible characters of the symmetric group, we try to explain what we can of the connection between the two topics, using ideas of White [8] and Stanton and White [7].

Section 2 investigates involution products from a purely combinatorial point of view. Section 3 discusses the relation to irreducible characters.

2. Products of Involutions in S_n

For a permutation $\sigma \in S_n$, we typically use cycle notation in this work. For example, $\sigma = (1236547)$ is the map $1 \mapsto 2 \mapsto 3 \mapsto 6 \mapsto 5 \mapsto 4 \mapsto 7 \mapsto 1$, while $\tau = (135)(26)(4)(7)$ maps $1 \mapsto 3 \mapsto 5 \mapsto 1$, $2 \mapsto 6 \mapsto 2$, and fixes 4 and 7. It is useful to draw our permutations as digraphs on the set $[n] := \{1, 2, \ldots, n\}$, in order to visualize the cycle structure. For example,

$\tau =$ 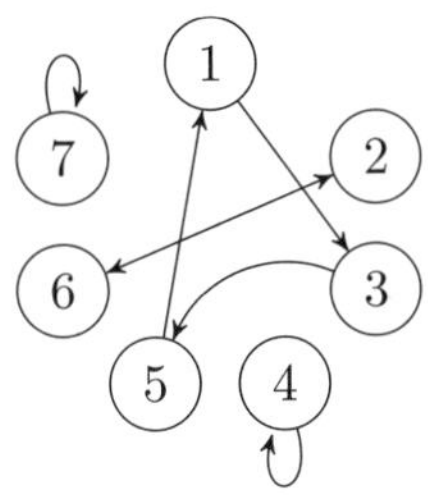

The cycle structure of a permutation is encoded by a *partition.* A partition $\lambda \vdash n$ is a collection of positive integers whose sum is n. We usually order the parts of $\lambda = (\lambda_1, \lambda_2, \ldots)$ in nonincreasing order; that is, $\lambda_1 \geq \lambda_2 \geq \cdots$ and $\sum \lambda_i = n$. An alternate notation is to let j_i denote the number of parts of size i in λ, and to write $\lambda = 1^{j_1} 2^{j_2} \cdots$, often suppressing any $j_i = 1$. For example, τ above has two one-cycles, one two-cycle, and one three-cycle; we encode this information with the partition $\lambda = (3, 2, 1, 1) = 1^2 2^1 3^1 = 1^2 23$.

The product of two permutations can be found by superimposing two digraphs of the form described above. For example, if we wish to compute the product of σ and τ from above, we draw the edges of σ and τ distinctly, say with solid and dashed edges, respectively. Then $\sigma\tau(i) = j$ if there is a directed path from i to $\tau(i)$

along a dashed edge, and from $\tau(i)$ to j along a solid edge. For example,

$\sigma\tau =$ 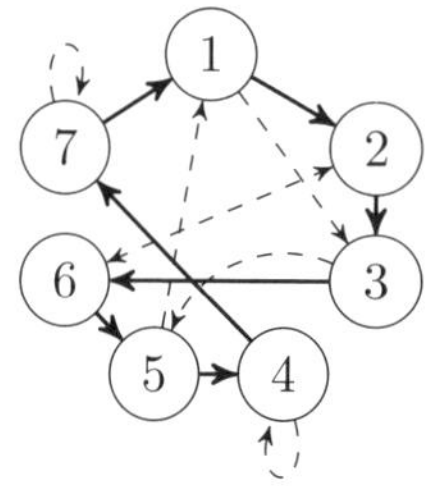

$=$ 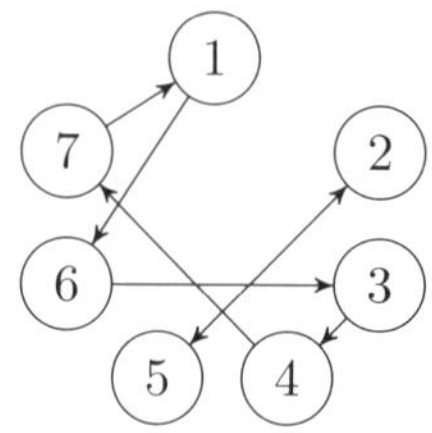

$= (16347)(25)$

We now consider the special case of *involutions*; that is, permutations whose squares are the identity. In an involution, each cycle must have size one or two, meaning that we can represent such elements graphically by *partial matchings*. For example,

$\iota = (14)(23)(57)(6) =$ 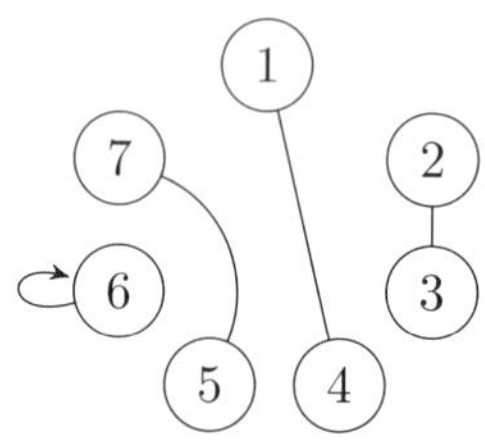

The product of two involutions in S_n can then be expressed as a decorated graph on $[n]$ (with loops) for which every vertex has precisely one edge of each decoration, counting a loop as a single edge. We call such a graph an *involution product graph*. For example, with the involution ι as above, and $\kappa = (1)(27)(35)(46)$, we have

$\iota\kappa =$ 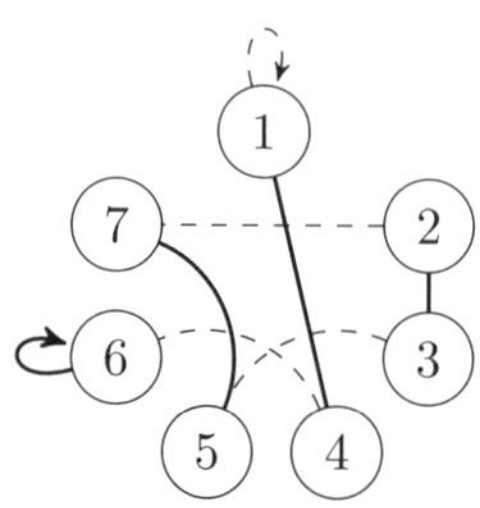

$=$ 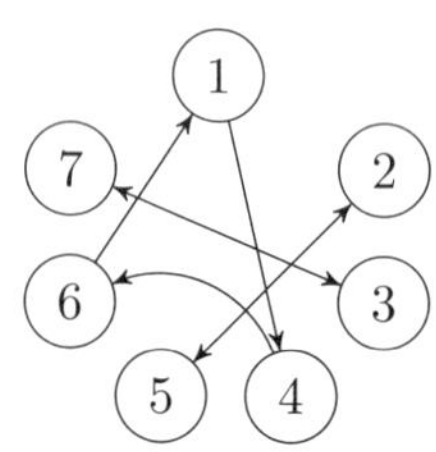

.

We see from this example that the set of involutions is not closed under multiplication. Indeed, it is well-known that any permutation can be written as a product of involutions (see [5, Exercise 10.1.17]), often in many different ways. Our goal is to describe, and to count, all the ways in which this can be done.

If an involution product graph defines a permutation ρ, we say it is an *involution graph for* ρ. In what follows, we identify the set of pairs of involutions whose product is ρ with the set of involution graphs for ρ. Let $N(\rho)$ denote the number of ways $\rho \in S_n$ can be written as a product of involutions. To be clear,

$$N(\rho) = |\{(\sigma,\tau) \in S_n^2 : \sigma^2 = \tau^2 = 1, \sigma\tau = \rho\}| = |\{\text{involution graphs for } \rho\}|.$$

Example 2.1. $N((123)(456)) = 12$, as depicted in Figure 1.

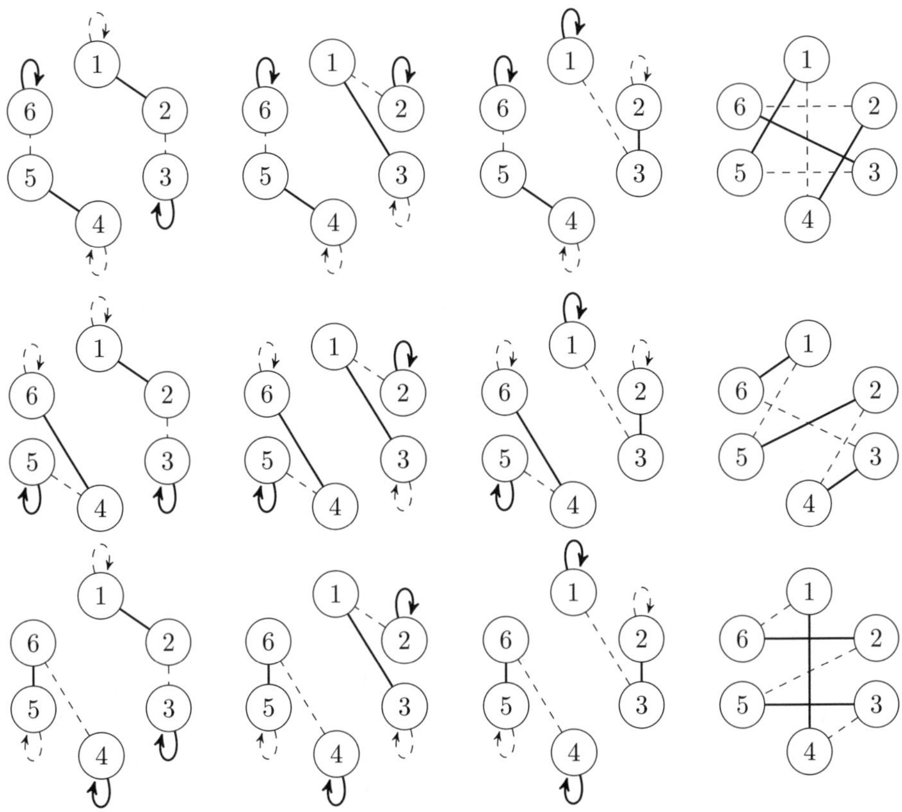

Figure 1: The involution product graphs depicting the twelve ways to write the permutation $(123)(456)$ as the product of two involutions, $\sigma\tau$, where σ is denoted with solid marks and τ with dashed marks.

Suppose $\rho' = g\rho g^{-1}$ for some $g \in S_n$. Let $\rho = \sigma\tau$ for some involutions σ and τ. Then $\sigma' = g\sigma g^{-1}$ and $\tau' = g\tau g^{-1}$ are also involutions, and we have $\rho' = \sigma'\tau'$. Hence, $N(\rho) = N(\rho')$, and we see that the number of ways to write an element of S_n as a product of involutions depends only on its conjugacy class (we could make the same observation in any finite group). Partitioning S_n by conjugacy class is equivalent to partitioning it by cycle type. Thus, for any partition $\lambda \vdash n$, we define $N(\lambda)$ to be the number of ways any particular permutation with cycle type λ can be written as a product of involutions. Carter [1, Theorem C] shows that any element in a finite Weyl group can be written as a product of two involutions, so one could study the same question for conjugacy classes of Weyl groups.

Define

$$R_m(k) = \sum_{0\le i\le m/2} \frac{k^{m-i}}{i!}\binom{m}{\underbrace{2,2,\dots,2}_{i}} = \sum_{0\le i\le m/2} \frac{k^{m-i}m!}{2^i i!(m-2i)!}. \tag{3}$$

We will prove the following formula for $N(\lambda)$.

Theorem 2.2. *Let $\lambda = 1^{j_1}2^{j_2}\cdots$. Then*

$$N(\lambda) = \prod_{i=1}^{n} R_{j_i}(i). \tag{4}$$

Before proving the theorem, we need some preliminary results.

Lemma 2.3. *Suppose an involution product graph for ρ has an alternating path of the form:*

$$1 \dashrightarrow v_1 \to 2 \dashrightarrow v_2 \to 3 \dashrightarrow v_3 \to \cdots \to k \dashrightarrow v_k \to 1,$$

where the first, third, fifth, … edges in this path are dashed and the second, fourth, sixth, … edges are solid. Then both

$$(12\cdots k) \quad \text{and} \quad (v_1\, v_k \cdots v_2)$$

are (possibly identical) k-cycles in ρ.

Proof. Fix an involution product graph of ρ with the alternating path described. Then obviously $(12\cdots k)$ is a k-cycle in ρ. Consider now $(v_1\, v_k\cdots v_2)$.

Since the involution graph has precisely one edge of each decoration at every vertex, we see that $v_i \neq v_j$ if $i \neq j$. That is, the path cannot have both $i \dashrightarrow v_i \to (i+1)$ and $j \dashrightarrow v_i \to (j+1)$ if $i \neq j$, where the first edges in each of these excerpts is dashed and the second is solid. Hence, $|\{v_1,\dots,v_k\}| = k$, and because all arrows in the product graph are reversible, we get another alternating path that shows $(v_1 v_k v_{k-1}\cdots v_2)$ is a k-cycle in ρ:

$$v_1 \dashrightarrow 1 \to v_k \dashrightarrow k \to v_{k-1} \dashrightarrow (k-1) \to \cdots \to v_2 \dashrightarrow 2 \to v_1.$$

□

If differently decorated cycles in an involution product graph form an alternating path as described in Lemma 2.3, then those cycles are *interlaced.* If a cycle does not interlace with any other, then that cycle is *isolated.*

A first consequence of Lemma 2.3 is the following.

Corollary 2.4. *The connected components of an involution product graph of a permutation ρ each either describe a cycle in ρ or two disjoint cycles of the same size. In particular,*

1. *three or more cycles cannot be interlaced in an involution product graph, and*
2. *if $k \neq k'$, then a k-cycle and a k'-cycle cannot be interlaced in an involution product graph.*

The next consequence is the key to our enumeration of involution product graphs.

Corollary 2.5. *For all $k \geq 3$,*

1. *There are precisely k ways to write a k-cycle as a product of involutions.*
2. *There are precisely k ways to write two disjoint k-cycles as a product of involutions, given that their elements are in the same connected component of the involution graph.*

Proof. If the cycles from Lemma 2.3 are fixed, then the designation of the label v_1 uniquely determines all the edges of this connected component of the involution graph. □

Proposition 2.6. *Suppose $\lambda = k^m = (k, k, \ldots, k) \vdash km$. Then,*

$$N(k^m) = R_m(k),$$

where $R_m(k)$ is defined in Equation (3).

Proof. Let ρ be a fixed permutation of cycle type λ. By Corollary 2.4, an involution graph for ρ either has all the k-cycles in disjoint connected components, or some pairs of the cycles are connected. For example, with $m = 7$ k-cycles, perhaps two pairs of them are interlaced, while the other three are isolated, as in the following picture.

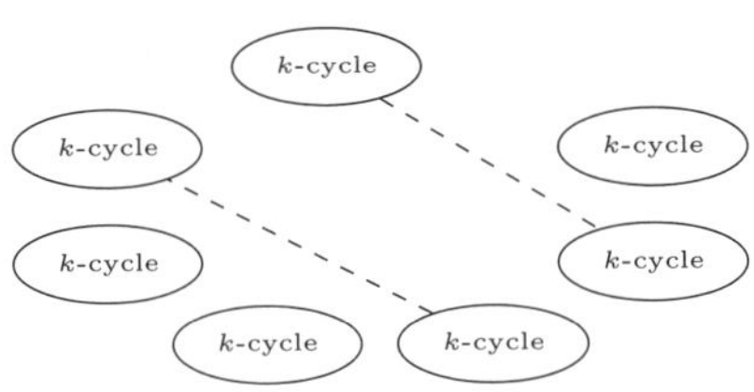

There are precisely

$$\frac{1}{i!}\binom{m}{2, 2, \ldots, 2, m-2i} = \frac{m!}{2^i i!(m-2i)!}$$

ways to choose a matching of i pairs of the k-cycles, leaving $m - 2i$ of the k-cycles isolated. Given such a matching, Corollary 2.5 tells us there are k ways to draw each connected component, giving $k^i \cdot k^{m-2i} = k^{m-i}$ choices. Hence, the number of involution product graphs for ρ that have i pairs of k-cycles is:

$$\frac{k^{m-i} m!}{2^i i!(m-2i)!}.$$

To count all possible involution graphs for ρ, we sum over all i to obtain the desired result:

$$N(\rho) = \sum_{0\le i\le m/2} \frac{k^{m-i}m!}{2^i i!(m-2i)!} = R_m(k).$$

□

We are now able to prove Theorem 2.2.

Proof of Theorem 2.2. Let ρ be a permutation of type $\lambda = 1^{j_1}2^{j_2}\cdots$. By Corollary 2.4, two cycles of ρ are in different connected components if they are of different size. Hence, for $k \neq k'$, the k-cycles and k'-cycles are mutually independent, and to count the number of involution product graphs for ρ, we simply multiply the number of possibilities for each cycle size:

$$N(\lambda) = N(1^{j_1}) \cdot N(2^{j_2}) \cdots = \prod_{j=1}^{n} R_{j_i}(i).$$

□

3. Connections with the Character Table of S_n

Now that we have enumerated the number of pairs of involutions with a fixed product, we will describe why understanding such involution products may be related to characters of the symmetric group. Our tone is primarily expository here, built upon substantial work of White [8] and Stanton and White [7]. (A more recent approach to many of the same topics can be found in [2].) Omitted details may be found in those papers.

To begin, we recall that the well-known Murnaghan-Nakayama rule provides the following formula for an irreducible character of the symmetric group S_n:

$$\chi_\mu^\lambda = \sum_T (-1)^{\mathrm{ht}(T)},$$

where the sum is taken over all *rim-hook tableaux* T of shape λ and content μ, and $\mathrm{ht}(T)$ is the sum of the heights of the hooks in T minus the number of hooks. In general, $(-1)^{\mathrm{ht}(T)}$ gives the *sign* of a tableau T. For example,

$$T = \begin{array}{|c|c|c|c|c|} \hline 1 & 1 & 1 & 1 & 3 \\ \hline 1 & 2 & 3 & 3 & 3 \\ \hline 2 & 2 & 4 & 6 \\ \cline{1-4} 2 & 4 & 4 \\ \cline{1-3} 5 \\ \cline{1-1} \end{array}$$

is a rim-hook tableau with content $\mu = (5,4,4,3,1,1)$, its height is calculated by $\mathrm{ht}(T) = 2+3+2+2+1+1-6 = 5$, and $(-1)^{\mathrm{ht}(T)} < 0$ so T is a negative tableau. See either [4, Sec. 4.10] or [6, Sec. 7.17] for precise definitions and further discussion.

This formula is lovely in that it gives an explicit combinatorial description for the characters. However, it involves tremendous cancellation. For instance, consider the character table for S_6 in Figure 2. There are 40 rim-hook tableau of content $(2,2,1,1)$, yet in the column indexed by $\mu = (2,2,1,1)$ we see that only 12 terms are non-cancelling; that is, the sum of the absolute values of the entries in this column is $1+1+1+2+1+2+1+1+1+1 = 12$.

$\lambda\backslash\mu$	6	51	42	411	33	321	3111	222	2211	21111	111111
6	1	1	1	1	1	1	1	1	1	1	1
51	-1	0	-1	1	-1	0	2	-1	1	3	5
42	0	-1	1	-1	0	0	0	3	1	3	9
411	1	0	0	0	1	-1	1	-2	-2	2	10
33	0	0	-1	-1	2	1	-1	-3	1	1	5
321	0	1	0	0	-2	0	-2	0	0	0	16
3111	-1	0	0	0	1	1	1	2	-2	-2	10
222	0	0	-1	1	2	-1	-1	3	1	-1	5
2211	0	-1	1	1	0	0	0	-3	1	-3	9
21111	1	0	-1	-1	-1	0	2	1	1	-3	5
111111	-1	1	1	-1	1	-1	1	-1	1	-1	1
$N(\mu)$	6	5	8	8	12	6	12	20	12	20	76

Figure 2: The character table of S_6, along with $N(\mu)$, the number of ways to write a permutation of cycle type μ as a product of involutions.

On the other hand, there are precisely 12 rim hook tableaux of content (3,3), and so there is no cancellation in this column. As we will see shortly, it follows from work of White [8] that, when all parts of μ are the same size, the Murnaghan-Nakayama formula for χ^λ_μ is *cancellation-free* (Corollary 3.6). Such a partition μ is called *rectangular*. Our feeling is that understanding involution products may lead to a similar cancellation-free formula for any content μ.

The idea begins with the following observation. For any λ, the Murnaghan-Nakayama rule gives

$$\chi^\lambda_{1^n} = |\{\text{standard Young tableaux of shape } \lambda\}|.$$

On the other hand, the Schensted insertion algorithm gives a bijection between the set of standard Young tableaux with n boxes and the set of involutions in S_n. If we let $I^\lambda = \{\sigma \in S_n : \sigma^2 = 1, \mathrm{sh}(\sigma) = \lambda\}$, where $\mathrm{sh}(\sigma)$ denotes the shape of the image of σ upon the Schensted insertion, then

$$\chi^\lambda_{1^n} = |I^\lambda|.$$

To generalize this idea to $\mu \neq 1^n$, we first need to discuss a more general insertion algorithm.

In [8], White defines a map Sch that generalizes the usual Schensted insertion algorithm. Given a pair of rim hook tableaux (P, Q) of the same shape and content, we say the pair is *positive* if P and Q have the same sign; the pair is *negative* otherwise. The map Sch provides a bijection that maps a positive pair (P, Q) of content μ to either a negative pair (P', Q') of tableaux of content μ or to a "hook permutation" of content μ.

Definition 3.1. Fix a positive integer m. A *hook permutation* $(H_1, H_2, \ldots, H_m)$ of *type* $\rho = (\rho_1, \rho_2, \ldots, \rho_m) = 1^{j_1}2^{j_2}\cdots k^{j_k}$ and shapes $(\tau^{(1)}, \tau^{(2)}, \ldots, \tau^{(m)})$ has the properties

- the H_i are each hook tableaux of shape $\tau^{(i)}$,
- $\rho_i = |\tau^{(i)}|$, and
- if $\rho_i > \rho_j$, then the content of H_i is less than the content of H_j.

Using the notation of Definition 3.1, define the intervals $I_1 = [1, j_k]$, $I_2 = [j_k + 1, j_k + j_{k-1}]$, $I_3 = [j_k + j_{k-1} + 1, j_k + j_{k-1} + j_{k-2}]$, and so on. Then a hook permutation maps each I_ℓ bijectively onto itself, and the image of $i \in I_\ell$ is assigned a hook consisting of $k + 1 - \ell$ boxes, whose shape is $\tau^{(i)}$. We can think of this as a sequence of m hooks, where the lengths of the hooks is nonincreasing, and the entries of the k-hooks are I_1, the entries of the $(k-1)$-hooks are I_2, and so on.

Example 3.2. The following is an example of a hook permutation of type $\rho = 1^3 3 4^3 5$.

$$\mathcal{H} = \left(\begin{array}{|c|c|c|c|}\hline 1&1&1&1\\\hline 1\\\cline{1-1}\end{array} \quad \begin{array}{|c|}\hline 4\\\hline 4\\\hline 4\\\hline 4\\\hline\end{array} \quad \begin{array}{|c|c|}\hline 2&2\\\hline 2\\\cline{1-1} 2\\\cline{1-1}\end{array} \quad \begin{array}{|c|}\hline 3\\\hline 3\\\hline 3\\\hline 3\\\hline\end{array} \quad \begin{array}{|c|c|}\hline 5&5\\\hline 5\\\cline{1-1}\end{array} \quad \begin{array}{|c|}\hline 8\\\hline\end{array} \quad \begin{array}{|c|}\hline 7\\\hline\end{array} \quad \begin{array}{|c|}\hline 6\\\hline\end{array} \right)$$

It is easy to see that the number of hook permutations of type $(\rho_1, \rho_2, \ldots, \rho_m) = 1^{j_1}2^{j_2}\cdots k^{j_k}$ is $1^{j_1} j_1! 2^{j_2} j_2! \cdots k^{j_k} j_k!$, since there are j_i different hooks of length i.

We will choose to think of Sch as a map from the set of hook permutations of a given content to the non-cancelling positive pairs of rim-hook tableaux (P, Q) having that same content. The algorithm involved in which turns a hook permutation $(H_1, \ldots, H_m)$ into a pair of rim hook tableaux (P, Q) builds the tableaux by successively inserting the hooks $H_1, \ldots, H_m$. The first tableau is the insertion tableau, and the second is the recording tableau. We briefly describe the process here, and the full details can be found in [8].

The first hook H_1 is inserted into a pair of empty tableaux (P_0, Q_0), yielding (P_1, Q_1). Subsequently, the hook H_i is inserted into (P_{i-1}, Q_{i-1}) to form (P_i, Q_i).

When inserting the hook H_i with content a and size ℓ, the tableau P_{i-1} is divided into an *inside part* consisting of all entries less than a, and an *outside part* consisting of all entries greater than a.

The hook H_i is first placed so that its upper left corner coincides with the upper left corner of the inside part of P_{i-1}, and it slides (changing shape, in the process) down the first column of P_{i-1}, ℓ units at a time, until it is completely beyond the inside part of the tableau. It then slithers up the outer rim of the inside part of P_{i-1}, ℓ units at a time, until it reaches a position so that the union of this rim hook and the inside part is a partition. This new rim hook is now considered part of the inside part of the tableau.

Now the rim hooks in the outside part of P_{i-1} successively move into the inside part, in increasing order of content. If the rim hook being added does not overlap the inside part at all, then it joins the inside part without changing position. If it overlaps the inside part but does not lie completely inside the inside part, then it joins the inside part after the overlapping entries each move one step southeast, and the non-overlapping entries do not move. Finally, if the rim hook being added, of length ℓ', overlaps the inside part completely, then it slithers up the outer rim of the inside part, ℓ' units at a time, until it reaches a position for which the union of this rim hook and the inside part is a partition. Once the entire outside part has been moved to the inside, the result is (P_i, Q_i), where Q_i records the newly added boxes. Finally, $(P, Q) = (P_m, Q_m)$.

Example 3.3. Given the rim hook permutation $\mathcal{H}$ of Example 3.2, the insertion algorithm produces the following sequence of tableaux:

$$P_1 = \begin{array}{cccc} 1 & 1 & 1 & 1 \\ 1 & & & \end{array}, \quad P_2 = \begin{array}{cccc} 1 & 1 & 1 & 1 \\ 1 & 4 & & \\ 4 & 4 & & \\ 4 & & & \end{array}, \quad P_3 = \begin{array}{cccc} 1 & 1 & 1 & 1 \\ 1 & 2 & 2 & \\ 2 & 2 & 4 & \\ 4 & 4 & 4 & \end{array}, \quad P_4 = \begin{array}{ccccc} 1 & 1 & 1 & 1 & 4 \\ 1 & 2 & 2 & 4 & 4 \\ 2 & 2 & 3 & 4 & \\ 3 & 3 & 3 & & \end{array},$$

$$P_5 = \begin{array}{ccccc} 1 & 1 & 1 & 1 & 4 \\ 1 & 2 & 2 & 4 & 4 \\ 2 & 2 & 3 & 4 & \\ 3 & 3 & 3 & & \\ 5 & 5 & 5 & & \end{array}, \quad P_6 = \begin{array}{ccccc} 1 & 1 & 1 & 1 & 4 \\ 1 & 2 & 2 & 4 & 4 \\ 2 & 2 & 3 & 4 & \\ 3 & 3 & 3 & & \\ 5 & 5 & 5 & & \\ 8 & & & & \end{array}, \quad P_7 = \begin{array}{ccccc} 1 & 1 & 1 & 1 & 4 \\ 1 & 2 & 2 & 4 & 4 \\ 2 & 2 & 3 & 4 & \\ 3 & 3 & 3 & & \\ 5 & 5 & 5 & & \\ 7 & 8 & & & \end{array},$$

$$P_8 = \begin{array}{ccccc} 1 & 1 & 1 & 1 & 4 \\ 1 & 2 & 2 & 4 & 4 \\ 2 & 2 & 3 & 4 & \\ 3 & 3 & 3 & & \\ 5 & 5 & 5 & & \\ 6 & 7 & 8 & & \end{array}$$

and thus the algorithm Sch produces the following pair of tableaux:

$$\mathsf{Sch}(\mathcal{H}) = \left(\begin{array}{|c|c|c|c|c|} \hline 1 & 1 & 1 & 1 & 4 \\ \hline 1 & 2 & 2 & 4 & 4 \\ \hline 2 & 2 & 3 & 4 \\ \cline{1-4} 3 & 3 & 3 \\ \cline{1-3} 5 & 5 & 5 \\ \cline{1-3} 6 & 7 & 8 \\ \cline{1-3} \end{array} \; , \; \begin{array}{|c|c|c|c|c|} \hline 1 & 1 & 1 & 1 & 4 \\ \hline 1 & 2 & 3 & 4 & 4 \\ \hline 2 & 2 & 3 & 4 \\ \cline{1-4} 2 & 3 & 3 \\ \cline{1-3} 5 & 5 & 5 \\ \cline{1-3} 6 & 7 & 8 \\ \cline{1-3} \end{array} \right)$$

The following facts describe the aspects of the Sch construction that are most relevant to our work here.

Proposition 3.4. *Fix a positive integer m and a composition $\rho = (\rho_1, \ldots, \rho_m) = 1^{j_1} \ldots k^{j_k}$. Let $\mathcal{H} = (H_1, \ldots, H_m)$ be a hook permutation of type ρ.*

(a) If the content of all the hooks $\{H_1, \ldots, H_i\}$ is less than the content of all the hooks $\{H_{i+1}, \ldots, H_m\}$, then the tableaux built from the hook permutation $(H_1, \ldots, H_i)$ are stable for the duration of the procedure. That is, if $\mathsf{Sch}(H_1, \ldots, H_i) = (P', Q')$, and $\mathsf{Sch}(H_1, \ldots, H_m) = (P, Q)$, then the rim hook tableau P' is a subtableau of the rim hook tableau P, and similarly Q' is a subtableau of Q.

(b) In Sch, the shape being constructed propagates down and to the right, never upward; that is, once the top row of the shape has been established, it is maintained as the top row for the duration of the procedure.

White [8] shows many of the properties of Sch are particularly nice in the special case of rectangular content; that is, when all rim hooks have the same length. In [7], Stanton and White consider this case in further detail.

Proposition 3.5 ([8], Corollaries 9 and 10)**.** *If $\mu = k^m$, then the inverse procedure Sch^{-1} is a bijection from hook permutations of content μ to the set of all pairs (P, Q) of rim hook tableaux of the same shape and content μ. In particular, every pair of tableaux (P, Q) is positive.*

The following is immediate.

Corollary 3.6. *Any two rim hook tableaux of the same shape and the same rectangular content have the same sign. Hence,*

$$|\chi^{\lambda}_{k^m}| = |\{P : P \text{ has shape } \lambda \text{ and content } k^m\}|.$$

In other words, there is no cancellation in the Murnaghan-Nakayama formula for χ^{λ}_{μ} when μ is rectangular.

We now show how this result is related to involution products.

Definition 3.7. A hook permutation $\mathcal{H}$ of type ρ is a *hook-block involution* if

- the underlying permutation is an involution, and
- the hooks associated to the letters in each cycle of the involution are the same.

Example 3.8. The hook permutation

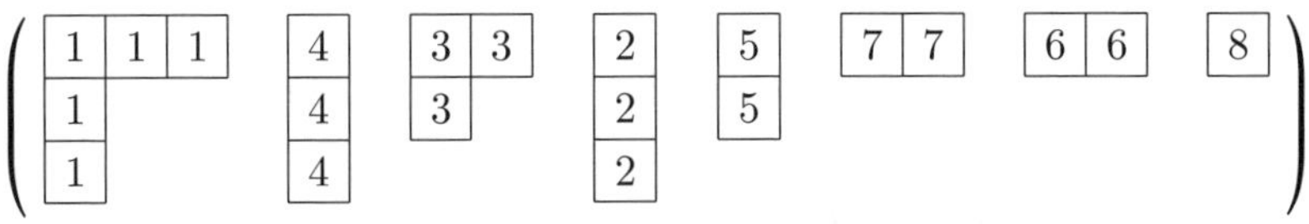

is a hook-block involution. In this case, the underlying permutation is the involution $(1)(24)(3)(5)(67)(8)$, and the shape equalities are $\tau^{(2)} = \tau^{(4)}$ and $\tau^{(6)} = \tau^{(7)}$.

To any pair of involutions we may assign a hook-block involution as follows. We look at all of the k-cycles in the involution product graph, in decreasing order of k. Suppose that we have examined the k'-cycles for all $k' > k$, and so far created a hook-block involution on the letters $\{1, \ldots, M\}$. Now suppose that there are r k-cycles in the involution product graph. Label these $\{M+1, \ldots, M+r\}$ in order of the smallest element appearing in each. We thus inherit an involution of $\{M+1, \ldots, M+r\}$, based on which k-cycles are interlaced. To each letter in this involution, we associate a hook, where the choice of v_1 in Lemma 2.3 determines the height of the first column in each hook of the 2-cycle: if v_1 is in the ith position when the cycle is written with the smallest letter in the first position, then the height of that first column in the hook is i. Note that any pair of letters that are each other's images under the involution get assigned the same hook shape, by Lemma 2.3. We demonstrate with an example.

Example 3.9.

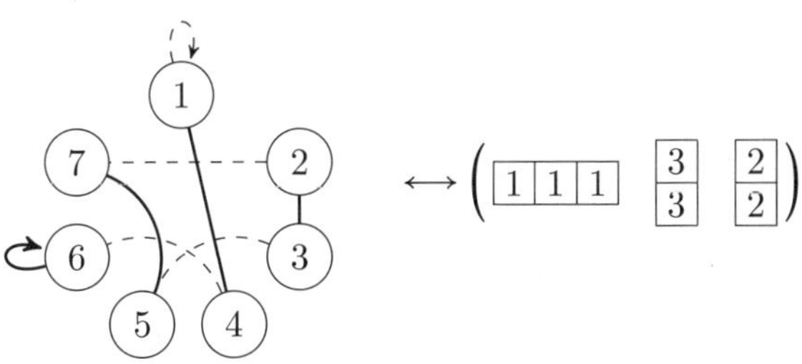

First of all, the product is $(146)(25)(37)$, which has cycle type $(3, 2, 2)$. Hence we know the hook-block involution will have a 3-hook and two 2-hooks. Since 1 is the smallest element of the 3-cycle, we see where its dashed arrow points to determine the height of the 3-hook. It points to itself, so we choose the hook of height 1. (If the 1 had had a dashed arrow to the 4, then the 3-hook would have had height

2. If it had pointed to the 6 it would have had height 3.) There are two 2-cycles, and they lie in the same connected component of the matching graph. Hence, the corresponding 2-hooks are transposed in the hook permutation. The shape of the hooks is determined by where the 2 is connected with its dashed edge. If it had been connected to the 3, then the 2-hooks would be horizontal. However, the dashed edge connects to the 7 (which is the second-largest element in its cycle), and so the hooks have height 2.

A hook-block involution in which all hooks are the same size k is a "k-partial involution" [7].

Example 3.10. 1. The hook permutation

$$\left(\begin{array}[t]{|c|}\hline 3\\\hline 3\\\hline 3\\\hline\end{array} \quad \begin{array}[t]{|c|c|}\hline 2&2\\\hline 2\\\cline{1-1}\end{array} \quad \begin{array}[t]{|c|}\hline 1\\\hline 1\\\hline 1\\\hline\end{array} \quad \begin{array}[t]{|c|c|}\hline 5&5\\\hline 5\\\cline{1-1}\end{array} \quad \begin{array}[t]{|c|c|}\hline 4&4\\\hline 4\\\cline{1-1}\end{array} \quad \begin{array}[t]{|c|c|c|}\hline 6&6&6\\\hline\end{array} \right)$$

is a 3-partial involution because the underlying permutation is $(13)(2)(45)(6)$, which is an involution, and it has the shape equalities $\tau^{(1)} = \tau^{(3)}$ and $\tau^{(4)} = \tau^{(5)}$.

2. The hook permutation

$$\left(\begin{array}[t]{|c|}\hline 3\\\hline 3\\\hline 3\\\hline\end{array} \quad \begin{array}[t]{|c|c|}\hline 5&5\\\hline 5\\\cline{1-1}\end{array} \quad \begin{array}[t]{|c|}\hline 1\\\hline 1\\\hline 1\\\hline\end{array} \quad \begin{array}[t]{|c|c|}\hline 2&2\\\hline 2\\\cline{1-1}\end{array} \quad \begin{array}[t]{|c|c|}\hline 4&4\\\hline 4\\\cline{1-1}\end{array} \quad \begin{array}[t]{|c|c|c|}\hline 6&6&6\\\hline\end{array} \right)$$

is not a 3-partial involution because the underlying permutation $(13)(254)(6)$ is not an involution.

3. The hook permutation

$$\left(\begin{array}[t]{|c|}\hline 3\\\hline 3\\\hline 3\\\hline\end{array} \quad \begin{array}[t]{|c|c|}\hline 2&2\\\hline 2\\\cline{1-1}\end{array} \quad \begin{array}[t]{|c|}\hline 1\\\hline 1\\\hline 1\\\hline\end{array} \quad \begin{array}[t]{|c|c|}\hline 5&5\\\hline 5\\\cline{1-1}\end{array} \quad \begin{array}[t]{|c|}\hline 4\\\hline 4\\\hline 4\\\hline\end{array} \quad \begin{array}[t]{|c|c|c|}\hline 6&6&6\\\hline\end{array} \right)$$

is not a 3-partial involution because, although the underlying permutation is the involution $(13)(2)(45)(6)$, this hook permutation has the shape inequality $\tau^{(4)} \neq \tau^{(5)}$.

Indeed, k-partial involutions have nice properties.

Theorem 3.11 ([7]). *A hook permutation $\mathcal{H}$ of type $\mu = k^m$ is a k-partial involution if and only if $\mathsf{Sch}(\mathcal{H}) = (P, P)$. That is, we have a bijection,*

$$\{\ k\text{-partial involutions}\ \} \xleftrightarrow{\mathsf{Sch}} \{\ \text{rim-hook tableaux } P \text{ of content } k^m\ \}.$$

Now let

$$I_k^\lambda = \{\mathcal{H} : \mathcal{H} \text{ is a } k\text{-partial involution and } \mathsf{Sch}(\mathcal{H}) \text{ has shape } \lambda\}.$$

Then, together with Corollary 3.6, we obtain the following.

Corollary 3.12. *Let k and m be positive integers. Then for all $\lambda \vdash n$,*

$$|\chi_{k^m}^\lambda| = |I_k^\lambda|.$$

Because $N(\mu)$ counts all hook-block involutions of content μ, Theorem 2.2 implies the following result.

Corollary 3.13. *Let k and m be positive integers. Then,*

$$\sum_{\lambda \vdash km} |\chi_{k^m}^\lambda| = N(k^m) = \sum_{0 \le i \le m/2} \frac{k^{m-i} m!}{2^i i! (m-2i)!}.$$

This leads one to wonder whether something similar is true for non-rectangular μ.

Question 1. *For which $\mu \vdash n$ is it true that*

$$\sum_{\lambda \vdash n} |\chi_\mu^\lambda| = N(\mu)? \tag{5}$$

Note that the sums of the characters themselves, not the absolute values of the characters, have been studied in other contexts, and have been shown in [2] to equal the number of certain hook-block involutions. Not taking the absolute values of characters allows the possibility of canceling terms, a possibility we are interested in avoiding.

Corollary 3.12 shows that Equation (5) holds for $\mu = k^m$ and by computer we have checked that it holds for all $\mu \vdash n$ when $n \le 7$. However, when $n = 8$, we begin to see some discrepancies. In each of these cases $N(\mu)$ is greater than $\sum_{\lambda \vdash n} |\chi_\mu^\lambda|$. See Table 1, where for $n \le 10$ we have listed all the shapes $\mu \vdash n$ for which Equation (5) does not hold.

If we check all shapes $\mu \vdash n$ with $n \le 16$, we find that there are indeed a number of cases in which Equation (5) holds. We summarize the number of agreements ($N(\mu) = \sum_{\lambda \vdash n} |\chi_\mu^\lambda|$) and discrepancies ($N(\mu) > \sum_{\lambda \vdash n} |\chi_\mu^\lambda|$) in Table 2. We have also included the number of rectangular partitions $\mu = k^m$ since in these cases we know that equality holds. For example, when $n = 16$ there are 64 shapes $\mu \vdash 16$ for which $N(\mu) = \sum_{\lambda \vdash n} |\chi_\mu^\lambda|$, yet we can only explain 5 of these: $1^{16}, 2^8, 4^4, 8^2$, and 16^1.

If we are to move beyond the rectangular case, we need to understand how to associate involution products, or hook-block involutions, to a collection of rim-hook tableaux. Theorem 3.11 does this in the case when μ is rectangular. In the more

μ	$N(\mu) - \sum_{\lambda \vdash n} \lvert\chi_\mu^\lambda\rvert$
2^24	4
1^224	4
1^44	4
1^22^3	4
1^42^2	8
1^62	4
234	4
1^234	4
12^24	4
1^324	4
1^54	4
1^33^2	8
2^33	4
1^22^23	4
1^423	4
1^63	20
1^32^3	4
1^52^2	8
1^72	4

μ	$N(\mu) - \sum_{\lambda \vdash n} \lvert\chi_\mu^\lambda\rvert$
235	4
1^235	4
1^325	4
1^55	8
2^34	8
1^22^24	8
1^424	8
1^64	8
2^23^2	8
1^43^2	8
1^32^23	8
1^523	16
1^73	56
1^22^4	8
1^42^3	28
1^62^2	48
1^82	32

Table 1: When Equation (5) does not hold.

n	≤ 7	8	9	10	11	12	13	14	15	16
Discrepancies:	0	6	13	17	24	46	60	83	114	167
Agreements:	all	16	17	25	32	31	41	52	62	64
Known agreements ($\mu = k^m$):	-	4	3	4	2	6	2	4	4	5
Partitions of n:	-	22	30	42	56	77	101	135	176	231

Table 2: Counting the number of shapes μ for which $N(\mu) = \sum_{\lambda \vdash n} \lvert\chi_\mu^\lambda\rvert$.

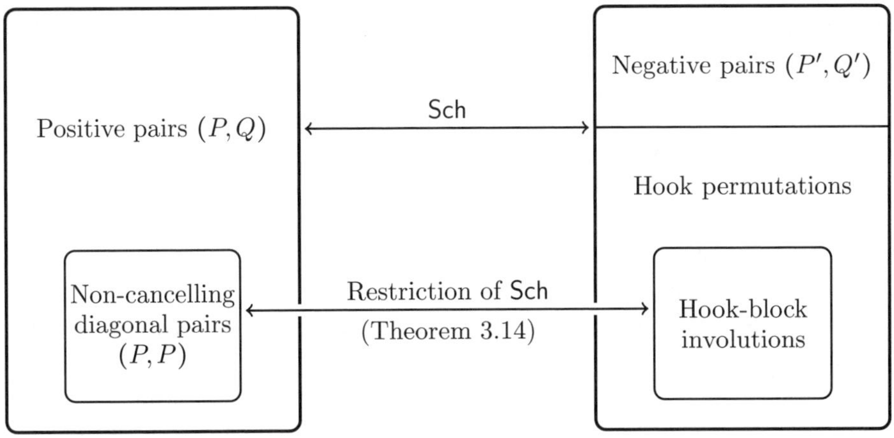

Figure 3: Schematic diagram for the Sch bijection with fixed shape λ and content μ.

general case, Theorem 3.14 below shows that Sch does something similar, giving a bijection between hook-block involutions and the set of tableaux P for which (P, P) is a non-cancelling pair. See Figure 3 for an illustration of these results and their scopes.

Theorem 3.14. *A hook permutation $\mathcal{H}$ of content μ is a hook-block involution if and only if $\mathsf{Sch}(\mathcal{H}) = (P, P)$. That is, the non-cancelling pairs (P, P) of content μ are in bijection with the set of hook-block involutions.*

Proof of Theorem 3.14. Fix an integer $m > 0$ and a composition $\mu = (\mu_1, \ldots, \mu_m) = 1^{j_1} \ldots k^{j_k}$. Let $\mathcal{H} = (H_1, \ldots, H_m)$ be a hook permutation of type μ.

Suppose that $\mathsf{Sch}(\mathcal{H}) = (P, P)$. By Theorem 3.11 and Proposition 3.4(a), the initial j_k letters and hooks in $\mathcal{H}$ must form a k-partial involution. Inductively, suppose now that all hooks $\{H_1, \ldots, H_r\}$ of length greater than ℓ have been inserted, forming the pair (P', P') of shape ν. Note that the entries of P' are all less than or equal to r. Let $\pi = \nu \cup \alpha$ be the shape obtained by appending ℓ sufficiently long rows, each of length at least $|\nu| + \ell j_\ell$, on top of ν, and such that $|\pi| = \ell t$ for some t.

Let $\widehat{P}$ be a rim hook tableau of shape π, and content ℓ^t, whose entries are $\{r - t + 1, r - t + 2, \ldots, r\}$ (allowing entries to be non-positive). By Proposition 3.5 and Theorem 3.11, $\mathsf{Sch}^{-1}(\widehat{P}, \widehat{P})$ equals some hook permutation $\widehat{\mathcal{H}} = (\widehat{H}_{r-t+1}, \widehat{H}_{r-t+2}, \ldots, \widehat{H}_r)$ which is in fact an ℓ-partial involution. (We are allowing permutations of sets that include non-positive values due to the content of $\widehat{P}$.)

Now define the hook permutation

$$\mathcal{H}^+ = (\widehat{H}_{r-t+1}, \widehat{H}_{r-t+2}, \ldots, \widehat{H}_r, H_{r+1}, H_{r+2}, \ldots, H_{r+j_\ell}),$$

where $\{H_{r+1}, \ldots, H_{r+j_\ell}\}$ are the hooks of size ℓ in $\mathcal{H}$. The point of Proposition 3.4(b) is that adding ℓ rows on top of the shape ν to form π does not affect the insertion of the ℓ-hooks $\{H_{r+1}, \ldots, H_{r+j_\ell}\}$ via Sch, except to allow us to create a shape with rectangular content. Because all content in the hooks $\{H_{r+j_\ell+1}, \ldots, H_m\}$ is less than the content of the hooks $\{H_{r+1}, \ldots, H_{r+j_\ell}\}$, we must have that $\mathsf{Sch}(\mathcal{H}^+) = (P^+, P^+)$ by Proposition 3.4(a). Moreover, the hooks with content $\{r+1, \ldots, r+j_\ell\}$ in P^+ without α, its top ℓ rows, occupy the same positions that they do in P. Theorem 3.11 implies that $\mathcal{H}^+$ must be an ℓ-partial involution. Because $\widehat{H}$ was an ℓ-partial involution as well, this means that $(H_{r+1}, H_{r+2}, \ldots, H_{r+j_\ell})$ itself is an ℓ-partial involution. Consequently, by the inductive hypothesis, $\mathcal{H}$ must be a hook-block involution.

Now suppose that $\mathsf{Sch}(\mathcal{H}) = (P, Q)$ where $P \neq Q$. Then the tableaux first diverge during the insertion of the hooks of length i for some i. A similar argument to the above shows that this means that the hooks of length i do not form a i-partial involution. Thus the full hook permutation $\mathcal{H}$ is not a hook-block involution. □

Theorem 3.14 shows it is possible to associate a collection of $N(\mu)$ rim-hook tableaux of content μ to the set of all hook-block involutions of content μ. Unfortunately, it is not obvious how to deduce when $\sum_{\lambda \vdash n} |\chi^\lambda_\mu|$ and $N(\mu)$ coincide from this result. Moreover, as the following example shows, even when $N(\mu)$ and $\sum_{\lambda \vdash n} |\chi^\lambda_\mu|$ are equal, applying Sch to the set of hook-block involutions does not necessarily yield a result like Corollary 3.12.

Example 3.15. While $N(2211) = 12$, applying Sch to the corresponding twelve hook-block involutions yields no tableaux of shape 6 or shape 51, while $\chi^{6}_{2211} = \chi^{51}_{2211} = 1$. Compare Figures 4 and 5 with the column corresponding to 2211 in Figure 2.

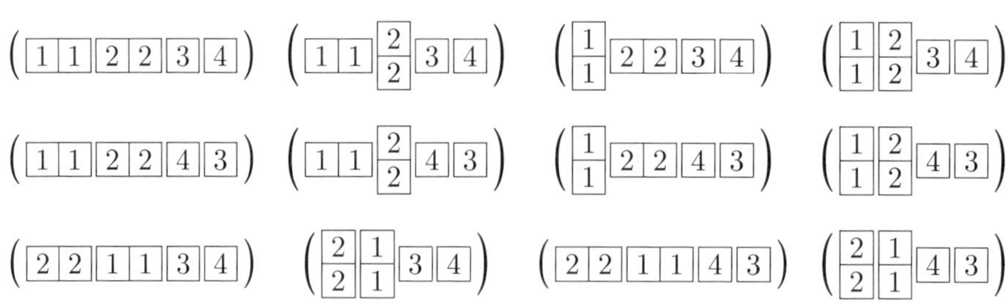

Figure 4: The hook-block involutions corresponding to the involution product graphs for $\sigma = (12)(34)(5)(6)$.

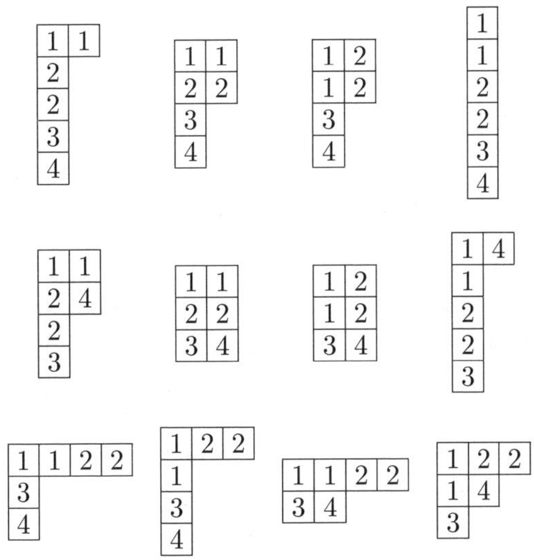

Figure 5: The image under Sch of the hook-block involutions of content $(2,2,1,1)$.

3.1. Connection with commuting pairs

As a final thought, we communicate an observation of Allan Trojan, shared with us via private communication, in response to an early draft of this paper.

For a positive integer n, let $C(n)$ denote the set of all ordered pairs of commuting elements of S_n,

$$C(n) = \{(\sigma,\tau) \in S_n^2 : \sigma\tau = \tau\sigma\}.$$

Define an equivalence relation on $C(n)$ by conjugacy acting diagonally; that is, say $(\sigma,\tau) \sim (g\sigma g^{-1}, g\tau g^{-1})$ for each $g \in S_n$. For $\mu \vdash n$, define

$$C(\mu) = \{(\sigma,\tau) \in S_n^2 : \sigma\tau = \tau\sigma \text{ and } \operatorname{sh}(\sigma) = \mu\} \subseteq C(n).$$

Let $c(\mu)$ denote the number of conjugacy classes in $C(\mu)$.

Example 3.16. There are five conjugacy classes of commuting pairs in S_4 whose first element has cycle type $(2,2)$, so $c((2,2)) = 5$. Taking $\sigma = (12)(34)$ in each class, they are:

$$(\sigma,1), (\sigma,(12)), (\sigma,(12)(34)), (\sigma,(13)(24)), (\sigma,(1324)).$$

Let $\operatorname{nz}(\mu)$ count the nonzero entries in column μ of the character table of S_n. Trojan observes that $c(\mu)$ is linked to $\operatorname{nz}(\mu)$ in much the same way that our numbers $N(\mu)$ are linked to the sum of absolute values of the entries in column μ. Namely,

- $c(\mu) = \operatorname{nz}(\mu)$ for all $\mu \vdash n$ where $n \leq 7$,

- if $\mu = k^m$ is rectangular, then $c(\mu) = \mathrm{nz}(\mu)$, and
- if $n = 8, 9$, or 10, the partitions μ for which $c(\mu)$ and $\mathrm{nz}(\mu)$ differ are precisely those listed in Table 1.

This certainly suggestions some interesting connections to explore.

Acknowledgments The authors wish to thank Allan Berele, Ira Gessel, and John Stembridge for conversations that helped prompt us to better understand characters of S_n. Thanks to Dennis White for helping us understand Sch, and thanks to Peter Winkler for helpful discussions related to the formula for $N(\mu)$. We are also grateful to Allan Trojan for sharing his comments, and to the suggestions of an anonymous referee. Lastly, we should mention the role played by the OEIS, without which we would never have guessed at a connection between involution products and characters of the symmetric group.

References

[1] R. W. Carter, Conjugacy classes in the Weyl group, *Compositio Math.* **25** (1972), 1–59.

[2] D. Gouyou-Beauchamps and P. Nadeau, Signed enumeration of ribbon tableaux: an approach through growth diagrams, *J. Algebr. Comb.* **36** (2012), 67–102.

[3] The On-Line Encyclopedia of Integer Sequences, available electronically at `oeis.org`.

[4] B. E. Sagan, *The symmetric group. Representations, combinatorial algorithms, and symmetric functions*, Second edition. Graduate Texts in Mathematics, 203. Springer-Verlag, New York, 2001.

[5] W. R. Scott, *Group Theory*, Prentice-Hall, Englewood Cliffs, New Jersey, 1964.

[6] R. P. Stanley, *Enumerative Combinatorics, vol. 2*, Cambridge Studies in Advanced Mathematics, no. 62, Cambridge University Press, Cambridge, 2001.

[7] D. W. Stanton and D. E. White, A Schensted algorithm for rim hook tableaux, *J. Combin. Theory Ser. A* **40** (1985), 211–247.

[8] D. E. White, A bijection proving orthogonality of the characters of S_n, *Adv. Math.* **50** (1983), 160–186.

ON THE HIGHER-DIMENSIONAL GENERALIZATION OF A PROBLEM OF ROTH

József Borbély
Department of Algebra and Number Theory, Eötvös Lóránd University, Budapest, Hungary

Received: 9/23/12, Revised: 3/12/13, Accepted: 8/7/13, Published: 9/26/13

Abstract

Long ago Roth conjectured that for any k-coloring of the positive integers the equation $x+x'=n, x \neq x'$ has a monochromatic solution in (x,x') for more than cM integers n up to M (where c is an absolute constant independent of k). Later Erdős, Sárközy and T. Sós proved this conjecture with $\frac{1}{2}-\varepsilon$ in place of c. In this paper we will prove a higher-dimensional generalization of this theorem by using a higher-dimensional extension of the well known Hilbert cube-lemma. We will also give bounds for the number of monochromatic solutions in higher dimension.

1. Introduction

K. F. Roth conjectured (see [2] and [6]) that for an arbitrary k-coloring of the positive integers there are more than cM integers $n \leq M$ such that the equation $x+x'=n, x \neq x'$, has a monochromatic solution in (x,x'). In [1] Erdős, Sárközy and T. Sós proved this conjecture in the following form:

Theorem 1. *For every $k \geq 2$ there exists a positive integer $M_0(k)$ such that for any $M \geq M_0(k)$ and an arbitrary k-coloring of the set $\mathbb{N}$, the number of positive integers $n \leq M$ for which there is a monochromatic solution of the equation $x+x'=n$, $x \neq x'$, is greater than $\frac{M}{2} - 3M^{1-2^{-k-1}}$*

The proof of this theorem was based on the density version of Hilbert's cube lemma (for the original coloring version see [7]). Szemerédi proved that if we consider a sequence of positive integers of positive density, then the sequence must contain a so-called Hilbert d-cube or affine d-cube, i.e., for every d a set of the form $u+\sum_{i=1}^{d}\varepsilon_i v_i$, where ε_i=0 or 1 for every i. In [7] Hilbert used the coloring version of this lemma in studying irreducibility of polynomials with integer coefficients. Later, Szemerédi gave the density version of this lemma (see [3] and [10]). This density version of Hilbert's cube lemma is generally called Szemerédi's cube lemma. Erdős, Sárközy and T. Sós used the following quantitative version of this lemma:

Lemma 1. *(Szemerédi's cube lemma): If H is a subset of (1,M) for M large enough and H has at least* $3M^{1-2^{-d}}$ *elements, then H contains a Hilbert-cube.*

Our first goal is to give a higher-dimensional generalization of Lemma 1. With this generalization we will prove the following result:

Theorem 2. *For fixed positive integers* r, s, k, *there is a positive integer* m_0 *with the following property: for any positive integer* $m > m_0$ *and for any* k*-coloring of the elements of the set* $(1,m)^r$ *there are at least* $\left[\frac{m}{s}\right]^r - 3\cdot\left(2^{r-1}m^r\right)^{1-2^{-ks+k-1}}$ *vectors* $\vec{x}$ *in* $(1,m)^r$*, such that one can find pairwise distinct vectors* $\vec{x_1}, \vec{x_2}, \cdots, \vec{x_s}$ *of the same color in* $(1,m)^r$*, whose sum is* $\vec{x}$.

The special case $r = 1$ and $s = 2$ in Theorem 2 gives the result of Theorem 1. After the proof of Theorem 2 we will study the number of solutions of the equation $\vec{x} + \vec{x'} = \vec{n}, \vec{x} \neq \vec{x'}$ for vectors $\vec{n}$ in $(1,m)^r$, where $\vec{x}$ and $\vec{x'}$ are monochromatic. We will get the following result:

Theorem 3. *For every positive real number* α *and* β *with the property* $\alpha^r + \beta^r \leq \frac{1}{2^{2r+1}k}$ *there is a positive integer* $m_{\alpha\beta}$*, such that for every* $m > m_{\alpha\beta}$ *and for every* k*-coloring of* $\mathbb{N}^r$ *the number of elements in* $(1,m)^r$ *having representations as a sum of two monochromatic distinct vectors in more than* $\frac{\beta^r}{2}m^r$ *ways is more than* $\alpha^r m^r$.

One can observe that the result of Theorem 2 is asymptotically independent of the number of the colors, if we fix r and s. In Theorem 3 we want to search for an arbitrary k-coloring a "large number" of vectors with a "large number of representations". Here we will see that these "large numbers" already depend on the number of the colors for fixed r and s. First we study only the case of two summands, later we show a way of studying the case of more summands.

2. The Generalization of Hilbert's Cube Lemma

Similarly to the original definition one can interprete d-cubes in the set of r-dimensional vectors.

Definition 1. If $\vec{u}, \vec{v_1}, \vec{v_2}, \cdots, \vec{v_d}$ are r-dimensional vectors, then the set of the sums in the form $\vec{u} + \sum_{i=1}^{d} \varepsilon_i \vec{v_i}$, where ε_i=0 or 1 for every i, is an affine d-cube or a d-dimensional Hilbert cube.

Lemma 2. *If H is a subset of distinct vectors of* $(1,m)^r$ *and the set H has at least* $3\cdot\left(2^{r-1}m^r\right)^{1-2^{-d}}$ *elements, then H contains an affine d-cube.*

Proof. Our proof is a generalization of the proof given in [1]. We will define sets $H_0, H_1, \cdots, H_d$ and r-dimensional vectors $\vec{v_1}, \vec{v_2}, \cdots, \vec{v_d}$ in the following way:

(i) $H_0 = H$

(ii) $H_j \cup \left\{\vec{b} + \vec{v_j} | \vec{b} \in H_j\right\}$ is a subset of H_{j-1} for every $j = 1, 2, \cdots, d$

(iii)$|H_j| \geq |H|^{2^j} \left(3 \cdot 2^{r-1} m^r\right)^{-\left(2^j - 1\right)}$

We construct $H_0, H_1, \cdots, H_d$ and $\vec{v_1}, \vec{v_2}, \cdots, \vec{v_d}$ recursively. Let $H_0 = H$. Assume that $0 \leq j \leq d-1$ and in the case $j > 0$ sets $H_0, H_1, \cdots, H_j$ and vectors $\vec{v_1}, \vec{v_2}, \cdots, \vec{v_j}$ have been defined. Let F be the set of the vectors in $(-(m-1), (m-1))^r$, whose first nonzero coordinate is positive. Denote by $f(H_j, \vec{h})$ the number of solutions of the equation $\vec{b} - \vec{b'} = \vec{h}$, where $\vec{b}, \vec{b'} \in H_j$ and $\vec{h} \in F$.

Let L be the maximum value of the numbers $f(H_j, \vec{h})$, where $\vec{h} \in F$ and $\vec{h} \neq \vec{v_1}, \vec{v_2}, \cdots, \vec{v_j}$. For all r-dimensional vectors $\vec{h}$ we have $f(H_j, \vec{h}) \leq |H_j|$. Clearly $\sum_{\vec{h} \in F} f(H_j, \vec{h}) = \binom{|H_j|}{2}$. We give an estimate for L. We can majorize $f(H_j, \vec{h})$ by $|H_j|$, if $\vec{h} \in \{\vec{v_1}, \vec{v_2}, \cdots, \vec{v_j}\}$ and by L otherwise. Thus we get the estimate

$$\binom{|H_j|}{2} \leq j\,|H_j| + \frac{(2m-1)^r - 1}{2} L < j\,|H_j| + 2^{r-1} m^r L.$$

Thus we have $L \geq \frac{1}{2^r m^r}\left(|H_j|^2 - |H_j| - 2j\,|H_j|\right) = \frac{|H_j|}{3 \cdot 2^{r-1} \cdot m^r}\left(\frac{3}{2}|H_j| - \frac{3}{2} - 3j\right)$. According to our assumption we have (for m large enough) the estimate

$$|H_j| \geq |H|^{2^j} \left(3 \cdot 2^{r-1} m^r\right)^{-\left(2^j - 1\right)} > \left(3 \cdot \left(2^{r-1} m^r\right)^{1 - 2^{-d}}\right)^{2^j} \left(3 \cdot 2^{r-1} m^r\right)^{-\left(2^j - 1\right)}$$

$$= 3 \cdot \left(2^{r-1} m^r\right)^{1 - 2^{j-d}} > 3 \cdot \left(2^{r-1} m^r\right)^{1 - 2^{-1}} > 3 + 6d > 3 + 6j.$$

So we have $\frac{|H_j|^2}{3 \cdot 2^{r-1} m^r} < L$. By (iii) we get $L > |H|^{2^{j+1}} \left(3 \cdot 2^{r-1} m^r\right)^{-\left(2^{j+1} - 1\right)}$. This means that the vector $\vec{h}$ can play the role of $\vec{v_{j+1}}$ and we are able to define set H_{j+1}, too. Thus indeed we can define sets $H_0, H_1, \cdots, H_d$ and r-dimensional vectors $\vec{v_1}, \vec{v_2}, \cdots, \vec{v_d}$ recursively. □

3. The Proof of the Generalization of Roth's Problem

Here we give a proof of Theorem 2.

Proof. Assume to the contrary that there is an appropriate k-coloring for infinitely many positive integers m, such that the number of vectors $\vec{x}$ with the given property in $(1, m)^r$ is less than $\left[\frac{m}{s}\right]^r - 3 \cdot \left(2^{r-1} m^r\right)^{1 - 2^{-ks+k-1}}$. We will get a contradiction via Lemma 2. Let S be the subset of $(1, m)^r$, in which all the r coordinates of the elements are divisible by s. Let S' denote the set of those elements of S, which do not have a representation in the given form. According to our assumption we have $3 \cdot \left(2^{r-1}m^r\right)^{1-2^{-ks+k-1}} < |S'|$. Now we can apply Lemma 2. In S' one can find an affine $(ks-k+1)$-cube, so that there are r-dimensional vectors $\vec{u}, \vec{v_1}, \vec{v_2}, \cdots, \vec{v_{ks-k+1}}$ in $(-(m-1),(m-1)))^r$, such that all the sums $\vec{u} + \sum_{i=1}^{ks-k+1} \varepsilon_i \vec{v_i}$ are in S',where ε_i=0 or 1 for every i. By the pigeonhole principle there are s vectors in the set $\left\{\frac{1}{s}\vec{u} + \vec{v_i}, i = 1, 2, \cdots, ks-k+1\right\}$ with the same color. We can assume without loss of generality, that these vectors are $\vec{v_1}, \vec{v_2}, \cdots, \vec{v_s}$. By the definition of the Hilbert-cube the vector $\vec{u} + \sum_{i=1}^{s} \vec{v_i} = \sum_{i=1}^{s} \left(\frac{1}{s}\vec{u} + \vec{v_i}\right)$ is an element of S', which contradicts the definition of S'. □

4. On the Number of Representations as the Sum of Two Monochromatic Distinct Vectors

In this section we study the number of representations, if the number of summands is $s = 2$. Our goal is to prove Theorem 3. The key will be the following lemma:

Lemma 3. *If for positive real numbers α and β there exist infinitely many positive integers m, such that there is a k-coloring of $\mathbb{N}^r$, for which at most $\alpha^r m^r$ elements of $(1,m)^r$ have representations as a sum of two monochromatic distinct vectors of $\mathbb{N}^r$ in more than $\frac{\beta^r}{2}m^r$ ways, then $\alpha^r + \beta^r > \frac{1}{2^{2r+1}k}$.*

Proof. Let be m a positive integer with the given property. Let a be the minimal and b the maximal positive integer such that $\frac{a}{m} \geq \alpha$ and $\frac{b}{m} \leq \beta$. In this case at most a^r elements of $(1,m)^r$ have representations as a sum of two monochromatic distinct vectors in more than $\frac{b^r}{2}$ ways. Let $f(\vec{x})$ denote the number of representations of $\vec{x} = (x_1, x_2, \cdots, x_r,)$ as the sum of two distinct vectors in $(1,x_1) \times (1,x_2) \times \cdots \times (1,x_r)$. Clearly $f(\vec{x}) = \left[\frac{\prod_{i=1}^{r}(x_i-1)}{2}\right]$, because one can order the vectors of this set (except at most one vector) into disjoint pairs such that each pair consists of two distinct vectors with sum $\vec{x}$. For all vectors $\vec{x}$ of the set $(1,m)^r$ let $g(\vec{x})$ denote the number of representations of $\vec{x}$ as the sum of two distinct monochromatic vectors of $(1,m)^r$. Let m_i be the number of vectors with the i-th color in $\left(1, \left[\frac{m}{2}\right]\right)^r$. Clearly $\sum_{i=1}^{k} m_i = \left[\frac{m}{2}\right]^r$ and $\sum_{i=1}^{k} \binom{m_i}{2} \leq \sum_{\vec{x} \in (1,m)^r} g(\vec{x})$. We give an upper estimate for the sum $\sum_{\vec{x} \in (1,m)^r} g(\vec{x})$, too. If one can write the vectors $\vec{v_1}, \vec{v_2}, \cdots, \vec{v_t} (t \leq a^r)$, in more than $\frac{b^r}{2}$ ways as the sum of two monochromatic distinct vectors, then

$$\sum_{i=1}^{t} g(\vec{v_i}) \leq \sum_{\vec{x} \in (m-a+1,m)^r} f(\vec{x}).$$

Hence we have the following upper estimate:

$$\sum_{\vec{x}\in(1,m)^r} g(\vec{x}) \le \frac{b^r}{2}(m)^r + \sum_{\vec{x}\in(m-a+1,m)^r} f(\vec{x})$$

$$\le \frac{b^r}{2}m^r + \frac{1}{2}\sum_{x_r=m-a+1}^{m}\cdots\sum_{x_1=m-a+1}^{m}(x_1-1)(x_2-1)\cdots(x_r-1)$$

$$= \frac{b^r}{2}m^r + \frac{1}{2}\sum_{x_r=m-a}^{m-1}\cdots\sum_{x_1=m-a}^{m-1}x_1x_2\cdots x_r = \frac{b^r}{2}m^r + \frac{1}{2}\left(am - \frac{a^2+a}{2}\right)^r$$

$$< \frac{b^r}{2}m^r + (am)^r.$$

Using the Cauchy-Schwarz-inequality we get

$$\sum_{i=1}^{k}\binom{m_i}{2} = \frac{1}{2}\sum_{i=1}^{k}m_i^2 - \frac{1}{2}\left[\frac{m}{2}\right]^r \ge \frac{1}{2}k\left(\frac{\left[\frac{m}{2}\right]^r}{k}\right)^2 - \frac{1}{2}\left[\frac{m}{2}\right]^r.$$

By dividing by m^{2r} we get

$$\frac{\frac{1}{2}k\left(\frac{\left[\frac{m}{2}\right]^r}{k}\right)^2 - \frac{1}{2}\left[\frac{m}{2}\right]^r}{m^{2r}} < \frac{\frac{1}{2}b^r m^r + \frac{1}{2}(am)^r}{m^{2r}}.$$

If m tends to infinity, then the left-hand side is asymptotically $\frac{1}{2^{2r+1}k}$ and the right-hand side is asymptotically $\frac{1}{2}\beta^r + \frac{1}{2}\alpha^r$. Clearly we have $\frac{1}{2}\beta^r + \frac{1}{2}\alpha^r < \alpha^r + \beta^r$, and hence $\alpha^r + \beta^r > \frac{1}{2^{2r+1}k}$. □

With Lemma 3 we can easily prove Theorem 3. The proof can be done in the following way:

Proof. Assume, for a contradiction, that there are positive numbers α and β such that $\alpha^r + \beta^r \le \frac{1}{2^{2r+1}k}$ and for infinitely many positive integers m there is a k-coloring of $\mathbb{N}^r$ such that the number of elements in $(1,m)^r$ having representations as a sum of two monochromatic distinct vectors in more than $\frac{\beta^r}{2}m^r$ ways is at most $\alpha^r m^r$. This contradicts Lemma 3. □

Remark. Let k' be the maximal odd integer such that $k \ge k'$. We color the elements of $\mathbb{N}^r$ with at most k colors in the following way: if the sum of the coordinates is 0 or 1 modulo k', then we color this vector by the first color; the other vectors are colored by the other $k'-1$ colors according to the other residue classes. In this case there are at most $3\cdot\left[\frac{m}{k'}\right]m^{r-1}$ vectors in $(1,m)^r$, for which the number of representations as a sum of two distinct monochromatic vectors is at least $2\cdot\left[\frac{m}{k'}\right]m^{r-1}$ (to see this, one should only observe the vectors having the sum of coordinates equal to 0, 1 or 2 modulo k'). Thus Theorem 2 cannot be improved significantly.

5. Further Remarks

We studied in the last section only the case $s = 2$ (i.e., the number of addends is two). We can do calculations in a similar way using the further asymptotic formulae (see [9]). The method we use will not be combinatorially new, one needs only the same technique, but with a little more work. Let us denote by $p(n, s)$ the number of partititons of n into exactly s not necessarily distinct parts and $q(n, s)$ the number of partititons of n into exactly s distinct parts. We formulate the relevant statements of [9] in Lemma 4.

Lemma 4. *$p(n, s) \sim \frac{n^{s-1}}{(1\cdot 2\cdot\dots\cdot(s-1))^2\cdot s}$ and $q(n, s) \sim \frac{n^{s-1}}{(1\cdot 2\cdot\dots\cdot(s-1))^2\cdot s}$ where $n >> 1$ and $s = O(1)$.*

We give only a sketch of how our methods generally work. Let $H(x_1, x_2, \cdots, x_r)$ denote the number of representations of the vector $\vec{x} = (x_1, x_2, \cdots, x_r)$ as a sum of s distinct vectors in $(1, x_1)\times(1, x_2)\times\cdots\times(1, x_r)$. Let m_i be the number of vectors with the i-th color in $\left(1, \left[\frac{m}{s}\right]\right)^r$. It is obvious that $\sum_{i=1}^{k} \binom{m_i}{s} \leq \sum_{\vec{x}\in(1,m)^r} T(\vec{x})$, where $T(\vec{x})$ is the number of ways vector $\vec{x}$ can be written as a sum of s monochromatic distinct vectors. A similar argument to that of Section 4 shows that we have to majorize $H(x_1, x_2, \cdots, x_r)$.

It is easy to verify that

$$q(x_1, s)\prod_{i=2}^{r}(1\cdot 2\cdot\dots\cdot s\cdot q(x_i, s)) \leq H(x_1, x_2, \cdots, x_r)$$
$$\leq p(x_1, s)\prod_{i=2}^{r}(1\cdot 2\cdot\dots\cdot s\cdot p(x_i, s)).$$

By Lemma 4 we get the conclusion that $H(x_1, x_2, \cdots, x_r)$ is asymptotically equal to

$$\frac{x_1^{s-1}}{(1\cdot 2\cdot\dots\cdot(s-1))^2\cdot s}\prod_{i=2}^{r}\left(1\cdot 2\cdots s\frac{x_i^{s-1}}{(1\cdot 2\cdot\dots\cdot(s-1))^2\cdot s}\right),$$

if s is fixed and $x_i \gg 1$. With further calculation one can achieve analogous results as formulated in Theorem 3.

References

[1] Paul Erdős, András Sárközy and Vera T. Sós, *On a conjecture of Roth and Some Related Problems I Irregularities of partitions*, Pap. Meet., (Fertőd, 1986), 4759, Algorithms Combin. Study Res. Texts, 8, Springer, Berlin, 1989.

[2] Paul Erdős, *Some unsolved problems*, Publ. Math. Inst. Hung. Acad. Sci., Ser. A 6, 221-254, 1961.

[3] R. L. Graham, B. L. Rotschild and J. H. Spencer, *Ramsey Theory*, Wiley, 1980.

[4] David S. Gunderson and Vojtech Rödl, *An alternate proof of Szemerédi's cube lemma using extremal hypergraphs*, Diskrete Strukturen in der Mathematik, Preprintreihe, 1995.

[5] David S. Gunderson and Vojtech Rödl, *Extremal problems for affine cubes of integers*, Combin. Probab. Comput **7**, no. 1, 65-79, 1998.

[6] R. K. Guy, *Unsolved Problems in Number Theory*, Springer-Verlag, 1980.

[7] David Hilbert, *Ueber die Irreduzibilitaet ganzer rationaler Functionen, mit ganzzahligen Koeffizienten*, Crelles J. Math., Volume 110, 104-129, 1892.

[8] Bruce M. Landman and Aaron Robertson, *Ramsey Theory on the Integers*, American Mathematical Soc., 2004.

[9] Charles Knessl and Joseph B. Keller, *Partititon asymptotics from recursion equations*, SIAM Journal on Applied Mathematics, Vol. 50, No. 2, Apr., 323-338. 1990.

[10] Endre Szemerédi, *On sets of integers containing no k elements in arithmetic progression*, Acta Arithmetica 27, 199-245, 1975.

INDEPENDENT DIVISIBILITY PAIRS ON THE SET OF INTEGERS FROM 1 TO N

Rosemary Sullivan
Dept. of Mathematics, West Chester University, West Chester, Pennsylvania
rsullivan@wcupa.edu

Neil Watling
Department of Mathematics, Widener University, Chester, Pennsylvania
nawatling@widener.edu

Received: 6/2/12, Revised: 8/6/13, Accepted: 8/30/13, Published: 9/26/13

Abstract

Let $\mathbb{S}_N$ denote the set of integers from 1 up to N and A_i be the event that a number selected from $\mathbb{S}_N$ is divisible by i. For the sample space $\mathbb{S}_N$, with the uniform probability measure, consider the question of the independence of the events A_i and A_j, $i \neq j$. We determine a characterization in terms of N, i and j. Using this we consider various situations and supplementary questions.

1. Introduction

Statistical independence is a fundamental concept in probability theory. Two events A and B are statistically independent if $P(A)P(B) = P(A \cap B)$; see Chung [3], Kac [5] or Ross [8] for example. Probabilists often illustrate statistical independence with examples from games of chance, whereas number theorists like to demonstrate the property using sets of positive integers. The natural density measure of a set A of natural numbers [2, 3, 5] is defined as,

$$D(A) = \lim_{n\to\infty} \frac{|A \cap \{1,2,\ldots,n\}|}{n},$$

where $|\cdot|$ denotes the cardinality of a set. For example, if $A = \{3,6,9,\ldots\}$ then $D(A) = \frac{1}{3}$. If the "divisibility event", A_{q_1}, is the set of natural numbers divisible by q_1 then $D(A_{q_1})D(A_{q_2}) = D(A_{q_1} \cap A_{q_2})$ if and only if q_1 and q_2 are coprime. Here $D(A_{q_1}) = 1/q_1$, $D(A_{q_2}) = 1/q_2$ and $D(A_{q_1} \cap A_{q_2}) = D(A_{q_1 q_2}) = 1/q_1 q_2$. In this sense the "event of divisibility by q_1" and the "event of divisibility by q_2" are independent on the set of natural numbers if and only if q_1 and q_2 are coprime.

However, it should be noted that $D(A)$ is not a probability measure since it is not countably additive [2]; so this is not true statistical independence.

Suppose instead we restrict to a finite set $\mathbb{S}_N = \{1, \ldots, N\}$, where N is a natural number, with the uniform probability measure. We can analogously define A_{q_i} to be the event that a number selected from $\mathbb{S}_N$ is divisible by q_i. Now which events A_{q_1} and A_{q_2} are independent? This is more complicated to answer since we have the additional parameter N. As a simple illustration, consider the events A_2 and A_3. It is a straightforward computation to show these events are not independent on $\mathbb{S}_{10}$, even though 2 and 3 are coprime, but are independent on $\mathbb{S}_{12}$.

As another illustration if $N = 24$, $P(A_3) = 8/24$, $P(A_5) = 4/24$ and $P(A_3 \cap A_5) = P(A_{15}) = 1/24 \neq 8/24 \cdot 4/24$. So once again being coprime is not sufficient but it is, as we shall see, necessary. Clearly the cardinality of the set, N, plays a role. Perhaps we should require q_1 and q_2 to be coprime factors of N. This is in fact not necessary. For example, if we keep $N = 24$ but consider A_3 and A_7 then $P(A_3) = 8/24$, $P(A_7) = 3/24$ and $P(A_3 \cap A_7) = P(A_{21}) = 1/24 = 8/24 \cdot 3/24$ so these two events are independent.

Define A_{q_1} and A_{q_2} to be an *independent divisibility pair on* $\mathbb{S}_N$ if $P(A_{q_1})P(A_{q_2}) = P(A_{q_1} \cap A_{q_2})$ where $P(\cdot)$ is the uniform distribution on $\mathbb{S}_N$. Theorem 2.1 gives the necessary and sufficient conditions on N, q_1 and q_2 for A_{q_1} and A_{q_2} to be an independent divisibility pair on $\mathbb{S}_N$.

With the conditions for independence known, it is possible to consider various supplementary questions. If we fix q_1 and q_2, for which N are the events A_{q_1} and A_{q_2} an independent divisibility pair on $\mathbb{S}_N$? If instead we fix N, which pairs (A_{q_1}, A_{q_2}) are an independent divisibility pair on $\mathbb{S}_N$? These questions are examined in Sections 3 and 4 respectively. In Section 5, we analyze the question of characterizing those N which possess a specified number of independent divisibility pairs. In particular, we prove that if N is the product of a Sophie Germain prime and its corresponding safe prime then there is a unique independent divisibility pair given by these two primes. We conclude in Section 6 with some conjectures and generalizations.

2. The Necessary and Sufficient Conditions for A_{q_1} and A_{q_2} to be Independent on $\mathbb{S}_N$

Theorem 2.1. *Let N be a natural number and $\mathbb{S}_N = \{1, \ldots, N\}$. Let q_1 and q_2 be two natural numbers, with $1 < q_1 < q_2 < N$, and let A_{q_i} be the event that a number selected from $\mathbb{S}_N$ is divisible by q_i. The events A_{q_1} and A_{q_2} are independent on $\mathbb{S}_N$ if and only if q_1 and q_2 are coprime and*

$$N = tq_1q_2 + r_1q_1 = q_1(tq_2 + r_1), \tag{1}$$

with t a natural number and r_1 a non-negative integer such that $r_1q_1 < q_2$.

Proof. Assume q_1 and q_2 are coprime and satisfy (1). It is straightforward to compute the respective probabilities for A_{q_1}, A_{q_2} and the intersection:

$$P(A_{q_1}) = \frac{tq_2 + r_1}{N} = \frac{1}{q_1}, \qquad P(A_{q_2}) = \frac{tq_1}{N} = \frac{t}{tq_2 + r_1},$$

$$P(A_{q_1} \cap A_{q_2}) = P(A_{q_1 q_2}) = \frac{t}{N} = \frac{t}{q_1(tq_2 + r_1)},$$

confirming they are indeed independent.

Now assume $1 < q_1 < q_2 < N$, A_{q_1} and A_{q_2} are independent on $\mathbb{S}_N$ and $N \geq \operatorname{lcm}(q_1, q_2)$, the least common multiple of q_1 and q_2. (If $q_2 < N < \operatorname{lcm}(q_1, q_2)$ then the intersection event is empty. So $P(A_{q_1})P(A_{q_2}) \neq P(A_{q_1} \cap A_{q_2})$ since the left-hand side is nonzero whereas the right-hand side is zero. If $q_1 < N < q_2$ then A_{q_2} is empty, so the events A_{q_1} and A_{q_2} are degenerately independent but the conditions on q_1, q_2 and N have been violated.) By the definition of independence,

$$\left\lfloor \frac{N}{q_1} \right\rfloor \cdot \left\lfloor \frac{N}{q_2} \right\rfloor = N \cdot \left\lfloor \frac{N}{\operatorname{lcm}(q_1, q_2)} \right\rfloor. \tag{2}$$

If we suppose $2 \leq \gcd(q_1, q_2)$, the greatest common divisor of q_1 and q_2, then

$$\left\lfloor \frac{N}{q_1} \right\rfloor \cdot \left\lfloor \frac{N}{q_2} \right\rfloor \leq \frac{N}{q_1} \cdot \frac{N}{q_2} = N \cdot \frac{N}{q_1 q_2} < N \cdot \left\lfloor \frac{N \gcd(q_1, q_2)}{q_1 q_2} \right\rfloor = N \cdot \left\lfloor \frac{N}{\operatorname{lcm}(q_1, q_2)} \right\rfloor$$

which contradicts (2), so q_1 and q_2 must be coprime. Hence $\operatorname{lcm}(q_1, q_2) = q_1 q_2$ and (2) becomes

$$\left\lfloor \frac{N}{q_1} \right\rfloor \cdot \left\lfloor \frac{N}{q_2} \right\rfloor = N \cdot \left\lfloor \frac{N}{q_1 q_2} \right\rfloor. \tag{3}$$

We may assume $N = t(q_1 q_2) + r$ where $t \geq 1$ and $0 \leq r < q_1 q_2$, and also $r = r_1 q_1 + s_1$ or $r = r_2 q_2 + s_2$ with $r_1, r_2 \geq 0$, $0 \leq s_1 < q_1$, $0 \leq s_2 < q_2$. Then (3) gives

$$(tq_2 + r_1) \cdot (tq_1 + r_2) = (t(q_1 q_2) + r) \cdot t,$$

which can be simplified to

$$r_2(tq_2 + r_1) = ts_1. \tag{4}$$

Now if $r_2 \geq 1$, then $r_2(tq_2 + r_1) \geq tq_2 > tq_1 > ts_1$ giving a contradiction to (4), so $r_2 = 0$. Hence, from (4), $s_1 = 0$, and thus, $r = r_1 q_1 = s_2 < q_2$. So $N = t(q_1 q_2) + r_1 q_1$ where $t \geq 1$ and $0 \leq r_1 q_1 < q_2$ as required. □

It should be noted that Theorem 2.1 is a special case of Theorem 1 in [4]. Their requirement that N is composite is clearly true here but we expect a more restrictive condition since we have very specific events.

Observe (1) is satisfied if N is a multiple of two coprime numbers q_1 and q_2. However N need only be a multiple of the smaller number q_1, not the larger q_2. Conversely, if q_1 is not a divisor of N the events A_{q_1} and A_{q_2} are dependent on $\mathbb{S}_N$.

3. The Natural Density Measure of Independent A_{q_1} and A_{q_2}

Recall A_2 and A_3 are independent on $\mathbb{S}_N$ if $N = 12$ but not if $N = 10$. In fact, given Theorem 2.1, it is clear A_2 and A_3 are independent on $\mathbb{S}_N$ if and only if $N \equiv 0$ or $2 \mod 6$. That is, "one third of the time they are independent" and "two thirds of the time they are dependent." More precisely if we define $I_{q_1,q_2} = \{N : A_{q_1}, A_{q_2}$ are independent on $\mathbb{S}_N\}$ then,

$$D(I_{2,3}) = \lim_{n\to\infty} \frac{|I_{2,3} \cap \{1, 2, \ldots, n\}|}{n} = \frac{1}{3}.$$

Below are two more examples using Theorem 2.1.

Example 3.1. If $q_1 = 5$ and $q_2 = 53$ then the events A_5 and A_{53} are independent on $\mathbb{S}_N$ if and only if $N \equiv 0, 5, 10, \ldots, 50 \mod 265$. So $D(I_{5,53}) = 11/265 \approx 0.0415$.

Example 3.2. If $q_1 = 7$ and $q_2 = 510$ then the events A_7 and A_{510} are independent on $\mathbb{S}_N$ if and only if $N \equiv 0, 7, 14, \ldots, 504 \mod 3570$. So $D(I_{7,510}) = 73/3570 \approx 0.020448$.

In general, for q_1 and q_2 coprime, if $q_2 = pq_1 + r$, $1 \le r \le q_1 - 1$, then the events A_{q_1} and A_{q_2} are independent on $\mathbb{S}_N$ if and only if $N \equiv 0, q_1, 2q_1, \ldots, pq_1 \mod q_1q_2$. So $D(I_{q_1,q_2}) = (p+1)/(q_1q_2)$. Using the inequalities,

$$\begin{aligned} 0 < \frac{1}{q_1^2} < \frac{1}{q_1(q_1 - \frac{1}{p+1})} &= \frac{p+1}{q_1((p+1)q_1 - 1)} \le \frac{p+1}{q_1q_2} \\ &= \frac{p+1}{q_1(pq_1 + r)} \le \frac{p+1}{q_1(pq_1 + 1)} \le \frac{2}{q_1(q_1 + 1)} \le \frac{1}{3}, \end{aligned}$$

we can construct a table of intervals for possible values of $D(I_{q_1,q_2})$:

q_1	2	3	4	5	6	7	q_1
$D(I_{q_1,q_2})$	$(\frac{1}{4}, \frac{1}{3}]$	$(\frac{1}{9}, \frac{1}{6}]$	$(\frac{1}{16}, \frac{1}{10}]$	$(\frac{1}{25}, \frac{1}{15}]$	$(\frac{1}{36}, \frac{1}{21}]$	$(\frac{1}{49}, \frac{1}{28}]$	$(\frac{1}{q_1^2}, \frac{2}{q_1(q_1+1)}]$

If we fix q_1 and let q_2 get large, then p gets large, and we approach from above the left-hand end point of the interval, $1/q_1^2$.

Conversely, for $q_2 = q_1 + 1$ we have $p = 1$, $r = 1$ and the density becomes $\frac{2}{q_1(q_1+1)}$, with the maximum value of $\frac{1}{3}$ occurring when $q_1 = 2$. So consecutive pairs have the more explicit table of densities:

(q_1, q_2)	$(2,3)$	$(3,4)$	$(4,5)$	$(5,6)$	$(6,7)$	$(7,8)$	(q_1, q_1+1)
$D(I_{q_1,q_2})$	$\frac{1}{3}$	$\frac{1}{6}$	$\frac{1}{10}$	$\frac{1}{15}$	$\frac{1}{21}$	$\frac{1}{28}$	$\frac{2}{q_1(q_1+1)}$

In both tables these densities are larger than one would intuitively expect. For the latter it appears to be double, and for the former the density is always larger than $1/q_1^2$.

4. The Independent Divisibility Pairs on $\mathbb{S}_N$ for a Particular N

If we choose a particular N, then which events A_{q_1} and A_{q_2} are an independent divisibility pair on $\mathbb{S}_N$? Equivalently, for which pairs of coprime natural numbers q_1, q_2 do N, q_1 and q_2 satisfy (1)?

Since $N = q_1(tq_2 + r)$ and $r \geq 0$, we have $q_2 \leq \frac{N}{q_1 t}$. Now $rq_1 < q_2$ so $N < q_1tq_2 + q_2$, hence $q_2 > \frac{N}{q_1t+1}$. Thus, bounds for q_2 in terms of N, t and q_1 are:

$$\frac{N}{q_1t+1} < q_2 \leq \frac{N}{q_1t}. \tag{5}$$

As $N = q_1(tq_2 + r)$ and $q_2 \geq q_1 + 1$, $t = \frac{1}{q_2}\left(\frac{N}{q_1} - r\right) \leq \frac{N}{q_1(q_1+1)}$, so bounds for t are:

$$1 \leq t \leq \frac{N}{q_1(q_1+1)}. \tag{6}$$

Also, $N \geq q_1(q_1 + 1)$ and consequently $q_1 \leq \frac{\sqrt{4N+1}-1}{2}$. Hence the bounds on q_1, a factor of N, are:

$$2 \leq q_1 \leq \frac{\sqrt{4N+1}-1}{2}. \tag{7}$$

Definition 4.1. Define (q_1, q_2) to be a valid pair on N if q_1, q_2 and N are natural numbers satisfying (1) and its conditions.

If (q_1, q_2) is a valid pair on N then the events A_{q_1} and A_{q_2} are an independent divisibility pair on $\mathbb{S}_N$.

Example 4.2. When $N = 24$ there are 8 valid pairs: $(2,3)$, $(2,5)$, $(2,9)$, $(2,11)$, $(3,4)$, $(3,7)$, $(3,8)$ and $(4,5)$.

The bounds (5), (6), and (7) enable us to write an algorithm that can be used to generate all possible valid pairs for a particular N. Start at $q_1 = 2$ and proceed through all possible values for q_1 given by (7). For each q_1 check if it is a factor of N. If so, use (6) to compute the range of values for t and (5) to compute the range of values for potential q_2. Now just increment through all these potential q_2's and at each stage check if the potential q_2 is coprime to q_1. If so then this value for q_2 would make (q_1, q_2) a valid pair on N.

Figure 1 has some annotated lines of pseudocode showing this algorithm.

Definition 4.3. Define the function $v(N)$ to be the number of valid pairs on N.

The algorithm in Figure 1 can be used to determine the values of $v(N)$. Figure 2 has a table listing N and $v(N)$ for $N \leq 100$.

```
Input N
count=0;
for q1=2 to (√(4*N+1) − 1)/2,                              % from (7)
  if N mod q1 == 0,                          % check q1 is a factor of N
   for t=1 to N/(q1*(q1+1)),                                % from (6)
    for potentialq2 = ceiling(N/(t*q1+1)) to N/(t*q1),      % from (5)
     if gcd(q1,potentialq2) == 1,                  % check q1, q2 coprime
      print(q1,potentialq2), count++              % if so then valid pair
return(count)                                     % number of valid pairs
```

Figure 1: Pseudocode to generate all valid pairs on N.

5. Bounds on $v(N)$ and the Characterization of those N with $v(N) < 5$

If we allow q_1, q_2 and N to vary, is there a characterization for those N with a particular number of valid pairs? Can we characterize those N for which $v(N) = 0$ or 1 for example? To begin we will give some simple bounds for $v(N)$.

Theorem 5.1. *Let $N = p_1^{n_1} p_2^{n_2} \dots p_r^{n_r}$ be the prime decomposition of N and $v(N)$ be the number of valid pairs on N. Then upper and lower bounds for $v(N)$ are respectively,*

$$v(N) < \frac{N \ln N}{2}, \tag{8}$$

and

$$v(N) \geq \frac{[(2n_1+1)\dots(2n_r+1)] - 2[(n_1+1)\dots(n_r+1)] + 1}{2}. \tag{9}$$

Proof. From (1) we have $q_2 = \frac{(N/q_1)-r_1}{t}$. But $t \geq 1$ and $r_1 \geq 0$, so $q_1 < q_2 \leq N/q_1$ and hence, for a given factor q_1 of N, the maximum number of valid q_2's is $(N/q_1) - q_1$. Since q_1 is a factor of N between 2 and $\lfloor\sqrt{N}\rfloor$, we have

$$v(N) < \sum_{q_1=2}^{\lfloor\sqrt{N}\rfloor} \left(\frac{N}{q_1} - q_1\right) < \sum_{q_1=2}^{\lfloor\sqrt{N}\rfloor} \frac{N}{q_1} < N \ln(\sqrt{N}) = \frac{N \ln N}{2}.$$

If we consider (1) and fix $r_1 = 0$, the number of choices for coprime factors q_1 and q_2, with $1 < q_1 < q_2$, will clearly give a lower bound for $v(N)$.

We can view this counting problem in terms of boxes and colored balls instead of factors and primes as follows: the two coprime values q_1 and q_2 correspond to two different boxes and the r distinct primes correspond to r different colors with n_i balls of color i, $i = 1, \dots, r$. We now employ the inclusion-exclusion principle.

The q_i are coprime factors of N so we do not need to use all the balls, but we cannot have the same colored ball in both boxes. For each color, i, there are

N	$v(N)$	N	$v(N)$	N	$v(N)$	N	$v(N)$	N	$v(N)$
1	0	21	1	41	0	61	0	81	5
2	0	22	3	42	13	62	12	82	11
3	0	23	0	43	0	63	10	83	0
4	0	24	8	44	12	64	11	84	32
5	0	25	0	45	6	65	4	85	3
6	1	26	4	46	6	66	17	86	13
7	0	27	3	47	0	67	0	87	10
8	1	28	4	48	16	68	12	88	17
9	0	29	0	49	0	69	7	89	0
10	1	30	10	50	9	70	17	90	35
11	0	31	0	51	6	71	0	91	2
12	3	32	7	52	9	72	25	92	20
13	0	33	3	53	0	73	0	93	8
14	3	34	5	54	14	74	14	94	15
15	2	35	2	55	1	75	12	95	6
16	1	36	9	56	12	76	14	96	29
17	0	37	0	57	3	77	2	97	0
18	4	38	6	58	8	78	21	98	17
19	0	39	4	59	0	79	0	99	13
20	5	40	9	60	25	80	21	100	21

Figure 2: Values of $v(N)$ for $N \leq 100$

$(2n_i+1)$ ways to place some or none of the i colored balls into the two boxes. Hence for all r colors we have $(2n_1+1)\dots(2n_r+1)$ total possibilities.

However, we cannot have an empty box, since neither q_1 nor q_2 can be 1; so we need to remove these possibilities from the above total. These situations correspond to putting the balls into only one box rather than two. Consequently, each color has (n_i+1) choices, and for all colors there are $(n_1+1)\dots(n_r+1)$ possibilities. Now observe that either the q_1 box or the q_2 box could be empty, so multiply this latter number by 2. However, the case where both boxes are simultaneously empty has now been counted once but excluded twice, hence correct this by adding back one possibility.

Finally, only half the remaining pairs satisfy the requirement $q_1 < q_2$, so we divide the expression by two. □

The next two theorems give relatively simple characterizations for those N with exactly zero, one, two, three or four valid pairs.

Theorem 5.2. *N has no valid pairs, that is $v(N) = 0$, if and only if N is a prime or the square of a prime.*

Proof. It is clear that if N is prime then it is impossible to decompose N in the form specified by (1). (In fact it is known that the set $\mathbb{S}_N$ is a dependent set, so it has no independent events whatsoever [4].) If N has at least two distinct prime factors, then choose these as q_1 and q_2. With $r_1 = 0$ and $t = N/(q_1q_2)$ equation (1) is satisfied demonstrating there is at least one valid pair. Consequently the only remaining cases are powers of a single prime. If $N = p^n$, $n \geq 3$, choose $q_1 = p$ and $q_2 = p^{(n-1)} - 1$, then q_1 and q_2 are coprime with $q_1 < q_2$ and $N = q_1q_2 + q_1$. However, if $N = p^2$, where p is prime, we must have $q_1 = p$ in which case it is impossible to choose q_2 satisfying $q_2 > q_1$ and $q_1q_2 \leq N$ simultaneously. □

Once again this is more restrictive than the general situation. It is known that if N is prime then $\mathbb{S}_N$ is a dependent set and any two events are dependent. However, if N is the square of a prime, it is possible to have independent events. For example, if $N = 9$ there are no independent sets of our divisibility type, but $A = \{1, 3, 5\}$ and $B = \{5, 7, 9\}$ are clearly independent.

Theorem 5.3. *Let $v(N)$ denote the number of valid pairs on N. Then*

(a) $v(N) = 1$ if and only if $N = 6 = 2 \cdot 3$, $N = 8 = 2^3$, $N = 16 = 2^4$ or $N = p_1p_2$, with p_1, p_2 prime and $p_2 = 2p_1 + 1$. The unique valid pair is $(2, 3)$, $(2, 3)$, $(2, 7)$ and $(p_1, 2p_1 + 1)$ respectively.

(b) $v(N) = 2$ if and only if $N = p_1p_2$ with p_1, p_2 prime, $2 < p_1$, and $p_1 + 2 \leq p_2 \leq 2p_1 - 1$.

(c) $v(N) = 3$ if and only if $N = 12 = 2^2 \cdot 3$, $N = 14 = 2 \cdot 7$, $N = 22 = 2 \cdot 11$, $N = 27 = 3^3$, $N = 57 = 3 \cdot 19$, $N = p_1p_2$ with p_1, p_2 prime and $p_2 = 3p_1 + 2$, or $N = p_1p_2$ with p_1, p_2 prime and $p_2 = 4p_1 + 1$, $p_1 \geq 5$.

(d) $v(N) = 4$ if and only if $N = 18 = 2 \cdot 3^2$, $N = 26 = 2 \cdot 13$, $N = 28 = 2^2 \cdot 7$, $N = 39 = 3 \cdot 13$, $N = p_1p_2$ with p_1, p_2 prime, $p_1 \geq 5$, $2p_1 + 3 \leq p_2 \leq 3p_1 - 2$, or $N = p_1p_2$ with p_1, p_2 prime, $p_1 \geq 5$, $p_2 = 4p_1 + 3$, with $2p_1 + 1$ not divisible by 3.

Proof. If N is the product of at least three distinct prime factors, then (9) gives $v(N) \geq 6$. Hence, we can restrict to N having one or two distinct prime factors.

If $N = p^n$, from Theorem 5.2, $n \geq 3$ in which case $(p, p^{n-1} - c)$, $1 \leq c \leq p - 1$ are valid pairs. So we need only consider $p \leq 5$. If $p = 5$, then $(5, 5^{n-1} - 1)$, $(5, 5^{n-1} - 2)$, $(5, 5^{n-1} - 3)$, $(5, 5^{n-1} - 4)$, and $(5, \frac{5^{n-1}-1}{2})$ are five valid pairs. If $p = 3$, then $(3, 3^{n-1} - 1)$, $(3, 3^{n-1} - 2)$, $(3, 3^{n-1} - 4)$, $(3, 3^{n-1} - 5)$, and $(3, 3^{n-1} - 7)$ are five valid pairs if $n \geq 5$. For $N = 3^3 = 27$ and $N = 3^4 = 81$ there are three and five valid pairs respectively (see Figure 2). If $p = 2$, then $(2, 2^{n-1} - 1)$, $(2, 2^{n-1} - 3)$, $(2, 2^{n-1} - 5)$, $(2, 2^{n-1} - 7)$, and $(2, 2^{n-1} - 9)$ are five valid pairs if $n \geq 6$. It is easy to check that $(2, 3)$ and $(2, 7)$ are the unique valid pair for $N = 2^3 = 8$ and $N = 2^4 = 16$ respectively. The case $N = 2^5 = 32$ has seven valid pairs (see Figure 2).

If $N = p_1^{n_1}p_2^{n_2}$, $n_1, n_2 \geq 1$, from (9) $v(N) \geq n_1n_2$. In particular there are the n_1n_2 valid pairs given by (p_1^a, p_2^b) (possibly in reverse order) where $1 \leq a \leq n_1$,

$1 \le b \le n_2$. So we need only consider when $n_1 n_2 \le 4$.

In the cases $(n_1, n_2) = (2, 2)$, $(n_1, n_2) = (4, 1)$, and $(n_1, n_2) = (1, 4)$ a fifth valid pair is given by $(p_1, p_1 p_2^2 - 1)$, $(p_1, p_1^3 p_2 - 1)$, and $(p_2, p_1 p_2^3 - 1)$ respectively.

If $(n_1, n_2) = (3, 1)$, then $(p_1, p_1^2 p_2 - 1)$ and $(p_1, p_1 p_2 - 1)$ give two additional valid pairs whereas for $(n_1, n_2) = (1, 3)$ they are $(p_2, p_1 p_2^2 - 1)$ and $(p_2, p_2^2 - 1)$.

If $(n_1, n_2) = (2, 1)$, then $(p_1, p_1 p_2 - c)$, with $1 \le c \le p_1 - 1$ are valid pairs. So there are at least four additional pairs if $p_1 \ge 5$. If $p_1 = 3$, then $(3, 3p_2 - 1)$, $(3, 3p_2 - 2)$, and $(3, \frac{3p_2-1}{2})$ are three additional valid pairs. If $p_1 = 2$, then $(2, 2p_2 - 1)$, $(2, 2p_2 - 3)$, and $(2, 2p_2 - 5)$ are three additional valid pairs if $p_2 \ge 11$. This leaves $N = 2^2 \cdot 3 = 12$, $N = 2^2 \cdot 5 = 20$ and $N = 2^2 \cdot 7 = 28$ which have 3, 5, and 4 valid pairs respectively (see Figure 2).

If $(n_1, n_2) = (1, 2)$, then $(p_2, p_1 p_2 - c)$, $1 \le c \le p_1 - 1$ are valid pairs. Hence there are at least four additional pairs if $p_1 \ge 5$. If $p_1 = 3$, then $(p_2, 3p_2 - 1)$, $(p_2, 3p_2 - 2)$, and $(p_2, \frac{3p_2-1}{2})$ are three additional valid pairs. If $p_1 = 2$, then $(2, p_2^2 - 2)$, $(2, p_2^2 - 4)$, and $(2, p_2^2 - 6)$ are three additional valid pairs if $p_2 \ge 5$. For $N = 2 \cdot 3^2 = 18$ there are 4 valid pairs (see Figure 2).

Hence we are reduced to the case where N is the product of two primes p_1 and p_2. First suppose that $p_1 = 2$ so $N = 2p_2$. Here $(2, p_2)$, $(2, p_2 - 2)$, $(2, p_2 - 4)$, $(2, p_2 - 6)$, and $(2, p_2 - 8)$ are valid pairs if $p_2 > 24$. So we are left with $p_2 = 3$, 5, 7, 11, 13, 17, 19, and 23. These have 1, 1, 3, 3, 4, 5, 6, and 6 valid pairs respectively (see Figure 2). Now suppose $p_1 = 3$, so $N = 3p_2$. In this case $(3, p_2)$ together with four of $(3, p_2 - 1)$, $(3, p_2 - 2)$, $(3, p_2 - 3)$, $(3, p_2 - 4)$, $(3, p_2 - 5)$ and $(3, p_2 - 6)$ are valid pairs provided $p_2 > 24$. This leaves $p_2 = 5$, 7, 11, 13, 17, 19 and 23. These have 2, 1, 3, 4, 6, 3 and 7 valid pairs respectively (see Figure 2).

Finally, assume $p_1 \ge 5$. So p_1 and p_2 are odd primes, $p_2 \ne c(p_1 + 1)$, $p_2 \ne cp_1$ and at most one of $p_2 - 1$, $p_2 - 2$, $p_2 - 3$, $p_2 - 4$ and $p_2 - 5$ is a multiple of p_1. Now suppose that (p_1, q_3) is a valid pair on $N = p_1 p_2$ and $(p_1, q_3) \ne (p_1, p_2)$. Then we must have $p_2 = tq_3 + r$ with $r \ge 1$, $t \ge 1$, q_3 coprime to p_1, and $q_3 > rp_1$. Hence $q_3 = \dfrac{p_2 - r}{t}$ and $p_2 > trp_1 + r$. So restrictions on the value of p_2 will generate restrictions on the possible values of r and t, which in turn give restrictions on the possible forms for q_3, and hence the possible valid pairs. Consequently, we have the following:

If $p_2 > 5(p_1 + 1)$, at least five of (p_1, p_2), $(p_1, p_2 - 1)$, $(p_1, (p_2 - 1)/2)$, $(p_1, p_2 - 2)$, $(p_1, p_2 - 3)$, $(p_1, p_2 - 4)$, and $(p_1, p_2 - 5)$ are valid pairs.

If $4(p_1 + 1) < p_2 < 5(p_1 + 1)$, then $p_2 - 1$ is coprime to p_1, so at least five of (p_1, p_2), $(p_1, p_2 - 1)$, $(p_1, (p_2 - 1)/2)$, $(p_1, p_2 - 2)$, $(p_1, p_2 - 3)$, and $(p_1, p_2 - 4)$ are valid pairs.

If $p_2 = 4p_1 + 3$, then (p_1, p_2), $(p_1, p_2 - 1)$, $(p_1, (p_2 - 1)/2)$, and $(p_1, p_2 - 2)$ are valid pairs. If $2p_1 + 1$ is divisible by 3, then $(p_1, (p_2 - 1)/3)$ is a fifth valid pair.

If $p_2 = 4p_1 + 1$, then (p_1, p_2), $(p_1, p_2 - 2)$, and $(p_1, p_2 - 3)$ are the only valid pairs.

If $3(p_1+1) < p_2 \leq 4p_1-1$, then (p_1,p_2), (p_1,p_2-1), $(p_1,(p_2-1)/2)$, (p_1,p_2-2), and (p_1,p_2-3) are valid pairs. If $2p_1-1$ is divisible by 3, then $(p_1,(p_2-1)/3)$ is a sixth valid pair.

If $p_2 = 3p_1+2$, then (p_1,p_2), (p_1,p_2-1), and $(p_1,(p_2-1)/2)$ are the only valid pairs.

If $2p_1+3 \leq p_2 \leq 3p_1-2$, then (p_1,p_2), (p_1,p_2-1), $(p_1,(p_2-1)/2)$, and (p_1,p_2-2) are the only valid pairs.

If $p_2 = 2p_1+1$, then (p_1,p_2) is the only valid pair.

If $p_1+2 \leq p_2 \leq 2p_1-1$, then (p_1,p_2) and (p_1,p_2-1) are the only valid pairs. □

6. Remarks, Conjectures, and Generalizations

Remark 6.1. Primes p such that $2p+1$ is also prime are called Sophie Germain primes with $2p+1$ called a safe prime. The numbers are listed in sequence A005384 in [6]. For more information see [7], [9], and [1]. It is interesting to note that the largest known Sophie Germain prime, found in April 2012, is $18543637900515 \cdot 2^{666667}-1$ which has 200,701 digits. It is conjectured there are infinitely many such primes.

Remark 6.2. Primes p such that $3p+2$ or $4p+1$ are also prime do not appear to be named, but the numbers are listed in sequences A023208 and A023212 in [6]. (Primes of the form $(4n+1)$ can be called Pythagorean, [6, A002144], since they correspond to the length of the hypotenuse of a right triangle with integer sides.) Figure 3 contains a table of all N values less than 50000 with exactly three valid pairs, together with the corresponding p_1 and p_2 values.

N	p_1	p_2	N	p_1	p_2	N	p_1	p_2
12	2	3	689	13	53	18023	67	269
14	2	7	901	17	53	18881	79	239
22	2	11	1121	19	59	20833	83	251
27	3	-	1633	23	71	21389	73	293
33	3	11	2581	29	89	23941	89	269
57	3	19	4181	37	113	25043	79	317
85	5	17	5513	37	149	28421	97	293
161	7	23	5633	43	131	32033	103	311
203	7	29	7439	43	173	37733	97	389
533	13	41	10561	59	179	48641	127	383

Figure 3: Values of $N < 50000$ with $v(N) = 3$

Given the lower bound, (9), it is clear that N exist for which $v(N) > M$ for any non-negative integer M.

Conjecture 6.3. N exist such that $v(N) = M$ for any non-negative integer M.

We have used Mathematica to confirm values of N for M up to 20,000.

Definition 6.4. Define the function $r(N)$ to be the ratio of the number of valid pairs on N to the value of N, that is, $r(N) = v(N)/N$.

Conjecture 6.5. For all N, the ratio of the number of valid pairs on N to the value of N itself is less than one half. That is, $r(N) < 0.5$.

Note that the conjecture is equivalent to $v(N) < N/2$, which is a better upper bound for $v(N)$ than (8). We can confirm this is true for $N \leq 1,000,000$ with the two largest values of $r(N)$ being 0.46317 and 0.465922 when $N = 360360$ and $N = 720720$ respectively. Figure 4 contains a table with some larger values of N.

N	Prime Factorization	$v(N)$	$r(N)$
360,360	$2^3.3^2.5.7.11.13$	166,908	0.46317
720,720	$2^4.3^2.5.7.11.13$	335,799	0.465922
12,252,240	$2^4.3^2.5.7.11.13.17$	5,782,591	0.471962
232,792,560	$2^4.3^2.5.7.11.13.17.19$	111,124,547	0.477354
5,354,228,880	$2^4.3^2.5.7.11.13.17.19.23$	2,576,614,559	0.48123
26,771,144,400	$2^4.3^2.5^2.7.11.13.17.19.23$	12,955,084,847	0.48392
80,313,433,200	$2^4.3^3.5^2.7.11.13.17.19.23$	39,012,925,759	0.485758

Figure 4: Values of $r(N)$ for select N

These N values have the property that they are the smallest number containing all factors from 2 to m, where m =15, 16, 18, 22, 24, 26 and 28 respectively. The next number in this sequence would be $N = 2^4.3^3.5^2.7.11.13.17.19.23.29$, containing all factors from 2 to 30. (These are the distinct entries in the integer sequence, A003418 [6], where a_n is the least common multiple of 1 up to n.)

Conjecture 6.6. The only values of N for which $r(N) > 0.45$ are multiples of 420, that is, have factors 2 through 7 inclusive.

Once again we can confirm this for $N \leq 1,000,000$.

Definition 6.7. Define the function $a(N)$ to be the average number of valid pairs from 1 up to N, that is, $a(N) = \dfrac{\sum_{i=1}^{N} v(i)}{N}$.

Figure 5 shows a plot of the values of $\frac{a(N)}{N}$ for $N \leq 1,000,000$.

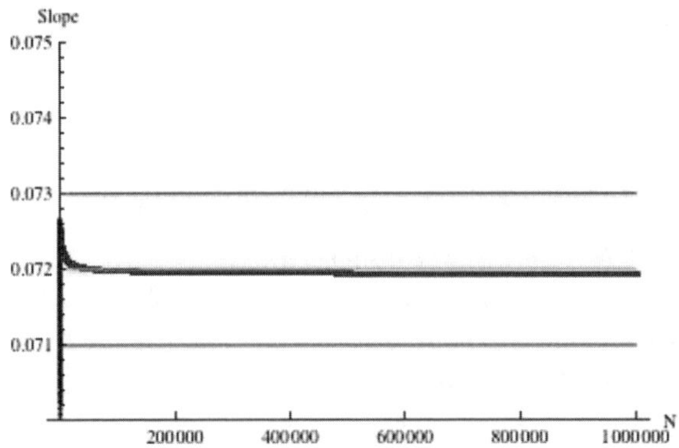

Figure 5: Plot of $\frac{a(N)}{N}$

Conjecture 6.8. $a(N) \approx 0.072N$ or $\sum_{i=1}^{N} v(i) \approx 0.072N^2$.

Definition 6.9. Define the function $\rho(N)$ to be the average ratio from 1 up to N, that is, $\rho(N) = \dfrac{\sum_{i=1}^{N} r(i)}{N}$.

Figure 6 shows a plot of the values of $\rho(N)$ for $N \leq 1,000,000$. The darker-colored curve corresponds to a model of the form $a\left(\dfrac{\ln(N)}{N}\right) + b$. The grey line on the graph corresponds to 0.143878 the value of b given by the model.

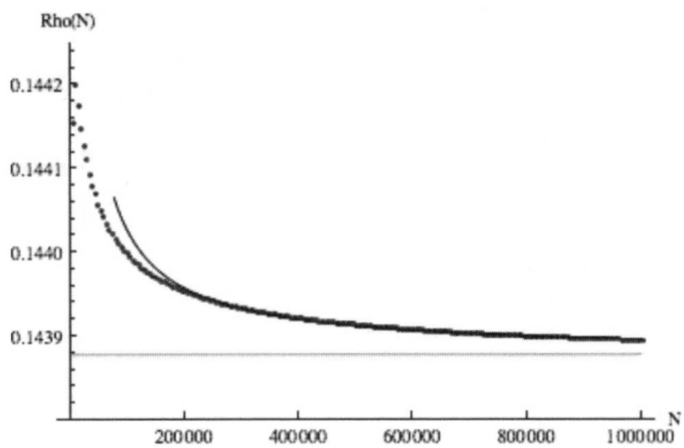

Figure 6: Plot of $\rho(N)$ and model $a\left(\dfrac{\ln(N)}{N}\right) + b$

Conjecture 6.10. The limiting value of $\rho(N)$ is between 0.14387 and 0.14389.

Remark 6.11. Statistical independence is also defined for more than two events, [3, 8], so it is possible to generalize these ideas for more than two divisibility events. In the case of three divisibility events we have the analog of Theorem 2.1:

Theorem 6.12. *Let N be a natural number and $\mathbb{S}_N = \{1, \ldots, N\}$. Let q_1, q_2, and q_3 be three natural numbers, with $1 < q_1 < q_2 < q_3 < N$, and let A_{q_i} be the event that a number selected from $\mathbb{S}_N$ is divisible by q_i. The events A_{q_1}, A_{q_2}, and A_{q_3} are independent on $\mathbb{S}_N$ if and only if q_1, q_2, and q_3 are coprime and*

$$N = tq_1q_2q_3 + r_1q_1q_2 = q_1q_2(tq_3 + r_1),$$

with t a natural number and r_1 a non-negative integer such that $r_1q_1q_2 < q_3$.

We can then consider the same questions as for pairs. The results, not surprisingly, are more complicated in this situation and are not presented here.

References

[1] Chris Caldwell, *The Prime Pages*, http://primes.utm.edu/

[2] Yung-Pin Chen, *On Primes, Density Measures, and Statistical Independence*, College Math. J. **36** (2005), 284–288.

[3] Kai Lai Chung, *Elementary Probability Theory with Stochastic Processes*, 3rd ed., Undergraduate Texts in Mathematics, Springer-Verlag, New York, 1979.

[4] Bennett Eisenberg and B. K. Ghosh, *Independent Events in a Discrete Uniform Probability Space*, Amer. Statist. **41** (1987), 52–56.

[5] Mark Kac, *Statistical Independence in Probability, Analysis, and Number Theory*, The Carus Mathematical Monographs, No. 12 (1959), The Mathematical Association of America.

[6] The On-Line Encyclopedia of Integer Sequences, published electronically at http://oeis.org, 2010.

[7] James E. Pommersheim, Tim K. Marks and Erica L. Flapan, *Number Theory*, John Wiley & Sons, 2010.

[8] Sheldon Ross, *A First Course in Probability*, 5th ed., Prentice Hall, 1998.

[9] Eric W. Weisstein, "Sophie Germain Prime." From MathWorld–A Wolfram Web Resource. http://mathworld.wolfram.com/SophieGermainPrime.html

CYCLES IN REPEATED EXPONENTIATION MODULO p^n

Lev Glebsky
Instituto de Investigación en Comunicación Óptica Universidad Autónoma de San Luis Potosí Av. Karakorum 1470, Lomas 4a 78210 San Luis Potosi, Mexico
glebsky@cactus.iico.uaslp.mx

Received: 4/13/12, Revised: 6/16/13, Accepted: 9/1/13, Published: 10/10/13

Abstract

We consider a dynamical system generated by exponentiation modulo r, that is, by the map $u \mapsto f_q(u)$, where $f_q(u) \equiv q^u \pmod r$ and $0 \le f_g(u) \le r-1$. The number of cycles is estimated from above in the case when $r = p^n$ with a prime integer p and $\gcd(q,p) = 1$. Also a more general class of functions is considered.

1. Introduction and Formulation of Results

Given a number $r \in \mathbb{N}$, we consider a dynamical system generated by exponentiation modulo r, that is, by the map $u \mapsto f_q(u)$ where $f_q(u) \equiv q^u \pmod r$ and $0 \le f_q(u) \le r-1$. In [2] the author with Igor Shparlinski considered the case when r is prime. We gave some estimates for the number of $1-, 2-, 3-$periodic points of f. We believe that our estimates are very far from being strict (it seems that better estimates are not known). But the strictness is not the main issue. The technique we use does not work even for 4-periodic points. So, it is a challenging problem to prove any nontrivial estimate for the number of k-periodic points with $k \ge 4$. Maybe one of the difficulties is that f is not an algebraic factor of q^x: if, for example, $\gcd(r, \phi(r)) = 1$ then one can choose a representative $y \equiv x \mod r$ such that q^y has any possible value modulo r. The situation with large $\gcd(r, \phi(r))$ may be more tractable. In that case, instead of considering the function f, one may consider the graph with the edges from $x \in \mathbb{Z}_r = \mathbb{Z}/r\mathbb{Z}$ to all $q^y \mod r$, $y \equiv x \mod r$. We show that it works very well at least for $r = p^n$ with a prime p. Also, "additively close" to q^x functions are considered, see Corollary 5.

The main results of the present paper are Theorem 1, Theorem 2, and Theorem 4. A weaker result than Theorem 2 is proved in the author's preprint [1]. Then independently, among other things, a result equivalent to Theorem 2 is established in [3]. The authors of [3] use p-adic analysis and in the present article we use combinatorial methods. As it is shown in [3] the algebraic reason to consider a graph instead of a function is that $f(x) = q^x$ has the m-valued continuous extension on p-adic numbers where m is the multiplicative order of q modulo p. The results formulated in Theorem 1 and Theorem 4 do not appear in [3].

In what follows we suppose that $\gcd(q,p) = 1$. Let $\Gamma_{p,n,q}$ be a directed graph with the set of vertices $V(\Gamma) = \mathbb{Z}_{p^n} = \mathbb{Z}/p^n\mathbb{Z}$ and the set of edges $E = \{(x, q^y \mod p^n) \mid x \in \mathbb{Z}_{p^n},\ y \in \mathbb{Z},\ y \equiv x \mod p^n\}$.

Theorem 1. *Let* $\gcd(q,p) = \gcd(g,p) = 1$ *and the multiplicative orders of* q *and* g *modulo* p^n *coincide. Then the graphs* $\Gamma_{p,n,q}$ *and* $\Gamma_{p,n,g}$ *are isomorphic.*

Given a graph Γ, a sequence $v_1, v_2, \dots, v_k, v_{k+1}$ with $v_j \in V(\Gamma)$, $(v_j, v_{j+1}) \in E(\Gamma)$, $v_1 = v_{k+1}$ is said to be a k-cycle (with initial vertex marked) in Γ. Let $C_{p,n,q}(k)$ be the number of k-cycles (with initial vertex marked) in $\Gamma_{p,n,q}$.

Theorem 2. *Let* m *be the multiplicative order of* q *modulo* p. *Then* $C_{p,n,q}(k) = m^k$ *and the out-degree of any edge of the graph* $\Gamma_{p,n,q}$ *is* m.

Corollary 3. *The number of* k*-periodic points for* $f(x) \equiv q^x \mod p^n$, $0 \le f(x) < p^n$, *is less than* m^k *where* m *is the multiplicative order of* q *modulo* p.

The same technique is used to estimate the number of k-cyclic points in "additive perturbations" of graph Γ. Precisely, let us define $\Gamma^r_{p,n,q}$ as follows: the set of vertices is $V(\Gamma) = \mathbb{Z}_{p^n}$ and the set of edges is $E = \{(x, q^y + c \mod p^n) \mid x \in \mathbb{Z}_{p^n},\ y \in \mathbb{Z},\ y \equiv x \mod p^n,\ c = -r, -r+1, \dots, r\}$. Let $C^r_{p,n,q}(k)$ be the number of k-cycles (with the initial vertex marked) in $\Gamma^r_{p,n,q}$.

Theorem 4. $C^r_{p,n,q}(k) \le p + 4krp[2(2r+1)(p-1)]^{k-1}(n-1)$.

So, $C^r_{p,n,q}$ grows no more than linearly in n (but the number of all vertices grows exponentially). Let us note that the estimate may be nontrivial even for $r > p$. For example, if k is fixed, $p > 2$, $r \sim p^{\epsilon n}$, and $\epsilon < 1/k$, then the estimate of the Theorem is nontrivial for large enough n.

Corollary 5. *Let* $f : \mathbb{Z}_{p^n} \to \mathbb{Z}_{p^n}$ *be such that* $f(x) \equiv q^x + g(x) \mod p^n$, *where* $g : \mathbb{Z}_{p^n} \to \{-r, \dots, r\}$. *Then the number of* k*-periodic points of* f *is less than* $p + 4krp[2(2r+1)(p-1)]^{k-1}(n-1)$.

One could also consider other additive perturbations. For example, with $E = \{(x, q^y \mod p^n) \mid y \equiv x + j \mod p^n,\ j = -r, \dots, r\}$. But it gives nothing new. Indeed, let Γ^r be the graph defined as $V(\Gamma^r) = \mathbb{Z}_{p^n}$ and $E(\Gamma^r) = \{(x, x + j \mod p^n),\ j = -r, \dots, r\}$. Then both additive perturbations are composition of Γ^r and $\Gamma_{p,n,q}$, but with the different order. It is clear that the order does not change the number of cycles. The same is true for the composition $\Gamma^{r_1} - \Gamma_{p,n,q} - \Gamma^{r_2}$ (two-sided additive perturbation).

2. Proof of Theorem 2

In what follows we identify $\mathbb{Z}_{p^n}$ with $\{0, 1, 2, \ldots, p^n - 1\}$. So, $y = (x \mod p^n)$ often means that $y \equiv x \mod p^n$ and $y \in \{0, 1, \ldots p^n - 1\}$.

Lemma 6. *Let $A_1, A_2, ..., A_r$ be elements of an associative (not necessarily commutative) ring $\mathcal{A}$. Let $M \in Mat_{n\times n}(\mathcal{A})$,*

$$M = \begin{pmatrix} A_1 & A_2 & \ldots & A_n \\ A_1 & A_2 & \ldots & A_n \\ \vdots & \vdots & \ldots & \vdots \\ A_1 & A_2 & \ldots & A_n \end{pmatrix}.$$

Then $trace(M^k) = (A_1 + A_2 + \cdots + A_r)^k$.

Proof.

$$M^k = \begin{pmatrix} 1 \\ 1 \\ \vdots \\ 1 \end{pmatrix} \left(\begin{pmatrix} A_1 & A_2 & \ldots & A_r \end{pmatrix} \begin{pmatrix} 1 \\ 1 \\ \vdots \\ 1 \end{pmatrix} \right)^{(k-1)} \begin{pmatrix} A_1 & A_2 & \ldots & A_r \end{pmatrix} =$$

$$(A_1 + A_2 + \cdots + A_r)^{k-1} \begin{pmatrix} A_1 & A_2 & \ldots & A_n \\ A_1 & A_2 & \ldots & A_n \\ \vdots & \vdots & \ldots & \vdots \\ A_1 & A_2 & \ldots & A_n \end{pmatrix}.$$

□

Lemma 7. *Let A_n be the adjacency matrix of $\Gamma_{p,n,q}$. Then*

1)

$$A_1 = \begin{pmatrix} 0 & a_1 & a_2 & \ldots & a_{p-1} \\ 0 & a_1 & a_2 & \ldots & a_{p-1} \\ \vdots & \vdots & \vdots & \ldots & \vdots \\ 0 & a_1 & a_2 & \ldots & a_{p-1} \end{pmatrix}$$

with $a_i \in \{0, 1\}$ and $a_1 + a_2 \cdots + a_{p-1} = m$, where m is the multiplicative order of q modulo p;

2) *for $n > 1$ one has*

$$A_n = \begin{pmatrix} B_1^n & B_2^n & \ldots & B_p^n \\ B_1^n & B_2^n & \ldots & B_p^n \\ \vdots & \vdots & \ldots & \vdots \\ B_1^n & B_2^n & \ldots & B_p^n \end{pmatrix},$$

where $B_j^n \in Mat_{p^{n-1}\times p^{n-1}}(\{0, 1\})$ and $B_1^n + B_2^n + ... + B_p^n = A_{n-1}$.

Proof. For Item 1 it suffices to note that q^x modulo p takes m different values and depends on x modulo $p-1$ only (not on x modulo p). Let us prove Item 2. First of all we represent $x \in \mathbb{Z}_{p^n} = \{0, 1, 2, \ldots, p^n - 1\}$ as $x = y + bp^{n-1}$, where $y \in \{0, 1, \ldots, p^{n-1} - 1\}$ and $b \in \{0, 1, \ldots, p-1\}$. The block structure of A_n corresponds to the above described representation, such that b's are indexing our blocks and y's are indexing the elements inside the blocks. Let $O^n(x) = \{y \in Z_{p^n} \mid (x, y) \in E(\Gamma_{p,n,q})\}$. Item 2 follows from the following facts

(i) $O^n(x)$ is independent of blocks. That is $O^n(x) = O^n(z)$ if $x \equiv z \mod p^{n-1}$.

(ii) Let $\psi : \mathbb{Z}_{p^n} \to \mathbb{Z}_{p^{n-1}}$ be defined as $\psi(x) \equiv x \mod p^{n-1}$. Then for any $y \in \{0, 1, 2, \ldots, p^{n-1} - 1\}$ ψ defines a bijection $O^n(y) \leftrightarrow O^{n-1}(y)$.

Fact **i)**. To find $q^z \mod p^n$ it suffices to know $z \mod (p-1)p^{n-1}$. Let $P_x = \{z \in \mathbb{Z}_{(p-1)p^{n-1}} \mid \exists a \in \mathbb{Z}\ a \equiv z \mod (p-1)p^{n-1}$ and $a \equiv x \mod p^n\}$. One has that $O^n(x) = \{q^z \mod p^n \mid z \in P_x\}$. By Chinese Remainder Theorem $P_x = P_y$ if and only if $x \equiv y \mod p^{n-1}$.

Fact **ii)** Observe that $p^n \equiv 1 \mod p-1$. So, $O^n(y) = \{q^y q^{bp^n} \mod p^n \mid b \in \{0, 1, \ldots, p-2\}\}$ and $O^{n-1}(y) = \{q^y q^{bp^{n-1}} \mod p^{n-1} \mid b \in \{0, 1, \ldots, p-2\}\}$. Now, $q^{bp^{n-1}} \equiv q^{bp^n} \mod p^n$. Indeed, $bp^{n-1} - bp^n \equiv 0 \mod (p-1)p^{n-1}$. So, $O^n(y) = \{q^y q^{bp^{n-1}} \mod p^n \mid b \in \{0, 1, \ldots, p-2\}\}$. It remains to prove that for $b_1, b_2 \in \{0, 1, \ldots, p-2\}$ the congruence

$$q^{b_1 p^{n-1}} \equiv q^{b_2 p^{n-1}} \mod p^{n-1} \tag{1}$$

implies the congruence

$$q^{b_1 p^{n-1}} \equiv q^{b_2 p^{n-1}} \mod p^n. \tag{2}$$

Let $q \equiv g^r \mod p^n$ for primitive g. The first congruence is equivalent to $(b_1 - b_2)rp^{n-1} \equiv 0 \mod (p-1)p^{n-2}$. It implies $(p-1) | (b_1 - b_2)r$. So, $(b_1 - b_2)rp^{n-1} \equiv 0 \mod (p-1)p^{n-1}$ and the second congruence follows. Now, Fact **(i)** implies that all block rows in A_n are equal. Fact **(ii)** implies that the sum of the blocks in the first row is A_{n-1}. □

Now it is easy to finish the proof of Theorem 2. Firstly, $C_{p,n,q}(k) = trace((A_n)^k)$. Using Lemma 6, Lemma 7 and compatibility of the trace and multiplication with the block structure we get

$$trace((A_n)^k) = trace((A_{n-1})^k) = \cdots = trace((A_1)^k) = (a_1 + a_2 + \cdots + a_{p-1})^k.$$

The second statement of the theorem is a simple consequence of Lemma 7.

3. Proof of Theorem 1

Adopting the notation of the previous section for different q, let $O^n_q(x) = \{y \in \mathbb{Z}_{p^n} \mid (x, y) \in E(\Gamma_{p,n,q})\}$.

Lemma 8. *Let M be the multiplicative order of q modulo p^n. Let r be the greatest r such that $p^r|M$. Then the following statements are equivalent:*

1) $O_q^n(x_1) \cap O_q^n(x_2) \neq \emptyset$;

2) $x_1 - x_2 \equiv 0 \mod p^r$;

3) $O_q^n(x_1) = O_q^n(x_2)$.

Proof. 1)$\Longrightarrow$2). Let $q^{y_1} \equiv q^{y_2} \mod p^n$ and $x_j \equiv y_j \mod p^n$, $j = 1, 2$. This implies that $y_1 \equiv y_2 \mod M$ and $x_1 \equiv x_2 \mod p^r$.
2)$\Longrightarrow$3). The values of $q^x \mod p^n$ depend on $x \mod (p-1)p^r$. So, the implication follows by the Chinese Reminder Theorem.
The last implication is trivial. □

Under our conditions the multiplicative orders of q and $g \mod p^k$, $k \leq n$, coincide. So, we inductively construct isomorphisms $h_k : \mathbb{Z}_{p^k} \to \mathbb{Z}_{p^k}$ of graphs $\Gamma_{p,k,q}$ and $\Gamma_{p,k,g}$. Let $\psi : \mathbb{Z}_{p^{k+1}} \to \mathbb{Z}_{p^k}$ be the natural projection (as in the previous section). We inductively keep the following properties:

1. h_k is an isomorphism of $\Gamma_{p,k,q}$ and $\Gamma_{p,k,g}$. In particular, it implies that h_k is a bijection.

2. The following diagram is commutative:

$$\begin{array}{ccc} \mathbb{Z}_{p^{k+1}} & \xrightarrow{h_{k+1}} & \mathbb{Z}_{p^{k+1}} \\ \downarrow \psi & & \downarrow \psi \\ Z_{p^k} & \xrightarrow{h_k} & Z_{p^k} \end{array}$$

Trivially, $h_1 = \text{id}$ (the identity map) satisfies our hypothesis. Assume that h_k is constructed. Let $x \in \{0, 1, \ldots, p^k - 1\} = \mathbb{Z}_{p^k}$. Since ψ and h_k are bijections on $O^k(x)$, there is a unique h_{k+1}^x making the following diagram commute:

$$\begin{array}{ccc} O_q^{k+1}(x) & \xrightarrow{h_{k+1}^x} & O_g^{k+1}(h_k(x)) \\ \downarrow \psi & & \downarrow \psi \\ O_q^k(x) & \xrightarrow{h_k} & O_g^k(h_k(x)) \end{array}$$

Let $X = \bigcup_{x \in \mathbb{Z}_{p^k}} O_q^{k+1}(x) = \bigcup_{x \in \mathbb{Z}_{p^{k+1}}} O_q^{k+1}(x)$ and $Y = \bigcup_{x \in \mathbb{Z}_{p^k}} O_g^{k+1}(x)$. Lemma 8 implies that $O_q^{k+1}(x)$, $x \in \mathbb{Z}_{p^k}$ forms a partition of X and $O_q^{k+1}(x_1) = O_q^{k+1}(x_2)$ if and only if $O_g^{k+1}(h_k(x_1)) = O_g^{k+1}(h_k(x_2))$. So, the set of functions h_{k+1}^x, $x \in$

$\mathbb{Z}_{p^k}$, define bijection $h_{k+1} : X \to Y$. By construction, the following diagram is commutative:

$$\begin{array}{ccc} X & \xrightarrow{h_{k+1}} & Y \\ \big\downarrow \psi & & \big\downarrow \psi \\ \psi(X) & \xrightarrow{h_k} & \psi(Y) \end{array}$$

So, we can define $h_{k+1} : \mathbb{Z}_{p^{k+1}} \setminus X \to \mathbb{Z}_{p^{k+1}} \setminus Y$ making the corresponding diagram commute. Taking into account that $O^{k+1}(x) = O^{k+1}(\psi(x))$ one can check that the h_{k+1} constructed satisfies properties 1. and 2..

4. Proof of Theorem 4

For $A, B \in Mat_{d\times d}(\{0,1\})$ we will write $A \preceq B$ if $A_{i,j} = 1$ implies $B_{i,j} = 1$.

Lemma 9. *Let A_n be the adjacency matrix of $\Gamma^{+r}_{p,n,q}$. Then*
(1)

$$A_1 \preceq \begin{pmatrix} 1 & 1 & 1 & \dots & 1 \\ 1 & 1 & 1 & \dots & 1 \\ \vdots & \vdots & \vdots & \dots & \vdots \\ 1 & 1 & 1 & \dots & 1 \end{pmatrix};$$

(2) for $n > 1$ one has

$$A_n \preceq \begin{pmatrix} B_1^n & B_2^n & \dots & B_p^n \\ B_1^n & B_2^n & \dots & B_p^n \\ \vdots & \vdots & \dots & \vdots \\ B_1^n & B_2^n & \dots & B_p^n \end{pmatrix} + X, \text{ where } B_j^n \in \underset{p^{n-1}\times p^{n-1}}{Mat}(\{0,1\}),$$

$B_1^n + B_2^n + \ldots + B_p^n = A_{n-1}$ and $X \in Mat_{p^n\times p^n}(\{0,1\})$ with less then $2rp$ non-zero columns;
(3) The number of 1s in a row of A_n is less than $2r(p-1)$.

Proof. Item (1) is trivial. The proof of Item (2) proceeds in the same way as the one of Lemma 7. As in Lemma 7 the block structure corresponds to the representation $x = y + bp^{n-1}$ with $0 \le y < p^{n-1}$ and $b \in \{0, 1, \ldots, p-1\}$. The neighborhood of a vertex x in $\Gamma^r_{p,n,q}$ is easy to construct using the neighborhood of x in $\Gamma_{p,n,q}$. So, let $O^n(x) = \{y \in \mathbb{Z}_{p^n} \mid (x, y) \in E(\Gamma_{p,n,q})\}$ and $O^n_r(x) = \{y \in \mathbb{Z}_{p^n} \mid (x, y) \in E(\Gamma^r_{p,n,q})\}$. By definition of $\Gamma^r_{p,n,q}$ one has that $y + bp^{n-1} \in O^n_r(x)$ if and only if $(y + bp^{n-1} + j \mod p^n) \in O^n(x)$ for some $j = -r, \ldots, r$. Observe that $y + j = (y + j \mod p^{n-1})$ for $r \le y < p^{n-1} - r$. So, $(y + j \mod p^{n-1}) + bp^{n-1} = (y + j + bp^{n-1} \mod p^n)$ for such y. It implies that

$$(B_0^n + B_1^n + \cdots + B_{p-1}^n)_{jy} = (A_{n-1})_{jy}$$

for $r \leq y < p^{n-1} - r$. (Recall that $O^n(x)$ satisfies items **(i)** and **(ii)** of the proof of Lemma 7.) So, there exist no more than $2r$ columns in each block where the equality does not hold. Item (2) of the lemma follows. Item (3) holds since the number of 1s in a row x is $|O_r^n(x)|$ but $|O_r^n(x)| \leq (2r+1)|O^n(x)| = (2r+1)(p-1)$. □

Now we are ready to prove Theorem 4. We have

$$C_{p,n,q}^r(k) = c_n = trace(A_n^k) \leq trace(A_{n-1}^k) + \Delta = c_{n-1} + \Delta,$$

where Δ is the sum of the traces of $2^k - 1$ matrices P_s, each of them being a product of k matrices containing X. Observe that $trace(P_s) \leq 2krp((2r+1)(p-1))^{k-1}$. Indeed, this is the number of k-periodic paths such that at least one edge of a path corresponds to the matrix X (X-edge). We have no more than $2rp$ ends of X-edges. Starting counting from these vertices we get no more than $2rp((2r+1)(p-1))^{k-1}$ cycles. Counting cycles with initial points we get our estimate. Observe that $c_1 \leq p$ and, by induction, $c_n \leq p + 4krp(2(2r+1)(p-1))^{k-1}(n-1)$.

Acknowledgments. The author thanks Igor Shparlinski, Edgardo Ugalde and Alexander Pyshchev for useful suggestions. The work was partially supported by PROMEP grant UASLP-CA21 and by CONACyT grant 50312.

References

[1] Lev Glebsky, Cycles in Repeated Exponentiation Modulo p^n, *Preprint* arXiv:1006.2500v1.

[2] Lev Glebsky and Igor E. Shparlinski, Short cycles in repeated exponentiation modulo a prime, *Des. Codes Cryptogr.* **56** (2010), no. 1, 35-42.

[3] Joshua Holden and Margaret M. Robinson, Counting fixed points, two-cycles, and collisions of the discrete exponential functions using p-adic methods, *J. Aust. Math. Soc.* **92** (2012), 163-178.

COMPUTATION OF AN IMPROVED LOWER BOUND TO GIUGA'S PRIMALITY CONJECTURE

Jonathan Borwein
Centre for Computer-assisted Research Mathematics and its Applications (CARMA), School of Mathematical and Physical Sciences, University of Newcastle, Australia
jonathan.borwein@newcastle.edu.au jborwein@gmail.com

Christopher Maitland
Mathematical Sciences Institute, Australian National University, Canberra, Australia
c3053540@gmail.com

Matthew Skerritt
Centre for Computer-assisted Research Mathematics and its Applications (CARMA), School of Mathematical and Physical Sciences, University of Newcastle, Australia
matt.skerritt@gmail.com

Received: 3/7/13, Accepted: 9/5/13, Published: 10/10/13

Abstract

Our most recent computations tell us that any counterexample to Giuga's 1950 primality conjecture must have at least 19,908 decimal digits. Equivalently, any number which is both a Giuga and a Carmichael number must have at least 19,908 decimal digits. This bound has not been achieved through exhaustive testing of all numbers with up to 19,908 decimal digits, but rather through exploitation of the properties of Giuga and Carmichael numbers. This bound improves upon the 1996 bound of one of the authors. We present the algorithm used, and our improved bound. We also discuss the changes over the intervening years as well as the challenges to further computation.

1. Introduction

In a 1950 paper [9] Giuseppe Giuga formulated his now well-known prime number conjecture:

Conjecture 1 (Giuga Conjecture). An integer, $n > 1$, is prime if and only if

$$s_n := \sum_{k=1}^{n-1} k^{n-1} \equiv -1 \pmod{n}.$$

In 1995, Takahashi Agoh [1] showed that this is equivalent to:

Conjecture 2 (Giuga-Agoh Conjecture)**.** An odd integer, $n > 1$, is prime if and only if the even Bernoulli number B_{n-1} satisfies

$$nB_{n-1} \equiv -1 \pmod{n}.$$

A more recent generalisation of Giuga's conjecture can be found in [12].

1.1. Giuga and Carmichael numbers

Herein, we present enough background material to facilitate understanding of the algorithm with which lower bounds to any possible counterexample were computed. For more details, including proofs of the theorems below, see [6, 7, 8].

We begin by observing that if p is prime, then $s_p \equiv p-1 \pmod{p}$ is an immediate consequence of Fermat's little theorem. The question, then, is whether there exists any composite n with $s_n \equiv n-1 \pmod{n}$. To this end, Giuga proved the following theorem.

Theorem 3. *An integer, $n > 1$, satisfies $s_n \equiv -1 \pmod{n}$ if and only if each prime divisor p of n satisfies:*

$$p \mid \left(\frac{n}{p} - 1\right), \tag{1}$$

$$(p-1) \mid \left(\frac{n}{p} - 1\right). \tag{2}$$

□

Any composite number satisfying (1) is known as a *Giuga number*. It is currently unknown whether there are infinitely many Giuga numbers, or whether there are any odd Giuga numbers. Recent papers on Giuga numbers include [10, 11]. Note that Giuga numbers must be square-free. To see this we suppose $n = p^2q$ and observe that (1) implies $p \mid (pq-1)$, which is impossible.

Giuga numbers may be characterised in the following manner, which is useful in the computation of exclusion bounds for counterexamples.

Theorem 4. *A composite number $n \in \mathbb{N}$ is a Giuga number if and only if it is square-free and satisfies*

$$\sum_{p \mid n} \frac{1}{p} - \prod_{p \mid n} \frac{1}{p} \in \mathbb{N}.$$

□

Composite, square-free numbers satisfying (2) are called *Carmichael numbers.* We may replace (2) with the equivalent formulation:

$$(p-1) \mid (n-1). \tag{3}$$

It has been shown that there are infinitely many Carmichael numbers [2]. Indeed, for any $\varepsilon > 0$, for sufficiently large x the number of Carmichael numbers $n \leq x$ exceeds $x^{\beta-\varepsilon}$ where $\beta := 0.33222... > 2/7$ is an explicit constant. More recently, various interesting density estimates have been given in [5, 14, 16]. For example, the number of Carmichael numbers $n \leq x$ that are not Giuga numbers exceeds $x^{\beta-\varepsilon}$ for sufficiently large x. It has been observed that $x^{\beta-\varepsilon}$ may be replaced by $x^{1/3}$ as a consequence of a result of Harman [13].

1.2. Normal sequences of primes

Note that if p is a prime divisor of a Carmichael number, n say, then for no $k \in \mathbb{N}$ is $kp+1$ a prime divisor of n. We can see this if we suppose that there is some prime p such that $p \mid n$, $kp+1$ is prime, and $(kp+1) \mid n$. Then, by (3), $(kp+1)-1 = kp \mid (n-1)$ which is impossible if $p \mid n$. A consequence of this observation is that $p = 2$ cannot be a prime divisor of a Carmichael number, and that therefore every Carmichael number must be odd.

We say that the prime divisors of a Carmichael number, n, are *normal* according to the following definition:

Definition 5. A set of odd primes, P, is *normal* if no $p, q \in P$ satisfies $p \mid (q-1)$.

The following extension of the idea of a normal set of primes is useful in the computation of exclusion bounds.

Definition 6. Let P be a normal set of primes. For any prime, p, we say *p is normal to P* if $P \cup \{p\}$ is normal.

In light of this we can reformulate Theorem 3 to say that a counterexample to Giuga's conjecture must be both a Giuga number and a Carmichael number. Giuga's conjecture then simply states that there is no such number. A detailed study on Giuga's conjecture and variants related to normal families of primes is made in [17].

2. Exclusion Bounds

We may compute exclusion bounds for counterexamples to Giuga's conjecture by exploiting the properties of Giuga and Carmichael numbers described in the previous section.

Notation 1 (Odd primes and sets of odd primes).

- Let q_k denote the kth odd prime.
- Let Q denote the collection of all odd primes $\{3, 5, 7, 11, \dots\}$

- Let Q_k denote the set $\{q_1, \ldots, q_{k-1}\}$ of odd primes strictly less than q_k
- Let $\mathcal{N}_d$ denote the set $\{\mathcal{N} \subset Q_d \mid \mathcal{N} \text{ is normal}\}$ of all finite normal sets whose maximal element is less than q_d.

To begin with, an easy lower bound can be found by observing that, by Theorem 4, the reciprocal sum of prime factors of any Giuga number must be greater than 1, and that any Carmichael number must be odd. We observe that the set $Q_{10} = \{3, 5, 7, 11, 13, 17, 19, 23, 29\}$ is the smallest set of odd primes whose reciprocal sum is greater than 1. Thus a counterexample must, because $|Q_{10}| = 9$, have at least 9 prime factors and, because $\prod_{p \in Q_{10}} p > 10^9$, must have at least 10 digits.

Observe that the above result does not fully take into account the fact that a counterexample must also be a Carmichael number. Doing so raises the bounds considerably.

We exploit a binary tree structure, where the vertices of the tree are normal sets of primes. We begin with the empty set as the root of the tree, and consider this to be depth 1 of the tree. Each vertex, $\mathcal{N}$ say, at depth d of the tree has itself as a child. Additionally, if q_d is normal to $\mathcal{N}$ then $\mathcal{N} \cup \{q_d\}$ is also a child of $\mathcal{N}$. This structure is shown in Figure 1. With this structure, the collection of all sets at depth d of this tree is precisely the set $\mathcal{N}_d$, defined above.

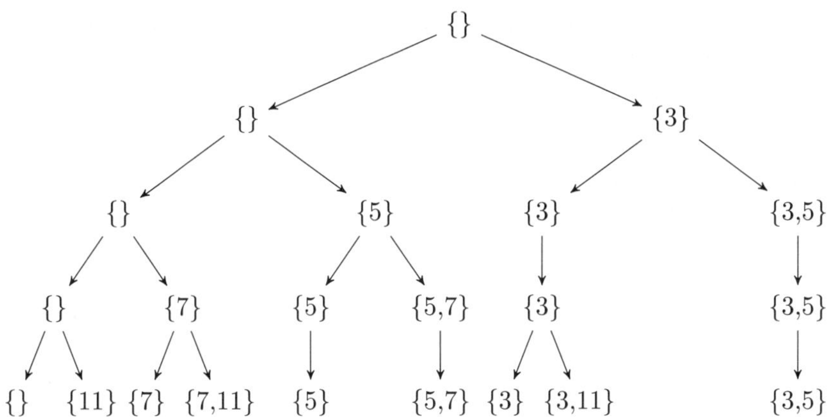

Figure 1: The first 5 levels of a tree of normal sets of primes.

For each set, $\mathcal{N}$, on level d of the tree we compute a new set which we call $T_d(\mathcal{N})$. We begin by assigning $T_d(\mathcal{N}) = \mathcal{N}$. We then look at the reciprocal sum of $T_d(\mathcal{N})$

$$r := \sum_{q \in T_d(\mathcal{N})} \frac{1}{q}$$

If $r \leq 1$ and $\mathcal{N} \cup \{q_d\}$ is normal then we set $T_d(\mathcal{N}) = T_d(\mathcal{N}) \cup \{q_d\}$, otherwise we leave it unchanged. We repeat this process with q_{d+1}, q_{d+2} and so on until $r > 1$, at which point we are done.

Definition 7 (j-value)**.** Let $\mathcal{N} \in \mathcal{N}_d$ be a vertex at depth d of the tree. Denote by $T_d(\mathcal{N})$ the set computed as described, above. Then we say that

$$j_d(\mathcal{N}) := |T_d(\mathcal{N})|$$

is the *j-value* or, equivalently, the *score* of $\mathcal{N}$.

Note that there is a special case when the reciprocal sum of $\mathcal{N}$ is greater than 1 to begin with. In this case, no primes are appended to $T_d(\mathcal{N})$, and we have $T_d(\mathcal{N}) = \mathcal{N}$ and so $j_d(\mathcal{N}) = |\mathcal{N}|$.

Example 8. The lower bound we computed at the beginning of this section is equivalent to computing $T_1(\{\}) = \{3, 5, 7, 11, 13, 17, 19, 23, 29\}$ and $j_1(\{\}) = 9$. We may also compute

$$T_2(\{\}) = \{5, 7, 11, 13, 17, \ldots, 107, 109\} \text{ so that } j_2(\{\}) = 27$$
$$T_2(\{3\}) = \{3, 5, 11, 17, 23, \ldots, 317, 347\} \text{ so that } j_2(\{3\}) = 32.$$

Observe that, in general, it is not true that $T_d(\mathcal{N})$ is a normal set; our criterion for inclusion of a prime, q, into $T_d(\mathcal{N})$ is that it is normal to $\mathcal{N}$, and not that it is normal to $T_d(\mathcal{N})$. We can see, above, that both $T_2(\{\})$ and $T_2(\{3\})$ contain both 5 an 11, and so are not normal. □

For a fixed depth, d, the values of $j_d(\mathcal{N})$ for all $\mathcal{N} \in \mathcal{N}_d$ allow us to compute a lower bound for the number of primes needed for a counterexample. For each $\mathcal{N}$, we consider a hypothetical family of all counterexamples whose set of prime factors less than q_d is precisely $\mathcal{N}$, and $j_d(\mathcal{N})$ is an approximation—specifically a lower bound—of the number of primes required for any such counterexample, should one exist.

To see that $j_d(\mathcal{N})$ is a lower bound, recall that $T_d(\mathcal{N})$ is not (in general) a normal set, but that the prime factors of a counterexample must form a normal set. As such, whenever $T_d(\mathcal{N})$ is not normal, some of its elements must be absent in any counterexample from the above hypothetical family. Each such absent element—because of the way $T_d(\mathcal{N})$ was constructed—must be replaced by at least one prime greater than $\max(T_d(\mathcal{N}))$ in order for the reciprocal sum to stay above 1. Note that if $T_d(\mathcal{N})$ is normal, then we have the smallest normal set containing $\mathcal{N}$ with a reciprocal sum greater than 1, although it is not necessarily the case that these normal primes are the factors of a counterexample. We may, nonetheless, conclude that $j_d(\mathcal{N})$ is a lower bound on the number of prime factors in any counterexample whose prime factors less than q_d are exactly the set $\mathcal{N}$.

Recall that $\mathcal{N}_d$ is the collection of all normal subsets of Q_d and that the leaves (or vertices) of our tree for a fixed depth, d, are precisely the sets contained in $\mathcal{N}_d$.

Any counterexample to Giuga's conjecture must have as its set of prime factors less than q_d exactly one of $\mathcal{N} \in \mathcal{N}_d$, and so if we compute $j_d(\mathcal{N})$ for all $\mathcal{N} \in \mathcal{N}_d$, then the smallest such value

$$j_d := \min\{j_d(\mathcal{N}) \mid N \in \mathcal{N}_d\}$$

must be a lower bound for the number of prime factors in any counterexample to Giuga's conjecture. From this we compute the corresponding lower bound

$$P_d := \prod_{k=1}^{j_d} q_k$$

as the product of the first j_d odd primes.

Example 9. Any counterexample to Giuga's conjecture will either have 3 as a prime factor or not, corresponding to the normal sets $\{\}$ and $\{3\}$ sitting at depth 2 of the tree in Figure 1. We have computed the values of $j_2(\{\})$ and $j_2(\{3\})$ in Example 8, above. In the former case there must be at least 32 prime factors; in the latter case there must be at least 27. We may conclude, then, that such a counterexample must have at least 27 prime factors—that is $j_2 = 27$—and must therefore be greater than $P_2 = \prod_{k=1}^{27} q_k > 10^{42}$. □

The algorithm, then, is simply to build the tree down to the desired depth, d, then to calculate the j-value for every vertex at that depth, so as to find the minimum j-value as the value of j_d. P_d may easily be computed once j_d is known. Regrettably the number of vertices grows exponentially with the depth of the tree, so computation of j_d for large d is slow and so the algorithm is of limited use in the form hitherto described. Fortunately a relatively simple observation leads to significant improvement in performance.

We observe that the j-values of a set in our tree do not decrease as we travel down the tree. That is, for a set $\mathcal{N}$ at depth d of the tree, the value of $j_d(\mathcal{N})$ is no larger than the value of $j_{d+1}(\mathcal{N}')$ so long as $\mathcal{N}'$ is a child of $\mathcal{N}$. This follows from the following theorem which is proved in [8].

Theorem 10. *Let $\mathcal{N} \in \mathcal{N}_d$ be a normal set of primes. Then $j_{d+1}(\mathcal{N}) \geq j_d(\mathcal{N})$ Furthermore, let $\mathcal{N}' = N \cup \{q_d\}$. If $\mathcal{N}'$ is normal, then $j_{d+1}(\mathcal{N}') \geq j_d(\mathcal{N})$.* □

A corollary of this theorem is that any normal superset of $\mathcal{N}$ whose reciprocal sum is greater than 1 must have at least $j_d(\mathcal{N})$ elements. That is, if $\mathcal{N} \subset N'$, $\mathcal{N}'$ is normal, and $\sum_{q\in\mathcal{N}'} 1/q > 1$, then $|\mathcal{N}'| \geq j_d(\mathcal{N})$.

We exploit Theorem 10 to significantly cut down the number of vertices (leaves) needed when computing j_d for a fixed d. Inasmuch as j_d is the minimum j-value of any vertex at depth d of the tree, the j-value of any vertex at that depth is an upper bound for j_d. Any vertex higher in the tree—which is to say at depth $d' < d$—whose j-value exceeds such an upper bound need not have any of its children examined.

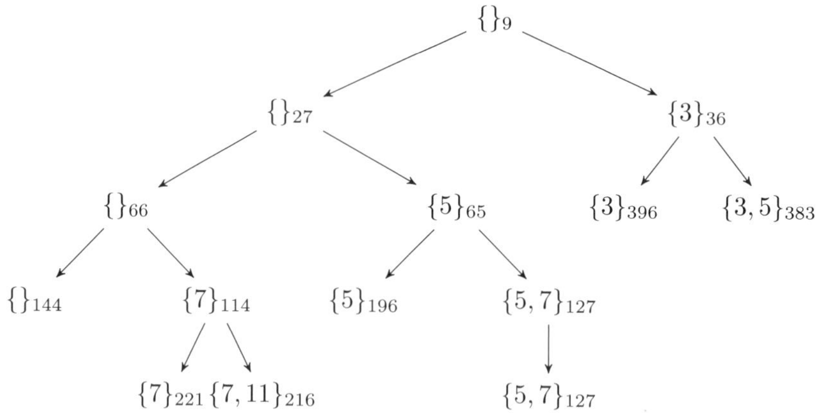

Figure 2: The first 5 levels of the trimmed tree of normal sets of primes.

In essence, we may trim the tree early. A trimmed tree, along with the j-values for each vertex can be seen in Figure 2.

Note that the algorithm still exhibits signs of exponential growth (see below), however the work saved by this method is enough to make the algorithm much more tractable and proved sufficient, in 1996, to make the difference between being able to calculate j_{19} and j_{100} (see [8]).

The amount of work saved by the above optimisation is entirely dependent on the choice of initial upper bound. An excellent candidate for a good such bound is given by the family of sets described as follows:

Definition 11 (*L*-sets)**.** Let $k \geq 5$ then

$$L_5 := \{5, 7\} \quad \text{and} \quad L_{k+1} := \begin{cases} L_k \cup q_k & \text{if this set is normal,} \\ L_k & \text{otherwise.} \end{cases}$$

In every computation we have performed for depths $d > 5$ the value of j_d has been given by these sets. That is to say that $j_d = j_d(L_d)$ for all $5 \leq d \leq 311$.

These L-sets lead to an interesting observation: the algorithm described here is exhaustible. The first of these sets to have a reciprocal sum greater than 1 is the set $L_{27{,}692}$, which has 8,135 elements. Note that the primes in $L_{27{,}692}$ are not the factors of a counterexample to Giuga's conjecture. Nonetheless, it must be the case that $j_d \leq 8{,}135$ for all $d \geq 27{,}692$, and so our algorithm can never compute an exclusion bound higher than 8,135 prime factors, even if it turns out that $j_{27{,}692} < j_{27{,}692}(L_{27{,}692})$. It is the sincerest hope of the authors to eventually exhaust this algorithm, although given the apparently exponential growth of even the trimmed tree, it is exceptionally unlikely to happen.

It was with this algorithm in 1996 that Borwein et al [6] were able to compute $j_{135} = 3{,}459$. Similarly, it is with this algorithm that the present authors have been able to compute $j_{311} = 4{,}771$ thus obtaining the following result.

Theorem 12. *Any counterexample to Giuga's primality conjecture is an odd square-free number with at least* 4,771 *prime factors and so must exceed* $10^{19{,}907}$. □

3. Computation Improvements, Implementation Details, and Technical Challenges

In 1996, Borwein et al [6] were able to compute j_{100} in *Maple* using the algorithm described above. Borwein reports in [8] that the computation required "a few CPU hours for each k" for values of k near 100. Using a C++ implementation (which took two months to produce) they were able to compute j_{135}, taking "303 CPU hours" [6]. The C++ implementation then crashed irrevocably before doing any extra work [8]. Thus, in 1996, the best results for exclusion bounds for Giuga's conjecture were

$$j_{135} = 3{,}459 \qquad\qquad P_{135} > 10^{13{,}886}$$

No further work on these computations was done by any of the present authors until mid-2009. Effort was spent to reduce the number of operations in the implementation, without deviating from the algorithm. In early 2011, j_{173} was computed in approximately thirty five hours using *Maple*, obtaining:

$$j_{173} = 3{,}795 \qquad\qquad P_{173} > 10^{15{,}408}$$

The main issue with the *Maple* code was long run times, although memory issues were beginning to creep in. An attempt was made to parallelise the algorithm using an early version of *Maple*'s inbuilt multithreading capabilities. At the time, *Maple* 13 was the current version. Unfortunately this attempt failed due to poor machine memory management, which was most likely a result of the infancy of the threading library.

Significant execution time improvements were achieved with a C++ implementation. In contrast to the authors' earlier experience with C++ (described above, and in [8]) the implementation was now easy and straightforward, took only a day or two, and worked extremely well. This implementation was able to compute j_{173} in a little under ten minutes. This implementation stored only the leaves of the tree and in the interests of simplicity of code, did not store the tree after the computation had completed.

Remark 13. We note here that a second, equivalent approach to exploiting Theorem 10 was experimented with at this time. The variant is to build the tree up from the root vertex $\{\}$ by iteratively computing only the immediate children of the vertex with the lowest j-value, favoring vertices closer to the root in the case of a tie. This variation has the advantage that it always processes the least number of vertices possible in any execution. In practice, it showed small (but essentially insignificant) speed improvements over the method described above, most likely due to the exceptionally good bounds provided by the L_d sets. Unfortunately, this variation did not easily lend itself to a threaded (parallel) implementation and has not been used since.

Further execution time improvements were achieved by threading the code. Threading was achieved via use of Apple's threading library "Grand Central Dispatch", which conceals the traditional low-level tools of threading (semaphores, threads, mutexes, etc) and instead presents a much more intuitive method of passing code blocks into work queues, allowing the library to do the rest (see [3, 4] for more information). This allowed for a quick and painless threading implementation and avoided nearly all of the hurdles normally associated with introducing threading to an existing program. It took approximately two days to implement the threading, and this new threaded implementation could compute j_{173} in approximately two minutes.

A grid implementation, running on an Apple X-grid cluster, was experimented with. Unfortunately the grid facilities lacked any easy interprocess communications and the problem did not divide into evenly sized sub-problems. This avenue was abandoned in favor of improving the already working single-machine threaded implementation.

3.1. Final Computations

With this implementation we were able to compute to depth 280 and obtained the following results.

$$j_{280} = 4{,}581 \qquad\qquad P_{280} > 10^{19{,}023}$$

During the computation to depth 280 our implementation was exceeding the memory of the computer it was running on. Computation speed drastically slowed due to excessive paging. Unfortunately, no detailed records were kept of this computation, however we know the machine had 12 or 14 gigabytes of memory. To fix the paging problem, it was realised that much of the data in memory consisted of leaves which had completed processing. The program was thus modified to write these completed leaves to disk, in the form of a Berkeley database file, to save on RAM. The Berkeley database format was chosen because it already incorporated caching, as well as sorting of the entries (see [15] for more information). The newly modified program was run and computed to depth 301 to obtain the following results:

$$j_{301} = 4{,}669 \qquad\qquad P_{301} > 10^{19{,}432}$$

The computation took approximately ninety five hours.

A side effect of the change was that, since the completed tree was now being written to disk, later computations could pick up where previous computations had left off[1]. Unfortunately, similar memory exhaustion and excessive paging problems had crept back in by the end of the computation. To fix this problem, the program was modified to write all leaves to disk and to only hold in memory those leaves which were currently being processed.

The new implementation was able to compute $j_{311} = 4{,}771$, taking approximately an extra six days[2] and producing an output file of approximately one hundred and twenty gigabytes.

$$j_{311} = 4{,}771 \qquad\qquad P_{311} > 10^{19{,}907}$$

Memory issues are still present, although the most recent implementation delayed them until approximately depth 308. The most recent memory issues appear to be caused not by large amounts of tree data, but by large numbers of jobs waiting to be executed by the threading API[3]. If this is the case, then we begin to see a cost of the easy implementation of threading thanks to the Grand Central Dispatch library. It may be the case, however, that the memory issues are caused by tree data being created faster than the existing data can be written to disk, combined with the serialised writing to disk. Efforts to identify and eradicate the memory problems are ongoing, and no further computations have yet been performed.

We should note the diminishing returns of these fixes to memory exhaustion; each fix has taken longer to implement and has yielded less improvement than the fix which preceded it. Growth of execution time and file output size is discussed in the next section.

4. Runtime Analysis

Due to the unpredictable nature of the placement of primes in general, and of normal pairs of primes in particular, it is very difficult to analyse the runtime and space complexity of the above algorithm algebraically. However, measurements of runtimes and file output sizes strongly suggest the trimmed tree still grows exponentially.

[1]Unfortunately, since this was the first time this implementation was being run, it was unable to exploit this feature and the quoted ninety five hours was to compute the tree in its entirety to depth 301.

[2]That is, the computation which started at depth 301 from the output of the previous implementation and ran through to depth 311 took six days. This brings the total execution time for depth 311 to over ten days.

[3]Tree data is stored on disk and not passed into processing functions. Instead, the first act of the function which processes the leaf is to retrieve the leaf from disk.

The following measurements were taken on a 2.8Ghz quad core i7 processor with 16 gigabytes of RAM, running Mac OS X 10.7 (Lion). The computations were performed by the most recent implementation as described in the previous section—the one used to compute j_{311}. The program computed j_d for all depths $1 \leq d \leq 311$. The ability of this implementation to begin a new computation starting from a previously completed computation was exploited to measure the size of the output file of the tree at each depth. Total computation time up to depth d was calculated by adding together the times for all computations from depth d' to $d' + 1$ for $d' < d$.

The granularity of the time measurements is 1 second and as such the total computation times for small d may be a little unreliable, as the sub-second execution times are rounded to 0 or 1 seconds. This should not affect our conclusions very much when we consider that the individual execution times reach into the minutes at around depth $d = 66$, hours at around depth $d = 95$, and eventually even days.

Of more concern is the timings for individual executions of depths $d > 301$. The computer these computations were performed on was the personal computer of one of the authors, and with the execution of a single step in depth taking close to a day (or more for the later ones), it was sometimes necessary to pause the computation for hours at a time to allow the computer to be used for other purposes. It is unclear whether these pauses were included in the timings or not. As such the timings for those computations are somewhat suspect and have not been used in any of the following analysis. Note that these pauses will not have affected file output sizes and so the file size data for these computations has been kept and used.

Note that for any $\mathcal{N} \in \mathcal{N}_d$ for which $\mathcal{N} \cup \{q_d\}$ is not normal, $\mathcal{N}$ only has one child in the tree (namely, itself), and $j_{d+1}(\mathcal{N}) = j_d(\mathcal{N})$ as a direct result the method by which the j-values are computed. So in the case that the vertex, $\mathcal{N}$, at depth d whose j-value realises the value j_d[4] (i.e., $\mathcal{N} \subset \mathcal{N}_d$ where $j_d(\mathcal{N}) = j_d$), if q_d is not normal to $\mathcal{N}$ we have that $j_{d+1} = j_d$ and no other computation for depth $d + 1$ needs to be performed. The program was modified to recognise this condition and to simply skip ahead to the next depth, $d' > d$ for which $\mathcal{N} \cup \{q_{d'}\}$ is normal. As such the data appears to "miss" some depths, but this is simply the program skipping these trivial tree depths.

When we plot the log of total execution time, or the log of output file size, against depth (see Figure 3) we see an essentially linear pattern, consistent with exponential growth. We expect fluctuation in the number of vertices processed due to the unpredictable nature of which primes will and will not be included in the computation of $j_d(\mathcal{N})$ for any given vertex $\mathcal{N}$, so the observed variation from linear behavior is not inconsistent. Furthermore, the particulars and vagaries of the output file format (a Berkeley database) might also be a source of some variation

[4] In principle there is nothing in the algorithm to prevent multiple sets realising the minimum j-value, but in practice this has not yet been observed.

in the file sizes.

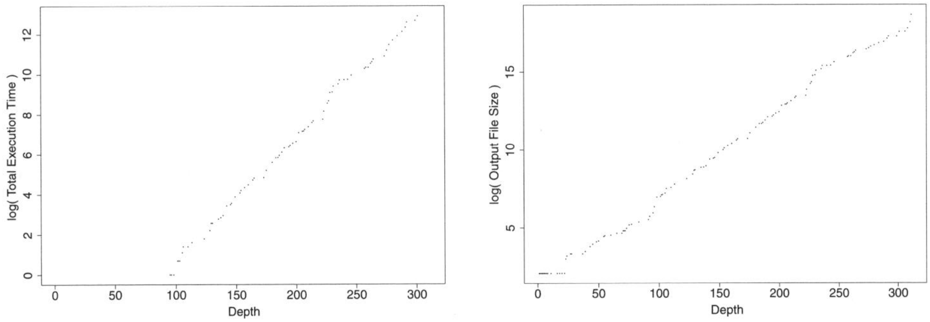

Figure 3: Log-linear plot of cumulative execution time in seconds (left) and output file size in kilobytes (right) against tree depth.

4.1. Predicted Future Performance

Applying a least-squares linear model to this data, we make the following 95% prediction intervals, for the current implementation of the algorithm running on the machine described, above. Total computation time for j_{350} is expected to take between 69 and 188 days and is expected to output a data file of between 490 gigabytes and 2.2 terabytes. We expect the total computation time for j_{400} to be between 4 and 12 years and expect an output file between 7.5 and 36 terabytes in size.

In addition, the following predictions are found by looking at the 95% prediction intervals for predictions of all depths 312 through to 400, and choosing those depths which contain the desired size or time respectively. We predict a terabyte sized output file at a depth between 337 and 364, and we predict a cumulative month long computation will be needed for depths between 321 and 337, and a cumulative year long computation for depths between 361 and 377.

5. Further Work

The current bottleneck in computations appears to be storage; both RAM and hard disk. However, the statistical predictions suggest that time will shortly become the bottleneck for further computations. It is the authors' intention to fix the RAM issues and to compress the output file to address this growth. It is hoped that doing so will delay memory growth enough to once again make execution time the bottleneck . It is also the authors' intention to re-explore grid computation, most probably with the use of the Open MPI API in the hopes that spreading the load across many computers will help alleviate both the time and the storage limitations.

Acknowledgements

The authors would like to thank Dr David Allingham for his support and advice regarding the CARMA computational infrastructure, and Dr Robert King for his advice in interpreting the statistical results.

References

[1] Takashi Agoh. On Giuga's conjecture. *Manuscripta Math.*, 87(4):501–510, 1995.

[2] W. R. Alford, Andrew Granville, and Carl Pomerance. There are infinitely many Carmichael numbers. *Ann. of Math. (2)*, 139(3):703–722, 1994.

[3] Apple Inc. Concurrency programming guide. `http://developer.apple.com/library/ios/#documentation/General/Conceptual/ConcurrencyProgrammingGuide/`, 2013. [Online; accessed 25-Jan-2013].

[4] Apple Inc. Grand central dispatch (gcd) reference. `http://developer.apple.com/library/ios/#documentation/Performance/Reference/GCD_libdispatch_Ref/`, 2013. [Online; accessed 25-Jan-2013].

[5] William D. Banks, C. Wesley Nevans, and Carl Pomerance. A remark on Giuga's conjecture and Lehmer's totient problem. *Albanian J. Math.*, 3(2):81–85, 2009.

[6] D. Borwein, J. M. Borwein, P. B. Borwein, and R. Girgensohn. Giuga's conjecture on primality. *Amer. Math. Monthly*, 103(1):40–50, 1996.

[7] J. M. Borwein and E. Wong. A survey of results relating to Giuga's conjecture on primality. In *Advances in mathematical sciences: CRM's 25 years (Montreal, PQ, 1994)*, volume 11 of *CRM Proc. Lecture Notes*, pages 13–27. Amer. Math. Soc., Providence, RI, 1997.

[8] Jonathan M Borwein, David H Bailey, and Roland Girgensohn. *Experimentation in Mathematics.* Computational Paths to Discovery. A K Peters Ltd, 2004.

[9] Giuseppe Giuga. Su una presumibile proprietà caratteristica dei numeri primi. *Ist. Lombardo Sci. Lett. Rend. Cl. Sci. Mat. Nat. (3)*, 14(83):511–528, 1950.

[10] José María Grau, Florian Luca, and Antonio M. Oller-Marcén. On a variant of Giuga numbers. *Acta Math. Sin. (Engl. Ser.)*, 28(4):653–660, 2012.

[11] José María Grau and Antonio M. Oller-Marcén. Giuga numbers and the arithmetic derivative. *J. Integer Seq.*, 15(4):Article 12.4.1, 5, 2012.

[12] Jos María Grau and Antonio M. Oller-Marcén. Generalizing Giuga's conjecture. *ArXiv e-prints*, March 2011.

[13] Glyn Harman. Watt's mean value theorem and Carmichael numbers. *Int. J. Number Theory*, 4(2):241–248, 2008.

[14] Florian Luca, Carl Pomerance, and Igor Shparlinski. On Giuga numbers. *Int. J. Mod. Math.*, 4(1):13–18, 2009.

[15] Oracle. Oracle berkeley db. `http://www.oracle.com/technetwork/products/berkeleydb/overview/index.html`, 2013. [Online; accessed 25-Jan-2013].

[16] Vicentiu Tipu. A note on Giuga's conjecture. *Canad. Math. Bull.*, 50(1):158–160, 2007.

[17] E. Wong. Computations on normal families of primes. *MSc, Simon Fraser University*, 1997.

CONVOLUTION SUMS INVOLVING LEGENDRE-JACOBI SYMBOL AND DIVISOR FUNCTIONS

Mohamed El Bachraoui
Dept. of Mathematical Sciences, United Arab Emirates University, Al-Ain, United Arab Emirates
melbachraoui@uaeu.ac.ae

Received: 5/10/12, Revised: 1/20/13, Accepted: 9/18/13, Published: 10/10/13

Abstract

We use elementary tools to evaluate a variety of convolution sums which involve the Legendre-Jacobi symbol and the divisor functions.

1. Introduction

Let $\mathbb{N} = \{1, 2, 3, \ldots\}$ and let $\mathbb{N}_0 = \{0, 1, 2, \ldots\}$. We will use the following definitions.

Definition 1. Let the function σ be defined on $\mathbb{Q}$ as follows: $\sigma(0) = 1$, if $n \in \mathbb{Q} \backslash \mathbb{N}_0$, then $\sigma(n) = 0$, and if $n \in \mathbb{N}$, then

$$\sigma(n) = \sum_{d|n} d.$$

Definition 2. Let $n, r \in \mathbb{N}_0$, let $m \in \mathbb{N}$, and let

$$\sigma_{r,m}(n) = \sum_{\substack{d|n \\ d \equiv r \bmod m}} d.$$

We will be using the basic fact that for all $m, n \in \mathbb{N}$

$$\sigma_{0,m}(n) = m\sigma(n/m). \tag{1}$$

Definition 3. Let the function σ^* be defined on $\mathbb{Q}$ as follows: if $n \in \mathbb{Q} \setminus \mathbb{N}$, then $\sigma^*(n) = 0$ and if $n \in \mathbb{N}$, then

$$\sigma^*(n) = \sum_{d|n,\ \frac{n}{d} \text{ odd}} d.$$

By [10, Theorem 3.4] we have for all $n \in \mathbb{N}$ that

$$\sigma^*(n) = \sigma(n) - \sigma(n/2). \tag{2}$$

For an integer n and an odd prime number p we shall use $(n\,|\,p)$ to denote the Legendre-Jacobi symbol of n with respect to p, see for instance Apostol [2] about properties of $(n\,|\,p)$. In this note we shall evaluate a variety of convolution sums involving the Legendre-Jacobi symbol and the divisor functions. Our main tool is the following theorem.

Theorem 1. [6] *Let $\alpha \in \mathbb{N}$. Let $A_1, A_2, \dots A_\alpha \subseteq \mathbb{N}$ and $\underline{A} = (A_1, A_2, \dots, A_\alpha)$ and let $f_i : A_i \to \mathbb{C}$ for $i = 1, 2, \dots, \alpha$ be arithmetic functions and let $\underline{f} = (f_1, f_2, \dots, f_\alpha)$. If both*

$$F_{\underline{A}}(q) = \prod_{i=1}^{\alpha} \prod_{n \in A_i} (1 - q^n)^{-\frac{f_i(n)}{n}} = \sum_{n=0}^{\infty} p_{\underline{A},\underline{f}}(n) q^n$$

and

$$\sum_{i=1}^{\alpha} \sum_{n \in A_i} \frac{f_i(n)}{n} q^n$$

converge absolutely and represent analytic functions in the unit disk $|q| < 1$, then

$$n p_{\underline{A},\underline{f}}(n) = \sum_{k=1}^{n} \left(p_{\underline{A},\underline{f}}(n-k) \sum_{i=1}^{\alpha} f_{i,A_i}(k) \right),$$

where $p_{\underline{A},\underline{f}}(0) = 1$ and

$$f_{i,A_i}(k) = \sum_{\substack{d|k \\ d \in A_i}} f_i(d).$$

We note that Theorem 1 for the special case $\alpha = 1$, has been given in Apostol [2] and in Robbins [9] to give formulas relating arithmetic functions to sums of divisors functions. The authors' key argument is that generating functions for the arithmetic functions they considered have the form of infinite products ranging over a single set of natural numbers. In our previous work [6] we used Theorem 1 to deal with arithmetic functions whose generating functions involve finitely many infinite products ranging over different sets of natural numbers and to derive a variety of inductive formulas for such functions. Among our results in [6] we were able by means of Theorem 1 to reproduce the following result of Liouville [7].

$$(-1)^n s(n) n = \frac{\sigma(n) - \sigma_{1,2}(n)}{2} + \sum_{k \geq 1} (-1)^{k+1} \Big(\sigma(n - k^2) + \sigma_{1,2}(n - k^2) \Big),$$

where

$$s(n) = \begin{cases} 1, & \text{if } n = m^2 \\ 0, & \text{otherwise.} \end{cases}$$

We note further that the infinite products along with their power series expansions which we will use in this paper are all found as theorems in the recent paper of Alaca, Alaca, and Williams [1]. Besides several new results on infinite products of Ramanujan type in [1], the authors reproduced a variety of other results including some due to Carlitz [5] and Ramanujan [8].

2. The Results

From now on we will suppose that n is a positive integer. While the corollaries in this section are given along with their proofs, proofs for the theorems are provided in the next section.

Theorem 2. *We have*

$$(a)\quad \sum_{j=1}^{n-1}\Big(\sum_{d|j} d^2\left(-3\,|\,d\right)\Big)\Big(\sigma(n-j)-\sigma\left(\frac{n-j}{3}\right)\Big)=\frac{1}{9}\sigma(n)-\frac{1}{9}\sigma\left(\frac{n}{3}\right)-\frac{n}{9}\sum_{d|n} d^2\left(-3\,|\,d\right).$$

$$(b)\quad \sum_{j=1}^{n-1}\Big(\sum_{d|j} d^2\left(-3\,|\,\frac{j}{d}\right)\Big)\Big(\sigma(n-j)-9\sigma\left(\frac{n-j}{3}\right)\Big)=\frac{n-1}{3}\sum_{d|n} d^2\left(-3\,|\,\frac{n}{d}\right).$$

Corollary 1. *If p is an odd prime, then*

$$(a)\quad \sum_{j=1}^{p-1}\Big(\sum_{d|j} d^2\left(-3\,|\,d\right)\Big)\Big(\sigma(p-j)-\sigma\left(\frac{p-j}{3}\right)\Big)=\begin{cases}0, & \text{if } p=3\\ \frac{1-p^3}{9}, & \text{if } p\equiv 1 \pmod 3\\ \frac{1+p^3}{9}, & \text{if } p\equiv -1 \pmod 3.\end{cases}$$

$$(b)\quad \sum_{j=1}^{p-1}\Big(\sum_{d|j} d^2\left(-3\,|\,\frac{j}{d}\right)\Big)\Big(\sigma(p-j)-9\sigma\left(\frac{p-j}{3}\right)\Big)=\begin{cases}6, & \text{if } p=3\\ \frac{(p-1)(p^2+1)}{3}, & \text{if } p\equiv 1 \pmod 3\\ \frac{(p-1)(p^2-1)}{3}, & \text{if } p\equiv -1 \pmod 3.\end{cases}$$

Proof. The case $p=3$ is easily checked in both parts. Suppose now that $p\neq 3$. Then as

$$\sigma(p)=p+1,\ \sigma(p/3)=0,\ \text{and}\ \left(-3\,|\,p\right)=\begin{cases}1, & \text{if } p\equiv 1 \pmod 3\\ -1, & \text{if } p\equiv -1 \pmod 3,\end{cases}$$

Theorem 2(a) yields

$$9\sum_{j=1}^{p-1}\Big(\sum_{d|j} d^2\left(-3\,|\,d\right)\Big)\Big(\sigma(p-j)-\sigma\left(\frac{p-j}{3}\right)\Big)=p+1-p\Big(\left(-3\,|\,1\right)+p^2\left(-3\,|\,p\right)\Big)$$

$$=\begin{cases}1-p^3, & \text{if } p\equiv 1 \pmod 3\\ 1+p^3, & \text{if } p\equiv -1 \pmod 3,\end{cases}$$

which after division by 9 gives part (a). Further by Theorem 2 (b) we have

$$\sum_{j=1}^{p-1}\Big(\sum_{d|j} d^2\left(-3\,|\,\tfrac{j}{d}\right)\Big)\Big(\sigma(p-j)-9\sigma\left(\frac{p-j}{3}\right)\Big)=\frac{p-1}{3}\Big(\left(-3\,|\,p\right)+p^2\left(-3\,|\,1\right)\Big)$$

$$=\begin{cases}\frac{(p-1)(1+p^2)}{3}, & \text{if } p\equiv 1 \pmod 3\\ \frac{(p-1)(-1+p^2)}{3}, & \text{if } p\equiv -1 \pmod 3),\end{cases}$$

which proves part (b). □

Theorem 3. *The following identities are true:*

(a) $\displaystyle\sum_{j=1}^{n-1}\Big(\sum_{d|j} d^2\left(-4\,|\,d\right)\Big)\Big(\sigma^*(n-j)+4\sigma^*\left(\frac{n-j}{2}\right)\Big)=\frac{\sigma^*(n)}{4}+\sigma^*\left(\frac{n}{2}\right)-\frac{n}{4}\sum_{d|n} d^2\left(-4\,|\,d\right).$

(b) $\displaystyle\sum_{j=1}^{n-1}\Big(\sum_{d|j} d^2\left(-4\,|\,\tfrac{j}{d}\right)\Big)\Big(\sigma(n-j)-3\sigma\left(\frac{n-j}{2}\right)-4\sigma\left(\frac{n-j}{4}\right)\Big)=\frac{n-1}{4}\sum_{d|n} d^2\left(-4\,|\,\frac{n}{d}\right).$

(c) $\displaystyle\sum_{j=1}^{n-1}\Big(\sum_{d|j}\left(-4\,|\,d\right)\Big)\Big(\sigma^*(n-j)-4\sigma^*\left(\frac{n-j}{2}\right)\Big)=-\frac{1}{4}\sigma^*(n)+\sigma^*\left(\frac{n}{2}\right)+\frac{n}{4}\sum_{d|n}\left(-4\,|\,d\right).$

Corollary 2. *If p is an odd prime, then ;*

(a) $$\sum_{j=1}^{p-1}\Big(\sum_{d|j} d^2\left(-4\,|\,d\right)\Big)\Big(\sigma^*(p-j)+4\sigma^*\left(\frac{p-j}{2}\right)\Big)=\begin{cases}\frac{1-p^3}{4}, & \textit{if } p\equiv 1 \pmod 4\\ \frac{1+p^3}{4}, & \textit{if } p\equiv -1 \pmod 4).\end{cases}$$

(b) $$\sum_{j=1}^{p-1}\Big(\sum_{d|j} d^2\left(-4\,|\,\tfrac{j}{d}\right)\Big)\Big(\sigma(p-j)-3\sigma\left(\frac{p-j}{2}\right)-4\sigma\left(\frac{p-j}{4}\right)\Big)$$

$$=\begin{cases}\frac{(p-1)(p^2+1)}{4}, & \textit{if } p\equiv 1 \pmod 4\\ \frac{(p-1)(p^2-1)}{4}, & \textit{if } p\equiv -1 \pmod 4).\end{cases}$$

(c) $$\sum_{j=1}^{p-1}\Big(\sum_{d|j}\left(-4\,|\,d\right)\Big)\Big(\sigma^*(p-j)-4\sigma^*\left(\frac{p-j}{2}\right)\Big)=\begin{cases}\frac{(p-1)}{4}, & \textit{if } p\equiv 1 \pmod 4\\ \frac{-p-1}{4}, & \textit{if } p\equiv -1 \pmod 4).\end{cases}$$

Proof. Suppose that p is an odd prime. Then Theorem 3 (a) implies

$$-4\sum_{j=1}^{p-1}\Big(\sum_{d|j} d^2\left(-4\,|\,d\right)\Big)\Big(\sigma^*(p-j)+\sigma^*\left(\frac{p-j}{2}\right)\Big)$$

$$=p\Big(\left(-4\,|\,1\right)+p^2\left(-4\,|\,p\right)\Big)-p-1$$

$$=\begin{cases}p^3-1, & \text{if } p\equiv 1 \pmod 4\\ -p^3-1, & \text{if } p\equiv -1 \pmod 4),\end{cases}$$

which upon division by -4 yields the required formula in part (a). Next by Theorem 3 (b) we get

$$\sum_{j=1}^{p-1}\Big(\sum_{d|j} d^2\,(-4\,|\,\tfrac{j}{d}\,)\Big)\Big(\sigma(p-j)-3\sigma\left(\frac{p-j}{2}\right)-4\sigma\left(\frac{p-j}{4}\right)\Big)$$
$$=\frac{p-1}{4}\Big((-4\,|\,p\,)+p^2\,(-4\,|\,1\,)\Big)$$
$$=\begin{cases}\frac{(p-1)(1+p^2)}{4}, & \text{if } p\equiv 1 \pmod 4\\ \frac{(p-1)(-1+p^2)}{4}, & \text{if } p\equiv -1 \pmod 4,\end{cases}$$

giving part (b). Further by Theorem 3 (c) we have

$$\sum_{j=1}^{p-1}\Big(\sum_{d|j}(-4\,|\,d\,)\Big)\Big(\sigma^*(p-j)-4\sigma^*\left(\frac{p-j}{2}\right)\Big)=\frac{p\Big(1+(-1\,|\,p\,)\Big)-p-1}{4}$$
$$=\begin{cases}\frac{p-1}{4}, & \text{if } p\equiv 1 \pmod 4\\ \frac{-p-1}{4}, & \text{if } p\equiv -1 \pmod 4,\end{cases}$$

which proves part (c). □

Theorem 4. *We have*

(*a*) $$\sum_{j=1}^{n-1}\Big(\sum_{d|j} d\,(5\,|\,d\,)\Big)\Big(\sigma(n-j)-\sigma\left(\frac{n-j}{5}\right)\Big)=\frac{1}{5}\sigma(n)-\frac{1}{5}\sigma\left(\frac{n}{5}\right)-\frac{n}{5}\sum_{d|n} d\,(5\,|\,d\,).$$

(*b*) $$\sum_{j=1}^{n-1}\Big(\sum_{d|j} d\,(5\,|\,\tfrac{j}{d}\,)\Big)\Big(\sigma(n-j)-25\sigma\left(\frac{n-j}{5}\right)\Big)=(n-1)\sum_{d|n} d\,(5\,|\,\tfrac{n}{d}\,).$$

Corollary 3. *If p is an odd prime, then*

(*a*) $$\sum_{j=1}^{p-1}\Big(\sum_{d|j} d\,(5\,|\,d\,)\Big)\Big(\sigma(p-j)-\sigma\left(\frac{p-j}{5}\right)\Big)=\begin{cases}0, & \textit{if } p=5\\ \frac{1-p^2}{5}, & \textit{if } p\equiv \pm 1 \pmod 5\\ \frac{1+p^2}{5}, & \textit{if } p\equiv \pm 2 \pmod 5.\end{cases}$$

(*b*) $$\sum_{j=1}^{p-1}\Big(\sum_{d|j} d\,(5\,|\,\tfrac{j}{d}\,)\Big)\Big(\sigma(p-j)-25\sigma\left(\frac{p-j}{5}\right)\Big)$$

$$=\begin{cases}17, & \textit{if } p=5\\ (p-1)(p+5), & \textit{if } p\equiv \pm 1 \pmod 5\\ (p-1)(p-5), & \textit{if } p\equiv \pm 2 \pmod 5.\end{cases}$$

Proof. The results are easily checked if $p = 5$. Now suppose that $p \neq 5$ is an odd prime. Then by Theorem 4(a) we find

$$\sum_{j=1}^{p-1}\Big(\sum_{d|j} d\,(5\,|\,d)\Big)\Big(\sigma(p-j)-\sigma\Big(\frac{p-j}{5}\Big)\Big) = \frac{-p\Big((5\,|\,1)+p\,(5\,|\,p)\Big)+p+1}{5}$$

$$= \begin{cases} \frac{1-p^2}{5}, & \text{if } p \equiv \pm 1 \pmod 5 \\ \frac{1+p^2}{5}, & \text{if } p \equiv \pm 2 \pmod 5, \end{cases}$$

which proves part (a). Part (b) follows similarly by Theorem 4(b). □

Theorem 5. *The following formulas hold:*

(a) $$\sum_{j=1}^{n-1}\Big(\sum_{d|j} d\,(8\,|\,d)\Big)\Big(\sigma^*(n-j)+2\sigma^*\Big(\frac{n-j}{2}\Big)+8\sigma^*\Big(\frac{n-j}{4}\Big)\Big)$$

$$= \frac{1}{2}\sigma^*(n)+\sigma^*\Big(\frac{n}{2}\Big)+4\sigma^*\Big(\frac{n}{4}\Big)-\frac{n}{2}\sum_{d|n} d\,(8\,|\,d)\,.$$

(b) $$\sum_{j=1}^{n-1}\Big(\sum_{d|j} d\,(8\,|\,\tfrac{j}{d})\Big)\Big(\sigma(n-j)-3\sigma\Big(\frac{n-j}{2}\Big)-2\sigma\Big(\frac{n-j}{4}\Big)-8\sigma\Big(\frac{n-j}{8}\Big)\Big)$$

$$= \frac{n-1}{2}\sum_{d|n} d\,(8\,|\,\tfrac{j}{d})\,.$$

(c) $$\sum_{j=1}^{n-1}\Big(\sum_{d|j} (-8\,|\,d)\Big)\Big(\sigma^*(n-j)-2\sigma^*\Big(\frac{n-j}{2}\Big)-8\sigma^*\Big(\frac{n-j}{4}\Big)\Big)$$

$$= -\sigma^*(n)+\sigma^*\Big(\frac{n}{2}\Big)+8\sigma^*\Big(\frac{n}{4}\Big)+n\sum_{d|n}(-8\,|\,d)\,.$$

Corollary 4. *If p is an odd prime, then*

(a) $$\sum_{j=1}^{p-1}\Big(\sum_{d|j} d\,(8\,|\,d)\Big)\Big(\sigma^*(p-j)+2\sigma^*\Big(\frac{p-j}{2}\Big)+8\sigma^*\Big(\frac{p-j}{4}\Big)\Big)$$

$$= \begin{cases} \frac{1-p^2}{2}, & \text{if } p \equiv \pm 1 \pmod 8 \\ \frac{1+p^2}{2}, & \text{if } p \equiv \pm 3 \pmod 8. \end{cases}$$

(b) $$\sum_{j=1}^{p-1}\Big(\sum_{d|j} d\,(8\,|\,\tfrac{j}{d})\Big)\Big(\sigma(p-j)-3\sigma\Big(\frac{p-j}{2}\Big)-2\sigma\Big(\frac{p-j}{4}\Big)-8\sigma\Big(\frac{p-j}{8}\Big)\Big)$$

$$= \begin{cases} \frac{p^2-1}{2}, & \text{if } p \equiv \pm 1 \pmod 8 \\ \frac{(p-1)^2}{2}, & \text{if } p \equiv \pm 3 \pmod 8. \end{cases}$$

$$(c)\quad \sum_{j=1}^{p-1}\Big(\sum_{d\mid j}(-8\,|\,d)\Big)\Big(\sigma^*(p-j)-2\sigma^*\Big(\frac{p-j}{2}\Big)-8\sigma^*\Big(\frac{p-j}{4}\Big)\Big)$$
$$=\begin{cases}\frac{p-1}{2}, & \textit{if } p\equiv 1,3 \pmod 8\\ \frac{-p-1}{2}, & \textit{if } p\equiv -1,-3 \pmod 8.\end{cases}$$

Proof. By Theorem 5(a) we get

$$\sum_{j=1}^{p-1}\Big(\sum_{d\mid j}d\,(8\,|\,d)\Big)\Big(\sigma^*(p-j)+2\sigma^*\Big(\frac{p-j}{2}\Big)+8\sigma^*\Big(\frac{p-j}{4}\Big)\Big)$$
$$=\frac{-1}{2}\Big(p\big((8\,|\,1)+p\,(8\,|\,p)\big)-p-1\Big)=\begin{cases}\frac{1-p^2}{2}, & \text{if } p\equiv \pm 1 \pmod 8\\ \frac{1+p^2}{2}, & \text{if } p\equiv \pm 3 \pmod 8,\end{cases}$$

which proves part (a). Parts (b) and (c) follow similarly by Theorem 5, parts (b) and (c). □

Theorem 6. *The following identities are valid:*

$$(a)\quad \sum_{j=1}^{n-1}\Big(\sum_{d\mid j}d\,(12\,|\,d)\Big)\Big(\sigma(n-j)+3\sigma\Big(\frac{n-j}{3}\Big)$$
$$+8\sigma\Big(\frac{n-j}{4}\Big)+12\sigma\Big(\frac{n-j}{6}\Big)-24\sigma\Big(\frac{n-j}{12}\Big)\Big)$$
$$=\sigma(n)+3\sigma\Big(\frac{n}{3}\Big)+8\sigma\Big(\frac{n}{4}\Big)+12\sigma\Big(\frac{n}{6}\Big)-24\sigma\Big(\frac{n}{12}\Big)-n\sum_{d\mid n}d\,(12\,|\,d).$$

$$(b)\quad \sum_{j=1}^{n-1}\Big(\sum_{d\mid j}d\,(12\,|\,\tfrac{j}{d})\Big)\Big(\sigma(n-j)-2\sigma\Big(\frac{n-j}{2}\Big)$$
$$-3\sigma\Big(\frac{n-j}{3}\Big)-2\sigma\Big(\frac{n-j}{4}\Big)-6\sigma\Big(\frac{n-j}{12}\Big)\Big)$$
$$=\frac{n-1}{2}\sum_{d\mid n}d\,(12\,|\,\tfrac{n}{d}).$$

Corollary 5. *If p is an odd prime, then*

$$(a)\quad \sum_{j=1}^{p-1}\Big(\sum_{d\mid j}d\,(12\,|\,d)\Big)\Big(\sigma(p-j)+3\sigma\Big(\frac{p-j}{3}\Big)$$
$$+8\sigma\Big(\frac{p-j}{4}\Big)+12\sigma\Big(\frac{p-j}{6}\Big)-24\sigma\Big(\frac{p-j}{12}\Big)\Big)$$
$$=\begin{cases}4, & \textit{if } p=3\\ 1-p^2, & \textit{if } p\equiv \pm 1 \pmod{12}\\ 1+p^2, & \textit{if } p\equiv \pm 5 \pmod{12}.\end{cases}$$

$$(b)\quad \sum_{j=1}^{p-1}\Big(\sum_{d|j} d\,(\,12\,|\,\tfrac{j}{d}\,)\Big)\Big(\sigma(p-j)-2\sigma\left(\frac{p-j}{2}\right)$$

$$-3\sigma\left(\frac{p-j}{3}\right)-2\sigma\left(\frac{p-j}{4}\right)-6\sigma\left(\frac{p-j}{12}\right)\Big)$$

$$=\begin{cases}2, & \textit{if}\ p=3\\ \frac{p^2-1}{2}, & \textit{if}\ p\equiv\pm1 \pmod{12}\\ \frac{(p-1)^2}{2}, & \textit{if}\ p\equiv\pm5 \pmod{12}.\end{cases}$$

Proof. Part (a) follows by Theorem 6(a) and part (b) follows by Theorem 6(b) using the same sort arguments as before. □

Theorem 7. *We have*

$$\sum_{j=1}^{n-1}\Big(\sum_{d|j}(\,-15\,|\,d\,)\Big)\Big(\sigma(n-j)-6\sigma\left(\frac{n-j}{3}\right)-10\sigma\left(\frac{n-j}{5}\right)+15\sigma\left(\frac{n-j}{15}\right)\Big)$$

$$=-\sigma(n)+6\sigma\left(\frac{n}{3}\right)+10\sigma\left(\frac{n}{5}\right)-15\sigma\left(\frac{n}{15}\right)+n\sum_{d|n}(\,-15\,|\,d\,)\,.$$

An immediate consequence is the next corollary.

Corollary 6. *If p is an odd prime, then we have*

$$\sum_{j=1}^{p-1}\Big(\sum_{d|j}(\,-15\,|\,d\,)\Big)\Big(\sigma(p-j)-6\sigma\left(\frac{p-j}{3}\right)-10\sigma\left(\frac{p-j}{5}\right)+15\sigma\left(\frac{p-j}{15}\right)\Big)$$

$$=\begin{cases}5, & \textit{if}\ p=3\\ 9, & \textit{if}\ p=5\\ p-1, & \textit{if}\ p\equiv 1,17,19,23,-7,-11,-13,-29 \pmod{60}\\ -p-1, & \textit{if}\ p\equiv 7,11,13,29,-1,-17,-19,-23 \pmod{60}.\end{cases}$$

Theorem 8. *We have*

$$\sum_{j=1}^{n-1}\Big(\sum_{d|j}(\,-20\,|\,d\,)\Big)\Big(\sigma(n-j)-2\sigma\left(\frac{p-j}{2}\right)-4\sigma\left(\frac{n-j}{4}\right)$$

$$-5\sigma\left(\frac{n-j}{5}\right)-10\sigma\left(\frac{n-j}{10}\right)+20\sigma\left(\frac{n-j}{20}\right)\Big)$$

$$=-\sigma(n)+2\sigma\left(\frac{n}{2}\right)+4\sigma\left(\frac{n}{4}\right)+5\sigma\left(\frac{n}{5}\right)+10\sigma\left(\frac{n}{10}\right)-20\sigma\left(\frac{n}{20}\right)+n\sum_{d|n}(\,-20\,|\,d\,)\,.$$

An immediate consequence is the next corollary.

Corollary 7. *If p is an odd prime, then*

$$\sum_{j=1}^{p-1}\Big(\sum_{d|j}(-20\,|\,d)\Big)\Big(\sigma(p-j)-2\sigma\left(\frac{p-j}{2}\right)-4\sigma\left(\frac{p-j}{4}\right)-5\sigma\left(\frac{p-j}{5}\right)-10\sigma\left(\frac{p-j}{10}\right)+20\sigma\left(\frac{p-j}{20}\right)\Big)$$
$$=\begin{cases}4, & \text{if } p=5\\ p-1, & \text{if } p\equiv 1,3,7,9 \pmod{20}\\ -p-1, & \text{if } p\equiv -1,-3,-7,-9 \pmod{20}.\end{cases}$$

Theorem 9. *We have*

$$\sum_{j=1}^{n-1}\Big(\sum_{d|j}(-24\,|\,d)\Big)\Big(\sigma(n-j)-2\sigma\left(\frac{n-j}{2}\right)-3\sigma\left(\frac{n-j}{3}\right)-8\sigma\left(\frac{n-j}{8}\right)-12\sigma\left(\frac{n-j}{12}\right)+24\sigma\left(\frac{n-j}{24}\right)\Big)$$
$$=-\sigma(n)+2\sigma\left(\frac{n}{2}\right)+3\sigma\left(\frac{n}{3}\right)+8\sigma\left(\frac{n}{8}\right)-12\sigma\left(\frac{n}{12}\right)-24\sigma\left(\frac{n}{24}\right)+n\sum_{d|n}(-24\,|\,d).$$

We have the following consequence.

Corollary 8. *If p is an odd prime, then we have*

$$\sum_{j=1}^{p-1}\Big(\sum_{d|j}(-24\,|\,d)\Big)\Big(\sigma(p-j)-2\sigma\left(\frac{p-j}{2}\right)-3\sigma\left(\frac{p-j}{3}\right)-8\sigma\left(\frac{p-j}{8}\right)-12\sigma\left(\frac{p-j}{12}\right)+24\sigma\left(\frac{p-j}{24}\right)\Big)$$
$$=\begin{cases}2, & \text{if } p=3\\ p-1, & \text{if } p\equiv 1,5,7,11 \pmod{24}\\ -p-1, & \text{if } p\equiv -1,-5,-7,-11 \pmod{24}.\end{cases}$$

3. Proofs of the Theorems

Throughout this section we shall suppose that $q\in\mathbb{C}$ such that $|q|<1$.

Proof of Theorem 2. The following two identities are due to Carlitz [5] and new proofs have been reproduced in Alaca *et al.* [1, Theorems 2.2 and 2.3].

$$\prod_{n=1}^{\infty}(1-q^n)^9(1-q^{3n})^{-3}=\sum_{n=0}^{\infty}a(n)q^n=1-9\sum_{n=1}^{\infty}\Big(\sum_{d\mid n}d^2\left(-3\,|\,d\right)\Big)q^n, \quad (3)$$

where

$$a(0)=1 \text{ and } a(n)=-9\sum_{d\mid n}d^2\left(-3\,|\,d\right).$$

$$\prod_{n=1}^{\infty}(1-q^{3n})^9(1-q^n)^{-3}=\sum_{n=0}^{\infty}b(n)q^n=\sum_{n=1}^{\infty}\Big(\sum_{d\mid n}d^2\left(-3\,|\,\tfrac{n}{d}\right)\Big)q^{n-1}, \quad (4)$$

where

$$b(0)=1 \text{ and } b(n)=\sum_{d\mid(n+1)}d^2\left(-3\,|\,\frac{n+1}{d}\right).$$

To prove part (a) we apply Theorem 1 to the identity (3) with the sets $A_1=\mathbb{N}$ and $A_2=3\mathbb{N}$ and the functions $f_1(k)=-9k$ on A_1 and $f_2(k)=3k$ on A_2 to get

$$\begin{aligned} na(n)&=-9\sigma(n)+3\sigma_{0,3}(n)+\sum_{j=1}^{n-1}a(j)\big(-9\sigma(n-j)+3\sigma_{0,3}(n-j)\big)\\ &=-9\sigma(n)+9\sigma(n/3)-9\sum_{j=1}^{n-1}a(j)\Big(\sigma(n-j)-\sigma\left(\frac{n-j}{3}\right)\Big),\end{aligned} \quad (5)$$

where the second identity follows by the relation (1). Now use the definition of $a(j)$ and divide by 9 to conclude the desired formula. As to part (b), apply Theorem 1 to the identity (4) for the sets $A_1=3\mathbb{N}$ and $A_2=\mathbb{N}$ and the functions $f_1(k)=-9k$ on A_1 and $f_2(k)=3k$ on A_2 to get

$$nb(n)=\sum_{j=0}^{n-1}b(j)\big(-9\sigma_{0,3}(n-j)+3\sigma(n-j)\big)=3\sum_{j=0}^{n-1}b(j)\Big(\sigma(n-j)-9\sigma\left(\frac{n-j}{3}\right)\Big). \quad (6)$$

Equivalently,

$$\sum_{j=1}^{n}\Big(\sum_{d\mid j}d^2\left(-3\,|\,\tfrac{j}{d}\right)\Big)\Big(\sigma(n+1-j)-9\sigma\left(\frac{n+1-j}{3}\right)\Big)=\frac{n}{3}\sum_{d\mid n+1}d^2\left(-3\,|\,\frac{n+1}{d}\right),$$

as desired. □

Proof of Theorem 3. The following two identities are due to Carlitz [5], see [1, Theorems 2.4 and 2.5] for new proofs.

$$\prod_{n=1}^{\infty}(1-q^n)^4(1-q^{2n})^6(1-q^{4n})^{-4}=\sum_{n=0}^{\infty}a(n)q^n=1-4\sum_{n=1}^{\infty}\Big(\sum_{d\mid n}d^2\left(-4\,|\,d\right)\Big)q^n, \quad (7)$$

where

$$a(0)=1 \text{ and } a(n)=-4\sum_{d|n} d^2\left(-4\,|\,d\right).$$

$$\prod_{n=1}^{\infty}(1-q^{2n})^6(1-q^{4n})^4(1-q^n)^{-4}=\sum_{n=0}^{\infty} b(n)q^n=\sum_{n=1}^{\infty}\Big(\sum_{d|n} d^2\left(-4\,|\,\frac{n}{d}\right)\Big)q^{n-1}, \quad (8)$$

where

$$b(0)=1 \text{ and } b(n)=\sum_{d|(n+1)} d^2\left(-4\,|\,\frac{n+1}{d}\right).$$

By virtue of Theorem 1 applied to the formula (7) and identity (2) we obtain

$$\begin{aligned} na(n) &= -4\sigma(n)-12\sigma(n/2)+16\sigma(n/4)+\\ &\sum_{j=1}^{n-1} a(j)\Big(-4\sigma(n-j)-12\sigma\left(\frac{n-j}{2}\right)+16\sigma\left(\frac{n-j}{4}\right)\Big) \\ &= -4\sigma^*(n)-16\sigma^*(n/2)-4\sum_{j=1}^{n-1} a(j)\Big(\sigma^*(n-j)+4\sigma^*\left(\frac{n-j}{2}\right)\Big), \end{aligned} \quad (9)$$

which is equivalent to

$$n\sum_{d|n} d^2\left(-4\,|\,d\right)=\sigma^*(n)+4\sigma^*(n/2)-4\sum_{j=1}^{n-1}\Big(\sum_{d|j} d^2\left(-4\,|\,d\right)\Big)\Big(\sigma^*(n-j)+4\sigma^*\left(\frac{n-j}{2}\right)\Big),$$

giving part (a). As to part (b), applying Theorem 1 to the relation (8) and proceeding as before we find

$$\sum_{j=1}^{n}\Big(\sum_{d|j} d^2\left(-4\,|\,\frac{j}{d}\right)\Big)\Big(\sigma(n+1-j)-3\sigma\left(\frac{n+1-j}{2}\right)-4\sigma\left(\frac{n+1-j}{4}\right)\Big)$$

$$=\frac{n}{4}\sum_{d|n+1} d^2\left(-4\,|\,\frac{n+1}{d}\right),$$

as desired. As to part (c), we have by [1, Theorem 4.5]

$$\prod_{n=1}^{\infty}(1-q^{2n})^{10}(1-q^n)^{-4}(1-q^{4n})^{-4}=1+4\sum_{n=1}^{\infty}\Big(\sum_{d|n}\left(-4\,|\,d\right)\Big)q^n, \quad (10)$$

which by Theorem 1 and the relation (2) translates into

$$16\sum_{j=1}^{n-1}\Big(\sum_{d|j}\left(-4\,|\,d\right)\Big)\Big(\sigma^*(n-j)-4\sigma^*\left(\frac{n-j}{2}\right)\Big)$$

$$= -4\sigma^*(n) + 16\sigma^*\left(\frac{n}{2}\right) + 4n\sum_{d|n}(-4\,|\,d);$$

yielding the desired identity. □

Proof of Theorem 4. The following two identities are due to Ramanujan, see Alaca *et al.* [1, Theorems 3.2 and 3.3] for proofs based on work by Bailey [3, 4]:

$$\prod_{n=1}^{\infty}(1-q^n)^5(1-q^{5n})^{-1} = 1 - 5\sum_{n=1}^{\infty}\Big(\sum_{d|n} d\,(5\,|\,d)\Big)q^n. \tag{11}$$

$$\prod_{n=1}^{\infty}(1-q^{5n})^5(1-q^n)^{-1} = \sum_{n=1}^{\infty}\Big(\sum_{d|n} d\,(5\,|\,\frac{n}{d})\Big)q^{n-1}. \tag{12}$$

By Theorem 1 applied to (11) we have

$$-5n\sum_{d|n} d\,(5\,|\,d) = -5\sigma(n)+5\sigma(n/5)+\sum_{j=1}^{n-1}\Big(-5\sum_{d|j} d\,(5\,|\,d)\Big)\Big(-5\sigma(n-j)+5\sigma\left(\frac{n-j}{5}\right)\Big),$$

or equivalently,

$$\sum_{j=1}^{n-1}\Big(\sum_{d|j} d\,(5\,|\,d)\Big)\Big(\sigma(n-j)-\sigma\left(\frac{n-j}{5}\right)\Big) = \frac{1}{5}\sigma(n) - \frac{1}{5}\sigma\left(\frac{n}{5}\right) - \frac{n}{5}\sum_{d|n} d\,(5\,|\,d),$$

which gives part (a). As to part (b), apply Theorem 1 to identity (12) to get

$$n\sum_{d|n+1} d\,(5\,|\,\frac{n+1}{d}) = \sum_{j=1}^{n}\Big(\sum_{d|j} d\,(5\,|\,\frac{j}{d})\Big)\Big(\sigma(n+1-j) - 25\sigma\left(\frac{n+1-j}{5}\right)\Big),$$

and proceed as in part (a). □

Proof of Theorem 5. We have the following two formulas of Carlitz [5] which appear as Theorem 3.4 and Theorem 3.5 respectively in Alaca *et al.* [1]:

$$\prod_{n=1}^{\infty}(1-q^n)^2(1-q^{2n})(1-q^{4n})^3(1-q^{8n})^{-2} = 1 - 2\sum_{n=1}^{\infty}\Big(\sum_{d|n} d\,(8\,|\,d)\Big)q^n, \tag{13}$$

$$\prod_{n=1}^{\infty}(1-q^{2n})^3(1-q^{4n})(1-q^{8n})^2(1-q^n)^{-2} = \sum_{n=1}^{\infty}\Big(\sum_{d|n} d\,(8\,|\,\frac{n}{d})\Big)q^{n-1}. \tag{14}$$

Part (a) follows by Theorem 1 applied to (13) and part (b) follows by Theorem 1 applied to (14). To prove part (c) use the same sort of argument with an application of Theorem 1 to the following identity which is Theorem 4.6 in [1]:

$$\prod_{n=1}^{\infty}(1-q^{2n})^3(1-q^{4n})^3(1-q^n)^{-2}(1-q^{8n})^{-2} = 1 + 2\sum_{n=1}^{\infty}\Big(\sum_{d|n}(-8\,|\,d)\Big)q^n.$$

□

Proof of Theorem 6. We proceed as in the previous proofs. We have the following two identities by Alaca *et al.* [1, Theorems 3.6 and 3.7]:

$$\prod_{n=1}^{\infty}(1-q^n)(1-q^{3n})(1-q^{4n})^2(1-q^{6n})^2(1-q^{12n})^{-2} = 1 - \sum_{n=1}^{\infty}\Big(\sum_{d|n} d\,(12\,|\,d)\Big)q^n,$$

$$\prod_{n=1}^{\infty}(1-q^{2n})^2(1-q^{3n})^2(1-q^{4n})(1-q^{12n})(1-q^n)^{-2} = \sum_{n=1}^{\infty}\Big(\sum_{d|n} d\,(12\,|\,\frac{n}{d})\Big)q^{n-1}.$$

Then application of Theorem 1 to the first identity yields part (a) and application of Theorem 1 to the second identity yields part (b). □

Proof of Theorem 7. Just as before, apply Theorem 1 to the following result of Alaca *et al.* [1, Theorem 4.2].

$$\prod_{n=1}^{\infty}(1-q^{3n})^2(1-q^{5n})^2(1-q^n)^{-1}(1-q^{15n})^{-1} = 1 + \sum_{n=1}^{\infty}\Big(\sum_{d|n}(-15\,|\,d)\Big)q^n.$$

□

Proof of Theorem 8. By [1, Theorem 4.3] we have

$$\prod_{n=1}^{\infty}(1-q^{2n})(1-q^{4n})(1-q^{5n})(1-q^{10n})(1-q^n)^{-1}(1-q^{20n})^{-1} = 1 + \sum_{n=1}^{\infty}\Big(\sum_{d|n}(-20\,|\,d)\Big)q^n,$$

which by virtue of Theorem 1 for an odd prime p yields the desired identity. □

Proof of Theorem 9. Combine Theorem 1 with the following formula [1, Theorem 4.4]

$$\prod_{n=1}^{\infty}(1-q^{2n})(1-q^{3n})(1-q^{8n})(1-q^{12n})(1-q^n)^{-1}(1-q^{24n})^{-1} = 1 + \sum_{n=1}^{\infty}\Big(\sum_{d|n}(-24\,|\,d)\Big)q^n.$$

□

Acknowledgment The author is grateful to the referee for valuable comments and interesting suggestions.

References

[1] A. Alaca, S. Alaca, and K. Williams, *Some infinite products of Ramanujan type*, Canad. Math. Bull. Vol. 52 (4), (2009), 481-492.

[2] T. M. Apostol, *Introduction to Analytic Number Theory*, Undergraduate Texts in Mathematics, Springer, first edition, 1976.

[3] W. N. Bailey, *A note on two of Ramanujan's formulae*, Quart. J. Math. Oxford Ser. (2) **3**(1952), 29-31.

[4] W. N. Bailey, *A further note on two of Ramanujan's formulae*, Quart. J. Math. Oxford Ser. (2) **3**(1952), 158-160.

[5] L. Carlitz, *Note on some partition formulae*, Quart. J. Math. Oxford Ser. (2) **4**(1953), 168-172.

[6] M. El Bachraoui, *Inductive formulas for some arithmetic functions*, International Journal of Number Theory 7, Number 8, (2011), 2261-2268.

[7] J. Liouville, *Sur quelques formules générales qui peuvent être utiles dans la théorie des nombres*, (7th article), J. Math. Pures Appl. 4, (1859), 1-8.

[8] S. Ramanujan, *The lost notebook and other unpublished papers*, Narosa Publishing House, New Delhi, 1988.

[9] N. Robbins, *Some identities connecting partition functions to other number theoretic functions*, Rocky Mountain Journal of Mathematics, Volume 29, Number 1, (1999), 335-345.

[10] K. S. Williams, *Number Theory in the Spirit of Liouville*, Cambridge University Press, New York, First edition, 2011.

OPTIMAL PRIMITIVE SETS WITH RESTRICTED PRIMES

William D. Banks
Department of Mathematics, University of Missouri, Columbia, Missouri
bankswd@missouri.edu

Greg Martin
Department of Mathematics, University of British Columbia, Vancouver, Canada
gerg@math.ubc.ca

Received: 1/8/13, Accepted: 9/28/13, Published: 10/10/13

Abstract

A set of natural numbers is *primitive* if no element of the set divides another. Erdős conjectured that if $\mathcal{S}$ is any primitive set, then

$$\sum_{n\in\mathcal{S}} \frac{1}{n\log n} \leqslant \sum_{p\in\mathbb{P}} \frac{1}{p\log p},$$

where $\mathbb{P}$ denotes the set of primes. In this paper, we make progress towards this conjecture by restricting the setting to smaller sets of primes. Let $\mathcal{P}$ denote any subset of $\mathbb{P}$, and let $\mathbb{N}(\mathcal{P})$ denote the set of natural numbers all of whose prime factors are in $\mathcal{P}$. We say that $\mathcal{P}$ is *Erdős-best among primitive subsets of* $\mathbb{N}(\mathcal{P})$ if the inequality

$$\sum_{n\in\mathcal{S}} \frac{1}{n\log n} \leqslant \sum_{p\in\mathcal{P}} \frac{1}{p\log p}$$

holds for every primitive set $\mathcal{S}$ contained in $\mathbb{N}(\mathcal{P})$. We show that if the sum of the reciprocals of the elements of $\mathcal{P}$ is small enough, then $\mathcal{P}$ is Erdős-best among primitive subsets of $\mathbb{N}(\mathcal{P})$. As an application, we prove that the set of twin primes exceeding 3 is Erdős-best among the corresponding primitive sets.

This problem turns out to be related to a similar problem involving multiplicative weights. For any real number $t > 1$, we say that $\mathcal{P}$ is *t-best among primitive subsets of* $\mathbb{N}(\mathcal{P})$ if the inequality

$$\sum_{n\in\mathcal{S}} n^{-t} \leqslant \sum_{p\in\mathcal{P}} p^{-t}$$

holds for every primitive set $\mathcal{S}$ contained in $\mathbb{N}(\mathcal{P})$. We show that if the sum on the right-hand side of this inequality is small enough, then $\mathcal{P}$ is t-best among primitive subsets of $\mathbb{N}(\mathcal{P})$.

1. Introduction

A nonempty set of natural numbers is called *primitive* if no element of the set divides another (for later convenience, we stipulate that the singleton set $\{1\}$ is not primitive). In 1935, Erdős [3] established the convergence of the sum of $1/(n \log n)$ over all elements n of a given primitive set; from this he deduced that the lower asymptotic density of a primitive set must equal 0 (in contrast to the upper density, which can be positive, as shown by Besikovitch [1]). Erdős actually proved that this sum of $1/(n \log n)$ is bounded by a universal constant:

$$\sup_{\mathcal{S} \text{ primitive}} \sum_{n \in \mathcal{S}} \frac{1}{n \log n} < \infty.$$

Noting that the set $\mathbb{P}$ of all primes is itself primitive and contains many small elements, Erdős proposed that the supremum on the left-hand side is attained when $\mathcal{S} = \mathbb{P}$.

Conjecture 1 (Erdős). For any primitive set $\mathcal{S}$, we have $\displaystyle\sum_{n \in \mathcal{S}} \frac{1}{n \log n} \leqslant \sum_{p \in \mathbb{P}} \frac{1}{p \log p}$.

This conjecture is still open, although it has been established for primitive sets $\mathcal{S}$ with additional properties (see for example [7]).

We consider a generalization of this problem, to primitive sets whose elements are restricted to have only certain prime factors. For a given set of primes $\mathcal{P}$, let $\mathbb{N}(\mathcal{P})$ denote the set of natural numbers divisible only by primes in $\mathcal{P}$ (the multiplicative semigroup generated by $\mathcal{P}$), that is,

$$\mathbb{N}(\mathcal{P}) = \{n \in \mathbb{N} \colon p \mid n \Rightarrow p \in \mathcal{P}\}.$$

We say that $\mathcal{P}$ is *Erdős-best among primitive subsets of* $\mathbb{N}(\mathcal{P})$ if the inequality

$$\sum_{n \in \mathcal{S}} \frac{1}{n \log n} \leqslant \sum_{p \in \mathcal{P}} \frac{1}{p \log p}$$

holds for every primitive set $\mathcal{S}$ contained in $\mathbb{N}(\mathcal{P})$. In this terminology, Conjecture 1 can be restated as the assertion that $\mathbb{P}$ is Erdős-best among primitive subsets of $\mathbb{N}$. A similar heuristic, together with some computational evidence, leads us to generalize the conjecture of Erdős to these restricted sets.

Conjecture 2. Any set of primes $\mathcal{P}$ is Erdős-best among primitive subsets of $\mathbb{N}(\mathcal{P})$.

Our first result shows that this conjecture holds if the sum of the reciprocals of the elements of $\mathcal{P}$ is small enough.

Theorem 3. *Let $\mathcal{P}$ be a set of primes such that*

$$\sum_{p \in \mathcal{P}} p^{-1} \leqslant 1 + \Big(1 - \sum_{p \in \mathcal{P}} p^{-2}\Big)^{1/2}, \tag{1}$$

Then $\mathcal{P}$ is Erdős-best among primitive subsets of $\mathbb{N}(\mathcal{P})$.

Remark. The square root on the right-hand side of the inequality (1) is always well-defined, as $\sum_{p\in\mathcal{P}} p^{-2} \leqslant \sum_{p\in\mathbb{P}} p^{-2} < 1$. In fact, the latter sum can be precisely evaluated using the rapidly converging series

$$\sum_{p\in\mathbb{P}} p^{-2} = \sum_{m=1}^{\infty} \frac{\mu(m)}{m} \log\zeta(2m) = 0.45224742\cdots. \tag{2}$$

We conclude that if $\mathcal{P}$ satisfies the inequality (1) then $\sum_{p\in\mathcal{P}} p^{-1} < 2$, while if $\sum_{p\in\mathcal{P}} p^{-1} \leqslant 1 + \left(1 - \sum_{p\in\mathbb{P}} p^{-2}\right)^{1/2} = 1.74010308\cdots$ then $\mathcal{P}$ satisfies (1).

The following application of Theorem 3 is quickly derived in Section 2.

Corollary 4. *Let $\mathcal{T}$ denote the set of twin primes exceeding 3, that is, the set of all primes $p > 3$ for which $p-2$ or $p+2$ is also prime. Then $\mathcal{T}$ is Erdős-best among primitive subsets of $\mathbb{N}(\mathcal{T})$.*

We find it amusing that we can identify the optimal primitive subset of $\mathbb{N}(\mathcal{T})$ without needing to determine whether that subset is finite or infinite!

We can further generalize this problem by demanding that the integers in our sets have at least a certain number of prime factors. For every natural number k, define

$$\mathbb{N}_k = \{n \in \mathbb{N} \colon \Omega(n) = k\} \quad \text{and} \quad \mathbb{N}_{\geqslant k} = \{n \in \mathbb{N} \colon \Omega(n) \geqslant k\},$$

where as usual $\Omega(n)$ denotes the number of prime factors of n counted with multiplicity; for example, $\mathbb{N}_1 = \mathbb{P}$ and $\mathbb{N}_{\geqslant 1} = \mathbb{N} \setminus \{1\}$. Note that each of the sets $\mathbb{N}_k$ is itself a primitive set. One step towards a proof of Conjecture 1 would thus be to establish the natural conjecture

$$\sum_{p\in\mathbb{P}} \frac{1}{p\log p} > \sum_{n\in\mathbb{N}_2} \frac{1}{n\log n} > \sum_{n\in\mathbb{N}_3} \frac{1}{n\log n} > \cdots,$$

but this is still an open problem: it was shown by Zhang [7] that the first sum over $\mathbb{P}$ is indeed larger than any of the other sums, but even this partial result is nontrivial. However, in the setting of restricted prime factors, we can establish the analogous chain of inequalities and in fact more. For any set $\mathcal{P}$ of primes, define

$$\mathbb{N}_k(\mathcal{P}) = \mathbb{N}(\mathcal{P}) \cap \mathbb{N}_k \quad \text{and} \quad \mathbb{N}_{\geqslant k}(\mathcal{P}) = \mathbb{N}(\mathcal{P}) \cap \mathbb{N}_{\geqslant k},$$

so that we are now simultaneously restricting the allowable prime factors $\mathcal{P}$ and the minimum number of prime factors k.

Theorem 5. *Let $\mathcal{P}$ be a set of primes for which the inequality (1) holds. Then for every natural number k, the set $\mathbb{N}_k(\mathcal{P})$ is Erdős-best among primitive subsets of $\mathbb{N}_{\geqslant k}(\mathcal{P})$.*

Manifestly, Theorem 3 is the special case $k = 1$ of Theorem 5. Since $\mathbb{N}_k(\mathcal{P}) \subset \mathbb{N}_{\geqslant j}(\mathcal{P})$ for all $k \geqslant j$, Theorem 5 implies in particular that when $\mathcal{P}$ is a set of primes satisfying the inequality (1), we have

$$\sum_{p\in\mathcal{P}} \frac{1}{p\log p} > \sum_{n\in\mathbb{N}_2(\mathcal{P})} \frac{1}{n\log n} > \sum_{n\in\mathbb{N}_3(\mathcal{P})} \frac{1}{n\log n} > \cdots. \tag{3}$$

We do not formulate Theorem 5 simply for its own sake: our proof of Theorem 3 requires comparing sets containing elements with different numbers of prime factors, and the chain of inequalities (3) is a stable yardstick upon which these comparisons can be made.

Finally, we modify the problem in yet a different way. Instead of counting an integer n with weight $1/(n \log n)$ in these sums, we may instead count it with weight n^{-t} for some fixed real number t. We can establish an analogue of Theorem 5 for these weights as well. We say that $\mathbb{N}_k(\mathcal{P})$ is *t-best among primitive subsets of* $\mathbb{N}_{\geqslant k}(\mathcal{P})$ if the inequality

$$\sum_{n\in\mathcal{S}} n^{-t} \leqslant \sum_{p\in\mathbb{N}_k(\mathcal{P})} n^{-t}$$

holds for every primitive set $\mathcal{S}$ contained in $\mathbb{N}_k(\mathcal{P})$.

Theorem 6. *Let $t > 1$ be a real number, and let $\mathcal{P}$ be a set of primes satisfying the inequality*

$$\sum_{p\in\mathcal{P}} p^{-t} \leqslant 1 + \Big(1 - \sum_{p\in\mathcal{P}} p^{-2t}\Big)^{1/2}. \tag{4}$$

Then for every natural number k, the set $\mathbb{N}_k(\mathcal{P})$ is t-best among primitive subsets of $\mathbb{N}_{\geqslant k}(\mathcal{P})$.

In fact, it suffices to establish Theorem 6 when $\mathcal{P}$ is finite, as we show in Section 2.

It can be verified that the function

$$\sum_{p\in\mathbb{P}} p^{-t} - 1 - \Big(1 - \sum_{p\in\mathbb{P}} p^{-2t}\Big)^{1/2}$$
$$= \sum_{m=1}^{\infty} \frac{\mu(m)}{m} \log\zeta(tm) - 1 - \Big(1 - \sum_{m=1}^{\infty} \frac{\mu(m)}{m} \log\zeta(2tm)\Big)^{1/2},$$

which is defined for $t > 1$, has a unique zero $\tau = 1.1403659\cdots$ and is positive for all $t > \tau$. Furthermore, by monotonicity, if the inequality (4) is satisfied for $\mathcal{P} = \mathbb{P}$ then it is satisfied for any set of primes. We therefore have the following corollary.

Corollary 7. *If $t > \tau$, then any set of primes $\mathcal{P}$ is t-best among all primitive subsets of* $\mathbb{N}(\mathcal{P})$.

This assertion does not necessarily hold for every t: in fact, if $\sum_{p\in\mathcal{P}} p^{-1}$ diverges, then one can establish, using equation (9) below, the existence of a number $\delta(\mathcal{P}) > 0$ such that $\sum_{p\in\mathcal{P}} p^{-t} < \sum_{n\in\mathbb{N}_2(\mathcal{P})} n^{-t}$ for all t between 1 and $1+\delta(\mathcal{P})$.

These new weighted sums $\sum_{n\in\mathcal{S}} n^{-t}$ are much easier to handle than the original sums $\sum_{n\in\mathcal{S}} 1/(n\log n)$, because n^{-t} is a multiplicative function of n. However, Theorem 6 is not merely analogous to Theorem 5: in the next section we actually derive the latter from the former. Once Section 2 is done, the only remaining task is to prove Theorem 6, which we accomplish in Section 3.

2. Quick Derivations

In this section we give the three quick derivations described in the introduction: first we demonstrate that the infinite case of Theorem 3 follows from the finite case, then we deduce Theorem 5 (of which Theorem 3 is a special case) from Theorem 6, and finally we derive Corollary 4 from Theorem 3. For convenience we introduce the notation

$$\Sigma_t(\mathcal{S}) = \sum_{n\in\mathcal{S}} n^{-t}$$

for any set $\mathcal{S} \subset \mathbb{N}$ and any $t > 1$.

Proof of Theorem 3 for infinite $\mathcal{P}$, assuming Theorem 3 for finite $\mathcal{P}$. Let $t > 1$ be a real number, let $\mathcal{P}$ be an infinite set of primes satisfying the inequality (4), and let $\mathcal{S}$ be any primitive subset of $\mathbb{N}(\mathcal{P})$; we want to show that $\Sigma_t(\mathcal{S}) \leqslant \Sigma_t(\mathcal{P})$. Enumerate $\mathcal{P}$ as $\mathcal{P} = \{p_1, p_2, \dots\}$, and for each natural number n, let $\mathcal{P}_n = \{p_1, \dots, p_n\}$. Also let $\mathcal{S}_n = \mathcal{S} \cap \mathbb{N}(\mathcal{P}_n)$, so that $\mathcal{S}_n$ is a primitive subset of $\mathbb{N}(\mathcal{P}_n)$. Note that the $\mathcal{P}_n$ form a nested sequence of sets whose union is $\mathcal{P}$, and similarly for $\mathcal{S}_n$ and $\mathcal{S}$.

Because the inequality (4) holds for $\mathcal{P}$, it also holds for each $\mathcal{P}_n$ by monotonicity:

$$\Sigma_t(\mathcal{P}_n) < \Sigma_t(\mathcal{P}) \leqslant 1 + \big(1 - \Sigma_{2t}(\mathcal{P})\big)^{1/2} < 1 + \big(1 - \Sigma_{2t}(\mathcal{P}_n)\big)^{1/2}.$$

Therefore we may apply Theorem 3 to the finite set $\mathcal{P}_n$ for each natural number n, concluding that $\Sigma_t(\mathcal{S}_n) \leqslant \Sigma_t(\mathcal{P}_n)$. Taking the limit as n tends to infinity (valid by the dominated convergence theorem, for example), we deduce that $\Sigma_t(\mathcal{S}) \leqslant \Sigma_t(\mathcal{P})$ as desired. □

Lemma 8. *Let $\mathcal{U}$ be a subset of $\mathbb{N}$, and let $\mathcal{S}^\star$ be a primitive subset of $\mathcal{U}$. If $\mathcal{S}^\star$ is t-best among all primitive subsets of $\mathcal{U}$ for every $t > 1$, then $\mathcal{S}^\star$ is also Erdős-best among all primitive subsets of $\mathcal{U}$.*

Proof. If $\mathcal{S}$ is any primitive subset of $\mathcal{U}$, then

$$\sum_{n\in\mathcal{S}} \frac{1}{n\log n} = \sum_{n\in\mathcal{S}} \int_1^\infty n^{-t}\,dt = \int_1^\infty \sum_{n\in\mathcal{S}} n^{-t}\,dt = \int_1^\infty \Sigma_t(\mathcal{S})\,dt.$$

(The leftmost sum is finite by Erdős's result [3], and the interchange of integral and sum is justified because all terms are positive.) The hypothesis that $\mathcal{S}^\star$ is t-best among all primitive subsets of $\mathcal{U}$ for every $t > 1$ means that $\Sigma_t(\mathcal{S}) \leqslant \Sigma_t(\mathcal{S}^\star)$ for every $t > 1$. It follows that

$$\sum_{n\in\mathcal{S}} \frac{1}{n\log n} = \int_1^\infty \Sigma_t(\mathcal{S})\,dt \leqslant \int_1^\infty \Sigma_t(\mathcal{S}^\star)\,dt = \sum_{n\in\mathcal{S}^\star} \frac{1}{n\log n},$$

as required. □

Proof of Theorem 5, assuming Theorem 6. If $\mathcal{P}$ is a set of primes for which the inequality (1) holds, then for any $t > 1$,

$$\sum_{p\in\mathcal{P}} p^{-t} < \sum_{p\in\mathcal{P}} p^{-1} \leqslant 1 + \Big(1 - \sum_{p\in\mathcal{P}} p^{-2}\Big)^{1/2} < 1 + \Big(1 - \sum_{p\in\mathcal{P}} p^{-2t}\Big)^{1/2}.$$

Theorem 6 implies that $\mathbb{N}_k(\mathcal{P})$ is t-best among primitive subsets of $\mathbb{N}_{\geqslant k}(\mathcal{P})$ for every natural number k and every $t > 1$. It follows from Lemma 8 that $\mathbb{N}_k(\mathcal{P})$ is also Erdős-best among primitive subsets of $\mathbb{N}_{\geqslant k}(\mathcal{P})$. □

Proof of Corollary 4, assuming Theorem 3. It suffices to verify that the inequality (1) is satisfied with the set $\mathcal{T} = \{5, 7, 11, 13, \ldots\}$ consisting of twin primes exceeding 3. On one hand, if B is the Brun constant defined by

$$B = \sum_{p\colon p+2\in\mathbb{P}} \Big(\frac{1}{p} + \frac{1}{p+2}\Big) = \Big(\frac{1}{3} + \frac{1}{5}\Big) + \Big(\frac{1}{5} + \frac{1}{7}\Big) + \Big(\frac{1}{11} + \frac{1}{13}\Big) + \cdots,$$

then the bound $B < 2.347$ has been given by Crandall and Pomerance [2, pp. 16–17] (for a proof, see Klyve [4, Chapter 3]), and therefore

$$\sum_{p\in\mathcal{T}} p^{-1} = B - \frac{1}{3} - \frac{1}{5} < 1.814. \tag{5}$$

On the other hand, we have

$$\sum_{p\in\mathcal{T}} p^{-2} < \sum_{n=1}^{\infty} \Big(\frac{1}{(6n-1)^2} + \frac{1}{(6n+1)^2}\Big) < \sum_{n=1}^{\infty} \Big(\frac{1}{6n(6n-3)} + \frac{1}{6n(6n+3)}\Big)$$
$$= \sum_{n=1}^{\infty} \frac{1}{9}\Big(\frac{1}{2n-1} - \frac{1}{2n+1}\Big) = \frac{1}{9}. \tag{6}$$

We conclude that

$$\sum_{p\in\mathcal{T}} p^{-1} < 1.814 < 1.9428 < 1 + \Big(1 - \frac{1}{9}\Big)^{1/2} < 1 + \Big(1 - \sum_{p\in\mathcal{T}} p^{-2}\Big)^{1/2},$$

and thus Theorem 3 can be applied to deduce that $\mathcal{T}$ is Erdős-best among primitive subsets of $\mathbb{N}(\mathcal{T})$. □

Let $\mathcal{T}_3 = \mathcal{T} \cup \{3\}$ be the set of all twin primes, including 3. It can be shown that $\sum_{p\in\mathcal{T}_3} p^{-2}$ is between 0.19725177 and 0.19725181. To show that $\mathcal{T}_3$ is Erdős-best among all primitive subsets of $\mathbb{N}(\mathcal{T}_3)$, it therefore suffices to establish the unconditional bound $B < 2.0959621$ on the Brun constant. The true value of Brun's constant is believed to be $B = 1.90216\cdots$ (see for example Sebah and Demichel [6]), and if this is the case, then it follows from Theorem 3 that $\mathcal{T}_3$ is indeed Erdős-best. Regrettably, the value $B = 1.90216\cdots$ is quoted in several places in the literature in a manner that suggests it has been rigorously established, but at the present time no bound better than $B < 2.347$ is known unconditionally.

3. Proof of Theorem 6

We now turn to the sole remaining task, namely, establishing Theorem 6 when the set of primes $\mathcal{P}$ is finite; we accomplish this task with the more detailed Proposition 11 stated below. As before, $\mathbb{N}(\mathcal{P})$ denotes the set of natural numbers all of whose prime factors lie in $\mathcal{P}$. We recall the previously defined notation

$$\mathbb{N}_k = \{n \in \mathbb{N} \colon \Omega(n) = k\} \quad \text{and} \quad \mathbb{N}_{\geqslant k} = \{n \in \mathbb{N} \colon \Omega(n) \geqslant k\}$$

(with $\mathbb{N}_0 = \{1\}$), as well as

$$\mathbb{N}_k(\mathcal{P}) = \mathbb{N}(\mathcal{P}) \cap \mathbb{N}_k \quad \text{and} \quad \mathbb{N}_{\geqslant k}(\mathcal{P}) = \mathbb{N}(\mathcal{P}) \cap \mathbb{N}_{\geqslant k}.$$

We also recall the notation $\Sigma_t(\mathcal{S}) = \sum_{n\in\mathcal{S}} n^{-t}$ for any set $\mathcal{S}$ of natural numbers.

Lemma 9. *Let $t > 1$ be a real number, and let $\mathcal{P}$ be a finite set of primes. Suppose that*

(i) for every proper subset $\mathcal{Q}$ of $\mathcal{P}$ and for every $k \in \mathbb{N}$, the primitive set $\mathbb{N}_k(\mathcal{Q})$ is t-best among all primitive subsets of $\mathbb{N}_{\geqslant k}(\mathcal{Q})$;

(ii) the inequality $\Sigma_t(\mathbb{N}_k(\mathcal{P})) \geqslant \Sigma_t(\mathbb{N}_{k+1}(\mathcal{P}))$ holds for all $k \in \mathbb{N}$.

Then $\mathbb{N}_k(\mathcal{P})$ itself is t-best among all primitive subsets of $\mathbb{N}_{\geqslant k}(\mathcal{P})$, for every $k \in \mathbb{N}$.

Proof. Fix $k \in \mathbb{N}$, and let $\mathcal{S}$ be a primitive subset of $\mathbb{N}_{\geqslant k}(\mathcal{P})$; we need to show that $\Sigma_t(\mathcal{S}) \leqslant \Sigma_t(\mathbb{N}_k(\mathcal{P}))$. Define $\ell = \min\{\Omega(n) \colon n \in \mathcal{S}\}$ (so that $\ell \geqslant k$), and fix a number $s \in \mathcal{S}$ with $\Omega(s) = \ell$. We proceed to partition both $\mathbb{N}_\ell(\mathcal{P})$ and $\mathcal{S}$ according to the greatest common divisor of their elements with s.

Let d denote any divisor of s. Notice that

$$\{n \in \mathbb{N}_\ell(\mathcal{P}) \colon (n, s) = d\} = d \cdot \left\{m \in \tfrac{1}{d}\mathbb{N}_\ell(\mathcal{P}) \cap \mathbb{N} \colon \left(m, \tfrac{s}{d}\right) = 1\right\} = d \cdot \mathbb{N}_{\ell-\Omega(d)}(\mathcal{Q}_d),$$

where $\mathcal{Q}_d$ is the set of primes in $\mathcal{P}$ that do not divide s/d; note that $\mathcal{Q}_d$ is a proper subset of $\mathcal{P}$ when $d \neq s$.

We define

$$\mathcal{S}_d = \tfrac{1}{d}\{n \in \mathcal{S}\colon (n,s)=d\} = \big\{m \in \tfrac{1}{d}\mathcal{S} \cap \mathbb{N}\colon \big(m, \tfrac{s}{d}\big) = 1\big\},$$

noting that $\mathcal{S}_d$ is a primitive subset of $\mathbb{N}_{\geqslant \ell-\Omega(d)}(\mathcal{Q}_d)$. With this notation, the sets $\mathbb{N}_\ell(\mathcal{P})$ and $\mathcal{S}$ can be decomposed as the disjoint unions

$$\mathbb{N}_\ell(\mathcal{P}) = \bigcup_{d\mid s} d\cdot \mathbb{N}_{\ell-\Omega(d)}(\mathcal{Q}_d) \quad \text{and} \quad \mathcal{S} = \bigcup_{d\mid s} d\cdot \mathcal{S}_d.$$

Therefore

$$\begin{aligned}\Sigma_t(\mathbb{N}_\ell(\mathcal{P})) - \Sigma_t(\mathcal{S}) &= \sum_{d\mid s} d^{-t}\big(\Sigma_t(\mathbb{N}_{\ell-\Omega(d)}(\mathcal{Q}_d)) - \Sigma_t(\mathcal{S}_d)\big)\\ &= \sum_{\substack{d\mid s\\ d\neq s}} d^{-t}\big(\Sigma_t(\mathbb{N}_{\ell-\Omega(d)}(\mathcal{Q}_d)) - \Sigma_t(\mathcal{S}_d)\big), \qquad (7)\end{aligned}$$

since $\mathbb{N}_{\ell-\Omega(s)}(\mathcal{Q}_s) = \mathcal{S}_s = \{1\}$.

However, $\mathcal{Q}_d$ is a proper subset of $\mathcal{P}$ when d is a proper divisor of s, and so hypothesis (i) tells us that $\mathbb{N}_{\ell-\Omega(d)}(\mathcal{Q}_d)$ is t-best among primitive subsets of $\mathbb{N}_{\geqslant \ell-\Omega(d)}(\mathcal{Q}_d)$. In particular, $\mathcal{S}_d$ is a primitive subset of $\mathbb{N}_{\geqslant \ell-\Omega(d)}(\mathcal{Q}_d)$, and so $\Sigma_t(\mathcal{S}_d) \leqslant \Sigma_t(\mathbb{N}_{\ell-\Omega(d)}(\mathcal{Q}_d))$ for every proper divisor d of s. Consequently, equation (7) demonstrates that $\Sigma_t(\mathcal{S}) \leqslant \Sigma_t(\mathbb{N}_\ell(\mathcal{P}))$. Finally, since $\ell \geqslant k$, hypothesis (ii) tells us that $\Sigma_t(\mathbb{N}_\ell(\mathcal{P})) \leqslant \Sigma_t(\mathbb{N}_k(\mathcal{P}))$; thus we have derived the required inequality $\Sigma_t(\mathcal{S}) \leqslant \Sigma_t(\mathbb{N}_k(\mathcal{P}))$. □

The proof of Proposition 11 relies upon one remaining statement, which has an elegant proof from the field of algebraic combinatorics. For all natural numbers k and m, define the polynomial

$$h_k(x_1,\dots,x_m) = \sum_{1\leqslant j_1\leqslant\cdots\leqslant j_k\leqslant m} x_{j_1}x_{j_2}\cdots x_{j_k}. \qquad (8)$$

Lemma 10. *Let $(x_1,\dots,x_m)$ be an arbitrary m-tuple of nonnegative real numbers. If $h_1(x_1,\dots,x_m) \geqslant h_2(x_1,\dots,x_m)$, then $h_k(x_1,\dots,x_m) \geqslant h_{k+1}(x_1,\dots,x_m)$ for all $k \in \mathbb{N}$.*

Proof. For ease of notation, we suppress the dependence of the polynomials h_k on the quantities $x_1,\dots,x_m$. For any $k \in \mathbb{N}$, the Jacobi–Trudi identity tells us that the determinant

$$\det\begin{pmatrix} h_{k+1} & h_k \\ h_{k+2} & h_{k+1}\end{pmatrix} = h_{k+1}^2 - h_k h_{k+2}$$

is equal to the Schur function s_λ corresponding to the partition $\lambda = (k{+}1, k{+}1)$; see, for example, Macdonald [5, Ch. I, §3, Eq. (4.3)]. Since the monomials comprising

s_λ have nonnegative coefficients (and we are evaluating at nonnegative real numbers $x_1, \ldots, x_m$), this determinant is nonnegative, which implies that $h_k/h_{k+1} \leqslant h_{k+1}/h_{k+2}$ for each $k \in \mathbb{N}$. However, our assumption is that $1 \leqslant h_1/h_2$, and therefore $1 \leqslant h_k/h_{k+1}$ for all $k \in \mathbb{N}$, as required. □

Proposition 11. *Let $t > 1$ be a real number, and let $\mathcal{P}$ be a finite set of primes satisfying the inequality $\Sigma_t(\mathcal{P}) \leqslant 1 + \sqrt{1 - \Sigma_{2t}(\mathcal{P})}$. Then*

(i) for every proper subset $\mathcal{Q}$ of $\mathcal{P}$ and for every $k \in \mathbb{N}$, the primitive set $\mathbb{N}_k(\mathcal{Q})$ is t-best among all primitive subsets of $\mathbb{N}_{\geqslant k}(\mathcal{Q})$;

(ii) the inequality $\Sigma_t(\mathbb{N}_k(\mathcal{P})) \geqslant \Sigma_t(\mathbb{N}_{k+1}(\mathcal{P}))$ holds for all $k \in \mathbb{N}$;

(iii) the primitive set $\mathbb{N}_k(\mathcal{P})$ is t-best among all primitive subsets of $\mathbb{N}_{\geqslant k}(\mathcal{P})$, for every $k \in \mathbb{N}$.

Remark. Note that the inequality $\Sigma_t(\mathcal{P}) \leqslant 1 + \sqrt{1 - \Sigma_{2t}(\mathcal{P})}$ is exactly the same as the hypothesis (4) of Theorem 6, while conclusion (iii) is the same as the conclusion of that theorem; hence this proposition implies Theorem 6. Note also that conclusions (i) and (ii) are the same as the hypotheses of Lemma 9, while conclusion (iii) is the same as the conclusion of that lemma. We feel that this redundancy makes clearer the structure of this proposition's proof.

Proof. We proceed by induction on the cardinality $\#\mathcal{P}$. Suppose first that $\#\mathcal{P} = 1$, so that $\mathcal{P} = \{p\}$ for some prime p. Conclusion (i) holds vacuously. Clearly $\mathbb{N}_k(\mathcal{P}) = \{p^k\}$ and $\mathbb{N}_{k+1}(\mathcal{P}) = \{p^{k+1}\}$, from which it follows that

$$\Sigma_t(\mathbb{N}_k(\mathcal{P})) = p^{-kt} > p^{-(k+1)t} = \Sigma_t(\mathbb{N}_{k+1}(\mathcal{P}))$$

for all $k \in \mathbb{N}$, establishing conclusion (ii). Finally, since $\mathcal{P}$ satisfies both (i) and (ii), Lemma 9 tells us that $\mathcal{P}$ satisfies conclusion (iii) as well.

Now suppose that $\#\mathcal{P} > 1$. First, let $\mathcal{Q}$ be any proper subset of $\mathcal{P}$. Since $\#\mathcal{Q} < \#\mathcal{P}$, the induction hypothesis is that the proposition holds for $\mathcal{Q}$; in particular, conclusion (iii) holds for $\mathcal{Q}$. Thus conclusion (i) holds for $\mathcal{P}$.

Turning now to (ii), we begin by treating the case $k = 1$, which requires us to establish the inequality $\Sigma_t(\mathcal{P}) \geqslant \Sigma_t(\mathbb{N}_2(\mathcal{P}))$. Note that

$$\Sigma_t(\mathcal{P})^2 = \Big(\sum_{p \in \mathcal{P}} p^{-t}\Big)^2 = 2 \sum_{\substack{n \in \mathbb{N}(\mathcal{P}) \\ \Omega(n)=2}} n^{-t} - \sum_{p \in \mathcal{P}} p^{-2t} = 2\Sigma_t(\mathbb{N}_2(\mathcal{P})) - \Sigma_{2t}(\mathcal{P}). \qquad (9)$$

The inequality $\Sigma_t(\mathcal{P}) \geqslant \Sigma_t(\mathbb{N}_2(\mathcal{P}))$ is therefore equivalent to $\Sigma_t(\mathcal{P})^2 - 2\Sigma_t(\mathcal{P}) + \Sigma_{2t}(\mathcal{P}) \leqslant 0$, which holds if and only if $1 - \sqrt{1 - \Sigma_{2t}(\mathcal{P})} \leqslant \Sigma_t(\mathcal{P}) \leqslant 1 + \sqrt{1 - \Sigma_{2t}(\mathcal{P})}$.

The second inequality is exactly the condition we have placed on $\mathcal{P}$, and so it remains only to prove the first inequality. First, note that $0 < \Sigma_{2t}(\mathcal{P}) < \Sigma_2(\mathcal{P}) \leqslant \Sigma_2(\mathbb{P}) < 1$ by equation (2); therefore $1 - \Sigma_{2t}(\mathcal{P}) \leqslant \sqrt{1 - \Sigma_{2t}(\mathcal{P})}$. Consequently,

$$1 - \sqrt{1 - \Sigma_{2t}(\mathcal{P})} \leqslant \Sigma_{2t}(\mathcal{P}) = \sum_{p \in \mathcal{P}} p^{-2t} < \sum_{p \in \mathcal{P}} p^{-t} = \Sigma_t(\mathcal{P}),$$

as required.

This argument establishes conclusion (ii) in the case $k = 1$. However, if we write $\mathcal{P} = \{p_1, \ldots, p_m\}$, note that

$$\Sigma_t(\mathbb{N}_k(\mathcal{P})) = \sum_{n \in \mathbb{N}_k(\mathcal{P})} n^{-t} = \sum_{1 \leqslant j_1 \leqslant \cdots \leqslant j_k \leqslant m} p_{j_1}^{-t} p_{j_2}^{-t} \cdots p_{j_k}^{-t} = h_k\big(p_1^{-t}, \ldots, p_m^{-t}\big)$$

using notation (8). We have just shown that $h_1\big(p_1^{-t}, \ldots, p_m^{-t}\big) \geqslant h_2\big(p_1^{-t}, \ldots, p_m^{-t}\big)$, and so Lemma 10 imples that $h_k\big(p_1^{-t}, \ldots, p_m^{-t}\big) \geqslant h_{k+1}\big(p_1^{-t}, \ldots, p_m^{-t}\big)$ for all $k \in \mathbb{N}$, which establishes conclusion (ii) in full. Finally, since $\mathcal{P}$ satisfies both (i) and (ii), Lemma 9 tells us that $\mathcal{P}$ satisfies conclusion (iii) as well, which completes the proof of the proposition. □

Acknowledgements. The authors thank David Speyer for suggesting the proof of Lemma 10. This work began during visits by the authors to Brigham Young University, whose hospitality and support are gratefully acknowledged.

References

[1] A. S. Besicovitch, 'On the density of certain sequences of integers,' *Math. Ann.* **110** (1934), 336–341.

[2] R. Crandall and C. Pomerance, *Prime numbers: a computational perspective.* Second edition. Springer, New York, 2005.

[3] P. Erdős, 'Note on sequences of integers no one of which is divisible by any other,' *J. London Math. Soc.* **10** (1935), 3–27.

[4] D. Klyve, *Explicit Bounds on Twin Primes and Brun's Constant.* Ph.D. thesis, Dartmouth College, Hanover NH, 2007.

[5] I. G. Macdonald, *Symmetric functions and Hall polynomials.* Second edition. Oxford Mathematical Monographs. The Clarendon Press, Oxford University Press, New York, 1995.

[6] P. Sebah and X. Gourdon, 'Introduction to twin primes and Brun's constant computation,' reported online at `http://numbers.computation.free.fr/Constants/constants.html`.

[7] Z. Zhang, 'On a problem of Erdős concerning primitive sequences,' *Math. Comp.* **60** (1993), 827–834.

A ZERO-SUM THEOREM OVER $\mathbb{Z}$

Marvin L. Sahs
Mathematics Department, Illinois State University, Normal, Illinois
marvinsahs@gmail.com

Papa A. Sissokho
Mathematics Department, Illinois State University, Normal, Illinois
psissok@ilstu.edu

Jordan N. Torf
Mathematics Department, Illinois State University, Normal, Illinois
jordan.torf@gmail.com

Received: 1/12/13, Accepted: 10/13/13, Published: 10/23/13

Abstract

A *zero-sum* sequence of integers is a sequence of nonzero terms that sum to 0. Let $k > 0$ be an integer and let $[-k, k]$ denote the set of all nonzero integers between $-k$ and k. Let $\ell(k)$ be the smallest integer ℓ such that any zero-sum sequence with elements from $[-k, k]$ and length greater than ℓ contains a proper nonempty zero-sum subsequence. In this paper, we prove a more general result which implies that $\ell(k) = 2k - 1$ for any $k > 1$.

1. Introduction

For any multiset S, let $|S|$ denote the number of elements in S, let $\max(S)$ denote the maximum element in S, and let $\Sigma S = \sum_{s \in S} s$. Let A and B be nonempty multisets of positive integers. The pair $\{A, B\}$ is said to be *irreducible* if $\Sigma A = \Sigma B$, and for every nonempty proper mutisubsets $A' \subsetneq A$ and $B' \subsetneq B$, $\Sigma A' \neq \Sigma B'$ holds. If $\{A, B\}$ fails to be irreducible, we say that it is *reducible*. It is easy to see that if $\{A, B\}$ is irreducible, then $A \cap B = \emptyset$ or $|A| = |B| = 1$.

We define the *length* of $\{A, B\}$ as

$$\ell(A, B) = |A| + |B|.$$

An irreducible pair $\{A, B\}$ is said to be *k-irreducible* if $\max(A \cup B) \leq k$. We define

$$\ell(k) = \max_{\{A,B\}} \ell(A, B), \tag{1}$$

where the maximum is taken over all k-irreducible pairs $\{A, B\}$.

For $k > 1$, let

$$A = \{\underbrace{k, \ldots, k}_{k-1}\} \text{ and } B = \{\underbrace{k-1, \ldots, k-1}_{k}\}. \tag{2}$$

Then $\{A, B\}$ is k-irreducible and $\ell(A, B) = 2k - 1$. This implies that $\ell(k) \geq 2k - 1$. El-Zanati, Seelinger, Sissokho, Spence, and Vanden Eynden introduced k-irreducible pairs in connection with their work on irreducible λ-fold partitions (e.g., see [2]). They also conjectured that $\ell(k) = 2k - 1$.

In our main theorem below, we prove a more general result which implies this conjecture.

Theorem 1. *If $\{A, B\}$ is an irreducible pair, then $|A| \leq \max(B)$ and $|B| \leq \max(A)$. Consequently, $\ell(k) = 2k - 1$ for any $k > 1$.*

One may naturally ask which k-irreducible pairs $\{A, B\}$ achieve the maximum possible length. We answer this question in the the following corollary.

Corollary 1. *Let $k > 1$ be an integer. A k-irreducible pair $\{A, B\}$ has (maximum possible) length $\ell(A, B) = 2k - 1$ if and only if $\{A, B\}$ is the pair shown in (2).*

A *zero-sum* sequence is a sequence of nonzero terms that sum to 0. A zero-sum sequence is said to be *irreducible* if it does not contain a proper nonempty zero-sum subsequence.

Let k be a positive integer, and let $[-k, k]$ denote the set of all nonzero integers between $-k$ and k. Given a zero-sum sequence τ with elements from $[-k, k]$, let A_τ be the multiset of all positive integers from τ, and let B_τ be the multiset containing the absolute values of all negative integers from τ. Then the sequence τ is irreducible if and only if the pair $\{A_\tau, B_\tau\}$ is irreducible. Moreover, the number $\ell(k)$, defined in (1), is also equal to the smallest integer ℓ such that any zero-sum sequence with elements from $[-k, k]$ and length greater than ℓ contains a proper nonempty zero-sum subsequence. It follows from Theorem 1 that $\ell(k) = 2k - 1$.

Let G be a finite (additive) abelian group of order n. The *Davenport constant* of G, denoted by $D(G)$, is the smallest integer m such that any sequence of elements from G with length m contains a nonempty zero-sum subsequence. Another key constant, $E(G)$, is the smallest integer m such that any sequence of elements from G with length m contains a zero-sum subsequence of length exactly n. The constant $E(G)$ was inspired by the well-known result of Erdös, Ginzburg, and Ziv [3], which states that $E(\mathbb{Z}/n\mathbb{Z}) = 2n - 1$. Subsequently, Gao [4] proved that $E(G) = D(G) + n - 1$. There is a large number of research papers dealing with the constants $D(G)$ and $E(G)$. We refer the interested reader to the survey papers of Caro [1] and Gao–Geroldinger [5] for further information.

Using the language of zero-sum sequence, we can view our main theorem as a zero-sum theorem. Whereas zero-sum sequences are traditionally studied for finite abelian groups such as $\mathbb{Z}/n\mathbb{Z}$, we consider in this paper zero-sum sequences over the infinite group $\mathbb{Z}$.

The rest of the paper is structured as follows. In Section 2, we prove our main results (Theorem 1 and Corollary 1), and in Section 3, we end with some concluding remarks.

2. Proofs of Theorem 1 and Corollary 1

Suppose we are given a k-irreducible pair $\{A, B\}$. We may assume that $A = \{x_1 \cdot a_1, \ldots, x_n \cdot a_n\}$ and $B = \{y_1 \cdot b_1, \ldots, y_m \cdot b_m\}$, where the a_i's and b_j's are all positive integers such that $1 \leq a_i, b_j \leq k$ for $1 \leq i \leq n$, $1 \leq j \leq m$. We also assume that the a_i's (resp. b_j's) are pairwise distinct. Moreover, $x_i > 0$ and $y_j > 0$ are the multiplicities of a_i and b_j respectively. For any pair (a_i, b_j), let

1. C be the multiset obtained from A by: (i) removing one copy of a_i, and (ii) introducing one copy of $a_i - b_j$ if $a_i > b_j$;

2. D be the multiset obtained from B by: (i) removing one copy of b_j, and (ii) introducing one copy of $b_j - a_i$ if $b_j > a_i$.

We say that $\{C, D\}$ is (a_i, b_j)-*derived* from $\{A, B\}$. We also call the above process an (a_i, b_j)-*derivation*. Consider the integers $p > 0$, $q > 0$, and $z_{ij} \geq 0$ for $p \leq i \leq q$ and $u \leq j \leq v$. We say that $\{C, D\}$ is $\prod_{i=p}^{q} \prod_{j=u}^{v} (a_i, b_j)^{z_{ij}}$-derived from $\{A, B\}$ if it is obtain by performing on $\{A, B\}$ an (a_i, b_j)-derivation z_{ij} times for each (i, j) pair. (If $z_{ij} = 0$, then we simply do not perform the corresponding (a_i, b_j)-derivation.)

We illustrate this operation with the following example. Let $A = \{3 \cdot 7, 2 \cdot 1\} = \{7, 7, 7, 1, 1\}$ and $B = \{3 \cdot 6, 5\} = \{6, 6, 6, 5\}$. Then $\{A, B\}$ is 7-irreducible. A $(7, 6)^2(7, 5)$-derivation of (A, B) yields the pair $\{C, D\}$, where $C = \{2, 1, 1, 1, 1\}$ and $D = \{6\}$. Note that $\{C, D\}$ is 6-irreducible (thus, 7-irreducible).

In general, the order in which the derivation is done makes a difference. For example, if $A = \{5, 5\}$ and $B = \{2, 2, 2, 2, 2\}$, then we can do a $(5, 2)$ derivation followed by a $(3, 2)$-derivation on $\{A, B\}$, but not in reverse order. However, all the derivation used in our proofs can be done in any order.

We will use the following lemma.

Lemma 1. *Let* $A = \{x_1 \cdot a_1, \ldots, x_n \cdot a_n\}$ *and* $B = \{y_1 \cdot b_1, \ldots, y_m \cdot b_m\}$ *be multisets, where the* a_i*'s and* b_i*'s are all positive integers such that* $1 \leq a_i, b_j \leq k$ *for* $1 \leq i \leq n$, $1 \leq j \leq m$. *Moreover,* $x_i > 0$ *and* $y_j > 0$ *are the multiplicities of* a_i *and* b_j *respectively. Suppose that* $\{A, B\}$ *is a* k*-irreducible pair with length* $|A| + |B| > 2$.

(i) If $\{C, D\}$ is (a_i, b_j)-derived from (A, B), then it is k-irreducible.
(ii) Let $p > 0$, $q > 0$, and $z_{ij} \geq 0$ for $p \leq i \leq q$ and $u \leq j \leq v$, be integers. Assume that $\sum_{j=u}^{v} z_{ij} \leq x_i$ and $\sum_{i=p}^{q} z_{ij} \leq y_j$. If $\{C, D\}$ is $\prod_{i=p}^{q} \prod_{j=u}^{v} (a_i, b_j)^{z_{ij}}$-derived from $\{A, B\}$, then it is k-irreducible.

Proof. We first prove (i). Without loss of generality, we may assume that $a_i > b_j$ since the proof is similar for $a_i < b_j$. Let

$$C = (A - \{a_i\}) \cup \{a_i - b_j\} \text{ and } D = B - \{b_j\}. \tag{3}$$

Since $\{A, B\}$ is irreducible, we have

$$\Sigma A = \Sigma B \Rightarrow \Sigma C = \Sigma A - a_i + (a_i - b_j) = \Sigma B - b_j = \Sigma D.$$

Then C and D are nonempty since C is clearly nonempty by (3). Assume that $\{C, D\}$ is reducible. Then, there exist nonempty proper multisubsets $C' \subsetneq C$ and $D' \subsetneq D$ such that $\Sigma C' = \Sigma D'$. Let $\overline{C}' = C - C'$ and $\overline{D}' = D - D'$. Then $\overline{C}' \subsetneq C$ and $\overline{D}' \subsetneq D$ are also nonempty proper multisubsets that satisfy $\Sigma\overline{C}' = \Sigma\overline{D}'$. However, it follows from the definition of C in (3) that either C' or $\overline{C}'$ is a proper multisubset of A, because there is a copy of the element $a_i - b_j$ which cannot be in both C' and $\overline{C}'$. It also follows from the definition of D in (3) that both D' and $\overline{D}'$ are proper multisubsets of B. Thus, either the pair $\{C', D'\}$ or $\{\overline{C}', \overline{D}'\}$ is a witness to the reducibility of $\{A, B\}$. This contradicts the fact that $\{A, B\}$ is irreducible. Hence, if $\{A, B\}$ is irreducible, then $\{C, D\}$ is also irreducible. In addition, it follows from (3) that $\max(C) \leq \max(A)$ and $\max(D) \leq \max(B)$. Hence, if $\{A, B\}$ is k-irreducible, then $\{C, D\}$ is also k-irreducible.

To prove (ii), observe that we can apply (i) recursively by performing (in any order) on $\{A, B\}$ an (a_i, b_j)-derivation z_{ij} times for each (i, j) pair. The conditions on the z_{ij}'s guarantee that there are enough pairs (a_i, b_j) in $A \times B$ to independently perform all the (a_i, b_j)-derivations for $p \leq i \leq q$ and $u \leq j \leq v$. □

We will also need the following basic lemma.

Lemma 2. *Let x_i and y_j be positive integers, where $1 \leq i \leq n$ and $1 \leq j \leq t + 1$. If*

$$\sum_{j=1}^{t} y_j \leq \sum_{i=1}^{n} x_i \text{ and } \sum_{j=1}^{t+1} y_j > \sum_{i=1}^{n} x_i,$$

then there exist integers $z_{ij} \geq 0$, $1 \leq i \leq n$ and $1 \leq j \leq t + 1$, such that

$$\sum_{i=1}^{n} z_{ij} = y_j \text{ for } 1 \leq j \leq t, \quad z_{i,t+1} = x_i - \sum_{j=1}^{t} z_{ij} \geq 0, \quad \text{and } y_{t+1} > \sum_{i=1}^{n} z_{i,t+1}.$$

Proof. For each j, $1 \le j \le t$, consider y_j marbles of color j. For each i, $1 \le i \le n$, consider a bin with capacity x_i (i.e., it can hold x_i marbles). Since $M = \sum_{j=1}^{t} y_j \le \sum_{i=1}^{n} x_i = C$, we can distribute all the M marbles into the n bins (with total capacity C) without exceeding the capacity of any given bin. Since $M + y_{t+1} = \sum_{j=1}^{t+1} y_j > C$, we can use some of the y_{t+1} marbles to top off the bins that were not already full.

For $1 \le i \le n$ and $1 \le j \le t$, we define z_{ij} to be the number of marbles in bin i that have color j. Then the numbers $z_{i,t+1} = x_i - \sum_{j=1}^{t} z_{ij} \ge 0$ are well defined for all $1 \le i \le n$. Hence the z_{ij}'s satisfy the required properties. □

We now prove our main theorem.

Proof of Theorem 1.

Let $\{A, B\}$ be a k-irreducible pair. We write $A = \{x_1 \cdot a_1, \ldots, x_n \cdot a_n\}$ and $B = \{y_1 \cdot b_1, \ldots, y_m \cdot b_m\}$, where the a_i's and b_i's are all positive integers such that $a_i, b_j \le k$ for $1 \le i \le n$ and $1 \le j \le m$. Moreover, $x_i > 0$ and $y_j > 0$ are the multiplicities of a_i and b_j respectively. Consequently, we may assume that the a_i's (resp. b_j's) are pairwise distinct. Without loss of generality, we may also assume that

$$a_1 > \ldots > a_n \text{ and } b_1 > \ldots > b_m. \tag{4}$$

We shall prove by induction on $r = \max(A) + \max(B) \ge 2$ that

$$|A| \le \max(B) \quad \text{and} \quad |B| \le \max(A). \tag{5}$$

If $r = 2$, then $k \le 2$ and the only possible irreducible pair is $\{\{1\}, \{1\}\}$. Thus, the inductive statement (5) is clearly true.

If $a_i = b_j$ for some pair (i, j), then $A = \{a_i\} = B$. (Otherwise, $A' = \{a_i\} \subsetneq A$ and $B' = \{a_i\} \subsetneq B$ are nonempty proper subsets satisfying $\Sigma A' = \Sigma B'$, which contradicts the irreducibility of $\{A, B\}$.) Moreover, $|A| = |B| = 1 \le a_i = \max(A) = \max(B)$ holds. Since $k > 1$, we further obtain $\ell(A, B) = |A| + |B| = 2 < 2k - 1$.

So we can assume that $A \cap B = \emptyset$. Without loss of generality, we may also assume that $a_1 = \max(A) > \max(B) = b_1$.

Suppose that the theorem holds for all k-irreducible pairs $\{C, D\}$ with $2 \le r' = \max(C) + \max(D) < r$. To prove the inductive step, we consider two parts.

Part I: In this part, we show $|A| \le \max(B)$. We consider two cases.

`Case 1:` $y_1 > x_1$.

Since $y_1 > x_1$, we can perform a $(a_1, b_1)^{x_1}$-derivation from $\{A, B\}$ to obtain (by Lemma 1) the k-irreducible pair $\{C, D\}$, where

$$C = \{x_1 \cdot (a_1 - b_1), x_2 \cdot a_2, \ldots, x_n \cdot a_n\}$$

and $D = \{(y_1 - x_1) \cdot b_1, y_2 \cdot b_2, \ldots, y_m \cdot b_m\}$.

Since $r' = \max(C) + \max(D) = \max\{a_1 - b_1, a_2\} + b_1 < r$, it follows from the induction hypothesis that

$$|C| = \sum_{i=1}^{n} x_i \le \max(D) = b_1. \tag{6}$$

It follows from (6) that $|A| = \sum_{i=1}^{n} x_i = |C| \le b_1$ as required.

`Case 2:` $y_1 \le x_1$.

Since $y_1 \le x_1$, we can perform a $(a_1, b_1)^{y_1}$-derivation from $\{A, B\}$ to obtain (by Lemma 1) the k-irreducible pair $\{C, D\}$, where

$$C = \{(x_1 - y_1) \cdot a_1, y_1 \cdot (a_1 - b_1), x_2 \cdot a_2, \ldots, x_n \cdot a_n\}$$

and $D = \{y_2 \cdot b_2, \ldots, y_m \cdot b_m\}$.

Since $r' = \max(C) + \max(D) \le a_1 + b_2 < r$, it follows from the induction hypothesis that

$$|C| = (x_1 - y_1) + y_1 + \sum_{i=2}^{n} x_i = \sum_{i=1}^{n} x_i \le \max(D) = b_2. \tag{7}$$

It follows from (7) that $|A| = \sum_{i=1}^{n} x_i = |C| \le b_2 < b_1$. This concludes the first part of the proof.

Part II: In this part, we show that $|B| \le \max(A) = a_1$. Assume that $|B| > a_1$. Then since $a_1 > b_1$ and $|A| \le b_1$ (by Part I), we obtain $|B| > |A|$. We now consider the cases $a_n > b_1$ and $b_1 > a_n$. (Recall that $b_1 \ne a_n$ since $A \cap B = \emptyset$.)

`Case 1:` $a_n > b_1$.

Then it follows from our general assumption (4) that

$$a_1 > \ldots > a_n > b_1 > \ldots > b_m.$$

We consider the following two subcases.

`Case 1.1:` $y_1 > \sum_{i=1}^{n} x_i$.

Then, we can perform a $\prod_{i=1}^{n} (a_i, b_1)^{x_i}$-derivation from $\{A, B\}$ to obtain (by Lemma 1) the k-irreducible pair $\{C, D\}$, where

$$C = \{x_1 \cdot (a_1 - b_1), x_2 \cdot (a_2 - b_1), \ldots, x_n \cdot (a_n - b_1)\},$$

and

$$D = \left\{\left(y_1 - \sum_{i=1}^{n} x_i\right) \cdot b_1, y_2 \cdot b_2, \ldots, y_m \cdot b_m\right\}.$$

Since $r' = \max(C) + \max(D) = (a_1 - b_1) + b_1 < r$, it follows from the induction hypothesis that

$$|C| = \sum_{i=1}^{n} x_i \le \max(D) \text{ and } |D| = \sum_{j=1}^{m} y_j - \sum_{i=1}^{n} x_i \le \max(C). \tag{8}$$

Thus, it follows from (8) that

$$|B| = \sum_{j=1}^{m} y_j = |C| + |D| \le \max(C) + \max(D) = (a_1 - b_1) + b_1 = a_1.$$

`Case 1.2:` $y_1 \le \sum_{i=1}^{n} x_i$.

Recall from the first paragraph in Part II that

$$\sum_{j=1}^{m} y_j = |B| > |A| = \sum_{i=1}^{n} x_i.$$

Consequently, the above inequality together with $y_1 \le \sum_{i=1}^{n} x_i$ imply that there exists an integer t, $1 \le t < m$, such that

$$\sum_{j=1}^{t} y_j \le \sum_{i=1}^{n} x_i \text{ and } \sum_{j=1}^{t+1} y_j > \sum_{i=1}^{n} x_i. \tag{9}$$

Then it follows from Lemma 2 that there exist integers $z_{ij} \ge 0$, $1 \le i \le n$ and $1 \le j \le t+1$, such that

$$\sum_{i=1}^{n} z_{ij} = y_j \text{ for } 1 \le j \le t, \quad z_{i,t+1} = x_i - \sum_{j=1}^{t} z_{ij} \ge 0, \text{ and } y_{t+1} > \sum_{i=1}^{n} z_{i,t+1}.$$

Thus, we can perform a $\prod_{i=1}^{n} \prod_{j=1}^{t+1} (a_i, b_j)^{z_{ij}}$-derivation from $\{A, B\}$ to obtain (by Lemma 1) the k-irreducible pair $\{C, D\}$, where

$$\begin{aligned} C = \{z_{11} \cdot (a_1 - b_1), \ldots, z_{1,t+1} \cdot (a_1 - b_{t+1}), \ldots,& \\ z_{i1} \cdot (a_i - b_1), \ldots, z_{i,t+1} \cdot (a_i - b_{t+1}), \ldots,& \\ z_{n1} \cdot (a_n - b_1), \ldots, z_{n,t+1} \cdot (a_n - b_{t+1})\},& \end{aligned}$$

and

$$D = \Big\{ \Big(y_{t+1} - \sum_{i=1}^{n} z_{i,t+1} \Big) \cdot b_{t+1}, y_{t+2} \cdot b_{t+2}, \ldots, y_m \cdot b_m \Big\}.$$

Since $a_1 > \ldots > a_n > b_1 > \ldots > b_m$, it follows that

$$\max(C) \le \max(A) - \min_{1 \le j \le t+1} b_j = a_1 - b_{t+1} \text{ and } \max(D) = b_{t+1}.$$

Thus, $r' = \max(C) + \max(D) \leq (a_1 - b_{t+1}) + b_{t+1} < r$ and it follows from the induction hypothesis that

$$|C| = \sum_{j=1}^{t}\sum_{i=1}^{n} z_{ij} + \sum_{i=1}^{n} z_{i,t+1} = \sum_{j=1}^{t} y_j + \sum_{i=1}^{n} z_{i,t+1} \leq \max(D), \tag{10}$$

and

$$|D| = (y_{t+1} - \sum_{i=1}^{n} z_{i,t+1}) + \sum_{j=t+2}^{m} y_j \leq \max(C). \tag{11}$$

From (10) and (11), we obtain

$$|B| = \sum_{j=1}^{m} y_j = |C| + |D| \leq \max(C) + \max(D) \leq a_1 - b_{t+1} + b_{t+1} = a_1,$$

as required.

Case 2: $b_1 > a_n$.

Let s be that smallest index such that $b_1 > a_s$. Since $a_1 > b_1 > a_n$, the integer s exists and $2 \leq s \leq n$. We consider the following two subcases.

Case 2.1: $y_1 \leq \sum_{i=s}^{n} x_i$.

Since $y_1 \leq \sum_{i=s}^{n} x_i$, there exist integers $z_i \geq 0$, $s \leq i \leq n$ such that $x_i \geq z_i$, and $y_1 = \sum_{i=s}^{n} z_i$. We can perform an $\prod_{i=s}^{n}(a_i, b_1)^{z_i}$-derivation from $\{A, B\}$ to obtain (by Lemma 1) the k-irreducible pair $\{C, D\}$, where

$$C = \{x_1 \cdot a_1, \ldots, x_{s-1} \cdot a_{s-1}, (x_s - z_s) \cdot a_s, \ldots, (x_n - z_n) \cdot a_n\},$$

and

$$D = \{z_s \cdot (b_1 - a_s), \ldots, z_n \cdot (b_1 - a_n), y_2 \cdot b_2, \ldots, y_m \cdot b_m\}.$$

Since $r' = \max(C) + \max(D) \leq a_1 + \max\{b_1 - a_n, b_2\} < r$, it follows from the induction hypothesis that

$$|D| = \sum_{i=s}^{n} z_i + \sum_{j=2}^{m} y_j = y_1 + \sum_{j=2}^{m} y_j = \sum_{j=1}^{m} y_j \leq \max(C) = a_1. \tag{12}$$

Thus, it follows from (12) that $|B| = \sum_{j=1}^{m} y_i = |D| \leq a_1$ as required.

Case 2.2: $y_1 > \sum_{i=s}^{n} x_i$.

Since $y_1 > \sum_{i=s}^{n} x_i$, we can perform a $\prod_{i=s}^{n}(a_i, b_1)^{x_i}$-derivation from $\{A, B\}$ to obtain (by Lemma 1) the k-irreducible pair $\{A', B'\}$, where

$$A' = \{x_1 \cdot a_1, x_2 \cdot a_2, \ldots, x_{s-1} \cdot a_{s-1}\},$$

and

$$B' = \Big\{ (y_1 - \sum_{i=s}^{n} x_i) \cdot b_1, x_s \cdot (b_1 - a_s), \ldots, x_n \cdot (b_1 - a_n), y_2 \cdot b_2, \ldots, y_m \cdot b_m \Big\}.$$

Note that $\max(B') = b_1$. We can now rename the distinct elements of the multiset B' as $b'_1, \ldots, b'_{m'}$ such that $\max(B') = b'_1 > \ldots > b'_{m'} = \min(B')$. Let y'_j be the multiplicity of b'_j for $1 \le j \le m'$. We also let $a'_i = a_i$ for $1 \le i \le s-1 = n'$.

Recall from Part I that $|A| \le \max(B) = b_1$. Hence,

$$|A'| = \sum_{i=1}^{s-1} x_i \le \sum_{i=1}^{n} x_i = |A| \le b_1.$$

If $|B'| \le |A'|$, then $|B| = \sum_{j=1}^{m} y_j = |B'| \le |A'| \le b_1 < a_1$, and we are done. So, we may assume that $|B'| > |A'|$. Since $a'_{n'} = a_{s-1} > b_1 = b'_1$ (owing to the definition of s and the fact that $A \cap B = \emptyset$), it follows that

$$a'_1 > \ldots > a'_{n'} > b'_1 > \ldots > b'_{m'}.$$

We can now proceed as in Part II (Case 1) to infer that

$$|B'| \le \max(A') \Longrightarrow |B| = \sum_{j=1}^{m} y_j = |B'| \le \max(A') = a_1.$$

This concludes the second part of the proof.

We conclude from Part I and Part II that

$$|A| \le \max(B) = b_1 \quad \text{ and } \quad |B| \le \max(A) = a_1.$$

Moreover, these inequalities imply that

$$\ell(A, B) = |A| + |B| \le b_1 + a_1 \le 2k - 1,$$

where the last inequality follows from the fact that $1 \le b_1 < a_1 \le k$. Finally, since $\ell(k) \ge 2k-1$ (see the example in (2) from Section 1), it follows that $\ell(k) = 2k-1$. □

We now prove the corollary.

Proof of Corollary 1. Let $A = \{x_1 \cdot a_1, \ldots, x_n \cdot a_n\}$ and $B = \{y_1 \cdot b_1, \ldots, y_m \cdot b_m\}$ be multisets, where the a_i's and b_i's are all positive integers such that $a_i, b_j \le k$ for $1 \le i \le n$ and $1 \le j \le m$. Moreover, $x_i > 0$ and $y_j > 0$ are the multiplicities of a_i and b_j respectively. We also assume that the a_i's (resp. b_j's) are pairwise distinct. Without loss of generality, we may also assume that

$$A \cap B = \emptyset;\ a_1 > \ldots > a_n;\ b_1 > \ldots > b_m;\ \text{and } a_1 > b_1.$$

Suppose that $\{A, B\}$ is a k-irreducible pair such that $\ell(A, B) = 2k - 1$. Then it follows from Theorem 1 (and the above setup) that

$$|A| = \max(B) = b_1 = k - 1 \text{ and } |B| = \max(A) = a_1 = k. \tag{13}$$

For a proof by contradiction assume that the pair $\{A, B\}$ is different from the pair $\{\{k \cdot (k-1)\}, \{(k-1) \cdot k\}\}$. We consider two cases.

Case 1: $x_1 \geq y_1$.
We perform a $(a_1, b_1)^{y_1}$-derivation from $\{A, B\}$ to obtain (by Lemma 1) the k-irreducible pair $\{C, D\}$, where

$$C = \{(x_1 - y_1) \cdot a_1, y_1 \cdot (a_1 - b_1), x_2 \cdot a_2, \ldots, x_n \cdot a_n\}$$

and $D = \{y_2 \cdot b_2, \ldots, y_m \cdot b_m\}$.

Since $a_1 > b_1$, $y_1 > 0$, and $\sum A = \sum B$, we have $m > 1$, so that $b_2 \in D$. Hence, C and D are both nonempty. We now use Theorem 1 on the irreducible pair $\{C, D\}$ to infer that

$$|C| = (x_1 - y_1) + y_1 + \sum_{i=2}^{n} x_i = \sum_{i=1}^{n} x_i \leq \max(D) = b_2. \tag{14}$$

It follows from (14) that $|A| = \sum_{i=1}^{n} x_i = |C| \leq b_2 < b_1$. This contradicts the fact that $|A| = b_1$ (see (13)).

Case 2: $y_1 > x_1$.
We perform a $(a_1, b_1)^{x_1}$-derivation from $\{A, B\}$ to obtain (by Lemma 1) the k-irreducible pair $\{C, D\}$, where

$$C = \{x_1 \cdot (a_1 - b_1), x_2 \cdot a_2, \ldots, x_n \cdot a_n\} = \{x_1 \cdot 1, x_2 \cdot a_2, \ldots, x_n \cdot a_n\},$$

and $D = \{(y_1 - x_1) \cdot b_1, y_2 \cdot b_2, \ldots, y_m \cdot b_m\}$.

If $n = 1$, then $x_1 = |A| = k - 1$. So $y_1 > x_1$ and $\ell(A, B) = 2k - 1$ imply $y_1 = k$, contradicting that $\{A, B\}$ is different from $\{\{k \cdot (k-1)\}, \{(k-1) \cdot k\}\}$. Thus we may assume that $n \geq 2$, that is, $a_2 \in C$.

Since $a_2 \neq b_1 = k - 1$, we must have $z = b_1 - a_2 > 0$. If $z < x_1$ also holds, then $C' = \{a_2, z \cdot 1\} \subsetneq C$ and $D' = \{b_1\} \subsetneq D$ form a witness for the reducibility of $\{C, D\}$, which is a contradiction. Thus, we must have $z = b_1 - a_2 \geq x_1$. We now use Theorem 1 on the irreducible pair $\{C, D\}$ to infer that

$$|D| = (y_1 - x_1) + \sum_{j=2}^{m} y_j = -x_1 + \sum_{j=1}^{m} y_j \leq \max(C) = a_2 \leq b_1 - x_1. \tag{15}$$

It follows from (15) that $|B| = \sum_{j=1}^{m} y_j = x_1 + |D| \leq b_1 = k - 1$. This contradicts the fact that $|B| = a_1 = k$ (see (13)). □

3. Concluding Remarks

One may wonder if our results can be extended to other infinite abelian groups. For instance, consider irreducible pairs $\{A, B\}$, where A and B are multisets of rational numbers. Is there a suitable (and interesting) definition of irreducibility that would guarantee the finiteness of $\ell(A, B)$?

Finally, we remark that Theorem 1 can be used to bound the number of λ-*fold vector space partitions* (e.g., see [2]). We shall address this application in a subsequent paper.

Acknowledgement The authors thank G. Seelinger, L. Spence, and C. Vanden Eynden for providing useful suggestions that led to an improved version of this paper. We also thank an anonymous referee for valuable comments.

References

[1] Y. Caro, Zero-sum problems – a survey, *Discrete Math.* **152** (1996), 93–113.

[2] S. El-Zanati, G. Seelinger, P. Sissokho, L. Spence, and C. Vanden Eynden, On Lambda-fold Partitions of Finite Vector Spaces and Duality. *Discrete Math.* **311/4** (2011), 307–318.

[3] P. Erdös, A. Ginzburg and A. Ziv, A theorem in additive number theory, *Bull. Res. Council Israel* **10F** (1961), 41–43.

[4] W. Gao, A Combinatorial Problem on Finite Abelian Groups, *J. Number Theory* **58** (1996), 100–103.

[5] W. Gao, A. Geroldinger, Zero-sum problems in finite abelian groups: A survey, *Expo. Math.* **24** (2006), 337–369.

ON THE DIVISIBILITY OF $a^n \pm b^n$ BY POWERS OF n

Salvatore Tringali[1]
Laboratoire Jacques-Louis Lions (LJLL), Université Pierre et Marie Curie (UPMC), Paris, France
tringali@ann.jussieu.fr

Received: 1/22/13, Revised: 7/31/13, Accepted: 10/21/13, Published: 11/11/13

Abstract

We determine all triples (a, b, n) of positive integers such that a and b are relatively prime and n^k divides $a^n + b^n$ (respectively, $a^n - b^n$), when k is the maximum of a and b (in fact, we answer a slightly more general question). As a by-product, we see that, for $m, n \in \mathbb{N}^+$ with $n \geq 2$, n^m divides $m^n + 1$ if and only if $(m, n) = (2, 3)$ or $(1, 2)$, which generalizes problems from the 1990 and 1999 editions of the International Mathematical Olympiad. The results are related to a conjecture by K. Győry and C. Smyth on the finiteness of the sets $R_k^{\pm}(a, b) := \{n \in \mathbb{N}^+ : n^k \mid (a^n \pm b^n)\}$, where a, b, k are fixed integers with $k \geq 3$, $\gcd(a, b) = 1$ and $|ab| \geq 2$; in particular, we find that the conjecture is true for $k \geq \max(|a|, |b|)$.

1. Introduction

It is a problem from the 1990 edition of the International Mathematical Olympiad (shortly, IMO) to find all integers $n \geq 2$ such that $n^2 \mid (2^n + 1)$. This appears as Problem 7.1.15 (p. 147) in [1], together with a solution by the authors (p. 323), who show that the only possible n is 3. On the other hand, Problem 4 in the 1999 IMO asks for all pairs (n, p) of positive integers such that p is a (positive rational) prime, $n \leq 2p$ and $n^{p-1} \mid ((p - 1)^n + 1)$. This is Problem 5.1.3 (p. 105) in the same book as above, whose solution by the authors (p. 105) is concluded with the remark that "With a little bit more work, we can even erase the condition $n \leq 2p$." Specifically, it is found that the required pairs are $(1, p)$, $(2, 2)$ and $(3, 3)$, where p is an arbitrary prime.

It is now fairly natural to ask whether similar conclusions can be drawn in relation to the more general problem of determining all pairs (m, n) of positive integers for which $n^m \mid (m^n + 1)$. In fact, the question is answered in the positive, and even in

[1]This research has been funded from the European Community's 7th Framework Programme (FP7/2007-2013) under Grant Agreement No. 276487 (project ApProCEM), and partially from the French ANR Project No. ANR-12-BS01-0011 (project CAESAR).

a stronger form, by Theorem 1 below, where the following observations are taken into account to rule out from the analysis a few trivial cases: Given $a, b \in \mathbb{Z}$ and $n, k \in \mathbb{N}^+$, we have that $1^k \mid (a^n \pm b^n)$ and $n^k \mid (a^n - a^n)$. Furthermore, $n^k \mid (a^n \pm b^n)$ if and only if $n^k \mid (b^n \pm a^n)$, and also if and only if $n^k \mid ((-a)^n \pm (-b)^n)$. Finally, $n^k \mid (a^n + (-a)^n)$ for n odd and $n^k \mid (a^n - (-a)^n)$ for n even.

Theorem 1. *Let a, b and n be integers such that $n \geq 2$, $a \geq \max(1, |b|)$ and $b \geq 0$ for n even, and set $\delta := \gcd(a, b)$, $\alpha := \delta^{-1}a$ and $\beta := \delta^{-1}b$.*

(1) Assume that $\beta \neq -\alpha$ when n is odd. Then, $n^a \mid (a^n + b^n)$ and $n^\alpha \mid (\alpha^n + \beta^n)$ if and only if $(a, b, n) = (2, 1, 3)$ or $(2^c, 2^c, 2)$ for $c \in \{0, 1, 2\}$.

(2) Let $\beta \neq \alpha$. Then, $n^a \mid (a^n - b^n)$ and $n^\alpha \mid (\alpha^n - \beta^n)$ if and only if $(a, b, n) = (3, 1, 2)$ or $(2, -1, 3)$.

Theorem 1 will be proved in Section 2. In fact, our proof is just the result of a meticulous refinement of the solutions already known for the IMO problems mentioned in the preamble. Thus, our only possible merit, if any at all, has been that of bringing into focus a clearer picture of (some of) their essential features.

Some comments on Theorem 1 are in order before proceeding. First, it would be interesting to extend the theorem, possibly at the expense of some extra solutions, by removing the assumption that $n^\alpha \mid (\alpha^n + \beta^n)$ or $n^\alpha \mid (\alpha^n - \beta^n)$ (the notation is the same as in the statement of the result), but at present we do not have great ideas for this. Secondly, three out of the six triples obtained by the present formulation come from the identity $2^3 + 1^3 = 3^2$. Lastly, the result yields a solution of the problems which have originally stimulated this work, as we have the following corollary (of which we omit the obvious proof):

Corollary 2. *Let $m, n \in \mathbb{N}^+$. Then $n^m \mid (m^n + 1)$ if and only if either $(m, n) = (2, 3)$, $(m, n) = (1, 2)$, or $n = 1$ and m is arbitrary.*

For the notation and terminology used herein without definition, as well as for material concerning classical topics in number theory, the reader should refer to [5]. In particular, we write $\mathbb{Z}$ for the integers, $\mathbb{N}$ for the nonnegative integers, and $\mathbb{N}^+$ for $\mathbb{N} \setminus \{0\}$, each of these sets being endowed with its ordinary addition $+$, multiplication $\cdot$ and total order $\leq$ (as is customary, $\geq$ will stand for the dual order of $\leq$). For $a, b \in \mathbb{Z}$ with $a^2 + b^2 \neq 0$, we denote by $\gcd(a, b)$ the greatest common divisor of a and b. Lastly, we let $\mathbb{P}$ be the set of all (positive rational) primes and, for $c \in \mathbb{Z} \setminus \{0\}$ and $p \in \mathbb{P}$, we use $e_p(c)$ for the greatest exponent $k \in \mathbb{N}$ such that $p^k \mid c$, which is extended to $\mathbb{Z}$ by $e_p(0) := \infty$.

We will make use at some point of the following lemma, which belongs to the folklore and is typically attributed to É. Lucas [6] and R. D. Carmichael [3] (the latter having fixed an error in Lucas' original work in the 2-adic case).

Lemma 3 (Lifting-the-exponent lemma). *For all $x, y \in \mathbb{Z}$, $\ell \in \mathbb{N}^+$ and $p \in \mathbb{P}$ such that $p \nmid xy$ and $p \mid (x-y)$, the following conditions are satisfied:*

(1) If $p \geq 3$, ℓ is odd, or $4 \mid (x-y)$, then $e_p(x^\ell - y^\ell) = e_p(x-y) + e_p(\ell)$.

(2) If $p = 2$, ℓ is even and $e_2(x-y) = 1$, then $e_2(x^\ell - y^\ell) = e_2(x+y) + e_2(\ell)$.

The study of the congruences $a^n \pm b^n \equiv 0 \bmod n^k$ has a very long history, dating back at least to Euler, who proved that, for all relatively prime integers a, b with $a > b \geq 1$, every primitive prime divisor of $a^n - b^n$ is congruent to 1 modulo n; see [2, Theorem I] for a proof and [2, §1] for the terminology. However, since there are so many results related to the question, instead of trying to summarize them here, we just refer the interested reader to the paper [4], whose authors provide a thorough account of the existing literature on the topic and characterize, for $a, b \in \mathbb{Z}$ and $k \in \mathbb{N}^+$, the set $R_k^+(a,b)$, respectively $R_k^-(a,b)$, of all positive integers n such that n^k divides $a^n + b^n$, respectively $a^n - b^n$ (note that no assumption is made about the coprimality of a and b), while addressing the problem of finding the exceptional cases when $R_1^-(a,b)$ and $R_2^-(a,b)$ are finite; see, in particular, [4, Theorems 1–2 and 18]. Nonetheless, the related question of determining, given $a, b \in \mathbb{Z}$ with $\gcd(a,b) = 1$, all positive integers n such that n^k divides $a^n + b^n$ (respectively, $a^n - b^n$), when k is the maximum of $|a|$ and $|b|$, does not appear to be considered in either [4] or the references therein.

On another hand, it is suggested in [4] that $R_k^+(a,b)$ and $R_k^-(a,b)$ are both finite provided that a, b and k are fixed integers with $k \geq 3$, $\gcd(a,b) = 1$ and $|ab| \geq 2$ (the authors point out that the question is probably a difficult one, even assuming the ABC conjecture). Although far from being an answer to this, Theorem 1 in the present paper implies that, under the same assumptions as above, $R_k^+(a,b)$ and $R_k^-(a,b)$ are finite for $k \geq \max(|a|, |b|)$.

2. Proofs

First, for the sake of exposition, we give a couple of lemmas.

Lemma 4. *Let $x, y, z \in \mathbb{Z}$ and $\ell \in \mathbb{N}^+$ such that $\gcd(x,y) = 1$ and $z \mid (x^\ell + y^\ell)$. Then xy and z are relatively prime, $q \nmid (x^\ell - y^\ell)$ for every integer $q \geq 3$ for which $q \mid z$, and $4 \nmid z$ provided that ℓ is even. Moreover, if there exists an odd prime divisor p of both z and ℓ such that $\gcd(\ell, p-1) = 1$, then $p \mid (x+y)$, ℓ is odd and $e_p(z) \leq e_p(x+y) + e_p(\ell)$.*

Proof. The first part is routine (we omit the details). As for the second, let p be an odd prime dividing both z and ℓ with $\gcd(\ell, p-1) = 1$. Also, considering that z and xy are relatively prime (by the above), denote by y^{-1} an inverse of y modulo p and by

ω the order of xy^{-1} modulo p, viz the smallest $k \in \mathbb{N}^+$ such that $(xy^{-1})^k \equiv 1 \bmod p$; cf. [5, §6.8]. Since $(xy^{-1})^{2\ell} \equiv 1 \bmod p$, we have $\omega \mid 2\ell$. It follows from Fermat's little theorem and [5, Theorem 88] that ω divides $\gcd(2\ell, p-1)$, whence we get $\omega \mid 2$, using that $\gcd(\ell, p-1) = 1$. This in turn implies that $p \mid (x^2 - y^2)$, so that either $p \mid (x-y)$ or $p \mid (x+y)$. But $p \mid (x-y)$ would give that $p \mid (x^\ell - y^\ell)$, which is impossible by the first part of the claim (since $p \geq 3$). So $p \mid (x+y)$, with the result that ℓ is odd: For, if $2 \mid \ell$, then $p \mid 2x^\ell$ (because $p \mid z \mid (x^\ell + y^\ell)$ and $y \equiv -x \bmod p$), which would lead to $\gcd(x, y) \geq p$ (again, using that p is odd), and thus to a contradiction. The rest is an immediate application of Lemma 3. □

Lemma 5. *Let $x, y, z \in \mathbb{Z}$ such that x, y are odd and $x, y \geq 0$. Then $x^2 - y^2 = 2^z$ if and only if $z \geq 3$, $x = 2^{z-2} + 1$ and $y = 2^{z-2} - 1$.*

Proof. Since x and y are odd, $x^2 - y^2$ is divisible by 8, namely $z \geq 3$, and there exist $i, j \in \mathbb{N}^+$ such that $i + j = z$, $x - y = 2^i$ and $x + y = 2^j$. It follows that $x = 2^{j-1} + 2^{i-1}$ and $y = 2^{j-1} - 2^{i-1}$, and then $j > i$ and $i = 1$ (otherwise x and y would be even). The rest is straightforward. □

Now, we are ready to write down the proof of the main result.

Proof of Theorem 1. Point (1): Assume that $n^a \mid (a^n + b^n)$, $n^\alpha \mid (\alpha^n + \beta^n)$, and $\beta \neq -\alpha$ when n is odd. Since α and β are coprime (by construction), it holds that $\beta \neq 0$, for otherwise $n \mid (\alpha^n + \beta^n)$ and $n \geq 2$ would give $\gcd(\alpha, \beta) \geq 2$. Also, $\alpha = |\beta|$ if and only if $\alpha = \beta = 1$ and $n = 2$ (as $\beta \geq 0$ for n even), hence 2^δ divides $2\delta^2$, which is possible if and only if $\delta \in \{1, 2, 4\}$ and gives $(a, b, n) = (1, 1, 2)$, $(2, 2, 2)$, or $(4, 4, 2)$. So, we are left with the case when

$$\alpha \geq 2 \quad \text{and} \quad \alpha > |\beta| \geq 1, \tag{1}$$

since $\alpha \geq \max(1, |\beta|)$. Considering that $4 \mid n^2$ for n even, it follows from Lemma 4 that n is odd and $\gcd(\alpha\beta, n) = 1$. Denote by p the smallest prime divisor of n. Again by Lemma 4, it is then found that p divides $\alpha + \beta$ and

$$\alpha - 1 \leq (\alpha - 1) \cdot e_p(n) \leq e_p(\alpha + \beta). \tag{2}$$

Furthermore, $\alpha + \beta \geq 1$ by equation (1), so that

$$\alpha + \beta = p^r s, \quad \text{with } r, s \in \mathbb{N}^+ \text{ and } p \nmid s. \tag{3}$$

Therefore, equations (1) and (3) yield that $2\alpha \geq p^r s + 1$. This implies by equation (2), since $r = e_p(\alpha + \beta)$, that $3^r s \leq p^r s \leq 2r + 1$, which is possible only if $p = 3$ and $r = s = 1$. Thus, by equations (2) and (3), we get $\alpha + \beta = 3$ and $\alpha = 2$, namely $(\alpha, \beta) = (2, 1)$. Also, $e_3(n) = 1$, and hence $n = 3t$ for some $t \in \mathbb{N}^+$ with $\gcd(6, t) = 1$. It follows that $t^2 \mid (\gamma^t + 1)$ for $\gamma = 2^3$.

So suppose, for the sake of contradiction, that $t \geq 2$ and let q be the least prime divisor of t. Then, another application of Lemma 4 gives $2e_q(t) \leq e_q(\gamma+1) + e_q(t)$, and accordingly $1 \leq e_q(t) \leq e_q(\gamma+1) = e_q(3^2)$, which is however absurd, due to the fact that $\gcd(3,t) = 1$. Hence $t = 1$, i.e., $n = 3$, and putting everything together we obtain the claim, because $2^3 + 1^3 = 3^2$ and $3^{2\delta}$ divides $\delta^2 \cdot (2^3 + 1^3)$ only if $\delta = 1$.

Point (2). Assume that $n^a \mid (a^n - b^n)$, $n^\alpha \mid (\alpha^n - \beta^n)$, and $\beta \neq \alpha$. Since $\gcd(\alpha, \beta) = 1$, we get as in the proof of point (1) that $\beta \neq 0$, while $\alpha = |\beta|$ only if $\alpha = 1$, $\beta = -1$, and n is odd (again, $\beta \geq 0$ for n even), which is however impossible, because it would give that $n \mid 2$ with $n \geq 3$. So, we can suppose from now on that α and β satisfy the same conditions as in equation (1), and write n as $2^r s$, where $r \in \mathbb{N}$, $s \in \mathbb{N}^+$ and $\gcd(2,s) = 1$. We have the following:

`Case 1:` $r = 0$, i.e., $n = s$. Then, n is odd, and hence $n^a \mid (a^n + (-b)^n)$ and $n^\alpha \mid (\alpha^n + (-\beta)^n)$, so by point (1) we get $(a,b,n) = (2,-1,3)$.

`Case 2:` $r \geq 1$. Since n is even and $\gcd(\alpha,\beta) = 1$, both α and β are odd, that is $8 \mid (\alpha^2 - \beta^2)$. It follows from point (1) of Lemma 3 that

$$e_2(\alpha^n - \beta^n) = e_2(\alpha^2 - \beta^2) + e_2(2^{r-1}s) = e_2(\alpha^2 - \beta^2) + r - 1. \tag{4}$$

(With the notation there, we apply the lemma with $x = \alpha^2$, $y = \beta^2$, $\ell = 2^{r-1}s$, and $p = 2$.) Also, $2^{r\alpha} \mid (\alpha^n - \beta^n)$, so equation (4) yields

$$(\alpha - 1) \cdot r \leq e_2(\alpha^2 - \beta^2) - 1. \tag{5}$$

There now exist $u, v \in \mathbb{N}^+$ with $u \geq 2$ and $\gcd(2,v) = 1$ such that $\alpha^2 - \beta^2 = 2^{u+1}v$, with the result that $\alpha > 2^{u/2}\sqrt{v}$. Hence, taking also into account that $2^x \geq x + 1$ for every $x \in \mathbb{R}$ with $x \geq 1$, we get by equation (5) that

$$\left(\frac{u}{2} + 1\right)\sqrt{v} \leq 2^{u/2}\sqrt{v} < \frac{u}{r} + 1, \tag{6}$$

which is possible only if $r = 1$ and $\sqrt{v} < 2$. Then $2^{u/2}\sqrt{v} < u + 1$, in such a way that $2 \leq u \leq 5$ and $v = 1$ (using that v is odd). In consequence of Lemma 5, all of this implies, at the end of the day, that $\alpha = 2^z + 1$, $b = 2^z - 1$ and $n = 2s$ (recall that we want the conditions in equation (1) to be satisfied and $\beta \geq 0$ for n even), where z is an integer between 1 and 4. But we need $2^z \leq z + 1$ by equation (5), so necessarily $z = 1$, i.e., $\alpha = 3$ and $\beta = 1$. Finally, we check that $(2s)^3 \mid (3^{2s} - 1^{2s})$ if and only if $s = 1$: For, if $s \geq 2$ and q is the smallest prime divisor of s, then $0 < 3e_q(s) \leq e_q(3^2 - 1)$ by Lemma 4, which is absurd since $\gcd(2,s) = 1$. This gives $(a,b,n) = (3,1,2)$, while it is trivially seen that $2^{3\delta}$ divides $\delta^2 \cdot (3^2 - 1^2)$ if and only if $\delta = 1$.

Putting all the pieces together, the proof is thus complete. □

Acknowledgements The author thanks Paolo Leonetti (Università Bocconi di Milano) and Carlo Sanna (Università di Torino) for useful discussions. He is also grateful to an anonymous referee for remarks that have helped to improve the overall quality of the paper.

References

[1] T. Andreescu and D. Andrica, *Number Theory - Structures, Examples, and Problems*, 1st ed., Birkhäuser, 2009.

[2] G. D. Birkhoff and H. S. Vandiver, *On the Integral Divisors of* $a^n - b^n$, Ann. of Math. **5** (Jul., 1904), no. 4, 173–180.

[3] R. D. Carmichael, *On the Numerical Factors of Certain Arithmetic Forms*, Amer. Math. Monthly **16** (1909), no. 10, 153–159.

[4] K. Győry and C. Smyth, *The divisibility of* $a^n - b^n$ *by powers of* n, Integers **10** (Jul., 2010), no. 3, 319–334.

[5] G. H. Hardy and E. M. Wright, *An Introduction to the Theory of Numbers*, 6th ed. (revised by D. R. Heath-Brown and J. H. Silverman), Oxford University Press, 2008.

[6] É. Lucas, *Théorie des Fonctions Numériques Simplement Periodiques*, Amer. J. Math. **1** (1878), 184–196, 197-240, 289–321.

LOWER BOUNDS FOR SUMSETS OF MULTISETS IN $\mathbb{Z}_p^2$

Greg Martin
Dept. of Mathematics, University of British Columbia, Vancouver, BC, Canada
gerg@math.ubc.ca

Alexis Peilloux
Université Pierre-et-Marie Curie, aculté de Mathématiques,Paris, France
alexis.peilloux@etu.upmc.fr

Erick B. Wong
Dept. of Mathematics, University of British Columbia, Vancouver, BC, Canada
erick.wong@alumni.ubc.ca

Received: 8/31/12, Revised: 10/7/13, Accepted: 10/31/13, Published: 11/11/13

Abstract

The classical Cauchy–Davenport theorem implies the lower bound $n+1$ for the number of distinct subsums that can be formed from a sequence of n elements of the cyclic group $\mathbb{Z}_p$ (when p is prime and $n < p$). We generalize this theorem to a conjecture for the minimum number of distinct subsums that can be formed from elements of a multiset in $\mathbb{Z}_p^m$; the conjecture is expected to be valid for multisets that are not "wasteful" by having too many elements in nontrivial subgroups. We prove this conjecture in $\mathbb{Z}_p^2$ for multisets of size $p+k$, when k is not too large in terms of p.

1. Introduction

Determining the number of elements in a particular abelian group that can be written as sums of given sets of elements is a topic that goes back at least two centuries. The most famous result of this type, involving the cyclic group $\mathbb{Z}_p$ of prime order p, was established by Cauchy in 1813 [1] and rediscovered by Davenport in 1935 [2, 3] (here $\#A$ denotes the cardinality of A):

Lemma 1.1 (Cauchy–Davenport Theorem). *Let A and B be subsets of $\mathbb{Z}_p$, and define $A+B$ to be the set of all elements of the form $a+b$ with $a \in A$ and $b \in B$. Then $\#(A+B) \geq \min\{p, \#A+\#B-1\}$.*

The lower bound is easily seen to be best possible by taking A and B to be intervals, for example. It is also easy to see that the lower bound of $\#A+\#B-1$ does not hold for general abelian groups $\mathbf{G}$ (take A and B to be the same nontrivial subgroup of

G). There is, however, a well-known generalization obtained by Kneser in 1953 [4], which we state in a slightly simplified form that will be quite useful for our purposes (see [8, Theorem 4.1] for an elementary proof):

Lemma 1.2 (Kneser's Theorem). *Let A and B be subsets of a finite abelian group $\mathbf{G}$, and let m be the largest cardinality of a proper subgroup of $\mathbf{G}$. Then $\#(A+B) \geq \min\{\#\mathbf{G}, \#A+\#B-m\}$.*

Given a sequence $A=(a_1,\ldots,a_k)$ of (not necessarily distinct) elements of an abelian group $\mathbf{G}$, a related result involves its *sumset* ΣA, which is the set of all sums of any number of elements chosen from A (not to be confused with $A+A$, which it contains but usually properly):

$$\Sigma A = \left\{ \sum_{j\in J} a_j : J \subseteq \{1,\ldots,k\} \right\}.$$

(Note that we allow J to be empty, so that the group's identity element is always an element of ΣA, even when A itself is empty) When $\mathbf{G}=\mathbb{Z}_p$, one can prove the following result by writing $\Sigma A = \{0,a_1\}+\cdots+\{0,a_k\}$ and applying the Cauchy–Davenport theorem inductively:

Lemma 1.3. *Let $A=(a_1,\ldots,a_k)$ be a sequence of nonzero elements of $\mathbb{Z}_p$. Then $\#\Sigma A \geq \min\{p,k+1\}$.*

This result can also be proved directly by induction on k, and in fact such a proof will discover why the order p of the cyclic group must be prime (intuitively, the sequence A could lie completely within a nontrivial subgroup). For a formal proof, see [6, Lemma 2]. Again the lower bound is easily seen to be best possible, by taking $a_1=\cdots=a_k$.

It is a bit misleading to phrase such results in terms of sequences, since the actual order of the elements in the sequence is irrelevant (given that we are considering only abelian groups). We prefer to use *multisets*, which are simply sets that are allowed to contain their elements with multiplicity. If we let m_x denote the multiplicity with which the element x occurs in the multiset A, then the definition of ΣA can be written in the form

$$\Sigma A = \left\{ \sum_{x\in\mathbf{G}} \delta_x x : 0 \leq \delta_x \leq m_x \right\},$$

where $\delta_x x$ denotes the group element $x+\cdots+x$ obtained by adding δ_x summands all equal to x.

When using multisets, we should choose our notation with care: the hypotheses of such results tend to involve the total number of elements of the multiset A counting multiplicity, while the conclusions involve the number of distinct elements of ΣA. Consequently, throughout this paper, we use the following notational conventions:

- $|S|$ denotes the total number of elements of the multiset S, counted with multiplicity;
- $\#S$ denotes the number of distinct elements of the multiset S, or equivalently the cardinality of S considered as a (mere) set.

In this notation, Lemma 1.3 can be restated as:

Lemma 1.4. *Let A be a multiset contained in $\mathbb{Z}_p$ such that $0 \notin A$. Then $\#\Sigma A \geq \min\{p, |A|+1\}$.*

The purpose of this paper is to improve, as far as possible, this lower bound for multisets contained in the larger abelian group $\mathbb{Z}_p^2$. We cannot make any progress without some restriction upon our multisets: if a multiset is contained within a nontrivial subgroup of $\mathbb{Z}_p^2$ (of cardinality p), then so is its sumset, in which case the lower bound $\min\{p, |A|+1\}$ from Lemma 1.4 is the best we can do. Therefore we restrict to the following class of multisets. We use the symbol $\mathbf{0} = (0,0)$ to denote the identity element of $\mathbb{Z}_p^2$.

Definition 1.5. A multiset A contained in $\mathbb{Z}_p^2$ is called *valid* if:

- $\mathbf{0} \notin A$; and
- every nontrivial subgroup contains fewer than p points of A, counting multiplicity.

The exact number p in the second condition has been carefully chosen: any nontrivial subgroup of $\mathbb{Z}_p^2$ is isomorphic to $\mathbb{Z}_p$, and so Lemma 1.4 applies to these nontrivial subgroups. In particular, any multiset A containing $p-1$ nonzero elements of a nontrivial subgroup will automatically have that entire subgroup contained in its sumset ΣA, so allowing p nonzero elements in a nontrivial subgroup would always be "wasteful".

We believe that the following lower bound should hold for sumsets of valid multisets:

Conjecture 1.6. Let A be a valid multiset contained in $\mathbb{Z}_p^2$ such that $p \leq |A| \leq 2p-3$. Then $\#\Sigma A \geq (|A|+2-p)p$. In other words, if $|A| = p+k$ with $0 \leq k \leq p-3$, then $\#\Sigma A \geq (k+2)p$.

It is easy to see that this conjectured lower bound would be best possible: if A is the multiset that contains the point $(1,0)$ with multiplicity $p-1$ and the point $(0,1)$ with multiplicity $k+1$, then the set ΣA is precisely $\{(s,t)\colon s \in \mathbb{Z}_p, 0 \leq t \leq k+1\}$, which has $(k+2)p$ distinct elements. Conjecture 1.6 is actually part of a larger assertion (Conjecture 4.3) concerning lower bounds for sumsets in $\mathbb{Z}_p^m$. We note that Conjecture 1.6 is vacuous for $p = 2$, and we shall see in Section 4 that the conditions of the more general conjecture render it similarly unremarkable when $p = 2$.

One of our results completely resolves the first two cases $k = 0$ and $k = 1$ of this conjecture:

Theorem 1.7. *Let p be a prime.*

a. If A is any valid multiset contained in $\mathbb{Z}_p^2$ with $|A| = p$, then $\#\Sigma A \geq 2p$.

b. Suppose that $p \geq 5$. If A is any valid multiset contained in $\mathbb{Z}_p^2$ with $|A| = p+1$, then $\#\Sigma A \geq 3p$.

It turns out that proving part (b) of the theorem requires a certain amount of computation for a finite number of primes (see the remarks following the proof of the theorem in Section 3). Extending the conjecture to larger values of k would require, by our methods, more and more computation to take care of small primes p as k grows. However, we are able to establish the conjecture when p is large enough with respect to k, or equivalently when k is small enough with respect to p:

Theorem 1.8. *Let p be a prime, and let $2 \leq k \leq \sqrt{p/(2\log p + 1)} - 1$ be an integer. If A is any valid multiset contained in $\mathbb{Z}_p^2$ with $|A| = p + k$, then $\#\Sigma A \geq (k+2)p$.*

A contrapositive version of Theorem 1.8 is also enlightening:

Corollary 1.9. *Let p be a prime, and let $2 \leq k \leq \sqrt{p/(2\log p + 1)} - 1$ be an integer. Let A be a multiset contained in $\mathbb{Z}_p^2 \setminus \{\mathbf{0}\}$ with $|A| = p + k$. If $\#\Sigma A < (k+2)p$, then there exists a nontrivial subgroup of $\mathbb{Z}_p^2$ that contains at least p points of A, counting multiplicity.*

Our methods of proof stem from two main ideas. First, we will obviously exploit the structure of $\mathbb{Z}_p^2$ as a direct sum of cyclic groups of prime order, within which we can apply the known Lemma 1.4 after using projections. Section 2 contains several elementary lemmas in this vein (see in particular Lemma 2.8). It is important for us to utilize the flexibility coming from the fact that $\mathbb{Z}_p^2$ can be decomposed as the direct sum of two subgroups in many different ways. Second, our methods work best when there exists a single subgroup that contains many elements of the given multiset; however, by selectively replacing pairs of elements with their sums, we can increase the number of elements in a subgroup in a way that improves our lower bounds upon the sumset (see Lemma 3.2). These methods, which appear in Section 3, combine to provide the proofs of Theorems 1.7 and 1.8. Finally, Section 4 contains a generalization of Conjecture 1.6 to higher-dimensional direct sums of $\mathbb{Z}_p$, together with examples demonstrating that the conjecture would be best possible.

Subsequent to the completion of this paper, K. Matomäsi [7] has given an elegant argument which reduces our Conjecture 4.3 to the single subcase (c), and thereby establishes Conjecture 1.6 in full generality, by a result of Peng [9].

2. Sumsets in Abelian Groups and Direct Products

All of the results in this section are valid for general finite abelian groups and have correspondingly elementary proofs, although the last two lemmas seem rather less standard than the first few. In this section, **G**, **H**, and **K** denote finite abelian groups, and e denotes a group's identity element.

Lemma 2.1. *Let $B_0, B_1, B_2, \ldots, B_j$ be multisets in* $\mathbf{G}$*, and set $A = B_0 \cup B_1 \cup \cdots \cup B_j$. For each $1 \leq i \leq j$, specify an element $x_i \in \Sigma B_i$, and set $C = B_0 \cup \{x_1, \ldots, x_j\}$. Then $\Sigma C \subseteq \Sigma A$.*

Proof. For each $1 \leq i \leq j$, choose a submultiset $D_i \subseteq B_i$ such that the sum of the elements of D_i equals x_i. By definition, every element y of ΣC equals the sum of the elements of some subset E of B_0, plus $\sum_{i \in I} x_i$ for some $I \subseteq \{1, \ldots, j\}$. But then y equals the sum of the elements of $E \cup \bigcup_{i \in I} D_i$, which is an element of ΣA since $E \cup \bigcup_{i \in I} D_i \subseteq B_0 \cup \bigcup_{1 \leq i \leq j} B_i = A$. □

Lemma 2.2. *Let $A_1, A_2, \ldots, A_j$ be multisets in* $\mathbf{G}$*, and set $A = A_1 \cup \cdots \cup A_j$. If m is the largest cardinality of a proper subgroup of* $\mathbf{G}$*, then either $\Sigma A = \mathbf{G}$ or $\#\Sigma A \geq (\sum_{i=1}^{j} \#\Sigma A_i) - (j-1)m$.*

Proof. Since $\Sigma A = \Sigma A_1 + \Sigma A_2 + \cdots + \Sigma A_j$ (viewed as ordinary sets), this follows immediately by inductive application of Kneser's theorem. □

For the remainder of this section, we will be dealing with groups that can be decomposed into a direct sum.

Definition 2.3. A subgroup **H** of **G** is called an *internal direct summand* if there exists a subgroup **K** of **G** such that **G** is the internal direct sum of **H** and **K**, or in other words, such that $\mathbf{H} \cap \mathbf{K} = \{e\}$ and $\mathbf{H} + \mathbf{K} = \mathbf{G}$. Equivalently, **H** is an internal direct summand of **G** if there exists a *projection homomorphism* $\pi_{\mathbf{H}} \colon \mathbf{G} \to \mathbf{H}$ that is the identity on **H**. Note that this projection homorphism does depend on the choice of **K** but is uniquely determined by $\pi_{\mathbf{H}}^{-1}(e) = \mathbf{K}$.

Lemma 2.4. *For any homomorphism $f \colon \mathbf{G} \to \mathbf{H}$, and any subset X of* $\mathbf{G}$*, we have $f(\Sigma X) = \Sigma(f(X))$. In particular, if* $\mathbf{H}$ *is an internal direct summand of* $\mathbf{G}$*, then $\pi_{\mathbf{H}}(\Sigma X) = \Sigma(\pi_{\mathbf{H}}(X))$ for any subset X of* $\mathbf{G}$.

Proof. Given $y \in f(\Sigma X)$, there exists $x \in \Sigma X$ such that $f(x) = y$. Hence we can find $x_1, \ldots, x_j \in X$ such that $x_1 + \cdots + x_j = x$, and so $f(x_1 + \cdots + x_j) = y$. But f is a homomorphism, and so $f(x_1) + \cdots + f(x_j) = y$, so that $y \in \Sigma(f(X))$. This shows that $f(\Sigma X) \subseteq \Sigma(f(X))$; the proof of the reverse inclusion is similar. □

Lemma 2.5. *Let* $\mathbf{G} = \mathbf{H} \oplus \mathbf{K}$*, and let* D *and* E *be multisets contained in* $\mathbf{H}$ *and* $\mathbf{K}$*, respectively. For any* $z \in \mathbf{G}$*,*

$$z \in \Sigma(D \cup E) \quad \text{if and only if} \quad \pi_{\mathbf{H}}(z) \in \Sigma D \text{ and } \pi_{\mathbf{K}}(z) \in \Sigma E.$$

Proof. Since $z = \pi_{\mathbf{H}}(z) + \pi_{\mathbf{K}}(z)$, the "if" direction is obvious. For the converse, note that

$$\pi_{\mathbf{H}}(z) \in \pi_{\mathbf{H}}\big(\Sigma(D \cup E)\big) = \Sigma\big(\pi_{\mathbf{H}}(D \cup E)\big)$$

by Lemma 2.4. On the other hand, $\pi_{\mathbf{H}}(D) = D$ and $\pi_{\mathbf{H}}(E) = \{e\}$, and so

$$\pi_{\mathbf{H}}(z) \in \Sigma\big(\pi_{\mathbf{H}}(D) \cup \pi_{\mathbf{H}}(E)\big) = \Sigma\big(D \cup \{e\}\big) = \Sigma D$$

(since the sumset is not affected by whether e is an allowed summand). A similar argument shows that $\pi_{\mathbf{K}}(z) \in \Sigma E$, which completes the proof of the lemma. □

Lemma 2.6. *Let* $\mathbf{H}$ *and* $\mathbf{K}$ *be subgroups of* $\mathbf{G}$ *satisfying* $\mathbf{H} \cap \mathbf{K} = \{e\}$*. Let* D *and* E *be multisets contained in* $\mathbf{H}$ *and* $\mathbf{K}$*, respectively. Then* $\#\Sigma(D \cup E) = \#\Sigma D \cdot \#\Sigma E$*.*

Proof. Notice that every element of $\Sigma(D \cup E)$ is contained in $\mathbf{H} + \mathbf{K}$; therefore we may assume without loss of generality that $\mathbf{G} = \mathbf{H} \oplus \mathbf{K}$. In particular, we may assume that $\mathbf{H}$ and $\mathbf{K}$ are internal direct summands of $\mathbf{G}$, so that the projection maps $\pi_{\mathbf{H}}$ and $\pi_{\mathbf{K}}$ exist and every element $z \in \mathbf{G}$ has a unique representation $z = x + y$ where $x \in \mathbf{H}$ and $y \in \mathbf{K}$; note that $x = \pi_{\mathbf{H}}(z)$ and $y = \pi_{\mathbf{K}}(z)$ in this representation.

To establish the lemma, it therefore suffices to show that $z = \pi_{\mathbf{H}}(z) + \pi_{\mathbf{K}}(z) \in \Sigma(D \cup E)$ if and only if $\pi_{\mathbf{H}}(z) \in \Sigma D$ and $\pi_{\mathbf{K}}(z) \in \Sigma E$; but this is exactly the statement of Lemma 2.5. □

The next lemma is a bit less standard yet still straightforward: in a direct product of two abelian groups, it characterizes the elements of a sumset that lie in a given coset of one of the direct summands.

Lemma 2.7. *Let* $\mathbf{H}$ *and* $\mathbf{K}$ *be subgroups of* $\mathbf{G}$ *satisfying* $\mathbf{H} \cap \mathbf{K} = \{e\}$*. Let* D *and* E *be multisets contained in* $\mathbf{H}$ *and* $\mathbf{K}$*, respectively. For any* $y \in \mathbf{K}$*:*

a. if $y \in \Sigma E$*, then* $(\mathbf{H} + \{y\}) \cap \Sigma(D \cup E) = \Sigma D + \{y\}$*;*

b. if $y \notin \Sigma E$*, then* $(\mathbf{H} + \{y\}) \cap \Sigma(D \cup E) = \emptyset$*.*

Proof. As in the proof of Lemma 2.6, we may assume without loss of generality that $\mathbf{G} = \mathbf{H} \oplus \mathbf{K}$. Suppose that z is an element of $(\mathbf{H}+\{y\}) \cap \Sigma(D \cup E)$. Since $z \in \mathbf{H}+\{y\}$, we may write $z = x+y$ for some $x \in \mathbf{H}$, whence $\pi_{\mathbf{K}}(z) = \pi_{\mathbf{K}}(x) + \pi_{\mathbf{K}}(y) = e+y = y$. On the other hand, since $z \in \Sigma(D \cup E)$, we see that $y \in \Sigma E$ by Lemma 2.5. In other words, the presence of any element $z \in (\mathbf{H} + \{y\}) \cap \Sigma(D \cup E)$ forces $y \in \Sigma E$, which establishes part (b) of the lemma.

We continue under the assumption $y \in \Sigma E$ to prove part (a). The inclusions $\Sigma D+\{y\} \subseteq \mathbf{H}+\{y\}$ and $\Sigma D+\{y\} \subseteq \Sigma(D\cup E)$ are both obvious, and so $\Sigma D+\{y\} \subseteq (\mathbf{H}+\{y\})\cap\Sigma(D\cup E)$. As for the reverse inclusion, let $z \in (\mathbf{H}+\{y\})\cap\Sigma(D\cup E)$ as above; then $\pi_{\mathbf{H}}(z) \in \Sigma D$ by Lemma 2.5, whence $z = \pi_{\mathbf{H}}(z)+\pi_{\mathbf{K}}(z) = \pi_{\mathbf{H}}(z)+y \in \Sigma D+\{y\}$ as required. □

Finally we can establish the lemma that we will make the most use of when we return to the setting $\mathbf{G} = \mathbb{Z}_p^2$ in the next section.

Lemma 2.8. *Let* $\mathbf{G} = \mathbf{H} \oplus \mathbf{K}$*, and let* C *be a multiset contained in* $\mathbf{G}$*. Let* $D = C \cap \mathbf{H}$*, let* $F = C \setminus D$*, and let* $E = \pi_{\mathbf{K}}(F)$*. Then* $\#\Sigma C \geq \#\Sigma D \cdot \#\Sigma E$*.*

Proof. Lemma 2.6 tells us that $\#\Sigma(D \cup E) = \#\Sigma D \cdot \#\Sigma E$, and so it suffices to show that $\#\Sigma C \geq \#\Sigma(D \cup E)$. We accomplish this by showing that

$$\#\big((\mathbf{H}+\{y\})\cap\Sigma C\big) \geq \#\big((\mathbf{H}+\{y\})\cap\Sigma(D\cup E)\big) \tag{1}$$

for all $y \in \mathbf{K}$.

For any $y \in \mathbf{K}\setminus\Sigma E$, Lemma 2.7 tells us that $(\mathbf{H}+\{y\})\cap\Sigma(D\cup E) = \emptyset$, in which case the inequality (1) holds trivially. For any $y \in \Sigma E$, Lemma 2.7 tells us that $(\mathbf{H}+\{y\})\cap\Sigma(D\cup E) = \Sigma D+\{y\}$, and so the right-hand side of the inequality (1) equals $\#\Sigma D$.

On the other hand, since $\Sigma E = \Sigma(\pi_{\mathbf{K}}(F)) = \pi_{\mathbf{K}}(\Sigma F)$ by Lemma 2.4, there exists at least one element $z \in \Sigma F$ satisfying $\pi_{\mathbf{K}}(z) = y$; as $\mathbf{G} = \mathbf{H}\oplus\mathbf{K}$, this is equivalent to saying that $z \in \mathbf{H}+\{y\}$. Since $\Sigma D \subseteq \mathbf{H}$, we have $\Sigma D+\{z\} \subseteq \mathbf{H}+\{y\}$ as well. But the inclusion $\Sigma D+\{z\} \subseteq \Sigma D+\Sigma F = \Sigma C$ is trivial, and therefore $\Sigma D+\{z\} \subseteq (\mathbf{H}+\{y\})\cap\Sigma C$; in particular, the left-hand side of the inequality (1) is at least $\#\Sigma D$. Combined with the observation that the right-hand side equals $\#\Sigma D$, this lower bound establishes the inequality (1) and hence the lemma. □

These lemmas might be valuable for studying sumsets in more general abelian groups. They will prove to be particularly useful for studying sumsets in $\mathbb{Z}_p^2$, however, essentially because there are many ways of writing $\mathbb{Z}_p^2$ as an internal direct sum of two subgroups (which are simply lines through $\mathbf{0}$).

3. Lower Bounds for Sumsets

In this section we establish Theorems 1.7 and 1.8; the proofs employ two combinatorial propositions which we defer to the next section. It would be possible to prove these two theorems at the same time, at the expense of a bit of clarity; however, we find it illuminating to give complete proofs of Theorem 1.7 (the cases $|A| = p$ and $|A| = p+1$) first, as the proofs will illustrate the methods used to prove the more general Theorem 1.8. Seeing the limitations of the proof of Theorem 1.7 will also motivate the formulation of our main technical tool, Lemma 3.2.

Throughout this section, A will denote a valid multiset contained in $\mathbb{Z}_p^2$. For any $x \in \mathbb{Z}_p^2$, we let $\langle x \rangle$ denotes the subgroup of $\mathbb{Z}_p^2$ generated by x (that is, the line passing through both the origin $\mathbf{0}$ and x), and we let m_x denote the multiplicity with which x appears in A, so that $|A| = \sum_{x \in \mathbb{Z}_p^2} m_x$. The fact that A is valid means that $m_{\mathbf{0}} = 0$ and $\sum_{t \in \langle x \rangle} m_t < p$ for every $x \in \mathbb{Z}_p^2 \setminus \{\mathbf{0}\}$.

Our first lemma quantifies the notion that we can establish sufficiently good lower bounds for the cardinality of ΣA if we know that there are enough elements of A lying in one subgroup of $\mathbb{Z}_p^2$. Naturally, the method of proof is to partition A into the elements lying in that subgroup and all remaining elements, project the remaining elements onto a complementary subgroup, and then use Lemma 1.4 in each subgroup separately.

Lemma 3.1. *Let A be any valid multiset contained in $\mathbb{Z}_p^2$. Suppose that for some $x \in \mathbb{Z}_p^2 \setminus \{\mathbf{0}\}$,*

$$\sum_{y \in \langle x \rangle} m_y \geq |A| - (p-1). \tag{2}$$

Then $\#\Sigma A \geq (|A| + 2 - p)p$.

Remark. The conclusion is trivial if $|A| < p - 1$; also, the fact that A is valid means that the left-hand side of equation (2) is at most $p - 1$, and so the lemma is vacuous if $|A| > 2p - 2$. Therefore in practice the lemma will be applied only to multisets A satisfying $p - 1 \leq |A| \leq 2p - 2$.

Proof. Let $D = A \cap \langle x \rangle$; note that $|D| \leq p - 1$ since A is a valid multiset, and note also that $|D| = \sum_{y \in \langle x \rangle} m_y \geq |A| - (p-1)$ by assumption. Set $F = A \setminus D$. Choose any nontrivial subgroup $\mathbf{K}$ of $\mathbb{Z}_p^2$ other than $\langle x \rangle$, and set $E = \pi_{\mathbf{K}}(F)$. Then by Lemma 2.8, we know that $\#\Sigma A \geq \#\Sigma D \cdot \#\Sigma E$. By Lemma 1.4 and the fact that $\mathbf{0} \notin D \cup E$, we obtain

$$\begin{aligned} \#\Sigma A &\geq \min\{p, 1 + |D|\} \cdot \min\{p, 1 + |E|\} \\ &= \min\{p, 1 + |D|\} \cdot \min\{p, 1 + |A| - |D|\}, \end{aligned} \tag{3}$$

since $|E| = |F| = |A| - |D|$. The inequalities $|D| \leq p - 1$ and $|A| - |D| \leq p - 1$ ensure that p is the larger element in both minima, and so we have simply

$$\#\Sigma A \geq (1 + |D|)(1 + |A| - |D|) = \tfrac{1}{4}|A|^2 + |A| + 1 - \left(|D| - \tfrac{1}{2}|A|\right)^2.$$

The pair of inequalities $|D| \leq p - 1$ and $|A| - |D| \leq p - 1$ is equivalent to the inequality $\left||D| - \frac{1}{2}|A|\right| \leq p - 1 - \frac{1}{2}|A|$; therefore

$$|\Sigma A| \geq \tfrac{1}{4}|A|^2 + |A| + 1 - \left(p - 1 - \tfrac{1}{2}|A|\right)^2 = (|A| + 2 - p)p,$$

as claimed. □

This lemma alone is sufficient to establish Theorem 1.7.

Proof of Theorem 1.7(a). When $|A| = p$, the right-hand side of the inequality (2) equals 1, and so the inequality holds for any $x \in A$. Therefore Lemma 3.1 automatically applies, yielding $\#\Sigma A \geq (|A|+2-p)p = 2p$ as desired. (In fact essentially the same proof gives the more general statement: if A is a multiset contained in $\mathbb{Z}_p^2 \setminus \{\mathbf{0}\}$ but not contained in any proper subgroup, and $|A| \geq p$, then $\#\Sigma A \geq 2|A|$.) □

Proof of Theorem 1.7(b). We are assuming that $|A| = p+1$. Suppose first that there exists a nontrivial subgroup of $\mathbb{Z}_p^2$ that contains at least two points of A (including possibly two copies of the same point). Choosing any nonzero element x in that subgroup, we see that the inequality (2) is satisfied, and so Lemma 3.1 yields $\#\Sigma A \geq (|A| + 2 - p)p = 3p$ as desired.

From now on we may assume that there does not exist a nontrivial subgroup of $\mathbb{Z}_p^2$ that contains at least two points of A. Since there are only $p+1$ nontrivial subgroups of $\mathbb{Z}_p^2$, it must be the case that A consists of exactly one point from each of these $p+1$ subgroups; in particular, the elements of A are distinct. We can verify the assertion for $p \leq 11$ by exhaustive computation (see the remarks after the end of this proof), so from now on we may assume that $p \geq 13$.

Suppose first that all sums of pairs of distinct elements from A are distinct. All these sums are elements of ΣA, and thus $\#\Sigma A \geq \binom{p+1}{2} > 3p$ since $p \geq 13$.

The only remaining case is when two pairs of distinct elements from A sum to the same point of $\mathbb{Z}_p^2$. Specifically, suppose that there exist $x_1, y_1, x_2, y_2 \in A$ such that $x_1 + y_1 = x_2 + y_2$. Partition $A = B_0 \cup B_1 \cup B_2$ where $B_1 = \{x_1, y_1\}$ and $B_2 = \{x_2, y_2\}$ and hence $B_0 = A \setminus \{x_1, y_1, x_2, y_2\}$; note that this really is a partition of A, as the fact that $x_1 + y_1 = x_2 + y_2$ forces all four elements to be distinct. Moreover, if we define $z = x_1 + y_1 = x_2 + y_2$, then we know that $z \neq \mathbf{0}$ since x_1 and y_1 are in different subgroups.

Define C to be the multiset $B_0 \cup \{z, z\}$; by Lemma 2.1, we know that $\#\Sigma A \geq \#\Sigma C$. Define $D = C \cap \langle z \rangle$; we claim that $|D| = 3$. To see this, note that A has exactly one point in every nontrivial subgroup, and in particular A has exactly one point in $\langle z \rangle$. Furthermore, that point cannot be x_1 for example, since then $y_1 = z - x_1$ would also be in that subgroup; similarly that point cannot be x_2, y_1, or y_2. We conclude that B_0 has exactly one point in $\langle z \rangle$, whence C has exactly three points in $\langle z \rangle$.

Now define $F = C \setminus D$, so that $|F| = |C| - |D| = (|B_0|+2) - 3 = (|A|-4+2) - 3 = p - 4$. Let $\mathbf{K}$ be any nontrivial subgroup other than $\langle z \rangle$, and set $E = \pi_{\mathbf{K}}(F)$. The lower bounds $\#\Sigma D \geq 4$ and $\#\Sigma E \geq p - 3$ then follow from Lemma 1.4. By Lemma 2.8, we conclude that $\#\Sigma C \geq \#\Sigma D \cdot \#\Sigma E = 4(p-3) > 3p$ since $p \geq 13$. □

Remark. The computation that verifies Theorem 1.7(b) for $p \leq 11$ should be done a little bit intelligently, since there are 10^{12} subsets A of $\mathbb{Z}_{11}^2$ (for example) consisting of exactly one nonzero element from each nontrivial subgroup. We describe the computation in the hardest case $p = 11$. Let us write the elements of $\mathbb{Z}_{11}^2$ as ordered pairs (s, t) with s and t considered modulo 11. By separately dilating the two coordinates of $\mathbb{Z}_{11}^2$ (which does not alter the cardinality of ΣA), we may assume without loss of generality that A contains both $(1, 0)$ and $(0, 1)$. We also know every such A contains a subset of the form $\{(i, i), (j, 2j), (k, 3k), (\ell, 4\ell)\}$ for some integers $1 \le i, j, k, \ell \le 10$. Therefore the cardinality of every such ΣA is at least as large as the cardinality of one of the subsumsets $\Sigma\big(\{(1, 0), (0, 1), (i, i), (j, 2j), (k, 3k), (\ell, 4\ell)\}\big)$.

There are 10^4 such subsumsets, and direct computation shows that all of them have more than 33 distinct elements except for the cases $\Sigma\big(\{(1, 0), (0, 1), \pm(1, 1), \pm(1, 2), \pm(1, 3), \pm(1, 4)\}\big)$, which each contain 32 distinct elements. It is then easily checked that any subsumset of the form $\Sigma\big(\{(1, 0), (0, 1), \pm(1, 1), \pm(1, 2), \pm(1, 3), \pm(1, 4), (m, 5m)\}\big)$ with $1 \le m \le 10$ contains more than 33 distinct elements. This concludes the verification of Theorem 1.7(b) for $p = 11$, and the cases $p \le 7$ are verified even more quickly.

We now foreshadow the proof of Theorem 1.8 by reviewing the structure of the proof of Theorem 1.7(b). In that proof, we quickly showed that the desired lower bound held if there were enough elements of A in the same subgroup. Also, the desired lower bound certainly held if there were enough distinct sums of pairs of elements of A. If however no subgroup contained enough elements of A and there were only a few distinct sums of pairs of elements of A, then we showed that we could find multiple pairs of elements summing to the same point in $\mathbb{Z}_p^2$. Replacing those elements in A with multiple copies of their joint sum, we found that the corresponding subgroup now contained enough elements to carry the argument through.

The following lemma quantifies the final part of this strategy, where we replace j pairs of elements of A with their joint sum and then use our earlier ideas to bound the cardinality of the sumset from below.

Lemma 3.2. *Let A be any valid multiset contained in $\mathbb{Z}_p^2$, and let $z \in \mathbb{Z}_p^2 \setminus \{\mathbf{0}\}$. For any integer j satisfying*

$$0 \le j \le \tfrac{1}{2} \sum_{t \in \mathbb{Z}_p^2 \setminus \langle z \rangle} \min\{m_t, m_{z-t}\}, \tag{4}$$

we have

$$\#\Sigma A \ge \min\Big\{p, 1 + j + \sum_{y \in \langle z \rangle} m_y\Big\} \min\Big\{p, 1 + |A| - 2j - \sum_{y \in \langle z \rangle} m_y\Big\}.$$

Remark. This can be seen as a generalization of Lemma 3.1, as equation (3) is the special case $j = 0$ of this lemma.

Proof. Partition $A = B_0 \cup B_1 \cup \cdots \cup B_j$, where for each $1 \le i \le j$, the multiset B_i has exactly two elements, neither contained in $\langle z \rangle$, that sum to z (the complementary submultiset B_0 is unrestricted). The upper bound (4) for j is exactly what is required for such a partition to be possible; the factor of $\frac{1}{2}$ arises because the sum on the right-hand side of (4) double-counts the pairs $(t, z-t)$ and $(z-t, t)$. Then set C equal to B_0 with j additional copies of z inserted. By Lemma 2.1, we know that $\#\Sigma A \ge \#\Sigma C$.

Now let D be the intersection of C with the subgroup $\langle z \rangle$, and let $F = C \setminus D$. Let $\mathbf{K}$ be any nontrivial subgroup other than $\langle z \rangle$, and set $E = \pi_{\mathbf{K}}(F)$. By Lemma 2.8, we know that $\#\Sigma C \ge \#\Sigma D \cdot \#\Sigma E$. However, the number of elements of D (counting multiplicity) is $j + |B_0 \cap \langle z \rangle|$; this is the same as $j + |A \cap \langle z \rangle|$ (since no elements of $B_1, \dots, B_j$ lie in $\langle z \rangle$), or in other words $j + \sum_{y \in \langle z \rangle} m_y$. Similarly, the number of elements of E (equivalently, of F) is equal to the number of elements of $B_0 \setminus \langle z \rangle$; this is the same as $|A \setminus \langle z \rangle| - 2j$, or in other words $|A| - 2j - \sum_{y \in \langle z \rangle} m_y$. The lower bounds $\#\Sigma D \ge \min\{p, 1 + j + \sum_{y \in \langle z \rangle} m_y\}$ and $\#\Sigma E \ge \min\{p, 1 + |A| - 2j - \sum_{y \in \langle z \rangle} m_y\}$ then follow from Lemma 1.4; the chain of inequalities $\#\Sigma A \ge \#\Sigma C \ge \#\Sigma D \cdot \#\Sigma E$ establishes the lemma. □

We are now ready to use Lemma 3.2 to establish Conjecture 1.6 when $|A| = p+k$, for all but finitely many primes p depending on k. Let $H_k = 1 + \frac{1}{2} + \cdots + \frac{1}{k}$ denote the kth harmonic number.

Theorem 3.3. *Let $k \ge 2$ be any integer, and let A be any valid multiset contained in $\mathbb{Z}_p^2$ such that $|A| = p + k$. If $p \ge 4(k+1)^2 H_k - 2k$, then $\#\Sigma A \ge (k+2)p$.*

Remark. Using the elementary bound $H_k \le \gamma + \log(k+1)$, where γ denotes the Euler–Mascheroni constant, we see that Theorem 3.3 holds as long as $p \ge 4(k+1)^2(\gamma + \log(k+1))$. Theorem 1.8 can thus be readily deduced from Theorem 3.3 as follows: If $k + 1 \le \sqrt{p/(2\log p + 1)}$ then $p \ge 4(k+1)^2(\frac{1}{4} + \frac{1}{2}\log p)$. In this case we have $p \ge (1 + 2\log 2)(k+1)^2$, whence $\log p \ge \frac{4}{5} + 2\log(k+1)$ and $\frac{1}{4} + \frac{1}{2}\log p \ge \gamma + \log(k+1)$.

Proof. If there are $k+1$ elements of A in some nontrivial subgroup, then we are done by Lemma 3.1. Therefore we may assume that there are at most k points in each subgroup; in particular, $m_x \le k$ for all $x \in \mathbb{Z}_p^2$. We now argue that if ΣA is small, then there must be lots of pairs of elements of A that add to the same element of $\mathbb{Z}_p^2$, at which point we will be able to invoke Lemma 3.2. We may assume that $\Sigma A \ne \mathbb{Z}_p^2$, for otherwise we are done immediately.

For each $1 \le i \le k$, we define the level set $A_i = \{x \in \mathbb{Z}_p^2 : m_x \ge i\}$. Notice that A can be written precisely as the multiset union $A_1 \cup A_2 \cup \cdots \cup A_k$, and so $\sum_{i=1}^k \#A_i = |A| = p + k$. Let B_i be the multiset formed by the sums of pairs of elements of A_i not in the same subgroup:

$$B_i = \{x + y \colon x, y \in A_i, \ \langle x \rangle \ne \langle y \rangle\}.$$

Note that $\mathbf{0} \notin B_i$ (the restriction $\langle x\rangle \neq \langle y\rangle$ ensures that $x \neq -y$) and that every element of B_i occurs with even multiplicity (the restriction $\langle x\rangle \neq \langle y\rangle$ ensures that $x \neq y$). It is not hard to estimate the relative sizes of $\#A_i$ and $|B_i|$: for each $x \in A_i$ there are at most k elements of A lying in the subgroup $\langle x\rangle$. Since each such x occurs with multiplicity at least i in A, there are at most k/i distinct values of y excluded by the condition $\langle x\rangle \neq \langle y\rangle$. Hence $|B_i| \geq \#A_i(\#A_i - k/i)$, which implies that

$$\#A_i \leq \frac{k}{i} + \sqrt{|B_i|}. \tag{5}$$

Since $\sum_{i=1}^{k} \#A_i$ is fixed, this shows that $|B_i|$ cannot be very small on average. At the same time, $\#B_i$ cannot get very large: if $\sum_{i=1}^{k} \#B_i \geq (2k+1)p$, then (under our assumption that $\Sigma A \neq \mathbb{Z}_p^2$) Lemma 2.2 already yields

$$\#\Sigma A \geq \sum_{i=1}^{k} \#\Sigma A_i - (k-1)p > \sum_{i=1}^{k} \#B_i - (k-1)p \geq (k+2)p.$$

where the middle inequality holds because $B_i \subseteq \Sigma A_i$. We may therefore assume henceforth that

$$\sum_{i=1}^{k} \#B_i < (2k+1)p. \tag{6}$$

Let us now introduce the weighted height parameter

$$\eta = \max_{1\leq i\leq k} \left\{ \frac{i|B_i|}{2\#B_i} : \#B_i > 0 \right\}. \tag{7}$$

We shall show shortly that $\eta > k+1$. Assuming so, then for some $1 \leq i \leq k$, we have

$$\frac{|B_i|}{2\#B_i} > \frac{k+1}{i},$$

so by the pigeonhole principle, there exists some $z \in B_i$ (in particular $z \neq \mathbf{0}$) occurring with multiplicity greater than $2(k+1)/i$; since this multiplicity is an even integer, it must be at least $2(k+2)/i$. For each solution $x+y=z$ corresponding to an occurrence of z in B_i, we have by construction that $x, y \notin \langle z\rangle$ and $m_x, m_y \geq i$, so for this particular choice of z,

$$\tfrac{1}{2} \sum_{t\in\mathbb{Z}_p^2\setminus\langle z\rangle} \min\{m_t, m_{z-t}\} \geq k+2.$$

Furthermore, $\sum_{y\in\langle z\rangle} m_y \leq k$ by assumption. Therefore we are free to apply Lemma 3.2 with $j = (k+2) - \sum_{y\in\langle z\rangle} m_y$, which gives the lower bound

$$\#\Sigma A \geq \min\{p, k+3\} \min\left\{p, p-k-3+\sum_{y\in\langle z\rangle} m_y\right\} \geq (k+3)(p-k-3) \geq (k+2)p$$

(the last step used the inequality $p \ge (k+3)^2$, which holds under the hypotheses of the theorem).

It remains only to verify that $\eta > k+1$. Summing the inequality (5) over all $1 \le i \le k$ yields

$$p+k = \sum_{i=1}^{k} \#A_i \le kH_k + \sum_{i=1}^{k} \sqrt{|B_i|} \le kH_k + \sqrt{2\eta} \sum_{i=1}^{k} \sqrt{\frac{\#B_i}{i}},$$

using the definition (7) of η. We estimate the rightmost sum using Cauchy–Schwarz together with the inequality (6):

$$\sum_{i=1}^{k} \sqrt{\frac{\#B_i}{i}} \le \left(\sum_{i=1}^{k} \#B_i\right)^{1/2} \left(\sum_{i=1}^{k} \frac{1}{i}\right)^{1/2} < \sqrt{(2k+1)pH_k}.$$

Combining the previous two inequalities gives $p+k-kH_k < \sqrt{\eta(4k+2)pH_k}$, so that

$$\eta > \frac{(p+k-kH_k)^2}{(4k+2)pH_k} > \frac{p(p+2(k-kH_k))}{(4k+2)pH_k} = \frac{(p+2k)-2kH_k}{(4k+2)H_k} \ge \frac{4(k+1)^2H_k - 2kH_k}{(4k+2)H_k}$$

by the hypothesis on the size of p. In other words,

$$\eta > \frac{2(k+1)^2 - k}{2k+1} = k+1+\frac{1}{2k+1},$$

which completes the proof of the theorem. □

4. A Wider Conjecture

As mentioned earlier, Conjecture 1.6 is just one part of a more far-reaching conjecture concerning sumsets of multisets in $\mathbb{Z}_p^m$. Before formulating that wider conjecture, we must expand the definition of a valid multiset to $\mathbb{Z}_p^m$.

Definition 4.1. Let p be an odd prime, and let m be a positive integer. A multiset A contained in $\mathbb{Z}_p^m$ is *valid* if:

- $\mathbf{0} \notin A$; and
- for each $1 \le d \le m$, every subgroup of $\mathbb{Z}_p^m$ that is isomorphic to $\mathbb{Z}_p^d$ contains fewer than dp points of A, counting multiplicity.

When $m = 1$, a multiset contained in $\mathbb{Z}_p$ is valid precisely when it does not contain 0; when $m = 2$ and $|A| < 2p$, this definition of valid agrees with Definition 1.5 for multisets contained in $\mathbb{Z}_p^2$. Note that in particular, Definition 4.1(b) implies that every valid multiset contained in $\mathbb{Z}_p^m$ has at most $mp-1$ elements, counting multiplicity. We now give an example showing that this upper bound $mp-1$ can in fact be achieved. Throughout this section, let $\{x_1,\ldots,x_m\}$ denote a generating set for $\mathbb{Z}_p^m$, and let $\mathbf{K}_d = \langle x_1,\ldots,x_d\rangle$ denote the subgroup of $\mathbb{Z}_p^m$ generated by $\{x_1,\ldots,x_d\}$, so that $\mathbf{K}_d \cong \mathbb{Z}_p^d$.

Example 4.2. Let A_1 be the multiset consisting of $p-1$ copies of x_1; for $2 \le j \le m$ let $A_j = \{x_j + ax_1 : 0 \le a \le p-1\}$; and define $B_m = \bigcup_{j=1}^m A_j$. Then $|B_m| = (p-1)+(m-1)p = mp-1$ and $\mathbf{0} \notin B_m$. To verify that B_m is a valid subset of $\mathbb{Z}_p^m$, let $\mathbf{H}$ be any subgroup of $\mathbb{Z}_p^m$ that is isomorphic to $\mathbb{Z}_p^d$; we need to show that B_m contains fewer than dp points of $\mathbf{H}$.

First suppose that $x_1 \notin \mathbf{H}$, which implies that $bx_1 \notin \mathbf{H}$ for every nonzero multiple bx_1 of x_1. Then for each $2 \le j \le m$, at most one of the elements of A_j can be in $\mathbf{H}$, since the difference of any two such elements is a nonzero multiple of x_1. Therefore $|B_m \cap \mathbf{H}| = \ell$ for some $1 \le \ell \le m-1$, and in fact all ℓ of these elements are of the form $x_j + ax_1$ for ℓ distinct values of j. Since no such element is in the subgroup spanned by the others, we conclude that $d \ge \ell$, and so the necessary inequality $|B_m \cap \mathbf{H}| = \ell \le d < dp$ is amply satisfied.

Now suppose that $x_1 \in \mathbf{H}$. Then for each $2 \le j \le m$, either all or none of the elements of A_j are in $\mathbf{H}$. By reindexing the x_i, we may choose an integer $1 \le \ell \le m$ such that $\mathbf{H}$ contains $A_1 \cup \cdots \cup A_\ell$ and is disjoint from $A_{\ell+1} \cup \cdots \cup A_m$. In particular, $|B_m \cap \mathbf{H}| = (p-1)+(\ell-1)p = \ell p - 1$. But $\mathbf{H}$ contains $\{x_1,\ldots,x_\ell\}$ and hence $d \ge \ell$, so that $\ell p - 1 \le dp - 1$ as required.

We may now state our wider conjecture; Conjecture 1.6 is the special case $q = 1$ of part (a) of this conjecture.

Conjecture 4.3. Let p be an odd prime. Let m be a positive integer, and let A be a valid multiset of $\mathbb{Z}_p^m$ with $|A| \ge p$. Write $|A| = qp + k$ with $0 \le k \le p-1$.

a. If $0 \le k \le p-3$, then $\#\Sigma A \ge (k+2)p^q$.

b. If $k = p-2$, then $\#\Sigma A \ge p^{q+1} - 1$.

c. If $k = p-1$, then $\#\Sigma A \ge p^{q+1}$.

In particular, if $|A| = mp-1$ then $\Sigma A = \mathbb{Z}_p^m$.

We remark that the quantity dp in Definition 4.1, bounding the number of elements in a valid multiset that can lie in a rank-d subgroup, has been carefully chosen in light of this conjecture: by Conjecture 4.3(c), any valid multiset A with at least $dp-1$ elements counting multiplicity must satisfy $\#\Sigma A \ge p^d$. In particular, if A is a valid multiset contained in a subgroup $\mathbf{H} < \mathbb{Z}_p^m$ that is isomorphic to $\mathbb{Z}_p^d$, then $|A| \ge dp-1$ implies that $\Sigma A = \mathbf{H}$. Therefore allowing dp elements in such a subgroup would always be "wasteful". Of course, the validity of Definition 4.1 for rank-d subgroups depends crucially upon the truth of Conjecture 4.3(c) for $q = d - 1$.

The conjecture is restricted to multisets A with $|A| \geq p$ because we already know the truth for smaller multisets, for which the definition of "valid" is simply the condition that $\mathbf{0} \notin A$: when $|A| \leq p - 1$, the best possible lower bound is $\#\Sigma A \geq |A| + 1$ as in Lemma 1.4. We remark that Peng [9, Theorem 2] has proved Conjecture 4.3(c) in the case $m = 2$ and $q = 1$, even under a slightly weaker hypothesis; in other words, he has shown that if A is a valid multiset contained in $\mathbb{Z}_p^2$ with $|A| = 2p - 1$, then $\Sigma A = \mathbb{Z}_p^2$. (Mann and Wou [5] have proved in the case that A is actually a set—that is, a multiset with distinct elements—that $\#A = 2p - 2$ suffices to force $\Sigma A = \mathbb{Z}_p^2$.) Peng considers the higher-rank groups $\mathbb{Z}_p^m$ as well, but the multisets he allows (see [10, Theorem 1]) form a much wider class than our valid multisets for $q \geq 2$, and so $|A|$ must be much larger than required by Conjecture 4.3(c) in order to imply $\Sigma A = \mathbb{Z}_p^m$ when $m \geq 3$. Finally, we mention that we have completely verified Conjecture 4.3 by exhaustive computation for the groups $\mathbb{Z}_p^2$ with $p \leq 7$ and also for the group $\mathbb{Z}_3^3$.

It is easy to see that all of the lower bounds in Conjecture 4.3(a), if true, would be best possible. Given $q \geq 1$ and $0 \leq k \leq p - 3$, let A' be any valid multiset contained in $\mathbf{K}_q$ with $|A'| = qp - 1$ (such as the one given in Example 4.2 with $m = q$), and let A be the union of A' with $k + 1$ copies of x_{q+1}. Then $\Sigma A = \{y + ax_{q+1} \colon y \in \Sigma A', 0 \leq a \leq k + 1\}$ and thus $\#\Sigma A' = (k + 2)\#\Sigma A \leq (k + 2)p^q$ since ΣA is contained in $\mathbf{K}_q$. Similarly, the fact that there exists a valid multiset contained in $\mathbf{K}_{q+1}$ with $qp + (q - 1) = (q + 1)p - 1$ elements (such as the one given in Example 4.2 with $m = q + 1$) shows that the lower bound in Conjecture 4.3(c) would be best possible, since the sumset of this multiset would still be contained in $\mathbf{K}_{q+1}$ and thus would have at most p^{q+1} distinct elements.

The lower bound in Conjecture 4.3(b) might seem counterintuitive, especially in comparison with the pattern established in Conjecture 4.3(a). However, we can give an explicit example showing that the lower bound $p^{q+1} - 1$ for $\#\Sigma A$ cannot be increased:

Example 4.4. When p is an odd prime, define B'_m to be the set B_m from Example 4.2 with one copy of x_1 removed, so that B'_m contains x_1 with multiplicity only $p - 2$. Since B_m is a valid multiset contained in $\mathbb{Z}_p^m$, so is B'_m. We have $|B'_m| = |B_m| - 1 = (mp - 1) - 1 = (m - 1)p + (m - 2)$, and we claim that $-x_1 \notin \Sigma B'_m$; this will imply that $\#\Sigma B'_m \leq p^m - 1$, and so the lower bound for $\#\Sigma A$ in Conjecture 4.3(b) cannot be increased. (In fact it is not hard to show that every other element of $\mathbb{Z}_p^m$ is in $\Sigma B'_m$, and so $\#\Sigma B'_m$ is exactly equal to $p^m - 1$.)

Suppose for the sake of contradiction that $-x_1 \in \Sigma B'_m$, and let C be a submultiset of B'_m such that $-x_1 = \sum_{y \in C} y$. For each $2 \leq j \leq m$, define $\ell_j = |C \cap A_j| =$

$\#\big(C \cap \{x_j + ax_1 \colon 0 \le a \le p-1\}\big)$. Then

$$-x_1 = \sum_{y\in C} y = tx_1 + \ell_2 x_2 + \ell_3 x_3 + \cdots + \ell_m x_m$$

for some integer t. It follows from this equation that each ℓ_j must equal either 0 or p. However, if $\ell_j = p$ then

$$\sum_{y\in C\cap A_j} y = \sum_{0\le a\le p-1} (x_j + ax_1) = px_j + \frac{p(p-1)}{2} x_1 = \mathbf{0}$$

(since p is odd). So in either case, if $s = |C \cap A_1|$ is the multiplicity with which x_1 appears in C, then

$$-x_1 = \sum_{y\in C} y = sx_1 + \sum_{j=2}^{m} \sum_{y\in C\cap A_j} y = sx_1 + \mathbf{0} + \cdots + \mathbf{0}.$$

This is a contradiction, however, since s must lie between 0 and $p-2$. Therefore $-x_1$ is indeed not an element of $\Sigma B'_m$, as claimed.

A naive application of Lemma 2.8 seems to relatively ineffective in extending our main theorem to partial results toward Conjecture 4.3 for the higher-rank groups $\mathbb{Z}_p^m$, say using the decomposition $\mathbb{Z}_p^m = \mathbb{Z}_p^{m-1} \oplus \mathbb{Z}_p$ inductively. If a multiset C is very well-distributed across subgroups, then we expect the cardinality of $D = C \cap \mathbb{Z}_p^{m-1}$ to be about p times smaller than that of C. Without exploiting the structure of C as in the proof of Theorem 1.8, we would thus require $|C| \gg p^{m-1}$ elements to obtain $\Sigma C = \mathbb{Z}_p^m$. This is comparable to the aforementioned results of Peng, rather than the linear growth in m which is predicted by Conjecture 4.3.

The line of questioning in this section turns out to be uninteresting when $p = 2$: when the multiset A does not contain $\mathbf{0}$, the condition that no rank-1 subgroup of $\mathbb{Z}_2^m$ contain 2 points of A is simply equivalent to A not containing any element with multiplicity greater than 1. It is easy to check that if A consists of any q points in $\mathbb{Z}_2^m$ that do not lie in any subgroup isomorphic to $\mathbb{Z}_2^{q-1}$, then ΣA fills out the entire rank-q subgroup generated by A. In other words, the analogous definition of "valid" for multisets in $\mathbb{Z}_2^m$ would simply be a set of q points that generate a rank-q subgroup of $\mathbb{Z}_2^m$, and we would always have $\#\Sigma A = 2^{|A|} = 2^{\#A}$ for valid (multi)sets in $\mathbb{Z}_2^m$.

Acknowledgments The collaboration leading to this paper was made possible thanks to Jean–Jacques Risler, Richard Kenyon, and especially Ivar Ekeland; the authors also thank the University of British Columbia and the Institut d'Études Politiques de Paris for their undergraduate exchange program. The first author thanks Andrew Granville for conversations that explored this topic and eventually led to the formulation of the conjectures herein.

References

[1] A. Cauchy, *Recherches sur les nombres*, Jour. Ecole Polytechn. **9** (1813), 99–116.

[2] H. Davenport, *On the addition of residue classes*, J. London Math. Soc. **10** (1935), 30–32.

[3] H. Davenport, *A historical note*, J. London Math. Soc. **22** (1947), 100–101.

[4] M. Kneser, *Abschätzungen der asymptotischen Dichte von Summenmengen*, Math. Z. **58** (1953), 459–484.

[5] H.B. Mann and Y.F. Wou, *An addition theorem for the elementary abelian group of type* (p,p), Monatsh. Math. **102** (1986), 273–308.

[6] G. Martin, *Dense Egyptian fractions*, Trans. Amer. Math. Soc. **351** (1999), no. 9, 3641–3657.

[7] K. Matomäki, *On sumsets of multisets in* $\mathbb{Z}_p^m$, Electr. J. Combin. **20** (2013), P30.

[8] M.B. Nathanson, *Additive Number Theory: Inverse Problems and the Geometry of Sumsets*, vol. 165 of Graduate Texts in Mathematics, Springer–Verlag, New York, 1996.

[9] C. Peng, *Addition theorems in elementary abelian groups, I*, J. Number Th. **27** (1987), 46–57.

[10] C. Peng, *Addition theorems in elementary abelian groups, II*, J. Number Th. **27** (1987), 58–62.

ON SEQUENCES WITHOUT GEOMETRIC PROGRESSIONS

Melvyn B. Nathanson
Department of Mathematics, Lehman College (CUNY), Bronx, New York
melvyn.nathanson@lehman.cuny.edu

Kevin O'Bryant
Dept. of Mathematics, College of Staten Island (CUNY), Staten Island, New York
obryant@mail.csi.cuny.edu

Received: 6/23/13, Accepted: 10/31/13, Published: 11/11/13

Abstract
An improved upper bound is obtained for the density of sequences of positive integers that contain no k-term geometric progression.

1. A Problem of Rankin

Let $k \geq 3$ be an integer. Let $r \neq 0, \pm 1$ be a real number. A *geometric progression of length k with common ratio r* is a sequence $(a_0, a_1, a_2, \ldots, a_{k-1})$ of nonzero real numbers such that

$$r = \frac{a_i}{a_{i-1}}$$

for $1, 2, \ldots, k-1$. For example, $(3/4, 3/2, 3, 6)$ and $(8, 12, 18, 27)$ are geometric progressions of length 4 with common ratios 2 and $3/2$, respectively. A *k-geometric progression* is a geometric progression of length k with common ratio r for some r. If the sequence $(a_0, a_1, a_2, \ldots, a_{k-1})$ is a k-geometric progression, then $a_i \neq a_j$ for $0 \leq i < j \leq k-1$.

A finite or infinite set of real numbers is *k-geometric progression free* if the set does not contain numbers $a_0, a_1, \ldots, a_{k-1}$ such that the sequence $(a_0, a_1, \ldots, a_{k-1})$ is a k-geometric progression. Rankin [3] introduced k-geometric progression free sets, and proved that there exist infinite k-geometric progression free sets with positive asymptotic density.[1] For example, the set Q of square-free positive integers, with asymptotic density $d(Q) = \pi^2/6$, contains no k-term geometric progression for $k \geq 3$.

[1] If $A(n)$ denotes the number of positive integers $a \in A$ with $a \leq n$, then the *upper asymptotic density* of A is $d_U(A) = \limsup_{n\to\infty} A(n)/n$, and the *asymptotic density* of A is $d(A) = \lim_{n\to\infty} A(n)/n$, if this limit exists.

Let A be a set of positive integers that contains no k-term geometric progression. Brown and Gordon [2] proved[2] that the upper asymptotic density of A, denoted $d_U(A)$, has the following upper bound:

$$d_U(A) \leq 1 - \frac{1}{2^k} - \frac{2}{5}\left(\frac{1}{5^{k-1}} - \frac{1}{6^{k-1}}\right).$$

Riddell [4] and Beiglböck, Bergelson, Hindman, and Strauss[1] proved that

$$d_U(A) \leq 1 - \frac{1}{2^k - 1}.$$

The purpose of this note is to improve these results.

2. An Upper Bound for Sets with No k-Term Geometric Progression

Theorem 1. *For integers $k \geq 3$ and $n \geq 2^{k-1}$, let $GPF_k(n)$ denote the set of subsets of $\{1, 2, \ldots, n\}$ that contain no k-term geometric progression. If $A \in GPF_k(n)$, then*

$$n - |A| \geq \left(\frac{1}{2^k - 1} + \frac{2}{5}\left(\frac{1}{5^{k-1}} - \frac{1}{6^{k-1}}\right) + \frac{4}{15}\left(\frac{1}{7^{k-1}} - \frac{1}{10^{k-1}}\right)\right) n + O\left(\frac{\log n}{k}\right).$$

Proof. Let

$$L = \left[\frac{\log 2n}{k \log 2}\right].$$

For $1 \leq \ell \leq L$ we have $2^{\ell k - 1} \leq n$. Let a be an odd positive integer such that

$$a \leq \frac{n}{2^{\ell k - 1}}.$$

The sequence

$$\left(2^{(\ell-1)k}a, 2^{(\ell-1)k+1}a, 2^{(\ell-1)k+2}a, \ldots, 2^{\ell k - 1}a\right)$$

is a geometric progression of length k with common ratio 2. If $A \in GPF_k(n)$, then A does not contain this geometric progression, and so at least one element in the set

$$X_\ell(a) = \left\{2^{(\ell-1)k}a, 2^{(\ell-1)k+1}a, 2^{(\ell-1)k+2}a, \ldots, 2^{\ell k - 1}a\right\}$$

is not an element of A. Because every nonzero integer has a unique representation as the product of an odd integer and a power of 2, it follows that, for integers $\ell = 1, \ldots, L$ and odd positive integers $a \leq 2^{1-\ell k}n$, the sets $X_\ell(a)$ are pairwise disjoint subsets of $\{1, 2, \ldots, n\}$.

[2] Brown and Gordon claimed a slightly stronger result, but their proof contains an (easily corrected) error.

For every real number $t \geq 1$, the number of odd positive integers not exceeding t is strictly greater than $(t-1)/2$. It follows that the cardinality of the set $\{1, 2, \ldots, n\} \setminus A$ is strictly greater than

$$\sum_{\ell=1}^{L} \frac{1}{2}\left(\frac{n}{2^{\ell k-1}} - 1\right) = \sum_{\ell=1}^{L}\left(\frac{n}{2^{\ell k}} - \frac{1}{2}\right) = n\sum_{\ell=1}^{L}\frac{1}{2^{\ell k}} + O\left(\frac{\log n}{k}\right)$$
$$= \frac{n}{2^k - 1} + O\left(\frac{\log n}{k}\right).$$

Note that if r is an odd integer and $r \in X_\ell(a)$, then $\ell = 1$ and $r = a$.

Let b be an odd integer such that

$$\frac{n}{6^{k-1}} < b \leq \frac{n}{5^{k-1}} \tag{1}$$

and b is not divisible by 5, that is,

$$b \equiv 1, 3, 7, \text{ or } 9 \pmod{10}. \tag{2}$$

We consider the following geometric progression of length k with ratio $5/3$:

$$(3^{k-1}b, 3^{k-2}5b, \ldots, 3^{k-1-i}5^i b, \cdots, 5^{k-1}b).$$

Every integer in this progression is odd, and

$$\frac{n}{2^{k-1}} < 3^{k-1}b < \cdots < 5^{k-1}b \leq n.$$

Let

$$Y(b) = \{3^{k-1}b, 3^{k-2}5b, \ldots, 3^{k-1-i}5^i b, \cdots, 5^{k-1}b\}.$$

It follows that $X_\ell(a) \cap Y(b) = \emptyset$ for all ℓ, a, and b. If the integers b and b' satisfy (1) and (2) with $b < b'$ and if $Y(b) \cap Y(b') \neq \emptyset$, then there exist integers $i, j \in \{0, 1, 2, \ldots, k-1\}$ such that $3^{k-1-i}5^i b = 3^{k-1-j}5^j b'$ or, equivalently,

$$5^{i-j}b = 3^{i-j}b'.$$

The inequality $b < b'$ implies that $0 \leq j < i \leq k-1$ and so $b' \equiv 0 \pmod 5$, which contradicts (2). Therefore, the sets $Y(b)$ are pairwise disjoint. The number of integers b satisfying inequality (1) and congruence (2) is

$$\frac{2}{5}\left(\frac{1}{5^{k-1}} - \frac{1}{6^{k-1}}\right)n + O(1).$$

Let c be an odd integer such that

$$\frac{n}{10^{k-1}} < c \leq \frac{n}{7^{k-1}} \tag{3}$$

and c is not divisible by 3 or 5, that is,

$$c \equiv 1, 7, 11, 13, 17, 19, 23, \text{ or } 29 \pmod{30}. \tag{4}$$

We consider the following geometric progression of length k with ratio $7/5$:

$$(5^{k-1}c, 5^{k-2}7c, \ldots, 5^{k-1-i}7^i c, \cdots, 7^{k-1}c).$$

Every integer in this progression is odd, and

$$\frac{n}{2^{k-1}} < 5^{k-1}c < \cdots < 7^{k-1}c \le n.$$

Let

$$Z(c) = \{5^{k-1}c, 5^{k-2}7c, \ldots, 5^{k-1-i}7^i c, \cdots, 7^{k-1}c\}.$$

It follows that $X_\ell(a) \cap Z(c) = \emptyset$ for all ℓ, a, and c. If c and c' satisfy (3) and (4) with $c < c'$ and if $Z(c) \cap Z(c') \neq \emptyset$, then there exist integers $i, j \in \{0, 1, 2, \ldots, k-1\}$ such that $5^{k-1-i}7^i c = 5^{k-1-j}7^j c'$ or, equivalently,

$$7^{i-j}c = 5^{i-j}c'.$$

The inequality $c < c'$ implies that $0 \le j < i \le k-1$ and so $c \equiv 0 \pmod 5$, which contradicts (4). Therefore, the sets $Z(c)$ are pairwise disjoint.

If b and c satisfy inequalities (1) and (3), respectively, then $c < b$. If $Y(b) \cap Z(c) \neq \emptyset$, then there exist integers $i, j \in \{0, 1, \ldots, k-1\}$ such that $5^{k-1-i}7^i c = 5^{k-1-j}3^j b$ or, equivalently,

$$5^j 7^i c = 5^i 3^j b.$$

Because $bc \not\equiv 0 \pmod 5$, it follows that $i = j$ and so

$$7^i c = 3^i b.$$

Because $c < b$, we must have $i \ge 1$ and so $c \equiv 0 \pmod 3$, which contradicts congruence (4). Therefore, $Y(b) \cap Z(c) = \emptyset$ and the sets $X_\ell(a)$, $Y(b)$, and $Z(c)$ are pairwise disjoint. The number of integers c satisfying inequality (3) and congruence (4) is

$$\frac{4}{15}\left(\frac{1}{7^{k-1}} - \frac{1}{10^{k-1}}\right) n + O(1).$$

Because A contains no k-term geometic progression, at least one element from each of the sets $X_\ell(a)$, $Y(b)$, and $Z(c)$ is not in A. This completes the proof. □

Corollary 1. *If A_k is a set of positive integers that contains no k-term geometric progression, then*

$$d_U(A_k) \le 1 - \frac{1}{2^k - 1} - \frac{2}{5}\left(\frac{1}{5^{k-1}} - \frac{1}{6^{k-1}}\right) - \frac{4}{15}\left(\frac{1}{7^{k-1}} - \frac{1}{10^{k-1}}\right).$$

Here is a table of upper bounds for $d_U(A)$ for various values of k:

k	3	4	5	6	7	10	17
$d_U(A_k) \leq$	0.84948	0.93147	0.96733	0.98404	0.99211	0.99902	0.99999

3. Open Problems

For every integer $k \geq 3$, let GPF_k denote the set of sets of positive integers that contain no k-term geometric progression. It would be interesting to determine precisely

$$\sup\{d_U(A) : A \in GPF_k\}$$

and

$$\sup\{d(A) : A \text{ has asymptotic density and } A \in GPF_k\}.$$

In the special case $k = 3$, Riddell [4, p. 145] claimed that if $A \in GPF_3$, then $d_U(A) < 0.8339$, but wrote, "The details are too lengthy to be included here."

An infinite sequence $A = (a_i)_{i=1}^{\infty}$ of positive integers is *syndetic* if it is strictly increasing with bounded gaps. Equivalently, A is syndetic if there is a number c such that $1 \leq a_{i+1} - a_i \leq c$ for all positive integers i. Beiglböck, Bergelson, Hindman, and Strauss [1] asked if every syndetic sequence must contain arbitrarily long finite geometric progressions.

References

[1] M. Beiglböck, V. Bergelson, N. Hindman, and D. Strauss, *Multiplicative structures in additively large sets*, J. Combin. Theory Ser. A **113** (2006), no. 7, 1219–1242.

[2] B. E. Brown and D. M. Gordon, *On sequences without geometric progressions*, Math. Comp. **65** (1996), no. 216, 1749–1754.

[3] R. A. Rankin, *Sets of integers containing not more than a given number of terms in arithmetical progression*, Proc. Roy. Soc. Edinburgh Sect. A **65** (1960/1961), 332–344 (1960/61).

[4] J. Riddell, *Sets of integers containing no n terms in geometric progression*, Glasgow Math. J. **10** (1969), 137–146.

INCLUSION-EXCLUSION POLYNOMIALS WITH LARGE COEFFICIENTS

Bartłomiej Bzdęga
Adam Mickiewicz University, Poznań, Poland
exul@amu.edu.pl

Received: 2/1/12, Revised: 10/17/13, Accepted: 11/3/13, Published: 11/11/13

Abstract

We prove that for every positive integer k there exists an inclusion-exclusion polynomial $Q_{\{q_1,q_2,\ldots,q_k\}}$ having height at least $d^{2^k}\prod_{j=1}^{k-2} q_j^{2^{k-j-1}-1}$, where d is a positive constant and $q_1 < q_2 < \ldots < q_k$ are pairwise coprime and arbitrarily large.

1. Introduction

The nth cyclotomic polynomial is the unique monic polynomial having as its simple roots all the nth primitive roots of unity. It can be shown that its coefficients are all integers.

It is well-known that if $n = p_1p_2\ldots p_k$, where $p_1, p_2, \ldots, p_k$ are distinct primes, then

$$\Phi_n(x) = \frac{(1-x^n)\cdot\prod_{1\le j_1<j_2\le k}\left(1-x^{n/(p_{j_1}p_{j_2})}\right)\cdot\ldots}{\prod_{1\le j_1\le k}(1-x^{n/p_{j_1}})\cdot\prod_{1\le j_1<j_2<j_3\le k}\left(1-x^{n/(p_{j_1}p_{j_2}p_{j_3})}\right)\cdot\ldots}. \qquad (1)$$

Bachman [1] defined a slightly more general class of polynomials, called inclusion-exclusion polynomials. The case $k = 3$ is considered in [2]. If we replace the primes $p_1, p_2, \ldots, p_k$ by pairwise coprime numbers $q_1, q_2, \ldots, q_k > 1$ in the formula above and put $m = q_1q_2\ldots q_k$ instead of n, then we arrive at the definition of the inclusion-exclusion polynomial Q_ρ, where $\rho = \{q_1, q_2, \ldots, q_k\}$.

We can expect that properties of inclusion-exclusion polynomials and cyclotomic polynomials are similar. In particular, we may use the same methods to bound the coefficients of polynomials of these both classes, as long as we use formula (1) only and do not make use of the assumption that the numbers $p_1, p_2, \ldots, p_k$ are prime. In the note we illustrate this by an example.

Throughout the paper we set $n = p_1p_2\ldots p_k$, $m = q_1q_2\ldots q_k$, and $\rho = \{q_1, q_2\ldots, q_k\}$. We also assume that $p_1 < p_2 < \ldots < p_k$ and $q_1 < q_2 < \ldots < q_k$.

Let A_n be the largest (in absolute value) coefficient of the polynomial Φ_n. Similarly we define A_ρ for the polynomial Q_ρ. Put

$$M_n = \prod_{j=1}^{k-2} p_j^{2^{k-j-1}-1} \quad \text{and} \quad M_\rho = \prod_{j=1}^{k-2} q_j^{2^{k-j-1}-1},$$

where both products equal 1 for $k < 3$.

We define C_k to be the smallest real number for which the inequality $A_n \leq C_k M_n$ holds for all sufficiently large $p_1 = \min\{p_1, p_2, \ldots, p_k\}$. In [4] it is proved that $C_k < (c + o(1))^{2^k}$, where $c \approx 0.9541$. In the proof the assumption that $p_1, p_2, \ldots, p_k$ are primes is never used, so this estimation is true for the inclusion-exclusion polynomials as well. More precisely, the following holds.

Theorem 1. *Let D_k be the smallest real number for which the inequality $A_\rho \leq D_k M_\rho$ holds for all sufficiently large $q_1 = \min\{q_1, q_2, \ldots, q_k\}$. We have $D_k < (c + o(1))^{2^k}$ with $c \approx 0.9541$.*

The aim of this paper is to construct for every positive integer k an infinite family of inclusion-exclusion polynomials Q_ρ with A_ρ at least $d^{2^k} M_\rho$, where d is a positive constant. In this way we prove the following theorem.

Theorem 2. *We have $D_k > (d + o(1))^{2^k}$ with $d \approx 0.5496$.*

2. Proof of Theorem 2

Bateman, Pomerance and Vaughan ([3], Lemma 5, p. 188) proved that if r is a positive integer and $p_j \equiv 2r \pm 1 \pmod{4r}$ for $j = 1, 2, \ldots, k$, then $A_n \geq (4r/\pi)^{2^{k-1}}/n$, where $n = p_1 p_2 \ldots p_k$. Their proof uses only formula (1) and does not require the assumption that $p_1, p_2, \ldots, p_k$ are primes. Therefore we deduce the following inclusion-exclusion polynomial version of this lemma.

Lemma 3. *Let r be a positive integer. If $q_j \equiv 2r \pm 1 \pmod{4r}$ for $j = 1, 2, \ldots, k$, then $A_\rho \geq (4r/\pi)^{2^{k-1}}/m$, where $m = q_1 q_2 \ldots q_k$.*

Now we are ready to prove the main result of this paper.

Proof of Theorem 2. Let N be a positive integer, $r = Nk!$ and $q_j = (4j-2)r+1$ for $j = 1, 2, \ldots, k$. First we check that the numbers $q_1, q_2, \ldots, q_k$ are pairwise coprime. By the Euclidean algorithm we have $(q_i, q_j) = ((4i-2)r+1, (4j-2)r+1) = (4(i-j)r, (4j-2)r+1) = 1$, because every prime divisor of $4(i-j)r$ divides N or is not greater than k, and the number $(4j-2)r+1$ has no such prime divisors.

By Lemma 3 we have

$$(A_\rho/M_\rho)^{2^{-k}} > \left(\frac{(4r/\pi)^{2^{k-1}}/m}{\prod_{j=1}^{k-2} q_j^{2^{k-j-1}-1}}\right)^{2^{-k}} = \frac{2}{\sqrt{\pi}}\left(\frac{r^{2^{k-1}}/q_k}{\prod_{j=1}^{k-1} q_j^{2^{k-j-1}}}\right)^{2^{-k}}$$
$$= \frac{2}{\sqrt{\pi}}\left(\frac{r}{q_k}\prod_{j=1}^{k-1}\left(\frac{r}{q_j}\right)^{2^{k-j-1}}\right)^{2^{-k}} = \frac{2}{\sqrt{\pi}}\prod_{j=1}^{\infty}(4j-2)^{-2^{-j-1}} + o(1) = d + o(1),$$

because the product is convergent. Numerical computations give $d \approx 0.5496$, which completes the proof. □

3. Discussion

Put $\overline{c} = \limsup_{k\to\infty}(C_k)^{2^{-k}}$ and $\underline{c} = \liminf_{k\to\infty}(C_k)^{2^{-k}}$ for cyclotomic polynomials, and similarly $\overline{d} = \limsup_{k\to\infty}(D_k)^{2^{-k}}$ and $\underline{d} = \liminf_{k\to\infty}(D_k)^{2^{-k}}$ for inclusion-exclusion polynomials. The following corollary is an immediate consequence of Theorems 1 and 2.

Corollary 4. *We have $d \le \underline{d} \le \overline{d} \le c$ with $d \approx 0.5496$ and $c \approx 0.9541$.*

It would be much more challenging to prove the analogous results for cyclotomic polynomials. Currently we know only that $\overline{c} \le c$, where $c \approx 0.9541$. We do not even know if $\overline{c} > 0$.

In the definition of C_k the assumption that $p_1 = \min\{p_1, p_2, \dots, p_k\}$ is sufficiently large is important. Without it we would have (see [4], concluding remarks)

$$\overline{c} \ge \underline{c} \ge \prod_{j=1}^{\infty} p_j^{-2^{3-j}} = c_1 \approx 0.0001442,$$

where p_j denotes the jth prime number.

Acknowledgments. The author is partially supported by NCN grant no. 2012/07/D/ST1/02111. He would like to thank Wojciech Gajda for his remarks on this note and Bartosz Naskręcki for his numerical computations of the constants c_1 and d.

References

[1] G. Bachman, *On ternary inclusion-exclusion polynomials*, Integers **10** (2010), 623–638.

[2] G. Bachman, P. Moree, *On a class of ternary inclusion – exclusion polynomials*, Integers **11** (2011), 77–91.

[3] P.T. Bateman, C. Pomerance, R.C. Vaughan, *On the size of the coefficients of cyclotomic polynomials*, Coll. Math. Soc. J. Bolyai **34** (1981), 171–202.

[4] B. Bzdęga, *On the height of cyclotomic polynomials*, Acta Arith. **152** (2012), 349–359.

AN EXPLICIT FORMULA FOR HIGHER ORDER BERNOULLI POLYNOMIALS OF THE SECOND KIND

Haiqing Wang
Dept. of Mathematics, Huizhou University, Huizhou, Guangdong, P. R. China

Guodong Liu[1]
Dept. of Mathematics, Huizhou University, Huizhou, Guangdong, P. R. China
gdliu@pub.huizhou.gd.cn

Received: 3/15/13, Accepted: 11/5/13, Published: 11/27/13

Abstract

In this paper, the authors establish a formula expressing the Bernoulli polynomials of the second kind and general order k, $b_n^{(k)}(x)$, in terms of those of first order, $b_n(x) = b_n^{(1)}(x)$.

1. Introduction and Results

The Bernoulli polynomials $b_n^{(k)}(x)$ of the second kind and of order k, for any integer k, may be defined by (see [2,4,10])

$$\left(\frac{t}{\log(1+t)}\right)^k (1+t)^x = \sum_{n=0}^{\infty} b_n^{(k)}(x)t^n, \quad |t| < 1. \tag{1}$$

The numbers $b_n^{(k)} = b_n^{(k)}(0)$ are the Bernoulli numbers of the second kind and of order k; $b_n^{(1)} = b_n$, $b_n^{(1)}(x) = b_n(x)$ are the ordinary Bernoulli numbers and polynomials of the second kind (see [1,4,5, 10-13]), and $C_n = n!b_n$ are the Cauchy numbers of the first kind (see [8, 13]). By (1.1), we have

$$b_n^{(k)}(x) = \sum_{\substack{v_1,\cdots,v_k\in\mathbb{N}_0 \\ v_1+\cdots+v_k=n}} b_{v_1}(x/k)b_{v_2}(x/k)\cdots b_{v_k}(x/k), \tag{2}$$

$$b_n^{(k)} = \sum_{\substack{v_1,\cdots,v_k\in\mathbb{N}_0 \\ v_1+\cdots+v_k=n}} b_{v_1}b_{v_2}\cdots b_{v_k}, \tag{3}$$

[1]Corresponding author.

and

$$b_n^{(k)}(x) = b_0^{(k)}\binom{x}{n} + b_1^{(k)}\binom{x}{n-1} + \cdots + b_{n-1}^{(k)}\binom{x}{1} + b_n^{(k)}. \tag{4}$$

where $n \in \mathbb{N}_0 = \mathbb{N} \cup \{0\}$, $\mathbb{N}$ being the set of positive integers.

The numbers b_n satisfy the recurrence relation (see [1, 5])

$$b_0 = 1, \quad \sum_{j=0}^{n} \frac{(-1)^j}{n-j+1} b_j = 0 \quad (n \geq 1), \tag{5}$$

so we find $b_1 = \frac{1}{2}, b_2 = -\frac{1}{12}, b_3 = \frac{1}{24}, b_4 = -\frac{19}{720}, b_5 = \frac{3}{160}, b_6 = -\frac{863}{60480}$.

The numbers b_n satisfy many interesting relations. For example (see [1,5,8])

$$b_n = \int_0^1 \binom{x}{n} dx, \quad 1 - \log 2 = \sum_{n=1}^{\infty} \frac{|b_n|}{n+1}, \quad \gamma = \sum_{n=1}^{\infty} \frac{|b_n|}{n}, \quad 1 + \sum_{n=1}^{\infty} \frac{|b_n| H_n}{n} = \frac{\pi^2}{6}, \tag{6}$$

where $\binom{x}{n} = \frac{x(x-1)(x-2)\cdots(x-n+1)}{n!}$, γ is the Euler constant and $H_n = \sum_{k=1}^{n} \frac{1}{k}$ is the n-th harmonic number.

The Bernoulli polynomials $B_n^{(k)}(x)$ of order k, for any integer k, may be defined by (see [3,4,6,7,10])

$$\left(\frac{t}{e^t - 1}\right)^k e^{xt} = \sum_{n=0}^{\infty} B_n^{(k)}(x) \frac{t^n}{n!}, \quad |t| < 2\pi. \tag{7}$$

The numbers $B_n^{(k)} = B_n^{(k)}(0)$ are the Bernoulli numbers of order k, and $B_n^{(1)} = B_n$ are the ordinary Bernoulli numbers. The numbers $B_n^{(n)}$ are called Nörlund numbers (see [3]), or Cauchy numbers of the second kind (see [8, 13]). Nörlund found the exponential generating function (see [10, p.150])

$$\frac{t}{(1+t)\log(1+t)} = \sum_{n=0}^{\infty} B_n^{(n)} \frac{t^n}{n!} \quad (|t| < 1). \tag{8}$$

These numbers b_n, $b_n^{(k)}$, $B_n^{(n)}$ and $B_n^{(k)}$ satisfy various identities. For example (see [1-4])

$$n! b_n = B_n^{(n)} + n B_{n-1}^{(n-1)}, \quad B_n^{(n)} = n! \sum_{j=0}^{n} (-1)^{n-j} b_j, \quad \text{and} \quad n! b_n^{(k)} = \frac{k}{k-n} B_n^{(n-k)}. \tag{9}$$

The paper's central result is a formula expressing the Bernoulli polynomials of the second kind and general order k, $b_n^{(k)}(x)$, in terms of those of first order, $b_n(x) = b_n^{(1)}(x)$. That is, we shall prove the following main conclusion.

Theorem. *Let $n, k \in \mathbb{N}$ and $n \geq k-1$. Then*

$$(-1)^{k-1}(k-1)!(n-k)!b_n^{(k)}(x) = \sum_{j=0}^{k-1}(n-1-j)!$$
$$\times \sum_{\substack{v_1,\cdots,v_{k-j}\in\mathbb{N}_0 \\ v_1+\cdots+v_{k-j}=j}} (n-j-1-x)^{v_1}(n-j-2-x)^{v_2}\cdots(n-k-x)^{v_{k-j}}b_{n-j}(x). \quad (10)$$

By taking $x=0$ in Equation (10), we can deduce the following.

Corollary 1. *Let $n, k \in \mathbb{N}$ and $n \geq k-1$. Then*

$$(-1)^{k-1}(k-1)!(n-k)!b_n^{(k)}$$
$$= \sum_{j=0}^{k-1}(n-1-j)! \sum_{\substack{v_1,\cdots,v_{k-j}\in\mathbb{N}_0 \\ v_1+\cdots+v_{k-j}=j}} (n-j-1)^{v_1}(n-j-2)^{v_2}\cdots(n-k)^{v_{k-j}}b_{n-j}. \quad (11)$$

Taking $k=2,3,4$ in (10) and (11), we immediately deduce the following expressions for the first few higher order Bernoulli polynomials and numbers of the second kind:

$$b_n^{(2)}(x) = (1-n)b_n(x) + (x+2-n)b_{n-1}(x) \quad (n \geq 1);$$
$$b_n^{(3)}(x) = \frac{1}{2}(n-1)(n-2)b_n(x)$$
$$+\frac{1}{2}(n-2)(2n-5-2x)b_{n-1}(x) + \frac{1}{2}(n-3-x)^2 b_{n-2}(x) \quad (n \geq 2);$$
$$b_n^{(4)}(x) = -\frac{1}{6}(n-1)(n-2)(n-3)b_n(x) - \frac{1}{6}(n-2)(n-3)(3n-9-3x)b_{n-1}(x)$$
$$-\frac{1}{6}(n-3)\left((n-3-x)^2 + (n-3-x)(n-4-x) + (n-4-x)^2\right)b_{n-2}(x)$$
$$-\frac{1}{6}(n-4-x)^3 b_{n-3}(x) \quad (n \geq 3);$$

$$b_n^{(2)} = (1-n)b_n + (2-n)b_{n-1} \quad (n \geq 1);$$
$$b_n^{(3)} = \frac{1}{2}(n-1)(n-2)b_n + \frac{1}{2}(n-2)(2n-5)b_{n-1} + \frac{1}{2}(n-3)^2 b_{n-2} \quad (n \geq 2);$$
$$b_n^{(4)} = -\frac{1}{6}(n-1)(n-2)(n-3)b_n - \frac{1}{6}(n-2)(n-3)(3n-9)b_{n-1}$$
$$-\frac{1}{6}(n-3)\left((n-3)^2 + (n-3)(n-4) + (n-4)^2\right)b_{n-2} - \frac{1}{6}(n-4)^3 b_{n-3}$$
$$(n \geq 3).$$

By (11), (3), and noting that $C_n = n!b_n$, we obtain an explicit formula for the sum involving Cauchy numbers of the first kind:

$$(-1)^{k-1}(k-1)!(n-k)! \sum_{\substack{v_1,\cdots,v_k\in\mathbb{N}_0\\ v_1+\cdots+v_k=n}} \frac{C_{v_1}C_{v_2}\cdots C_{v_k}}{v_1!v_2!\cdots v_k!}$$

$$=\sum_{j=0}^{k-1} \sum_{\substack{v_1,\cdots,v_{k-j}\in\mathbb{N}_0\\ v_1+\cdots+v_{k-j}=j}} (n-j-1)^{v_1}(n-j-2)^{v_2}\cdots(n-k)^{v_{k-j}}\frac{C_{n-j}}{n-j}. \quad (12)$$

Corollary 2. *Let $n, k \in \mathbb{N}_0$ and $m \in \mathbb{N}$. Then*

$$(-1)^k k! \int_0^m b_n^{(k)}(x)dx = \sum_{i=1}^{n+1}\binom{m}{i}\sum_{j=0}^{k}\frac{(n-i-j)!}{(n-i-k)!}$$

$$\times \sum_{\substack{v_1,\cdots,v_{k+1-j}\in\mathbb{N}_0\\ v_1+\cdots+v_{k+1-j}=j}} (n-i-j)^{v_1}(n-i-j-1)^{v_2}\cdots(n-i-k)^{v_{k+1-j}}b_{n+1-i-j}. \quad (13)$$

By taking $m = 1$ in (13), we can deduce the following:

$$(-1)^k k!(n-k-1)! \int_0^1 b_n^{(k)}(x)dx = \sum_{j=0}^{k}(n-1-j)!$$

$$\times \sum_{\substack{v_1,\cdots,v_{k+1-j}\in\mathbb{N}_0\\ v_1+\cdots+v_{k+1-j}=j}} (n-j-1)^{v_1}(n-j-2)^{v_2}\cdots(n-k-1)^{v_{k+1-j}}b_{n-j}. \quad (14)$$

Taking $k = 0, 1$; $m = 1, 2, 3$ in (13) and noting that $b_n^{(0)}(x) = \binom{x}{n}$, we have

$$\int_0^1 \binom{x}{n}dx = b_n, \quad \int_0^2 \binom{x}{n}dx = 2b_n + b_{n-1}, \quad \int_0^3 \binom{x}{n}dx = 3b_n + 3b_{n-1} + b_{n-2}.$$

and

$$\int_0^1 b_n(x)dx = (1-n)b_n + (2-n)b_{n-1},$$

$$\int_0^2 b_n(x)dx = 2(1-n)b_n + 3(2-n)b_{n-1} + (3-n)b_{n-2},$$

$$\int_0^3 b_n(x)dx = 3(1-n)b_n + 6(2-n)b_{n-1} + 4(3-n)b_{n-2} + (4-n)b_{n-3}.$$

2. Proof of Theorem

In this section, we shall complete the proof of the theorem. First, the following lemma (see [2,10]) is crucial to the proof of the theorem. To be more self-contained, we present a simpler proof here.

Lemma. *Let* $n, k \in \mathbb{N}_0$ *and* $m \in \mathbb{N}$. *Then*

$$b_n^{(k+1)}(x) = \frac{n-k}{-k} b_n^{(k)}(x) + \frac{n-k-1-x}{-k} b_{n-1}^{(k)}(x). \tag{15}$$

Proof. By (1), we have

$$\begin{aligned}\sum_{n=0}^{\infty}(n-k)b_n^{(k)}(x)t^{n-k-1} &= \frac{d}{dt}\left(\frac{1}{\log(1+t)}\right)^k (1+t)^x \\ &= -k\left(\frac{1}{\log(1+t)}\right)^{k+1}(1+t)^{x-1} + x\left(\frac{1}{\log(1+t)}\right)^k (1+t)^{x-1},\end{aligned} \tag{16}$$

i.e.,

$$\begin{aligned}&\sum_{n=1}^{\infty}(n-k-1)b_{n-1}^{(k)}(x)t^{n-k-1} \\ &= -kt\left(\frac{1}{\log(1+t)}\right)^{k+1}(1+t)^{x-1} + xt\left(\frac{1}{\log(1+t)}\right)^k (1+t)^{x-1}.\end{aligned} \tag{17}$$

By (16) and (17), we have

$$\begin{aligned}&\sum_{n=0}^{\infty}(n-k)b_n^{(k)}(x)t^{n-k-1} + \sum_{n=1}^{\infty}(n-k-1)b_{n-1}^{(k)}(x)t^{n-k-1} \\ &= -k\left(\frac{1}{\log(1+t)}\right)^{k+1}(1+t)^x + x\left(\frac{1}{\log(1+t)}\right)^k (1+t)^x \\ &= -k\sum_{n=0}^{\infty} b_n^{(k+1)}(x)t^{n-k-1} + x\sum_{n=0}^{\infty} b_n^{(k)}(x)t^{n-k}.\end{aligned} \tag{18}$$

Comparing the coefficient of t^{n-k-1} on both sides of (18), we get

$$b_n^{(k+1)}(x) = \frac{n-k}{-k} b_n^{(k)}(x) + \frac{n-k-1-x}{-k} b_{n-1}^{(k)}(x).$$

This proves the lemma. □

Now we complete the proof of the theorem by using mathematical induction and the method of coefficients (see [9]).

Proof of theorem. First note that (10) holds for $k = 1, 2$, by (15). Now suppose (10) is true for some natural number k and all $n \geq k - 1$. By superposition of (15), we have

$$
\begin{aligned}
&(-1)^k k!(n-k-1)! b_n^{(k+1)}(x) \\
&= (-1)^{k-1}(k-1)!(n-k)! b_n^{(k)}(x) + (-1)^{k-1}(k-1)!(n-k-1)!(n-k-1-x) b_{n-1}^{(k)}(x) \\
&= \sum_{j=0}^{k-1} (n-1-j)! \\
&\quad \times \sum_{\substack{v_1,\cdots,v_{k-j}\in\mathbb{N}_0 \\ v_1+\cdots+v_{k-j}=j}} (n-j-1-x)^{v_1}(n-j-2-x)^{v_2}\cdots(n-k-x)^{v_{k-j}} b_{n-j}(x) \\
&\quad +(n-k-1-x)\sum_{j=0}^{k-1}(n-2-j)! \\
&\quad \times \sum_{\substack{v_1,\cdots,v_{k-j}\in\mathbb{N}_0 \\ v_1+\cdots+v_{k-j}=j}} (n-j-2-x)^{v_1}(n-j-3-x)^{v_2}\cdots(n-k-1-x)^{v_{k-j}} b_{n-1-j}(x) \\
&= \sum_{j=0}^{k-1} (n-1-j)! \\
&\quad \times \sum_{\substack{v_1,\cdots,v_{k-j}\in\mathbb{N}_0 \\ v_1+\cdots+v_{k-j}=j}} (n-j-1-x)^{v_1}(n-j-2-x)^{v_2}\cdots(n-k-x)^{v_{k-j}} b_{n-j}(x) \\
&\quad +(n-k-1-x)\sum_{j=1}^{k}(n-1-j)! \\
&\quad \times \sum_{\substack{v_1,\cdots,v_{k+1-j}\in\mathbb{N}_0 \\ v_1+\cdots+v_{k+1-j}=j-1}} (n-j-1-x)^{v_1}(n-j-2-x)^{v_2}\cdots(n-k-1-x)^{v_{k+1-j}} b_{n-j}(x) \\
&= \sum_{j=0}^{k} (n-1-j)! \\
&\quad \times \sum_{\substack{v_1,\cdots,v_{k+1-j}\in\mathbb{N}_0 \\ v_1+\cdots+v_{k+1-j}=j}} (n-j-1-x)^{v_1}(n-j-2-x)^{v_2}\cdots(n-k-1-x)^{v_{k+1-j}} b_{n-j}(x),
\end{aligned}
$$

which shows that (10) is also true for the natural number $k+1$. The theorem follows by induction. □

Now we complete the proof of Corollary 2.

Proof of Corollary 2. By (11), (4), and noting that $\frac{d}{dx} b_n^{(k)}(x) = b_{n-1}^{(k-1)}(x)$ (see [11]), we have

$$(-1)^k k! \int_0^m b_n^{(k)}(x)dx = (-1)^k k! \left(b_{n+1}^{(k+1)}(m) - b_{n+1}^{(k+1)}\right) = (-1)^k k! \sum_{i=1}^{n+1} \binom{m}{i} b_{n+1-i}^{(k+1)}$$

$$= \sum_{i=1}^{n+1} \binom{m}{i} \sum_{j=0}^{k} \frac{(n-i-j)!}{(n-i-k)!}$$

$$\times \sum_{\substack{v_1,\cdots,v_{k+1-j}\in\mathbb{N}_0 \\ v_1+\cdots+v_{k+1-j}=j}} (n-i-j)^{v_1}(n-i-j-1)^{v_2}\cdots(n-i-k)^{v_{k+1-j}} b_{n+1-i-j}.$$

This completes the proof. □

Acknowledgement. The authors would like to thank the referee for his/her very useful suggestions. This work was supported by the Guangdong Provincial Natural Science Foundation (No. 8151601501000002).

References

[1] T. Agoh and K. Dilcher, Recurrence relations for Nörlund numbers and Bernoulli numbers of the second kind, *Fibonacci Quart.* **48.1** (2010), 4-12.

[2] L. Carlitz, A note on Bernoulli and Euler polynomials of the second kind, *Scripta Math.* **25** (1961), 323-330.

[3] F. T. Howard, Nörlund's number $B_n^{(n)}$, *Applications of Fibonacci Numbers*, Volume 5, Dordrecht: Kluwer, 1993, pp. 355-366.

[4] F. T. Howard, Congruences and recurrences for Bernoulli numbers of higher order, *Fibonacci Quart.* **32** (1994), 316-328.

[5] C. Jordan, Calculus of Finite Differences, New York: Chelsea, 1965.

[6] G.-D. Liu, Some computational formulas for Nörlund numbers, *Fibonacci Quart.* **45** (2007), 133-137.

[7] G.-D. Liu and H. M. Srivastava, Explicit formulas for the Nörlund polynomials $B_n^{(x)}$ and $b_n^{(x)}$, *Comput. Math. Appl.* **51** (2006), 1377-1384.

[8] D. Merlini, R. Sprugnoli and M. C. Verri, The Cauchy numbers, *Discrete Math.* **306** (2006), 1906-1920.

[9] D. Merlini, R. Sprugnoli and M. C. Verri, The method of coefficients, *Amer. Math. Monthly* **114** (2007), 40-57.

[10] N. E. Nörlund, Vorlesungenüber Differenzenrechnung, Springer-Verlag, Berlin 1924; Reprinted by Chelsea Publishing Company, Bronx, New York, 1954.

[11] T. R. Prabhakar, Sharda Gupta, Bernoulli polynomials of the second kind and general order, *Indian J. pure appl. Math.* **11** (1980), 1361-1368.

[12] M. Wu and H. Pan, Sums of products of Bernoulli numbers of the second kind, *Fibonacci Quart.* **45** (2007), 146-150.

[13] F.-Z. Zhao, Sums of products of Cauchy numbers, *Discrete Math.* **309** (2009), 3830-3842.

A NOTE ON A CONJECTURE OF ERDŐS, GRAHAM, AND SPENCER

Zhongfeng Zhang
School of Mathematics and Information Science, Zhaoqing University, Zhaoqing, China
bee2357@163.com

Pingzhi Yuan
School of Mathematics, South China Normal University, Guangzhou, China
yuanpz@scnu.edu.cn

Received: 2/3/13, Revised: 9/6/13, Accepted: 11/12/13, Published: 11/27/13

Abstract

Let $1 \le a_1 \le a_2 \le \cdots \le a_s$ be integers with $\sum_{i=1}^{s} 1/a_i \ge n + 9/31$. In this paper, we prove that this sum can be decomposed into n parts so that all partial sums are greater than or equal to 1.

1. Introduction

Erdős, Graham and Spencer [1] posed the conjecture that if $1 \le a_1 \le a_2 \le \cdots \le a_s$ are integers with $\sum_{i=1}^{s} 1/a_i < n - 1/30$, then this sum can be decomposed into n parts so that all partial sums are less than or equal to 1. A counterexample given by Guo [4], as shown by $a_1 = 2, a_2 = a_3 = 3, a_4 = 4, a_5 = \cdots = a_{11n-12} = 11$, tells us that we should replace $1/30$ by $5/132$ or a larger quantity.

On the other hand, Sándor [3] proved that the Erdős-Graham-Spencer conjecture is true for $\sum_{i=1}^{s} 1/a_i \le n - 1/2$, and recently, Chen [5], Fang and Chen [2] replace $1/2$ by $1/3$ and $2/7$, respectively.

In this paper, instead of improving the bound, we consider the following similar problem.

Problem 1. Find the least positive number $\eta^+ = \eta^+(n)$ such that when $1 \le a_1 \le a_2 \le \cdots \le a_s$ are integers with $\sum_{i=1}^{s} 1/a_i \ge n + \eta^+$, then this sum can be decomposed into n parts so that all partial sums are ≥ 1.

We get the following result.

Theorem 2. *Let n given, and let $\eta^+ = \eta^+(n)$ be defined as in Problem 1. Then*

$$\frac{5}{156} < \eta^+ \leq \frac{9}{31}.$$

Let $\eta^- = \eta^-(n)$ be the least positive number such that when $1 \leq a_1 \leq a_2 \leq \cdots \leq a_s$ are integers with $\sum_{i=1}^{s} 1/a_i \leq n - \eta^-$, then this sum can be decomposed into n parts so that all partial sums are less than or equal to 1. By the results of Guo [4], Fang and Chen [2] we know $\frac{5}{132} < \eta^- \leq \frac{2}{7}$. We have the following problem.

Problem 3. Is there any relationship between η^- and η^+?

In order to prove the theorem, we only need to consider those sequences such that each term is more than 1 and no partial sum (of two or more terms) is the inverse of a positive integer; otherwise, we may replace the partial sum by the inverse of the integer. We call a sequence $1 \leq a_1 \leq a_2 \leq \cdots \leq a_s$ *primitive* if no partial sum of $\sum_{i=1}^{s} 1/a_i$ is the inverse of a positive integer. In this paper, we consider multisets (i.e., sets with repetitions allowed) of positive integers. Let A be a multiset, and $T(A) = \sum_{i=1}^{s} 1/a_i$, then A is primitive if $1 \notin A$ and there is no multisubset A_1 of A with the cardinality of $A_1 \geq 2$ and $T(A_1)^{-1}$ being an integer.

2. Notation

For a multiset A and a positive real number x, let $m_A(a)$ denote the multiplicity of a in A, let $m(A)$ denote the cardinality of A and let

$$A(x) = \{a : a \in A, a < x\}.$$

For example, if $A = \{2, 3, 3, 4, 5, 5\}$, $B = \{3, 4, 5\}$, then $m_A(1) = 0$, $m_A(2) = 1$, $m_A(3) = 2$, $m_A(4) = 1$, $m_A(5) = 2$, $m(A) = 6$, and

$$A(4) = \{2, 3, 3\}, \quad A \setminus B = \{2, 3, 5\}.$$

With this notation, we say that A has an n^+-*quasiunit partition* if A can be decomposed into n multisubsets $A_1, A_2, \cdots, A_n$, with $T(A_i) \geq 1$ $(1 \leq i \leq n)$ and $m_{A_1}(a) + m_{A_2}(a) + \cdots + m_{A_n}(a) = m_A(a)$ for all integers a. In the following discussion, if we write $A = \cup_{i=1}^{n} A_i$, we mean that $\sum_{i=1}^{n} m_{A_i}(a) = m_A(a)$ for every $a \in A$, and, without loss of generality, we assume $n \geq 2$.

3. Preliminaries

Similar to Lemma 2 of [5], we have the following lemma.

Lemma 4. *Let η be a positive real number and n a positive integer. If for any positive integer $k \le n$, any finite primitive multiset A with $T(A) \ge k+\eta$ has a k^+-quasiunit partition, then any finite multiset A with $T(A) \ge n+\eta$ has an n^+-quasiunit partition.*

Proof. Let A be a finite multiset of positive integers. From Lemma 1 of [5] we know there exists an effective constructible finite primitive multiset A' and a nonnegative integer k such that $T(A) = k + T(A')$, and then the lemma follows. □

Lemma 5. *Let η be a positive real number and let A be a finite multiset with $T(A) = n+\eta$ and $A((n-1)/\eta) = B_1 \cup B_2 \cup \cdots \cup B_n$, and such that $\sum_{i=1}^{n} m_{B_i}(a) = m_A(a)$ for all integers a. Then A has an n^+- quasiunit partition if one of the following conditions holds:*

(i) $T(B_i) \ge 1$ for $1 \le i \le n$;

(ii) $T(B_i) \le 1 + \frac{\eta}{n-1}$ for $1 \le i \le n$;

(iii) $T(A(\frac{n-1}{\eta})) \le n - \frac{2}{7}$.

Proof. (i) It is obvious in this case.

(ii) If for every $1 \le i \le n$, one has $T(B_i) \ge 1$, then it is just case (i). If there exists $1 \le j \le n$, such that $T(B_j) < 1$, then

$$T\left(\cup_{i=1}^{s} B_i\right) = \sum_{i=1}^{s} T(B_i) < n + \frac{(n-1)\eta}{n-1} = n + \eta = T(A).$$

Thus $A \setminus A((n-1)/\eta) \ne \varnothing$. Add $a \in A \setminus A((n-1)/\eta)$ to B_j. If $T(B_j) \ge 1$, we have $1 \le T(B_j) < 1 + \eta/(n-1)$. Otherwise, repeat this process for $A \setminus A((n-1)/\eta) \setminus \{a\}$. Since A is a finite set, we may get $T(B_i) \ge 1$ after finite steps, and then the result follows.

(iii) It follows from (ii) and the result of [2]. □

Let $B = \{b_1, b_2, \cdots, b_r\}$, $C = \{c_1, c_2, \cdots, c_s\}$ be multisets. We write $F(B) \ge F(C)$ if $r \ge s$ and $c_i \ge b_i$ for $1 \le i \le s$.

Lemma 6. *Let B, C be multisets with $F(B) \ge F(C)$ and $\mu_i > 0$ for $1 \le i \le n$. If $B = \cup_{i=1}^{n} B_i$ and $T(B_i) \le \mu_i$, then C can be decomposed into $C = \cup_{i=1}^{n} C_i$ with $T(C_i) \le \mu_i$.*

Proof. Let $B = \{b_1, b_2, \cdots, b_r\}$, $C = \{c_1, c_2, \cdots, c_s\}$ be multisets and $c_i \ge b_i$ for $1 \le i \le s$. Let $B_i \cap \{b_1, b_2, \cdots, b_s\} = \{b_{i_1}, b_{i_2}, \cdots, b_{i_k}\}$ and put $C_i = \{c_{i_1}, c_{i_2}, \cdots, c_{i_k}\}$. Then $T(C_i) \le T(B_i) \le \mu_i$. □

4. The Proof

We are now ready to give the proof of Theorem 2.

Proof. We begin by proving $\eta^+ > 5/156$. Let $n \geq 2$ and

$$a_1 = 2, \quad a_2 = a_3 = 3, \quad a_4 = 4, \quad a_5 = \cdots = a_{13n-14} = 13;$$

then

$$\sum_{i=1}^{13n-14} \frac{1}{a_i} = n + \frac{5}{156}.$$

Let $A = \{a_1, a_2, \cdots, a_{13n-14}\} = B_1 \cup B_2 \cup \cdots \cup B_n$. Then there exists $1 \leq j \leq n$ such that $T(B_j) < 1$, which yields $\eta^+ > 5/156$. In fact, suppose that $T(B_j) \geq 1$ for every $1 \leq j \leq n$. Without loss of generality, we assume $2 \in B_1$.

If $\{3, 3\} \subset B_1$, then

$$\frac{1}{2} + \frac{1}{3} + \frac{1}{3} = 1 + \frac{1}{6} > 1 + \frac{5}{156},$$

which is impossible.

If $3 \in B_1$, then

$$\frac{1}{2} + \frac{1}{3} + \frac{1}{4} = 1 + \frac{1}{12} > 1 + \frac{5}{156},$$

and we know that $4 \notin B_1$; thus

$$T(B_1) = \frac{1}{2} + \frac{1}{3} + \frac{r}{13},$$

which is impossible since

$$\frac{1}{2} + \frac{1}{3} + \frac{2}{13} = \frac{77}{78} < 1, \quad \frac{1}{2} + \frac{1}{3} + \frac{3}{13} = 1 + \frac{5}{78} > 1 + \frac{5}{156}.$$

Therefore we have

$$T(B_1) = \frac{1}{2} + \frac{r}{13},$$

which is also impossible since

$$\frac{1}{2} + \frac{6}{13} = \frac{25}{26} < 1, \quad \frac{1}{2} + \frac{7}{13} = 1 + \frac{1}{26} > 1 + \frac{5}{156}.$$

We have proved the left inequality in Theorem 2, and we proceed to prove the right inequality in it, that is, $\eta^+ \leq 9/31$.

By Lemma 4, we assume A is primitive with $T(A) \geq n + 9/31$. By Lemmas 3-5 in [2] and the proof of the main theorem in the same paper, we know that $T(A(31(n-1)/9)) \leq T(A(7(n-1)/2)) \leq n - 1/3$ for $n = 2, 3, 4$ and $n \geq 11$. Then from Lemma 5(iii), A has an n^+- quasiunit partition.

To finish the proof, we treat the cases $5 \le n \le 10$. First we have the following equalities:

$$\frac{1}{3}+\frac{1}{6}=\frac{1}{2},\quad \frac{1}{4}+\frac{1}{12}=\frac{1}{3},\quad \frac{2}{5}+\frac{1}{10}=\frac{1}{2},\quad \frac{1}{5}+\frac{2}{15}=\frac{1}{3},$$

$$\frac{1}{5}+\frac{1}{20}=\frac{1}{4},\quad \frac{3}{7}+\frac{1}{14}=\frac{1}{2},\quad \frac{2}{7}+\frac{1}{21}=\frac{1}{3},\quad \frac{1}{8}+\frac{1}{24}=\frac{1}{6},$$

$$\frac{1}{9}+\frac{1}{18}=\frac{1}{6},\quad \frac{5}{11}+\frac{1}{22}=\frac{1}{2},\quad \frac{1}{15}+\frac{1}{30}=\frac{1}{10},\quad \frac{6}{13}+\frac{1}{26}=\frac{1}{2}.$$

Since A is primitive, the two fractions in the left-hand side of each equality above cannot exist in $T(A)$ at the same time.

Below, the notation $b(v)$ in the multiset B means that $m_B(b) = v$.

(1) $n = 5$

Let $B = \{2, 3(2), 4, 5(4), 7(6), 8, 9(2), 11(10), 13(12)\}$; then $F(B) \ge F(A(124/9))$. By Lemma 5(ii) and Lemma 6, A has a 5^+- quasiunit partition since $B = \cup_{i=1}^{5} B_i$ with $B_1 = \{2, 3, 9(2)\}$, $B_2 = \{3, 4, 5(2), 13\}$, $B_3 = \{5(2), 7(4), 11\}$, $B_4 = \{7(2), 8, 11(7)\}$, $B_5 = \{11(2), 13(11)\}$ and $T(B_i) < 1 + \frac{9}{124}$ for $1 \le i \le 5$.

(2) $n = 6$

Let $B = \{2, 3(2), 4, 5(4), 7(6), 8, 9(2), 11(10), 12, 13(12), 15, 16, 17(16)\}$; then $F(B) \ge F(A(155/9))$. Let $B = \cup_{i=1}^{6} B_i$ with $B_1 = \{2, 3, 9(2)\}$, $B_2 = \{3, 4, 17(8)\}$, $B_3 = \{5(4), 8, 15, 16\}$, $B_4 = \{7(6), 13, 17(2)\}$, $B_5 = \{11(9), 17(4)\}$, $B_6 = \{11, 13(11), 17(2)\}$. Then $T(B_i) < 1 + \frac{9}{155}$ for $1 \le i \le 6$, which implies that A has a 6^+- quasiunit partition by Lemma 5 and Lemma 6.

(3) $n = 7, 8$

Let $B = \{2, 3(2), 4, 5(4), 7(6), 8, 9(2), 11(10), 13(12), 15, 16, 17(16), 19(18)\}$, $C = B \cup \{23(22)\}$. Then $B = \cup_{i=1}^{7} B_i$, $C = \cup_{i=1}^{8} C_i$ with $B_1 = \{2, 13(7)\}$, $B_2 = \{3(2), 4\}$, $B_3 = \{5(4), 11, 13(2)\}$, $B_4 = \{7(6), 9, 13\}$, $B_5 = \{11(9), 9, 19(2)\}$, $B_6 = \{8, 13(2), 15, 16, 17(2), 19\}$, $B_7 = \{17(4), 19(15)\}$, $C_1 = B_1$, $C_2 = B_2$, $C_3 = \{5(4), 13(3)\}$, $C_4 = \{7(6), 11(2)\}$, $C_5 = \{9(2), 11(8), 23(2)\}$, $C_6 = B_6$, $C_7 = B_7$, $C_8 = \{19(2), 23(20)\}$ and $T(B_i) < 1+\frac{3}{62}$ for $1 \le i \le 7$, $T(C_i) \le 1+\frac{9}{217}$ for $1 \le i \le 8$. It is easy to see that $F(T(A(62/3))) \le F(B)$ and $F(T(A(217/9))) \le F(C)$. From Lemma 5(ii) and Lemma 6, we know that A has a 7^+- quasiunit partition for $n = 7$ and an 8^+- quasiunit partition for $n = 8$.

(4) $n = 9, 10$

Let $B = \{2, 3(2), 4, 5(4), 7(6), 8, 9(2), 11(10), 13(12), 15, 16, 17(16), 19(18), 23(22), 25(4), 27(2)\}$, $C = B \cup \{28, 29(28)\}$. Then $B = \cup_{i=1}^{9} B_i$, $C = \cup_{i=1}^{10} C_i$ with $B_1 = \{2, 4, 8, 27(2)\}$, $B_2 = \{3(2), 9(2)\}$, $B_3 = \{5(4), 15\}$, $B_4 = \{7(6), 16\}$, $B_5 = \{11(10), 25(2)\}$, $B_6 = \{13(12), 25\}$, $B_7 = \{17(16), 25\}$, $B_8 = \{19(18)\}$, $B_9 = \{23(22)\}$, $C_1 = B_1 \cup \{28)\}$, $C_i = B_i$ for $2 \le i \le 9$, $C_{10} = \{29(28)\}$, and for all i, $T(B_i), T(C_i)$ are less than 1. Since $F(T(A(248/9))) \le F(B)$, $F(T(A(31))) \le F(C)$, from Lemma 5(ii) and Lemma 6, we get $\eta^+ \le \frac{9}{31}$ for $n = 9, 10$. □

Acknowledgments. This work was supported by the Guangdong Provincial Natural Science Foundation (No. S2012040007653) and NSF of China (No. 11271142). We would like to thank the referee and the editor for their valuable suggestions.

References

[1] P. Erdős, R. Graham, *Old and New Problems and Results in Combinatorial Number Theory*, Enseign. Math. (2), vol. 28, Enseignement Math., Geneva, 1980.

[2] J. Fang and Y. Chen, *On a conjecture of Erdős, Graham and Spencer, II*, Discrete Appl. Math. 156 (2008), 2950-2958.

[3] C. Sándor, *On a problem of Erdős*, J. Number Theory 63 (1997), 203-210.

[4] S. Guo, *A counterexample to a conjecture of Erdős, Graham and Spencer*, Electron. J. Comb. 15 (2008), #N43, 3 pp.

[5] Y. Chen, *On a conjecture of Erdős, Graham and Spencer*, J. Number Theory 119 (2006), 307-314.

SQUARES AND DIFFERENCE SETS IN FINITE FIELDS

C. Bachoc[1]
Univ Bordeaux, Institut de Mathématiques de Bordeaux, Talence, France
bachoc@math.u-bordeaux1.fr

M. Matolcsi[2]
Alfréd Rényi Institute of Mathematics, Hungarian Academy of Sciences, Budapest, Hungary
matolcsi.mate@renyi.mta.hu

I. Z. Ruzsa[3]
Alfréd Rényi Institute of Mathematics, Hungarian Academy of Sciences, Budapest, Hungary
ruzsa.z.imre@mta.renyi.hu

Received: 4/29/13, Accepted: 11/17/13, Published: 11/27/13

Abstract

For infinitely many primes $p = 4k + 1$ we give a slightly improved upper bound for the maximal cardinality of a set $B \subset \mathbb{Z}_p$ such that the difference set $B - B$ contains only quadratic residues. Namely, instead of the "trivial" bound $|B| \leq \sqrt{p}$ we prove $|B| \leq \sqrt{p} - 1$, under suitable conditions on p. The new bound is valid for approximately three quarters of the primes $p = 4k + 1$.

1. Introduction

Let q be a prime-power, say $q = p^k$. We will be interested in estimating the maximal cardinality $s(q)$ of a set $B \subset \mathbb{F}_q$ such that the difference set $B - B$ contains only squares. While our main interest is in the case $k = 1$, we find it instructive to compare the situation for different values of k.

This problem makes sense only if -1 is a square; to ensure this we assume $q \equiv 1 \pmod 4$. The universal upper bound $s(q) \leq \sqrt{q}$ can be proved by a pigeonhole argument or by simple Fourier anlysis, and it has been re-discovered several times (see [8, Theorem 3.9], [12, Problem 13.13], [4, Proposition 4.7], [3, Chapter XIII, Theorem 14], [11, Theorem 31.3], [10, Proposition 4.5], [7, Section 2.8] for various

[1]This study has been carried out with financial support from the French State, managed by the French National Research Agency (ANR) in the frame of the Investments for the future Programme IdEx Bordeaux (ANR-10-IDEX-03-02).

[2]Supported by ERC-AdG 228005, and OTKA Grants No. K81658 and the Bolyai Scholarship.

[3]Supported by ERC-AdG 228005, and OTKA Grants No. K81658. All authors supported by the CNRS-MTA "Haagacant" project.

proofs). For even k we have equality, since $\mathbb{F}_{p^k}$ can be constructed as a quadratic extension of $\mathbb{F}_{p^{k/2}}$, and then every element of the embedded field $\mathbb{F}_{p^{k/2}}$ will be a square. It is known that every case of equality can be obtained by a linear transformation from this one, [2].

Such problems and results are often formulated in terms of the Paley graph P_q, which is the graph with vertex set $\mathbb{F}_q$ and an edge between x and y if and only if $x - y = a^2$ for some non-zero $a \in \mathbb{F}_q$. Paley graphs are self-complementary, vertex and edge transitive, and $(q, (q-1)/2, (q-5)/4, (q-1)/4)$-strongly regular (see [3] for these and other basic properties of P_q). Paley graphs have received considerable attention over the past decades because they exhibit many properties of random graphs $G(q, 1/2)$ where each edge is present with probability 1/2. Indeed, P_q form a family of *quasi-random* graphs, as shown in [5].

With this terminology $s(q)$ is the *clique number* of P_q. The general lower bound $s(q) \geq (\frac{1}{2} + o(1)) \log_2 q$ is established in [6], while it is proved in [9] that $s(p) \geq c \log p \log \log \log p$ for infinitely many primes p. The "trivial" upper bound $s(p) \leq \sqrt{p}$ is notoriously difficult to improve, and it is mentioned explicitly in the selected list of problems [7]. The only improvement we are aware of concerns the special case $p = n^2 + 1$ for which it is proved in [13] that $s(p) \leq n - 1$ (the same result was proved independently by T. Sanders – unpublished, personal communication). It is more likely, heuristically, that the lower bound is closer to the truth than the upper bound. Numerical data [16, 15] up to $p < 10000$ suggest (very tentatively) that the correct order of magnitude for the clique number of P_p is $c \log^2 p$ (see the discussion and the plot of the function $s(p)$ at [17]).

In this note we prove the slightly improved upper bound $s(p) \leq \sqrt{p} - 1$ for the *majority* of the primes $p = 4k + 1$ (we will often suppress the dependence on p, and just write s instead of $s(p)$).

We will denote the set of nonzero quadratic residues by Q, and that of nonzero non-residues by NQ. Note that $0 \notin Q$ and $0 \notin NQ$.

2. The Improved Upper Bound

Theorem 2.1. *Let q be a prime-power, $q = p^k$, and assume that k is odd and $q \equiv 1 \pmod 4$. Let $s = s(q)$ be the maximal cardinality of a set $B \subset \mathbb{F}_q$ such that the difference set $B - B$ contains only squares.*
(i) If $[\sqrt{q}]$ is even then $s^2 + s - 1 \leq q$; *(ii) if $[\sqrt{q}]$ is odd then $s^2 + 2s - 2 \leq q$.*

Proof. The claims hold if $s < [\sqrt{q}]$. Hence we may assume that $s \geq [\sqrt{q}]$.

Lemma 2.2. *Let $D \subset \mathbb{F}_q$ be a set such that $D \subset NQ$, $D - D \subset Q \cup \{0\}$. With $r = |D|$ we have*

$$s(q) \leq 1 + \frac{q-1}{2r}. \tag{1}$$

Proof. Let B be a maximal set such that $B - B \subset Q \cup \{0\}$, $|B| = s(q) = s$. Consider the equation $b_1 - b_2 = zd$, $b_1, b_2 \in B$, $d \in D$, $z \in NQ$. This equation has exactly $s(s-1)r$ solutions; indeed, every pair of distinct $b_1, b_2 \in B$ and a $d \in D$ determines z uniquely. On the other hand, given b_1 and z, there can be at most one pair b_2 and d to form a solution. Indeed, if there were another pair b_2', d', then by substracting the equations $b_1 - b_2 = zd$, $b_1 - b_2' = zd'$ we get $(b_2' - b_2) = z(d - d')$, a contradiction, as the left hand side is a square and the right hand side is not. This gives $s(s-1)r \le s(q-1)/2$ as wanted. □

We try to construct such a set D in the form $D = (B - t) \cap NQ$ with a suitable t. The required property then follows from $D - D \subset B - B$.

Let χ denote the quadratic multiplicative character, i.e., $\chi(t) = 1$ according to whether $t \in Q$ or $t \in NQ$ (and $\chi(0) = 0$). Let

$$\varphi(t) = \sum_{b \in B} \chi(b - t). \tag{2}$$

Clearly $\varphi(t) = |(B - t) \cap Q| - |(B - t) \cap NQ|$, and hence for $t \notin B$ we have

$$|(B - t) \cap NQ| = \frac{s - \varphi(t)}{2}.$$

To find a large set in this form we need to find a negative value of φ.

We list some properties of this function. For $t \in B$ we have $\varphi(t) = s - 1$, and otherwise $\varphi(t) \le s - 2$, $\varphi(t) \equiv s \pmod 2$ (the inequality expresses the maximality of B). Furthermore, $\sum_t \varphi(t) = 0$, and, since translations of the quadratic character have the quasi-orthogonality property

$$\sum_t \chi(t + a)\chi(t + b) = -1$$

for $a \neq b$, we conclude that

$$\sum_t \varphi(t)^2 = s(q-1) - s(s-1) = s(q-s).$$

By subtracting the contribution of $t \in B$ we obtain

$$\sum_{t \notin B} \varphi(t) = -s(s-1); \qquad \sum_{t \notin B} \varphi(t)^2 = s(q-s) - s(s-1)^2 = s(q - s^2 + s - 1).$$

These formulas assume an even nicer form by introducing the function $\varphi_1(t) = \varphi(t) + 1$:

$$\sum_{t \notin B} \varphi_1(t) = q - s^2, \tag{3}$$

$$\sum_{t \notin B} \varphi_1(t)^2 = (s+1)(q - s^2). \tag{4}$$

As a byproduct, the second equation shows the familiar estimate $s \leq \sqrt{q}$, so we have $s = [\sqrt{q}] < \sqrt{q}$ (recall that we assume that $s \geq [\sqrt{q}]$, the theorem being trivial otherwise).

Now we consider separately the cases of odd and even s. If s is even, then, since $\sum_{t \notin B} \varphi(t) < 0$ and each summand is even, we can find a t with $\varphi(t) \leq -2$. This gives us an r with $r \geq (s+2)/2$, and on substituting this into (1) we obtain the first case of the theorem.

If s is odd, we claim that there is a t with $\varphi(t) \leq -3$. Otherwise we have $\varphi(t) \geq -1$, that is, $\varphi_1(t) \geq 0$ for all $t \notin B$. We also know $\varphi(t) \leq s-2$, $\varphi_1(t) \leq s-1$ for $t \notin B$. Consequently

$$\sum_{t \notin B} \varphi_1(t)^2 \leq (s-1) \sum_{t \notin B} \varphi_1(t) = (s-1)(q-s^2),$$

a contradiction to (4). (Observe that to reach a contradiction we need that $q - s^2$ is strictly positive. In case of an even k it can happen that $q = s^2$ and the function φ_1 vanishes outside B.)

This t provides us with a set D with $r \geq (s+3)/2$, and on substituting this into (1) we obtain the second case of the theorem. □

Remark 2.3. An alternative proof for the case $q = p$ and s being odd is as follows. Assume by contradiction that φ_1 is even-valued and nonnegative. Then by (3) it must be 0 for at least

$$q - |B| - \frac{q-s^2}{2} = \frac{q+s^2-2s}{2}$$

values of t. Let $\tilde{\chi}, \tilde{\varphi}, \tilde{\varphi}_1$ denote the images of χ, φ, φ_1 in $\mathbb{F}_q$ (i.e., the functions are evaluated mod p). By the previous observation $\tilde{\varphi}_1$ has at least $(q+s^2-2s)/2$ zeroes. On the other hand, we have $\tilde{\chi}(x) = x^{\frac{q-1}{2}}$, and hence $\tilde{\varphi}_1$ is a polynomial of degree $(q-1)/2$; its leading coefficient is $s = [\sqrt{q}] \neq 0 \bmod p$ (This last fact may fail if $q = p^k$, even if k is odd. Therefore this proof is restricted in its generality. Nevertheless we include it here, because we believe that it has the potential to lead to stronger results if $q = p$.) Consequently $\tilde{\varphi}_1$ can have at most $(q-1)/2$ zeros, a contradiction. In the case of even k we can have $s = \sqrt{q} \equiv 0 \pmod{p}$ and so the polynomial $\tilde{\varphi}_1$ can vanish, as it indeed does when B is a subfield.

Remark 2.4. It is clear from (1) that any improved lower bound on r will lead to an improved upper bound on s. If one thinks of elements of $\mathbb{Z}_p$ as being quadratic residues randomly with probability 1/2, then we expect that $r \geq \frac{s}{2} + c\sqrt{s}$. This would lead to an estimate $s \leq \sqrt{p} - cp^{1/4}$. This seems to be the limit of this method. In order to get an improved lower bound on r one can try to prove non-trivial upper bounds on the third moment $\sum_{t \in \mathbb{Z}_p} \varphi^3(t)$. To do this, we would need that the distribution of numbers $\frac{b_1-b_2}{b_1-b_3}$ is approximately uniform on Q as b_1, b_2, b_3 ranges over B. This is plausible because if $s \approx \sqrt{p}$ then the distribution of $B - B$

must be close to uniform on NQ. However, we could not prove anything rigorous in this direction.

Remark 2.5. Theorem 2.1 gives the bound $s \leq [\sqrt{p}] - 1$ for about three quarters of the primes $p = 4k+1$. Indeed, part *(ii)* gives this bound for almost all p such that $n = [\sqrt{p}]$ is odd, with the only exception when $p = (n+1)^2 - 3$. Part *(i)* gives the improved bound $s \leq n-1$ if $n^2 + n - 1 > p$. This happens for about half of the primes $p = 4k+1$ for which n is even. To make these statements rigorous we note that $\sqrt{p}/2$ is uniformly distributed modulo one, when p ranges over primes of the form $p = 4k+1$: this is a special case of a result of Balog, [1, Theorem 1].

Acknowledgment The authors are grateful to Péter Csikvári for insightful comments regarding the prime-power case.

References

[1] A. Balog, *On the distribution of p^θ mod 1*, Acta Math. Hungar. 45 (1985), no. 1-2, 179-199.

[2] A. Blokhuis, *On subsets of $GF(q^2)$ with square differences*, Indag. Math., vol. 87, no. 4, pp. 369-372, (1984).

[3] B. Bollobás, Random Graphs, (second ed.), Cambridge University Press, Cambridge, 2001.

[4] P. J. Cameron, *Automorphism groups in graphs*, in: R. J. Wilson, L. W. Beineke (Eds.), Selected Topics in Graph Theory, vol. 2, Academic Press, NewYork, (1983), pp. 89-127.

[5] F. R. K. Chung, R. L. Graham, R. M. Wilson, *Quasi-random graphs*, Combinatorica, Volume 9, Issue 4, (1989), pp 345-362.

[6] S. D. Cohen, *Clique numbers of Paley graphs*, Quaest. Math. **11**, (2) (1988), 225-231.

[7] E. Croot, V. Lev, *Open problems in additive combinatorics*, Additive combinatorics CRM Proc. Lecture Notes Amer. Math. Soc., Providence, RI, 43, (2007), 207-233.

[8] P. Delsarte, *An algebraic approach to the association schemes of coding theory*, Philips Res. Rep. Suppl. 10 (1973).

[9] S. Graham, C. Ringrose, *Lower bounds for least quadratic non-residues*, Analytic Number Theory (Allterton Park, IL, 1989), 269-309.

[10] M. Krivelevich, B. Sudakov, *Pseudo-random graphs*, in: More Sets, Graphs and Numbers, Bolyai Society Mathematical Studies 15, Springer, (2006), 199-262.

[11] J. H. van Lint, R. M. Wilson, *A Course in Combinatorics*, Cambridge University Press, Cambridge, 1992 (2nd edition in 2001).

[12] L. Lovász, *Combinatorial Problems and Exercises*, North-Holland, Amsterdam, 1979 (2nd edition in 1993).

[13] E. Maistrelli, D. B. Penman, *Some colouring problems for Paley graphs*, Discrete Math. **306** (2006) 99-106.

[14] M. Matolcsi, I. Z. Ruzsa, *Difference sets and positive exponential sums I. General properties*, J. Fourier Anal. Appl., to appear.

[15] Web-page of Geoffrey Exoo with clique numbers of Paley graphs for $7000 < p < 10000$, http://ginger.indstate.edu/ge/PALEY/

[16] Web-page of J. B. Shearer with clique numbers of Paley graphs for $p < 7000$, http://www.research.ibm.com/people/s/shearer/indpal.html

[17] Web-page discussion of clique numbers and plot of the function $s(p)$ for $p < 10000$, http://mathoverflow.net/questions/48591/cliques-paley-graphs-and-quadratic-residues

A FAMILY OF FIBONACCI-LIKE CONDITIONAL SEQUENCES

Daniel Panario
School of Mathematics and Statistics, Carleton University, Ottawa, Canada
daniel@math.carleton.ca

Murat Sahin
Department of Mathematics, Ankara University, Ankara, Turkey
msahin@ankara.edu.tr

Qiang Wang [1]
School of Mathematics and Statistics, Carleton University, Ottawa, Canada
wang@math.carleton.ca

Received: 3/4/12, Revised: 8/12/13, Accepted: 11/18/13, Published: 12/13/13

Abstract

Let $\{a_{i,j}\}$ be real numbers for $0 \leqslant i \leqslant r-1$ and $1 \leqslant j \leqslant 2$, and define a sequence $\{v_n\}$ with initial conditions v_0, v_1 and conditional linear recurrence relation $v_n = a_{t,1}v_{n-1} + a_{t,2}v_{n-2}$ where $n \equiv t \pmod r$ $(n \geqslant 2)$. The sequence $\{v_n\}$ can be viewed as a generalization of many well-known integer sequences, such as Fibonacci, Lucas, Pell, Jacobsthal, etc. We find explicitly a linear recurrence equation which is satisfied by $\{v_n\}$, generating functions, matrix representations and extended Binet's formulas for $\{v_n\}$ in terms of a generalized continuant.

1. Introduction

The Fibonacci sequence is a sequence of integers in which each subsequent term is the sum of the two preceding it starting with 0 and 1. The so-called Fibonacci numbers appeared in the solution of a problem by Leonardo Pisano, in his book Liber Abaci (1202), concerning reproduction patterns of rabbits. Now, this sequence appears in many areas of mathematics. The Fibonacci sequence has been generalized in several directions, namely, by changing the initial values, by mixing two sequences, by demanding that the numbers in the sequences not be integers, and by having more than two parameters; see [6, 8, 10, 11, 12, 13, 14, 15, 16, 20, 21].

[1]The authors are partially supported by NSERC of Canada. Qiang Wang is also partially supported by National Natural Science Foundation of China (Grant No. 61170289).

Yet another extension of the Fibonacci sequence is given in [4] as follows. Let $a_0, a_1, \ldots, a_{r-1}$ be real numbers; define a sequence $\{q_n\}$, with initial values $q_0 = 0$ and $q_1 = 1$, and for $n \geqslant 2$,

$$q_n = \begin{cases} a_0 q_{n-1} + q_{n-2}, & \text{if } n \equiv 0 \pmod{r}, \\ a_1 q_{n-1} + q_{n-2}, & \text{if } n \equiv 1 \pmod{r}, \\ \vdots & \vdots \\ a_{r-1} q_{n-1} + q_{n-2}, & \text{if } n \equiv r-1 \pmod{r}. \end{cases}$$

It remained as an open problem in [4] to find a closed form for the generating function, and to obtain a Binet-like formula for $\{q_n\}$. Later on, the sequence $\{q_n\}$ was studied in [3], where it was called *k-periodic Fibonacci sequences* (here k corresponds to our r), and the open problem was solved. Independently, in [17] and [19], one of the authors of this paper solved this open problem by using continuants while also introducing a matrix representation for the sequence $\{q_n\}$.

We now introduce a new sequence $\{v_n\}$ that is a generalization of the sequence $\{q_n\}$. Let $\{a_{i,j}\}$ be real numbers for $0 \leqslant i \leqslant r-1$ and $1 \leqslant j \leqslant 2$, and define a sequence $\{v_n\}$ with initial conditions v_0, v_1, and for $n \geqslant 2$,

$$v_n = \begin{cases} a_{0,1} v_{n-1} + a_{0,2} v_{n-2}, & \text{if } n \equiv 0 \pmod{r}, \\ a_{1,1} v_{n-1} + a_{1,2} v_{n-2}, & \text{if } n \equiv 1 \pmod{r}, \\ \vdots & \vdots \\ a_{r-1,1} v_{n-1} + a_{r-1,2} v_{n-2}, & \text{if } n \equiv r-1 \pmod{r}. \end{cases} \tag{1}$$

We call $\{v_n\}$ a *Fibonacci-like conditional sequence.*

We can get a great number of different sequences by providing the values of $r, a_{i,j}$ and the initial values in the sequence $\{v_n\}$. The following are some examples of sequences which are special cases of $\{v_n\}$. We note that some of the following examples are given with the a reference number to Sloane's *On-Line Encyclopedia of Integer Sequences*. Let us assume initial values $v_0 = 0$ and $v_1 = 1$.

1. If we take $a_{i,j} = 1$ for $0 \leqslant i \leqslant r-1$ and $1 \leqslant j \leqslant 2$, we get the classical Fibonacci numbers.

2. If we take $a_{i,1} = 1$ and $a_{i,2} = 1$ for $0 \leqslant i \leqslant r-1$, we get Pell's sequence.

3. If we take $a_{i,1} = k$ and $a_{i,2} = 1$ for $0 \leqslant i \leqslant r-1$, we get k-Fibonacci numbers.

4. If we take $r = 2$, $a_{0,1} = a_{1,2} = 1$ and $a_{1,1} = a_{0,2} = 2$, we get [*A*005824]

5. If we take $r = 2$, $a_{0,1} = 2$ and $a_{0,2} = a_{1,1} = a_{1,2} = 1$, we get [*A*048788]. The terms of the sequence are numerators of the continued fraction convergent to $\sqrt{3} - 1$.

6. If we take $r = 2$, $a_{0,1} = a_{0,2} = 1$ and $a_{1,1} = a_{1,2} = 2$, we get [$A001045$] (Jacobsthal sequence).

7. If we take $r = 3$ and $(a_{0,1}, a_{0,2}, a_{1,1}, a_{1,2}, a_{2,1}, a_{2,2}) = (1, 1, 0, 1, 1, 1)$, we get [$A097564$].

8. If we take $r = 3$ and $(a_{0,1}, a_{0,2}, a_{1,1}, a_{1,2}, a_{2,1}, a_{2,2}) = (1, 1, 1, 0, 0, 0)$, we get [$A117567$] which gives Riordan arrays.

9. If we take $r = 3$ and $(a_{0,1}, a_{0,2}, a_{1,1}, a_{1,2}, a_{2,1}, a_{2,2}) = (0, 1, 1, 1, 1, 1)$, we get [$A092550$] which is the two-steps-forward and one-step backwards Fibonacci-based switched sequence inspired by Per Bak's sand piles.

10. If we take $r = 3$ and $(a_{0,1}, a_{0,2}, a_{1,1}, a_{1,2}, a_{2,1}, a_{2,2}) = (2, 0, 2, 1, 2, 0)$, we get [$A004647$].

11. If we take $r = 3$ and $(a_{0,1}, a_{0,2}, a_{1,1}, a_{1,2}, a_{2,1}, a_{2,2}) = (2, 0, 1, 1, 1, 1)$, we get [$A133335$].

12. If we take $r = 2$ with $a_{0,2} = a_{1,2} = 1$ and any nonzero real numbers $a_{0,1}$ and $a_{1,1}$ we get *generalized Fibonacci sequences* [4].

13. If we take $a_{i,2} = 1$ for $0 \leqslant i \leqslant r - 1$ with any nonzero real numbers $a_{i,1}$, we get the sequences studied in [3, 17, 19].

14. If we take $r = 2$ and real numbers $a_{i,j}$, not all zeros, we get the sequences defined in [18].

Next we give the structure of the paper. In Section 2, we introduce the concept of the generalized continuant and the generalized continued fraction, we make a connection between them and we find explicitly a linear recurrence relation which is satisfied by $\{v_n\}$ for any given integer $r \geq 2$. In Sections 3 and 4, we find generating functions, matrix representations and extended Binet's formulas for $\{v_n\}$ in terms of a generalized continuant. Conclusions and further work are given in Section 5.

2. Linear Recurrence of $\{v_n\}$

Continuants are a sequence of polynomials extensively studied by Euler. They are defined by the following recurrence

$$\begin{aligned} K_0() &= 1; \\ K_1(x_1) &= x_1; \\ K_n(x_1, \ldots, x_n) &= K_{n-1}(x_1, \ldots, x_{n-1})x_n + K_{n-2}(x_1, \ldots, x_{n-2}). \end{aligned}$$

It is well known that continuants are the key to the study of continued fractions like

$$a_n + \cfrac{1}{a_{n-1} + \cfrac{1}{a_{n-2} + \cfrac{1}{\ddots + \vdots \cfrac{1}{a_0}}}} .$$

Indeed, the above continued fraction is equal to

$$\frac{K(a_n, a_{n-1}, \ldots, a_0)}{K(a_{n-1}, a_{n-2}, \ldots, a_0)} .$$

We refer to pages 301-309 in [5] for detailed information about the continuant.

Let us consider a generalization of continued fractions of the form

$$a_{n,1} + \cfrac{a_{n,2}}{a_{n-1,1} + \cfrac{a_{n-1,2}}{a_{n-2,1} + \cfrac{a_{n-2,2}}{\ddots + \vdots \cfrac{a_{1,2}}{a_{0,1}}}}}$$

and denote it by $[a_{n,1}, a_{n-1,1}, \ldots, a_{0,1}]_{\overrightarrow{b}}$ where $\overrightarrow{b} = [a_{n,2}, a_{n-1,2}, \ldots, a_{1,2}]$ for any positive integer n. If we take $\overrightarrow{b} = [1, 1, \ldots, 1]$ then we get the classical continued fraction. For example, let us take $[1, 2, 1, 1, 2, 1]_{\overrightarrow{b}}$ where $\overrightarrow{b} = [1, 1, 2, 1, 1]$; then

$$[1, 2, 1, 1, 2, 1]_{\overrightarrow{b}} = 1 + \cfrac{1}{2 + \cfrac{1}{1 + \cfrac{2}{1 + \cfrac{1}{2 + \cfrac{1}{1}}}}} = \frac{17}{12} .$$

We associate this generalized continued fraction with the following Fibonacci-like sequences

$$A_n = \begin{cases} A_{n-1} + 2A_{n-2}, & \text{if } n \equiv 0 \pmod 3, \\ 2A_{n-1} + A_{n-2}, & \text{if } n \equiv 1 \pmod 3, \\ A_{n-1} + A_{n-2}, & \text{if } n \equiv 2 \pmod 3, \end{cases} \qquad A_0 = 1 \text{ and } A_1 = 3,$$

and

$$B_n = \begin{cases} B_{n-1} + B_{n-2}, & \text{if } n \equiv 0 \pmod 3, \\ B_{n-1} + 2B_{n-2}, & \text{if } n \equiv 1 \pmod 3, \\ 2B_{n-1} + B_{n-2}, & \text{if } n \equiv 2 \pmod 3, \end{cases} \qquad B_0 = 1 \text{ and } B_1 = 1.$$

Here, $[1, 2, 1, 1, 2, 1]_{\overrightarrow{b}} = \frac{17}{12} = \frac{A_5}{B_5}$ where $\overrightarrow{b} = [1, 1, 2, 1, 1]$. The sequences $\{A_n\}$ and $\{B_n\}$ correspond to a generalized continuant; explanation of the mathematical background of this example is forthcoming.

In order to study the general conditional sequences introduced in Section 1, we now introduce a generalized continuant with an input of two-dimensional arrays $\{a_{i,j} \mid i, j \geq 0\}$.

Definition 1 (Generalized Continuant). Let $K() = 1, K(a_{i,j}) = a_{i,j}$, where i, j are nonnegative integers, and for $n \geqslant 2$,

$$\begin{aligned} &K(a_{n,j}, a_{n-1,j}, \ldots, a_{i+1,j}, a_{i,j}) \\ &\quad = a_{i,j}K(a_{n,j}, a_{n-1,j}, \ldots, a_{i+1,j}) + a_{i+1,j+1}K(a_{n,j}, a_{n-1,j}, \ldots, a_{i+2,j}). \end{aligned}$$

For example, let us suppose that

$$[a_{0,1}, \ldots, a_{5,1}] = [1, 2, 1, 1, 2, 1] \quad \text{and} \quad [a_{0,2}, \ldots, a_{5,2}] = [2, 1, 1, 2, 1, 1],$$

then we can obtain $K(a_{5,1}, \ldots, a_{0,1}) = 34$ and $K(a_{4,1}, \ldots, a_{0,1}) = 24$. We emphasize that the definition of K depends on both input sequences. For example, let $[b_{0,1}, \ldots, b_{5,1}] = [1, 2, 1, 1, 2, 1]$ and $[b_{0,2}, \ldots, b_{5,2}] = [2, 1, 1, 1, 2, 2]$. Although $[b_{5,1}, b_{4,1}, b_{3,1}] = [b_{2,1}, b_{1,1}, b_{0,1}] = [1, 2, 1]$, we still have $K(b_{5,1}, b_{4,1}, b_{3,1}) = 6 \neq K(b_{2,1}, b_{1,1}, b_{0,1}) = 4$ because $[b_{5,2}, b_{4,2}, b_{3,2}] \neq [b_{2,2}, b_{1,2}, b_{0,2}]$.

We note that this generalized continuant satisfies the following identity which can serve as an alternative definition.

Theorem 2. *For a positive integer $n \geqslant 1$, and $0 \leq k \leq n-1$, the generalized continuant satisfies*

$$K(a_{n,1}, a_{n-1,1}, \ldots, a_{k,1}) = a_{n,1}K(a_{n-1,1}, \ldots, a_{k,1}) + a_{n,2}K(a_{n-2,1}, \ldots, a_{k,1}).$$

Proof. We shall prove the formula by induction on the length of the generalized continuant. By the definition of the continuant (Definition 1), we can write

$$\begin{aligned} K(a_{k+1,1}, a_{k,1}) &= a_{k,1}K(a_{k+1,1}) + a_{k+1,2} \\ &= a_{k,1}a_{k+1,1} + a_{k+1,2} = a_{k+1,1}K(a_{k,1}) + a_{k+1,2}. \end{aligned}$$

Hence it is true for the base case (length 2), that is simply the case $n = k+1$. Let us assume now that the formula holds for the generalized continuant of the above form of length less than or equal to m (that is, $n = m + k - 1$). If we use the definition of continuant and the hypothesis of induction, we get

$$\begin{aligned}
&K(a_{m+k,1}, a_{m+k-1,1}, \ldots, a_{k,1}) \\
&= a_{k,1}K(a_{m+k,1}, a_{m+k-1,1}, \ldots, a_{k+1,1}) \\
&\quad + a_{k+1,2}K(a_{m+k,1}, a_{m+k-1,1}, \ldots, a_{k+2,1}) \\
&= a_{k,1}(a_{m+k,1}K(a_{m+k-1,1}, \ldots, a_{k+1,1}) + a_{m+k,2}K(a_{m+k-2,1}, \ldots, a_{k+1,1})) \\
&\quad + a_{k+1,2}(a_{m+k,1}K(a_{m+k-1,1}, \ldots, a_{k+2,1}) + a_{m+k,2}K(a_{m+k-2,1}, \ldots, a_{k+2,1})) \\
&= a_{m+k,1}(a_{k,1}K(a_{m+k-1,1}, \ldots, a_{k+1,1}) + a_{k+1,2}K(a_{m+k-1,1}, \ldots, a_{k+2,1})) \\
&\quad + a_{m+k,2}(a_{k,1}K(a_{m+k-2,1}, \ldots, a_{k+1,1}) + a_{k+1,2}K(a_{m+k-2,1}, \ldots, a_{k+2,1})) \\
&= a_{m+k,1}K(a_{m+k-1,1}, \ldots, a_{k,1}) + a_{m+k,2}K(a_{m+k-2,1}, \ldots, a_{k,1}).
\end{aligned}$$

□

Similar to the classical continuant, by mathematical induction we can easily establish the following connection between the generalized continuant and the generalized continued fraction.

Theorem 3. *For any positive integer n,*

$$[a_{n,1}, a_{n-1,1}, \ldots, a_{0,1}]_{\overrightarrow{b}} = \frac{K(a_{n,1}, a_{n-1,1}, \ldots, a_{0,1})}{K(a_{n-1,1}, a_{n-2,1}, \ldots, a_{0,1})},$$

where $\overrightarrow{b} = [a_{n,2}, a_{n-1,2}, \ldots, a_{1,2}]$.

Proof. First we observe that the result is true for $n = 0$ and $n = 1$,

$$a_{0,1} = \frac{K(a_{0,1})}{K()} \quad \text{and} \quad a_{1,1} + \frac{a_{1,2}}{a_{0,1}} = \frac{K(a_{1,1}, a_{0,1})}{K(a_{0,1})}$$

by the definition of the generalized continuant. Assume that the statement of the theorem holds for n. By using hypothesis of the induction and Theorem 2 (taking $k = 0$), we get

$$\begin{aligned}
&[a_{n+1,1}, a_{n,1}, \ldots, a_{0,1}]_{\overrightarrow{b_1}} \\
&= a_{n+1,1} + \frac{a_{n+1,2}}{[a_{n,1}, a_{n-1,1}, \ldots, a_{0,1}]_{\overrightarrow{b_2}}} \\
&= a_{n+1,1} + \frac{a_{n+1,2}}{\dfrac{K(a_{n,1}, a_{n-1,1}, \ldots, a_{0,1})}{K(a_{n-1,1}, a_{n-2,1}, \ldots, a_{0,1})}} \\
&= \frac{a_{n+1,1}K(a_{n,1}, a_{n-1,1}, \ldots, a_{0,1}) + a_{n+1,2}K(a_{n-1,1}, a_{n-2,1}, \ldots, a_{0,1})}{K(a_{n,1}, a_{n-1,1}, \ldots, a_{0,1})} \\
&= \frac{K(a_{n+1,1}, a_{n,1}, \ldots, a_{0,1})}{K(a_{n,1}, a_{n-1,1}, \ldots, a_{0,1})}
\end{aligned}$$

where $\overrightarrow{b_1} = [a_{n+1,2}, a_{n,2}, \ldots, a_{1,2}]$ and $\overrightarrow{b_2} = [a_{n,2}, a_{n-1,2}, \ldots, a_{1,2}]$. □

For a continued fraction $[a_{n,1}, a_{n-1,1}, \ldots, a_{0,1}]_{\overrightarrow{b}}$ with $\overrightarrow{b} = [a_{n,2}, a_{n-1,2}, \ldots, a_{1,2}]$, the convergents $\dfrac{K(a_{n,1}, a_{n-1,1}, \ldots, a_{0,1})}{K(a_{n-1,1}, a_{n-2,1}, \ldots, a_{0,1})}$ are

$$\frac{a_{0,1}}{1}, \frac{a_{0,1}a_{1,1}+a_{1,2}}{a_{0,1}}, \frac{a_{2,1}(a_{0,1}a_{1,1}+a_{1,2})+a_{2,2}(a_{0,1})}{a_{1,1}(a_{0,1})+a_{1,2}}, \ldots.$$

If successive convergents are found, with numerators $A_0, A_1, A_2, \ldots$ and denominators $B_0, B_1, B_2, \ldots$, then the relevant recursive relations are

$$A_0 = a_{0,1}, \;\; A_1 = a_{1,1}a_{0,1} + a_{1,2} \;\text{ and }\; A_n = a_{n,1}A_{n-1} + a_{n,2}A_{n-2} \;\text{ for }\; n \geq 2,$$

and

$$B_0 = 1, \;\; B_1 = a_{0,1} \;\text{ and }\; B_n = a_{n-1,1}B_{n-1} + a_{n-1,2}B_{n-2} \;\text{ for }\; n \geq 2.$$

Let us define, for given positive integer r,

$$\begin{aligned} K_1^{(i)} &= K(a_{i,1}, a_{i-1,1}, a_{i-2,1} \ldots, a_{1,1}, a_{0,1}, a_{r-1,1}, \ldots, a_{i+2,1}, a_{i+1,1}), \quad \text{and} \\ K_2^{(i)} &= K(a_{i-1,1}, a_{i-2,1}, a_{i-3,1}, \ldots, a_{1,1}, a_{0,1}, a_{r-1,1}, \ldots, a_{i+2,1}), \end{aligned}$$

where as usual i is taken modulo r.

Theorem 4. *We have, for* $i = 0, \ldots, r-1$,

$$K_1^{(i)} + a_{i+1,2}K_2^{(i)} = K_1^{(i+1)} + a_{i+2,2}K_2^{(i+1)}.$$

Proof. By Theorem 2,

$$\begin{aligned} K_1^{(i+1)} &= K(a_{i+1,1}, a_{i,1}, \ldots, a_{i+3,1}, a_{i+2,1}) \\ &= a_{i+1,1}K(a_{i,1}, \ldots, a_{i+3,1}, a_{i+2,1}) + a_{i+1,2}K(a_{i-1,1}, a_{i-2,1}, \ldots, a_{i+2,1}). \end{aligned}$$

Since $K_2^{(i+1)} = K(a_{i,1}, a_{i-1,1}, \ldots, a_{i+4,1}, a_{i+3,1})$, we get

$$\begin{aligned} & K_1^{(i+1)} + a_{i+2,2}K_2^{(i+1)} \\ =\; & a_{i+1,1}K(a_{i,1}, \ldots, a_{i+3,1}, a_{i+2,1}) + a_{i+1,2}K(a_{i-1,1}, a_{i-2,1}, \ldots, a_{i+2,1}) \\ & + a_{i+2,2}K(a_{i,1}, a_{i-1,1}, \ldots, a_{i+4,1}, a_{i+3,1}). \end{aligned} \tag{2}$$

By the definition of the continuant,

$$\begin{aligned} K_1^{(i)} &= K(a_{i,1}, a_{i-1,1}, \ldots, a_{i+2,1}, a_{i+1,1}) \\ &= a_{i+1,1}K(a_{i,1}, a_{i-1,1}, \ldots, a_{0,1}, a_{r-1,1}, \ldots, a_{i+2,1}) \\ &\quad + a_{i+2,2}K(a_{i,1}, a_{i-1,1}, \ldots, a_{0,1}, a_{r-1,1}, \ldots, a_{i+3,1}), \end{aligned}$$

and since $K_2^{(i)} = K(a_{i-1,1}, a_{i-2,1}, \ldots, a_{i+3,1}, a_{i+2,1})$, we get

$$\begin{aligned} & K_1^{(i)} + a_{i+1,2}K_2^{(i)} \\ = \; & a_{i+1,1}K(a_{i,1}, a_{i-1,1}, \ldots, a_{i+2,1}) + a_{i+2,2}K(a_{i,1}, a_{i-1,1}, \ldots, a_{i+3,1}) \\ & + a_{i+1,2}K(a_{i-1,1}, a_{i-2,1}, \ldots, a_{i+3,1}, a_{i+2,1}). \end{aligned} \tag{3}$$

Thus, using (2) and (3), we get $K_1^{(i)} + a_{i+1,2}K_2^{(i)} = K_1^{(i+1)} + a_{i+2,2}K_2^{(i+1)}$, for $i = 0, \ldots, r-1$. □

Let us define $K_1 = K_1^{(0)}$ and $K_2 = K_2^{(0)}$. By Theorem 4, we can write

$$K_1^{(i)} + a_{i+1,2}K_2^{(i)} = K_1 + a_{1,2}K_2, \quad \text{for } i = 0, \ldots, r-1.$$

The following theorem extends results in [2]. The author of [2] provides a combinatorial description; here, we give our results in terms of the generalized continuant.

Theorem 5. *For any $n \geq 2r$, the sequence $\{v_n\}$ satisfies the following $2r$-order linear recurrence*

$$v_n = (K_1 + a_{1,2}K_2)\, v_{n-r} + (-1)^{r+1} (a_{0,2}a_{1,2}a_{2,2} \ldots a_{r-1,2})\, v_{n-2r}.$$

Proof. By the definition of the sequence $\{v_n\}$, we have

$$v_{mr+i} = a_{i,1}v_{mr+i-1} + a_{i,2}v_{mr+i-2} = K(a_{i,1})\, v_{mr+i-1} + a_{i,2}K()\, v_{mr+i-2}.$$

If we substitute $v_{mr+i-1} = a_{i-1,1}v_{mr+i-2} + a_{i-1,2}v_{mr+i-3}$ in the above equality and rearrange it, we get

$$\begin{aligned} v_{mr+i} &= (a_{i,1}a_{i-1,1} + a_{i,2})v_{mr+i-2} + a_{i-1,2}a_{i,1}v_{mr+i-3} \\ &= K(a_{i,1}, a_{i-1,1})\, v_{mr+i-2} + a_{i-1,2}K(a_{i,1})\, v_{mr+i-3}. \end{aligned}$$

If we proceed in this way, we obtain

$$\begin{aligned} v_{mr+i} &= K(a_{i,1})\, v_{mr+i-1} + a_{i,2}K()\, v_{mr+i-2} \\ &= K(a_{i,1}, a_{i-1,1})\, v_{mr+i-2} + a_{i-1,2}K(a_{i,1})\, v_{mr+i-3} \\ &= K(a_{i,1}, a_{i-1,1}, a_{i-2,1})\, v_{mr+i-3} + a_{i-2,2}K(a_{i,1}, a_{i-1,1})\, v_{mr-4} \\ &\vdots \quad \vdots \\ &= K(a_{i,1}, a_{i-1,1}, \ldots, a_{i+2,1}, a_{i+1,1})\, v_{mr+i-r} \\ &\quad + a_{i+1,2}K(a_{i,1}, a_{i-1,1}, \ldots, a_{i+2,1})\, v_{mr+i-r-1} \\ &= K_1^{(i)} v_{mr+i-r} + a_{i+1,2}K(a_{i,1}, a_{i-1,1}, \ldots, a_{i+2,1})\, v_{mr+i-r-1}. \end{aligned} \tag{4}$$

Now by the definition of the sequence, since $mr + i - 2r + 2 \equiv i + 2 \pmod{r}$, we have

$$\begin{aligned} a_{i+2,2}v_{mr+i-2r} &= v_{mr+i-2r+2} - a_{i+2,1}v_{mr+i-2r+1} \\ &= K()\, v_{mr+i-2r+2} - K(a_{i+2,1})\, v_{mr+i-2r+1}. \end{aligned}$$

Multiplying the above equation by $a_{i+3,2}$, substituting

$$a_{i+3,2}v_{mr+i-2r+1} = v_{mr+i-2r+3} - a_{i+3,1}v_{mr+i-2r+2}$$

in the above equality and arranging it, we obtain

$$\begin{aligned} & a_{i+2,2}a_{i+3,2}v_{mr+i-2r} \\ = \ & a_{i+3,2}v_{mr+i-2r+2} - a_{i+2,1}a_{i+3,2}v_{mr+i-2r+1} \\ = \ & -K(a_{i+2,1})\, v_{mr+i-2r+3} + (a_{i+2,1}a_{i+3,1} + a_{i+3,2})\, v_{mr+i-2r+2} \\ = \ & -K(a_{i+2,1})\, v_{mr+i-2r+3} + K(a_{i+3,1}, a_{i+2,1})\, v_{mr+i-2r+2}. \end{aligned}$$

If we proceed in this way, we get

$$\begin{aligned} a_{i+2,2}v_{mr+i-2r} &= K()\, v_{mr+i-2r+2} - K(a_{i+2,1})\, v_{mr+i-2r+1} \\ a_{i+2,2}a_{i+3,2}v_{mr+i-2r} &= -K(a_{i+2,1})\, v_{mr+i-2r+3} \\ &\quad +K(a_{i+3,1}, a_{i+2,1})\, v_{mr+i-2r+2} \\ a_{i+2,2}a_{i+3,2}a_{i+4,2}v_{mr+i-2r} &= K(a_{i+3,1}, a_{i+2,1})\, v_{mr+i-2r+4} \\ &\quad -K(a_{i+4,1}, a_{i+3,1}, a_{i+2,1})\, v_{mr+i-2r+3} \end{aligned}$$

and so on up to

$$\begin{aligned} & a_{i+2,2}a_{i+3,2}\dots a_{i+r,2}v_{mr+i-2r} \\ = \ & (-1)^r K(a_{i-1,1}, a_{i-2,1}, \dots, a_{i+3,1}, a_{i+2,1})\, v_{mr+i-r} \\ & +(-1)^{r+1} K(a_{i,1}, a_{i-1,1}, \dots, a_{i+3,1}, a_{i+2,1})\, v_{mr+i-r-1}. \end{aligned}$$

Multiplying the last equation by $(-1)^r a_{i+1,2}$, reducing the first subscript of the coefficients $a_{i,j}$ modulo r and rearranging the expression, we get

$$\begin{aligned} & (-1)^r a_{0,2}a_{1,2}a_{2,2}\dots a_{r-1,2}v_{mr+i-2r} && (5) \\ = \ & a_{i+1,2}K(a_{i-1,1}, a_{i-2,1}, \dots, a_{i+3,1}, a_{i+2,1})\, v_{mr+i-r} \\ & -a_{i+1,2}K(a_{i,1}, a_{i-1,1}, \dots, a_{i+3,1}, a_{i+2,1})\, v_{mr+i-r-1} \\ = \ & a_{i+1,2}K_2^{(i)} v_{mr+i-r} \\ & -a_{i+1,2}K(a_{i,1}, a_{i-1,1}, \dots, a_{i+3,1}, a_{i+2,1})\, v_{mr+i-r-1}. && (6) \end{aligned}$$

Let $n = mr + i$; summing (4) and (6), we get

$$v_n = \left(K_1^{(i)} + a_{1,2}K_2^{(i)}\right) v_{n-r} + (-1)^{r+1}(a_{0,2}a_{1,2}a_{2,2}\dots a_{r-1,2})\, v_{n-2r}.$$

By Theorem 4, we get the result. □

3. The Generating Function of the Sequence $\{v_n\}$

The generating function for $\{v_n\}$ is

$$G(x) = \sum_{m=0}^{\infty} v_m x^m.$$

If we define

$$G_i(x) = \sum_{m=0}^{\infty} v_{mr+i} x^{mr+i}, \quad i = 0, 1, \ldots, r-1,$$

then we can write

$$G(x) = \sum_{i=0}^{r-1} G_i(x). \tag{7}$$

Theorem 6. *The generating function of the sequence* $\{v_n\}$ *is*

$$G(x) = \frac{\sum_{i=0}^{r-1} \left(v_i x^i + (v_{r+i} - A v_i) x^{r+i}\right)}{1 - Ax^r + (-1)^r B x^{2r}},$$

where

$$A = K_1 + a_{1,2} K_2 \quad \text{and} \quad B = \prod_{i=0}^{r-1} a_{i,2}.$$

Proof. Multiplying $G_i(x)$ by $1 - Ax^r + (-1)^r Bx^{2r}$ for $i = 0, \ldots, r-1$ and using Theorem 5 and Theorem 25.5.1 in [1], we get for $i = 0, 1, \ldots, r-1$,

$$G_i(x) = \frac{v_i x^i + (v_{r+i} - A v_i) x^{r+i}}{1 - Ax^r + (-1)^r B x^{2r}}. \tag{8}$$

Hence, using (7) we get the desired result. □

Example 7. For $r = 3$, the sequence $\{v_n\}$ satisfies $v_0 = 0, v_1 = 1$ and for $n \geq 2$

$$v_n = \begin{cases} a_{0,1} v_{n-1} + a_{0,2} v_{n-2}, & \text{if } n \equiv 0 \pmod{3}, \\ a_{1,1} v_{n-1} + a_{1,2} v_{n-2}, & \text{if } n \equiv 1 \pmod{3}, \\ a_{2,1} v_{n-1} + a_{2,2} v_{n-2}, & \text{if } n \equiv 2 \pmod{3}. \end{cases}$$

Let us find the generating function of the sequence $\{v_n\}$ for the given integer $r = 3$. Since $r = 3$, using Definition 1 we have

$$\begin{aligned} K_1 &= K(a_{0,1}, a_{2,1}, a_{1,1}) = a_{1,1} K(a_{0,1}, a_{2,1}) + a_{2,2} K(a_{0,1}) \\ &= a_{1,1}\left(a_{2,1} K(a_{0,1}) + a_{0,2} K()\right) + a_{2,2} K(a_{0,1}) \\ &= a_{0,1} a_{1,1} a_{2,1} + a_{0,2} a_{1,1} + a_{0,1} a_{2,2}, \end{aligned}$$

and

$$K_2 = K(a_{2,1}) = a_{2,1}.$$

Thus, we have

$$A = K_1 + a_{1,2} K_2 = a_{0,1} a_{1,1} a_{2,1} + a_{0,2} a_{1,1} + a_{0,1} a_{2,2} + a_{1,2} a_{2,1}.$$

By Theorem 6, we get

$$G(x) = \frac{\sum_{i=0}^{2} \left(v_i x^i + (v_{3+i} - A v_i) x^{3+i}\right)}{1 - Ax^3 + (-1)^3 B x^6}.$$

Since $v_0 = 0$, $v_1 = 1$, $v_2 = a_{2,1}$, $v_3 = a_{0,1}a_{2,1} + a_{0,2}$, $v_4 = a_{0,1}a_{1,1}a_{2,1} + a_{0,2}a_{1,1} + a_{1,2}a_{2,1}$ and $v_5 = a_{0,1}a_{1,1}a_{2,1}^2 + a_{0,2}a_{1,1}a_{2,1} + a_{1,2}a_{2,1}^2 + a_{0,1}a_{2,1}a_{2,2} + a_{0,2}a_{2,2}$, we finally obtain

$$G(x) = \frac{x + a_{2,1}x^2 + (a_{0,1}a_{2,1} + a_{0,2})x^3 - a_{0,1}a_{2,2}x^4 + a_{0,2}a_{2,2}x^5}{1 - (a_{0,1}a_{1,1}a_{2,1} + a_{0,2}a_{1,1} + a_{0,1}a_{2,2} + a_{1,2}a_{2,1})x^3 - a_{0,2}a_{1,2}a_{2,2}x^6}.$$

4. The Extended Binet's Formula of the Sequence $\{v_n\}$

By Theorem 5, we can consider the conditional sequence $\{v_n\}$ as a constant coefficient $2r$-order linear recurrence for any positive integer s. So, we can find the matrix representation of the sequence $\{v_n\}$ and we can use methods for linear recurrences to obtain the extended Binet's formula for the conditional sequence $\{v_n\}$.

Let us define the $2r \times 2r$ matrix

$$N = \begin{bmatrix} 0 & 1 & 0 & \dots & 0 & \dots & 0 & 0 \\ 0 & 0 & 1 & \dots & 0 & \dots & 0 & 0 \\ 0 & 0 & 0 & \dots & 0 & \dots & 0 & 0 \\ \vdots & \vdots & \vdots & & \vdots & \dots & \vdots & \vdots \\ 0 & 0 & 0 & \dots & 0 & \dots & 0 & 1 \\ (-1)^r \prod_{i=0}^{r-1} a_{i,2} & 0 & 0 & \dots & (K_1 + a_{1,2}K_2) & \dots & 0 & 0 \end{bmatrix},$$

where $K_1 + a_{1,2}K_2$ is the entry in the $(2r)$-th row and $(r+1)$-th column. In fact, N is the companion matrix for the polynomial

$$q(x) = x^{2r} - (K_1 + a_{1,2}K_2)x^r - (-1)^r \prod_{i=0}^{r-1} a_{i,2}.$$

The polynomial $q(x)$ is the characteristic polynomial of N. By using an inductive argument, we can give the matrix representation of the conditional sequence $\{v_n\}$:

$$N^n \begin{bmatrix} v_0 \\ v_1 \\ \vdots \\ v_{2r-1} \end{bmatrix} = \begin{bmatrix} v_n \\ v_{n+1} \\ \vdots \\ v_{n+2r-1} \end{bmatrix}. \tag{9}$$

This matrix representation is important since it may be used to derive many interesting properties of the conditional sequence $\{v_n\}$.

Let $\mu_1, \mu_2, \dots, \mu_{2r}$ be the eigenvalues of the matrix N. Let us assume that the polynomial $q(x)$ has no multiple roots, so these eigenvalues are all distinct. In this case, N can be diagonalized using the Vandermonde matrix

$$V(\mu_1, \mu_2 \ldots, \mu_{2r}) = \begin{bmatrix} 1 & 1 & \cdots & 1 & 1 \\ \mu_1 & \mu_2 & \cdots & \mu_{2r-1} & \mu_{2r} \\ \mu_1^2 & \mu_2^2 & \cdots & \mu_{2r-1}^2 & \mu_{2r}^2 \\ \vdots & \vdots & \vdots & \vdots & \vdots \\ \mu_1^{2r-1} & \mu_2^{2r-1} & \cdots & \mu_{2r-1}^{2r-1} & \mu_{2r}^{2r-1} \end{bmatrix}.$$

We can give the following theorem using the matrix methods in [9].

Theorem 8. *Assume $q(x)$ has roots $\mu_1, \mu_2, \ldots, \mu_{2r}$ and also assume the polynomial $q(x)$ has no multiple roots. Then, the Binet-like formula for the conditional sequence $\{u_n\}$ is $v_n = \sum_{i=1}^{2r} \frac{\mu_i^n}{q'(\mu_i)}$, where $q'$is the derivative of q.*

We can also give the following general extended Binet's formula for the sequence $\{v_n\}$ independent of whether $q(x)$ has multiple roots or not.

Theorem 9. *The terms of the sequence $\{v_n\}$ satisfy*

$$v_{mr+i} = (-1)^{r(m+1)} \left[\frac{\alpha^m - \beta^m}{\alpha - \beta} v_{r+i} - B \frac{\alpha^{m-1} - \beta^{m-1}}{\alpha - \beta} v_i \right],$$

where

$$\alpha = \frac{(-1)^r A + \sqrt{A^2 - 4(-1)^r B}}{2} \quad \text{and} \quad \beta = \frac{(-1)^r A - \sqrt{A^2 - 4(-1)^r B}}{2},$$

that is, α and β are the roots of the polynomial $p(z) = z^2 - (-1)^r Az + (-1)^r B$, where $A = K_1 + a_{1,2}K_2$ and $B = \prod_{i=0}^{r-1} a_{i,2}$.

Proof. By (8), the generating function for the subsequence $\{v_{mr+i}\}$ is given by

$$G_i(x) = \frac{v_i x^i + (v_{r+i} - Av_i) x^{r+i}}{1 - Ax^r + (-1)^r Bx^{2r}}.$$

Using the identities $\alpha + \beta = (-1)^r A$, $\alpha - \beta = \sqrt{A^2 - 4(-1)^r B}$, $\alpha\beta = (-1)^r B$, $(-1)^r A\alpha^m - \alpha^{m+1} = \beta\alpha^m$ and $(-1)^r A\beta^m - \beta^{m+1} = \alpha\beta^m$, we get

$$\begin{aligned} G_i(x) &= \frac{v_i x^i + (v_{r+i} - Av_i) x^{r+i}}{1 - Ax^r + (-1)^r Bx^{2r}} \\ &= \frac{x^i [v_i + (v_{r+i} - Av_i) x^r]}{\alpha - \beta} \left[\frac{\alpha}{1 - (-1)^r \alpha x^r} - \frac{\beta}{1 - (-1)^r \beta x^r} \right] \end{aligned}$$

$$
\begin{aligned}
&= x^i\left[v_i+(v_{r+i}-Av_i)\,x^r\right]\sum_{m=0}^{\infty}\frac{(-1)^{mr}\left(\alpha^{m+1}-\beta^{m+1}\right)x^{mr}}{\alpha-\beta}\\
&= x^i\left[\sum_{m=0}^{\infty}\frac{(-1)^{mr}\left(\alpha^{m+1}-\beta^{m+1}\right)v_i}{\alpha-\beta}x^{mr}\right.\\
&\qquad\left.+\sum_{m=0}^{\infty}\frac{(-1)^{mr}\left(\alpha^{m+1}-\beta^{m+1}\right)(v_{r+i}-Av_i)}{\alpha-\beta}x^{mr+r}\right]\\
&= x^i\left[v_i+\sum_{m=1}^{\infty}\frac{(-1)^{mr}\left(\alpha^{m+1}-\beta^{m+1}\right)v_i}{\alpha-\beta}x^{mr}\right.\\
&\qquad\left.+(-1)^r\sum_{m=1}^{\infty}\frac{(-1)^{mr}\left(\alpha^{m}-\beta^{m}\right)(v_{r+i}-Av_i)}{\alpha-\beta}x^{mr}\right]\\
&= x^iv_i+\sum_{m=1}^{\infty}(-1)^{mr}\left[(-1)^r\frac{(\alpha^m-\beta^m)}{\alpha-\beta}v_{r+i}\right.\\
&\qquad\left.-\frac{\left((-1)^rA\alpha^m-\alpha^{m+1}\right)-\left((-1)^rA\beta^m-\beta^{m+1}\right)}{\alpha-\beta}v_i\right]x^{mr+i}\\
&= \sum_{m=0}^{\infty}(-1)^{mr}\left[(-1)^r\frac{(\alpha^m-\beta^m)}{\alpha-\beta}v_{r+i}-\frac{\beta\alpha^m-\alpha\beta^m}{\alpha-\beta}v_i\right]x^{mr+i}\\
&= \sum_{m=0}^{\infty}(-1)^{mr+r}\left[\frac{(\alpha^m-\beta^m)}{\alpha-\beta}v_{r+i}-B\frac{\alpha^{m-1}-\beta^{m-1}}{\alpha-\beta}v_i\right]x^{mr+i}.
\end{aligned}
$$

Thus, we obtain

$$
v_{mr+i}=(-1)^{r(m+1)}\left[\frac{\alpha^m-\beta^m}{\alpha-\beta}v_{r+i}-B\frac{\alpha^{m-1}-\beta^{m-1}}{\alpha-\beta}v_i\right].
$$

□

Corollary 10. *If we take the values* $a_{0,2}=a_{1,2}=\cdots=a_{r-1,2}=1$ *in the sequence* $\{v_n\}$ *we get a special case the* k-periodic Fibonacci sequence *in [3]. Also, substituting* $B=1$ *in Theorem 6 and 9, we get Theorem 13 and Theorem 16 in [3], respectively.*

5. Conclusion

In this paper we consider the Fibonacci-like conditional sequences $\{v_n\}$ as given in Equation (1). We define the concept of generalized continuant and use it to show that $\{v_n\}$ satisfies a linear recurrence relation. We also derive generating functions, matrix representations and extended Binet's formulas for $\{v_n\}$ in terms of the generalized continuant. It would be interesting to find more useful applications of the generalized continuant and generalized fractions.

Acknowledgement We thank the referee for helpful suggestions to improve the presentation of this paper.

References

[1] N. Biggs, Discrete Mathematics, Oxford University Press (2002).

[2] C. Cooper, An identity for period k second order linear recurrence systems, Congr. Numer. 200 (2010), 95–106.

[3] M. Edson, S. Lewis and O. Yayenie, The k-periodic Fibonacci sequence and an extended Binet's formula, Integers 11 (2011), 739–751.

[4] M. Edson and O. Yayenie, A new generalization of Fibonacci sequence and extended Binet's formula, Integers 9 (2009), 639–654.

[5] R. L. Graham, D. E. Knuth and O. Patashnik, Concrete Mathematics: A Foundation for Computer Science, 2nd ed. Reading, MA: Addison-Wesley (1994).

[6] A. F. Horadam, A generalized Fibonacci sequence, Amer. Math. Monthly 68 (1961), 455–459.

[7] F. Howard, Applications of Fibonacci Numbers, Volume 9, Kluwer Academic Publishers (2004).

[8] D. V. Jaiswal, On a generalized Fibonacci sequence, Labdev J. Sci. Tech. Part A 7 (1969), 67–71.

[9] D. Kalman, Generalized Fibonacci numbers by matrix methods, Fibonacci Quart. 20 (1982), 73–76.

[10] S. T. Klein, Combinatorial representation of generalized Fibonacci numbers, Fibonacci Quart. 29 (1991), 124–131.

[11] A. T. Krassimir, A. C. Liliya, S. D. Dimitar, A new perspective to the generalization of the Fibonacci sequence, Fibonacci Quart. 23 (1985), 21–28.

[12] G. Y. Lee, S. G. Lee, H. G. Shin, On the k-generalized Fibonacci matrix Q_k, Linear Algebra Appl. 251 (1997), 73–88.

[13] G. Y. Lee, S. G. Lee, J. S. Kim, H. K. Shin, The Binet formula and representations of k-generalized Fibonacci numbers, Fibonacci Quart. 39 (2001), 158–164.

[14] D. H. Lehmer, An extended theory of Lucas functions. Ann. of Math. 31 (1930), 419–448.

[15] S. P. Pethe, C. N. Phadte, A generalization of the Fibonacci sequence, Applications of Fibonacci numbers, Vol. 5 (St. Andrews, 1992), 465–472.

[16] J. C. Pond, Generalized Fibonacci summations, Fibonacci Quart. 6 (1968), 97–108.

[17] M. Sahin, The generating function of a family of sequences in terms of the continuant, Appl. Math. Comput. 217 (12), (2011), 5416–5420.

[18] M. Sahin, The Gelin-Cesàro identity in some conditional sequences, Hacet. J. Math. Stat. 40 (6) (2011), 855–861.

[19] M. Sahin, The Binet-like formula of a family of the conditional sequences by matrix methods, Ars Combinatoria 102 (2011), 393–398.

[20] G. Sburlati, Generalized Fibonacci sequences and linear congruences, Fibonacci Quart. 40 (2002), 446–452.

[21] J. E. Walton, A. F. Horadam, Some further identities for the generalized Fibonacci sequence $\{H_n\}$, Fibonacci Quart. 12 (1974), 272–280.

ON THE RESIDUE CLASSES OF $\pi(n)$ MODULO t

Ping Ngai Chung
Department of Mathematics, Massachusetts Institute of Technology, Cambridge, Massachusetts
briancpn@mit.edu

Shiyu Li [1]
Dept of Mathematics, University of California - Berkeley, Berkeley, California
jjl2357@berkeley.edu

Received: 8/22/13, Accepted: 11/21/13, Published: 12/13/13

Abstract

The prime number theorem is one of the most fundamental theorems of analytic number theory, stating that the prime counting function, $\pi(x)$, is asymptotic to $x/\log x$. However, it says little about the parity of $\pi(n)$ as an arithmetic function. Using Selberg's sieve, we give, for any fixed integers r and t, a positive lower bound on the proportion of positive integers n such that $\pi(n)$ is congruent to r modulo t. Moreover, we generalize this to the counting function of any set of primes with positive density.

1. Introduction and Statement of Results

The prime counting function $\pi(x)$ has been a topic of interest for mathematicians throughout history. One of the first main results about this function is the prime number theorem, which states that $\pi(x) \sim \frac{x}{\log x}$. This asymptotic carries deep implications for the distribution of prime numbers. However, there are many problems about primes which remain unsolved in spite of this theorem. Two examples are the Goldbach conjecture, stating that every even number greater than 2 is the sum of two primes, and the twin prime conjecture, which claims that there are infinitely many pairs of primes that differ by 2.

In spectacular fashion, there has been a spate of recent advances concerning the gaps between prime numbers. In 2005, Goldston, Pintz and Yıldırım [4] proved that there exist infinitely many consecutive primes which are much closer than average. Recent work by Zhang [12] and the polymath project [10] shows that there are

[1]The authors are grateful to the NSF's support of the REU at Emory University.

infinitely many pairs of primes differing by at most an absolute constant. These advances depend critically on sifting techniques, particularly Selberg's sieve.

In this paper, we provide another application of Selberg's sieve, one which does not seem to have been considered previously in the literature. We treat $\pi(n)$ as an arithmetic function, and consider its distribution mod t as n varies. The prime number theorem alone says little about the parity of $\pi(n)$ or the proportion of positive integers n for which $\pi(n)$ is even. Nevertheless, numerical evidence shows that this seems to be the case about half the time:

x	100	1000	10^4	10^5	10^6	10^7
$\#\{n < x \mid \pi(n) \text{ is even}\}/x$	0.510	0.523	0.486	0.502	0.498	0.499

Define $T_{r,t}(x)$ to be the number of positive integers $n < x$ such that $\pi(n) \equiv r \bmod t$. In this paper, we prove the following.

Theorem 1.1. *For any $0 \le r < t$, we have that*

$$\liminf_{x\to\infty} \frac{T_{r,t}(x)}{x} \ge \frac{1}{16t^2}.$$

If we assume the Hardy-Littlewood prime k-tuple conjecture, we can replace 16 *by* 2 *above.*

Remark 1. Recall that the Hardy-Littlewood prime k-tuple conjecture states that given $\mathfrak{a} = (\alpha_1, ..., \alpha_k)$, where $\alpha_1, \dots, \alpha_k$ are distinct integers which do not cover all residue classes to any prime modulus, the number $\pi(x; \mathfrak{a})$ of positive integers $m < x$ for which $m - \alpha_1, ..., m - \alpha_k$ are all prime satisfies $\pi(x;\mathfrak{a}) \sim Bx(\log x)^{-k}$, where B is a certain nonzero infinite product over primes. With this estimate, it is possible to improve Theorem 2.3 by a factor of $2^k k!$. In our application, $k = 2$, explaining the improvement of the lower bound by a factor of 8.

Remark 2. Theorem 1.1 was obtained independently and almost simultaneously by Alboiu [1].

Theorem 1.1 is a special case of a more general theorem. We say that a set $\mathfrak{S}$ of primes has density δ if

$$\frac{\#\{p < x : p \in \mathfrak{S}\}}{\#\{p < x : p \text{ prime}\}} \to \delta \qquad \text{as } x \to \infty.$$

Define $T_{r,t}(x; \mathfrak{S})$ to be the number of positive integers $n < x$ such that $|\mathfrak{S}(n)| \equiv r \bmod t$, where $|\mathfrak{S}(n)|$ denotes the number of primes in $\mathfrak{S}$ less than n.

Theorem 1.2. *If we assume the notation above and that $\mathfrak{S}$ has positive density δ, then*

$$\liminf_{x\to\infty} \frac{T_{r,t}(x;\mathfrak{S})}{x} \ge \frac{\delta^2}{16t^2}.$$

If we assume the Hardy-Littlewood prime k-tuple conjecture, we can replace 16 *by* 2 *above.*

Examples. Two examples of sets of primes with positive density follow.

1. Let $K/\mathbb{Q}$ be a finite Galois extension with Galois group G, and let C be a conjugacy class in G. Then the set of unramified primes p for which the Frobenius automorphism σ_p is in C has density $\#C/\#G$ by the Chebotarëv density theorem.

2. Given a positive irrational $\alpha > 1$, the set of primes p of the form $\lfloor \alpha n \rfloor$ for some integer n has density α^{-1} by [8, Thm. 2].

In particular, we may consider $T_{r,t}(x; q, a)$ to be the number of positive integers $n < x$ such that $\pi(n; q, a) \equiv r \bmod t$, where $\pi(x; q, a)$ is the counting function of the primes congruent to a mod q. In this case, we have the following stronger result.

Theorem 1.3. *For any* $0 \le r < t$, $0 \le a < q$, *where* $\gcd(a, q) = 1$, *we have that*

$$\liminf_{x \to \infty} \frac{T_{r,t}(x; q, a)}{x} \ge \frac{1}{16\phi(q)t^2},$$

where $\phi(q)$ *is Euler's totient function.*

Remark. Results similar to those in Theorems 1.1, 1.2, and 1.3 can be obtained by making use of Brun's sieve. However, the constants will be weaker.

To prove our theorem, we will rely on a judicious application of Selberg's sieve. In particular, using either Selberg's sieve or the prime k-tuple conjecture, we bound the number of primes which have a particular "small" gap, thus also bounding the number of primes which have any small gap. This lets us say that not all gaps occurring at primes p_{nt+r} can be small, and thus $T_{r,t}(x)$ cannot be too small.

2. Selberg's Sieve

We now introduce the main tool, Selberg's sieve. In this, as well as many other sieve methods (see [2], [7, Ch. 6]), we are given a finite set $\mathcal{A}$ and a set of primes $\mathcal{P}$, and we wish to estimate the size of the set

$$S(\mathcal{A}, \mathcal{P}) := \{a \in \mathcal{A} \mid (a, p) = 1 \text{ for all } p \in \mathcal{P}\}.$$

To do this, we use the inclusion-exclusion principle. For each squarefree integer d, let $\mathcal{A}_d$ be the set of elements in $\mathcal{A}$ divisible by d. Then

$$|S(\mathcal{A}, \mathcal{P})| = \sum_{d \mid P} \mu(d) |\mathcal{A}_d|,$$

where P is the product of all the primes in $\mathcal{P}$. Sieve methods are most useful when the elements of $\mathcal{A}_d$ distribute approximately evenly in $\mathcal{A}$, in which case one would expect $|\mathcal{A}_d| = g(d)X + R_d$, where X is approximately the size of $\mathcal{A}$, $g(d)$ is a multiplicative function that is the density of $\mathcal{A}_d$ in $\mathcal{A}$, so that $0 < g(d) < 1$, and R_d is a small error term. Typically, a sieve method gives an estimate for $|S(\mathcal{A}, \mathcal{P})|$ in the following form

$$|S(\mathcal{A}, \mathcal{P})| = X \prod_{p \in \mathcal{P}} (1 - g(p)) + \text{ error term.}$$

The main term is what one would expect if sifting by the primes in $\mathcal{P}$ are pairwise independent events, and the error term varies from different sieve methods. However, when the primes in $\mathcal{P}$ are fairly large, this independence condition fails. Hence, most sieves can only obtain an upper bound for the sifted quantity in question.

The key idea of Selberg's sieve is to replace the Möbius function in the formula of $|S(\mathcal{A}, \mathcal{P})|$ by an optimally chosen quadratic form so that the resulting estimates are minimal. Formally, it is stated in the following theorem.

Theorem 2.1. *(Selberg's sieve) [7, Thm. 6.5] With notation as above, let $h(d)$ be the multiplicative function given by $h(p) = g(p)(1 - g(p))^{-1}$ and set*

$$H(D) = \sum_{d < \sqrt{D}, d | P} h(d)$$

for any $D > 1$. If we assume that

$$\begin{aligned} |R_d| &\leq g(d)d, && (1)\\ g(d)d &\geq 1 \qquad \text{if } d|P, && (2)\\ \sum_{y \leq p \leq x} g(p) \log p &\ll \log(2x/y) \qquad \text{for all } 2 \leq y \leq x, && (3) \end{aligned}$$

then

$$S(\mathcal{A}, \mathcal{P}) \leq \frac{X}{H(D)} + O\left(\frac{D}{\log^2 D}\right).$$

With further restrictions on $g(d)$, one may obtain an estimate for $H(D)$ which is frequently useful in practice.

Lemma 2.2. *[6, Lemmas 5.3, 5.4] Suppose the conditions in Theorem 2.1 hold. Moreover, suppose there exist positive real numbers κ, A_1, A_2, L such that*

$$0 \leq g(p) \leq 1 - \frac{1}{A_1}; \qquad (4)$$

$$-L \leq \sum_{w \leq p \leq z} g(p) \log p - \kappa \log \frac{z}{w} \leq A_2 \qquad \text{for all } 2 \leq w \leq z. \qquad (5)$$

If P is the product of all primes $p < \sqrt{D}$, then

$$H(D) = C(\log \sqrt{D})^{\kappa} \left(1 + O\left(\frac{L}{\log D}\right)\right),$$

where

$$C = \frac{1}{\Gamma(\kappa+1)} \prod_p (1-g(p))^{-1} \left(1-\frac{1}{p}\right)^{\kappa}.$$

The implied constant depends only on A_1, A_2, and κ. In particular, it is independent of L.

Remark. This is a combination of Lemmas 5.3 and 5.4 of Halberstam and Richert's book [6]. Note that exactly this formulation appears in [5].

In our work, we will find the following theorem, which is proved as a direct application of Selberg's sieve, to be useful.

Theorem 2.3. *[7, Thm. 6.7] Let $\mathfrak{a} = (\alpha_1, \ldots, \alpha_k)$ be distinct integers which do not cover all residue classes to any prime modulus and $\alpha_i \le c \log x$ for some absolute constant c for all i. Then the number $\pi(x;\mathfrak{a})$ of positive integers $m \le x$ for which $m-\alpha_1, \ldots, m-\alpha_k$ are all prime satisfies*

$$\pi(x;\mathfrak{a}) \le 2^k k! Bx(\log x)^{-k} \left(1 + O\left(\frac{\log\log x}{\log x}\right)\right),$$

where

$$B = \prod_p \left(1-\frac{\nu(p)}{p}\right)\left(1-\frac{1}{p}\right)^{-k}$$

and $\nu(p)$ is the number of roots of the polynomial $f(m) = (m-\alpha_1)\cdots(m-\alpha_k)$ modulo p. The implied constant depends only on k and c.

Proof. We apply Selberg's sieve (Theorem 2.1). Take $\mathcal{A}$ to be the set of polynomial values

$$f(m) = (m-\alpha_1)\cdots(m-\alpha_k)$$

for positive integers $m \le x$. Take $g(p) = \nu(p)p^{-1}$ and $X = x$. Note that $\nu(p) \le k$ for all primes p and $\nu(p) = k$ for sufficiently large p; thus, conditions (1) - (5) hold for $\kappa = k$, $A_1 = k+1$, $A_2 = 5k$, $L = (k-1)\log(\max \alpha_i) = O_{k,c}(\log\log x)$. By Lemma 2.2, we have that

$$\begin{aligned} H(D)^{-1} &= k! \prod_p \left(1-\frac{\nu(p)}{p}\right)\left(1-\frac{1}{p}\right)^{-k} \left(\frac{\log D}{2}\right)^{-k} \left(1 + O_k\left(\frac{L}{\log D}\right)\right) \\ &= 2^k k! B (\log D)^{-k} \left(1 + O_k\left(\frac{L}{\log D}\right)\right), \end{aligned}$$

hence, by Theorem 2.1 with $D = x \log^{-(k-1)} x$, we have

$$\pi(x;\mathfrak{a}) \le S(\mathcal{A},\mathcal{P}) + \sqrt{D} \le 2^k k! Bx(\log x)^{-k} \left(1 + O_k\left(\frac{\max\{L, \log\log x\}}{\log x}\right)\right).$$

The result follows. □

Remark 1. The prime k-tuple conjecture predicts that $\pi(x;\mathfrak{a}) \sim Bx(\log x)^{-k}$.

Remark 2. For our purpose, it is important that the implied constant depends only on k and c but not the individual values of α_i since we will take α_i to infinity as $x \to \infty$ in the proof of Lemma 3.1.

3. Proof of Theorem 1.2

We are now ready to begin the proof of the main theorem. For each positive integer k, let S_k be the number of positive integers n such that $|\mathfrak{S}(n)| = k$. We label the elements in $\mathfrak{S}$ by $q_1, q_2 \ldots$ in increasing order, and we note that $S_k = q_{k+1} - q_k$. In other words, S_k is just the gap between q_k and q_{k+1}. Observe that for each prime q_m,

$$T_{r,t}(q_m; \mathfrak{S}) = \sum_{\substack{k \le m \\ k \equiv r (\text{mod } t)}} S_k.$$

By a simple application of the prime number theorem, we see that, for all $A > 1$,

$$T_{r,t}(x, \mathfrak{S}) = \sum_{\substack{q_k < x, \\ k \equiv r (\text{mod } t)}} (q_{k+1} - q_k) + O_t\left(\frac{x}{\log^A x}\right), \tag{6}$$

where O_t indicates that the implied constant may depend on t.

The proof of Theorem 1.2 relies on the following key lemma, which gives an upper bound on the number of small gaps.

Lemma 3.1. *Let $f(x)$ tend to infinity as $x \to \infty$ and $f(x) = O(\log x)$. Then the number of pairs of primes (p_i, p_j) with $p_i, p_j < x$ such that*

$$0 < p_i - p_j < f(x)$$

is at most

$$\frac{8x}{\log^2 x} f(x)(1 + o(1)).$$

To prove this lemma, we start with an auxiliary estimate, which follows from Theorem 2.3.

Lemma 3.2. *Let a be a positive integer such that $a \le c \log x$ for some positive constant c. Then the number of prime solutions $p_i, p_j < x$ to $p_i - p_j = a$ is at most*

$$\frac{8x}{\log^2 x} B(a) \left(1 + O\left(\frac{\log\log x}{\log x}\right)\right)$$

where

$$B(a) := \prod_{p|a}\left(1-\frac{1}{p}\right)^{-1}\prod_{p\nmid a}\left(1-\frac{2}{p}\right)\left(1-\frac{1}{p}\right)^{-2}$$

and the implied constant depends only on c. Assuming the prime k-tuple conjecture, the upper bound can be improved by a factor of 8.

Proof. In Theorem 2.3, take $k = 2$, $\mathfrak{a} = (0, -a)$. Note that $\nu(p) = 1$ if $p|a$ and $\nu(p) = 2$ if $p \nmid a$. Then the result follows. □

We also need the following estimate.

Lemma 3.3. *Assuming the notation in Lemma 3.2, we have that*

$$\sum_{\substack{a\le f(x),\\ 2|a}} B(a) = f(x)(1+o(1)).$$

Proof.

$$\begin{aligned}\sum_{\substack{a\le f(x),\\ 2|a}} B(a) &= \sum_{\substack{a\le f(x),\\ 2|a}}\prod_{p|a}\left(1-\frac{1}{p}\right)^{-1}\prod_{p\nmid a}\left(1-\frac{2}{p}\right)\left(1-\frac{1}{p}\right)^{-2}\\ &= 2\prod_{p\ge 3}\left(1-\frac{2}{p}\right)\left(1-\frac{1}{p}\right)^{-2}\sum_{\substack{a\le f(x),\\ 2|a}}\psi(a),\end{aligned}$$

where

$$\psi(a) := \prod_{p|a, p\ge 3}\left(1+\frac{1}{p-2}\right).$$

Note that $\psi(a)$ is a multiplicative function, so

$$\begin{aligned}L(s,\psi) &= \sum_{n=1}^{\infty}\frac{\psi(n)}{n^s}\\ &= \prod_p\left(1+\frac{\psi(p)}{p^s}+\frac{\psi(p^2)}{p^{2s}}+\dots\right)\\ &= \prod_p\left(1+\psi(p)\left(\frac{1}{p^s}+\frac{1}{p^{2s}}+\dots\right)\right)\\ &= \left(1-\frac{1}{2^s}\right)^{-1}\prod_{p\ge 3}\left(1+\frac{p-1}{(p^s-1)(p-2)}\right).\end{aligned}$$

Observe that $L(s,\psi)=\zeta(s)A(s)$, where

$$A(s)=\prod_{p\geq 3}\left(1+\frac{1}{p^s(p-2)}\right)$$

is absolutely convergent in $\mathrm{Re}(s)>0$. Hence by the Wiener-Ikehara Tauberian theorem [9, Cor. 8.8],

$$\begin{aligned}\sum_{\substack{a\leq f(x),\\ 2|a}}\psi(a)=\sum_{a\leq f(x)/2}\psi(a)&=A(1)\frac{f(x)}{2}(1+o(1))\\ &=\prod_{p\geq 3}\left(1+\frac{1}{p(p-2)}\right)\frac{f(x)}{2}(1+o(1)).\end{aligned}$$

Therefore

$$\begin{aligned}\sum_{\substack{a\leq f(x),\\ 2|a}}B(a)&=2\prod_{p\geq 3}\left[\left(1-\frac{2}{p}\right)\left(1-\frac{1}{p}\right)^{-2}\left(1+\frac{1}{p(p-2)}\right)\right]\frac{f(x)}{2}(1+o(1))\\ &=f(x)(1+o(1)).\end{aligned}$$

□

Using Lemma 3.2 and 3.3, we can give an upper bound on the number of pairs of primes with prime gap in a given range.

Proof of Lemma 3.1. By Lemmas 3.2 and 3.3, the number of such pairs (p_i,p_j) is bounded by

$$\begin{aligned}&\sum_{a\leq f(x)}\frac{8x}{\log^2 x}B(a)\left(1+O\left(\frac{\log\log x}{\log x}\right)\right)\\ &=\frac{8x}{\log^2 x}\left(1+O\left(\frac{\log\log x}{\log x}\right)\right)\sum_{\substack{a\leq f(x),\\ 2|a}}B(a)+O(f(x))\\ &=\frac{8x}{\log^2 x}f(x)(1+o(1)).\end{aligned}$$

□

If we take $f(x)=\alpha\log x$ with α small, then we can give an upper bound on the number of consecutive primes with an unusually small prime gap. From this regularity result, we will now deduce the main theorem.

Proof of Theorem 1.2. Recall from (6) that

$$T_{r,t}(x;\mathfrak{S}) = \sum_{\substack{q_k<x,\\ k\equiv r(\mathrm{mod}\ t)}} (q_{k+1}-q_k) + O_t\left(\frac{x}{\log^A x}\right).$$

Given a function $f(x)$ which tends to ∞ as $x \to \infty$ and $f(x) = O(\log x)$, by Lemma 3.1, we know that

$$\#\{p_n < x \mid p_{n+1}-p_n < f(x)\} \leq \frac{8x}{\log^2 x} f(x)(1+o(1)). \tag{7}$$

Note that

$$\#\{q_n < x \mid n \equiv r(\mathrm{mod}\ t)\} = \frac{\delta\pi(x)}{t}(1+o(1)) \geq \frac{\delta x}{t\log x}(1+o(1))$$

by the prime number theorem and the estimate in [11, Cor. 1, P.69]. Hence

$$\#\{q_n < x \mid q_{n+1}-q_n \geq f(x) \text{ and } n \equiv r(\mathrm{mod}\ t)\} \geq \left(\frac{\delta x}{t\log x} - \frac{8x}{\log^2 x} f(x)\right)(1+o(1))$$

In particular, for $f(x) \leq \delta \log x/(8t)$, the main term is nonnegative. Hence

$$\begin{aligned}
T_{r,t}(x;\mathfrak{S}) &= \sum_{\substack{q_k<x,\\ k\equiv r(\mathrm{mod}\ t)}} (q_{k+1}-q_k) + O_t\left(\frac{x}{\log^A x}\right) = \sum_{y=1}^{\infty} \sum_{\substack{q_k<x,\\ k\equiv r(\mathrm{mod}\ t),\\ q_{k+1}-q_k=y}} \sum_{z=1}^{y} 1 + O_t\left(\frac{x}{\log^A x}\right) \\
&= \sum_{z=1}^{\infty}\sum_{y=z}^{\infty} \sum_{\substack{q_k<x,\\ k\equiv r(\mathrm{mod}\ t),\\ q_{k+1}-q_k=y}} 1 + O_t\left(\frac{x}{\log^A x}\right) \geq \sum_{z=1}^{\delta\log x/(8t)} \sum_{\substack{q_k<x,\\ k\equiv r(\mathrm{mod}\ t),\\ q_{k+1}-q_k\geq z}} 1 + O_t\left(\frac{x}{\log^A x}\right) \\
&\geq \sum_{z=1}^{\delta\log x/(8t)} \left(\frac{\delta x}{t\log x} - \frac{8x}{\log^2 x} z\right)(1+o(1)) + O_t\left(\frac{x}{\log^A x}\right) \\
&= \left(\frac{\delta x}{t\log x}\frac{\delta \log x}{8t} - \frac{8x}{2\log^2 x}\left(\frac{\delta\log x}{8t}\right)^2\right)(1+o(1)) \\
&= \frac{\delta^2 x}{16t^2}(1+o(1)).
\end{aligned}$$

Therefore, we have

$$\frac{T_{r,t}(x;\mathfrak{S})}{x} \geq \frac{\delta^2}{16t^2}(1+o(1)),$$

and so

$$\liminf_{x\to\infty} \frac{T_{r,t}(x;\mathfrak{S})}{x} \geq \frac{\delta^2}{16t^2}.$$

If we assume the prime k-tuple conjecture, the upper bound in Lemma 3.2 can be improved by a factor of 8. Proceeding analogously in our above proof, we obtain

$$\liminf_{x\to\infty} \frac{T_{r,t}(x;\mathfrak{S})}{x} \geq \frac{\delta^2}{2t^2}.$$

□

Proof of Theorem 1.3. Theorem 1.3 follows *mutatis mutandis* by considering only prime gaps divisible by q and primes congruent to a mod q in Lemma 3.2, in which case we obtain an extra factor of $\phi(q)$ in the denominator of the upper bound in (7). We also recall that $\pi(x;q,a) \sim \pi(x)/\phi(q)$ by Dirichlet's theorem on arithmetic progressions. □

Acknowledgements The authors would like to thank Ken Ono for running the Emory REU in Number Theory and making valuable suggestions. We would also like to thank Robert Lemke Oliver for his guidance in both research and writing.

References

[1] M. Alboiu, *On the parity of the prime counting function and related problems.* Preprint.

[2] A. Cojocaru and M. Murty, *An introduction to sieve methods and their applications.* London Math. Soc. Stud. Texts, 2006.

[3] S. R. Finch, *Mathematical Constants.* Encyclopedia Math. Appl., vol. 94, Cambridge Univ. Press, 2003.

[4] D. Goldston, J. Pintz and C. Yıldırım, *Primes in tuples. I.* Ann. of Math. (2) 170 (2009), no. 2, 819-862.

[5] D. Goldston, S. Graham, J. Pintz and C. Yıldırım, *Small gaps between products of two primes.* Proc. Lond. Math. Soc. (3) 98 (2009), no. 3, 741-774.

[6] H. Halberstam and H.-E. Richert, *Sieve methods.* London Math. Soc. Monogr. Ser., 1974.

[7] H. Iwaniec and E. Kowalski, *Analytic number theory.* Amer. Math. Soc. Colloq. Publ., 2004.

[8] D. Leitmann and D. Wolke, *Primzahlen der Gestalt* $[f(n)]$. (German) Math. Z. 145 (1975), no. 1, 81-92.

[9] H. Montgomery, R. Vaughan, *Multiplicative Number Theory I. Classical Theory.* Cambridge Stud. Adv. Math., 2007.

[10] Polymath, *Improving the bounds for small gaps between primes.* `http://michaelnielsen.org/polymath1/index.php?title=Bounded_gaps_between_primes`.

[11] J. B. Rosser and L. Schoenfeld, *Approximate formulas for some functions of prime numbers.* Illinois J. Math. Volume 6, Issue 1 (1962), 64-94.

[12] Y. Zhang, *Bounded gaps between primes.* Ann. of Math., to appear.

S-DIOPHANTINE QUADRUPLES WITH TWO PRIMES CONGRUENT TO 3 MODULO 4

László Szalay
Institute of Mathematics, University of West Hungary, Sopron, Hungary
laszalay@emk.nyme.hu

Volker Ziegler[1]
Johann Radon Institute for Computational and Applied Mathematics (RICAM)
Austrian Academy of Sciences, Linz, Austria
volker.ziegler@ricam.oeaw.ac.at

Received: 3/12/13, Revised: 11/8/13, Accepted: 11/29/13, Published: 12/13/13

Abstract
Let S be a fixed set of primes and let $a_1, \ldots, a_m$ denote distinct positive integers. We call the m-tuple $(a_1, \ldots, a_m)$ an S-Diophantine tuple if the integers $a_i a_j + 1 = s_{i,j}$ are S-units for all $i \neq j$. In this paper, we show that if $S = \{p, q\}$ and $p, q \equiv 3 \pmod 4$, then no S-Diophantine quadruple exists.

1. Introduction

It is an old problem to find m-tuples $(a_1, \ldots, a_m)$ of distinct positive integers such that

$$a_i a_j + 1 = \square \tag{1}$$

for $i \neq j$. Such m-tuples are called Diophantine m-tuples and have been studied since ancient times by several authors. Most notable is Dujella's result [3] that no Diophantine sextuple exists and that there are only finitely many quintuples. Even more is believed to be true. A folklore conjecture states that there exist no quintuples at all.

Beside Diophantine m-tuples, various variants have also been considered. For instance, Bugeaud and Dujella [1] examined m-tuples, where $\square$ in (1) is replaced by a k-th power, and Dujella and Fuchs [4] investigated a polynomial version. Later Fuchs, Luca and the first author [5] replaced $\square$ by terms of a given binary recurrence sequence, in particular, the Fibonacci sequence [6]. Recently the authors

[1]The second author was supported by the Austrian Science Found (FWF) under the project P 24801-N26

substituted □ by S-units [7]. For complete overview we suggest Dujella's web page on Diophantine tuples [2].

In this paper, we continue our research on S-Diophantine m-tuples. Let S be a fixed set of primes. Then we call the m-tuple $(a_1, \ldots, a_m)$ with positive distinct integers a_i $(1 \leq i \leq m)$ an S-Diophantine m-tuple, if we have $a_i a_j + 1 = s_{i,j}$, $1 \leq i < j \leq n$, are S-units.

In a recent paper [7] the authors showed that if $S = \{p, q\}$ and $C(\xi) < p < q < p^{\xi}$ for some $\xi > 1$ and for some explicitly computable constant $C(\xi)$, then no S-Diophantine quadruple exists. This result and numerical experiments (see [7, Lemma 9], where we found no quadruples with $1 \leq a < b < c < d \leq 1000$) raise the question of whether or not an S-Diophantine quadruple with $|S| = 2$ exists. In this connection we have the following conjecture.

Conjecture 1.1. There is no pair of primes (p, q) such that a $\{p, q\}$-Diophantine quadruple exists.

Unfortunately, we can prove only the following weaker statement which is the main result of this paper.

Theorem 1.2. *Let $S = \{p, q\}$ with primes $p, q \equiv 3 \pmod 4$. Then no S-Diophantine quadruple exists.*

The proof of Theorem 1.2 is organized as follows. In the next section we provide two auxiliary results which enable us to prove Theorem 1.2 partially in Section 3. Later the proof is completed in the last section of the paper. Here we note that Lemma 2.2 is the only place where we used the assertion p, q congruent to 3 modulo 4, so the technique we applied might be useful in the proof of the conjecture.

2. Auxiliary Results

We start with a very useful lemma (see [7, Lemma 2]) which excludes some divisibility relations for S-Diophantine triples.

Lemma 2.1. *Assume that (a, b, c) is an S-Diophantine triple with $1 \leq a < b < c$. If $ac + 1 = s$ and $bc + 1 = t$, then $s \nmid t$.*

This lemma is exactly Lemma 2 in [7]. The proof is short and, since we intend to keep this paper independent and self-contained, we repeat the proof here.

Proof. Assume that $s \mid t$. Then

$$m = \frac{bc+1}{ac+1} = \frac{b}{a} + \frac{a-b}{a^2c+a} = \frac{b}{a} + \frac{\theta}{a^2} \in \mathbb{Z}$$

holds if $|\theta| < 1$. Therefore m is an integer if and only if $\theta = 0$. Thus, we have $a = b$ which is a contradiction to $a < b$. □

Now we deduce a few restrictions on the exponents appearing in the prime factorization of the S-units $s_{i,j}$.

Lemma 2.2. *Let $S = \{p, q\}$ with $p, q \equiv 3 \pmod 4$ and let (a, b, c) be an S-Diophantine triple. Further assume that*

$$ab + 1 = p^{\alpha_1} q^{\beta_1}, \qquad ac + 1 = p^{\alpha_2} q^{\beta_2}, \qquad bc + 1 = p^{\alpha_3} q^{\beta_3}.$$

Then at least one of $\alpha_1, \alpha_2, \alpha_3$ is zero and at least one of $\beta_1, \beta_2, \beta_3$ is equal to zero.

Proof. Using the notation of the lemma we have

$$(abc)^2 = \left(p^{\alpha_1} q^{\beta_1} - 1\right)\left(p^{\alpha_2} q^{\beta_2} - 1\right)\left(p^{\alpha_3} q^{\beta_3} - 1\right).$$

If α_1, α_2 and α_3 are all positive, then $(abc)^2 \equiv -1 \pmod p$ and we arrive at a contradiction since the Legendre symbol $(-1/p) = -1$. Similarly, at least one of β_1, β_2 and β_3 must be zero. □

3. Proof of Theorem 1.2

For the rest of the paper we assume that $S = \{p, q\}$ and $p, q \equiv 3 \pmod 4$.

Suppose now that (a, b, c, d) is an S-Diophantine quadruple. Therefore there are non-negative integers α_i, β_i, $i = 1, \ldots, 6$, such that

$$\begin{aligned} ab + 1 &= p^{\alpha_1} q^{\beta_1}, & bc + 1 &= p^{\alpha_4} q^{\beta_4}, \\ ac + 1 &= p^{\alpha_2} q^{\beta_2}, & bd + 1 &= p^{\alpha_5} q^{\beta_5}, \\ ad + 1 &= p^{\alpha_3} q^{\beta_3}, & cd + 1 &= p^{\alpha_6} q^{\beta_6} \end{aligned}$$

hold. Since (a, b, c) is an S-Diophantine triple, according to Lemma 2.2, at least one of α_1, α_2 and α_4 is zero. Let us assume for the moment that all of them vanish, i.e., $\alpha_1 = \alpha_2 = \alpha_4 = 0$. Without loss of generality, we may suppose $a < b < c$. Thus, $ac + 1 \mid bc + 1$ and Lemma 2.1 yields a contradiction. Subsequently, at least one of α_1, α_2 and α_4 is non-zero, and the same is true for β_1, β_2 and β_4.

Proposition 3.1. *If exactly one of α_1, α_2 and α_4 is zero, or exactly one of β_1, β_2 and β_4 is zero, then (a, b, c, d) cannot be an S-Diophantine quadruple.*

Proof. By switching p and q if necessary, and by rearranging the quadruple (a, b, c, d) we may assume that $\alpha_1 = 0$ and α_2, α_4 are positive. Note that (b, c, d) is also an S-Diophantine triple. In view of Lemma 2.2, one of α_5 and α_6 must be zero.

Letting $\alpha_5 = 0$, it implies $\alpha_6 = 0$. Indeed, consider the S-Diophantine triple (a, c, d) and the corresponding equations $ac + 1 = p^{\alpha_2} q^{\beta_2}$, $ad + 1 = p^{\alpha_3} q^{\beta_3}$ and $cd + 1 = p^{\alpha_6} q^{\beta_6}$. By Lemma 2.2, one of α_3 and α_6 vanishes. But $\alpha_3 = 0$ leads to a contradiction because it provides $ab+1=q^{\beta_1}$, $ad+1=q^{\beta_3}$ and $bd+1=q^{\beta_5}$, which contradicts Lemma 2.1. Hence $\alpha_5=\alpha_6=0$.

Thus, the following lemma completes the proof of the proposition.

Lemma 3.2. *There exists no S-Diophantine quadruple (a,b,c,d) with $\alpha_1=\alpha_6=0$.*

The proof of this lemma is long and technical; we postpone the proof to the forthcoming section. However, assuming this lemma, the proof of our proposition is complete. □

By virtue of Proposition 3.1, at least two of α_1,α_2 and α_4 are zero, and similarly, at least two of β_1,β_2 and β_4 equal zero. Therefore one pair fulfills $(\alpha_i,\beta_i)=(0,0)$ with $i\in\{1,2,4\}$. But, this is impossible since all of $ab+1$, $ac+1$ and $bc+1$ are at least 3.

Hence, up to the proof of Lemma 3.2, we have proved Theorem 1.2.

4. Proof of Lemma 3.2

In view of the assumptions of Lemma 3.2, we have to study the system

$$\begin{aligned} ab+1&=q^{\beta_1}, & bc+1&=p^{\alpha_4}q^{\beta_4},\\ ac+1&=p^{\alpha_2}q^{\beta_2}, & bd+1&=p^{\alpha_5}q^{\beta_5},\\ ad+1&=p^{\alpha_3}q^{\beta_3}, & cd+1&=q^{\beta_6}.\end{aligned}$$

Consider the triple (a,b,c). By Lemma 2.2, we deduce that either $\beta_2=0$ or $\beta_4=0$. There is no restriction in assuming $\beta_2=0$ (switch a and b as well as the corresponding exponents, if necessary). Thus, we obtain the system

$$\begin{aligned} ab+1&=q^{\beta_1}, & bc+1&=p^{\alpha_4}q^{\beta_4},\\ ac+1&=p^{\alpha_2}, & bd+1&=p^{\alpha_5}q^{\beta_5},\\ ad+1&=p^{\alpha_3}q^{\beta_3}, & cd+1&=q^{\beta_6}.\end{aligned}$$

Subsequently,

$$ab\cdot cd=\left(q^{\beta_1}-1\right)\left(q^{\beta_6}-1\right)=\left(p^{\alpha_2}-1\right)\left(p^{\alpha_5}q^{\beta_5}-1\right)=ac\cdot bd.$$

Assuming $\beta_5>0$ we get

$$1\equiv 1-p^{\alpha_2} \pmod q. \tag{2}$$

Note that the positivity of a,b,c and d entails that β_1 and β_6 are also positive integers. However, equation (2) yields the contradiction $q|p^{\alpha_2}$. Thus, we have

$\beta_5 = 0$, and the system

$$ab + 1 = q^{\beta_1}, \qquad bc + 1 = p^{\alpha_4} q^{\beta_4},$$
$$ac + 1 = p^{\alpha_2}, \qquad bd + 1 = p^{\alpha_5},$$
$$ad + 1 = p^{\alpha_3} q^{\beta_3}, \qquad cd + 1 = q^{\beta_6}$$

is valid. Consider now the equation

$$ac \cdot bd = (p^{\alpha_2} - 1)(p^{\alpha_5} - 1) = \left(p^{\alpha_3} q^{\beta_3} - 1\right)\left(p^{\alpha_4} q^{\beta_4} - 1\right) = ad \cdot bc$$

and its consequence

$$p^{\alpha_2+\alpha_5} - p^{\alpha_2} - p^{\alpha_5} = p^{\alpha_3+\alpha_4} q^{\beta_3+\beta_4} - p^{\alpha_3} q^{\beta_3} - p^{\alpha_4} q^{\beta_4}. \tag{3}$$

By switching a, b and c, d simultaneously, we may assume that $\alpha_5 \geq \alpha_2$. Moreover, the p-adic valuation of the left and right hand side of (3) coincide, hence the least two of $\alpha_2, \alpha_3, \alpha_4$ and α_5 must be equal. In particular, we have the following three cases: $\alpha_2 = \alpha_3 \leq \alpha_4$, $\alpha_2 = \alpha_4 \leq \alpha_3$ and $\alpha_3 = \alpha_4 < \alpha_2$. Note that with $\alpha_2 = \alpha_5$ at least one further exponent is necessarily minimal.

Similarly, we can arrive at the equation

$$q^{\beta_1+\beta_6} - q^{\beta_1} - q^{\beta_6} = p^{\alpha_3+\alpha_4} q^{\beta_3+\beta_4} - p^{\alpha_3} q^{\beta_3} - p^{\alpha_4} q^{\beta_4},$$

where we may assume that $\beta_6 \geq \beta_1$. Thus, the least two of $\beta_1, \beta_3, \beta_4$ and β_6 must coincide. Hence, in total, we have 9 possibilities which will be treated successively (see Table 1).

α	β
$\alpha_2 = \alpha_3 \leq \alpha_4$	$\beta_1 = \beta_3 \leq \beta_4$
	$\beta_1 = \beta_4 \leq \beta_3$
	$\beta_3 = \beta_4 < \beta_1$
$\alpha_2 = \alpha_4 \leq \alpha_3$	$\beta_1 = \beta_3 \leq \beta_4$
	$\beta_1 = \beta_4 \leq \beta_3$
	$\beta_3 = \beta_4 < \beta_1$
$\alpha_3 = \alpha_4 < \alpha_2$	$\beta_1 = \beta_3 \leq \beta_4$
	$\beta_1 = \beta_4 \leq \beta_3$
	$\beta_3 = \beta_4 < \beta_1$

Table 1: List of cases

4.1. The Case $\alpha_2 = \alpha_3 \le \alpha_4$ and $\beta_1 = \beta_3 \le \beta_4$

Consider the triple (a, b, c) with

$$ab + 1 = q^{\beta_1}, \qquad ac + 1 = p^{\alpha_2}, \qquad bc + 1 = p^{\alpha_4} q^{\beta_4}.$$

The assumption $\beta_1 \le \beta_4$ implies $ab < bc$, i.e., $a < c$ immediately. Similarly, $a < b$ is concluded from $\alpha_2 \le \alpha_4$. Hence either $ab+1 \mid bc+1$ with $a < c < b$ or $ac+1 \mid bc+1$ with $a < b < c$ holds. But both cases contradict Lemma 2.1.

4.2. The Case $\alpha_2 = \alpha_3 \le \alpha_4$ and $\beta_1 = \beta_4 \le \beta_3$

We clone the treatment of the previous case. Consider the triple (a, b, c) and deduce $a < c$ and $a < b$. Then either $ab + 1 \mid bc + 1$ with $a < c < b$ or $ac + 1 \mid bc + 1$ with $a < b < c$, and we arrive again at a contradiction.

4.3. The Case $\alpha_2 = \alpha_3 \le \alpha_4$ and $\beta_3 = \beta_4 < \beta_1$

For simplicity we omit certain subscripts by writing $\beta := \beta_3 = \beta_4$ and $\alpha := \alpha_2 = \alpha_3$. Comparing $ac + 1$ with $bc + 1$ and $ad + 1$ we obtain $a < b$ and $c < d$, and therefore $\alpha_4 < \alpha_5$. Moreover, by the properties of triple (a, b, c) we have $c < b$, otherwise a contradiction to Lemma 2.1 occurs.

Now consider the equation

$$ad \cdot bc = \left(p^{\alpha} q^{\beta} - 1\right)\left(p^{\alpha_4} q^{\beta} - 1\right) = \left(p^{\alpha} - 1\right)\left(p^{\alpha_5} - 1\right) = ac \cdot bd$$

modulo p^{α_4}. We get $p^{\alpha} q^{\beta} - 1 \equiv p^{\alpha} - 1 \pmod{p^{\alpha_4}}$ and then

$$q^{\beta} \equiv 1 \pmod{p^{\alpha_4 - \alpha}}. \tag{4}$$

This yields $p^{\alpha_4 - \alpha} \mid q^{\beta} - 1$, i.e., $p^{\alpha_4 - \alpha} \le q^{\beta} - 1$.

At this point we distinguish the two cases: $\beta_1 \ge 2\beta$ and $\beta_1 < 2\beta$. Taking

$$ad \cdot bc = \left(p^{\alpha} q^{\beta} - 1\right)\left(p^{\alpha_4} q^{\beta} - 1\right) = \left(q^{\beta_6} - 1\right)\left(q^{\beta_1} - 1\right) = ab \cdot cd$$

modulo $q^{2\beta}$, the congruence

$$q^{\beta}\left(p^{\alpha_4} + p^{\alpha}\right) \equiv 0 \pmod{q^{2\beta}}$$

follows. This yields $q^{\beta} \mid p^{\alpha_4 - \alpha} + 1$. Thus, $p^{\alpha_4 - \alpha} \ge q^{\beta} - 1$, which together with (4) implies $p^{\alpha_4 - \alpha} = q^{\beta} - 1$. Clearly, the two sides have opposite parity.

In case of $2\beta > \beta_1$ consider

$$ad \cdot bc = \left(p^{\alpha} q^{\beta} - 1\right)\left(p^{\alpha_4} q^{\beta} - 1\right) = \left(q^{\beta_1} - 1\right)\left(q^{\beta_6} - 1\right) = ab \cdot cd$$

modulo q^{β_1}. Thus,

$$q^{\beta}\left(p^{\alpha_4} + p^{\alpha}\right) \equiv 0 \pmod{q^{\beta_1}}.$$

Consequently, $q^{\beta_1-\beta} \mid p^{\alpha_4-\alpha}+1$.

A simple calculation shows that

$$\frac{b}{a}=\frac{bc}{ac}=\frac{p^{\alpha_4}q^{\beta}-1}{p^{\alpha}-1}=p^{\alpha_4-\alpha}q^{\beta}+\frac{p^{\alpha_4-\alpha}q^{\beta}-1}{p^{\alpha}-1}>p^{\alpha_4-\alpha}q^{\beta}.$$

If we assume $q^{\beta_1-\beta}\neq p^{\alpha_4-\alpha}+1$, then $2q^{\beta_1-\beta}\le p^{\alpha_4-\alpha}+1$ follows. Thus, we have

$$\frac{b}{a}>2q^{\beta_1}-q^{\beta}>\frac{5}{3}q^{\beta_1}>q^{\beta_1}>b,$$

which is a contradiction. Then $q^{\beta_1-\beta}=p^{\alpha_4-\alpha}+1$ holds, and it contradicts the fact that $q^{\beta_1-\beta}$ and $p^{\alpha_4-\alpha}+1$ have opposite parity.

4.4. The Case $\alpha_2=\alpha_4\le\alpha_3$ and $\beta_1=\beta_3\le\beta_4$

Similar to the case 4.1 ($\alpha_2=\alpha_3\le\alpha_4$ and $\beta_1=\beta_3\le\beta_4$), consider the triple (a,b,c) to find a contradiction to Lemma 2.1.

4.5. The Case $\alpha_2=\alpha_4\le\alpha_3$ and $\beta_1=\beta_4\le\beta_3$

Again, similar to the case 4.1, the triple (a,b,c) contradicts Lemma 2.1.

4.6. The Case $\alpha_2=\alpha_4\le\alpha_3$ and $\beta_3=\beta_4<\beta_1$

Write $\beta:=\beta_3=\beta_4$ and $\alpha:=\alpha_2=\alpha_4$. Comparing $ac+1$ with $bc+1$ we obtain $a<b$. Since $p^{\alpha_3}q^{\beta}-1=ad<bd=p^{\alpha_5}-1$ we have $\alpha_3\le\alpha_5$. The equations

$$ad\cdot bc=\left(p^{\alpha_3}q^{\beta}-1\right)\left(p^{\alpha}q^{\beta}-1\right)=\left(p^{\alpha}-1\right)\left(p^{\alpha_5}-1\right)=ac\cdot bd$$

modulo p^{α_3} admit $p^{\alpha}q^{\beta}-1\equiv p^{\alpha}-1 \pmod{p^{\alpha_3}}$. Therefore, $q^{\beta}\equiv 1 \pmod{p^{\alpha_3-\alpha}}$, and hence $p^{\alpha_3-\alpha}\mid q^{\beta}-1$. We also have $c\mid c(b-a)=p^{\alpha}(q^{\beta}-1)$. Thus, $c\mid q^{\beta}-1$ and $c<q^{\beta}$ follow. Since c and p are coprime (note that $ac+1=p^{\alpha_3}$), $c\mid \frac{q^{\beta}-1}{p^{\alpha_3-\alpha}}$. Clearly, $bc+1=p^{\alpha}q^{\beta}$ implies $p^{\alpha}q^{\beta}\le\frac{q^{\beta}-1}{p^{\alpha_3-\alpha}}b+1$ and then

$$b\ge\frac{p^{\alpha_3}q^{\beta}-p^{\alpha_3-\alpha}}{q^{\beta}-1}\ge p^{\alpha_3}-\frac{p^{\alpha_3}}{p^{\alpha}q^{\beta}}\ge p^{\alpha_3}\left(1-\frac{1}{p^{\alpha}}\right).$$

On the other hand, $b\mid b(a-c)=q^{\beta}(q^{\beta_1-\beta}-p^{\alpha})$. Thus, $b\mid q^{\beta_1-\beta}-p^{\alpha}$. The assumption $b<p^{\alpha}$ leads to the contradiction $bc+1<p^{\alpha}q^{\beta}$. Therefore, we necessarily obtain $q^{\beta_1-\beta}>p^{\alpha}$. Hence $b\le q^{\beta_1-\beta}-p^{\alpha}$. But $ab+1=q^{\beta_1}\le(q^{\beta_1-\beta}-p^{\alpha})a+1$, and we have

$$a\ge\frac{q^{\beta_1}-1}{q^{\beta_1-\beta}-p^{\alpha}}.$$

For the moment, assume that $d > b$. Then we have

$$p^{\alpha_3}q^{\beta} = ad+1 > ab > p^{\alpha_3}\left(1-\frac{1}{p^{\alpha}}\right)\frac{q^{\beta_1}-1}{q^{\beta_1-\beta}-p^{\alpha}} = p^{\alpha_3}q^{\beta}\left(1-\frac{1}{p^{\alpha}}\right)\frac{q^{\beta_1}-1}{q^{\beta_1}-p^{\alpha}q^{\beta}} > p^{\alpha_3}q^{\beta}.$$

Indeed,

$$\frac{p^{\alpha}-1}{p^{\alpha}}\cdot\frac{q^{\beta_1}-1}{q^{\beta_1}-p^{\alpha}q^{\beta}} > 1$$

is implied by $p^{\alpha}+q^{\beta_1} < p^{2\alpha}q^{\beta}$, which is coming from

$$q^{\beta}p^{2\alpha} = (ac+1)(bc+1) > ab+1+ac+1 = q^{\beta_1}+p^{\alpha}.$$

Hence $d < b$. But this relation, together with $c < a$ leads to

$$cd+1 = q^{\beta_6} < q^{\beta_1} = ab+1,$$

which contradicts the assumption $\beta_1 \leq \beta_6$.

4.7. The Case $\alpha_3 = \alpha_4 < \alpha_2$ and $\beta_1 = \beta_3 \leq \beta_4$

By switching p, q and b, c, we arrive at the case $\alpha_2 = \alpha_3 \leq \alpha_4$ and $\beta_3 = \beta_4 < \beta_1$.

4.8. The Case $\alpha_3 = \alpha_4 < \alpha_2$ and $\beta_1 = \beta_4 \leq \beta_3$

This is equivalent to the case $\alpha_2 = \alpha_4 \leq \alpha_3$ and $\beta_3 = \beta_4 < \beta_1$ by exchanging p and q, respectively b and c.

4.9. The Case $\alpha_3 = \alpha_4 < \alpha_2$ and $\beta_3 = \beta_4 < \beta_1$

First suppose $c < a$. From $cd+1 = q^{\beta_6} \geq q^{\beta_1} = ab+1$ we deduce $d > b$. Then $ad+1 = p^{\alpha}q^{\beta} = bc+1$, contradicting $c < a$ and $b < d$.

Now assume that $b < a$. Then $bd+1 = p^{\alpha_5} \geq p^{\alpha_2} = ac+1$, and therefore $d > c$ follows. Thus, we again arrive at a contradiction to $ad+1 = p^{\alpha}q^{\beta} = bc+1$.

Consequently, $a < b$, $a < c$, and

$$b \mid b(c-a) = q^{\beta}(p^{\alpha}-q^{\beta_1-\beta}) \quad\text{and}\quad c \mid c(b-a) = p^{\alpha}(q^{\beta}-p^{\alpha_2-\alpha}).$$

Hence $b < q^{\beta}$ and $c < p^{\alpha}$ follow, and $p^{\alpha}q^{\beta} < bc+1 = p^{\alpha}q^{\beta}$ shows the final contradiction.

References

[1] Y. Bugeaud and A. Dujella. On a problem of Diophantus for higher powers. *Math. Proc. Cambridge Philos. Soc.*, 135(1): 1–10, 2003.

[2] A. Dujella. On `http://web.math.hr/~duje/dtuples.html`.

[3] A. Dujella. There are only finitely many Diophantine quintuples. *J. Reine Angew. Math.*, 566: 183–214, 2004.

[4] A. Dujella and C. Fuchs. Complete solution of the polynomial version of a problem of Diophantus. *J. Number Theory*, 106(2): 326–344, 2004.

[5] C. Fuchs, F. Luca, and L. Szalay. Diophantine triples with values in binary recurrences. *Ann. Sc. Norm. Super. Pisa Cl. Sci. (5)*, 7(4): 579–608, 2008.

[6] F. Luca and L. Szalay. Fibonacci Diophantine triples. *Glas. Mat. Ser. III*, 43(63)(2): 253–264, 2008.

[7] L. Szalay and V. Ziegler. On an S-unit variant of Diophantine m-tuples. *Publ. Math. Debrecen*, 83(1-2): 97–121, 2013.

THE AVERAGE LARGEST PRIME FACTOR

Eric Naslund
Department of Mathematics, Princeton University, Princeton, New Jersey
naslund@math.princeton.edu

Received: 2/18/13, Revised: 7/17/13, Accepted:12/5/13 , Published: 12/13/13

Abstract

Let $P(n)$ denote the largest prime factor of an integer n. In this paper we look at the average of $P(n)$, and prove that

$$\frac{1}{x}\sum_{n\leq x} P(n) = \mathrm{li}_g(x) + O_\epsilon\left(xe^{-c(\log x)^{3/5-\epsilon}}\right),$$

where $\mathrm{li}_g(x) = \int_2^x \frac{t}{x}\frac{[x/t]}{\log t}dt$. This improves a result of De Koninck and Ivíc and allows us to deduce their asymptotic expansion

$$\frac{1}{x}\sum_{n\leq x} P(n) = c_0\frac{x}{\log x} + c_1\frac{1!x}{(\log x)^2} + \cdots + c_{k-1}\frac{(k-1)!x}{(\log x)^k} + O\left(\frac{x}{(\log x)^{k+1}}\right)$$

as a corollary, with the advantage that we can give the constants explicitly as

$$c_n = \frac{1}{2^{n+1}}\sum_{j=0}^{n}\frac{2^j(-1)^j\zeta^{(j)}(2)}{j!}.$$

1. Introduction

Let $P(n)$ denote the largest prime factor of an integer n. The moments of $P(n)$ were first looked at in 1976 by Knuth and Pardon [4], and concerning the average, they proved that

$$\sum_{n\leq x} P(n) \sim \frac{\pi^2}{12}\frac{x^2}{\log x}. \tag{1}$$

The above asymptotic also appeared in Erdős and Alladi's 1977 paper [1] on additive arithmetic functions. Later, De Koninck and Ivić [2] proved that $\sum_{n\leq x} P(n)$ has an asymptotic expansion of the form

$$\sum_{n\leq x} P(n) = x^2\left(\frac{d_1}{\log x} + \frac{d_2}{\log^2 x} + \cdots + \frac{d_m}{\log^m x} + O\left(\frac{1}{\log^{m+1} x}\right)\right)$$

where the constants d_i are computable, but not given explicitely. This expansion appears again in [3], where Ivić finds a similar formula for the kth largest prime factor. In this paper, we calculate the average of $P(n)$ up to an error of the form $O_\epsilon\left(xe^{-c(\log x)^{3/5-\epsilon}}\right)$, like that of the prime number theorem. This allows us to deduce De Koninck and Ivić's expansion as a corollary, as well as give the constants d_i explicitely. Our main result is:

Theorem 1. *Letting* $li_g(x) = \int_2^x \frac{t}{x}\frac{[x/t]}{\log t}dt$, *we have that*

$$\frac{1}{x}\sum_{n\leq x} P(n) = li_g(x) + O_\epsilon\left(xe^{-c(\log x)^{3/5-\epsilon}}\right). \tag{2}$$

For any integer k, $li_g(x)$ *has the asymptotic expansion*

$$li_g(x) = c_0\frac{x}{\log x} + c_1\frac{1!x}{(\log x)^2} + \cdots + c_{k-1}\frac{(k-1)!x}{(\log x)^k} + O\left(\frac{x}{(\log x)^{k+1}}\right) \tag{3}$$

where

$$c_n = \frac{1}{2^{n+1}}\sum_{j=0}^{n}\frac{2^j(-1)^j\zeta^{(j)}(2)}{j!}. \tag{4}$$

These constants c_n *satisfy* $c_n = 1 + O\left(\frac{n}{2^n}\right)$.

Notice in particular that $c_0 = \frac{\pi^2}{12}$, so we deduce that

$$\sum_{n\leq x} P(n) = \frac{\pi^2}{12}\frac{x^2}{\log x} + O\left(\frac{x^2}{(\log x)^2}\right).$$

Although this expansion has previously been presented in [2], and again in [3], there are some advantages to the result above. We are able to give the constants c_n explicitely in terms of the zeta function without much additional work, and our proof is basic requiring only an application of the hyperbola method. Furthermore, we give the explicit integral form which has an error term like that of the prime number theorem, paralleling how $\pi(x)$ is approximated by the function $\mathrm{li}(x)$.

2. The Main Theorem

For each integer $n \leq x$, there is at most one prime $p > \sqrt{x}$ such that $p|n$, and for each prime p there will be exactly $\left[\frac{x}{p}\right]$ integers less than x which are divisible by p. Combining these two facts, we see that

$$\sum_{\sqrt{x}<p\leq x} p\left[\frac{x}{p}\right] \leq \sum_{n\leq x} P(n) \leq \sum_{p\leq x} p\left[\frac{x}{p}\right],$$

and since $\sum_{p\leq\sqrt{x}} p\left[\frac{x}{p}\right] \leq x\sum_{p\leq\sqrt{x}} 1 = O\left(\frac{x^{3/2}}{\log x}\right)$, it follows that

$$\sum_{n\leq x} P(n) = \sum_{p\leq x} p\left[\frac{x}{p}\right] + O\left(\frac{x^{3/2}}{\log x}\right). \tag{5}$$

By writing $\left[\frac{x}{p}\right] = \sum_{n\leq x:\ p|n} 1$, and rearranging the order of summation, we obtain

$$\sum_{p\leq x} p\left[\frac{x}{p}\right] = \sum_{p\leq x} p \sum_{n\leq x:\ p|n} 1 = \sum_{n\leq x}\sum_{p\leq \frac{x}{n}} p. \tag{6}$$

Letting $\mathcal{K}(x) = \sum_{p\leq x} p$, we have the estimate

$$\mathcal{K}(x) = \int_2^x \frac{t}{\log t}dt + O_\epsilon\left(x^2 e^{-c(\log x)^{3/5-\epsilon}}\right), \tag{7}$$

which follows from the prime number theorem (See [5]) along with partial summation. Using (7) along with equations (5), and (6), we have that

$$\begin{aligned}
\sum_{n\leq x} P(n) &= \sum_{n\leq x}\int_2^{\frac{x}{n}} \frac{t}{\log t}dt + \sum_{n\leq x} O_\epsilon\left(\frac{x^2}{n^2}e^{-c\left(\log\frac{x}{n}\right)^{3/5-\epsilon}}\right) + O\left(\frac{x^{\frac{3}{2}}}{\log x}\right)\\
&= \int_2^x \frac{t\left[\frac{x}{t}\right]}{\log t}dt + O_\epsilon\left(x^2 e^{-c(\log x)^{3/5-\epsilon}}\right) &(8)\\
&= x\mathrm{li}_g(x) + O_\epsilon\left(x^2 e^{-c(\log x)^{3/5-\epsilon}}\right), &(9)
\end{aligned}$$

which proves (2). To recover the asymptotic expansion in (3), we turn our attention to this integral function and make the substitution $t = \frac{x}{u}$ to obtain

$$\mathrm{li}_g(x) = \frac{1}{x}\int_2^x \frac{t\left[\frac{x}{t}\right]}{\log t}dt = x\int_1^{x/2} \frac{[u]}{u^3\log\left(\frac{x}{u}\right)}du. \tag{10}$$

We can rewrite the integral above as

$$\int_1^{x/2} \frac{[u]}{u^3\log\left(\frac{x}{u}\right)}du = \frac{1}{\log x}\int_1^{x/2}\frac{[u]}{u^3}\left(1-\frac{\log u}{\log x}\right)^{-1} du,$$

and for any integer $k\geq 1$, by the geometric series expansion

$$\left(1-\frac{\log u}{\log x}\right)^{-1} = 1 + \frac{\log u}{\log x} + \cdots + \left(\frac{\log u}{\log x}\right)^{k-1} + \left(\frac{\log u}{\log x}\right)^{k}\left(1-\frac{\log u}{\log x}\right)^{-1},$$

we have that

$$\begin{aligned}
\int_1^{x/2} \frac{[u]}{u^3\log\left(\frac{x}{u}\right)}du &= \sum_{n=0}^{k-1}\frac{1}{(\log x)^{n+1}}\int_1^{x/2}\frac{[u]}{u^3}(\log u)^n\, du\\
&\quad + \frac{1}{(\log x)^{k+1}}\int_1^{x/2}\frac{[u]}{u^3}(\log u)^k\left(1-\frac{\log u}{\log x}\right)^{-1} du.
\end{aligned}$$

Each of these integrals is absolutely convergent on $[1, \infty)$, and so the last term contributes an error term of at most $O_k\left(\frac{1}{\log^{k+1} x}\right)$. Setting $c_k = \frac{1}{k!}\int_1^\infty \frac{[u]}{u^3}(\log u)^k\, du$, and applying the bound

$$\int_{x/2}^\infty \frac{[u]}{u^3}(\log u)^n\, du \le \int_{x/2}^\infty \frac{(\log u)^n}{u^2} du = O\left(\frac{(\log x)^n}{x}\right),$$

it follows that

$$\mathrm{li}_g(x) = \frac{x}{\log x}\sum_{n=0}^{k-1}\frac{n! c_n}{(\log x)^n} + O_k\left(\frac{x}{\log^{k+1} x}\right). \tag{11}$$

To evaluate these constants c_n, we use exponential generating series. Integration by parts tells us that for $s > 1$

$$\zeta(s) = \sum_{n=1}^\infty n^{-s} = \int_1^\infty x^{-s} d[x] = s\int_1^\infty [x]\, x^{-s-1} dx,$$

and so

$$\sum_{k=0}^\infty c_k z^k = \int_1^\infty [x]\, x^{z-3} dx = \frac{\zeta(2-z)}{2-z}.$$

Multiplying the power series expansions for $\zeta(2-z)$ and $\frac{1}{2-z}$ yields

$$\left(\sum_{j=0}^\infty (-1)^j \zeta^{(j)}(2)\frac{z^j}{j!}\right)\left(\frac{1}{2}\sum_{k=0}^\infty \frac{z^k}{2^k}\right) = \frac{1}{2}\sum_{n=0}^\infty z^n\left(\sum_{k+j=n}(-1)^j\frac{\zeta^{(j)}(2)}{j!}\frac{1}{2^k}\right),$$

and hence

$$c_n = \frac{1}{2^{n+1}}\sum_{j=0}^n \frac{2^j(-1)^j\zeta^{(j)}(2)}{j!}.$$

To find the size of c_n we note that

$$\begin{aligned}\frac{(-1)^j\zeta^{(j)}(2)}{j!} &= \frac{1}{j!}\sum_{k=1}^\infty \frac{(\log k)^j}{k^2}\\ &= \frac{1}{j!}\int_1^\infty \frac{(\log x)^j}{x^2}dx + \frac{1}{j!}\int_1^\infty \frac{(j - 2\log x)\log^{j-1} x}{x^3}\{x\}\, dx.\end{aligned}$$

Substituting $x = e^u$, the first integral becomes $\Gamma(j+1)$, and the second is $O\left(\frac{j!}{2^j}\right)$, which implies that

$$\frac{(-1)^j}{j!}\zeta^{(j)}(2) - 1 = O\left(\frac{1}{2^j}\right),$$

and so $c_n = 1 + O\left(\frac{n}{2^n}\right)$. This establishes all of Theorem 1.

Acknowledgment I am grateful to Greg Martin for his suggestions and encouragement.

References

[1] K. Alladi and P. Erdős. On an additive arithmetic function. *Pacific J. Math.* 71(2):275–294, 1977.

[2] Jean-Marie De Koninck and Aleksandar Ivić. The distribution of the average prime divisor of an integer. *Arch. Math. (Basel)* **43(1)** (1984), 37–43.

[3] Aleksandar Ivić. On the kth prime factor of an integer. *Zb. Rad. Prirod.-Mat. Fak. Ser. Mat.* **20(1)** (1990), 63–73.

[4] Donald E. Knuth and Luis Trabb Pardo. Analysis of a simple factorization algorithm. *Theoret. Comput. Sci.* **3(3)** (1976/77), 321–348.

[5] Hugh L. Montgomery and Robert C. Vaughan. *Multiplicative number theory. I. Classical theory*, volume 97 of *Cambridge Studies in Advanced Mathematics*. Cambridge University Press, Cambridge, 2007.